calculus with
analytic geometry

3rd edition

EDWIN J. PURCELL
University of Arizona

calc analytic

Prentice-Hall, Inc., Englewood Cliffs, New Jersey   07632

ulus with geometry

Library of Congress Cataloging in Publication Data

PURCELL, EDWIN JOSEPH, (date)
 Calculus with analytic geometry.

 Includes index.
 1. Calculus. 2. Geometry, Analytic. I. Title.
QA303.P99 1978   515'.15   77-7977
ISBN 0-13-112052-2

CALCULUS WITH ANALYTIC GEOMETRY, 3rd edition
Edwin J. Purcell

Cover Illustration: "*Color Ribbon*" by Joanna Pinsky

Printed in the United States of America

10 9 8 7 6 5 4 3 2

Prentice-Hall International, Inc., *London*
Prentice-Hall of Australia Pty. Limited, *Sydney*
Prentice-Hall of Canada, Ltd., *Toronto*
Prentice-Hall of India Private Limited, *New Delhi*
Prentice-Hall of Japan, Inc., *Tokyo*
Prentice-Hall of Southeast Asia Pte. Ltd., *Singapore*
Whitehall Books Limited, *Wellington, New Zealand*

contents

3

limits and continuity

77

4

the derivative

112

5

formulas for differentiation of algebraic functions. applications

135

6

further applications of the derivative

175

Contents

7   antiderivatives

8   the definite integral

9   applications of the definite integral

10 transcendental functions

11 techniques of integration

12 conics. polar coordinates

13 indeterminate forms. improper integrals

521

14 infinite series

547

19 multiple integrals   769

20 differential equations   823

appendix

857

answers to odd-numbered exercises

900

index

939

preface

This book presents a first course in calculus and analytic geometry. The author has tried to write it in a simple and straightforward style, with ample explanations, an abundance of illustrative examples, and carefully graded exercise sets, so that it will be unusually suitable for a reader of average ability to study alone or with minimum help from a teacher. There is enough material for three semesters.

Starting with simple first principles, each new concept is motivated by a natural, intuitive introduction. Seven basic concepts are stressed: function, limit of a function, continuity, derivative, antiderivative, definite integral, and infinite series. Effort is made to impress on the reader that a mastery of these ideas is indispensable in acquiring a genuine understanding of calculus.

At the same time, there is an abundance of material dealing with the degree of accuracy of computed results and with other aspects of the computational work that is so important for progress in science and technology.

Since ϵ, δ methods are necessary for a proper definition of the limit of a function, a very thorough treatment of inequalities and absolute values precedes it. However, the use of ϵ, δ methods is minimized in later work by utilizing limit theorems wherever possible.

Set notation is introduced early. It is employed when clearly advantageous, but not slavishly.

Vectors in two- and three-dimensional space are presented with a firm mathematical basis and are applied widely. Vectors do not supplant a sound foundation in Cartesian plane analytic geometry but complement it and make possible a more concise formulation of some of the theorems that were first derived in the classical

Cartesian manner. In three-dimensional analytic geometry, vectors are used from the outset. This avoids duplication of effort and contributes to a better understanding of both subjects.

To make this third edition more readable for the increasing numbers of students with less preparation for calculus than formerly, the following improvements have been made.

Many new illustrative examples with complete solutions clarify the theorems, definitions, and techniques. Students who study them should be able to do most of the exercises.

The exercise sets have been reworked and excessive algebraic manipulation has been eliminated. Each set now starts with easy variations of the illustrative examples, progresses through exercises of increasing difficulty, and concludes with some to challenge the stronger students.

Each chapter begins with an intuitive preview of the main ideas to be discussed and their relation to what has gone before. This helps the reader to see the developing calculus as a whole rather than as a series of isolated processes.

At the end of each chapter there is a set of review exercises. A student often finds it easy to work the exercises in a particular section because the method has just been explained. But a set of miscellaneous review exercises based on the material of an entire chapter causes the student to review the chapter and gain a better understanding of it.

Many proofs have been simplified and some of the more tedious ones have been moved to the appendix.

Stronger students will find that the logical development of this third edition and the careful statement of its theorems maintain the integrity of earlier editions. The organization has been improved by moving infinite series forward to Chapter 14, so that the first fifteen chapters now constitute a two semester course in single variable calculus. Chapters 16 to 19 treat the calculus of two or more variables.

The preliminary material has been shortened in order to start the actual calculus sooner. The chapter on the definite integral has been rewritten; it is simpler, more direct, and easier to understand. The treatment of trigonometric functions has been much improved.

There are new sections on Lagrange multipliers and on surface area.

Throughout this book the principal definitions and theorems are prominently labeled, numbered, and displayed, both for easy reference and to keep the main structure of the material before the reader's eyes. The number 7.3.4, for example, refers to the fourth numbered definition or theorem in Section 3 of Chapter 7. The number 14.6 refers to Section 6 of Chapter 14. Fig. 11-5 indicates the fifth figure in Chapter 11.

The most used theorems and definitions are printed in color to enable the student to concentrate his efforts to advantage.

I wish to thank many users of the earlier editions, both faculty and students, for their comments, criticism and encouragement. My thanks are also due to my wife, Bernice Lee Purcell, who made the index and typed the manuscript.

EDWIN J. PURCELL
University of Arizona

preliminaries

For students who are familiar with the preliminary concepts discussed in this chapter, a careful reading of most of it will suffice. However, *inequalities* and *absolute value* are so important in calculus that a mastery of Sections 1.5 and 1.6 is indispensable for success in studying this book. These topics are seldom covered adequately in high school, and a large proportion of the exercises in Sections 1.5 and 1.6 should be worked.

1.1 INTRODUCTION

Prior to the seventeenth century, algebra and geometry were studied as separate, unconnected subjects. The Greeks had perfected elementary geometry two thousand years ago, and in the centuries that followed the Hindus and Arabs cultivated algebra. Their algebra dealt with numbers, whereas Euclidian geometry was concerned with points, lines, planes, and the like.

There seemed little connection between algebra and geometry until the seventeenth century when two French mathematicians, René Descartes (1596–1650), who was also a philosopher, and Pierre de Fermat (1601–1655), invented a method, now called *analytic geometry*, that uses algebraic operations and equations to solve geometric problems; their method also shed new light on algebra by exhibiting its equations as geometric curves.

The basis for analytic geometry was Descartes' coordinate system, which associated the numbers of algebra with the points of geometry. By means of Cartesian

1

coordinates, large parts of algebra and geometry were seen to be two aspects of the same thing, somewhat as two different languages may express the same meaning. For instance the algebraic statement "Two distinct equations of the first degree in two variables have a single common solution or none" is equivalent to the geometric theorem "Two distinct lines in the same plane intersect in a single point or are parallel."

The names generally associated with the invention of *calculus* are Isaac Newton (1642–1727) and Gottfried Wilhelm Leibniz (1646–1716). Newton, an Englishman, developed calculus as a tool for his investigations in physics and astronomy. The German, Leibniz, was a universal genius who, independently of Newton and almost simultaneously, also developed calculus.

Calculus is based on the properties of numbers, and by using a Cartesian coordinate system, much of calculus can be presented in geometric terms. Thus the recently discovered analytic geometry was an ideal prelude to the invention of calculus.

Calculus, unlike the mathematics that preceded it, is the study of change and growth. The two basic processes of calculus are *differentiation* and *integration*. Differentiation gives the instantaneous rate of change of a varying quantity, and integration measures the total effect of continuous change. The key to Newton's and Leibniz's success in developing the calculus was their insight into the intimate relation between differentiation and integration as inverse processes, somewhat as multiplication and division of numbers are inverse operations.

Many scattered ideas from calculus were known to predecessors of Newton and Leibniz, even as far back as Archimedes (287–212 B.C.), who, without any algebra, succeeded in finding the areas of circles and regions under a parabola. For circles, he computed the areas of inscribed regular polygons of more and more sides. As the number of sides increased, the areas of the polygons increased and approached the area of the circle as a limit (Fig. 1-1). This is an example of integration.

The area of a circle is the limit of the area of an inscribed regular polygon of *n* sides as the number of sides, *n*, increases indefinitely

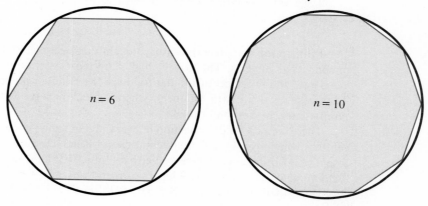

$n = 6$ $n = 10$

Figure 1-1

In the generation just before Newton, the problem of finding maximum and minimum values of a function was solved for some individual cases by finding the points on its graph where the tangent line is horizontal (Fig. 1-2). This led to a method for determining the direction of the tangent line to a curve at any point on the curve.

Let P be an arbitrarily chosen point on a curve and draw the secant line through P and a neighboring point Q on the curve (Fig. 1-3). Draw the vertical and horizontal

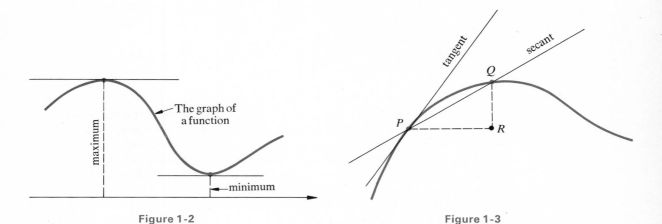

Figure 1-2   Figure 1-3

line segments, RQ and PR; the ratio of their lengths, RQ/PR, is a measure of the steepness of ascent of a point on the secant and hence of the direction of the secant. Keeping P fixed on the curve, allow Q to approach P along the curve. This causes the secant PQ to rotate about P and approach the position of the tangent line at P. The limit of the ratio RQ/PR as Q approaches P along the curve gives the direction of the tangent line at P. This is an example in differentiation.

Notice that in both examples the word *limit* was used. Limit is the most important concept in calculus and is what distinguishes calculus from all previous mathematics.

Thus Newton and Leibniz were not the first to differentiate or integrate. In particular, Isaac Barrow, Newton's teacher at Cambridge, understood the area problem and the tangent problem and probably knew that they were inverse to each other. The importance of Newton and Leibniz in calculus resulted from their consolidation of the known fragments into a general method, incorporating what is now known as the fundamental theorem, which is applicable to very large classes of functions, both algebraic and transcendental. Leibniz also devised a good notation, much of which is still being used.

Today calculus is essential in engineering and the physical sciences, and is being used more and more in biology and such social sciences as economics, sociology, and psychology. Without calculus, one could not design radar systems or cyclotrons, to name just a few. Calculus is used to determine the orbits of earth satellites and the paths for space travel.

Calculus is generally considered to be one of the greatest intellectual achievements of mankind.

1.2 SETS

A **set** is a collection of things. Some examples of sets are the letters in our alphabet, all American citizens, the positive integers, and the positions on a baseball team.

The **elements** of a set are the objects belonging to the set; they may or may not be material. In this book we shall be chiefly concerned with sets of real numbers and sets of points. The statement *a* **is an element of the set** *S* is symbolized by

$$a \in S,$$

and *a* **is not an element of** *S* is symbolized by $a \notin S$.

A set is **defined** when its description is sufficient to enable us to determine whether any arbitrary object belongs to the set. For instance, if S is the set of all integers greater than $\frac{4}{3}$, then $7 \in S$, $\frac{9}{4} \notin S$, and $-3 \notin S$. It is essential that if a is any object whatever, the definition of a set will enable us to give the unqualified answer "Yes" or "No" to the question "Does a belong to the set?" Thus "all beautiful women" fails to define a set because the decision of membership would be a matter of opinion.

When the number of elements of a set is finite, we can define the set by listing its elements. For example, the set consisting of the numbers π, $\sqrt{2}$, and 8 can be written $\{\pi, \sqrt{2}, 8\}$. Other sets are $\{a, b, c, d\}$ and $\{2, 3, 5, 7, 11\}$.

A different kind of set is

$$S = \{\{-1, 6\}, \{8, 16, 24\}, \{z, w\}\}.$$

This is a **set of sets** (or a collection of sets) whose three elements are the sets $\{-1, 6\}$, $\{8, 16, 24\}$, and $\{z, w\}$. Notice that $-1 \notin S$, although $\{-1, 6\} \in S$.

If the number of elements of a set is not finite, or if it is not convenient to list all the elements of a set, some rule that enables us to determine whether any given object belongs to the set will suffice. To illustrate, "the set of all numbers that can be expressed in the form $2n$, where n is an integer" defines the set of even integers.

The symbol

$$\{x \mid \cdots\}$$

means "the set of elements x such that"; the three dots here stand for some statement or statements about the elements of the set that clearly define the set. For example, $\{x \mid x$ is a real number and $2x^2 - 5x - 3 = 0\}$ is the set of real numbers x such that $2x^2 - 5x - 3 = 0$ is true; in other words, it is the set consisting of the real roots of $2x^2 - 5x - 3 = 0$, which is the set $\{3, -\frac{1}{2}\}$. Again, $\{x \mid x$ is a positive integer and x is less then 10$\}$ is the set $\{1, 2, 3, 4, 5, 6, 7, 8, 9\}$. Another example is $\{x \mid x$ is a negative integer and $x^2 - x - 6 = 0\}$, which is the set $\{-2\}$.

Notice that $\{x \mid x$ is a real number and $x^2 + 1 = 0\}$ contains no elements at all. It is the **empty set** and is represented by $\varnothing$. Thus $\varnothing = \{x \mid x$ is a real number, $x^2 + 1 = 0\}$, $\varnothing = \{\theta \mid \sin \theta = 1.5\}$, and $\varnothing = \{y \mid y$ is a living person and y signed the Declaration of Independence$\}$.

Two sets, A and B, are **equal**, written

$$A = B,$$

if and only if every element of A is an element of B and every element of B is an element of A; that is, $A = B$ if and only if A and B are two labels for the same set.

The sets $\{2, 1, 3\}$, $\{1, 2, 3\}$, and $\{\frac{6}{2}, \frac{19}{19}, 5 - 3\}$ are all equal. Moreover, $\{2, 1, 3\}$ $= \{2, 1, 2, 3, 3, 3\}$; in both sets the elements are the three numbers 1, 2, and 3.

A set A is said to be a subset of a set B, written

$$A \subseteq B,$$

if and only if every element of A is an element of B (Fig. 1-4).

For instance, $\{2, 7, 6\} \subseteq \{1, 2, 3, 4, 5, 6, 7\}$, and the set of all right triangles is a subset of the set of all triangles.

Notice that every set is a subset of itself, and that the empty set is a subset of every set. Symbolically, if A is a set, then $A \subseteq A$ and $\varnothing \subseteq A$.

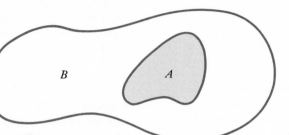

$$A \subseteq B$$

Figure 1-4

As another example, the set $\{a, b, c\}$ has eight subsets—namely, $\{a, b, c\}$, $\{a, b\}$, $\{a, c\}$, $\{b, c\}$, $\{a\}$, $\{b\}$, $\{c\}$, and $\varnothing$.

When proving that two sets A and B are equal, we usually try to show that $A \subseteq B$ and $B \subseteq A$.

If A is a subset of B and B is not a subset of A—that is, if every element of A is an element of B, but there is at least one element of B that is not an element of A—then A is said to be a **proper subset** of B. This situation is expressed by

$$A \subset B.$$

Thus the set of squares is a proper subset of the set of rectangles, and $\{5, 2\} \subset \{1, 2, 3, 4, 5, 6\}$.

By the **union** of two sets A and B, written

$$A \cup B,$$

we mean the set composed of every element of A and every element of B (Fig. 1-5). Symbolically,

$$A \cup B = \{x \mid x \in A \text{ or } x \in B\}.$$

The statement "$x \in A$ or $x \in B$" means that x is an element of A or of B or of both A and B, and we shall always use the word "or" in this inclusive sense. As an example of union, if $A = \{4, -1, 6\}$ and $B = \{6, 1, \pi\}$, then $A \cup B = \{4, -1, 6, 1, \pi\}$.

By the **intersection** of two sets A and B, written

$$A \cap B,$$

we mean the set composed of the elements that are in both A and B (Fig. 1-6). That is,

$$A \cap B = \{x \mid x \in A \text{ and } x \in B\}.$$

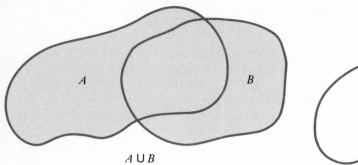

$A \cup B$

Figure 1-5

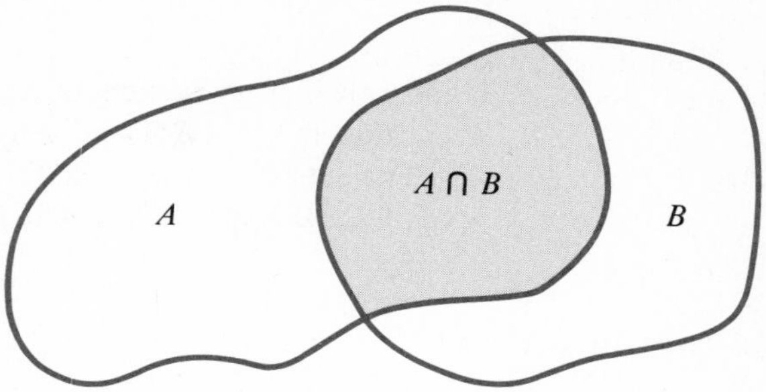

$A \cap B$

Figure 1-6

For example, the intersection of the sets $A = \{4, -1, 6\}$ and $B = \{6, 1, \pi\}$ is $A \cap B = \{6\}$.

As illustrations, if $N = \{1, 2, 3, \cdots\}$ is the set of all positive integers (or natural numbers), $A = \{x \mid x = 2n - 1, n \in N\}$, the set of odd positive integers, and $B = \{x \mid x = 2n, n \in N\}$, the set of even positive integers, then $A \cap B = \emptyset$ and $A \cup B = N$.

Exercises

1. Let $A = \{1, 2, 3, 4, 5\}$, $B = \{1, 4, 5, 6\}$, and $C = \{2, 3, 5\}$. Tell whether each of the following statements is true or false.

(a) $3 \in A$; (b) $1 \in C$;
(c) $2 \notin C$; (d) $3 \notin B$;
(e) $B \subseteq A$; (f) $C \subseteq A$;
(g) $C \subset A$; (h) $4 \in A \cup C$;
(i) $6 \in B \cup C$; (j) $4 \in A \cap C$;
(k) $6 \in B \cap C$; (l) $A \cap C = C$;
(m) $B \cap C = A$; (n) $A \cup C = A$.

2. Let $A = \{a, c, e\}$, $B = \{a, b, c, d, e\}$, and $C = \{b, c, d\}$. Tell whether each of the statements below is true or false.

(a) $c \in A$; (b) $a \in C$;
(c) $e \notin C$; (d) $b \notin A$;
(e) $C \subseteq B$; (f) $C \subset B$;
(g) $\emptyset \subset A$; (h) $d \in A \cup C$;
(i) $c = A \cap C$; (j) $c \in A \cap C$;
(k) $A \cup C = B$; (l) $A \cup B = B$;
(m) $A \cap C = B$; (n) $A \cap B = A$.

3. Rewrite each of the sets shown by listing its elements.

(a) $\{x \mid x = 2n + 1, n = 1, 2, 3, 4\}$;
(b) $\{x \mid x = n^3, n = 2, 4, 5\}$;

(c) $\{x \mid x$ is a positive integer, $x^2 - 3x + 2 = 0\}$;
(d) $\{x \mid x^3 + 2x^2 - 8x = 0\}$.

4. Rewrite each of these sets by listing its elements.

(a) $\{x \mid x = 2n, n = 1, 2, 3, 4, 5\}$;
(b) $\{x \mid x = 2n^2 - 1, n = 1, 2, 3, 4\}$;
(c) $\{x \mid x^2 - 9 = 0\}$;
(d) $\{x \mid 2x^3 - 3x^2 - 2x = 0\}$.

5. Use the notation $\{x \mid \ldots\}$ to describe each of the following sets.

(a) $\{3, 5, 7, 9, 11\}$;
(b) $\{4, 7, 10, 13, 16\}$;
(c) $\{2, 4, 8, 16, 32, \cdots\}$; (*Hint:* Let N be the set of all positive integers.)
(d) $\{3, 9, 27, 81, 243, \cdots\}$;
(e) $\{-2, 5\}$; (*Hint:* What equation in x has the solutions -2 and 5?)
(f) $\{-1, 0, 7\}$.

6. Let N be the set of positive integers. Use the notation $\{x \mid \ldots\}$ to describe each of these sets.

(a) the set of even positive integers;
(b) the set of odd positive integers;
(c) $\{1, 4, 9, 16, 25, \cdots\}$;
(d) $\{2, 5, 10, 17, 26, \cdots\}$;

(e) {2, 9}. (*Hint:* See Exercise 5(e));
(f) {0, 3, 4}.

7. Let $A = \{a, c, d, e, g\}$, $B = \{b, c, d, f\}$, and $C = \{d, e, g\}$. Find
(a) $A \cup C$;   (b) $B \cup C$;
(c) $A \cap C$;   (d) $B \cap C$.

8. Let $A = \{1, 3, 4, 5\}$, $B = \{1, 2, 4, 6\}$, and $C = \{2, 3, 5\}$. Find
(a) $A \cup (B \cap C)$;
(b) $(A \cup B) \cap (A \cup C)$;
(c) $A \cap (B \cup C)$;
(d) $(A \cap B) \cup (A \cap C)$.

9. Let $A = \{a, b, c, e, g\}$, $B = \{b, c, d, f\}$, and $C = \{a, d, e, g\}$. Find
(a) $A \cup (B \cap C)$;
(b) $(A \cup B) \cap (A \cup C)$;
(c) $A \cap (B \cup C)$;
(d) $(A \cap B) \cup (A \cap C)$.

10. If A and B are sets, prove that always
(a) $A \cap (A \cup B) = A$;
(b) $A \cup (A \cap B) = A$.

11. Tell whether each of the following statements is true or false; and if false, explain why.
(a) If S is a set, then $\varnothing \subseteq S$.
(b) $\{3\} = 3$;   (c) $0 = \varnothing$;
(d) $\{0\} = \varnothing$;   (e) $0 \in \varnothing$.

12. If A, B, and C are nonempty sets, which of the statements below are true and which false? If false, give a counterexample—that is, an example in which the statement in false.
(a) If $5 \in A$ and $A \subseteq B$, then always $5 \in B$;
(b) If $5 \in A$ and $A \in B$, then always $5 \in B$;
(c) $(A \cap B) \cap C = A \cap (B \cap C)$;
(d) $(A \cup B) \cup C = A \cup (B \cup C)$;
(e) $(A \cap B) \cup C = (A \cup B) \cap C$.

1.3 REAL NUMBERS

Calculus is based on the real number system and its properties. We begin by recalling the kinds of real numbers.

The **integers** are the elements of the set

$$\{\cdots, -3, -2, -1, 0, 1, 2, 3, 4, 5, \cdots\}.$$

A **rational number** is any number that can be expressed in the form p/q, where p and q are integers and $q \neq 0$.

They are called *rational* because p/q is a *ratio*. The numbers $\frac{5}{13}$, $-\frac{1}{2}$, $0.25 = \frac{1}{4}$, and $-2.3 = -\frac{23}{10}$ are rational.

The real numbers that are not rational are called **irrational** (that is, not rational). Such familiar numbers as $\sqrt{2}$, $\sqrt{3}$, $\sqrt[3]{7}$, π, and e are irrational numbers.

The ancient Greeks, two thousand years ago, knew that $\sqrt{2}$ was not a rational number and Euclid proved it. In his proof he assumed the contrary to be true—namely, that $\sqrt{2}$ *is a rational number*—and then proceeded to show that this assumption leads to a contradiction. The details of the proof are as follows.

Assume that $\sqrt{2}$ is a rational number. Then

(1)
$$\sqrt{2} = \frac{p}{q},$$

where p and q are integers and $q \neq 0$. Since common factors in the numerator and denominator can be canceled, we can, without loss of generality, assume that *p and q have no common factor* other than 1. Squaring both members of (1), we get $2 = p^2/q^2$;

that is,

$$(2) \qquad 2q^2 = p^2.$$

Since p^2 is expressed as 2 times an integer, p^2 is an even integer. Whenever the square of an integer is even, the integer itself is even. Thus *p is even* and can be written $p = 2k$, where k is an integer. Substituting this result in (2), we obtain $2q^2 = (2k)^2$, or

$$(3) \qquad q^2 = 2k^2.$$

Thus q^2 is even and so *q is even.* Since p and q are both even, they contain the common factor 2. But this contradicts our original assumption. Therefore it is not true that $\sqrt{2}$ is a rational number. ∎*

It is also easy to prove that $\sqrt{3}$ and $\sqrt[3]{7}$ are irrational (that is, not rational), but it is difficult to prove that π and e are irrational. The latter are a more complicated kind of irrational number called **transcendental** numbers.

Throughout this book *the set of positive integers will be denoted by N, the set of all integers by Z, the set of rational numbers* (quotients of integers) *by Q,* and *the set of all real numbers by R.* Clearly,

$$N \subset Z \subset Q \subset R.$$

Any rational number can be written as a decimal, since by definition it can always be expressed as a ratio of integers; and if we divide the denominator into the numerator by long division, we obtain a decimal. For example, $\frac{1}{2} = 0.5$, $\frac{1}{25} = 0.04$, $\frac{8}{3} = 2.666666 \cdots$, $\frac{1}{11} = 0.09090909 \cdots$, $\frac{3}{7} = 0.428571\ 428571\ 428571 \cdots$.

Irrational numbers, too, can be expressed as decimals; for instance, $\sqrt{2} = 1.41421\ 35624 \cdots$, $\sqrt{3} = 1.73205\ 08076 \cdots$, $\pi = 3.14159\ 26535\ 89793 \cdots$, and $e = 2.71828\ 18284\ 59045 \cdots$.

The decimal representation of a **rational number** either terminates (that is, consists entirely of zeros to the right of some digit) or else repeats in regular cycles forever (see the examples above). A little experimenting with the long division process will show why. Conversely, any terminating or repeating decimal is a rational number. This is obvious in the case of the terminating decimal (for instance, $3.137 = 3137/1000$) and is easy to prove in the case of repeating decimals. Suppose, for example, that we have the repeating decimal $n = 0.136\ 136\ 136 \cdots$. Then $1000n = 136.136\ 136\ 136 \cdots$ and $1000n - n = 136$. Thus $n = 136/999$. This method applies to any repeating decimal.

It follows that no decimal that is nonterminating and nonrepeating is a rational number. The **irrational numbers** are the numbers whose decimal representations are nonterminating and nonrepeating. The **real numbers** consist of all numbers that have a decimal representation—that is, the rational numbers and the irrational numbers.

There are still other numbers, which the reader has encountered in previous courses, that are not *real*. For example, the roots of the equation $x^2 + 1 = 0$ are $\pm\sqrt{-1}$, and they are called **imaginary numbers**. Other imaginary numbers are $\sqrt{-7}$,

---

*The symbol ∎ will be used to denote the end of a proof.

$\sqrt[4]{-6}$, $2 - \sqrt{-1}$, and $13/\sqrt[8]{-5}$. All these numbers involve an indicated even root of a negative number. In fact, every imaginary number can be expressed in the form $a + b\sqrt{-1}$, where a and b are real numbers and $b \neq 0$. If $b = 0$, we, of course, have the real numbers. The **complex numbers** include the real and the imaginary numbers.

Unless there is a clear indication to the contrary, *the word "number" will always mean "real number"* in this book.

Calculus can be derived from a set of axioms for the real number system and from suitable definitions stated in terms of real numbers. We state briefly such a set of axioms. It is unnecessary to memorize any of them; generally, they are properties of real numbers that have long been familiar from one's experience in arithmetic and algebra.

Any two real numbers, x and y, may be added to form their **sum**, $x + y$, and multiplied to form their **product**, $x \cdot y$ (usually abbreviated xy).

The set of real numbers, R, and the two operations, **addition** and **multiplication**, are said to form a **field** because they obey the following seven **field axioms** (or axioms of arithmetic).

Let x, y, and z be real numbers.

F1 **Closure Laws.** $x + y$ and xy are unique real numbers.

F2 **Commutative Laws.** $x + y = y + x$ and $xy = yx$.

F3 **Associative Laws.** $x + (y + z) = (x + y) + z$ and $x(yz) = (xy)z$.

F4 **Distributive Law.** $x(y + z) = xy + xz$.

F5 **Identity Elements.** There are two distinct real numbers, 0 and 1, such that for each real number x, $x + 0 = x$ and $x \cdot 1 = x$.

F6 **Negatives.** For each real number x there is a real number $-x$ (read "the negative of x") such that $x + (-x) = 0$.

[*Caution.* The negative of x is not necessarily a negative number. For example, if $x = -4$, the negative of x is $-(-4) = 4$.]

F7 **Reciprocals.** For each real number except 0, there exists a real number x^{-1} such that $xx^{-1} = 1$.

The operation of **subtraction** is defined in terms of addition:

$$x - y = x + (-y).$$

Here $x - y$ is read "x minus y" and is the result of subtracting y from x.

The operation of **division** is defined in terms of multiplication. If $y \neq 0$,

$$\frac{x}{y} = xy^{-1}.$$

The left member ("x divided by y") is often written as $x \div y$ or x/y.

Notice that *division by zero* is undefined, and thus *never permissible*.

Many of the rules of arithmetic and elementary algebra can be derived from the preceding seven field axioms.

Example 1. Prove the cancellation law for addition—namely, if a, b, and c are real numbers and $a + b = a + c$, then $b = c$.

Proof. Assume that $a + b = a + c$. By substitution,

$$-a + (a + b) = -a + (a + c),$$

or

$$(-a + a) + b = (-a + a) + c, \qquad \text{(by F3)}$$

or

$$[a + (-a)] + b = [a + (-a)] + c, \qquad \text{(by F2)}$$

or

$$0 + b = 0 + c, \qquad \text{(by F6)}$$

or

$$b = c. \qquad \text{(F2 and F5)}$$

Example 2. Let $a, b \in R$. Then $ab = 0$ if and only if $a = 0$ or $b = 0$.

Proof. (i) Let $b = 0$. Then $a + ab = a \cdot 1 + a \cdot 0 = a(1 + 0) = a \cdot 1 = a = a + 0$ (by F4, F5, and substitution). Thus $a + ab = a + 0$ and therefore $ab = 0$ (by cancellation). The proof is similar if $a = 0$.

(ii) Let $ab = 0$. If $b \neq 0$, $b^{-1} \in R$ and $(ab)b^{-1} = a(bb^{-1}) = a \cdot 1 = a$ (by F3 and F7). Also, $(ab)b^{-1} = 0 \cdot b^{-1} = b^{-1} \cdot 0 = 0$ [by substitution, F2, and (i)]. Therefore $a = 0$.

Similarly, if $ab = 0$ and $a \neq 0$, then $b = 0$.

We cannot yet say that one number is greater (or less) than another number because the field axioms alone do not imply any ordering of the real numbers.

However, the real number system is also an **ordered** field because it contains a proper subset P (soon to be called the positive numbers) that obeys the following three **order axioms**.

O1 $1 \in P$ and $0 \notin P$.

O2 If $x, y \in P$, then $x + y$ and xy are numbers of P.

O3 If x is a real number, then one and only one of the following three statements is true:

$$x \in P, \qquad x = 0, \qquad -x \in P.$$

The numbers belonging to P are the **positive numbers**. A real number x is a **negative number** if and only if $-x \in P$. Zero is neither positive nor negative.

1.3.1 Definition. Let a and b be real numbers. Then $a < b$, read "a is **less than** b," if and only if $b - a$ is a positive number; and $a > b$, read "a is **greater than** b," if and only if $b - a$ is a negative number.

By way of illustration, $2 < 5$ because $5 - 2 = 3$ is a positive number; $-11 < -3$ because $-3 - (-11) = 8$ is a positive number; $4 > -1$ because $-1 - 4 = -5$ is a negative number; and $\frac{5}{9} > \frac{1}{3}$ because $\frac{1}{3} - \frac{5}{9} = -\frac{2}{9}$ is a negative number.

Clearly, $a < b$ *if and only if* $b > a$.

From 1.3.1 and O3, we have the following important corollary.

1.3.2 Law of Trichotomy. If a and b are real numbers, then one and only one of the following statements is true:
$$a < b, \qquad a = b, \qquad a > b.$$

The next theorem enables us to use the quickly written symbols "$a > 0$" instead of the longer "a is a positive number", and "$a < 0$" instead of "a is a negative number."

1.3.3 Theorem. $a > 0$ if and only if a is a positive number; and $a < 0$ if and only if a is a negative number.

Proof. $a > 0$ if and only if $0 < a$. But $0 < a$ if and only if $a - 0 = a$ is a positive number. Therefore $a > 0$ if and only if a is a positive number. The proof of the second part is analogous. ∎

The basic definition 1.3.1 can now be restated in briefer form.

1.3.4 Corollary. *Let* $a, b \in R$. Then $a < b$ if and only if $b - a > 0$, and $a > b$ if and only if $b - a < 0$.

We have seen that the real number system is an ordered field because it obeys the seven field axioms and the three order axioms. However, this fact is not enough to characterize the real number system completely. Its proper subset, the rational numbers, constitutes an ordered field that does not include such real numbers as $\sqrt{2}$ and π.

One more axiom is needed, the **axiom of completeness.** This will be a new concept for most of those using this book, and a thorough discussion of it will be postponed until needed (Section 13.6).

An ordered field whose elements obey the axiom of completeness is called a **complete ordered field.**

1.3.5 Definition. The real number system is a complete ordered field.

Exercises

1. Which of these numbers are rational? $\sqrt{4}$, 2.718, $1 + \sqrt{2}$, $(1 + \sqrt{3})^2$, $(1 + \sqrt{5})$, $(1 - \sqrt{5})$, 7/13, $\sqrt{2}/(3\sqrt{2})$, 0.3333 $\cdots$, $-2.6666 \cdots$, 1.034 034 034 $\cdots$, 6.19407 34820 90206 $\cdots$.

2. Express each of the following rational numbers as a quotient of integers. You can prove your answer, if you wish to, by dividing its denominator into its numerator.
(a) 0.7777 $\cdots$;
(b) 0.13 13 13 $\cdots$;
(c) $-0.027\ 027\ 027 \cdots$;
(d) 0.037 037 037 $\cdots$;
(e) 21. 62 037 037 037 $\cdots$; (*Hint:* Since 21.62 = 2162/100, the given number is equal to 2162/100 $+ x/100$, where $x = 0.037\ 037\ 037 \cdots$.)
(f) 2.179 40 40 40 $\cdots$.

3. Tell whether each of the statements below is true or false and use 1.3.5 to show why.
(a) $2 < -20$;
(b) $1 > -39$;
(c) $-3 < \dfrac{5}{9}$;
(d) $-4 > -16$;
(e) $\dfrac{6}{7} < \dfrac{34}{39}$;
(f) $-\dfrac{5}{7} < -\dfrac{44}{59}$.

4. Do the set Z (of all integers) and the operations of addition and multiplication constitute a field? If not, tell which field axioms Z fails to satisfy.

5. Find the value of each of the following; if undefined, say so.
(a) $0 + 0$;
(b) $0 \cdot 0$;
(c) $\dfrac{0}{0}$;
(d) $\dfrac{0}{8}$;
(e) $\dfrac{8}{0}$;
(f) $0 - 0$.

6. Let $a, b, c, d \in Z$. Show that if $b \neq 0$ and $d \neq 0$, then
(a) $\left(\dfrac{a}{b} + \dfrac{c}{d}\right) \in Q$;
(b) $\left(\dfrac{a}{b}\right)\left(\dfrac{c}{d}\right) \in Q$.
Is $\left(\dfrac{a}{b} \div \dfrac{c}{d}\right) \in Q$?

7. Prove that -1, the negative of 1, is a negative number.

8. Show by examples that if $x \in R$ and $x \neq 0$, then $-x$ (the negative of x) can be either a positive number or a negative number, depending on x.

1.4 THE COORDINATE LINE. INTERVALS

We can give geometric interpretation to all that has been said about real numbers by associating them with the points of a line.

Choose any two distinct points on a straight line and label them 0 (zero) and 1. The point 0 (zero) is called the **origin**, and we sometimes mark the origin with the letter O as well as with zero. We define the **positive direction** on the line to be *from* the point O *to* the point 1, and the length of the line segment terminated by O and 1 to be the **unit of length**; that is, we define the length of this segment to be 1.

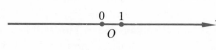

Figure 1-7

The line can be in any position, and the points O and 1 can be any two distinct points on it. However, for the present we shall follow the usual custom of depicting the line as horizontal with its positive direction to the right, as indicated by the arrowhead (Fig. 1-7).

One unit to the right of 1 we put the number 2; specifically, this means that we mark a point 2 on the line to the right of 1 so that the line segment between zero and 1 is congruent to the line segment between 1 and 2.

One unit to the right of 2 we put the number 3; and we continue in this way so that each positive integer, 1, 2, 3, 4, 5, $\cdots$, is the *coordinate* (label) of a unique point on the line.

One unit to the left of O we put the number -1; one unit to the left of -1 we put -2, and so on. In this way each integer—positive, negative, or zero—is the coordinate of a unique point on the line (Fig. 1-8).

Now consider any rational number, say $\frac{4}{5}$. Divide the line segment between 0 and 1 into five equal segments (Fig. 1-9). Lay off four such segments to the right of 0, and label the right-hand endpoint of the fourth segment with the coordinate $\frac{4}{5}$.

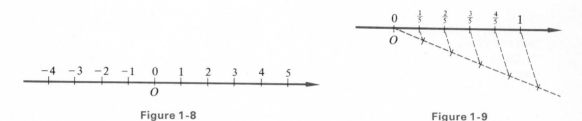

Figure 1-8

Figure 1-9

Similarly, each rational number becomes the coordinate of a unique point on the line.

This is a very extensive set of points. For, midway between any two of these points, no matter how close together, there is always another point with a rational number as label (Section 1.8, Exercise 23). It follows that between any two *rational points* on the line (that is, points with rational coordinates) there is an infinity of other rational points. We express this situation by saying that the set of rational points on the line is **everywhere dense**.

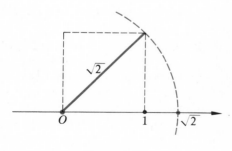

Figure 1-10

Yet despite the density of this set of points many points on the line are still without labels. For example, if we construct a square on the line segment between O and 1 (Fig. 1-10) and then draw a circular arc with center O and radius equal to the length of the diagonal of the square, the arc intersects the line in a point that cannot have a rational coordinate because the distance between O and this point is $\sqrt{2}$ (by the Pythagorean Theorem) and $\sqrt{2}$ is not a rational number.

To "fill the gaps" in the coordinate line—that is, to assign real numbers to the points on the line that do not have rational coordinates—we define a *coordinate line*.

First, however, some mathematical language must be explained. A **one-to-one correspondence** between the real numbers and the points on a line means that to each point on the line there corresponds one real number, its coordinate, and each real number is the coordinate of one and only one point on the line.

The symbol $P{:}(x)$ means the point P whose coordinate is x. When several points of the line are under discussion, we can designate them by $P_1{:}(x_1)$, $P_2{:}(x_2)$, and so on; read "P-one with coordinate x-one," and so on.

1.4.1 **Definition.** Consider a straight line in the Euclidean sense, and choose two distinct points, O and U, on it. Call the direction from O to U the positive direction on the line. This line is a **coordinate line** if it has the following properties.

(i) There is a one-to-one correspondence between the numbers of R and the points of the line, in which zero corresponds to the point O and 1 corresponds to the point U.

(ii) By the **distance** between any two points, $P_1{:}(x_1)$ and $P_2{:}(x_2)$, of the line, or the **length** of the line segment P_1P_2, is meant the nonnegative number

$$|P_1P_2| = |x_2 - x_1|.^*$$

(iii) Segments of the line that are geometrically congruent have equal lengths.

(iv) The direction from a point $P_1{:}(x_1)$ to a point $P_2{:}(x_2)$ on the line is positive if and only if $x_1 < x_2$.

If the coordinate line is horizontal with its positive direction to the right, $P_1{:}(x_1)$ is to the left of $P_2{:}(x_2)$ if and only if $x_1 < x_2$. Thus $P_1{:}(\frac{1}{2})$ is to the left of $P_2{:}(\sqrt{2})$ because $\frac{1}{2} < \sqrt{2}$ (Fig. 1-11).

*The **directed distance** from $P_1{:}(x_1)$ to $P_2{:}(x_2)$ is defined by*

$$\overline{P_1P_2} = x_2 - x_1.$$

It follows from 1.4.1(iv) that $\overline{P_1P_2}$ is positive if and only if the direction on the coordinate line *from P_1 to P_2* is positive. Moreover,

$$\overline{P_1P_2} = -\overline{P_2P_1}.$$

Let $P_1{:}(x_1)$ and $P_2{:}(x_2)$ be two points on a coordinate line; $P{:}(x)$ is the **midpoint** of the segment P_1P_2 if and only if

(1) $$\overline{P_1P} = \overline{PP_2}$$

(Fig. 1-12). But the directed distance $\overline{P_1P} = x - x_1$, and $\overline{PP_2} = x_2 - x$. Thus (1) is equivalent to $x - x_1 = x_2 - x$, or

$$x = \frac{x_1 + x_2}{2}.$$

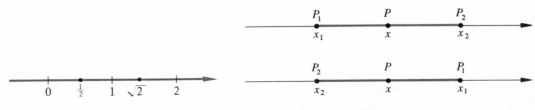

Figure 1-11

Figure 1-12

*The reader who is not familiar with the concept of *absolute value* is referred to 1.6.1.

1.4.2 Theorem.   The coordinate of the midpoint, $P:(x)$, of the line segment whose endpoints are $P_1:(x_1)$ and $P_2:(x_2)$ is

$$x = \frac{x_1 + x_2}{2}.$$

We conclude this section with a brief discussion of **intervals**.

Notation.   Let $a, b, c \in R$.
$a \leq b$ means that either $a < b$ or else $a = b$.
$a < b < c$ means that $a < b$ and $b < c$.
$a \leq b < c$ means that $a \leq b$ and $b < c$.

The number b is said to be **between** the numbers a and c if and only if either $a < b < c$ or $c < b < a$.

1.4.3 Definition.   Let $a, b \in R$ with $a < b$.
    (i)   The **open interval** (a, b) is defined by

$$(a, b) = \{x \mid a < x < b\}.$$

    (ii)   The **closed interval** $[a, b]$ is

$$[a, b] = \{x \mid a \leq x \leq b\}.$$

    (iii)   The **half-open interval** $[a, b)$ is

$$[a, b) = \{x \mid a \leq x < b\},$$

and similarly for $(a, b]$.

Thus $\{x \mid x \in R, -9 \leq x < 4\}$, for example, can be written more briefly as $[-9, 4)$.

Intervals can be represented geometrically on a coordinate line (Fig. 1-13). The open interval (a, b) can be pictured as the set of all points *between* the points whose coordinates are a and b but not including those endpoints. Incidentally, *we shall often say "the point x" when we mean either the point whose coordinate is x, or the real number x itself.* Similarly, we shall often refer to a set of real numbers as a set of points (on a coordinate line).

Thus, the closed interval $[a, b]$ may be said to consist of the endpoints a and b, and all points between a and b (Fig 1-13). The half-open interval $[a, b)$ contains the point a and all the points between a and b, but not the endpoint b.

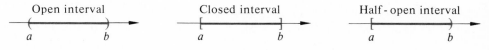

| Open interval | Closed interval | Half-open interval |

Figure 1-13

It will be convenient to denote by $[a, \infty)$ the set of real numbers greater than or equal to a; that is

$$[a, \infty) = \{x \,|\, x \in R,\ x \geq a\}.$$

A half-line

Figure 1-14

The "interval" $[a, \infty)$ starts with the point a and extends indefinitely to the right (Fig. 1-14). It is called a **half-line**.

Similarly, $(-\infty, a] = \{x \,|\, x \leq a\}$ is a half-line extending indefinitely to the left.

1.5   INEQUALITIES

We recall from 1.3.3 that $a > 0$ may be read "a is a positive number" and $a < 0$ may be read "a is a negative number."

For convenient reference in reading this section, we repeat the following definition and corollary from Section 1.3.

1.5.1   Definition.   Let $a, b \in R$. Then $a < b$ if and only if $b - a > 0$, and $a > b$ if and only if $b - a < 0$.

1.5.2   Law of Trichotomy.   If a and b are real numbers, then one and only one of the following statements is true:

$$a < b, \qquad a = b, \qquad a > b.$$

Statements in the form $a < b$ and $c > d$ are called **inequalities**. For reasons that will become apparent in Chapter 3, calculus has been called a science of inequalities. This is probably an overstatement, but it is certainly true that we must have some facility in working with inequalities.

The inequality $2x - 3 < 10$ is true for some real numbers x and false for others. For example, if 4 is substituted for x, the inequality is $8 - 3 < 10$, which is a true statement. But if 9 is substituted for x, the resulting inequality, $18 - 3 < 10$, is false.

The **solution set** of an inequality in x, say, is the set of all real numbers that, when substituted for x, make the given inequality a true statement.

For instance, the solution set of the inequality $x - 1 > 4$ is the set of all numbers x such that $x > 5$.

To **solve an inequality** is to find its solution set. In order to solve inequalities, we need rules comparable to the transposition and cancellation laws for equations. The most basic of such rules are given in the next four theorems.

1.5.3   Transitive Law of Inequality.   Let a, b, and c be real numbers. If $a < b$ and $b < c$, then $a < c$.

Proof. Since $a < b$ and $b < c$, both $b - a$ and $c - b$ are positive numbers. The sum of two positive numbers is a positive number, and so $(b - a) + (c - b) > 0$, or $c - a > 0$. Therefore $a < c$. Hence if $a < b$ and $b < c$, then $a < c$. ∎

1.5.4   Theorem. Let $a, b, c \in R$. Then $a < b$ if and only if $a + c < b + c$.

Proof. $a < b$ if and only if $b - a > 0$ (by 1.5.1). But $b - a = (b + c) - (a + c)$. Therefore $a < b$ if and only if $(b + c) - (a + c) > 0$—that is, if and only if $a + c < b + c$. ∎

This theorem enables us to add any number to, or subtract any number from, both members of an inequality without changing the direction of the inequality. To illustrate, $2 < 7$ implies $2 + 5 < 7 + 5$, and vice versa.

1.5.5   Theorem. Let $a, b, c, d \in R$. If $a < b$ and $c < d$, then $a + c < b + d$.

Proof. $b - a > 0$ and $d - c > 0$. Thus $(b - a) + (d - c) > 0$. But this may be written $(b + d) - (a + c) > 0$, which implies that $a + c < b + d$ (by 1.5.1). ∎

1.5.6   Theorem. Let $a, b, c \in R$.
(i)   When c is a positive number, $a < b$ if and only if $ac < bc$.
(ii)   When c is a negative number, $a < b$ if and only if $ac > bc$.

In other words, we can multiply or divide both members of an inequality by the same *positive* number without changing the direction of the inequality; but if we multiply both members of an inequality by the same *negative* number, we reverse the direction of the inequality. As an illustration, $-2 < 3$ if and only if $-2(11) < 3(11)$; but $-2 < 3$ if and only if $-2(-5) > 3(-5)$.

Proof of 1.5.6(i).   Assume that $c > 0$ and $a < b$. Then both c and $b - a$ are positive, and so $c(b - a) > 0$. But $c(b - a) = bc - ac$. Therefore $bc - ac > 0$ and $ac < bc$. We have proved that if $c > 0$ and $a < b$, then $ac < bc$.

Next, assume that $c > 0$ and $ac < bc$. If $a = b$, then $ac = bc$; and if $a > b$, then $ac > bc$. Since both conclusions contradict our assumption that $ac < bc$, neither $a = b$ nor $a > b$ is true. Therefore, by 1.5.2, $a < b$. ∎

The proof of 1.5.6(ii) is similar and is left for the reader.

In addition to the preceding four basic rules, a few more, given below, are worth knowing. Their proofs are easy and are also left for the reader.

1.5.7   Theorem. If $a \neq 0$, then $a^2 > 0$.

1.5.8   Theorem. If $a \neq 0$, $1/a$ has the same sign as a.

1.5.9 Theorem. If a and b have the same sign, and if $a < b$, then $1/a > 1/b$.

1.5.10 Theorem. Let $a < b$. Then, if a and b are both positive, $a^2 < b^2$; and, if a and b are both negative, $a^2 > b^2$.

Example 1. Solve the inequality $2x - 7 < x + 4$.

Solution. Solving this inequality means finding its solution set—that is, finding the set of all numbers that, when substituted for x, make the given inequality a true statement.

Let x be any number for which $2x - 7 < x + 4$. By adding 7 to both members of this inequality, we obtain $2x < x + 11$, and this latter inequality is true for a number x if and only if the given inequality is true for that same number x (1.5.4). Now add $-x$ to both members of $2x < x + 11$. We get $x < 11$, and this inequality is true for a number x if and only if the original inequality is true for that same number x (1.5.4). Therefore the solution set for the given inequality is $(-\infty, 11)$. Its graph is shown in Fig. 1-15.

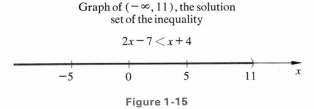

Graph of $(-\infty, 11)$, the solution
set of the inequality

$$2x - 7 < x + 4$$

Figure 1-15

Notation. If A and B are statements, the symbols $A \implies B$ mean "if A is true, then B is true" or, what is the same thing, "A implies B." The symbols $B \impliedby A$ mean "B is implied by A" or "B is true if A is true." The symbols $A \iff B$ mean "A is true if and only if B is true" or "A implies B and B implies A."

Two statements, A and B, are said to be **equivalent** if and only if A implies B and B implies A. So $A \iff B$ can also be read "A is equivalent to B."

In Example 1 above we can rewrite the statement "the inequality $2x - 7 < x + 4$ is true for a number x if and only if $x < 11$" in the shorter form "$2x - 7 < x + 4 \iff x < 11$." This statement is usually read "$2x - 7 < x + 4$ if and only if $x < 11$" or "$2x - 7 < x + 4$ is equivalent to $x < 11$."

Example 2. Solve the inequalities $-5 < 2x + 6 < 4$.

Solution. $-5 < 2x + 6 < 4 \iff -5 < 2x + 6$ and $2x + 6 < 4$.
But

$$-5 < 2x + 6 \iff -11 < 2x \iff \frac{-11}{2} < x;$$

and

$$2x + 6 < 4 \iff 2x < -2 \iff x < -1.$$

Therefore

$$-5 < 2x + 6 < 4 \iff \frac{-11}{2} < x \text{ and } x < -1.$$

Thus the solution set of the given inequalities is $(-\frac{11}{2}, \infty) \cap (-\infty, -1) = (-\frac{11}{2}, -1)$. It consists of all numbers between $-\frac{11}{2}$ and -1 (Fig. 1-16).

The open interval $(-11/2, -1)$

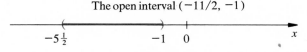

Figure 1-16

Example 3.   Find the solution set of the inequalities

$$-3 < \frac{2}{3x - 9} < -1, \qquad x \neq 3.$$

Solution.   There are two cases to be considered: $x > 3$ and $x < 3$.
     (a) Assume that $x > 3$. Then $3x - 9 > 0$, and

$$-3 < \frac{2}{3x - 9} < -1 \iff -3(3x - 9) < 2 < -(3x - 9)$$

$$\iff -9x + 27 < 2 \quad \text{and} \quad 2 < -3x + 9$$

$$\iff \frac{25}{9} < x \quad \text{and} \quad x < \frac{7}{3}$$

$$\iff \frac{25}{9} < x < \frac{7}{3},$$

which is impossible, since $\frac{25}{9} > \frac{7}{3}$.
     (b) Assume that $x < 3$. Then $3x - 9 < 0$, and

$$-3 < \frac{2}{3x - 9} < -1 \iff -3(3x - 9) > 2 > -(3x - 9)$$

$$\iff -9x + 27 > 2 \quad \text{and} \quad 2 > -3x + 9$$

$$\iff x < \frac{25}{9} \quad \text{and} \quad \frac{7}{3} < x$$

$$\iff \frac{7}{3} < x < \frac{25}{9}.$$

Therefore the solution set of the given inequalities is $(\frac{7}{3}, \frac{25}{9})$.

Example 4.   Solve the quadratic inequality $3x^2 - x - 2 < 0$.

Solution. $3x^2 - x - 2 < 0 \iff (x - 1)(3x + 2) < 0$. Thus $x - 1$ and $3x + 2$ must differ in sign.
     (a) *Assume that $x - 1 > 0$ and $3x + 2 < 0$.*
     $x - 1 > 0$ and $3x + 2 < 0 \iff x > 1$ and $x < -\frac{2}{3}$. But there exists no number x such that x is greater than 1 and at the same time less than $-\frac{2}{3}$.
     (b) *Assume that $x - 1 < 0$ and $3x + 2 > 0$.*
     $x - 1 < 0$ and $3x + 2 > 0 \iff x < 1$ and $x > -\frac{2}{3}$. Therefore the solution set for the given quadratic inequality is the open interval $(-\frac{2}{3}, 1)$. See Fig. 1-17.

The open interval $(-2/3, 1)$

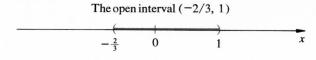

Figure 1-17

Exercises

1. Show by a drawing each of the following intervals or half-lines.
(a) $(-4, 1)$; (b) $[-4, 1]$;
(c) $(-4, 1]$; (d) $[-4, 1)$;
(e) $[1, \infty)$; (f) $(-\infty, -4]$.

2. Use the notation of Exercise 1 to describe the intervals below.

(a)

(b)

(c)

(d)

In Exercises 3 to 30, find the solution set of the given inequalities and sketch its graph.

3. $4x - 7 < 3x + 5$.

4. $2x + 16 < x + 25$.

5. $7x - 1 < 10x + 4$.

6. $6x - 10 > 5x - 16$.

7. $10x + 1 > 8x + 5$.

8. $3x + 5 > 7x + 17$.

9. $-6 < 2x + 3 < -1$.

10. $-3 < 4x - 9 < 11$.

11. $-2 < 1 - 5x < 3$.

12. $4 < 5 - 3x < 7$.

13. $2 + 3x < 5x + 1 < 16$.

14. $2x - 4 < 6 - 7x < 3x + 6$.

15. $3x + 6 < x + 4 < 2x - 1$.

16. $-x - 8 < 4x + 1 < 3x - 3$.

17. $\dfrac{1}{x} < 5$.

18. $\dfrac{7}{2x} < 3$.

19. $\dfrac{1}{3x - 2} < 4$.

20. $\dfrac{3}{x + 5} > 2$.

21. $-4 < \dfrac{1}{2x + 5} < -1$.

22. $2 < \dfrac{1}{x - 3} < 8$.

23. $-3 < 4 - \dfrac{1}{x} < 2$.

24. $4 < \dfrac{3}{7x - 1} < 8$.

25. $3x^2 - 11x - 4 < 0$.

26. $2x^2 + 7x - 15 > 0$.

27. $x^2 - 5x + 6 > 0$.

28. $2x^2 + 7x + 3 < 0$.

29. $\dfrac{1}{2x} + \dfrac{9}{6x + 4} < 3$.

30. $\dfrac{1}{x} - \dfrac{1}{3x + 1} < 2$.

31. Prove 1.5.7.

32. Prove 1.5.8.

33. Prove 1.5.9.

1.6 ABSOLUTE VALUE

The concept of *absolute value* is extremely useful in calculus and the reader should acquire skill in working with it.

1.6.1 Definition. The **absolute value** of a real number x, denoted by $|x|$, is defined by

$$|x| = x \quad \text{if} \quad x \geq 0,$$
$$|x| = -x \quad \text{if} \quad x < 0.$$

For instance, $|5| = 5$, $|0| = 0$, and $|-2| = -(-2) = 2$. For each real number x, $0 \leq |x|$; and either $x = |x|$ or $x = -|x|$.

1.6.2 Corollary. If $x \in R$, then

$$-|x| \leq x \leq |x|.$$

The notation for absolute value was used in our definition, 1.4.1(ii), of the **length** of a line segment whose endpoints are $P_1:(x_1)$ and $P_2:(x_2)$—namely,

$$|P_1 P_2| = |x_2 - x_1|.$$

We recall from algebra that the symbol $\sqrt{a}$, where $a \geq 0$, means the *nonnegative* number whose square is a. Thus $\sqrt{4} = 2$, *not* ± 2; and $-\sqrt{4} = -2$.

1.6.3 Corollary. For any real number x,

$$\sqrt{x^2} = |x| \quad \text{and} \quad |x|^2 = x^2.$$

To illustrate, $\sqrt{6^2} = |6| = 6$, $\sqrt{(-6)^2} = |-6| = 6$, and $|-7|^2 = (-7)^2 = 49$.

Some basic properties of the absolute value concept are given in the theorems that follow.

1.6.4 Theorem. If a and b are real numbers,

$$|a| \cdot |b| = |ab|.$$

Proof. By 1.6.3, $|a| \cdot |b| = \sqrt{a^2} \sqrt{b^2} = \sqrt{a^2 b^2} = |ab|$. ∎

If $b \neq 0$, $|b| \cdot |a/b| = |b(a/b)| = |a|$. Thus 1.6.4 has the following corollary.

1.6.5 Corollary. If $a, b \in R$ and $b \neq 0$, then

$$\frac{|a|}{|b|} = \left| \frac{a}{b} \right|.$$

The next theorem and its corollary will be used frequently.

1.6.6 Theorem. Let $a > 0$. Then

$$|x| < a \iff -a < x < a,$$

and

$$|x| > a \iff x > a \quad \text{or} \quad x < -a.$$

For example, $|x| < 4 \iff -4 < x < 4$, which is the open interval $(-4, 4)$ whose midpoint is the origin [Fig. 1-18(a)]; and $|x| > 4 \iff x < -4$ or $x > 4$, which consists of the half-lines $(-\infty, -4)$ and $(4, \infty)$ [Fig. 1-18(b)].

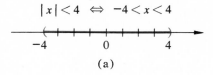

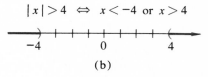

(a)

(b)

Figure 1-18

Proof of 1.6.6. We shall prove the first part of 1.6.6—namely, that $|x| < a \iff -a < x < a$—and leave the proof of the second part for the reader.

We must show that $|x| < a \implies -a < x < a$, and that $-a < x < a \implies |x| < a$. It will then follow that $|x| < a \iff -a < x < a$.

(i) *Assume that* $|x| < a$. Since $a > 0$, $-a < 0$. If $0 \leq x$, then $|x| = x$ and $-a < 0 \leq x < a$; thus $-a < x < a$.

If $x < 0$, then $|x| = -x$, and $-a < 0 < -x < a$—from which $-a < -x < a$ or, if we multiply through by -1, $a > x > -a$. But this is equivalent to $-a < x < a$.

Therefore $|x| < a \implies -a < x < a$.

(ii) *Assume that* $-a < x < a$.

If $0 \leq x$, then $|x| = x < a$. Thus $|x| < a$.

If $x < 0$, then $|x| = -x$, and since $-a < x$, we have $-x < a$, or $|x| < a$.

Therefore, for any x, $-a < x < a \implies |x| < a$. ∎

1.6.7 Corollary. If $a > 0$ and if b and x are real numbers, then

$$|x - b| < a \iff -a < x - b < a$$

or, equivalently,

$$|x - b| < a \iff b - a < x < b + a.$$

The inequality $|x - b| < a$ says the "x differs from the number b by less than a."

Thus (Fig. 1-19) the set $\{x \mid |x - b| < a\}$ is the open interval $(b - a, b + a)$ whose midpoint is b; in particular, when $b = 0$, $\{x \mid |x| < a\}$ is the open interval $(-a, a)$ whose midpoint is the origin. It then follows from 1.4.1(ii) that on a coordinate line

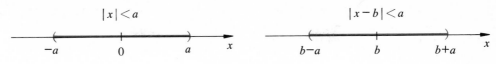

Figure 1-19

$|x - b|$ is the distance between the point x and the point b, and that $|x|$ is the distance between x and the origin.

The preceding theorem and its corollary (1.6.7) enable us to replace $|x - 3| < 0.4$, say, by $3 - 0.4 < x < 3 + 0.4$ or $2.6 < x < 3.4$, and vice versa, whenever it suits our purpose. The importance of this fact will be seen in the proof of the next theorem and in the solutions of the seven examples that conclude this section.

1.6.8 **Theorem** (*Triangle Inequality*).   If a and b are any real numbers,
$$|a + b| \le |a| + |b|.$$
Proof.   From 1.6.2, $-|a| \le a \le |a|$ and $-|b| \le b \le |b|$. Hence (by 1.5.5)
$$-(|a| + |b|) \le a + b \le (|a| + |b|),$$
or (by 1.6.6)
$$|a + b| \le |a| + |b|. \quad \blacksquare$$

1.6.9 **Corollary.**   If a and b are any real numbers, then
$$|a - b| \le |a| + |b|.$$

Proof. $|a - b| = |a + (-b)| \le |a| + |-b| = |a| + |b|. \quad \blacksquare$

1.6.10 **Corollary.**   If a and b are any real numbers, then
$$|a| - |b| \le |a - b|.$$

Proof. $|a| = |(a - b) + b| \le |a - b| + |b|.$ Therefore
$$|a| - |b| \le |a - b|. \quad \blacksquare$$

Example 1.   Solve the inequality $|x - 4| < 2$.

Solution.   By 1.6.7, $|x - 4| < 2 \iff -2 < x - 4 < 2 \iff 2 < x < 6$. Thus the solution set of the given inequality is $\{x \mid 2 < x < 6\}$—that is, the open interval $(2, 6)$. Its graph is shown in Fig. 1-20.

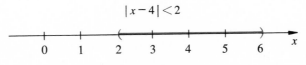

Figure 1-20

Example 2. Solve the inequality $|2x + 3| < 5$.

Solution. By 1.6.7, $|2x + 3| < 5 \iff -5 < 2x + 3 < 5 \iff -8 < 2x < 2 \iff -4 < x < 1$. Therefore the solution set is the open interval $(-4, 1)$.

1.6.11 Theorem. If $a > 0$ and if b and x are real numbers, then

$$|x - b| > a \iff x - b < -a \quad \text{or} \quad x - b > a,$$

which can be written

$$|x - b| > a \iff x < b - a \quad \text{or} \quad x > b + a.$$

Proof. From the law of trichotomy (1.5.2), for each real number x either $|x - b| > a$ or $|x - b| \le a$. Thus all real numbers *not* in the solution set of $|x - b| \le a$ form the solution set of $|x - b| > a$. Since

$$|x - b| \le a \iff b - a \le x \le b + a$$

(by 1.6.7), the solution set of $|x - b| \le a$ is the closed interval

$$[b - a, b + a].$$

Therefore the solution set of $|x - b| > a$ is

$$(-\infty, b - a) \cup (b + a, \infty).$$

But $(-\infty, b - a)$ is the solution set of the inequality $x < b - a$, and $(b + a, \infty)$ is the solution set of $x > b + a$. Therefore $|x - b| > a \iff x < b - a$ or $x > b + a$. ∎

Example 3. Solve the inequality $|3x - 5| > 1$.

Solution. By 1.6.11,

$$|3x - 5| > 1 \iff 3x - 5 < -1 \text{ or } 3x - 5 > 1.$$

But $3x - 5 < -1 \iff x < \frac{4}{3}$, and $3x - 5 > 1 \iff x > 2$. Therefore the solution set of $|3x - 5| > 1$ is

$$(-\infty, \tfrac{4}{3}) \cup (2, \infty)$$

(see Fig. 1-21).

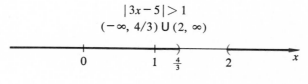

Figure 1-21

Two sets of real numbers whose intersection is $\varnothing$ and whose union is R are said to be **complements** of each other with respect to R. Thus the solution set of $|3x - 5| > 1$ is the complement with respect to R of the solution set of $|3x - 5| \le 1$.

Example 4. Show that

$$|x - 4| < 0.2 \iff 11 - 0.4 < 2x + 3 < 11 + 0.4.$$

Solution. By 1.6.7,

$$|x - 4| < 0.2 \iff -0.2 < x - 4 < 0.2$$
$$\iff 4 - 0.2 < x < 4 + 0.2.$$

But

$$4 - 0.2 < x < 4 + 0.2 \iff 8 - 0.4 < 2x < 8 + 0.4$$
$$\iff 11 - 0.4 < 2x + 3 < 11 + 0.4$$

by 1.5.6 and 1.5.4. Therefore

$$|x - 4| < 0.2 \iff 11 - 0.4 < 2x + 3 < 11 + 0.4.$$

Example 5. Let ϵ (epsilon) be a positive number. Show that

$$|x - 2| < \frac{\epsilon}{5} \implies 6 - \epsilon < 5x - 4 < 6 + \epsilon.$$

Solution. By 1.6.7,

$$|x - 2| < \frac{\epsilon}{5} \iff -\frac{\epsilon}{5} < x - 2 < \frac{\epsilon}{5}$$
$$\iff 2 - \frac{\epsilon}{5} < x < 2 + \frac{\epsilon}{5}.$$

But

$$2 - \frac{\epsilon}{5} < x < 2 + \frac{\epsilon}{5} \iff 10 - \epsilon < 5x < 10 + \epsilon$$
$$\iff 6 - \epsilon < 5x - 4 < 6 + \epsilon$$

by 1.5.6 and 1.5.4. Therefore

$$|x - 2| < \frac{\epsilon}{5} \iff 6 - \epsilon < 5x - 4 < 6 + \epsilon.$$

Example 6. Let ϵ be an arbitrarily chosen positive number. Find a positive number δ (delta) such that

$$|x - 2| < \delta \implies 2 - \epsilon < 4x - 6 < 2 + \epsilon.$$

Preliminary Analysis. By using 1.5.4, 1.5.6, and 1.6.7, we can write

$$2 - \epsilon < 4x - 6 < 2 + \epsilon \iff 8 - \epsilon < 4x < 8 + \epsilon$$
$$\iff 2 - \frac{\epsilon}{4} < x < 2 + \frac{\epsilon}{4}$$
$$\iff |x - 2| < \frac{\epsilon}{4},$$

which suggests that if we let $\delta = \epsilon/4$ and reverse the sequence of steps in this analysis, we will show that

$$|x - 2| < \delta = \frac{\epsilon}{4} \implies 2 - \epsilon < 4x - 6 < 2 + \epsilon.$$

Solution to Example 6. Let $\delta = \epsilon/4$. Then

$$|x - 2| < \delta \iff |x - 2| < \frac{\epsilon}{4}$$
$$\iff 2 - \frac{\epsilon}{4} < x < 2 + \frac{\epsilon}{4}$$
$$\iff 8 - \epsilon < 4x < 8 + \epsilon$$
$$\iff 2 - \epsilon < 4x - 6 < 2 + \epsilon.$$

Therefore, if $\delta = \epsilon/4$,

$$|x - 2| < \delta \iff 2 - \epsilon < 4x - 6 < 2 + \epsilon.$$

Example 7. Solve the inequality $|3x + 1| < 2|x - 6|$.

Solution. Since the absolute value of a number is always nonnegative, $0 \le |3x + 1| < 2|x - 6|$. Thus $|x - 6| > 0$, and the given inequality can be written

$$\left|\frac{3x + 1}{x - 6}\right| < 2.$$

By 1.6.6, this result is equivalent to

(1) $$-2 < \frac{3x + 1}{x - 6} < 2.$$

Because $|x - 6| > 0$, $x \ne 6$. Thus two cases must be considered, depending on whether $x > 6$ or $x < 6$.

(a) *Assume that $x > 6$.* Then $x - 6 > 0$, and by multiplying all three members of (1) by $x - 6$, we obtain

$$-2x + 12 < 3x + 1 < 2x - 12,$$

which is equivalent to

$$-2x + 12 < 3x + 1 \quad \text{and} \quad 3x + 1 < 2x - 12,$$

or

$$x > \tfrac{11}{5} \quad \text{and} \quad x < -13.$$

But this result is impossible, and so x cannot be greater than 6.

(b) *Assume that $x < 6$.* Then $x - 6 < 0$, and by multiplying (1) through by $x - 6$, we obtain

$$-2(x - 6) > 3x + 1 > 2(x - 6),$$

from which

$$x < \tfrac{11}{5} \quad \text{and} \quad -13 < x \quad \text{and} \quad x < 6,$$

which may be rewritten

$$(-\infty, \tfrac{11}{5}) \cap (-13, \infty) \cap (-\infty, 6).$$

But $(-\infty, \tfrac{11}{5}) \cap (-\infty, 6) = (-\infty, \tfrac{11}{5})$. Therefore the solution set of the given inequality is

$$(-\infty, \tfrac{11}{5}) \cap (-13, \infty) = (-13, \tfrac{11}{5}).$$

Exercises

In Exercises 1 to 12, find the solution sets of the given inequalities.

1. $|x + 1| < 4$.
2. $|x - 2| < 5$.
3. $|3x + 4| < 8$.
4. $|2x - 7| < 3$.
5. $\left|\frac{x}{3} - 2\right| < 6$.
6. $\left|\frac{3x}{5} + 1\right| < 4$.

7. $|2x - 7| > 3$. (*Hint:* Find the solution set of $|2x - 7| \le 3$. Its complementary set with respect to R is the solution set of $|2x - 7| > 3$.)
8. $|5x - 6| > 1$.
9. $|4x + 2| > 10$.
10. $\left|\frac{x}{2} + 7\right| > 2$.
11. $\left|2 + \frac{5}{x}\right| > 1$.
12. $\left|\frac{1}{x} - 3\right| > 6$.

13. Show that $|x - 2| < 0.5 \implies 7 - 1.5 < 3x + 1 < 7 + 1.5$.

14. Show that $|x - 3| < 2 \implies 1 - 4 < 2x - 5 < 1 + 4$.

15. Show that $|x - 4| < 0.06 \implies 7 - 0.03 < \frac{1}{2}x + 5 < 7 + 0.03$.

16. Show that $|x - 6| < 0.1 \implies 3 - 0.5 < 5x - 27 < 3 + 0.5$.

17. Let ϵ (epsilon) be a positive number. Show that

$$|x - 5| < \frac{\epsilon}{10} \implies 14 - \epsilon < 10x - 36 < 14 + \epsilon.$$

18. Let ϵ be an arbitrarily chosen positive number. Show that

$$|x - 1| < \frac{\epsilon}{2} \implies 3 - \epsilon < 2x + 1 < 3 + \epsilon.$$

19. Find a positive number δ (delta) such that

$$|x + 2| < \delta \implies 4.8 < 5x + 15 < 5.2.$$

(*Hint:* $4.8 < 5x + 15 < 5.2 \iff 5 - 0.2 < 5x + 15 < 5 + 0.2$)

20. Find a positive number δ such that

$$|x - 3| < \delta \implies 6.9 < 4x - 5 < 7.1.$$

21. Let ϵ be a positive number. Find a positive number δ such that

$$|x - 5| < \delta \implies 3 - \epsilon < 2x - 7 < 3 + \epsilon.$$

22. Let ϵ be a positive number. Find a positive number δ such that

$$|x - 4| < \delta \implies 18 - \epsilon < 3x + 6 < 18 + \epsilon.$$

23. Solve the inequality $|x - 2| < 3|x + 7|$.

24. Solve the inequality $|2x - 5| < |x + 4|$.

25. Solve the inequality $2|2x - 3| < |x + 10|$.

26. If a, b, and c are any real numbers, prove that

$$|a + b + c| \le |a| + |b| + |c|.$$

1.7 INDUCTION

We saw in Section 1.3 that the set of positive integers (natural numbers) is

$$N = \{1, 2, 3, 4, \cdots\},$$

in which 1 is the number postulated in F5, $2 = 1 + 1$, $3 = 2 + 1$, $4 = 3 + 1$, $\cdots$.

The reader has undoubtedly used a method of proof called **mathematical induction** in several previous courses. The idea behind it was as follows. In order to prove that a formula (or theorem) about positive integers was true for all positve integers, we first verified (usually by actual substitution) that the formula was true for the number 1. Then we proved that whenever the formula was true for a positive integer n, it must also be true for $n + 1$. Since it was verified to be true for 1, it had to be true for 2; since it was true for 2, it had to be true for 3; and so on for all positive integers.

The validity of this method of proof by mathematical induction is based on the following axiom.

1.7.1   Axiom of Induction.   If S is a subset of the set of positive integers, N, such that

   (i) $1 \in S$, and
   (ii) $k \in S$ implies $(k + 1) \in S$, then $S = N$.

Its use in proving theorems that involve positive integers is shown in the following examples. The reader should fix the *pattern* of these proofs in his mind.

Example 1. Use mathematical induction to prove that the formula

(1)
$$1 + 2 + 3 + \cdots + n = \frac{n(n+1)}{2}$$

is true for every positive integer n.

Proof. Let S be the set of all positive integers for which the formula (1) is true. Clearly, $1 \in S$ because $1 = 1(1+1)/2 = 1$ is true.

If $k \in S$—that is, if (1) is true for a positive integer k—then

$$1 + 2 + 3 + \cdots + k = \frac{k(k+1)}{2}.$$

By adding $k + 1$ to both members of this equation, we obtain

$$1 + 2 + 3 + \cdots + k + (k+1) = \frac{k(k+1)}{2} + (k+1)$$

$$= \frac{k(k+1) + 2(k+1)}{2} = \frac{(k+1)(k+2)}{2},$$

which is formula (1) when $n = (k+1)$. Thus if (1) is true for $n = k$, it is also true for $n = (k+1)$. That is, $k \in S$ implies that $(k+1) \in S$. Therefore (by 1.7.1) $S = N$. This proves that formula (1) is true for every positive integer n.

Example 2. Prove that the formula

(2)
$$1^2 + 2^2 + 3^2 + \cdots + n^2 = \frac{n(n+1)(2n+1)}{6}$$

is true for every positive integer n.

Proof. Let S be the set of positive integers for which the formula (2) is true. Then $1 \in S$ because $1^2 = 1(1+1)(2 \cdot 1 + 1)/6 = 1$ is true.

If $k \in S$,

$$1^2 + 2^2 + 3^2 + \cdots + k^2 = \frac{k(k+1)(2k+1)}{6}.$$

By adding $(k+1)^2$ to both members of this equation, we have

$$1^2 + 2^2 + 3^2 + \cdots + k^2 + (k+1)^2 = \frac{k(k+1)(2k+1)}{6} + (k+1)^2$$

$$= \frac{k(k+1)(2k+1) + 6(k+1)^2}{6}$$

$$= \frac{(k+1)(2k^2 + 7k + 6)}{6} = \frac{(k+1)(k+2)(2k+3)}{6},$$

which is formula (2) when $n = k + 1$. Thus if (2) is true for $n = k$, it is also true for $n = k + 1$. Therefore $k \in S \implies (k+1) \in S$ and, by 1.7.1, $S = N$. That is, formula (2) is true for all positive integers n.

Example 3. Prove that the formula

(3)
$$1 \cdot 2 + 2 \cdot 3 + 3 \cdot 4 + \cdots + n \cdot (n+1) = \tfrac{1}{3}n(n+1)(n+2)$$

is true for every positive integer n.

Proof. Let S be the set of positive integers for which formula (3) is true. Then $1 \in S$ because $1 \cdot 2 = (\tfrac{1}{3})(1)(1+1)(1+2) = 2$.

If $k \in S$,
$$1 \cdot 2 + 2 \cdot 3 + \cdots + k(k + 1) = \tfrac{1}{3}k(k + 1)(k + 2).$$

By adding $(k + 1)(k + 2)$ to both members, we obtain
$$1 \cdot 2 + 2 \cdot 3 + \cdots + k(k + 1) + (k + 1)(k + 2) = \tfrac{1}{3}k(k + 1)(k + 2) + (k + 1)(k + 2)$$
$$= \tfrac{1}{3}(k + 1)(k + 2)(k + 3),$$

which is formula (3) when $n = (k + 1)$. Thus if (3) is true for $n = k$, it is also true for $k + 1$. Therefore $k \in S \implies k + 1 \in S$ and, by 1.7.1, $S = N$. That is, formula (3) is true for all positive integers n.

Exercises

1. Prove by mathematical induction that
$$1 + 3 + 5 + \cdots + (2n - 1) = n^2$$
is true for every positive integer n.

2. Use mathematical induction to prove that
$$1^3 + 2^3 + 3^3 + \cdots + n^3 = \left[\frac{n(n + 1)}{2}\right]^2$$
is true for every positive integer n.

3. Prove by mathematical induction the formula for the sum of the first n terms of an *arithmetic progression*,
$$a + (a + d) + (a + 2d) + \cdots + [a + (n - 1)d]$$
$$= \frac{n[2a + (n - 1)d]}{2},$$
where $a \in R$, d is the common difference between successive terms, and n is any positive integer.

4. Prove the formula for the sum of the first n terms of a *geometric progression*,
$$a + ar + ar^2 + \cdots + ar^{n-1} = \frac{a(1 - r)^n}{1 - r},$$

where $a \in R$, $r \neq 1$ is the common ratio of any two successive terms, and n is any positive integer.

5. Prove that if a and b are real numbers, then
$$a^n - b^n = (a - b)(a^{n-1} + a^{n-2}b + a^{n-3}b$$
$$+ \cdots + ab^{n-2} + b^{n-1})$$
is true for all positive integers $n > 1$. [*Hint:* In the second step of proof by mathematical induction, when showing that $k \in S \implies k + 1 \in S$, use the obvious identity $a^{k+1} - b^{k+1} = a(a^k - b^k) + b^k(a - b)$.]

6. Use mathematical induction to prove that if a and b are positive numbers and $a < b$, then $a^n < b^n$ is true for all positive integers n.

7. Prove that $2^n < n!$ for all positive integers $n > 3$. [*Hint:* Verify that $4 \in S$, and then show that $k \in S$ and $k > 3$ imply $(k + 1) \in S$.]

8. Prove that if $a > -1$ and $a \neq 0$, then
$$(1 + a)^n > 1 + na$$
for every positive integer $n > 1$. This is known as Bernoulli's inequality.

1.8   REVIEW EXERCISES

In each of Exercises 1 through 20, find the solution set of the given inequality and sketch its graph.

1. $6x + 3 > 2x - 5$.

2. $|3x - 4| < 6$.

3. $\left|\dfrac{x + 3}{2x - 5}\right| < 7$.

4. $|8 - 3x| \geq |2x|$.

5. $1 < |x + 2| < 4$.

6. $4 - x < 5 + 2x$.

7. $\dfrac{3}{1 - x} \leq 2$.

8. $2 < |x - 3| < 5$.

9. $|4 + x| < |3 - 2x|$.

10. $\dfrac{3}{x} < \dfrac{5}{8}$.

11. $\dfrac{2}{x} - 3 > \dfrac{3}{x} - 8$.

12. $x^2 - x - 6 < 0$.

13. $4 \le 7 - 3x < 13$.

14. $2x - 3 < \dfrac{x}{4} + \dfrac{1 - x}{2}$.

15. $-2 < 4 - x < 9$.

16. $|5 - x| \ge 6$.

17. $2x^2 - 9x < 5$.

18. $|2x + 1| > 3$.

19. $\left| \dfrac{4 - 3x}{2 + x} \right| \le \dfrac{3}{4}$.

20. $\dfrac{2x - 1}{3 - x} < \dfrac{x}{4 + x}$.

21. Prove that $x < y \implies x < (x + y)/2 < y$.

22. Prove that if a and b are rational numbers, then $a + b$ and ab are rational numbers. (*Hint:* Use the definition of a rational number, Section 1.3.)

23. Let $P_1:(x_1)$ and $P_2:(x_2)$ be points on a coordinate line. Prove that if $x_1, x_2 \in R$, the midpoint of the line segment $P_1 P_2$ has a rational coordinate.

24. Prove the *binomial theorem.* If $a, b \in R$ and $n \in N$, then

$$(a + b)^n = a^n + \frac{n}{1!} a^{n-1}b + \frac{n(n - 1)}{2!} a^{n-2}b^2$$

$$+ \frac{n(n - 1)(n - 2)}{3!} a^{n-3}b^3 + \cdots$$

$$+ \frac{(n - 1)n}{2!} a^2 b^{n-2} + \frac{n}{1!} ab^{n-1} + b^n,$$

in which the rth term in the right-hand member is

$$\frac{n(n - 1)(n - 2) \cdots (n - r + 2)}{(r - 1)!} a^{n-r+1}b^{r-1},$$

$$r = 1, 2, 3, \cdots, n + 1.$$

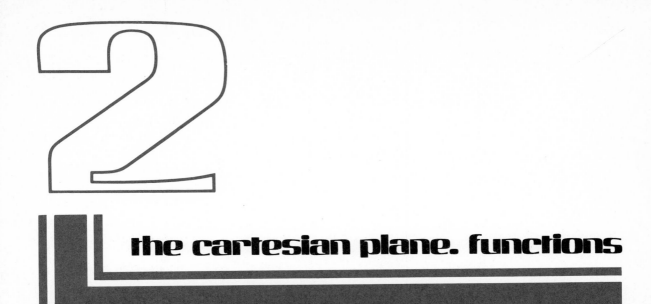

the cartesian plane. functions

The plane analytic geometry needed by calculus students is presented in the early part of the chapter. It leads to graph sketching for equations and inequalities in two variables. The chapter concludes with a careful treatment of functions, their definition, inverses, and combinations. This work on functions is essential preparation for the study of calculus.

2.1 RECTANGULAR CARTESIAN COORDINATES

Consider two mutually perpendicular coordinate lines (1.4.1) that intersect at their zero points and have the same unit of length. Although these perpendicular lines may be oriented in any way we please, they are usually pictured as horizontal and vertical, with their positive directions to the right and upward (Fig. 2-1).

The horizontal coordinate line, Ox, is called the **x axis** and the vertical line, Oy, is the **y axis**. Their intersection O is called the **origin**.

Let P be an arbitrarily chosen point in the plane of the coordinate axes. The **projection** of the point P on a line l means the point of intersection of l with a line through P perpendicular to l (Fig. 2-2).

With any point P in the plane of the coordinate lines, we associate two numbers, its x coordinate and its y coordinate. The **x coordinate** (or **abscissa**) of P is the coor-

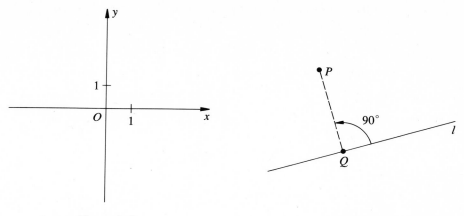

Figure 2-1   Figure 2-2

dinate x of the projection of P on the x axis; and the y **coordinate** (or **ordinate**) of P is the coordinate y of the projection of P on the y axis (Fig. 2-3).

The coordinates of P are written (x, y), always in that order. Thus $(2, -3)$ are the coordinates of the point that is two units to the right of the y axis and three units below the x axis (Fig. 2-4). The point whose coordinates are $(-1, 4)$ is one unit to the left of the y axis and four units above the x axis. The point $(5, 0)$ is on the x axis, five units to the right of the origin.

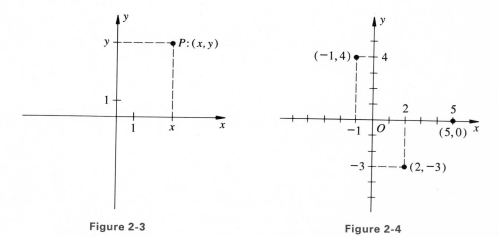

Figure 2-3   Figure 2-4

The symbol $P{:}(x, y)$ means the point P whose coordinates are (x, y). When several points are being discussed, it is convenient to designate them by $P_1{:}(x_1, y_1)$, $P_2{:}(x_2, y_2)$, $P_3{:}(x_3, y_3)$, and so forth.

Let $P_1{:}(x_1, y_1)$ and $P_2{:}(x_2, y_2)$ be any two points in the plane. If the line segment P_1P_2 is not parallel to a coordinate axis, we denote by Q the point of intersection of the line through P_1 that is parallel to the x axis and the line through P_2 that is parallel to the y axis (Fig. 2-5). By the Pythagorean theorem,

$$|P_1P_2|^2 = |P_1Q|^2 + |QP_2|^2.$$

But $|P_1Q| = |x_2 - x_1|$ and $|QP_2| = |y_2 - y_1|$, by 1.4.1. Thus

(1) $$|P_1P_2| = \sqrt{|x_2 - x_1|^2 + |y_2 - y_1|^2}.$$

If the line segment P_1P_2 is parallel to the x axis (Fig. 2-6), then $y_1 = y_2$. Moreover, by 1.6.1,

$$|P_1P_2| = |x_2 - x_1|.$$

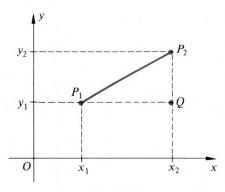

Figure 2-5   Figure 2-6

But this is just what is given by formula (1) when $y_1 = y_2$. Similarly, if P_1P_2 is parallel to the y axis,

$$|P_1P_2| = |y_2 - y_1|.$$

2.1.1 Definition. The (undirected) **distance** between any two points in the plane, $P_1:(x_1, y_1)$ and $P_2:(x_2, y_2)$, or the **length** of the line segment P_1P_2, is

$$|P_1P_2| = \sqrt{(x_2 - x_1)^2 + (y_2 - y_1)^2}.$$

In using this **distance formula**, it makes no difference which of the given points we call P_1 and which we call P_2, since $(x_2 - x_1)^2 = (x_1 - x_2)^2$.

We shall often say "the point (x, y)" instead of "the point whose coordinates are (x, y)."

Example. The distance between the points $(2, 5)$ and $(4, 1)$ is $\sqrt{(4 - 2)^2 + (1 - 5)^2} = 2\sqrt{5}$; and the length of the line segment whose endpoints are $(-3, -6)$ and $(-5, 1)$ is

$$\sqrt{[-5 - (-3)]^2 + [1 - (-6)]^2} = \sqrt{53}.$$

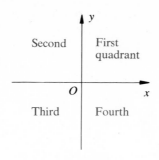

Figure 2-7

Coordinate axes separate the plane into four regions called **quadrants**. The numbering of the quadrants is indicated in Fig. 2-7.

Pairs of numbers like the Cartesian coordinates of P, in which one number is designated the first number and the other the second, are called **ordered pairs** of numbers. Two ordered pairs of numbers, (a, b) and (c, d), are **equal** if and only if $a = c$ and $b = d$. Thus $(2, 5) \neq (5, 2)$; they are different ordered pairs, and, if they are coordinates of points in the plane, the points are distinct.

Let A and B be sets of numbers. The **Cartesian product** of A and B is the set of ordered pairs defined by

$$A \times B = \{(x, y) \mid x \in A, y \in B\}.$$

As an illustration, if $A = \{1, -6, 4\}$ and $B = \{2, 3\}$, then

$$A \times B = \{(1, 2), (1, 3), (-6, 2), (-6, 3), (4, 2), (4, 3)\}.$$

If R is the set of real numbers, the Cartesian product $R \times R$ is indicated by R^2. Thus

$$R^2 = R \times R = \{(x, y) \mid x \in R, y \in R\}.$$

The **Cartesian plane** is the set of points

$$E^2 = \{P{:}(x, y) \mid (x, y) \in R^2, |P_1P_2| = \sqrt{(x_2 - x_1)^2 + (y_2 - y_1)^2}\}.$$

A Cartesian coordinate system in the plane establishes a **one-to-one correspondence** between the points of the Cartesian plane and the ordered pairs of numbers belonging to R^2. This means that to each point of the plane there corresponds a unique ordered pair of real numbers, and each ordered pair of real numbers gives the coordinates of a unique point in the plane.

Exercises

1. Draw a pair of coordinate axes and mark the points whose coordinates are $(4, 1)$, $(-2, 3)$, $(-1, -4)$, $(5, -5)$, $(0, 6)$, and $(-5, 0)$. Label each point with its coordinates.

2. Sketch the line segments terminated by each of the following pairs of points, and find their lengths.
(a) $(1, 2)$ and $(2, 4)$;
(b) $(3, 1)$ and $(6, -2)$;
(c) $(-1, -4)$ and $(2, -3)$;
(d) $(-5, 3)$ and $(-5, -2)$.

3. Sketch the line segments terminated by each of the following pairs of points, and find their lengths.
(a) $(2, 1)$ and $(6, 4)$;
(b) $(-2, -1)$ and $(2, -2)$;
(c) $(4, -3)$ and $(4, 1)$;
(d) $(-6, -2)$ and $(-1, 1)$.

4. Sketch the triangle whose vertices are $P_1{:}(0, 1)$, $P_2{:}(2, 5)$, and $P_3{:}(-1, 4)$, and show that it is isoceles.

5. Show that the triangle whose vertices are $P_1{:}(-3, 2)$, $P_2{:}(0, -1)$, and $P_3{:}(5, 4)$ is a right triangle. Make a sketch.

6. Use the distance formula to determine whether the point $P_1{:}(4, -2)$ is on the line segment whose endpoints are $P_2{:}(-5, -6)$ and $P_3{:}(10, 1)$. Make a sketch.

7. Draw the quadrilateral whose vertices are $(-2, 6)$, $(4, 3)$, $(1, -3)$, and $(-5, 0)$, and prove that it is a square.

8. Express by an equation in x and y the statement that a point $P:(x, y)$ is always at a distance of four units from the point $(5, 3)$. On what geometric figure must all such points lie? Make a sketch.

9. Express in analytic language (that is, by means of an equation in x and y) the statement that a point P is on the perpendicular bisector of the line segment whose endpoints are $A:(1, -5)$ and $B:(-2, 2)$. Make a sketch. (*Hint:* The distances $|AP|$ and $|BP|$ must be equal for all positions of P.)

10. What geometric figure is defined by each of the following sets of points?
(a) $\{P:(x, y) \mid x = -3\}$;
(b) $\{P:(x, y) \mid y = 0\}$;

(c) $\{P:(x, y) \mid \sqrt{(x - 1)^2 + (y - 4)^2} = 5\}$;
(d) $\{P:(x, y) \mid \sqrt{x^2 + y^2} = 3\}$.

11. Name and sketch the geometric figure formed by the set of points
$$\{P:(x, y) \mid \sqrt{(x - 1)^2 + (y - 3)^2} = \sqrt{(x - 5)^2 + (y - 5)^2}\}.$$

12. Find the coordinates of the point on the x axis that is equidistant from the points $(-5, 3)$ and $(2, 4)$. Make a sketch.

13. Express by an equation in x and y the statement that a point $P:(x, y)$ is equidistant from the y axis and the point $(4, 0)$. After sketching the coordinate axes, mark a few such points P and draw a smooth curve through them. This curve is called a *parabola*.

14. Write the Cartesian product of the sets $\{-5, 0, 2\}$ and $\{1, -4, -3\}$.

2.2 SLOPE AND MIDPOINT OF A LINE SEGMENT

In the Cartesian plane, all lines parallel to a coordinate axis have the same positive direction as that axis. On other lines the positive direction is optional.

It follows (Section 1.4) that if two points, $P_1:(x_1, y_1)$ and $P_2:(x_2, y_1)$, are on a line that is parallel to the x axis, the **directed distance** from P_1 to P_2 is

$$\overline{P_1P_2} = x_2 - x_1;$$

and if P_1 and P_2 are on a line parallel to the y axis, then

$$\overline{P_1P_2} = y_2 - y_1$$

(Fig. 2-8).

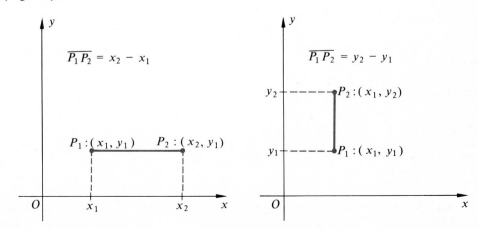

Figure 2-8

Consider a nonvertical line segment P_1P_2 in the Cartesian plane, and, for the moment, let $P_1:(x_1, y_1)$ be its *left* endpoint. Its other endpoint is $P_2:(x_2, y_2)$ (Fig. 2-9). A line through P_1 that is parallel to the x axis intersects a line through P_2, parallel to the y axis, in a point $Q:(x_2, y_1)$. Then $\overline{QP_2} = y_2 - y_1$ and $\overline{P_1Q} = x_2 - x_1$ are directed distances. Their ratio,

(1) $$\frac{\overline{QP_2}}{\overline{P_1Q}} = \frac{y_2 - y_1}{x_2 - x_1}$$

is called the **slope** of the line segment P_1P_2 and is designated by m.

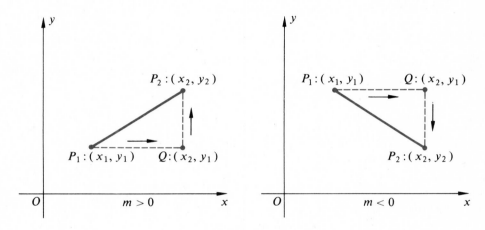

Figure 2-9

The slope of a line segment is a measure of its steepness. If $m > 0$, P_1P_2 points upward to the right; and, if $m < 0$, the segment points downward to the right (Fig. 2-9). In (1), $\overline{P_1Q}$ is sometimes called the *run* and $\overline{QP_2}$ the *rise* (or *fall*).

Notice that the value of m is the same no matter which endpoint is labeled P_1 and which P_2, because

$$\frac{y_2 - y_1}{x_2 - x_1} = \frac{y_1 - y_2}{x_1 - x_2}.$$

2.2.1 Definition. The **slope** of the line segment whose endpoints are $P_1:(x_1, y_1)$ and $P_2:(x_2, y_2)$ is

$$m = \frac{y_2 - y_1}{x_2 - x_1}, \qquad x_1 \neq x_2.$$

If $y_1 = y_2$, the line segment P_1P_2 is parallel to the x axis (Fig. 2-8) and its slope is zero. If $x_1 = x_2$, then P_1P_2 is parallel to the y axis and *has no slope*.

Example 1. The slope of the line segment whose endpoints are $(-2, 3)$ and $(4, 1)$ is $(1 - 3)/[4 - (-2)]$ $= -\frac{1}{3}$. The slope of the line segment terminated by $(5, 0)$ and $(1, -3)$ is $(-3 - 0)/(1 - 5) =$ $\frac{3}{4}$. The line segment whose endpoints are $(-4, 1)$ and $(-4, -4)$ is vertical and has no slope. The reader should sketch each of these line segments.

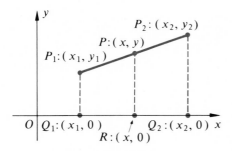

Figure 2-10

Let $P{:}(x, y)$ be the midpoint of the line segment P_1P_2 (Fig. 2-10). Project P_1, P, and P_2 onto the x axis; their projections are $Q_1{:}(x_1, 0)$, $R{:}(x, 0)$, and $Q_2{:}(x_2, 0)$. Since parallel lines intercept proportional segments on any two transversals, P is the midpoint of the line segment P_1P_2 if and only if R is the midpoint of Q_1Q_2—that is, if and only if $x = (x_1 + x_2)/2$ (by 1.4.2). Similarly, $y = (y_1 + y_2)/2$.

It is easy to verify that these formulas are consistent with 1.4.2 when P_1P_2 is parallel to a coordinate axis.

2.2.2 Theorem. The coordinates of the **midpoint** of the line segment whose endpoints are $P_1{:}(x_1, y_1)$ and $P_2{:}(x_2, y_2)$ are

$$x = \frac{x_1 + x_2}{2}, \qquad y = \frac{y_1 + y_2}{2}.$$

Example 2. Prove analytically (that is, by algebra) that the length of the line segment joining the midpoints of the nonparallel sides of any trapezoid is equal to half the sum of the lengths of the parallel sides.

Solution. Our proof must be valid for every trapezoid, not merely some particular one. Observe that a wise choice of position for the coordinate axes will often simplify the algebra.

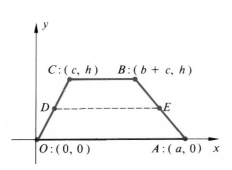

Figure 2-11

Draw a general trapezoid. A good position for the coordinate axes is shown in Fig. 2-11. If we denote the lengths of the parallel sides by a and b, the altitude by h, and the x coordinate of C by c, the vertices will be $O{:}(0, 0)$, $A{:}(a, 0)$, $B{:}(b + c, h)$, and $C{:}(c, h)$. We wish to prove that the length of the line segment joining the midpoints, D and E, of OC and AB is equal to $\frac{1}{2}(a + b)$.

By the midpoint formulas (2.2.2), the coordinates of D are $(c/2, h/2)$ and the coordinates of E are $((a + b + c)/2, h/2)$. Since the y coordinates of D and E are equal, DE is horizontal and, by 1.4.1,

$$|DE| = |\tfrac{1}{2}(a + b + c) - \tfrac{1}{2}c| = \tfrac{1}{2}(a + b).$$

The symbols a, b, and h represent any positive numbers and c is any real number. Thus we have proved the theorem for all trapezoids, not just for a particular one.

Exercises

Make a sketch for each exercise.

1. Determine from a sketch the slope of a line segment P_1P_2 if

(a) P_2 is five units to the right of and two units above P_1;

(b) P_2 is seven units to the left of and six units above P_1.

2. By means of a sketch find the slope of a line segment P_1P_2 if

(a) P_2 if four units to the right of and nine units below P_1;

(b) P_2 is eight units to the left of and three units below P_1.

3. Use 2.2.1 and 2.2.2 to find the slope and the midpoint of the line segment whose endpoints are

(a) $(-1, 2)$ and $(3, 4)$;

(b) $(6, -4)$ and $(2, 2)$;

(c) $(-2, -5)$ and $(3, -5)$;

(d) $(-5, -1)$ and $(-2, -2)$.

4. Use 2.2.1 and 2.2.2 to find the slope and the midpoint of the line segment whose endpoints are

(a) $(2, 4)$ and $(6, 2)$;

(b) $(-3, 2)$ and $(-1, -8)$;

(c) $(-4, -1)$ and $(-4, -5)$;

(d) $(7, 0)$ and $(0, -2)$.

5. The slope of a line segment is 3 and one endpoint is $(-2, 5)$. If the other endpoint is on the x axis, what are its coordinates?

6. The slope of a line segment is $-\frac{1}{4}$ and one endpoint is $(5, 1)$. If the other endpoint is on the y axis, what are its coordinates?

7. Find the slopes of the sides of the triangle whose vertices are $(-7, 1)$, $(-4, 5)$, and $(3, -2)$.

8. Find the slopes of the sides of the triangle whose vertices are $(-3, -7)$, $(8, -4)$, and $(5, 4)$.

9. Using slopes only, determine whether the point $(2, 3)$ is on the line segment whose endpoints are $(-5, 1)$ and $(7, 4)$.

10. By means of the slope formula, determine whether the point $(-1, 0)$ is on the line segment terminated by $(-9, -2)$ and $(11, 3)$.

11. If one endpoint of a line segment is $(-3, -4)$ and its midpoint is $(2, -1)$, find the coordinates of its other endpoint.

12. One end of a diameter of a circle is $(-2, -7)$ and its center is $(3, 0)$. Find the coordinates of the other end of the diameter.

13. The endpoints of a line segment are $P_1:(x_1, y_1)$ and $P_2:(x_2, y_2)$. Find the coordinates of the point $P:(x, y)$ on the line segment that is one-third of the way from P_1 to P_2. (*Hint:* See the derivation of 2.2.2.)

14. The endpoints of a line segment are $P_1:(x_1, y_1)$ and $P_2:(x_2, y_2)$. Find the coordinates of the point $P:(x, y)$ on the line segment whose distance from P_1 is $1/n$ of the directed distance $\overline{P_1P_2}$, where n is a positive integer.

15. By means of the slope formula, put into analytic language (that is, express by means of an equation in x and y) the statement that the slope of a line segment joining the point $(2, 5)$ to the variable point $P:(x, y)$ is always equal to 3. On what geometric figure must all such points P lie?

16. Use the slope formula to put into analytic language the statement that the slope of a line segment joining the point $(-3, 2)$ to a point $P:(x, y)$ is equal to -1. On what geometric figure must all such points P lie?

17. Prove analytically that two medians of any isosceles triangle are equal.

2.3 SUBSETS OF THE CARTESIAN PLANE

By the **solution set** of an equation in x and y, say, is meant the set of all ordered pairs of numbers such that when the first number in a pair is substituted for x and the second for y, the equation becomes a true statement.

Thus the solution set of the equation $2x - y - 4 = 0$ is

$$S = \{(x, y) \,|\, 2x - y - 4 = 0\}.$$

Clearly, $(1, -2)$ belongs to this solution set because $2(1) - (-2) - 4 = 0$ is a true statement, and $(5, 6) \in S$, since $2(5) - (6) - 4 = 0$ is true. But $(2, 3) \notin S$ because $2(2) - (3) - 4 \neq 0$. This is often expressed by saying that $(1, -2)$ and $(5, 6)$ **satisfy the equation** but $(2, 3)$ does not satisfy it.

Probably the simplest way to discover elements of the solution set of $2x - y - 4 = 0$ is to rewrite the equation as $y = 2x - 4$, substitute numbers for x in the right-hand member, and compute the corresponding values of y. Thus if -3 is substituted for x, we find $y = -10$, and $(-3, -10)$ belongs to the solution set. Such results are included in the accompanying *table of values*.

x	-3	-2	-1	0	1	2	3	4	5	6
y	-10	-8	-6	-4	-2	0	2	4	6	8

By the **graph of an equation** in x and y, we mean the subset of E^2 (the Cartesian plane) consisting of those points, and only those points, whose coordinates form the solution set of the equation.

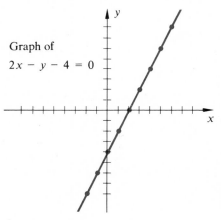

Graph of

$2x - y - 4 = 0$

Figure 2-12

It follows that a point belongs to (or is on) the graph of an equation if and only if the coordinates of the point satisfy the equation.

To **sketch the graph** of an equation, draw a pair of coordinate axes and plot a sufficient number of points of the graph so that when a smooth curve is drawn through them, the unplotted points on the sketch will have coordinates that approximately satisfy the equation. A sketch of the graph of the equation $2x - y - 4 = 0$, discussed above, is shown in Fig. 2-12. It is a straight line. We shall establish in Section 2.4 that every equation of the first degree in x and y has for its Cartesian graph a straight line and vice versa.

The graph of the equation $x^2 + y^2 = 16$ is

$$\{P:(x, y) \,|\, x^2 + y^2 = 16\}.$$

By solving the equation for y in terms of x, we obtain

$$y = \pm\sqrt{16 - x^2}.$$

It is now easy to substitute values of x in the right member of this equation and compute the corresponding values of y. Using this procedure, we find the following table of values.

x	-4	-3	-2	-1	0	1	2	3	4
y	0	$\pm\sqrt{7}$	$\pm2\sqrt{3}$	$\pm\sqrt{15}$	±4	$\pm\sqrt{15}$	$\pm2\sqrt{3}$	$\pm\sqrt{7}$	0

Notice in this table that a reading like $x = -2$, $y = \pm2\sqrt{3}$ means that the *two* points, $(-2, 2\sqrt{3})$ and $(-2, -2\sqrt{3})$, are on the graph.

When all these points are plotted and a smooth curve is drawn through them, the graph of $x^2 + y^2 = 16$ is seen to be a circle with its center at the origin and having radius 4 (Fig. 2-13).

There are equations in x and y that have no graphs because their solution sets are empty. Thus $\{P:(x, y)\,|\,x^2 + y^2 = -4\} = \varnothing$ because Cartesian coordinates are real numbers and the square of a real number cannot be negative.

Two graphs in the same Cartesian plane may or may not have points in common. A straight line may intersect a circle in two points, be tangent to it, or miss it completely.

The **intersection** of two graphs means the set of points common to the two graphs (sets). To find the points of intersection of the graphs of two equations, solve the equations simultaneously. The real pairs of solutions (if any) are the coordinates of the points of intersection of the two graphs.

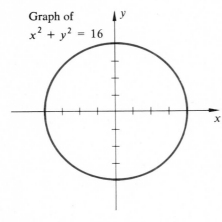

Graph of
$x^2 + y^2 = 16$

Figure 2-13

Example. Find $\{P:(x, y)\,|\,2x - y - 4 = 0\} \cap \{P:(x, y)\,|\,x^2 + y^2 = 16\}$; that is, find the set of points of intersection of the line $2x - y - 4 = 0$ and the circle $x^2 + y^2 = 16$ (Fig. 2-14).

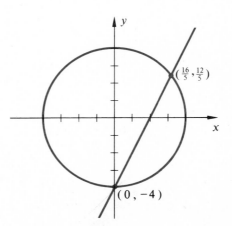

$\left(\frac{16}{5}, \frac{12}{5}\right)$

$(0, -4)$

Figure 2-14

Solution. By solving the given equations simultaneously, we find the set of points of intersection of the line and circle to be $\{P:(0, -4), P:(\frac{16}{5}, \frac{12}{5})\}$.

Exercises

In Exercises 1 to 16, make a table of values, plot the corresponding points in a Cartesian plane, and draw a smooth curve through them. If no pair of real coordinates satisfies the given equation, state that the equation has no graph.

1. $x + 2y - 4 = 0$.

2. $x - 4y + 3 = 0$.

3. $2x + y + 4 = 0$.

4. $7x - 3y + 21 = 0$.

5. $x = 2$.

6. $y = -5$.

7. $y = 0$.

8. $x = -1$.

9. $x^2 + y^2 = 9$.

10. $x^2 + y^2 = -9$.

11. $(x - 2)^2 + y^2 = 4$.

12. $x^2 + (y + 3)^2 = 9$.

13. $y^2 = 2x$. (Parabola.)

14. $y = x^2 + 1$. (Parabola.)

15. $25x^2 + 9y^2 - 175 = 0$. (Ellipse.)

16. $9x^2 + 16y^2 - 144 = 0$. (Ellipse.)

In each of Exercises 17 to 22, sketch the graphs of the given pair of equations and find the coordinates of their points of intersection.

17. $x - 4y + 8 = 0, 3x + y - 15 = 0$.

18. $4x - 5y + 13 = 0, x + 2y = 0$.

19. $2x - y = 0, x^2 + y^2 = 25$.

20. $x - 2y + 4 = 0, x^2 + y^2 = 16$.

21. $x + y - 2 = 0, y = x^2$.

22. $x - 2y^2 = 0, x^2 - 4y = 0$.

2.4 THE STRAIGHT LINE

Euclid did not define a **straight line** but left it as one of the undefined elements whose existence he assumed. From his axioms, however, several characteristic properties of a line can be deduced.

Let $P_1:(x_1, y_1)$ be an arbitrarily chosen fixed point in the Cartesian plane. A line through P_1 is either vertical (that is, parallel to the y axis) or is not.

If the line through P_1 is vertical, all points on it have the same x coordinate, x_1 (Fig. 2-15).

The vertical line through $P_1:(x_1, y_1)$ is

(1) $$\{P:(x, y) \mid x = x_1\}.$$

In particular, the y axis is the graph of the equation $x = 0$.

All nonvertical lines through P_1 have slopes. A characteristic property of a non-vertical line l through $P_1:(x_1, y_1)$ is that any other point $P:(x, y)$ is on l if and only if the slope of the line segment P_1P is m, where m is constant for the given line l (Fig. 2-16). But, by 2.2.1, the slope of P_1P is $(y - y_1)/(x - x_1)$. Therefore P is on l if and

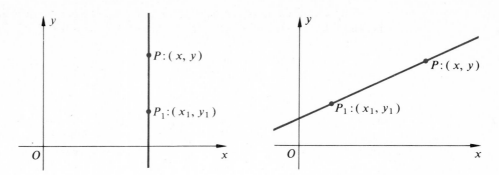

Figure 2-15   Figure 2-16

only if $(y - y_1)/(x - x_1) = m$. If we rewrite this equation as

(2) $$y - y_1 = m(x - x_1),$$

then the coordinates of P_1 also satisfy it. This is called the **point-slope form** of the equation of a line.

The line through the point $P_1:(x_1, y_1)$, having slope m, is

(3) $$\{P:(x, y) \,|\, y - y_1 = m(x - x_1)\}.$$

Example 1. Find the equation of the line through the point $(-2, 3)$ with slope $\frac{1}{2}$(Fig. 2-17).

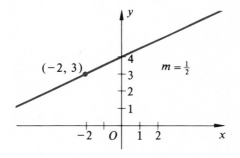

Figure 2-17

Solution. By substituting $(-2, 3)$ for (x_1, y_1) and $\frac{1}{2}$ for m in (2), we find $y - 3 = \frac{1}{2}(x + 2)$, or $x - 2y + 8 = 0$.

Example 2. Find the equation of the line through the two points $(1, 4)$ and $(3, -2)$.

Solution. By 2.2.1, the slope of the line is

$$m = \frac{4 - (-2)}{1 - 3} = -3.$$

From (3), the equation of the line through $(1, 4)$ with slope -3 is $y - 4 = -3(x - 1)$, or $3x + y - 7 = 0$.

The y coordinate of the point of intersection of a nonvertical line and the y axis is called the **y intercept** of the line. An analogous definition holds for the **x intercept** of a line.

Thus a line whose y intercept is b intersects the y axis in the point $(0, b)$. By (3), the equation of the line through $(0, b)$ and having slope m is $y - b = m(x - 0)$ or

$$(4) \qquad\qquad y = mx + b.$$

This is known as the **slope-intercept form** of the equation of a line; the slope of the line is m and its y intercept is b. Every nonvertical line has an equation (2) and hence an equation of the form (4).

Since every line in the Cartesian plane is either vertical or has a slope, every line has an equation $x = k$ or $y = mx + b$, where k, m, and b are constants. But both equations can be written in the form

$$(5) \qquad\qquad Ax + By + C = 0,$$

where not both A and B are zero. Thus every line in the Cartesian plane has an equation of the form (5). This is known as the **general form** of the equation of a line in the Cartesian plane.

If $B = 0$, then $A \neq 0$ and (5) can be written $x = -C/A$, the graph of which is a vertical line. If $B \neq 0$, (5) can be written $y = (-A/B)x + (-C/B)$, which is in the form (4); thus its graph is a line whose *slope* is $-A/B$ and whose y intercept is $-C/B$. Therefore the Cartesian graph of (5) is always a straight line.

These properties of Euclid's (undefined) straight lines suggest the following definition of a straight line in the Cartesian plane.

2.4.1 Definition.   A **straight line** in the Cartesian plane is

$$\{P{:}(x, y) \mid Ax + By + C = 0,\ A, B, C \in R,\ A \neq 0 \text{ or } B \neq 0\}.$$

Consequently, an equation of the first degree in the variables is frequently called a **linear equation.**

We shall often say "the line $Ax + By + C = 0$" instead of "the line whose equation is $Ax + By + C = 0$."

2.4.2 Corollary.   If $B \neq 0$, the slope of the line $Ax + By + C = 0$ is

$$m = \frac{-A}{B}.$$

Example 3.   Find the x intercept, the y intercept, and the slope of the line $3x - 5y + 15 = 0$. Then sketch the line.

Solution.   To find the x intercept of the line, we substitute $y = 0$ in $3x - 5y + 15 = 0$ and obtain $x = -5$. Thus -5 is the x intercept. By substituting $x = 0$ in the given equation, we get $y = 3$; so the y intercept is 3. By 2.4.2, the slope of the line is $-A/B = \frac{3}{5}$. The graph is the

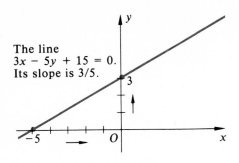

The line
$3x - 5y + 15 = 0$.
Its slope is 3/5.

Figure 2-18

line whose x and y intercepts are -5 and 3 (Fig. 2-18).

Another method is to write the given equation in the slope intercept form

$$y = \tfrac{3}{5}x + 3,$$

from which we read the slope as $\tfrac{3}{5}$ and the y intercept as 3. By substituting $y = 0$ in this equation, we find the x intercept to be -5.

It should be noted that the graph of $Ax + By + C = 0$ is identical with the graph of $k(Ax + By + C) = 0$ if $k \neq 0$. For if $P_1:(x_1, y_1)$ is an arbitrarily chosen point on the first line, then $Ax_1 + By_1 + C = 0$. Thus $k(Ax_1 + By_1 + C) = k \cdot 0 = 0$, and so P_1 also lies on the graph of $k(Ax + By + C) = 0$. Similarly, every point on the second graph is also on the first. In other words, both members of an equation of a line can be multiplied by any nonzero constant without affecting the graph.

Exercises

Make a sketch for each exercise.

1. Find an equation of the line through the point $(-5, 1)$ with slope -1.

2. Find an equation of the line through the point $(4, -2)$ with slope 3.

3. Find an equation of the horizontal line whose y intercept is 5.

4. Find an equation of the vertical line whose x intercept is -4.

5. Find an equation of the line whose slope is -2 and whose x intercept is 3.

6. Find an equation of the line whose slope is $\tfrac{1}{2}$ and whose y intercept is -7.

7. Find an equation of the line through the points $(2, -1)$ and $(-5, 4)$.

8. Find an equation of the line through the points $(-2, -3)$ and $(6, -1)$.

9. Find an equation of the line whose x intercept is -3 and whose y intercept is 2.

10. Find an equation of the line whose x intercept is 4 and whose y intercept is -1.

11. Find the intercepts of the line

$$\frac{x}{a} + \frac{y}{b} = 1, \qquad ab \neq 0.$$

This is called the **intercept form** of the equation of a line.

12. Show algebraically that the lines $2x - 3y + 4 = 0$ and $4x - 6y - 9 = 0$ do not intersect. Then find their slopes.

13. The endpoints of a line segment are $P_1:(-3, -1)$ and $P_2:(5, 3)$. Find the equation of the line through the midpoint of the segment P_1P_2 and perpendicular to P_1P_2. [*Hint:* A point $P:(x, y)$ is on the line if and only if $|P_1P| = |P_2P|$.]

14. The endpoints of a line segment are $(-2, 4)$ and $(6, -2)$. Find the equation of the line through the midpoint of the line segment and perpendicular to it.

15. Find the coordinates of the point on the line $x - 2y + 4 = 0$ that is equidistant from the points $A:(3, 2)$ and $B:(-1, -4)$. (*Hint:* Find the point of intersection of the given line and the perpendicular bisector of the line segment AB.)

16. Find the coordinates of the center of the circle through the points $A:(0, 0)$, $B:(2, 4)$, and $C:(-4, 6)$. (*Hint:* Find the intersection of the perpendicular bisectors of the line segments AB and BC.)

Sec. 2.5 / Parallel and Perpendicular Lines. Circles

45

17. Sketch the graph of $x^2 - y^2 = 0$. [*Hint:* The given equation can be rewritten $(x - y)(x + y) = 0$; and a point is on the graph if and only if its coordinates make the expression in *either* parentheses zero.]

18. Sketch the graph of $x^2 - 4xy - 8x = 0$. (*Hint:* Factor the left-hand member.)

2.5 PARALLEL AND PERPENDICULAR LINES. CIRCLES

By a **vertical line** in the Cartesian plane is meant

$$\{P:(x, y) \mid x = k, k \in R\}.$$

2.5.1 Definition. All vertical lines are **parallel** to each other. Two nonvertical lines in the Cartesian plane are **parallel** if and only if their slopes are equal (Fig. 2-19).

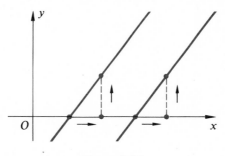

Figure 2-19

From 2.4.2 and 2.5.1 we have the following corollary.

2.5.2 Corollary. Two nonvertical lines, $A_1x + B_1y + C_1 = 0$ and $A_2x + B_2y + C_2 = 0$, are parallel if and only if

$$\frac{A_1}{B_1} = \frac{A_2}{B_2}.$$

Every line parallel to $Ax + By + C = 0$ has an equation of the form $Ax + By + D = 0$.

Example 1. Find an equation of the line through the point $(2, 1)$ and parallel to the line $x + 5y + 5 = 0$ (Fig. 2-20).

Solution. By 2.5.2, the desired equation is $x + 5y + D = 0$, where the value of D is still to be determined. Since the line is to go through the point $(2, 1)$, the coordinates of this point must satisfy the equation of the line. Thus $(2) + 5(1) + D = 0$, from which $D = -7$. Therefore the equation of the line through the point $(2, 1)$ and parallel to $x + 5y + 5 = 0$ is $x + 5y - 7 = 0$.

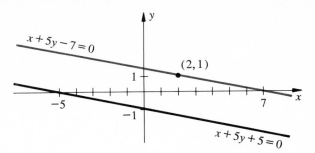

Figure 2-20

Consider next two nonvertical perpendicular lines, l_1 and l_2, and denote their slopes by m_1 and m_2, respectively. Through the origin draw a line l_1', parallel to l_1, and a line l_2' that is parallel to l_2 (Fig. 2-21). By 2.5.1, the slope of l_1' is m_1 and the slope of l_2' is m_2. Let $P_1{:}(x_1, y_1)$ be an arbitrarily chosen point on l_1', distinct from O, and let $P_2{:}(x_2, y_2)$ be a point on l_2', also distinct from O.

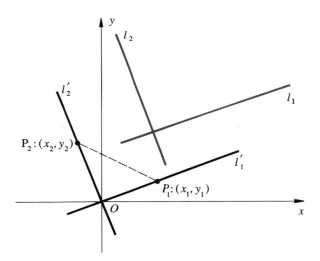

Figure 2-21

Since l_1 and l_2 are mutually perpendicular in the Euclidean sense, then l_1' and l_2' are mutually perpendicular, and P_1OP_2 is a right triangle. By the Pythagorean theorem,

$$|OP_1|^2 + |OP_2|^2 = |P_1P_2|^2,$$

from which

$$(x_1^2 + y_1^2) + (x_2^2 + y_2^2) = [(x_2 - x_1)^2 + (y_2 - y_1)^2].$$

After expansion, cancellation, and transposition of terms, this equation becomes $2x_1x_2 + 2y_1y_2 = 0$, or

(1)
$$\frac{y_1}{x_1} = -\frac{x_2}{y_2}.$$

But $y_1/x_1 = m_1$, the slope of l_1', and $y_2/x_2 = m_2$, the slope of l_2'. Thus (1) is equivalent to

$$m_1 = -\frac{1}{m_2}.$$

Therefore, if l_1 and l_2 are perpendicular in the Euclidean sense, $m_1 = -1/m_2$. Since the order of the preceding steps is reversible, the converse is also valid.

2.5.3   Definition.   Let m_1 and m_2 be the slopes of two intersecting (nonvertical) lines in the Cartesian plane. The lines are **mutually perpendicular** if and only if

$$m_1 = -\frac{1}{m_2}.$$

Example 2.   Which of the following lines are parallel and which are mutually perpendicular?
  (a) $x + 2y - 4 = 0.$ (b) $3x + 6y + 5 = 0.$
  (c) $2x - y + 6 = 0.$ (d) $2x + 4y - 8 = 0.$

Solution.   Since (d) can be written $2(x + 2y - 4) = 0$, (a) and (d) are equations of the same line.
  By 2.4.2, the slope of (a) is $m_1 = -\frac{1}{2}$; the slope of (b) is $m_2 = -\frac{3}{6} = -\frac{1}{2}$; and the slope of (c) is $m_3 = 2$.
  Consequently, (2.5.1 and 2.5.3) lines (a) and (b) are parallel, lines (a) and (c) are mutually perpendicular, and lines (b) and (c) are mutually perpendicular.
  The reader should sketch these lines.

Example 3.   Find an equation of the line through the point (2, 1) that is perpendicular to the line $x + 5y + 5 = 0$ (Fig. 2-22).

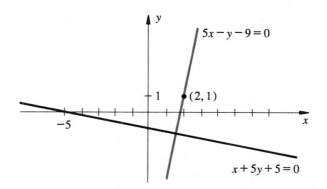

Figure 2-22

Solution.   Since the slope of the given line is $-\frac{1}{5}$, the slope of any line perpendicular to it is 5 (by 2.4.2 and 2.5.3).
  Using the point-slope form of the equation of a line, the line through the point (2, 1), having slope 5, is $y - 1 = 5(x - 2)$, or

$$5x - y - 9 = 0.$$

Let $C:(a, b)$ be an arbitrarily chosen point in the Cartesian plane, and let r be a nonnegative real number. The set of points $P:(x, y)$ in the plane such that $|CP| = r$ is called a **circle**. The point C is the **center** of the circle, and the number r is its **radius.**

From the distance formula, $|CP| = \sqrt{(x - a)^2 + (y - b)^2}$. Thus an equation of the circle is $\sqrt{(x - a)^2 + (y - b)^2} = r$, or more simply,

$$(2) \qquad (x - a)^2 + (y - b)^2 = r^2.$$

This is called the **center-radius form** of the equation of a circle because it exhibits the coordinates (a, b) of the center and the radius r.

When the center of the circle is the origin of coordinates, equation (2) has its simplest form:

$$(3) \qquad x^2 + y^2 = r^2.$$

Example 4.   Find the equation of the circle whose center is $(2, -3)$ and whose radius is 4.

Solution.   By substituting $a = 2, b = -3$, and $r = 4$ in (2), we obtain $(x - 2)^2 + (y + 3)^2 = 16$ (Fig. 2-23).

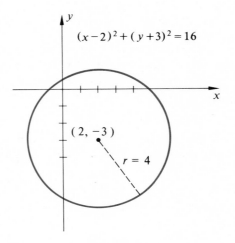

$$(x - 2)^2 + (y + 3)^2 = 16$$

$(2, -3)$

$r = 4$

Figure 2-23

By expanding and collecting terms, equation (2) can be rewritten

$$x^2 + y^2 - 2ax - 2by + (a^2 + b^2 - r^2) = 0.$$

This is in the form

$$(4) \qquad x^2 + y^2 + Dx + Ey + F = 0,$$

where D, E, and F are real numbers. Equation (4) is known as the **general form** of the equation of a circle. Since every circle has a center and a radius, every circle has an equation in the center-radius form (2). Therefore, all circles are included in the graphs of (4) for various selections of the constants D, E, and F.

But is the graph of the general equation (4) always a circle? This equation can be written

$$(x^2 + Dx) + (y^2 + Ey) = -F.$$

If we "complete the squares" in the parentheses by adding $D^2/4$ and $E^2/4$ to both members of the equation, we obtain

$$\left(x^2 + Dx + \frac{D^2}{4}\right) + \left(y^2 + Ey + \frac{E^2}{4}\right) = \frac{D^2}{4} + \frac{E^2}{4} - F$$

or

(5) $$\left(x + \frac{D}{2}\right)^2 + \left(y + \frac{E}{2}\right)^2 = \frac{D^2 + E^2 - 4F}{4}.$$

By comparing (5) with (2), we see that if $D^2 + E^2 - 4F > 0$, the graph of (5) is a circle whose center is $(-D/2, -E/2)$ and whose radius is $\sqrt{D^2 + E^2 - 4F}/2$. If $D^2 + E^2 - 4F = 0$, (5) becomes

$$\left(x + \frac{D}{2}\right)^2 + \left(y + \frac{E}{2}\right)^2 = 0,$$

the graph of which is a single point, $(-D/2, -E/2)$; this is often called a **point-circle**. It may be thought of as the limiting form of a circle whose radius decreases and approaches zero. If $D^2 + E^2 - 4F < 0$, equation (5) has no graph because its left-hand member is the sum of two squares, and the square of a real number cannot be negative. The following theorem summarizes this discussion.

2.5.4   Theorem.   The graph of the equation

$$x^2 + y^2 + Dx + Ey + F = 0$$

is a circle if $D^2 + E^2 - 4F > 0$ and a point-circle if $D^2 + E^2 - 4F = 0$. If $D^2 + E^2 - 4F < 0$, the equation has no graph.

In sketching a circle whose general equation is given, the reader is advised to use the method of "completing the squares" to find the radius and the coordinates of the center. It is unnecessary to memorize the formulas for the radius and the center.

Example 5.   Sketch the graph of the equation $4x^2 + 4y^2 + 16x - 8y - 5 = 0$.

Solution.   Since the given equation can be rewritten

$$x^2 + y^2 + 4x - 2y - \tfrac{5}{4} = 0,$$

its graph, if any, will be a circle or a point-circle (by 2.5.4). If we collect the terms in x and the terms in y, and transpose the constant term, the result is

$$(x^2 + 4x) + (y^2 - 2y) = \tfrac{5}{4}.$$

Completing the squares in the parentheses by adding 4 and 1 to both members of this equation, we obtain

$$(x^2 + 4x + 4) + (y^2 - 2y + 1) = 4 + 1 + \tfrac{5}{4}$$

or

$$(x + 2)^2 + (y - 1)^2 = \tfrac{25}{4}.$$

Comparison of this equation with (2) shows that the center of the circle is $(-2, 1)$ and that its radius is $\frac{5}{2}$. It is now easy to sketch the circle (Fig. 2-24).

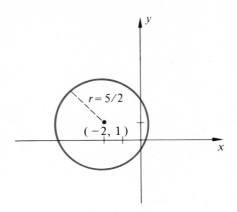

Figure 2-24

Exercises

Make a sketch for each exercise.

1. Find an equation of the line through the point $(2, 3)$ and parallel to the line $x + 3y - 3 = 0$.

2. Find an equation of the line through the point $(2, -\frac{9}{2})$ and parallel to the line $7x - 5y + 35 = 0$.

3. Find an equation of the line through the point $(-1, -4)$ and perpendicular to the line $x - 2y + 2 = 0$.

4. Find an equation of the line through the point $(2, 1)$ and perpendicular to the line $x + 2y + 4 = 0$.

5. Find an equation of the line through the origin and perpendicular to the line joining $(-5, 1)$ and $(2, 4)$.

6. Find an equation of the line through the point $(-1, 4)$ and perpendicular to the line whose x intercept is 3 and y intercept is -2.

7. Given the points A:$(1, 3)$ and B:$(4, -1)$, find the equation of the line through the midpoint of the line segment AB and perpendicular to AB.

8. Find the equation of the perpendicular bisector of the line segment joining the points $(-2, -1)$ and $(5, 5)$.

9. Find the coordinates of the point on the line

$3x - y + 3 = 0$ that is equidistant from the points A:$(2, 4)$ and B:$(6, -2)$. (*Hint:* Find the point of intersection of the given line and the perpendicular bisector of the line segment AB.)

10. Show that the points $(-3, 2)$, $(0, 3)$, $(1, 0)$, and $(-2, -1)$ are vertices of a square.

11. Write the center-radius form and the general form of the equation of the circle with center at $(3, -6)$ and radius 6.

12. Give the center-radius form and the general form of the equation of the circle whose center is $(-2, 1)$ and radius is 5.

13. Find an equation of the circle whose center is $(-3, 7)$ and that goes through the origin.

14. Find an equation of the circle whose center is $(3, 2)$ and that passes through the point $(-4, -5)$.

15. Find an equation of the circle through the points $(-4, 8)$, $(4, -4)$, and $(3, -9)$.

16. Find an equation of the circle through the four points given in Exercise 10.

17. Find the center and the radius of the circle $x^2 + y^2 + 14x - 4y + 52 = 0$. Then sketch the circle.

18. Show that $x^2 + y^2 - 4x - 12y + 40 = 0$ is an equation of a point-circle. Graph it.

19. Find the equation of the line that is tangent to the circle $x^2 + y^2 + 6x + 10y + 5 = 0$ at the point $(2, -3)$. Make a sketch.

20. Find the equation of the line that is tangent to the circle $x^2 + y^2 + 2x - 4y - 8 = 0$ at the point $(2, 4)$. Make a sketch.

21. Find the equation of the circle that is concentric with the circle $x^2 + y^2 - 8x + 10y + 5 = 0$ and that goes through the point $(-1, 3)$. Make a sketch.

22. Find the equation of the circle that is concentric with the circle $x^2 + y^2 - 4x + 10y = 0$ and that passes through the point $(-2, 3)$. Make a sketch.

2.6 SKETCHING GRAPHS OF EQUATIONS

The labor involved in sketching the graph of an equation can often be considerably reduced by making a preliminary analysis of the equation. Although calculus is the most powerful aid in graph sketching (Section 6.12), we should start using three preliminary tests now before plotting any points on the graph. These tests are for intercepts, symmetry, and excluded regions.

Intercepts.

The easiest points to locate when sketching the graph of an equation are usually its points of intersection, if any, with the coordinate axes. A point is on the x axis if and only if its y coordinate is zero. Therefore, to find the **x intercepts** of the graph of an equation (that is, the x coordinates of the points of intersection of the graph and the x axis), substitute $y = 0$ in the equation of the graph and find the corresponding values of x. Similarly, to find the **y intercepts** of a graph, put $x = 0$ in the equation of the graph and find the corresponding values of y.

Example 1. Find the intercepts of the graph of $y^2 - x + y - 6 = 0$.

Solution. Putting $y = 0$ in the given equation, we have $x = -6$, and so the x intercept is -6. Putting $x = 0$ in the equation, we find $y^2 + y - 6 = 0$ or $(y + 3)(y - 2) = 0$; the y intercepts are -3 and 2 (Fig. 2-25).

Two points, P and P', are said to be **symmetric** with respect to a line l if l is the perpendicular bisector of the segment PP' (Fig. 2-26). Each of the points, P and P', is

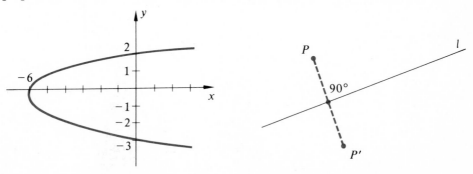

Figure 2-25   Figure 2-26

said to be the **reflection** of the other about the line l. A graph is **symmetric with respect to a line** l if the reflection about l of every point of the graph is also on the graph (Fig. 2-27). Here l is said to be a **line of symmetry** for the graph.

A graph is **symmetric with respect to a point** Q if corresponding to each point P of the graph there is a point P', also on the graph, such that Q is the midpoint of the line segment PP' (Fig. 2-28).

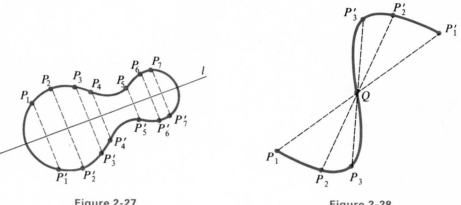

Figure 2-27   Figure 2-28

Consider a point $P{:}(x, y)$ in the Cartesian plane (Fig. 2-29). Its reflection with respect to the x axis is $P'{:}(x, -y)$; its reflection with respect to the y axis is $P'{:}(-x, y)$; and $P{:}(x, y)$ and $P'{:}(-x, -y)$ are symmetric with respect to the origin.

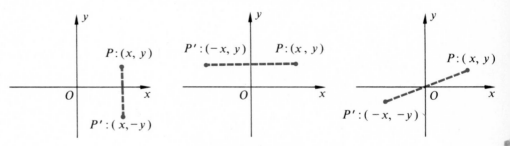

Figure 2-29

It follows that a graph is symmetric with respect to the x axis if, whenever the coordinates (x, y) satisfy the equation, $(x, -y)$ also satisfy the equation; a graph is symmetric with respect to the y axis if, whenever (x, y) satisfies the equation, $(-x, y)$ also satisfies the equation; and a graph is symmetric with respect to the origin if, whenever (x, y) satisfies the equation, $(-x, -y)$ also satisfies the equation.

Two equations are said to be **equivalent** if they have the same solution set. As in the equation of a line, both members of any equation in x and y can be multiplied by the same nonzero constant without affecting the solution set. In particular, if we change the sign of every term in an equation, we obtain an equivalent equation.

Symmetry.

The foregoing discussion implies the following tests for symmetry of a graph with respect to the coordinate axes and the origin.

The graph of an equation is

1. **symmetric** with respect to the **x axis** if whenever (x, y) satisfies the equation, then $(x, -y)$ also satisfies the equation;

2. **symmetric** with respect to the **y axis** if whenever (x, y) satisfies the equation, $(-x, y)$ also satisfies the equation;

3. **symmetric** with respect to the **origin** if whenever (x, y) satisfies the equation, $(-x, -y)$ also satisfies the equation.

Example 2. Show that the graph of $2y^2 - 3x + 1 = 0$ is symmetric with respect to the x axis (Fig. 2-30).

Solution. Let $P_1:(x_1, y_1)$ be any point on the graph of $2y^2 - 3x + 1 = 0$. That is, let

(1) $$2y_1^2 - 3x_1 + 1 = 0.$$

Then $P_1':(x_1, -y_1)$, the reflection of P_1 with respect to the x axis, is also on the graph because

$$2(-y_1)^2 - 3x_1 + 1 = 0$$

is equivalent to (1).

Example 3. Show that the graph of $y = x^3$ is symmetric with respect to the origin (Fig. 2-31).

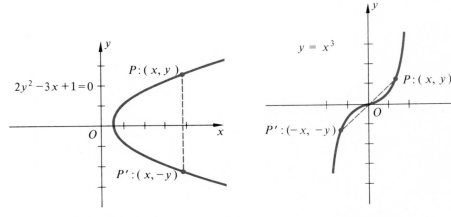

Figure 2-30   Figure 2-31

Solution. If x and y are replaced by $-x$ and $-y$ in the given equation, we obtain

$$(-y) = (-x)^3,$$

which is equivalent to $y = x^3$.

Excluded Regions.

By solving the equation $xy^2 - y^2 - x = 0$ for y in terms of x, we obtain

(2) $$y = \pm \sqrt{\frac{x}{x - 1}}.$$

Clearly, all values of x between 0 and 1 cause the expression under the radical sign to be negative and thus cause the corresponding values of y to be imaginary. For instance, if $x = \frac{1}{2}$, then $y = \pm\sqrt{-1}$. Moreover, if $x = 1$, the denominator in (2) is zero and so $x = 1$ is ruled out. As a result, the graph cannot exist in the vertical band $0 < x \le 1$. Knowledge of such excluded regions saves unnecessary labor in graph sketching.

Consequently, in order to find the **excluded vertical regions** for a polynomial equation, solve the given equation for y in terms of x; if there is a radical of even degree in the right-hand member, exclude all values of x that cause the expression under the radical sign to be negative; also exclude any values of x that make the denominator zero.

Similarly, to find **excluded horizontal regions,** solve the given equation for x in terms of y. If there is a radical of even degree in the right member, exclude all values of y that cause the expression under the radical sign to be negative. Also exclude any value of y that makes the denominator zero.

Example 4. Analyze the equation $x^2 - y^2 + 1 = 0$ and draw its graph.

Solution. Substituting zero for y in the given equation gives $x = \pm\sqrt{-1}$, which are imaginary numbers. Therefore the graph has no x intercept.

Substitution of zero for x in the given equation yields $y = \pm 1$. And so the graph intersects the y axis in the points $(0, 1)$ and $(0, -1)$.

If we substitute $-x$ for x throughout the given equation, it remains unchanged. Thus the graph is symmetric with respect to the y axis. When we substitute $-y$ for y throughout the equation, the equation is unchanged. Therefore the graph is symmetric with respect to the x axis.

The graph does not exist in the horizontal band where $-1 < y < 1$. For by solving the equation for x in terms of y, we obtain $x = \pm\sqrt{y^2 - 1}$; any value of y for which $|y| < 1$ makes $y^2 - 1$ negative and $\sqrt{y^2 - 1}$ imaginary. So no real value of x corresponds to any value of y between -1 and 1, and the graph has no points in the horizontal band between the lines $y = -1$ and $y = 1$.

Having all this information at our disposal, we need only plot a few points, say $(0, 1)$, $(1, \sqrt{2})$, and $(2, \sqrt{5})$. By symmetry, it follows that the points $(0, -1)$, $(1, -\sqrt{2})$, $(-1, \sqrt{2})$, $(-1, -\sqrt{2})$, $(2, -\sqrt{5})$, $(-2, \sqrt{5})$, and $(-2, -\sqrt{5})$ are also on the graph (Fig. 2-32). This graph is called a *hyperbola*.

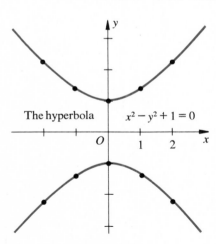

The hyperbola $\qquad x^2 - y^2 + 1 = 0$

Figure 2-32

Exercises

In Exercises 1 to 14, write an analysis of the given equation and then sketch its graph.

1. $3x^2 - 4y = 0$. (Parabola.)

2. $y^2 - 4x - 4 = 0$. (Parabola.)

3. $4x^2 + 9y^2 - 36 = 0$. (Ellipse.)

4. $16x^2 + y^2 - 16 = 0$. (Ellipse.)

5. $9x^2 - 4y^2 - 36 = 0$. (Hyperbola.)

6. $xy - 2 = 0$. (Hyperbola.)

7. $x^3 - y = 0$.

8. $x^3 - y^2 = 0$.

9. $x^4 + y^4 - 16 = 0$.

10. $x^2y - 4x + y = 0$.

11. $x^2y + 4y - 8 = 0$.

12. $x^2y - 4x + 2y = 0$.

13. $y = x(x - 3)(x - 5)$.

14. $x^3 + 3x^2 - y^2 = 0$.

15. Sketch the graph of each of the following equations.

(a) $y = \sqrt{3x}$;   (b) $y = -\sqrt{3x}$;

(c) $y^2 = 3x$.

16. Sketch the graph of each of the following equations.

(a) $2x - 5y = 0$;   (b) $2x + 5y = 0$;

(c) $4x^2 - 25y^2 = 0$.

2.7 GRAPHS OF INEQUALITIES

Section 2.5 showed that the Cartesian graph of the *equation*

(1) $$x^2 + y^2 = 9$$

is the set of points three units distant from the origin—that is, the circle whose center is the origin and whose radius is 3.

Similarly, the graph of the inequality

(2) $$x^2 + y^2 < 9$$

is the **open disk** consisting of all points *inside* the circle $x^2 + y^2 = 9$ [Fig. 2-33(a)]. The graph of

(3) $$x^2 + y^2 \leq 9$$

is the **closed disk** formed by the set of points *inside or on* the circle [Fig. 2-33(b)].

The graph of (2) is said to be an **open** region because it includes none of its

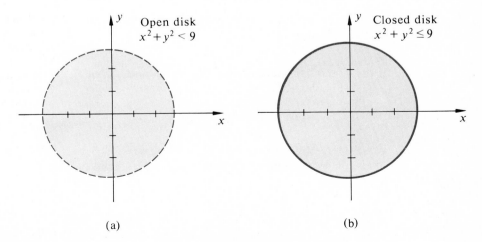

(a)   (b)

Figure 2-33

boundary points, and the graph of (3) is a **closed** region because it contains all of its boundary points (the circle itself).

The graph of

$$x^2 + y^2 > 9$$

is the open region consisting of all points in the plane that are outside the circle (1) (Fig. 2-34).

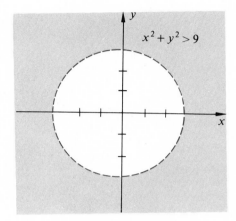

Figure 2-34

Just as the graph in the Cartesian plane of the equation

$$x = 5$$

is a vertical line five units to the right of the y axis, so the graph of the inequality

$$x > 5$$

is the open **half-plane** consisting of all points to the right of the line $x = 5$ [Fig. 2-35(a)]. The graph of

$$x \geq 5$$

is the closed half-plane consisting of all points on, or to the right of, the vertical line $x = 5$ [Fig. 2-35(b)].

The set

(4) $$\{P{:}(x, y) \mid x > 5, y < 2\}$$

is the intersection of the two open half-planes $x > 5$ and $y < 2$; that is, (4) is the set of points to the right of the vertical line $x = 5$ and below the horizontal line $y = 2$ (Fig. 2-36).

The graph of

$$-3 < x < 4$$

is the open vertical strip *between* the vertical lines $x = -3$ and $x = 4$ (Fig. 2-37). It is the intersection of the two open half-planes $x > -3$ and $x < 4$.

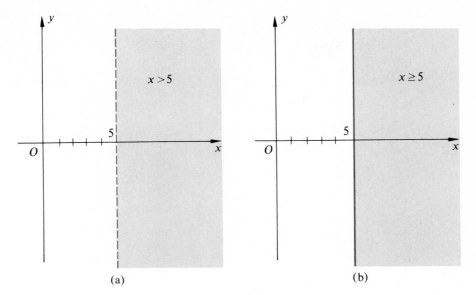

(a)   (b)

Figure 2-35

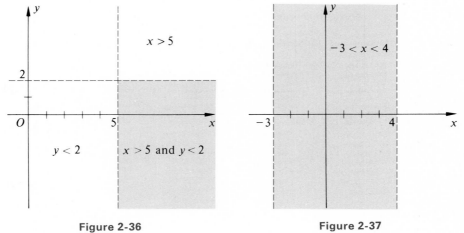

Figure 2-36   Figure 2-37

The set of points

$$\{P{:}(x, y)\,|-1 < x < 2, 3 < y < 5\}$$

is the open rectangular region shown in Fig. 2-38; it is the intersection of the open vertical strip $-1 < x < 2$ and the open horizontal strip $3 < y < 5$.

The **solution set of an inequality** in two variables, say x and y, one of which may be absent, is the set of ordered pairs of real numbers such that when the first number of a pair is substituted for x and the second for y, the inequality becomes a true statement.

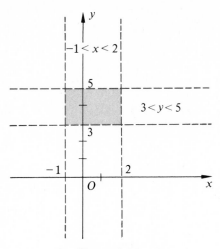

Figure 2-38

The Cartesian **graph of an inequality in x and y** is the set of points $P:(x, y)$ whose coordinates form the solution set of the inequality.

Example 1. Consider the line $2x + 3y - 6 = 0$. Show that a point $P_1:(x_1, y_1)$ is
      (a) *above* the line if and only if $2x_1 + 3y_1 - 6 > 0$,
      (b) *on* the line if and only if $2x_1 + 3y_1 - 6 = 0$,
      (c) *below* the line if and only if $2x_1 + 3y_1 - 6 < 0$.

Solution. The vertical line through $P_1:(x_1, y_1)$ is $x = x_1$ (Fig. 2-39). It intersects the given line in the point $Q:(x_1, (6 - 2x_1)/3)$. Then P_1 is above, on, or below the given line, depending on whether the directed distance $\overline{QP_1} > 0$, $\overline{QP_1} = 0$, or $\overline{QP_1} < 0$—that is, depending on whether $y_1 - (6 - 2x_1)/3 > 0$, $= 0$, or < 0. But this is equivalent to $2x_1 + 3y_1 - 6 > 0$, $= 0$, or < 0.

Proving the following theorem is left as an exercise for the reader.

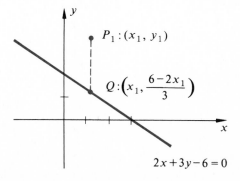

Figure 2-39

2.7.1   Theorem.   A point P_1:(x_1, y_1) is above, on, or below the nonvertical line $Ax + By + C = 0$, $B > 0$, depending on whether $Ax_1 + By_1 + C > 0$, $Ax_1 + By_1 + C = 0$, or $Ax_1 + By_1 + C < 0$.

Example 2.   Sketch the closed region

$$\{P:(x, y) \mid x^2 + (y - 1)^2 \leq 1, \ -x + 2y - 2 \geq 0\}.$$

Solution.   The given set of points is the intersection of the closed disk, with center $(0, 1)$ and radius 1, and the closed half-plane above or on the line $-x + 2y - 2 = 0$ (Fig. 2-40).

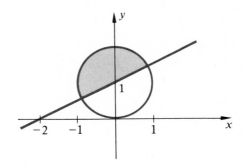

Figure 2-40

Exercises

1.   Sketch the closed region $\{P:(x, y) \mid x^2 + y^2 \leq 5\}$.

2.   Sketch the region $\{P:(x, y) \mid x^2 + y^2 > 5\}$. Is this region open or closed?

3.   Sketch the open half-plane $\{P:(x, y) \mid x + 5y - 10 > 0\}$.

4.   Sketch the closed half-plane $\{P:(x, y) \mid x - 2y - 4 \geq 0\}$.

5.   The equation of any nonvertical line can be written in the form $Ax + By + C = 0$, with $B > 0$. Show that the point P_1:(x_1, y_1) is *above, on,* or *below* the line, depending on whether $Ax_1 + By_1 + C > 0$, $Ax_1 + By_1 + C = 0$, or $Ax_1 + By_1 + C < 0$. (*Hint:* See Example 1.)

6.   Consider the circle whose center is C:(a, b) and whose radius is r. Show that a point P:(x, y) is
(a)   *inside* the circle if and only if $(x - a)^2 + (y - b)^2 < r^2$;
(b)   *on* the circle if and only if $(x - a)^2 + (y - b)^2 = r^2$;
(c)   *outside* the circle if and only if $(x - a)^2 + (y - b)^2 > r^2$.

(*Hint:* A point P is said to be **inside** the circle if and only if $|CP| < r$.)

7.   By means of two inequalities, define the set of all points above the line $5x + 4y - 20 = 0$ and to the right of the y axis. Sketch the graph of the set.

8.   By means of three inequalities in x and y, define the open region inside the triangle whose vertices are $(3, -1)$, $(10, 2)$, and $(7, 4)$.

9.   Sketch the open region $\{P:(x, y) \mid x^2 + y^2 < 3, y > 0\}$.

10.   Sketch the region $\{P:(x, y) \mid x^2 + y^2 < 4, x + y \geq 0\}$.

11.   Define by four inequalities in x and y the closed region whose boundaries are the triangle with vertices $(0, 5)$, $(-6, 0)$, and $(1, -3)$, and the circle with center at $(-2, 1)$ and radius 1. Sketch the region.

12.   Sketch the region

$$\{P:(x, y) \mid x^2 + y^2 - 6x - 8y + 24 > 0,$$
$$x^2 + y^2 - 12x - 10y + 25 < 0\}.$$

2.8 FUNCTIONS

The concept of *function* is one of the most basic in all mathematics, and it plays an indispensable role in calculus.

The table

x	-5	-4	-3	-2	-1	0	1	2	3	4	5
$y = 2x + 1$	-9	-7	-5	-3	-1	1	3	5	7	9	11

establishes a correspondence between the numbers of two sets,

$$\{-5, -4, -3, -2, -1, 0, 1, 2, 3, 4, 5\}$$

and

$$\{-9, -7, -5, -3, -1, 1, 3, 5, 7, 9, 11\},$$

such that to each element x of the first set there corresponds one and only one number $y \, (= 2x + 1)$ of the second set. This correspondence is an example of a function.

The graph (Fig. 2-41) of $y = x^2$, $-2 \leq x \leq 2$, sets up a correspondence between the numbers x of the closed interval $[-2, 2]$ and the numbers y of the closed interval $[0, 4]$, such that to each $x \in [-2, 2]$ there corresponds exactly one number $y \in [0, 4]$. Thus to $x = -2$ corresponds $y = x^2 = (-2)^2 = 4$; to $x = \frac{1}{2}$ corresponds $y = (\frac{1}{2})^2 = \frac{1}{4}$; and to $x = \frac{5}{3}$ corresponds $y = (\frac{5}{3})^2 = \frac{25}{9} = 2\frac{7}{9}$. This correspondence is another example of a function.

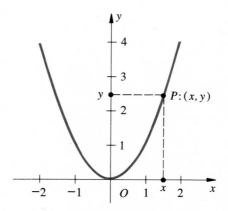

The graph of $y = x^2$, $-2 \leq x \leq 2$

Figure 2-41

2.8.1 Definition. A **function** is a rule of correspondence between two nonempty sets of elements, called the **domain** and the **range** of the function, such that to each element of the domain there corresponds one and only one element of the range, and each element of the range is the correspondent of at least one element of the domain. A function is often called a **mapping** and is said to **map** its domain onto its range.

Example 1. The formula,

$$V = l^3,$$

for the volume V of a cube in terms of the length l of an edge, determines a correspondence between the set of all positive numbers l and the set of all values of V. This correspondence is a function because to each positive number l there corresponds *one and only one* number V. The formula $V = l^3$ is the rule of correspondence between the domain, which is the set of positive numbers l, and the range, which is the set of positive numbers V.

Example 2. The equation

$$y = 2x^2 + 5, \qquad x \in R,$$

defines a function having domain R (the set of all real numbers) and range $\{y \mid y \in R, y \geq 5\}$. To each real number x there corresponds a *unique* number y in the range $[5, \infty)$. As an illustration, to $x = 3$ in the domain corresponds $y = 2(3)^2 + 5 = 23$ in the range; to -1 in the domain corresponds $2(-1)^2 + 5 = 7$ in the range; to $\frac{1}{2}$ in the domain corresponds $2(\frac{1}{2})^2 + 5 = 5\frac{1}{2}$ in the range.

Example 3. The equation

$$y = \sqrt{x - 7}, \qquad x \geq 7,$$

defines a function whose domain is the set $\{x \mid x \geq 7\}$ and whose range is the set $\{y \mid y \geq 0\}$. To each number x in the domain there corresponds *one and only one* number y in the range. As illustration, to $x = 7$ corresponds $y = 0$; to $x = 8$ corresponds $y = 1$; and to $x = 11$ corresponds $y = 2$. Notice that $x = 3$ could not be in the domain of this function because $y = \sqrt{3 - 7} = \sqrt{-4}$, which is imaginary.

Example 4. A package not exceeding 12 ounces can be sent by first-class mail in this country. For a parcel weighing w ounces, the cost c in cents to mail it is shown in the following table.

w	$0 < w \leq 1$	$1 < w \leq 2$	$2 < w \leq 3$	$3 < w \leq 4$	$\cdots$	$11 < w \leq 12$
c	10	20	30	40		120

This table is a rule of correspondence between the set $A = \{w \mid 0 < w \leq 12\}$ and the set $B = \{10, 20, 30, 40, \ldots, 110, 120\}$, such that to each number w in A there corresponds *one and only one* number c in B, and each number c in B is the correspondent of at least one number w in A. Thus the correspondence is a function whose domain is the set A and whose range is the set B.

A function, then, involves three things: a nonempty set of elements called the *domain* of the function, a nonempty set of elements called the *range* of the function, and a *rule of correspondence* that enables us to determine precisely which element of the range corresponds to each element of the domain. Of course, if we are given the domain of a function and its rule of correspondence, the range of the function is uniquely determined and need not be specified. The rule of correspondence may have the form of a formula or a table of corresponding values, or any other adequate form.

Although the elements of the domain or range of a function need not be numbers, it is to be understood that when they *are* numbers, they will be *real numbers* unless the contrary is expressly stated.

Functions are often represented by such letters as f, g, F, ϕ, ψ. If x is an element in the domain of a function f, then $f(x)$, read "f of x," is used to represent the corresponding element of the range. The element $f(x)$ of the range is called the **value** of the function f for the element x in the domain. The beginner is cautioned against thinking of $f(x)$ as "f times x"; $f(x)$ is a short way of writing the phrase "the value of the function f at the number x" or, more briefly, "f of x."

Thus if f is the function defined by

$$f(x) = \frac{x^2 - 5}{x},$$

with domain consisting of $\{x \mid x \in R, x \neq 0\}$, then $f(1) = (1^2 - 5)/1 = -4$, $f(-6) = [(-6)^2 - 5]/(-6) = -31/6$, and $f(\pi) = (\pi^2 - 5)/\pi$.

If no domain is specified when a function is defined, it will be understood that the domain consists of all those real numbers for which the value of the function exists and is real. For example, if g is the function defined by $g(u) = \sqrt{u^2 - 9}$ and if no domain is specified, its domain will be understood to be the set of all real numbers u such that $|u| \geq 3$. Some values of g are $g(5) = \sqrt{25 - 9} = 4$, $g(3) = \sqrt{9 - 9} = 0$, and $g(-4) = \sqrt{(-4)^2 - 9} = \sqrt{7}$; but $g(1) = \sqrt{1 - 9} = 2\sqrt{-2}$ does not exist because it is not a real number, and so 1 is not in the domain of the function.

When a function f is defined by an equation $y = f(x)$, x is often called the **independent variable** and y the **dependent variable**. Any element of the domain may be chosen for a value of the independent variable x, but that choice completely determines the corresponding value of the dependent variable y. Thus the value of y in $y = f(x)$ is dependent on the freely chosen value of x.

The domain of a function f will be symbolized by $\mathfrak{D}_f$ and its range by $\mathfrak{R}_f$. Thus for the function f defined by $f(x) = -\sqrt{x}$, $\mathfrak{D}_f$ is the set of nonnegative real numbers and $\mathfrak{R}_f$ is the set of nonpositive real numbers. In the symbolism of sets, $\mathfrak{D}_f = \{x \mid 0 \leq x\}$ and $\mathfrak{R}_f = \{f(x) \mid f(x) \leq 0\}$ for this function. Similarly, the domain of a function g is $\mathfrak{D}_g$ and its range is $\mathfrak{R}_g$.

Example 5. The function F, defined by

$$F(x) = \frac{1}{x^2 + 1},$$

has domain $\mathfrak{D}_F = \{x \mid x \in R\}$ and range $\mathfrak{R}_F = \{F(x) \mid F(x) \in R, 0 < F(x) \leq 1\}$. Some values of F are $F(0) = 1/(0^2 + 1) = 1$, $F(-2) = 1/[(-2)^2 + 1] = 1/5$, $F(6) = 1/(6^2 + 1) = 1/37$, and $F(0.2) = 1/[(0.2)^2 + 1] = 1/1.04 = 25/26$.

Example 6. If G is the function defined by

$$G(t) = \sqrt{16 - t^2}, \qquad \mathfrak{D}_G = \{t \mid -4 \leq t \leq 4\},$$

find the values of G at

(a) -1;   (b) 2;   (c) 0;   (d) $-\dfrac{3}{2}$;   (e) 4.

Solution. (a) The value of the function G at $t = -1$ is $G(-1) = \sqrt{16 - (-1)^2} = \sqrt{15}$;
(b) $G(2) = \sqrt{16 - 2^2} = 2\sqrt{3}$;
(c) $G(0) = \sqrt{16 - 0^2} = 4$;
(d) $G\left(-\dfrac{3}{2}\right) = \sqrt{16 - \left(-\dfrac{3}{2}\right)^2} = \dfrac{\sqrt{55}}{2}$;
(e) $G(4) = \sqrt{16 - 4^2} = 0$.

Example 7. Let $f(x) = 1/x$, $x \neq 0$. Then

$$f(t - 1) = \frac{1}{t - 1}, \qquad f(4 - a) = \frac{1}{4 - a},$$

$$f(3x) = \frac{1}{3x}, \qquad f(a^2) = \frac{1}{a^2},$$

$$f(x^2) = \frac{1}{x^2}, \qquad f(1/x) = 1/(1/x)$$

$$= x.$$

Example 8. If $f(x) = 1/x$, find and simplify

$$\frac{f(a + h) - f(a)}{h}, \qquad ah \neq 0, \quad (a + h) \neq 0.$$

Solution. Since

$$f(a + h) - f(a) = \frac{1}{a + h} - \frac{1}{a} = \frac{a - (a + h)}{(a + h)a} = \frac{-h}{a^2 + ah},$$

then

$$\frac{f(a + h) - f(a)}{h} = \frac{-h}{a^2 + ah} \cdot \frac{1}{h} = \frac{-1}{a^2 + ah}.$$

A function f, as defined above, is often called a **mapping** of its domain, $\mathfrak{D}_f$, onto its range, $\mathfrak{R}_f$. And $f(x)$, the value of a function f at x, is also called the **image** of x in the mapping f. So if f is the mapping defined by $f(x) = \sqrt{1 - x} + 3$, f maps its domain, $(-\infty, 1]$, onto its range, $[3, \infty)$. In particular, the image of 0 is 4 in this mapping, and the image of -8 is 6.

A function gives rise to a set of ordered pairs such that in each pair the first element belongs to the domain of the function and the second element is the corresponding element of the range. Therefore, for any function f, the set of such ordered pairs is

$$\{(x, f(x)) \mid x \in \mathfrak{D}_f\}.$$

This discussion suggests an alternative definition of a function that is equivalent to 2.8.1.

2.8.2 Definition. A **function** is a nonempty set of ordered pairs such that no two distinct pairs have the same first element. Thus if (x, y) and (x, z) are ordered pairs belonging to the same function, then $y = z$. The set of first elements of the ordered pairs is the **domain** of the function and the set of second elements is the **range**.

To illustrate, $\{(1, 3), (0, 5), (-1, 2)\}$ is a function with domain $\{-1, 0, 1\}$ and range $\{2, 3, 5\}$; also, $\{(x, 3x^2) \mid x \in R\}$ is a function, having domain R and range $[0, \infty)$. But $\{(1, 3), (0, 5), (1, 2)\}$ is not a function. (Why?)

When the elements of these ordered pairs are numbers, each ordered pair can be interpreted as the Cartesian coordinates of a point in the plane. The totality of such points for any given function is the Cartesian **graph** of that function.

Since to each element of the domain of a function there corresponds exactly one element of the range, *no vertical line can intersect the Cartesian graph of a function in more than one point* (see Figs. 2-41 and 2-42).

Example 9. The function ϕ, defined by $\phi(x) = \sqrt{16 - x^2}$, has for its graph the semicircle shown in Fig. 2-42. For this function, $\mathfrak{D} = \{x \mid |x| \le 4\}$ and $\mathfrak{R} = \{\phi(x) \mid 0 \le \phi(x) \le 4\}$.

Recall from Section 2.1 that the projection of a point P onto a straight line l is the point of intersection of l with a line through P perpendicular to l (Fig. 2-2).

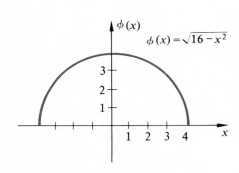

Figure 2-42

The **projection of a graph** onto a line is the set of projections of the points of the graph onto the line. As an illustration, the projection of the semicircle in Fig. 2-42 onto the x axis is the closed interval $[-4, 4]$.

When the domain of a function is a set of numbers, $\{x \mid x \in \mathfrak{D}\}$, we shall often speak of the domain as the corresponding set of *points* on the x axis. Thus we can refer to the domain of a function as the projection of its Cartesian graph onto the horizontal axis; similarly, its range is its projection onto the vertical axis (see Example 9, above, and Fig. 2-42).

Exercises

1. If $f(x) = x^2 - 1$, find
(a) $f(1)$;   (b) $f(-2)$;   (c) $f(0)$;
(d) $f(k)$;   (e) $f(-b)$;   (f) $f\left(\frac{1}{2}\right)$;
(g) $f(2t)$;   (h) $f(3x)$;   (i) $f\left(\frac{1}{x}\right)$.

2. If $F(x) = 3x^3 + x$, find
(a) $F(-6)$;   (b) $F\left(\frac{1}{2}\right)$;   (c) $F(-0.2)$;
(d) $F(\sqrt{3})$;   (e) $F(\pi)$;   (f) $F\left(\frac{1}{x}\right)$.

3. If $G(y) = 1/(y - 1)$, find
(a) $G(0)$;   (b) $G(0.999)$;   (c) $G(101)$;
(d) $G(y^2)$;   (e) $G(-x)$;   (f) $G\left(\frac{1}{x^2}\right)$.

4. If $\phi(t) = \sqrt{t}/(1 + t^2)$, find
(a) $\phi(0)$;   (b) $\phi\left(\frac{1}{4}\right)$;   (c) $\phi(x^3)$;
(d) $\phi(x + 2)$;   (e) $\phi(-t)$;   (f) $\phi\left(\frac{1}{z^4}\right)$.

5. Sketch the graph of each of the following sets

of ordered pairs. Tell which are functions and state their largest possible domains.

(a) $\{(x, y) \mid x^2 + y^2 = 9\}$;
(b) $\{(x, y) \mid x^2 + y^2 = 9, x \geq 0\}$;
(c) $\{(x, y) \mid x^2 + y^2 = 9, y \geq 0\}$;
(d) $\{(x, y) \mid x^2 + y^2 = 0\}$.

6. Sketch the graph of each of the sets of ordered pairs shown. Tell which are functions and state their largest possible domains.

(a) $\{(x, y) \mid y^2 - 4x = 0\}$;
(b) $\{(x, y) \mid y^2 - 4x = 0, y \geq 0\}$;
(c) $\{(x, y) \mid x^2 - 4y = 0\}$;
(d) $\{(x, y) \mid x^2 - 4y = 0, x < 0\}$.

7. Let $f(x) = 2x^2 - 1$. Find $f(a)$, $f(a + h)$, and $f(a + h) - f(a)$. Then find

$$\frac{f(a + h) - f(a)}{h}$$

and simplify your result.

8. If $F(t) = 4t^3$, find and simplify

$$\frac{F(a + h) - F(a)}{h}.$$

9. State the largest possible domain and range for each of the functions F, g, ψ, and H, whose rules of correspondence are

(a) $F(z) = \sqrt{2z + 3}$; (b) $g(v) = \dfrac{1}{4v - 1}$;

(c) $\psi(x) = \sqrt{x^2 - 9}$;
(d) $H(y) = -\sqrt{625 - y^4}$.

10. Find the largest possible domain and range for each of the functions f, G, ϕ, and F, whose rules of correspondence are

(a) $f(x) = \dfrac{4 - x^2}{x + 2}$; (b) $G(y) = \sqrt{(y + 1)^{-1}}$;

(c) $\phi(u) = |2u - 3|$; (d) $F(t) = t^{2/3} - 4$.

Recall that when a function f is defined by an equation $y = f(x)$, we call x the *independent variable* and y the *dependent variable*.

11. Denote by p the perimeter of an equilateral triangle. Express the area of an equilateral triangle as the value $A(p)$ of a function A whose independent variable is p.

12. A right circular cylinder of radius r is inscribed in a sphere of radius $2r$. Express the volume $V(r)$ of the cylinder as the value of a function V whose independent variable is the radius r, of the cylinder.

13. A right circular cone is inside a cube of edge l. The base of the cone is inscribed in one face of the cube and its vertex is in the opposite face. Express the volume $V(l)$ of the region between the cone and the cube as the value of a function V whose independent variable is the length, l, of an edge of the cube.

14. An open box is formed from a rectangular piece of cardboard by cutting out, and discarding, equal squares from the four corners and then bending up the sides. If the dimensions of the original piece of cardboard were 15 inches by 21 inches, express the volume of the box as the value of a function whose independent variable is the length x of an edge of the squares cut out.

15. An open box is to be made from a rectangular piece of sheet metal, 10 inches by 16 inches, by cutting out, and discarding, equal squares from each corner and folding up the sides. Express the volume $V(x)$ of the completed box as the value of a function V whose independent variable is the length, x, of a side of the squares cut out. What is the domain of the function? Draw the graph of the function, and from it estimate the size of the cut-out squares that makes the volume of the box a maximum.

16. Express the amount $E(x)$ by which a number x exceeds its cube, as the value of a function E whose independent variable is x. Draw the graph of the function and from the graph estimate the positive number that exceeds its cube by the maximum amount.

17. Graph the function F, defined by

$$F(x) = \begin{cases} 1 & \text{for } x \geq 0, \\ -1 & \text{for } x < 0. \end{cases}$$

18. The Heaviside unit function, H, is defined by

$$H(x) = \begin{cases} 1 & \text{if } x > 0, \\ \frac{1}{2} & \text{if } x = 0, \\ 0 & \text{if } x < 0. \end{cases}$$

Draw the graph of
(a) $H(t)$; (b) $H(t - 3)$;
(c) $H(t^2 - 1)$; (d) $H(t) - H(t + 1)$.

2.9 SPECIAL FUNCTIONS

A function f is called a **constant function** if its range consists of a single number k—that is, if $f(x) = k$ for all numbers x in its domain. For instance, if $f(x) = 4$ for all real numbers x, the function f is a constant function. The graph of this constant function is a horizontal straight line, four units above the x axis (Fig. 2-43). If $f(n) = 3$ for $n = 1, 2, 3, \cdots$, the graph of this particular constant function is the set of isolated points whose coordinates are $(1, 3), (2, 3), (3, 3), (4, 3), \cdots$.

The **identity function** is defined by the equation $f(x) = x$. If we let its domain be the set of all real numbers, its Cartesian graph is the straight line through the origin with slope 1 (Fig. 2-44). Some values of this function are $f(-3) = -3$, $f(\frac{1}{3}) = \frac{1}{3}$, $f(0.005) = 0.005$.

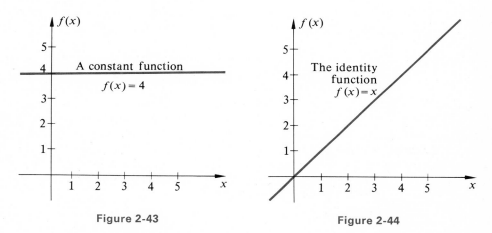

Figure 2-43   Figure 2-44

The function f, defined by

$$f(x) = a_0 x^n + a_1 x^{n-1} + \cdots + a_{n-1} x + a_n,$$

where n is a nonnegative integer and $a_0, a_1, \cdots, a_n$, are real constants, is called a **polynomial function**. If $a_0 \neq 0$, the **degree** of the polynomial function is n; and if $n = 0$, so that $f(x) = a_0$ for all numbers x, the polynomial function is a constant function. Thus a constant function whose value is not zero may be thought of as a polynomial function of degree zero.

If $a_0 = 0$ and $n = 0$, the degree of the polynomial function is not defined; this polynomial function without degree is called the **zero function** because its value is $f(x) = 0$ for all x.

A **linear function** is a polynomial function of degree 1. It is therefore defined by

$$f(x) = ax + b,$$

where a and b are constants and $a \neq 0$. Its graph is a straight line with slope a and y intercept b.

When the degree of a polynomial function is 2, we have a **quadratic function;** it is defined by the rule of correspondence

$$f(x) = ax^2 + bx + c,$$

where a, b, and c are constants and $a \neq 0$. Its graph is a parabola that is symmetric with respect to the vertical line $x = -b/2a$, and it opens upward if $a > 0$ and downward if $a < 0$ (Fig. 2-45).

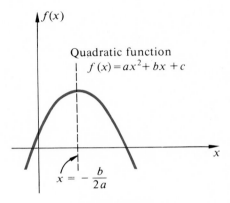

Quadratic function
$f(x) = ax^2 + bx + c$

$x = -\dfrac{b}{2a}$

Figure 2-45

Example. The rule of correspondence $f(x) = x^4 - 2x^3 + 7$ defines a polynomial function of degree 4; $f(x) = -2x^2 + x - 3$ defines a quadratic polynomial function; $f(x) = 12$ defines a polynomial function of degree zero; and the function defined by $f(x) = 0$ *for all* x is a polynomial function without degree.

A **rational function** is a function that can be expressed as the quotient of two polynomial functions. It can be defined by

$$f(x) = \frac{a_0 x^n + a_1 x^{n-1} + \cdots + a_n}{b_0 x^m + \cdots + b_m}$$

for values of x for which the denominator is not zero.

The **square root function** is defined by

$$f(x) = \sqrt{x}.$$

Recall that $\sqrt{x}$ is the nonnegative number whose square is x. The graph of the square root function is shown in Fig. 2-46.

An **algebraic function** is defined by $y = f(x)$ if and only if $y = f(x)$ satisfies identically an equation of the form

$$p_0(x)y^n + p_1(x)y^{n-1} + \cdots + p_{n-1}(x)y + p_n(x) = 0,$$

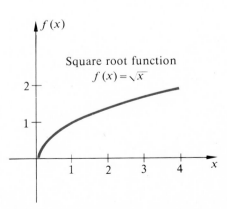

Square root function
$f(x) = \sqrt{x}$

Figure 2-46

where $p_0, p_1, \cdots, p_n$ are polynomials and n is a positive integer. A function that can be formed by a finite number of algebraic operations (addition, subtraction, multiplication, division, and root extraction) on the identity function and constant functions is algebraic. For instance, the function f defined by

$$f(x) = \frac{(2x^2 + 3x)\sqrt{x - 2}}{4x - (19x^5 + 25)^{2/3}}$$

is an algebraic function. However, not every algebraic function can be formed in this way.

A **transcendental function** is any function that is not algebraic. Examples of elementary transcendental functions are the trigonometric functions, the inverse trigonometric functions, the logarithmic functions, and the exponential functions.

An interesting special function is the **greatest integer function**. Its value $f(x)$, denoted by $[\![x]\!]$, is defined to be the greatest integer that is less than or equal to x. Thus $[\![\sqrt{2}]\!] = 1$, $[\![-\tfrac{4}{3}]\!] = -2$, and $[\![\pi]\!] = 3$. The graph of the greatest integer function is shown in Fig. 2-47.

Still another special function is the **absolute value function**. It is defined by

$$f(x) = |x|,$$

and its graph is shown in Fig. 2-48.

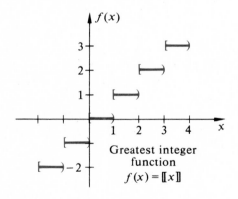

Greatest integer
function
$f(x) = [\![x]\!]$

Figure 2-47

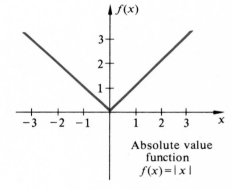

Absolute value
function
$f(x) = |x|$

Figure 2-48

Exercises

In Exercises 1 through 12, sketch the graph of the function whose rule of correspondence is given and state its domain and range. Specify which of the terms—algebraic, rational, irrational, polynomial, and constant—apply to the function.

1. $f(x) = -4$.

2. $f(x) = 0$.

3. $F(x) = 2x + 1$.

4. $F(x) = 3x - \sqrt{2}$.

5. $g(x) = 3x^2 + 2x - 1$.

6. $g(u) = \dfrac{u^3}{8}$.

7. $\phi(z) = \dfrac{2z + 1}{z - 1}$.

8. $G(t) = \dfrac{3t^2 + 2t - 1}{t + 1}$.

9. $f(w) = \sqrt{w - 1}$.

10. $h(x) = \sqrt{x^2 + 4}$.

11. $\Phi(x) = \dfrac{1}{\sqrt{x + 3}}$.

12. $f(x) = \sqrt{\dfrac{3x}{x^2 - 4}}$.

In Exercises 13 to 20, sketch the graph of the function whose rule of correspondence is given.

13. $f(x) = |2x|$.

14. $F(t) = -|x + 3|$.

15. $g(u) = \left[\!\!\left[\dfrac{x}{2}\right]\!\!\right]$.

16. $G(x) = [\![2x - 1]\!]$.

17. $\phi(x) = -[\![x]\!]$.

18. $\Phi(x) = [\![x^2]\!]$.

19. $g(t) = \begin{cases} 1 & \text{when } t \le 0, \\ t + 1 & \text{when } 0 < t < 2, \\ t^2 - 1 & \text{when } t \ge 2. \end{cases}$

20. $F(t) = \begin{cases} [\![t]\!] & \text{when } [\![t]\!] \text{ is even,} \\ 2t - [\![t + 1]\!] & \text{when } [\![t]\!] \text{ is odd.} \end{cases}$

21. Let x be any positive number. In each of the following, use the greatest integer notation to write a single equation defining the function N, when
(a) $N(x)$ is the number of even positive integers less than or equal to x.

(b) $N(x)$ is the number of odd positive integers less than or equal to x.

A function f is said to be an **even function** if $f(-x) = f(x)$ for all x in the domain of f; moreover, f is an **odd function** if $f(-x) = -f(x)$ for all x in the domain of f. It follows at once from the tests for symmetry in Section 2.6 that *the graph of an even function is symmetric with respect to the y axis and that the graph of an odd function is symmetric with respect to the origin.*

22. Give an example of
(a) an even function;
(b) an odd function;
(c) a function that is neither even nor odd.

23. Each of the following defines a function; tell whether the function is even, odd, or neither.
(a) $f(x) = 5$;
(b) $g(x) = 3x$;
(c) $F(x) = 12x^7 - 5x^3$;
(d) $F(x) = x^5 - 2x^3 + 4$;
(e) $G(x) = \dfrac{1}{x}, x \ne 0$;
(f) $\psi(x) = |x|$;
(g) $f(x) = 3|x| - 1$;
(h) $f(x) = \begin{cases} 1 & \text{if } x \text{ is rational,} \\ -1 & \text{if } x \text{ is irrational.} \end{cases}$
(i) $F(x) = [\![x]\!]$. (*Hint:* Sketch the graph.)

2.10 OPERATIONS ON FUNCTIONS

Functions are not numbers. But just as numbers may be added, subtracted, and so on, to get other numbers, we will define certain operations on functions and call the functions resulting from such operations the sum, product, etc., of the original functions.

2.10.1 Definition. If f and g are functions with domains $\mathfrak{D}_f$ and $\mathfrak{D}_g$, their **sum**, indicated by $f + g$, their **difference**, $f - g$, their **product**, $f \cdot g$, and their **quotient**, f/g, are the functions defined by

$$(f + g)(x) = f(x) + g(x),$$
$$(f - g)(x) = f(x) - g(x),$$
$$(f \cdot g)(x) = f(x) \cdot g(x),$$
$$(f/g)(x) = f(x)/g(x),$$

respectively; in each case, the domain consists of all values of x common to $\mathfrak{D}_f$ and $\mathfrak{D}_g$, —that is,

$$\mathfrak{D}_{f+g} = \mathfrak{D}_{f-g} = \mathfrak{D}_{f \cdot g} = \mathfrak{D}_{f/g} = \mathfrak{D}_f \cap \mathfrak{D}_g,$$

except that the values of x for which $g(x) = 0$ are excluded from the domain $\mathfrak{D}_{f/g}$.

Example 1. If f and g are functions defined by $f(x) = 1/(x - 3)$ and $g(x) = \sqrt{x}$, then their sum, $f + g$, is the function defined by $(f + g)(x) = f(x) + g(x) = 1/(x - 3) + \sqrt{x}$. The domain of $f + g$ is $\mathfrak{D}_{f+g} = \{x \mid x \geq 0 \text{ and } x \neq 3\}$.

Example 2. If f is the function defined by $f(x) = x/(x + 5)$ and g is the function defined by $g(x) = -2x^3$, then $f \cdot g$ is the function defined by $(f \cdot g)(x) = f(x) \cdot g(x) = -2x^4/(x + 5)$, and its domain is $\mathfrak{D}_{f \cdot g} = \{x \mid x \neq -5\}$.

We symbolize $f \cdot f$, the product of a function f by itself, by f^2. Thus if f is defined by $f(x) = -4x^3$, then f^2 is the function defined by $f^2(x) = f(x) \cdot f(x) = (-4x^3)(-4x^3) = 16x^6$.

Example 3. If $f(x) = \sqrt{x + 6}$ and $g(x) = 2x/(x^2 - 4)$, then

$$(f/g)(x) = f(x)/g(x) = (x^2 - 4)\sqrt{x + 6}/(2x),$$

and the domain of this quotient function is any real number not less than -6, except zero and ± 2.

Another, and very important, way to construct a new function from other functions is by *composition*. Let f and g be functions defined by $f(x) = 1/x$, $x \neq 0$, and $g(x) = 2x - 7$. Then the function F, defined by $F(x) = f(g(x)) = f(2x - 7) = 1/(2x - 7)$, is called the *composition of f with g*. The domain of F is the set of all real numbers x for which $2x - 7 \neq 0$—that is, all real numbers except $x = \frac{7}{2}$.

2.10.2 Definition. Let f and g be functions. The **composite function** $f \circ g$ (read "f circle g") is defined by

$$(f \circ g)(x) = f(g(x)).$$

Its domain consists of all numbers x in the domain of g for which $g(x)$ is in the domain of f.

The function $f \circ g$ is often called the **composition** of f with g or *the function f of the function g.*

Example 4. If f and g are functions defined by $f(x) = \sqrt{x}$ and $g(x) = 3x - 4$, find $F(x)$ when $F = f \circ g$, the composition of f with g. What is the domain of $f \circ g$?

Solution. $F(x) = (f \circ g)(x) = f(g(x)) = f(3x - 4) = \sqrt{3x - 4}$. Since the domain of f is the set of all nonnegative numbers, the domain of $F = f \circ g$ is the set of all real numbers x for which $g(x) = 3x - 4 \geq 0$—that is, $\mathfrak{D}_F = \{x \mid x \in R, x \geq \frac{4}{3}\}$.

Example 5. Let f and g be the functions defined by $f(x) = 6x/(x^2 - 9)$ and $g(x) = \sqrt{3x}$. Then the composite function $F = f \circ g$ is defined by

$$F(x) = (f \circ g)(x) = f(g(x)) = f(\sqrt{3x})$$

$$= \frac{6\sqrt{3x}}{(\sqrt{3x})^2 - 9} = \frac{6\sqrt{3x}}{3x - 9} = \frac{2\sqrt{3x}}{x - 3}.$$

The domain of $F = f \circ g$ is $[0, 3) \cup (3, \infty)$.

Exercises

In Exercises 1, 2, and 3, $f(x) = \sqrt{x^2 - 1}$ and $g(x) = 2/x$.

1. Write expressions defining
(a) $f + g$,   (b) $f - g$,
(c) $f \cdot g$,   (d) f/g,
and state their domains.

2. Form $f \circ g$, the composition of f with g, and tell its domain.

3. Form $g \circ f$, the composition of g with f, and state its domain.

In Exercises 4, 5, and 6, $f(x) = 1/(2x - 5)$ and $g(x) = \sqrt{x - 1}$.

4. Write expressions defining
(a) $f + g$,   (b) $f \cdot g$,   (c) f/g,
and give their domains.

5. Write an expression that defines, $f \circ g$, the composition of f with g. What is its domain?

6. Form $g \circ f$, the composition of g with f, and tell its domain.

In Exercises 7, 8, and 9, $f(x) = \sqrt{x - 4}$ and $g(x) = |x|$.

7. Find
(a) $(f + g)(x)$,   (b) $(f \cdot g)(x)$,   (c) $(f/g)(x)$,
and state their domains.

8. Find $(f \circ g)(x)$. What is the domain of $f \circ g$?

9. Find $(g \circ f)(x)$. What is the domain of $g \circ f$?

10. If $f(x) = [\![x]\!]$ and $g(x) = |x|$, write an expression defining $f \circ g$, the composition of f with g, and state its domain. Sketch the graph of $f \circ g$.

11. If $f(x) = |x|$ and $g(x) = [\![x]\!]$, construct $f \circ g$, and state its domain. Sketch the graph of $f \circ g$.

12. By means of a counterexample, show that the operation of composition of functions is not, in general, commutative; that is, find two functions f and g, such that $f \circ g$ exists and yet $f \circ g \neq g \circ f$.

13. Sketch the graph of the function F that is defined by $F(x) = |x| - [\![x]\!]$, and state its domain.

14. Sketch the graph of the function ϕ defined by $\phi(x) = |x|/[\![x]\!]$. State the domain and range of ϕ.

15. Tell whether the function defined by each of the following rules of correspondence is an even function, an odd function, or neither. (See page 69.)
(a) $f(t) = [\![t]\!]$;   (b) $F(s) = [\![s^2]\!]$.

16. Is the function f, defined by $f(x) = x - [\![x]\!]$ for $-2 \leq x \leq 2$, odd, even, or neither? Sketch its graph.

17. Is the function F, defined by $F(x) = x^2 - [\![x^2]\!]$ for $-2 \leq x \leq 2$, odd, even, or neither? Sketch its graph.

18. Let x and y be any real numbers. For the greatest integer function, prove that either $[\![x + y]\!] = [\![x]\!] + [\![y]\!]$ or $[\![x + y]\!] = [\![x]\!] + [\![y]\!] + 1$.

19. State whether each of the following is an odd function, an even function, or neither. Prove your statements.
(a) The sum of two even functions.
(b) The sum of two odd functions.
(c) The product of two even functions.
(d) The product of two odd functions.
(e) The product of an even function and an odd function.

20. Prove that a polynomial function can always be expressed as the sum of an odd function and an even function.

21. Let F be any function whose domain contains $-x$ whenever it contains x. Prove that
(a) $F(x) - F(-x)$ defines an odd function.
(b) $F(x) + F(-x)$ defines an even function.
(c) F can always be expressed as the sum of an odd function and an even function.

22. Show that if f and g are odd functions, $f \circ g$ is an odd function.

2.11 INVERSE OF A FUNCTION

We know (2.8.1) that a function is a correspondence between two sets of elements, the domain $\mathfrak{D}$ and range $\mathfrak{R}$, such that to each element in the domain there corresponds one and only one element of the range, and each element of the range is the correspondent of *at least one* element of the domain.

There is nothing in the definition of a function to prohibit several elements of the domain from having the same correspondent in the range. The quadratic function defined by $f(x) = x^2$, $-2 \leq x \leq 2$, has domain $\mathfrak{D}_f = [-2, 2]$ and range $\mathfrak{R}_f = [0, 4]$ (see Fig. 2-49). To each number x in $\mathfrak{D}_f$ corresponds one and only one number x^2 in $\mathfrak{R}_f$. But each positive number in the range $\mathfrak{R}_f$ is the correspondent of two distinct numbers, x and $-x$, in the domain $\mathfrak{D}_f$. Thus to the number 9/4 (in the range) there correspond both 3/2 and $-3/2$, since $f(3/2) = f(-3/2) = 9/4$.

When a function has the additional property that no two distinct elements of the domain have the same correspondent in the range, the function is said to be *one-to-one*.

For example (Fig. 2-50), the cubic function defined by $F(x) = x^3$, $-2 \leq x \leq 2$, is one-to-one because distinct numbers in the domain $\mathfrak{D}_F = [-2, 2]$ always have distinct cubes in the range $\mathfrak{R}_F = [-8, 8]$.

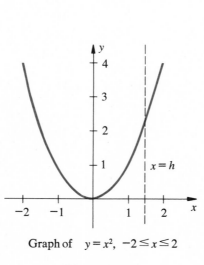

Graph of $y = x^2$, $-2 \leq x \leq 2$.

Figure 2-49

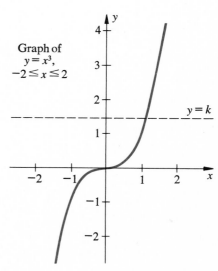

Graph of
$y = x^3$,
$-2 \leq x \leq 2$

$y = k$

Figure 2-50

2.11.1 Definition. A function f, with domain $\mathfrak{D}$ is said to be **one-to-one** if and only if

$$x_1 \in \mathfrak{D}, \quad x_2 \in \mathfrak{D}, \quad x_1 \neq x_2 \quad \text{implies} \quad f(x_1) \neq f(x_2).$$

Let f be a function with domain $\mathfrak{D}$ and range $\mathfrak{R}$. We have seen that any vertical line $x = h$, $h \in \mathfrak{D}$, intersects the graph of f in one and only one point (Fig. 2-49).

When f is one-to-one, it is also true that any horizontal line, $y = k$, $k \in \mathfrak{R}$, intersects the graph of the function in exactly one point (Fig. 2-50).

If a function f is one-to-one, its rule of correspondence, $y = f(x)$, defines a second function whose typical ordered pair of numbers is (y, x) instead of (x, y). This new function is called the *inverse* of f. Just as the set of first numbers x in the ordered pairs (x, y) of f constitutes the domain of f, so also the set of first numbers y in the ordered pairs (y, x) of the inverse of f forms the domain of the inverse. Thus the domain of the inverse function is the range of f, and the range of the inverse function is the domain of f.

2.11.2 Definition. If f is a one-to-one function, then the function f^{-1}, defined by

$$x = f^{-1}(y) \qquad \text{if and only if} \qquad y = f(x),$$

is called the **inverse** of f. The domain of f^{-1} is the range of f, and the range of f^{-1} is the domain of f.

Example 1. If f is the one-to-one function defined by $f(x) = 2x - 5$, its inverse can be found by solving $y = 2x - 5$ for x in terms of y. This procedure gives $x = \frac{1}{2}(y + 5)$. Thus $f^{-1}(y) = \frac{1}{2}(y + 5)$. The domain of f^{-1} is the range of f, which is the set R of all real numbers.

Example 2. Let f be the one-to-one function defined by $f(x) = \sqrt{x - 2}$; its domain is $[2, \infty)$ and its range is $[0, \infty)$. By solving $y = \sqrt{x - 2}$ for x in terms of y, we obtain $x = y^2 + 2$, $y \geq 0$. Thus f^{-1}, the inverse of f, is defined by $f^{-1}(y) = y^2 + 2$, $y \geq 0$. The domain of f^{-1} is $[0, \infty)$, which is the same as the range of f; and the range of f^{-1} is $[2, \infty)$, which is the domain of f.

If we eliminate y between the two equations in 2.11.2, we obtain $x = f^{-1}(f(x))$; and if we eliminate x between the same two equations, the result is $y = f(f^{-1}(y))$.

2.11.3 Theorem. If f is a one-to-one function with inverse f^{-1}, then f^{-1} is a one-to-one function with inverse f. Thus

$$f^{-1}(f(x)) = x \qquad \text{for} \qquad x \in \mathfrak{D}_f,$$
$$f(f^{-1}(y)) = y \qquad \text{for} \qquad y \in \mathfrak{D}_{f^{-1}}.$$

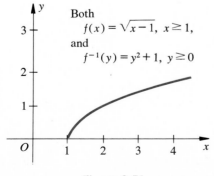

Both
$$f(x) = \sqrt{x - 1}, \ x \geq 1,$$
and
$$f^{-1}(y) = y^2 + 1, \ y \geq 0$$

Figure 2-51

Consider the one-to-one function f defined by

$$f(x) = \sqrt{x - 1}, \qquad \mathfrak{D}_f = [1, \infty), \qquad \mathfrak{R}_f = [0, \infty).$$

If we write $y = \sqrt{x - 1}$, the equation defining f^{-1} can be found by solving for x in terms of y:

$$f^{-1}(y) = y^2 + 1, \qquad \mathfrak{D}^{f^{-1}} = [0, \infty),$$
$$\mathfrak{D}_f = [1, \infty).$$

It is easy to verify that the single curve in Fig. 2-51 pictures the graph of *both* f and f^{-1}, but with one essential difference—namely, that the x axis is the axis of the independent variable for f

and the y axis is the axis of the independent variable for f^{-1}. That is, the domain of f is the interval $[1, \infty)$ *on the x axis*, and the domain of f^{-1} is the interval $[0, \infty)$ *on the y axis*.

Since we are accustomed to using the x axis for the independent variable, and since the choice of a letter to represent a variable is arbitrary, it is customary to interchange the letters x and y in $x = f^{-1}(y)$, obtaining $y = f^{-1}(x)$. Now x is the independent variable in both f and f^{-1}, and the domains of *both f and f*$^{-1}$ are on the x axis. So in our example above, $f(x) = \sqrt{x - 1}$ and $f^{-1}(x) = x^2 + 1$, $x \geq 0$ (Fig. 2-52).

The points P_1:(a, b) and P_2:(b, a) have their coordinates interchanged (Fig. 2-53). Since the line segment $P_1 P_2$ joining them is bisected perpendicularly by the line $y = x$

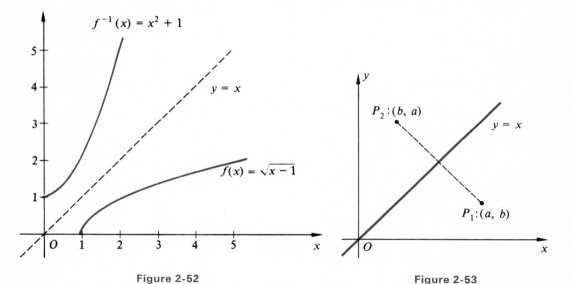

Figure 2-52   Figure 2-53

(Exercise 15), we say that each of the points P_1 and P_2 is the **reflection** of the other with respect to the line $y = x$. If we interchange x and y in the equation $y = f(x)$, the graph of the resulting equation $x = f(y)$ is said to be the **reflection of the graph** of $y = f(x)$ with respect to the line $y = x$. But the curve $x = f(y)$ is the same as the curve $y = f^{-1}(x)$. Thus the graph of $y = f^{-1}(x)$ is the reflection of the graph of $y = f(x)$ with respect to the line $y = x$.

This discussion suggests an easy way to draw the graph of f^{-1} from the graph of f even when it is inconvenient or impossible to solve $y = f(x)$ for x in terms of y. Simply *reflect the graph of* $y = f(x)$ *with respect to the line* $y = x$ *to obtain the graph of* $y = f^{-1}(x)$ (see Fig. 2-52).

Example 3. If f is the one-to-one function defined by $f(x) = x^3/2 + 1$, then $f^{-1}(x)$ can be found by

   (a) solving $y = x^3/2 + 1$ for x in terms of y to obtain $x = \sqrt[3]{2(y - 1)}$, and

(b) interchanging x and y in the latter equation to get $y = \sqrt[3]{2(x-1)}$. Then the inverse function f^{-1} is defined by

$$f^{-1}(x) = \sqrt[3]{2(x-1)}.$$

The graph of f^{-1} can be sketched from this equation, but it is easier to reflect the graph of f with respect to the line $y = x$ (Fig. 2-54).

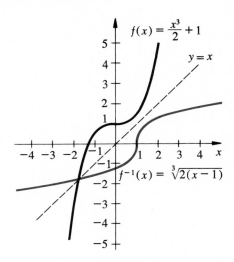

Figure 2-54

Exercises

In Exercises 1 through 10, find the inverse of f and prove your result by means of the relationship $f^{-1}(f(x)) = x$. State the domain and range of f^{-1}. Sketch the graphs of $y = f(x)$ and $y = f^{-1}(x)$ on the same axes.

1. $f(x) = 3x - 1$.

2. $f(x) = \dfrac{-x}{4} + 5$.

3. $f(x) = \sqrt{2x + 5}$.

4. $f(x) = -\sqrt{2 - x}$.

5. $f(x) = \dfrac{1}{x - 5}$.

6. $f(x) = \dfrac{1}{\sqrt{x - 3}}$.

7. $f(x) = x^{3/2},\ 0 \leq x \leq 4$.

8. $f(x) = (x + 1)^2$.

9. $f(x) = (x - 4)^3$.

10. $f(x) = \dfrac{1}{x}$.

In Exercises 11 to 14, sketch the graph of $y = f^{-1}(x)$ by reflecting the graph of $y = f(x)$ about the line $y = x$.

11. $f(x) = \dfrac{2x^2 + x - 36}{6},\ x \geq 0$.

12. $f(x) = x^3 - 3x + 1,\ -1 \leq x \leq 1$.

13. $f(x) = \sqrt{x^3 + x},\ 0 \leq x \leq 3$.

14. $f(x) = (x^3 + x)^{-1/2},\ 1 \leq x \leq 6$.

15. Show that the line $y = x$ is the perpendicular bisector of the line segment having endpoints P_1: (a, b) and P_2:$(b, a),\ a \neq b$.

2.12 REVIEW EXERCISES

1. Sketch the graph of each of the following equations. Tell which define y as a function of x, which define x as a function of y, and which define a one-to-one function.
(a) $x^2 + y^2 = 4$;
(b) $x^2 + y^2 = 4, 0 \le x \le 2$;
(c) $x^2 + y^2 = 4, 0 \le y \le 2$;
(d) $x^2 + y^2 = 4, 0 \le x \le 2, 0 \le y \le 2$.

2. Find an equation of the line through the point $P:(-3, 2)$ that is
(a) parallel to the line $x - 2y - 2 = 0$.
(b) perpendicular to the line $x - 2y - 2 = 0$;
Make a sketch.

3. Find the shortest distance between the point $(-3, 2)$ and the line $x - 2y - 2 = 0$. Make a sketch.

4. Name and sketch the graphs of $x^2 + y^2 - 6y = 0$ and $x^2 + y^2 - 8x = 0$. Without finding the coordinates of the two points of intersection of their graphs, show that the line $(x^2 + y^2 - 6y) - (x^2 + y^2 - 8x) = 0$ (which simplifies to $4x - 3y = 0$) goes through those points of intersection. [*Hint:* The coordinates of a point common to the graphs of the first two equations make the left member of each equation zero. What does that do to the line $(x^2 + y^2 - 6y) - (x^2 + y^2 - 8x) = 0$?]

5. Find the equation of the line through the points of intersection of the graphs of $x^2 + y^2 - 4 = 0$ and $x^2 + y^2 - 6x = 0$. Make a sketch. (*Hint:* See Exercise 4.)

6. The circles $x^2 + y^2 - 8y + 8 = 0$ and $x^2 + y^2 - 8x + 8 = 0$ have only one point in common. Find an equation of the line that is tangent to both circles at their common point. Make a sketch.

In each of Exercises 7 to 12, find the inverse of the function defined by the given equation and sketch the graphs of the function and its inverse on the same pair of axes.

7. $f(x) = x - 3$.

8. $g(x) = \dfrac{x^2}{2}, x \ge 0$.

9. $F(x) = \dfrac{x^2}{2}, x \le 0$.

10. $s(t) = 16t^2 + 3, t \ge 0$.

11. $\phi(w) = -w^3$.

12. $G(t) = (t - 1)^3 - 2, -1 \le t \le 3$.

13. State whether the given set of points is the graph of a function or not; if so, tell its domain and sketch its graph.
(a) $\{P:(x, y) \,|\, y = (2x^2 - 5x)^{-1}\}$;
(b) $\{P:(u, v) \,|\, u^2 + 4v^2 = 4\}$.

14. State whether the given set of points is the graph of a function; if so, give its domain and sketch its graph.
(a) $\{P:(s, t) \,|\, t^2 - 4s = 0\}$;
(b) $\{P:(x, y) \,|\, (x - 2)^3 - y - 1 = 0\}$.

15. Sketch the graph of the function whose rule of correspondence is $f(x) = |2x + 4|$, and state its domain and range.

16. Sketch the graph of the function whose rule of correspondence is $g(x) = [\![x + 2]\!]$, and give its domain and range.

17. Let x be any positive number. Use the greatest integer notation to write a single equation defining the function N, when
(a) $N(x)$ is the number of positive integers whose squares are less than or equal to x.
(b) $N(x)$ is the number of even positive integers whose squares are less than or equal to x.
(c) $N(x)$ is the number of odd positive integers whose cubes are less than or equal to x.

18. Let $f(x) = \sqrt{4 - x}$ and $g(x) = 4 - \sqrt{x}$.
(a) Find $f^{-1}(x)$ and state the domain and range of f^{-1}.
(b) Find $g^{-1}(x)$ and state the domain and range of g^{-1}.
(c) Find $f \circ g$ and give its domain and range.
(d) Find $(f \circ g)^{-1}$, the inverse of $f \circ g$, and tell its domain and range.
(e) Find $g^{-1} \circ f^{-1}$ and tell its domain and range.
(f) Show that $(f \circ g)^{-1} = g^{-1} \circ f^{-1}$.

19. Sketch the graph $\{P:(x, y) \,|\, x \in R, y > -x^2\}$.

Sketch the graph $\{P:(x, y) \,|\, -2 \le x \le 2, x^2 \le y \le 4\}$.

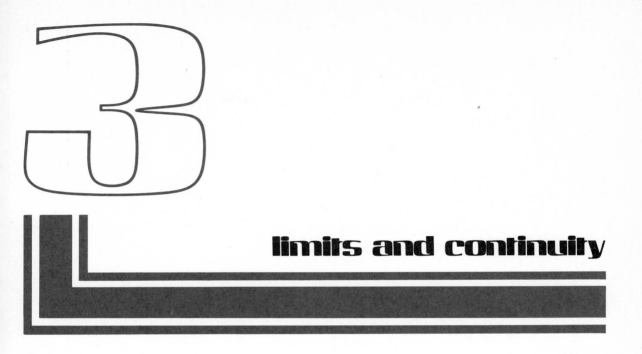

limits and continuity

The concept of *limit* is the most important one in calculus. It is what distinguishes calculus from all the mathematics learned earlier. While the idea of limit is simple, it is somewhat subtle and beginners often have trouble with it. The reader is advised to review these sections from time to time because it is impossible to really *understand* calculus without mastering the concept of limit.

The closely related idea of *continuity* is more easily grasped. Intuitively, a function is continuous if its graph has no breaks or sudden jumps and can be sketched without lifting the pencil from the paper. Although we can often decide whether a function is continuous by looking at a sketch of its graph, this test is not reliable and a precise definition of continuity is desirable.

Most of the functions studied in the early years of the calculus were continuous and there seemed no need to make the idea precise. But in time, as calculus was applied more and more frequently to scientific investigations, functions appeared that were not continuous. In the nineteenth century the French mathematician Augustin-Louis Cauchy and the German, Karl Weierstrass, formulated the precise definitions of limit and continuity that are used today.

3.1 INTRODUCTION TO LIMITS

We start with an intuitive discussion of limits. When a function f has a limit L as x approaches a number c, the value of the function, $f(x)$, gets closer and closer to the number L as x represents numbers that are closer and closer to c.

Example 1. Consider the function f whose rule of correspondence is $f(x) = 2x - 4$. Letting x represent numbers closer and closer to 1, we calculate the table below.

x	0.5	0.8	0.9	0.99	0.999		1.001	1.01	1.1	1.2	1.5
$f(x)$	-3	-2.4	-2.2	-2.02	-2.002		-1.998	-1.98	-1.8	-1.6	-1

It is evident from the table and from Fig. 3-1 that when x is a number very close to 1, $f(x)$ is very close to -2. We say that *the limit of the function f at 1 is -2.*

Example 2. Let g be the function defined by $g(x) = x^2 - 1$, and let x represent numbers closer and closer to 2, as shown in the following table. This table and Fig. 3-2 indicate that if x is a number

x	1	1.5	1.8	1.9	1.99	1.999		2.001	2.01	2.1	2.2	2.5	3
$g(x)$	0	1.25	2.24	2.61	2.96	2.996		3.004	3.04	3.41	3.84	5.25	8

very close to 2, $g(x)$ is very close to 3. *In fact, we can make $g(x)$ as close to 3 as we please by assigning to x a number sufficiently close to 2. The limit of g at 2 is 3.*

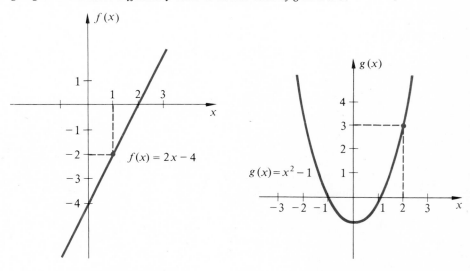

Figure 3-1   Figure 3-2

In view of the fact that $f(1) = -2$ in Example 1 and $g(2) = 3$ in Example 2, the reader may wonder why we bother with values of x *close* to 1 in Example 1 instead of letting x be 1 and why in Example 2 we are concerned with values of x close to 2 instead of letting x be 2. The answer is that the limit of a function at a particular number is not always the same as the value of the function at that number. Indeed, a function F may have a limit at a number c even though $F(c)$ does not exist; this situation is illustrated in the next three examples.

Example 3. Let F be the function defined by $F(x) = 3^{-1/x^2} + 1$. It may be seen from the graph (Fig. 3-3) or from the table below that when x represents numbers very close to zero, $F(x)$ is very close to 1. It can be proved that *the limit of F at 0 is* 1. Yet $F(0)$ does not exist, since division by zero is meaningless.

x	± 2	± 1	± 0.5	± 0.2
$F(x)$	1.7599 ...	1.3333 ...	1.012345679 ...	1.000 000 000 001 ...

Example 4. Let f be the function defined by $f(x) = (\sin x)/x$, where x is measured in radians (Fig. 3-4). As shown in the table below, if x gets closer and closer to zero, the value $f(x)$ becomes closer and closer to 1. It will be proved later that the limit of this function f at zero is 1. Notice, however, that the function f does not exist at $x = 0$, since the denominator of $f(x) = (\sin x)/x$ would be zero there.

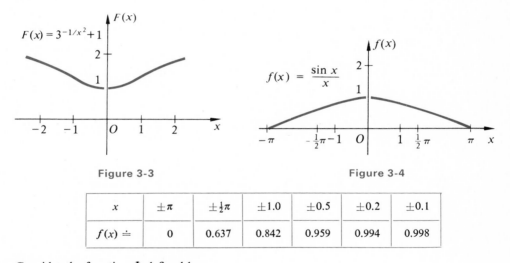

Figure 3-3   Figure 3-4

x	$\pm \pi$	$\pm \frac{1}{2}\pi$	± 1.0	± 0.5	± 0.2	± 0.1
$f(x) \doteq$	0	0.637	0.842	0.959	0.994	0.998

Example 5. Consider the function Φ defined by

$$\Phi(x) = \frac{x - 1}{\sqrt{x} - 1}.$$

Its domain is $\mathcal{D}_\Phi = \{x \mid x \geq 0, x \neq 1\}$. Notice that, for $x \geq 0$ and $x \neq 1$,

$$(1) \qquad \frac{x - 1}{\sqrt{x} - 1} = \sqrt{x} + 1.$$

Thus the graph of Φ is the same as the graph of the function defined by $\sqrt{x} + 1$ except that $\Phi(1)$ does not exist (Fig. 3-5). This fact is helpful in calculating values of $\Phi(x)$. As suggested by the graph and the accompanying table of values, the limit of the function Φ at $x = 1$ is 2 even though Φ does not exist at 1.

x	0	0.5	0.8	0.9	0.99	0.999	1.001	1.01	1.1	1.2	1.5
$\Phi(x)$	1	1.7071	1.8944	1.9487	1.9950	1.9995	2.0005	2.0050	2.0488	2.0954	2.2247

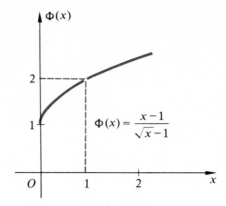

Figure 3-5

The expressions "closer and closer to" and "very close to" that we have been using are entirely too vague for the purposes of mathematics. Let us examine more carefully the table and figure in, say, Example 5.

If it is demanded that $\Phi(x) = (x - 1)/(\sqrt{x} - 1)$ differ from the limit 2 by less than 0.1, we need simply confine x to numbers in the interval (0.9, 1.1), $x \neq 1$, as may be seen from the table. If we wish $\Phi(x)$ to differ from the limit 2 by less than 0.01, we can keep x in the interval (0.99, 1.01), but $x \neq 1$. If it is demanded that $\Phi(x)$ differ from the limit 2 by less than 0.001, this can be accomplished by confining $x \neq 1$ to the interval (0.999, 1.001).

In fact, *we can make the value $\Phi(x)$ differ from the limit 2 by as little as we please by confining x to a sufficiently small interval about 1 (but $x \neq 1$). Such is the very essence of the concept of the limit of a function.*

Example 6. Let f be the function defined by

$$f(x) = \frac{x^2 - 9}{x - 3}, \qquad x \neq 3.$$

The domain of f is $(-\infty, 3) \cup (3, \infty)$. Notice that $(x^2 - 9)/(x - 3) = x + 3$ for all values of x different from 3. It can be shown that the limit of f at $x = 3$ is 6 even though $f(3)$ does not exist.

Can we make the value $f(x)$ differ from the limit 6 by as little as we please by confining x to a sufficiently small interval about 3 (but $x \neq 3$)? The answer is yes.

Suppose that we are asked to find an interval I about 3 on the x axis such that when $x \in I$, but $x \neq 3$, $f(x)$ will differ from its limit 6 by less than 1. We can proceed as follows.

The value of $f(x) = x + 3$ will differ from 6 by less than 1 if and only if $5 < f(x) < 7$—that is, if and only if $5 < x + 3 < 7$. Since

$$5 < x + 3 < 7 \Longleftrightarrow 2 < x < 4,$$

we see that when x is in the interval $I = (2, 4)$, $x \neq 3$, $f(x)$ will be in the interval (5, 7) and so will differ from its limit 6 by less than 1 (see Fig. 3-6).

Again, let us find an interval I_2 about 3 so that when $x \in I_2$, $x \neq 3$, the value of $f(x) = x + 3$ will differ from its limit 6 by less than 0.01—that is, $5.99 < x + 3 < 6.01$. Since

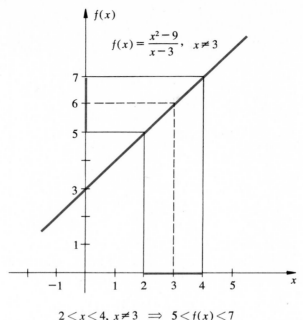

$$2 < x < 4, \; x \neq 3 \; \Longrightarrow \; 5 < f(x) < 7$$

Figure 3-6

$$5.99 < x + 3 < 6.01 \iff 2.99 < x < 3.01,$$

we see that $I_2 = (2.99, 3.01)$ will do. If $x \in I_2$ and $x \neq 3$, then $5.99 < f(x) < 6.01$, and so $f(x)$ will differ from the limit 6 by less than 0.01.

Clearly, we can always use this method to find a sufficiently small interval, I, about 3 so that when $x \in I$, $x \neq 3$, $f(x)$ will differ from 6 by as little as we please.

Example 7.   The value of the greatest integer function f at any real number x, symbolized by $f(x) = [\![x]\!]$ was defined (Section 2.9) to be the greatest integer less than or equal to x. Its graph for $1 \leq x < 3$ is shown in Fig. 3-7. Does f have a limit at 2? The answer is no; for all numbers x

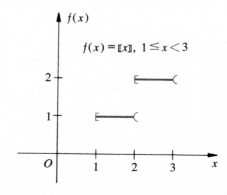

Figure 3-7

in the interval $[1, 3)$ and greater than 2, $[\![x]\!] = 2$, and for all numbers x in the interval $[1, 3)$ and less than 2, $[\![x]\!] = 1$. Thus no matter what number is proposed for the limit of the greatest integer function at 2, $f(x) = [\![x]\!]$ will always differ from the proposed limit by at least $\frac{1}{2}$ for all numbers x on one side of 2.

Exercises

In each of Exercises 1 through 9, a function f is defined and a number c is given. By computing $f(x)$ for values of x closer and closer to c (but not equal to c) and by drawing a careful graph of f for x close to c, try to discover whether f has a limit at c and, if so, the value of that limit.

1. $f(x) = \dfrac{2}{x} + 1, \qquad c = 2.$

2. $f(x) = \dfrac{4 - x^2}{2 - x}, \qquad c = 2.$

[*Hint:* $f(x)$ is not defined for $x = 2$. But

$$\frac{4 - x^2}{2 - x} = x + 2 \qquad \text{for all } x \neq 2.$$

Thus $f(x) = x + 2$, $x \neq 2$. This makes it easier to compute $f(x)$ for numbers x close to 2, but different from 2, and thus easier to sketch the graph of f. See Example 6.]

3. $f(x) = \dfrac{x^2 - 2x - 3}{x - 3}, \qquad c = 3.$

[*Hint:*

$$\frac{x^2 - 2x - 3}{x - 3} = \frac{(x - 3)(x + 1)}{x - 3} = x + 1, \qquad x \neq 3.$$

See Exercise 2.]

4. $f(x) = \dfrac{2x^2 + 5x - 3}{x + 3}, \qquad c = -3.$

[*Hint:*

$$\frac{2x^2 + 5x - 3}{x + 3} = \frac{(x + 3)(2x - 1)}{x + 3} = 2x - 1,$$
$$x \neq -3.$$

See Exercise 2.]

5. $f(x) = \dfrac{x^3 - 16x}{x^2 + 4x}, \qquad c = 0.$

[*Hint:*

$$\frac{x^3 - 16x}{x^2 + 4x} = \frac{x(x + 4)(x - 4)}{x(x + 4)} = x - 4,$$
$$x \neq 0, \ x \neq -4.$$

See Exercise 2.]

6. $f(x) = \dfrac{x - 9}{\sqrt{x} - 3}, \qquad c = 9.$

[*Hint:*

$$\frac{x - 9}{\sqrt{x} - 3} = \sqrt{x} + 3, \qquad x \neq 9.$$

See Exercise 2.]

7. $f(x) = \dfrac{\sin x}{\tan x}$, where $\sin x$ and $\tan x$ are the sine and tangent of an angle whose measure is x radians; $c = 0.$ $\left(\textit{Hint: } \dfrac{\sin x}{\tan x} = \cos x, x \neq 0.\right)$

8. $f(x) = \dfrac{\cos x - 1}{x}$, where $\cos x$ is the cosine of an angle whose measure is x radians; $c = 0.$

9. $f(x) = (1 + x)^{1/x}, \qquad c = 0.$

10. Let F be the function defined by

$$F(x) = \frac{4 - x^2}{2 - x}, \qquad x \neq 2.$$

It can be shown that the limit of F at 2 is 4, even though F is not defined at 2. Find an interval I on the x axis, having center 2, such that whenever $x \in I$, $x \neq 2$, $F(x)$ will differ from 4 by less than 1. (*Hint:* See Example 6 and Exercise 2.)

11. For the function F defined in Exercise 10, find an interval I_2, having 2 for its center, such that when $x \in I_2$, $x \neq 2$, $F(x)$ will differ from the limit 4 by less than 0.3. (*Hint:* See Example 6.)

12. Let F be the function defined in Exercise 10. Find an interval I_3, with 2 as a center, such that whenever $x \in I_3$, $x \neq 2$, then $F(x)$ will differ from its limit 4 by less than 0.05. (*Hint:* See Example 6.)

13. Let G be the function defined by

$$G(x) = \frac{2x^2 + 5x - 3}{x + 3}, \qquad x \neq -3.$$

The limit of G at -3 is -7, although $G(-3)$ does not

exist. Find an interval I on the x axis, having -3 for center, such that whenever $x \in I$, $x \neq -3$, $G(x)$ will differ from -7 by less than

(a) 0.8; (b) 0.2; (c) 0.06.

(*Hint:* See Example 6 and Exercise 4.)

3.2 DEFINITION OF THE LIMIT OF A FUNCTION

It is customary to use the Greek letters ϵ (epsilon) and δ (delta) in defining the limit of a function.

Recall from 1.6.7 that the statement "$f(x)$ differs from the number L by less than 0.3" can be expressed by

$$| f(x) - L | < 0.3$$

or, equivalently, by the inequalities

$$L - 0.3 < f(x) < L + 0.3.$$

As indicated in Fig. 3-8(a), this means that $f(x)$ is some number in the open interval $(L - 0.3, L + 0.3)$ on the vertical $f(x)$ axis.

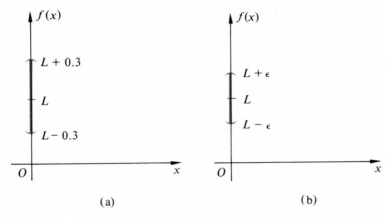

(a) (b)

Figure 3-8

More generally, if ϵ (epsilon) is a positive number, the statement $f(x)$ differs from the number L by less than ϵ can be written

$$| f(x) - L | < \epsilon$$

or, equivalently,

$$L - \epsilon < f(x) < L + \epsilon$$

[see Fig. 3-8(b)].

We next turn our attention to finding a simple, precise symbolism for the statement "x is confined to a small interval about c, but $x \neq c$."

Let δ (delta) be a positive number. Then x will be confined to the open interval $(c - \delta, c + \delta)$, having c for its midpoint and half-length δ (Fig. 3-9), if and only if

$$c - \delta < x < c + \delta,$$

which (by 1.6.6) can be written in the more compact form

$$|x - c| < \delta.$$

As an illustration, when $c = 2$ and $\delta = 0.5$, x is in the open interval $(2 - 0.5, 2 + 0.5) \Longleftrightarrow (1.5, 2.5)$ if and only if $|x - 2| < 0.5$.

To exclude the possibility of x being equal to c when $c - \delta < x < c + \delta$, we simply write

$$0 < |x - c| < \delta.$$

Now x is confined to $(c - \delta, c) \cup (c, c + \delta)$ (Fig. 3-10).

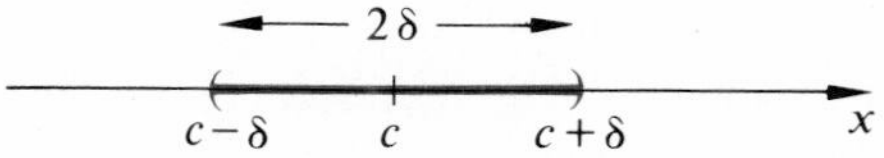

$$0 < |x - c| < \delta \Longleftrightarrow$$
$$x \epsilon \, (c - \delta, \, c + \delta) \text{ and } x \neq c$$

Figure 3-9

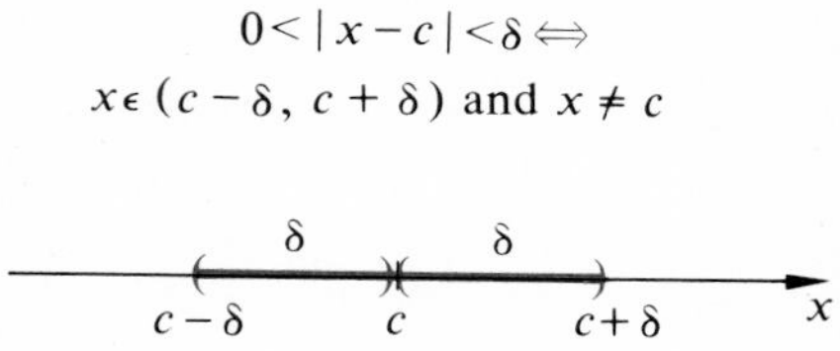

Figure 3-10

For example, let $c = 3$ and $\delta = 1$. Then x is confined to $(3 - 1, 3 + 1)$ with $x \neq 3$, which can be written $x \in (2, 3) \cup (3, 4)$ if and only if $0 < |x - 3| < 1$.

We are now ready for a precise definition of what is meant by saying that "a function f has a limit L at a point c."

The essence of this concept is that the value, $f(x)$, of the function can be made to differ from the limit L by as small a positive number ϵ as we please by confining the independent variable x to numbers in a sufficiently small interval about c, and yet $x \neq c$. (The stipulation $x \neq c$ is needed because f sometimes has a limit at c even though f is not defined at c.)

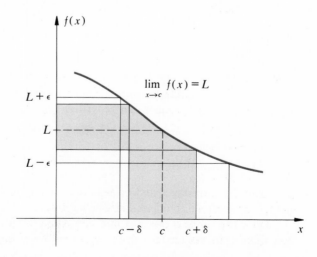

Figure 3-11

We can express this idea by saying that the function f has a limit L at c if and only if for *each* positive number ϵ, no matter how small, there is *some* positive number δ such that whenever x is a number in $(c - \delta, c + \delta)$ and $x \neq c$, then $|f(x) - L| < \epsilon$ (see Fig. 3-11).

3.2.1 **Definition.** Let c be a point in some open interval I and let f be a function that is defined at every point of I except possibly at c. The **limit of the function** f **at** c **is** L, written

$$\lim_{x \to c} f(x) = L,$$

if corresponding to each positive number ϵ (no matter how small) there exists some positive number δ such that whenever

$$0 < |x - c| < \delta,$$

then f will differ from L by less than ϵ; in symbols,

$$0 < |x - c| < \delta \implies |f(x) - L| < \epsilon$$

or, equivalently,

$$x \in (x - \delta, x + \delta), x \neq c \implies L - \epsilon < f(x) < L + \epsilon.$$

This is the most important definition in the calculus.

It is to be emphasized that *the number ϵ is given first*. Then we try to find *some* number δ that fulfills the conditions of the definition for the given ϵ. The value of δ depends on the value of ϵ; and usually the smaller the given number ϵ is, the smaller the number δ must be.

The definition of the limit of a function (3.2.1) requires that for *each* $\epsilon > 0$ we must be able to find some corresponding $\delta > 0$ such that

$$0 < |x - c| < \delta \implies |f(x) - L| < \epsilon$$

or, equivalently,

$$0 < |x - c| < \delta \implies L - \epsilon < f(x) < L + \epsilon.$$

Thus if a person A wants to convince another person B that $\lim_{x \to c} f(x) = L$, B can challenge A with any particular positive number ϵ he pleases and demand that A produce a sufficiently small $\delta > 0$, such that for every x that satisfies $0 < |x - c| < \delta$, $f(x)$ will differ from L by less than the selected ϵ.

Moreover, since the definition (3.2.1) says "each $\epsilon > 0$" (not "some $\epsilon > 0$"), after A has met B's first ϵ-challenge with an adequate δ-response, B can again challenge A with a new $\epsilon > 0$, smaller than the first, and demand that A produce a new $\delta > 0$ that is adequate for the new ϵ. We repeat that generally the smaller the value of ϵ, the smaller the corresponding δ must be.

It is clear from the terms of the definition (3.2.1) that two such ϵ-challenges with two adequate δ-responses are not sufficient to *prove* that $\lim_{x \to c} f(x) = L$; nor, for that matter, would five suffice, or even a million. Certainly the definition of the limit of a function (3.2.1) must be used in some other way to prove that a given function has a limit L at a selected point c. Such a method will now be illustrated by examples.

Example 1. A function F is defined by $F(x) = 5x - 3$ and has for its domain R, the set of all real numbers. Prove that $\lim_{x \to 4} (5x - 3) = 17$.

Preliminary Analysis. Corresponding to each $\epsilon > 0$, we must find some $\delta > 0$ such that

(1)
$$0 < |x - 4| < \delta \implies |(5x - 3) - 17| < \epsilon.$$

We start by finding an inequality involving $|x - 4|$ that is equivalent to the desired $|(5x - 3) - 17| < \epsilon$.

$$|(5x - 3) - 17| < \epsilon \iff 17 - \epsilon < 5x - 3 < 17 + \epsilon$$
$$\iff 20 - \epsilon < 5x < 20 + \epsilon$$
$$\iff 4 - \frac{\epsilon}{5} < x < 4 + \frac{\epsilon}{5}$$
$$\iff |x - 4| < \frac{\epsilon}{5}.$$

Thus

(2)
$$|x - 4| < \frac{\epsilon}{5} \iff |(5x - 3) - 17| < \epsilon.$$

It is important to understand that statement (2) is true for *each and every positive number* ϵ.

A comparison of (1) and (2) suggests that, in applying 3.2.1, we choose $\delta = \epsilon/5$ to correspond to the given ϵ.

Proof. Let ϵ be any positive number. Since the foregoing steps are reversible, we can write

$$0 < |x - 4| < \frac{\epsilon}{5} \implies |x - 4| < \frac{\epsilon}{5}$$
$$\iff 4 - \frac{\epsilon}{5} < x < 4 + \frac{\epsilon}{5}$$
$$\iff 20 - \epsilon < 5x < 20 + \epsilon$$
$$\iff 17 - \epsilon < 5x - 3 < 17 + \epsilon$$
$$\iff |(5x - 3) - 17| < \epsilon.$$

Thus

(3)
$$0 < |x - 4| < \frac{\epsilon}{5} \implies |(5x - 3) - 17| < \epsilon.$$

By choosing $\delta = \epsilon/5$ to correspond to the given ϵ, (3) becomes

$$0 < |x - 4| < \delta \implies |(5x - 3) - 17| < \epsilon.$$

Therefore (by 3.2.1), $\lim_{x \to 4} (5x - 3) = 17$.

Observe that what we have just done was find a *formula for δ in terms of ϵ*, so that no matter what positive number ϵ is given, we automatically have a corresponding $\delta > 0$ that satisfies the requirements of the definition (3.2.1). In Example 1 above, the formula was $\delta = \epsilon/5$. Thus if $\epsilon = 0.1$, $\delta = 0.02$ will do; if $\epsilon = 0.005$, $\delta = 0.001$ will do, and so on.

Example 2. Let f be the function defined by

$$f(x) = \frac{1 - 4x^2}{2x + 1}, \qquad x \neq -\frac{1}{2}.$$

Prove that $\lim_{x \to -1/2} f(x) = 2$.

Preliminary Analysis. For each positive number ϵ, we must find a corresponding positive number δ such that

(4) $$0 < \left| x - \frac{-1}{2} \right| < \delta \implies \left| \frac{1 - 4x^2}{2x + 1} - 2 \right| < \epsilon.$$

The given function f is not defined at $x = -\frac{1}{2}$ because division by zero is impossible. The domain of f is $(-\infty, -\frac{1}{2}) \cup (-\frac{1}{2}, \infty)$. But notice that $(1 - 4x^2)/(2x + 1) = 1 - 2x$ for all values of x *different from* $-\frac{1}{2}$ (see Fig. 3-12). Thus (4) can be written

(5) $$0 < |x - (-1/2)| < \delta \implies |(1 - 2x) - 2| < \epsilon,$$

for $x \neq -\frac{1}{2}$.

From the inequality $|(1 - 2x) - 2| < \epsilon$, $x \neq -\frac{1}{2}$, we shall find an equivalent inequality involving $|x - (-\frac{1}{2})|$. When $x \neq -\frac{1}{2}$,

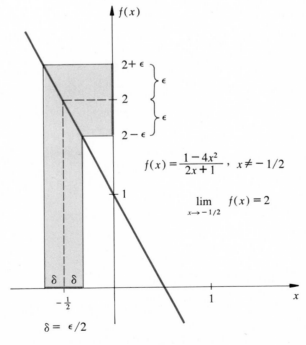

$$f(x) = \frac{1 - 4x^2}{2x + 1}, \quad x \neq -1/2$$

$$\lim_{x \to -1/2} f(x) = 2$$

$$\delta = \epsilon/2$$

Figure 3-12

$$|(1 - 2x) - 2| < \epsilon \iff 2 - \epsilon < 1 - 2x < 2 + \epsilon$$
$$\iff 1 - \epsilon < -2x < 1 + \epsilon$$
$$\iff -1 - \epsilon < 2x < -1 + \epsilon$$
$$\iff \frac{-1}{2} - \frac{\epsilon}{2} < x < \frac{-1}{2} + \frac{\epsilon}{2}$$
$$\iff \left| x - \frac{-1}{2} \right| < \frac{\epsilon}{2}.$$

Therefore

(6) $$\left| x - \frac{-1}{2} \right| < \frac{\epsilon}{2} \iff |(1 - 2x) - 2| < \epsilon.$$

A comparison of (5) and (6) suggests that we choose $\delta = \epsilon/2$ to correspond to the arbitrarily chosen ϵ.

Proof. Let ϵ be any positive number. By reversing the order of the preceding steps, we can write

$$0 < \left| x - \frac{-1}{2} \right| < \frac{\epsilon}{2} \implies \left| x - \frac{-1}{2} \right| < \frac{\epsilon}{2}$$
$$\iff \frac{-1}{2} - \frac{\epsilon}{2} < x < \frac{-1}{2} + \frac{\epsilon}{2}$$
$$\iff -1 - \epsilon < 2x < -1 + \epsilon$$
$$\iff 1 - \epsilon < -2x < 1 + \epsilon$$
$$\iff 2 - \epsilon < 1 - 2x < 2 + \epsilon$$
$$\iff |(1 - 2x) - 2| < \epsilon$$
$$\iff \left| \frac{1 - 4x^2}{2x + 1} - 2 \right| < \epsilon$$

for $x \neq -1/2$. Thus

(7) $$0 < \left| x - \frac{-1}{2} \right| < \frac{\epsilon}{2} \implies \left| \frac{1 - 4x}{2x + 1} - 2 \right| < \epsilon.$$

If we choose $\delta = \epsilon/2$ to correspond to ϵ, (7) becomes

$$0 < \left| x - \frac{-1}{2} \right| < \delta \implies \left| \frac{1 - 4x^2}{2x + 1} - 2 \right| < \epsilon.$$

Therefore (by 3.2.1)

$$\lim_{x \to -1/2} \frac{1 - 4x^2}{2x + 1} = 2.$$

We conclude this section by using 3.2.1 to establish the limits of some simple, often used functions.

3.2.2 **Theorem.** Let f be the constant function defined by $f(x) = k$, where k is a constant. Then its limit at any number c is k; that is,

$$\lim_{x \to c} f(x) = k \quad \text{or, equivalently,} \quad \lim_{x \to c} k = k.$$

Proof. Let ϵ be any positive number. Then for any $\delta > 0$ whatsoever,

$$0 < |x - c| < \delta \implies k - \epsilon < k < k + \epsilon$$

because $k - \epsilon < k < k + \epsilon$ is always true, quite independently of the value of x. Therefore, by 3.2.1, $\lim\limits_{x \to c} k = k$. ∎

As an illustration (Fig. 3-13), 5 is the limit at 2 of the constant function whose rule of correspondence is $f(x) = 5$; that is, $\lim\limits_{x \to 2} 5 = 5$.

3.2.3 Theorem. The limit of the identity function at any number c is c; that is,

$$\lim_{x \to c} x = c.$$

Proof. Let f be the identity function defined by $f(x) = x$ and let ϵ be any positive number. We seek a number $\delta > 0$ such that

$$0 < |x - c| < \delta \implies |x - c| < \epsilon.$$

Clearly, we can choose $\delta = \epsilon$. Then

$$0 < |x - c| < \delta = \epsilon \implies |x - c| < \epsilon.$$

Therefore (by 3.2.1)

$$\lim_{x \to c} f(x) = \lim_{x \to c} x = c. \quad ∎$$

As an illustration (Fig. 3-14), $\lim\limits_{x \to 6} x = 6$.

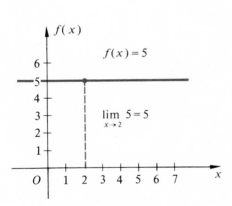

Figure 3-13

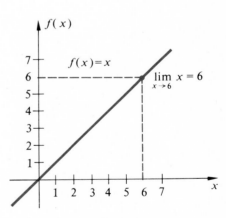

Figure 3-14

3.2.4 Note. If, for a given $\epsilon > 0$, we find a particular positive number δ_1 such that

(8) $$0 < |x - c| < \delta_1 \implies |f(x) - L| < \epsilon,$$

then δ_1 can be replaced by any smaller positive number δ_2. For if $0 < \delta_2 < \delta_1$, then from (8)

$$0 < |x - c| < \delta_2 < \delta_1 \implies |f(x) - L| < \epsilon.$$

This fact will be useful in proving the next theorem.

3.2.5 Theorem. The limit of the principal square root function at any positive number c is $\sqrt{c}$; that is,

$$\lim_{x \to c} \sqrt{x} = \sqrt{c}, \qquad c > 0.$$

Proof. Since the domain of the principal square root function is the set of all non-negative real numbers, we assume in what follows that $x \geq 0$ and $x \neq c$.

Let ϵ be an arbitrary positive number. We seek a $\delta > 0$ such that

$$0 < |x - c| < \delta \implies |\sqrt{x} - \sqrt{c}| < \epsilon.$$

From the inequality $|\sqrt{x} - \sqrt{c}| < \epsilon$ we shall find an equivalent inequality involving $|x - c|$.

$$|\sqrt{x} - \sqrt{c}| < \epsilon \iff |\sqrt{x} - \sqrt{c}| \cdot \frac{\sqrt{x} + \sqrt{c}}{\sqrt{x} + \sqrt{c}} < \epsilon$$

$$\iff \frac{|x - c|}{\sqrt{x} + \sqrt{c}} < \epsilon$$

$$\iff |x - c| < \epsilon(\sqrt{x} + \sqrt{c}).$$

Since these steps are reversible, we have shown that

(9) $$|x - c| < \epsilon(\sqrt{x} + \sqrt{c}) \iff |\sqrt{x} - \sqrt{c}| < \epsilon.$$

Because $\sqrt{x} \geq 0$, we have $\epsilon\sqrt{c} \leq \epsilon(\sqrt{x} + \sqrt{c})$. Thus

(10) $$0 < |x - c| < \epsilon\sqrt{c} \implies |x - c| < \epsilon(\sqrt{x} + \sqrt{c}).$$

If we choose $\delta = \epsilon\sqrt{c}$, it follows from (9) and (10) that

$$0 < |x - c| < \delta \implies |\sqrt{x} - \sqrt{c}| < \epsilon$$

(see Fig. 3-15). Therefore, (by 3.2.1)

$$\lim_{x \to c} \sqrt{x} = \sqrt{c}. \quad \blacksquare$$

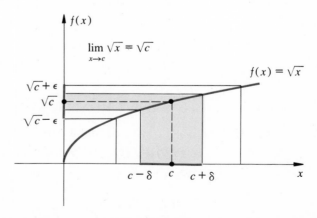

Figure 3-15

Exercises

In each of Exercises 1 through 9, we are given $\lim_{x \to c} f(x) = L$ and thus $f(x)$, c, and L. Corresponding to each $\epsilon > 0$, find a $\delta > 0$ (in terms of ϵ) such that

$$0 < |x - c| < \delta \implies |f(x) - L| < \epsilon.$$

Make a preliminary analysis.

1. $\lim_{x \to 2} (8x - 3) = 13$.

2. $\lim_{x \to -1} (2x + 7) = 5$.

3. $\lim_{x \to 5} \dfrac{x^2 - 25}{x - 5} = 10$.   (*Hint:* See Example 2.)

4. $\lim_{x \to 3/2} \dfrac{4x^2 - 9}{2x - 3} = 6$.

5. $\lim_{x \to 0} \dfrac{3x^2 - 4x}{x} = -4$.

6. $\lim_{x \to 1} \dfrac{2x^3 + 3x^2 - 2x - 3}{x^2 - 1} = 5$.

(*Hint:*
$$\frac{2x^3 + 3x^2 - 2x - 3}{x^2 - 1} = 2x + 3,$$
$$x \neq 1, \quad x \neq -1.)$$

7. $\lim_{x \to 9} \sqrt{x} = 3$.   (*Hint:* See the proof of 3.2.5.)

8. $\lim_{x \to 2} \sqrt{2x} = 2$.

9. $\lim_{x \to 5} \sqrt{x - 1} = 2$.

10. Prove that if $\lim_{x \to c} f(x) = L$ exists, then L is unique. [*Hint:* Show that the assumption that $\lim_{x \to c} f(x) = L_1$, $\lim_{x \to c} f(x) = L_2$, and $L_1 \neq L_2$ leads to a contradiction.]

11. Let F and G be functions such that $0 \leq F(x) \leq G(x)$ for all x in some open interval I, except possibly at one point a in I. Prove that if $\lim_{x \to a} G(x) = 0$, then also $\lim_{x \to a} F(x) = 0$.

3.3 THEOREMS ON LIMITS

Establishing the limits of even such simple functions as those in the preceding section can be a tedious task if we use only the definition of limit. Moreover, the difficulty increases with the complexity of the functions.

We shall now discuss a general theorem that enables us to express the limits of many complicated functions in terms of the known limits of simple functions.

3.3.1 Theorem. If the limits of the functions f and g exist at a number c, then

(i) $$\lim_{x \to c} [f(x) + g(x)] = \lim_{x \to c} f(x) + \lim_{x \to c} g(x);$$

(ii) $$\lim_{x \to c} [f(x) \cdot g(x)] = [\lim_{x \to c} f(x)][\lim_{x \to c} g(x)];$$

(iii) $$\lim_{x \to c} \frac{f(x)}{g(x)} = \frac{\lim_{x \to c} f(x)}{\lim_{x \to c} g(x)} \quad \text{if} \quad \lim_{x \to c} g(x) \neq 0;$$

(iv) $$\lim_{x \to c} \sqrt[n]{f(x)} = \sqrt[n]{\lim_{x \to c} f(x)},$$

where n is any positive integer, provided that $\lim_{x \to c} f(x) > 0$ when n is even.

This theorem is often remembered as

The limit of the sum of two functions is the sum of the limits of the functions; the limit of the product of two functions is the product of the limits of the functions; the limit of the quotient of two functions is the quotient of the limits of the functions, provided that the limit of the function in the denominator is not zero; the limit of the nth root of a function is the nth root of the limit of the function.

Of course, it must be understood in the preceding statements that the two functions have limits at a number c and that all limits mentioned are at c.

We will now prove part (i). The proofs of the other parts are more difficult and may be found in the Appendix (A.1).

Proof of 3.3.1(i).   Let $\lim\limits_{x \to c} f(x) = L$ and $\lim\limits_{x \to c} g(x) = M$. If ϵ is any positive number, then $\epsilon/2$ is positive. Since $\lim\limits_{x \to c} f(x) = L$, there exists some positive number δ_1 such that

$$(1) \qquad 0 < |x - c| < \delta_1 \implies L - \frac{\epsilon}{2} < f(x) < L + \frac{\epsilon}{2}.$$

Since $\lim\limits_{x \to c} g(x) = M$, there exists some positive number δ_2 such that

$$(2) \qquad 0 < |x - c| < \delta_2 \implies M - \frac{\epsilon}{2} < g(x) < M + \frac{\epsilon}{2}.$$

Let $\delta = \min\{\delta_1, \delta_2\}$.* Then both (1) and (2) are true if the left-hand members of the implications are replaced by $0 < |x - c| < \delta$. From this it follows that

$$0 < |x - c| < \delta \implies \left(L - \frac{\epsilon}{2}\right) + \left(M - \frac{\epsilon}{2}\right) < f(x) + g(x) < \left(L + \frac{\epsilon}{2}\right)$$
$$+ \left(M + \frac{\epsilon}{2}\right)$$

$$\implies (L + M) - \epsilon < f(x) + g(x) < (L + M) + \epsilon$$
$$\implies |f(x) + g(x) - (L + M)| < \epsilon.$$

Therefore, by 3.2.1,

$$\lim_{x \to c} [f(x) + g(x)] = L + M = \lim_{x \to c} f(x) + \lim_{x \to c} g(x). \quad \blacksquare$$

Example 1.   Prove that $\lim\limits_{x \to c} x^2 = c^2$.

Proof.   By 3.3.1(ii) and 3.2.3,

$$\lim_{x \to c} x^2 = \lim_{x \to c} [x \cdot x] = (\lim_{x \to c} x)(\lim_{x \to c} x)$$
$$= c \cdot c = c^2.$$

*$\delta = \min\{\delta_1, \delta_2\}$ means that if $\delta_1 \ne \delta_2$, δ is the smaller of δ_1 and δ_2; and if $\delta_1 = \delta_2$, then $\delta = \delta_1 = \delta_2$.

Example 2. Prove that $\lim\limits_{x \to -2} x^3 = -8$.

Proof. By 3.3.1(ii),

$$\lim_{x \to -2} x^3 = \lim_{x \to -2} [x \cdot x^2] = (\lim_{x \to -2} x)(\lim_{x \to -2} x^2)$$

$$= (\lim_{x \to -2} x)(\lim_{x \to -2} (x \cdot x)) = (\lim_{x \to -2} x)(\lim_{x \to -2} x)(\lim_{x \to -2} x).$$

But $\lim\limits_{x \to -2} x = -2$ (by 3.2.3). Therefore

$$\lim_{x \to -2} x^3 = (-2)^3 = -8.$$

The following corollary to 3.3.1(ii) is easily proved by induction (Section 1.7).

3.3.2 Corollary. If $\lim\limits_{x \to c} f(x)$ exists and n is any positive integer, then

$$\lim_{x \to c} [f(x)]^n = [\lim_{x \to c} f(x)]^n.$$

By taking $g(x) = k$, we have another corollary to 3.3.1(ii).

3.3.3 Corollary. If $\lim\limits_{x \to c} f(x)$ exists and k is any constant,

$$\lim_{x \to c} [k \cdot f(x)] = k \lim_{x \to c} f(x).$$

Example 3. Show that $\lim\limits_{x \to 4} (3x^2 - 2x) = 40$.

Solution. By using 3.3.1(i), 3.3.2, and 3.3.3, we obtain

$$\lim_{x \to 4} (3x^2 - 2x) = \lim_{x \to 4} 3x^2 + \lim_{x \to 4} (-2x)$$

$$= 3 \lim_{x \to 4} x^2 - 2 \lim_{x \to 4} x = 3(4^2) - 2(4) = 40.$$

Example 4. Show that $\lim\limits_{x \to 2} [5x^5 - 13x^2 + 10] = 118$.

Solution. By 3.3.1(i), 3.3.2, 3.3.3, and 3.2.2, we have

$$\lim_{x \to 2} [5x^5 - 13x^2 + 10] = \lim_{x \to 2} (5x^5) + \lim_{x \to 2} (-13x^2) + \lim_{x \to 2} 10$$

$$= 5 \lim_{x \to 2} x^5 - 13 \lim_{x \to 2} x^2 + \lim_{x \to 2} 10$$

$$= 5(2)^5 - 13(2)^2 + 10 = 160 - 52 + 10 = 118.$$

Example 5. Show that

$$\lim_{x \to 3} \frac{x^3 - 6}{2x + 1} = 3.$$

Solution. Using 3.3.1(i) and (iii), 3.3.2, and 3.3.3, we have

$$\lim_{x \to 3} \frac{x^3 - 6}{2x + 1} = \frac{\lim\limits_{x \to 3}(x^3 - 6)}{\lim\limits_{x \to 3}(2x + 1)} = \frac{\lim\limits_{x \to 3}(x^3) + \lim\limits_{x \to 3}(-6)}{\lim\limits_{x \to 3} 2x + \lim\limits_{x \to 3} 1}$$

$$= \frac{(3^3) - 6}{2 \lim\limits_{x \to 3} x + 1} = \frac{3^3 - 6}{2(3) + 1} = \frac{21}{7} = 3.$$

Example 6. Find $\lim\limits_{x \to 4} \sqrt{x^2 + 9}$.

Solution. By 3.3.1(i) and (iv), 3.3.2, and 3.3.3,

$$\lim_{x \to 4} \sqrt{x^2 + 9} = \sqrt{\lim_{x \to 4}(x^2 + 9)} = \sqrt{\lim_{x \to 4} x^2 + \lim_{x \to 4} 9}$$

$$= \sqrt{(4^2) + 9} = \sqrt{25} = 5.$$

Example 7. Find $\lim\limits_{x \to -3} [\sqrt{x^2 - 2}/(3x^3 - 17)]$.

Solution. By applying parts of 3.3.1 and corollaries 3.3.2 and 3.3.3, we have

$$\lim_{x \to -3} \frac{\sqrt{x^2 - 2}}{3x^3 - 17} = \frac{\lim\limits_{x \to -3} \sqrt{x^2 - 2}}{\lim\limits_{x \to -3}(3x^3 - 17)} = \frac{\sqrt{\lim\limits_{x \to -3}(x^2 - 2)}}{\lim\limits_{x \to -3}(3x^3) - \lim\limits_{x \to -3} 17}$$

$$= \frac{\sqrt{\lim\limits_{x \to -3} x^2 - \lim\limits_{x \to -3} 2}}{3 \lim\limits_{x \to -3} x^3 - \lim\limits_{x \to -3} 17} = \frac{\sqrt{(-3)^2 - 2}}{3(-3)^3 - 17} = \frac{\sqrt{7}}{-98}.$$

A *polynomial function f* was defined in Section 2.9 by

$$f(x) = a_0 x^n + a_1 x^{n-1} + \cdots + a_{n-1} x + a_n,$$

where n is a nonnegative integer and $a_0, a_1, \cdots, a_n$, are constants. It is an immediate consequence of theorems 3.2.2, 3.2.3, 3.3.1, 3.3.2, and 3.3.3 that if c is any real number,

$$\lim_{x \to c} f(x) = a_0 c^n + a_1 c^{n-1} + \cdots + a_{n-1} c + a_n = f(c).$$

Thus *the limit of any polynomial function at any chosen number c may be obtained by simply substituting c for x throughout the polynomial.*

Example 8. Let F be the polynomial function defined by $F(x) = 7x^5 - 10x^4 - 13x + 6$. Then

$$\lim_{x \to 2} F(x) = 7(2)^5 - 10(2)^4 - 13(2) + 6 = 44.$$

Because of its usefulness in finding many limits, we state it formally for easy reference.

3.3.4 Corollary. Let f be the polynomial function defined by

$$f(x) = a_0 x^n + a_1 x^{n-1} + \cdots + a_{n-1} x + a_n,$$

where n is a nonnegative integer and $a_0, a_1, \cdots, a_n$ are constants. Then the limit of f at any number c is

$$\lim_{x \to c} f(x) = f(c).$$

Any rational function can be expressed as the quotient of two polynomial functions (Section 2.9). Its domain is the set of all real numbers at which the denominator polynomial is not zero.

It follows from 3.3.4 and 3.3.1(iii) that the limit of any rational function, at any number c in its domain, may be obtained by substituting c for x throughout the two polynomials.

Example 9. Let f be the rational function defined by

$$f(x) = \frac{2x^7 - 9x^6 + 11x^2 - 2}{4x^9 - 10x^8 - 6x + 7}.$$

Then

$$\lim_{x \to -1} f(x) = \frac{2(-1)^7 - 9(-1)^6 + 11(-1)^2 - 2}{4(-1)^9 - 10(-1)^8 - 6(-1) + 7} = 2.$$

3.3.5 Corollary. If f is a **rational function** and c is any number in its domain, then

$$\lim_{x \to c} f(x) = f(c).$$

We conclude this section with some simple but important properties of the limit of a function that will be needed later on.

3.3.6 Theorem. $\lim_{x \to c} f(x) = L$ if and only if $\lim_{x \to c} [f(x) - L] = 0$.

Proof. Assume that $\lim_{x \to c} f(x) = L$. By 3.2.2 and 3.3.1(i),

$$\lim_{x \to c} [f(x) - L] = \lim_{x \to c} f(x) + \lim_{x \to c} (-L) = L - L = 0.$$

Conversely, assume that $\lim_{x \to c} [f(x) - L] = 0$. Since $f(x) = [(fx) - L] + L$, it follows from 3.2.2 and 3.3.1(i) that

$$\lim_{x \to c} f(x) = \lim_{x \to c} \{[f(x) - L] + L\}$$

$$= \lim_{x \to c} [f(x) - L] + \lim_{x \to c} L = 0 + L = L. \quad \blacksquare$$

3.3.7 Theorem. If $\lim_{x \to c} g(x) > 0$, then there exists an interval $(c - \delta, c + \delta)$, $\delta > 0$, such that $g(x) > 0$ for all $x \in (c - \delta, c + \delta)$, $x \neq c$.

That is, if x stays sufficiently close to (but distinct from) c, then $g(x)$ will stay so close to its positive limit that $g(x)$ cannot be negative (Fig. 3-16).

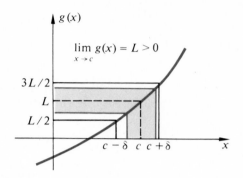

Figure 3-16

Proof of 3.3.7. We are given that $\lim\limits_{x \to c} g(x) = L > 0$. Let $\epsilon = L/2$. By the definition of the limit of a function (3.2.1) there exists some positive number δ such that

$$(3) \qquad 0 < |x - c| < \delta \implies L - \frac{L}{2} < g(x) < L + \frac{L}{2} \implies \frac{L}{2} < g(x) < \frac{3L}{2}$$

Since $0 < |x - c| < \delta \iff c - \delta < x < c + \delta$ and $x \neq c$, it follows from (3) that

$$0 < \frac{L}{2} < g(x)$$

for all $x \neq c$ in $(c - \delta, c + \delta)$. ∎

Similarly, we have

3.3.8 Theorem. If $\lim\limits_{x \to c} f(x) < 0$, there is an interval $(c - \delta, c + \delta)$, $\delta > 0$, such that $f(x) < 0$ for all x in $(c - \delta, c + \delta)$, $x \neq c$.

The next theorem is sometimes very useful.

3.3.9 Theorem. Let f, g, and h be functions such that $f(x) \leq g(x) \leq h(x)$ for all numbers x in some open interval I except possibly at one point $c \in I$. If $\lim\limits_{x \to c} f(x) = \lim\limits_{x \to c} h(x) = L$, then $\lim\limits_{x \to c} g(x) = L$.

Proof. Let $\epsilon > 0$. Since $\lim\limits_{x \to c} f(x) = \lim\limits_{x \to c} h(x) = L$, there exist positive numbers δ_1 and δ_2 such that, for $x \in I$,

$$0 < |x - c| < \delta_1 \implies L - \epsilon < f(x) < L + \epsilon$$

and
$$0 < |x - c| < \delta_2 \implies L - \epsilon < h(x) < L + \epsilon.$$

If $\delta = \min\{\delta_1, \delta_2\}$, then, for $x \in I$,

(4) $0 < |x - c| < \delta \implies L - \epsilon < f(x) < L + \epsilon$ and $L - \epsilon < h(x) < L + \epsilon.$

Since $f(x) \le g(x) \le h(x)$ for all $x \ne c$ in I, it follows from (4) that
$$0 < |x - c| < \delta \implies L - \epsilon < f(x) \le g(x) \le h(x) < L + \epsilon.$$

Thus for each $\epsilon > 0$ there is some $\delta > 0$ such that
$$0 < |x - c| < \delta \implies L - \epsilon < g(x) < L + \epsilon.$$

Therefore (by 3.2.1)
$$\lim_{x \to c} g(x) = L. \quad\blacksquare$$

This squeeze theorem (3.3.9) provides a useful technique for proving the limit of a function g at c. If we know, or can find, simpler approximating functions f and h with a common limit at c, and if $g(x)$ is "sandwiched" between $f(x)$ and $h(x)$ for all x close to c, then g is "squeezed" to the same limit as f and h at c.

3.3.10   Theorem.   If A and B are numbers and $|A - B| < \epsilon$ for all $\epsilon > 0$, then $A = B$.

Proof.   Assume that $A \ne B$. Then $|A - B| > 0$. Denote $|A - B|$ by C. Since ϵ can be any positive number, let $\epsilon = C/2$. Then $0 < |A - B| < \epsilon \implies 0 < C < C/2$, which is impossible. Therefore $A = B$. $\quad\blacksquare$

Exercises

Prove the statements in Exercises 1 through 10 by means of the theorems and corollaries of this section. In each case, indicate the theorems and corollaries used by giving their numbers.

1. $\lim_{x \to 3} (7x - 4) = 17.$

2. $\lim_{x \to -1} (2x^3 - 5x) = 3.$

3. $\lim_{x \to 2} [(x^2 + 1)(3x - 11)] = -25.$

4. $\lim_{x \to 0} [(4x^2 - 3)(7x^3 + 2x)] = 0.$

5. $\lim_{x \to 4} \dfrac{2x}{3x^3 - 16} = \dfrac{1}{22}.$

6. $\lim_{x \to -2} \dfrac{3x^4 - 8}{x^3 + 24} = \dfrac{5}{2}.$

7. $\lim_{x \to 3} \sqrt{3x - 5} = 2.$

8. $\lim_{x \to -3} \sqrt{5x^2 + 2x} = \sqrt{39}.$

9. $\lim_{x \to 4} (x^2 - 6x + 1)^3 = -343.$

10. $\lim_{x \to -2} (2x^3 + 15)^{13} = -1.$

11. Find $\lim_{x \to 2} (2x^2 + 3x - 7).$

12. Find $\lim_{x \to -3} (x^5 - 6x^3 + x^2 + 12).$

13. Find $\lim_{t \to 3} \dfrac{3t + 5}{2t^2 - 11}.$

14. Find $\lim_{t \to 4} \dfrac{t^3 - 7t}{2t^2 - 8}.$

15. Find $\lim_{w \to -2} \sqrt{3w^3 + 7w^2}.$

16. Find $\lim_{y \to 3} \dfrac{\sqrt{6y}}{y^2 - 2y}.$

17. Find $\lim\limits_{y\to 2}\left(\dfrac{4y^3 + 8y}{y + 4}\right)^{1/3}$.

18. Find $\lim\limits_{w\to 5}(2w^4 - 9w^3 + 19)^{-1/2}$.

In each of Exercises 19 to 24, an $f(x)$ and a number c are given. Use the theorems of this section to evaluate $\lim\limits_{x\to c} G(x)$, where

$$G(x) = \frac{f(x) - f(c)}{x - c}, \qquad x \neq c.$$

For instance, if $f(x) = 2x^2 - 1$ and $c = 3$, then $G(x) = [(2x^2 - 1) - 17]/(x - 3)$, $x \neq 3$. Notice that 3.3.5 does not apply to this rational function G because $c = 3$ is not in the domain of G. However, the definition (3.2.1) of the limit of the function G at 3 is concerned only with values of x close to 3 but *different* from 3. Since $G(x) = [(2x^2 - 1) - 17]/(x - 3) = 2(x^2 - 9)/(x - 3) = 2(x + 3)$ for all $x \neq 3$, we have $\lim\limits_{x\to 3} G(x) = \lim\limits_{x\to 3}(2x + 6) = 12$.

19. $f(x) = 5x^2$, $c = 2$.

20. $f(x) = 3x^2 - 5$, $c = -3$.

21. $f(x) = x^3$, $c = 5$.

22. $f(x) = 2x^3 - x + 1$, $c = -1$.

23. $f(x) = \dfrac{1}{x}$, $x \neq 0$; $c = 3$.

[*Hint:*
$$G(x) = \frac{1/x - 1/3}{x - 3} = \frac{3 - x}{3x(x - 3)} = \frac{-1}{3x},$$
$$x \neq 0, \quad x \neq 3.]$$

24. $f(x) = \dfrac{3}{x^2}$, $x \neq 0$; $c = -2$.

25. Prove that if $\lim\limits_{x\to c} f(x)$ exists, then
$$\lim_{x\to c}[f(x)]^2 = [\lim_{x\to c} f(x)]^2.$$

26. Prove 3.3.2 by induction (Section 1.7).

3.4 ONE-SIDED LIMITS

The definition (3.2.1) of the limit of a function f at c requires that f be defined in some *open* interval containing c, although not necessarily at c itself. Thus f must be defined in some $(a, c) \cup (c, b)$.

The principal square root function, defined by $f(x) = \sqrt{x}$, has for its domain $[0, \infty)$. Since $\sqrt{x}$ does not exist at any point to the left of 0, we cannot find $\lim\limits_{x\to 0}\sqrt{x}$.

But if, in our definition of the limit of a function f at a number c, we stipulate that $x > c$, so that x "approaches c from the right," we have what is known as a *right-hand limit*; it will be designated by $\lim\limits_{x\to c^+} f(x)$.

3.4.1 Definition. Let f be a function that is defined (at least) on an open interval (c, b). The **right-hand limit** of f at c is L, written

$$\lim_{x\to c^+} f(x) = L,$$

if corresponding to each positive number ϵ there is some positive number δ such that

$$0 < x - c < \delta \implies |f(x) - L| < \epsilon.$$

Notice that $0 < x - c < \delta$ implies that $x > c$, and so x *must be to the right of c.*

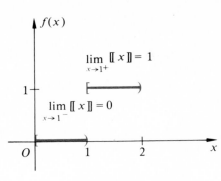

Figure 3-17

The definition of the **left-hand limit**, $\lim\limits_{x \to c^-} f(x) = L$, is similar, except that f must be defined in some open interval (a, c) and

$$0 < c - x < \delta \implies |f(x) - L| < \epsilon.$$

Here $0 < c - x < \delta$ implies that $x < c$; so the numbers x *must be to the left of c*.

As an illustration, although the greatest integer function (Fig. 3-17) does not have a limit at 1, it does have both a right-hand limit and a left-hand limit at 1—namely,

$$\lim_{x \to 1^+} [\![x]\!] = 1 \quad \text{and} \quad \lim_{x \to 1^-} [\![x]\!] = 0.$$

Example 1. Prove that $\lim\limits_{x \to 0^+} \sqrt{x} = 0$.

Solution. Since the domain of the principal square root function is $[0, \infty)$, we assume in what follows that $x \geq 0$.

Let ϵ be a positive number. We seek a $\delta > 0$ such that $0 < x - 0 < \delta \implies |\sqrt{x} - 0| < \epsilon$; that is, we seek a $\delta > 0$ such that

$$0 < x < \delta \implies \sqrt{x} < \epsilon.$$

Since $\epsilon > 0$,

$$0 < x < \epsilon^2 \implies 0 < \sqrt{x} < \epsilon \implies \sqrt{x} < \epsilon.$$

Therefore by choosing $\delta = \epsilon^2$, we have

$$0 < x < \delta \implies \sqrt{x} < \epsilon,$$

and so (by 3.4.1) $\lim\limits_{x \to 0^+} \sqrt{x} = 0$.

Proof of the following theorem is easy and is left as an exercise for the reader.

3.4.2   Theorem.   Let a function f be defined on an interval I having c as an interior point. Then $\lim\limits_{x \to c} f(x) = L$ if and only if $\lim\limits_{x \to c^+} f(x) = \lim\limits_{x \to c^-} f(x) = L$.

The basic theorems for finding limits in the preceding section—namely, 3.3.1, 3.3.2, and 3.3.4—are also true when all the limits are right-hand limits or all the limits are left-hand limits.

Example 2. Find the limit (if it exists) of the absolute value function at zero (see Fig. 2-47).

Solution. The absolute value function is defined by $f(x) = |x|$. Its domain is R.

For $x \geq 0$, $|x| = x$. Thus $\lim\limits_{x \to 0^+} |x| = \lim\limits_{x \to 0^+} x = 0$. When $x < 0$, $|x| = -x$, and $\lim\limits_{x \to 0^+} |x| = \lim\limits_{x \to -} (-x) = 0$.

Since $\lim\limits_{x \to 0^+} |x| = \lim\limits_{x \to 0^-} |x| = 0$, it follows from 3.4.2 that $\lim\limits_{x \to 0} |x| = 0$.

Example 3. Let G be the function defined by

$$G(x) = \begin{cases} 2 - x, & x > 1, \\ x^2, & x < 1, \end{cases}$$

(Fig. 3-18). Find $\lim_{x \to 1} G(x)$.

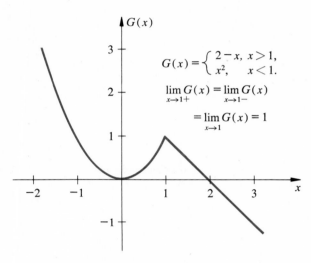

$G(x) = \begin{cases} 2-x, & x>1, \\ x^2, & x<1. \end{cases}$

$\lim_{x \to 1^+} G(x) = \lim_{x \to 1^-} G(x)$

$= \lim_{x \to 1} G(x) = 1$

Figure 3-18

Solution. $\lim_{x \to 1^+} G(x) = \lim_{x \to 1^+} (2 - x) = 1$. Also, $\lim_{x \to 1^-} G(x) = \lim_{x \to 1^-} x^2 = 1$. Therefore (by 3.4.2) $\lim_{x \to 1} G(x) = 1$.

Example 4.

$$\lim_{x \to 2^-} \frac{\sqrt{4 - x^2}}{2x^3} = \left[\lim_{x \to 2^-} \sqrt{4 - x^2} \right] \div \left[\lim_{x \to 2^-} (2x^3) \right] = \frac{0}{16} = 0.$$

Example 5. $\lim_{x \to 2^+} x[\![x]\!] = \left[\lim_{x \to 2^+} x \right]\left[\lim_{x \to 2^+} [\![x]\!] \right] = 2 \cdot 2 = 4.$

Exercises

In each of Exercises 1 to 10, find the indicated one-sided limit and make a sketch.

1. $\lim_{x \to 2^+} \sqrt{x - 2}.$

2. $\lim_{x \to 3^-} \sqrt{3 - x}.$

3. $\lim_{x \to 2^+} [\![x]\!].$

4. $\lim_{x \to 2^-} [\![x]\!].$

5. $\lim_{x \to 2^-} [\![x/2]\!].$

6. $\lim_{x \to 2^+} [\![x/2]\!].$

7. $\lim_{x \to 0^+} [\![-x]\!].$

8. $\lim_{x \to 2^-} x[\![x]\!].$

9. $\lim_{x \to -3^+} \frac{[\![x]\!]}{x}.$

10. $\lim\limits_{x \to -1+} \dfrac{|x|}{[\![x]\!]}$.

In Exercises 11 to 14, make use of 3.4.2.

11. Find $\lim\limits_{x \to 3} |x - 3|$. Make a sketch.

12. Find $\lim\limits_{x \to 2} \left| \dfrac{x}{2} - 1 \right|$. Make a sketch.

13. Sketch the graph of the function Φ defined by

$$\Phi(x) = \begin{cases} 2x, & x > 1, \\ x^3 + 1, & x < 1. \end{cases}$$

Then find $\lim\limits_{x \to 1} \Phi(x)$.

14. Sketch the graph of the function F defined by

$$F(x) = \begin{cases} \sqrt{x}, & x > 1, \\ x^2, & x < 1. \end{cases}$$

Find $\lim\limits_{x \to 1} F(x)$.

15. Write a definition of the left-hand limit, $\lim\limits_{x \to c^-} f(x) = L$, analogous to 3.4.1.

16. Prove 3.4.2.

3.5 CONTINUITY

We have seen that the limit of a function f at a number c can often be found by computing $f(c)$, the value of the function at c. For instance, $\lim\limits_{x \to -2} (4x^3 - 2x + 5) = 4(-2)^3 - 2(-2) + 5 = -23$. In the case of some other functions, however, this procedure is not always possible (see Examples 3, 4, 5, 6, and 7 of Section 3.1). A function f whose limit at a number c is $f(c)$ is said to be *continuous* at c.

3.5.1 Definition. A function f is **continuous** at a number c if

$$\lim\limits_{x \to c} f(x) = f(c).$$

Thus the continuity of f at c requires three things:

1. $\lim\limits_{x \to c} f(x) = L$ exists;
2. $f(c)$ exists (that is, f is defined at c); and
3. $L = f(c)$.

In Example 3 of Section 3.1, $\lim\limits_{x \to 0} F(x) = \lim\limits_{x \to 0} (3^{-1/x^2} + 1) = 1$, and so (1) is satisfied; but (2) is not satisfied because $F(0) = 3^{-1/0} + 1$ does not exist. Therefore the function F is not continuous at zero. (It is, however, continuous at every other number.)

A function is said to be **discontinuous** at any number where it is not continuous.

Example 1. Consider the function g defined by

$$g(x) = \frac{x^2 - 4}{x - 2}, \qquad x \neq 2,$$

and

$$g(2) = 1.$$

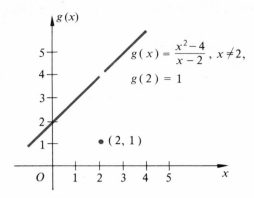

Figure 3-19

Its graph (Fig. 3-19) consists of the isolated point (2, 1) and every point on the line $y = x + 2$ except (2, 4). Since $g(x) = (x - 2)(x + 2)/(x - 2) = x + 2$ for $x \neq 2$, $\lim_{x \to 2} g(x) = 4$ and (1) is satisfied at 2; and because $g(2) = 1$ exists, (2) is satisfied. But $\lim_{x \to 2} g(x) = 4 \neq 1 = g(2)$, and so (3) fails. Therefore g is discontinuous at 2, although it is continuous at all other numbers.

Example 2. Let G be the function defined by $G(x) = (x^2 - 4)/(x - 2)$ for $x \neq 2$, and $G(2) = 4$ (compare with the function g in Example 1). Since $G(x) = x + 2$ for $x \neq 2$, $\lim_{x \to 2} G(x) = 4$ and (1) is satisfied. Because $G(2) = 4$ by the definition of G, (2) is satisfied. Moreover, $\lim_{x \to 2} G(x) = 4 = G(2)$, and so (3) is satisfied. Consequently, G is continuous at 2. It is also continuous at all other numbers. The graph of G is the line $y = x + 2$.

From 3.2.5, 3.3.4, 3.3.5, 3.5.1, and Example 2 of Section 3.4, we have the following corollary. The reader can save much labor by remembering it.

3.5.2 Corollary. All **polynomial functions** are continuous at every real number. Any **rational function** is continuous at all numbers in its domain. The **absolute value function** is continuous at every real number. The principal **square root function** is continuous at all positive numbers.

The next theorem is a simple consequence of the theorem on limits (3.3.1) and the definition of continuity.

3.5.3 Theorem. Let f and g be functions that are continuous at c. Then $f + g$ and fg are continuous at c, and, if $g(c) \neq 0$, f/g is continuous at c.

This theorem can be proved by merely replacing $\lim_{x \to c} f(x)$ and $\lim_{x \to c} g(x)$ by $f(c)$ and $g(c)$, respectively, in 3.3.1.

Example 3. Denote by F the function defined by

$$F(x) = \frac{|2x - 7|}{\sqrt{x} + x^2}.$$

At what numbers is F continuous?

Solution. Let $f(x) = |2x - 7|$. Since the absolute value function is continuous at all real numbers, f is continuous at all real numbers.

Let $g(x) = \sqrt{x} + x^2$. The principal square root function is continuous at all positive numbers and a polynomial function is continuous everywhere. Thus g is continuous at all positive numbers. Moreover, $g(x) \neq 0$ for any positive number x.

Therefore $F = f/g$ is continuous at every positive number (by 3.5.3).

3.5.4 Theorem. If g is continuous at c and f is continuous at $g(c)$, then the composite function, $f \circ g$, is continuous at c.

Proof. Let $\epsilon > 0$. Since f is continuous at $g(c)$, $\lim_{t \to g(c)} f(t) = f(g(c))$. So, corresponding to the $\epsilon > 0$, there is a $\delta_1 > 0$, such that

(1) $$|t - g(c)| < \delta_1 \implies |f(t) - f(g(c))| < \epsilon.$$

Let $t = g(x)$ in (1). Then

(2) $$|g(x) - g(c)| < \delta_1 \implies |f(g(x)) - f(g(c))| < \epsilon.$$

Because g is continuous at c, $\lim_{x \to c} g(x) = g(c)$. Thus, corresponding to the positive number δ_1 there is a $\delta_2 > 0$ such that

(3) $$|x - c| < \delta_2 \implies |g(x) - g(c)| < \delta_1.$$

From (2) and (3) we see that corresponding to each $\epsilon > 0$, there is some $\delta_2 > 0$ such that

$$|x - c| < \delta_2 \implies |f(g(x)) - f(g(c))| < \epsilon.$$

Therefore

$$\lim_{x \to c} f(g(x)) = f(g(c)),$$

or, in different notation,

$$\lim_{x \to c} (f \circ g)(x) = (f \circ g)(c),$$

which proves that $f \circ g$ is continuous at c. ∎

Example 4. Show that the function F, defined by

$$F(x) = \left[\frac{3x^2 + 1}{(x - 3)^2}\right]^{1/2}, \qquad x \neq 3,$$

is continuous at every real number except 3.

Solution. $F = f \circ g$, where

$$f(x) = x^{1/2} = \sqrt{x} \qquad \text{and} \qquad g(x) = \frac{3x^2 + 1}{(x - 3)^2}.$$

The principal square root function, f, is continuous at all positive numbers, and the rational function, g, is continuous at all numbers except 3. Moreover, $g(x) > 0$ for all $x \neq 3$. Thus (by 3.5.4) the composite function $F = f \circ g$ is continuous at every real number except 3.

3.5.5 Definition.   A function is **continuous on an open interval** if and only if it is continuous at every point in the open interval.

So if $F(x) = \sqrt{16 - x^2}$, F is continuous on the open interval $(-4, 4)$ (see Fig. 2-42). Again, if $f(x) = [\![x]\!]$, $1 \leq x < 3$, f is continuous on the two open intervals, $(1, 2)$ and $(2, 3)$ (see Fig. 3-7).

Just as a function may have a one-sided limit, it may have *one-sided continuity* at a point. Its definition is just what might be expected.

3.5.6 Definition.   A function f is **continuous from the right** at a number c if and only if

$$\lim_{x \to c^+} f(x) = f(c).$$

A function f is **continuous from the left** at c if and only if

$$\lim_{x \to c^-} f(x) = f(c).$$

For instance, if $F(x) = \sqrt{x - 1}$, F is continuous from the right at 1 because $\lim_{x \to 1^+} F(x) = \sqrt{1 - 1} = 0 = F(1)$.

As with limits (3.4.2), a function f is continuous at c if and only if

$$\lim_{x \to c^+} f(x) = \lim_{x \to c^-} f(x) = f(c).$$

It is customary to define continuity on a *closed interval* in the following manner.

3.5.7 Definition.   A function f is said to be **continuous on a closed interval** $[a, b]$ if and only if f is continuous on the open interval (a, b), continuous from the right at a, and continuous from the left at b.

To illustrate (Fig. 2-42), the function ϕ, defined by $\phi(x) = \sqrt{16 - x^2}$, is continuous on the closed interval $[-4, 4]$ because it is continuous at every number in the open interval $(-4, 4)$, continuous from the right at -4, and continuous from the left at 4. Notice, however, that although ϕ is continuous *from the right* at -4, ϕ is *not continuous* at -4. It is not defined to the left of -4 and so is not continuous from the left at -4. Similarly, ϕ is continuous *from the left* at 4 but is *not continuous* at 4.

Throughout this book we shall be particularly interested in functions that are continuous on some interval, and most of the theorems will concern such functions.

The reader should note that the graph of a function is unbroken wherever the function is continuous. There are no sudden jumps in the values of the function where it is continuous, and its graph can be drawn without lifting the pencil from the paper.

The following basic theorem for continuous functions establishes these facts; it is a theorem that is easy to understand but not so easy to prove. Its proof may be found in the Appendix (A.2.7).

3.5.8 The Intermediate Value Theorem. If f is continuous on a closed interval $[a, b]$ and if W is a number between $f(a)$ and $f(b)$, then there exists a number c between a and b such that $f(c) = W$.

In other words, a function f that is continuous on a closed interval $[a, b]$, takes on every value between $f(a)$ and $f(b)$ as x assumes all numbers between a and b.

If Fig. 3-20 shows the graph of a function, the function is continuous on the open intervals $(-\infty, 0)$ and $(0, r)$, continuous on the closed interval $[r, s]$, and continuous on the half-open interval $[t, \infty)$.

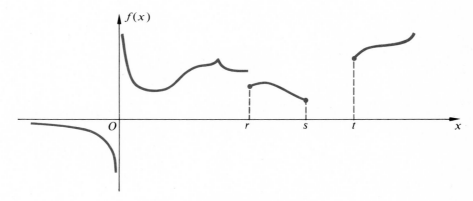

Figure 3-20

Exercises

In each of Exercises 1 through 7, state whether the indicated function is continuous at 2; if not, tell why.

1. $f(x) = 4x^5 - 2x^2 + 12$.

2. $F(x) = \dfrac{8}{x - 2}$

3. $g(x) = \dfrac{3x^2}{x - 1}$

4. $G(x) = \sqrt{x - 1}$.

5. $H(x) = \sqrt{x - 3}$.

6. $\Phi(x) = |3 - 5x^3|$.

7. $\phi(x) = [\![x]\!]$.

8. Is the function f, defined by

$$f(x) = \frac{x^2 - 9}{x - 3},$$

continuous at 3? Why? Sketch the graph of f.

9. Is the function g, defined by

$$g(x) = \frac{x^2 - 9}{x - 3} \quad \text{for} \quad x \neq 3 \quad \text{and} \quad g(3) = 4,$$

continuous at $x = 3$? Why?

10. Is the function F, defined by

$$F(x) = \frac{x^2 - 9}{x - 3} \quad \text{for} \quad x \neq 3 \quad \text{and} \quad F(3) = 6,$$

continuous at $x = 3$? Why?

11. Let H be the function defined by

$$H(x) = \frac{9x^2 - 4}{3x + 2}, \qquad x \neq -\frac{2}{3}.$$

What value assigned to $H(-\frac{2}{3})$ would make H continuous at $-\frac{2}{3}$?

12. Let Φ be the function defined by

$$\Phi(x) = \frac{x - 1}{\sqrt{x} - 1}, \qquad x \neq 1.$$

What value assigned to $\Phi(1)$ would make Φ continuous at 1? Why?

13. Let f be the function defined by

$$f(x) = \frac{\sin x}{x} \qquad x \neq 0.$$

What value assigned to $f(0)$ would make f continuous at 0? No proof is required. (*Hint:* See Example 4, Section 3.1)

14. Let ϕ be the function defined by

$$\phi(x) = \frac{x^4 + 4x - 1}{x - 2}, \qquad x \neq 2.$$

Is there a value that can be assigned to $\phi(2)$ to make ϕ continuous at 2?

15. Sketch the graph of the function G, defined by $G(x) = |x - 4| + 1$. Use one-sided limits to show that G is continuous at 4.

16. Let $f(x) = |x + 1|$ for $x \neq -1$, and $f(-1) = 3$. Sketch the graph of f and tell why it is discontinuous at -1.

In each of Exercises 17 to 20, sketch the graph of the indicated function and tell what intervals it is continuous on.

17. $F(x) = \sqrt{x - 2}$.

18. $G(x) = \dfrac{1}{\sqrt{x - 2}}$.

19. $f(x) = \sqrt{x^2 + 1}$.

20. $\phi(x) = \begin{cases} \dfrac{x + 2}{2} & \text{for } x > 0, \\ 0 & \text{for } x = 0, \\ 1 - x^2 & \text{for } x < 0. \end{cases}$

In each of Exercises 21 to 24, sketch the graph of the indicated function. Find all numbers at which the function is discontinuous and tell why it is discontinuous there.

21. $g(x) = \dfrac{x + 2}{2x^2 + x - 6}$

22. $H(x) = \dfrac{x + 2}{2x^2 + x - 6}$ for $x \neq -2$ and $x \neq \dfrac{3}{2}$;

$$H(-2) = -7, \qquad H\left(\frac{3}{2}\right) = 3.$$

23. $\Phi(x) = \dfrac{x + 2}{2x^2 + x - 6}$ for $x \neq -2$ and $x \neq \dfrac{3}{2}$;

$$\Phi(-2) = -\frac{1}{7}, \qquad \Phi\left(\frac{3}{2}\right) = -6.$$

24. $F(x) = 1$ for $x > 0$, $F(0) = \dfrac{1}{2}$, and $F(x) = 0$ for $x < 0$.

25. Define a function whose domain is R, and that is
(a) continuous on $(-\infty, 2) \cup (2, \infty)$ but discontinuous at 2;
(b) continuous from the right at 2 but not continuous from the left at 2;
(c) continuous on $[0, \infty)$ and on $(-\infty, 0)$ but discontinuous at 0.

26. Use the intermediate value theorem (3.5.8) to prove that the equation $x^3 + 3x - 2 = 0$ has a root between $x = 0$ and $x = 1$.

27. Prove the theorem: Let f be continuous at c. If $f(c) > 0$, there is an open interval I containing c such that $f(x) > 0$ for all $x \in I$. If $f(c) < 0$, there is an open interval I containing c such that $f(x) < 0$ for all $x \in I$. (*Hint:* See 3.3.7, 3.3.8, and 3.5.1.)

28. Prove that f is continuous at a number c if and only if $\lim_{t \to 0} f(t + c) = f(c)$. [*Hint:* In 3.2.1, let $L = f(c)$. By substituting $t + c$ for x in $|x - c| < \delta \implies |f(x) - f(c)| < \epsilon$, we obtain $|t - 0| < \delta \implies |f(t + c) - f(c)| < \epsilon$, from which $\lim_{t \to 0} f(t + c) = f(c)$.]

3.6 INCREMENTS

If x_1 and x_2 are two numbers and the value of a variable x changes from x_1 to x_2, then $x_2 - x_1$, the change in the value of x, is called an **increment of x** and is commonly denoted by Δx (read "delta x"). That is,

$$\Delta x = x_2 - x_1.$$

The symbol Δx never means "delta times x"; the Δ is a part of the symbol Δx, which denotes an increment of x (a change in the value of x). Thus Δx can be any number whatever.

For instance, if the value of x changes from $x_1 = 4.1$ to $x_2 = 5.7$, then $\Delta x = x_2 - x_1 = 1.6$; if the value of x changes from 4.1 to -0.3, then $\Delta x = (-0.3) - (4.1) = -4.4$.

Similarly, Δy means an increment of y—that is, a change in the value of y; Δu means an increment of u, and Δt means an increment of t.

Let x_1 and $x_2 = x_1 + \Delta x$ be numbers in the domain of a function f, and let $y = f(x)$. If y_1 is the value of the dependent variable y that corresponds to the value x_1 of the independent variable x, so that

(1) $$y_1 = f(x_1),$$

then, as the value of x changes from x_1 to $x_1 + \Delta x$, y changes from y_1 to $y_1 + \Delta y$, so that

(2) $$y_1 + \Delta y = f(x_1 + \Delta x).$$

By subtracting (1) from (2), member by member, we obtain

$$\Delta y = f(x_1 + \Delta x) - f(x_1).$$

Thus Δy is the change in the value of the function f as x changes from x_1 to $x_1 + \Delta x$.

3.6.1 Definition. If $y = f(x)$ and if x_1 and $x_1 + \Delta x$ are two numbers in the domain of f, then

$$\Delta y = f(x_1 + \Delta x) - f(x_1)$$

is the **increment of the dependent variable** y that corresponds to the increment Δx of the independent variable x at x_1. Equivalently,

$$\Delta f = f(x_1 + \Delta x) - f(x_1)$$

is the **increment of the function** f corresponding to the increment Δx of the independent variable x at x_1.

It is important to notice that $\Delta y (= \Delta f)$ depends for its value on the *two* numbers x_1 and Δx. It is the amount of change in the dependent variable brought about by changing the value of the independent variable from x_1 to $x_1 + \Delta x$.

Example 1. Let $f(x) = x^2 + 1$. If the value of x at $x_1 = 3$ takes on the increment $\Delta x = 5$, the *corresponding* increment of the function is

$$\Delta f = f(x_1 + \Delta x) - f(x_1) = [(3 + 5)^2 + 1] - [(3)^2 + 1] = 55.$$

Again, if $x_1 = 1.9$ and $\Delta x = 5$, then

$$\Delta f = [(1.9 + 5)^2 + 1] - [(1.9)^2 + 1] = 44.$$

If $x_1 = 8$ and $\Delta x = -0.3$, then

$$\Delta f = [(8 - 0.3)^2 + 1] - [8^2 + 1] = -4.71.$$

Example 2. If the correct length of an edge of a certain cube is 13.2 inches and a maximum possible error of ± 0.04 inch occurs in measuring it, what is the maximum possible error introduced in the computed volume?

Solution. The volume of a cube is given by

$$V = x^3,$$

where x is the length of an edge. We are given that $x_1 = 13.2$ and $\Delta x = \pm 0.04$; we seek the corresponding increment ΔV of V. It is

$$\Delta V = (x_1 + \Delta x)^3 - x_1^3 = (13.2 \pm 0.04)^3 - (13.2)^3.$$

Therefore the maximum possible error in computing the volume is $\Delta V = 20.972224$ cubic inches or $\Delta V = -20.845504$ cubic inches.

In Fig. 3-21, $P_1:(x_1, y_1)$ and P_2 are two points on the graph of $y = f(x)$. In going from P_1 to P_2 along the graph, the initial value x_1 of the independent variable x at P_1 takes on the increment $\Delta x = \overline{P_1 Q}$, and y_1 takes on the *corresponding* increment $\Delta y = \overline{QP_2}$. Recall from Section 1.4 that $\overline{P_1 Q}$ means the directed distance from P_1 to Q and that $\overline{QP_2}$ is the directed distance from Q to P_2. Thus the coordinates of P_2 are $(x_1 + \Delta x, y_1 + \Delta y)$, where $\Delta y = f(x_1 + \Delta x) - f(x_1)$.

Notice that the *slope of the secant $P_1 P_2$* in Fig. 3-21 is

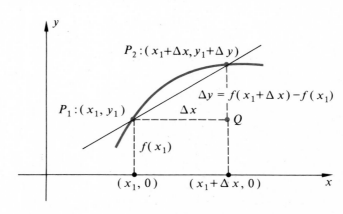

Figure 3-21

$$\frac{\Delta y}{\Delta x} = \frac{f(x_1 + \Delta x) - f(x_1)}{\Delta x}.$$

We shall use this result in the next chapter.

Exercises

In each of Exercises 1 to 4, a function f is defined. Find the value of $\Delta f = f(x_1 + \Delta x) - f(x_1)$ that corresponds to the given values of x_1 and Δx.

1. $f(x) = 3x^2 - 5$, $x_1 = 2$, and $\Delta x = 1$.

2. $f(x) = \dfrac{1}{x}$, $x_1 = -1$, and $\Delta x = 0.3$.

3. $f(x) = \dfrac{1}{2x - 1}$, $x_1 = 3$, and $\Delta x = -0.5$.

4. $f(x) = \sqrt{3x - 6}$, $x_1 = 5$, and $\Delta x = 0.02$.

5. The radius of a circle is measured and found to be 21.83 inches with a maximum possible error of 0.04 inch. What is the maximum possible error in the computed area of the circle?

6. If the inside of a closed cubicle container measures 16 inches each way, how much volume is lost if the container is lined—top, bottom, and sides— with cork that is 0.3 inch thick? [*Hint:* $V = x^3$. Find $\Delta V = (x_1 + \Delta x)^3 - x_1^3$, where $x_1 = 16$ and $\Delta x = -0.3$.]

7. How much metal is used to make a hollow spherical shell that is 0.2 inch thick if the inside diameter is 40 inches?

8. A cylindrical tank, open at the top, has a circular copper bottom and vertical steel sides. How much asphalt is needed to coat the inside steel sides to a uniform thickness of 0.4 inch if the inside dimensions of the tank before coating are radius of base $= 20$ inches, and height $= 75$ inches?

In Exercises 9 to 12, $f(x)$, x_1, and Δx are given. Sketch the graph of f for values of x near x_1, using a conveniently large scale. Find $\Delta y = \Delta f = f(x_1 + \Delta x) - f(x_1)$, and write the coordinates of the points $P_1:(x_1, y_1)$ and $P_2:(x_1 + \Delta x, y_1 + \Delta y)$ on the curve. Draw the secant $P_1 P_2$, find its slope, $\Delta y / \Delta x$, and label the appropriate horizontal and vertical line segments Δx and Δy (see Fig. 3-21).

9. $f(x) = x^2 - x$, $x_1 = 1$, and $\Delta x = 0.6$.

10. $f(x) = \dfrac{1}{x}$, $x_1 = -2$, and $\Delta x = 1.5$.

11. $f(x) = \dfrac{2}{x^2}$, $x_1 = 2$, and $\Delta x = -\dfrac{4}{5}$.

12. $f(x) = \sqrt{x}$, $x_1 = 0.25$, and $\Delta x = 0.75$.

In each of Exercises 13 and 14, the equation of a curve, $y = f(x)$, is given. Let x_1 and Δx be arbitrary numbers.

(a) Find $\Delta y = \Delta f = f(x_1 + \Delta x) - f(x_1)$.
(b) Form $\Delta y / \Delta x$, the slope of the secant through the two points, $P_1:(x_1, f(x_1))$ and $P_2:(x_1 + \Delta x, f(x_1 + \Delta x))$, on the curve.
(c) Keep x_1 fixed and denote by S the function of Δx defined by

$$S(\Delta x) = \frac{f(x_1 + \Delta x) - f(x_1)}{\Delta x}.$$

Remembering that x_1 is a constant, find the limit of S at $\Delta x = 0$—that is, find $\lim\limits_{\Delta x \to 0} (\Delta y / \Delta x)$. (See Exercises 19 to 24, Section 3.3.)
(d) Use (c) to write the equation of the line through $P_1:(x_1, f(x_1))$ whose slope is $\lim\limits_{\Delta x \to 0} (\Delta y / \Delta x)$.
(e) For $x_1 = 1$ and $\Delta x = 1$, make a careful sketch showing the curve, the two points P_1 and P_2 on it, the horizontal and vertical line segments $P_1 Q$ and $Q P_2$ whose lengths are Δx and Δy, and the secant line $P_1 P_2$.
(f) Substitute $x_1 = 1$ in the equation of the line found in (d) and carefully draw the graph of the resulting equation on the sketch of the curve in (e). How would you describe this line geometrically?

13. $y = x^2$.

14. $y = x^3$.

3.7 REVIEW EXERCISES

Use the (ϵ, δ)-definition of the limit of a function (3.2.1) to prove the statement in each of Exercises 1 through 4.

1. $\lim\limits_{x \to 2} 3x = 6$.

2. $\lim\limits_{x \to -1} (2x - 5) = -7$.

3. $\lim\limits_{x \to 3} \dfrac{x^2 - 9}{x - 3} = 6$.

4. $\lim\limits_{x \to 2} \sqrt{2x} = 2$.

In Exercises 5 to 12, use the theorems on limits to evaluate the indicated limits.

5. $\lim\limits_{x \to 1} (4x^3 - 2x^2 + 1)$.

6. $\lim\limits_{t \to -2} (t^6 + 8t^5 - t^2 + 5)$.

7. $\lim\limits_{w \to 5} \dfrac{w}{w^2 - 5}$.

8. $\lim\limits_{x \to 4} \dfrac{4x^3 - 256}{x - 4}$.

9. $\lim\limits_{x \to 3+} \dfrac{x^2 - 9}{\sqrt{x - 3}}$.

10. $\lim\limits_{x \to 1} \dfrac{\sqrt{x} - 1}{x - 1}$.

11. $\lim\limits_{w \to -2} \sqrt{\dfrac{3w - 10}{w - 7}}$.

12. $\lim\limits_{x \to 5} \dfrac{4 - \sqrt{1 + 3x}}{5 - x}$.

In Exercises 13 to 16, tell what intervals each indicated function is continuous on.

13. $f(x) = 13x^9 - 2x^6 + x^5 - 11x + 8$.

14. $F(x) = \dfrac{5x^5 + 17x - 4}{2x^2 + 9x - 5}$.

15. $G(x) = \sqrt{\dfrac{5 - x}{x - 2}}$.

16. $g(x) = \sqrt{x + 2} + \sqrt{9 - x}$.

In Exercises 17 to 20, sketch the graph of the indicated function and state what intervals it is continuous on.

17. $f(x) = |x - 2|$.

18. $G(x) = \dfrac{x - 1}{|x - 1|}$ for $x \neq 1$, and $G(1) = 1$.

19. $\Phi(t) = \dfrac{t^2}{1 - t^2}$.

20. $F(w) = \dfrac{w}{\sqrt{w^2 - 9}}$.

21. Let a function f be defined by
$$f(x) = \begin{cases} x - 2 & \text{for } x \neq 3, \\ 2 & \text{for } x = 3. \end{cases}$$
Find $\lim\limits_{x \to 3} f(x)$ and show that it is different from $f(3)$. Then sketch the graph of f.

22. Let F be defined by
$$F(x) = \begin{cases} 2x^2 - 3 & \text{for } x \neq -2, \\ 4 & \text{for } x = -2. \end{cases}$$
Find $\lim\limits_{x \to -2} F(x)$ and show that $\lim\limits_{x \to -2} F(x) \neq F(-2)$. Then sketch the graph of F.

23. Sketch the graph of the function ϕ, defined by $\phi(x) = 1/(x^2 - 1)$. On what intervals is ϕ continuous?

24. Let $f(x) = |x|/x$ for $x \neq 0$, and $f(0) = 0$. Let $g(x) = |x|/(-x)$ for $x \neq 0$, and $g(0) = 0$. Use one-sided limits to show that f and g are discontinuous at 0. Then show that $f + g$ is continuous at zero. Sketch the graphs of $f, g,$ and $f + g$.

25. Let $G(x) = \sqrt{x^2 - 5}$. Use 3.5.2 and 3.5.4 to prove that G is continuous at $x = 3$.

26. Is the function F, defined by $F(x) = [\![x]\!]$, continuous at 0? Why?

27. Is the function H, defined by $H(x) = [\![|x|]\!]$, continuous at 0? Why?

28. Let $f(x) = x^2 + 1$ if $x \leq 1$, and $f(x) = 3 - x$

if $x > 1$. Sketch the graph of f. Find $\lim\limits_{x \to 1^+} f(x)$, $\lim\limits_{x \to 1^-} f(x)$, and $\lim\limits_{x \to 1} f(x)$.

29. Let

$$g(x) = \begin{cases} \sqrt{4 - x^2} & \text{if } x \leq 0, \\ \sqrt{16 - x^2} - 2 & \text{if } x > 0. \end{cases}$$

On what interval or intervals is g continuous? Sketch the graph of g.

30. Use the intermediate value theorem (3.5.8) to prove that the equation $x^5 - 4x^3 - 3x + 1 = 0$ has at least one root between $x = 2$ and $x = 3$.

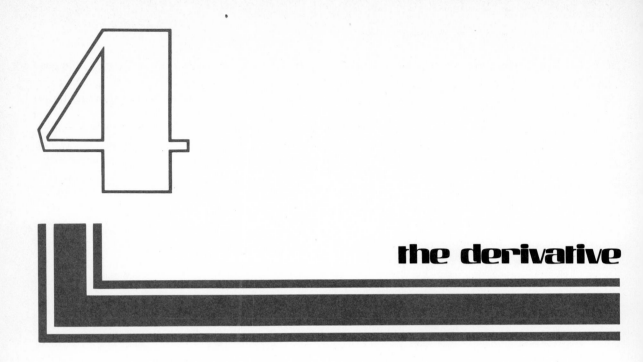

the derivative

The development of differential calculus in the seventeenth century was greatly influenced by two problems: (a) finding the slope of the tangent line to a given curve at a given point on the curve, and (b) finding the instantaneous velocity of a particle moving along a straight line at varying speeds. We shall see in the next two sections how these problems were solved, and how their solution led to the concept of the *derivative*. Finding derivatives is the basic process of the differential caluclus.

The derivative of a function is a limit. Furthermore, it will be shown that when this limit exists, the function is continuous. Thus function, limit, and continuity are intimately linked to the derivative.

To obtain the full benefit of this chapter, the reader should find all derivatives from their definition, as shown in the illustrative examples. Short formulas for derivatives will be developed in the next chapter, and can then be memorized and used freely.

4.1 TANGENT LINE TO A CURVE

We noticed in plane geometry that a straight line intersects a circle in two points, or is tangent to the circle, or fails to intersect the circle at all. Consequently, we might be tempted to define a tangent to a circle as a line that intersects the circle in one and only one point.

But such a definition would not do for most other curves. For instance, the tangent line to the graph of $y = x^3$ at the point $(1, 1)$ intersects the curve again at the point $(-2, -8)$ (Fig. 4-1). This situation indicates that a different approach is needed.

Since we can readily write the equation of a line through a given point if we know the slope of the line (Section 2.4), our task is to formulate a definition of the *slope of the tangent line* to a curve that will apply not only to circles but to other curves as well.

Consider a circle with center C, and let P be an arbitrarily chosen fixed point on the circle (Fig. 4-2). The tangent line to the circle at P is the line t through P that is perpendicular to CP.

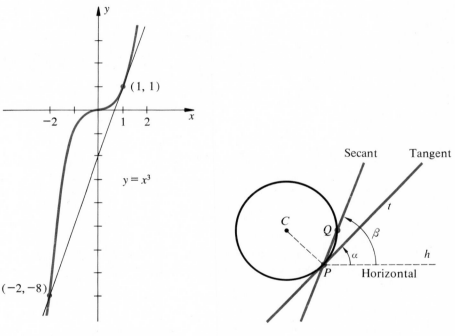

Figure 4-1   Figure 4-2

Let Q be any point on the circle distinct from P. The line through P and Q is a *secant* of the circle. Denote by h the horizontal half-line extending to the right from P. Indicate by α and β the smallest positive angles from h to the tangent line t and to the secant PQ, respectively (Fig. 4-2).

It is intuitively evident that we can make β differ from α by as little as desired by placing Q on the curve close enough to P. *Thus we can make the slope of the secant PQ differ from the slope of the tangent line t by as little as we please by choosing Q on the curve sufficiently close to P (but distinct from P).*

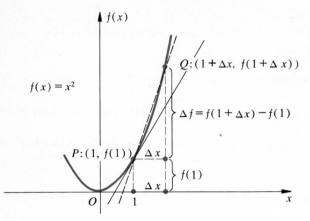

Figure 4-3

Let us use this idea to define the slope of the tangent line to the graph of $f(x) = x^2$ at some particular point on the curve, say at $P:(1, f(1))$ (Fig. 4-3). Let Q be any other point on the graph distinct from P, with coordinates $(1 + \Delta x, f(1 + \Delta x))$, where $\Delta x \neq 0$.

For the given fixed point $P:(1, f(1))$, the slope of the secant PQ is a function S of Δx, defined by

$$(1) \qquad S(\Delta x) = \frac{\Delta f}{\Delta x} = \frac{f(1 + \Delta x) - f(1)}{\Delta x}.$$

The point $Q:(1 + \Delta x, f(1 + \Delta x))$ on the curve can be made as close to $P:(1, f(1))$ as we like by taking Δx sufficiently close to zero. This fact, and our discussion of the tangent line to the circle, above, suggest that we *define* the slope of the tangent line to the graph of $f(x) = x^2$ at $P:(1, f(1))$ to be the limit of the slope of the secant (1) as $\Delta x \longrightarrow 0$, that is,

$$(2) \qquad \lim_{\Delta x \to 0} S(\Delta x) = \lim_{\Delta x \to 0} \frac{\Delta f}{\Delta x} = \lim_{\Delta x \to 0} \frac{f(1 + \Delta x) - f(1)}{\Delta x}.$$

To find this limit, we notice that since $f(x) = x^2$, the expression (1) for the slope of PQ can be simplified as follows:

$$S(\Delta x) = \frac{f(1 + \Delta x) - f(1)}{\Delta x} = \frac{(1 + \Delta x)^2 - 1^2}{\Delta x}$$

$$= \frac{2\,\Delta x + (\Delta x)^2}{\Delta x} = 2 + \Delta x,$$

for all numbers $\Delta x \neq 0$. Since S is a function of Δx and the definition (3.2.1) of the limit of S as $\Delta x \longrightarrow 0$ specifically excludes $\Delta x = 0$, then

$$\lim_{\Delta x \to 0} S(\Delta x) = \lim_{\Delta x \to 0} \frac{\Delta f}{\Delta x} = \lim_{\Delta x \to 0} (2 + \Delta x) = 2.$$

So the slope of the tangent line to the graph of $f(x) = x^2$ at $(1, 1)$ is 2. Using the point-slope form of the equation of a line, we find the equation of the tangent line to the curve at $P:(1, 1)$ to be $y - 1 = 2(x - 1)$, or

$$2x - y - 1 = 0.$$

We now state a formal definition of the slope of the tangent line to the graph of a function at a given point on the graph.

4.1.1   Definition.   If f is a function and $P:(c, f(c))$ is a point on the graph of f, the **slope of the tangent line** to the graph at $P:(c, f(c))$ is

$$\lim_{\Delta x \to 0} \frac{\Delta f}{\Delta x} = \lim_{\Delta x \to 0} \frac{f(c + \Delta x) - f(c)}{\Delta x},$$

provided that this limit exists.

Example 1. Find the slope of the tangent line to the curve $y = 3x^2 - 5$ at the point $P_1:(-1, -2)$ on the curve, and write the equation of that tangent line (Fig. 4-4).

Solution. By 4.1.1, the slope of the tangent line to the graph of $f(x) = 3x^2 - 5$ at $P_1:(-1, f(-1))$ $= P_1:(-1, -2)$ is

$$\begin{aligned}
\lim_{\Delta x \to 0} \frac{\Delta f}{\Delta x} &= \lim_{\Delta x \to 0} \frac{f(-1 + \Delta x) - f(-1)}{\Delta x} \\
&= \lim_{\Delta x \to 0} \frac{[3(-1 + \Delta x)^2 - 5] - [3(-1)^2 - 5]}{\Delta x} \\
&= \lim_{\Delta x \to 0} \frac{-6\,\Delta x + 3(\Delta x)^2}{\Delta x} \\
&= \lim_{\Delta x \to 0} (-6 + 3\,\Delta x) = -6.
\end{aligned}$$

Therefore the tangent line at $P_1:(-1, -2)$ has slope -6.

By substituting this slope in $y - y_1 = m(x - x_1)$, the point-slope form of the equation of a line, we find the equation of the tangent line to the curve $y = 3x^2 - 5$ at the point $(-1, -2)$ to be $y + 2 = -6(x + 1)$, or $6x + y + 8 = 0$.

Example 2. Find the slope of the tangent line to the curve $y = 1/(2x)$ at the point $P_1:(\frac{1}{2}, 1)$ on the curve, and write the equation of the tangent line (Fig. 4-5).

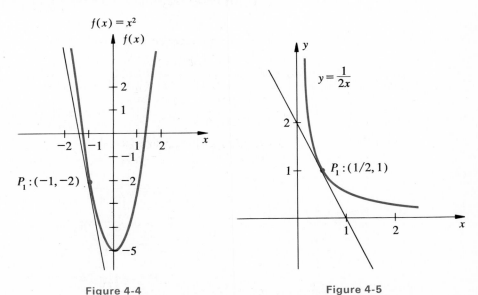

Figure 4-4   Figure 4-5

Solution. From 4.1.1, the slope of the tangent line to the graph of $f(x) = 1/(2x)$ at $P_1 = (\frac{1}{2}, f(\frac{1}{2})) = P_1 = (\frac{1}{2}, 1)$ is

$$\lim_{\Delta x \to 0} \frac{\Delta f}{\Delta x} = \lim_{\Delta x \to 0} \frac{f(1/2 + \Delta x) - f(1/2)}{\Delta x}$$

$$= \lim_{\Delta x \to 0} \frac{\dfrac{1}{2(1/2 + \Delta x)} - \dfrac{1}{2(1/2)}}{\Delta x}$$

$$= \lim_{\Delta x \to 0} \frac{\dfrac{1}{1 + 2\,\Delta x} - 1}{\Delta x}$$

$$= \lim_{\Delta x \to 0} \frac{1 - (1 + 2\,\Delta x)}{\Delta x\,(1 + 2\,\Delta x)}$$

$$= \lim_{\Delta x \to 0} \frac{-2}{1 + 2\,\Delta x} = -2.$$

We now substitute this slope ($m = -2$) and the coordinates of the point $P_1{:}(\frac{1}{2}, 1)$ in the point-slope form of the equation of a line, $y - y_1 = m(x - x_1)$, to obtain $y - 1 = -2(x - \frac{1}{2})$, or $2x + y - 2 = 0$. This is the equation of the tangent line to the curve $y = 1/(2x)$ at $(\frac{1}{2}, 1)$.

Example 3. Find the slope of the tangent line to the graph of $f(x) = \sqrt{3x - 2}$ at the point $P_1{:}(2, 2)$. Write the equation of the tangent line at P_1 on the curve.

Solution. By 4.1.1, the slope of the tangent line to the graph of $f(x) = \sqrt{3x - 2}$ at $P_1{:}(2, f(2))$ is

$$\lim_{\Delta x \to 0} \frac{\Delta f}{\Delta x} = \lim_{\Delta x \to 0} \frac{f(2 + \Delta x) - f(2)}{\Delta x}$$

$$= \lim_{\Delta x \to 0} \frac{\sqrt{3(2 + \Delta x) - 2} - \sqrt{3(2) - 2}}{\Delta x}$$

$$= \lim_{\Delta x \to 0} \frac{\sqrt{4 + 3\,\Delta x} - 2}{\Delta x}.$$

To simplify the latter fraction, we multiply its numerator and its denominator by $\sqrt{4 + 3\,\Delta x} + 2$, the *conjugate* of the numerator, obtaining

$$\frac{\sqrt{4 + 3\,\Delta x} - 2}{\Delta x} = \frac{\sqrt{4 + 3\,\Delta x} - 2}{\Delta x} \cdot \frac{\sqrt{4 + 3\,\Delta x} + 2}{\sqrt{4 + 3\,\Delta x} + 2}$$

$$= \frac{4 + 3\,\Delta x - 4}{\Delta x\,(\sqrt{4 + 3\,\Delta x} + 2)}$$

$$= \frac{3}{\sqrt{4 + 3\,\Delta x} + 2}.$$

Then

$$\lim_{\Delta x \to 0} \frac{\Delta f}{\Delta x} = \lim_{\Delta x \to 0} \frac{\sqrt{4 + 3\,\Delta x} - 2}{\Delta x}$$

$$= \lim_{\Delta x \to 0} \frac{3}{\sqrt{4 + 3\,\Delta x} + 2}$$

$$= \frac{3}{\sqrt{4 + 0} + 2} = \frac{3}{4}.$$

Therefore the slope of the tangent line to the graph of $f(x) = \sqrt{3x - 2}$ at P_1:(2, 2) is $\frac{3}{4}$, and the equation of that tangent line is $y - 2 = (\frac{3}{4})(x - 2)$, or $3x - 4y + 2 = 0$.

Exercises

In each of the following exercises, the equation of a curve and the coordinates of a point P_1:(x_1, y_1) on the curve are given. Use the definition 4.1.1 to find the slope of the tangent line to the curve at P_1 and write the equation of that tangent line. Make a sketch showing the tangent line and the curve.

1. $y = x^2$, P_1:(2, 4).
2. $y = x^2$, P_1:(−1, 1).
3. $y = 2x^2 + 1$, P_1:(0, 1).
4. $y = 2x^2 + 1$, P_1:(1, 3)
5. $y = -x^2 - 4x - 2$, P_1:(−1, 1).
6. $y = -x^2 - 4x - 2$, P_1:(−2, 2).

7. $y = \dfrac{2}{x}$, P_1:(1, 2).

8. $y = \dfrac{2}{x}$, P_1:(−2, −1).

9. $y = \dfrac{3}{2x - 1}$, P_1:(2, 1).

10. $y = \dfrac{3}{2x - 1}$, P_1:(0, −3).

11. $y = \sqrt{2x}$, P_1:(2, 2).
12. $y = \sqrt{2x}$, P_1:(1, $\sqrt{2}$).
13. $y = \sqrt{4x - 3}$, P_1:(1, 1).
14. $y = \sqrt{4x - 3}$, P_1:(3, 3).

4.2 INSTANTANEOUS VELOCITY

If we drive an automobile from one town to another 80 miles distant in 2 hours, our average velocity is 40 miles per hour. That is, the distance from the first position to the second position divided by the elapsed time is the average velocity.

But during our trip the speedometer reading was often different from 40 miles per hour; it registered zero to start with and at times touched 57 miles per hour. Just what do we mean by instantaneous velocity?

Consider the more precise example of an object falling in a vacuum. Experiment shows that if it starts from rest, the object falls approximately $16t^2$ feet in t seconds. Thus it falls 16 feet in the first second and 64 feet during the first 2 seconds; clearly, it falls faster and faster as time goes on (Fig. 4-6).

The *position* on the coordinate scale (Fig. 4-6) of the falling object at time t_1 is $16(t_1)^2$, and its position at another time $(t_1 + \Delta t)$ is $16(t_1 + \Delta t)^2$. Its displacement from time t_1 to time $t_1 + \Delta t$ is $16(t_1 + \Delta t)^2 - 16(t_1)^2$, and the elapsed time is Δt. Since **average velocity** is displacement divided by elapsed time, the average velocity of the falling object from time t_1 to time $t_1 + \Delta t$ is

$$\text{Average velocity} = \frac{16(t_1 + \Delta t)^2 - 16(t_1)^2}{\Delta t} \text{ feet per second.}$$

During the first second the average velocity is

$$\frac{16(0 + 1)^2 - 16(0)^2}{1} = 16 \text{ feet per second,}$$

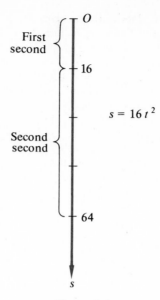

Figure 4-6

and during the next second its average velocity is

$$\frac{16(1+1)^2 - 16(1)^2}{1} = 48 \text{ feet per second.}$$

What is the *instantaneous* velocity of the falling object at the end of 1 second? There is no point in attempting to use the formula for average velocity to find instantaneous velocity because the time elapsed during an "instant" is zero and the distance traveled during an "instant" is also zero.

But if we find the average velocity of the falling object during a very short interval of time Δt, starting at the end of 1 second, we intuitively feel that it should approximate the instantaneous velocity of the object at the end of 1 second.

For example, the time elapsed from $t = 1$ to $t = 1.5$ is $\Delta t = 0.5$ and the average velocity during that interval is

$$\frac{16(1.5)^2 - 16(1)^2}{0.5} = 40 \text{ feet per second.}$$

The average velocity of the falling object from $t = 1$ to $t = 1.25$ is

$$\frac{16(1.25)^2 - 16(1)^2}{0.25} = 36 \text{ feet per second.}$$

Similarly, its average velocity from $t = 1$ to $t = 1.1$ is

$$\frac{16(1.1)^2 - 16(1)^2}{0.1} = 33.6 \text{ feet per second.}$$

In general, the *average velocity* of the falling object from $t = 1$ to $t = 1 + \Delta t$ is given by

$$(1) \qquad V(\Delta t) = \frac{16(1 + \Delta t)^2 - 16(1)^2}{\Delta t}.$$

Intuitively, we feel that whatever the instantaneous velocity at the end of 1 second is, we must be getting closer and closer to it as we compute the average velocity for shorter and shorter intervals of time Δt following 1 second. Accordingly, we *define* the instantaneous velocity at $t = 1$ as the limit of the average velocity from time 1 to time $1 + \Delta t$ as Δt approaches 0. That is, we define the instantaneous velocity at $t = 1$ to be

$$\lim_{\Delta t \to 0} V(\Delta t) = \lim_{\Delta t \to 0} \frac{16(1 + \Delta t)^2 - 16(1)^2}{\Delta t}.$$

In finding this limit, we notice that

$$\frac{16(1 + \Delta t)^2 - 16(1)^2}{\Delta t} = \frac{16[2(\Delta t) + (\Delta t)^2]}{\Delta t} = 32 + 16(\Delta t)$$

for all numbers $\Delta t \neq 0$. Since the definition of $\lim_{\Delta t \to 0} V(\Delta t)$ expressly excludes $\Delta t = 0$,

$$\lim_{\Delta t \to 0} V(\Delta t) = \lim_{\Delta t \to 0} \frac{16(1 + \Delta t)^2 - 16(1)^2}{\Delta t} = \lim_{\Delta t \to 0} [32 + 16(\Delta t)] = 32.$$

So the instantaneous velocity of the falling object at the end of 1 second is 32 feet per second.

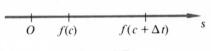

Figure 4-7

More generally, if f is a position function that gives the coordinate $s = f(t)$ at time t of a particle moving along a coordinate line (Fig. 4-7), then the displacement of the particle from time c to time $(c + \Delta t)$ is $f(c + \Delta t) - f(c)$ and the time consumed during this displacement is Δt. And so the average velocity of the moving particle from time c to time $(c + \Delta t)$ is

$$V(\Delta t) = \frac{\Delta f}{\Delta t} = \frac{f(c + \Delta t) - f(c)}{\Delta t}.$$

Its instantaneous velocity at time c is given by the following definition.

4.2.1 **Definition.** If f is a function such that the coordinate of a particle, moving along a coordinate line, at the end of t units of time is $s = f(t)$, then the **instantaneous velocity** of the moving particle at the end of c units of time is

$$\lim_{\Delta t \to 0} \frac{\Delta f}{\Delta t} = \lim_{\Delta t \to 0} \frac{f(c + \Delta t) - f(c)}{\Delta t}.$$

provided that this limit exists.

Example 1. If an object, starting from rest, falls approximately $16t^2$ feet in t seconds, what is its instantaneous velocity at the end of 3.8 seconds?

Solution. Let $f(t) = 16t^2$ and $c = 3.8$. Using the definition 4.2.1, we find the instantaneous velocity at the end of $c = 3.8$ seconds to be

$$\varsigma r = \lim_{\Delta t \to 0} \frac{\Delta f}{\Delta t} = \lim_{\Delta t \to 0} \frac{f((3.8 + \Delta t) - f(3.8)}{\Delta t}$$

$$= \lim_{\Delta t \to 0} \frac{16(3.8 + \Delta t)^2 - 16(3.8)^2}{\Delta t}$$

$$= \lim_{\Delta t \to 0} \frac{16[2(3.8) \Delta t + (\Delta t)^2]}{\Delta t}$$

$$= \lim_{\Delta t \to 0} 16(7.6 + \Delta t) = 121.6.$$

Consequently, the instantaneous velocity of the falling object at the end of 3.8 seconds is 121.6 feet per second.

Example 2. A particle moves along a coordinate line and s, its directed distance in feet from the origin at the end of t seconds, is given by $s = \sqrt{5t + 1}$. Find the instantaneous velocity of the particle at the end of 3 seconds.

Solution. In 4.2.1, let $f(t) = \sqrt{5t + 1}$ and $c = 3$. Then the desired instantaneous velocity is

$$\lim_{\Delta t \to 0} \frac{\Delta f}{\Delta t} = \lim_{\Delta t \to 0} \frac{f(3 + \Delta t) - f(3)}{\Delta t}$$

$$= \lim_{\Delta t \to 0} \frac{\sqrt{5(3 + \Delta t) + 1} - \sqrt{5(3) + 1}}{\Delta t}$$

$$= \lim_{\Delta t \to 0} \frac{\sqrt{16 + 5 \Delta t} - 4}{\Delta t}$$

To simplify this fraction, we multiply its numerator and denominator by the conjugate of the numerator—namely, $\sqrt{16 + 5 \Delta t} + 4$—and obtain

$$\frac{\sqrt{16 + 5 \Delta t} - 4}{\Delta t} \cdot \frac{\sqrt{16 + 5 \Delta t} + 4}{\sqrt{16 + 5 \Delta t} + 4} = \frac{5 \Delta t}{\Delta t \, (\sqrt{16 + 5 \Delta t} + 4)}$$

$$= \frac{5}{\sqrt{16 + 5 \Delta t} + 4}, \qquad \Delta t \neq 0.$$

Therefore

$$\lim_{\Delta t \to 0} \frac{\Delta f}{\Delta t} = \lim_{\Delta t \to 0} \frac{5}{\sqrt{16 + 5 \Delta t} + 4} = \frac{5}{8},$$

and so the instantaneous velocity of the moving particle at the end of 3 seconds is $\frac{5}{8}$ foot per second.

Example 3. Find how long it takes the falling object in Example 1 to reach an instantaneous velocity of 112 feet per second.

Solution. Let $f(t) = 16t^2$. By definition (4.2.1), the instantaneous velocity of the falling object at the end of c seconds is

$$\lim_{\Delta t \to 0} \frac{\Delta f}{\Delta t} = \lim_{\Delta t \to 0} \frac{f(c + \Delta t) - f(c)}{\Delta t}$$

$$= \lim_{\Delta t \to 0} \frac{16(c + \Delta t)^2 - 16c^2}{\Delta t}$$

$$= \lim_{\Delta t \to 0} (32t + 16 \Delta t) = 32c.$$

We seek the value of c at which the instantaneous velocity, $32c$, is 112 feet per second. From $32c = 112$, we find $c = 112/32 = 3.5$. Thus in 3.5 seconds the object has an instantaneous velocity of 112 feet per second.

Exercises

In Exercises 1 through 6, assume that an object, starting from rest, falls s feet in t seconds, where $s = 16t^2$. Use the definition 4.2.1 to find the instantaneous velocity of the falling object at the end of c seconds, for the given value of c.

1. $c = 5$.

2. $c = 1.25$.

3. $c = 2.1$.

4. $c = 3.5$.

5. $c = 0.75$.

6. $c = 4.2$.

7. How many seconds does it take the falling object in the preceding exercises to attain a velocity of
(a) 128 feet per second?
(b) 72 feet per second?
(c) 60 feet per second?

In Exercises 8 to 14, a particle moves along a coordinate line, and s, its directed distance from the origin at the end of t seconds, is measured in feet.

Find the instantaneous velocity of the particle at the end of c seconds.

8. $s = 3t - 2$, $c = 10$.

9. $s = 2t^2 + 5$, $c = 1.7$.

10. $s = t^3 + 1$, $c = 2$.

11. $s = \dfrac{1}{2t}$, $c = 0.5$.

12. $s = \dfrac{1}{t + 3}$, $c = 4$.

13. $s = \sqrt{t - 3}$, $c = 7$.

14. $s = \sqrt{t^2 + 7}$, $c = 3$.

15. A particle moves along a coordinate line. Its directed distance s, in feet, from the origin at the end of t seconds is given by $s = 3t^2 - 2$. How long does it take the particle to attain a velocity of 10 feet per second?

16. An object moves along a coordinate line and its directed distance from the origin at the end of t seconds is $s = -t^2 + 4t$ feet. In how many seconds after $t = 0$ does the object come to a stop?

4.3 THE DERIVATIVE

By comparing our definition of the slope of a tangent to a curve (4.1.1) with our definition of the instantaneous velocity of a particle moving on a straight line (4.2.1), the reader will see that they are formally the same.

Since the limit involved, namely,

$$\lim_{\Delta x \to 0} \frac{f(c + \Delta x) - f(c)}{\Delta x},$$

has two such striking interpretations, it is reasonable to suspect that it may have many others. Such, indeed, is the case.

The preceding limit is called the **value** of the derivative of the function f at c.

4.3.1 Definition. The **derivative** of a function f is another function f' whose value at any point c in the domain of f is

$$f'(c) = \lim_{\Delta x \to 0} \frac{f(c + \Delta x) - f(c)}{\Delta x},$$

provided that this limit exists.

If this limit does exist, we say that f is **differentiable** at c. The domain of f' is a subset of the domain of f.

Finding the derivative of a function is called **differentiation**; it is the basic process in differential calculus.

Another symbol for f' is $D_x f$ (read "the derivative of f with respect to x"); it indicates that x is the independent variable in the function f. If we let $y = f(x)$, other commonly used symbols for the derivative are $\frac{dy}{dx}$ and $D_x y$ (read "the derivative of y with respect to x").

Example 1. Let f be the function defined by $f(x) = 13x - 6$. Find $f'(4)$, the value of the derivative of f at 4.

Solution. By definition,

$$f'(4) = \lim_{\Delta x \to 0} \frac{f(4 + \Delta x) - f(4)}{\Delta x} = \lim_{\Delta x \to 0} \frac{[13(4 + \Delta x) - 6] - [13(4) - 6]}{\Delta x}$$

$$= \lim_{\Delta x \to 0} \frac{13(\Delta x)}{\Delta x}.$$

Since both numerator and denominator of the fraction $13(\Delta x)/\Delta x$ have limit zero as $\Delta x \to 0$, we cannot evaluate this limit by mere substitution. But

$$\frac{13(\Delta x)}{\Delta x} = 13$$

for all $\Delta x \neq 0$. Since the definition of limit as $\Delta x \to 0$ expressly excludes $\Delta x = 0$, we have

$$\lim_{\Delta x \to 0} \frac{13(\Delta x)}{\Delta x} = \lim_{\Delta x \to 0} 13 = 13.$$

Thus 13 is the value of the derivative of f at 4.

Example 2. If f is defined by $f(x) = x^3 + 7x$, find the derivative of f.

Solution. Let c be a number in the domain of f. Then

$$f'(c) = \lim_{\Delta x \to 0} \frac{f(c + \Delta x) - f(c)}{\Delta x} = \lim_{\Delta x \to 0} \frac{[(c + \Delta x)^3 + 7(c + \Delta x)] - (c^3 + 7c)}{\Delta x}$$

$$= \lim_{\Delta x \to 0} \frac{3c^2(\Delta x) + 3c(\Delta x)^2 + (\Delta x)^3 + 7(\Delta x)}{\Delta x}.$$

As in Example 1, the numerator and the denominator of this last fraction both approach zero as $\Delta x \to 0$. *Such is always the case in finding a derivative from its definition.* But

$$\frac{3c^2(\Delta x) + 3c(\Delta x)^2 + (\Delta x)^3 + 7(\Delta x)}{\Delta x} = 3c^2 + 3c(\Delta x) + (\Delta x)^2 + 7$$

for all $\Delta x \neq 0$. Therefore

$$\lim_{\Delta x \to 0} \frac{3c^2(\Delta x) + 3c(\Delta x)^2 + (\Delta x)^3 + 7(\Delta x)}{\Delta x} = \lim_{\Delta x \to 0} [3c^2 + 3c(\Delta x) + (\Delta x)^2 + 7]$$

$$= 3c^2 + 7.$$

That is, the value of the derivative of f at any point c is $3c^2 + 7$.

Thus f', the derivative of f, is the function defined by $f'(x) = 3x^2 + 7$. Its domain, like that of f, is R, the set of all real numbers.

Example 3. Find the derivative of the function f defined by $f(x) = 1/x$.

Solution. The value of the derivative at any number $c \neq 0$ is (by 4.3.1)

$$f'(c) = \lim_{\Delta x \to 0} \frac{f(c + \Delta x) - f(c)}{\Delta x} = \lim_{\Delta x \to 0} \frac{\dfrac{1}{(c + \Delta x)} - \dfrac{1}{c}}{\Delta x}.$$

As usual, both numerator and denominator of this latter quotient approach zero as $\Delta x \longrightarrow 0$. In order to find the limit of the quotient, we must change its form, but not its value, for all $\Delta x \neq 0$. Since

$$\frac{\dfrac{1}{(c + \Delta x)} - \dfrac{1}{c}}{\Delta x} = \frac{\dfrac{c - (c + \Delta x)}{(c + \Delta x)c}}{\Delta x} = \frac{-\Delta x}{(c + \Delta x)c} \cdot \frac{1}{\Delta x} = \frac{-1}{(c + \Delta x)c}$$

for all $\Delta x \neq 0$, then

$$\lim_{\Delta x \to 0} \frac{\dfrac{1}{(c + \Delta x)} - \dfrac{1}{c}}{\Delta x} = \lim_{\Delta x \to 0} \frac{-1}{(c + \Delta x)c} = -\frac{1}{c^2}.$$

Thus f' is the function defined by

$$f'(x) = -\frac{1}{x^2},$$

where x is any real number different from zero.

Example 4. Find the derivative of the function F defined by $F(x) = 1/\sqrt{x}$.

Solution. The domain of F is $(0, \infty)$. By 4.3.1, the value of the derivative of F at any positive number c is

$$F'(c) = \lim_{\Delta x \to 0} \frac{F(c + \Delta x) - F(c)}{\Delta x} = \lim_{\Delta x \to 0} \frac{\dfrac{1}{\sqrt{c + \Delta x}} - \dfrac{1}{\sqrt{c}}}{\Delta x}$$

$$= \lim_{\Delta x \to 0} \frac{\sqrt{c} - \sqrt{c + \Delta x}}{\Delta x \sqrt{c + \Delta x} \sqrt{c}}.$$

As always, when finding a derivative from its definition, both the numerator and the denominator of this latter fraction approach zero as Δx approaches zero. To find the limit, we must change the form of the quotient, without changing its value, for all positive numbers $c + \Delta x$, $\Delta x \neq 0$. This step is accomplished by multiplying both the numerator and the denominator of the quotient by $\sqrt{c} + \sqrt{c + \Delta x}$, the conjugate of the numerator.

$$\frac{\sqrt{c} - \sqrt{c + \Delta x}}{\Delta x \sqrt{c + \Delta x} \sqrt{c}} = \frac{\sqrt{c} - \sqrt{c + \Delta x}}{\Delta x \sqrt{c + \Delta x} \sqrt{c}} \cdot \frac{\sqrt{c} + \sqrt{c + \Delta x}}{\sqrt{c} + \sqrt{c + \Delta x}}$$

$$= \frac{c - (c + \Delta x)}{\Delta x \sqrt{c + \Delta x} \sqrt{c}(\sqrt{c} + \sqrt{c + \Delta x})}$$

$$= \frac{-1}{\sqrt{c + \Delta x} \sqrt{c}(\sqrt{c} + \sqrt{c + \Delta x})}$$

Thus

$$F'(c) = \lim_{\Delta x \to 0} \frac{-1}{\sqrt{c + \Delta x} \sqrt{c}(\sqrt{c} + \sqrt{c + \Delta x})}$$

$$= \frac{-1}{\sqrt{c} \sqrt{c}(\sqrt{c} + \sqrt{c})} = \frac{-1}{2c\sqrt{c}} = \frac{-1}{2c^{3/2}},$$

for all $c > 0$. Therefore F', the derivative of F, is the function defined by $F'(x) = -1/(2x^{3/2})$. Its domain is $(0, \infty)$.

Example 5. If $f(x) = \sqrt{x}, x > 0$, show that

$$f'(x) = \frac{1}{2\sqrt{x}}, \qquad x > 0.$$

Solution. Let c be an arbitrarily chosen positive number. By the definition of the derivative,

$$f'(c) = \lim_{\Delta x \to 0} \frac{\sqrt{c + \Delta x} - \sqrt{c}}{\Delta x}, \qquad c + \Delta x > 0.$$

But

$$\frac{\sqrt{c + \Delta x} - \sqrt{c}}{\Delta x} = \frac{\sqrt{c + \Delta x} - \sqrt{c}}{\Delta x} \cdot \frac{\sqrt{c + \Delta x} + \sqrt{c}}{\sqrt{c + \Delta x} + \sqrt{c}}$$

$$= \frac{(c + \Delta x) - c}{\Delta x(\sqrt{c + \Delta x} + \sqrt{c})}$$

$$= \frac{1}{\sqrt{c + \Delta x} + \sqrt{c}}.$$

Therefore

$$f'(c) = \lim_{\Delta x \to 0} \frac{1}{\sqrt{c + \Delta x} + \sqrt{c}} = \frac{1}{2\sqrt{c}}.$$

Since c can be any positive number,

$$f'(x) = \frac{1}{2\sqrt{x}}, \qquad x > 0.$$

Exercises

In Exercises 1 to 10, use the definition of the derivative, 4.3.1, to find $f'(x)$.

1. $f(x) = 5x - 4$.

2. $f(x) = ax + b$ (a, b constants).

3. $f(x) = 8x^2 - 1$.

4. $f(x) = ax^2 + bx + c$ (a, b, c constants).

5. $f(x) = 3x^3$.

6. $f(x) = 4x^3 + 3x^2 - 2$.

7. $f(x) = x^3 - 2x + 15$.

8. $f(x) = ax^3 + bx^2 + cx + d$ (a, b, c, d constants).

9. $f(x) = x^4$.

10. $f(x) = 2x^4 - 7x + 1$.

In Exercises 11 to 14, find $D_x F$ from 4.3.1. What is the domain of the derivative?

11. $F(x) = \dfrac{3}{5x}.$

12. $F(x) = \dfrac{2}{3x - 4}.$

13. $F(x) = \dfrac{6}{2x^2 + 3}.$

14. $F(x) = \dfrac{1}{x^2 + 2x - 3}.$

Use the definition of the derivative, 4.3.1, to find $\dfrac{dy}{dx}$ in Exercises 15 to 18.

15. $y = \dfrac{4}{x^3}.$

17. $y = \dfrac{1}{x^4}.$

16. $y = \dfrac{1}{7x^3 - 5}.$

18. $y = \dfrac{2}{3x^4 + 1}.$

In Exercises 19 to 24, use the definition, 4.3.1, to find the derivatives of the indicated functions.

19. $g(x) = \dfrac{1}{\sqrt{3x}}.$

20. $H(y) = \dfrac{3}{\sqrt{y - 2}}.$

21. $F(t) = \sqrt{t^2 + 4}.$
[*Hint:*
$$\dfrac{\sqrt{(c + \Delta t)^2 + 4} - \sqrt{c^2 + 4}}{\Delta t}$$
$$= \dfrac{\sqrt{(c + \Delta t)^2 + 4} - \sqrt{c^2 + 4}}{\Delta t} \cdot \dfrac{\sqrt{(c + \Delta t)^2 + 4} + \sqrt{c^2 + 4}}{\sqrt{(c + \Delta t)^2 + 4} + \sqrt{c^2 + 4}}$$
$$= \dfrac{2c + \Delta t}{\sqrt{(c + \Delta t)^2 + 4} + \sqrt{c^2 + 4}}.]$$

22. $\phi(z) = \sqrt{2z^2 - 1}.$

23. $f(u) = \dfrac{1}{\sqrt{u^2 + 9}}.$

24. $G(x) = \dfrac{5}{\sqrt{2x}}.$

If the right-hand limit (3.4.1),
$$\lim_{\Delta x \to 0+} \dfrac{f(c + \Delta x) - f(c)}{\Delta x},$$
exists, it is called the value of the **right-hand derivative** of f at c. Similarly, the value of the **left-hand derivative** of a function f at c is the left-hand limit,
$$\lim_{\Delta x \to 0-} \dfrac{f(c + \Delta x) - f(c)}{\Delta x},$$
provided that this limit exists.

25. Let f be a function that is defined on an interval I and let c be an interior point of I. Prove that $f'(c)$ exists if and only if both the left- and right-hand derivatives of f exist at c and their values are the same at c. (*Hint:* Use 3.4.2.)

26. Draw the graph of $F(x) = |x|$.
(a) Find the right-hand derivative of F at 0.
(b) Find the left-hand derivative of F at 0.
(c) Use Exercise 25 to show that $F'(0)$ does not exist.

27. Draw the graph of $f(x) = |x^2 - 1|$.
(a) Show that $f'(1)$ and $f'(-1)$ do not exist.
(b) Find $f'(x)$ for $x > 1$ and for $x < -1$.
(c) Find $f'(x)$ for $-1 < x < 1$.

28. Draw the graph of $F(t) = |t^3 + 1|$.
(a) Does $F'(-1)$ exist? If not, tell why; if so, find $F'(-1)$.
(b) Find $F'(t)$ for $t > -1$.
(c) Find $F'(t)$ for $t < -1$.

4.4 APPLICATIONS OF THE DERIVATIVE. RATE OF CHANGE

The quotient

(1)
$$\dfrac{f(c + \Delta x) - f(c)}{\Delta x},$$

which appears in the definition of the derivative (4.3.1), is called a **difference quotient.** (Notice that it defines a function of Δx, because c is a constant and Δx is a variable.) Its denominator is the difference between two values of x (namely, c and $c + \Delta x$) and its numerator is the difference between the corresponding values of the function. In other words, the denominator is a change in the value of x and the numerator is the corresponding change in the value of f. Thus (1) measures an *average rate of change of $f(x)$* with respect to the independent variable x, starting at $x = c$.

We define *the* **instantaneous rate of change** *of $f(x)$ with respect to the independent variable x, at $x = c$, as the limit of this average rate of change.* Consequently, the instantaneous rate of change is given by the value of the derivative at $x = c$,

$$f'(c) = \lim_{\Delta x \to 0} \frac{f(c + \Delta x) - f(c)}{\Delta x};$$

and, conversely, *the value of the derivative f' at any number c in its domain can always be interpreted as the instantaneous rate of change of $f(x)$ with respect to x at $x = c$.*

For instance, in the motion of a particle on a coordinate line (Section 4.2), the difference quotient

$$\frac{f(c + \Delta t) - f(c)}{\Delta t}$$

represents the change in the distance of the particle from the origin divided by the change in time. Thus it is an average rate of change of position with respect to time, starting with $t = c$, which is average velocity.

Then

$$\lim_{\Delta t \to 0} \frac{f(c + \Delta t) - f(c)}{\Delta t}$$

gives the instantaneous rate of change of position with respect to time, at $t = c$ (that is, the instantaneous velocity).

Again, on the Cartesian graph of $y = f(x)$ the difference quotient (1) is the change in the value of the function f divided by the change in the value of the independent variable x as a point moves along the curve from $(c, f(c))$ to $(c + \Delta x, f(c + \Delta x))$; that is, it is an average rate of change of y with respect to x (see Fig. 4-3). Then $f'(c)$ gives the instantaneous rate of change of y with respect to x at $x = c$.

Thus the slope of the tangent line to a curve at a given point measures the instantaneous rate of change of y with respect to x at the point. If $f'(c) > 0$, the tangent line points upward to the right, the curve is rising as x increases, and $f(x)$ is increasing at $x = c$. If $f'(c) < 0$, the tangent line points downward to the right, the curve is falling as x increases, and $f(x)$ is decreasing at $x = c$ (Fig. 4-8).

Note. In learning to *use* calculus, it is helpful to remember that *the derivative of a function f always gives the instantaneous rate of change of $f(x)$ with respect to the independent variable x, at any point in the domain of f'.*

Example 1. Find the rate of change of the volume of a cube with respect to the length of an edge of the cube.

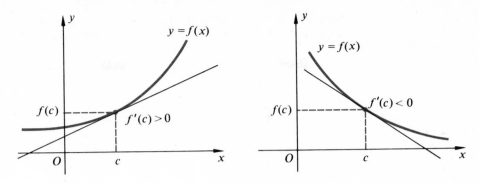

Figure 4-8

Solution. If x represents the length of an edge of a cube, then the volume V of the cube is given by $V = x^3$. We seek the rate of change of V with respect to x. That is, we seek $D_x V$, the derivative of V with respect to x. Let c be an arbitrary value of x. Then, at $x = c$,

$$D_x V = \lim_{\Delta x \to 0} \frac{(c + \Delta x)^3 - c^3}{\Delta x}$$

$$= \lim_{\Delta x \to 0} \frac{3c^2(\Delta x) + 3c(\Delta x)^2 + (\Delta x)^3}{\Delta x}$$

$$= \lim_{\Delta x \to 0} [3c^2 + 3c(\Delta x) + (\Delta x)^2] = 3c^2.$$

Since c can be any value of x, the rate of change of the volume of a cube with respect to the length of an edge is $3x^2$, where x is the length of an edge.

Example 2. If a small lead ball is projected directly upward with an initial velocity of 80 feet per second, its approximate height at the end of t seconds is $h = 80t - 16t^2$ feet. How fast is the ball rising at the end of 2 seconds? In how many seconds does the ball reach its maximum height? What is that maximum height?

Solution. Let c be an arbitrarily chosen positive number. The velocity of the ball at the end of c seconds is

$$V(c) = h'(c) = \lim_{\Delta t \to 0} \frac{\Delta h}{\Delta t} = \lim_{\Delta t \to 0} \frac{h(c + \Delta t) - h(c)}{\Delta t}$$

$$= \lim_{\Delta t \to 0} \frac{[80(c + \Delta t) - 16(c + \Delta t)^2] - (80c - 16c^2)}{\Delta t}$$

$$= \lim_{\Delta t \to 0} \frac{(80 - 32c)\,\Delta t - 16(\Delta t)^2}{\Delta t}$$

$$= \lim_{\Delta t \to 0} (80 - 32c - 16\,\Delta t) = 80 - 32c.$$

At the end of 2 seconds, $c = 2$ and the velocity of the ball is $V(2) = 80 - 32(2) = 16$ feet per second.

When the ball reaches its greatest height, its velocity is zero. By setting $V(c)$ equal to zero, we get $80 - 32c = 0$, from which $c = \frac{80}{32} = \frac{5}{2}$. Thus, the ball attains its greatest height in $2\frac{1}{2}$ seconds.

The greatest height is found by substituting $t = \frac{5}{2}$ in $h = 80t - 16t^2$. This step gives the maximum height of the ball as $80(\frac{5}{2}) - 16(\frac{5}{2})^2 = 100$ feet.

Exercises

1. Let $y = x^2/2 - 1$. Find the average rate of change of y with respect to x
(a) from $x = 2$ to $x = 2.5$;
(b) from $x = 2$ to $x = 2.1$.
What is the instantaneous rate of change of y with respect to x at $x = 2$? Sketch the graph of $y = x^2/2 - 1$. What is the slope of its tangent line at $(2, 1)$? Draw the tangent line to the graph at $(2, 1)$.

2. If $y = 1/x$, find the average rate of change of y with respect to x
(a) from $x = 1$ to $x = 2$;
(b) from $x = 1$ to $x = 1.2$.
Find the instantaneous rate of change of y with respect to x at $x = 1$. Sketch the graph of $y = 1/x$. What is the slope of its tangent line at $(1, 1)$? Draw the tangent line to the graph at $(1, 1)$.

3. Find the instantaneous rate of change of the area of a square with respect to the length of a side.

4. Show that the instantaneous rate of change of the area of a circle with respect to its radius is equal to the circumference. (*Hint:* $A = \pi r^2$. Find $D_r A$.)

5. The volume of a sphere of radius r is given by $V = 4\pi r^3/3$. Find the instantaneous rate of change of the volume with respect to the radius when $r = 6$.

6. Find the instantaneous rate of change of the volume of a right circular cone
(a) with respect to the radius of its base if the altitude is constant;
(b) with respect to the altitude if the radius of the base is constant.

7. Find the instantaneous rate of change of the area of an equilateral triangle with respect to the length of a side.

8. The hypotenuse of a right triangle has constant length 12, and sides of varying lengths x and y. Make a sketch showing the side having length x as horizontal and the the side of length y as vertical. Find the instantaneous rate of change of y with re-

spect to x when $x = 6$. (*Hint:* See Exercise 21 of Section 4.3.)

9. A ladder, 12 feet long, is leaning against a wall. If the foot of the ladder slides away from the wall along level ground, what is the instantaneous rate of change of the top of the ladder with respect to the distance of the foot of the ladder from the wall when the foot is 4 feet from the wall? (*Hint:* See Exercise 8.)

10. What is the instantaneous rate of change of the reciprocal of the fourth power of a number with respect to the number, when the number is 3?

11. Two electric charges s units apart repel each other according to the formula $r = k/s^2$, where $k > 0$ is a constant. Find the instaneous rate of change of r with respect to s.

12. If the temperature of a confined gas is kept constant, its pressure p and its volume V are related by $pV = k$, where k is a constant. Find the instantaneous rate of change of the volume of the gas with respect to its pressure when the pressure is 5 pounds per square foot.

13. Neglecting friction, if a stone is thrown directly upward with an initial velocity of 40 feet per second, its approximate height in feet at the end of t seconds is given by $h = 40t - 16t^2$. How fast is the stone rising at the end of 1 second?

14. In how many seconds does the stone in Exercise 13 reach the top of its flight? How high does it go? (*Hint:* At the top of its flight, the velocity of the stone is zero.)

15. Neglecting friction, a stone thrown directly downward from the edge of a cliff, with an initial velocity of 20 feet per second, will fall approximately $s = 16t^2 + 20t$ feet in t seconds. How fast is it falling at the end of 3.4 seconds?

16. Two lead balls are dropped from a high cliff into the sea. If one ball is dropped 2 seconds after the other, how fast is the distance between them increasing

(a) 1 second after the last ball is released?

(b) 2 seconds after the last ball is released?

[*Hint: t* seconds after the second ball is released, it has fallen $16t^2$ feet, the first ball has fallen $16(t + 2)^2$ feet, and the distance between them is $S(t) = 16(t + 2)^2 - 16t^2$. Find $S'(1)$ and $S'(2)$.]

17. A ball rolls down a long inclined plane and its distance from its starting point is $s = 4.5t^2$ feet at the end of t seconds. When will its velocity be 30 feet per second? (*Hint:* Find the instantaneous rate of change of s with respect to t at the end of c seconds. Then find the value of c that makes the rate 30 feet per second.)

18. A tank, in the form of an inverted right circular cone whose altitude is 18 feet and the radius of whose base is 6 feet, is being filled with water. Find the instantaneous rate of change of the volume of water in the tank with respect to the depth of the water when the depth is 12 feet. (*Hint:* Express V, the volume of water in the tank, as a function of its depth, h.)

19. Water is pouring into an upright cylindrical tank, the radius of whose circular base is 5 feet, at the rate of 30 cubic feet per minute. How fast is the water level rising? (*Hint:* Express the depth of the water, h, as a function of time, t, and find $D_t h$ when $t = c$ minutes.)

4.5 THE DERIVATIVE AND CONTINUITY

An alternate form of the definition of the derivative (4.3.1) is sometimes more convenient to work with.

4.5.1 Definition (*Alternate Form*).   The **derivative** of a function f is another function f' whose value at any point c in the domain of f is

$$f'(c) = \lim_{x \to c} \frac{f(x) - f(c)}{x - c},$$

provided that this limit exists.

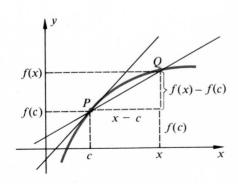

Figure 4-9

Figure 4-9 shows a geometric interpretation of this second form of the definition of the derivative. Let $P:(c, f(c))$ be a fixed point on the curve $y = f(x)$ and let $Q:(x, f(x))$, where $x \neq c$, be a neighboring point on the same curve. Then the difference quotient

$$\frac{f(x) - f(c)}{x - c}$$

is the slope of the secant PQ, and

$$\lim_{x \to c} \frac{f(x) - f(c)}{x - c}$$

is the slope of the tangent to the curve at P.

It is easy to show that the two forms of the definition of the derivative are equivalent. If

$$\lim_{\Delta x \to 0} \frac{f(c + \Delta x) - f(c)}{\Delta x} = L,$$

then (by 3.2.1) to each number $\epsilon > 0$ there corresponds a number $\delta > 0$ such that

(1) $\qquad \left| \dfrac{f(c + \Delta x) - f(c)}{\Delta x} - L \right| < \epsilon \qquad$ whenever $\qquad 0 < |\Delta x| < \delta.$

Letting $\Delta x = x - c$ in (1), we have

(2) $\qquad \left| \dfrac{f(x) - f(c)}{x - c} - L \right| < \epsilon \qquad$ whenever $\qquad 0 < |x - c| < \delta.$

But (2) means that

$$\lim_{x \to c} \frac{f(x) - f(c)}{x - c} = L \qquad \text{(by 3.2.1)},$$

which is the alternate form of the definition of the derivative (4.5.1).

Since these steps are reversible, each form of the definition of the derivative implies the other, and the two forms are equivalent.

Example 1. Let f be the function defined by $f(x) = 2x^2 - x + 5$. Use the alternate form of the definition of the derivative (4.5.1) to find f', the derivative of f.

Solution. The domain of f is R. Let c be an arbitrarily chosen real number. By 4.5.1, for $x \neq c$,

$$f'(c) = \lim_{x \to c} \frac{f(x) - f(c)}{x - c} = \lim_{x \to c} \frac{(2x^2 - x + 5) - (2c^2 - c + 5)}{x - c}$$

$$= \lim_{x \to c} \frac{2(x^2 - c^2) - (x - c)}{x - c} = \lim_{x \to c} \frac{(x - c)[2(x + c) - 1]}{x - c}$$

$$= \lim_{x \to c} (2x + 2c - 1) = 4c - 1.$$

Since c can be any real number, f' is the function defined by $f'(x) = 4x - 1$.

Example 2. If $G(x) = 2/\sqrt{3x}$, use 4.5.1 to find $G'(x)$.

Solution. Let c be an arbitrarily chosen positive number. By 4.5.1,

$$G'(c) = \lim_{x \to c} \frac{G(x) - G(c)}{x - c} = \lim_{x \to c} \frac{(2/\sqrt{3x}) - (2/\sqrt{3c})}{x - c}$$

$$= \lim_{x \to c} \frac{2(\sqrt{3c} - \sqrt{3x})}{\sqrt{3x}\sqrt{3c}\,(x - c)} = \lim_{x \to c} \frac{-2\sqrt{3}(\sqrt{x} - \sqrt{c})}{3\sqrt{cx}\,(x - c)}$$

$$= \lim_{x \to c} \frac{-2\sqrt{3}}{3\sqrt{cx}(\sqrt{x} + \sqrt{c})} = \frac{-2\sqrt{3}}{3\sqrt{c^2}(2\sqrt{c})} = \frac{-1}{c\sqrt{3c}}.$$

Since c can be any positive number,

$$G'(x) = \frac{-1}{x\sqrt{3x}}, \qquad x > 0.$$

The alternate form of the definition of the derivative (4.5.1) is useful in proving our next theorem.

4.5.2 Theorem. If f is a function and $f'(c)$ exists, then f is continuous at c.

Proof. Let

$$f'(c) = \lim_{x \to c} \frac{f(x) - f(c)}{x - c}$$

exist. Now

(3)
$$f(x) - f(c) = \frac{f(x) - f(c)}{x - c} \cdot (x - c)$$

identically for all $x \neq c$ in the domain of f. Taking the limit of both sides of (3) as $x \longrightarrow c$, and recalling that the limit of the product of two functions is equal to the product of the limits of the functions, we obtain

$$\lim_{x \to c} [f(x) - f(c)] = \left[\lim_{x \to c} \frac{f(x) - f(c)}{x - c} \right] \cdot \lim_{x \to c} (x - c) = f'(c) \cdot 0 = 0.$$

Since $\lim_{x \to c} [f(x) - f(c)] = 0$, it follows from 3.3.6 that $\lim_{x \to c} f(x) = f(c)$. Therefore f is continuous at c (by 3.5.1). ∎

This theorem tells us that if the derivative of a function exists at a point, the function must be continuous there. However, *the converse is not true*. There are functions that are continuous at certain points and yet do not possess a derivative there, as the following example shows.

Example 3. Show that the absolute value function, defined by $f(x) = |x|$, is continuous at $x = 0$ but is not differentiable there.

Solution. The graph of the function (Fig. 4-10) suggests that it is continuous at $(0, 0)$, since there is no break there; it also suggests that the graph has no (unique) tangent there and hence that the function is not differentiable at $x = 0$. We shall now prove this to be true.

By 3.5.2, the absolute value function is continuous at every real number; in particular, it is continuous at $x = 0$.

But the absolute value function is not differentiable at $x = 0$. From the alternate form of the definition of the derivative, $f'(0)$ exists if and only if

$$\lim_{x \to 0} \frac{f(x) - f(0)}{x - 0} = \lim_{x \to 0} \frac{|x| - |0|}{x} = \lim_{x \to 0} \frac{|x|}{x}$$

exists.

For $x \geq 0, |x| = x$, and the right-hand limit (3.4.1)

$$\lim_{x \to 0^+} \frac{f(x) - f(0)}{x - 0} = \lim_{x \to 0^+} \frac{|x|}{x} = \lim_{x \to 0^+} \frac{x}{x} = 1.$$

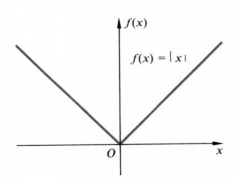

Figure 4-10

Again, for $x < 0, |x| = -x$, and the left-hand limit

$$\lim_{x \to 0^-} \frac{f(x) - f(0)}{x - 0} = \lim_{x \to 0^-} \frac{|x|}{x} = \lim_{x \to 0^-} \frac{-x}{x} = -1.$$

Since the right-hand limit and the left-hand limit of $[f(x) - f(0)]/(x - 0)$ at zero are not the same,

$$f'(0) = \lim_{x \to 0} \frac{f(x) - f(0)}{x - 0}$$

does not exist.

This example shows that a function may be continuous at a point and yet not be differentiable there. Continuity of a function is a *necessary condition* for the existence of its derivative, but it is not a *sufficient condition*. That is, if a function f has a derivative at a point c, then f is necessarily continuous at c; but the continuity of f at a point b is not sufficient to guarantee that $f'(b)$ exists.

We close this section with an alternate form of the definition of continuity.

4.5.3   Definition (*Alternate Form*).   Let f be a function that is defined throughout an interval containing c and $c + \Delta x$; then the function f is **continuous** at c if

$$\lim_{\Delta x \to 0} f(c + \Delta x) = f(c).$$

It is easy to show, as was done for 4.5.1, that this alternate form is equivalent to our earlier definition of continuity (3.5.1). The details are left as an exercise.

The following corollary is an immediate consequence of 4.5.3 and 3.3.1.

4.5.4   Corollary.   If f is continuous at c, and if $\Delta f = f(c + \Delta x) - f(c)$, then

$$\lim_{\Delta x \to 0} \Delta f = 0.$$

Exercises

In each of Exercises 1 through 8, use 4.5.1 to find the derivative of the indicated function and state the domain of the derivative.

1. $f(x) = 5x + 3$.

2. $F(x) = 2x^2$.

3. $G(t) = t^2 - 7$.

4. $\phi(w) = 4w^2 + 2w - 8$.

5. $g(x) = x^3$.

6. $f(u) = \dfrac{3u}{u^2 - 4}$.

7. $\Phi(t) = \sqrt{t^3 - 1}$.

8. $F(y) = \sqrt{4 - 9y^2}$.

Sketch the graph of each of the functions defined in Exercises 9 to 16. At what numbers is each function discontinuous?

9. $f(x) = \dfrac{2x^2 + 5x - 3}{2x - 1}$.

10. $\phi(t) = \dfrac{3t + 7}{3t^2 - 8t - 35}$.

11. $F(w) = \sqrt{16 - w^2}$.

12. $g(x) = \dfrac{x}{|2x|}$.

13. $G(z) = [\![z - 1]\!]$.

14. $f(y) = y[\![y]\!]$.

15. $H(u) = 2u + [\![u]\!]$.

16. $f(x) = \dfrac{x - 6}{\sqrt{|x - 6|}}$.

17. If a function f is discontinuous at a number c and $f(c)$ can be defined (or redefined) so that f becomes continuous at c, then f is said to have a **removable discontinuity** at c (see Figs. 3-3 to 3-6, 3-12, and 3-19). Which of the discontinuities in Exercises 9 to 16 are removable? How would you remove them?

18. If $\lim_{x \to c^+} f(x) = L_1$ and $\lim_{x \to c^-} f(x) = L_2$, and $L_1 \neq L_2$, then f is said to have a **jump discontinuity** at c (see Figs. 3-7 and 3-17). Which of the discontinuities in Exercises 9 to 16 are jump discontinuities?

19. Sketch the graph of the function F defined by $F(x) = |x|(x^2 - 1)$. Prove that F is continuous at 0

but is not differentiable there. (*Hint:* See 3.4.2, and Exercise 25 of Section 4.3.)

20. Show that the greatest integer function is discontinuous at every integer n. (*Hint:* Use 3.4.2.)

21. Prove that the greatest integer function does not have a derivative at $x = n$, if $n \in Z$. (*Hint:* Use Exercise 20 and 4.5.2.)

22. Prove that 4.5.3 and 3.5.1 are equivalent—that is, prove that 4.5.3 $\implies$ 3.5.1 and 3.5.1 $\implies$ 4.5.3. (*Hint:* See the proof in this section that 4.5.1 and 4.3.1 are equivalent.)

23. Let Φ be the function defined by

$$\Phi(x) = 2x^2 \qquad \text{for} \quad x \geq 0,$$
$$\Phi(x) = x^3 \qquad \text{for} \quad x < 0.$$

(a) Sketch its graph.
(b) Show that Φ is continuous and differentiable at all real numbers, and find $\Phi'(0)$.

(*Hint:* Consider separately the three cases, $x > 0$, $x < 0$, and $x = 0$. At $x = 0$, use Exercise 25 of Section 4.3.)

4.6 REVIEW EXERCISES

1. From memory, state the Δx form of the definition of the derivative of a function f. Use it to find $f'(x)$ when $f(x) = ax + b$, in which a and b are constants.

2. State from memory the alternate form of the definition of the derivative of a function F. Use it to find $F'(x)$ when $F(x) = Ax^2 + Bx + C$, in which A, B, and C are constants.

3. Use the Δx form of the definition of the derivative to find
(a) $f'(x)$ if $f(x) = x^2 - 5x + 2$.
(b) $D_x f$ if $f(x) = \dfrac{1}{x - 2}$,
(c) $\dfrac{dy}{dx}$ if $y = \sqrt{9 - x}$.

4. Use the alternate form of the definition of the derivative to do Exercise 3.

5. Find the equation of the tangent line to the curve $y = x^2 - 4x$ at the point $(1, -3)$. Sketch the curve and the tangent line.

6. Let $f(x) = (x - 2)^2$. Find $f'(c)$, where c is an arbitrarily chosen number.

7. Sketch the curve $y = (x - 2)^2$ and the line $2x - y + 2 = 0$. Find the coordinates of the point P on the curve at which its tangent line is parallel to the given line. Write the equation of the tangent line and include its graph in your sketch. [*Hint:* In the answer to Exercise 6, determine the value of c that makes $f'(c)$ equal to the slope of the given line, $2x - y + 2 = 0$.]

8. Find the coordinates of the point Q on the curve $y = (x - 2)^2$ at which its tangent line is perpendicular to the line $2x - y + 2 = 0$. Write the equation of that tangent line. Make a sketch. (*Hint:* See Exercises 6 and 7.)

9. A spherical balloon is expanding from the sun's heat. Find the instantaneous rate of change of the volume of the balloon with respect to its radius when the radius is 5.

10. The surface area of the balloon in Exercise 9 is given by $A = 4\pi r^2$, where r is its radius. Find the instantaneous rate of change of the surface area of the balloon with respect to the radius when the radius is 5.

11. Let $f(x) = (x + 1)/(x - 1)$. Use either form of the definition of the derivative to find $f'(2)$.

12. Use either form of the definition of the derivative to show that
(a) if $f(x) = x^2$, then $f'(x) = 2x$.
(b) if $f(x) = x^3$, then $f'(x) = 3x^2$.
(c) if $f(x) = x^4$, then $f'(x) = 4x^3$.
If $f(x) = x^5$, guess what $f'(x)$ is; then verify your answer. Finally, if $f(x) = x^n$, where n is a positive integer, guess what $f'(x)$ is.

13. Let f and g be functions that are differentiable at a number c. How are the tangent lines to their graphs at c related if
(a) $f'(c) = g'(c)$?   (b) $f'(c) = \dfrac{-1}{g'(c)}$?

14. Sketch the graph of each of the following indicated functions. Is the function continuous at zero? Is it differentiable at zero? If either answer is no, state your reason.

(a) $f(x) = 2$ for $x \geq 0$; $f(x) = x$ for $x < 0$.

(b) $g(x) = x^2$ for $x > 0$; $g(x) = x^3$ for $x \leq 0$.

(c) $F(x) = -x$ for $x \geq 0$; $F(x) = x^2$ for $x < 0$.

(d) $G(x) = x^2 - x$ for $x > 0$; $G(x) = |x|$ for $x \leq 0$.

15. Find the rate of change of the reciprocal of the cube of a number with respect to the number, when the number is 2.

16. A tank in the form of a right circular cylinder is full of water. The tank is 10 feet high and the radius of its circular base is 6 feet. If water is drained from the tank at a constant rate of 54 cubic feet per minute, how fast is the water level falling? (*Hint:* Express the depth of the water, h, as a function of the time, t, and find $D_t h$ when $t = c$ minutes.)

17. An object is projected directly upward from the ground with an initial velocity of 128 feet per second. Its height at the end of t seconds is approximately $s = 128t - 16t^2$ feet.

(a) Find its (instantaneous) velocity at the end of 2 seconds and tell whether it is moving up or down.

(b) Find its velocity at the end of 5 seconds and state whether it is moving up or down.

(c) What is the maximum height attained by the object?

(d) In how many seconds will it return to the ground?

18. Let $f(x) = x^2 + |x|$.

(a) Find the value of the right-hand derivative at zero.

(b) Find the value of the left-hand derivative at zero.

(c) Does $f'(0)$ exist?

19. A particle moves on a coordinate line. If its directed distance from the origin at the end of t seconds is given by $s = t^3 - 6t^2 + 9t$, find its instantaneous velocity, $v(t)$, at the end of $t = c$ seconds. For what values of c does the particle come to a stop? When is it moving to the right ($v > 0$) and when is it moving to the left ($v < 0$)?

20. (a) Find the value of the right-hand derivative of the greatest integer function at 1.

(b) Does the left-hand derivative of the greatest integer function at 1 exist? If so, find it; if not, tell why.

5

formulas for differentiation of algebraic functions. applications

The process of finding the derivative of a function directly from the definition of the derivative, by setting up a difference quotient and evaluating its limit, can be time consuming and tedious. Fortunately, there is a quicker way.

We shall apply this process, once and for all, to functions of certain basic types involving arbitrary constants, and memorize the results as formulas for differentiation. These **standard formulas for differentiation**, as they are sometimes called, enable us to set down the derivatives of many seemingly complicated functions as rapidly as we can write. For example, we shall be able to write, without hesitation, the derivative of $6\sqrt{x^7} - 0.01/x$ as $21\sqrt{x^5} + 0.01/x^2$.

Since f', the derivative of a function f, is also a function, it can often be differentiated to obtain f'', the *second derivative* of f. The second derivative gives the instantaneous rate of change of the first derivative and has important applications. Perhaps the simplest application of the second derivative is to motion. If an object is moving on a straight line and $s(t)$ gives its position at the end of t units of time, we know that $s'(t) = v(t)$ gives the velocity of the object at time t. Then $s''(t) = v'(t)$ gives the instantaneous rate of change of the velocity of the object. This is called the *acceleration* of the object at time t, and is denoted by $a(t)$.* Thus $s''(t) = v'(t) = a(t)$. When $a(t) > 0$, the velocity of the moving object is increasing; when $a(t) < 0$, the velocity is decreasing.

The second derivative, f'', of f, also being a function, can often be differentiated to give f''', the third derivative of f. The results of such successive differentiations are called **higher derivatives.** Higher derivatives will be discussed later in this chapter.

*This topic will be discussed more fully in Section 6.4.

5.1 DERIVATIVE OF A POLYNOMIAL FUNCTION

Since the graph of $f(x) = k$, where k is a constant, is a horizontal line whose slope is everywhere zero (see Fig. 3-13), it is natural to expect that $f'(x) = 0$ for all values of x—that is, that *the value of the derivative of a constant function is everywhere zero*.

5.1.1   Theorem.   If k is a constant and $f(x) = k$, then $D_x f = 0$ or

$$D_x k = 0.$$

Proof.   Let c be any real number. Since $f(x) = k$ for all values of x, $f(c) = k$ and $f(c + \Delta x) = k$. Therefore (by 4.3.1)

$$f'(c) = \lim_{\Delta x \to 0} \frac{f(c + \Delta x) - f(c)}{\Delta x} = \lim_{\Delta x \to 0} \frac{k - k}{\Delta x} = \lim_{\Delta x \to 0} 0 = 0,$$

or

$$D_x k = 0. \quad \blacksquare$$

The graph of $y = x$ is a line through the origin with slope 1 (see Fig. 3-14); so we would expect the value of the derivative of the identity function, defined by $f(x) = x$, to be 1 for every value of x.

5.1.2   Theorem.   If f is the identity function, defined by $f(x) = x$, then

$$D_x x = 1.$$

Proof.   Let c be any number. Then (by 4.3.1)

$$f'(c) = \lim_{\Delta x \to 0} \frac{f(c + \Delta x) - f(c)}{\Delta x} = \lim_{\Delta x \to 0} \frac{(c + \Delta x) - c}{\Delta x} = \lim_{\Delta x \to 0} \frac{\Delta x}{\Delta x} = \lim_{\Delta x \to 0} 1 = 1.$$

Thus

$$D_x x = 1. \quad \blacksquare$$

5.1.3   Theorem (*Power Rule*).   If f is the function defined by $f(x) = x^n$, where n is a positive integer, then

$$D_x x^n = n x^{n-1}.$$

Proof.   By 5.1.2, the theorem is valid when $n = 1$. Assume that n is an integer greater than 1 and that c is an arbitrarily chosen real number. Then

$$f'(c) = \lim_{\Delta x \to 0} \frac{f(c + \Delta x) - f(c)}{\Delta x} = \lim_{\Delta x \to 0} \frac{(c + \Delta x)^n - c^n}{\Delta x}.$$

By applying the binomial theorem to $(c + \Delta x)^n$, we obtain

$$(c + \Delta x)^n - c^n = c^n + n c^{n-1} \Delta x + \frac{n(n - 1)}{2} c^{n-2} \Delta x^2 + \cdots + (\Delta x)^n - c^n.$$

Thus

$$\frac{(c + \Delta x)^n - c^n}{\Delta x} = nc^{n-1} + \frac{n(n-1)}{2}c^{n-2} \Delta x + \cdots + (\Delta x)^{n-1},$$

in which every term after the first contains Δx as a factor. Therefore

$$f'(c) = \lim_{\Delta x \to 0} \frac{(c + \Delta x)^n - c^n}{\Delta x} = nc^{n-1}.$$

Since c can be any real number,

$$D_x x^n = nx^{n-1}. \quad \blacksquare$$

For instance,

$$D_x x^2 = 2x, \qquad D_x x^5 = 5x^4, \qquad D_x x^{18} = 18x^{17}.$$

5.1.4 Theorem. If k is a constant and f is a differentiable function, then

$$D_x[k \cdot f(x)] = k \cdot D_x f(x).$$

Proof. Let F be the function defined by $F(x) = k \cdot f(x)$, where f is differentiable at a number c. Then

$$F'(c) = \lim_{\Delta x \to 0} \frac{F(c + \Delta x) - F(c)}{\Delta x} = \lim_{\Delta x \to 0} \frac{k \cdot f(c + \Delta x) - k \cdot f(c)}{\Delta x}$$

$$= k \cdot \lim_{\Delta x \to 0} \frac{f(c + \Delta x) - f(c)}{\Delta x} \qquad \text{(by 3.3.3)}$$

$$= kf'(c).$$

Since c is any number in the domain of f', we have proved that

$$D_x[k \cdot f(x)] = k \cdot D_x f(x). \quad \blacksquare$$

As illustrations,

$$D_x(4x^2) = 4(D_x x^2) = 4(2x) = 8x;$$
$$D_x(-7x^9) = -7(D_x x^9) = -7(9x^8) = -63x^8.$$

5.1.5 Theorem. The derivative of the algebraic sum of two differentiable functions is equal to the algebraic sum of their derivatives:

$$D_x(f \pm g) = D_x f \pm D_x g.$$

Proof. Let f and g be differentiable at c, and let $H(x) = f(x) + g(x)$. Then

$$H'(c) = \lim_{\Delta x \to 0} \frac{[f(c + \Delta x) + g(c + \Delta x)] - [f(c) + g(c)]}{\Delta x}$$

$$= \lim_{\Delta x \to 0} \frac{[f(c + \Delta x) - f(c)] + [g(c + \Delta x) - g(c)]}{\Delta x}$$

$$= \lim_{\Delta x \to 0} \frac{f(c + \Delta x) - f(c)}{\Delta x} + \lim_{\Delta x \to 0} \frac{g(c + \Delta x) - g(c)}{\Delta x} \qquad \text{(by 3.3.1)}$$

$$= f'(c) + g'(c).$$

That is,
$$D_x[f(x) + g(x)] = D_x f(x) + D_x g(x).$$

Since

$$f(x) - g(x) = f(x) + (-1)g(x),$$

and

$$D_x[(-1)g(x)] = -D_x g(x) \qquad \text{(by 5.1.4)},$$

we have

$$D_x[f(x) - g(x)] = D_x f(x) - D_x g(x). \quad \blacksquare$$

It follows from 5.1.5 that
$$\begin{aligned}
D_x[f(x) + g(x) + h(x)] &= D_x[(f(x) + g(x)) + h(x)] \\
&= D_x[f(x) + g(x)] + D_x h(x) \\
&= D_x f(x) + D_x g(x) + D_x h(x).
\end{aligned}$$

By mathematical induction we can extend this result to the sum of any finite number of functions.

5.1.6   Corollary.   The derivative of the sum of any finite number of differentiable functions is equal to the sum of their derivatives.

The preceding theorems enable us to write down, rapidly and easily, the derivative of any polynomial function, as shown in the following examples.

Example 1.   Find the derivative of the polynomial function defined by
$$f(x) = 5x^3 + 7.$$

Solution.   By 5.1.5,
$$D_x(5x^3 + 7) = D_x(5x^3) + D_x(7).$$

Using 5.1.4, 5.1.3, and 5.1.1, in that order, we get
$$\begin{aligned}
D_x(5x^3) + D_x(7) &= 5(D_x x^3) + D_x(7) \\
&= 5(3x^2) + 0 = 15x^2.
\end{aligned}$$

Example 2.   Differentiate the polynomial function defined by
$$f(x) = 4x^6 - 3x^5 - 10x^2 + 5x + 16.$$

Solution.   Using 5.1.1, 5.1.3, 5.1.4, and 5.1.6, we obtain
$$\begin{aligned}
D_x(4x^6 - 3x^5 - 10x^2 + 5x + 16) &= D_x(4x^6) - D_x(3x^5) - D_x(10x^2) + D_x(5x) + D_x(16) \\
&= 4D_x x^6 - 3D_x x^5 - 10D_x x^2 + 5D_x x + D_x 16 \\
&= 4(6x^5) - 3(5x^4) - 10(2x) + 5(1) + 0 \\
&= 24x^5 - 15x^4 - 20x + 5.
\end{aligned}$$

Exercises

Use the formulas for differentiation to find $D_x y$ in Exercises 1 to 10.

1. $y = x^3$.

2. $y = 6x^4$.

3. $y = -3x^6$.

4. $y = 5x + 11$.

5. $y = 4x^2 - 3$.

6. $y = 2x^2 + 3x - 1$.

7. $y = 4x^3 - 7x + 2$.

8. $y = 8x^3 - 3x^2 + 5$.

9. $y = 11x^4 + x - 18$.

10. $y = 3x^5 - 4x^3 + 7$.

In Exercises 11 to 16, use the formulas for differentiation to find $\dfrac{dy}{dx}$.

11. $y = 7x^7 + 2x^6 - 6x^2 + 15$.

12. $y = 3x^8 + 5x^7 - 4x^5 + x^2 - 1$.

13. $y = 2x^{11} - 3x^8 + 11x^4 + 6x - 33$.

14. $y = 5x^{57} - 2x^{19}$.

15. $y = 4x^5 + 6x^4 - 3x^3 - 7x^2 - 5x + 14$.

16. $y = -x^6 - 4x^5 - 8x^4 + 2x^3 + 11x^2 - 12$.

5.2 DERIVATIVE OF A PRODUCT OR QUOTIENT OF FUNCTIONS

The derivative of a product of functions is *not* equal to the product of the derivatives of the functions.

5.2.1 Theorem. The derivative of the product of two differentiable functions is equal to the first function times the derivative of the second plus the second function times the derivative of the first:

$$D_x(u \cdot v) = u \cdot D_x v + v \cdot D_x u.$$

Proof. Let u and v be functions that are differentiable at a number c, and let $F = u \cdot v$. Then

$$F'(c) = \lim_{\Delta x \to 0} \frac{F(c + \Delta x) - F(c)}{\Delta x} = \lim_{\Delta x \to 0} \frac{u(c + \Delta x) \cdot v(c + \Delta x) - u(c) \cdot v(c)}{\Delta x}$$

$$= \lim_{\Delta x \to 0} \frac{u(c + \Delta x) \cdot v(c + \Delta x) - u(c + \Delta x) \cdot v(c) + u(c + \Delta x) \cdot v(c) - u(c) \cdot v(c)}{\Delta x}$$

$$= \lim_{\Delta x \to 0} \left[u(c + \Delta x) \cdot \frac{v(c + \Delta x) - v(c)}{\Delta x} + v(c) \cdot \frac{u(c + \Delta x) - u(c)}{\Delta x} \right]$$

$$= \left[\lim_{\Delta x \to 0} u(c + \Delta x) \right] \left[\lim_{\Delta x \to 0} \frac{v[c + \Delta x) - v(c)}{\Delta x} \right]$$

$$\quad + \left[\lim_{\Delta x \to 0} v(c) \right] \left[\lim_{\Delta x \to 0} \frac{u(c + \Delta x) - u(c)}{\Delta x} \right]$$

$$= \left[\lim_{\Delta x \to 0} u(c + \Delta x) \right] \cdot v'(c) + \left[\lim_{\Delta x \to 0} v(c) \right] \cdot u'(c).$$

Since u is differentiable at c, it is continuous there (by 4.5.2), and thus $\lim_{\Delta x \to 0} u(c + \Delta x) = u(c)$, by 4.5.3. Moreover, $v(c)$ is a constant, and so $\lim_{\Delta x \to 0} v(c) = v(c)$. Thus

$$F'(c) = u(c) \cdot v'(c) + v(c) \cdot u'(c)$$

or

$$D_x(u \cdot v) = u \cdot D_x v + v \cdot D_x u. \quad \blacksquare$$

The derivative of the product of a finite number of differentiable functions may be found by repeated use of this formula.

Example 1. If f is the function defined by $f(x) = (2x + 6)(5x^2 - x + 4)$, find $D_x f$.

Solution. Of course, we could multiply the two factors together and then differentiate $f(x)$ as a polynomial. But in order to illustrate the product rule, 5.2.1, let $u(x) = 2x + 6$ and $v(x) = 5x^2 - x + 4$. Then $D_x u = 2$ and $D_x v = 10x - 1$. Thus

$$D_x f(x) = u(x) \cdot D_x v(x) + v(x) \cdot D_x u(x)$$
$$= (2x + 6)(10x - 1) + (5x^2 - x + 4)(2)$$
$$= 30x^2 + 56x + 2.$$

Example 2. If $f(x) = (3x^2 - 5)(7x^3 + 6x + 1)$, find $D_x f$.

Solution.

$$D_x f(x) = (3x^2 - 5) \cdot D_x(7x^3 + 6x + 1) + (7x^3 + 6x + 1) \cdot D_x(3x^2 - 5)$$
$$= (3x^2 - 5)(21x^2 + 6) + (7x^3 + 6x + 1)6x = 105x^4 - 51x^2 + 6x - 30.$$

5.2.2 Theorem. If v is a differentiable function, then

$$D_x \left(\frac{1}{v} \right) = -\frac{D_x v}{v^2},$$

provided that $v(x) \neq 0$.

Proof. Let $F = 1/v$. Then

$$F'(c) = \lim_{\Delta x \to 0} \frac{F(c + \Delta x) - F(c)}{\Delta x} = \lim_{\Delta x \to 0} \frac{\dfrac{1}{v(c + \Delta x)} - \dfrac{1}{v(c)}}{\Delta x}$$

$$= \lim_{\Delta x \to 0} \frac{v(c) - v(c + \Delta x)}{\Delta x \cdot [v(c + \Delta x) \cdot v(c)]} = \lim_{\Delta x \to 0} \left[\frac{-\dfrac{v(c + \Delta x) - v(c)}{\Delta x}}{v(c + \Delta x) \cdot v(c)} \right]$$

$$= \frac{-\lim_{\Delta x \to 0} \dfrac{v(c + \Delta x) - v(c)}{\Delta x}}{\lim_{\Delta x \to 0} [v(c + \Delta x) \cdot v(c)]} = \frac{-D_x v}{v^2}.$$

That is,

$$D_x \left(\frac{1}{v} \right) = -\frac{D_x v}{v^2}. \quad \blacksquare$$

Example 3. If $f(x) = 1/(3x^2 - 5)$, find $D_x f$.

Solution. Using 5.2.2, we obtain

$$D_x f(x) = D_x\left(\frac{1}{3x^2 - 5}\right) = -\frac{D_x(3x^2 - 5)}{(3x^2 - 5)^2} = -\frac{6x}{(3x^2 - 5)^2}.$$

From 5.2.1 and 5.2.2 it is easy to establish the following important rule for differentiating a quotient of functions.

5.2.3 **Theorem.** The derivative of the quotient of two differentiable functions is equal to the denominator times the derivative of the numerator minus the numerator times the derivative of the denominator, all divided by the square of the denominator:

$$D_x\left(\frac{u}{v}\right) = \frac{v \cdot D_x u - u \cdot D_x v}{v^2} \qquad \text{when} \quad v(x) \neq 0.$$

Proof. Using 5.2.1 and 5.2.2, we can write

$$D_x\left(\frac{u}{v}\right) = D_x\left(u \cdot \frac{1}{v}\right) = u \cdot D_x\left(\frac{1}{v}\right) + \frac{1}{v} D_x u$$

$$= -\frac{u \cdot D_x v}{v^2} + \frac{D_x u}{v} = \frac{v \cdot D_x u - u \cdot D_x v}{v^2}. \quad \blacksquare$$

Example 4. Find $D_x f$ when

$$f(x) = \frac{3x - 5}{x^2 + 7}.$$

Solution. From 5.2.3,

$$D_x\left(\frac{3x - 5}{x^2 + 7}\right) = \frac{(x^2 + 7) \cdot D_x(3x - 5) - (3x - 5) \cdot D_x(x^2 + 7)}{(x^2 + 7)^2}$$

$$= \frac{(x^2 + 7)(3) - (3x - 5)(2x)}{(x^2 + 7)^2}$$

$$= \frac{-3x^2 + 10x + 21}{x^4 + 14x^2 + 49}.$$

Example 5. If $y = (2x^4 - 13x + 4)/(5x^3 + x)$, find $D_x y$.

Solution.

$$D_x y = \frac{(5x^3 + x) \cdot D_x(2x^4 - 13x + 4) - (2x^4 - 13x + 4) \cdot D_x(5x^3 + x)}{(5x^3 + x)^2}$$

$$= \frac{(5x^3 + x)(8x^3 - 13) - (2x^4 - 13x + 4)(15x^2 + 1)}{(5x^3 + x)^2}$$

$$= \frac{10x^6 + 6x^4 + 130x^3 - 60x^2 - 4}{25x^6 + 10x^4 + x^2}.$$

Example 6. From 5.2.2 we have

$$D_x x^{-1} = D_x\frac{1}{x} = -\frac{1}{x^2} = -x^{-2};$$

$$D_x x^{-2} = D_x\frac{1}{x^2} = -\frac{2x}{x^4} = -2x^{-3};$$

$$D_x x^{-3} = D_x\frac{1}{x^3} = \frac{-3x^2}{x^6} = -3x^{-4}.$$

Exercises

In Exercises 1 through 10, find $D_x f(x)$.

1. $f(x) = (2x - 7)(3x^2 + 4x - 1)$.

2. $f(x) = (x^2 - 5x + 3)(4x + 5)$.

3. $f(x) = (5x^2 - 7)(2x^2 + 6x - 7)$.

4. $f(x) = (x^3 + 2x - 3)(4x^2 + x + 2)$.

5. $f(x) = (x^2 - 5x + 1)(x^2 - 5x + 1)$.

6. $f(x) = (2x^3 + x^2 - 4)^2$.

7. $f(x) = \dfrac{1}{3x^2 + 2}$.

8. $f(x) = \dfrac{1}{5x^8}$.

9. $f(x) = \dfrac{1}{4x^4 - 3x^3 + 2}$.

10. $f(x) = \dfrac{x^4 - 2x^3 - 6x}{4x}$.

In Exercises 11 to 14, use the formulas for differentiation to find dy/dx.

11. $y = \dfrac{2x^2 - 1}{3x + 5}$.

12. $y = \dfrac{5x - 4}{3x^2 + 2}$.

13. $y = \dfrac{2x^3 - x + 4}{x^2 + 3x - 5}$.

14. $y = \dfrac{3x^3 - 4x + 10}{x^2 + 2x - 11}$.

In Exercises 15 and 16, use the formulas for differentiation to find $G'(t)$.

15. $G(t) = \dfrac{2t^2 + 9t - 3}{t^3 + 6t^2 - t + 2}$.

16. $G(t) = \dfrac{6t^3}{(2t^2 - 3)^2}$.

17. Let u and v be differentiable functions. If $u(0) = 4$, $u'(0) = -1$, $v(0) = -3$, and $v'(0) = 5$, find the value at $x = 0$ of

(a) $D_x(u \cdot v)$;   (b) $D_x\left(\dfrac{u}{v}\right)$.

18. Let f and g be differentiable functions. If $f(3) = 7$, $f'(3) = 2$, $g(3) = 6$, and $g'(3) = -10$, find the value at $x = 3$ of

(a) $(f \cdot g)'$;   (b) $\left(\dfrac{f}{g}\right)'$.

19. Show that if $y = x^{-n}$, where n is a positive integer, then

$$\frac{dy}{dx} = -nx^{-n-1}, \qquad x \neq 0.$$

(*Hint:* Let $v = x^n$ in 5.2.2.)

20. By means of 5.2.1, show that $D_x[f(x)]^2 = 2 \cdot f(x) \cdot f'(x)$.

21. Show that

$$D_x[f(x) \cdot g(x) \cdot h(x)] = f(x) \cdot g(x) \cdot h'(x)$$
$$+ f(x) \cdot g'(x) \cdot h(x)$$
$$+ f'(x) \cdot g(x) \cdot h(x).$$

(*Hint:* $f(x) \cdot g(x) \cdot h(x) = [f(x) \cdot g(x)] \cdot h(x)$.)

22. Use Exercise 21 to prove that $D_x[f(x)]^3 = 3[f(x)]^2 \cdot f'(x)$.

5.3 CHAIN RULE FOR DIFFERENTIATING COMPOSITE FUNCTIONS

We have developed formulas and techniques for the rapid differentiation of many types of functions. But should we attempt to use those formulas to find the derivative of, let us say, the function F defined by $F(x) = (2x^2 - 4x + 1)^{60}$, we would first have to multiply together the 60 quadratic factors, $(2x^2 - 4x + 1)$, and then differentiate the resulting polynomial of degree 120.

Fortunately, there is a far easier way to differentiate F, by using what is known as the **chain rule.** The chain rule enables us to write down the correct result, $F'(x) =$

$60(2x^2 - 4x + 1)^{59}(4x - 4)$, as fast as we can move our pencil. Moreover, the chain rule enables us to differentiate much more complicated functions—for instance, the function g defined by

$$g(x) = \sqrt[7]{(32x^{15} - 11x + 10)/(9x^4 - 3x^2 - 1)}.$$

In fact, the chain rule is so important that the reader will seldom again differentiate any function without using it.

We now state the chain rule and show by examples what it means and how it is used. Its proof will be discussed toward the end of this section.

5.3.1 **Theorem** (*Chain Rule*). Let $y = f(u)$ and $u = g(x)$, where f and g are functions. If g is differentiable at x and f is differentiable at $u = g(x)$, then the composite function defined by $y = f(g(x))$ is differentiable at x and

$$\frac{dy}{dx} = \frac{dy}{du} \cdot \frac{du}{dx}$$

or, equivalently,

$$D_x f(g(x)) = f'(u) \cdot g'(x).$$

The first form of the chain rule given above is in the Leibniz* notation, which is easier to remember.

Example 1. If $y = (x^3 - 10)^9$, find dy/dx.

Solution. Let $u = x^3 - 10$. Then $y = u^9$, $dy/du = 9u^8$, and $du/dx = 3x^2$. By the chain rule,

$$\frac{dy}{dx} = \frac{dy}{du} \cdot \frac{du}{dx} = 9u^8(3x^2)$$

$$= 9(x^3 - 10)^8(3x^2)$$

$$= 27x^2(x^3 - 10)^8.$$

Example 2. If $F(x) = (2x^2 - 4x + 1)^{60}$, find $D_x F(x)$.

Solution. Let $f(u) = u^{60}$ and let $u = g(x) = 2x^2 - 4x + 1$. Then $F(x) = f(u) = f(g(x)) = (2x^2 - 4x + 1)^{60}$. Since $f'(u) = 60u^{59}$ and $g'(x) = 4x - 4$, it follows from the chain rule that

$$D_x F(x) = D_x f(g(x)) = f'(u) \cdot g'(x)$$

$$= 60u^{59}(4x - 4)$$

$$= 60(2x^2 - 4x + 1)^{59}(4x - 4).$$

Example 3. If $y = \dfrac{1}{(2x^5 - 7)^3}$, find $\dfrac{dy}{dx}$.

*The German, Gottfried Wilhelm Leibniz (1646–1716), and the Englishman, Isaac Newton (1642–1727), independently and almost simultaneously established the calculus as a general symbolic method, performed by formal rules without the necessity of geometric imagery. Leibniz was equally famous as a philosopher and was, in fact, a universal genius. Starting as a jurist, he served as a diplomat, a theologian, historian, librarian, economist, and linguist. Leibniz understood the great importance of well-chosen symbols in mathematics, and much of the notation we use today is due to him.

Solution.   Let $u = 2x^5 - 7$. Then

$$y = \frac{1}{u^3}, \qquad \frac{dy}{du} = \frac{-3}{u^4}$$

(see Example 6, Section 5.2), and

$$\frac{du}{dx} = 10x^4.$$

Substituting these results in the chain rule gives

$$\frac{dy}{dx} = \frac{dy}{du} \cdot \frac{du}{dx} = \frac{-3}{u^4} \cdot (10x^4) = \frac{-30x^4}{(2x^5 - 7)^4}.$$

Example 4.   If $y = \left(\dfrac{3x^3 - 5x}{4x + 17} \right)^{13}$, find $\dfrac{dy}{dx}$.

Solution.   Let $u = (3x^3 - 5x)/(4x + 17)$. Then $y = u^{13}$ and, by the chain rule,

(1) $$\frac{dy}{dx} = \frac{dy}{du} \frac{du}{dx} = (13u^{12}) \frac{d}{dx} \left(\frac{3x^3 - 5x}{4x + 17} \right).$$

From 5.2.3,

$$\frac{d}{dx} \left(\frac{3x^3 - 5x}{4x + 17} \right) = \frac{(4x + 17)(9x^2 - 5) - (3x^3 - 5x)4}{(4x + 17)^2}$$

$$= \frac{24x^3 + 153x^2 - 85}{(4x + 17)^2}.$$

Substituting this result and $u = (3x^3 - 5x)/(4x + 17)$ in (1), we obtain

$$\frac{dy}{dx} = \frac{13(3x^3 - 5x)^{12}(24x^3 + 153x^2 - 85)}{(4x + 17)^{14}}.$$

Example 5.   If $y = \dfrac{\sqrt{x^2 + 1}}{1 + \sqrt{x^2 + 1}}$, find $\dfrac{dy}{dx}$.

Solution.   Let $u = \sqrt{v}$ and $v = x^2 + 1$. Then

$$y = \frac{u}{1 + u} = \frac{\sqrt{v}}{1 + \sqrt{v}} = \frac{\sqrt{x^2 + 1}}{1 + \sqrt{x^2 + 1}}.$$

Two applications of the chain rule give

$$\frac{dy}{dx} = \frac{dy}{du} \frac{du}{dx} \qquad \text{and} \qquad \frac{du}{dx} = \frac{du}{dv} \frac{dv}{dx},$$

or

(2) $$\frac{dy}{dx} = \frac{dy}{du} \frac{du}{dv} \frac{dv}{dx}.$$

Now

$$\frac{dy}{du} = \frac{(1 + u)(1) - (u)(1)}{(1 + u)^2} = \frac{1}{(1 + u)^2},$$

$$\frac{du}{dv} = \frac{1}{2\sqrt{v}} \qquad \text{(Example 5, Section 4.3)},$$

$$\frac{dv}{dx} = 2x.$$

Substituting in (2), we have

$$\frac{dy}{dx} = \frac{1}{(1+u)^2} \cdot \frac{1}{2\sqrt{v}} \cdot (2x)$$

or, since $u = \sqrt{v} = \sqrt{x^2 + 1}$,

$$\frac{dy}{dx} = \frac{1}{(1 + \sqrt{x^2 + 1})^2} \cdot \frac{1}{2\sqrt{x^2 + 1}} \cdot (2x)$$

$$= \frac{x}{\sqrt{x^2 + 1}(1 + \sqrt{x^2 + 1})^2}.$$

Although it is useful, when learning to use the chain rule, to write down specifically what is chosen for u, v, and so on in each individual problem, the reader will soon be able to do so mentally and thus speed up the work.

Let us now discuss a restricted proof of the chain rule. If f and g are functions defined by $y = f(u)$ and $u = g(x)$, then $y = f(g(x))$ defines a composite function whose independent variable is x. It is assumed that $f'(u_1)$ and $g'(x_1)$ exist, where $u_1 = g(x_1)$.

Let x_1 and $x_1 + \Delta x$ be numbers in the domain of g for which the corresponding numbers $u_1 = g(x_1)$ and $u_1 + \Delta u = g(x_1 + \Delta x)$ are in the domain of f. Then

$$\Delta u = g(x_1 + \Delta x) - g(x_1)$$

is the increment of the variable u that corresponds to the increment Δx of the independent variable x at x_1. If we write $y_1 = f(u_1)$ and $y_1 + \Delta y = f(u_1 + \Delta u)$, then

$$\Delta y = f(u_1 + \Delta u) - f(u_1)$$

is the increment of y that corresponds to the increment Δu of u at u_1, which, in turn, corresponds to the increment Δx of x at x_1. Thus Δy is the increment of the composite function that corresponds to the increment Δx of the independent variable x at x_1. That is,

$$\Delta y = f(g(x_1 + \Delta x)) - f(g(x_1)).$$

Then, by definition, the value of the derivative of the composite function at x_1 is

$$\frac{dy}{dx} = \lim_{\Delta x \to 0} \frac{\Delta y}{\Delta x} = \lim_{\Delta x \to 0} \frac{f(g(x_1 + \Delta x)) - f(g(x_1))}{\Delta x}.$$

For most functions g encountered in beginning calculus, there exists an open interval I containing x_1 such that at all points $x_1 + \Delta x$ in I, $\Delta u = g(x_1 + \Delta x) - g(x_1) \neq 0$ if $\Delta x \neq 0$. *For such functions only,*

$$(3) \qquad\qquad \frac{\Delta y}{\Delta x} = \frac{\Delta y}{\Delta u} \frac{\Delta u}{\Delta x}, \qquad \Delta x \neq 0,$$

is an identity in Δx. Since $u = g(x)$ is differentiable at x_1, it is continuous at x_1; so $\Delta u \longrightarrow 0$ as $\Delta x \longrightarrow 0$. By taking the limits of both members of (3) as $\Delta x \longrightarrow 0$, we obtain

$$\lim_{\Delta x \to 0} \frac{\Delta y}{\Delta x} = \lim_{\Delta u \to 0} \frac{\Delta y}{\Delta u} \cdot \lim_{\Delta x \to 0} \frac{\Delta u}{\Delta x}.$$

That is,

$$\frac{dy}{dx} = \frac{dy}{du}\frac{du}{dx}$$

at x_1.

However, there are other functions g such that in every interval I that contains x_1, no matter how small, there exist points $x_1 + \Delta x$, with $\Delta x \neq 0$, for which the corresponding increment Δu is zero. For these functions, (3) is no longer an identity, since division by zero is ruled out, and the foregoing derivation fails.

A *general* proof that the chain rule is true, even for such exceptional functions, may be found in the Appendix (page 865). It is more difficult.

Exercises

In Exercises 1 to 8, find $\frac{dy}{dx}$.

1. $y = (2 - 9x)^{15}$.

2. $y = (4x + 7)^{23}$.

3. $y = (5x^2 + 2x - 8)^5$.

4. $y = (3x^3 - 11x)^7$.

5. $y = (x^3 - 4x^2 + 2x - 1)^8$.

6. $y = (6x^5 - 3x^4 + 10)^6$.

7. $y = (ax^2 + bx + c)^n$, $n \in N$; a, b, c constants.

8. $y = (ax^3 + bx^2 + cx + d)^n$, $n \in N$; a, b, c, d constants.

In Exercises 9 to 12, find $D_x F$.

9. $F(x) = (4x - 7)^2(2x + 3)$.

10. $F(x) = (5x + 6)^2(x - 13)^3$.

11. $F(x) = (2x - 1)^3(x^2 - 3)^2$.

12. $F(x) = (3x^2 + 5)^2(x^3 - 11)^4$.

Differentiate the functions defined in Exercises 13 through 29.

13. $y = \dfrac{1}{(x + 2)^2}$.

14. $y = \dfrac{1}{(2x^2 - 5)^3}$.

15. $y = \dfrac{1}{(x^3 - 4)^2}$.

16. $y = \dfrac{1}{(7x^2 - 2x + 6)^2}$.

17. $f(x) = (3x^4 + x - 8)^{-3}$

18. $f(x) = \dfrac{1}{(2x^4 - 3x^2 + 1)^9}$.

19. $G(x) = \dfrac{(x + 1)^2}{3x - 4}$.

20. $G(x) = \dfrac{2x - 3}{(x^2 + 4)^2}$.

21. $F(x) = \dfrac{(3x^2 + 2)^2}{(2x^2 - 5)}$.

22. $F(x) = \dfrac{(x^2 - 1)^3}{(4x^3 + 5)^2}$.

23. $y = \left(\dfrac{2x + 3}{4x - 5}\right)^2$.

24. $F(u) = \left(\dfrac{4u + 3}{5 - 7u}\right)^3$.

25. $F(t) = \left(\dfrac{t^2 + 4}{5t - 9}\right)^4$.

26. $y = \left(\dfrac{x^2 - 6}{3x^2 + 1}\right)^5$.

27. $h(s) = \left(\dfrac{4s^3 + 5}{2s^5 - 3}\right)^{-7}$.

28. $y = \dfrac{(2x^2 - 1)^5}{1 + (2x^2 - 1)^5}$.

[*Hint*: Let $u = (2x^2 - 1)^5$.]

29. $y = \dfrac{(x^3 + 7)^4 - 5}{6 + (x^3 + 7)^4}$.

5.4 DERIVATIVE OF ANY RATIONAL POWER OF A FUNCTION

It was shown in 5.1.3 that the formula

(1) $$D_x x^n = n x^{n-1}$$

is true when n is any *positive integer*. In the present section we shall prove that the formula (1) is true when n is any rational number whatever.

But first consider $x^{1/q}$, where q is an integer greater than 1. For each number x that makes $x^{1/q}$ real,

(2) $$D_x x^{1/q} = \lim_{\Delta x \to 0} \frac{(x + \Delta x)^{1/q} - x^{1/q}}{\Delta x} = \lim_{\Delta x \to 0} \frac{(x + \Delta x)^{1/q} - x^{1/q}}{(x + \Delta x) - x}.$$

It is easy to prove by mathematical induction (Exercise 5, Section 1.7) that if q is any integer greater than 1,

$$a^q - b^q = (a - b)(a^{q-1} + a^{q-2}b + a^{q-3}b^2 + \cdots + ab^{q-2} + b^{q-1}).$$

When $a \neq b$, this identity can be written

(3) $$\frac{a - b}{a^q - b^q} = \frac{1}{a^{q-1} + a^{q-2}b + a^{q-3}b^2 + \cdots + ab^{q-2} + b^{q-1}},$$

where there are q terms in the denominator of the right-hand member. If we let $a = (x + \Delta x)^{1/q}$ and $b = x^{1/q}$, $\Delta x \neq 0$, (3) becomes

(4) $$\frac{(x + \Delta x)^{1/q} - x^{1/q}}{(x + \Delta x) - x}$$

$$= \frac{1}{(x + \Delta x)^{(q-1)/q} + (x + \Delta x)^{(q-2)/q}x^{1/q} + \cdots + (x + \Delta x)^{1/q}x^{(q-2)/q} + x^{(q-1)/q}}.$$

By substituting (4) in (2), and using 3.3.1, we obtain

$$D_x x^{1/q}$$

$$= \lim_{\Delta x \to 0} \frac{1}{(x + \Delta x)^{(q-1)/q} + (x + \Delta x)^{(q-2)/q}x^{1/q} + \cdots + (x + \Delta x)^{1/q}x^{(q-2)/q} + x^{(q-1)/q}}$$

$$= \frac{1}{q[x^{(q-1)/q}]} = \frac{1}{q} x^{1/q - 1}.$$

Thus

(5) $$D_x x^{1/q} = (1/q)x^{1/q - 1}.$$

Now let n be any positive rational number; that is, let $n = p/q$, where p and q are positive integers. Since $q = 1$ has been considered in (1), we assume that q is an integer greater than 1.

Using the chain rule and (1) and (5), we write

$$D_x x^n = D_x x^{p/q} = D_x (x^{1/q})^p = p(x^{1/q})^{p-1} D_x x^{1/q}$$

$$= p(x^{1/q})^{p-1}(1/q)x^{1/q - 1} = (p/q)x^{p/q - 1} = n x^{n-1}.$$

Therefore the formula (1) *is valid for any positive rational number n* and any value of x that makes x^{n-1} real.

If n is a *negative* rational number, write $n = -r$, where r is a positive rational number. By letting $x^r = v$ in 5.2.2, we obtain

$$D_x x^n = D_x x^{-r} = D_x \frac{1}{x^r} = \frac{-1}{(x^r)^2} \cdot (rx^{r-1}) = -rx^{-r-1} = nx^{n-1}.$$

Consequently, the formula (1) is also true when *n is a negative rational number*, and x is any number for which x^{n-1} is real.

Finally, if $n = 0$ and $x \neq 0$, $D_x x^n = D_x x^0 = D_x 1 = 0 = 0/x = 0 \cdot x^{0-1} = nx^{n-1}$; so the formula (1) is valid if $n = 0$ and $x \neq 0$.

This completes the proof of the following theorem.

5.4.1 **Theorem.** Let n be any rational number. Then

$$D_x x^n = nx^{n-1},$$

where x is any number for which x^{n-1} is real.

For instance,

$$D_x x^{3/4} = (3/4)x^{3/4-1} = \frac{3x^{-1/4}}{4};$$

and $D_x x^{-7} = -7(x^{-7-1}) = -7x^{-8}$.

Applying the chain rule to 5.4.1 gives the following important result.

5.4.2 **Theorem.** If n is a rational number and u is a differentiable function of x, then

$$D_x u^n = nu^{n-1} \cdot D_x u,$$

for values of x that make $[u(x)]^{n-1}$ real.

As an illustration, if $u(x) = 2x^3 - 7x + 9$ and $n = \frac{1}{5}$, then

$$D_x(2x^3 - 7x + 9)^{1/5} = \tfrac{1}{5}(2x^3 - 7x + 9)^{1/5-1} \cdot D_x(2x^3 - 7x + 9)$$
$$= \tfrac{1}{5}(2x^3 - 7x + 9)^{-4/5}(6x^2 - 7).$$

It will be proved later that 5.4.2 is also valid for *all real numbers n* when u is a positive-valued, differentiable function of x.

Example 1. By 5.4.1,

(a) $D_x(\sqrt{x}) = D_x(x^{1/2}) = \dfrac{1}{2}x^{-1/2} = \dfrac{1}{2x^{1/2}} = \dfrac{1}{2\sqrt{x}}$;

(b) $D_x(\sqrt[3]{x}) = D(x^{1/3}) = \dfrac{1}{3}x^{-2/3} = \dfrac{1}{3x^{2/3}} = \dfrac{1}{3\sqrt[3]{x^2}}$;

(c) $D_x\left(\dfrac{1}{\sqrt{x}}\right) = D_x(x^{-1/2}) = -\dfrac{1}{2}x^{-3/2} = -\dfrac{1}{2x^{3/2}} = -\dfrac{1}{2\sqrt{x^3}}$.

Example 2. From 5.4.2,

$$D_t\sqrt{t^4 - 3t + 17} = D_t(t^4 - 3t + 17)^{1/2}$$

$$= \frac{1}{2}(t^4 - 3t + 17)^{-1/2} \cdot D_t(t^4 - 3t + 17)$$

$$= \frac{1}{2}(t^4 - 3t + 17)^{-1/2}(4t^3 - 3)$$

$$= \frac{4t^3 - 3}{2\sqrt{t^4 - 3t + 17}}.$$

Example 3. If $y = 1/(x^4\sqrt{2x - 6})$, find $D_x y$.

Solution. Using 5.2.1 and 5.4.2, we can write

$$D_x y = D_x[x^{-4}(2x - 6)^{-1/2}]$$

$$= x^{-4} \cdot D_x(2x - 6)^{-1/2} + (2x - 6)^{-1/2} \cdot D_x x^{-4}$$

$$= x^{-4}\left[-\frac{1}{2}(2x - 6)^{-3/2} \cdot D_x(2x - 6)\right] + (2x - 6)^{-1/2}(-4x^{-5})$$

$$= x^{-4}\left[-\frac{1}{2}(2x - 6)^{-3/2}(2)\right] + (2x - 6)^{-1/2}(-4x^{-5})$$

$$= \frac{-1}{x^4(2x - 6)^{3/2}} + \frac{-4}{x^5(2x - 6)^{1/2}}$$

$$= \frac{-x - 4(2x - 6)}{x^5(2x - 6)^{3/2}} = \frac{24 - 9x}{x^5(2x - 6)^{3/2}}.$$

Exercises

Differentiate the indicated functions in Exercises 1 through 28.

1. $y = x^{2/3}$.

2. $y = x^{-13}$.

3. $y = (2x + 3)^{1/2}$.

4. $y = 5(3x - 7)^{-1}$.

5. $y = \dfrac{1}{x^2 + 4}$.

6. $y = \sqrt{4x - 5}$.

7. $y = \sqrt{2x^3 + 7}$.

8. $y = \sqrt{4 - x^2}$.

9. $y = \sqrt[3]{x + 2}$.

10. $y = \sqrt[5]{5x^2 - 4}$.

11. $s = \dfrac{1}{(5 - 2t)^3}$.

12. $s = \dfrac{1}{(t^2 - 6)^2}$.

13. $f(u) = \dfrac{\sqrt{9 + u^2}}{u}$.

14. $F(w) = w\sqrt{(w^2 - 2)^3}$.

15. $\phi(t) = 3t\sqrt[3]{8 - t^3}$.

16. $f(x) = (x - 2)^5 + \dfrac{1}{(x - 2)^5}$.

17. $s = t^{1/2} + t^{-1/2}$.

18. $G(w) = \left(\dfrac{1 + w}{1 - w}\right)^{1/2}$.

19. $y = \sqrt{2z - 1} - \dfrac{1}{\sqrt{2z - 1}}$.

20. $y = (a^{2/3} - x^{2/3})^{3/2}$, where a is a constant.

21. $y = \sqrt{\dfrac{7 - x}{12 + x}}$.

22. $f(x) = \sqrt[9]{7x^3 - 2x + 11}$.

23. $v = \left(\dfrac{2p - 3}{3p - 2}\right)^{2/3}$.

24. $u = \sqrt{\dfrac{3v - 1}{v^2 + 3}}$.

25. $F(x) = \left(\dfrac{x^2}{2 - x}\right)^{1/3}$

26. $F(z) = \left(\dfrac{\sqrt{6z}}{17z - 5}\right)^{3/2}$.

27. $g(u) = \sqrt{5 - \sqrt{2u}}$.

28. $y = \sqrt{1 - \sqrt{1 + x}}$.

29. Recalling from algebra the formula for the sum of a finite geometric series, we see that

$$1 + t + t^2 + t^3 + \cdots + t^n = \frac{1 - t^{n+1}}{1 - t}, \qquad t \neq 1,$$

is an identity in t. Prove that

$$1 + 2t + 3t^2 + 4t^3 + \cdots + nt^{n-1} = \frac{nt^{n+1} - (n + 1)t^n + 1}{(t - 1)^2}.$$

30. From the identity

$$\frac{1}{x} + \frac{1}{x^2} + \frac{1}{x^3} + \cdots + \frac{1}{x^n} = \frac{x^n - 1}{x^{n+1} - x^n}, \qquad x \neq 0 \text{ and } x \neq 1,$$

prove that

$$\frac{1}{x^2} + \frac{2}{x^3} + \frac{3}{x^4} + \cdots + \frac{n}{x^{n+1}} = \frac{x^{n+1} - (n + 1)x + n}{x^{n+1}(x - 1)^2}.$$

5.5 DERIVATIVES OF HIGHER ORDER

Since the derivative of a function is also a function (4.3.1), it can often be differentiated.

Let f be a function and f' be its derivative; then the derivative of f', if it exists, is called the **second derivative** of f and is symbolized by f'' (read "f double prime").

Example 1.   If f is the function defined by $f(x) = 2x^3 + x - 5$, then its (first) derivative is the function f' defined by $f'(x) = 6x^2 + 1$, and its second derivative is the function f'' defined by $f''(x) = 12x$.

Similarly, the **third derivative** of f is the first derivative of f'' and is designated by f''' (read "f triple prime").

When the first derivative of f is indicated by $D_x f$, it is customary to write $D_x^2 f$ for the second derivative of f. This symbol is suggested by thinking of $D_x f$ as the differentiation *operator* D_x applied to f. Then the second derivative would be represented by two applications of the operator D_x, and $D_x(D_x f)$ is customarily written as $D_x^2 f$. The symbolism $D_x(D_x f) = D_x^2 f$ has nothing to do with ordinary multiplication of numbers; it simply means that the function f is differentiated twice with respect to x.

In the same way, if $y = f(x)$ and the value of the first derivative is given by dy/dx, then the value of the second derivative is indicated by d^2y/dx^2 to signify two applications of the operator d/dx to y.

If n is a positive integer greater than 2, the **nth derivative** of a function is the (first) derivative of the $(n-1)$st derivative of the function. Symbols for the nth derivative of f are

$$f^{(n)}, \quad D_x^n f, \quad \text{and} \quad \frac{d^n f}{dx^n},$$

and when considering the successive derivatives of f, it is often convenient to represent f itself by $f^{(0)}$. If $y = f(x)$, the value of the nth derivative of f is often represented by $d^n y/dx^n$ or by $D_x^n y$.

Example 2. Let $y = 5x^4 - x^2 + 2$. Then

$$y' = \frac{dy}{dx} = D_x y = 20x^3 - 2x,$$

$$y'' = \frac{d^2 y}{dx^2} = D_x^2 y = 60x^2 - 2,$$

$$y''' = \frac{d^3 y}{dx^3} = D_x^3 y = 120x,$$

$$y^{(4)} = \frac{d^4 y}{dx^4} = D_x^4 y = 120,$$

$$y^{(5)} = \frac{d^5 y}{dx^5} = D_x^5 y = 0,$$

$$y^{(n)} = \frac{d^n y}{dx^n} = D_x^n y = 0, \quad \text{for } n = 5, 6, 7, \cdots.$$

Example 3. If $y = \dfrac{2x}{1-2x}$, then

$$y' = \frac{dy}{dx} = D_x y = \frac{2}{(1-2x)^2} = 2(1-2x)^{-2},$$

$$y'' = \frac{d^2 y}{dx^2} = D_x^2 y = 2(-2)(-2)(1-2x)^{-3},$$

$$y''' = \frac{d^3 y}{dx^3} = D_x^3 y = 2(-2)^2[(-2)(-3)](1-2x)^{-4},$$

$$y^{(4)} = \frac{d^4 y}{dx^4} = D_x^4 y = 2(-2)^3[(-2)(-3)(-4)](1-2x)^{-5},$$

$$\cdot \quad \cdot \quad \cdot$$

$$y^{(n)} = \frac{d^n y}{dx^n} = D_x^n y = \frac{2^n(n!)}{(1-2x)^{n+1}}.$$

Higher derivatives have many important applications. To illustrate, if an object moves in a straight line and if, t seconds after it starts, it is $s(t)$ feet from its starting position, its velocity is $v(t) = s'(t)$. Its *acceleration* is defined as the instantaneous rate of change of the velocity with respect to time and is therefore $a(t) = v'(t) = s''(t)$.

Another application of the second derivative is to curve sketching. It will be shown in Section 6.10 that the graph of $y = f(x)$ is concave upward when $f''(x) > 0$ and concave downward when $f''(x) < 0$.

Exercises

Find $f''(x)$ in Exercises 1 to 12.

1. $f(x) = 2x^2 - 7$.

2. $f(x) = 5x^3 + 1$.

3. $f(x) = x^3 + 3x^2 - 5x + 4$.

4. $f(x) = 3x^4 + 5x^2 - x + 11$.

5. $f(x) = x^7 - 2x^6 + 5x^3$.

6. $f(x) = ax^3 + bx^2 + cx + d$, where a, b, c, d are constants.

7. $f(x) = \dfrac{1}{x}$.

8. $f(x) = \dfrac{1}{2x - 5}$.

9. $f(x) = \sqrt{x}$.

10. $f(x) = \sqrt{x^2 + 8}$.

11. $f(x) = \sqrt{2x^2 - 13}$.

12. $f(x) = x + \dfrac{1}{x}$.

In Exercises 13 to 18, find $D_x^3 y$.

13. $y = x^3 + 3x^2 - 2x - 8$.

14. $y = 2x^5 - x^4$.

15. $y = (x^2 + 5)^2$.

16. $y = \dfrac{1}{x - 3}$.

17. $y = \dfrac{x}{2x + 1}$.

18. $y = \sqrt{3 - x}$.

19. Carefully sketch the graph of $f(x) = x^3 - 3x^2 + 1$ for $-1 \le x \le 3$. Find the points on the curve at which $f'(x) = 0$, and draw the tangents to the curve at these points. For what value of x is $f''(x) = 0$? Find the interval on which $f''(x) < 0$ and the interval on which $f''(x) > 0$. What is peculiar about the curve on each of these intervals?

20. Sketch the graph of $f(x) = (x - 2)^3 + 1$ for $0 \le x \le 4$. Find the point on the curve where $f'(x) = 0$, and draw the tangent to the curve at this point. Find the point on the curve where $f''(x) = 0$. On what interval is the curve concave downward

$[f''(x) < 0]$? On what interval is the curve concave upward $[f''(x) > 0]$?

In Exercises 21 and 22, find $D_x y$, $D_x^2 y$, and $D_x^3 y$, and from them conjecture the form of the nth derivative $(n \in N)$; then prove your result by mathematical induction.

21. $y = \dfrac{1}{x^2}$.

22. $y = \dfrac{1}{2x + 3}$.

23. Consider the three points (x_1, y_1), (x_2, y_2), and (x_3, y_3), with $x_1 < x_2 < x_3$. Let $A(x) = (x - x_1)(x - x_2)(x - x_3)$, $A_i(x) = A(x)/(x - x_i)$, and $L_i(x) = A_i(x)/A'(x_i)$ for $i = 1, 2, 3$. Show that the polynomial $P(x)$, of degree 2, whose graph contains the three given points is $P(x) = y_1 L_1(x) + y_2 L_2(x) + y_3 L_3(x)$.

24. Use Exercise 23 to find the polynomial of degree 2 whose graph goes through the three points $(2, -3)$, $(4, 2)$, and $(7, -1)$.

25. (a) Find the polynomial $P(x)$ of degree 3 whose graph goes through the four points (x_1, y_1), (x_2, y_2), (x_3, y_3), and (x_4, y_4).
   (b) Find the polynomial function P whose values at the numbers $-1, 2, 4,$ and 7 are $5, -3, 2,$ and 1, respectively.

26. The Lagrange Interpolation Formula. Let (x_1, y_1), (x_2, y_2), (x_3, y_3), $\cdots$, (x_n, y_n), with $x_1 < x_2 < x_3 < \cdots < x_n$, be n points in the plane.
(a) Show that the polynomial of degree $n - 1$ whose graph contains the n given points is

$$P(x) = y_1 L_1(x) + y_2 L_2(x) + y_3 L_3(x) + \cdots + y_n L_n(x),$$

where

$$L_i(x) = \frac{A_i(x)}{A'(x_i)},$$

$$A(x) = (x - x_1)(x - x_2)(x - x_3) \cdots (x - x_n),$$

and

$$A_i(x) = \frac{A(x)}{x - x_i}, \qquad i = 1, 2, 3, \cdots, n.$$

(b)   Show that the polynomial $P(x)$ is unique.

Note: The polynomial $P(x)$ is called the **Lagrange interpolation formula** for n given points, and the $L_i(x)$, $i = 1, 2, 3, \cdots, n$, are the Lagrange interpolation coefficients. This polynomial is used to approximate a function f whose graph also passes through the n points. If the nth derivative of f exists throughout an interval containing $x_1, x_2, x_3, \cdots, x_n$, the interpolation polynomial $P(x)$ will be a good approximation to $f(x)$ for values of x in the interval.

5.6 IMPLICIT DIFFERENTIATION

Most of the functions discussed so far were defined by an expression or expressions involving only one variable. For instance, a function f is defined by $f(x) = 3x^3 - 2x + 5$ with $\mathfrak{D}_f = R$.

But an equation like

(1) $$x^2 + y^2 = 1$$

also defines a function of x if we specify that to each number x_1 in the closed interval $[-1, 1]$ there corresponds the number $y_1 = \sqrt{1 - x^2}$. We say that equation (1) defines a function of x implicitly or that (1) defines an **implicit function** of x.

Actually, of course, equation (1) defines many functions of x, some of which are given by $f(x) = \sqrt{1 - x^2}$ with $\mathfrak{D}_f = [-1, 1]$, $g(x) = -\sqrt{1 - x^2}$ with $\mathfrak{D}_g = [-1, 1]$, and

$$h(x) = \begin{cases} \sqrt{1 - x^2} & \text{if } -1 \le x \le 1 \text{ and } x \text{ is rational,} \\ -\sqrt{1 - x^2} & \text{if } -1 \le x \le 1 \text{ and } x \text{ is irrational.} \end{cases}$$

The graphs of f and g are shown in Fig. 5-1.

It is easy to show that f and g are continuous on the closed interval $[-1, 1]$ and differentiable throughout the open interval $(-1, 1)$; but h is discontinuous everywhere and possesses no derivative anywhere.

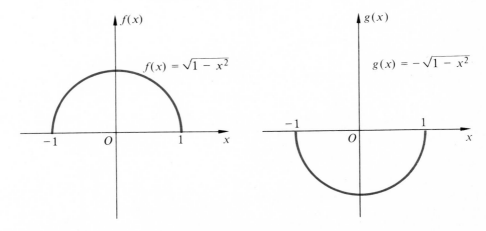

Figure 5-1

Although (1) is readily solvable for y in terms of x, an equation like

(2) $$2y^5 - 4y^4 + 5y^3 - 10y^2 + 3y + x = 0$$

cannot be solved for y in terms of x. Yet (2) also defines many functions of x implicitly. For if any real number is substituted for x in (2), the left-hand member of the resulting equation is a fifth-degree polynomial in y with real coefficients. Therefore the resulting equation has at least one real root. Moreover, there are numbers x that make (2) have *only* one real root, and the set of such numbers is the domain of one of the functions of x defined implicitly by (2).

If we are given an equation involving two variables, perhaps not solvable for one variable in terms of the other, we would like to know whether any functions are defined implicitly by the given equation and, if so, whether any such functions are differentiable. A complete discussion of this topic is better left for advanced calculus, and we shall assume in what follows that *the given equation in two variables, x and y, defines implicitly at least one function of x that is differentiable.*

A method for finding the derivative of a differentiable implicit function, without solving the defining equation (in two variables) for the dependent variable in terms of the independent variable, is explained in the following examples. This method is called **implicit differentiation.**

Example 1. Assuming that $x^2 + 5y^2 = 1$ defines implicitly a differentiable function of x, find $D_x y$.

Solution. If we differentiate every term of the given equation with respect to x, we can write

(3) $$D_x(x^2) + 5D_x(y^2) = D_x(1).$$

Keeping in mind that y is assumed to be a function of x, we differentiate y (with respect to x) by the power rule *and the chain rule* to obtain $D_x(y^2) = 2y(D_x y)$. Thus (3) becomes

$$2x + 5(2y)D_x y = 0.$$

Solving this equation for $D_x y$ gives

$$D_x y = \frac{-2x}{10y} = \frac{-x}{5y}, \qquad y \neq 0.$$

Example 2. Assuming that $x^3 + x^2 y - 10y^4 = 0$ defines implicitly a differentiable function of x, find $D_x y$.

Solution. Differentiating term by term with respect to x, we have

$$D_x(x^3) + D_x(x^2 y) - D_x(10y^4) = D_x(0).$$

Remembering that y is assumed to be a function of x, we use the power rule, the product rule, and the chain rule to obtain

$$3x^2 + [x^2 \cdot D_x y + y \cdot D_x(x^2)] - 10(4y^3 \cdot D_x y) = 0$$

or

$$3x^2 + x^2 \cdot D_x y + 2xy - 40y^3 \cdot D_x y = 0.$$

Solving this equation for $D_x y$, we have

$$D_x y = \frac{3x^2 + 2xy}{40y^3 - x^2}, \qquad 40y^3 - x^2 \neq 0.$$

Example 3. Find the equation of the tangent to the hyperbola $4x^2 - 9y^2 = 36$ at the point $(6, 2\sqrt{3})$.

Solution. We assume that the given equation defines implicitly a differentiable function of x. For brevity, we denote the derivative of y with respect to x by y'. If we differentiate the given equation term by term, with respect to x, we get

$$8x - 18yy' = 0,$$

from which we obtain $y' = 4x/9y$. Hence, at $(6, 2\sqrt{3})$ the slope of the tangent is $4/(3\sqrt{3})$ and the equation of the tangent is

$$4x - 3\sqrt{3}\,y - 6 = 0.$$

Example 4. Assuming that $y^2 = 2x^3$ defines implicitly a function of x, find $D_x^2 y$ by implicit differentiation.

Solution. Differentiating both sides of the given equation with respect to x, we obtain

$$(4) \qquad\qquad 2yD_x y = 6x^2.$$

We now differentiate each side of (4) with respect to x; since each of the factors, $2y$ and $D_x y$, in the left member of (4) defines a function of x, we use the formula for the derivative of a product of functions and obtain

$$(2y) \cdot D_x(D_x y) + (D_x y) \cdot D_x(2y) = 12x,$$

or

$$2yD_x^2 y + 2(D_x y)(D_x y) = 12x,$$

or

$$(5) \qquad\qquad yD_x^2 y + (D_x y)^2 = 6x.$$

Solving (4) for $D_x y$ gives $D_x y = 6x^2/(2y) = 3x^2/y$, $y \neq 0$; and if we substitute this result in (5), we have

$$yD_x^2 y + \left(\frac{3x^2}{y}\right)^2 = 6x,$$

from which

$$D_x^2 y = \frac{3x(2y^2 - 3x^3)}{y^3}, \qquad y \neq 0.$$

Since $y^2 = 2x^3$, this equation can be rewritten

$$D_x^2 y = \frac{3x}{2y}, \qquad y \neq 0.$$

Example 5. If $x^3 - y^3 = a^3$, in which a is a constant, defines implicitly a differentiable function of x, find y''.

Solution. Differentiating with respect to x, we get

$$(6) \qquad\qquad 3x^2 - 3y^2 y' = 0.$$

Again differentiating with respect to x, we obtain

$$(7) \qquad\qquad 2x - [y^2(y'') + (2yy')y'] = 0.$$

Solving (6) for y' and substituting it in (7), we have, for $y \neq 0$,

$$2x - \left[y^2 y'' + 2y\left(\frac{x^2}{y^2}\right)^2\right] = 0,$$

from which

$$y'' = \frac{-2x(x^3 - y^3)}{y^5},$$

or, since $x^3 - y^3 = a^3$,

$$y'' = \frac{-2a^3x}{y^5}.$$

Exercises

Assuming that each equation in Exercises 1 through 12 defines an implicit function of x, find $D_x y$ by implicit differentiation.

1. $x^2 - y^2 = 9$.

2. $4x^2 + 9y^2 = 36$.

3. $xy = 4$.

4. $b^2x^2 + a^2y^2 = a^2b^2$, where a, b are constants.

5. $xy^2 - x + 16 = 0$.

6. $x^3 - 3x^2y + 19xy = 0$.

7. $4x^3 + 11xy^2 - 2y^3 = 0$.

8. $\sqrt{xy} + 3y = 10x$.

9. $6x - \sqrt{2xy} + xy^3 = y^2$.

10. $\dfrac{y^2}{x^3} - 1 = y^{3/2}$.

11. $x^{2/3} + y^{2/3} = a^{2/3}$, where a is a constant.

12. $\dfrac{4x^2}{3y^3} - 6 = y^{2/3}$.

In Exercises 13 to 20, find $D_x^2 y$ by implicit differentiation.

13. $2y^2 = x^3$.

14. $xy^3 = 12$.

15. $x^2y^2 - y = 7$.

16. $x^3 - 4y^2 + 3 = 0$.

17. $2x^3 + 2y^3 = 1$.

18. $2x^2y - 4y^3 + 1 = 0$.

19. $x^{1/2} + y^{1/2} = a^{1/2}$, a is constant.

20. $x^{1/3} + y^{1/3} = a^{1/3}$, a is constant.

5.7 TANGENTS AND NORMALS

Now that we can find derivatives easily, without having to resort to the definition of the derivative, we return briefly to the *tangent line* (or *tangent*) to the graph of a function.

5.7.1 Definition. Let f be a function and let $f'(c)$ exist. Then the **tangent** to the graph of f at the point $(c, f(c))$ on the graph is the line through $(c, f(c))$ having slope $f'(c)$. Its equation is

$$y - f(c) = f'(c)(x - c).$$

This definition is consistent with our definition of the slope of the tangent (4.1.1) and the definition of the derivative (4.3.1).

The **slope of a graph** at a given point on the graph means the slope of the tangent to the graph at that point. Thus $f'(c)$ is the slope of the graph of the function f at the point $(c, f(c))$. In Fig. 5-2, the slope of the graph of $f(x) = (x - 1)^3/6 + 1$ at any point $(x, f(x))$ is $f'(x) = (x - 1)^2/2$. At $(-1, -\frac{1}{3})$ the slope is $f'(-1) = 2$, at $(1, 1)$ the slope is $f'(1) = 0$, and at $(2, \frac{7}{6})$ the slope is $f'(2) = \frac{1}{2}$.

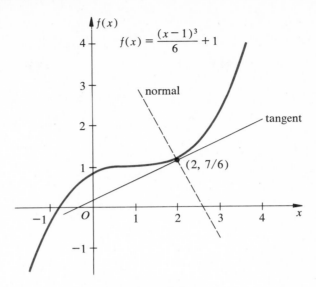

$$f(x) = \frac{(x-1)^3}{6} + 1$$

normal

tangent

$(2, 7/6)$

Figure 5-2

If the graph of a function f has a tangent at $(c, f(c))$, the line through the point $(c, f(c))$, perpendicular to the tangent, is called the **normal** to the graph at $(c, f(c))$. Since the slope of the tangent is $f'(c)$, it follows from 2.5.3 that the slope of the normal is $-1/f'(c)$, provided that $f'(c) \neq 0$.

5.7.2 Theorem. If f is a function and $f'(c)$ exists and is not zero, then the equation of the normal to the graph of f at the point $(c, f(c))$ on the graph is

$$y - f(c) = -\frac{1}{f'(c)}(x - c).$$

Example 1. Find the equations of the tangent and normal to the graph of

$$f(x) = \frac{(x - 1)^3}{6} + 1$$

at the point $(2, \frac{7}{6})$. Sketch the curve and its tangent and normal at the given point (Fig. 5-2).

Solution. We differentiate $f(x)$ to get $f'(x) = (x - 1)^2/2$. Then the slope of the curve at $(2, \frac{7}{6})$ is $f'(2) = \frac{1}{2}$, which is also the slope of the tangent to the curve at that point. From 5.7.1, the equation of the tangent is $y - \frac{7}{6} = \frac{1}{2}(x - 2)$, or $3x - 6y + 1 = 0$.

The slope of the normal to the curve at $(2, \frac{7}{6})$ is $-1/f'(2) = -1/(\frac{1}{2}) = -2$, and the equation of the normal is $y - \frac{7}{6} = -2(x - 2)$, or $12x + 6y - 31 = 0$.

Example 2. Find the equation of the normal to the curve $y = 2\sqrt{9 - x^2}/3$ that is parallel to the line $3x - 2y - 12 = 0$ (Fig. 5-3).

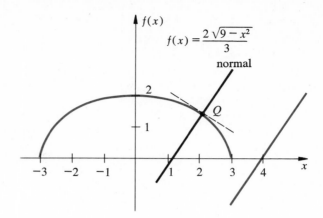

Figure 5-3

Solution. The slope of the given line is $\frac{3}{2}$. Writing the equation of the curve in the form $y = (\frac{2}{3})(9 - x^2)^{1/2}$, we find $dy/dx = (\frac{2}{3})(\frac{1}{2})(9 - x^2)^{-1/2}(-2x)$, or

$$\frac{dy}{dx} = \frac{-2x}{3\sqrt{9 - x^2}}, \qquad -3 < x < 3.$$

This gives the slope of the tangent to the curve at any point (x, y) on it, $-3 < x < 3$. Then the slope of the normal at (x, y) is $3\sqrt{9 - x^2}/(2x)$. We seek the value of x that makes this slope equal to the slope of the given line—namely, $\frac{3}{2}$. Setting

$$\frac{3\sqrt{9 - x^2}}{2x} = \frac{3}{2}$$

and solving for x, we find $x = 3/\sqrt{2} = 3\sqrt{2}/2$. This is the x coordinate of the point Q on the curve at which the normal has slope $\frac{3}{2}$. The y coordinate of Q, obtained by substituting this value of x in the given equation of the curve, is $\sqrt{2}$. Hence the coordinates of Q are $(3\sqrt{2}/2, \sqrt{2})$.

By substituting the coordinates of Q and the slope $\frac{3}{2}$ into the point-slope form of the equation of a line, we find the equation of the normal at Q to be

$$3x - 2y - \frac{5}{\sqrt{2}} = 0.$$

Exercises

In Exercises 1 to 6, find the equation of the tangent and the normal to the graph of $y = f(x)$ at the given point P. Make a sketch.

1. $y = x^2 - 3x + 4$; P:(3, 4).

2. $y = 2x^2 - 4x - 1$; P:(0, −1).

3. $y = x^3 - x^2 - 2x + 4$; P:(0, 4).

4. $y = \frac{1}{3}x^3 - x^2 + 1$; P:(3, 1).

5. $y = \frac{2}{x} + \frac{x}{2}$; P:$\left(1, \frac{5}{2}\right)$,

6. $y = \frac{1}{4}x^4 - 2x^2$; P:$\left(1, -\frac{7}{4}\right)$.

In Exercises 7 and 8, find $D_x y$ by implicit differentiation and write the equation of the tangent to the graph of the given equation at the given point P.

7. $x^2 - y^2 = 36$; P:$(6\sqrt{2}, 6)$.

8. $x^3 - 9x^2 - y^2 + 27x - 2y - 28 = 0$; $P:(4, -2)$.

9. Find the coordinates of the point on the graph of $y = x^2 + 2x - 2$ at which the slope of the graph is 2. Make a sketch.

10. Find the coordinates of the points on the curve $y = x^3 - 4x^2$ at which the slope of the curve is -5. Make a sketch.

11. Find the equation of the tangent line to the curve $y = x^4$ that is parallel to the line $x - 2y + 5 = 0$. Make a sketch.

12. Find the equations of the tangent lines to the curve $3y = x^3 - 3x^2 + 6x + 4$ that are parallel to the line $2x - y + 3 = 0$. Make a sketch.

13. Find the equations of the tangent lines to the graph of $y = x^4 - 4x^3 - 2x^2 + 8x$ that are parallel to the line $4x + y - 7 = 0$. Make a sketch.

14. Find the equations of the tangent lines to the graph of $x^3 - 9x^2 + 8y - 4 = 0$ that are parallel to the line $3x - y + 5 = 0$. Make a sketch.

15. Let P and Q be any two points on the curve $xy = 1$. Show that the area of the triangle formed by the coordinate axes and the tangent to the curve at P is equal to the area of the triangle formed by the coordinate axes and the tangent to the curve at Q. What is the value of this constant area?

16. Find the equations of the lines through the point $(\frac{1}{4}, -4)$ that are tangent to the curve $y = 2x^2 - 3$. Make a sketch.

5.8 RELATED RATES

If t represents time and $y = f(t)$ is differentiable, the instantaneous rate of change of y with respect to time is given at once by dy/dt (Section 4.4).

But there are many problems in which x and y are related by an equation not involving t, and it is implied that x and y are *unknown* functions of time. Often it is possible in such problems to find dx/dt or dy/dt, the time rate of change of x or y, without expressing either x or y directly as a function of t. The method is to differentiate both sides of the given equation implicitly with respect to t by means of the chain rule, as shown in the following examples.

Example 1. A small balloon is released at a point 150 feet away from an observer who is on level ground. If the balloon goes straight up at the rate of 8 feet per second, how fast is the distance from the observer to the balloon increasing when the balloon is 50 feet high?

Solution. Let t denote the number of seconds after the balloon is released.

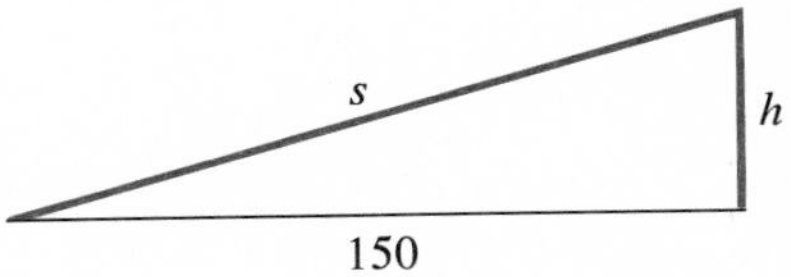

Figure 5-4

We first draw a diagram that is valid for all $t > 0$, not just at a particular instant (Fig. 5-4).

Since the vertical side of the triangle indicates the height of the balloon at any time $t > 0$, it varies in length and we label it with the letter h, a variable. Similarly, the hypotenuse, representing the distance from the observer to the balloon at any time t, also varies and is labeled with the variable s. But the base of the triangle, which is the distance between the observer and the point of release of the balloon, remains unchanged as t increases, and so is labeled with its given constant value 150. We stress again that Fig. 5-4 is valid at all instants, $t > 0$.

Next we ask three questions about the variables s and h:

(a) What is *given* concerning them?

(b)   What do we *want to know* about any of them?

(c)   *When* (at what particular instant as *t* increases) do we want to know it?

The answers to (a) and (b) are derivatives with respect to *t*. We are given that *h* increases at 8 feet per second, that is, we are given that $dh/dt = 8$. We want to know how fast *s* is increasing when $h = 50$ feet, that is, we want to know the value of ds/dt at the particular instant when $h = 50$.

We next write an equation, suggested by the diagram, that relates the variables *s* and *h*, and that is true for all $t > 0$, not just at some particular instant:

(1) $$s^2 = h^2 + (150)^2.$$

The variables *s* and *h* change as *t* increases, and are implicit functions of *t*. So we can differentiate equation (1) implicitly with respect to *t*, obtaining

$$2s\frac{ds}{dt} = 2h\frac{dh}{dt},$$

or

(2) $$s\frac{ds}{dt} = h\frac{dh}{dt}.$$

It is important to keep in mind that equation (2), like equation (1), is valid for all $t > 0$, not just at some particular instant.

Now, and *not before now*, we turn to the particular situation when $h = 50$. From (1) we find that when $h = 50$, $s = \sqrt{(50)^2 + (150)^2} = 50\sqrt{10}$. Moreover, we are given that $dh/dt = 8$. By substituting these results in equation (2), we find that when $h = 50$,

$$50\sqrt{10}\,\frac{ds}{dt} = 50(8),$$

or

$$\frac{ds}{dt} = \frac{8}{\sqrt{10}} \text{ feet per second.}$$

Thus, when the balloon is 50 feet high, its distance from the observer is increasing at the rate of $8/\sqrt{10} \doteq 2.53$ feet per second.

Note.   The reader is cautioned against the common mistake of substituting the given particular values before differentiating with respect to *t*. The relation (1) between *h* and *s* holds good throughout the problem, and should be differentiated with respect to *t* *before* the given particular values are substituted.

Example 2.   Water is pouring into a conical cistern at the rate of 8 cubic feet per minute. If the height of the inverted cone is 12 feet and the radius of its circular opening is 6 feet, how fast is the water level rising when the water is 4 feet deep?

Solution.   Denote the depth of the water in the cistern at any time *t* by the variable *h*, and denote the radius of the circular surface of the water by the variable *r* (Fig. 5-5).

We are *given* that the volume, *V*, of water in the cistern is increasing at the rate of 8 cubic feet per minute—that is, that $dV/dt = 8$. We *want to know* how fast the water level is rising, that is, dh/dt, *when* $h = 4$ feet.

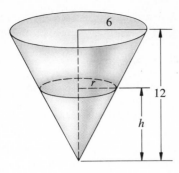

Figure 5-5

First let us find an equation relating the variables V and h for all $t > 0$. The formula for the volume of water in the cistern, $V = \frac{1}{3}\pi r^2 h$, contains the unwanted variable r. But from similar triangles in Fig. 5-5, we have $r/h = 6/12$, or $r = h/2$. By substituting this in $V = \frac{1}{3}\pi r^2 h$, we obtain

$$(3) \qquad V = \frac{\pi h^3}{12},$$

the desired equation relating the variables V and h for all $t > 0$.

The values of the variables V and h in (3) depend on the value of t (until the cistern is full), and are implicit functions of t. It we differentiate both members of (3) implicitly with respect to t, we get

$$\frac{dV}{dt} = \frac{3\pi h^2}{12}\frac{dh}{dt},$$

or

$$(4) \qquad \frac{dV}{dt} = \frac{\pi h^2}{4}\frac{dh}{dt},$$

which is true for all values of $t > 0$.

At this point, and not earlier, we turn our attention to the particular situation at the instant when $h = 4$. Substituting $h = 4$ and $dV/dt = 8$ in (2), we obtain

$$8 = \frac{\pi(4)^2}{4}\frac{dh}{dt},$$

from which

$$\frac{dh}{dt} = \frac{2}{\pi} \text{ foot per minute.}$$

And so when the depth of the water is 4 feet, the water level is rising at the rate of $2/\pi \doteq 0.637$ foot per minute.

Examples 1 and 2 suggest a procedure for the solution of related-rate problems.

STEP 1. Let t denote time. Draw a diagram that is valid for all $t > 0$. Label those lines whose lengths do not change, as t increases, with their given constant values. Assign letters to the quantities that vary with t, and label the appropriate lines of the figure with these variables.

STEP 2. State what is *given* about the variables and what information is *wanted* about them. This information will be in the form of derivatives with respect to t.

STEP 3. Write an equation relating the variables that is valid at all times $t > 0$, not just at some particular instant.

STEP 4. Differentiate the equation found in Step 3 implicitly with respect to t. The resulting equation, containing derivatives with respect to t, is true for all $t > 0$.

STEP 5.   Substitute in the equation just found in Step 4 all data that are valid *at the particular instant* for which the answer to the problem is required.

We illustrate this procedure in the next example.

Example 3. A plane flying north at 640 miles per hour passes over a certain town at noon, and a second plane going east at 600 miles per hour is directly over the same town 15 minutes later. If the planes are flying at the same altitude, how fast will they be separating at 1 : 15 P.M. ?

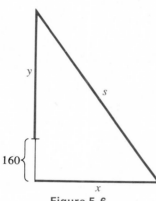

Figure 5-6

STEP 1.   Let t denote the number of hours after 12 : 15 A.M. Figure 5-6 shows the situation for all $t > 0$. The distance in miles from the town to the northbound plane when $t = 0$ (12 : 15 A.M.) is labeled with the constant $640/4 = 160$. For any $t > 0$, the distance in miles flown by the northbound plane (after 12 : 15 A.M.) is labeled with the variable y, the distance flown by the eastbound plane is labeled with the variable x, and the distance between the planes is indicated by the variable s.

STEP 2.   We are given that for all $t > 0$, $dy/dt = 640$, and $dx/dt = 600$. We want to know the value of ds/dt (how fast the planes are separating) at the particular instant when $t = 1$, that is, at 1 : 15 P.M.

STEP 3.   An equation relating the variables x, y, and s, valid for all $t > 0$, is suggested by the right triangle in the figure:

(5)
$$s^2 = x^2 + (y + 160)^2.$$

STEP 4.   Differentiating both members of (5) implicitly with respect to t, we have

$$2s\frac{ds}{dt} = 2x\frac{dx}{dt} + 2(y + 160)\frac{dy}{dt},$$

or

(6)
$$s\frac{ds}{dt} = x\frac{dx}{dt} + (y + 160)\frac{dy}{dt},$$

which is true for all $t > 0$.

STEP 5.   For all $t > 0$, $dx/dt = 600$ and $dy/dt = 640$. At the particular instant when $t = 1$, $x = 600(1) = 600$, $y = 640(1) = 640$, and, from (5), $s = \sqrt{(600)^2 + (640 + 160)^2} = 1000$. By substituting all these data in (6), we obtain

$$1000\frac{ds}{dt} = 600(600) + (640 + 160)(640),$$

from which

$$\frac{ds}{dt} = 872 \text{ miles per hour.}$$

Thus at 1 : 15 P.M. the planes are separating at the rate of 872 miles per hour.

Exercises

1. An edge of a variable cube is increasing at the rate of 3 inches per second. How fast is the volume of the cube increasing when an edge is 10 inches long?

2. Assuming that a soap bubble retains its spherical shape as it expands, how fast is its radius increasing when its radius is 2 inches, if air is blown into it at the rate of 4 cubic inches a second?

3. An airplane, flying horizontally at an altitude of 1 mile, passes directly over an observer. If the constant speed of the plane is 240 miles per hour, how fast is its distance from the observer increasing 30 seconds later? [*Hint:* In 30 seconds ($\frac{1}{120}$ hour) the plane goes 2 miles.]

4. A student is using a straw to drink from a conical paper cup, whose axis is vertical, at the rate of 3 cubic inches a second. If the height of the cup is 10 inches and the diameter of its opening is 6 inches, how fast is the level of the liquid falling when the depth of the liquid is 5 inches?

5. An airplane, flying west at 400 miles per hour, goes over a certain town at 11:30 A.M., and a second plane at the same altitude, flying south at 500 miles per hour, goes over the town at noon. How fast are they separating at 1:00 P.M.? (*Hint:* Let $t = 0$ at noon.)

6. A man on a dock is pulling in a rope that is fastened to the bow of a small boat. If the man's hands are 12 feet higher than the point where the rope is attached to the boat, and if he is retrieving the rope at the rate of 3 feet per second, how fast is the boat approaching the dock when 20 feet of rope are still out?

7. A 20-foot ladder is leaning against a wall. If the bottom of the ladder is pulled along the level pavement, directly away from the wall, at 2 feet per second, how fast is the top of the ladder moving down the wall when the foot of the ladder is 4 feet from the wall?

8. A man throws a stone into a still millpond, causing a circular ripple to spread. If the radius of the circle increases at the constant rate of 1.5 feet

per second, how fast is the enclosed area increasing at the end of 2 seconds?

9. Sand is pouring from a pipe at the rate of 16 cubic feet per second. If the falling sand forms a conical pile on the ground whose altitude is always $\frac{1}{4}$ the diameter of the base, how fast is the altitude increasing when the pile is 4 feet high?

10. A child is flying a kite. If the kite is 90 feet high and the wind is blowing it on a horizontal course at 5 feet per second, how fast is the child paying out cord when 150 feet of cord is out?

11. A swimming pool is 40 feet long, 20 feet wide, 8 feet deep at the deep end, and 3 feet deep at the shallow end, the bottom being rectangular (Fig. 5-7). If the pool is filled by pumping water into it at the rate of 40 cubic feet per minute, how fast is the water level rising when it is 3 feet deep at the deep end?

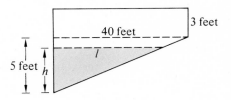

Figure 5-7

12. A particle P is moving along the graph of $y = \sqrt{x^2 - 4}$, $x \geq 2$, so that the x coordinate of P is increasing at the rate of five units per second. How fast is the y coordinate of P increasing when $x = 3$?

13. A metal disk expands during heating. If its radius increases at the rate of 0.02 inch per second, how fast is the area of one of its faces increasing when its radius is 8.1 inches?

14. Two ships sail from the same island port, one going north at 24 knots (that is, at 24 nautical miles per hour) and the other east at 30 knots. The northbound ship departed at 9:00 A.M. and the eastbound ship at 11:00 A.M. How fast is the distance between them increasing at 2:00 P.M.? (*Hint:* Let $t = 0$ at 11:00 A.M.)

15. A long, level highway bridge passes over a railroad track that is 100 feet below it and at right angles to it. If an automobile traveling 45 miles per hour (= 66 feet per second) is directly above a train going 60 miles per hour (= 88 feet per second), how fast will they be separating 10 seconds later?

16. Denote the perpendicular sides of a certain right triangle by a and b. If the length of a is increasing at the rate of 3 inches per second and the rate of b is decreasing at the rate of 2 inches per second, how fast is the area increasing when the length of a is 10

inches and the length of b is 13 inches? [*Hint:* Although three variables are involved, A (area), a, and b, all are implicit functions of t.]

17. The volume of a pyramid is increasing at the constant rate of 30 cubic inches per second, and the area of the base is increasing at the constant rate of 5 square inches per second. When the area of the base is 100 square inches and the altitude is 8 inches, how fast is the altitude increasing? (See the Hint in Exercise 16.)

5.9 DIFFERENTIALS

Consider a differentiable function f, and let $y = f(x)$. We have sometimes used the Leibnitz notation dy/dx for the derivative of y with respect to x. Up to now we have not thought of it as a quotient but simply as another commonly used symbol for the derivative. We are about to define a new concept, **differential**, which will give meaning to dy and dx separately and, among other things, permit us to think of dy/dx either as a symbol for the derivative or as the quotient of two differentials.

5.9.1 Definition. Let f be a function that is differentiable at a certain value x_1 of its independent variable x and let $y = f(x)$. Then

(i) the **differential**, dx, **of the independent variable** x is an arbitrary increment of x; that is,

$$dx = \Delta x;$$

(ii) the **differential** dy **of the dependent variable** y [or $df(x)$, the **differential of the function**] at the point x_1 is

$$dy = f'(x_1)\, dx.$$

Example 1. If f is the function defined by $f(x) = 2x^3 - 1$ and if $y = f(x)$, then $f'(x_1) = 6x_1^2$ and the differential of y at x_1 is $dy = 6x_1^2\, dx$.

If we let $x_1 = 5$ and $dx = 3$, say, then $dy = 6(25)(3) = 450$; if $x_1 = -1$ and $dx = 0.04$, $dy\ (= df(x)) = 6(-1)^2(0.04) = 0.24$.

If f is the identity function and $y = f(x) = x$, then $f'(x) = 1$ and [by 5.9.1(ii)] $dy = 1 \cdot dx = dx$. Since both parts, (i) and (ii), of 5.9.1 give $d(x) = dx$, the definitions 5.9.1 are consistent.

It should be recognized that $dx\ (= \Delta x)$ is another independent variable and that dy depends for its value on the *two* independent variables x_1 and dx. This is an example of a function of two independent variables, a concept that will be explained thoroughly in later chapters. For the present we will simply say that dy is a function of two independent variables, x_1 and dx, because to each ordered pair of numbers (x_1, dx) in its domain there corresponds one and only one number dy, given by $dy = f'(x_1)\,dx$. The domain of this function of two variables is the set of ordered pairs (x_1, dx), where x_1 is any number in the domain of f' and dx is any number whatever.

If we divide both members of $dy = f'(x_1)\,dx$ by the differential dx, we get

$$\frac{dy}{dx} = f'(x_1), \qquad dx \neq 0,$$

where dy/dx in the left member is the quotient of the two differentials, dy by dx. Thus dy/dx may be thought of as representing the quotient of two differentials or simply as a value of the derivative of y with respect to x, whichever suits our convenience in a particular situation.

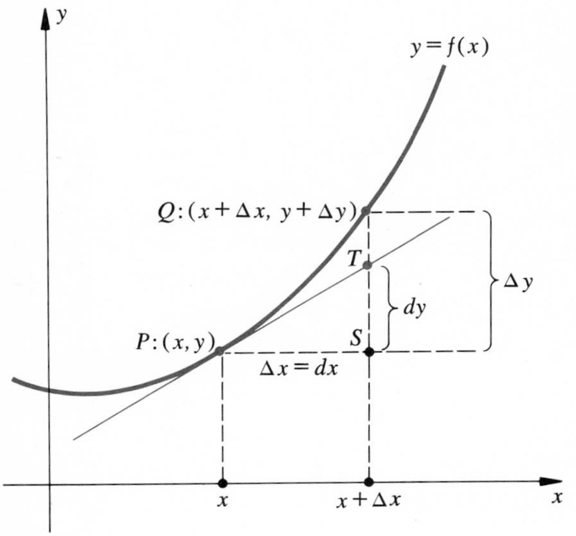

Figure 5-8

While the differential dx of the independent variable is an increment Δx, the differential dy of the dependent variable is not, in general, equal to the corresponding increment Δy because $\Delta y = f(x + \Delta x) - f(x)$ and $dy = f'(x)\,dx$. In Fig. 5-8, the difference between Δy and dy, for the same values of x and dx, is the directed distance $\overline{TQ}$.

Let $P{:}(x, y)$ and $Q{:}(x + \Delta x, y + \Delta y)$ be neighboring points on the graph of $y = f(x)$ (see Fig. 5-8). Since the slope of the tangent line at P is $f'(x) = \overline{ST}/\overline{PS}$ and $\overline{PS} = dx$, we have $\overline{ST} = f'(x)\,dx$, or $\overline{ST} = dy$. But $\Delta y = \overline{SQ}$. Hence the difference between Δy and dy is $\overline{TQ}$.

Figure 5-8 suggests that, for a fixed x, $|TQ|$ can be made as small as we please by taking $|dx| = |\Delta x|$ sufficiently small. This will be proved in the next section. It follows that the smaller we make $|dx|\ (\neq 0)$, the closer dy will approximate Δy.

Formulas for differentials are readily obtained by multiplying the corresponding formulas for derivatives by the differential of the independent variable. As an illustration, if u and v are differentiable functions of x, then

$$\frac{d(uv)}{dx} = u\frac{dv}{dx} + v\frac{du}{dx}.$$

If we multiply both sides of this equation by dx and treat $d(uv)/dx$, du/dx, and dv/dx as quotients of differentials, then

$$d(uv) = u\,dv + v\,du.$$

In words, *the differential of the product of two differentiable functions is equal to the first function times the differential of the second plus the second function times the differential of the first.*

Listed below in the first column are formulas for derivatives, written in the

Leibniz notation; opposite each, in the second column, are the corresponding formulas for differentials.

I $\quad \dfrac{dc}{dx} = 0$ $\qquad\qquad\qquad$ I' $\quad dc = 0$

II $\quad \dfrac{d(cu)}{dx} = c\dfrac{du}{dx}$ $\qquad\qquad$ II' $\quad d(cu) = c\,du$

III $\quad \dfrac{d(u + v)}{dx} = \dfrac{du}{dx} + \dfrac{dv}{dx}$ $\qquad$ III' $\quad d(u + v) = du + dv$

IV $\quad \dfrac{d(uv)}{dx} = u\dfrac{dv}{dx} + v\dfrac{du}{dx}$ $\qquad$ IV' $\quad d(uv) = u\,dv + v\,du$

V $\quad \dfrac{d(u/v)}{dx} = \dfrac{v(du/dx) - u(dv/dx)}{v^2}$ $\qquad$ V' $\quad d\left(\dfrac{u}{v}\right) = \dfrac{v\,du - u\,dv}{v^2}$

VI $\quad \dfrac{d(u^n)}{dx} = nu^{n-1}\dfrac{du}{dx}$ $\qquad$ VI' $\quad d(u^n) = nu^{n-1}\,du$

Since each formula for differentials can be obtained from the corresponding formula for derivatives by multiplying by dx, memorization of them is optional.

In $y = f(x)$, if x is not an independent variable but $x = g(t)$ for some function g, $\mathfrak{R}_g \subseteq \mathfrak{D}_f$, then y is also a function of the independent variable t, since $y = f(g(t)) = F(t)$. Now

(1) $\qquad\qquad\qquad dy = F'(t)\,dt \qquad$ and $\qquad dx = g'(t)\,dt.$

By applying the chain rule to $F(t) = f(g(t))$, we obtain

(2) $\qquad\qquad\qquad\qquad F'(t) = f'(x) \cdot g'(t).$

Multiplying both members of (2) by $dt \neq 0$ gives

$$F'(t)\,dt = f'(x) \cdot [g'(t)\,dt]$$

or, using (1),

$$dy = f'(x)\,dx.$$

Thus $dy = f'(x)\,dx$ even though t is the independent variable, not x.

5.9.2 Corollary. Let $y = f(x)$. Then $dy = f'(x)\,dx$ whether x is an independent variable or is dependent on still another variable.

Example 2. If $y = f(x) = 4x^3 - 6x + 1$, find dy.

Solution. Since $f'(x) = 12x^2 - 6$, $dy = f'(x)\,dx = (12x^2 - 6)\,dx$.

Example 3. If $y = 3x/\sqrt{x^2 + 1}$, find dy.

Solution. Using formula V' above, we obtain

$$dy = \frac{\sqrt{x^2 + 1} \cdot d(3x) - 3x \cdot d(\sqrt{x^2 + 1})}{x^2 + 1}$$

$$= \frac{3\sqrt{x^2 + 1}\,dx - 3x[\tfrac{1}{2}(x^2 + 1)^{-1/2} \cdot d(x^2 + 1)]}{x^2 + 1}$$

$$= \frac{3\sqrt{x^2 + 1}\, dx - 3x(2x\, dx)/[2\sqrt{x^2 + 1}]}{x^2 + 1}$$

$$= \frac{3\, dx}{(x + 1)^{3/2}}.$$

Example 4. If $x^2 - 3xy + 4y^2 = 10$ defines y implicitly as a function of x, find dy/dx.

Solution. It is often convenient to find the derivative of an implicit function by means of differentials. If we take differentials of both members of the given equation, we get

$$2x\, dx - (3x\, dy + 3y\, dx) + 8y\, dy = 0$$

or

$$(3x - 8y)\, dy = (2x - 3y)\, dx.$$

Thus

$$\frac{dy}{dx} = \frac{2x - 3y}{3x - 8y}, \qquad 3x - 8y \neq 0.$$

Exercises

1. Let $y = f(x) = x^3$. Find the value of dy when
(a) $x = 0.5$, $dx = 1$;
(b) $x = -1$, $dx = 0.75$.
Make a careful drawing of the graph of f for $-1.5 \le x \le 1.5$, and the tangents to the curve at $x = 0.5$ and $x = -1$; on this drawing label dy and dx for each of the given sets of data in (a) and (b).

2. Let $y = 1/x$. Find the value of dy when
(a) $x = 1$, $dx = 0.5$;
(b) $x = -2$, $dx = 0.75$.
Make a large scale drawing, as in Exercise 1, for $-3 \le x < 0$ and $0 < x \le 3$.

In Exercises 3 to 8, find dy.

3. $y = 2x^2 - 3x + 5$.

4. $y = 7x^3 - 3x^2 + 4$.

5. $y = (3 + 2x^3)^{-4}$.

6. $y = \dfrac{13x}{5x^2 + 2}$.

7. $y = \sqrt{4x^5 + 2x^4 - 5}$.

8. $y = (6x^8 - 11x^5 + x^2 + 2x)^{-2/3}$.

9. If $s = \sqrt[5]{(t^2 - 3)^2}$, find ds.

10. If $F(x) = (5x^2 + 1)^2(x - 7)^5$, find dF.

11. If $\phi(x) = \dfrac{1}{(x - 3)^2(x^2 + 14)}$, find $d\phi$.

12. If $G(w) = \dfrac{6w^3 - 5}{(w + 3)^4}$, find dG.

Assuming that each equation in Exercises 13 to 16 defines an implicit function of x, use differentials to find dy/dx.

13. $x^2 + 5xy - 2y^2 = 4$.

14. $2x^2 - 3xy + y^2 - 5x + 2 = 0$.

15. $x^4 + y^4 = 1$.

16. $x^{1/2} + y^{1/2} = a^{1/2}$.

5.10 DIFFERENTIALS AS LINEAR APPROXIMATIONS

Let f be a differentiable function and let $y = f(x)$. Suppose that we know $f(c)$ and $f'(c)$, the values of the function and its derivative at some point $x = c$, and we want to know $f(c + \Delta x)$, the value of the function at some point $c + \Delta x$ near c (Fig. 5-9).

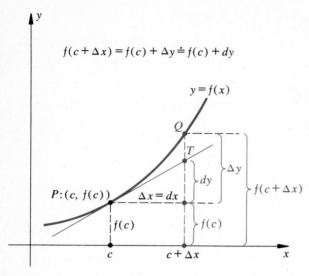

$$f(c+\Delta x) = f(c) + \Delta y \doteq f(c) + dy$$

$y = f(x)$

Q

T

Δy

dy

$P:(c, f(c))$ $\Delta x = dx$ $f(c+\Delta x)$

$f(c)$ $f(c)$

c $c + \Delta x$ x

Figure 5-9

It often happens that $f(c + \Delta x)$ is difficult or tedious to compute by direct substitution. Since $\Delta y = f(c + \Delta x) - f(c)$, $f(c + \Delta x) = \Delta y + f(c)$. So a good approximation for Δy would enable us to get a good approximation of $f(c + \Delta x)$. It will be shown below that the easily calculated differential, $dy = f'(c)\, dx$, is an excellent approximation of Δy when $dx (= \Delta x)$ is small. Thus

(1) $$f(c + \Delta x) \doteq dy + f(c).$$

This approximation has a simple geometric interpretation (Fig. 5-9). Since $f(c + \Delta x)$ is the y coordinate of the point Q on the curve and $dy + f(c)$ is the y coordinate of the point T on the tangent line, the approximation (1) means that we are replacing the curve $y = f(x)$ by its tangent line for values of x close to c. For this reason, $dy + f(c)$ is said to be a **linear approximation** of $f(c + \Delta x)$, the value of the function f at $c + \Delta x$.

To show that dy is a good approximation of Δy, we recall (3.3.6) that $\lim\limits_{\Delta x \to 0} (\Delta y / \Delta x) = f'(c)$ is equivalent to

$$\lim_{\Delta x \to 0} \left[\frac{\Delta y}{\Delta x} - f'(c) \right] = 0.$$

It now follows from the definition of limit that for each $\epsilon > 0$ there is a $\delta > 0$ such that

$$|\Delta x| < \delta \implies \left| \frac{\Delta y}{\Delta x} - f'(c) \right| < \epsilon.$$

But this result can be written

$$|\Delta x| < \delta \implies |\Delta y - f'(x)\,\Delta x| < \epsilon\,|\Delta x|$$

or since $\Delta x = dx$ and $f'(c)\,dx = dy$,

$$|dx| < \delta \implies |\Delta y - dy| < \epsilon\,|dx|.$$

Thus the smaller $|dx|$ is, the closer dy approximates Δy.

Example 1. If $y = x^2 - 3x + 4$, compare Δy and dy when $x = -1$ and $dx = 0.07$.

Solution. Substituting $x = -1$ and $dx = \Delta x = 0.07$ in $\Delta y = f(x + \Delta x) - f(x)$, we have

$$\Delta y = f(-1 + 0.07) - f(-1) = f(-0.93) - f(-1)$$
$$= [(-0.93)^2 - 3(-0.93) + 4] - [(-1)^2 - 3(-1) + 4] = -0.3451.$$

Also, $dy = f'(x)\,dx = (2x - 3)\,dx$; thus for $x = -1$ and $dx = 0.07$,

$$dy = [2(-1) - 3](0.07) = -0.35.$$

Hence $\Delta y = -0.3451$ and $dy = -0.35$.

Example 2. Use differentials to approximate $\sqrt{8}$.

Solution. We know that $\sqrt{9} = 3$ and that 8 is close to 9, which suggests that we use $y = f(x) = \sqrt{x}$ and let $x = 9$, $\Delta x = -1$. We seek $\sqrt{8} = \sqrt{9 + \Delta x}$.

Since $\Delta y = f(x + \Delta x) - f(x) = \sqrt{9 + \Delta x} - \sqrt{9}$, we can write $\sqrt{9 + \Delta x} = \sqrt{9} + \Delta y$, or

$$(2) \qquad \sqrt{8} = 3 + \Delta y.$$

We shall approximate Δy by the easily computed $dy = f'(x)\,dx$, and replace (2) by the approximation

$$(3) \qquad \sqrt{8} \doteq 3 + dy.$$

Now $f(x) = \sqrt{x} = x^{1/2}$, $f'(x) = \tfrac{1}{2}x^{-1/2} = \tfrac{1}{2}/\sqrt{x}$, and $f'(9) = \tfrac{1}{6}$. Moreover, $dx = \Delta x = -1$. Thus $dy = f'(9)\,dx = \tfrac{1}{6}(-1) = -\tfrac{1}{6}$, and (3) becomes

$$\sqrt{8} \doteq 3 + (-\tfrac{1}{6}) = \tfrac{8}{3} \doteq 2.833.$$

The exact value of $\sqrt{8}$, to three decimal places, is 2.828.

We have used the descriptive statement that "dy is a good approximation to Δy at a given point if $|dx| = |\Delta x|$ is sufficiently small."

But in any approximation method it is desirable to be able to find bounds for the possible error. In the present instance, we must make sure that $|\Delta y - dy|$, the error introduced in using the differential dy in place of the increment Δy in a particular calculation, is not so great as to be unacceptable.

It can be shown that if $f''(x)$ exists on the closed interval I having c and $c + \Delta x$ as endpoints, and if M is a number such that $|f''(x)| \leq M$ for all $x \in I$, then Δy differs from dy by not more than $\tfrac{1}{2}(\Delta x)^2 M$. That is,

5.10.1
$$|\Delta y - dy| \leq \tfrac{1}{2}(\Delta x)^2 M.$$

Example 3. Use differentials to find the approximate increase in the area of a soap bubble when its radius increases from 3 inches to 3.025 inches, and find bounds for the exact increase in area, ΔA.

Solution. The area of a (spherical) soap bubble is $A = 4\pi r^2$. The exact change in A is ΔA. We shall use dA as an approximation of ΔA when $r = 3$ and $\Delta r = dr = 0.025$:

$$dA = 8\pi r\,dr = 8\pi(3)(0.025) = 1.885 \text{ square inches.}$$

Since $d^2A/dr^2 = 8\pi$, a positive constant for all values of r, we take $M = 8\pi$. Substituting in $|\Delta A - dA| \leq \tfrac{1}{2}(\Delta r)^2 M$, we have

$$|\Delta A - 1.885| \leq \tfrac{1}{2}(0.025)^2(8\pi) < 0.007855,$$

from which

$$-0.007855 < \Delta A - 1.885 < 0.007855.$$

Therefore

$$1.877 < \Delta A < 1.893.$$

Example 4. If $y = 1/\sqrt{x^2 + 1}$, and if x increases from 2 to 2.04, use differentials to find the approximate change in y. By means of 5.10.1, estimate the error, $|\Delta y - dy|$, resulting from the use of dy instead of Δy.

Solution. The exact change in y is Δy. We shall use dy as an approximation to Δy when $x = 2$ and $\Delta x = dx = 0.04$.

From $y = (x^2 - 1)^{-1/2}$,

$$dy = -\frac{1}{2}(x^2 + 1)^{-3/2}2x\, dx = \frac{-x\, dx}{(x^2 + 1)^{3/2}}.$$

Thus for $x = 2$ and $dx = 0.04$, we have

$$dy = \frac{-2(0.04)}{5^{3/2}} = (-0.0032)\sqrt{5} \doteq -0.007155.$$

Since

$$f''(x) = \frac{2x^2 - 1}{(x + 1)^{5/2}} < \frac{2x^2}{(x^2)^{5/2}} = \frac{2}{x^3} \qquad \text{for} \quad x > 0,$$

the largest value of f'' on the closed interval [2, 2.04] is at $x = 2$—namely, $f''(2) = \frac{2}{8} = \frac{1}{4}$. By taking $M = \frac{1}{4}$, we have $f''(x) \leq M$ for $2 \leq x \leq 2.04$. Then (from 5.10.1)

$$|\Delta y - dy| \leq \tfrac{1}{2}(\Delta x)^2 M = \tfrac{1}{2}(0.02)^2(\tfrac{1}{4}) = 0.00005.$$

Exercises

1. If $y = x^2 - 3$, find and compare the values of Δy and dy when
(a) $x = 2$ and $dx = \Delta x = 0.5$;
(b) $x = 2$ and $dx = \Delta x = 0.2$;
(c) $x = 2$ and $dx = \Delta x = 0.1$;
(d) $x = 2$ and $dx = \Delta x = 0.01$.
Make a large scale drawing of the graph of $y = x^2 - 3$ for $1 \leq x \leq 3$ and draw the tangent to the curve at (2, 1). Then label appropriate line segments to illustrate parts (a) and (b). (See Fig. 5-8.)

2. If $y = x^4$, find and compare the values of Δy and dy when
(a) $x = -0.8$ and $dx = \Delta x = -0.3$;
(b) $x = -0.8$ and $dx = \Delta x = -0.1$.
Make a very large scale drawing of the graph of $y = x^4$ for $-1.2 \leq x \leq 0$ and draw the tangent to the curve at $(-\frac{3}{4}, \frac{81}{256})$. Then label appropriate line segments to illustrate parts (a) and (b). (See Fig. 5-8.)

3. Let $y = 5x^3 - 2x^2 + 6$. Use differentials to find the approximate change in y as x changes from 1 to 1.03 and bound the error by means of 5.10.1.

4. All six sides of a cubical metal box are 0.25 inch thick, and the volume of the interior of the box is 40 cubic inches. Use differentials to find the approximate volume of metal used to make the box.

5. The outside diameter of a thin spherical shell is 12 feet. If the shell is 0.3 inch thick, use differentials to approximate the volume of the interior of the shell.

6. The interior of an open cylindrical tank is 12 feet in diameter and 8 feet deep. The bottom is copper and the sides are steel. Use differentials to find approximately how many gallons of waterproofing paint are needed to apply a 0.05-inch coat to the steel part of the inside of the tank. (One gallon $\doteq$ 231 cubic inches.)

7. Use differentials to approximate $\sqrt{402}$.

8. Use differentials to approximate $\sqrt{35.9}$.

9. Use differentials to approximate the cube root of 28.

10. Use differentials to approximate the fourth root of 15.

11. Use differentials to approximate $2.9\sqrt{24.41}$.

12. Use differentials to approximate $4.01\sqrt{25.0801}$.

13. Assuming that the equator is a circle whose radius is approximately 4000 miles, how much longer than the equator would a concentric, coplanar circle be if each point on it were 2 feet above the equator?

14. The period of a simple pendulum of length L feet is given by $T = 2\pi\sqrt{L/g}$ seconds. We assume that g, the acceleration due to gravity on (or very near) the surface of the earth, is 32 feet per second per second. If the pendulum is that of a clock that keeps good time when $L = 4$ feet, how much time will the clock gain in 24 hours if the length of the pendulum is decreased to 3.97 feet?

5.11 DERIVATIVE OF THE INVERSE OF A FUNCTION

If $y = f(x)$ and $x = f^{-1}(y)$ define inverse functions (Section 2.11) that are differentiable, their derivatives are reciprocals. That is, $dx/dy = 1/(dy/dx)$.

For instance, if $y = f(x) = x^3$ and $x = f^{-1}(y) = y^{1/3}$, then $dy/dx = 3x^2$ and $dx/dy = \frac{1}{3}y^{-2/3} = \frac{1}{3}(x^3)^{-2/3} = 1/(3x^2)$.

These results are hardly surprising. For dy/dx can always be interpreted as the instantaneous rate of change of y with respect to x, and dx/dy as the instantaneous rate of change of x with respect to y. So in $y = f(x)$, if y is changing twice as fast as x, then x is changing half as fast as y.

5.11.1 Theorem. Let $y = f(x)$ define a function f that is differentiable on an interval I and let x_1 be a point in I at which $f'(x_1) \neq 0$. If f has an inverse f^{-1} that is continuous on an open interval J containing the corresponding point $y_1 = f(x_1)$, then $x = f^{-1}(y)$ is differentiable at y_1 and the value of its derivative dx/dy at y_1 is

$$\frac{dx}{dy} = \frac{1}{\dfrac{dy}{dx}}.$$

Proof. Since $y_1 = f(x_1)$, $f^{-1}(y_1) = f^{-1}(f(x_1)) = x_1$ (by 2.11.3). That is,

(1) $$x_1 = f^{-1}(y_1).$$

In the equation $x = f^{-1}(y)$, y is the independent variable. Let Δy be an increment of y such that $y_1 + \Delta y$ is in J. Then $x_1 + \Delta x = f^{-1}(y_1 + \Delta y)$ (Fig. 5-10). From it and from (1),

(2) $$\Delta x = f^{-1}(y_1 + \Delta y) - f^{-1}(y_1).$$

Because f^{-1} is continuous at y_1, $\lim_{\Delta y \to 0} [f^{-1}(y_1 + \Delta y) - f^{-1}(y_1)] = 0$. Thus, in (2), $\lim_{\Delta y \to 0} \Delta x = 0$. That is, $\Delta x \longrightarrow 0$ if $\Delta y \longrightarrow 0$. Therefore

(3) $$\lim_{\Delta y \to 0} \frac{\Delta y}{\Delta x} = \lim_{\Delta x \to 0} \frac{\Delta y}{\Delta x} = \frac{dy}{dx}.$$

If $\Delta y \neq 0$ in (2), $\Delta x \neq 0$, for otherwise $\Delta x = 0$ and $y_1 = f(x_1) = f(x_1 + \Delta x) = y_1 + \Delta y$, from which $\Delta y = 0$, a contradiction. Thus for $\Delta y \neq 0$,

$$\frac{\Delta x}{\Delta y} = \frac{1}{\Delta y/\Delta x}$$

(4)

is an identity.

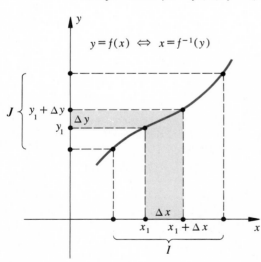

$$y = f(x) \iff x = f^{-1}(y)$$

Figure 5-10

From (3) and (4),

$$\frac{dx}{dy} = \lim_{\Delta y \to 0} \frac{\Delta x}{\Delta y} = \frac{1}{\displaystyle\lim_{\Delta y \to 0} \frac{\Delta y}{\Delta x}} = \frac{1}{\dfrac{dy}{dx}}. \quad \blacksquare$$

Example 1. A function f is defined by $y = f(x) = x^2$, $x > 0$. Find the derivative of the inverse function f^{-1}, defined by $x = f^{-1}(y)$.

Solution. We seek dx/dy. Using the preceding theorem, we first find $dy/dx = 2x$ and then write $dx/dy = 1/2x$, $x > 0$.

Of course, in this example it is very simple to solve $y = x^2$ for $x = y^{1/2}$ and then write $dx/dy = \frac{1}{2}y^{-1/2} = 1/2y^{1/2} = 1/2x$, $x \neq 0$. But for many other functions it is difficult or impossible to solve $y = f(x)$ for x in terms of y.

Example 2. Given $y = x/(x + 1)$, $x \neq -1$, find dx/dy.

Solution. $dy/dx = 1/(x + 1)^2$, $x \neq -1$. Using 5.11.1, we can write $dx/dy = 1/(dy/dx) = (x + 1)^2$, $x \neq -1$.

As in Example 1, it is not difficult to verify this result by solving the given equation, $y = x/(x + 1)$, $x \neq -1$, for x in terms of y. If we multiply both members of this equation by $(x + 1)$ and collect the terms involving x, we obtain $x(y - 1) = -y$, or $x = y/(1 - y)$, $y \neq 1$. Then $dx/dy = 1/(1 - y)^2$. To express this in terms of x, we use the original equation to write $1 - y = 1 - x/(x + 1) = 1/(x + 1)$. Then $dx/dy = 1/(1 - y)^2 = (x + 1)^2$, $x \neq -1$, which verifies our earlier result.

Example 3. From $y = x^3 + x$, find dx/dy.

Solution. $dy/dx = 3x^2 + 1$. Then $dx/dy = 1/(dy/dx) = 1/(3x^2 + 1)$, $x \in R$.

Exercises

In each of Exercises 1 through 10, the indicated function has a continuous inverse on the given interval. Use 5.11.1 to find dx/dy. Then check your result by finding the inverse function explicitly (by solving for x in terms of y), differentiating it with respect to y, and then expressing dx/dy in terms of x.

1. $y = 3x - 1$, $x \in R$.

2. $y = 4x^2$, $x > 0$.

3. $y = 9x^2 - 4$, $x > 0$.

4. $y = x^2 + 6x$, $x > 0$.

5. $y = x^3 - 2$, $x > 0$.

6. $y = \sqrt{2 - x}$, $x < 2$.

7. $y = x^{3/2}$, $x > 0$.

8. $y = (x + 1)^2$, $x > -1$.

9. $y = (x - 4)^3$, $x \neq 4$.

10. $y = \dfrac{2x}{x - 3}$, $x > 3$.

In Exercises 11 to 14, the indicated function is one-to-one on the given interval and has a continuous inverse. Find dx/dy *in terms of* y.

11. $y = 4 - x^2$, $x > 0$.

12. $y = \dfrac{1}{x}$, $x > 0$.

13. $y = \dfrac{x - 1}{x + 1}$, $x \in R$.

14. $y = \sqrt{x^2 + 4x}$, $x > 0$.

In Exercises 15 to 18, $y = f(x)$ and $x = f^{-1}(y)$ define functions that are inverse to each other.

15. If $y = f(0) = 5$ and $dy/dx = \frac{1}{3}$ at $x = 0$, find dx/dy at $y = 5$.

16. If $f(-2) = 4$ and $f'(-2) = 7$, find dx/dy at $y = 4$.

17. If $f(\frac{1}{4}) = -3$ and $f'(\frac{1}{4}) = \frac{2}{11}$, find dx/dy when $y = -3$.

18. If $f(1) = -8$ and $f'(1) = \sqrt{2}$, find dx/dy when $y = -8$.

In Exercises 19 to 22, the function f, defined by $y = f(x)$ on the given interval, has a continuous inverse. Find dx/dy at y_1 and at y_2.

19. $y = x^3 + 4x$, $x \geq 0$. $y_1 = 0$ and $y_2 = 5$. (*Hint:* To use 5.11.1 here, we must find the values of x that make $y = y_1$ and $y = y_2$ in the given equa-

tion. Try substituting some simple nonnegative numbers for x in the given equation.)

20. $y = 2x^3 + x - 5$, $x \in R$. $y_1 = -5$ and $y_2 = 13$.

21. $y = 2x^4 - 4$, $x > 0$. $y_1 = -2$ and $y_2 = 28$.

22. $y = \dfrac{x + 2}{x^2}$, $x > 0$. $y_1 = 3$ and $y_2 = 1$.

23. Let a differentiable function f have a continuous inverse f^{-1}, Show that if the graph of $y = f(x)$ has slope $m \neq 0$ at the point (c, d), then the graph of $y = f^{-1}(x)$ has slope $1/m$ at the point (d, c). (*Hint:* Each curve is the reflection of the other with respect to the line $y = x$. See Section 2.11.)

5.12 REVIEW EXERCISES

1. Find dy/dx if
(a) $y = (x - 2)^3(x + 7)^4$,
(b) $y = (1 - 2x^2)^2(x - 3)^5$.

2. Find $D_x y$ if $y = (x^2 - x + 2)^4(x^2 + 6x - 9)^3$.

3. Find $f'(t)$ if
(a) $f(t) = \dfrac{(3t - 1)^2}{(t + 4)^3}$,
(b) $f(t) = \dfrac{(1 + t^2)^5}{(1 - 5t)^4}$.

4. Find $F'(w)$ if $F(w) = \dfrac{(w^2 - w + 2)^3}{(w^2 - 2w - 1)^5}$.

5. Find dy/dx if
(a) $y = \sqrt{2x^2 - 3}$,
(b) $y = \sqrt{\dfrac{1 - x}{1 + x}}$.

6. Find the points on the curve $4y = 2x^3 - 3x^2 - 12x + 8$ at which the tangent lines are horizontal. Sketch the curve for $-3 \leq x \leq 3$, and draw the horizontal tangents.

7. Assume that a soap bubble retains its spherical shape as air is being blown into it at the rate of 3 cubic inches per second. Find
(a) how fast the radius is increasing when the radius is 2 inches.
(b) how fast the surface area of the bubble is increasing when the radius is 2 inches. [*Hint:* Use the answer to (a) in finding (b).]

8. Find the first five derivatives of
(a) $f(x) = 3x^2 - x + 5$;
(b) $f(x) = x^3 + 2x^2 - 4x - 1$;
(c) $f(x) = 2x^5 - 3x^3 + 8$.

9. What is the 20th derivative of
(a) $F(x) = 13x^{19} - 2x^{12} - 6x^5 + 18$?
(b) $G(x) = \dfrac{1}{x}$?

10. Find dy/dx from $x^3 + y^3 = x^3y^3$.

11. Find dy/dx if $x^{2/3} + y^{2/3} = 1$.

12. If $F(t) = (1 - t)/(1 + t)$,
(a) find $F'(t)$, $F''(t)$, and $F'''(t)$;
(b) then conjecture what $F^{(n)}(t)$ is.

13. Find the equations of the tangent lines to the curve $3y = x^3 + 3x^2 + 6x - 3$ that have slope 2. Sketch the curve for $-3 \leq x \leq 2$, and draw the required tangents.

14. Find dx/dy if
(a) $y = x^3 - 6x + 3$;
(b) $y = \dfrac{2x - 7}{3x + 1}$, $x \neq -\dfrac{1}{3}$.

15. A trough, 12 feet long, has a cross section in the form of an isoceles triangle 4 feet deep and 6 feet across. If water is filling the trough at the rate of 9 cubic feet per minute, how fast is the water level rising when the water is 3 feet deep?

16. If $x^5 + y^5 = (x + y)^2 - (x - y)^2$, find $D_x y$.

17. Show that the tangents to the curves $y^2 = 4x^3$ and $2x^2 + 3y^2 = 14$. at $x = 1$ are perpendicular to each other. Make a sketch. (*Hint:* Use implicit differentiation.)

18. A one-to-one function has an inverse (Section 2.11). The graph of a one-to-one function f is cut by any vertical line $x = h$, $h \in \mathfrak{D}_f$, in exactly one point, and it is also cut in exactly one point by any horizontal line $y = k$, $k \in \mathfrak{R}_f$. Sketch a curve that has these two properties on some interval $a \leq x \leq b$.

19. If $w = 8(u - 5)(u^3 - 2u + 6)$, find dw.

20. If $2xy^2 = 3x - 1$ defines an implicit function of x, use differentials to find dy/dx.

21. Given $x^3 + y^3 = 1$, use differentials to find dx/dy.

22. The altitude of a right circular cylinder is 10 inches. If the radius of the base changes from 2 inches to 2.06 inches, use differentials to compute the approximate corresponding change in the volume of the cylinder and find bounds for the error.

23. If $\Phi(t) = |t^2 - 4|$, find $\Phi'(t)$. What are the domains of Φ and Φ'? Sketch the graph of Φ for $-3 \leq t \leq 3$.

24. Show that if f is a polynomial function and $f(x) = 0$ has a k-fold root at $x = a$, $k \in N$, then $f'(x) = 0$ has a $(k - 1)$-fold root at a. [*Hint:* $f(x) = (x - a)^k g(x)$, where $g(x)$ is a polynomial.]

6

further applications of the derivative

From the formulas developed in the preceding chapter and the experience gained in using them, we can now differentiate very many functions with comparative ease. It is time to apply that skill.

We know that the derivative of a function gives the instantaneous rate of change of the function with respect to the independent variable; and we have seen that two interpretations of this are the slope of the tangent to the graph of the function and the velocity of a particle moving on a straight line. Both of these applications of the derivative will be explored further.

One of the more interesting and useful applications of the derivative to be discussed in this chapter is finding maximum and minimum values of a function. In many practical problems in engineering, science, geometry, and economics, the proper conditions for some varying quantity to be a maximum or minimum are sought. As a simple example consider finding the dimensions for a cylindrical can of given volume that require the least material. As another example, where should a book be placed on a table, between two lamps of different intensities, for maximum illumination? Again, in the manufacture of a certain article, how should stated controllable conditions be adjusted for maximum profit?

An accurately drawn graph of a function can tell us much about the behavior of the function. But plotting a sufficiently large number of points to ensure an accurate curve can be a boring, tedious task. In this chapter we shall see that the first and second derivatives of a function are powerful aids in sketching its graph. Before plotting a single point, they enable us to know where the curve is rising and where it is falling, where it is concave upward and where it is concave downward, and at what points the tops of the "waves" and the bottoms of the "troughs" occur. Furthermore, by an extension of the definition of limit, we shall be able to find any asymptotes that the curve may have. With all that information at our disposal, it is seldom necessary to plot more than a few selected points to draw an accurate graph.

For these and other purposes, several far-reaching theorems on continuous functions will first be established, among them Rolle's theorem and the mean value theorem. Both of these theorems depend on the concept of the maximum and minimum values of a function on an interval.

6.1 ROLLE'S THEOREM

6.1.1 Definition.   Let f be a function that is defined on a set S containing a point c.

(i) $f(c)$ is the **maximum** of f on S if $f(c) \geq f(x)$ for all points x in S.
(ii) $f(c)$ is the **minimum** of f on S if $f(c) \leq f(x)$ for all $x \in S$.
(iii) $f(c)$ is an **extreme value** of f on S if either (i) or (ii) holds.

Some functions have a maximum or a minimum on a set, others do not. The function defined by $f(x) = x^3$ has the maximum 8 on the closed interval [0, 2] and the minimum 0; yet on the open interval (0, 2) it has neither a maximum nor a minimum (Fig. 6-1). The following fundamental theorem guarantees the existence of a maximum and a minimum if the function is *continuous* and if the set is a *closed interval*.

6.1.2 Maximum–Minimum Theorem.   If a function f is continuous on a closed interval [a, b], there are numbers x_1 and x_2 in [a, b] such that $f(x_1)$ is the maximum value of f on [a, b] and $f(x_2)$ is the minimum value of f on [a, b].

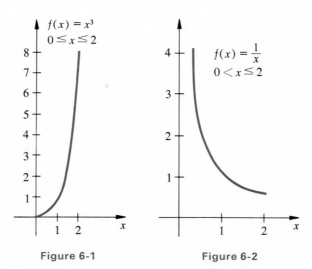

Figure 6-1   Figure 6-2

A proof of this theorem is given in the Appendix (A.2.4 and A.2.5). Notice that the conclusion of this theorem need not hold if the interval is not closed. To illustrate, the function whose value is given by $1/x$ is continuous on the half-open interval $(0, 1]$ but has no maximum value there (Fig. 6-2).

Example 1. Let f be the function defined by $f(x) = -2x^3 + 3x^2$. On the closed interval $[-\frac{1}{2}, \frac{3}{2}]$, its maximum value is 1 and its minimum value is zero (Fig. 6-3). The maximum value 1 occurs at both $x = -\frac{1}{2}$ and $x = 1$, and the minimum occurs at $x = 0$ and $x = \frac{3}{2}$. So, while the maximum–minimum theorem guarantees the existence of *at least* one point, x_1, in the closed interval at which the maximum value occurs, there may be several points in the interval at which the function has the (same) maximum value.

Example 2. The function F, defined by $F(x) = x^{2/3}$, is continuous everywhere On the closed interval $[-1, 2]$ the maximum value of F is $F(2) = \sqrt[3]{4}$ and the minimum is $F(0) = 0$ (Fig. 6-4).

Notice in Example 1 (Fig. 6-3) that when an extreme value occurs at an *interior* point of $[-\frac{1}{2}, \frac{3}{2}]$, the tangent to the graph is horizontal there: thus $f'(0) = 0$ and $f'(1) = 0$. In Example 2, however, at the interior point of the interval $[-1, 2]$ at which the extreme value occurs, the tangent is vertical and thus $F'(0)$ fails to exist (Fig. 6-4). These examples illustrate the following basic theorem, which is due to Fermat.

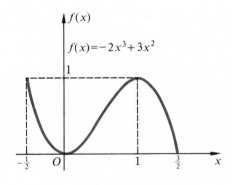

Figure 6-3

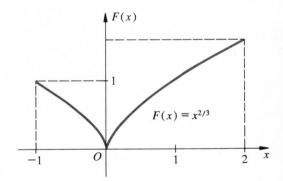

Figure 6-4

6.1.3 **Theorem.** Let f be defined on an interval I and let c be an **interior point** of I. If $f(c)$ is the maximum or minimum of f on I, then either $f'(c) = 0$ or $f'(c)$ fails to exist.

Proof. If $f(c)$ is the maximum value of f on I, then (by 6.1.1) $f(c) \geq f(x)$ for all $x \in I$; that is,

$$f(x) - f(c) \leq 0$$

for all $x \in I$.

For $x < c$, $x - c < 0$ and

$$\frac{f(x) - f(c)}{x - c} \geq 0.$$

Therefore, by 3.3.8,

(1) $$\lim_{x \to c} \frac{f(x) - f(c)}{x - c}$$

cannot be negative. For $x > c$, $x - c > 0$ and

$$\frac{f(x) - f(c)}{x - c} \le 0;$$

therefore, by 3.3.7, the limit (1) cannot be positive.

Since the limit (1) cannot be negative and cannot be positive, either (1) is zero or it fails to exist. That is, $f'(c) = 0$ or else $f'(c)$ does not exist.

The proof is analogous when $f(c)$ is the minimum of f on I. ∎

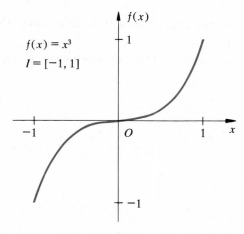

$f(x) = x^3$

$I = [-1, 1]$

Figure 6-5

The converse of 6.1.3 is not true. A function f can be defined on an interval I and $f'(c)$ can be zero for some interior point c of I, and yet $f(c)$ may not be an extreme value of f on I. This case is shown by $f(x) = x^3$ (Fig. 6-5) on $I = [-1, 1]$; $f'(0) = 0$ and 0 is an interior point of I, and yet $f(0)$ is neither a maximum nor a minimum of f on $[-1, 1]$. Thus 6.1.3 furnishes a *necessary* condition for a function, defined on an interval, to have an extreme value at an interior point of the interval but not a *sufficient* condition.

We are now prepared for a theorem on continuous functions that is basic in our development of the calculus. It is called Rolle's theorem in honor of its discoverer, the French mathematician, Michel Rolle (1652–1719).

In geometric terms, Rolle's theorem records the simple fact that if a continuous function f possesses a derivative at each point between $x = a$ and $x = b$ and if $f(a) = f(b)$, then there is at least one point on the graph between $x = a$ and $x = b$ where the tangent is horizontal (Fig. 6-6).

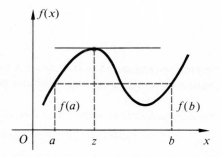

Figure 6-6

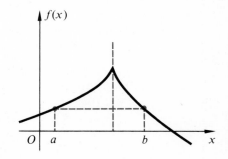

Figure 6-7

Of course, if the derivative of the function fails to exist at some point between $x = a$ and $x = b$, there may be no horizontal tangent even though the function is continuous and $f(a) = f(b)$ (Fig. 6-7).

6.1.4 Rolle's Theorem. If a function f is continuous on a closed interval $[a, b]$ and is differentiable on the open interval (a, b), and if $f(a) = f(b)$, then there exists a number z between a and b, $a < z < b$, such that $f'(z) = 0$.

Proof. If $f(x) = f(a)$ for all x in $[a, b]$, then f is a constant function and $f'(z) = 0$ for every number z in (a, b).

If $f(x) > f(a)$ for some x in (a, b), then the maximum of the continuous function f on the closed interval $[a, b]$, guaranteed to exist by 6.1.2, cannot be at either of the endpoints, a or b. Thus there exists some *interior* point z of $[a, b]$ such that $f(z)$ is the maximum of f on $[a, b]$ (Fig. 6-6). Since f is differentiable on (a, b) by hypothesis, and z is an interior point of $[a, b]$, $f'(z) = 0$ (by 6.1.3).

If $f(x)$ is not greater than $f(a)$ for any x in (a, b), but $f(x) < f(a)$ for some x in (a, b), the proof is analogous to the above. ∎

Example 3. Let f be defined by $f(x) = x^2 - x - 2$, $-2 \leq x \leq 3$ (see Fig. 6-8). Then $f'(x) = 2x - 1$. Since f and f' are polynomials, f is continuous on closed $[-2, 3]$ and f' exists on open $(-2, 3)$. Moreover, $f(-2) = f(3) = 4$. Thus the hypotheses of Rolle's theorem are satisfied by f on $[-2, 3]$, and there must be a number z between -2 and 3 such that $f'(z) = 0$. To verify this, we set $f'(x) = 2x - 1 = 0$ and find that $f'(\frac{1}{2}) = 0$. Therefore $z = \frac{1}{2}$.

The disarming simplicity of Rolle's theorem gives little indication of its importance, but we shall use it to obtain some very important results. That each of the three

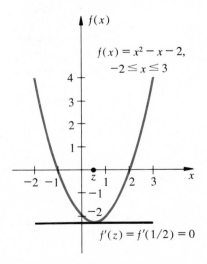

Figure 6-8

hypotheses stated in Rolle's theorem is essential will be shown in Exercises 11, 12, and 13.

Example 4. Use Rolle's theorem to find the number of real roots of the equation $f(x) = 3 - x - x^2 = 0$, and locate each nonintegral root between successive integers.

Solution. $f'(x) = -1 - 2x$. Since f and f' are polynomial functions, they are continuous everywhere. Also, $f'(x) = 0$ only at $x = -\frac{1}{2}$; so $f'(x)$ cannot be zero at any number in the open intervals $(-\infty, -\frac{1}{2})$ and $(-\frac{1}{2}, \infty)$.

It then follows from Rolle's theorem that $f(x) = 0$ cannot have more than one root in either of the open intervals $(-\infty, -\frac{1}{2})$ and $(-\frac{1}{2}, \infty)$. Otherwise there would be two numbers a and b in one of the intervals, say in $(-\infty, -\frac{1}{2})$, such that $f(a) = f(b) = 0$ and thus, by Rolle's theorem, there would be a number z between a and b in $(-\infty, -\frac{1}{2})$ such that $f'(z) = 0$, a contradiction.

By trial we find that $f(-3) = -3$ and $f(-2) = 1$; so it follows from the intermediate value theorem (3.5.8) that there is a number x between -3 and -2 for which $f(x) = 0$. Therefore there is one and only one root of $f(x) = 0$ in $(-\infty, -\frac{1}{2})$.

Similarly, since $f(1) = 1$ and $f(2) = -3$, there is a root of $f(x) = 0$ between 1 and 2. This is the only root in $(-\frac{1}{2}, \infty)$.

Finally, $f(-\frac{1}{2}) = \frac{13}{4}$. So $f(x) = 0$ has exactly two real roots, one between -3 and -2, the other between 1 and 2 (Fig. 6-9).

Of course, this information about a quadratic equation could have been obtained more easily from the quadratic formula, but the method explained here can be used on many equations more complicated than a quadratic.

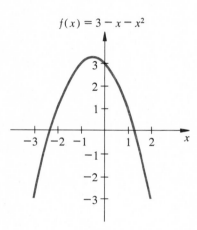

$f(x) = 3 - x - x^2$

Figure 6-9

Example 5. Find the number of real roots of the equation $2x^3 - 9x^2 + 1 = 0$ and locate each nonintegral root between two successive integers.

Solution. Let $f(x) = 2x^3 - 9x^2 + 1$. Then

$$f'(x) = 6x^2 - 18x = 6x(x - 3).$$

Since f and f' are polynomial functions, they are continuous everywhere. Moreover, $f'(x) = 0$ only for $x = 0$ and $x = 3$. Thus $f'(x)$ cannot be zero at any point in the open intervals $(-\infty, 0)$, $(0, 3)$, and $(3, \infty)$.

It follows from this, and from Rolle's theorem, that $f(x) = 0$ cannot have more than one root in any one of the open intervals $(-\infty, 0)$, $(0, 3)$, and $(3, \infty)$. For otherwise there would be two numbers, a and b, in one of the three intervals, say in $(-\infty, 0)$, such that $f(a) = f(b) = 0$, and thus, by Rolle's theorem, there would be a number z between a and b, and consequently in $(-\infty, 0)$, such that $f'(z) = 0$, a contradiction.

Since $f(0) = 1$ and $f(3) = -26$, it follows from the intermediate value theorem (3.5.8) that there is a value of x between 0 and 3 for which $f(x) = 0$. Therefore there is one and only one root of $f(x) = 0$ in $[0, 3]$.

Since $f(-1) = -10$ and $f(0) = 1$, the intermediate value theorem ensures that $f(x) = 0$ for some x between -1 and 0. Thus $f(x) = 0$ has one and only one root in $(-\infty, 0]$. Similarly, $f(3) = -26$ and $f(5) = 26$. Thus $f(x) = 0$ for some number x between 3 and 5.

This analysis shows that the given equation has exactly three real roots. By substituting successive integers for x in $f(x)$ and using the intermediate value theorem, we find the roots to be in $(-1, 0)$, $(0, 1)$,and $(4, 5)$.

Exercises

In each of Exercises 1 to 4, a function is defined and a closed interval I is given. Find the extreme values of the function on I. Afterward, sketch the graph of the given function on I.

1. $f(x) = -x^2 + 4x - 1$, $I = [0, 3]$.

2. $f(x) = x^2 + 3x$, $I = [-2, 1]$.

3. $G(x) = \frac{1}{3}(2x^3 + 3x^2 - 12x)$, $I = [-3, 3]$.

4. $h(w) = 4w^3 + 3w^2 - 6w + 1$, $I = [-2, 1]$.

5. Let $F(x) = x^3 - 6x^2 + 12x - 8$. Find all numbers x for which $F'(x) = 0$. Use 6.1.3 to determine whether F has any extreme value on the *open* interval $(0, 3)$. If not, explain.

6. The function f, defined by $f(x) = 2 - |x|$, exists for all $x \in R$. It has one extreme value on the open interval $(-2, 2)$. Use 6.1.3 to find that extreme value. Then make a sketch. [*Hint:* Find $f'(x)$ for $-2 < x < 0$ and for $0 < x < 2$. Does $f'(x) = 0$ for any x in $(-2, 0) \cup (0, 2)$? Does $f'(0)$ exist?]

7. Let $F(x) = |x|^{1/2}$. The function F is defined for all $x \in R$ and has an extreme value on the *open* interval $(-2, 4)$. Use 6.1.3 to find that extreme value. Then make a sketch.

In Exercises 8, 9, and 10, a function is defined and a closed interval is given. Verify that the given function satisfies the three conditions for Rolle's theorem on the given interval and find all possible values of z. Make a sketch.

8. $f(x) = 2x^2 - 4x + 3$, $[-1, 3]$.

9. $\phi(x) = x^2 - x - 2$, $[-1, 2]$.

10. $g(x) = x^3 - 9x$, $[-3, 3]$.

The three hypotheses stated in Rolle's theorem are

  (i) f is continuous on a closed interval $[a, b]$,
  (ii) f' exists on the open interval (a, b), and
  (iii) $f(a) = f(b)$.

In each of Exercises 11, 12, and 13, a function f is defined on a closed interval $[a, b]$ and satisfies two of the foregoing three hypotheses. Is there a number z between a and b such that $f'(z) = 0$? If not, tell which of these hypotheses f fails to satisfy. Make a sketch.

11. A function f is defined on $[a, b] = [0, 2]$ by $f(x) = 1/x$ for $x \in (0, 2]$ and $f(0) = \frac{1}{2}$. [*Hint:* See 3.5.7.]

12. $f(x) = x^{2/3}$, $x \in [-1, 1] = [a, b]$.

13. $f(x) = x^2$, $x \in [1, 2] = [a, b]$.

In Exercises 14 to 17, tell how many real roots the given equation has, and locate each nonintegral root between successive integers.

14. $3x^2 - 24x + 40 = 0$.

15. $x^2 + 2x - 2 = 0$.

16. $2x^3 - 3x^2 - 12x + 1 = 0$.

17. $2x^5 + 5x^4 + 1 = 0$.

18. Prove the following theorem. Let f be a function that has a continuous derivative on $[a, b]$. If $f(a)$ and $f(b)$ have opposite signs and if $f'(x) \neq 0$ for all numbers x between a and b, $a < x < b$, then the equation $f(x) = 0$ has one and only one root between a and b. (*Hint:* Use the intermediate value theorem and Rolle's theorem.)

6.2 THE MEAN VALUE THEOREM

The celebrated **mean value theorem** is an extension of Rolle's theorem, or, what is the same thing, Rolle's theorem is a special case of the mean value theorem.

In geometric terms, the mean value theorem states that if the graph of a continuous function has a tangent at every point between A and B (Fig. 6-10), then there

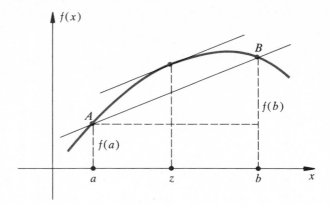

Figure 6-10

is at least one point on the curve between A and B at which the tangent is parallel to the line AB.

Since $[f(b) - f(a)]/(b - a)$ is the slope of the line connecting the points A and B in Fig. 6-10, and $f'(x)$ is the slope of the tangent to the curve at any point $(x, f(x))$ on the curve between A and B, an analytic formulation of the preceding statement is that there is a point z in the open interval (a, b) such that

$$\frac{f(b) - f(a)}{b - a} = f'(z)$$

or

$$f(b) - f(a) = (b - a) \cdot f'(z).$$

This makes it easy to remember the mean value theorem but gives no indication of its monumental importance in calculus. In time, the indispensable part that the mean value theorem plays in the development of the calculus will become evident to the reader.

Notice that when $f(a) = f(b)$, the mean value theorem becomes Rolle's theorem, since the line AB is horizontal.

6.2.1 **Mean Value Theorem.** If f is continuous on the closed interval $[a, b]$ and differentiable on the open interval (a, b), then there exists a number z between a and b, $a < z < b$, such that

(1) $$f(b) - f(a) = (b - a)f'(z).$$

Proof. We define a function F with domain $[a, b]$ by

$$F(x) = f(x) - f(a) - \frac{f(b) - f(a)}{b - a}(x - a).$$

Notice that F is continuous on the closed interval $[a, b]$ and differentiable on the open interval (a, b) because f is, and also that $F(a) = F(b) = 0$. Thus F fulfills all the conditions of Rolle's theorem, and so there exists a number z between a and b such that

$$F'(z) = f'(z) - \frac{f(b) - f(a)}{b - a} = 0.$$

Consequently,

$$f'(z) = \frac{f(b) - f(a)}{b - a}$$

or

$$f(b) - f(a) = (b - a)f'(z). \quad \blacksquare$$

Example. Let $f(x) = 2\sqrt{x}$; then $f'(x) = 1/\sqrt{x}$. Since f is continuous, and f' exists, for all $x > 0$, f satisfies the conditions for the mean value theorem on the closed interval $[1, 4]$. Thus there is a number z between 1 and 4 such that

$$f(4) - f(1) = [4 - 1]\frac{1}{\sqrt{z}}$$

or

$$4 - 2 = 3\left(\frac{1}{\sqrt{z}}\right),$$

from which $\sqrt{z} = \frac{3}{2}$ and $z = \frac{9}{4}$.

In Fig. 6-11, $[f(4) - f(1)]/(4 - 1) = \frac{2}{3}$ is the slope of the line AB, and $f'(\frac{9}{4}) = \frac{2}{3}$ is the slope of the tangent to the curve at T.

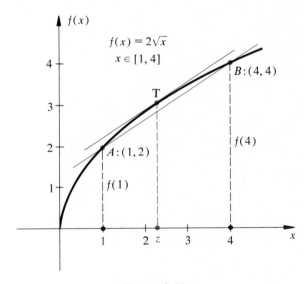

Figure 6-11

In reading the proof of the mean value theorem for the first time, an alert reader may wonder how the seemingly artificial function F, which so neatly fits the conditions of Rolle's theorem, was found. What suggested the function F? Exercises 13 to 16 below provide an answer.

If we let $b - a = \Delta x$ in 6.2.1 and notice that $a < z < b$ is equivalent to $z = a + \lambda \Delta x$, where $0 < \lambda < 1$, then $f(b) - f(a) = (b - a)f'(z)$ can be expressed as $f(a + \Delta x) - f(a) = \Delta x \cdot f'(a + \lambda \Delta x)$. This leads to the following **alternate form** of the mean value theorem, which will be used later on.

6.2.2 Alternate Mean Value Theorem. If f is continuous on the closed interval I whose endpoints are a and $a + \Delta x$, $\Delta x \neq 0$, and if f is differentiable at every interior point of I, then there is a number λ between 0 and 1, $0 < \lambda < 1$, such that

$$f(a + \Delta x) - f(a) = \Delta x \cdot f'(a + \lambda \Delta x).$$

This alternate form of the mean value theorem *is also true when* $\Delta x < 0$. Proof of this statement is left as an exercise.

Exercises

In each of Exercises 1 through 10, a function is defined and a closed interval is given. Decide whether the mean value theorem applies to the given function on the given interval; if so, find all possible values of z; if not, state the reason. In each exercise, sketch the graph of the given function on the given interval.

1. $f(x) = x^2 + 2x$, $[-2, 2]$.
2. $f(x) = x^2 + 3x - 1$, $[-3, 1]$.
3. $g(x) = \dfrac{x^3}{3}$, $[-2, 2]$.
4. $g(x) = \dfrac{1}{3}(x^3 + x - 4)$, $[-1, 2]$.
5. $F(t) = \dfrac{t + 3}{t - 3}$, $[-1, 4]$.
6. $F(x) = \dfrac{x + 1}{x - 1}$, $\left[\dfrac{3}{2}, 5\right]$.
7. $h(x) = x^{2/3}$, $[0, 2]$.
8. $h(x) = x^{2/3}$, $[-2, 2]$.
9. $\phi(x) = x + \dfrac{1}{x}$, $\left[-1, \dfrac{1}{2}\right]$.
10. $\phi(x) = x + \dfrac{1}{x}$, $\left[\dfrac{1}{2}, \dfrac{3}{2}\right]$.

11. Show that if f is the quadratic function defined by $f(x) = cx^2 + dx + e$, $c \neq 0$, then the z of the mean value theorem is always the midpoint of any given closed interval $[a, b]$.

12. Prove that the alternate form of the mean value theorem (6.2.2) is also valid when Δx is negative.

(*Hint:* The closed interval is $[a + \Delta x, a]$ when $\Delta x < 0$.)

Exercises 13 to 16 provide a *motivation* for the choice of the function F in the proof of the mean value theorem.

13. In the proof of the mean value theorem, $A:(a, f(a))$ and $B:(b, f(b))$ are two points on the graph of f (Fig. 6-12). Show that the constant

$$\frac{f(b) - f(a)}{b - a}$$

is the slope of the line AB.

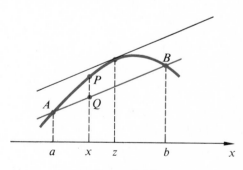

Figure 6-12

14. Show that the equation of the line AB that passes through $A:(a, f(a))$ and has the slope

$$\frac{f(b) - f(a)}{b - a}$$

can be written

$$y = f(a) + \frac{f(b) - f(a)}{b - a}(x - a),$$

where (x, y) is any point on AB.

15. Let $P:(x, f(x))$ be any point on the graph of f such that $a \le x \le b$ (Fig. 6-12). Drop a perpendicular from P to the x axis, intersecting the line AB in some point Q. Show that the directed distance from Q to P is

$$\overline{QP} = f(x) - f(a) - \frac{f(b) - f(a)}{b - a}(x - a)$$

and thus that the function F, defined in the proof of the mean value theorem, has for its value at any point x in $[a, b]$ the vertical distance $\overline{QP}$ from the line AB to the graph of f. [*Hint:* The y coordinate of P is $f(x)$ and, from Exercise 14, the y coordinate of Q is

$$f(a) + \frac{f(b) - f(a)}{b - a}(x - a).]$$

16. Figure 6-12 suggests that the tangent to the graph of f at P will be parallel to the secant AB when $\overline{QP} = F(x)$ is a maximum. But F is continuous on $[a, b]$ and differentiable on (a, b) because f is, and thus a necessary condition for F to have an extreme value at an interior point z of $[a, b]$ is $F'(z) = 0$ (by 6.1.3). Now notice in Fig. 6-12 that $F(a) = F(b) = 0$. So F fulfills the conditions for Rolle's theorem, and the existence of the desired $z \in (a, b)$ such that $F'(z) = 0$ is ensured. This completes a *motivation* for the choice of the function F in the proof of the mean

value theorem. Incidentally, there are other simple geometric motivations.

17. This extension of the mean value theorem is sometimes useful.

Extended Mean Value Theorem. If a function f and its first derivative f' are continuous on a closed interval $[a, b]$, and if f' is differentiable on open (a, b), then there is a number z between a and b such that

$$f(b) = f(a) + (b - a)f'(a) + \frac{(b - a)^2}{2}f''(z).$$

Complete its proof by filling in the details of the following outline.

Outline of Proof. Define a constant K by

$$(1) \quad f(b) = f(a) + (b - a)f'(a) + \frac{(b - a)^2}{2}K,$$

and let F be the function defined on $[a, b]$ by

$$F(x) = f(b) - f(x) - (b - x)f'(x) - \frac{(b - x)^2}{2}K.$$

Then verify that

$$(2) \qquad F'(x) = -(b - x)f''(x) + (b - x)K.$$

Show that F fulfills the conditions of Rolle's theorem and hence that there exists a number z between a and b such that $F'(z) = 0$. From this, and from (2), $F'(z) = -(b - z)f''(z) + (b - z)K = 0$; thus $K = f''(z)$. Conclude the proof by substituting $K = f''(z)$ in (1).

6.3 INCREASING FUNCTIONS AND DECREASING FUNCTIONS

Consider the function f whose graph is shown in Fig. 6-13. If we think of a particle P as moving along this curve from A to B, the particle rises. Moving from B to C along the curve, the particle falls; and from C to D the particle rises again. Thus the function f is increasing to the right from A to B, decreasing to the right from B to C, and increasing to the right from C to D.

The essence of the concept of a function f increasing on an interval I is that the larger we choose the number x in I, the greater the value of the function, $f(x)$, will be. That is, if x_1 and x_2 are any two numbers in I and $x_1 < x_2$, then $f(x_1) < f(x_2)$ (see Fig. 6-14).

For instance, the quadratic function, $f(x) = x^2$, is increasing on the interval

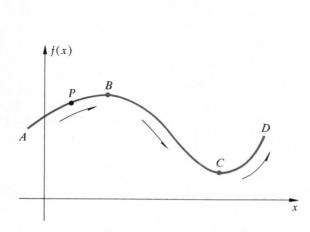

Figure 6-13

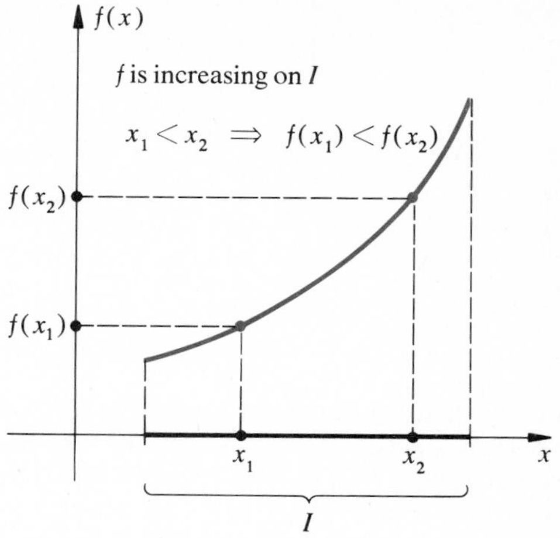

Figure 6-14

$[0, \infty)$ because for any two nonnegative numbers, x_1 and x_2, $x_1 < x_2 \Longrightarrow x_1^2 < x_2^2$ (by 1.5.10); that is, $0 \leq x_1 < x_2 \Longrightarrow f(x_1) < f(x_2)$ (see Fig. 4-3).

6.3.1 Definition. Let f be a function that is defined on an interval I (open, closed, or neither). Then (Fig. 6-15)

(i) f is **increasing** on I if and only if for every pair of numbers, x_1 and x_2, in I,
$$x_1 < x_2 \Longrightarrow f(x_1) < f(x_2).$$

(ii) f is **decreasing** on I if and only if for every pair of numbers, x_1 and x_2, in I,
$$x_1 < x_2 \Longrightarrow f(x_1) > f(x_2).$$

(iii) f is **nondecreasing** on I if and only if for every pair of numbers, x_1 and x_2, in I,
$$x_1 < x_2 \Longrightarrow f(x_1) \leq f(x_2).$$

(iv) f is **nonincreasing** on I if and only if for every pair of numbers, x_1 and x_2, in I,
$$x_1 < x_2 \Longrightarrow f(x_1) \geq f(x_2).$$

If any of these four properties holds, f is **monotonic** on I. If f is increasing on I or if f is decreasing on I, f is **strictly monotonic** on I.

We saw in 2.11.1 that a function f is *one-to-one* on an interval I if for every pair of numbers x_1 and x_2 in I, $x_1 \neq x_2$ implies $f(x_1) \neq f(x_2)$. Clearly, from 6.3.1, if a function is increasing on I, it is one-to-one on I; and if a function is decreasing on I, it is one-to-one on I.

6.3.2 Corollary. A function that is increasing on an interval I is one-to-one on I, and a function that is decreasing on I is one-to-one on I.

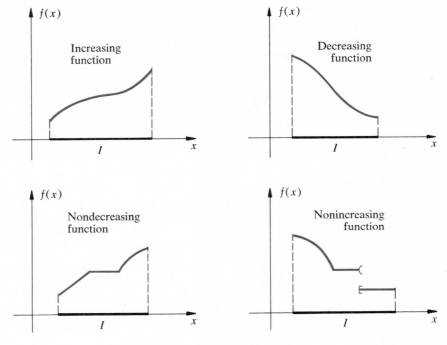

Figure 6-15

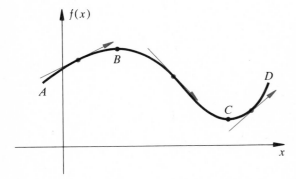

Figure 6-16

When a function is differentiable on an interval, there is a simple test for finding where it is increasing and where it is decreasing. Let Fig. 6-16 represent the graph of a differentiable function f. Since f is differentiable, its graph has a tangent line at each point on it. We observe that on AB and CD, where f is increasing, the tangent line always has a positive slope (that is, the tangent points upward to the right), and on BC, where f is decreasing, its tangent line has a negative slope (it points downward to the right). Since the slope of the tangent to the curve at any point $(x, f(x))$ is $f'(x)$, this observation suggests that f is increasing where $f'(x) > 0$ and decreasing where $f'(x) < 0$.

6.3.3 Theorem. Let f be a function that is continuous on an interval I (open, closed, or neither) and differentiable at every interior point of I.

(i)   If $f'(x) > 0$ for all x interior to I, f is **increasing** on I.

(ii) If $f'(x) < 0$ for all interior points x of I, f is **decreasing** on I.

Proof of (i).   Assume that $f'(x) > 0$ for all points x interior to I. Let x_1 and x_2 be any two numbers of I such that $x_1 < x_2$. Then by the mean value theorem (6.2.1) there is some number z between x_1 and x_2, $x_1 < z < x_2$, such that

(1) $$f(x_2) - f(x_1) = (x_2 - x_1) \cdot f'(z).$$

But $f'(z) > 0$ because z is an interior point of I, and $x_1 < x_2$ implies that $x_2 - x_1 > 0$. So it follows from (1) that $f(x_2) - f(x_1) > 0$ or, equivalently, $f(x_1) < f(x_2)$. Since x_1 and x_2 are any two numbers in I with $x_1 < x_2$, f is increasing on I (by 6.3.1).
      The proof of (ii) is analogous. ∎

Example 1.   If $f(x) = 3 - x - x^2$, find the interval on which f is increasing and the interval on which f is decreasing.

Solution. $f'(x) = -1 - 2x$. Thus $-\frac{1}{2}$ is the only number x for which $f'(x) = 0$. The polynomial function f is continuous everywhere and, in particular, on the intervals $(-\infty, -\frac{1}{2}]$ and $[-\frac{1}{2}, \infty)$. Let us examine the behavior of $f'(x)$ at the interior points of these intervals.
      Since $x < -\frac{1}{2} \iff 2x < -1 \iff 0 < -1 - 2x \iff -1 - 2x > 0$, it follows that $f'(x) = -1 - 2x > 0$ for all $x < -\frac{1}{2}$. Therefore, by 6.3.3, f is increasing on the interval $(-\infty, -\frac{1}{2}]$ (see Fig. 6-9).
      Similarly, $x > -\frac{1}{2} \iff -1 - 2x < 0$; so $f'(x) = -1 - 2x < 0$ for all $x > -\frac{1}{2}$. Thus (6.3.3) f is decreasing on $[-\frac{1}{2}, \infty)$.

Example 2.   Let $F(x) = 2x^3 - 3x^2 - 12x + 7$. Find where F is increasing and where F is decreasing.

Solution. $F'(x) = 6x^2 - 6x - 12 = 6(x + 1)(x - 2)$. Thus $F'(x) = 0$ only at $x = -1$ and $x = 2$. The given function F is continuous on each of the three intervals $(-\infty, -1]$, $[-1, 2]$, and $[2, \infty)$. Let us examine $F'(x)$ for $x < -1$, for $-1 < x < 2$, and for $x > 2$ (Fig. 6-17).

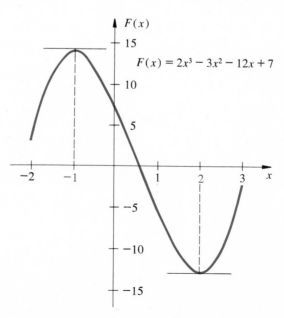

Figure 6-17

Clearly, for $x < -1$, we have $(x + 1) < 0$ and $(x - 2) < 0$; so

$$F'(x) = 6(x + 1)(x - 2) > 0$$

for all $x < -1$. Thus F is increasing on $(-\infty, -1]$.

For $-1 < x < 2$, $(x + 1) > 0$ and $(x - 2) < 0$. Consequently,

$$F'(x) = 6(x + 1)(x - 2) < 0,$$

and F is decreasing on $[-1, 2]$.

Finally, for $x > 2$, $(x + 1) > 0$ and $(x - 2) > 0$; so $F'(x) = 6(x + 1)(x - 2) > 0$ for all $x > 2$. Therefore F is increasing on $[2, \infty)$.

Many of the differentiable functions studied in a first course in calculus have *continuous* first derivatives. The following theorem and its corollary, which are easy to prove, can reduce considerably the labor involved in finding where such a function is increasing and where it is decreasing.

6.3.4 **Theorem.** If f' is continuous on an interval I (open, closed, or neither) and if $f'(x) \neq 0$ for all interior points x of I, then either f is increasing throughout I or else f is decreasing on I.

Proof. We shall prove that either $f'(x) > 0$ for all interior points x of I or else $f'(x) < 0$ for all interior points x of I.

Assume the contrary—namely, that there are two interior points, x_1 and x_2, of I, with $x_1 < x_2$, such that $f'(x_1)$ and $f'(x_2)$ have opposite signs. Since f' is continuous on $[x_1, x_2]$, it follows from the intermediate value theorem (3.5.8) that there is a number c between x_1 and x_2, $x_1 < c < x_2$, for which $f'(c) = 0$. But this is contrary to the hypothesis. Therefore there is no pair of numbers, x_1 and x_2, interior to I, such that $f'(x_1)$ and $f'(x_2)$ have opposite signs.

Since $f'(x)$ exists and is nonzero for all interior points x of I, either $f'(x) > 0$ at all interior points of I or $f'(x) < 0$ at all interior points of I. So f is increasing throughout I or f is decreasing on all of I. ∎

From the proof of 6.3.4 we have the following useful corollary.

6.3.5 **Corollary.** Let f' be continuous on an interval I (open, closed, or neither) and let $f'(x) \neq 0$ at all interior points of I. Then f is increasing throughout I if $f'(x) > 0$ at a single interior point x of I, and f is decreasing on I if $f'(x) < 0$ at a single interior point of I.

Example 3. Use 6.3.5 to solve Example 1.

Solution. $f(x) = 3 - x - x^2$ and $f'(x) = -1 - 2x$. Thus the polynomial function f' is continuous everywhere and has the value zero only at $x = -\frac{1}{2}$.

Since f' is continuous on $(-\infty, -\frac{1}{2}]$ and $f'(x) \neq 0$ for all x in the open interval $(-\infty, -\frac{1}{2})$, we choose some convenient number in $(-\infty, -\frac{1}{2})$, say -1, and find that $f'(-1) = 1 > 0$. It now follows from 6.3.5 that f is increasing throughout $(-\infty, -\frac{1}{2}]$.

Similarly, $f'(0) = -1 < 0$; so f is decreasing on $[-\frac{1}{2}, \infty)$.

Example 4. Solve Example 2 by finding the two numbers x for which $F'(x) = 0$ and the values of F at exactly three other points.

Solution. From $F(x) = 2x^3 - 3x^2 - 12x + 7$, we have $F'(x) = 6x^2 - 6x - 12 = 6(x + 1)(x - 2)$. F' is a polynomial function, and $F'(x) = 0$ only for $x = -1$ and $x = 2$. Thus F is continuous on the three intervals $(-\infty, -1]$, $[-1, 2]$, and $[2, \infty)$; and $F'(x) \neq 0$ at all interior points of these intervals.

Choose any convenient number in $(-\infty, -1)$, say -2. Since $F'(-2) = 24 > 0$, it follows from 6.3.5 that F is increasing on $(-\infty, -1]$.

Again, $F'(0) = -12 < 0$; so F is decreasing on $[-1, 2]$. Finally, $F'(3) = 24 > 0$, and F is increasing on $[2, \infty)$.

Exercises

In each of Exercises 1 to 6, use 6.3.5 to find all intervals on which the given function is increasing and all intervals on which it is decreasing. Afterward sketch the graph of the function and draw its horizontal tangents.

1. $f(x) = x^2 - 4x + 2$.

2. $f(x) = 2x - x^2$.

3. $F(x) = x^3 - 1$.

4. $F(x) = \frac{1}{2}(2x^3 + 9x^2 - 13)$.

5. $\phi(x) = x^4 + 4x$.

6. $\phi(t) = \frac{t^4}{2} - \frac{4t^2}{3}$.

7. Show that if F is increasing on $[a, b]$ and also on $[b, c]$, then F is increasing on $[a, c]$. (*Hint:* Use 6.3.1.)

8. Show that if $\phi(t) = 2t^5 - 15t^4 + 30t^3 - 6$, ϕ is increasing everywhere.

9. If $F(x) = (2 - x)/x^2$, find where F is increasing and where F is decreasing. Sketch its graph and the horizontal tangent.

10. If $f(x) = (x^2 - 6x)/(x + 1)^2$, find where f is increasing and where f is decreasing.

11. If $H(w) = 2w - 1/w$, find where H is increasing and where H is decreasing.

12. A one-to-one function f, defined on an interval $I = [a, b]$, has an inverse f^{-1} defined on the corresponding interval $J = f(I)$. The graph of f is cut by any vertical line $x = h$, $h \in I$, in exactly one point, and it is also cut in exactly one point by any horizontal line $y = k$, $k \in J$. Sketch a curve on an interval $[a, b]$ that has these two properties and such that

(a) its left endpoint is lower than its right endpoint;

(b) its left endpoint is higher than its right endpoint.

What can you say of each curve?

6.4 VELOCITY AND ACCELERATION IN RECTILINEAR MOTION

Consider a particle moving on a coordinate line. If $s = f(t)$ defines the **position function** f that gives the directed distance s of the particle from the origin at the end of t units of time, then the instantaneous **velocity** of the particle at time t was defined (see 4.2.1) to be

$$v = f'(t).$$

The **speed** of the particle is $|v|$, the absolute value of its velocity.

The **acceleration** *of a moving particle is the time rate of change of its velocity; in rectilinear motion it is given by*

$$a = \frac{dv}{dt} = f''(t).$$

Example 1. The position of a particle, moving on a horizontal coordinate line, at time t is given by $s = t^3 - 9t^2 + 24t$, where s is measured in feet and t in seconds. Describe the motion for the particle for $t \geq 0$.

Solution. The velocity of the particle at any time t is $v = s'(t) = 3t^2 - 18t + 24$, and its acceleration is $a = v'(t) = s''(t) = 6t - 18$. By setting $v = 3t^2 - 18t + 24 = 0$ and solving for t, we find that $v = 0$ when $t = 2$ or 4. Similarly, $a = 0$ when $t = 3$.

When $v = s'(t) > 0$, s is increasing and the particle is moving to the right; when $v = s'(t) < 0$, s is decreasing and the particle is moving to the left. When $a = v'(t) > 0$, the velocity of the particle is increasing; when $a = v'(t) < 0$, the velocity of the particle is decreasing.

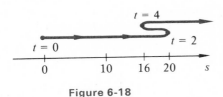

Figure 6-18

The motion, shown schematically in Fig. 6-18, is as follows. When $t = 0$, the particle is at the origin, moving to the right at 24 feet per second and slowing down. When 2 seconds have elapsed, it comes to a stop, 20 feet to the right of the origin. It then starts moving to the left, increasing its speed for 1 second and then slowing down for the next second, coming to a stop at the end of 4 seconds, 16 feet to the right of the origin. Finally, it starts moving to the right, going faster and faster forever.

Example 2. A stone thrown directly upward with a speed of 80 feet per second reaches a height of $s = 80t - 16t^2$ feet in t seconds. Neglecting air resistance, how high does it go? What is its acceleration at the end of 1 second? At the top of its flight?

Solution. $v = D_t s = 80 - 32t$. At the top of the stone's flight, $v = 0$. Letting $80 - 32t = 0$, we get $t = \frac{80}{32} = 2\frac{1}{2}$; thus the stone attains its greatest height at the end of $2\frac{1}{2}$ seconds, and that height is $s = 80(\frac{5}{2}) - 16(\frac{25}{4}) = 100$ feet.

The acceleration is given by $a = D_t v = -32$ feet per second per second. This constant acceleration is due to gravity, and it is the same at the end of 1 second as at the top of the flight—namely, -32 feet per second per second.

Exercises

In Exercises 1 and 2, the position of a moving particle on a coordinate line is given by $s = f(t)$, where s is measured in feet and t in seconds ($t \geq 0$). Find

(a)   where it starts from.
(b)   when it is moving in the positive direction.
(c)   when it comes to a stop.
(d)   when it is moving in the negative direction.

(e)   when it is back at its starting point.

1. $s = 12t - 2t^2$.

2. $s = t^3 - 6t^2$.

In each of Exercises 3 through 9, the position of a moving particle on a coordinate line is given by $s = f(t)$, where s is measured in feet and t in seconds.

Find an expression for the velocity and the acceleration in terms of t and describe the motion of the particle for $t \geq 0$. Make a schematic drawing, as in Fig. 6-18.

3. $s = 2t^2 - 4t - 5$.

4. $s = -t^2 + 6t + 1$.

5. $s = 3t^2 - 12t + 4$.

6. $s = 2t^3 - 6t + 5$.

7. $s = t^3 - 9t^2 + 24t - 1$.

8. $s = t^3 - 9t^2 + 15t + 2$.

9. $s = t^2 + \dfrac{16}{t}$, $t > 0$.

10. If $s = \frac{1}{2}t^4 - 5t^3 + 12t^2$, find the velocity of the moving particle when its acceleration is zero.

11. If $s = \frac{1}{10}(t^4 - 14t^3 + 60t^2)$, find the velocity of the moving particle when its acceleration is zero.

12. Two particles move along a coordinate line. At the end of t seconds their directed distances from the origin, in feet, are given by $s_1 = 4t - 3t^2$ and $s_2 = t^2 - 2t$, respectively.
(a) When do they have the same velocity?
(b) When do they have the same speed? (The *speed* of a particle is the absolute value of its velocity.)
(c) When do they have the same position?

13. The positions of two particles, P_1 and P_2, on a coordinate line, at the end of t seconds, is given by $s_1 = 3t^3 - 12t^2 + 18t + 5$ and $s_2 = -t^3 + 9t^2 - 12t$. When do the two particles have the same velocity?

14. A projectile is fired directly upward with an initial velocity of v_0 feet per second. Its height in t seconds is given by $s = v_0 t - 16t^2$ feet. What must its initial velocity be for the projectile to reach a height of 1 mile?

15. An object thrown directly upward with an initial velocity of 48 feet per second is approximately $s = 48t - 16t^2$ feet high at the end of t seconds.
(a) What is the maximum height attained?
(b) How fast is it moving, and in which direction, at the end of 1 second?
(c) How long does it take to return to its original position?

16. An object thrown directly downward from the top of a cliff with an initial velocity of v_0 feet per second falls approximately $s = v_0 t + 16t^2$ feet in t seconds. If it strikes the ocean below in 3 seconds with a velocity of 140 feet per second, how high is the cliff?

17. A block of wood, sliding down an inclined plane, travels $s = 6t^2$ feet in t seconds. How long does it take the block to attain a velocity of 50 feet per second?

18. If the block in Exercise 17 is pushed, at the start, so that its initial velocity is 9 feet per second, it will slide $s = 6t^2 + 9t$ feet in t seconds. When will the velocity of the block be five times its initial velocity?

19. Find a position function for a particle that always moves toward the right on a horizontal coordinate line, and yet has zero velocity at one and only one point. (*Hint:* See Fig. 6-5.)

6.5 LOCAL MAXIMA AND MINIMA

We recall, from 6.1.1, that the maximum and the minimum of a function on a set are its largest and smallest values on the set, provided that such values exist. Moreover, we saw in the maximum-minimum theorem (6.1.2) that if the function is *continuous* and if the set is a *closed interval*, the function always has a maximum and a minimum on the interval. The reader should now review 6.1.1 carefully before proceeding.

Let f be a continuous function whose graph is shown in Fig. 6-19; the domain of f is the interval $[a, b]$. Clearly, A is the highest point on the graph and M is the lowest; thus the ordinate of A, which is $f(a)$, is the maximum of f on the closed interval $[a, b]$, and the ordinate of M, which is $f(b)$, is the minimum.

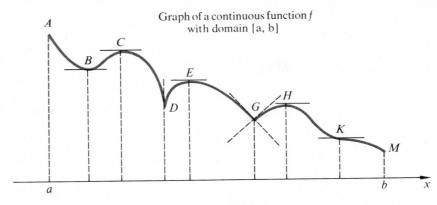

Graph of a continuous function f
with domain $[a, b]$

Figure 6-19

Although A and M are endpoints of the graph of f, there are interior points, C, E, and H, which are higher than all other points on the graph very close to them (Fig. 6-19). The ordinates of such points are called **local maximum** values of the function. Also, points B, D, and G are lower than all other points on the graph very close to them, and the ordinates of these points are called **local minimum** values of the function.

6.5.1 Definition.   Let f be a function.
(i) $f(c)$ is a **local maximum** of f if $f(c)$ is the maximum of f on some interval having c as an interior point.
(ii) $f(c)$ is a **local minimum** of f if $f(c)$ is the minimum value of f on some interval having c as an interior point.
(iii) $f(c)$ is a **local extreme value** of f if either (i) or (ii) holds.

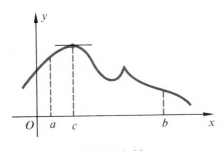

Figure 6-20

Notice that the maximum of a function on an interval (6.1.1) need not be a local maximum. In Fig. 6-19, the maximum of the function on the interval $[a, b]$ is $f(a)$, and the minimum is $f(b)$. But neither $f(a)$ nor $f(b)$ is a local extreme value because a and b are not interior points of the domain of f. However, (Fig. 6-20), if $f(c)$ is the maximum of f on $[a, b]$ and if c is an *interior* point of $[a, b]$, then $f(c)$ is also a local maximum of f.

From the definition of a local extreme value (6.5.1) and from theorem 6.1.3, we have an immediate corollary.

6.5.2 Corollary.   If $f(c)$ is a local extreme value, then $f'(c) = 0$ or $f'(c)$ fails to exist.

A number c at which a function f is defined and for which either $f'(c) = 0$ or $f'(c)$ does not exist is called a **critical number** for f.

It follows from 6.5.2 that the local extreme values of a function, if there are any, occur at critical numbers of the function. But not every critical number c of a function makes $f(c)$ a local maximum or a local minimum. As an illustration, the function defined by $f(x) = x^3$ (Fig. 6-5) has a critical number 0, since $f'(x) = 3x^2$ and $f'(0) = 0$. But $f(x) < f(0)$ for all $x < 0$ and $f(x) > 0$ for all $x > 0$. Thus $f(0)$ is not a local extreme value of f (by 6.5.1).

The first step in finding the local extreme values of a given function is to find its critical numbers. To decide whether the function has a local extreme value at a particular critical number, the following test is often useful.

6.5.3   **First Derivative Test for Local Extreme Values.**   Let f be continuous on an open interval (a, b) that contains a critical number c and let f' be defined at all points of (a, b) except possibly at c.

(i)   If $f'(x) > 0$ for all x in (a, c) and if $f'(x) < 0$ for all x in (c, b), then $f(c)$ is a **local maximum** of f (Fig. 6-21).

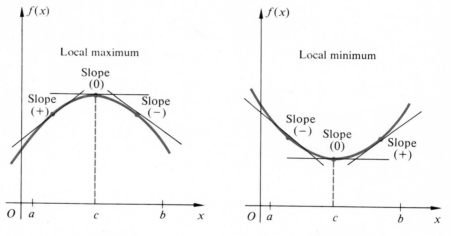

Figure 6-21

Figure 6-22

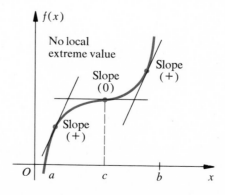

Figure 6-23

(ii)   If $f'(x) < 0$ for all $x \in (a, c)$ and if $f'(x) > 0$ for all x in (c, b) then $f(c)$ is a **local minimum** of f (Fig. 6-22).

(iii)   If $f'(x)$ has the same sign for all x in both (a, c) and (c, b), then $f(c)$ is not a local extreme values of f (Fig. 6-23).

Proof of (i).   Since $f'(x) > 0$ for all x in (a, c), f is increasing on $(a, c]$ (by 6.3.2). Thus $f(x) < f(c)$ for all x in $(a, c]$ (by 6.3.1). Again, since $f'(x) < 0$ for all $x \in (c, b)$, f is decreasing on $[c, b)$ and $f(c) > f(x)$ for all $x \in [c, b)$.

Thus $a < c < b$, and $f(c) > f(x)$ for all x in (a, b). Therefore $f(c)$ is a local maximum of f (6.1.1 and 6.5.1).

The proof of (ii) is analogous to that of (i), and the proof of (iii) is left as an exercise. ∎

To determine the local extreme values of a given function f, we may

1.   find the critical numbers for f; they are
     (i)   the roots of $f'(x) = 0$, and
     (ii)   the numbers for which $f'(x)$ fails to exist;
2.   test each critical number by means of the first derivative test (6.5.3).

Example 1.   Find the local extreme values of the function defined by $f(x) = x^2 - 6x + 5$.

Solution.   The polynomial function f is continuous everywhere, and its derivative, $f'(x) = 2x - 6$, exists for all x. Thus the only critical number for f is the single root of $f'(x) = 0$—namely, $x = 3$.

Since $f'(x) = 2x - 6 < 0$ for $x < 3$, f is decreasing on $(-\infty, 3]$; and because $f'(x) = 2x - 6 > 0$ for $x > 3$, f is increasing on $[3, \infty)$. Therefore, by the first derivative test, $f(3) = -4$ is a local minimum of f. Since 3 is the only critical number for f, there is no other extreme value (Fig. 6-24).

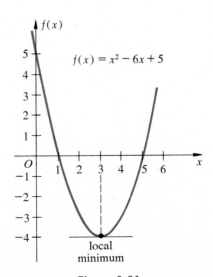

Figure 6-24

Example 2. Find the local maximum and the local minimum of the function f defined by $f(x) = \frac{1}{3}x^3 - x^2 - 3x + 4$.

Solution. $f'(x) = x^2 - 2x - 3 = (x + 1)(x - 3)$. Since f and f' are polynomial functions, they are continuous everywhere. Thus the only critical numbers for f are the roots of

$$f'(x) = (x + 1)(x - 3) = 0$$

—that is, -1 and 3. We shall use the first derivative test to show that $f(-1)$ and $f(3)$ are local extreme values of f.

Since f' is continuous on $(-\infty, -1]$ and is nonzero on $(-\infty, -1)$, we choose any convenient number in $(-\infty, -1)$, say -2, and find that $f'(-2) = 5 > 0$. Thus (by 6.3.5) f is increasing on $(-\infty, -1]$. Also, f' is continuous on $[-1, 3]$ and nonzero on $(-1, 3)$; so $f'(0) = -3 < 0$ implies that f is decreasing on $[-1, 3]$. Therefore, by the first derivative test, $f(-1) = 5\frac{2}{3}$ is a local maximum of f (Fig. 6-25).

Finally, since f' is continuous on $[3, \infty)$ and nonzero on $(3, \infty)$, $f'(4) = 5 > 0$ ensures that f is increasing on $[3, \infty)$. Thus $f(3) = -5$ is a local minimum of f (Fig. 6-25).

Example 3. Find the local extreme value of $f(x) = x^{2/3}$ (Fig. 6-26).

$f(x) = \frac{1}{3}x^3 - x^2 - 3x + 4$

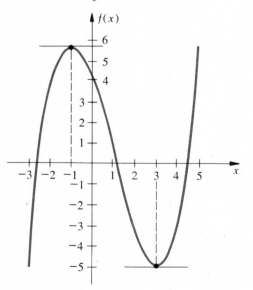

Figure 6-25

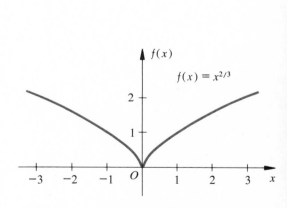

Figure 6-26

Solution. $f'(x) = 2/(3x^{1/3})$, $x \neq 0$. There is no number x for which $f'(x) = 0$, and there is just one number, 0, for which $f'(x)$ fails to exist. So 0 is the only critical number for f.

Since $f'(x) = 2/(3x^{1/3}) < 0$ for all $x < 0$ and $f'(x) > 0$ for all $x > 0$, f is decreasing on $(-\infty, 0)$ and increasing on $(0, \infty)$. Therefore, by the first derivative test, $f(0) = 0$ is a local minimum of f.

Exercises

In each of the following exercises, a function is defined. Find (a) its critical numbers, and (b) its local extreme values. Then sketch the graph of the function and draw its horizontal tangents.

1. $f(x) = x^2 - 8x + 7$.

2. $f(x) = x^2 + 2x$.

3. $F(x) = -x^2 + 4x - 3$.

4. $F(x) = x^2 - 2x + 3$.

5. $f(x) = x^3 - 3x - 1$.

6. $f(x) = x^3 - 3x^2 + 2$.

7. $G(x) = \frac{1}{4}x^3 - 3x - 1$.

8. $G(x) = \frac{1}{3}x^3 + x^2 - 3x$.

9. $\phi(x) = \frac{1}{4}x^4 + 1$.

10. $\phi(x) = 3x^4 - 4x^3$.

11. $g(t) = 2 - (t - 1)^{2/3}$.

12. $g(t) = \left(1 - \frac{3}{t}\right)^2, 1 \leq t \leq 5$.

13. $H(x) = (x + 2)^{2/3} - 4$.

14. $H(t) = t^2 + \frac{2}{t}, t > 0$.

15. $\Phi(x) = \dfrac{x^2}{\sqrt{x^2 + 1}}$.

16. $\Phi(x) = 2x - \frac{1}{x^2}, x \neq 0$.

17. Show that if
$$f(x) = x + \frac{1}{x},$$
the local minimum of f is greater than the local maximum of f. Sketch the graph of f.

6.6 SECOND DERIVATIVE TEST FOR LOCAL EXTREME VALUES

There is another test for local maxima and minima that is sometimes easier to apply. Just as the first derivative measures the rate of change of the function, the second derivative measures the rate of change of the first derivative. Thus the second derivative gives the rate of change of the slope of the tangent to the curve. When the second derivative is positive at a number c, it means that the first derivative is increasing there.

If $f'(c) = 0$ and $f''(c) > 0$, then $f'(x)$ is increasing from negative values to positive values as x increases through c. Therefore $f(c)$ is a local minimum of f (by 6.5.3). Interpreted geometrically, since $f'(x)$ is the slope of the tangent to the graph of f at $(x, f(x))$, then for values of x a little less than c the tangent points downward toward the right; for $x = c$ the tangent is horizontal; and for x slightly greater than c the tangent points upward to the right (Fig. 6-22).

Similarly, if $f'(c) = 0$ and $f''(c) < 0$, then $f'(x)$ decreases from positive values to negative values as x increases through c; this means (6.5.3) that $f(c)$ is a local maximum of f. Thus the slope, $f'(x)$, of the tangent to the graph of f changes from positive to zero to negative as x increases through c. For numbers x close to c but less than c the tangent to the graph at $(x, f(x))$ points upward toward the right; at $(c, f(c))$ the tangent is horizontal; and for x a little greater than c the tangent points down to the right (Fig. 6-21).

6.6.1 **Second Derivative Test for Local Extreme Values.** Let f' and f'' exist at every point of an open interval (a, b) containing c, and let $f'(c) = 0$.
   (i) If $f''(c) < 0$, $f(c)$ is a **local maximum** of f.
   (ii) If $f''(c) > 0$, $f(c)$ is a **local minimum** of f.

Proof of (i). Since, by definition and hypothesis,

$$f''(c) = \lim_{x \to c} \frac{f'(x) - f'(c)}{x - c} < 0,$$

there exists an open interval $(a, b) \subseteq I$ containing c such that

(1)
$$\frac{f'(x) - f'(c)}{x - c} < 0$$

for all $x \neq c$ in (a, b) (by 3.3.8).

When $a < x < c$, $x - c < 0$, and thus $f'(x) - f'(c) > 0$ in (1). Since $f'(c) = 0$, this becomes $f'(x) > 0$ for all x in (a, c).

Similarly, if $c < x < b$, $x - c > 0$ and $f'(x) - f'(c) < 0$ in (1). Since $f'(c) = 0$, $f'(x) < 0$ for all x in (c, b).

Therefore, by the first derivative test (6.5.3), $f(c)$ is a relative maximum of f.
The proof of (ii) is similar. ▮

Example 1. Given $f(x) = x^2 - 6x + 5$, use the second derivative test to find the local extreme values of f.

Solution. Since $f'(x) = 2x - 6 = 2(x - 3)$, the only critical number for f is 3. Now $f''(x) = 2$ for all x, and, in particular, $f''(3) = 2 > 0$. Consequently, by the second derivative test, $f(3) = -5$ is a local minimum of f (see Fig. 6-24).

Example 2. Find the local maximum and the local minimum of the function F defined by $F(x) = 2x^3 - 3x^2 - 12x + 5$.

Solution. $F'(x) = 6x^2 - 6x - 12 = 6(x + 1)(x - 2)$, and $F''(x) = 12x - 6$. Thus $F'(-1) = F'(2) = 0$; so -1 and 2 are critical numbers for F.

Since $F''(-1) = -18 < 0$, then $F(-1) = 12$ is a local maximum of F; and since $F''(2) = 18 > 0$, $F(2) = -15$ is a local minimum of F.

It sometimes happens that at a critical number $x = c$ the second derivative $f''(c)$, as well as the first derivative, is zero. In this case, the preceding theorem fails to give any information; $f(c)$ may be a local extreme value or it may not.

6.7 FINDING EXTREME VALUES

In the two preceding sections we discussed *local* extreme values of a function and tests for them. It is helpful, when sketching the graph of a function, to know where its local extreme values are. But in many problems we need to know the largest or the smallest value that a function can have on a given interval, sometimes on the entire domain of the function. Such extreme values may be among its local extreme values and thus be at interior points of the interval, *but often an extreme value is at an endpoint of the interval.* It is important to keep in mind the possibility of **endpoint extreme values** when seeking the maximum or minimum of a function.

Procedure for **finding the extreme values** of a continuous function f on a closed interval $[a, b]$:

1.   Find the critical numbers for f. They are the points in the open interval (a, b) at which either $f'(x) = 0$ or $f'(x)$ fails to exist.

2.   Test each critical number c to determine whether $f(c)$ is a local extreme value. When applicable, it is often easier to use the second derivative test for this purpose than the first derivative test.

3.   Compare the local maxima with $f(a)$ and $f(b)$, the endpoint values of f. The largest of all these is the maximum of f on $[a, b]$. Compare the local minima with $f(a)$ and $f(b)$; the smallest of these is the minimum of f on $[a, b]$.

Example 1.   Find the extreme values of the function defined by $f(x) = x^2 - 2x - 1$ on the closed interval $[-1, 2]$.

Solution.   The polynomial function f is continuous on the closed interval $[-1, 2]$ and therefore has a maximum there (6.1.2).

Since $f'(x) = 2x - 2 = 2(x - 1)$, the only critical number is 1. Now $f''(x) = 2$ for all x in $(-1, 2)$. Thus $f''(1) = 2 > 0$, and it follows from the second derivative test that $f(1) = -2$ is a local minimum of f. Comparing this with the values of f at the endpoints, $f(-1) = 2$ and $f(2) = -1$, we see that the minimum of f on $[-1, 2]$ is -2 and the maximum is 2 (Fig. 6-27).

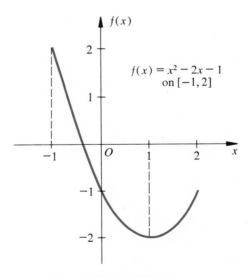

$$f(x) = x^2 - 2x - 1$$
$$\text{on } [-1, 2]$$

Figure 6-27

Example 2.   If $f(x) = x^3$, $-1 \leq x \leq 1$, determine the maximum and minimum of f on $[-1, 1]$.

Solution. $f'(x) = 3x^2$, and so the only critical number is 0. Since $f''(x) = 6x$, $f''(0) = 0$, and thus the second derivative test fails to give any information.

But $f'(x) = 3x^2 > 0$ for all numbers x in the open intervals $(-1, 0)$ and $(0, 1)$; it follows from the first derivative test that $f(0)$ is not a local extreme value of f.

We turn to the endpoints of $[-1, 1]$ and find that $f(-1) = -1$ and $f(1) = 1$. Therefore -1 is the minimum of f on $[-1, 1]$ and 1 is the maximum (Fig. 6-5).

Example 3. Find the maximum and the minimum of the function defined by $F(x) = 6x^{1/2} - 3x$ on the closed interval $[0, 4]$.

Solution. Since

$$F'(x) = \frac{3}{x^{1/2}} - 3 = 3\left(\frac{1 - x^{1/2}}{x^{1/2}}\right), \qquad x > 0,$$

the only critical number for F is 1.

$$F''(x) = -\frac{3}{2x^{3/2}}, \qquad x > 0;$$

and

$$F''(1) = -\frac{3}{2} < 0.$$

This shows that $F(1) = 3$ is a local maximum (by the second derivative test).

The value of the function at the endpoints of $[0, 4]$ is $F(0) = F(4) = 0$. Comparing it with the local maximum, $F(1) = 3$, we see that the maximum of F on $[0, 4]$ is 3 and the minimum is 0 (Fig. 6-28).

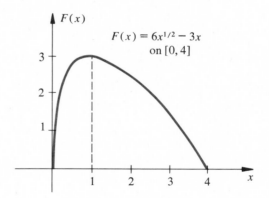

Figure 6-28

Exercises

In each exercise, find the extreme values of the indicated function on the given interval I. Then sketch the graph and draw all horizontal tangents.

1. $f(x) = 4 - x^2$; $I = [-1, 3]$.

2. $f(x) = (x - 3)^2$; $I = [0, 5]$.

3. $F(x) = (x - 3)^3 + 4$; $I = [1, 4]$.

4. $F(x) = \frac{1}{4}(x^3 - 12x)$; $I = [-3, 4]$.

5. $g(x) = x^3 - 3x^2 + 1$; $I = [-1, 3]$.

6. $g(x) = \frac{1}{4}(2x^3 - 3x^2 - 12x + 8)$; $I = (-3, 4]$.

7. $H(x) = (x + 1)^4$; $I = [-2, 1]$.

8. $H(x) = \sqrt{x} + \frac{1}{\sqrt{x}}$; $I = \left[\frac{1}{4}, 9\right]$.

9. $f(x) = \left(\frac{x - 2}{x - 5}\right)^2$; $I = [-4, 4]$.

10. $f(x) = x^2 + \frac{1}{x^2}$; $I = (0, \infty)$.

11. Let f be the quadratic function defined by $f(x)$

$= Ax^2 + Bx + C$, $A \neq 0$. Show that $f(x) \geq 0$ for all $x \in R$ if and only if $A > 0$ and $B^2 - 4AC \leq 0$.

[*Hint:* $f(x) \geq 0$ for all $x \in R$ if and only if the minimum of f is greater than or equal to zero.]

6.8 APPLIED PROBLEMS IN MAXIMA AND MINIMA

As noted earlier, many situations in science, engineering, geometry, and economics involve the determination of the maximum or minimum value of some varying quantity. Let us consider some problems, not requiring extensive knowledge of other subjects, that illustrate a technique for translating a problem from words to a mathematical equation and then finding the desired maximum or minimum.

The method may be summarized as follows.

1. By reading the problem, decide which quantity is to be a maximum (or minimum) and, after assigning a letter to it, express it as a function of only one independent variable. Sometimes two equations involving three variables are obtained; eliminate one of the variables so as to express the variable to be maximized (or minimized) in terms of just one independent variable.

2. Find the derivative of the function in step 1 and set it equal to zero. The roots of the resulting equation are critical numbers for the function. Other critical numbers are any isolated values of the independent variable at which the derivative fails to exist. By inspection or by one of the tests developed in the preceding sections, determine at which of the critical numbers the function has local extreme values.

3. Find the maximum (or minimum) of the function by comparing the local maxima (or minima) with each other and with the value of the function at any endpoint that its domain of definition may have.

Example 1. A handbill is to contain 50 square inches of printed matter, with 4-inch margins at top and bottom and 2-inch margins on each side. What dimensions for the page would use the least paper?

Solution. Let x be the width and y the height of the handbill (Fig. 6-29). Its area is

$$(1) \qquad A = xy.$$

We wish A to be a minimum.

In (1), A is expressed in terms of two variables, x and y. We will find an equation connecting x and y so that either x or y can be eliminated from (1). The dimensions of the printed part are $x - 4$ and $y - 8$ and its area is 50 square inches; so $(x - 4)(y - 8) = 50$ or

$$(2) \qquad y = \frac{50}{x - 4} + 8.$$

Substituting from (2) in (1), we get

$$(3) \qquad A = \frac{50x}{x - 4} + 8x, \qquad x \neq 4.$$

Now A is expressed as a function of one variable x. To find the critical numbers for this function, we

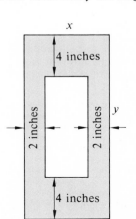

Figure 6-29

differentiate A with respect to x, obtaining

$$D_x A = \frac{(x-4)50 - 50x}{(x-4)^2} + 8,$$

or

(4)
$$D_x A = \frac{8(x-4)^2 - 200}{(x-4)^2}$$

Although $D_x A$ does not exist for $x = 4$, this is not a critical number for A because A itself, as given in (3), does not exist for $x = 4$. The critical numbers for A will be found by setting $D_x A$ equal to zero:

(5)
$$\frac{8(x-4)^2 - 200}{(x-4)^2} = 0.$$

The roots of (5) are $x = 9$ and $x = -1$, but the latter cannot be the width of a rectangle. From the nature of the problem we conclude that $x = 9$ is the width that makes A a minimum. However, it is easy to confirm this by the second derivative test. Rewriting $D_x A$ as $D_x A = -200(x-4)^{-2} + 8$, we find

$$D_x^2 A = \frac{400}{(x-4)^3},$$

which clearly is positive for $x = 9$. Therefore $x = 9$ in (3) makes A a local minimum.

To see that $x = 9$ also makes A the minimum, we notice in (4) that $D_x A < 0$ for $4 < x < 9$, and $D_x A > 0$ for $x > 9$. Thus A is decreasing for $4 < x < 9$ and increasing for $x > 9$, and the local minimum for A at $x = 9$ is also its minimum.

The height, $y = 18$, is found by substituting $x = 9$ in (2). So the dimensions for the handbill that will use the least paper are 9 inches by 18 inches.

Example 2. A rectangular beam is to be cut from a log with circular cross section. If the strength of the beam is proportional to its width and the square of its depth, find the dimensions that give the strongest beam.

Solution. Denote the diameter of the log by a, and the width and depth of the beam by w and d, respectively (Fig. 6-30). Let S be the strength of the beam. We wish to maximize S; that is, we seek the width and depth that make S a maximum.

From the conditions of the problem,

(6)
$$S = kwd^2,$$

where k is a constant of proportionality. Here S depends on both w and d. We wish to express S as a function of one variable only; so we seek a relation between w and d. Since a is the hypotenuse of a right triangle, $w^2 + d^2 = a^2$, or

(7)
$$d^2 = a^2 - w^2.$$

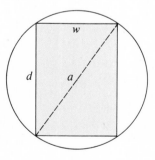

Figure 6-30

By substituting (7) in (6), we obtain

(8)
$$S = k(a^2 w - w^3),$$

which expresses S as a function of w. Now

$$\frac{dS}{dw} = k(a^2 - 3w^2).$$

Setting this result equal to zero, we have $a^2 - 3w^2 = 0$, or

$$(9) \qquad\qquad w = \frac{a}{\sqrt{3}},$$

and by substituting (9) in (7), we find

$$(10) \qquad\qquad d = \frac{\sqrt{2}\,a}{\sqrt{3}}.$$

Therefore when w and d are as in (9) and (10), we have the strongest beam.

We can express this result more concisely if we divide (10) by (9), member by member. Doing so gives

$$d = \sqrt{2}\,w$$

as the proportion that gives the strongest beam.

In step 1 of the procedure outlined earlier for solving problems in maxima and minima, it was suggested that when two equations involving three variables are obtained from the given data, we eliminate one of the variables so as to express the quantity to be maximized or minimized as a function of one variable only. Sometimes it is better not to eliminate the third variable but to *differentiate the two equations implicitly*. This is particularly true when relative proportions are wanted instead of specific quantities.

Example 3. A cylindrical glass jar has a metal top. If the metal costs three times as much as the glass, per unit of area, find the proportions of the least costly jar that holds a given amount.

Solution. Let C denote the cost of the jar, V its (constant) volume, r the radius of its base, and h its height. Let a be the cost of the glass per unit of area; then $3a$ is the cost of the metal per unit of area.

From the given data, $C = 3a(\pi r^2) + a(\pi r^2 + 2\pi rh)$, or

$$(11) \qquad\qquad C = 4a\pi r^2 + 2a\pi rh,$$

and

$$(12) \qquad\qquad V = \pi r^2 h.$$

In these equations V and π are constants, and C, r, and h are variables. By differentiating (11) and (12) implicitly with respect to r, we get

$$(13) \qquad\qquad \frac{dC}{dr} = 8a\pi r + 2a\pi\left(r\frac{dh}{dr} + h\right)$$

and

$$(14) \qquad\qquad 0 = \pi\left(r^2\frac{dh}{dr} + 2rh\right),$$

since V is a constant. From (14) we have

$$(15) \qquad\qquad \frac{dh}{dr} = -\frac{2h}{r},$$

and if we substitute this result in (13) we obtain

$$(16) \qquad\qquad \frac{dC}{dr} = 8a\pi r - 2a\pi h.$$

In order to minimize C, we set $dC/dr = 0$ and obtain $8a\pi r - 2a\pi h = 0$, or

(17)
$$h = 4r.$$

To see whether (17) makes C a local minimum, we will use the second derivative test. Accordingly, we differentiate (16) with respect to r and then use (15), obtaining

(18)
$$\frac{d^2C}{dr^2} = 8a\pi - 2a\pi \frac{dh}{dr} = 8a\pi + 4a\pi \frac{h}{r}.$$

For $h = 4r$, (18) becomes

$$\frac{d^2C}{dr^2} = 8a\pi + 16a\pi = 24a\pi > 0.$$

Therefore (17) makes C a minimum.

The most economical jar is one whose height is four times the radius of the base.

Example 4. Find the dimensions of the right circular cylinder of greatest volume that can be inscribed in a given right circular cone.

Solution. Let a be the altitude and b the radius of the base of the given right circular cone (Fig. 6-31). Denote by h, r, and V the altitude, the radius of the base, and the volume of an inscribed cylinder. Here V, r, and h are variables, whereas a and b are constants.

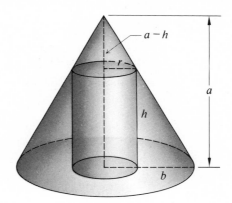

Figure 6-31

The volume of the cylinder is

(19)
$$V = \pi r^2 h.$$

From similar triangles,

(20)
$$\frac{r}{a - h} = \frac{b}{a}, \quad \text{or} \quad r = b - \frac{bh}{a}.$$

Consider h to be the independent variable. To maximize V, we differentiate (19) and (20) with respect to h and set $D_h V = 0$. From (19) we have

(21)
$$D_h V = \pi r^2 + 2\pi rh \cdot D_h r = 0, \quad \text{or} \quad D_h r = -\frac{r}{2h},$$

and from (20)

(22)
$$D_h r = -\frac{b}{a}.$$

From (21) and (22),

(23)
$$r = \frac{2b}{a}h.$$

By combining (23) with (20), we obtain

(24)
$$h = \frac{a}{3}, \qquad r = \frac{2b}{3}.$$

Rather than test the critical number formally for a maximum or minimum, we need only note that the volume is zero for the extreme values $r = b$ and $r = 0$, and positive for intermediate values. Consequently, there is a maximum value of the volume for the one critical number $r = \frac{2}{3}b$.

Thus, the volume of the inscribed cylinder will be greatest when its altitude is $\frac{1}{3}$ of the altitude of the cone and the radius of its base is $\frac{2}{3}$ of the radius of the base of the cone.

Exercises

1. Find the two numbers whose sum is 10 and whose product is the maximum.

2. What number exceeds its square by the maximum amount?

3. Find two numbers whose product is -12 and the sum of whose squares is the minimum.

4. What number exceeds its principal square root by the least amount?

5. Find the volume of the largest box that can be made from a piece of cardboard, 24 inches square, by cutting equal squares from each corner and turning up the sides.

6. An open box is to be made from a rectangular piece of sheet metal, 20 inches by 32 inches, by cutting equal squares from each corner and folding up the sides. Find the volume of the box with greatest capacity that can be so constructed.

7. Find the points on the hyperbola $x^2 - 4y^2 - 4 = 0$ that are closest to the point $(5, 0)$.

8. Find the maximum area of a rectangle inscribed in a semicircle of radius r.

9. A piece of wire, 16 inches long, is cut into two pieces; one piece is bent to form a square and the other is bent to form a circle. Where should the cut be made in order that the sum of the areas of the square and circle be a minimum?

10. A window is made in the shape of a rectangle surmounted by a semicircle whose diameter is equal to the width of the rectangle. If the perimeter of the window is 16 feet, what dimensions will admit the most light?

11. What is the shape of the rectangular field of given area that requires the least amount of fencing?

12. An open box is needed with a capacity of 36,000 cubic inches. If the box must be twice as long as it is wide, what dimensions would require the least material?

13. A concrete cistern with a square base is to be constructed to hold 12,000 cubic feet of water. If the metal top costs twice as much per square foot as the concrete sides and floor, what are the most economical dimensions for the cistern?

14. The cross section of a proposed open wooden spillway is a rectangle. For the spillway to have a given cross-sectional area, what relation between the width and depth will require the least lumber?

15. Find the maximum volume that a right circular cylinder can have when it is inscribed in a right circular cone, if the altitude of the cone is 20 inches and the radius of its base is 6 inches.

16. A metal rain gutter is to have 3-inch sides and a 3-inch horizontal bottom, the sides making equal angles with the bottom. How wide should the opening across the top be for maximum carrying capacity?

17. Find the greatest volume that a right circular cone can have if it is inscribed in a sphere of radius r.

18. Find the relative dimensions of the right circular cone with greatest curved surface area that can be inscribed in a given sphere.

19. What are the relative dimensions of the right circular cylinder with greatest curved surface area that can be inscribed in a given sphere?

20. A right circular cone is to be inscribed in another right circular cone of given volume, with the same axis and with the vertex of the inner cone touching the base of the outer cone. What must be the ratio of their altitudes for the inscribed cone to have maximum volume?

21. A small island is 2 miles from the nearest point P on the straight shoreline of a large lake. If a man on the island can row his boat 3 miles per hour and can walk 4 miles per hour, where should he land his boat in order to arrive in the shortest time at a town 10 miles down the shore from P?

22. If the strength of a rectangular beam is proportional to its width and the square of its depth, find the dimensions of the strongest beam that can be cut from a log whose cross section has the form of the ellipse $9x^2 + 8y^2 = 72$.

23. The illumination at a point is inversely proportional to the square of the distance of the point from the light source and directly proportional to the intensity of the light source. If two light sources are s feet apart, and their intensities are I_1 and I_2, respectively, at what point between them will the sum of their illuminations be a minimum?

24. At 7:00 A.M. one ship was 60 miles due east from a second ship. If the first ship sailed west at 20 miles per hour and the second ship sailed southeast at 30 miles per hour, when were they closest together?

25. A powerhouse is located on one bank of a straght river that is w feet wide. A factory is situated on the opposite bank of the river, l feet downstream from the point A directly opposite the power house. What is the most economical path for a cable connecting the powerhouse to the factory if it costs a dollars per foot to lay the cable under water and b dollars on land $(a > b)$?

26. Show that the rectangle with maximum perimeter that can be inscribed in a circle is a square.

27. An open irrigation ditch of given cross-sectional area is to be lined with concrete to prevent seepage. If the two equal sides are perpendicular to the flat bottom, find the relative dimensions that require the least concrete.

28. A closed cylindrical can is to have a specified surface area. Find the relative dimensions of such a can when the volume is a maximum.

29. Find the dimensions of the rectangle of greatest area that can be inscribed in the region bounded by the curve $y = (a^3 - x^3)^{1/3}$ and the coordinate axes (a is a positive constant). Make a sketch.

30. Find the equation of the line that is tangent to the ellipse $b^2x^2 + a^2y^2 = a^2b^2$ in the first quadrant and that forms with the coordinate axes the triangle with smallest possible area. (a and b are positive constants.)

31. A closed box in the form of a rectangular parallelepiped with a square base is to have a given volume. If the material used in the bottom costs 20% more per square inch than the material in the sides, and the top costs 50% more per square inch than the sides, find the most economical proportions for the box.

32. An observatory is to be in the form of a right circular cylinder surmounted by a hemispherical dome. If the hemispherical dome costs twice as much per square foot as the cylindrical wall, what are the most economical proportions for a given volume?

33. A proposed tunnel of given cross-sectional area is to to have a horizontal floor, vertical side walls of equal height, and a semicircular ceiling. If the ceiling costs twice as much per square yard to build as the vertical sides and floor, find the most economical ratio of the diameter of the semicircular cylinder to the height of the vertical side walls.

34. Find the most economical proportions for a conical paper cup with given volume.

6.9 APPLICATIONS TO ECONOMICS

Economists use two terms to express the variation of a variable y with respect to another variable x, *average* and *marginal*.

The *average variation* of y with respect to x corresponds to the average rate of change $\Delta y/\Delta x$, and is expressed by a difference quotient

$$\frac{\Delta y}{\Delta x} = \frac{f(x + \Delta x) - f(x)}{\Delta x}.$$

Here Δx can be comparatively large.

On the other hand, the concept of *marginal variation* is concerned with the variation of y with respect to x when Δx is very small; it is defined to be the limit of $\Delta y/\Delta x$ as $\Delta x \longrightarrow 0$. Thus the marginal variation of y with respect to x is the derivative

$$\frac{dy}{dx} = \lim_{\Delta x \to 0} \frac{\Delta y}{\Delta x} = \lim_{\Delta x \to 0} \frac{f(x + \Delta x) - f(x)}{\Delta x}.$$

Let x be the number of units of a particular commodity produced and marketed. The total cost, y, of producing and selling x units is given by

$$y = C(x),$$

where C is a **total cost function.**

The **average cost** per unit is

$$\frac{y}{x} = \frac{C(x)}{x}.$$

If the level of output x is increased by an amount Δx, the total cost will be increased by an amount Δy. The **marginal cost** is defined to be

$$\lim_{\Delta x \to 0} \frac{\Delta y}{\Delta x} = \lim_{\Delta x \to 0} \frac{C(x + \Delta x) - C(x)}{\Delta x} = C'(x);$$

it is the rate of increase in cost per unit increase in output.

Example 1. Let $C(x) = 8300 + 3.25x - 0.0002x^2$ define a total cost function, where x denotes the number of units produced and marketed. The average cost per unit is

$$\frac{C(x)}{x} = \frac{8300}{x} + 3.25 - 0.0002x;$$

and the marginal cost is

$$C'(x) = 3.25 - 0.0004x.$$

Thus if $x = 2000$ units are produced and sold each week, the average cost per unit is

$$\frac{C(2000)}{2000} = \frac{8300}{2000} + 3.25 - 0.0002(2000) = 7.00.$$

The marginal cost of producing and selling one more unit is

$$C'(2000) = 3.25 - (0.0004)(2000) = 2.45;$$

and the marginal cost for six more units is approximately

$$6(2.45) = 14.70.$$

Let x be the number of units of a particular commodity produced and marketed. The **total revenue function** R has for its value, $R(x)$, the total revenue from the sale of x units. As an illustration, if the selling price per unit is 4.50 and 7100 units are sold, the total revenue is $R(x) = 7100(4.50) = 31,950.00$.

The **profit**, $P(x)$, in producing and selling x units is the total revenue less the total cost; that is,

$$(1) \qquad\qquad P(x) = R(x) - C(x).$$

The price per unit at which a commodity can be sold is a function of the *demand*. When the market is flooded with too many units of a particular commodity, the price must be lowered in order to sell them all; if too few units are available to meet the demand, the price tends to rise.

The **demand function** p has for its value, $p(x)$, the price per unit at which x units of a particular commodity can be sold under prevailing market conditions.

Thus the total revenue function R is related to the demand function p by

$$(2) \qquad\qquad R(x) = x \cdot p(x).$$

Example 2. A manufacturer estimates that he can sell 1000 units per week if he sets the unit price at \$3.00, and that his sales will rise 100 with each 10-cent decrease in price. Find
    (a)   the demand function;
    (b)   the number of units that he should sell each week to obtain the maximum possible revenue;
    (c)   the maximum weekly revenue.

Solution. Let x be the number of units sold each week, $x \geq 1000$. The demand function, p, is given by

$$p(x) = 3.00 - \frac{(0.10)(x - 1000)}{100}, \qquad x \geq 1000,$$

and the weekly revenue is

$$R(x) = x \cdot p(x) = 3x - \frac{x^2 - 1000x}{1000}.$$

Then

$$R'(x) = 3 - \frac{2x - 1000}{1000} = \frac{4000 - 2x}{1000},$$

and the only critical number for R is $x = 2000$.

Since $R'(x) > 0$ for $1000 \leq x < 2000$ and $R'(x) < 0$ for $x > 2000$, $R(2000) = \$4000$ is the maximum possible weekly revenue. It is obtained by selling 2000 units per week.

The **marginal revenue**, $R'(x)$, is the rate of increase in revenue per unit increase in sale. Similarly, $p'(x)$ is the **marginal price**, and $P'(x)$ is the **marginal profit**. By differentiating both members of (1) with respect to x, we obtain

$$(3) \qquad\qquad P'(x) = R'(x) - C'(x),$$

which says that *the marginal profit is the marginal revenue minus the marginal cost.*

It often happens that the profit, $P(x)$, is a maximum when $P'(x) = 0$ and that there is just one value of x for which $P'(x) = 0$ (that is, just one critical number for the profit function P). Under such circumstances it follows from (3) that the total profit is a maximum when that number of units are produced and sold that make the marginal revenue equal to the marginal cost.

Example 3. In manufacturing and selling x units of a certain commodity, the demand function p and the cost function C are given by

$$p(x) = (5.00 - 0.002x) \text{ dollars}$$

and

$$C(x) = (300 + 1.10x) \text{ dollars}.$$

Find the number, x_1, of units that would make the profit a maximum, and find the maximum profit. Also find the marginal revenue and the marginal cost when x_1 units are produced and sold.

Solution. From (1) and (2), $P(x) = x \cdot p(x) - C(x)$. Thus

$$P(x) = x(5 - 0.002x) - (300 + 1.10x)$$
$$= -300 + 3.90x - 0.002x^2.$$

Now $P'(x) = 3.90 - 0.004x$, and so P has just one critical number, $x = 975$. By the first derivative test, $P(975) = \$1601.25$ is the maximum profit. Thus $x_1 = 975$.

The marginal revenue is $R'(x) = x \cdot p'(x) + p(x) = 5 - 0.004x$; thus $R'(975) = \$1.10$. The marginal cost is $C'(975) = \$1.10$.

Exercises

1. The total cost of producing and selling x units of a certain commodity per month is $C(x) = 1200 + (3.25)x - (0.0002)x^2$. If the production level is 1800 units per month, find the average cost, $C(x)/x$, of each unit and the marginal cost.

2. The total cost of producing and selling x units of a certain commodity per week is $C(x) = 1100 + x^2/1200$. Find the average cost, $C(x)/x$, of each unit and the marginal cost, at a production level of 900 units per week.

3. The total cost of producing and marketing x units of a certain commodity is given by

$$C(x) = \frac{80,000x - 400x^2 + x^3}{40,000}.$$

For what number x is the *average* cost a minimum?

4. The total cost of producing and selling $100x$ units of a particular commodity per week is

$$C(x) = 1000 + 33x - 9x^2 + x^3.$$

Find (a) the level of production at which the *marginal* cost is a minimum, and (b) the minimum marginal cost.

5. A demand function, p, is defined by

$$p(x) = 20 + 4x - \frac{x^2}{3},$$

where $x \geq 0$ is the number of units.
(a) Find the total revenue function and the marginal revenue function.
(b) On what interval is the total revenue increasing?
(c) For what number x is the marginal revenue a maximum?

6. For the demand function defined by $p(x) = (182 - x/36)^{1/2}$, find the number of units x_1 that makes the total revenue a maximum and state the maximum possible revenue. What is the marginal revenue when the optimum number of units, x_1, are sold?

7. For the demand function given by $p(x) = 800/(x + 3) - 3$, find the number of units x_1 that makes the total revenue a maximum and state the maximum possible revenue. What is the marginal revenue when the optimum number of units, x_1, are sold?

8. A river boat company offers a Fourth-of-July excursion to a fraternal organization with the understanding that there will be at least 400 passengers. The price of each ticket will be $12.00, and the company agrees to refund to every passenger $0.20 for each 10 passengers in excess of 400. Write an expression for $p(x)$ and find the number x_1 of passengers that makes the total revenue a maximum.

9. A merchant finds that he can sell 4000 yards of a particular fabric each month if he prices it at $6.00 per yard and that his monthly sales will increase by 250 yards for each $0.15 reduction in the price per yard. Write an expression for $p(x)$ and find the price per yard that would bring maximum revenue.

10. A manufacturer estimates that he can sell 500

articles per week if his unit price is $20.00 and that his weekly sales will rise by 50 with each $0.50 reduction in price. The cost of producing and selling x articles a week is $C(x) = 4200 + 5.10x + 0.0001 x^2$. Find

(a) the demand function;
(b) the level of weekly production for maximum profit;
(c) the price per article at the optimum level of production;
(d) the marginal price at that level of production.

11. The monthly overhead of a manufacturer of a certain commodity is $6000 and the cost of material is $1.00 per unit. If he manufacturers not more than 4500 units per month, his labor cost is $0.40 per unit; but for each unit over 4500 he must pay "time and a half" for labor. He can sell 4000 units per month at $7.00 per unit, and he estimates that his monthly sales will rise by 100 for each $0.10 reduction in price. Find (a) the total cost function, (b) the demand function, and (c) the number of units that he should produce each month for maximum profit.

6.10 CONCAVITY. POINTS OF INFLECTION

Consider a function f that is differentiable on an interval I. If the slope of the tangent to the graph of f increases as x increases on I, the form of the graph is cupped upward [Fig. 6-32(a)]. If the slope of the tangent to the graph decreases as x increases, the graph is cupped downward [Fig. 6-32(b)]. Since $f'(x)$ gives the slope of the tangent, we say that the graph of f is *concave upward* when f' is increasing and *concave downward* when f' is decreasing.

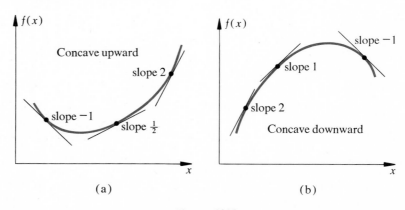

(a)   (b)

Figure 6-32

6.10.1   **Definition.**   Let f be a function that is differentiable on an interval I.

(i)   The graph of f is said to be **concave upward** on I if and only if f' is increasing on I.

(ii)   The graph of f is **concave downward** on I if and only if f' is decreasing on I.

When the second derivative of a function f exists, the sign of $f''(x)$ provides an easy test for concavity.

We recall (6.2.3) that if $f'(x) > 0$, the function f is increasing. It follows, then, that if $f''(x) > 0$, the derived function f' is increasing. But this means that the graph of f is concave upward. Similarly, when $f''(x) < 0$, the function f' is decreasing and the graph of f is concave downward.

6.10.2   **Test for Concavity.**   Let f be a differentiable function and let f'' exist at every point of an open interval (a, b).

(i)   If $f''(x) > 0$ for all x in (a, b), the graph of f is **concave upward** on (a, b).

(ii)   If $f''(x) < 0$ for all x in (a, b), the graph of f is **concave downward** on (a, b).

Example 1.   If $f(x) = x^2 - 6x + 5$, find where the graph of f is concave upward and where it is concave downward.

Solution. $f'(x) = 2x - 6$ and $f''(x) = 2$. Since $f''(x) = 2 > 0$ for all $x \in R$, the graph of f is concave upward everywhere (Fig. 6-24).

Example 2.   If f is defined by $f(x) = \frac{1}{3}x^3 - x^2 - 3x + 4$, find the intervals on which the graph of f is concave upward, and the intervals on which it is concave downward.

Solution.   We have $f'(x) = x^2 - 2x - 3$, and $f''(x) = 2x - 2 = 2(x - 1)$. Since $x - 1 < 0$ for all $x < 1$ and $x - 1 > 0$ for all $x > 1$, then $f''(x) < 0$ for $x < 1$ and $f''(x) > 0$ for $x > 1$. Thus the graph of f is concave downward on $(-\infty, 1)$ and concave upward on $(1, \infty)$ (see Fig. 6-25).

Often the graph of a function has several intervals of concavity. When an interval on which the graph is concave upward is adjacent to an interval on which it is concave downward, the point separating them is of particular interest in curve sketching.

6.10.3   **Definition.**   Let f be continuous at c. The point $(c, f(c))$ is called a **point of inflection** of the graph of f if there is an open interval (a, b) containing c such that f is concave upward on (a, c) and concave downward on (c, b), or vice versa (Fig. 6-33).

When the graph of a function has a tangent line at a point of inflection, the graph crosses its tangent there.

The following theorem is often useful in finding the points of inflection of a given function.

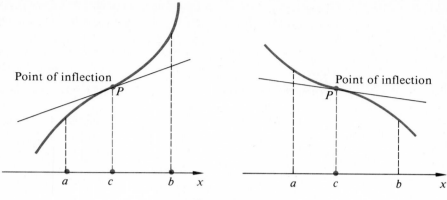

Figure 6-33

6.10.4 Theorem. Let f be differentiable on an interval I having c as an interior point, and let f'' be continuous at c. If $(c, f(c))$ is a point of inflection of the graph of f, then necessarily $f''(c) = 0$.

Proof. Since $f''(c)$ exists, it is positive, negative, or zero. We will show that it cannot be positive or negative.

Assume that $f''(c) > 0$. Since f'' is continuous at c, there exists an open interval $(a, b) \subseteq I$ containing c such that $f''(x) > 0$ for all $x \in (a, b)$ (see Exercise 27, Section 3.5). But this means that f is concave upward on (a, c) and also on (c, b), which (by 6.10.3) contradicts the fact that $(c, f(c))$ is a point of inflection of the graph of f. Therefore $f''(c)$ cannot be positive.

Similarly, $f''(c)$ cannot be negative. Therefore $f''(c) = 0$. ∎

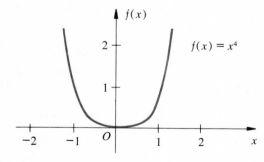

Figure 6-34

Although this theorem is a primary aid in finding points of inflection, $f''(c) = 0$ does not guarantee that $(c, f(c))$ is a point of inflection of the graph of f. To illustrate, consider the function defined by $f(x) = x^4$. Since $f''(x) = 12x^2$, $f''(0) = 0$. But $(0, 0)$ cannot be a point of inflection because $f''(x) > 0$ for all $x \neq 0$, which means that the graph is concave upward everywhere except at $(0, 0)$ (Fig. 6-34).

Contrariwise, $(c, f(c))$ may be a point of inflection even when $f''(c)$ does not exist. This situation is shown in Examples 4 and 5 below.

Nevertheless, 6.10.4 is often useful in finding points of inflection. Just as in searching for local extreme values of a function f we first found the critical numbers for f—that is, the numbers x for which $f'(x) = 0$ or $f'(x)$ did not exist—and then proceeded to test each critical number separately, so also in seeking the points of inflection of the graph of f we first find the numbers x for which $f''(x) = 0$ or $f''(x)$

fails to exist. We then test each such number by determining the concavity of the graph on both sides of the number. Remember that in order for P to be a point of inflection on a graph, *it is essential that the concavity of the graph change from upward to downward, or vice versa*, at P.

This procedure for finding the points of inflection of a graph will be used in the following examples.

Example 3. Find any points of inflection the graph of $f(x) = \frac{1}{6}x^3 - 2x$ may have.

Solution. From $f'(x) = \frac{1}{2}x^2 - 2$ and $f''(x) = x$, we see that $f''(0) = 0$ and that $f''(x)$ exists and is different from zero for all $x \neq 0$. So we turn our attention to the point $(0, 0)$ as the only possible point of inflection.

Since $f''(x) = x$, we have $f''(x) < 0$ for all $x < 0$ and $f''(x) > 0$ for all $x > 0$. Thus the graph of f is concave downward on $(-\infty, 0)$ and concave upward on $(0, \infty)$. Therefore $(0, 0)$ is a point of inflection on the graph of f.

From $f'(x) = \frac{1}{2}x^2 - 2$, we find that the graph has a tangent at its point of inflection, $(0, 0)$, with slope $f'(0) = -2$. The graph crosses its tangent at the point of inflection (Fig. 6-35).

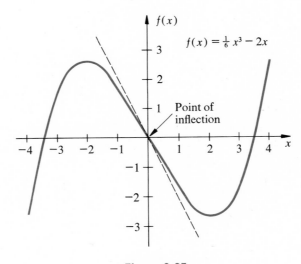

Figure 6-35

Example 4. Show that the graph $F(x) = x^{1/3} + 2$ has a point of inflection at $(0, 2)$ even though $F''(0)$ does not exist.

Solution. We find that

$$F'(x) = \frac{1}{3x^{2/3}} \quad \text{and} \quad F''(x) = \frac{-2}{9x^{5/3}}.$$

$F''(0)$ does not exist, but $F''(x)$ exists and is different from zero for all $x \neq 0$. We investigate the possibility that $(0, 2)$ might be a point of inflection.

For $x < 0$, $x^{5/3} < 0$, and so $F''(x) > 0$. Thus the graph is concave upward on $(-\infty, 0)$. When $x > 0$, $x^{5/3} > 0$ and $F''(x) < 0$; consequently, the graph is concave downward on $(0, \infty)$. Therefore $(0, 2)$ is a point of inflection of the graph of the continuous function F.

Although $F'(0)$ does not exist, the graph is tangent to the y axis at $(0, 2)$. Because this tangent line is vertical, it has no slope. However, the graph crosses its (vertical) tangent at the point of inflection $(0, 2)$ (Fig. 6-36).

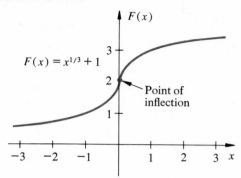

Figure 6-36

Example 5. Consider the function F defined by

$$F(x) = \frac{1}{2}|x^2 - 4| + 1, \qquad x \geq 0.$$

For $x \geq 2$, $|x^2 - 4| = x^2 - 4$; and for $0 \leq x \leq 2$, $|x^2 - 4| = -(x^2 - 4) = 4 - x^2$. Thus $F(x)$ can be written

(1)
$$F(x) = \begin{cases} 3 - \dfrac{x^2}{2}, & 0 \leq x \leq 2, \\[2mm] \dfrac{x^2}{2} - 1, & 2 \leq x. \end{cases}$$

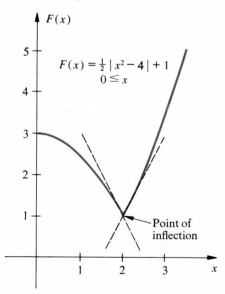

Figure 6-37

Then

(2)
$$F'(x) = \begin{cases} -x, & 0 < x < 2, \\ x, & 2 < x, \end{cases}$$

and

(3)
$$F''(x) = \begin{cases} -1, & 0 < x < 2, \\ 1, & 2 < x. \end{cases}$$

From (3), the graph is concave downward on $(0, 2)$ and concave upward on $(2, \infty)$; and from (1), F is continuous at $x = 2$. Therefore the point $(2, 1)$ is a point of inflection of the graph of F.

Perhaps the most interesting fact about this graph is that it does not have any tangent, vertical or otherwise, at its point of inflection. From (2), the value of the left-hand derivative at 2 is -2, and the value of the right-hand derivative at 2 is 2. Therefore (Exercise 25, Section 4.3) the graph of F has no tangent at $(2, 1)$ (see Fig. 6-37).

Exercises

In Exercises 1 through 8, find where the graph of the indicated function is concave upward and where it is concave downward. Find all points of inflection. No sketch is required.

1. $f(x) = (x - 3)^2$.

2. $f(x) = 4 - x^2$.

3. $F(x) = x^3 - 12x$.

4. $F(x) = (x - 3)^3 + 4$.

5. $g(x) = 2x^3 - 3x^2 - 12x + 8$.

6. $g(x) = x^4 - 6x^3 - 24x^2 + x + 2$.

7. $f(x) = 3x^2 - \dfrac{1}{x^2}$.

8. $f(x) = x^3 + 9x^2 + 6x - 1$.

In Exercises 9 to 12, find where the graph is concave upward and where it is concave downward. Find the points of inflection. Then sketch the graph.

9. $g(x) = (x + 1)^4$.

10. $g(x) = \sqrt{x} + \dfrac{1}{\sqrt{x}}$.

11. $F(x) = \sqrt{9 - x^2}$.

12. $F(x) = \dfrac{x\sqrt{x - 2}}{6}$.

In each of Exercises 13 to 16, find the intervals on which the graph is concave upward and the intervals on which it is concave downward. Make a sketch and explain why the graph has no point of inflection.

13. $f(x) = \dfrac{1}{x}$.

14. $f(x) = \dfrac{x}{x + 1}$.

15. $G(x) = 2x + \dfrac{1}{2x}$.

16. $G(x) = \dfrac{2x^2}{4 - x^2}$.

6.11 LIMITS AS $x \longrightarrow \infty$. ASYMPTOTES

A nonempty set S of real numbers has an **upper bound** u, or is **bounded above**, if there is a real number u such that no member of the set S is greater than u.

Thus 4 is an upper bound for the set of all negative numbers, and so is $\frac{1}{2}$; in fact, any nonnegative number is an upper bound for the set of negative numbers.

As an exercise, the reader should write definitions of **lower bound** and **bounded below** analogous to what is stated above.

We now extend the basic concept of the limit of a function (3.2.1). Consider the function f defined by $f(x) = 1/x$. Then $f(100) = 0.01$, $f(10,000) = 0.0001$, and $f(1,000,000) = 0.000\,001$. It is clear that we can make $f(x) = 1/x$ as close to zero as we please by taking x sufficiently large. We say that the limit of f is zero as x increases without bound, and we symbolize it by $\lim\limits_{x \to \infty} (1/x) = 0$.

6.11.1 **Definition.** The limit of f is L as x increases without bound (or as x approaches infinity), written

$$\lim_{x \to \infty} f(x) = L \qquad \text{or} \qquad \lim_{x \to +\infty} f(x) = L,$$

if the domain of f has no upper bound and if for each $\epsilon > 0$ there exists a number M such that

$$x > M \quad \text{and} \quad x \in \mathcal{D}_f \Longrightarrow |f(x) - L| < \epsilon.$$

Similarly,

6.11.2 **Definition.** The limit of f is L as x decreases without bound (or as x approaches negative infinity), written

$$\lim_{x \to -\infty} f(x) = L,$$

if the domain of f has no lower bound and if for each $\epsilon > 0$ there exists a number M such that

$$x < M \quad \text{and} \quad x \in \mathcal{D}_f \Longrightarrow |f(x) - L| < \epsilon.$$

Example 1. Show that $\lim\limits_{x \to \infty} (1/x) = 0$.

Preliminary Analysis. By 6.11.1, we must show that for each $\epsilon > 0$ there is a number M such that $x > M \Longrightarrow |1/x - 0| < \epsilon$. Let ϵ be an arbitrarily chosen positive number. Then, for $x > 0$,

$$\left| \frac{1}{x} - 0 \right| < \epsilon \iff -\epsilon < \frac{1}{x} < \epsilon \iff 0 < \frac{1}{x} < \epsilon \iff 0 < 1 < x \cdot \epsilon$$

$$\iff 0 < \frac{1}{\epsilon} < x \iff \frac{1}{\epsilon} < x.$$

Thus

$$x > \frac{1}{\epsilon} \iff \left| \frac{1}{x} - 0 \right| < \epsilon.$$

Solution. Let $M = 1/\epsilon$. Since

$$x > M = \frac{1}{\epsilon} \Longrightarrow \left| \frac{1}{x} - 0 \right| < \epsilon,$$

it follows from 6.11.1 that

$$\lim_{x \to \infty} \frac{1}{x} = 0.$$

The next example, although somewhat similar to Example 1, gives a useful result.

Example 2.   Prove that if p is a positive integer,

$$\lim_{x \to \infty} \frac{1}{x^p} = 0.$$

Preliminary   We wish to show that for each $\epsilon > 0$ there is some number M such that $x > M \Longrightarrow$
Analysis. $|1/x^p - 0| < \epsilon$.

Let $x > 0$. Then

$$\left| \frac{1}{x^p} - 0 \right| < \epsilon \iff -\epsilon < \frac{1}{x^p} < \epsilon \iff 0 < \frac{1}{x^p} < \epsilon \impliedby 0 < \frac{1}{x} < \epsilon^{1/p}$$

$$\impliedby 0 < 1 < x(\epsilon^{1/p}) \iff \frac{1}{\epsilon^{1/p}} < x,$$

in which we used Exercise 6, Section 1.7. Thus $x > 1/\epsilon^{1/p} \Longrightarrow |1/x^p - 0| < \epsilon$.

Solution.   Let ϵ be an arbitrarily chosen positive number and let $M = 1/\epsilon^p$. Then

$$x > M = \frac{1}{\epsilon^p} \Longrightarrow \left| \frac{1}{x^p} - 0 \right| < \epsilon,$$

as found above. Therefore (6.11.1)

$$\lim_{x \to \infty} \frac{1}{x^p} = 0.$$

Our much-used limit theorem [3.3.1(i) to (iv)] is also true for limits as $x \longrightarrow \infty$ or as $x \longrightarrow -\infty$, and the proof of this new limit theorem is similar to that of the earlier one.

Example 3.   Find $\lim\limits_{x \to \infty} (2 - 3x + x^2)/(7 + 4x - 5x^2)$.

Solution.   Let $x \neq 0$. Dividing numerator and denominator by the highest power of x, namely x^2, and using the limit theorem and Example 2 above, we can write

$$\lim_{x \to \infty} \frac{2 - 3x + x^2}{7 + 4x - 5x^2} = \lim_{x \to \infty} \frac{(2/x^2) - (3/x) + 1}{(7/x^2) + (4/x) - 5}$$

$$= \frac{\lim\limits_{x \to \infty} [(2/x^2) - (3/x) + 1]}{\lim\limits_{x \to \infty} [(7/x^2) + (4/x) - 5]}$$

$$= -\frac{1}{5}.$$

Now consider $1/(x - 3)$ as x approaches 3 *from the right*. Since $x > 3$, $1/(x - 3)$ is positive; and as x becomes sufficiently close to 3, $1/(x - 3)$ increases without bound. This situation is indicated by

$$\lim_{x \to 3^+} \frac{1}{x - 3} = \infty.$$

6.11.3   **Definition.**   If to each positive number M there corresponds a $\delta > 0$ such that the interval $(c, c + \delta)$ is in the domain of f and

$$0 < x - c < \delta \Longrightarrow f(x) > M,$$

we say that $f(x)$ **increases without bound as** x **approaches** c **from the right**, and we symbolize it by

$$\lim_{x \to c^+} f(x) = \infty \qquad \text{or} \qquad \lim_{x \to c^+} f(x) = +\infty.$$

This is sometimes expressed by saying that $f(x)$ becomes positively infinite as x approaches c from the right, but the reader should realize that f does not have any limit as $x \longrightarrow c^+$. As used in calculus, ∞ is not a number. The preceding symbols mean that f does not have a right-hand limit at c *because f increases without bound* as $x \longrightarrow c^+$.

Similar definitions hold for $\lim\limits_{x \to c^-} f(x) = \infty$, $\lim\limits_{x \to c^-} f(x) = -\infty$, and $\lim\limits_{x \to c^+} f(x) = -\infty$.

Finally, we have

6.11.4   Definition.

$$\lim_{x \to \infty} f(x) = \infty \qquad \text{or} \qquad \lim_{x \to +\infty} f(x) = +\infty$$

if the domain of f has no upper bound and if for each positive number M there corresponds a positive number N such that

$$x > N \quad \text{and} \quad x \in \mathfrak{D}_f \Longrightarrow f(x) > M.$$

The definitions for $\lim\limits_{x \to -\infty} f(x) = +\infty$, $\lim\limits_{x \to -\infty} f(x) = -\infty$, and $\lim\limits_{x \to \infty} f(x) = -\infty$ are analogous. The reader should write them out.

Let $f(x) = 1/(x - 3)$, $x > 3$. We saw earlier that

$$\lim_{x \to 3^+} \frac{1}{x - 3} = \infty.$$

The graph of f, for $x > 3$, is shown in Fig. 6-38. As $x \longrightarrow 3^+$, a point $P:(x, f(x))$ on the graph of f rises without bound and becomes arbitrarily close to the vertical line $x = 3$. We say that the line $x = 3$ is a *vertical asymptote* of the graph of f.

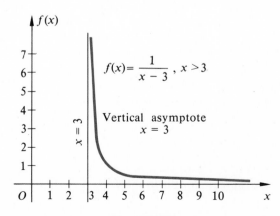

$$f(x) = \frac{1}{x - 3}, \ x > 3$$

Vertical asymptote
$x = 3$

Figure 6-38

6.11.5 Definition. The line $x = c$ is a **vertical asymptote** of the graph of the function f if any of the following four statements is true:

(i) $\lim\limits_{x \to c^+} f(x) = \infty$;   (ii) $\lim\limits_{x \to c^+} f(x) = -\infty$;

(iii) $\lim\limits_{x \to c^-} f(x) = \infty$;   (iv) $\lim\limits_{x \to c^-} f(x) = -\infty$.

Returning to Fig. 6-38, the graph of f becomes indefinitely close ot the x axis as x increases without bound. That is,

$$\lim_{x \to \infty} \frac{1}{x - 3} = 0.$$

We say that the x axis is a *horizontal asymptote* of the graph of f.

6.11.6 Definition. If either

$$\lim_{x \to \infty} f(x) = b \qquad \text{or} \qquad \lim_{x \to -\infty} f(x) = b,$$

the line $y = b$ is a **horizontal asymptote** of the graph of the function f.

Example 4. Find the vertical and horizontal asymptotes of the graph of f, if

(1) $$f(x) = \sqrt{\frac{x}{x - 1}}.$$

Solution. Since

$$\lim_{x \to 1^+} \sqrt{\frac{x}{x - 1}} = \infty,$$

the line $x = 1$ is a vertical asymptote of the graph of f (Fig. 6-39).

Returning to (1), if we divide the numerator and the denominator of the fraction under the radical sign by x, we obtain

$$f(x) = \sqrt{\frac{x}{x - 1}} = \frac{1}{\sqrt{1 - 1/x}}, \qquad x \neq 0, \quad x \neq 1.$$

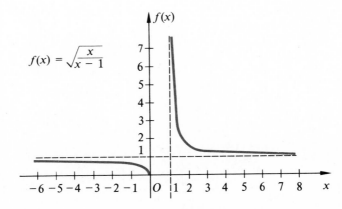

$$f(x) = \sqrt{\frac{x}{x - 1}}$$

Figure 6-39

Thus

$$\lim_{x \to \infty} f(x) = \frac{1}{\sqrt{1 - \lim_{x \to \infty} (1/x)}} = 1$$

and

$$\lim_{x \to -\infty} f(x) = \frac{1}{\sqrt{1 - \lim_{x \to -\infty} (1/x)}} = 1.$$

Therefore (by 6.11.6), $y = 1$ is a horizontal asymptote *in both directions.*

6.11.7 Definition. The line $y = ax + b$ is an **asymptote** of the graph of the function f if either

$$\lim_{x \to \infty} [f(x) - (ax + b)] = 0 \qquad \text{or} \qquad \lim_{x \to -\infty} [f(x) - (ax + b)] = 0.$$

Example 5. Find the nonvertical asymptote of the graph of the function f defined by

(2) $$f(x) = \frac{2x^4 + 3x^3 - 2x - 4}{x^3 - 1}.$$

Solution. By dividing the denominator into the numerator in its right-hand member, (2) can be rewritten

$$f(x) = 2x + 3 - \frac{1}{x^3 - 1}.$$

Since

$$\lim_{x \to \infty} \left[\left(2x + 3 - \frac{1}{x^3 - 1} \right) - (2x + 3) \right] = \lim_{x \to \infty} \left[-\frac{1}{x^3 - 1} \right] = 0,$$

the line $y = 2x + 3$ is an asymptote of the graph of f (by 6.11.7).

Exercises

Evaluate the limits in Exercises 1 through 8.

1. $\lim\limits_{x \to \infty} \dfrac{3 - 2x}{x + 5}.$

2. $\lim\limits_{x \to -\infty} \dfrac{5x + 1}{x - 1}.$

3. $\lim\limits_{x \to -\infty} \dfrac{2x^2 - x + 5}{5x^2 + 6x - 1}.$

4. $\lim\limits_{x \to \infty} \dfrac{2x + 7}{x^2 - x}.$

5. $\lim\limits_{y \to -\infty} \dfrac{9y^3 + 1}{y^2 - 2y + 2}.$ (*Hint:* Divide numerator and denominator by y^2.)

6. $\lim\limits_{x \to \infty} \dfrac{a_0 x^n + a_1 x^{n-1} + \cdots + a_{n-1} x + a_n}{b_0 x^n + b_1 x^{n-1} + \cdots + b_{n-1} x + b_n}$, where $a_0 \neq 0$, $b_0 \neq 0$, and $n \in N$.

7. $\lim\limits_{x \to \infty} (\sqrt{2x^2 + 3} - \sqrt{2x^2 - 5})$. (*Hint:* Multiply and divide by $\sqrt{2x^2 + 3} + \sqrt{2x^2 - 5}$.)

8. $\lim\limits_{x \to \infty} \dfrac{3x + 7}{\sqrt{4x^2 - 1}}.$

9. Write a definition of $\lim\limits_{x \to c^-} f(x) = -\infty$. (*Hint:* See 6.11.3.)

10. Prove that if $\lim\limits_{x \to \infty} f(x) = L$ and $\lim\limits_{x \to \infty} g(x) = M$, then $\lim\limits_{x \to \infty} [f(x) + g(x)] = L + M$. [*Hint:* See proof of 3.3.1(i).]

In Exercises 11 and 12, find the indicated limits.

11. $\lim\limits_{t \to 3^+} \dfrac{3 + t}{3 - t}.$

12. $\lim\limits_{t\to3^-} \dfrac{3+t}{3-t}.$

13. Write a definition of $\lim\limits_{x\to\infty} f(x) = -\infty$. (*Hint:* See 6.11.4.)

14. Find $\lim\limits_{x\to0^+} \dfrac{[\![x]\!]}{x}$ and $\lim\limits_{x\to0^-} \dfrac{[\![x]\!]}{x}$.

In Exercises 15 to 20, find the horizontal and vertical asymptotes of the graphs of the indicated functions, and sketch the graphs.

15. $f(x) = \dfrac{3}{x+1}.$

16. $f(x) = \dfrac{1}{x^2}.$

17. $F(x) = \dfrac{2x}{x-3}.$

18. $F(x) = \dfrac{5-x^3}{9-x^2}.$

19. $\phi(x) = \dfrac{2x}{\sqrt{x^2+5}}.$

20. $\phi(x) = \dfrac{14}{2x^2+7}.$

In Exercises 21 to 24, find all the asymptotes of the graphs of the indicated functions. Sketch the graphs.

21. $f(x) = \dfrac{x^2+1}{x}.$

22. $f(x) = \dfrac{x^5+1}{x^4}.$

23. $F(x) = \dfrac{x^2}{2x-4}.$

24. $F(x) = \dfrac{2x^3+3}{x^2}.$

6.12 GRAPH SKETCHING

As Section 2.6 showed, graph sketching can frequently be shortened by first testing the equation for symmetry with respect to the coordinate axes or the origin. When convenient, one also tries to find the x and y intercepts, at least approximately.

But calculus has now provided us with much more powerful aids for sketching the graph of a differentiable function. The domain and range of the function may confine its graph to a limited region of the plane. It is very helpful to know where the local extreme values of the function occur and the location of any points of inflection. A study of the sign of the first derivative provides information on where the graph is rising and where it is falling; and the sign of the second derivative tells where the graph is concave upward and where it is concave downward. Knowledge of asymptotes is particularly helpful in graph sketching. Finally, points where f, f', or f'' fails to exist should be noted and interpreted.

Often a very good sketch of the graph can be made from the data gathered in these preliminary tests, with very little actual point-plotting.

Procedure for Sketching the Graph of a Differentiable Function.

1. Make a *preliminary analysis* as follows.
    (a) From the domain and range of the function, find *excluded regions* of the plane.
    (b) Test for *symmetry* with respect to the coordinate axes or the origin, and find *intercepts* when convenient.
    (c) Use the first derivative to find the intervals on which the graph is *increasing* and the intervals on which it is *decreasing*.

(d) Find the critical numbers for the function, and use the first or second derivative test to find the *local extreme values*.

(e) From the second derivative, find the intervals on which the graph is *concave upward* and the intervals on which it is *concave downward*.

(f) Find the *points of inflection* of the graph.

(g) Find the *asymptotes*.

2. From this analysis, and by plotting a few more points, draw the graph.

Example 1. Analyze and sketch the graph of

$$f(x) = x + \frac{4}{x}, \qquad x \neq 0.$$

Solution. Since $f(-x) = -f(x)$, f is an odd function and its graph is symmetric with respect to the origin. For $x > 0$, $f(x) > 0$; and for $x < 0$, $f(x) < 0$. Thus the graph of f is confined to the first and third quadrants; and because of the symmetry of the graph with respect to the origin, we can concentrate our attention on the first quadrant (Fig. 6-40).

The equation $f(x) = x + 4/x = (x^2 + 4)/x = 0$ has no real solution; so the graph of f has no x intercept. Neither can f have a y intercept, because $x = 0$ is not in the domain of f.

Now

$$f'(x) = 1 - \frac{4}{x^2} = \frac{x^2 - 4}{x};$$

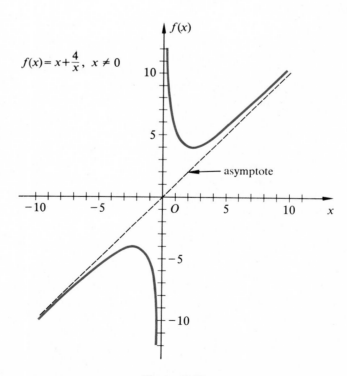

$$f(x) = x + \frac{4}{x}, \quad x \neq 0$$

asymptote

Figure 6-40

so the critical numbers for f are 2 and -2. For $0 < x < 2$, $f'(x) < 0$; and for $x > 2$, $f'(x) > 0$. Thus f is decreasing on $(0, 2]$ and increasing on $[2, \infty)$. At the critical number 2, $f'(2) = 0$. Therefore $f(2) = 4$ is a local minimum of f.

Since $f''(x) = 8/x^3$, $f''(x) > 0$ for $x > 0$. Consequently, the graph of f is concave upward on $(0, \infty)$, and has no point of inflection in the first quadrant.

Because $\lim_{x \to 0^+} f(x) = \infty$, the y axis is a vertical asymptote for the graph of f. There is no horizontal asymptote. But

$$\lim_{x \to \infty} \left[\left(x + \frac{4}{x}\right) - x\right] = \lim_{x \to \infty} \frac{4}{x} = 0,$$

and so $y = x$ is also an asymptote.

By plotting the point $(2, 4)$, where the function has a local minimum, and possibly one or two more points, say $(\frac{1}{2}, \frac{17}{2})$ and $(6, \frac{20}{3})$, the part of the graph in the first quadrant can be sketched easily from the information obtained in the preliminary analysis. The remainder of the graph, in the third quadrant, can be sketched by symmetry (Fig. 6-40).

Example 2.   After making a preliminary analysis, sketch the graph of

$$f(x) = \frac{3x^5 - 20x^3}{32}.$$

Solution.   The domain of f is R and the range of f is R; so there are no excluded regions. Since $f(-x) = -f(x)$, f is an odd function and its graph is symmetric with respect to the origin. The x intercepts are 0 and $\pm\sqrt{20/3} \doteq \pm 2.6$; the y intercept is 0.

We differentiate to obtain

$$f'(x) = \frac{15x^4 - 60x^2}{32} = \frac{15x^2(x^2 - 4)}{32},$$

from which we see that the critical numbers for f are -2, 0, and 2. For $x < -2$, $f'(x) > 0$, and so f is increasing on $(-\infty, -2)$. For $-2 < x < 0$ and $0 < x < 2$, $f'(x) < 0$; thus f is decreasing on $(-2, 0)$ and $(0, 2)$. For $x > 2$, $f'(x) > 0$; so f is increasing on $(2, \infty)$. Moreover, $f'(-2) = f'(2) = 0$. Thus $f(-2) = 2$ is a local maximum of f and $f(2) = -2$ is a local minimum.

Again differentiating, we get

$$f''(x) = \frac{15x(x^2 - 2)}{8}.$$

Thus $f''(-\sqrt{2}) = f''(0) = f''(\sqrt{2}) = 0$.

Since $f''(x) < 0$ for $x < -\sqrt{2}$, $f''(x) > 0$ for $-\sqrt{2} < x < 0$, $f''(x) < 0$ for $0 < x < \sqrt{2}$, and $f''(x) > 0$ for $x > \sqrt{2}$, it follows that the graph of f is concave downward on $(-\infty, -\sqrt{2})$, concave upward on $(-\sqrt{2}, 0)$, concave downward on $(0, \sqrt{2})$, and concave upward on $(\sqrt{2}, \infty)$. Therefore the points of inflection are $(-\sqrt{2}, 7\sqrt{2}/8) \doteq (-1.4, 1.2)$, $(0, 0)$, and $(\sqrt{2}, -7\sqrt{2}/8) \doteq (1.4, -1.2)$.

There are no asymptotes. Making use of all this information and plotting a few more points, say $(1, -0.5)$ and $(3, 5.9)$, the graph is easy to sketch (Fig. 6-41).

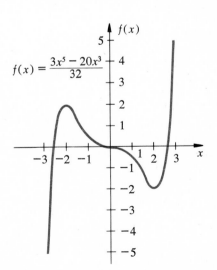

$$f(x) = \frac{3x^5 - 20x^3}{32}$$

Figure 6-41

Example 3. After analyzing the equation

$$F(x) = \frac{\sqrt{x}\,(x-5)^2}{4},$$

sketch its graph.

Solution. The domain of F is $[0, \infty)$ and the range is $[0, \infty)$; so the graph of F is confined to the first quadrant. The x intercepts are 0 and 5, and the y intercept is 0.

From

$$F'(x) = \frac{5(x-1)(x-5)}{8\sqrt{x}}, \qquad x > 0,$$

we find the critical numbers to be 1 and 5. Since $F'(x) > 0$ for $0 < x < 1$, $F'(x) < 0$ for $1 < x < 5$, and $F'(x) > 0$ for $x > 5$, the function is increasing on $[0, 1]$, decreasing on $[1, 5]$, and increasing on $[5, \infty)$. Moreover, at the critical numbers, $F'(1) = 0$ and $F'(5) = 0$; thus $F(1) = 4$ is a local maximum and $F(5) = 0$ is a local minimum.

So far it has been clear sailing. But on calculating the second derivative, we obtain

(1)
$$F''(x) = \frac{5(3x^2 - 6x - 5)}{16x^{3/2}}, \qquad x > 0.$$

Since F'' exists (and is continuous) for all $x > 0$, the only candidates for points of inflection have x coordinates satisfying $3x^2 - 6x - 5 = 0$, $x > 0$. The roots of this quadratic equation are $x = 1 \pm 2\sqrt{6}/3$. But $1 - 2\sqrt{6}/3 < 0$; so the only number in the domain of F that makes $F''(x) = 0$ is $x = 1 + 2\sqrt{6}/3 \doteq 2.6$.

The right-hand member of (1) is too complicated to determine *by simple inspection* the interval on which $F''(x) < 0$ and the interval on which $F''(x) > 0$. However, since F'' is continuous for all $x > 0$ and $F''(x) = 0$ only for $x = 1 + 2\sqrt{6}/3$, $F''(x)$ must be positive throughout one of the intervals $(0, 1 + 2\sqrt{6}/3)$, $(1 + 2\sqrt{6}/3, \infty)$, and negative on the other. Thus the graph of F is concave upward on one of these intervals and concave downward on the other. Since it is concave downward at $x = 1$, where it has a local maximum, the graph of F is concave downward on the interval $(0, 1 + 2\sqrt{6}/3)$. At $x = 5$, F has a local minimum which means that the graph of F is concave upward there. Therefore the graph of F is concave upward on the interval $(1 + 2\sqrt{6}/3, \infty)$. It now follows that the point $(1 + 2\sqrt{6}/3, F(1 + 2\sqrt{6}/3)) \doteq (2.6, 2.3)$ is the only point of inflection of the graph.

The graph has no asymptotes. But by using all this information, plotting the local maximum and minimum points, $(1, 4)$ and $(5, 0)$, the (approximate) point of inflection $(2.6, 2.3)$, and a few other points, say $(0, 0)$, $(4, \frac{1}{2})$, $(7, \sqrt{7}) \doteq (7, 2.6)$, and $(9, 12)$, it is easy to sketch the graph (Fig. 6-42).

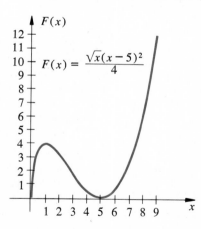

$$F(x) = \frac{\sqrt{x}(x-5)^2}{4}$$

Figure 6-42

Exercises

In each of Exercises 1 to 8, make a preliminary analysis as outlined above, and then sketch the graph.

1. $f(x) = x^3 - 4x$.

2. $f(x) = x^3 - 3x^2 + 3x - 3$.

3. $F(x) = \dfrac{x^4 - 18x^2 + 20}{20}$.

4. $F(x) = \dfrac{x^4(x^2 - 6)}{8}$.

5. $g(x) = \dfrac{(x + 2)^2}{x}$.

6. $g(x) = \dfrac{4x}{x^2 + 2}$.

7. $\phi(x) = 2x\sqrt{x + 3}$.

8. $\phi(x) = \dfrac{x^2}{x^2 - 4}$.

9.   Sketch a possible graph of a function f that has all the following properties:
(a) f is everywhere continuous;
(b) $f(2) = -3, f(6) = 1$;
(c) $f'(2) = 0, f'(x) > 0$ for $x \neq 2, f'(6) = 3$;
(d) $f''(6) = 0, \; f''(x) > 0$ for $2 < x < 6, \; f''(x) < 0$ for $x > 6$.

10.   Sketch a possible graph of a function f that has all the following properties:
(a) f is everywhere continuous;
(b) $f(-3) = 1$;

(c) $f'(x) < 0$ for $x < -3, f'(x) > 0$ for $x > -3$, $f''(x) < 0$ for $x \neq -3$.

11.   Sketch a possible graph of a function f that has all the following properties:
(a) f is everywhere continuous;
(b) $f(-4) = -3, f(0) = 0, f(3) = 2$;
(c) $f'(-4) = 0, f'(3) = 0, f'(x) > 0$ for $x < -4$, $f'(x) > 0$ for $-4 < x < 3, \; f'(x) < 0$ for $x > 3$;
(d) $f''(-4) = 0, f''(0) = 0, f''(x) < 0$ for $x < -4$, $f''(x) > 0$ for $-4 < x < 0, \; f''(x) < 0$ for $x > 0$.

In each of Exercises 12, 13, and 14, a function H is defined.
(a)   Find H' and state its domain.
(b)   Find H'' and state its domain.
(c)   Find the extrema and points of inflection, if any, of H.
(d)   Draw the graph of H.

12. $H(x) = x\,|x|$.

13. $H(x) = x^2\,|x|$.

14. $H(x) = x^3\,|x|$.

6.13   NEWTON'S METHOD FOR DETERMINING THE ROOTS OF $f(x) = 0$

In mathematics we are often faced with the problem of finding the roots of an equation $f(x) = 0$. To be sure, if $f(x)$ is a linear or quadratic polynomial, formulas for writing the exact solutions exist. But for other algebraic equations and for transcendental equations, there are usually no formulas for finding exact solutions. In such cases, we resort to approximation processes, and one of the best is **Newton's method.***

Consider a differentiable function f. The (real) roots of the equation $f(x) = 0$ are, of course, the x intercepts of the graph of $y = f(x)$.

Denote by r an unknown real root of $f(x) = 0$. If we can find an approximation x_1 to r by graphing $y = f(x)$ or by any other means, then, under circumstances to be detailed below, a better approximation to r is given by the intersection x_2 of the tangent to $y = f(x)$ at $(x_1, f(x_1))$ with the x axis (Fig. 6-43).

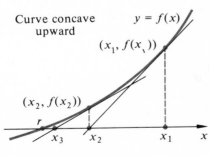

Curve concave upward

Figure 6-43

*Isaac Newton (1642–1727), one of the greatest mathematicians and physicists who ever lived, was educated at Cambridge University and was 23 years old when he developed the calculus, stated his law of universal gravitation, and derived from it (by means of his calculus) the laws of planetary motion that had taken Kepler 20 years of intensive effort to deduce from masses of empirical data.

Since x_2 is also an approximation to r, the tangent to $y = f(x)$ at $(x_2, f(x_2))$ intersects the x axis in a still better approximation to r, x_3 (Fig. 6-43).

By a sufficient number of applications of this process, the real root r of $f(x) = 0$ can be found to as many decimal places of accuracy as desired. This method is due to Isaac Newton.

If $f'(x)$ exists and $f'(x) \neq 0$ throughout a sufficiently large interval containing x_1 and r, and either $f''(x) > 0$ throughout the interval or $f''(x) < 0$ for all values of x in the interval, then the nth approximation, x_n, can be made as close as we please to r by taking n sufficiently large. We know, from 6.3.3 and 6.10.2, that the geometric significance of these two conditions is that the graph is either rising throughout the interval or is falling throughout the interval, and the curve will be concave upward throughout the whole interval (Fig. 6-43) or concave downward throughout the interval (Fig. 6-44).

Let x_1 be a first approximation to a root r of $f(x) = 0$. The tangent to the graph of $y = f(x)$ at $(x_1, f(x_1))$ is

$$y - f(x_1) = f'(x_1)(x - x_1),$$

and its x intercept x_2 is found by setting $y = 0$ in the preceding equation and solving for x. We obtain

(1)
$$x_2 = x_1 - \frac{f(x_1)}{f'(x_1)}.$$

Similarly, the next approximation x_3 is the x intercept of the tangent to the curve $y = f(x)$ at $(x_2, f(x_2))$:

(2)
$$x_3 = x_2 - \frac{f(x_2)}{f'(x_2)}.$$

The nth approximation x_n is (Fig. 6-45)

(3)
$$x_n = x_{n-1} - \frac{f(x_{n-1})}{f'(x_{n-1})},$$

where n is any positive integer greater than 1.

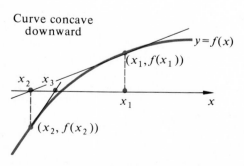

Figure 6-44

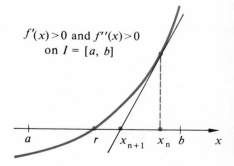

Figure 6-45

Observe in (3) that if $x_n = x_{n-1}$ for some n, then $f(x_{n-1}) = 0$, and thus $x_{n-1} = x_n$ is a root of $f(x) = 0$.

The iterative process of Newton's method is well adapted for programming on a computer. It can be arranged so that the process ends if $x_n = x_{n-1}$ for some n.

Example 1. Approximate the real root of the equation $x^3 + 3x - 6 = 0$.

Solution. Here, $f(x) = x^3 + 3x - 6 = 0$. Since $f(1) = -2$ and $f(2) = 8$, the graph of $y = x^3 + 3x - 6$ crosses the x axis at a point $(r, 0)$ between 1 and 2, and r is a root of $x^3 + 3x - 6 = 0$. Moreover, r is probably nearer 1 than 2, so we let our first approximation to r be $x_1 = 1$.

Since $f'(x) = 3x^2 + 3$, we have $f'(x_1) = f'(1) = 6$. Using formula (1), we find our second approximation x_2 to r to be

$$x_2 = 1 - \frac{(-2)}{6} = \frac{4}{3}.$$

From formula (2), with $x_2 = \frac{4}{3}$, $f(\frac{4}{3}) = \frac{10}{27}$, and $f'(\frac{4}{3}) = \frac{25}{3}$, we obtain x_3, our third approximation to r:

$$x_3 = \frac{4}{3} - \frac{\frac{10}{27}}{\frac{25}{3}} = \frac{58}{45} \doteq 1.289.$$

Actually, the value of r correct to three decimal places is 1.288. By continuing to apply Newton's method, we can find r correct to as many decimal places as needed.

In any approximation process it is important to know how good the approximation is. The following theorem provides a bound for the error at any stage of Newton's process; it also establishes sufficient conditions for Newton's method to give better and better approximations, x_n, to an (unknown) exact root of $f(x) = 0$, as n increases. Proof of this theorem will be deferred until the end of the section.

6.13.1 Theorem. Let r be an (unknown) root of $f(x) = 0$ and x_n an nth approximation to r in Newton's method. Assume that I is an interval containing r and x_n such that

    (i) $f'(x) \neq 0$ for all points x in I.

    (ii) either $f''(x) > 0$ for all x in I or else $f''(x) < 0$ for all x in I.

    (iii) there are positive numbers m and M such that $m \leq |f'(x)|$ and $|f''(x)| \leq M$ for all x in I (Fig. 6-45).

Then

$$|r - x_{n+1}| \leq \frac{M}{2m} |r - x_n|^2.$$

The error in the nth approximation is $|r - x_n|$, and the error in the $(n + 1)$st approximation is $|r - x_{n+1}|$. If $(M/2m) < 2$ for the interval I, it follows from the foregoing theorem that the number of decimal places of accuracy in each approximation is at least double that in the preceding approximation. As an illustration, if x_2 is accurate to two decimal places (so that its absolute error is less than 0.005), then the error in x_3 will be less than $2(0.005)^2 = 0.00005$.

Example 2. The equation $x^4 + 4x^3 + 1 = 0$ has a root between -1 and 0. Find that root, correct to three decimal places.

Solution. Let $f(x) = x^4 + 4x^3 + 1$. The function f is continuous everywhere and has two critical numbers, 0 and -3. There is a point of inflection (with a horizontal tangent) at $(0, 1)$, and $f(-3) = -26$ is the minimum. The graph of $f(x) = 0$ crosses the x axis between -4 and -3, and also between -1 and 0. The reader can easily sketch the graph from these data.

We seek the root, r, of $f(x) = 0$ that is between -1 and 0. By trial, we find that $f(-0.7) = -0.1319$ and $f(-0.6) = 0.2656$. Thus r is between -0.7 and -0.6.

We choose $x_1 = -0.7$ as our first approximation to r in Newton's method. Let us use 6.13.1 to determine how many steps of Newton's method are needed to find r correct to three decimal places. Since $-0.7 < r < -0.6$, then $|r - (-0.7)| = |r - x_1| < 0.1$.

Let m be the minimum of $|f'(x)|$ on $I = [-0.7, -0.6]$. Since $f'(x) = 4x^3 + 12x^2 = 4x^2(x + 3) > 0$ for $-3 < x < 0$, $|f'(x)| < f'(x)$ on I. Moreover, $f''(x) = 12x^2 + 24x = 12x(x + 2) < 0$ for $-2 < x < 0$, and so f' is decreasing on I. Therefore $f'(-0.6) = 3.456$ is the minimum of $f'(x) = |f'(x)|$ on I, and $m = 3.456 \leq |f'(x)|$ on I.

Let M be the maximum of $|f''(x)|$ on I. Since $f''(x) < 0$ on $I = [-0.7, -0.6]$, the maximum of $|f''(x)|$ is the minimum of $f''(x)$ on that interval. Because $f'''(x) = 24x + 24 = 24(x + 1) > 0$ for $x > -1$, f'' is increasing on I. Thus $f''(-0.7) = 10.82$ is the minimum of $f''(x)$ and the maximum of $|f''(x)|$ on I, and $|f''(x)| \leq M = 10.82$ on I.

We find

$$\frac{M}{2m} = \frac{10.82}{6.912} = 1.565 \cdots < 1.6.$$

Then (by 6.13.1)

$$|r - x_2| < (1.6)|r - x_1|^2 = (1.6)(0.1)^2 = 0.016 < 0.05.$$

Thus x_2, the second approximation to r in Newton's method, will be correct to one decimal place.

Again, $|r - x_3| < (1.6)|r - x_2|^2 = (1.6)(0.016)^2 = 0.0004096 < 0.0005$, from which we see that the third approximation, x_3, will have the desired accuracy to three decimal places.

Let us now use Newton's method to calculate x_3. We chose $x_1 = -0.7$. Thus

$$x_2 = x_1 - \frac{f(x_1)}{f'(x_1)} = -0.7 - \frac{f(-0.7)}{f'(-0.7)}$$

$$= -0.7 - \frac{-0.1319}{4.508} = -0.670741 \doteq -0.67;$$

and

$$x_3 = x_2 - \frac{f(x_2)}{f'(x_2)} = -0.67 - \frac{-0.00154079}{4.183748} = -0.669963175 \cdots.$$

The value of r, correct to three decimal places, is -0.670.

Proof of 6.13.1 In Newton's method the $(n + 1)$th approximation is

$$x_{n+1} = x_n - \frac{f(x_n)}{f'(x_n)}.$$

By subtracting r from both members of this equation, we obtain

(4) $$x_{n+1} - r = (x_n - r) - \frac{f(x_n)}{f'(x_n)}.$$

From the extended mean value theorem (Exercise 17, Section 6.2),

$$f(r) = f(x_n) + (r - x_n)f'(x_n) + \frac{(r - x_n)^2}{2}f''(z),$$

where z is some number between x_n and r. Since $f(r) = 0$, this equation may be written

(5)
$$-f(x_n) = (r - x_n)f'(x_n) + \frac{(r - x_n)^2}{2}f''(z).$$

By substituting the right member of (5) for $-f(x_n)$ in (4), we have

$$(x_{n+1} - r) = \frac{(r - x_n)^2}{2}\frac{f''(z)}{f'(x_n)}.$$

Thus

(6)
$$|r - x_{n+1}| = \frac{|r - x_n|^2}{2}\left|\frac{f''(z)}{f'(x_n)}\right|.$$

But $0 < m \le |f'(x_n)|$ and $|f''(z)| \le M$. Therefore

$$|r - x_{n+1}| \le \frac{M}{2m}|r - x_n|^2. \quad\blacksquare$$

Exercises

In Exercises 1 to 4, use Newton's method to find a decimal representation of the given irrational number, correct to three decimal places.

1. $\sqrt{7}$. (*Hint:* Find an approximation to the positive root of $x^2 - 7 = 0$ by Newton's method.)

2. $\sqrt[3]{6}$.

3. $\dfrac{1}{\sqrt[3]{71}}$.

4. $\sqrt[4]{47}$.

5. Find the largest root of $x^3 + 6x^2 + 9x + 2 = 0$, correct to three decimal places

6. Find the x intercept of the graph of $y = 7x^3 + x - 6$, correct to three decimal places.

7. Use Newton's method to find an approximation to the positive root of $x^4 + x^3 + x^2 - 2x - 3 = 0$ that is correct to three decimal places.

8. The function f, defined by $f(x) = 2x^3 + 9x^2 + 12x + 6$, has one and only one zero. Use Newton's method to approximate the value of x that makes $f(x) = 0$.

9. Use Newton's method to approximate the coordinates of the point of intersection of the graphs of $y = 1/x$ and $y = x^2 - 4x$. [*Hint:* Eliminate y between the given equations and approximate the (only) real root of the resulting equation in x.]

10. Sketch the graph of $y = x^{1/3}$. Obviously, its only x intercept is zero. But assuming that we are using Newton's method to approximate its x intercept, show that for all $x_1 \ne 0$, $|x_2| > |x_1|$. Explain this failure of Newton's method.

11. Can Newton's method be used to approximate the x intercept of the graph of $y = |x|^{1/2}$? Why? (*Hint:* Show that $|x_2| = |x_1|$ for all $x_1 \ne 0$.)

12. Approximate to four decimal places of accuracy the critical number for f, if $f(x) = \frac{1}{4}x^4 + 2x^2 - x + 1$.

13. Show that $f(x) = \frac{1}{10}x^5 + x^3 - 6x^2 + 1$ has one and only one point of inflection, and approximate its x coordinate, correct to four decimal places.

14. Approximate to four decimal places of accuracy the number x that makes $f(x) = x^5 + 2x^3 - 4x^2 + 2$ a relative minimum.

6.14 REVIEW EXERCISES

1. Sketch a possible graph of a function f that has all the following properties:
(a) f is continuous everywhere;
(b) $f(-1) = 1$;
(c) $f'(-1) = 0$;
(d) $f''(x) > 0$ for $x \in R$.

2. Sketch a possible graph of a function F that has all the following properties:
(a) F is everywhere continuous;
(b) $F(-2) = 3$, $F(2) = -1$;
(c) $F'(x) = 0$ for $x > 2$;
(d) $F''(x) < 0$ for $x < 2$.

3. A particle moves along a coordinate line. If $s = \frac{1}{2}t^4 - 5t^3 + 12t^2$ gives its directed distance, in feet, from the origin at the end of t seconds, $t \geq 0$, find the velocity of the moving particle when its acceleration is zero.

4. Find where the function f, defined by $f(x) = x^2(x - 4)$, is increasing and where it is decreasing. Find the local extreme values of f. Find the point of inflection. Sketch the graph.

5. A long sheet of metal, 16 inches wide, is to be turned up at both sides to make a horizontal gutter with vertical sides. How many inches should be turned up at each side for maximum carrying capacity?

6. A metal water trough with equal semicircular ends and open top is to have a given capacity (Fig. 6-46). What relative dimensions require the least material?

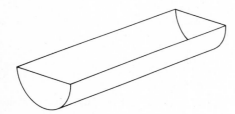

Figure 6-46

7. A particle moves along a coordinate line. If $s = (t - 1)^2(t - 4)$ gives its directed distance from the origin, measured in feet, at the end of t seconds, $t \geq 0$, find
(a) where the particle starts and with what initial velocity;
(b) where and when the particle comes to a stop;
(c) when it is moving in the positive direction and how far;
(d) when it is moving in the negative direction and how far;
(e) when it is back at its starting point;

(f) when its acceleration is negative and when it is positive.

8. Find the maximum and the minimum, if they exist, of the function defined by

$$f(x) = \frac{4}{x^2 + 1} + 2.$$

Find the points of inflection, and sketch the graph.

9. For each of the following functions, decide whether the mean value theorem applies on the indicated interval I. If so, find all possible values of z; if not, tell why. Make a sketch.
(a) $f(x) = \frac{x^3}{3}$; $I = [-3, 3]$.
(b) $F(x) = x^{3/5} + 1$; $I = [-1, 1]$.
(c) $G(x) = x^2[\![x^2 + 1]\!]$; $I = [-1, 1]$.

10. Sketch a possible graph of a function f that has all the following properties:
(a) f is continuous everywhere;
(b) $f(-3) = 2$, $f(1) = 4$;
(c) $f'(x) > 0$ for $x < -3$, $f'(x) > 0$ for $x > 1$;
(d) $f''(x) < 0$ for $x < -3$, $f''(x) = 0$ for $-3 < x < 1$, $f''(x) < 0$ for $x > 1$.

11. Sketch a possible graph of a function F that has all the following properties:
(a) F is everywhere continuous;
(b) $F(-1) = 6$, $F(3) = -2$;
(c) $F'(x) < 0$ for $x < -1$, $F'(-1) = F'(3) = -2$, $F'(7) = 0$;
(d) $F''(x) < 0$ for $x < -1$, $F''(x) = 0$ for $-1 < x < 3$, $F''(x) > 0$ for $x > 3$.

12. A fence, 8 feet high, is parallel to the wall of a building and 1 foot from the building. What is the shortest plank that can go over the fence, from the level ground, to prop the wall?

13. A page of a book is to contain 27 square inches of print. If the margins at the top, bottom, and one side are 2 inches, and the margin at the other side is 1 inch, what size page would use the least paper?

14. Sketch a possible graph of a function F that has all the following properties:
(a) F is everywhere continuous;
(b) $F(-3) = 5$, $F(0) = 2$, $F(4) = 5$, $F(6) = 0$;
(c) $F''(x) = 0$ for $x < 0$, $F''(x) > 0$ for $0 < x < 4$, $F''(x) = 0$ for $x > 4$.

15. Find the maximum and the minimum of the function defined on the closed interval $[-2, 2]$ by

$$f(x) = \begin{cases} \frac{1}{4}(x^2 + 6x + 8), & -2 \le x \le 0, \\ -\frac{1}{6}(x^2 + 4x - 12), & 0 \le x \le 2. \end{cases}$$

Find where the graph is concave upward and where it is concave downward. Sketch the graph.

16. Sketch a possible graph of a function G that has all the following properties:

(a) G is continuous on its domain, which is $\mathcal{D}_G$
  $= (-\infty, -2) \cup (-2, 2) \cup (2, \infty)$;
(b) $G(0) = 0$;
(c) $G''(x) > 0$ for all $x \in \mathcal{D}_G$;
(d) $\lim\limits_{x \to \infty} G(x) = 0$, $\lim\limits_{x \to -\infty} G(x) = 0$;
(e) $\lim\limits_{x \to -2^-} G(x) = \lim\limits_{x \to -2^+} G(x) = \infty$, $\lim\limits_{x \to 2^-} G(x) = \lim\limits_{x \to 2^+} G(x) = \infty$.

17. Sketch a possible graph of a function g that has all the following properties:

(a) g is continuous on its domain, which is $\mathcal{D}_g = (-\infty, 0) \cup (0, \infty)$;
(b) $g(-2) = 0$, $g(1) = 2$;
(c) $g'(1) = 0$;
(d) $g''(x) > 0$ for $x \in \mathcal{D}_g$;
(e) $\lim\limits_{x \to 0^-} g(x) = \lim\limits_{x \to 0^+} g(x) = \infty$;
(f) $\lim\limits_{x \to \infty} [g(x) - x] = 0$, $\lim\limits_{x \to -\infty} [g(x) - x] = 0$.

18. Sketch a possible graph of a function H that has all the following properties:

(a) H is continuous on its domain, which is $\mathcal{D}_H = (1, \infty)$;
(b) $H(3) = 1$, $H(5) = 4$, $H(7) = 7$, $H(9) = 5$;
(c) $H'(3) = 0$, $H'(7) = 0$;
(d) $H''(x) > 0$ for $1 < x < 5$, $H''(5) = 0$, $H''(x) < 0$ for $5 < x < 9$, $H''(9) = 0$, $H''(x) > 0$ for $x > 9$;
(e) $\lim\limits_{x \to 1^+} H(x) = \infty$, $\lim\limits_{x \to \infty} H(x) = 0$.

19. Let $s = 100t - 200\sqrt{t+1} + 200$ define the position function of a particle moving on a horizontal coordinate line, where s is its directed distance (in feet) from the origin at the end of t seconds, $t \ge 0$. Show that the particle starts from rest at the origin, moves to the right with increasing speed forever, and yet never quite attains a velocity of 100 feet per second.

20. Sketch the graph of the unique function f that has all the following properties:

(a) f is everywhere continuous;
(b) $f(-6) = 4$;
(c) $f'(x) = 1$ for $x < -6$, $f(x) = -2$ for $-6 < x < -4$, $f'(x) = 0$ for $-4 < x < 0$, $f'(x) = 2x$ for $x > 0$.

[*Hint:* What function on the interval $(0, \infty)$ can be differentiated to yield $f'(x) = 2x$?]

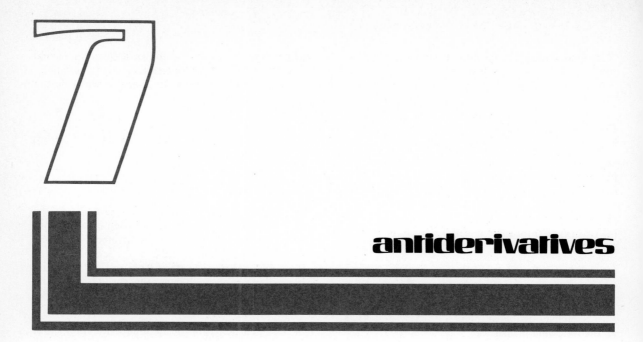

antiderivatives

This chapter forms a transition from differential calculus to integral calculus. It provides the link, exploited so successfully by Newton and Leibniz, between those two great concepts, the derivative and the definite integral.

The reader has been familiar with *inverse operations* for a long time. Addition and subtraction are operations that are inverse to each other, as are multiplication and division. Each operation in a pair of inverse operations undoes what the other does. We are now about to discuss what will temporarily be called *antidifferentiation*. Differentiation and antidifferentiation are almost inverse operations.

7.1 ANTIDERIVATIVES

Often in calculus we are given the derivative of an unknown function and it is necessary to discover the function.

Example 1. Find the equation of a curve whose tangent at any point (x, y) on the curve has its slope equal to $4x^3$.

Solution. We seek an equation $y = f(x)$ such that

$$D_x y = 4x^3.$$

From our experience with differentiation we recognize $4x^3$ as the result of differentiating x^4. Hence

$$y = x^4$$

is the equation of a curve whose tangent has slope $4x^3$ at every point (x, y) on the curve.

It should be noted that this problem has more than one solution. In fact, it has infinitely many correct solutions. For if C is any constant whatever, we see that any equation of the form $y = x^4 + C$ is a solution, since here, too, $D_x y = 4x^3$.

7.1.1 **Definition.** If f is the derivative of a function F, that is, if $f = F'$, then F is an **antiderivative** of f.

As an illustration, consider $F(x) = x^2$ and $f(x) = 2x$. Since $F'(x) = 2x = f(x)$, f is the derivative of F and F is an antiderivative of f. Again, if $F(x) = x^3$ and $f(x) = 3x^2$, we have $F'(x) = 3x^2 = f(x)$; so f is the derivative of F and F is an antiderivative of f.

Thus differentiation and antidifferentiation strongly resemble inverse processes. But while a function has at most one derivative, its derivative can have many anti-derivatives, one of which is the original function. We have just seen that $F(x) = x^3$ defines an antiderivative F of the function f whose value is $f(x) = 3x^2$; yet $x^3 - 6$ and $x^3 + \sqrt{2}$ also define antiderivatives of f, because $D_x(x^3 - 6) = D_x(x^3 + \sqrt{2})$ $= 3x^2 = f(x)$. In fact, if $F'(x) = 3x^2$, then $F(x) = x^3 + C$ (where C is an arbitrary constant) defines a whole family of antiderivatives of f.

We shall prove that $F(x) + C$, where F is any particular antiderivative of f and C is an arbitrary constant, defines all possible antiderivatives of the function f. Let us first establish a consequence of the mean value theorem.

7.1.2 **Lemma.** If $H'(x) = 0$ for all x in an interval (a, b), then $H(x) = C$ throughout this interval, where C is a constant.

Proof. Let x_1 be an arbitrarily selected point in (a, b) and let x be any other point in (a, b). Since H' exists for all points in (a, b) by hypothesis, then H' exists and H is continuous on the closed interval whose endpoints are x_1 and x. Thus the function H satisfies the conditions for the mean value theorem (6.2.1) on the closed interval whose endpoints are x_1 and x, and there exists a number z between x_1 and x such that

$$H(x) - H(x_1) = H'(z)(x - x_1).$$

But $H'(z) = 0$ by hypothesis. Therefore $H(x) - H(x_1) = 0$, or

$$H(x) = H(x_1) = C$$

for all x in (a, b). ∎

It is now easy to prove the following theorem.

7.1.3 **Theorem.** If $F'(x) = G'(x)$ for all x in (a, b), then

$$F(x) = G(x) + C,$$

where C is a constant; that is, if the derivatives of two functions are equal, the functions differ by an additive constant (which may be zero).

Proof. Let H be the function defined on (a, b) by $H = F - G$. Since

$$H'(x) = F'(x) - G'(x) = 0$$

for all x in (a, b), then $H(x) = C$, where C is a constant (7.1.2). That is, $F(x) - G(x) = C$, or

$$F(x) = G(x) + C$$

for all x in (a, b). ∎

When F is a particular antiderivative of f, we have the following corollary.

7.1.4 Corollary. The most general antiderivative of f is defined by

$$F(x) + C,$$

where F is any particular antiderivative of f and C is an arbitrary constant; all anti-derivatives of f may be obtained from this by giving C particular values.

7.2 FINDING ANTIDERIVATIVES

From the beginning of differentiation, we have used three interchangeable symbols for the derivative of a function f—namely, the operator notation $D_x f$, the Lagrange symbol f', and the Leibniz notation dy/dx when $y = f(x)$. Each has its advantages, depending on the circumstances, and by now the reader should be accustomed to working comfortably with all of them. In the present chapter the D_x symbol will predominate, because it suggests a simple notation for antiderivatives.

Just as D_x means the derivative with respect to x, the symbol A_x will be used, for the present, to indicate the most general antiderivative (with respect to x). Thus $A_x(3x^2) = x^3 + C$, where C is an arbitrary constant, is read "the antiderivatives of $3x^2$ are $x^3 + C$." It would be more precise to say "the antiderivatives of the function whose value is given by $3x^2$ is the function whose value is given by $x^3 + C$," but we shall often use the briefer language in this chapter for simplicity and clarity.

Let F be an antiderivative of f, so that $A_x f(x) = F(x) + C$. By differentiating both members of this equation with respect to x, we obtain

$$D_x[A_x f(x)] = D_x[F(x) + C].$$

Since F is an antiderivative of f, $D_x F(x) = F'(x) = f(x)$, by 7.1.1, and $D_x C = 0$. Therefore $D_x A_x f(x) = f(x)$, uniquely.

7.2.1 Theorem. The derivative of an antiderivative of a function f is f itself; in symbols,

$$D_x A_x f(x) = f(x).$$

But if the order of the two operators, D_x and A_x, is reversed, the result is not unique. For $A_x[D_x F(x)] = A_x F'(x) = A_x f(x) = F(x) + C$, from 7.1.1 and 7.1.4.

Consequently,

$$A_x D_x F(x) = F(x) + C,$$

where C is an arbitrary constant.

Our familiarity with the power formula for differentiation suggests that

$$A_x(4x^3) = x^4, \qquad A_x x^3 = \frac{x^4}{4}, \qquad A_x x^4 = \frac{x^5}{5},$$

and

$$A_x x^n = \frac{x^{n+1}}{n+1} \qquad \text{if} \qquad n \neq -1.$$

7.2.2 Theorem. If n is any rational number except -1, then

$$A_x x^n = \frac{x^{n+1}}{n+1} + C, \qquad n \neq -1,$$

where C is an arbitrary constant.

It follows from 7.2.1 that this formula, like every formula for antidifferentiation, is readily proved by showing that the derivative of the right member is the same as the expression following the antiderivative sign in the left member. So *in order to verify a formula for antidifferentiation, we simply differentiate both members of the formula.*

Example 1. From 7.2.2,

$$A_x x^{10} = \frac{x^{11}}{11} + C;$$

$$A_t \left(\frac{1}{t^3}\right) = A_t t^{-3} = \frac{t^{-2}}{-2} + C = -\frac{1}{2t^3} + C;$$

$$A_x \sqrt{x} = A_x x^{1/2} = \frac{x^{3/2}}{\frac{3}{2}} + C = \frac{2x\sqrt{x}}{3} + C.$$

It should be emphasized that every formula for differentiation has a counterpart for antidifferentiation that can be proved by using 7.2.1. Examples of this are given in the next theorem.

7.2.3 Theorem.

(i) The antiderivatives of the sum of two functions are the sum of the antiderivatives of the functions:

$$A_x[f(x) + g(x)] = A_x f(x) + A_x g(x).$$

(ii) The antiderivatives of a constant times a function are the constant times the antiderivatives of the function:

$$A_x[k \cdot f(x)] = k[A_x f(x)].$$

Proof of (i). Since the derivative of the sum of two functions is equal to the sum of the derivatives of the functions, we can differentiate the right member of the above formula and use 7.2.1 to obtain

$$D_x[A_x f(x) + A_x g(x)] = D_x A_x f(x) + D_x A_x g(x)$$
$$= f(x) + g(x),$$

and the theorem follows (7.1.1).

A proof of part (ii) is similar, and is left as an exercise. ∎

It follows from the second part of this theorem that *a constant factor may be moved across the antiderivative sign.* But a factor that is not constant cannot be moved across the antiderivative sign.

Example 2. Find $A_x(4x^2 - 2x + 7)$.

Solution. From 7.2.2 and 7.2.3,

$$A_x(4x^2 - 2x + 7) = A_x(4x^2) + A_x(-2x) + A_x(7)$$
$$= 4A_x x^2 - 2A_x x + 7A_x x^0$$
$$= \frac{4x^3}{3} - \frac{2x^2}{2} + \frac{7x^1}{1} + C$$
$$= \frac{4x^3}{3} - x^2 + 7x + C.$$

Example 3. Find $A_x(17x^5 - 10x^2 + 1/x^6 + 2\sqrt{x})$.

Solution. By 7.2.2 and 7.2.3,

$$A_x\left(17x^5 - 10x^2 + \frac{1}{x^6} + 2\sqrt{x}\right) = A_x(17x^5) + A_x(-10x^2) + A_x(x^{-6}) + A_x(2x^{1/2})$$
$$= 17A_x x^5 - 10A_x x^2 + A_x x^{-6} + 2A_x x^{1/2}$$
$$= \frac{17x^6}{6} - \frac{10x^3}{3} + \frac{x^{-5}}{-5} + \frac{2x^{3/2}}{\frac{3}{2}} + C$$
$$= \frac{17x^6}{6} - \frac{10x^3}{3} - \frac{1}{5x^5} + \frac{4x\sqrt{x}}{3} + C.$$

Example 4. Find $A_w[(5w^3 - w + 15)/w^3]$.

Solution.

$$A_w\frac{5w^3 - w + 15}{w^3} = A_w(5 - w^{-2} + 15w^{-3})$$
$$= 5w - \frac{w^{-1}}{-1} + \frac{15w^{-2}}{-2} + C$$
$$= 5w + \frac{1}{w} - \frac{15}{2w^2} + C.$$

Exercises

Find the antiderivatives of each of the functions defined by the following expressions.

1. $6x^2 - 6x + 1$.

2. $3x^2 + 10x - 7$.

3. $18x^8 - 25x^4 + 3x^2$.

4. $x^2(20x^7 - 7x^4 + 6)$.

5. $\dfrac{4}{x^5}$.

6. $\dfrac{4x^6 + 3x^5 - 8}{x^5}$.

In each of Exercises 7 to 12, find the most general function f that satisfies the given equation.

7. $f'(x) = 20x^3 - 6x^2 + 17$.

8. $f'(x) = 28x^{13} + 15x^4 - 8x + 7$.

9. $f'(x) = \dfrac{4}{x^3} - 8x^3 - 5$.

10. $f'(x) = \dfrac{(x^2 - 4)^2}{2x^2}$.

11. $f'(x) = \sqrt{5x^3}$.

12. $f'(x) = \sqrt[5]{32x^3}$.

In each of Exercises 13 to 16, find the most general function f that satisfies the given equation. (*Hint:* Each antidifferentiation produces a new arbitrary constant.)

13. $f''(x) = 12x - 6$.

14. $f''(x) = 10x^2 + 3$.

15. $f''(x) = 45\sqrt{x} + 6x$.

16. $f''(x) = \dfrac{3}{\sqrt{x}}$.

Verify the statements in Exercises **17** to **21** by differentiating the right member.

17. $A_x[(7x - 3)^5(7)] = \dfrac{(7x - 3)^6}{6} + C$.

18. $A_x[(x^2 - 5)^4(2x)] = \dfrac{(x^2 - 5)^5}{5} + C$.

19. $A_x[(2x^2 - x + 4)^2(4x - 1)] = \dfrac{(2x^2 - x + 4)^3}{3} + C$.

20. $A_x[(5x^3 - 7)^{1/2}(15x^2)] = \dfrac{(5x^3 - 7)^{3/2}}{\frac{3}{2}} + C$.

21. $A_x[(x^5 - 2x + 11)^{-1/2}(5x^4 - 2)]$
    $= \dfrac{(x^5 - 2x + 11)^{1/2}}{\frac{1}{2}} + C$.

22. In each of Exercises 17 to 21, what is the relation of the expression in the second parentheses to the expression inside the first parentheses? What formula for an antiderivative of $[f(x)]^k[f'(x)]$ do these exercises suggest? Verify your answer by differentiation.

7.3 GENERALIZED POWER FORMULA FOR ANTIDERIVATIVES

Since every formula for differentiation gives rise to a formula for antidifferentiation, we might expect that the powerful chain rule for finding derivatives would have interesting counterparts for antiderivatives.

Let f be a differentiable function and let $u = f(x)$. If we differentiate $u^{n+1}/(n + 1)$, $n \neq -1$, *with respect to x* by means of the chain rule, we obtain

$$D_x\left[\frac{u^{n+1}}{n+1}\right] = \left[D_u\frac{u^{n+1}}{n+1}\right] \cdot D_x u = \frac{(n+1)u^n}{n+1} \cdot D_x u = u^n D_x u,$$

or

(1) $$D_x\left[\frac{u^{n+1}}{n+1}\right] = u^n D_x u, \qquad n \neq 1.$$

Therefore, by the definition of an antiderivative with respect to x (7.1.1),

$$A_x[u^n D_x u] = \frac{u^{n+1}}{n+1} + C, \qquad n \neq -1.$$

Since $u = f(x)$, this can also be written

$$A_x\{[f(x)]^n \cdot f'(x)\} = \frac{[f(x)]^{n+1}}{n+1} + C, \qquad n \neq -1.$$

7.3.1   Theorem.   If f is a differentiable function and $u = f(x)$, then

$$A_x[u^n \cdot D_x u] = \frac{u^{n+1}}{n+1} + C, \qquad n \neq -1,$$

or

$$A_x\{[f(x)]^n \cdot f'(x)\} = \frac{[f(x)]^{n+1}}{n+1} + C, \qquad n \neq -1.$$

Example 1.   Find $A_x[6x(3x^2 - 1)^5]$.

Solution.   If we let $u = 3x^2 - 1$, then $D_x u = 6x$, and we can write

$$A_x[6x(3x^2 - 1)^5] = A_x[u^5 \cdot D_x u] = \frac{u^6}{6} + C = \frac{(3x^2 - 1)^6}{6} + C$$

(by 7.3.1).

This can also be done without using u as follows. Let $f(x) = 3x^2 - 1$; then $f'(x) = 6x$, and

$$A_x[6x(3x^2 - 1)^5] = A_x[(3x^2 - 1)^5 \cdot D_x(3x^2 - 1)] = \frac{(3x^2 - 1)^6}{6} + C$$

(by 7.3.1).

Example 2.   Find $A_x[x\sqrt{10x^2 + 1}]$.

Solution.   Let $f(x) = 10x^2 + 1$; then $f'(x) = 20x$. We can write

$$A_x[x\sqrt{10x^2 + 1}] = \frac{1}{20} A_x[(10x^2 + 1)^{1/2} \cdot 20x],$$

since a constant factor, such as $\frac{1}{20}$, can be moved across the antiderivative sign [see 7.2.3(ii)]. Thus

$$\frac{1}{20} A_x[(10x^2 + 1)^{1/2} \cdot 20x] = \frac{1}{20} A_x[(10x^2 + 1)^{1/2} \cdot D_x(10x^2 + 1)]$$

$$= \frac{1}{20} \frac{(10x^2 + 1)^{3/2}}{\frac{3}{2}} + C = \frac{(10x^2 + 1)^{3/2}}{30} + C$$

(by 7.3.1). Finally,

$$A_x[x\sqrt{10x^2 + 1}] = \frac{(10x^2 + 1)^{3/2}}{30} + C.$$

The reader should verify this result by differentiating the right member with respect to x.

It is important to notice in Example 2 that although we can supply a needed *constant* factor after the antiderivative sign and compensate for it by putting its reciprocal in front of the antiderivative sign because of 7.2.3(ii), we *cannot* supply an expression involving the variable, since a variable cannot be moved across the antiderivative sign. Thus

(2) $$A_x[(10x^2 + 1)^{1/2}] \neq \frac{2(10x^2 + 1)^{3/2}}{3} + C,$$

because

$$D_x\left[\frac{2(10x^2 + 1)^{3/2}}{3}\right] = (10x^2 + 1)^{1/2} \cdot D_x(10x^2 + 1)$$

$$= (10x^2 + 1)^{1/2} \cdot 20x \neq (10x^2 + 1)^{1/2},$$

and we cannot supply the needed factor, x, in the left member of (2).

Example 3.   Find $A_x\left[\dfrac{x^2}{\sqrt{x^3 + 6}}\right].$

Solution.

$$A_x\left[\frac{x^2}{\sqrt{x^3 + 6}}\right] = A_x[(x^3 + 6)^{-1/2}x^2].$$

If we let $x^3 + 6 = f(x)$, then $f'(x) = 3x^2$. So if we insert the factor 3 just before the x^2 and compensate for it by putting the factor $\frac{1}{3}$ before the antiderivative sign, we have

$$A_x\left[\frac{x^2}{\sqrt{x^3 + 6}}\right] = \frac{1}{3}A_x[(x^3 + 6)^{-1/2}(3x^2)] = \frac{1}{3}A_x[(x^3 + 6)^{-1/2} \cdot D_x(x^3 + 6)]$$

$$= \frac{1}{3}\frac{(x^3 + 6)^{1/2}}{\frac{1}{2}} = \frac{2}{3}\sqrt{x^3 + 6} + C.$$

Exercises

In each of Exercises 1 through 16, find the anti-derivatives of the indicated function and verify your result by differentiation.

1. $3(3x + 1)^4.$

2. $2x(x^2 - 4)^3.$

3. $(5x^3 - 18)^7(15x^2).$

4. $(x^2 - 3x + 2)^2(2x - 3).$

5. $3x^4(2x^5 + 9)^3.$

6. $6x\sqrt{3x^2 + 7}.$

7. $(5x^2 + 1)(5x^3 + 3x - 8)^6.$

8. $(5x^2 + 1)\sqrt{5x^3 + 3x - 2}.$

9. $3x\sqrt[3]{2x^2 - 11}.$

10. $\dfrac{3x}{\sqrt{2x^2 + 5}}.$

11. $\dfrac{x - 2}{x^3 - 3x - 2}.$ (*Hint:* Factor the denominator.)

12. $\dfrac{8x^6 - 22x^2 + 3}{2x^2}.$ (*Hint:* Divide denominator into numerator.)

13. $\dfrac{3x^3 + 6x^2 + 3x + 1}{x^2 + 2x + 1}.$

14. $\dfrac{x}{9x^4 + 6x^2 + 1}.$

15. $\sqrt{\dfrac{5x^2}{3x^2 + 7}}.$

16. $(x^3 - 6)^3(2x^5 - 12x^2).$

In Exercises 17 and 18, a is a nonzero constant. Find $A_x f(x)$ and verify your results by differentiation.

17. $f(x) = \dfrac{x}{a^2\sqrt{a^2 - x^2}}.$

18. $f(x) = \dfrac{x - a}{a^2\sqrt{2ax - x^2}}.$

19. Let $f(x) = x/(x + 1)$, $x \neq -1$.
(a) Find $f''(x)$.
(b) Find $A_x(A_x f''(x))$.
(c) Is your answer to (b) the same as the original $f(x)$? Explain.

20. Let f and g be functions such that they and their first derivatives all have a common domain.

Prove the antidifferentiation formula

$$A_x\{f(x)g'(x) + f'(x)g(x)\} = f(x)g(x) + C.$$

In Exercises 21 and 22, use the formula from Exercise 20.

21. Find $A_x\left[\dfrac{x^2}{2\sqrt{x-1}} + 2x\sqrt{x-1}\right]$.

22. Find $A_x\left[\dfrac{-x^3}{(2x+5)^{3/2}} + \dfrac{3x^2}{\sqrt{2x+5}}\right]$.

23. For all values of x, find
    (a) $A_x|x|$; (b) $A_x(x|x|)$;
    (c) $A_x(x^2|x|)$.

7.4 SOME APPLICATIONS OF ANTIDERIVATIVES. DIFFERENTIAL EQUATIONS

Example 1. Find the equation of the curve that passes through the point $(-1, 2)$ and whose slope at any point on the curve is equal to twice the abscissa of that point.

Solution. Since the slope of a curve at a point (x, y) on the curve is $D_x y$ and the abscissa of that point is x, we have

(1) $$D_x y = 2x.$$

Equation (1) is an example of what is known as a **differential equation** (although "derivative equation" might have been more appropriate). To *solve* the differential equation (1) means to find a function f such that $y = f(x)$ satisfies (1) identically.

If we take antiderivatives with respect to x of both members of (1), we obtain

$$A_x D_x y = A_x(2x),$$

or $y = x^2 + C$. Thus all the solutions of the differential equation (1) are given by

(2) $$y = x^2 + C,$$

where C is an arbitrary constant.

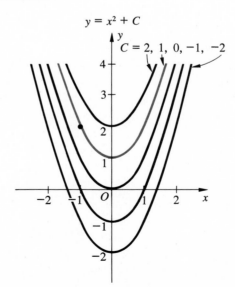

$$y = x^2 + C$$

Figure 7-1

The graph of (2) is a family of "parallel" curves, one curve for each value assigned to C (Fig. 7-1). By "parallel" curves we mean that, for any particular value of x, the slopes of all curves of the family are the same.

To find the curve of the family that goes through $(-1, 2)$, we determine the value of C by substituting $x = -1$ and $y = 2$ in (2), obtaining $C = 1$. So the equation of the desired curve is

$$y = x^2 + 1.$$

Example 2. Near the earth's surface an object is allowed to fall freely, starting from rest. Neglecting air resistance, find formulas for its velocity at the end of t seconds and for the distance traveled in t seconds.

Solution. Let s represent the number of feet that the object falls in t seconds, and choose an s axis with its origin at the point where the object starts its fall and with its positive direction downward (Fig. 7-2).

On or near the surface of the earth, the acceleration of the object, due to gravity, is approximately 32 feet per second per second and is directed toward the center of the earth. Thus

$$a = 32.$$

Since $a = D_t v$, this can be written

$$(3) \qquad D_t v = 32.$$

To solve this differential equation, we take antiderivatives with respect to t of both members:

$$A_t D_t v = A_t (32)$$

or

$$(4) \qquad v = 32t + C_1.$$

Figure 7-2

Since the object falls from rest, the initial conditions are that $v = 0$ when $t = 0$. Substituting $v = 0$ and $t = 0$ in (4), we find $C_1 = 0$. Therefore

$$(5) \qquad v = 32t$$

is the desired formula for the velocity of the falling object at the end of t seconds.

Now $v = D_t s$, and (5) may be rewritten

$$(6) \qquad D_t s = 32t.$$

Solving (6) by antidifferentiating both members, we get $A_t D_t s = A_t(32t)$, or

$$(7) \qquad s = 16t^2 + C_2.$$

But $s = 0$ when $t = 0$. By substituting these initial conditions in (7), we find that $C_2 = 0$. Thus the equation that gives the distance traveled by the falling object in t seconds is

$$s = 16t^2.$$

Compare this example with Example 2 of Section 6.4.

Exercises

1. Find the equation of the curve through the origin whose slope at any point on it is three times the directed distance of that point from the line $x = -2$. Sketch the curve.

2. Find the equation of the curve whose slope at any point on it is equal to four times the cube of the abscissa of that point, and that goes through the point $(-1, 5)$. Sketch the curve.

3. Find the equation of the curve whose slope is five more than the slope of the curve $3y = x^3 + 3x^2 - 24x$ for each value of x, and that goes through the point $(3, 5)$. Sketch both curves on the same axes.

4. Find the equation of the curve through $(-2, -\frac{1}{3})$, if its slope is the negative reciprocal of the slope of $xy = 2$ for each nonzero value of x. Graph both curves on the same axes.

In Exercises 5 to 8, the acceleration, a, of a particle moving on the x axis is given as a function of the time t, where x is measured in feet and t in seconds. If the initial velocity and the initial position of the moving particle are the given values of v_0 and x_0, find the velocity and position of the particle at the given time t.

5. $a = 7$, $v_0 = -3$, $x_0 = 4$, $t = 6$.

6. $a = -6t$, $v_0 = 3$, $x_0 = 6$, $t = 2$.

7. $a = 15\sqrt{t} + 8$, $v_0 = -6$, $x_0 = -44$, $t = 4$.

8. $a = \dfrac{72}{(2t + 3)^3}$, $v_0 = 2$, $x_0 = -6$, $t = 3$.

In Exercises 9 and 10, assume that the acceleration due to gravity at, or near, the earth's surface is 32 feet per second per second and neglect friction. Use antidifferentiation, and do not refer to the formulas developed in Example 2.

9. From what height must a ball be dropped so as to strike the ground with a velocity of 136 feet per second?

10. A ball is thrown directly downward from a tower 364 feet high, with an initial velocity of 48 feet per second. In how many seconds will it strike the ground and with what velocity?

11. If the brakes of a car, when fully applied, produce a constant deceleration of 11 feet per second per second, what is the shortest distance in which the car can be braked to a halt when it is traveling 60 miles per hour?

12. A block slides down an inclined plane with a constant acceleration of 8 feet per second per second. If the inclined plane is 75 feet long and the block reaches the bottom in 3.75 seconds, what was the initial velocity of the block?

13. On the moon, the acceleration due to gravity is approximately $\frac{1}{6}$ of what it is on the earth. If a high jumper can clear 7 feet on earth, how high should he be able to jump on the moon?

14. Compare the velocity at impact of an object dropped from a height of 200 feet above the moon's surface with the velocity at impact of an object dropped from the same height above the earth's surface.

15. A particle starts from the origin with an initial velocity of 3.5 linear units per second and moves along the x axis. If its acceleration at the end of t seconds is given by

$$a(t) = \frac{-9}{4(3t + 1)^{3/2}},$$

find an expression $x(t)$ for its position in t seconds, $t \geq 0$.

16. What constant acceleration will cause a car to increase its velocity from 45 to 60 miles per hour in 10 seconds?

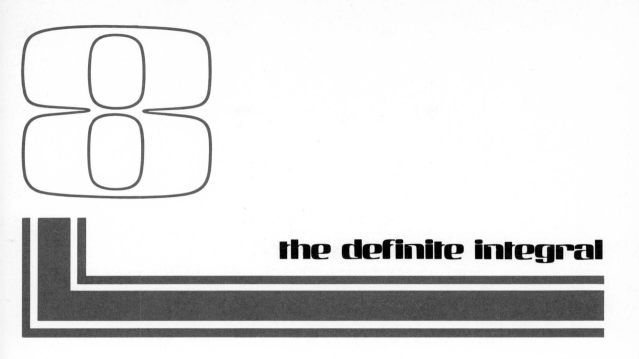

the definite integral

We have been studying differential calculus for some time. In the present chapter we start *integral calculus*. Unlike the derivative, which was conceived in the seventeenth century, the idea of the *definite integral* stems from antiquity. Archimedes, in the third century B.C., succeeded in finding the areas of a few plane regions bounded wholly, or in part, by curves. His method, which we are about to explain, is basic in the concept of the definite integral.

Although we shall start with a discussion of area in order to motivate the concept of the definite integral, the subsequent definition of the definite integral (8.3.4) and the proofs of its properties will be based entirely on real numbers, not on geometric intuition.

Just as the derivative was introduced by discussing the problem of finding tangents to curves and later was shown to have many practical applications unrelated to geometry, the definite integral will be shown in the next chapter to have many important applications that are very different from area, among them *work*, *fluid pressure*, and *moments of inertia*.

8.1 INTRODUCTION TO AREA

Every rectangle and every triangle has a number associated with it called its *area*. The area of a rectangle is defined to be the product of its length and width, and the area of a triangle is one-half the product of the length of its base by its altitude.

Since a polygon can always be decomposed into triangles (Fig. 8-1), its area is defined as the sum of the areas of the composing triangles.

Yet how can we find the area of a plane region enclosed by a curve, or whose boundary includes a segment of a curve (Fig. 8-2)? Indeed, what do we mean by such an "area"?

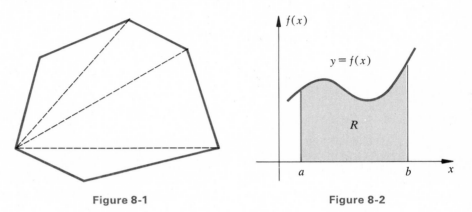

Figure 8-1   Figure 8-2

In light of our experience with rectilinear figures, we would expect an acceptable definition of the area of a plane region to imply the following properties.

8.1.1   Properties of the Area of a Plane Region.
   (i)   The area of a plane region is a nonnegative (real) number.
   (ii)   Congruent regions have equal areas.
   (iii)   If a region is contained in a second region (that is, if every point of the first region is also a point of the second region), then the area of the first region is less than or equal to the area of the second region.
   (iv)   The area of the union of two nonoverlapping regions is equal to the sum of the areas of the two regions.
   (v)   The area of a rectangle is the product of the lengths of two adjacent sides.

Two thousand years ago the Greek mathematician Archimedes assumed that the area of a circle was a number that could be approximated more and more closely by computing the areas of regular *inscribed* polygons of more and more sides, and that it could also be approximated more and more closely by finding the areas of regular *circumscribed* polygons of more and more sides.

Thus the area of a given circle would be a number C lying between the area, $a(4)$, of an inscribed square and the area, $A(4)$, of a circumscribed square (Fig. 8-3); that is,

$$a(4) < C < A(4).$$

The discrepancy between C and the area of the inscribed square is the area of the shaded region in Fig. 8-3(a), and the discrepancy between C and the area of the circumscribed square is the area of the shaded region in Fig. 8-3(b).

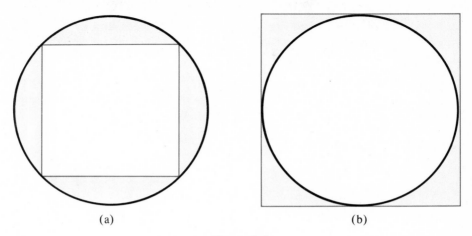

(a) (b)

Figure 8-3

A second, and much better, approximation to the area of the circle would be obtained by calculating the area, $a(8)$, of a regular inscribed octagon and the area, $A(8)$, of a regular circumscribed octagon (Fig. 8-4). Again

$$a(8) < C < A(8),$$

and the discrepancies between C and $a(8)$ and between C and $A(8)$ are indicated by the shaded regions in Fig. 8-4.

By continuing this process of inscribing and circumscribing regular polygons of 2^{n+1} sides, $n = 1, 2, 3, \cdots$, and assuming that for each value of n the area of the circle is a number between the areas of the inscribed and circumscribed polygons, the area of the circle can be approximated as accurately as desired. This will now be illustrated for a circle of radius 1.

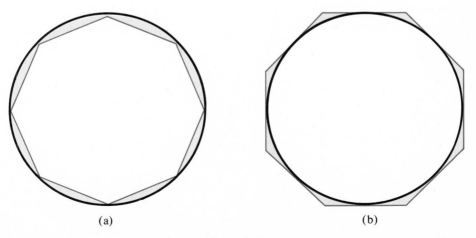

(a) (b)

Figure 8-4

If we denote by C the "area" of a circle of radius 1, by $a(n)$ the approximate area of an inscribed regular polygon of 2^{n+1} sides, and by $A(n)$ the approximate area of a circumscribed regular polygon of 2^{n+1} sides, and if we assume that

$$a(n) < C < A(n)$$

for $n = 1, 2, 3, \cdots$, then*

n	$a(n)$	C	$A(n)$
1	2.0000 $<$	C $<$	4.0000
2	2.8284 $<$	C $<$	3.3137
3	3.0615 $<$	C $<$	3.1826
4	3.1214 $<$	C $<$	3.1517
5	3.1365 $<$	C $<$	3.1441

We shall prove later that the "area" of a circle with radius r is given by πr^2. Then $C = \pi(1^2) = \pi$, and the above table indicates that π is a number between 3.1365 and 3.1441.

Let us now see how Archimedes' method has been adapted in the integral calculus. Consider the region R bounded by the parabola $f(x) = x^2$, the x axis, and the vertical

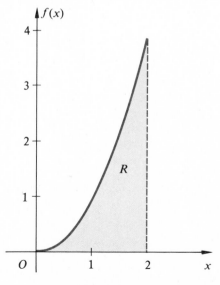

Figure 8-5

*Methods for computing such approximating areas may be found in trigonometry books. See, for instance, M. Richardson, *Plane and Spherical Trigonometry*, New York, Macmillan, 1950, Sec. 65.

line $x = 2$ (Fig. 8-5). We seek an area A of R that will have the properties listed in 8.1.1. Archimedes' method suggests that we proceed as follows.

Partition the interval $[0, 2]$ into n subintervals

$$[x_0, x_1], [x_1, x_2], [x_2, x_3], \cdots, [x_{n-1}, x_n],$$

where $0 = x_0 < x_1 < x_2 < x_3 < \cdots < x_{n-1} < x_n = 2$, and, for convenience in calculating, take the n intervals of the partition to be of equal length $\Delta x = 2/n$ (Fig. 8-6). Thus

(1)
$$x_0 = 0, \quad x_1 = \Delta x, \quad x_2 = 2\,\Delta x, \quad x_3 = 3\,\Delta x, \cdots,$$

$$x_i = i\,\Delta x, \cdots, \quad x_n = n\,\Delta x = 2,$$

where $\Delta x = 2/n$.

Since f is increasing on $[0, 2]$, it is increasing on every subinterval $[x_{i-1}, x_i]$ of the above partition, for $i = 1, 2, 3, \cdots, n$. So $f(x_{i-1})$ is the minimum value of f on $[x_{i-1}, x_i]$ and $f(x_i)$ is the maximum there (Fig. 8-7).

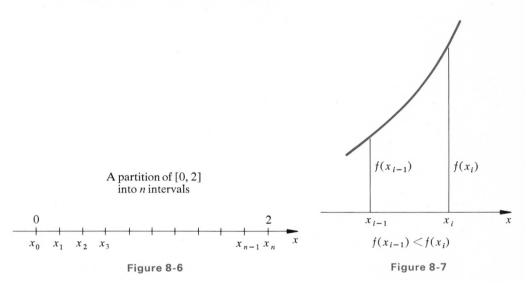

A partition of $[0, 2]$
into n intervals

$f(x_{i-1}) < f(x_i)$

Figure 8-6　　　　　　　　　　　　**Figure 8-7**

The union of the rectangles having base $[x_{i-1}, x_i]$ and altitude $f(x_{i-1})$, $i = 1, 2, 3, \cdots, n$, is called an **inscribed polygon** for the region R [Fig. 8-8(a)]. The area of this inscribed polygon is the sum of the areas of the rectangles and will be denoted $a(n)$. Thus

(2)
$$a(n) = f(x_0)\,\Delta x + f(x_1)\,\Delta x + f(x_2)\,\Delta x + \cdots$$
$$+ f(x_{n-1})\,\Delta x,$$

where $\Delta x = 2/n$ is the length of each of the equal subintervals $[x_{i-1}, x_i]$.

The union of the rectangles with base $[x_{i-1}, x_i]$ and altitude $f(x_i)$, $i = 1, 2, 3, \cdots, n$, is called a **circumscribed polygon** for the region R [Fig. 8-8(b)]. The area of the circumscribed polygon is the sum, $A(n)$, of the areas of the rectangles that constitute it.

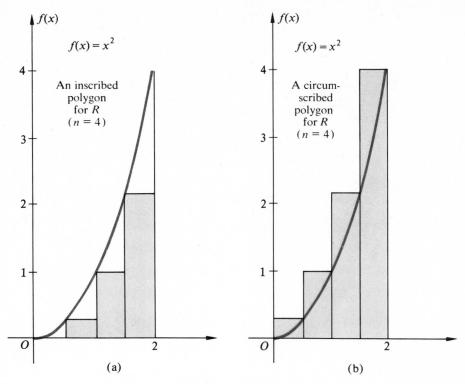

Figure 8-8

That is,

$$(3) \qquad A(n) = f(x_1)\,\Delta x + f(x_2)\,\Delta x + \cdots + f(x_n)\,\Delta x$$

is the area of the circumscribed polygon.

Denote by A the "area" (as yet undefined) of the region R (Fig. 8-5). For each partition of $[0, 2]$, the associated inscribed polygon is contained in R [Fig. 8-8(a)], and R is contained in the associated circumscribed polygon [Fig. 8-8(b)]. Thus the number A must satisfy the inequalities

$$(4) \qquad a(n) \le A \le A(n)$$

for every positive integer n.

Since $f(x) = x^2$, we have $f(x_{i-1}) = x_{i-1}^2$; and it follows from (1) and from $\Delta x = 2/n$ that

$$f(x_{i-1})\,\Delta x = x_{i-1}^2\,\Delta x = (i-1)^2(\Delta x)^3 = (i-1)^2\left(\frac{2}{n}\right)^3 = \frac{8(i-1)^2}{n^3}.$$

Then the area of the inscribed polygon is

$$a(n) = 0 + \frac{8}{n^3}(1^2) + \frac{8}{n^3}(2^2) + \frac{8}{n^3}(3^2) + \cdots + \frac{8}{n^3}(n-1)^2$$

$$= \frac{8}{n^3}[1^2 + 2^2 + 3^2 + \cdots + (n-1)^2]$$

$$= \frac{8}{n^3}\left[\frac{(n-1)n(2n-1)}{6}\right]$$

(by Example 2, Section 1.7). Consequently,

$$(5) \qquad a(n) = \frac{4}{3}\left(2 - \frac{3}{n} + \frac{1}{n^2}\right), \qquad n = 1, 2, 3, \cdots.$$

Again from (1), $x_i = i(\Delta x)$, $f(x_i) = x_i^2 = [i(\Delta x)]^2$, and $f(x_i)\,\Delta x = [i(\Delta x)]^2\,\Delta x = i^2(\Delta x)^3 = 8i^2/n^3$; and so the area of the circumscribed polygon is

$$A(n) = \frac{8}{n^3}(1^2) + \frac{8}{n^3}(2^2) + \frac{8}{n^3}(3^2) + \cdots + \frac{8}{n^3}(n^2)$$

$$= \frac{8}{n^3}(1^2 + 2^2 + 3^2 + \cdots + n^2)$$

$$= \frac{8}{n^3}\left[\frac{n(n+1)(2n+1)}{6}\right]$$

(by Example 2, Section 1.7). That is,

$$(6) \qquad A(n) = \frac{4}{3}\left(2 + \frac{3}{n} + \frac{1}{n^2}\right), \qquad n = 1, 2, 3, \cdots.$$

It follows from (4), (5), and (6) that

$$a(n) = \frac{4}{3}\left(2 - \frac{3}{n} + \frac{1}{n^2}\right) \le A \le \frac{4}{3}\left(2 + \frac{3}{n} + \frac{1}{n^2}\right) = A(n).$$

By letting the number, n, of intervals in our partition of $[0, 2]$ increase indefinitely, we have

$$\lim_{n\to\infty} a(n) = \lim_{n\to\infty} \frac{4}{3}\left(2 - \frac{3}{n} + \frac{1}{n^2}\right) \le A \le \lim_{n\to\infty} \frac{4}{3}\left(2 + \frac{3}{n} + \frac{1}{n^2}\right) = \lim_{n\to\infty} A(n).$$

But

$$(7) \qquad \lim_{n\to\infty} \frac{4}{3}\left(2 - \frac{3}{n} + \frac{1}{n^2}\right) = \frac{8}{3} \qquad \text{and} \qquad \lim_{n\to\infty} \frac{4}{3}\left(2 + \frac{3}{n} + \frac{1}{n^2}\right) = \frac{8}{3}.$$

Therefore

$$(8) \qquad \lim_{n\to\infty} a(n) = \frac{8}{3} = A = \frac{8}{3} = \lim_{n\to\infty} A(n).$$

The fact that $\lim_{n\to\infty} a(n) = \lim_{n\to\infty} A(n) = A$ in (8) suggests that we *define* the **area of the region R** to be

$$A = \lim_{n\to\infty} [f(w_1)\,\Delta x + f(w_2)\,\Delta x + \cdots + f(w_n)\,\Delta x],$$

where $f(w_i)$ is either the minimum value of f on $[x_{i-1}, x_i]$ for all $i = 1, 2, 3, \cdots, n$ or the maximum value of f on $[x_{i-1}, x_i]$ for $i = 1, 2, 3, \cdots, n$. In either case, *the area of the region R is $2\frac{2}{3}$ square units.*

A few words about limits like (7):

$$\lim_{n\to\infty} a(n) = L,$$

where $a(n)$ is defined only for positive *integral* values of n [see (2) above], means that corresponding to each positive number ϵ there is a number N such that $|a(n) - L| < \epsilon$ whenever $n > N$ and *n is a positive integer*. The reader should compare this definition with 6.11.1.

Example 1.   Consider the plane region R bounded by $f(x) = x + 3$, the x axis, and the lines $x = -2$ and $x = 3$ [Fig. 8-9(a)].

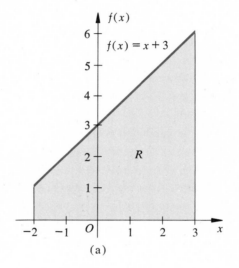

(a)

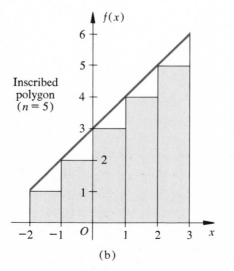

(b)

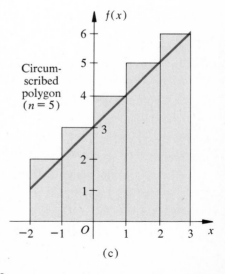

(c)

Figure 8-9

Partition the interval $[-2, 3]$ into five equal subintervals, each of length $\Delta x = 1$. Form the inscribed polygon that is the union of the five nonoverlapping rectangles having altitudes $f(-2) = 1$, $f(-1) = 2$, $f(0) = 3$, $f(1) = 4$, $f(2) = 5$, and bases $[-2, -1]$, $[-1, 0]$, $[0, 1]$, $[1, 2]$, $[2, 3]$, respectively [Fig. 8-9(b)].

Also form the circumscribed polygon that is the union of the five rectangles with altitudes $f(-1) = 2$, $f(0) = 3$, $f(1) = 4$, $f(2) = 5$, $f(3) = 6$, and bases as before $[-2, -1]$, $[-1, 0]$, $[0, 1]$, $[1, 2]$, and $[2, 3]$, respectively [Fig. 8-9(c)].

The area $a(n)$ of the inscribed polygon is the sum of the areas of its five rectangles: $a(5) = 1 \cdot 1 + 2 \cdot 1 + 3 \cdot 1 + 4 \cdot 1 + 5 \cdot 1 = 15$. Similarly, the area $A(n)$ of the circumscribed polygon is $A(5) = 2 \cdot 1 + 3 \cdot 1 + 4 \cdot 1 + 5 \cdot 1 + 6 \cdot 1 = 20$.

Since the inscribed polygon is entirely contained in R, and R is contained in the circumscribed polygon, we assume that the area A of the region R is bounded by the inequalities

$$a(5) = 15 \leq A \leq 20 = A(5).$$

Had we partitioned $[-2, 3]$ into ten equal subintervals, each of length $\Delta x = \frac{1}{2}$, we would have found the area $a(10)$ of the inscribed polygon to be $a(10) = 16.25$ and the area $A(10)$ of the circumscribed polygon to be $A(10) = 18.75$. Then the area A would be bounded more closely by

$$a(10) = 16.25 \leq A \leq 18.75 = A(10).$$

The greater the number of the equal subintervals in the partition of $[-2, 3]$, the closer A is bounded.

Exercises

In each of the following exercises, $f(x)$, a, b, and n are given.

(a) Sketch the region R bounded by $y = f(x)$, the x axis, and the lines $x = a$ and $x = b$. Partition $[a, b]$ into n equal subintervals, each of length $(b - a)/n$. On this drawing, sketch the inscribed polygon and compute its area, $a(n)$.

(b) On another drawing of R, sketch the circumscribed polygon and find its area, $A(n)$.

(c) Using the numbers $a(n)$ and $A(n)$, bound the area A of R by the inequality $a(n) \leq A \leq A(n)$.

1. $f(x) = x + 1$, $a = 0$, $b = 3$, $n = 3$.

2. $f(x) = \frac{1}{2}x$, $a = 1$, $b = 7$, $n = 6$.

3. $f(x) = 2x + 3$, $a = -1$, $b = 2$, $n = 3$.

4. $f(x) = -\frac{1}{2}x + 6$, $a = -4$, $b = 4$, $n = 8$.

5. $f(x) = -x + 3$, $a = -3$, $b = 2$, $n = 10$.

6. $f(x) = 3x - 2$, $a = 1$, $b = 3$, $n = 4$.

7. $f(x) = x^2 + 2$, $a = 0$, $b = 3$, $n = 3$.

8. $f(x) = 2x^2 + 1$, $a = -1$, $b = 3$, $n = 4$.

9. $f(x) = \frac{1}{3}x^2 + 3$, $a = 0$, $b = 4$, $n = 4$.

10. $f(x) = x^2 + x$, $a = 0$, $b = 2$, $n = 4$.

8.2 THE SUMMATION NOTATION

In our development of the definite integral we shall often have occasion to write long sums like

$$a_1 + a_2 + a_3 + \cdots + a_n.$$

A brief, convenient way of expressing such sums will now be explained. As soon as this new notation becomes familiar, the reader will find that it not only saves much labor in writing the sums, but also makes their manipulation simpler and clearer.

The sum $a_1 + a_2 + a_3 + \cdots + a_n$ is compactly indicated by $\sum_{i=1}^{n} a_i$, which is called the **summation notation.** The Greek letter $\sum$ (sigma) corresponds to our S and suggests the word *sum.* Other examples of the summation notation are

$$\sum_{i=1}^{5} S_i = S_1 + S_2 + S_3 + S_4 + S_5,$$

$$\sum_{j=1}^{n} \frac{1}{j} = \frac{1}{1} + \frac{1}{2} + \frac{1}{3} + \cdots + \frac{1}{n},$$

$$\sum_{k=1}^{4} \frac{k}{k^2 + 1} = \frac{1}{1^2 + 1} + \frac{2}{2^2 + 1} + \frac{3}{3^2 + 1} + \frac{4}{4^2 + 1},$$

$$\sum_{i=3}^{6} \frac{\sqrt{i + 1}}{2i} = \frac{\sqrt{3 + 1}}{2(3)} + \frac{\sqrt{4 + 1}}{2(4)} + \frac{\sqrt{5 + 1}}{2(5)} + \frac{\sqrt{6 + 1}}{2(6)}.$$

More generally $\sum_{i=m}^{n} F(i)$, where m and n are integers and $m \leq n$, represents the sum of $n - m + 1$ terms, the first of which is obtained by replacing i by m in $F(i)$, the second by replacing i by $m + 1$ in $F(i)$, $\cdots$, and the last term is $F(n)$, obtained by letting $i = n$ in $F(i)$:

$$\sum_{i=m}^{n} F(i) = F(m) + F(m + 1) + \cdots + F(n).$$

As an illustration,

$$\sum_{t=2}^{5} \frac{4t^2}{t - 1} = \frac{4(2^2)}{2 - 1} + \frac{4(3^2)}{3 - 1} + \frac{4(4^2)}{4 - 1} + \frac{4(5^2)}{5 - 1}.$$

In the summation notation, our definition of the area of R (page 249) may be written

$$A = \lim_{n \to \infty} \sum_{i=1}^{n} f(w_i) \, \Delta x.$$

Some useful formulas, written in the summation notation, are

(1) $$\sum_{i=1}^{n} i = 1 + 2 + 3 + \cdots + n = \frac{n(n + 1)}{2};$$

(2) $$\sum_{i=1}^{n} i^2 = 1^2 + 2^2 + 3^2 + \cdots + n^2 = \frac{n(n + 1)(2n + 1)}{6};$$

(3) $$\sum_{i=1}^{n} i^3 = 1^3 + 2^3 + 3^3 + \cdots + n^3 = \left[\frac{n(n + 1)}{2} \right]^2;$$

(4) $$\sum_{i=1}^{n} i^4 = 1^4 + 2^4 + 3^4 + \cdots + n^4 = \frac{n(n + 1)(6n^3 + 9n^2 + n - 1)}{30}.$$

The first three were proved in Section 1.7 by mathematical induction.

If all the c_i in $\sum_{i=1}^{n} c_i$ are the same constant c, then $\sum_{i=1}^{n} c_i = c + c + c + \cdots + c = nc$. As a result, we adopt the convention

$$\sum_{i=1}^{n} c = nc,$$

where c is a constant. To illustrate,

$$\sum_{i=1}^{5} 1 = 5(1) = 5, \qquad \sum_{i=1}^{7} 8 = 7(8) = 56, \qquad \sum_{i=1}^{10} -2 = 10(-2) = -20.$$

Five basic properties of the summation notation are stated in the following theorem. Their easy proofs are left as exercises.

8.2.1 Theorem. If c is a constant,

(i) $$\sum_{i=1}^{n} ca_i = c \sum_{i=1}^{n} a_i;$$

(ii) $$\sum_{i=1}^{n} (a_i + b_i) = \sum_{i=1}^{n} a_i + \sum_{i=1}^{n} b_i;$$

(iii) $$\sum_{i=1}^{n} (a_i - b_i) = \sum_{i=1}^{n} a_i - \sum_{i=1}^{n} b_i;$$

(iv) $$\sum_{i=1}^{n} (a_i - a_{i-1}) = a_n - a_0;$$

(v) If $\Delta x_i = x_i - x_{i-1}$, $\sum_{i=1}^{n} \Delta x_i = x_n - x_0$.

Example. Find the value of $\sum_{i=1}^{n} 2i(i - 5)$.

Solution. By use of 8.2.1(i) and (iii), and formulas (1) and (2), above, we can write

$$\sum_{i=1}^{n} 2i(i - 5) = \sum_{i=1}^{n} (2i^2 - 10i) = 2\sum_{i=1}^{n} i^2 - 10\sum_{i=1}^{n} i$$

$$= \frac{2n(n + 1)(2n + 1)}{6} - \frac{10n(n + 1)}{2}$$

$$= \frac{n(n + 1)(2n + 1)}{3} - \frac{15n(n + 1)}{3}$$

$$= \frac{n(n + 1)[(2n + 1) - 15]}{3}$$

$$= \frac{2n(n + 1)(n - 7)}{3}.$$

Exercises

In each of Exercises 1 to 6, find the value of the indicated sum.

1. $\sum_{i=1}^{10} (3i - 1)$.

2. $\sum_{i=1}^{6} 5i^2$.

3. $\sum_{i=4}^{6} \frac{i - 1}{3i + 1}$.

4. $\sum_{i=1}^{5} (-1)^{i+1} 3^{i-1}$.

5. $\sum_{j=3}^{7} (j + 2)^2$.

6. $\sum_{k=6}^{8} \frac{(-1)^k}{k(2k + 1)}$.

In Exercises 7 to 12, evaluate the indicated sums by using the properties of the summation notation, 8.2.1, and formulas (1), (2), and (3), above.

7. $\sum_{i=1}^{n} (3i - 2)$.

10. $\sum_{i=1}^{n} [(i - 1)(4i + 3)]$.

8. $\sum_{i=1}^{n} (4i + 3)$.

11. $\sum_{i=1}^{n} 5i^2(i + 4)$.

9. $\sum_{i=1}^{n} i(3i - 2)$.

12. $\sum_{i=1}^{n} (2i - 3)^2$.

13. Prove 8.2.1(i) by mathematical induction.

14. Prove 8.2.1(ii) by mathematical induction.

15. Prove 8.2.1(iii) by mathematical induction.

16. Prove 8.2.1(iv) by mathematical induction.

17. Prove $\sum_{i=1}^{n} i^4 = n(n + 1)(6n^3 + 9n^2 + n - 1)/30$.

[*Hint:* Since $i^5 - (i - 1)^5 = 5i^4 - 10i^3 + 10i^2 - 5i + 1$, then $\sum_{i=1}^{n} [i^5 - (i - 1)^5] = \sum_{i=1}^{n} (5i^4 - 10i^3 + 10i^2 - 5i + 1)$. Now use

8.2.1(iv) on the left member of the latter equation, and (1), (2), and (3) on the right member.]

18. If

$$A = \sum_{i=1}^{n} a_i^2, \qquad B = \sum_{i=1}^{n} a_i b_i, \qquad \text{and} \qquad C = \sum_{i=1}^{n} b_i^2,$$

where the a_i and b_i are real numbers, $i = 1, 2, 3, \cdots, n$, prove that

(1) $$At^2 + 2Bt + C \geq 0$$

for all real numbers t. [*Hint:* $At^2 + 2Bt + C = \sum_{i=1}^{n} (a_i t + b_i)^2$.]

19. Prove the **Cauchy-Schwarz inequality,**

$$\left(\sum_{i=1}^{n} a_i b_i \right)^2 \leq \left(\sum_{i=1}^{n} a_i^2 \right) \left(\sum_{i=1}^{n} b_i^2 \right),$$

where the a_i and b_i are real numbers, $i = 1, 2, 3, \cdots, n$. [*Hint:* In the notation of Exercise 18, we wish to prove that $B^2 - AC \leq 0$. Use Exercise 11, Section 6.7, on (1) in Exercise 18 above.]

8.3 THE DEFINITE INTEGRAL

We first encountered the limit

$$\lim_{\Delta x \to 0} \frac{\Delta f}{\Delta x}$$

as the slope of the tangent to a curve. But it had so many other applications that we studied it for its own sake and called it the derivative of f at x.

It turns out that the limit

$$\lim_{n \to \infty} \sum_{i=1}^{n} f(w_i) \, \Delta x,$$

which appeared in our introduction to area, also has many other important interpretations, such as the volume of a solid, the area of a surface of revolution, the length of a curve, the work done by a variable force, and centers of gravity. Quite apart from its applications, this limit is a simple example of what is known as a *Riemann integral* or a *definite integral*, which is symbolized by $\int_{a}^{b} f(x) \, dx$ and read *the definite integral of f from a to b.*

Thus in 8.1, where the interval $[a, b]$ was $[0, 2]$, our definition of the area of R (page 249) can be written

$$\lim_{n \to \infty} \sum_{i=1}^{n} f(w_i) \, \Delta x = \int_{a}^{b} f(x) \, dx.$$

The concept of the Riemann integral is much broader than we might be led to believe from our brief discussion of area. The subintervals into which $[a, b]$ is divided need not have the same length; the point w_i in each subinterval $[x_{i-1}, x_i]$, $i = 1, 2, 3, \cdots, n$, can be any point in the subinterval, not just the point w_i at which $f(w_i)$ is the minimum value of f on $[x_{i-1}, x_i]$ or the point w_i for which $f(w_i)$ is the maximum value of f on $[x_{i-1}, x_i]$; $f(x)$ can be positive, negative, or zero for points x in $[a, b]$; and the function f does not have to be continuous at all points in $[a, b]$, although we shall usually assume that f is continuous on $[a, b]$ when we speak of its definite integral $\int_{a}^{b} f(x) \, dx$ in this book.

The limit involved in the more general Riemann integral is somewhat different from the limits studied so far. To describe it precisely, some new terms are needed.

8.3.1   **Definition.**   In the closed interval $[a, b]$ let $x_1, x_2, x_3, \cdots, x_{n-1}$ be $n - 1$ arbitrarily chosen points between a and b, so that $a < x_1 < x_2 < \cdots x_{n-1} < b$, and let $x_0 = a$ and $x_n = b$, where n is any positive integer. Then the set of n closed subintervals

$$[x_0, x_1], [x_1, x_2], [x_2, x_3], \cdots, [x_{n-1}, x_n]$$

is called a **partition** of $[a, b]$ and is symbolized by

$$P = \{x_0, x_1, x_2, \cdots, x_n\}$$

(Fig. 8-10). The points $x_1, x_2, \cdots, x_{n-1}$ are the **points of division** of the partition P.

Two partitions of the interval $[-3, 2]$

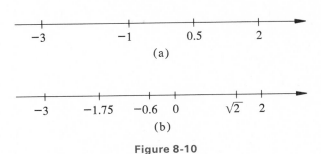

(a)

(b)

Figure 8-10

Example 1.   Consider the closed interval $[-3, 2]$. One partition of the interval $[-3, 2]$ is

$$P = \{-3, -1, 0.5, 2\},$$

which consists of the three subintervals

$$[-3, -1], [-1, 0.5], [0.5, 2]$$

[Fig. 8-10(a)]. Another partition of $[-3, 2]$,

$$P = \{-3, -1.75, -0.6, 0, \sqrt{2}, 2\},$$

consists of the five subintervals

$$[-3, -1.75], [-1.75, -0.6], [-0.6, 0], [0, \sqrt{2}], [\sqrt{2}, 2]$$

[Fig. 8-10(b)]. There are indefinitely many possible partition of $[-3, 2]$, including the interval itself when $n = 1$.

Denote the length of the first interval of a partition P of $[a, b]$ by Δx_1; that is, let $\Delta x_1 = x_1 - x_0$. Denote the length of the second interval by $\Delta x_2 = x_2 - x_1$; and continue in this way so that the length of the ith interval is

$$\Delta x_i = x_i - x_{i-1}, \qquad i = 1, 2, 3, \cdots, n.$$

The length of the longest interval in a partition P is called the **norm** *of the partition P and is symbolized by* $|P|$. As an illustration, the norm of the partition of $[-3, 2]$ shown in Fig. 8-10(a) is $|P| = -1 - (-3) = 2$, and the norm of the partition shown in Fig. 8-10(b) is $|P| = \sqrt{2} - 0 = \sqrt{2}$.

Consider a partition P of an interval $[a, b]$ and a function f that is defined on $[a, b]$ [Fig. 8-11]. In the first interval $[x_0, x_1]$ of the partition, choose any point w_1 so that $x_0 \leq w_1 \leq x_1$; in the second interval, select any point w_2 so that $x_1 \leq w_2 \leq x_2$; and continue in the same way, choosing one point in each of the n intervals, so that

$$x_{i-1} \leq w_i \leq x_i, \qquad i = 1, 2, 3, \cdots, n.$$

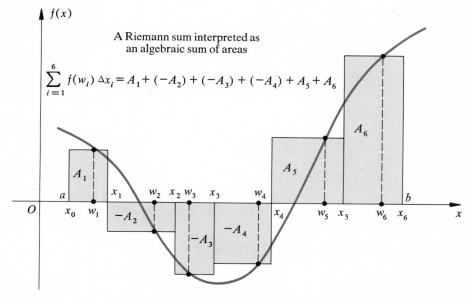

Figure 8-11

Then

$$\sum_{i=1}^{n} f(w_i)\,\Delta x_i = f(w_1)\,\Delta x_1 + f(w_2)\,\Delta x_2 + \cdots + f(w_n)\,\Delta x_n$$

is called a *Riemann sum* for f on $[a, b]$.

8.3.2 Definition. Let f be a function that is defined on a closed interval $[a, b]$ and let $P = \{x_0, x_1, x_2, \cdots, x_n\}$ be any partition of $[a, b]$. If w_i is an arbitrarily chosen point in the ith interval $[x_{i-1}, x_i]$ of P and $\Delta x_i = x_i - x_{i-1}$ is the length of that interval, $i = 1, 2, 3, \cdots, n$, then

$$\sum_{i=1}^{n} f(w_i)\,\Delta x_i$$

is called a **Riemann sum** for f on $[a, b]$.

Example 2. Let $f(x) = x^2 - 1$, and let $P = \{-2, -\frac{1}{2}, \frac{3}{4}, 2\}$. If $w_1 = -\frac{3}{2}$, $w_2 = 0$, and $w_3 = \frac{7}{4}$, find the Riemann sum for f on the partition P. Make a sketch showing this Riemann sum as an algebraic sum of areas (Fig. 8-12).

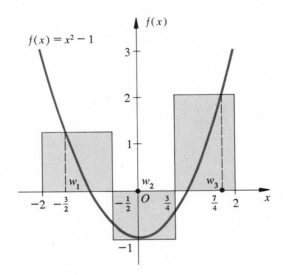

Figure 8-12

Solution. $f(w_1) = (-\frac{3}{2})^2 - 1 = \frac{5}{4}$, $f(w_2) = 0^2 - 1, = -1$, and $f(w_3) = (\frac{7}{4})^2 - 1 = \frac{33}{16}$. Also, $\Delta x_1 = -\frac{1}{2} - (-2) = \frac{3}{2}$, $\Delta x_2 = \frac{3}{4} - (-\frac{1}{2}) = \frac{5}{4}$, and $\Delta x_3 = 2 - \frac{3}{4} = \frac{5}{4}$. Thus the Riemann sum is

$$\sum_{i=1}^{3} f(w_i)\,\Delta x_i = \left(\frac{5}{4}\right)\left(\frac{3}{2}\right) + (-1)\left(\frac{5}{4}\right) + \left(\frac{33}{16}\right)\left(\frac{5}{4}\right) = \frac{205}{64}.$$

For a given interval $[a, b]$ and a given positive number δ, there are indefinitely many partitions P with norm $|P| < \delta$. Moreover, for each such partition we can choose the points w_i in an indefinitely great number of ways. Thus for a particular function f defined on a particular interval $[a, b]$, there are infinitely many Riemann sums for which the norm $|P|$ of the associated partition P is less than δ.

8.3.3 Definition. Let f be defined on a closed interval $[a, b]$ and let L be a number. The symbols

$$\lim_{|P| \to 0} \sum_{i=1}^{n} f(w_i) \, \Delta x_i = L$$

mean that corresponding to each $\epsilon > 0$ there is a $\delta > 0$ such that

$$\left| \sum_{i=1}^{n} f(w_i) \, \Delta x_i - L \right| < \epsilon$$

is true for all Riemann sums $\sum_{i=1}^{n} f(w_i) \, \Delta x_i$ for f on $[a, b]$ for which the norm, $|P|$, of the associated partition P satisfies $|P| < \delta$. In this case, we say that

$$\lim_{|P| \to 0} \sum_{i=1}^{n} f(w_i) \, \Delta x_i$$

exists and that its value is L.

In other words, $\lim\limits_{|P| \to 0} \sum\limits_{i=1}^{n} f(w_i) \, \Delta x_i = L$ means that for the function f, defined on the interval $[a, b]$, the values of all of its Riemann sums on $[a, b]$ will be as close as we please to the number L if we make the norms $|P|$ of the associated partitions P sufficiently small.

We are now ready for a definition of the definite integral.

8.3.4 Definition. Let f be a function that is defined on a closed interval $[a, b]$. If

$$\lim_{|P| \to 0} \sum_{i=1}^{n} f(w_i) \, \Delta x_i$$

exists (8.3.3), then its value is called the **definite integral** (or **Riemann* integral**) of f from a to b, and it is denoted by $\int_a^b f(x) \, dx$. *That is,*

$$\int_a^b f(x) \, dx = \lim_{|P| \to 0} \sum_{i=1}^{n} f(w_i) \, \Delta x_i.$$

A function that possesses a definite integral on a given interval is said to be **integrable** on that interval.

In the symbol $\int_a^b f(x) \, dx$ for the definite integral, $f(x)$ is called the **integrand;**

---

*This definition is due to G. B. F. Riemann, a famous German mathematician of the nineteenth century.

and it is customary to call a the *lower limit of integration* and b the *upper limit of integration*. The use of the word "limit" here is unfortunate because it has nothing to do with the highly specialized meaning of the word as used elsewhere in calculus; a and b are simply the endpoints of the closed interval in the definition of a definite integral, above.

It is not difficult to prove that *when the definite integral exists, its value is unique* (see Appendix A.4.1.).

In the definition (8.3.4) of the definite integral, $\int_a^b f(x)\,dx$, we assumed that $a < b$. To dispense with this restriction, we make the following two definitions.

8.3.5   Definition.   If $a > b$, then

$$\int_a^b f(x)\,dx = -\int_b^a f(x)\,dx,$$

provided that $\int_b^a f(x)\,dx$ exists.

8.3.6   Definition.   If f is defined at a number a, then

$$\int_a^a f(x)\,dx = 0.$$

Now that we have the definition of a definite integral, two questions arise: (a) what functions, if any, are integrable, and (b) how do we find the value of a definite integral?

A partial answer to the first question is given in the following theorem, which is proved in more advanced books.

8.3.7   Theorem.   If a function f is continuous on a closed interval $[a, b]$, then the definite integral

$$\int_a^b f(x)\,dx = \lim_{|P| \to 0} \sum_{i=1}^n f(w_i)\,\Delta x_i$$

exists.

Notice that in this theorem the interval $[a, b]$ is *closed*. Certain functions that are continuous in an open interval, or even in a half-open interval, are not integrable there. For example, $\int_0^2 (1/x)\,dx$ does not exist, although $f(x) = 1/x$ is continous in the half-open interval $(0, 2]$.

Continuity of a function on a closed interval is a *sufficient* condition for the existence of its definite integral there, but it is not a *necessary* condition. That is, if f is continuous on $[a, b]$, we are sure that $\int_a^b f(x)\,dx$ exists; but sometimes this integral exists even if the function is discontinuous at some points in $[a, b]$.

A partition of $[a, b]$ in which all the subintervals are of equal length $\Delta x = (b - a)/n$ is called a **regular partition** of $[a, b]$. Regular partitions were used in our brief discussion of area (8.1).

If f is integrable on $[a, b]$, then

$$\lim_{|P| \to 0} \sum_{i=1}^{n} f(w_i) \, \Delta x_i$$

exists and is equal to $\int_{a}^{b} f(x) \, dx$ independently of how the partitions P are formed and how the w_i are selected in each partition (8.3.3 and 8.3.4). It follows that we can form Riemann sums in any simple way that serves our purpose, such as by using regular partitions and letting the w be the left endpoints in each subinterval, or the right endpoints of each subinterval, or the midpoints. *No matter how simply the Riemann sums* $\sum_{i=1}^{n} f(w_i) \, \Delta x_i$ *are formed, their values will be as close as we please to* $\int_{a}^{b} f(x) \, dx$, *provided that the norms* $|P|$ *of the partitions are sufficiently small.* This fact is of considerable importance in the application of definite integrals to engineering, the physical sciences, geometry, and economics.

Example 3. If $f(x) = x + 3$, find the value of the definite integral $\int_{-2}^{3} (x + 3) \, dx$.

Solution. (Fig. 8-9). Partition the interval $[-2, 3]$ into n equal subintervals, each of length $\Delta x = 5/n$. In each subinterval $[x_{i-1}, x_i]$ of the partition, let $w_i = x_i$, the right endpoint of the subinterval, $i = 1, 2, 3, \cdots, n$. We seek

$$\lim_{|P| \to 0} \sum_{i=1}^{n} f(w_i) \, \Delta x_i.$$

Now $x_0 = -2$, $x_1 = -2 + \Delta x = -2 + (5/n)$, $x_2 = -2 + 2(\Delta x) = -2 + 2(5/n)$, $\cdots$, $x_i = -2 + i(\Delta x) = -2 + i(5/n)$, $\cdots$, $x_n = -2 + n(5/n)$. Moreover, $w_i = x_i = -2 + i(5/n)$, and $f(w_i) = w_i + 3 = [-2 + i(5/n)] + 3 = 1 + i(5/n)$. Then the typical Riemann sum may be written

$$\sum_{i=1}^{n} f(w_i) \, \Delta x_i = \sum_{i=1}^{n} \left[1 + i\left(\frac{5}{n}\right) \right] \frac{5}{n}$$

$$= \frac{5}{n} \sum_{i=1}^{n} 1 + \frac{25}{n^2} \sum_{i=1}^{n} i$$

$$= \frac{5}{n} (n) + \frac{25}{n^2} \left[\frac{n(n + 1)}{2} \right]$$

$$= 5 + \frac{25}{2} \left(1 + \frac{1}{n} \right).$$

Since all the partitions of $[-2, 3]$ are regular partitions, $|P| \longrightarrow 0$ is equivalent to $n \longrightarrow \infty$. Thus

$$\lim_{|P| \to 0} \sum_{i=1}^{n} f(w_i) \, \Delta x_i = \lim_{n \to \infty} \left[5 + \frac{25}{2} \left(1 + \frac{1}{n} \right) \right]$$

$$= 5 + \frac{25}{2} + \frac{25}{2} \lim_{n \to \infty} \frac{1}{n}$$

$$= \frac{35}{2} = 17\tfrac{1}{2}.$$

Therefore

$$\int_{-2}^{3} (x + 3) \, dx = 17\tfrac{1}{2}.$$

Compare with Example 1, Section 8.1.

Example 4. Find $\int_{-1}^{5} (3x^2 - 2) \, dx$.

Solution. Let P be a regular partition of $[-1, 5]$ into n equal subintervals, each of length $\Delta x = 6/n$. In each subinterval $[x_{i-1}, x_i]$ of P, choose w_i to be the right endpoint, so that $w_i = x_i$, $i = 1, 2, 3, \cdots, n$. Since $x_i = -1 + i(\Delta x) = -1 + 6i/n$, $f(w_i) = 3w_i^2 - 2 = 3(-1 + 6i/n)^2 - 2$, and

$$f(w_i) \, \Delta x = \left[3\left(-1 + \frac{6i}{n} \right)^2 - 2 \right] \frac{6}{n} = \frac{6}{n} - \frac{216i}{n^2} + \frac{648i^2}{n^3}.$$

Then

$$\sum_{i=1}^{n} f(w_i) \, \Delta x = \sum_{i=1}^{n} \left[\frac{6}{n} - \frac{216i}{n^2} + \frac{648i^2}{n^3} \right]$$

$$= \frac{6}{n} \sum_{i=1}^{n} 1 - \frac{216}{n^2} \sum_{i=1}^{n} i + \frac{648}{n^3} \sum_{i=1}^{n} i^2$$

$$= \frac{6}{n} (n) - \frac{216}{n^2} \left[\frac{n(n+1)}{2} \right] + \frac{648}{n^3} \left[\frac{n(n+1)(2n+1)}{6} \right]$$

$$= 6 - 108\left(1 + \frac{1}{n} \right) + 108\left(2 + \frac{3}{n} + \frac{1}{n^2} \right)$$

$$= 114 + \frac{216}{n} + \frac{108}{n^2}.$$

Therefore

$$\int_{-1}^{5} (3x^2 - 2) \, dx = \lim_{|P| \to 0} \sum_{i=1}^{n} f(w_i) \, \Delta x$$

$$= \lim_{n \to \infty} \left(114 + \frac{216}{n} + \frac{108}{n^2} \right)$$

$$= 114.$$

Exercises

In each of Exercises 1 to 5, write out the Riemann sum $\sum_{i=1}^{n} f(w_i) \, \Delta x_i$ for the given data and find its value. Graph the given function and show the Riemann sum as the algebraic sum of areas of rectangles, as in Fig. 8-11.

1. $f(x) = x - 1$, $[a, b] = [3, 7]$, $P = \{3, 3.75, 4.25, 5.5, 6, 7\}$; $w_1 = 3$, $w_2 = 4$, $w_3 = 4.75$, $w_4 = 6$, $w_5 = 6.5$.

2. $f(x) = -\frac{x}{2} + 3$, $[a, b] = [-3, 2]$, $P = \{-3, -1.3, 0, 0.9, 2\}$; $w_1 = -2$, $w_2 = -0.5$, $w_3 = 0$, $w_4 = 2$.

3. $f(x) = \frac{x^2}{2} + 1$, $[a, b] = [-1, 2]$, $P = \{-1, -0.25, 0.75, 1.5, 2\}$; $w_1 = -1$, $w_2 = 0$, $w_3 = 1$, $w_4 = 2$.

4. $f(x) = \dfrac{1}{x}$, $[a, b] = [1, 3]$, $P = \{1, 1\frac{7}{8}, 2\frac{1}{4}, 3\}$;

$w_1 = 1\frac{1}{2}$, $w_2 = 2$, $w_3 = 2\frac{1}{2}$.

5. $f(x) = \dfrac{x^3}{3} + 1$, $[a, b] = [0, 2]$, $P = \left\{0, \dfrac{2}{3}, 1,\right.$

$\left.\dfrac{5}{4}, 2\right\}$; $w_1 = \frac{1}{2}$, $w_2 = \frac{3}{4}$, $w_3 = 1$, $w_4 = \frac{3}{2}$.

In each of Exercises 6 to 11, proceed as in Examples 3 and 4, above, and find the value of the given definite integral.

6. $\displaystyle\int_1^7 \frac{1}{2} x \, dx$.

7. $\displaystyle\int_{-1}^2 (2x + 3) \, dx$.

8. $\displaystyle\int_0^3 (x^2 + 2) \, dx$.

9. $\displaystyle\int_2^4 (x^2 - 1) \, dx$. $\left(Hint: w_i = 2 + \dfrac{2i}{n}.\right)$

10. $\displaystyle\int_0^4 \left(\frac{1}{3} x^2 + 3\right) dx$.

11. $\displaystyle\int_{-1}^5 (x^2 - 2x + 3) \, dx$. $\left(Hint: w_i = -1 + \dfrac{6i}{n}.\right)$

8.4 THE FUNDAMENTAL THEOREM OF CALCULUS

We have been able to evaluate a few definite integrals directly from the definition (8.3.4) because we had formulas for the sums $1 + 2 + 3 + \cdots + n$ and $1^2 + 2^2 + 3^2 + \cdots + n^2$. But calculating definite integrals from Riemann sums is tedious in the simplest cases, and impossible for most of the integrals commonly encountered. Fortunately there is another way that makes no use of Riemann sums.

As noted, Isaac Newton and Gottfried Leibniz are credited with the simultaneous, but independent, discovery of calculus. Yet the concepts of the derivative as the slope of a tangent to a curve, and area as a definite integral, were known to others who preceded them. Why, then do Newton and Leibniz figure so prominently in the history of calculus? They do so because they understood and exploited the importance of an intimate relation between antiderivatives and definite integrals; a relation that enables us to compute easily the exact values of very many definite integrals without using Riemann sums. This discovery is so important that it is called **the fundamental theorem of calculus**.

Before discussing the fundamental theorem, two properties of definite integrals are needed. They are easy to understand and remember. Their proofs, being somewhat time consuming, may be found in the Appendix (A.4.3 and A.4.4).

8.4.1   **Theorem** (*Interval Additive Property*).   If f is continuous on an interval containing the three points a, b, and c, then

$$\int_a^b f(x) \, dx + \int_b^c f(x) \, dx = \int_a^c f(x) \, dx.$$

8.4.2   **Theorem.**   If f is continuous on $[a, b]$ and m and M denote the minimum and maximum of f on $[a, b]$, then

$$m(b - a) \le \int_a^b f(x) \, dx \le M(b - a).$$

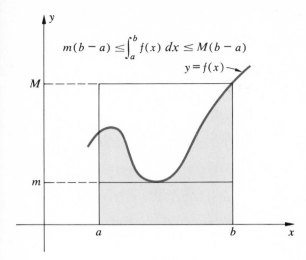

$$m(b-a) \leq \int_a^b f(x)\,dx \leq M(b-a)$$

$$y = f(x)$$

Figure 8-13

These inequalities are illustrated in Fig. 8-13, where the area of the shaded region is $\int_a^b f(x)\,dx$, the area of the smaller rectangle having base $(b-a)$ and altitude m is $m(b-a)$, and the area of the larger rectangle, with base $(b-a)$ and altitude M, is $M(b-a)$.

Let f be a function that is integrable on $[a, b]$. From the definition of a definite integral (8.3.4) and the discussion that preceded it, it is evident that the value of the definite integral $\int_a^b f(x)\,dx$ depends only on the function f and the limits of integration a and b and not at all on the particular letter used to represent the independent variable. For this reason, the x appearing in $\int_a^b f(x)\,dx$ is called a **dummy variable**; *it can be replaced by any other letter not already in use without affecting the value of the definite integral.* Thus

$$\int_a^b f(x)\,dx = \int_a^b f(t)\,dt = \int_a^b f(w)\,dw.$$

Now consider a function f that is continuous on a closed interval $[a, b]$. The definite integral $\int_a^b f(t)\,dt$ is a unique number that always exists when f is continuous on closed $[a, b]$ (by 8.3.7).

Let x be a number in $[a, b]$. Since f is continuous on $[a, b]$, it is continuous on the closed interval $[a, x]$, where $a \leq x \leq b$, and so

$$(1) \qquad\qquad \int_a^x f(t)\,dt$$

is a unique number. Corresponding to each number x in $[a, b]$, integral (1) has one and only one value. Thus it defines a function G whose domain is $[a, b]$ and whose value at any point x in $[a, b]$ is given by

$$G(x) = \int_a^x f(t)\,dt.$$

We call $\int_a^x f(t)\,dt$ an **integral with a variable upper limit.**

Now we are prepared for the fundamental theorem of calculus, that all-important connecting link between differential calculus and integral calculus.

8.4.3 **The Fundamental Theorem of Calculus.**   Let f be a function that is continuous on the closed interval $[a, b]$. Then
   (i)   The function G defined by

$$G(x) = \int_a^x f(t)\, dt, \qquad a \leq x \leq b,$$

is differentiable on $[a, b]$ and is an antiderivative of f; that is,

$$G'(x) = D_x \int_a^x f(t)\, dt = f(x)$$

for all $x \in [a, b]$.
   (ii)   If F is any particular antiderivative of f so that $F'(x) = f(x)$, then

$$\int_a^b f(x)\, dx = F(b) - F(a).$$

As an illustration of the use of part (ii), let $f(x) = 3x^2 - 2$. Since $D_x(x^3 - 2x) = 3x^2 - 2$, $F(x) = x^3 - 2x$ is an antiderivative of $f(x) = 3x^2 - 2$. Therefore

$$\int_a^b f(x)\, dx = \int_{-1}^5 (3x^2 - 2)\, dx = F(5) - F(-1)$$

$$= [5^3 - 2(5)] - [(-1)^3 - 2(-1)] = 114.$$

Contrast this almost trivial calculation with the tedious evaluation of the same definite integral by Riemann sums in Example 4, Section 8.3!

Proof of (i).   Let G be the function defined by

$$G(x) = \int_a^x f(t)\, dt$$

on the closed interval $[a, b]$ and let x_1 and $x_1 + \Delta x$ be arbitrary numbers in $[a, b]$. We wish to show that

$$G'(x_1) = \lim_{\Delta x \to 0} \frac{G(x_1 + \Delta x) - G(x_1)}{\Delta x} = f(x_1).$$

From the definition of G,

(2) $$G(x_1 + \Delta x) - G(x_1) = \int_a^{x_1 + \Delta x} f(t)\, dt - \int_a^{x_1} f(t)\, dt.$$

Since a, x_1, and $x_1 + \Delta x$ are points of $[a, b]$, it follows from 8.4.1 that

$$\int_a^{x_1} f(t)\, dt + \int_{x_1}^{x_1 + \Delta x} f(t)\, dt = \int_a^{x_1 + \Delta x} f(t)\, dt,$$

or $$\int_a^{x_1 + \Delta x} f(t)\, dt - \int_a^{x_1} f(t)\, dt = \int_{x_1}^{x_1 + \Delta x} f(t)\, dt.$$

Thus (2) can be rewritten

(3) $$G(x_1 + \Delta x) - G(x_1) = \int_{x_1}^{x_1 + \Delta x} f(t)\, dt.$$

In the remainder of this proof we shall think of the arbitrarily chosen number x_1 in $[a, b]$ as fixed and keep $\Delta x \neq 0$ at our disposal.
   Denote by I_Δ the closed interval with endpoints x_1 and $x_1 + \Delta x$. Since f is continuous on I_Δ, there are numbers u and v in I_Δ such that $f(u) = m$ is the minimum of

f on I_Δ and $f(v) = M$ is the maximum (6.1.2). Because of the continuity of f on I_Δ, $\lim_{\Delta x \to 0} f(x) = f(x_1)$ for all numbers x in I_Δ. Thus $\lim_{\Delta x \to 0} f(u) = f(x_1)$ and $\lim_{\Delta x \to 0} f(v) = f(x_1)$, which may be written

(4)
$$\lim_{\Delta x \to 0} m = f(x_1) = \lim_{\Delta x \to 0} M.$$

From 8.4.2, when $\Delta x > 0$

$$m(\Delta x) \le \int_{x_1}^{x_1 + \Delta x} f(t)\, dt \le M(\Delta x),$$

and when $\Delta x < 0$, the inequalities are reversed. In either case, using (3),

(5)
$$m \le \frac{G(x_1 + \Delta x) - G(x_1)}{\Delta x} \le M.$$

It follows from (4), (5) and the squeeze theorem (3.3.9) that

$$\lim_{\Delta x \to 0} \frac{G(x_1 + \Delta x) - G(x_1)}{\Delta x} = f(x_1).$$

That is, $G'(x_1) = f(x_1)$. ∎

Proof of (ii). Since F is an antiderivative of f by hypothesis, and $G(x) = \int_a^x f(t)\, dt$ also defines an antiderivative of f, $F(x)$ and $\int_a^x f(t)\, dt$ differ by an additive constant (by 7.1.3). That is,

(6)
$$\int_a^x f(t)\, dt = F(x) + C.$$

To evaluate C, we set $x = a$ in (6) and obtain $\int_a^a f(t)\, dt = F(a) + C$. But $\int_a^a f(t)\, dt = 0$. Thus $F(a) + C = 0$, from which $C = -F(a)$; and (6) becomes

(7)
$$\int_a^x f(t)\, dt = F(x) - F(a).$$

Therefore

$$\int_a^b f(t)\, dt = F(b) - F(a),$$

or, since t is a dummy variable,

$$\int_a^b f(x)\, dx = F(b) - F(a). \quad ∎$$

Part (ii) of the fundamental theorem enables us to find the value of the definite integral of any continuous function if we know an antiderivative of the integrand. For if F is an antiderivative of f, so that $F'(x) = f(x)$ for all x in $[a, b]$, then the fundamental theorem states that

$$\int_a^b f(x)\, dx = F(b) - F(a).$$

It will be convenient to symbolize $F(b) - F(a)$ by $F(x) \Big]_a^b$ or by $\left[F(x) \right]_a^b$. In this notation, the preceding equation can be written

$$\int_a^b f(x)\, dx = F(x) \Big]_a^b.$$

The reader should now review Section 7.2 and its illustrative examples.

Example 1.   Evaluate $\int_{1}^{5} x^4\, dx$.

Solution.   By 7.2.2, the most general antiderivative of x^4 is $x^5/5 + C$. Consequently,

$$\int_{1}^{5} x^4\, dx = \left(\frac{x^5}{5} + C\right)\Big]_{1}^{5} = \left(\frac{5^5}{5} + C\right) - \left(\frac{1^5}{5} + C\right) = 624\tfrac{4}{5}.$$

Notice that C, the constant of antidifferentiation, appears only as $C - C$ in evaluating a definite integral. Accordingly, in applying the fundamental theorem, we shall let $C = 0$ when choosing a particular antiderivative of the integrand.

Example 2.   Find $\int_{-2}^{4} (4x^3 - 5x + 12)\, dx$.

Solution.   From 7.2.2 and 7.2.3, an antiderivative of the integrand is

$$A_x(4x^3 - 5x + 12) = 4A_x x^3 - 5A_x x + 12A_x 1$$

$$= \frac{4x^4}{4} - \frac{5x^2}{2} + 12x$$

$$= \frac{1}{2}(2x^4 - 5x^2 + 24x).$$

Therefore

$$\int_{-2}^{4} (4x^3 - 5x + 12)\, dx = \frac{1}{2}(2x^4 - 5x^2 + 24x)\Big]_{-2}^{4}$$

$$= \frac{1}{2}[2(4)^4 - 5(4)^2 + 24(4)]$$

$$- \frac{1}{2}[2(-2)^4 - 5(-2)^2 + 24(-2)]$$

$$= 282.$$

Example 3.   Find $\int_{0}^{3} \frac{1}{\sqrt{s+1}}\, ds$.

Solution. $1/\sqrt{s+1}$ may be written $(s+1)^{-1/2}$. Since $D_s[2(s+1)^{1/2}] = (s+1)^{-1/2}$, $2\sqrt{s+1}$ is an antiderivative of $1/\sqrt{s+1}$. Therefore

$$\int_{0}^{3} \frac{1}{\sqrt{s+1}}\, ds = 2\sqrt{s+1}\Big]_{0}^{3} = 2\sqrt{3+1} - 2\sqrt{0+1}$$

$$= 2\sqrt{4} - 2\sqrt{1} = 2.$$

Example 4.   Let $G(x) = \int_{-3}^{x} \sqrt{t+4}\, dt$, $-3 \leq x$, an integral with a variable upper limit, define a function G. Use part (i) of the fundamental theorem to find $G'(x)$, and then antidifferentiate the result to get $G(x) + C$.

Solution.   Since the function f defined by $f(x) = \sqrt{x+4}$ is continuous on $[-3, \infty)$, the function G, defined by

(8) $$G(x) = \int_{-3}^{x} \sqrt{t+4}\, dt, \qquad -3 \leq x,$$

is differentiable on $[-3, \infty)$, and

$$G'(x) = D_x \int_{-3}^{x} \sqrt{t + 4} \, dt = \sqrt{x + 4}$$

by part (i) of the fundamental theorem. So $G'(x) = (x + 4)^{1/2}$, $-3 \leq x$.

In finding G, we use 7.2.2 to obtain an antiderivative of G',

$$(9) \qquad\qquad G(x) = \frac{(x + 4)^{3/2}}{\frac{3}{2}} + C = \frac{2(x + 4)^{3/2}}{3} + C.$$

To evaluate the constant of integration C in this problem, we notice that in (8)

$$G(-3) = \int_{-3}^{-3} \sqrt{t + 4} \, dt = 0.$$

Using this result in (9), we have

$$G(-3) = \frac{2(-3 + 4)^{3/2}}{3} + C = \frac{2}{3} + C = 0,$$

from which $C = -\frac{2}{3}$. Therefore

$$G(x) = \frac{2(x + 4)^{3/2}}{3} - \frac{2}{3}, \qquad -3 \leq x.$$

Exercises

In each of Exercises 1 through 10, use part (ii) of the fundamental theorem to find the value of the given definite integral.

1. $\displaystyle\int_{0}^{2} x^3 \, dx.$

2. $\displaystyle\int_{-3}^{5} (6x^2 - 2x - 3) \, dx.$

3. $\displaystyle\int_{-5}^{2} (4x^3 + 7) \, dx.$

4. $\displaystyle\int_{1}^{4} (7z^6 - 6z^5 - 22) \, dz.$

5. $\displaystyle\int_{1}^{4} \frac{2}{5w^2} \, dw.$

6. $\displaystyle\int_{-2}^{3} \frac{1}{(t + 3)^2} \, dt.$

7. $\displaystyle\int_{2}^{10} \sqrt{y - 1} \, dy.$

8. $\displaystyle\int_{-4}^{-2} \left(y^2 + \frac{1}{y^3}\right) dy.$

9. $\displaystyle\int_{1}^{4} \frac{8 - s^4}{2s^2} \, ds.$ (*Hint:* Divide denominator into numerator.)

10. $\displaystyle\int_{-2}^{2} \frac{2x^3 + 11x^2 + 12x - 9}{(x + 3)^2} \, dx.$ (*Hint:* Divide denominator into numerator.)

In each of Exercises 11 to 16, a function G is defined on an interval by means of an integral with a variable upper limit x. Use part (i) of the fundamental theorem to find $G'(x)$ and then antidifferentiate this result to get $G(x) + C$. Finally, determine the value of the constant of antidifferentiation, C, from the fact that $G(a) = \int_{a}^{a} f(t) \, dt = 0$.

11. $G(x) = \displaystyle\int_{-6}^{x} (2t - 1) \, dt, -6 \leq x.$

12. $G(x) = \displaystyle\int_{1}^{x} (6t^2 + 5) \, dt, 1 \leq x.$

13. $G(x) = \displaystyle\int_{-5}^{x} (4t^3 + 6t^2 - 7) \, dt, -5 \leq x.$

14. $G(x) = \displaystyle\int_{-1}^{x} \sqrt{t + 2} \, dt, -1 \leq x.$

15. $G(x) = \displaystyle\int_{-10}^{x} \frac{1}{(t - 3)^2} \, dt, -10 \leq t < 3.$

16. $G(x) = \displaystyle\int_{1}^{x} \left(3t^2 + \frac{1}{\sqrt{t}}\right) dt, 1 \leq x.$

8.5 PROPERTIES OF DEFINITE INTEGRALS

Six basic properties of definite integrals are given in the theorems that follow. They are important for working with integrals.

8.5.1 Theorem (*Homogeneous Property*).   If a function f is continuous on a closed interval $[a, b]$ and k is a constant, then

$$\int_a^b kf(x)\,dx = k\int_a^b f(x)\,dx.$$

Thus a *constant* factor can be moved across the integral sign. (But a variable factor cannot be moved across the integral sign.)

8.5.2 Theorem (*Integrand Additive Property*).   If the functions f and g are continuous on $[a, b]$, then

$$\int_a^b [f(x) + g(x)]\,dx = \int_a^b f(x)\,dx + \int_a^b g(x)\,dx.$$

That is, an integral of a sum of functions is equal to the sum of the integrals of the functions.

8.5.3 Theorem.   If f is continuous on $[a, b]$ and m_1 and M_1 are any numbers such that $m_1 \leq f(x) \leq M_1$ for all $x \in [a, b]$, then

$$m_1(b - a) \leq \int_a^b f(x)\,dx \leq M_1(b - a).$$

Notice that m_1 and M_1 in this theorem are not necessarily the minimum and maximum of f on $[a, b]$.

8.5.4 Theorem.   If f is continuous on $[a, b]$ and $f(x) \geq 0$ for all $x \in [a, b]$, then

$$\int_a^b f(x)\,dx \geq 0.$$

In other words, if a continuous function is nonnegative, its definite integral is nonnegative.

8.5.5 Theorem (*Comparison Property*).   If f and g are continuous on $[a, b]$ and $f(x) \leq g(x)$ for $a \leq x \leq b$, then

$$\int_a^b f(x)\,dx \leq \int_a^b g(x)\,dx.$$

This is a simple extension of 8.5.4 because $g(x) - f(x)$ is continuous and nonnegative.

8.5.6   Theorem (*Interval Additive Property*).   If f is continuous on an interval containing the three points a, b, and c, then

$$\int_a^b f(x)\,dx + \int_b^c f(x)\,dx = \int_a^c f(x)\,dx,$$

no matter what the order of the points a, b, and c.

This is just a restatement of 8.4.1. It is placed here with the other five properties for completeness.

It is very easy to prove the first five properties. In proving 8.5.1 and 8.5.2, both parts of the fundamental theorem are used. We shall prove 8.5.2 and leave the proof of 8.5.1 as an exercise.

Proof of 8.5.2.   Since f and g are continuous on $[a, b]$, they have antiderivatives F and G such that $F'(x) = f(x)$ and $G'(x) = g(x)$, $a \le x \le b$. Because $D_x[F(x) + G(x)] = F'(x) + G'(x) = f(x) + g(x)$, we see that $[F(x) + G(x)]$ is an antiderivative of $[f(x) + g(x)]$ on $[a, b]$. Therefore

$$\int_a^b [f(x) + g(x)]\,dx = [F(x) + G(x)]\Big]_a^b = [F(b) + G(b)] - [F(a) + G(a)]$$

$$= [F(b) - F(a)] + [G(b) - G(a)]$$

$$= \int_a^b f(x)\,dx + \int_a^b g(x)\,dx. \quad \blacksquare$$

Proof of 8.5.3.   Denote by m and M the minimum and the maximum of f on $[a, b]$. Since $m_1 \le m \le f(x) \le M \le M_1$, it follows from 8.4.2 that $m_1(b - a) \le m(b - a) \le \int_a^b f(x)\,dx \le M(b - a) \le M_1(b - a)$. $\quad \blacksquare$

Proof of 8.5.4.   Since f is continuous on $[a, b]$ and $f(x) \ge 0$, f has an antiderivative F on $[a, b]$ and $F'(x) = f(x) \ge 0$, $a \le x \le b$. Thus F is a nondecreasing function on $[a, b]$, from which $F(b) \ge F(a)$. Therefore $\int_a^b f(x)\,dx = F(b) - F(a) \ge 0$. $\quad \blacksquare$

Proof of 8.5.5. $g(x) - f(x)$ is continuous and nonnegative on $[a, b]$. Thus, by 8.5.4, $\int_a^b [g(x) - f(x)]\,dx \ge 0$. Since

$$\int_a^b [g(x) - f(x)]\,dx = \int_a^b g(x)\,dx - \int_a^b f(x)\,dx \ge 0,$$

then

$$\int_a^b f(x)\,dx \le \int_a^b g(x)\,dx. \quad \blacksquare$$

The reader should now review Section 7.3 and its illustrative examples before proceeding.

Example 1. Find the value of $\int_1^3 (2x^2 - 8)^3(4x)\, dx$.

Solution. To use the fundamental theorem, we need an antiderivative of $(2x^2 - 8)^3(4x)$, the integrand. From the generalized power formula for antiderivatives (7.3.1),

$$A_x\{[f(x)]^n \cdot f'(x)\} = \frac{[f(x)]^{n+1}}{n + 1}.$$

Let $f(x) = 2x^2 - 8$. Then $f'(x) = 4x$, and an antiderivative of the integrand is $A_x\{(2x^2 - 8)^3 (4x)\} = (2x^2 - 8)^4/4$. Thus

$$\int_1^3 (2x^2 - 8)^3(4x)\, dx = \frac{(2x^2 - 8)^4}{4}\bigg]_1^3$$

$$= \frac{[2(3)^2 - 8]^4}{4} - \frac{[2(1)^2 - 8]^4}{4}$$

$$= 2176.$$

Example 2. Evaluate $\int_0^2 (2x^2)\sqrt{x^3 + 1}\, dx$.

Solution. Using 8.5.1, we can rewrite the given integral,

$$(1) \qquad\qquad \int_0^2 (2x^2)\sqrt{x^3 + 1}\, dx = 2\int_0^2 (x^3 + 1)^{1/2}(x^2)\, dx.$$

Let $f(x) = (x^3 + 1)$; then $f'(x) = 3x^2$. Notice that the integrand in the right-hand member or equation (1) is not quite in the form $[f(x)]^n f'(x) = (x^3 + 1)^{1/2}(3x^2)$ needed to apply the generalized power formula for antiderivatives. But if we multiply and divide this integrand by the constant 3, we obtain

$$2\int_0^2 (x^3 + 1)^{1/2}(x^2)\, dx = \frac{2}{3}\int_0^2 (x^3 + 1)^{1/2}(3x^2)\, dx = \frac{2}{3}\frac{(x^3 + 1)^{3/2}}{\frac{3}{2}}\bigg]_0^2$$

$$= \frac{4}{9}[(2^3 + 1)^{3/2} - (0 + 1)^{3/2}] = \frac{104}{9}.$$

In Example 2 we were able to supply a needed *constant* factor in the integrand and compensate for it by putting its reciprocal in front of the integral sign, because of 8.5.1. We repeat, however, that we cannot supply a factor involving the variable, since a variable cannot be moved across the integral sign.

Example 3. Find $\int_{-1}^2 \frac{(w^2 + 6w)}{\sqrt{w^3 + 9w^2 + 1}}\, dw$.

Solution. Let $f(w) = w^3 + 9w^2 + 1$; then $f'(w) = 3w^2 + 18w$. So

$$\int_{-1}^2 \frac{(w^2 + 6w)}{\sqrt{w^3 + 9w^2 + 1}}\, dw = \frac{1}{3}\int_{-1}^2 (w^3 + 9w^2 + 1)^{-1/2}(3w^2 + 18w)\, dw$$

$$= \frac{2}{9}(w^3 + 9w^2 + 1)^{3/2}\bigg]_{-1}^2 = \frac{2}{9}[(45)^{3/2} - (9)^{3/2}]$$

$$= 6(5\sqrt{5} - 1).$$

Example 4. Find $\displaystyle\int_2^4 \frac{x^3 + 4x^2 + 4x + 1}{(x + 2)^2}\, dx$.

Solution.   The given integral may be written

$$\int_2^4 \frac{x^3 + 4x^2 + 4x + 1}{x^2 + 4x + 4}\, dx.$$

When the integrand is the quotient of two polynomials and *the degree of the numerator is equal to or greater than the degree of the denominator*, always divide the denominator into the numerator. Here

$$\frac{x^3 + 4x^2 + 4x + 1}{x^2 + 4x + 4} = x + \frac{1}{x^2 + 4x + 4}$$

$$= x + \frac{1}{(x + 2)^2}.$$

Thus the given integral is equivalent to

$$\int_2^4 \left[x + \frac{1}{(x + 2)^2} \right] dx = \int_2^4 x\, dx + \int_2^4 (x + 2)^{-2}\, dx$$

$$= \frac{x^2}{2} \Big]_2^4 + \frac{(x + 2)^{-1}}{-1} \Big]_2^4 = \frac{x^2}{2} \Big]_2^4 - \frac{1}{x + 2} \Big]_2^4$$

$$= \left(\frac{16}{2} - \frac{4}{2} \right) - \left(\frac{1}{6} - \frac{1}{4} \right) = 6\tfrac{1}{12}.$$

Exercises

In each of Exercises 1 through 12, find the value of the definite integral.

1. $\displaystyle\int_5^8 \sqrt{3x + 1}\, dx.$

2. $\displaystyle\int_0^7 \frac{1}{\sqrt{2x + 2}}\, dx.$

3. $\displaystyle\int_{-6}^{-1} \left(x + \frac{1}{x^2} \right) dx.$

4. $\displaystyle\int_3^7 x\sqrt{2x^2 - 17}\, dx.$

5. $\displaystyle\int_{-1}^0 \frac{5s}{\sqrt{3 - s^2}}\, dx.$

6. $\displaystyle\int_{-1}^2 x^3\sqrt{3x^4 + 1}\, dx.$

7. $\displaystyle\int_{-3}^3 \frac{5t}{\sqrt{3 + 2t^2}}\, dt.$

8. $\displaystyle\int_1^3 \frac{x^2 + 1}{\sqrt{x^3 + 3x}}\, dx.$

9. $\displaystyle\int_{-2}^1 \frac{2w^3 - 12w^2 + 18w + 5}{9 - 6w + w^2}\, dw.$ (*Hint:* Divide denominator into numerator.)

10. $\displaystyle\int_1^3 \frac{4x^3 + 8x^2 - 11x + 4}{(2x - 1)^2}\, dx.$

11. $\displaystyle\int_{-4}^1 \frac{1 - s^5}{2s^2}\, ds.$

12. $\displaystyle\int_a^{8a} (a^{1/3} - x^{1/3})^3\, dx.$

A function f is said to be an **even function** if $f(-x) = f(x)$ for all x in the domain of f. A function f is an **odd function** if $f(-x) = -f(x)$ for all x in the domain of f (see Exercises, Section 2.9).

13. Prove that if f is an odd function, integrable on $[-a, a]$, then

$$\int_{-a}^a f(x)\, dx = 0.$$

14. If f is an even function, integrable on $[-a, a]$,

show that

$$\int_{-a}^{a} f(x)\,dx = 2\int_{0}^{a} f(x)\,dx.$$

15. Find $\displaystyle\int_{-5}^{5} (x^3 - 5x)^{1/3}\,dx$.

16. Find $\displaystyle\int_{-3}^{3} |x|\,dx$.

17. Find $\displaystyle\int_{-4}^{4} |x^3|\,dx$.

18. Prove that if f is continuous on $[a, b]$, then

$$\left|\int_{a}^{b} f(x)\,dx\right| \le \int_{a}^{b} |f(x)|\,dx.$$

[*Hint:* Consider separately the two cases

(a) $\displaystyle\int_{a}^{b} f(x)\,dx \ge 0$ and (b) $\displaystyle\int_{a}^{b} f(x)\,dx < 0$.]

19. If the functions f and g are continuous on $[a, b]$, show that for any real constant t

(a) $\displaystyle\int_{a}^{b} [tf(x) + g(x)]^2\,dx$ exists;

(b) $\displaystyle t^2 \int_{a}^{b} [f(x)]^2\,dx + 2t \int_{a}^{b} f(x)\,g(x)\,dx$

$$+ \int_{a}^{b} [g(x)]^2\,dx \ge 0.$$

[*Hint:* Notice that (a) and the left member of (b) are identical; then apply 8.5.4 to (a).]

20. Let f and g be functions that are continuous on $[a, b]$. Prove the Cauchy-Schwarz inequality for integrals,

$$\left(\int_{a}^{b} f(x)\,g(x)\,dx\right)^2 \le \int_{a}^{b} [f(x)]^2\,dx \int_{n}^{b} [g(x)]^2\,dx.$$

(*Hint:* See Exercise 19, Section 8.2.)

8.6 THE MEAN VALUE THEOREM FOR INTEGRALS

Consider a function f that is continuous on a closed interval $[a, b]$ and, for the moment, let $f(x) \ge 0$ for $a \le x \le b$ (Fig. 8-14). Let R denote the region $ABEG$ bounded by the curve $y = f(x)$, the x axis, and the vertical lines $x = a$ and $x = b$.

The figure suggests that there is a point μ between a and b such that the area of the shaded rectangle $ABCD$, with altitude $f(\mu)$, is equal to the area of the region R. The area of the rectangle $ABCD$ is $f(\mu)(b - a)$, and the area of the region R is $\displaystyle\int_{a}^{b} f(x)\,dx$;

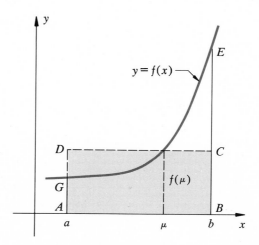

Figure 8-14

thus

$$\int_a^b f(x)\,dx = f(\mu)(b - a).$$

This suggests (but does not prove) the following theorem.

8.6.1 Mean Value Theorem for Integrals. If f is continuous on the closed interval $[a, b]$, there is a number μ between a and b such that

$$\int_a^b f(x)\,dx = f(\mu)(b - a).$$

Proof. Since f is continuous on the closed interval $[a, b]$, it follows from the maximum-minimum theorem (6.1.2) that there are numbers x_1 and x_2 in $[a, b]$ such that $f(x_1) = m$ is the minimum of f on $[a, b]$ and $f(x_2) = M$ is the maximum. Thus

$$m \le f(x) \le M$$

for $a \le x \le b$. Therefore, (by 8.5.3)

$$m(b - a) \le \int_a^b f(x)\,dx \le M(b - a)$$

or

$$f(x_1) \le \frac{\int_a^b f(x)\,dx}{b - a} \le f(x_2).$$

Consequently, by the intermediate value theorem (3.5.8), there is a number μ between x_1 and x_2 such that

$$f(\mu) = \frac{\int_a^b f(x)\,dx}{b - a},$$

or

$$\int_a^b f(x)\,dx = f(\mu)(b - a). \quad \blacksquare$$

This theorem does not offer an easy way to find the value of $\int_a^b f(x)\,dx$ because we do not know the exact value of μ, but it will be important in proving later theorems.

8.6.2 Definition. If f is continuous on $[a, b]$, the number

$$f(\mu) = \frac{\int_a^b f(x)\,dx}{b - a}$$

is called the **average value** of f on $[a, b]$.

Example. If $f(x) = (x + 1)^3$, find the average value of f on $[-3, 3]$.

Solution.   Since

$$\int_{-3}^{3} (x + 1)^3 \, dx = \frac{(x + 1)^4}{4} \Bigg]_{-3}^{3} = 60,$$

the average value of f on $[-3, 3]$ is

$$\frac{\displaystyle\int_{-3}^{3} (x + 1)^3 \, dx}{3 - (-3)} = \frac{60}{6} = 10.$$

8.7 APPROXIMATE INTEGRATION BY THE TRAPEZOIDAL RULE

We know that if f is continuous on a closed interval $[a, b]$, the definite integral $\int_{a}^{b} f(x) \, dx$ exists. To evaluate a definite integral by means of the fundamental theorem, we must first find an antiderivative of the integrand. Although we shall soon learn how to find the antiderivatives of many functions, there are still other (continuous) functions whose antiderivatives cannot be expressed in terms of *elementary functions*— that is, in terms of the functions studied in a first course in calculus. A few examples of such troublesome functions are

$$f(x) = \sqrt{1 + x^4}, \qquad f(x) = \frac{1}{\sqrt{1 + x^3}}, \qquad f(x) = \frac{\sin x}{x}$$

(see Exercises 9 to 14, Section 11.15).

Nevertheless, there are methods for approximate integration that enable us to compute the values of definite integrals correctly to as many decimal places as are needed in practical applications. Although a definite integral has many interpretations not connected with geometry, it can always be interpreted as an algebraic sum of areas if doing so suits our purpose. Hence if f is a function that is integrable on $[a, b]$, any method for approximating the area of the region bounded by the graph of $y = f(x)$, the x axis and the vertical lines $x = a$ and $x = b$ will give an approximation to the value of the definite integral $\int_{a}^{b} f(x) \, dx$. A simple method known as the **trapezoidal rule** will now be explained.

Let f be a function that is nonnegative and integrable on an interval $[a, b]$. Partition $[a, b]$ into n equal subintervals, each of length Δx, by a regular partition $P = \{x_0, x_1, x_2, \cdots, x_n\}$, and draw the ordinates $f(a) = f(x_0), f(x_1), f(x_2), \cdots, f(x_n) = f(b)$ [see Fig. 8-15 in which n is taken as 4]. Denote the points $(x_i, f(x_i))$ by Q_i, $i = 0, 1, 2, \cdots, n,$ and draw the line segments $Q_{i-1}Q_i$. Since the area A_i of each of the trapezoids thus formed is equal to the product of one-half the sum of the lengths of the parallel sides by the distance between them, we have

$$A_i = \tfrac{1}{2}[f(x_{i-1}) + f(x_i)] \, \Delta x;$$

and the sum of these areas is an approximation to the area A of the region bounded

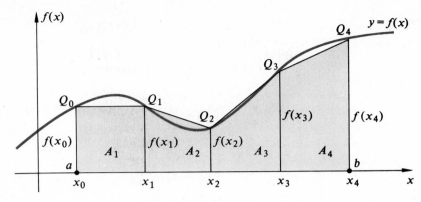

Figure 8-15

by the x axis, the graph of $y = f(x)$, and the vertical lines $x = a$ and $x = b$. That is,

$$A \doteq \tfrac{1}{2}[f(x_0) + f(x_1)]\,\Delta x + \tfrac{1}{2}[f(x_1) + f(x_2)]\,\Delta x$$
$$+ \tfrac{1}{2}[f(x_2) + f(x_3)]\,\Delta x + \cdots + \tfrac{1}{2}[f(x_{n-1}) + f(x_n)]\,\Delta x.$$

We therefore write

8.7.1   Trapezoidal Rule.   If f is continuous on the closed interval $[a, b]$, then

$$\int_a^b f(x)\,dx = \frac{\Delta x}{2}[f(x_0) + 2f(x_1) + 2f(x_2) + \cdots$$
$$+ 2f(x_{n-1}) + f(x_n)],$$

where $\Delta x = (b - a)/n$ and $x_i = a + i\,\Delta x$, $i = 0, 1, 2, \cdots, n$.

By 8.3.4, this sum can be made as close as we please to the value of $\int_a^b f(x)\,dx$ by taking Δx sufficiently small, and this is accomplished by taking n sufficiently large because the partition is regular.

Example 1.   Find an approximate value of $\int_1^3 x^4\,dx$ by the trapezoidal rule, taking $n = 8$.

Solution.   When $n = 8$, $\Delta x = \tfrac{2}{8} = 0.25$. Since $f(x) = x^4$,

$$
\begin{aligned}
x_0 &= 1.00, & f(x_0) &= (1.00)^4 = 1.0000 \\
x_1 &= 1.25, & f(x_1) &\doteq (1.25)^4 \doteq 2.4414 \\
x_2 &= 1.50, & f(x_2) &= (1.50)^4 = 5.0625 \\
x_3 &= 1.75, & f(x_3) &\doteq (1.75)^4 \doteq 9.3789 \\
x_4 &= 2.00, & f(x_4) &= (2.00)^4 = 16.0000 \\
x_5 &= 2.25, & f(x_5) &\doteq (2.25)^4 \doteq 25.6289 \\
x_6 &= 2.50, & f(x_6) &= (2.50)^4 = 39.0625
\end{aligned}
$$

$$x_7 = 2.75, \quad f(x_7) = (2.75)^4 \doteq 57.1914$$

$$x_8 = 3.00, \quad f(x_8) = (3.00)^4 = 81.0000$$

By the trapezoidal rule, with $\Delta x/2 = \frac{1}{8}$,

$$\int_1^3 x^4 \, dx \doteq \tfrac{1}{8}[1.0000 + 2(2.4414)$$
$$+ 2(5.0625) + 2(9.3789)$$
$$+ 2(16.0000) + 2(25.6289)$$
$$+ 2(39.0625) + 2(57.1914)$$
$$+ 81.0000] = 48.9414.$$

Of course, it is easy to evaluate this integral exactly by the fundamental theorem, since we know that an antiderivative of x^4 is $x^5/5$. Thus

$$\int_1^3 x^4 \, dx = \frac{x^5}{5}\bigg]_1^3 = \frac{3^5 - 1^5}{5} = \frac{242}{5} = 48.4.$$

Example 2. Approximate the value of $\displaystyle\int_1^2 \frac{1}{x} \, dx.$

Solution. Since we cannot, as yet, find an antiderivative of $1/x = x^{-1}$, we shall use the trapezoidal rule. By taking $n = 10$, $\Delta x = (2 - 1)/10 = 0.1$; and

$$x_0 = 1.0, \quad f(x_0) = \frac{1}{1.0} = 1.0000$$

$$x_1 = 1.1, \quad f(x_1) = \frac{1}{1.1} = 0.9091$$

$$x_2 = 1.2, \quad f(x_2) = \frac{1}{1.2} = 0.8333$$

$$x_3 = 1.3, \quad f(x_3) = \frac{1}{1.3} = 0.7692$$

$$x_4 = 1.4, \quad f(x_4) = \frac{1}{1.4} = 0.7143$$

$$x_5 = 1.5, \quad f(x_5) = \frac{1}{1.5} = 0.6667$$

$$x_6 = 1.6, \quad f(x_6) = \frac{1}{1.6} = 0.6250$$

$$x_7 = 1.7, \quad f(x_7) = \frac{1}{1.7} = 0.5882$$

$$x_8 = 1.8, \quad f(x_8) = \frac{1}{1.8} = 0.5556$$

$$x_9 = 1.9, \quad f(x_9) = \frac{1}{1.9} = 0.5623$$

$$x_{10} = 2.0, \quad f(x_{10}) = \frac{1}{2.0} = 0.5000.$$

From the trapezoidal rule, with $\Delta x/2 = 0.05$,

$$\int_1^2 \frac{1}{x}\,dx \doteq 0.05[1.0000 + 2(0.9091) + 2(0.8333) + 2(0.7692)$$
$$+\, 2(0.7143) + 2(0.6667) + 2(0.6250) + 2(0.5882)$$
$$+\, 2(0.5556) + 2(0.5263) + 0.5000] = 0.694.$$

Later on we shall see that the value of this integral, correct to three decimal places, is 0.693.

When using an approximation process, it is important to know how good the approximation is. It can be shown* that if f'' exists on $[a, b]$, then the error E_n introduced in using the trapezoidal rule to determine the (approximate) value of a definite integral $\int_b^b f(x)\,dx$ is bounded by the inequality

(1) $$|E_n| \le \frac{M(b-a)^3}{12n^2},$$

where n is the number of (equal) intervals in the regular partition of $[a, b]$ and M is any number such that $|f''(x)| \le M$ for $a \le x \le b$.

Example 3. In using the trapezoidal rule to approximate the value of the definite integral in Example 2 correct to three decimal places, into how many equal subintervals should we partition $[1, 2]$?

Solution. From $f(x) = 1/x$, we find $f''(x) = 2/x^3$. Since f'' is a decreasing function on $[1, 2]$, $f''(1) = 2$ is the maximum of f'' on $[1, 2]$. This suggests that we let $M = 2$ in the error formula (1), above, and obtain

$$|E_n| \le \frac{2(2-1)}{12n^2} = \frac{1}{6n^2}.$$

For the approximate value of our definite integral to be accurate to three decimal places, we must have $|E_n| < 0.0005$. Thus we seek the smallest positive integer n that satisfies

$$\frac{1}{6n^2} < 0.0005.$$

To solve this inequality for n, we rewrite it as

$$n^2 > \frac{1}{6(0.0005)} = \frac{1000}{3},$$

from which $n > \sqrt{1000/3} \doteq 18.26$. So $n = 19$ will do.

Exercises

In Exercises 1 to 4, find an approximate value of the given definite integral by the trapezoidal rule, using the given value of n. Keep four decimal places in your calculations, and round off your answers to three decimal places.

1. $\int_0^2 x^4\,dx$, $n = 10$.

2. $\int_1^2 \sqrt{x}\,dx$, $n = 5$.

*See, for example, J. M. H. Olmsted, *Advanced Calculus*, Appleton-Century-Crofts, 1961, pp. 118, 119.

3. $\int_{1}^{2} \frac{1}{x} dx, n = 8.$

4. $\int_{1}^{2} \frac{1}{x^2} dx, n = 4.$

In Exercises 5 to 10, use the trapezoidal rule to find an approximate value of the given definite integral correct to two decimal places. In each exercise, first find a value of n that will ensure this accuracy.

5. $\int_{0}^{1} \frac{1}{x^2 + 2} dx.$

6. $\int_{2}^{3} x\sqrt{x^2 + 4} \, dx.$

7. $\int_{1}^{2} \sqrt{1 + x^3} \, dx.$

8. $\int_{2}^{3} \sqrt{x^4 + 2} \, dx.$

9. $\int_{1}^{4} \frac{2}{\sqrt{1 + x}} dx.$

10. $\int_{0}^{2} \frac{1}{\sqrt{1 + x^2}} dx.$

8.8 REVIEW EXERCISES

1. Find the value of $\int_{0}^{4} \frac{x}{\sqrt{x^2 + 9}} dx.$

2. Find $\int_{2}^{4} y^2 \sqrt{y^3 - 1} \, dy.$

3. Find $\int_{2}^{8} x\sqrt{2x} \, dx.$

4. Let P be a regular partition of the interval $[0, 2]$ into four equal subintervals, and let $f(x) = x^2 - 1$. Write out the Riemann sum for f on P, in which w_i is the right endpoint of each subinterval of P, $i = 1$, 2, 3, 4. Find the value of this Riemann sum, and make a sketch.

5. If $f(x) = \int_{-2}^{x} \frac{1}{t + 3} dt, -2 \le x,$ find $f'(7)$.

6. Find $\int_{0}^{3} (2 - \sqrt{x + 1})^2 \, dx.$

7. If $f(x) = 2x^2 \sqrt{x^3 - 4}$, find the average value of f on $[2, 5]$.

8. Find $\int_{2}^{4} \frac{5x^2 - 1}{x^2} dx.$

9. Evaluate $\sum_{i=1}^{n} (3^i - 3^{i-1}).$

10. Evaluate $\sum_{i=1}^{n} 2i(4i - 3).$

11. Find $\int_{1}^{4} \frac{2 + \sqrt{x}}{x^2} dx.$

12. If $G(x) = \int_{-4}^{x} \frac{t}{\sqrt{3t^2 + 1}} dt, -4 \le x,$ find $G'(x)$; then antidifferentiate the result to get $G(x) + C$. Finally, evaluate C for this problem.

13. Given $G(x) = \int_{0}^{x} 3t(2t^2 - 1)^5 \, dt,$ express $G(x)$ without an integral sign.

14. Let $f(x) = x/\sqrt{1 + x}, x \ge 0$. If the average value of f on $[3, 8]$ is $2\frac{2}{15}$, find $\int_{3}^{8} (x/\sqrt{1 + x}) \, dx.$

15. Find $\int_{1}^{4} \frac{x^2 - 2x + 5}{\sqrt{x}} dx.$

16. Find $\int_{-1}^{2} \frac{3x - 1}{(3x^2 - 2x + 1)^2} dx.$

17. Find $\int_{1}^{5} |x - 2| \, dx.$ (*Hint:* On $[1, 2]$, $|x - 2|$ $= 2 - x$; and on $[2, 5]$, $|x - 2| = x - 2$. Use 8.5.6.)

18. Find $\int_{-5}^{5} 2x(x^4 - 1)^{10} \, dx.$ (*Hint:* See Exercise 13, Section 8.5)

19. If $f(x) = \sqrt{x^2 + 1}/x^3$, f is an odd function. Can you find $\int_{-2}^{2} (\sqrt{x^2 + 1}/x^3) \, dx$? If so, what is its value? If not, explain why.

20. Find $\int_{0}^{2} (2x^3 - 2x^2 - 2x - 3)^3(3x + 1)$ $(x - 1) \, dx.$

9

applications
of the definite integral

The definite integral is much richer in practical applications than the derivative. For a start, we shall see in this chapter some of the ways that the definite integral is used in geometry, engineering, physics, and economics.

In particular, the definite integral will be used to find areas of plane regions, surface areas of solids of revolution, and volumes of solids of revolution. We shall also use the definite integral to find the total force of water pressure on a submerged surface, to determine the work done by a variable force on an object moving along a straight line, to find the length of an arc of a plane curve, to find the moment of mass of a lamina about a line, to find the center of gravity of a lamina and of a solid of revolution, and to find the total profit in manufacturing and selling a commodity under certain varying conditions.

9.1 PLANE AREAS

The brief discussion of "area" in Section 8.1 served to *motivate* our development of the concept of the definite integral. But, as pointed out earlier, the definition of the definite integral (8.3.4) and the proofs of its properties were based entirely on real numbers, not on geometric intuition. It is therefore not circular for us to define the area of a plane region by means of a definite integral.

Consider a function f that is continuous on a closed interval $[a, b]$, with $f(x) \geq 0$ for $a \leq x \leq b$. We wish to define the area of the region

$$R = \{P:(x, y) \mid a \leq x \leq b, 0 \leq y \leq f(x)\}$$

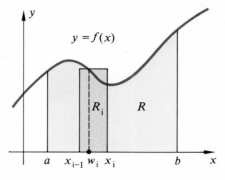

Figure 9-1

bounded by the curve $y = f(x)$, the x axis, and the vertical lines $x = a$ and $x = b$ (Fig. 9-1). Let $P = \{x_0, x_1, x_2, \cdots, x_n\}$ be a partition of $[a, b]$ into n subintervals

$$[x_0, x_1], [x_1, x_2], \cdots, [x_{i-1}, x_i], \cdots, [x_{n-1}, x_n],$$

and let w_i be an arbitrarily chosen point in the ith subinterval $[x_{i-1}, x_i]$, $i = 1, 2, \cdots, n$. Form n *approximating rectangles*, a typical one being R_i (Fig. 9-1) whose base $[x_{i-1}, x_i]$ has length $\Delta x_i = x_i - x_{i-1}$ and whose altitude is $f(w_i)$. The area of the rectangle R_i is $f(w_i)\,\Delta x_i$, and the sum of the areas of the n approximating rectangles,

$$\sum_{i=1}^{n} f(w_i)\,\Delta x_i,$$

is a Riemann sum that approximates the area of R more and more closely as the norm $|P|$ of the partition decreases. We define the area of the region R to be

$$\lim_{|P| \to 0} \sum_{i=1}^{n} f(w_i)\,\Delta x_i.$$

Since f is continuous on $[a, b]$, this limit exists and is the definite integral $\int_a^b f(x)\,dx$.

So, if f is continuous and nonnegative valued on the closed interval $[a, b]$, the *area A* of the region R bounded by the curve $y = f(x)$, the x axis, and the vertical lines $x = a$ and $x = b$ is defined to be

$$(1) \qquad A = \int_a^b f(x)\,dx.$$

Example 1. Find the area of the region R bounded by the curve $y = \frac{1}{3}x^2 + 1$, the x axis, and the lines $x = -2$ and $x = 3$ (Fig. 9-2).

Solution. The function f, defined by $f(x) = \frac{1}{3}x^2 + 1$, is continuous and positive valued on the closed interval $[-2, 3]$. Thus, by (1), the area of R is

$$\text{Area} = \int_{-2}^{3} \left(\frac{x^2}{3} + 1\right) dx = \left[\frac{x^3}{9} + x\right]_{-2}^{3}$$

$$= \left(\frac{(3)^3}{9} + 3\right) - \left(\frac{(-2)^3}{9} + (-2)\right)$$

$$= 6 - \left(\frac{-26}{9}\right) = 8\frac{8}{9} \text{ square units.}$$

Example 2. Find the area of the region R bounded by the curve $y = x^4 - 2x^3 + 2$, the x axis, and the lines $x = -1$ and $x = 2$ (Fig. 9-3).

Solution. The function f, defined by $f(x) = x^4 - 2x^3 + 2$, is continuous and positive valued on the given interval $[-1, 2]$. From (1), the area of R is

$$\text{Area} = \int_{-1}^{2} (x^4 - 2x^3 + 2)\,dx = \left[\frac{x^5}{5} + \frac{2x^4}{4} + 2x\right]_{-1}^{2}$$

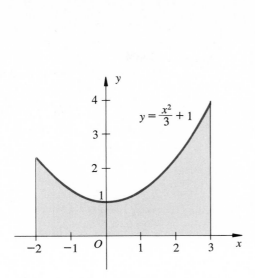

Figure 9-2

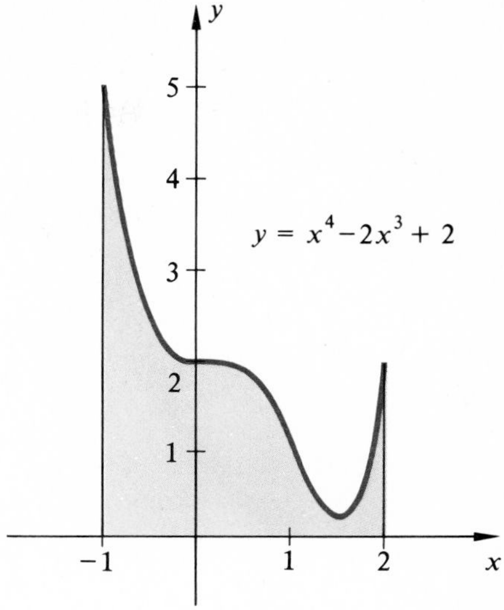

Figure 9-3

$$= \left(\frac{32}{5} - \frac{32}{4} + 4\right) - \left(-\frac{1}{5} - \frac{2}{4} - 2\right)$$

$$= \frac{12}{5} - \left(-\frac{27}{10}\right) = \frac{51}{10} \text{ square units.}$$

We said, in 8.1.1, that areas of plane regions should be defined so that they would have the following five characteristics:

1. The area of a plane region is a nonnegative number.
2. Congruent regions have equal areas.
3. If a region is contained in a second region (that is, if every point of the first region is also a point of the second region), then the area of the first region is less than or equal to the area of the second region.
4. The area of the union of two nonoverlapping regions is equal to the sum of the areas of two regions.
5. The area of a rectangle is the product of the lengths of two adjacent sides.

It can be shown from the properties of definite integrals that our definition (1) of area fulfills these requirements.

Now consider a function f that is continuous on $[a, b]$, but with $f(x) \leq 0$ for $a \leq x \leq b$. What is the area of the region below the x axis, above the curve $y = f(x)$, and between the lines $x = a$ and $x = b$ (Fig. 9-4)? From property 1, the area must

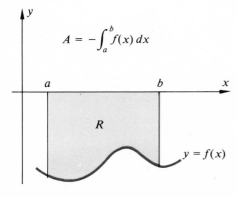

$$A = -\int_a^b f(x)\, dx$$

Figure 9-4

be a nonnegative number. But when $f(x) \leq 0$ on $[a, b]$, $\int_a^b f(x)\, dx \leq 0$ by the comparison property (8.5.5). So we define the area of this region to be

(2) $$A = -\int_a^b f(x)\, dx.$$

The following definition covers both of the preceding situations and also the case where $f(x)$ is sometimes positive and sometimes negative for $a \leq x \leq b$.

9.1.1 Definition.   Let f be continuous on the closed interval $[a, b]$. The **area** of the region bounded by the curve $y = f(x)$, the x axis and the lines $x = a$ and $x = b$ is

$$A = \int_a^b |f(x)|\, dx.$$

Example 3.   Find the area of the region bounded by the curve $y = x^3 - 3x^2 - x + 3$, the segment of the x axis from -1 to 2, and the line $x = 2$ (Fig. 9-5).

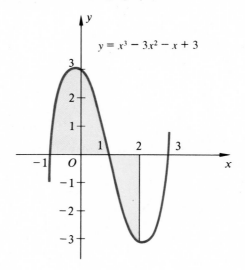

$$y = x^3 - 3x^2 - x + 3$$

Figure 9-5

Solution.   Since $x^3 - 3x^2 - x + 3 = (x + 1)(x - 1)(x - 3)$, the graph of $y = x^3 - 3x^2 - x + 3$ crosses the x axis at -1, 1, and 3. The graph is above the x axis between -1 and 1 and below the x axis between 1 and 2 (Fig. 9-5). To apply the definition 9.1.1, we separate the integral into two parts and write

$$A = \int_{-1}^{2} |x^3 - 3x^2 - x + 3|\, dx$$

$$= \int_{-1}^{1} (x^3 - 3x^2 - x + 3)\, dx - \int_{1}^{2} (x^3 - 3x^2 - x + 3)\, dx$$

$$= \left[\frac{x^4}{4} - x^3 - \frac{x^2}{2} + 3x \right]_{-1}^{1} - \left[\frac{x^4}{4} - x^3 - \frac{x^2}{2} + 3x \right]_{1}^{2}$$

$$= 5\tfrac{3}{4} \text{ square units.}$$

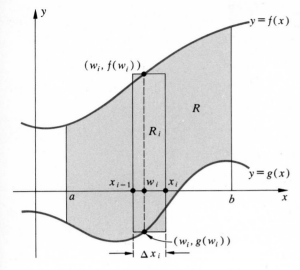

Figure 9-6

Let f and g be functions that are continuous on a closed interval $[a, b]$, with $g(x) \leq f(x)$ for $a \leq x \leq b$. We wish to define the area of the region

$$R = \{P:(x, y) \mid a \leq x \leq b, \, g(x) \leq y \leq f(x)\}$$

bounded by the curves $y = f(x)$ and $y = g(x)$ and the vertical lines $x = a$ and $x = b$ (Fig. 9-6). Partition the interval $[a, b]$ into n subintervals $[x_{i-1}, x_i]$ and let w_i be an arbitrarily chosen point in $[x_{i-i}, x_i]$, $i = 1, 2, 3, \cdots, n$. Form n approximating rectangles, a typical one being R_i (Fig. 9-6) whose altitude is $f(w_i) - g(w_i)$, the directed distance from the point $(w_i, g(x_i))$ to the point $(w_i, f(w_i))$, and whose base has length $\Delta x_i = x_i - x_{i-1}$. The area of R_i is $[f(w_i) - g(x_i)] \, \Delta x_i$, and the sum of the areas of the n approximating rectangles,

$$\sum_{i=1}^{n} [f(w_i) - g(w_i)] \, \Delta x_i,$$

is a Riemann sum. The area of the region R is defined to be

$$(3) \qquad \lim_{|P| \to 0} \sum_{i=1}^{n} [f(w_i) - g(w_i)] \, \Delta x_i.$$

Since f and g are continuous on $[a, b]$, the function $f - g$ is continuous on $[a, b]$. Therefore the limit (3) exists and is equal to the definite integral $\int_a^b [f(x) - g(x)] \, dx$.

9.1.2 Definition. If f and g are continuous on $[a, b]$ and $g(x) \leq f(x)$ for $a \leq x \leq b$, then the **area** of the region R, bounded by the curves $y = f(x)$ and $y = g(x)$, and the lines $x = a$ and $x = b$, is defined by

$$A = \int_a^b [f(x) - g(x)] \, dx.$$

Here, too, the definition of area can be shown to satisfy the five requirements of area listed earlier.

Example 4. Find the area of the region bounded by the curves $y = x^4$ and $y = 2x - x^2$ (Fig. 9-7).

Solution. The given curves intersect at $(0, 0)$ and $(1, 1)$. From 9.1.2, the area of the region between the curves, from $x = 0$ to $x = 1$, is

$$A = \int_0^1 [(2x - x^2) - x^4] \, dx$$

$$= \left[x^2 - \frac{x^3}{3} - \frac{x^5}{5} \right]_0^1$$

$$= \left(1 - \frac{1}{3} - \frac{1}{5} \right) - \left(0 - \frac{0}{3} - \frac{0}{5} \right)$$

$$= \frac{7}{15} \text{ square unit.}$$

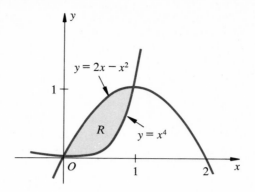

Figure 9-7

Sometimes it is easier to find an area by taking y as the independent variable, as the next example illustrates.

Example 5.   Find the area of the region bounded by the parabola $y^2 = 4x$ and the line $4x - 3y - 4 = 0$ (Fig. 9-8).

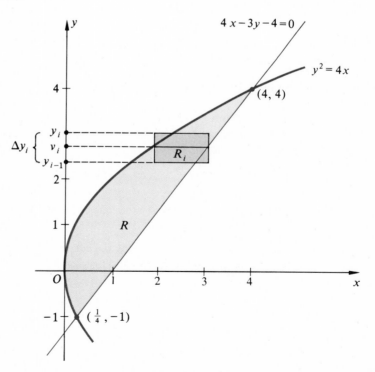

Figure 9-8

Solution. The parabola and the line intersect at $(4, 4)$ and $(\frac{1}{4}, -1)$.

Partition the interval $[-1, 4]$ *on the y axis* into n subintervals $[y_{i-1}, y_i]$, and let v_i be an arbitrarily chosen number in $[y_{i-1}, y_i]$, $i = 1, 2, \cdots, n$. Form n approximating rectangles R_i, a typical one of which is shown in Fig. 9-8. The area of R_i is

$$\left[\frac{3v_i + 4}{4} - \frac{v_i^2}{4}\right] \Delta y_i,$$

and the area of the given region is

$$A = \lim_{|P| \to 0} \sum_{i=1}^{n} \left[\frac{3v_i + 4}{4} - \frac{v_i^2}{4}\right] \Delta y_i$$

$$= \int_{-1}^{4} \left[\frac{3y + 4}{4} - \frac{y^2}{4}\right] dy = 5\frac{5}{24} \text{ square units.}$$

Of course, the same result is obtained by solving the given equations for x in terms of y, and using 9.1.2 with x and y interchanged.

Exercises

In Exercises 1 through 8, sketch the region bounded by the given curve, the x axis, and the given vertical lines. Find its area.

1. $y = 4 - \frac{1}{3}x^2$, $x = 0$, $x = 3$.

2. $y = 4x - x^2$, $x = 1$, $x = 3$.

3. $y = x^2 - 2x - 3$, $x = 0$, $x = 2$.

4. $y = \frac{1}{2}(x^2 - 10)$, $x = -2$, $x = 3$.

5. $y = x^3$, $x = -1$, $x = 2$.

6. $y = x^2 - 2x - 3$, $x = -2$, $x = 1$.

7. $y = \frac{1}{4}(x^3 - x^2 - 12x)$, $x = -2$, $x = 2$.

8. $y = \sqrt[3]{x}$, $x = -1$, $x = 8$.

9. Sketch the region bounded by the curve $y = \sqrt{x - 4}$, the x axis, and the line $x = 8$; then find its area.

10. Sketch the region bounded by the curve $y = x^2 - 4x + 3$ and the line $x - y - 1 = 0$; find its area.

11. Find the area of the region bounded by the curve $y = x^2$ and the line $x - y + 2 = 0$. Make a sketch.

12. Find the area of the region bounded by the curve $y = 2\sqrt{x}$, the line $2x - y - 4 = 0$, and the y axis. Make a sketch.

13. Find the area of the region bounded by the curves $y = x^2 - 4x$ and $y = -x^2$. Make a sketch.

14. Find the area of the region bounded by the curves $y = x^2 - 2$ and $y = 2x^2 + x - 4$. Make a sketch.

15. Make a large-scale sketch of the region bounded by the curves $y = x^3$ and $y = -\frac{1}{2}(x^2 - 2x - 1)$. Then find its area.

16. Make a large-scale sketch of the region bounded by the curves $y = x^{2/3}$ and $y = x^2$ and find its area.

In each of Exercises 17 to 20, use the method of Example 5 to find the area of the region bounded by the graphs of the given equations. Sketch the region and show a typical horizontal approximating rectangle, R_i, as in Fig. 9-8.

17. $y^2 + x - 4 = 0$, $x + y - 2 = 0$.

18. $y^2 - x - 3y = 0$, $x - y + 3 = 0$.

19. $y^2 - 2x = 0$, $y^2 + 4x - 12 = 0$.

20. $x = y^4$, $x = 2 - y^4$.

9.2 VOLUME OF A SOLID OF REVOLUTION

When a plane region, lying entirely on one side of a fixed line in its plane, is revolved about that line, it generates a **solid of revolution**. The fixed line is called the **axis** of the solid of revolution.

As an illustration, if the region bounded by a semicircle and its diameter is revolved about that diameter, it sweeps out a spherical solid. If the region inside a right triangle is revolved about one of its legs, it generates a conical solid. When a circular disk is revolved about a line in its plane that does not intersect the disk, it sweeps out a torus.

All plane sections of a solid of revolution that are perpendicular to its axis are circular disks or regions bounded by two concentric circles.

We seek the volume of a solid of revolution. But first we must define what is meant by the "volume" of a solid of revolution.

Just as in our discussion of a plane area (9.1) we assumed that the area of a rectangle is the product of its length and width, we start our investigation of volumes of solids of revolution by assuming that the volume of a right circular cylinder is $\pi r^2 h$, where r is the radius of its circular base and h is its altitude.

Let f be a function that is continuous on the closed interval $[a, b]$, with $f(x) \geq 0$ for $a \leq x \leq b$. We wish to define the volume of the solid of revolution generated by revolving about the x axis the region R that is bounded by the curve $y = f(x)$, the x axis, and the vertical lines $x = a$ and $x = b$ (Fig. 9-9).

Subdivide the interval $[a, b]$ into n subintervals by a partition P, and choose n points w_i, one in each subinterval. Draw n approximating rectangles with base $[x_{i-1}, x_i]$ and altitude $f(w_i)$, $i = 1, 2, 3, \cdots, n$; a typical one is shown in Fig. 9-9 as $EFGH$.

When the region R is revolved about the x axis to generate a solid of revolution, the n rectangles sweep out n right circular cylinders. The cylinder swept out by the typical rectangle ($EFGH$ in Fig. 9-9) is shown in Fig. 9-10; since the radius of its base is $f(w_i)$ and its altitude is Δx_i, its volume is

$$\Delta V_i = \pi [f(w_i)]^2 \, \Delta x_i.$$

The sum of the volumes of the n cylinders is the Riemann sum

(1)
$$\sum_{i=1}^{n} \Delta V_i = \sum_{i=1}^{n} \pi [f(w_i)]^2 \, \Delta x_i.$$

This Riemann sum approximates what we would like to think of as the "volume" of the solid of revolution, and this approximation improves as the norm of the partition, $|P|$, gets smaller.

We *define* the volume of the solid of revolution under consideration to be

(2)
$$V = \lim_{|P| \to 0} \sum_{i=1}^{n} \Delta V_i = \lim_{|P| \to 0} \sum_{i=1}^{n} \pi [f(w_i)]^2 \, \Delta x_i.$$

Since f is continuous on the closed interval $[a, b]$, the function $f^2 = f \cdot f$ is also

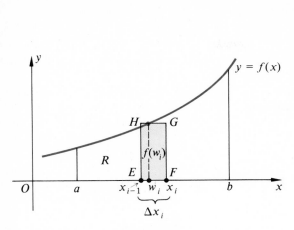

Figure 9-9

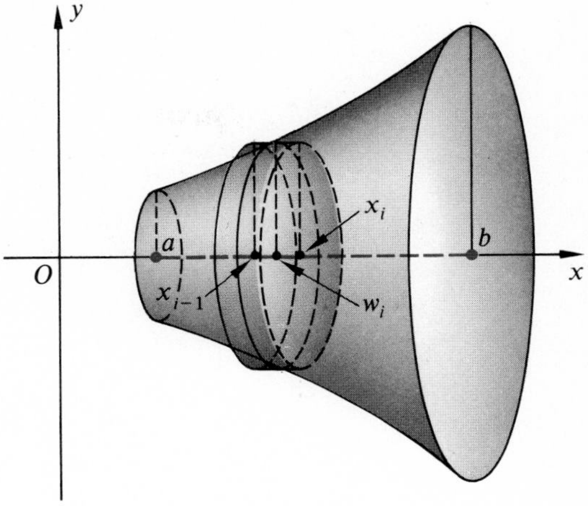

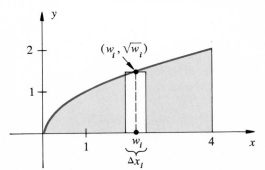

Figure 9-10

continuous on $[a, b]$; therefore the limit (2) exists and is equal to the definite integral

$$\pi \int_a^b [f(x)]^2 \, dx.$$

9.2.1 Definition. If f is a function that is continuous on the closed interval $[a, b]$, with $f(x) \geq 0$ for $a \leq x \leq b$, then the **volume** of the solid of revolution generated by revolving about the x axis the region bounded by the curve $y = f(x)$, the x axis, and the lines $x = a$ and $x = b$ is

$$V = \lim_{|P| \to 0} \sum_{i=1}^n \pi[f(w_i)]^2 \, \Delta x_i = \pi \int_a^b [f(x)]^2 \, dx.$$

Example 1. Find the volume of the solid of revolution generated by revolving about the x axis the plane region, R, bounded by the curve $y = \sqrt{x}$, the x axis, and the line $x = 4$.

Solution. The region R and a typical approximating rectangle are shown in Fig. 9-11. When the region R is revolved about the x axis, the typical rectangle generates an *element of volume*, a cylinder whose volume is $\Delta v_i = \pi(\sqrt{w_i})^2 \, \Delta x_i$. Thus

$$V = \lim_{|P| \to 0} \sum_{i=1}^n \pi(\sqrt{w_i})^2 \, \Delta x_i = \pi \int_0^4 (\sqrt{x})^2 \, dx$$

$$= \pi \int_0^4 x \, dx = \pi \left[\frac{x^2}{2} \right]_0^4 = 8\pi \text{ cubic units.}$$

Similarly, if g is continuous on the closed interval $[c, d]$ of the y axis and $g(y) \geq 0$ for $c \leq y \leq d$, the volume of the solid of revolution generated by revolving about the y axis the region bounded

Figure 9-11

by $x = g(y)$, the y axis, and the horizontal lines $y = c$ and $y = d$ is defined to be

$$V = \lim_{|P| \to 0} \sum_{i=1}^{n} \pi[g(v_i)]^2 \, \Delta y_i = \pi \int_c^d [g(y)]^2 \, dy,$$

where v_i is a point in the ith subinterval $[y_{i-1}, y_i]$ of a partition P of $[c, d]$.

Example 2. Find the volume of the solid generated by revolving about the y axis the region bounded by the curve $y = x^3$, the y axis, and the line $y = 3$ (Fig. 9-12).

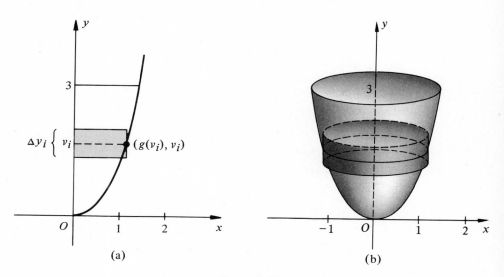

(a)

(b)

Figure 9-12

Solution. Here $g(y) = \sqrt[3]{y}$. Thus

$$V = \lim_{|P| \to 0} \sum_{i=1}^{n} \pi[\sqrt[3]{v_i}]^2 \, \Delta y_i = \pi \int_0^3 y^{2/3} \, dy$$

$$= \frac{3}{5} \pi y^{5/3} \Big]_0^3 = \frac{9\sqrt[3]{9}}{5} \pi \text{ cubic units.}$$

Let f and g be continuous on $[a, b]$ and let $f(x) \geq g(x) \geq 0$ for $a \leq x \leq b$. We shall now extend the definition of volume to the solid of revolution S generated by revolving about the x axis the region $R = \{P : (x, y) \mid a \leq x \leq b, 0 \leq g(x) \leq y \leq f(x)\}$, bounded by the curves $y = f(x)$ and $y = g(x)$ and the two lines $x = a$ and $x = b$ (Fig. 9-13).

Partition the interval $[a, b]$ into n subintervals $[x_{i-1}, x_i]$ and let w_i be a point in $[x_{i-1}, x_i]$, $i = 1, 2, 3, \cdots, n$. Form n approximating rectangles R_i and revolve them about the x axis to form n approximating *washer-shaped* solids S_i, a typical one of which is shown in Fig. 9-13. If, for convenience, we use the notation $f(x)^2$ as the equivalent of $[f(x)]^2$, the volume of S_i is $\Delta V_i = \pi[f(w_i)^2 - g(w_i)^2] \, \Delta x_i$.

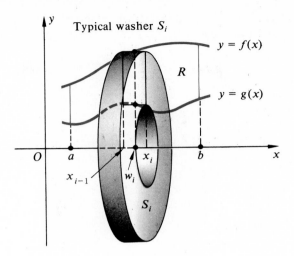

Figure 9-13

9.2.2 **Definition.** Let the functions f and g be continuous on $[a, b]$, with $f(x) \geq g(x) \geq 0$ for $a \leq x \leq b$. The **volume** V of the solid of revolution generated by revolving the region $R = \{P{:}(x, y) \mid a \leq x \leq b, 0 \leq g(x) \leq y \leq f(x)\}$ about the x axis is

$$V = \lim_{|P| \to 0} \sum_{i=1}^{n} \pi [f(w_i)^2 - g(w_i)^2] \, \Delta x_i = \pi \int_a^b [f(x)^2 - g(x)^2] \, dx.$$

This definition becomes the previous one (9.2.1) if $g(x) = 0$ for $a \leq x \leq b$.

Example 3. Find the volume generated by revolving about the x axis the region bounded by the parabolas $y = x^2$ and $y^2 = 8x$ (Fig. 9-14).

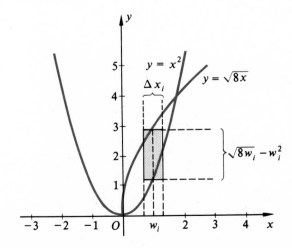

Figure 9-14

Solution. The given curves intersect at $(0, 0)$ and $(2, 4)$. The washer-shaped element of volume, S_i, has volume $\Delta V_i = \pi(\sqrt{8w_i})^2 \, \Delta x_i - \pi(w_i^2)^2 \, \Delta x_i = \pi(8w_i - w_i^4) \, \Delta x_i$. Then

$$V = \lim_{|P| \to 0} \sum_{i=1}^{n} \pi(8w_i - w_i^4) \, \Delta x_i$$

$$= \pi \int_0^2 (8x - x^4) \, dx = \pi \left[\frac{8x^2}{2} - \frac{x^5}{5} \right]_0^2 = \frac{48\pi}{5}.$$

Similarly, if f and g are functions of y that are continuous on $[c, d]$, with $0 \le g(y) \le f(y)$ for $c \le y \le d$, the volume of the solid generated by revolving about the y axis the region R, bounded by the curves $x = f(y)$ and $x = g(y)$ from $y = c$ to $y = d$ is

$$V = \lim_{|P| \to 0} \sum_{i=1}^{n} \pi[f(v_i)^2 - g(v_i)^2] \, \Delta y_i$$

$$= \pi \int_c^d [f(y)^2 - g(y)^2] \, dy,$$

where v_i is a point in the ith subinterval $[y_{i-1}, y_i]$ of a partition P of $[c, d]$.

Example 4. Find the volume of the solid generated by revolving about the line $y = -1$ the region R bounded by the parabolas $y = x^2$ and $y^2 = 8x$ (Fig. 9-15).

Solution. The parabolas intersect at $(0, 0)$ and $(2, 4)$.

Partition the interval $[0, 2]$ into n subintervals $[x_{i-1}, x_i]$ and let w_i be a point in $[x_{i-1}, x_i]$, $i = 1, 2, 3, \cdots, n$. As above, form n approximating rectangles for the region R; a typical one, R_i, is shown in Fig. 9-15.

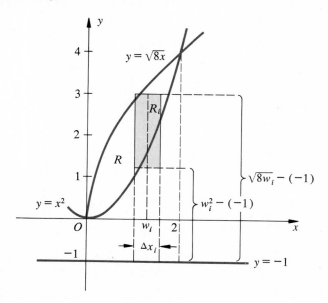

Figure 9-15

As R_i is revolved about the horizontal line $y = -1$, it sweeps out a washer-shaped solid whose volume is

$$\Delta V_i = \pi[\sqrt{8w_i} - (-1)]^2 \, \Delta x_i - \pi[w_i^2 - (-1)]^2 \, \Delta x_i$$
$$= \pi(-w_i^4 - 2w_i^2 + 8w_i + 4\sqrt{2} \sqrt{w_i}) \, \Delta x_i.$$

The volume of the solid of revolution generated by revolving the region R about the line $y = -1$ is

$$V = \lim_{|P| \to 0} \sum_{i=1}^{n} \pi(-w_i^4 - 2w_i^2 + 8w_i + 4\sqrt{2} \sqrt{w_i}) \, \Delta x_i$$

$$= \pi \int_0^2 (-x^4 - 2x^2 + 8x + 4\sqrt{2} x^{1/2}) \, dx$$

$$= \frac{224\pi}{15} \text{ cubic units.}$$

Exercises

In each of Exercises 1 to 6, sketch the region R bounded by the graphs of the given equations, showing a typical approximating rectangle. Then find the volume of the solid generated by revolving R about the x axis.

1. $y = \dfrac{x^2}{4}$, $x = 4$, $y = 0$.

2. $y = x^3$, $x = 2$, $y = 0$.

3. $y = \dfrac{1}{x}$, $x = 1$, $x = 4$, $y = 0$.

4. $y = x^{3/2}$, $x = 1$, $x = 3$, $y = 0$.

5. $y = \sqrt{4 - x^2}$, $x = -1$, $x = 2$, $y = 0$.

6. $y = x^{2/3}$, $x = 1$, $x = 8$, $y = 0$.

In Exercises 7 to 12, sketch the region R bounded by the graphs of the given equations and show a typical horizontal approximating rectangle. Find the volume of the solid generated by revolving R about the y axis.

7. $x = y^2$, $x = 0$, $y = 2$.

8. $x = \dfrac{2}{y}$, $y = 1$, $y = 6$, $x = 0$.

9. $x = \sqrt{y}$, $y = 4$, $x = 0$.

10. $x = y^{2/3}$, $y = 8$, $x = 0$.

11. $x = y^{3/2}$, $y = 4$, $x = 0$.

12. $x = \sqrt{9 - y^2}$, $x = 0$.

13. Find the volume of the solid generated by revolving about the x axis the region bounded by the upper half of the ellipse

$$\frac{x^2}{a^2} + \frac{y^2}{b^2} = 1$$

and the x axis and thus find the volume of a prolate spheroid. Here a and b are positive constants, with $a > b$.

14. Find the volume of the solid generated by rotating about the x axis the region bounded by the line $y = 4x$ and the parabola $y = 4x^2$. Make a sketch.

15. Find the volume of the solid generated by revolving about the x axis the region bounded by the line $x - 2y = 0$ and the parabola $y^2 - 2x = 0$. Make a sketch.

16. Find the volume of the solid generated by revolving about the x axis the region in the first quadrant bounded by the circle $x^2 + y^2 = r^2$, the x axis, and the line $x = r - h$, $0 < h < r$, and thus find the volume of a spherical segment of height h, radius of sphere r.

17. Find the volume of the solid generated by revolving about the y axis the region bounded by the line $y = 4x$ and the parabola $y = 4x^2$. Make a sketch.

18. Find the volume of the solid generated by revolving about the line $y = 2$ the region in the first

quadrant bounded by the parabolas $3x^2 - 16y + 48 = 0$ and $x^2 - 16y + 80 = 0$, and the y axis. Make a sketch.

19. Find the volume of the solid generated by revolving the region in the first quadrant bounded by the curve $y^2 = x^3$, the line $x = 4$, and the x axis
(a) about the line $x = 4$;
(b) about the line $y = 8$.

Make a sketch.

20. Find the volume of the solid generated by revolving the region bounded by the curve $y^2 = x^3$, the line $y = 8$, and the y axis
(a) about the line $x = 4$;
(b) about the line $y = 8$.
Make a sketch.

9.3 VOLUME BY CYLINDRICAL SHELLS

Again consider a region R in the first quadrant, bounded by the graph of a continuous function f, the x axis, and the vertical lines $x = a$ and $x = b$ [Fig. 9-16(a)]. If this region is revolved about the y axis, it generates a solid of revolution whose axis is the y axis [Fig. 9-16(b)].

Sometimes it is possible to find this volume by rewriting 9.2.2 with y as the independent variable, but often it is easier to proceed as follows.

Partition $[a, b]$ into n subintervals and let the points w_i consist of the *midpoints* of the n subintervals. Construct n approximating rectangles with base $[x_{i-1}, x_i]$ and altitude $f(w_i)$, $i = 1, 2, \cdots, n$; a typical rectangle is shown in Fig. 9-16(a). When the region R is revolved about the y axis, the rectangles sweep out hollow cylindrical shells of thickness Δx_i and height $f(w_i)$. The volume of the cylindrical shell swept out by revolving the typical rectangle about the y axis [Fig. 9-16(c)] is equal to the volume of the outer cylinder minus the volume of the inner cylinder; that is, the element of volume is

$$(1) \qquad \pi x_i^2 f(w_i) - \pi x_{i-1}^2 f(w_i) = \pi(x_i^2 - x_{i-1}^2) f(w_i)$$
$$= \pi(x_i - x_{i-1})(x_i + x_{i-1}) f(w_i).$$

Since $x_i - x_{i-1} = \Delta x$ and $(x_i + x_{i-1})/2$ is the midpoint w_i of the subinterval $[x_{i-1}, x_i]$, the volume (1) of the typical cylindrical shell can be written

$$(2) \qquad \Delta V_i = 2\pi w_i f(w_i) \Delta x_i.$$

The sum of the volumes of the n cylindrical shells is the Riemann sum

$$(3) \qquad \sum_{i=1}^{n} \Delta V_i = \sum_{i=1}^{n} 2\pi w_i f(w_i) \Delta x_i,$$

and the volume of the solid of revolution is defined to be

$$\lim_{|P| \to 0} \sum_{i=1}^{n} 2\pi w_i f(w_i) \Delta x_i.$$

Since f is continuous on $[a, b]$, this limit exists and is the definite integral $2\pi \int_a^b x f(x) \, dx$.

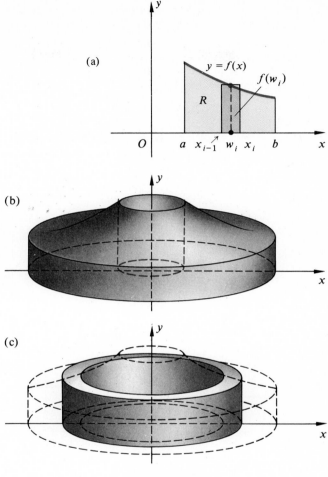

Figure 9-16

9.3.1 Definition.   Let f be continuous on $[a, b]$. If the region bounded by $y = f(x)$, the x axis, and the lines $x = a$ and $x = b$ lies in the first quadrant, the **volume** of the solid of revolution generated by revolving this region about the y axis is

$$V = 2\pi \lim_{|P| \to 0} \sum_{i=1}^{n} w_i f(w_i) \, \Delta x_i = 2\pi \int_a^b x f(x) \, dx.$$

For those solids of revolution to which both 9.2.2 (with x and y interchanged) and 9.3.1 apply, it can be shown that the two definitions give consistent results. In fact, both 9.2.2 and 9.3.1 are special cases of a more general definition of volume to be given in Chapter 19.

Example 1. Find the volume of the solid of revolution generated by revolving about the y axis the region bounded by the curve $y = (x - 1)^3$, the x axis, and the line $x = 2$ (Fig. 9-17).

Solution. The only x intercept of the given curve is 1. By 9.3.1, the desired volume is

$$2\pi \int_1^2 x(x - 1)^3 \, dx = 2\pi \int_1^2 (x^4 - 3x^3 + 3x^2 - x) \, dx$$

$$= 2\pi \left[\frac{x^5}{5} - \frac{3x^4}{4} + x^3 - \frac{x^2}{2} \right]_1^2 = \frac{9\pi}{10} \text{ cubic units.}$$

Example 2. Find the volume of the solid generated by revolving about the y axis the region in the first quadrant that is above the parabola $y = x^2$ and below the parabola $y = 2 - x^2$ (Fig. 9-18).

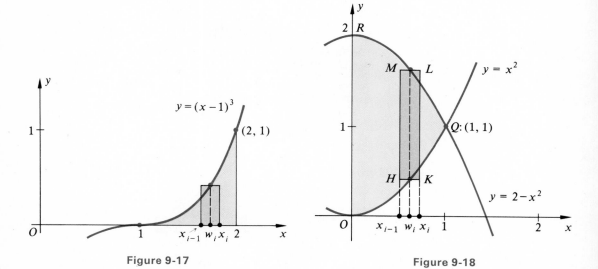

Figure 9-17   Figure 9-18

Solution. The two curves intersect at Q:(1, 1); and the region to be revolved about the y axis is OQR (Fig. 9-18). The element of volume is the cylindrical shell swept out by revolving the typical rectangle $HKLM$ about the y axis; its altitude is $\overline{KL} = (2 - w_i^2) - w_i^2 = 2(1 - w_i^2)$, where w_i is the midpoint of the subinterval $[x_{i-1}, x_i]$, and its volume is

$$\Delta V_i = \pi[x_i^2 - x_{i-1}^2][2(1 - w_i^2)] = 4\pi w_i(1 - w_i^2) \, \Delta x_i.$$

The volume of the solid of revolution is

$$\lim_{|P| \to 0} \sum_{i=1}^n 4\pi w_i(1 - w_i^2) \, \Delta x_i = 4\pi \int_0^1 (x - x^3) \, dx = 4\pi \left[\frac{x^2}{2} - \frac{x^4}{4} \right]_0^1 = \pi \text{ cubic units.}$$

What is the volume V of the solid generated by revolving about the vertical line $x = h$ the region $R = \{P:(x, y) \mid h \le a \le x \le b, 0 \le y \le f(x)\}$ (Fig. 9-19)?

Proceeding as above when $x = h$ was the y axis, we find that the volume ΔV_i of the hollow cylindrical shell of thickness Δx_i, height $f(w_i)$, and central axis $x = h$, which is swept out by revolving the ith approximating rectangle about $x = h$, is equal

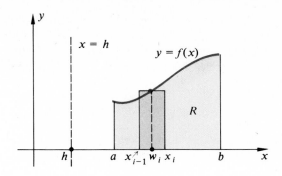

Figure 9-19

to the volume of the outer cylinder, $\pi(x_i - h)^2 f(w_i)$, minus the volume of the inner cylinder, $\pi(x_{i-1} - h)^2 f(w_i)$; thus

$$\begin{aligned}
\Delta V_i &= \pi(x_i - h)^2 f(w_i) - \pi(x_{i-1} - h)^2 f(w_i) \\
&= \pi[(x_i - h)^2 - (x_{i-1} - h)^2] f(w_i) \\
&= \pi(x_{i-1} + x_i - 2h)(x_i - x_{i-1}) f(w_i) \\
&= 2\pi[\tfrac{1}{2}(x_{i-1} + x_i) - h]\, \Delta x_i\, f(w_i) \\
&= 2\pi(w_i - h) f(w_i)\, \Delta x_i.
\end{aligned}$$

We define the volume V to be

$$\begin{aligned}
V &= 2\pi \lim_{|P| \to 0} \sum_{i=1}^{n} (w_i - h) f(w_i)\, \Delta x_i \\
&= 2\pi \int_a^b (x - h) f(x)\, dx.
\end{aligned}$$

9.3.2 **Definition.** Let f be continuous and nonnegative valued on $[a, b]$. Denote by R the region bounded by the curve $y = f(x)$, the x axis, and the lines $x = a$ and $x = b$. The **volume** of the solid generated by revolving the region R about a vertical line $x = h$ is

$$V = 2\pi \int_a^b |x - h|\, f(x)\, dx,$$

where either $h \le a < b$ or $a < b \le h$.

Example 3. Find the volume of the solid generated by revolving about the vertical line $x = -2$ the region bounded by the curve $y = x^2 - x + 2$, the lines $x = 1$ and $x = 3$, and the x axis.

Solution. Using $h = -2$ in 9.3.2, we find

$$\begin{aligned}
V &= 2\pi \int_1^3 (x + 2)(x^2 - x + 2)\, dx \\
&= 2\pi \int_1^3 (x^3 + x^2 + 4)\, dx = \frac{220\pi}{3}.
\end{aligned}$$

When f is a function of y, continuous and nonnegative valued on $[c, d]$, the formula in 9.3.2, with x, a, b, and h replaced respectively by y, c, d, and k, defines the volume of the solid generated by revolving about the horizontal line $y = k$ the region bounded by $x = f(y)$, $y = c$, $y = d$, and the y axis.

Exercises

Use the cylindrical shell method to find the volumes of the following solids of revolution. Sketch the given regions and show a typical approximating rectangle.

1. The solid generated by revolving about the y axis the region bounded by the curve $y = 4/x$, the lines $x = 1$ and $x = 4$, and the x axis.

2. The solid generated by revolving the region bounded by the curve $y = x^2$, the x axis, and the line $x = 2$, about (a) the y axis; (b) the line $x = 2$.

3. The solid generated by revolving the region bounded by $y = \sqrt{x}$, the x axis, and the line $x = 4$, about (a) the y axis; (b) the line $x = 4$.

4. The solid generated by revolving the region in the first quadrant that is bounded by the parabola $y = 4 - x^2$ and the coordinate axes, about (a) the y axis; (b) the line $x = 2$.

5. The solid generated by revolving the region bounded by the curve $y = \frac{1}{4}x^3 + 2$ and the lines $y = 2 - x$ and $x = 2$, about (a) the y axis; (b) the line $x = -1$.

6. The solid generated by revolving the region bounded by the parabola $y = x^2$ and the line $y = 2x$ about (a) the y axis; (b) the line $x = 3$.

7. The solid generated by revolving the region bounded by the curve $x = y^2$ and the lines $y = 2$ and $x = 0$, about (a) the x axis; (b) the line $y = 2$.

8. The solid generated by the region in the first quadrant bounded by the curve $x = \sqrt{2y} + 1$, the line $y = 2$, and the coordinate axes, about (a) the x axis; (b) the line $y = 3$.

9. The solid generated by revolving about the x axis the region in the first quadrant bounded by the circle $x^2 + y^2 + 6x - 16 = 0$ and the coordinate axes.

10. The solid generated by revolving the region bounded by the curve $x = 8/y^3$, the lines $y = 1$ and $y = 4$, and the y axis, about (a) the x axis; (b) the line $y = 4$.

11. The solid generated by revolving the region bounded by the curves $x = y^{1/2}$ and $x = y^3/32$, about (a) the x axis; (b) the line $y = 4$.

12. The solid generated by revolving the region given in Exercise 11 about (a) the y axis; (b) the line $x = 2$.

9.4 WORK

An application of the definite integral that does not involve areas or volumes is in finding the work done by a variable force when it is applied to an object that is moving along a straight line. In this section, the direction of the force is either the same as the direction of the motion or opposite the direction of motion.

If a *constant* force F acts along the x axis on an object that moves from a point a to a point b on the x axis, then the **work** done by F is defined in physics to be

$$(1) \qquad\qquad\qquad W = F(b - a);$$

that is, *work = force times displacement*. If the force is measured in pounds and the displacement in feet, the work done in (1) is given in foot-pounds.

But if the force F along the x axis is not constant, but is a continuous function of the position x of the object on which it acts, what do we *mean* by the "work" done by the variable force F as the object moves from a point a to a point b on the x axis?

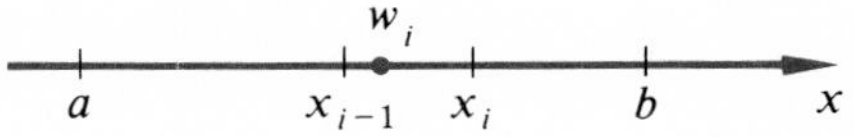

Partition the interval $[a, b]$ into n subintervals $[x_{i-1}, x_i]$, $i = 1, 2, 3, \cdots, n$, and let w_i be an arbitrarily chosen point in the ith subinterval (Fig. 9-20). If we assume that the force is constant throughout the subinterval $[x_{i-1}, x_i]$ and equal to $F(w_i)$ there, the work done on the object as it moves from x_{i-1} to x_i is [by (1)]

Figure 9-20

(2) $\Delta W_i = F(w_i)\,\Delta x_i,$

where $\Delta x_i = x_i - x_{i-1}$ is the displacement. We define the total work done by the variable force $F(x)$ as the object moves from a to b as

(3) $W = \lim_{|P| \to 0} \sum_{i=1}^{n} \Delta W_i = \lim_{|P| \to 0} \sum_{i=1}^{n} F(w_i)\,\Delta x_i.$

Since $x_{i-1} \leq w_i \leq x_i$, (3) is the limit of a Riemann sum; and because F was assumed continuous on $[a, b]$, this limit exists and is equal to $\int_a^b F(x)\,dx$.

9.4.1   Definition.   Let F be continuous on $[a, b]$. If $F(x)$ is the value at a point x on $[a, b]$ of a force acting along the x axis, then the **work** done by this force on an object as it moves from a to b is

$$W = \lim_{|P| \to 0} \sum_{i=1}^{n} \Delta W_i = \lim_{|P| \to 0} \sum_{i=1}^{n} F(w_i)\,\Delta x_i = \int_a^b F(x)\,dx.$$

Example 1.   If the natural length of a helical spring is 10 inches, and if it takes a force of 3 pounds to keep it extended 2 inches, find the work done in stretching the spring from its natural length to a length of 15 inches (Fig. 9-21).

Solution.   By *Hooke's law*, the force $F(x)$ necessary to keep a spring stretched x units beyond its natural length is

(4) $F(x) = k \cdot x,$

where k is the **spring constant**. To evaluate k for this particular spring, we know that $F(2) = 3$. Substituting this in (1), we find $3 = k \cdot 2$, or $k = \frac{3}{2}$. Thus

$$F(x) = \tfrac{3}{2}x.$$

When the spring is 10 inches long, $x = 0$; and when it is 15 inches long, $x = 5$. Therefore the work done in stretching the spring from its normal length of 10 inches to a length of 15 inches is

$$W = \int_0^5 \frac{3}{2}x\,dx = \frac{3}{2}\frac{x^2}{2}\Big]_0^5 = \frac{3}{4}(25 - 0) = 18\tfrac{3}{4} \text{ inch-pounds.}$$

Example 2. A tank in the form of an inverted right circular cone is full of water (Fig. 9-22). If the height of the tank is 10 feet and the radius of its top lip is 4 feet, find the work done in pumping all the water over the lip of the top of the tank.

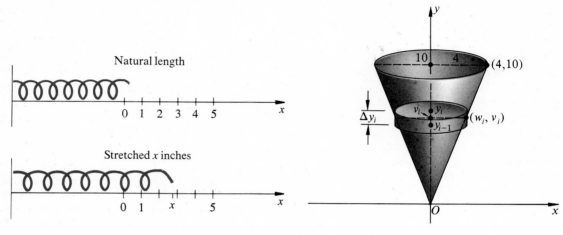

Natural length

Stretched x inches

Figure 9-21   **Figure 9-22**

Solution. Place the coordinate axes as shown in Fig. 9-22. Partition the interval [0, 10] on the y axis into n subintervals $[y_{i-1}, y_i]$ and denote by v_i an arbitrarily chosen point in $[y_{i-1}, y_i]$, $i = 1, 2, 3, \cdots, n$. Let w_i be the x coordinate of the point of intersection of the horizontal line $y = v_i$ and the line joining the origin to the point (4, 10).

A typical element of volume is the cylinder $\Delta V_i = \pi w_i^2 \, \Delta y_i$, as shown in Fig. 9-22. Since (w_i, v_i) is a point on the line joining the origin to (4, 10), $w_i = 4v_i/10$, and the element of volume can be written $\Delta V_i = (4\pi v_i^2/25) \, \Delta y_i$. The weight of this volume of water is $\delta(4\pi v_i^2/25) \, \Delta y_i$, which is equal to the force necessary to lift ΔV_i. Here δ is the density of water, which is approximately 62.4 pounds per cubic foot. Since the displacement of ΔV_i in moving it to the top of the cone is approximately $10 - v_i$, the work done in lifting ΔV_i to the top of the tank is approximately

$$\Delta W_i = \delta\left(\frac{4\pi v_i^2}{25}\right)(10 - v_i) \, \Delta y_i.$$

So the total work done in pumping all the water out of the tank at the top is

$$W = \delta\left(\frac{4\pi}{25}\right) \lim_{|P| \to 0} \sum_{i=1}^{n} (10 - v_i)v_i^2 \, \Delta y_i = \delta\left(\frac{4\pi}{25}\right) \int_0^{10} (10 - y)y^2 \, dy$$

$$= \frac{400\pi\delta}{3} \doteq 26{,}125 \text{ foot-pounds.}$$

Exercises

1. A force of 8 pounds is required to keep a certain spring stretched $\frac{1}{2}$ foot beyond its normal length. Find the value of the spring constant and find the work done in stretching the spring this much.

2. For the spring in Exercise 1, how much work is done in stretching the spring 1 foot?

3. For any spring obeying Hook's law, show that

the work done in stretching a spring a distance d is given by $W = \frac{1}{2}kd^2$.

4. For a certain type of nonlinear spring, the force required to stretch the spring a distance s is given by the formula $F = ks^{4/3}$. For a certain such spring, the force required to stretch it 8 inches is 1 pound. How much work is done in stretching this spring 27 inches?

5. A spring is such that the force required to stretch it s feet is given by $F = 11s$ pounds. How much work is done in compressing it 2 feet?

6. Two similar springs S_1 and S_2, each 3 feet long, are such that the force required to stretch either of them a distance of s feet is $F = 8s$ pounds. One end of one spring is fastened to one end of the other, and the combination is stretched between the walls of a room 12 feet wide (Fig. 9-23). What work is done in moving the midpoint, P, 1 foot to the right?

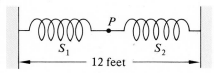

Figure 9-23

7. A volume v of gas is confined in a cylinder, one end of which is closed by a movable piston. If A is the area in square inches of the face of the piston and x is the distance in inches from the cylinder head to the piston, then $v = Ax$. The pressure of the confined gas is a continuous function p of the volume, and $p(v) = p(Ax)$ will be denoted by $f(x)$. Show that the work done by the piston in compressing the gas from a volume $v_1 = A(x_1)$ to a volume $v_2 = A(x_2)$ is

$$W = A \int_{x_1}^{x_2} f(x)\, dx.$$

[*Hint:* The total force on the face of the piston is $p(v) \cdot A = p(Ax) \cdot A = A \cdot f(x)$.]

8. A cylinder and piston, whose cross-sectional area is 1 square inch, contains 16 cubic inches of gas under a pressure of 40 pounds per square inch. If the pressure and the volume of the gas are related adiabatically (that is, without loss of heat) by the law $pv^{1.4} = $ constant, how much work is done by the piston in compressing the gas to 2 cubic inches?

9. If the area of the face of the piston in Exercise 8 is 2 square inches, find the work done by the piston.

10. One cubic foot of air under a pressure of 80 pounds per square inch expands adiabatically to 4 cubic feet according to the law $pv^{1.4} = c$. Find the work done by the gas.

11. A tank in the form of an inverted cone has depth 8 feet and radius of top 5 feet. If water in the tank is 6 feet deep, how much work is done in pumping all the water over the edge of the top of the tank?

12. A spherical tank, of radius 12 feet, is full of water. Find the work done in pumping all the water out of the tank through the top.

13. If the maximum depth of the water in the spherical tank of Exercise 12 is 18 feet, how much work is done in pumping all the water to a point 10 feet above the top of the tank?

14. According to Coulomb's law, two like electrical charges repel each other with a force that is inversely proportional to the distance between them. If the force of repulsion is 10 dynes when they are 2 centimeters apart, find the work done in bringing the charges from 5 centimeters apart to 1 centimeter apart. (An **erg** is a unit of work in the metric system and is equal to one dyne-centimeter.)

9.5 LIQUID PRESSURE

It is shown in physics that the force on a horizontal plate submerged in a liquid is equal to the weight of the column of liquid above it. If the weight of a cubic foot of a liquid is δ pounds and the horizontal plate is h feet below the surface of the liquid, the force on a square foot of the plate is δh pounds. But force per unit of area is called *pressure.* Thus *the pressure p at a depth of h feet in a liquid is*

$$p = \delta h$$

pounds per square foot, where δ is the weight in pounds of a cubic foot of the liquid.

Another principle of physics (attributed to Pascal) is that *the pressure at a given point in a liquid is the same in all directions.*

To find the total force on a submerged *horizontal* plate, we need only multiply the pressure at that depth by the area of the plate. But when the submerged plate is *vertical,* the total force on its face is more complicated. The difficulty is that the pressure on the plate varies with the depth, being greater near the bottom of the plate than near the top.

Let f and g be functions, that are continuous on a closed interval $[c, d]$, with $g(y) \leq f(y)$ for $c \leq y \leq d$. Denote by R the plane region bounded by $x = g(y)$, $x = f(y)$, $y = c$, and $y = d$ (Fig. 9-24), and let the line $y = k$, $k \geq d$, be along the surface of the liquid. We wish to define the *force* of liquid pressure on the vertical submerged region R.

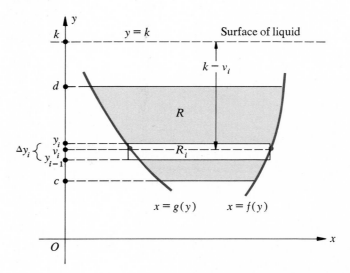

Figure 9-24

Partition the interval $[c, d]$ on the y axis into n subintervals $[y_{i-1}, y_i]$ and let v_i be an arbitrarily chosen point such that $y_{i-1} \leq v_i \leq y_i$ $(i = 1, 2, 3, \cdots, n)$. Form n approximating rectangles R_i, a typical one of which is shown in Fig. 9-24. The area of the typical rectangle R_i is $\Delta A_i = [f(v_i) - g(v_i)] \Delta y_i$.

The pressure on the rectangle R_i is $\delta(k - y_i)$ at the top and $\delta(k - y_{i-1})$ at the bottom. Since $y_{i-1} \leq v_i \leq y_i$, we have $\delta(k - y_i) \leq \delta(k - v_i) \leq \delta(k - y_{i-1})$. Thus $\delta(k - v_i)$ is an approximation to the pressure at any point on R_i. If we assume that the pressure on R_i is constant and equal to $\delta(k - v_i)$, then the force on the typical rectangle R_i is $\delta(k - v_i)\,\Delta A_i$. So the Riemann sum

$$\sum_{i=1}^{n} \delta(k - v_i)\,\Delta A_i = \sum_{i=1}^{n} \delta(k - v_i)[f(v_i) - g(v_i)]\,\Delta y_i$$

approximates the "total force" on the submerged vertical region R, and this approximation improves as the norm of the partition gets smaller.

9.5.1 Definition. The **force** of liquid pressure on the vertical submerged region R, described above, is

$$F = \lim_{|P|\to 0} \sum_{i=1}^{n} \delta(k - v_i)[f(v_i) - g(v_i)]\,\Delta y_i$$

$$= \delta \int_{c}^{d} (k - y)[f(y) - g(y)]\,dy,$$

where δ is the density of the liquid.

In applying this definition, it is essential that the *long* sides of the approximating rectangles be *parallel to the surface of the liquid*.

Example. Find the force on one end of a horizontal cylindrical tank of diameter 12 feet if it is half full of water (Fig. 9-25).

Solution. Place the coordinate axes with the origin at the center of the circular end of the tank (Fig. 9-25). Then the equation of the circle is $x^2 + y^2 = 36$.

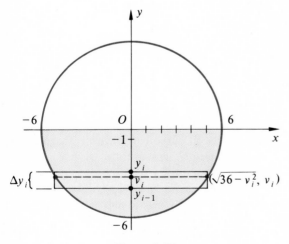

A typical approximating rectangle is shown in Fig. 9-25; its area is $\Delta A_i = 2\sqrt{36 - v_i^2}\,\Delta y_i$; and because we have put the x axis at the surface of the liquid, the force on the typical approximating rectangle is $\delta(0 - v_i)\,\Delta A_i = -\delta v_i(2\sqrt{36 - v_i^2})\,\Delta y_i$. Thus the total force on the end of the horizontal cylindrical tank is

$$F = \lim_{|P|\to 0} \sum_{i=1}^{n} -\delta\sqrt{36 - v_i^2}\,2v_i\,\Delta y_i$$

$$= \delta \int_{-6}^{0} (36 - y^2)^{1/2}(-2y)\,dy$$

$$= \left[\frac{2(36 - y^2)^{3/2}}{3}\right]_{-6}^{0} = 144\delta.$$

Since a cubic foot of water weighs approximately 62.4 pounds,

$$F = 144\delta = 144(62.4) = 8985.6 \text{ pounds.}$$

Figure 9-25

Exercises

Assume that a cubic foot of water weighs approximately 62.4 pounds; that is, assume that $\delta \doteq 62.4$ pounds per cubic foot.

1. Water reaches the top of a rectangular dam that is 50 feet wide and 20 feet deep. Find the total force on the dam.

2. Water reaches the top of a vertical rectangular dam that is w feet wide and h heet high. Show that the total force on this dam equals its area times the pressure halfway down.

3. A dam is in the form of an inverted isosceles triangle 40 feet wide and 30 feet deep at its deepest point (that is, the base of the triangle is at the top). Find the total force on the dam when the lake is full.

4. Show that the total force on the triangular dam in Exercise 3 will be the same irrespective of the shape of the triangle, provided only that the base (on top) equals 40 feet and the altitude (greatest depth) is 30 feet. (*Hint:* The width at any depth is to 30 minus the depth as 40 is to 30. Use similar triangles.)

5. Find the total force on a dam in the form of an inverted triangle of base b and altitude (maximum depth) h.

6. Find the total force on a semicircular dam, 20 feet in diameter.

7. Find the total force on the semielliptical dam whose face is the region bounded below by the curve $x = \pm 2\sqrt{100 - y^2}$ and bounded above by the x axis.

8. The face of a vertical dam is the parabolic region bounded by the curve $3x^2 = 10y$ and the line $y = 30$. If the water level is $y = 30$, what is the total force on the dam?

9. A dam contains a submerged rectangular gate, 5 feet wide and 4 feet high; the water level is 30 feet above the top of the gate. Find the total force on the gate.

10. A dam has a submerged gate bounded by an equilateral triangle, 2 feet on a side, with the horizontal base nearest the surface of the water and 10 feet below it. Find the total force on the gate.

11. What would be the force on the triangular gate in Exercise 10 if the base is horizontal but the opposite vertex is closer than the base to the surface of the water and 10 feet below the surface?

12. Show that if a dam in the form of a vertical rectangle is divided in half by means of a diagonal, the force on one-half the dam is twice the force on the other half.

13. What is the total force on all sides of a cube if each edge is 1 foot long, and if its top is horizontal and 100 feet below the surface of a lake?

9.6 ARC LENGTH

A function f is said to be **smooth** on an interval I *if it has a derivative f' that is continuous on I.* Thus the slope of the tangent to the graph of a smooth function makes no sudden changes, and, consequently, a smooth curve can have no sharp corners. Most of the functions we have been considering are smooth, at least on some interval.

Let f be a smooth function on a closed interval $[a, b]$, so that f' is continuous on $[a, b]$. Consider the arc AB of the graph of $y = f(x)$, where A and B are the points on the graph having coordinates $(a, f(a))$ and $(b, f(b))$, respectively (Fig. 9-26).

Partition the closed interval $[a, b]$ into n subintervals by a partition P and denote the length of the ith subinterval $[x_{i-1}, x_i]$ by $\Delta x_i = x_i - x_{i-1}$, where $i = 1, 2, 3, \cdots, n$. Let $|P|$ be the norm of the partition, so that $\Delta x_i \leq |P|$. Denote by Q_i the point on the curve with coordinates $(x_i, f(x_i))$, and let the point $A{:}(a, f(a))$ be Q_0.

By the distance formula, the length of a typical chord $Q_{i-1}Q_i$ is

$$|Q_{i-1}Q_i| = \sqrt{(\Delta x_i)^2 + (\Delta y_i)^2},$$

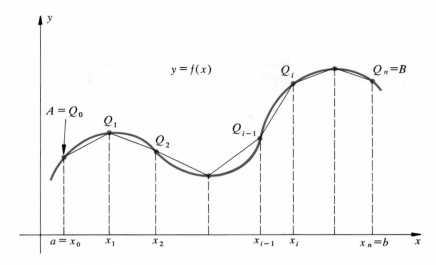

Figure 9-26

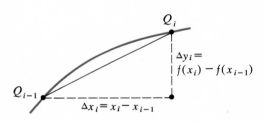

Figure 9-27

where $\Delta x_i = x_i - x_{i-1}$ and $\Delta y_i = f(x_i) - f(x_{i-1})$ (Fig. 9-27), or

(1) $|Q_{i-1}Q_i| = \sqrt{1 + \left(\dfrac{\Delta y_i}{\Delta x_i}\right)^2}\, \Delta x_i.$

Since f is smooth on $[a, b]$, f' exists on the closed subinterval $[x_{i-1}, x_i]$, and the mean value theorem applies. Thus there exists a number z_i between x_{i-1} and x_i such that

$$f(x_i) - f(x_{i-1}) = f'(z_i)(x_i - x_{i-1})$$

or

$$\frac{\Delta y_i}{\Delta x_i} = f'(z_i).$$

So (1) can be rewritten

(2) $|Q_{i-1}Q_i| = \sqrt{1 + [f'(z_i)]^2}\, \Delta x_i,$

where z_i is some point in the ith subinterval $[x_{i-1}, x_i]$.

The sum of the lengths of the chords $Q_{i-1}Q_i$ ($i = 1, 2, 3, \cdots, n$) is the Riemann sum

$$\sum_{i=1}^{n} \sqrt{1 + [f'(z_i)]^2}\, \Delta x_i,$$

which approximates what we would like to think of as the length of the arc AB. Moreover, this approximation improves as $|P|$, the norm of the partition, decreases.

We define the length of the arc AB as

(3) $$\lim_{|P|\to 0} \sum_{i=1}^{n} \sqrt{1 + [f'(z_i)]^2}\, \Delta x_i,$$

provided that this limit exists. Since f' is continuous on $[a, b]$, the function g, defined by $g(x) = \sqrt{1 + [f'(x)]^2}$, is continuous on $[a, b]$. Therefore the limit (3) exists and is equal to $\int_a^b \sqrt{1 + [f'(x)]^2}\, dx$.

9.6.1 Definition. If a function f is smooth on the closed interval $[a, b]$, the **length of the arc** of the graph of $y = f(x)$ from the point $A:(a, f(a))$ to the point $B:(b, f(b))$ is

$$L = \int_a^b \sqrt{1 + \left(\frac{dy}{dx}\right)^2}\, dx.$$

Similarly,

9.6.2 Definition. If g is a smooth function on the closed interval $[c, d]$, the **length of the arc** of $x = g(y)$ from $y = c$ to $y = d$ is

$$L = \int_c^d \sqrt{1 + \left(\frac{dx}{dy}\right)^2}\, dy.$$

Example. Find the length of the arc of the curve $y = x^{3/2}$ from the point $(1, 1)$ to the point $(4, 8)$ (Fig. 9-28).

Solution. Since $dy/dx = 3x^{1/2}/2$, the length of the arc from $(1, 1)$ to $(4, 8)$ is

(4) $$\int_1^4 \sqrt{1 + \left(\frac{3x^{1/2}}{2}\right)^2}\, dx = \int_1^4 \left(1 + \frac{9x}{4}\right)^{1/2} dx.$$

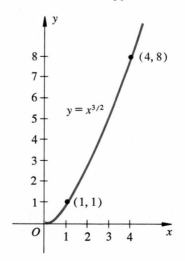

Figure 9-28

Let $u = [1 + (9x/4)]$; then $du/dx = 9/4$ and the integral (4) can be rewritten

$$\frac{4}{9} \int_1^4 \left(1 + \frac{9x}{4}\right)^{1/2} \left(\frac{9}{4}\right) dx.$$

By 7.3.1, an antiderivative of this integrand is

$$A_x\left[\left(1 + \frac{9x}{4}\right)^{1/2}\left(\frac{9}{4}\right)\right] = \frac{2}{3}\left(1 + \frac{9x}{4}\right)^{3/2}.$$

Therefore

$$\frac{4}{9} \int_1^4 \left(1 + \frac{9x}{4}\right)^{1/2} \left(\frac{9}{4}\right) dx = \frac{4}{9}\left[\frac{2}{3}\left(1 + \frac{9x}{4}\right)^{3/2}\right]_1^4$$

$$= \frac{8}{27}\left(10^{3/2} - \frac{13^{3/2}}{8}\right) \doteq 7.6.$$

Let the function f be smooth on $[a, b]$, and let s be the function defined on $[a, b]$ by

$$s(x) = \int_a^x \sqrt{1 + [f'(t)]^2}\, dt,$$

where the dummy variable has been designated t to avoid confusion with the variable upper limit of integration, x (see Section 8.4). Then $s(x)$ gives the length of the arc of $y = f(x)$ from the point $(a, f(a))$ to the point $(x, f(x))$, where x is any number in $[a, b]$. By the fundamental theorem (8.4.3),

$$s'(x) = \frac{ds}{dx} = \sqrt{1 + [f'(x)]^2},$$

which may be written

$$ds = \sqrt{1 + \left(\frac{dy}{dx}\right)^2}\, dx \qquad \text{or} \qquad ds = \sqrt{1 + \left(\frac{dx}{dy}\right)^2}\, dy.$$

This proves the following theorem.

9.6.3   **Theorem.**   Let f be a smooth function on $[a, b]$. Then the **differential of arc**, ds, for the curve $y = f(x)$ is given by

$$ds = \sqrt{1 + \left(\frac{dy}{dx}\right)^2}\, dx = \sqrt{1 + \left(\frac{dx}{dy}\right)^2}\, dy.$$

By squaring both members of

$$ds = \sqrt{1 + \left(\frac{dy}{dx}\right)^2}\, dx.$$

we obtain the useful and easily remembered relation $(ds)^2 = (dx)^2 + (dy)^2$, which is usually written

9.6.4

$$ds^2 = dx^2 + dy^2.$$

Since the formulas in 9.6.3 for the differential of arc length, ds, are so easily obtained from $ds^2 = dx^2 + dy^2$, most students memorize only 9.6.4.

Exercises

1. Find the length of the segment of the line $y = 3x + 5$ from $x = 1$ to $x = 4$. Check by the distance formula.

2. Find the length of the line $4x - 3y + 2 = 0$ from $x = -1$ to $x = 3$. Check by the distance formula.

(Notice in Exercises 1 and 2 that the arc length method for finding the distance between points on a given line is sometimes more expeditious than the use of the distance formula, since the slope of a line, and hence dy/dx, can be found by inspection.)

3. Find the length of the curve $y = 2x^{3/2}$ from $x = \frac{1}{3}$ to $x = 7$.

4. Find the length of the arc of the graph of $y = \frac{2}{3}(x^2 + 1)^{3/2}$ from $x = 1$ to $x = 4$.

5. Find the length of the arc of the curve $y = (4 - x^{2/3})^{3/2}$ from $x = 1$ to $x = 8$.

6. Find the length of the segment of the curve $y = (x^4 + 3)/6x$ from $x = 1$ to $x = 4$.

7. Find the length of the arc of the curve $x = \frac{2}{3}(y - 1)^{3/2}$ from $y = 1$ to $y = 4$.

In Exercises 8, 9, and 10, write the definite integrals for the indicated arc lengths, but wait until Chapter 11 to evaluate them.

8. The length of the arc of the parabola $y = x^2/2$ from $x = 0$ to $x = 2$.

9. The length of the arc of the semicircle $y = \sqrt{9 - x^2}$ from $x = -1$ to $x = 2$.

10. The length of the cubic curve $3y = x^3$ from $x = 0$ to $x = a$.

9.7 AREA OF A SURFACE OF REVOLUTION

Consider a function f that is smooth on a closed interval $[a, b]$, with $f(x) \geq 0$ for $a \leq x \leq b$. If the arc of the curve $y = f(x)$, from the point $(a, f(a))$ to the point $(b, f(b))$, is revolved about the x axis, a surface of revolution S is swept out (Fig. 9-29). The purpose of the following discussion is to motivate the definition (9.7.1) of the area of S, given below.

As in 9.6, partition $[a, b]$ into n intervals $[x_{i-1}, x_i]$, $i = 1, 2, 3, \cdots, n$. Let Q_i be the point on the curve whose coordinates are $(x_i, f(x_i))$, and denote the point $(a, f(a))$ by Q_0 (Fig. 9-30).

If the broken line formed by the n chords $Q_{i-1}Q_i$ of the curve is revolved about the x axis, it sweeps out a surface (Fig. 9-30) that approximates S, and this approximation improves as the norm $|P|$ of the partition decreases.

The lateral area of the frustum of a cone, having slant height s and radii of its bases r_1 and r_2, is $\pi(r_1 + r_2)s$ (Appendix A.6.1). Thus each chord $Q_{i-1}Q_i$, as it revolves about the x axis, sweeps out the lateral surface of a frustum of a cone whose area is

$$\pi[f(x_{i-1}) + f(x_i)]|Q_{i-1}Q_i|.$$

By equation (2) of 9.6, this can be rewritten as

$$\pi[f(x_{i-1}) + f(x_i)]\sqrt{1 + f'(z_i)^2}\,\Delta x_i,$$

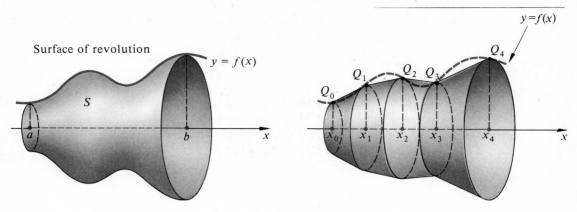

Figure 9-29

Figure 9-30

in which z_i is some number between x_{i-1} and x_i.

Thus

$$(1) \qquad \pi \sum_{i=1}^{n} [f(x_{i-1}) + f(x_i)]\sqrt{1 + f'(z_i)^2}\, \Delta x_i$$

is the area of the approximating surface swept out by the broken line formed by the n chords $Q_{i-1}Q_i$ (Fig. 9-30).

It can be shown that because of the continuity of f on the closed interval $[a, b]$ and because $x_{i-1} < z_i < x_i$, the sum $f(x_{i-1}) + f(x_i)$ can be made to differ from $2f(z_i)$ by as little as we please by taking the norm $|P|$ of the partition sufficiently small. By substituting $2f(z_i)$ for $f(x_{i-1}) + f(x_i)$ in (1), we obtain

$$(2) \qquad 2\pi \sum_{i=1}^{n} f(z_i)\sqrt{1 + f'(z_i)^2}\, \Delta x_i.$$

Since f and f' are continuous on $[a, b]$, the function F defined by

$$F(x) = f(x)\sqrt{1 + f'(x)^2}$$

is continuous on $[a, b]$. So (2) is a Riemann sum for the function F on $[a, b]$, and, by the fundamental theorem,

$$2\pi \lim_{|P| \to 0} \sum_{i=1}^{n} f(z_i)\sqrt{1 + f'(z_i)^2}\, \Delta x_i = 2\pi \int_{a}^{b} f(x)\sqrt{1 + f'(x)^2}\, dx$$

exists.

This suggests the following definition of the area of the surface of revolution S.

9.7.1 Definition. Let f be a smooth function on $[a, b]$, with $f(x) \geq 0$ for $a \leq x \leq b$. The **area of the surface of revolution** swept out by revolving about the x axis the segment of the curve $y = f(x)$, from the point $(a, f(a))$ to the point $(b, f(b))$, is

$$2\pi \int_{a}^{b} f(x)\sqrt{1 + f'(x)^2}\, dx.$$

Example.   Find the area of the surface of revolution generated by revolving about the x axis the segment of the curve $y = \sqrt{x}$ from $(1, 1)$ to $(4, 2)$.

Solution.   By substituting $f(x) = \sqrt{x}$ and $f'(x) = 1/(2\sqrt{x})$ in 9.7.1, we have

$$2\pi \int_1^4 \sqrt{x} \sqrt{1 + \frac{1}{4x}} \, dx = \pi \int_1^4 \sqrt{4x + 1} \, dx = \frac{\pi}{4} \int_1^4 (4x + 1)^{1/2}(4) \, dx$$

$$= \frac{\pi}{6} \Big[(4x + 1)^{3/2} \Big]_1^4 = \frac{\pi(17^{3/2} - 5^{3/2})}{6}.$$

It is left as an exercise for the reader to motivate, in a similar manner, the definition of the area of the surface of revolution swept out when a segment of a curve $y = f(x)$ is revolved *about the y axis*.

9.7.2   Definition.   Let f be a smooth function on $[a, b]$, with $a \geq 0$. The **area of the surface of revolution** swept out by revolving about the y axis the segment of the curve $y = f(x)$ from $(a, f(a))$ to $(b, f(b))$ is

$$2\pi \int_a^b x\sqrt{1 + f'(x)^2} \, dx.$$

To make the formulas in 9.7.1 and 9.7.2 easier to remember, it is worth noticing that they can be written as

$$2\pi \int_a^b y \, ds \qquad \text{and} \qquad 2\pi \int_a^b x \, ds,$$

respectively, where $ds = \sqrt{1 + f'(x)^2} \, dx$ is the differential of arc as given in 9.6.3.

Exercises

1.   Use 9.7.1 to find the area of the lateral surface of the right circular cone generated by revolving about the x axis the segment of the line $y = ax$ from $x = 0$ to $x = h$ $(a > 0, h > 0)$.

2.   Use 9.7.1 to show that the lateral surface area of a frustum of a right circular cone is $\pi(r_1 + r_2)s$, where s is its slant height and r_1 and r_2 are the radii of its bases. (This will show that the definition 9.7.1 is consistent with the assumption about the lateral area of a frustrum of a cone used in motivating 9.7.1.)

3.   Find the area of the surface generated by revolving about the x axis the segment of the semicircle $y = \sqrt{25 - x^2}$ from $x = -2$ to $x = 3$.

4.   Find the area of the surface generated by revolving about the x axis the segment of the curve $y = x^3/3$ from $x = 1$ to $x = \sqrt{7}$.

5.   Find the area of the surface generated by revolving about the y axis the segment of the parabola $y = x^2$ from $x = 0$ to $x = 2\sqrt{3}$.

6.   Find the area of the surface generated by revolving about the y axis the segment of the parabola $y = \frac{1}{2}x^2 - 1$ from $x = 0$ to $x = 2\sqrt{2}$.

7.   Find the area of the surface generated by revolving about the x axis the segment of the curve $y = (x^6 + 2)/(8x^2)$ from $x = 1$ to $x = 3$.

8.   Find the area of the surface generated by revolving the curve segment of Exercise 7 about the y axis.

9.8 CENTROID OF A PLANE REGION

When two boys, weighing w_1 and w_2 pounds, respectively, sit on a seesaw at distances of d_1 feet and d_2 feet from the center [Fig. 9-31(a)], the seesaw will balance if and only if

$$(1) \qquad w_1 d_1 = w_2 d_2.$$

So, if the first boy weighs twice as much as the second boy, the first boy must be half as far from the center as the second boy in order to balance the seesaw.

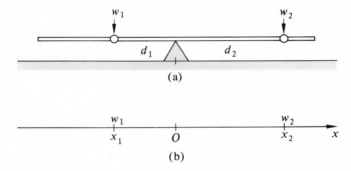

(a)

(b)

Figure 9-31

If the seesaw is replaced by a horizontal coordinate line whose origin is at the center of the seesaw and whose unit of length is 1 foot, then w_1 and w_2 will be at points whose coordinates are $x_1 = -d_1$ and $x_2 = d_2$ [Fig. 9-31(b)]. The condition (1) for "balance" now becomes

$$(2) \qquad w_1 x_1 + w_2 x_2 = 0.$$

In physics the product of the mass of a particle by its *directed* distance from a line is called the **moment** of the particle about the line. The condition (2) for the system of two weights to balance is simply that *the sum of their moments about the y axis is zero.*

Consider next a finite number n of particles, with masses $m_1, m_2, m_3, \cdots, m_n$, situated at the points $(x_1, y_1), (x_2, y_2), \cdots, (x_n, y_n)$, respectively, in the coordinate plane. The **moment** of this system of n masses about the y axis is defined to be

$$M_y = m_1 x_1 + m_2 x_2 + \cdots + m_n x_n = \sum_{i=1}^{n} m_i x_i,$$

and the **moment** of this system about the x axis is

$$M_x = m_1 y_1 + m_2 y_2 + \cdots + m_n y_n = \sum_{i=1}^{n} m_i y_i.$$

Example 1. If five particles, having masses 1, 4, 2, 3, and 2 units, are located at the points $(6, -1)$, $(2, 3)$, $(-4, 2)$, $(-7, 4)$, and $(2, -2)$, respectively (Fig. 9-32), then the moment of this system about y axis is

$$M_y = \sum_{i=1}^{5} m_i x_i = 1(6) + 4(2) + 2(-4) + 3(-7) + 2(2) = -11,$$

and its moment about the x axis is

$$M_x = \sum_{i=1}^{5} m_i y_i = 1(-1) + 4(3) + 2(2) + 3(4) + 2(-2) = 23.$$

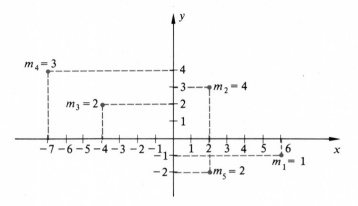

Figure 9-32

Returning to the system of n masses in the plane, the total **mass** of this system is

$$m = \sum_{i=1}^{n} m_i.$$

The point $(\bar{x}, \bar{y})$, such that

(3) $$m\bar{x} = M_y \quad \text{and} \quad m\bar{y} = M_x,$$

is called the **center of mass** or the **center of gravity** of the system of n masses.

Thus the center of mass $(\bar{x}, \bar{y})$ is a point such that if the total mass m of the system were concentrated there, its moments $m\bar{x}$ and $m\bar{y}$ about the y and x axes would be the same as the moments M_y and M_x of the system of n particles about those same axes.

If we think of the n particles as a rigid system in a horizontal plane, the system will balance at the center of mass $(\bar{x}, \bar{y})$. For if $(\bar{x}, \bar{y})$ is the origin of coordinates, then $\bar{x} = \bar{y} = 0$ and the moments M_y and M_x of the system about the coordinate axes are zero [by (3)]. So the system will "balance" at $(\bar{x}, \bar{y})$, and for this reason $(\bar{x}, \bar{y})$ is often called the **center of gravity** of the system.

Example 2. The total mass m of the system of five masses in Example 1 is

$$m = 1 + 4 + 2 + 3 + 2 = 12,$$

and the center of mass of that system is the point $(\bar{x}, \bar{y})$, when

$$\bar{x} = \frac{M_y}{m} = \frac{-11}{12} \quad \text{and} \quad \bar{y} = \frac{M_x}{m} = \frac{23}{12}.$$

Let us turn next to homogeneous *laminas* (or thin sheets). A lamina is said to be **homogeneous** if two pieces of it have equal weights whenever their areas are equal. Since the density of a homogeneous lamina is its mass per unit of area, the mass of a homogeneous lamina of density δ and area A is

$$m = \delta A.$$

We wish to define, in a manner consistent with our experience with systems of n particles, the *moment* of a homogeneous lamina about a line in its plane, and the *center of mass* (or center of gravity) of the lamina.

Since a rigid *rectangular* sheet, having uniform thickness and consistency, can be balanced on its geometric center [Fig. 9-33(a)], it is natural to define its center of gravity to be its geometric center [Fig. 9-33(b)].

Center of gravity of a homogeneous
rectangular lamina

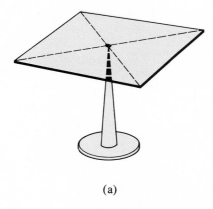

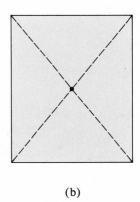

(a)   (b)

Figure 9-33

9.8.1 Definition. Let R be a homogeneous rectangular lamina, having density δ and area A, and let l be a line in the plane of R. The **moment of R about** l is

$$M_l(R) = (\delta A)d,$$

where δA is the mass of R and d is the directed distance from l to the center of R (Fig. 9-34).

Our discussion of moments of systems of n particles suggests the following **additive assumption** about the moments of laminas.

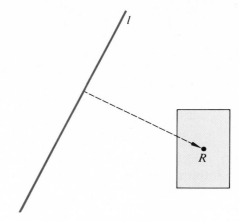

Figure 9-34

If a homogeneous lamina A is the union of two nonoverlapping laminas B and C, and if l is a line in their plane, then the moment of A about the line l is the sum of the moments of B and C about l:

$$M_l(A) = M_l(B) + M_l(C).$$

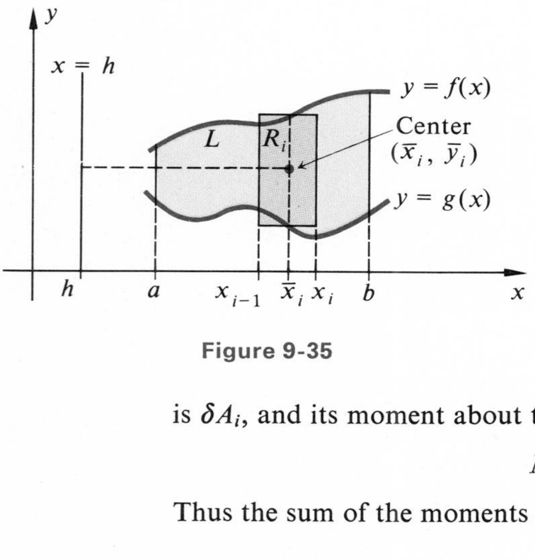

Let f and g be continuous functions on $[a, b]$ with $g(x) \leq f(x)$ for $a \leq x \leq b$. Consider the homogeneous lamina L of density δ and bounded by the curves $y = f(x)$ and $y = g(x)$ and the lines $x = a$ and $x = b$ (Fig. 9-35).

The purpose of this discussion is to motivate the definition (to be given below) of the moment of the lamina L about a vertical line $x = h$.

Partition $[a, b]$ into n intervals $[x_{i-1}, x_i]$, $i = 1$, 2, 3, $\cdots$, n, by a partition P, and denote by $\bar{x}_i$ the midpoint of each $[x_{i-1}, x_i]$. Form n approximating rectangular laminas R_i, a typical one of which is shown in Fig. 9-35. The area of R_i is A_i, its mass is δA_i, and its moment about the line $x = h$ is

$$M_{x=h}(R_i) = (\bar{x}_i - h)\delta A_i.$$

Figure 9-35

Thus the sum of the moments of the n rectangular laminas R_i about the line $x = h$ is

(4) $$\sum_{i=1}^{n} (\bar{x}_i - h)\delta A_i.$$

Since the union of the n rectangular laminas R_i approximates the given lamina L and this approximation improves as the norm $|P|$ of the partition decreases, it is natural to *define* the moment of L about $x = h$ to be

(5)
$$M_{x=h}(L) = \lim_{|P| \to 0} \sum_{i=1}^{n} (\bar{x}_i - h)\delta A_i,$$

provided that this limit exists.

Now, the altitude of R_i is $f(\bar{x}_i) - g(\bar{x}_i)$ and the length of its base is $x_i - x_{i-1} = \Delta x_i$, so the area of R_i is $A_i = [f(\bar{x}_i) - g(\bar{x}_i)]\,\Delta x_i$. Thus (4) can be written

$$\delta \sum_{i=1}^{n} (\bar{x}_i - h)[f(\bar{x}_i) - g(\bar{x}_i)]\,\Delta x_i$$

which is a Riemann sum for the continuous function that is defined by $(x - h)[f(x) - g(x)]$ on $[a, b]$. Consequently, the limit in (5) exists and is equal to

$$\delta \int_a^b (x - h)[f(x) - g(x)]\,dx.$$

9.8.2   Definition.   Let f and g be functions that are continuous on $[a, b]$ with $g(x) \leq f(x)$ for $a \leq x \leq b$, and let L be the homogeneous lamina of density δ whose face is the region

$$R = \{P{:}(x, y) \,|\, a \leq x \leq b,\, g(x) \leq y \leq f(x)\}.$$

The **moment of L about the vertical line** $x = h$ is

$$M_{x=h}(L) = \delta \int_a^b (x - h)[f(x) - g(x)]\,dx.$$

Similarly, the moment of the rectangular lamina R_i about a horizontal line $y = k$ is

(6)
$$M_{y=k}(R_i) = (\bar{y}_i - k)\delta A_i,$$

where $\bar{y}_i$ is the y coordinate of the center of R_i (Fig. 9-35). Since $\bar{y}_i = \frac{1}{2}[f(\bar{x}_i) + g(\bar{x}_i)]$, equation (6) can be rewritten as

$$M_{y=k}(R_i) = \tfrac{1}{2}\delta[f(\bar{x}_i) + g(\bar{x}_i) - 2k][f(\bar{x}_i) - g(\bar{x}_i)]\,\Delta x_i,$$

and we define the **moment about the horizontal line** $y = k$ of the lamina L to be

$$M_{y=k}(L) = \tfrac{1}{2}\delta \lim_{|P| \to 0} \sum_{i=1}^{n} [f(\bar{x}_i) + g(\bar{x}_i) - 2k][f(\bar{x}_i) - g(\bar{x}_i)]\,\Delta x_i,$$

or

$$M_{y=k}(L) = \tfrac{1}{2}\delta \int_a^b [f(x) + g(x) - 2k][f(x) - g(x)]\,dx.$$

9.8.3   Definition.   The **moment about the horizontal line** $y = k$ of the homogeneous lamina L, described in 9.8.2, is

$$M_{y=k}(L) = \tfrac{1}{2}\delta \int_a^b [f(x) + g(x) - 2k][f(x) - g(x)]\,dx.$$

When $h = k = 0$, the definitions 9.8.2 and 9.8.3 give the moments of L about the coordinate axes. It is customary to write M_y instead of $M_{x=0}$ for a moment about the y axis and M_x in place of $M_{y=0}$ for a moment about the x axis.

9.8.4 **Corollary.** Let f and g be continuous on $[a, b]$, with $g(x) \leq f(x)$ for $a \leq x \leq b$. Denote by L the homogeneous lamina, of density δ, that is bounded by the curves $y = f(x)$ and $y = g(x)$, and the vertical lines $x = a$ and $x = b$. Then

    (i)   the **moment about the y axis** of L is

$$M_y(L) = \delta \int_a^b x[f(x) - g(x)]\, dx.$$

    (ii) the **moment about the x axis** of L is

$$(M_x L) = \tfrac{1}{2}\delta \int_a^b [f(x)^2 - g(x)^2]\, dx.$$

A similar theorem holds when y is the independent variable and the given interval is $[c, d]$ on the y axis; it may be obtained by interchanging x and y in 9.8.4 and replacing a and b by c and d, respectively.

Example 3. Find the moments about the y axis and the x axis of the homogeneous lamina L, of density δ, that is bounded by the curve $y = x^2/3 + 1$, the x axis, and the lines $x = -2$ and $x = 3$ (Fig. 9-2).

Solution. In using 9.8.4, $f(x) = x^2/3 + 1$, and $g(x) = 0$, the x axis. The interval $[a, b]$ is $[-2, 3]$. Then

$$M_y(L) = \delta \int_{-2}^3 x\left[\frac{1}{3}x^2 + 1 - 0\right] dx$$

$$= \delta \int_{-2}^3 \left(\frac{x^3}{3} + x\right) dx = \delta\left[\frac{x^4}{12} + \frac{x^2}{2}\right]_{-2}^3$$

$$= \frac{95\delta}{12}.$$

Also,

$$M_x(L) = \frac{\delta}{2} \int_{-2}^3 \left[\left(\frac{x^2}{3} + 1\right)^2 - 0\right] dx$$

$$= \frac{\delta}{2} \int_{-2}^3 \left(\frac{x^4}{9} + \frac{2x^2}{3} + 1\right) dx$$

$$= \frac{\delta}{2}\left[\frac{x^5}{45} + \frac{2x^3}{9} + x\right]_{-2}^3 = \frac{85\delta}{9}.$$

Example 4. Find the moments about the coordinate axes of the homogeneous lamina L, of density δ, that is bounded by the parabola $y = \tfrac{1}{2}(-x^2 - 2x + 3)$, the line $x = 2$, and the x axis (Fig. 9-36).

Solution. In applying 9.8.4, the interval $[-3, 2]$ will be broken up into $[-3, 1]$ and $[1, 2]$. On $[-3, 1]$, the $f(x)$ of 9.8.4 is the given parabola and $g(x)$ is the x axis; and on $[1, 2]$ the x axis is the $f(x)$ of 9.8.4 and the parabola is $g(x)$. Thus

$$M_y(L) = \delta \int_{-3}^1 \frac{1}{2}x(-x^2 - 2x + 3)\, dx + \delta \int_1^2 -\frac{1}{2}x(-x^2 - 2x + 3)\, dx$$

$$= \frac{1}{2}\delta \int_{-3}^1 (-x^3 - 2x^2 + 3x)\, dx - \frac{1}{2}\delta \int_1^2 (-x^3 - 2x^2 + 3x)\, dx$$

$$= -\frac{27\delta}{8}.$$

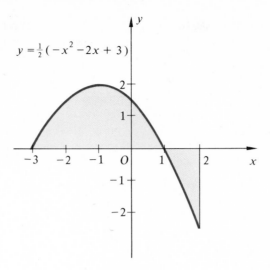

Figure 9-36

$$M_x(L) = \frac{1}{2}\delta \int_{-3}^{1} \frac{1}{4}(-x^2 - 2x + 3)^2 \, dx + \frac{1}{2}\delta \int_{1}^{2} -\frac{1}{4}(-x^2 - 2x + 3)^2 \, dx$$

$$= \frac{1}{8}\delta \int_{-3}^{1} (x^4 + 4x^3 - 2x^2 - 12x + 9) \, dx$$

$$\quad - \frac{1}{8}\delta \int_{1}^{2} (x^4 + 4x^3 - 2x^2 - 12x + 9) \, dx$$

$$= \frac{133\delta}{40}.$$

9.8.5 Definition. The **center of mass** (or **center of gravity**) of the homogeneous lamina L, described in 9.8.2, is the point $(\bar{x}, \bar{y})$ such that the moment of L about each of the lines $x = \bar{x}$ and $y = \bar{y}$ is zero.

Thus the lamina L will balance on $(\bar{x}, \bar{y})$; and for many problems in mechanics we can consider the mass of L to be concentrated at its center of mass $(\bar{x}, \bar{y})$.

From 9.8.2 the moment of L about the line $x = \bar{x}$ is

$$\delta \int_{a}^{b} (x - \bar{x})[f(x) - g(x)] \, dx,$$

which (by 9.8.5) is equal to zero. Therefore

(7) $$\delta \int_{a}^{b} x[f(x) - g(x)] \, dx - \bar{x} \, \delta \int_{a}^{b} [f(x) - g(x)] \, dx = 0.$$

Since the **mass**, $m(L)$, of a *homogeneous* lamina L is its density δ times its area,

(8) $$m(L) = \delta \int_{a}^{b} [f(x) - g(x)] \, dx.$$

So (7) can be written as

$$M_y(L) - \bar{x}\, m(L) = 0,$$

and the x coordinate of the center of mass $(\bar{x}, \bar{y})$ of L is $\bar{x} = M_y(L)/m(L)$. Similarly, $\bar{y} = M_x(L)/m(L)$.

These results are collected in the following corollary to 9.8.5.

9.8.6 Corollary. Let f and g be continuous on $[a, b]$, with $g(x) \le f(x)$ for $a \le x \le b$. If L is the homogeneous lamina of density δ bounded by $y = f(x)$, $y = g(x)$, $x = a$, and $x = b$, then the **center of mass** (or **center of gravity**) of L is the point $(\bar{x}, \bar{y})$, where

$$\bar{x} = \frac{M_y(L)}{m(L)} = \frac{\int_a^b x[f(x) - g(x)]\, dx}{\int_a^b [f(x) - g(x)]\, dx}.$$

and

$$\bar{y} = \frac{M_x(L)}{m(L)} = \frac{\frac{1}{2}\int_a^b [f(x)^2 - g(x)^2]\, dx}{\int_a^b [f(x) - g(x)]\, dx}.$$

Example 5. Find the center of gravity of the homogeneous lamina L of Example 3 (Fig. 9-2)

Solution. We found in Example 3 that the moments of L about the y axis and x axis were $M_y(L) = 95\delta/12$ and $M_x(L) = 85\delta/9$. By equation (8), above, and Example 1 of Section 9.1, the mass of L is

$$m(L) = \delta \int_{-2}^{3} \left(\frac{x^2}{3} + 1\right) dx = \frac{80\delta}{9}.$$

Therefore, by 9.8.6,

$$\bar{x} = \frac{M_y(L)}{m(L)} = \frac{95\delta/12}{80\delta/9} = \frac{57}{64}.$$

$$\bar{y} = \frac{M_x(L)}{m(L)} = \frac{85\delta/9}{80\delta/9} = \frac{17}{16}.$$

Example 6. Find the center of mass of the homogeneous lamina L of Example 4 (Fig. 9-36).

Solution. By (8), the mass of L is

$$m(L) = \delta \int_{-3}^{1} \frac{1}{2}(-x^2 - 2x + 3)\, dx + \delta \int_{1}^{2} -\frac{1}{2}(-x^2 - 2x + 3)\, dx = \frac{13\delta}{2}.$$

From 9.8.6 and Example 4,

$$\bar{x} = \frac{M_y(L)}{m(L)} = \frac{-27\delta/8}{13\delta/2} = -\frac{27}{52},$$

$$\bar{y} = \frac{M_x(L)}{m(L)} = \frac{133\delta/40}{13\delta/2} = \frac{133}{260}.$$

Notice that in the formulas for $\bar{x}$ and $\bar{y}$ in 9.8.6 the constant density factor δ, which appeared in both the numerator and denominator, has been canceled and the denominator is simply the area of the lamina. Consequently, we often speak of the point $(\bar{x}, \bar{y})$ in 9.8.6 as the **centroid** of the plane region bounded by $y = f(x)$, $y = g(x)$, $x = a$, and $x = b$.

We define the moments of this *plane region* about the x and y axes as follows.

9.8.7   Definition.   Let f and g be continuous on $[a, b]$, with $g(x) \leq f(x)$ for $a \leq x \leq b$. If R is the plane region

$$R = \{P{:}(x, y) \,|\, a \leq x \leq b, \; g(x) \leq y \leq f(x)\},$$

then the **moments of** R about the x and y axes are

$$M_x(R) = \frac{1}{2} \int_a^b [f(x)^2 - g(x)^2]\, dx,$$

$$M_y(R) = \int_a^b x[f(x) - g(x)]\, dx,$$

and the **centroid of** R, $(\bar{x}, \bar{y})$, is given by

$$\bar{x} = \frac{M_y(R)}{A(R)}, \qquad \bar{y} = \frac{M_x(R)}{A(R)},$$

where $A(R) = \int_a^b [f(x) - g(x)]\, dx$ is the area of R.

Analogous definitions, with x and y interchanged, apply to a region bounded by the curves $x = f(y)$ and $x = g(y)$, and the horizontal lines $y = c$ and $y = d$, where f and g are continuous on $[c, d]$ and $g(y) \leq f(y)$ for $c \leq y \leq d$.

Example 7.   Find the centroid of the region $R = \{P{:}(x, y) \,|\, 0 \leq x \leq 1$ and $0 \leq x^3 \leq y \leq \sqrt{x}\}$ (Fig. 9-37).

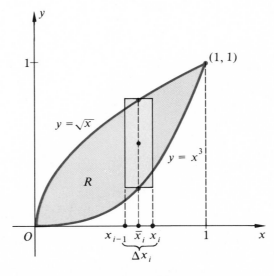

Figure 9-37

Solution. The area of the typical approximating rectangle shown in Fig. 9-37 is

$$\Delta A_i = (\sqrt{\bar{x}_i} - \bar{x}_i^3)\, \Delta x_i,$$

where $\bar{x}_i$ is the midpoint of $[x_{i-1}, x_i]$. Thus the area of the given region is

$$A = \lim_{|P|\to 0} \sum_{i=1}^{n} (\sqrt{\bar{x}_i} - \bar{x}_i^3)\, \Delta x_i = \int_0^1 (x^{1/2} - x^3)\, dx = \frac{5}{12}.$$

The centroid of the typical approximating rectangle is $(\bar{x}_i, \frac{1}{2}(\sqrt{\bar{x}_i} + \bar{x}_i^3))$, and its moments about the x and y axes are

$$\frac{1}{2}(\sqrt{\bar{x}_i} + x_i^3)\, \Delta A_i = \frac{1}{2}(\bar{x}_i - \bar{x}_i^6)\, \Delta x_i \qquad \text{and} \qquad \bar{x}_i\, \Delta A_i = \bar{x}_i(\sqrt{\bar{x}_i} - \bar{x}_i^3)\, \Delta x_i,$$

respectively.

The moments of the given region about the x and y axes are

$$M_x = \lim_{|P|\to 0} \sum_{i=1}^{n} \frac{1}{2}(\bar{x}_i - \bar{x}_i^6)\, \Delta x_i = \frac{1}{2}\int_0^1 (x - x^6)\, dx = \frac{5}{28},$$

and

$$M_y = \lim_{|P|\to 0} \sum_{i=1}^{n} \bar{x}_i(\sqrt{\bar{x}_i} - \bar{x}_i^3)\, \Delta x_i = \int_0^1 x(x^{1/2} - x^3)\, dx = \frac{1}{5}.$$

The centroid of the given region is $(\bar{x}, \bar{y})$, where

$$\bar{x} = \frac{M_y}{A} = \frac{12}{25} \qquad \text{and} \qquad \bar{y} = \frac{M_x}{A} = \frac{3}{7}.$$

A region R is **symmetric with respect to a line** l if corresponding to each point P in R (but not on l) there is another point P' in R such that l is the perpendicular bisector of the line segment PP' (Fig. 9-38).

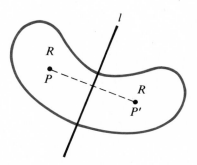

Figure 9-38

9.8.8 Theorem. Let f and g be continuous on $[a, b]$, with $g(x) \le f(x)$ for $a \le x \le b$. If the plane region $R = \{P:(x, y)\,|\,a \le x \le b,\ g(x) \le y \le f(x)\}$ has a vertical or horizontal line of symmetry, the centroid of R is on that line (or those lines) of symmetry.

For example, the centroid of a region bounded by a circle, an ellipse, or a rectangle, is its geometric center.

Proof of 9.8.8 is left as an exercise.

We conclude this section with a useful theorem connecting centroids of plane regions and volumes of solids of revolution. It was stated (in a more general form) by the Greek mathematician Pappus of Alexandria toward the end of the third century A.D.

9.8.9 Pappus' Theorem.   Let f and g be continuous on $[a, b]$, with $g(x) \leq f(x)$ for $a \leq x \leq b$. The volume of the solid of revolution generated by revolving the region $R = \{P:(x, y) \,|\, a \leq x \leq b, g(x) \leq y \leq f(x)\}$ about a horizontal or vertical line, in the plane of R but not intersecting R, is equal to the area of R times the circumference of the circle described by the centroid of R.

Proof of this theorem is easy and is left as an exercise.

Exercises

1. The masses and coordinates of a system of particles in the coordinate plane are given by the following: 3, (1, 1); 2, (7, 1); 4, (−2, −5); 6, (−1, 0); 2, (4, 6). Find the moments of this system with respect to the coordinate axes, and find the coordinates of the center of mass.

2. The masses and coordinates of a system of particles are given by the following: 3, (−3, 2); 6, (−2, −2); 2, (3, 5); 5, (4, 3); 1, (7, −1). Find the moments of this system with respect to the coordinate axes, and find the coordinates of the center of mass.

In each of Exercises 3 to 6, divide the indicated plane region into rectangular regions and use 9.8.1 and the additive assumption to find the moments about the coordinate axes of the given region. Then find its centroid.

3.
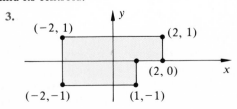

4.
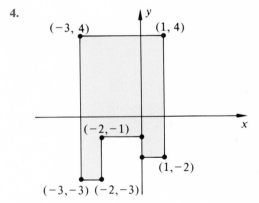

5.
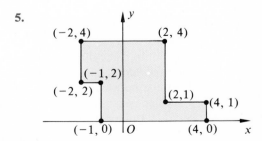

6.

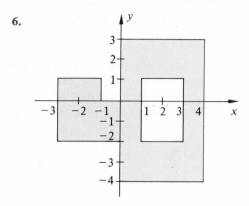

7. Find the centroid of the region R bounded by the curve $y = 4 - x^2$ and the x axis. Make a sketch. (*Hint:* Look for symmetry.)

8. Find the centroid of the region $R = \{P:(x, y)\,|\,0 \le x \le 3, 0 \le y \le 3x - x^2\}$. Make a sketch.

In Exercises 9 to 15, find M_x, M_y, and the centroid of the given region R. Make a sketch.

9. $R = \{P:(x, y)\,|\,0 \le x \le 4, 0 \le y \le \frac{1}{2}x^2\}$.

10. $R = \{P:(x, y)\,|\,0 \le x \le 3, 0 \le y \le 4 - \frac{1}{3}x^2\}$. (*Hint:* See Exercise 1, Section 9.1.)

11. $R = \{P:(x, y)\,|\,0 \le x \le 2, 0 \le y \le x^3\}$.

12. $R = \{P:(x, y)\,|\,1 \le x \le 3, 0 \le y \le 4x - x^2\}$. (*Hint:* See Exercise 2, Section 9.1.)

13. $R = \{P:(x, y)\,|\,-2 \le x \le 3, \frac{1}{2}(x^2 - 10) \le y \le 0\}$. (*Hint:* See Exercise 4, Section 9.1.)

14. $R = \{P:(x, y)\,|\,0 \le x \le 4, 2x - 4 \le y \le 2\sqrt{x}\,\}$. (*Hint:* See Exercise 12, Section 9.1.)

15. $R = \{P:(x, y)\,|\,-1 \le x \le 2, x^2 \le y \le x + 2\}$. (*Hint:* See Exercise 11, Section 9.1.)

16. Find the moment of the region $R = \{P:(x, y)\,|\,0 \le x \le 3, 0 \le y \le 3x - x^2\}$ about the line $x = -3$.

17. Find the moment of the region given in Exercise 16 about the line $y = 4$.

18. Find the moment of the region $R = \{P:(x, y)\,|\,0 \le x \le 4, 2x - 4 \le y \le 2\sqrt{x}\,\}$ about the line $x = 6$.

19. Find the centroid of the region $R = \{P:(x, y)\,|\,y^2 - 3y - 4 \le x \le -y - 1, -1 \le y \le 3\}$.

20. Use Pappus' theorem to find the volume generated by revolving about the line $y = -2$ the region R in Exercise 19.

21. Prove Pappus' theorem (9.8.9). (*Hint:* Use 9.2.2 and 9.8.6, also 9.3.2 and 9.8.6.)

22. Verify that the following definition is 9.8.7 with x and y interchanged, a replaced by c, and b replaced by d.

Definition. Let f and g be continuous on $[c, d]$, with $g(y) \le f(y)$ for $c \le y \le d$. If

$$R = \{P:(x, y)\,|\,g(y) \le x \le f(y), c \le y \le d\},$$

then

$$M_y(R) = \frac{1}{2} \int_c^d [f(y)^2 - g(y)^2]\,dy,$$

$$M_x(R) = \int_c^d y[f(y) - g(y)]\,dy,$$

and $(\bar{x}, \bar{y})$, the **centroid** of R, is given by

$$\bar{x} = \frac{M_y(R)}{A(R)} \quad \text{and} \quad \bar{y} = \frac{M_x(R)}{A(R)},$$

where $A(R) = \int_c^d [f(y) - g(y)]\,dy$ is the area of R.

9.9 CENTROID OF A SOLID OF REVOLUTION

In three-dimensional analytic geometry (Chapter 16), three coordinate axes are used: the familiar x and y axes, and a new one, the z axis, which is perpendicular to the other two. *The plane through the origin, perpendicular to the x axis, is called the yz* **plane** because it contains the y axis and the z axis (Fig. 9-39).

In motivating the definition (to be given below) of the **moment with respect to the** yz **plane** of a homogeneous solid of revolution, we start with two assumptions.

1.   If a homogeneous solid has a plane of symmetry, the center of mass (center of gravity) of the solid is a point in that plane of symmetry.

2.   If a homogeneous solid S is thought of as divided into two solids, S_1 and S_2, that have no interior points in common, then the moment of S with respect to a given plane is equal to the sum of the moments of S_1 and S_2 with respect to the given plane.

It follows from (1) that the center of mass of a homogeneous solid of revolution is on the axis of revolution, since every plane containing that axis is a plane of symmetry for the solid. In particular, the center of mass of a homogeneous right circular cylindrical solid is on its axis of revolution midway between the parallel circular faces (Fig. 9-40).

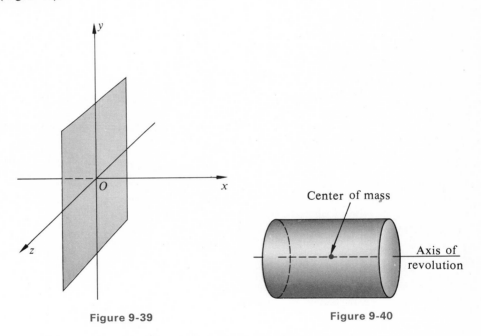

Figure 9-39   **Figure 9-40**

Let f be a function that is continuous on $[a, b]$, and let $f(x) \geq 0$ for $a \leq x \leq b$. Denote by R the region $R = \{P{:}(x, y) \mid a \leq x \leq b, 0 \leq y \leq f(x)\}$ (Fig. 9-41). Partition $[a, b]$ into n subintervals $[x_{i-1}, x_i]$, $i = 1, 2, 3, \cdots, n$, and denote by w_i the midpoint of each of the subintervals $[x_{i-1}, x_i]$. Form n approximating rectangles with altitudes $f(w_i)$ and bases of length $\Delta x_i = x_i - x_{i-1}$. A typical such approximating rectangle is shown in Fig. 9-41.

If the region R is revolved about the x axis to generate a solid of revolution S, the n approximating rectangles sweep out n approximating right circular cylinders, the sum of whose volumes approximates the volume of the solid of revolution S. A typical one of these approximating cylinders is shown in Fig. 9-42, its volume is $\pi[f(w_i)]^2 \Delta x_i$ and its mass is $\pi\delta[f(w_i)]^2 \Delta x_i$, where δ is the density of the homogeneous

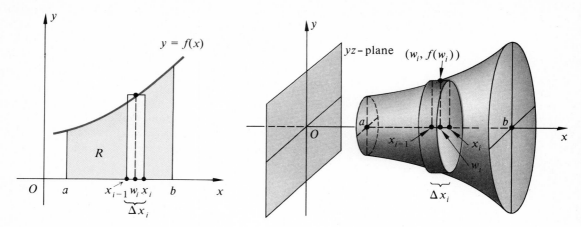

Figure 9-41   Figure 9-42

solid. The center of mass of this typical approximating cylinder is the center of the cylinder $(w_i, 0)$, and its moment with respect to the yz plane is defined to be $w_i[\pi\delta\{f(w_i)\}^2\,\Delta x_i]$. The sum of the moments of the n approximating cylinders with respect to the yz plane is

$$\sum_{i=1}^{n} \pi\delta w_i[f(w_i)]^2\,\Delta x_i.$$

We define the moment of the homogeneous solid of revolution S with respect to the yz plane to be

$$\lim_{|P|\to 0}\sum_{i=1}^{n} \pi\delta w_i[f(w_i)]^2\,\Delta x_i = \pi\delta\int_a^b x[f(x)]^2\,dx.$$

9.9.1 Definition. Let f be continuous on $[a, b]$ with $f(x) \geq 0$ for $a \leq x \leq b$. If S is the homogeneous solid of revolution generated by revolving the plane region $R = \{P{:}(x, y)\,|\,a \leq x \leq b,\; 0 \leq y \leq f(x)\}$ about the x axis, then the **moment** of S with respect to the yz plane is

$$M_{yz} = \pi\delta\int_a^b x[f(x)]^2\,dx,$$

where δ is the constant density of S.

Since the volume of this solid of revolution S is $V = \pi\int_a^b [f(x)]^2\,dx$ (by 9.2.1), and S is homogeneous, the **mass** of S is

$$m = \delta V = \delta\pi\int_a^b [f(x)]^2\,dx,$$

where δ is the constant density of S.

9.9.2 Definition. The **center of mass** of the solid of revolution S (described in 9.9.1) is that point on the x axis whose x coordinate is

$$\bar{x} = \frac{M_{yz}}{m},$$

where M_{yz} is the moment of S with respect to the yz plane and m is the mass of S; that is,

$$\bar{x} = \frac{\int_a^b x[f(x)]^2 \, dx}{\int_a^b [f(x)]^2 \, dx}.$$

The density constant δ does not appear in the latter formula for $\bar{x}$. For this reason, the center of mass of a *homogeneous* solid depends only on the geometric properties of size, shape, and position of the solid and is often called the **centroid** of the solid.

Example 1. Find the centroid of the solid of revolution S generated by revolving about the x axis the plane region $R = \{P{:}(x, y) \,|\, 0 \le x \le 4, 0 \le y \le \sqrt{x}\}$ (Fig. 9-11).

Solution. By 9.9.2, the centroid of this solid of revolution, S, is the point on the x axis whose x coordinate is

$$\bar{x} = \frac{M_{yz}}{m}.$$

To find M_{yz}, the moment of S with respect to the yz plane, we use 9.9.1 to write

$$M_{yz} = \pi\delta \int_a^b x\,[f(x)]^2 \, dx = \pi\delta \int_0^4 x(\sqrt{x})^2 \, dx$$

$$= \pi\delta \int_0^4 x^2 \, dx = \pi\delta\left[\frac{x^3}{3}\right]_0^4 = \frac{64\pi\delta}{3}.$$

From Example 1, Section 9.2, the volume of S is $V = 8\pi$; so the mass of S is $m = 8\pi\delta$. Therefore

$$\bar{x} = \frac{64\pi\delta/3}{8\pi\delta} = \frac{8}{3},$$

and the centroid of S is the point $(\frac{8}{3}, 0)$.

Example 2. Find the centroid of the solid of revolution generated by revolving about the x axis the region in first quadrant bounded by the curve $y^2 = 4x^2(1 - x)$ and the x axis (Fig. 9-43).

Solution. Using 9.9.1 and 9.2.1, we find

$$M_{yz} = \pi\delta \int_0^1 x[4x^2(1 - x)] \, dx = 4\pi\delta \int_0^1 (x^3 - x^4) \, dx = \frac{\pi\delta}{5}$$

and

$$V = \pi \int_0^1 4x^2(1 - x) \, dx = \frac{\pi}{3}.$$

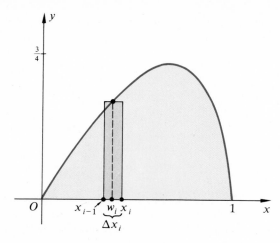

Figure 9-43

By 9.9.2, the centroid of this solid of revolution is the point on the x axis whose x coordinate is

$$\bar{x} = \frac{M_{yz}}{\delta V} = \frac{3}{5}.$$

Thus the centroid is the point $(\frac{3}{5}, 0)$.

The following theorem is a direct consequence of assumption (1) and 9.9.1.

9.9.3 Theorem. Let f and g be continuous on $[a, b]$ with $f(x) \geq g(x) \geq 0$ for $a \leq x \leq b$. If S is the homogeneous solid of revolution generated by revolving the plane region $R = \{P:(x, y) \,|\, a \leq x \leq b,\ g(x) \leq y \leq f(x)\}$ about the x axis, then the **moment** of S with respect to the yz plane is

$$M_{yz} = \pi\delta \int_a^b x[f(x)^2 - g(x)^2]\, dx,$$

where δ is the constant density of S.

The volume of the above homogeneous solid of revolution S is

$$V = \pi \int_a^b [f(x)^2 - g(x)^2]\, dx$$

(by 9.2.2), and so the **mass** of S is

$$m = \delta V = \delta\pi \int_a^b [f(x)^2 - g(x)^2]\, dx.$$

9.9.4 Definition. The **centroid** of the solid of revolution S (described in 9.9.3) is the point on the x axis whose x coordinate is

$$\bar{x} = \frac{M_{yz}}{m},$$

where M_{yz} is the moment of S with respect to the yz plane and m is the mass of S; that is,

$$\bar{x} = \frac{\int_a^b x[f(x)^2 - g(x)^2]\, dx}{\int_a^b [f(x)^2 - g(x)^2]\, dx}.$$

In engineering, it is often convenient to think of the mass of an object as concentrated at its centroid. This is illustrated by the following example.

Example 3. A tank in the form of an inverted right circular cone is full of water (Fig. 9-22). If the height, h, of the tank is 10 feet and the radius of its circular top lip is 4 feet, find the work done in pumping all the water over the edge of the top of the tank.

Solution. We shall consider the total mass of the water to be concentrated at the centroid of the inverted cone, and then find the work done in raising this mass from the centroid to the top of the tank.

The volume of the cone is $\pi r^2 h/3 = 160\pi/3$, and the mass of the water it contains is $\delta(160\pi/3)$. From Exercise 1, below, the y coordinate of the centroid of the conical body of water is $\bar{y} = 3h/4 = 15/2$. Thus the distance from the centroid to the top of the tank is $10 - 15/2 = 5/2$ feet. Then the work done in raising this mass of water to the top of the tank is

$$W = \delta\left(\frac{160\pi}{3}\right)\left(\frac{5}{2}\right) = \frac{400\pi\delta}{3}$$

$$\doteq 26{,}125 \text{ foot-pounds},$$

since δ, the weight of a cubic foot of water, is approximately 62.4 pounds.

The reader should compare this solution to that given in Example 2, Section 9.4, for the same problem.

Exercises

In Exercises 1 through 8, find the x coordinate of the centroid of the solid of revolution generated by revolving about the x axis the given plane region R. Make a sketch.

1. $R = \left\{ P:(x, y)\,\middle|\, 0 \le x \le h,\, 0 \le y \le \frac{rx}{h} \right\}.$

2. $R = \left\{ P:(x, y)\,\middle|\, 0 \le x \le 5,\, 0 \le y \le 3\!\left(1 - \frac{x}{5}\right) \right\}.$

3. $R = \{ P:(x, y)\,|\, 0 \le x \le 4,\, 0 \le y \le \sqrt{16 - x^2} \}.$

4. $R = \{ P:(x, y)\,|\, 0 \le x \le 4,\, 0 \le y \le 2\sqrt{x} \}.$

5. $R = \{ P:(x, y)\,|\, -2 \le x \le 1,\, 0 \le y \le 4 - x^2 \}.$

6. $R = \left\{ P:(x, y)\,\middle|\, -3 \le x \le -1,\, 0 \le y \le \frac{-x^3}{3} \right\}.$

7. $R = \{ P:(x, y)\,|\, 0 \le x \le 4,\, x^{3/2} \le y \le 2x \}.$

8. $R = \{ P:(x, y)\,|\, 0 \le x \le 2,\, x^3 \le y \le 4\sqrt{2x} \}.$

In Exercises 9 and 10, find the y coordinate of the centroid of the solid of revolution generated by revolving about the y axis the given plane region R. Make a sketch.

9. $R = \left\{ P:(x, y)\,\middle|\, \frac{y}{4} \le x \le \frac{\sqrt{y}}{2},\, 0 \le y \le 4 \right\}.$

10. $R = \{ P:(x, y)\,|\, (4 - y^{2/3})^{3/2} \le x \le (64 - y^2)^{1/2},\, 0 \le y \le 8 \}.$

11. Use the method shown in Example 3 to do Exercise 11 of Section 9.4.

12. Use the method of Example 3 to do Exercise 12 of Section 9.4.

9.10 APPLICATION TO ECONOMICS

We recall from Section 6.9 that the marginal revenue function R' is the derivative of the total revenue function R, and that the marginal cost function C' is the derivative of the total cost function C. The marginal revenue, $R'(x)$, is the rate of increase in total revenue per unit increase in output, and the marginal cost, $C'(x)$, is the rate of increase of total cost per unit increase in output.

We also saw that the profit function P is defined by

(1) $$P(x) = R(x) - C(x);$$

that is, the total profit, $P(x)$, in producing and selling x units of a commodity is the total revenue less the total cost. We differentiated both members of (1) to obtain the useful relation

(2) $$P'(x) = R'(x) - C'(x).$$

In many problems in economics we know the marginal revenue function and the marginal cost function, and we would like to know the output x that makes the total profit, $P(x)$, a maximum. The values of x that make $P'(x) = 0$ are critical numbers for the profit function P, and often there is one and only one such number x that makes $P(x)$ a maximum. We can use equation (2) to find that critical number by setting $R'(x) = C'(x)$ and solving for x. The solution, $x = k$, is the output that maximizes the profit $P(x)$.

To find the total profit of the operation from output zero to output k, we evaluate the definite integral $\int_0^k P'(x)\, dx$. Since we know $R'(x)$ and $C'(x)$, we can use (2) to write

$$\int_0^k P'(x)\, dx = \int_0^k [R'(x) - C'(x)]\, dx.$$

This gives the total profit as output is increased from zero to the optimum number k.

Example 1. In the manufacture of a certain article, the marginal revenue function and the marginal cost function are given in thousands of dollars by

$$R'(x) = 22 - 5x - 4x^2$$

and

$$C'(x) = 10 - 6x - 3x^2,$$

where x is in hundreds of articles manufactured and sold. Find the output that maximizes the profit, and also find the total profit as output increases from zero to that optimum number.

Solution. We must first find the output x that maximizes the total profit $P(x)$. Since

$$P'(x) = R'(x) - C'(x)$$
$$= (22 - 5x - 4x^2) - (10 - 6x - 3x^2)$$
$$= 12 + x - x^2 = (3 + x)(4 - x),$$

the critical numbers for $P(x)$ are $x = -3$ and $x = 4$. In the present context, only $x = 4$ is pertinent; so we test $x = 4$ by the second derivative test. $P''(x) = 1 - 2x$, and $P''(4) = 1 - 8 < 0$. Therefore $x = 4$ makes the total profit P a maximum.

The total profit from output 0 to output 4 is given by

$$\int_0^4 P'(x)\,dx = \int_0^4 (12 + x - x^2)\,dx$$
$$= \left[12x + \frac{x^2}{2} - \frac{x^3}{3}\right]_0^4$$
$$= 34\tfrac{2}{3}.$$

Thus the total profit from the time of startup until the output reaches 400 is $34,666.67.

The same method can frequently be used in somewhat similar situations not involving marginal revenue or marginal cost.

Example 2. A manufacturing company decides to increase its number of salesmen. If the cost, in units of $10,000, of employing x additional salesmen is given by

$$C(x) = 4\sqrt{\frac{2x}{3}},$$

and the additional revenue, in units of $10,000, resulting from their employment is

$$R(x) = 2\sqrt{x + 3} + 2,$$

find the optimum number of new salesmen to employ and the total resulting *net* revenue.

Solution. For each additional salesman hired, the total cost increases and the resulting revenue increases. At first the revenue greatly exceeds the cost, as can be verified by a sketch of the graphs of the functions R and C on the same coordinate axes. But eventually, as the number of new salesmen increases, their total cost exceeds the revenue that they bring in. The company will stop hiring new salesmen at this point.

To determine how many new salesmen to hire, we set $R(x) = C(x)$ and solve for x. Thus

$$2\sqrt{x + 3} + 2 = 4\sqrt{\frac{2x}{3}}.$$

By squaring both members of this equation, we obtain

$$(x + 3) + 2\sqrt{x + 3} + 1 = \frac{8x}{3},$$

or

$$\sqrt{x + 3} = \frac{5x - 12}{6}.$$

If we now square both members of this latter equation and collect terms, we get

$$25x^2 - 156x + 36 = 0$$

or

$$(x - 6)(25x - 6) = 0.$$

So the optimum number of new salesmen is $x = 6$.

Net revenue is the revenue minus the cost—that is, $R(x) - C(x)$. The *total* net revenue from this enlargement of the sales force is

$$\int_0^6 \left[2\sqrt{x+3} + 2 - 4\sqrt{\frac{2x}{3}} \right] dx$$

$$= 2 \int_0^6 (x+3)^{1/2}\, dx + 2 \int_0^6 dx - 4\sqrt{\frac{2}{3}} \int_0^6 x^{1/2}\, dx$$

$$= \frac{4}{3} \Big[(x+3)^{3/2} \Big]_0^6 + 2\Big[x \Big]_0^6 - \frac{8}{3}\sqrt{\frac{2}{3}} \Big[x^{3/2} \Big]_0^6$$

$$= 16 - 4\sqrt{3} \doteq 16 - 4(1.7321) = 9.0716.$$

The total net revenue is approximately \$90,716.

Exercises

1. In manufacturing and selling a certain commodity, the marginal revenue function, R', and the marginal cost function, C', are given (in thousands of dollars) by

$$R'(x) = 230 - 10x,$$
$$C'(x) = x^2 - 20x + 155,$$

where x is the output in units of a hundred. Find the output that maximizes the profit, and the total profit from startup ($x = 0$) to that optimum output.

2. If the marginal revenue function and the marginal cost function in a certain manufacturing operation are

$$R'(x) = 42 - 2x,$$
$$C'(x) = 6 + (x + 6)$$

find the output x that makes the profit a maximum, and also find the total profit from output zero to that optimum output.

3. Find the optimum output (that is, the output x that makes profit a maximum) and the total profit in manufacturing an article, if the marginal revenue function is

$$R'(x) = 5 + 4x - x^2$$

and the marginal cost function is given by

$$C'(x) = -x^2 + 6x - 3.$$

4. If the marginal revenue function and the marginal cost function are given by

$$R'(x) = 400 - 15x,$$
$$C'(x) = 3x^2 - 60x + 400,$$

find the profit-maximizing output x, and also the total profit from zero output to the optimum output.

5. A manufacturing company has purchased an additional machine, the earnings from which, in units of \$1000, are given by

$$E(t) = 96 - t^2,$$

where t is the number of years that the machine is used. The cost of maintenance and repair of the machine, in units of \$1000, is

$$M(t) = \frac{t^2}{2}.$$

If the company disposes of the machine when the cost of its maintenance and repair equals its earnings, how many years will the machine be used, and what will its total net earnings (earnings less maintenance and repairs) be during that time?

6. A shipping company is considering the purchase of a certain computer. The resulting savings in wages and increase in business due to faster shipping are estimated to be worth

$$E(t) = \sqrt{t} + 7$$

at time t, where $E(t)$ is in units of \$5000 and t is in years. The cost of servicing the machine at time t is given by

$$S(t) = \frac{t^2}{9} + 1,$$

where $S(t)$ is in units of \$5000. If the computer is used until its cost of servicing equals its related earnings, how many years will it be used, and what will its total *net* earnings be during the entire time that it is used? [Net earnings at time $t = E(t) - S(t)$.]

7. A company has decided to enlarge its sales force. If the cost, in units of \$12,000, of employing x

new salesmen is

$$C(x) = \frac{2x}{3},$$

and the resulting additional revenue, in units of $12,000, is estimated to be

$$R(x) = 2\sqrt{x},$$

what is the optimum number of new salesmen to employ, and what will the total resulting *net* revenue be?

8. A telephone solicitation company is consider-

ing increasing its number of solicitors. The cost, in units of $10,000, of employing x new solicitors is

$$C(x) = \frac{x}{2},$$

and the resulting additional revenue, in units of $10,000, is estimated to be

$$R(x) = (2x)^{1/3}.$$

How many new solicitors should be employed for greatest profit and what is the total associated net revenue?

9.11 REVIEW EXERCISES

In Exercises 1 to 7, a plane region R, bounded by the curve $y = 4x - x^2$ and the x axis is given.

1. Find the area of the region R. Sketch the region and a rectangular element of area.

2. Find the volume of the solid of revolution S_1 generated by revolving the region R about the x axis. Make a sketch.

3. Use the cylindrical shell method to find the volume of the solid S_2 generated by revolving the region R about the y axis. Make a sketch.

4. Use the washer method to find the volume of the solid S_2 of Exercise 3. Sketch the region R and a horizontal element of area.

5. Find the coordinates of the centroid of the region R.

6. Use Pappus' theorem and Exercises 1 and 5 to find the volume of the solid S_1 in Exercise 2.

7. Use Pappus' theorem and Exercises 1 and 5 to find the volume of the solid S_2 in Exercise 3.

8. The natural length of a certain spring is 16 inches, and a force of 8 pounds is required to keep it stretched 4 inches. Find the work done
(a) in stretching it from a length of 18 inches to a length of 24 inches;
(b) in compressing it from its natural length to a length of 14 inches.

9. An upright cylindrical tank is 10 feet in diameter and 10 feet high. If water in the tank is 8 feet deep, how much work is done in pumping all the water over the edge of the top of the tank?

10. An object weighing 300 pounds is suspended from the top of a building by a uniform cable. If the cable is 100 feet long and weighs 120 pounds, how much work is done in pulling the object to the top?

11. A region R is bounded by the line $y = 3x$ and the parabola $y = x^2$. Find the area of R by
(a) taking x as the independent variable;
(b) taking y as the independent variable.
Sketch the region R, and show a vertical element of area in (a) and a horizontal element of area in (b).

12. Find the moments, $M_x(R)$ and $M_y(R)$, about the coordinate axes of the region R given in Exercise 11. Then find the centroid of R.

13. Use Pappus' theorem and Exercises 11 and 12 to find the volume of the solid of revolution generated by revolving the region R about the x axis.

14. Use Pappus' theorem and Exercises 11 and 12 to find the volume of the solid generated by revolving the region R about the y axis.

15. Each end of a tank is an inverted isosceles triangle 10 feet wide and 8 feet deep [Fig. 9-44(a)].

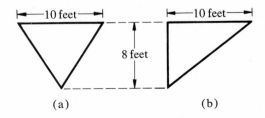

(a) (b)

Figure 9-44

What is the total force on one end when the tank is full of water?

16. Each end of a tank is an inverted right triangle with its 10-foot horizontal base at the top and the adjacent vertical side 8 feet long [Fig. 9-44(b)]. Find the force on one end when the tank is full of water.

17. Find the length of the arc of the curve $y = x^3/3 + 1/(4x)$ from $x = 1$ to $x = 3$.

18. Find the area of the surface generated by revolving about the x axis the segment of the curve $y = x^3/6 + 1/(2x)$ from $x = 1$ to $x = 3$.

19. A rectangular swimming pool is 30 feet long and 15 feet wide. If the depth of the water in the pool varies constantly from 3 feet at one end to 8 feet at the other end, find the total force of the water on a side of the pool.

20. Find the area of the surface generated by revolving about the y axis the arc of the curve $y = 9 - x^2$ from $x = 0$ to $x = \sqrt{6}$.

21. Find the centroid of the solid generated by revolving about the x axis the plane region

$$R = \left\{ P{:}(x, y) \,\middle|\, 0 \le x \le 3, \frac{x^2}{3} \le y \le 3 \right\}.$$

Make a sketch.

22. Find the centroid of the solid of revolution generated by revolving about the x axis the plane region

$$R = \left\{ P{:}(x, y) \,\middle|\, 0 \le x \le 2\sqrt{2}, \frac{x^2}{4} \le y \le \frac{16 - y^2}{4} \right\}.$$

Make a sketch.

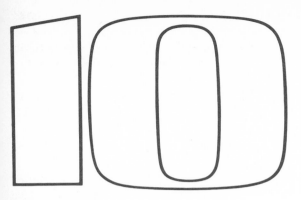

transcendental functions

We have developed a considerable body of calculus in the preceding chapters, but up to now it has been applied only to algebraic functions. A function that is not algebraic is called *transcendental*. In the present chapter we extend the calculus to the *elementary transcendental functions*—namely, the logarithm function, the exponential function, the trigonometric functions, and the inverse trigonometric functions.

The reader became accustomed to working with exponents and logarithms in his precalculus algebra course, and learned the laws of exponents:

$$a^M a^N = a^{M+N},$$
$$(a^M)^N = a^{MN},$$
$$a^0 = 1.$$

Most elementary algebra books define the logarithm to base a of a number N to be the exponent of the power to which a must be raised to equal N. That is, for $a > 0$ and $a \neq 1$,

(1) $$x = \log_a N \qquad \text{if and only if} \qquad N = a^x.$$

The principal use of logarithms in algebra and trigonometry was for computation. The properties of logarithms that made them such powerful tools in computation are readily derived from the laws of exponents. They are

(2)
$$\log_a MN = \log_a M + \log_a N,$$
$$\log_a \frac{M}{N} = \log_a M - \log_a N,$$
$$\log_a M^N = N \log_a M,$$
$$\log_a 1 = 0.$$

To understand the definition (1) of a logarithm, we must know what a^x means. This is easy if x is a rational number p/q, where p and q are integers and $q \neq 0$. By definition,

$$a^{p/q} = (\sqrt[q]{a})^p.$$

But what does a^x mean when x is irrational? For example, what does $7^{\sqrt{3}}$ mean, and what does $(\sqrt{13})^\pi$ mean?

It was assumed, but not proved, in elementary algebra that the equivalent equations

$$x = \log_a y \qquad \text{and} \qquad y = a^x$$

$a > 0$ and $a \neq 1$, have a unique solution x for each positive number y. For computational purposes, this assumption is acceptable. But in calculus we are interested in the logarithmic and exponential *functions*, defined by $y = \log_a x$ and $y = a^x$, $a > 0$ and $a \neq 1$; and to ensure the continuity and differentiability of these functions, we must be certain that they are defined for *all* numbers x of an interval, not just for rational numbers.

Although it is possible to continue with the approach used in elementary algebra and define these functions for irrational value of x, doing so would be difficult. Moreover, subsequent proofs that the logarithmic and exponential functions have the characteristic properties (2) for irrational numbers would be long and tedious.

Fortunately, the calculus provides a simple and elegant way to define the logarithmic and exponential functions, and to derive their familiar characteristic properties. In this approach we define the logarithm function first and then obtain the exponential function as the inverse of the logarithm function.

We ask the student to read this presentation with a fresh mind. No use will be made of any previous knowledge of logarithms. But it will soon be shown that the logarithm function and the exponential function, as defined by calculus for *all* numbers of their interval domains (not just for rational numbers), possess all the characteristic properties familiar from elementary algebra, where those functions were defined only for rational numbers.

In computation, the best base for a system of logarithms is $a = 10$. The logarithms to base 10, used in algebra and trigonometry, are called *common logarithms*. Another base, $e \doteq 2.718$, is used for the system of *natural logarithms*, so important in calculus, as we shall see.

10.1 THE NATURAL LOGARITHM FUNCTION

Since the rational function t^n, where n is any selected integer, is continuous for all $t > 0$, the integral with variable upper limit

(1)
$$y = \int_1^x t^n \, dt \qquad (n \text{ an integer, and } x > 0)$$

defines a function whose domain is the set of all positive numbers x (see Section 8.4).

When $n \neq -1$, an antiderivative of t^n is $t^{n+1}/(n+1)$. Thus the value of the function (1) at any positive number x is

$$\int t^n \, dt = \frac{t^{n+1}}{n+1}\bigg]_1^x = \frac{x^{n+1}}{n+1} - \frac{1}{n+1},$$

provided that $n \neq -1$.

It is natural to inquire what function (1) defines when $n = -1$.

10.1.1 Definition. The **natural logarithm function** is defined by

$$\ln x = \int_1^x \frac{1}{t} \, dt, \qquad x > 0.$$

Its domain is the set of all positive numbers.

The notation $\ln x$ is read "the natural logarithm of x" or, briefly, "logarithm x". The derivative of the natural logarithm function is

$$D_x \ln x = D_x \int_1^x \frac{1}{t} \, dt = \frac{1}{x}$$

(by the fundamental theorem). That is,

10.1.2
$$D_x \ln x = \frac{1}{x} \qquad x > 0.$$

Since this derivative exists for all $x > 0$, the natural logarithm function is continuous throughout its domain (4.5.2)

If u is a positive-valued, differentiable function of x, then (by the chain rule) $D_x \ln u = (D_u \ln u)(D_x u) = (1/u)D_x u$.

10.1.3
$$D_x \ln u = \frac{1}{u} D_x u, \qquad u(x) > 0.$$

Notice that 10.1.2 is a special case of the more general formula 10.1.3.

Example 1. Find $D_x \ln \sqrt{x}$.

Solution. Applying 10.1.3, with $u = \sqrt{x} = x^{1/2}$, we get

$$D_x \ln \sqrt{x} = \frac{1}{x^{1/2}} \cdot D_x x^{1/2}$$

$$= \frac{1}{x^{1/2}} \cdot \frac{1}{2x^{1/2}} = \frac{1}{2x}.$$

Example 2. Find $D_x \ln (x^2 - x - 2)$.

Solution. Let $u = x^2 - x - 2$. Since $x^2 - x - 2 = (x + 1)(x - 2)$, $u = 0$ for $x = -1$ and $x = 2$. Furthermore, $u > 0$ for $x < -1$, $u < 0$ for $-1 < x < 2$, and $u > 0$ for $x > 2$. Since $\ln u$

is defined only for those values of x that make $u > 0$, the domain of $\ln (x^2 - x - 2)$ is $(-\infty, -1) \cup (2, \infty)$.

Using formula 10.1.3, we find

$$D_x \ln (x^2 - x - 2) = \frac{1}{x^2 - x - 2} \cdot D_x(x^2 - x - 2)$$

$$= \frac{2x - 1}{x^2 - x - 2},$$

for $x \in (-\infty, -1) \cup (2, \infty)$.

If $u(x) < 0$, then $-u(x) > 0$ and 10.1.3 gives $D_x \ln(-u) = [1/(-u)]D_x(-u) = (1/u)D_x u$. This result and 10.1.3 are combined in $D_x \ln|u| = (1/u) D_x u$, $u \neq 0$. By antidifferentiating both members of the latter equation, we obtain the important formula

10.1.4
$$A_x\left(\frac{1}{u}D_x u\right) = \ln|u| + C, \quad u \neq 0.$$

Example 3. If $f(x) = \dfrac{5}{2x + 7}$, find the antiderivatives of f.

Solution. We can write

$$\frac{5}{2x + 7} = \frac{5}{2}\left(\frac{1}{2x + 7}\right) \cdot (2) = \frac{5}{2}\left[\frac{1}{2x + 7} \cdot D_x(2x + 7)\right].$$

By letting $u = 2x + 7$ in 10.1.4, we obtain

$$A_x\left(\frac{5}{2x + 7}\right) = \frac{5}{2} A_x\left[\frac{1}{2x + 7} \cdot D_x(2x + 7)\right]$$

$$= \frac{5}{2} \ln|2x + 7| + C.$$

Example 4. Find $\displaystyle\int_{-1}^{3} \frac{x}{10 - x^2}\, dx$.

Solution. The given integral can be written

$$\int_{-1}^{3} \frac{x}{10 - x^2}\, dx = -\frac{1}{2} \int_{-1}^{3} \frac{1}{10 - x^2} (-2x)\, dx$$

$$= -\frac{1}{2} \int_{-1}^{3} \frac{1}{10 - x^2} D_x(10 - x^2)\, dx.$$

Setting $10 - x^2 = u$ in 10.1.4 gives

$$A_x\left[\frac{1}{10 - x^2} \cdot D_x(10 - x^2)\right] = \ln (10 - x^2) + C$$

for $10 - x^2 > 0$, that is, for $-\sqrt{10} < x < \sqrt{10}$. Therefore

$$-\frac{1}{2}\int_{-1}^{3}\frac{1}{10 - x^2}\, D_x(10 - x^2)\, dx = -\frac{1}{2}\Big[\ln (10 - x^2)\Big]_{-1}^{3}$$

$$= -\frac{1}{2}(\ln 1 - \ln 9) = \frac{1}{2}\ln 9 - \frac{1}{2}\ln 1.$$

It will be shown below that $\ln 1 = 0$ and $\frac{1}{2}\ln 9 = \ln 9^{1/2} = \ln 3$, which is the final answer.

The properties of logarithms that made them so useful in earlier courses will now be proved for the natural logarithm function.

10.1.5 Theorem. If a and b are positive numbers and n is any rational number, then

  (i) $\ln 1 = 0$;

  (ii) $\ln ab = \ln a + \ln b$;

  (iii) $\ln \dfrac{a}{b} = \ln a - \ln b$;

  (iv) $\ln a^n = n \ln a$.

Proof.

  (i) $\ln 1 = \displaystyle\int_{1}^{1}\frac{1}{t}\, dt = 0.$

  (ii) Let $x > 0$. From 10.1.3,

$$D_x \ln ax = \left(\frac{1}{ax}\right)a = \frac{1}{x}$$

and, by 10.1.2,

$$D_x \ln x = \frac{1}{x}.$$

Since their derivatives are the same, $\ln ax$ and $\ln x$ differ only by an additive constant (7.1.3). That is,

(2) $\ln ax = \ln x + C, \qquad x > 0.$

To evaluate C, we substitute $x = 1$ in (2) and use (i), obtaining $\ln a = 0 + C$. Thus (2) becomes

(3) $\ln ax = \ln a + \ln x,$

which is true for all $x > 0$, in particular for $x = b$. Therefore

$$\ln ab = \ln a + \ln b.$$

  (iii) Replace a by $1/b$ in (ii). This gives

$$\ln \frac{1}{b} + \ln b = \ln \left(b \cdot \frac{1}{b}\right) = \ln 1 = 0,$$

so that

$$\ln \frac{1}{b} = -\ln b.$$

If we let $x = 1/b$ in equation (3), we obtain

$$\ln \left(a \cdot \frac{1}{b} \right) = \ln a + \ln \frac{1}{b} = \ln a - \ln b,$$

or

$$\ln \frac{a}{b} = \ln a - \ln b.$$

(iv)   By 10.1.3,

$$D_x \ln x^n = \frac{1}{x^n} \cdot nx^{n-1} = \frac{n}{x},$$

and by 10.1.2,

$$D_x(n \ln x) = \frac{n}{x}.$$

Since their derivatives are the same, $\ln x$ and $n \ln x$ differ by a constant. That is,

$$\ln x^n = n \ln x + C.$$

We evaluate C by letting $x = 1$. The result is

$$\ln 1 = n \ln 1 + C,$$

and since $\ln 1 = 0$, it follows that $C = 0$. Therefore

$$\ln x^n = n \ln x, \qquad x > 0,$$

and by letting $x = a$,

$$\ln a^n = n \ln a. \quad \blacksquare$$

Example 5.   Find dy/dx if

$$y = \ln \sqrt[3]{\frac{x - 1}{x^2}}.$$

Solution.   Differentiation of logarithm functions can often be simplified by first using the properties of logarithms (10.1.5). Thus, from properties (iii) and (iv),

$$y = \ln \left(\frac{x - 1}{x^2} \right)^{1/3} = \frac{1}{3} \ln \left(\frac{x - 1}{x^2} \right)$$

$$= \frac{1}{3}[\ln (x - 1) - \ln x^2] = \frac{1}{3}[\ln (x - 1) - 2 \ln x].$$

Therefore

$$\frac{dy}{dx} = \frac{1}{3} \left(\frac{1}{x - 1} - \frac{2}{x} \right) = \frac{2 - x}{3x^2 - 3x}.$$

The labor of differentating expressions involving quotients, products, or powers can often be substantially reduced by first applying the logarithm function to the expression and using the properties of logarithms. This method, called **logarithmic differentiation**, is illustrated in the next two examples.

Example 6. Differentiate $y = \dfrac{\sqrt{1 - x^2}}{(x + 1)^{2/3}}$.

Solution. By applying the natural logarithm function to both members of the given equation and using properties of logarithms, we obtain

$$\ln y = \frac{1}{2} \ln (1 - x^2) - \frac{2}{3} \ln (x + 1).$$

Differentiating implicitly with respect to x, we get

$$\frac{1}{y} \frac{dy}{dx} = \frac{(-2x)}{2(1 - x^2)} - \frac{2}{3(x + 1)} = \frac{-(x + 2)}{3(1 - x^2)}.$$

Thus

$$\frac{dy}{dx} = \frac{-y(x + 2)}{3(1 - x^2)} = \frac{-\sqrt{1 - x^2}\,(x + 2)}{3(x + 1)^{2/3}\,(1 - x^2)}$$

$$= \frac{-(x + 2)}{3(x + 1)^{2/3}(1 - x^2)^{1/2}}.$$

Example 7. Differentiate $y = x^x$.

Solution. Applying the natural logarithm function to both sides of the given equation and using properties of logarithms, we obtain

$$\ln y = x \ln x.$$

If we differentiate this implicitly with respect to x, we obtain

$$\frac{1}{y} \frac{dy}{dx} = x\left(\frac{1}{x}\right) + (1) \ln x = 1 + \ln x.$$

Therefore

$$\frac{dy}{dx} = y(1 + \ln x) = x^x(1 + \ln x).$$

In order to gain a better understanding of the natural logarithm function, let us draw the graph of

$$(4) \qquad\qquad y = \ln x, \qquad x > 0,$$

Since $1/t > 0$ for all $t > 0$,

$$\ln x = \int_1^x \frac{1}{t}\, dt$$

is positive for $x > 1$, zero for $x = 1$, and negative for $0 < x < 1$ (by 8.5.4, 8.3.6, 8.3.5). Thus the graph of $y = \ln x$ intersects the x axis only at $(1, 0)$; to the right of this point the graph is above the x axis; and between the origin and $(1, 0)$ the graph is below the x axis.

Moreover, by 10.1.2,

$$\frac{dy}{dx} = \frac{1}{x}, \qquad x > 0;$$

so the derivative of $y = \ln x$ exists and is positive valued for all $x > 0$. Therefore the natural logarithm function is continuous and increasing throughout its domain (4.5.2 and 6.3.2), and its graph always rises as x increases.

Furthermore,

$$\frac{d^2y}{dx^2} = -\frac{1}{x^2}$$

is negative for all $x > 0$, and thus the graph of $y = \ln x$ is everywhere concave downward (6.10.2).

We shall now calculate approximate values of $y = \ln x$ for a few numbers x. In Example 2 of Section 8.7, we used the trapezoidal rule to calculate

$$\ln 2 = \int_1^2 \frac{1}{x}\, dx \doteq 0.694.$$

From this, and from the properties of the natural logarithm function, we have

$$\ln 4 = \ln 2^2 = 2\ln 2 \doteq 2(0.694) = 1.388,$$
$$\ln 8 = \ln 2^3 = 3\ln 2 \doteq 3(0.694) = 2.082,$$
$$\ln 16 = \ln 2^4 = 4\ln 2 \doteq 4(0.694) = 2.776,$$
$$\ln\left(\tfrac{1}{2}\right) = \ln 2^{-1} = -\ln 2 \doteq -0.694,$$
$$\ln\left(\tfrac{1}{4}\right) = \ln 2^{-2} = -2\ln 2 \doteq -2(0.694) = -1.388.$$

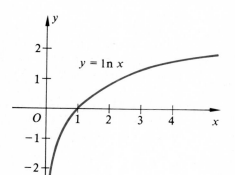

Figure 10-1

Plotting these points and remembering that the curve is everywhere continuous, rising, and concave downward, we see that the graph of $y = \ln x$ resembles Fig. 10-1. It is left for the reader to show that the graph of $y = \ln x$ is asymptotic to the negative y axis and rises without bound as x increases indefinitely (Exercises 37 and 38), and thus that *the range of the natural logarithm function is $(-\infty, \infty)$*.

A far better method than the trapezoidal rule for computing values of the natural logarithm function will be explained in Chapter 14. A table of values of the natural logarithm function is given in the Appendix (A.8.5).

Notice in the table of natural logarithms that

$$\ln 2.7 \doteq 0.993,$$
$$\ln 2.8 \doteq 1.030.$$

Since the natural logarithm function is continuous on $(0, \infty)$, it follows from the intermediate value theorem (3.5.8) that there is a number x between 2.7 and 2.8 such that $\ln x = 1$. Moreover, the natural logarithm function is increasing throughout its domain and so is a one-to-one function (6.3.2). Therefore the number x, such that $\ln x = 1$, is unique. This number is so important in mathematics that it has been given the special symbol e (after Leonard Euler, 1707–1783, who established the use of e and π in modern mathematics with their present-day meanings).

10.1.6 Definition. The unique solution of the equation $\ln x = 1$ is denoted by e; that is,

$$\ln e = 1.$$

This number e hás the approximate value

$$e \doteq 2.718281828459 \cdots.$$

It is neither a terminating decimal nor a repeating decimal and therefore e is not a rational number. Like π, e is a special kind of irrational number called a *transcendental number*.

From $\ln e = 1$ and the properties of logarithms,

(5) $$\ln e^r = r \ln e = r,$$

when r is a *rational* number. We would like to give a meaning to e^s when s is an *irrational* number. Since $\ln$ is a one-to-one continuous function on $(0, \infty)$, there is a unique number x such that $\ln x = s$. We *define* e^s to be that number x. That is,

(6) $$\ln e^s = s$$

for any irrational number s. Now e^x has meaning for all real numbers x.

10.1.7 Definition. Let x be any real number. Then e^x is the unique number for which

$$\ln e^x = x.$$

Thus the values of the natural logarithm function as defined in 10.1.1 are the logarithms to base e, briefly encountered by the reader in elementary algebra.

Since it turns out that $\ln x$ is much more important in calculus than $\log_{10} x$ or logarithms to any other base, it is to be understood that, *when speaking of the* **logarithm function**, *we mean the function whose value is* $\ln x$.

Exercises

In each of Exercises 1 to 6, state the domain of the given logarithm function f and find its derivative.

1. $f(x) = \ln (x - 5)^4$.

2. $f(x) = \ln (2x^3)$.

3. $f(x) = \ln (x^2 - 5x + 6)$.

4. $f(x) = \ln \sqrt{3x - 25}$.

5. $f(x) = \ln \left(\dfrac{2 - x}{x^3}\right)$.

6. $f(x) = \ln (\sqrt{x + 1} - \sqrt{x})$.

7. Find $D_x(x \ln x)$.

8. Find $D_x [\ln (\ln x)]$. (*Hint:* Let $u = \ln x$ in 10.1.3.)

In Exercises 9 to 12, use the properties of logarithms to simplify the given logarithm function before differentiating.

9. If $f(x) = \ln \sqrt[5]{\dfrac{x - 2}{x^4}}$, find $f'(x)$. (*Hint:* See Example 5.)

10. If $f(x) = \ln \dfrac{\sqrt{x^2 - 4}}{\sqrt[3]{(x - 2)^2}}$, find $f'(x)$.

11. Find $\dfrac{dy}{dx}$ if $y = \ln (x^2 \sqrt{4 - x^3})$.

12. Find $D_x \ln (\sqrt{x^2 - 1} \sqrt{2x - 6})$.

In Exercises 13 to 20, find the antiderivatives of the indicated functions.

13. $f(x) = \dfrac{3}{x - 3}$.

14. $f(x) = \dfrac{4x}{x^2 + 10}$.

15. $y = \dfrac{x}{a^2 - x^2}$.

16. $y = \dfrac{x^2 + 2x + 4}{x^3 + 3x^2 + 12x + 10}$.

17. $f(x) = \dfrac{3}{2(3x + 11)}$.

18. $f(x) = \dfrac{8x + 18}{(2x - 1)(x + 5)}$.

19. $y = \ln x$. (*Hint:* Write $\ln x = 1 + \ln x - 1$, and use Exercise 7.)

20. $y = \dfrac{1}{x \ln x}$. (*Hint:* Let $u = \ln x$ in 10.1.4.)

Evaluate the definite integrals in Exercises 21 to 24.

21. $\displaystyle\int_1^4 \dfrac{1}{3x - 2}\, dx$.

22. $\displaystyle\int_3^5 \dfrac{x}{x^2 - 1}\, dx$.

23. $\displaystyle\int_1^3 \dfrac{x - 1}{x + 1}\, dx$. (*Hint:* First divide denominator into numerator.)

24. $\displaystyle\int_1^4 \dfrac{\ln x}{x}\, dx$. (*Hint:* Use 7.3.1.)

25. Sketch the region bounded by the coordinate axes, the curve $y = 6/(x + 2)$, and the vertical line $x = 4$, and find its area.

26. Sketch the region bounded by the curve $xy = 4$ and the line $x + y - 5 = 0$; then find its area.

27. A particle moving on a straight line has an acceleration $a = 4/(5 - t)^2$ feet per second per second. If $v = 0$ and $s = 2$ when $t = 0$, what is the position of the particle when $t = 4$?

28. Find the coordinates of the point on the graph

of $y = \ln \sqrt{x}$ from which the tangent to the curve passes through the origin. Make a sketch.

In Exercises 29 to 34, use logarithmic differentiation to find dy/dx.

29. $y = \dfrac{x + 11}{\sqrt{x^3 - 4}}$

30. $y = (x^2 + 3x)(x - 2)(x^2 + 1)$.

31. $y = \dfrac{\sqrt{x + 13}}{(x - 4)\sqrt[3]{2x + 1}}$

32. $y = (x - 1) \ln x$.

33. $y = x^{(x^2)}$.

34. $y = \dfrac{2x^2 + 1}{\ln (2x^2 + 1)}$.

35. Without using the trapezoidal rule, show that
$$\tfrac{1}{2} \le \ln 2 \le 1.$$
(*Hint:* $1/2 \le 1/t \le 1$ for all t in $[1, 2]$.)

36. Show that for every positive rational number K,
$$\ln 4^K > K.$$
(*Hint:* Since $\ln 4 \doteq 1.388$, $\ln 4 > 1$ and $K \ln 4 > K$.)

37. Show that
$$\lim_{x \to \infty} \ln x = \infty,$$
and thus that the increasing function $\ln x$ has no upper bound. [*Hint:* By 6.11.4, it will be sufficient to prove that for every positive rational number K there is a positive number x such that $\ln x > K$. Use Exercise 36 and the fact that $\ln x$ is increasing on $(0, \infty)$.]

38. Show that
$$\lim_{x \to 0^+} \ln x = -\infty,$$
and thus that the curve $y = \ln x$ is asymptotic to the negative y axis (6.11.5). [*Hint:* $\ln x = -\ln (1/x)$, $x > 0$.]

10.2 INVERSE OF A FUNCTION*

A *function* is a correspondence between two sets of elements, its domain $\mathfrak{D}$ and its range $\mathfrak{R}$, such that to each element of the domain there corresponds one and only one element of the range, and each element of the range is the image (that is, the correspondent) of *at least one* element of the domain.

*It is suggested that the reader review Sections 2.11 and 5.11 before proceeding.

Of course, different elements of the domain of a function may have the same image in the range. For instance, the equation $y = x^2 - 1$ defines a function having domain R and range $[-1, \infty)$. To an arbitrarily chosen real number, say $x = 10$, there corresponds one and only one number, 99, in the range. But another number, -10, from the domain also has 99 for its image in the range, since $(-10)^2 - 1 = 99$.

When a function has the additional property that no two distinct elements of the domain have the same image in the range, the function is said to be *one-to-one*. In this case, each element of the range is the image of exactly one element of the domain.

Let f be a one-to-one function with domain $\mathfrak{D}$ and range $\mathfrak{R}$. Because f is a function, any vertical line $x = h$, $h \in \mathfrak{D}$, intersects the graph of f in one and only one point; and because f is one-to-one, any horizontal line $y = k$, $k \in \mathfrak{R}$, intersects the graph of f in exactly one point (Fig. 2-50).

The rule of correspondence $y = f(x)$ of a one-to-one function f, defines a second function, f^{-1}, having the rule of correspondence $x = f^{-1}(y)$. The second function f^{-1} is the *inverse* of f. A typical ordered pair of numbers for the inverse function f is (y, x) instead of (x, y). Just as the set of the first numbers x in the ordered pairs (x, y) of f is the domain of f, the set of first numbers y in the ordered pairs (y, x) of f^{-1} forms the domain of the inverse f^{-1}. The domain of the inverse function f^{-1} is the range of f, and the range of the inverse function is the domain of f. To recapitulate,

10.2.1 If f is a one-to-one function, its inverse f^{-1}, defined by

$$x = f^{-1}(y) \qquad \text{if and only if} \qquad y = f(x),$$

is also a one-to-one function and has f for its inverse. Thus

$$f^{-1}(f(x)) = x \qquad \text{for} \quad x \in \mathfrak{D}_f,$$
$$f(f^{-1}(y)) = y \qquad \text{for} \quad y \in \mathfrak{D}_{f^{-1}}.$$

The domain of f^{-1} is the range of f and the range of f^{-1} is the domain of f.

Recall from Section 6.3 that a function that is increasing on an interval I is one-to-one on I; also, a function that is decreasing on I is one-to-one on I.

10.2.2 Inverse Function Theorem. If a function f is continuous and increasing on a closed interval $[a, b]$, then f has an inverse f^{-1} that is continuous and increasing on $[f(a), f(b)]$.

Proof. By 6.3.2, f is a one-to-one function on $[a, b]$. Therefore f has an inverse f^{-1} that is defined on $[f(a), f(b)]$.

Let y_1 and y_2 be any two numbers in $[f(a), f(b)]$ such that $y_1 < y_2$. We wish to show that $f^{-1}(y_1) < f^{-1}(y_2)$, and therefore that f^{-1} is increasing on $[f(a), f(b)]$. Assume the contrary—namely, that $y_1 < y_2$ and $f^{-1}(y_2) \leq f^{-1}(y_1)$. Since f is increasing, $f^{-1}(y_2) \leq f^{-1}(y_1) \implies f(f^{-1}(y_2)) \leq f(f^{-1}(y_1))$. But $f(f^{-1}(y_2)) = y_2$ and

$f(f^{-1}(y_1)) = y_1$. Thus $y_2 \leq y_1$, which is a contradiction. Therefore

$$f(a) \leq y_1 < y_2 \leq f(b) \implies f^{-1}(y_1) < f^{-1}(y_2),$$

and f^{-1} is increasing on $[f(a), f(b)]$.

It remains to prove that f^{-1} is continuous on $[f(a), f(b)]$. Let y_1 be any number in the open interval $(f(a), f(b))$. We shall prove that $\lim_{y \to y_1} f^{-1}(y) = f^{-1}(y_1)$ and thus that f^{-1} is continuous on $(f(a), f(b))$.

Denote $f^{-1}(y_1)$ by x_1. Then $a < x_1 < b$ because f^{-1} is increasing on $[f(a), f(b)]$.

Let ϵ be any positive number, but small enough so that $x_1 - \epsilon$ and $x_1 + \epsilon$ are both in $[a, b]$, and denote $f(x_1 - \epsilon)$ and $f(x_1 + \epsilon)$ by c and d, respectively (Fig. 10-2).

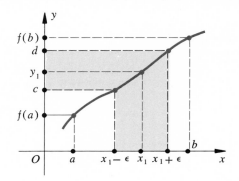

Figure 10-2

Since $a \leq x_1 - \epsilon < x_1 < x + \epsilon \leq b$, we have

$$f(a) \leq c < y_1 < d \leq f(b)$$

because f is increasing on $[a, b]$. Denote by δ the smaller of the two positive numbers $y_1 - c$ and $d - y_1$. Then $\delta \leq y_1 - c$ and $\delta \leq d - y_1$, from which $c \leq y_1 - \delta$ and $y_1 + \delta \leq d$. Thus

$$c \leq y_1 - \delta < y < y_1 + \delta \implies x_1 - \epsilon < f^{-1}(y) < x_1 + \epsilon$$

for all numbers y in the open interval $(y_1 - \delta, y_1 + \delta)$, or

$$y \in (y_1 - \delta, y_1 + \delta) \implies x_1 - \epsilon < f^{-1}(y) < x_1 + \epsilon.$$

But $x_1 = f^{-1}(y_1)$; so this can be written

(1) $$y \in (y_1 - \delta, y_1 + \delta) \implies f^{-1}(y_1) - \epsilon < f^{-1}(y) < f^{-1}(y_1) + \epsilon.$$

We have shown that for each $\epsilon > 0$ there is a positive number δ such that (1) holds. Therefore

$$\lim_{y \to y_1} f^{-1}(y) = f^{-1}(y_1),$$

and f^{-1} is continuous at y_1. Since y_1 was any number in $(f(a), f(b))$, f^{-1} is continuous on the open interval $(f(a), f(b))$.

That f^{-1} is also continuous at the endpoints $f(a)$ and $f(b)$ can be proved similarly, using one-sided limits. ∎

An inverse function theorem for *decreasing* functions can be proved in the same manner.

10.2.3 Theorem. If f is continuous and decreasing on $[a, b]$, then f has an inverse f^{-1} that is continuous and decreasing on $[f(a), f(b)]$.

A function is said to be **strictly monotonic** on an interval I if it is either increasing throughout I or decreasing throughout I.

The following important theorem is an immediate consequence of the inverse function theorems and theorem 5.11.1.

10.2.4 Theorem. Let a function f be continuous and strictly monotonic on $[a, b]$, and let $y = f(x)$. If dy/dx exists and is nonzero at a point x in $[a, b]$, then the inverse function $x = f^{-1}(y)$ is differentiable at the corresponding point y in $[f(a), f(b)]$ and the value of its derivative dx/dy at y is

$$\frac{dx}{dy} = \frac{1}{\dfrac{dy}{dx}}.$$

It is customary to use the horizontal axis for the independent variable in graphing a function. Since the choice of a letter to represent a variable is arbitrary, we shall write $y = f^{-1}(x)$ instead of $x = f^{-1}(y)$. Now x is the independent variable for both f and its inverse f^{-1}. This provides an easy way to draw the graph of f^{-1} from the graph of f, even when it is difficult or impossible to solve the equation $y = f(x)$ for x in terms of y. We simply *reflect the graph of $y = f(x)$ with respect to the line $y = x$ to obtain the graph of $y = f^{-1}(x)$* (Fig. 2-54).

10.3 THE EXPONENTIAL FUNCTION

We saw, in 10.1.7, that for each real number x, e^x has one and only one value. Thus e^x defines a function having domain $(-\infty, \infty)$.

10.3.1 Definition. The function defined by

$$y = e^x, \qquad x \in R,$$

is called the **exponential function**. Its domain is $(-\infty, \infty)$.

By 10.1.7, $\ln e^x = x$ for all real numbers x. If $y = e^x$, then $\ln y = \ln e^x = x$. Thus

(1) $$y = e^x \implies x = \ln y.$$

Moreover, if $x = \ln y$, then $\ln e^x = \ln y$ because $x = \ln e^x$ (10.1.7). Since the logarithm function, ln, is one-to-one, it follows from $\ln e^x = \ln y$ that $e^x = y$. So

(2) $\qquad\qquad\qquad\qquad x = \ln y \implies y = e^x$

From (1) and (2) we have

10.3.2 $\qquad\qquad\qquad\qquad y = e^x \qquad$ if and only if $\qquad x = \ln y.$

Therefore (10.2.1) *the exponential function is the inverse of the logarithm function.*

It follows that the **range** of the exponential function is $(0, \infty)$ because the domain of its inverse, the logarithm function, is $(0, \infty)$. Also, the graph of the exponential function, $y = e^x$, can be obtained by reflecting the graph of $y = \ln x$ about the line $y = x$ (Fig. 10-3).

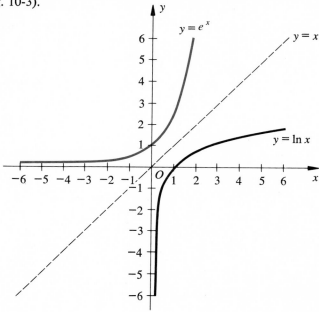

Figure 10-3

Since the logarithm function is one-to-one and its inverse is the exponential function, the exponential function is one-to-one and its inverse is the logarithm function (10.2.1). Moreover,

(3)
$$\ln (e^x) = x \qquad \text{for} \quad x \in R,$$
$$e^{(\ln x)} = x \qquad \text{for} \quad x > 0.$$

The exponential function has the following familiar properties.

10.3.3 **Theorem.** Let a and b be any real numbers. Then
  (i) $e^0 = 1$;
  (ii) $e^a e^b = e^{a+b}$;
  (iii) $(e^a)^b = e^{ab}$.

Proof. (i)   Since $\ln 1 = 0$, $e^0 = e^{(\ln 1)} = 1$, by (3).

(ii)   Let $\alpha = e^a$ and $\beta = e^b$. Then $\ln \alpha = a$ and $\ln \beta = b$, by 10.3.2, and

$$a + b = \ln \alpha + \ln \beta = \ln (\alpha\beta).$$

From $a + b = \ln (\alpha\beta)$ we have $\alpha\beta = e^{a+b}$ (by 10.3.2). But $\alpha\beta = e^a e^b$, Therefore

$$e^a e^b = e^{a+b}.$$

(iii)   To prove property (iii):

$$(e^a)^b = e^{\ln [(e^a)^b]} \qquad \text{[by (3)]}$$
$$= e^{b \ln (e^a)}$$
$$= e^{ba} \qquad \text{[by (3)].} \quad \blacksquare$$

It is a surprising and remarkable fact that the exponential function is its own derivative.

The **derivative of the exponential function** is given by

10.3.4 $$D_x e^x = e^x, \qquad x \in R.$$

For if we let $y = e^x$, then $x = \ln y$ and $dx/dy = 1/y = 1/e^x$. But, by 10.2.4, $dy/dx = 1/(dx/dy)$. Therefore $dy/dx = 1(1/e^x) = e^x$. That is, $D_x e^x = e^x$.

It follows that $D_x(Ce^x) = Ce^x$, where C is any constant (including zero). It is important to note that the function defined by $f(x) = Ce^x$ is the only function that is identical with its derivative, indeed, identical with each of its higher derivatives.

A more general differentiation formula than 10.3.4 is obtained by using the chain rule.

10.3.5   **Theorem.**   If u is a differentiable function of x, then

$$D_x e^u = e^u D_x u.$$

10.3.6   **Corollary.**   If u is a differentiable function of x, the **antiderivatives** with respect to x of $e^u D_x u$ are given by

$$A_x[e^u D_x u] = e^u + C.$$

Example 1.   Find $D_x(e^{\sqrt{x}})$.

Solution.   Using 10.3.5 with $u = \sqrt{x}$, we obtain

$$D_x(e^{\sqrt{x}}) = (e^{\sqrt{x}}) \frac{1}{2\sqrt{x}} = \frac{e^{\sqrt{x}}}{2\sqrt{x}}.$$

Example 2.   If $y = e^{x^2 \ln x}$, find dy/dx.

Solution.   From 10.3.5, in which $u = x^2 \ln x$, we get

$$\frac{dy}{dx} = e^{x^2 \ln x} \frac{d}{dx}(x^2 \ln x)$$
$$= e^{x^2 \ln x}\left(x^2 \cdot \frac{1}{x} + 2x \ln x\right)$$
$$= xe^{x^2 \ln x}(1 + \ln x^2).$$

Example 3. Let $f(x) = xe^{x/2}$, $x \in R$. Find the intervals on which f is increasing and the intervals on which f is decreasing. Find where the graph of f is concave upward and where it is concave downward. Find any extreme values and points of inflection. Then sketch the graph of f.

Solution. From $f(x) = xe^{x/2}$, we find

$$f'(x) = \frac{xe^{x/2}}{2} + e^{x/2} = e^{x/2}\left(\frac{x+2}{2}\right)$$

and

$$f''(x) = \frac{xe^{x/2}}{4} + \frac{e^{x/2}}{2} + \frac{e^{x/2}}{2} = e^{x/2}\left(\frac{x+4}{4}\right).$$

Keeping in mind that $e^{x/2} > 0$ for all numbers x, it is easy to see that $f'(x) < 0$ for $x < -2$, $f'(-2) = 0$, and $f'(x) > 0$ for $x > -2$. Thus f is decreasing on $(-\infty, -2)$, increasing on $(-2, \infty)$, and has its minimum value $f(-2) = -2/e \doteq -0.7$ at $x = -2$.

Also, $f''(x) < 0$ for $x < -4$, $f''(-4) = 0$, and $f''(x) > 0$ for $x > -4$; so the graph of f is concave downward on $(-\infty, -4)$, concave upward on $(-4, \infty)$, and has a point of inflection at $(-4, -4e^{-2}) \doteq (-4, -0.54)$. (To make this approximation of $-4e^{-2}$, we used table A.8.6 in the Appendix.)

Clearly, the curve $y = xe^{x/2}$ intersects the x axis only at the origin; it is below the negative x axis and above the positive x axis.

Having all this information, we need calculate the approximate coordinates of only a few points in order to sketch the graph (Fig. 10-4).

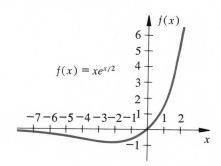

Figure 10-4

Example 4. Find the antiderivatives with respect to x of e^{-4x}.

Solution. In formula 10.3.6, let $u = -4x$. Then $D_x(-4x) = -4$, and

$$A_x[e^{-4x}] = -\frac{1}{4}A_x[e^{-4x}(-4)]$$

$$= -\frac{1}{4}A_x[e^{-4x}D_x(-4x)]$$

$$= -\frac{1}{4}e^{-4x} + C.$$

Example 5. If

$$f(x) = \frac{x^2}{e^{x^3}},$$

find the antiderivatives with respect to x of f.

Solution.   We can write $x^2/(e^{x^3}) = x^2 e^{-x^3}$. Let $u = -x^3$; then $D_x(-x)^2 = -3x^2$. Thus

$$A_x[x^2 e^{-x^3}] = -\frac{1}{3} A_x[e^{-x^3}(-3x^2)]$$

$$= -\frac{1}{3} A_x[e^{-x^3} D_x(-x^3)]$$

$$= -\frac{1}{3} e^{-x^3} + C$$

$$= \frac{-1}{3e^{x^3}} + C.$$

Example 6.   Find the value of $\displaystyle\int_1^3 xe^{-3x^2}\, dx$.

Solution.   In order to apply the fundamental theorem, we must find an antiderivative of xe^{-3x^2}. Let $u = -3x^2$ in 10.3.6. We have $D_x(-3x^2) = -6x$, and

$$A_x(xe^{-3x^2}) = -\frac{1}{6} A_x[e^{-3x^2}(-6x)]$$

$$= -\frac{1}{6} A_x[e^{-3x^2} D_x(-3x^2)]$$

$$= -\frac{1}{6} e^{-3x^2} + C.$$

Then

$$\int_1^3 xe^{-3x^2}\, dx = -\frac{1}{6} \int_1^3 e^{-3x^2}(-6x)\, dx$$

$$= -\frac{1}{6} \int_1^3 e^{-3x^2} D_x(-3x^2)\, dx$$

$$= -\frac{1}{6} \Big[e^{-3x^2}\Big]_1^3$$

$$= -\frac{1}{6}(e^{-27} - e^{-3}) = \frac{e^{-3} - e^{-27}}{6}$$

$$\doteq 0.000\ 000\ 747.$$

Exercises

In each of Exercises 1 through 16, differentiate the given expression.

1. $y = e^{2x+1}$.

2. $y = e^{3x^2-x}$.

3. $y = e^{\sqrt{x+1}}$.

4. $y = e^{1/x^3}$.

5. $y = \frac{1}{2}(e^x + e^{-x})$.

6. $y = \dfrac{e^x - e^{-x}}{e^x + e^{-x}}$

7. $y = e^{\ln x}$.

8. $y = e^{(\ln x)/x}$.

9. $y = x^2 e^x$.

10. $y = e^{x^2} \ln x$.

11. $y = \sqrt{x-1}\, e^{\sqrt{x-1}}$.

12. $y = \dfrac{e^{\sqrt{x+2}}}{\sqrt{x+2}}$.

13. $y = x^3 e^{x^3}$.

14. $y = \dfrac{\ln x}{e^x}$

15. $y = \dfrac{e^x}{\ln x}$.

16. $y = \dfrac{2}{e^x + e^{-x}}$.

In each of Exercises 17 to 24, find the antiderivatives with respect to x of the indicated functions.

17. $f(x) = e^{3x+1}$.

18. $f(x) = xe^{x^2-3}$.

19. $F(x) = \dfrac{e^{\ln 2x}}{x}$.

20. $F(x) = (x + 3)e^{x^2+6x}$.

21. $g(x) = \dfrac{e^x}{e^x - 1}$.

22. $g(x) = \dfrac{1}{x^2 e^{1/x}}$.

23. $\phi(x) = e^{x+e^x}$.

24. $\phi(x) = \dfrac{5e^{\sqrt{x+5}}}{\sqrt{x+5}}$.

25. We saw that the most general function f for which $D_t f(t) = f(t)$ is given by $f(t) = Ce^t$, where C is any constant. Find the most general function F for which $D_t F(t) = k \cdot F(t)$, where k is a factor of proportionality.

26. Assuming that a *population* (the number of people in a country, the number of bacteria in a colony, etc.) increases at a rate that is proportional to the number of individuals present, show that the population y in t years is given by

$$y = Ae^{kt},$$

where k is a factor of proportionality and A is the population when $t = 0$. (*Hint:* See Exercise 25.)

27. Determine the value of the constant k in $y = Ae^{kt}$ for the population to increase 4% in one year. (*Hint:* What would the value of y be when $t = 1$?)

28. In how many years would a population double if it increased at a constant rate of 4% a year?

29. The **half-life** of a radioactive substance is the time it takes a given amount of the substance to lose one-half of its mass. If the half-life of radium is 1690 years, how long does it take to lose $\frac{1}{3}$ of its mass? (*Hint:* The **law of decay** for radioactive substances is similar to the law of growth of populations except that k turns out to be negative.)

30. A substance in a chemical reaction is used up at a rate that is proportional to the amount of the substance present at any time. If $\frac{3}{4}$ of the substance is used up in 2 hours, how much of the substance is left at the end of 1 hour and 30 minutes?

31. Prove that for all positive numbers x

$$e^x > 1 + x.$$

[*Hint:* Consider $f(x) = e^x - x - 1$, $x > 0$.]

10.4 EXPONENTIAL AND LOGARITHMIC FUNCTIONS WITH BASES OTHER THAN e

Let a be any positive number and let

(1) $$b = \ln a.$$

Then

(2) $$a = e^b$$

by 10.3.2. Eliminating b between (1) and (2) gives

(3) $$a = e^{\ln a}.$$

If we raise both members of (3) to the power x, we obtain

(4) $$a^x = (e^{\ln a})^x = e^{x \ln a}.$$

Although a^x was defined only for *rational* exponents x in elementary algebra, we defined $e^{x \ln a}$ (in 10.1.7) for all real numbers x, rational and *irrational*. Equation (4) suggests that we define a^x when x is irrational (or rational) as being equal to $e^{x \ln a}$.

10.4.1 Definition. If a is any positive number and x is any real number, then

$$a^x = e^{x \ln a}.$$

We are now prepared to define the exponential function to base a.

10.4.2 Definition. If a is a positive number and x is any real number, the function f defined by

$$f(x) = a^x = e^{x \ln a}$$

is called the **exponential function to base** a.

The familiar properties of exponentials are now easy to prove.

10.4.3 Theorem. If a and b are positive numbers and x and y are any real numbers, then

(i) $a^0 = 1$; (ii) $a^1 = a$;

(iii) $a^x a^y = a^{x+y}$; (iv) $a^{-x} = \dfrac{1}{a^x}$;

(v) $\dfrac{a^x}{a^y} = a^{x-y}$; (vi) $(a^x)^y = a^{xy}$;

(vii) $a^x b^x = (ab)^x$.

To prove (iii) and (vi), we use 10.4.2 and 10.3.3, obtaining

$$a^x a^y = e^{x \ln a} e^{y \ln a} = e^{x \ln a + y \ln a} = e^{(x+y)(\ln a)} = a^{x+y},$$

which is (iii), and

$$(a^x)^y = (e^{x \ln a})^y = e^{xy \,(\ln a)} = a^{xy},$$

which is (vi). The proofs of the remaining parts of 10.4.3 are left for the reader.

A formula for the derivative of a^x is found by using the chain rule to differentiate both members of $a^x = e^{x \ln a}$ (10.4.1) with respect to x:

$$D_x a^x = D_x e^{x \ln a} = e^{x \ln a} D_x(x \ln a) = e^{x \ln a} \ln a = a^x \ln a.$$

That is,

(5) $$D_x a^x = a^x \ln a.$$

From (5) and the chain rule, we have the more general formula

10.4.4 Theorem. If a is a positive number and u is a differentiable function of x, then

$$D_x a^u = a^u \ln a \cdot D_x u.$$

Of course, each differentiation formula gives rise to an antidifferentiation formula.

10.4.5 Corollary. If a is a positive number and u is a differentiable function of x, the antiderivatives of $a^u D_x u$ are given by

$$A_x[a^u D_x u] = \frac{a^u}{\ln a} + C.$$

The inverse of the exponential function to base a is called the **logarithmic function to base** a.

10.4.6 Definition. If a is a positive number different from 1, the **logarithmic function to base** a is the inverse of the exponential function to base a; its rule of correspondence is

$$y = \log_a x \qquad \text{if and only if} \qquad a^y = x.$$

The most commonly used base in computations is $a = 10$, and logarithms to base 10 are known as **common logarithms** (or Briggsian logarithms). **Natural logarithms** (to base $e \doteq 2.7182818 \cdots$) are also called Napierian logarithms.

Logarithms to base a obey the same laws (10.1.5) as natural logarithms. The reader is asked to prove this statement in Exercise 25.

Let $y = \log_a x$. Then $a^y = x$ and $\ln (a^y) = \ln x$, or $y \ln a = \ln x$. Thus $y = \ln x/\ln a$; and since $y = \log_a x$, we have

10.4.7
$$\log_a x = \frac{\ln x}{\ln a}.$$

Differentiating both members of 10.4.7 gives

(6)
$$D_x \log_a x = \frac{1}{\ln a} D_x \ln x = \frac{1}{x \ln a}.$$

If we let $x = e$ in 10.4.7, we have

$$\log_a e = \frac{\ln e}{\ln a} = \frac{1}{\ln a}.$$

By substituting this in (6), we obtain

(7)
$$D_x \log_a x = \frac{\log_a e}{x}.$$

10.4.8 Theorem. If a is a positive number different from 1, and u is a positive-valued, differentiable function of x, then

$$D_x \log_a u = \frac{\log_a e}{u} D_x u.$$

Before leaving exponentials, one loose end remains to be tied up. In 5.4.2 we proved that

$$D_x u^n = nu^{n-1} D_x u$$

when n was any *rational* number and u was a differentiable function of x (for values of x that made u^{n-1} real). We promised to extend this formula to include *irrational* values of n, provided that u is positive. We shall now do so.

10.4.9   Theorem. If u is a positive-valued differentiable function of x and n is any real constant, then

$$D_x u^n = n u^{n-1} D_x u.$$

Proof.   Let $y = u^n$. By 10.4.1, for each (positive) value of u, $u^n = e^{n \ln u}$, or

$$(8) \qquad\qquad y = e^{n \ln u}.$$

If we apply the natural logarithm function to both members of (8), we get

$$(9) \qquad\qquad \ln y = n \ln u.$$

By differentiating both sides of (9) with respect to x, we obtain

$$\frac{1}{y} D_x y = \frac{n}{u} D_x u,$$

or

$$D_x y = \frac{ny}{u} D_x u = \frac{n u^n}{u} D_x u = n u^{n-1} D_x u. \quad \blacksquare$$

Example 1.   Differentiate $y = 7^{(x^3 - 4)}$.

Solution.   If we let $a = 7$ and $u = x^3 - 4$ in 10.4.4, we have

$$D_x y = 7^{(x^3 - 4)} (\ln 7) D_x (x^3 - 4)$$
$$= 7^{(x^3 - 4)} (\ln 7)(3x^2) = (3 \ln 7) x^2 7^{(x^3 - 4)}.$$

Example 2.   Differentiate $y = \log_{10} (x^4 + 13)$.

Solution.   Using 10.4.8, we obtain

$$D_x y = \frac{\log_{10} e}{x^4 + 13} \cdot (4x^3) = \frac{4x^3 \log_{10} e}{x^4 + 13}.$$

Example 3.   Differentiate $y = (x^5 - 1)^\pi,\ x > 1$.

Solution.   By 10.4.9,

$$\frac{dy}{dx} = \pi(x^5 - 1)^{\pi - 1}(5x^4) = 5\pi x^4 (x^5 - 1)^{\pi - 1}.$$

Example 4.   Find the antiderivatives of $f(x) = \dfrac{3^{\sqrt{x}}}{\sqrt{x}}$.

Solution.   Let $a = 3$ and $u = \sqrt{x}$ in formula 10.4.5. Then $D_x u = 1/(2\sqrt{x})$, and

$$A_x \left(\frac{3^{\sqrt{x}}}{\sqrt{x}} \right) = 2 A_x \left(3^{\sqrt{x}} \cdot \frac{1}{2\sqrt{x}} \right)$$
$$= 2 A_x (3^{\sqrt{x}} D_x \sqrt{x}) = \frac{2(3^{\sqrt{x}})}{\ln 3} + C.$$

Example 5. Find the antiderivatives of $f(x) = 10^{x \ln x}(2 + \ln x^2)$.

Solution. In 10.4.5 let $a = 10$ and $u = x \ln x$; then $D_x u = 1 + \ln x$. Since $2 + \ln x^2 = 2 + 2 \ln x = 2(1 + \ln x)$,

$$A_x[10^{x \ln x}(2 + \ln x^2)] = 2A_x[10^{x \ln x}(1 + \ln x)]$$
$$= \frac{2[10^{x \ln x}]}{\ln 10} + C.$$

Example 6. Evaluate $\int_0^3 2^x \, dx$.

Solution. An antiderivative of the integrand is $2^x/\ln 2$ by 10.4.5. Thus

$$\int_0^3 2^x \, dx = \left[\frac{2^x}{\ln 2}\right]_0^3 = \frac{2^3 - 2^0}{\ln 2} = \frac{7}{\ln 2}.$$

Example 7. Evaluate $\int_{1/2}^1 x^{-2} 5^{1/x} \, dx$.

Solution. To find an antiderivative of the integrand, $x^{-2} 5^{1/x}$, let $a = 5$ and $u = 1/x$ in 10.4.5. Then $D_x u = -1/x^2$; and we can write $x^{-2} 5^{1/x} = -5^{1/x}(-1/x^2) = -5^{1/x} D_x(1/x)$. Thus

$$A_x[x^{-2} 5^{1/2}] = -A_x\left[5^{1/x} D_x \frac{1}{x}\right]$$
$$= -\frac{5^{1/x}}{\ln 5} + C.$$

Therefore, by the fundamental theorem,

$$\int_{1/2}^1 x^{-2} 5^{1/x} \, dx = -\int_{1/2}^1 5^{1/x} \frac{-1}{x^2} \, dx$$
$$= -\int_{1/2}^1 5^{1/2} D_x \frac{1}{x} \, dx$$
$$= -\left[\frac{5^{1/2}}{\ln 5}\right]_{1/2}^1 = -\frac{1}{\ln 5}(5 - 5^2)$$
$$= \frac{20}{\ln 5}.$$

Exercises

In Exercises 1 through 10, differentiate the indicated functions.

1. $y = 5^{x^3}$.

2. $y = 3^{2x^4 - 4x}$.

3. $y = 7^{\sqrt{x-1}}$.

4. $y = \log_{10} e^x$.

5. $y = x \, 2^x$.

6. $y = 2^x \ln (x + 5)$.

7. $y = \sqrt{\log_{10} x}$.

8. $y = \log_3 (x^3 + 3)$.

9. $y = \log_2 \frac{(x^3 - 27)}{(x^3 + 27)}$.

10. $y = \log_{10} (\log_{10} x)$.

11. Let $y = u^v$, where u and v are differentiable functions of x and u is positive valued. By taking the

natural logarithm of both members, show that

$$D_x u^v = v u^{v-1} D_x u + u^v \ln u \, D_x v.$$

12. Find $\dfrac{dy}{dx}$ if $y = x^{(x^z)}$.

In each of Exercises 13 to 20, find the antiderivatives of f.

13. $f(x) = x 2^{x^2}$.

14. $f(x) = 10^{5x-1}$.

15. $f(x) = \dfrac{e^{\sqrt{x-1}}}{\sqrt{x-1}}$

16. $f(x) = \dfrac{10^{\sqrt{2x+1}}}{\sqrt{2x+1}}$.

17. $f(x) = 8^{(x^2+2x)}(x+1)$.

18. $f(x) = (\ln x + 1)10^{x \ln x}$.

19. $f(x) = \dfrac{5^{(\ln x)/x}(1 - \ln x)}{x^2}$

20. $f(x) = 3^{1/(2x)} x^{-2}$.

21. Find the value of $\displaystyle\int_1^2 10^{2x-1} \, dx$.

22. Find the value of $\displaystyle\int_1^4 \dfrac{5^{\sqrt{x}}}{\sqrt{x}} \, dx$.

23. Evaluate $\displaystyle\int_0^2 x \, 3^{-x^2} \, dx$.

24. Evaluate $\displaystyle\int_0^1 (10^{3x} + 10^{-3x}) \, dx$.

25. Prove the theorem. If a, x, and y are positive numbers, $a \neq 1$, and if n is any real number, then
   (i) $\log_a 1 = 0$;
   (ii) $\log_a xy = \log_a x + \log_a y$;
   (iii) $\log_a \dfrac{x}{y} = \log_a x - \log_a y$;
   (iv) $\log_a x^n = n \log_a x$.

10.5 TRIGONOMETRIC FUNCTIONS

We assume that the reader has studied trigonometry and is familiar with the definitions of the trigonometric functions based on angle measure, and the principal trigonometric identities. In calculus, however, we study trigonometric functions of numbers—that is, trigonometric functions whose domains are sets of real numbers. Let us start by defining these functions.

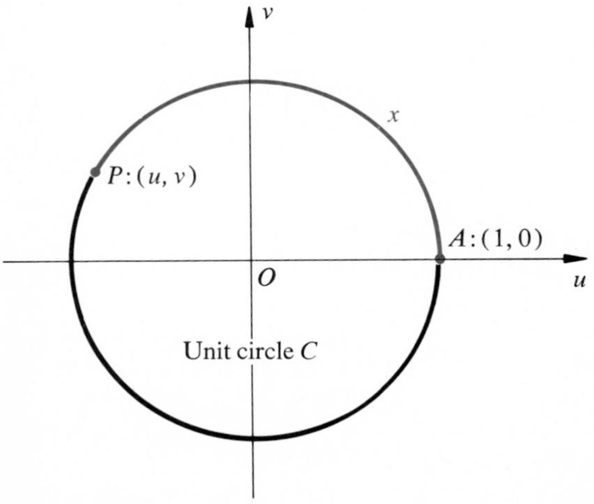

Consider a unit circle C with its center at the origin of a uv coordinate system (Fig. 10-5). Its equation is $u^2 + v^2 = 1$. The graph of $v = \sqrt{1 - u^2}$, $-1 < u < 1$, is the upper semicircle between the points $(1, 0)$ and $(-1, 0)$. Since its derivative,

$$\frac{dv}{du} = \frac{-u}{\sqrt{1 - u^2}}, \qquad -1 < u < 1,$$

is continuous, any arc of C smaller than a semicircle is a smooth curve and has length (9.6.1). Any larger arc of a circle can be thought of as the union of smaller arcs, and its length is the sum of the lengths of the smaller arcs.

Denote by A the point $(1, 0)$ on the unit circle C, and let x be an arbitrarily chosen positive number (Fig. 10-5). Then there is exactly one point $P{:}(u, v)$ on C such that the length of the arc $\overset{\frown}{AP}$, measured

Figure 10-5

counterclockwise from A along the unit circle, is x. The circumference of C is 2π; so if $x > 2\pi$, it will take more than a complete circuit of the unit circle to trace the arc $\overparen{AP}$. If $x = 0$, $P = A$.

Similarly if $x < 0$, there is exactly one point $P{:}(u, v)$ on the unit circle C such that the length of the arc $\overparen{AP}$, measured *clockwise* on C, is $|x|$.

This establishes a correspondence between the real numbers and the points of a unit circle C. To each real number x there corresponds one and only one point P of C. However, each point of C is the correspondent of indefinitely many real numbers.

For if P is a point of C, any number of complete circuits of C that start at P will finish at P. Since the circumference of C is 2π, the unique point P on C that corresponds to a given number x also corresponds to the numbers $x + 2\pi$, $x + 4\pi$, $x - 2\pi$, $\cdots$. In fact, *when a point $P{:}(u, v)$ on C corresponds to a real number x, the same point P corresponds to all the numbers $x + n \cdot 2\pi$, where n is any integer.*

We now use this correspondence to define the trigonometric functions sine, cosine, tangent, cotangent, secant, and cosecant, abbreviated sin, cos, tan, cot, sec, and csc, respectively.

10.5.1 Definition. Let x be a real number and let $P{:}(u, v)$ be the unique point of a unit circle C that corresponds to x, as described above. Then (Fig. 10-6)

$$\sin x = v, \qquad\qquad \csc x = \frac{1}{v}, \qquad v \neq 0,$$

$$\cos x = u, \qquad\qquad \sec x = \frac{1}{u}, \qquad u \neq 0,$$

$$\tan x = \frac{v}{u}, \qquad u \neq 0, \qquad \cot x = \frac{u}{v}, \qquad v \neq 0.$$

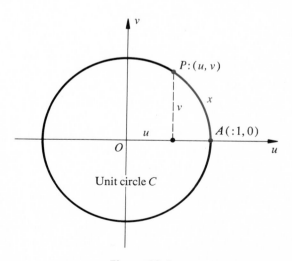

Figure 10-6

It follows from this definition that

$$\tan x = \frac{\sin x}{\cos x}, \qquad \csc x = \frac{1}{\sin x}, \qquad \sec x = \frac{1}{\cos x},$$

$$\cot x = \frac{1}{\tan x} = \frac{\cos x}{\sin x}$$

whenever the denominator is not zero.

We saw, above, that the point $P:(u, v)$ of the unit circle C that corresponds to the number x also corresponds to each of the numbers $x + n \cdot 2\pi$, where n is any integer. Thus

$$\sin x = v = \sin (x + n \cdot 2\pi)$$

$$\cos x = u = \cos (x + n \cdot 2\pi),$$

$n = \pm 1, \pm 2, \pm 3, \cdots$, and similarly for all the other trigonometric functions. The values of the trigonometric functions repeat in successive intervals of 2π. Because of this, the trigonometric functions are said to be *periodic*.

A function f is **periodic** if there is a positive number p such that $f(x + p) = f(x)$ for every x in the domain of f. The smallest such positive number p is called the **period** of the function.

The sine, cosine, secant, and cosecant functions have the period 2π. The tangent and cotangent have the period π, as we shall see.

It is this property of periodicity that makes the trigonometric functions so important in the study of recurring natural (and manmade) phenomena, such as vibrations, planetary motions, sound waves, alternating current, tides, and breathing.

The relation of trigonometric functions of real numbers to the familiar trigonometric functions of angle measure is simple. In the unit circle C, draw the line segment OP, and denote the central angle $\angle AOP$ by θ (Fig. 10-7).

We recall that the **radian measure** of a central angle θ of any circle is defined to be

$$\theta = \frac{s}{r},$$

where s is the length of the subtended arc, $0 \leq s < 2\pi$, and r is the radius of the circle.

If the circle is a unit circle, $r = 1$ and $s = x$ (Fig. 10-8), and

$$\theta = \frac{s}{r} = \frac{x}{1} = x.$$

So, in a unit circle the radian measure of a central angle is equal to the measure of its arc.

From the familiar angle definitions of sine and cosine,

$$\sin \theta = \frac{v}{1} = v, \qquad \cos \theta = \frac{u}{1} = u$$

(Fig. 10-7). But $\sin x = v$ and $\cos x = u$, where x is a real number (10.5.1). Consequently, *if x is any real number, $\sin x$ and $\cos x$ are equal to the sine and cosine of an*

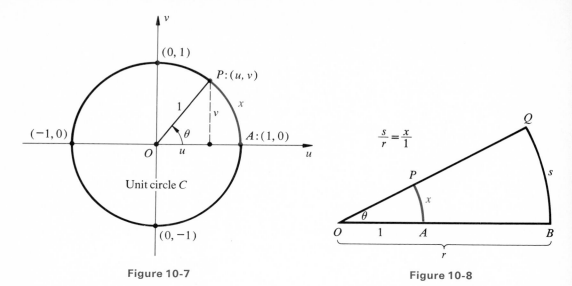

Figure 10-7   Figure 10-8

angle θ whose radian measure is x. The same statement holds for the other trigonometric functions.

Because of this relation between the trigonometric functions of a number x and the trigonometric functions of an angle θ, all the identities that were established for functions of angles in trigonometry are also valid for the trigonometric functions of numbers. The more important of these identities are listed at the end of this section.

The reader is probably more accustomed to the degree measure of angles than to radian measure, but the latter is used in calculus. Since the circumference of a unit circle is 2π, the radian measure of a straight angle is π. Thus π radians equals 180°. If d is the degree measure of a given angle and θ is the radian measure of the same angle, then

(1) $$\frac{d}{180°} = \frac{\theta}{\pi}.$$

This is the basic formula for converting degrees to radians, and vice versa. From it we find that

$$1 \text{ radian} \doteq 57.3°,$$

although we shall not have occasion to use this fact in calculations.

Example. (a) Express 165° in radians, and (b) express $\frac{3}{5}\pi$ radians in degrees.

Solution. (a) By substituting $d = 165$ in (1), we find

$$\frac{165}{180} = \frac{\theta}{\pi},$$

or $\theta = \frac{11}{12}\pi$. Thus, an angle of 165° is equal to an angle of $\frac{11}{12}\pi$ radians.
   (b) If $\theta = \frac{3}{5}\pi$ in (1), then

$$\frac{d}{180} = \frac{3\pi}{5} \cdot \frac{1}{\pi},$$

or $d = 108$. Therefore, an angle of $\frac{3}{5}\pi$ radians is equal to an angle of 108°.

The radian and degree measures of some often used angles are given in the following table.

Radians	$\frac{\pi}{6}$	$\frac{\pi}{4}$	$\frac{\pi}{3}$	$\frac{\pi}{2}$	$\frac{2\pi}{3}$	$\frac{3\pi}{4}$	$\frac{5\pi}{6}$	π
Degrees	30	45	60	90	120	135	150	180

It is easy to read some values of the trigonometric functions from Fig. 10-7 and from the triangles in Fig. 10-9. Notice that the signs of these values are governed by the signs of u and v in the four quadrants of the unit circle.

x	$\sin x = v$	$\cos x = u$	$\tan x = \dfrac{u}{v}$
0	0	1	0
$\frac{\pi}{6}$	$\frac{1}{2}$	$\frac{\sqrt{3}}{2}$	$\frac{\sqrt{3}}{3}$
$\frac{\pi}{4}$	$\frac{\sqrt{2}}{2}$	$\frac{\sqrt{2}}{2}$	1
$\frac{\pi}{3}$	$\frac{\sqrt{3}}{2}$	$\frac{1}{2}$	$\sqrt{3}$
$\frac{\pi}{2}$	1	0	Not defined
$\frac{2\pi}{3}$	$\frac{\sqrt{3}}{2}$	$-\frac{1}{2}$	$-\sqrt{3}$
$\frac{3\pi}{4}$	$\frac{\sqrt{2}}{2}$	$\frac{-\sqrt{2}}{2}$	-1
$\frac{5\pi}{6}$	$\frac{1}{2}$	$\frac{-\sqrt{3}}{2}$	$\frac{-\sqrt{3}}{3}$
π	0	-1	0

Instead of memorizing these values, most people make quick sketches like the triangles in Fig. 10-9, and remember that the ratios of the lengths of their sides are $1:2:\sqrt{3}$ and $1:1:\sqrt{2}$.

Graphs of the sine, cosine, and tangent functions are shown in Fig. 10-10. Notice that the sine and cosine have the period 2π, but that the tangent function has the period π. More information about the graphs of these functions will be obtained from their derivatives (Section 10.7).

The more important of the trigonometric identities are listed here for reference.

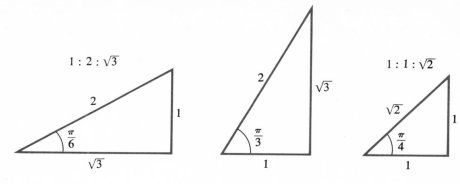

$$1 : 2 : \sqrt{3} \qquad\qquad 1 : 1 : \sqrt{2}$$

Figure 10-9

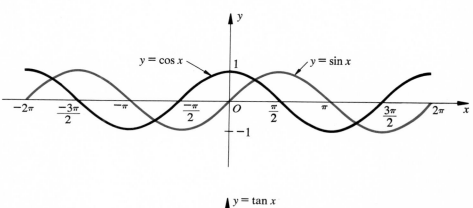

$y = \cos x \qquad y = \sin x$

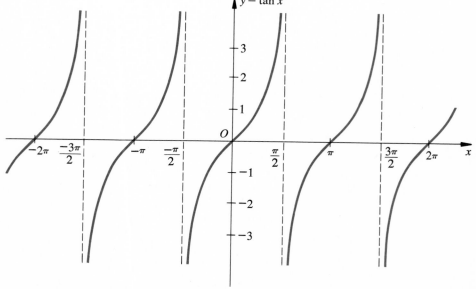

$y = \tan x$

Figure 10-10

10.5.2   In each of the following *trigonometric identities*, x and y are any real numbers for which all the functions involved are defined.

Reduction formulas.

$$\sin(-x) = -\sin x. \qquad \sin\left(\frac{\pi}{2} \pm x\right) = \cos x.$$

$$\cos(-x) = \cos x. \qquad \cos\left(\frac{\pi}{2} \pm x\right) = \mp\sin x.$$

$$\tan(-x) = -\tan x. \qquad \tan\left(\frac{\pi}{2} \pm x\right) = \mp\cot x.$$

Pythagorean identities.

$$\sin^2 x + \cos^2 x = 1.$$
$$1 + \tan^2 x = \sec^2 x.$$
$$1 + \cot^2 x = \csc^2 x.$$

Addition identities.

$$\sin(x + y) = \sin x \cos y + \cos x \sin y.$$
$$\cos(x + y) = \cos x \cos y - \sin x \sin y.$$
$$\tan(x - y) = \frac{\tan x - \tan y}{1 + \tan x \tan y}.$$

Double-angle identities.

$$\sin 2x = 2 \sin x \cos x.$$
$$\cos 2x = \cos^2 x - \sin^2 x$$
$$= 2 \cos^2 x - 1 = 1 - 2 \sin^2 x.$$

Half-angle identities.

$$\sin^2 \frac{x}{2} = \frac{1 - \cos x}{2}. \qquad \cos^2 \frac{x}{2} = \frac{1 + \cos x}{2}.$$

Sums and products.

$$\sin x + \sin y = 2 \sin\left(\frac{x + y}{2}\right) \cos\left(\frac{x - y}{2}\right).$$

$$\cos x + \cos y = 2 \cos\left(\frac{x + y}{2}\right) \cos\left(\frac{x - y}{2}\right).$$

The reduction formulas are easily read from quick sketches, and this procedure is advised. The other identities, however, should be memorized.

The Pythagorean identities follow from the right triangle in Fig. 10-7. All later identities in 10.5.2 may be obtained from the first two addition identities. Their derivation can be found in any trigonometry book.

Exercises

1. Change to radian measure.
(a) 210°,   (b) −60°,   (c) 315°,
(d) −630°,   (e) 300°,   (f) −540°.

2. Change to degree measure.
(a) $\frac{3\pi}{2}$,   (b) $\frac{-5\pi}{4}$,   (c) $\frac{-11\pi}{6}$,
(d) $\frac{17\pi}{12}$,   (e) $\frac{-3\pi}{4}$,   (f) $\frac{4\pi}{3}$,

3. As in Fig. 10-7, locate the points P on a unit circle that correspond to the numbers x given below, and find the uv coordinates of each point.
(a) $x = \frac{3\pi}{2}$,   (b) $x = \frac{3\pi}{4}$,   (c) $x = \frac{-\pi}{3}$,
(d) $x = -\pi$,   (e) $x = \frac{13\pi}{6}$,   (f) $x = \frac{7\pi}{6}$.

4. Evaluate $\sin x$ and $\cos x$ for each of the numbers x in Exercise 3.

5. At which of the numbers x in Exercise 3 is the tangent function undefined? Find the value of $\tan x$ at each of the others.

6. Find all numbers x for which
(a) $\sin x = 0$,   (b) $\sin x = 1$,
(c) $\sin x = -1$,   (d) $\sin x = \frac{1}{2}$,
(e) $\sin x = -\frac{1}{2}$.

7. Find all numbers x for which
(a) $\cos x = 0$,   (b) $\cos x = 1$,
(c) $\cos x = -1$,   (d) $\cos x = \frac{1}{2}$,
(e) $\cos x = -\frac{1}{2}$.

8. Find all numbers x for which
(a) $\tan x$ is undefined,   (b) $\tan x = 0$,
(c) $\tan x = 1$,   (d) $\tan x = -1$.

9. Use the addition identities and reduction identities to show that $\sin(\pi - x) = \sin x$ and $\cos(\pi - x) = -\cos x$.

10. What are the maximum and minimum values of (a) the sine function, and (b) the cosine function? (*Hint:* See Fig. 10-7.)

10.6 SOME TRIGONOMETRIC LIMITS

In order to find formulas for the derivatives of the trigonometric functions, we must first establish a few trigonometric limits.

Recall from plane geometry that the area of a sector of a circle is $A = \frac{1}{2}r^2\theta$, where r is the radius of the circle and θ is the radian measure of the central angle. If the circle is a unit circle and the length of the subtended arc is x, then the area of the sector is

$$A = \frac{x}{2}.$$

Consider a unit circle C, having the equation $u^2 + v^2 = 1$ (Fig. 10-11). Denote by A the point $(1, 0)$, and let $P:(u, v)$ be a point of C such that the length of the arc $\overset{\frown}{AP}$ is x, $0 < x < \pi/2$. The right triangles OAT and OBP are similar, and $\overline{AT}/1 = v/u$. Thus the coordinates of T are $(1, v/u)$.

The triangle OAP is contained in the circular sector OAP, and the circular sector OAP is contained in the triangle OAT (Fig. 10-11). Consequently (8.1.1),

(1)   Area $\triangle OAP \leq$ area sector $OAP \leq$ area $\triangle OAT$.

Since $u = \cos x$ and $v = \sin x$,

$$\text{Area } \triangle OAP = \frac{1}{2}|OA| \cdot |BP| = \frac{1}{2}(1)v = \frac{1}{2}\sin x,$$

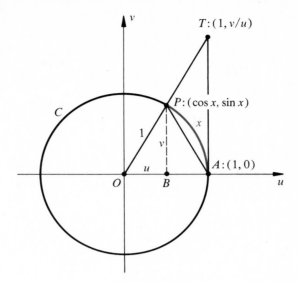

Figure 10-11

$$\text{Area sector } OAP = \frac{1}{2}x,$$

$$\text{Area } \triangle OAT = \frac{1}{2}|OA| \cdot |AT| = \frac{1}{2}\frac{v}{u} = \frac{1}{2}\frac{\sin x}{\cos x}.$$

By substituting this in (1) and multiplying through by 2, we get

$$\sin x \leq x \leq \frac{\sin x}{\cos x}, \qquad 0 < x < \frac{\pi}{2}.$$

But $\sin x = v > 0$ when $0 < x < \pi/2$. Therefore

(2) $$0 < \sin x \leq x \leq \frac{\sin x}{\cos x}, \qquad 0 < x < \frac{\pi}{2}.$$

We shall first use (2) to prove that the sine and cosine functions are continuous at 0. From (2),

(3) $$0 < \sin x \leq x$$

if $0 < x < \pi/2$. Since $\lim\limits_{x \to 0^+} 0 = 0$ and $\lim\limits_{x \to 0^+} x = 0$, it follows from the "squeeze theorem" (3.3.9) that

(4) $$\lim_{x \to 0^+} \sin x = 0.$$

If $-\pi/2 < x < 0$, then $0 < -x < \pi/2$ and, from (3), $0 < \sin(-x) \leq -x$. Since $\sin(-x) = -\sin x$,

$$0 < \sin(-x) \leq -x \iff 0 < -\sin x \leq -x$$

$$\iff x \leq \sin x < 0.$$

But $\lim\limits_{x\to 0^-} x = 0$ and $\lim\limits_{x\to 0^-} 0 = 0$. Thus

$$(5) \qquad \lim_{x\to 0^-} \sin x = 0.$$

From (4) and (5), $\lim\limits_{x\to 0} \sin x = 0$. Since $\sin 0 = 0$, $\lim\limits_{x\to 0} \sin x = \sin 0$, which proves that *the sine function is continuous at* 0.

For $-\pi/2 \le x \le \pi/2$, $\cos x = \sqrt{1 - \sin^2 x}$. Thus

$$\lim_{x\to 0} \cos x = \lim_{x\to 0} \sqrt{1 - \sin^2 x}$$

$$= \sqrt{1 - (\lim_{x\to 0} \sin x)^2}$$

$$= 1 = \cos 0.$$

Therefore *the cosine function is continuous at* 0.

We now use (2) to establish an important limit.

10.6.1   Theorem.   If x is a real number,

$$\lim_{x\to 0} \frac{\sin x}{x} = 1.$$

Proof.   From (2),

$$(6) \qquad 0 < \sin x \le x \le \frac{\sin x}{\cos x}$$

for $0 < x < \pi/2$. If we divide inequalities (6) by $\sin x$, the result is

$$1 \le \frac{x}{\sin x} \le \frac{1}{\cos x}.$$

By taking reciprocals and reversing the direction of these inequalities, we obtain

$$(7) \qquad \cos x \le \frac{\sin x}{x} \le 1, \qquad 0 < x < \frac{\pi}{2}.$$

If $-\pi/2 < x < 0$, then $0 < -x < \pi/2$ and, from (7),

$$\cos(-x) \le \frac{\sin(-x)}{-x} \le 1.$$

But $\cos(-x) = \cos x$, $\sin(-x) = -\sin x$, and $\sin(-x)/(-x) = (\sin x)/x$. Thus

$$\cos x \le \frac{\sin x}{x} \le 1,$$

for $-\pi/2 < x < \pi/2$, $x \ne 0$. Since cosine is continuous at 0, $\lim\limits_{x\to 0} \cos x = 1$, and it follows from the "squeeze" theorem that

$$\lim_{x\to 0} \frac{\sin x}{x} = 1. \quad \blacksquare$$

Another important limit is given in the following theorem.

10.6.2 Theorem. If x is a real number, then

$$\lim_{x \to 0} \frac{1 - \cos x}{x} = 0.$$

Proof. For $-\pi/2 < x < \pi/2$, $x \neq 0$,

$$\frac{1 - \cos x}{x} = \frac{1 - \cos x}{x} \cdot \frac{1 + \cos x}{1 + \cos x} = \frac{1 - \cos^2 x}{x(1 + \cos x)}$$

$$= \frac{\sin^2 x}{x(1 + \cos x)} = \frac{\sin x}{x} \cdot \frac{\sin x}{1 + \cos x}.$$

Thus

$$\lim_{x \to 0} \frac{1 - \cos x}{x} = \lim_{x \to 0} \left[\frac{\sin x}{x} \cdot \frac{\sin x}{1 + \cos x} \right]$$

$$= \left[\lim_{x \to 0} \frac{\sin x}{x} \right] \cdot \frac{\displaystyle\lim_{x \to 0} \sin x}{\displaystyle\lim_{x \to 0} (1 + \cos x)}$$

$$= 1 \cdot \frac{0}{1 + 1} = 0. \quad \blacksquare$$

Example. Find $\lim\limits_{x \to 0} (\sin 2x \cot x)$.

Solution. For $0 < |x| < \pi/2$, we have

$$\sin 2x \cot x = (2 \sin x \cos x) \left(\frac{\cos x}{\sin x} \right)$$

$$= 2 \cos^2 x.$$

Since the cosine function is continuous at 0 and $\cos 0 = 1$, $\lim\limits_{x \to 0} \cos x = 1$. Thus

$$\lim_{x \to 0} (\sin 2x \cot x) = \lim_{x \to 0} (2 \cos^2 x)$$

$$= 2(\lim_{x \to 0} \cos x)^2 = 2.$$

10.7 DERIVATIVES OF THE TRIGONOMETRIC FUNCTIONS

The basic formula for differentiating the trigonometric functions is the formula for the derivative of the sine. Once it is established, the others follow easily.

10.7.1 Theorem. If u is a differentiable function of x and n is any integer, then

$$D_x \sin u = \cos u \, D_x u.$$

$$D_x \cos u = -\sin u \, D_x u.$$

$$D_x \tan u = \sec^2 u \, D_x u, \qquad u \neq \frac{(2n-1)\pi}{2}.$$

$$D_x \cot u = -\csc^2 u \, D_x u, \qquad u \neq n\pi.$$

$$D_x \sec u = \sec u \tan u D_x u, \qquad n \neq \frac{(2n-1)\pi}{2}.$$

$$D_x \csc u = -\csc u \cot u \, D_x u, \qquad u \neq n\pi.$$

Proof. Let x be a real number. Using the addition identity and the trigonometric limits of the preceding section, we can write

$$D_x \sin x = \lim_{\Delta x \to 0} \frac{\sin(x + \Delta x) - \sin x}{\Delta x} = \lim_{\Delta x \to 0} \frac{\sin x \cos \Delta x + \cos x \sin \Delta x - \sin x}{\Delta x}$$

$$= \lim_{\Delta x \to 0} \left(-\sin x \frac{1 - \cos \Delta x}{\Delta x} + \cos x \frac{\sin \Delta x}{\Delta x} \right)$$

$$= -\sin x \lim_{\Delta x \to 0} \frac{1 - \cos \Delta x}{\Delta x} + \cos x \lim_{\Delta x \to 0} \frac{\sin \Delta x}{\Delta x}$$

$$= (-\sin x) \cdot 0 + (\cos x) \cdot 1 = \cos x.$$

Thus

$$D_x \sin x = \cos x.$$

If u is a differentiable function of x, we use the chain rule to obtain

$$D_x \sin u = \cos u \, D_x u.$$

The formula for the derivative of the cosine follows quickly from this result. Since $\cos u = \sin(\pi/2 - u)$, we have

$$D_x \cos u = D_x \sin\left(\frac{\pi}{2} - u \right)$$

$$= \cos\left(\frac{\pi}{2} - u \right) D_x \left(\frac{\pi}{2} - u \right)$$

$$= \sin u \,(-D_x u) = -\sin u \, D_x u,$$

which proves that

$$D_x \cos u = -\sin u \, D_x u.$$

As to the tangent function, if $u \neq (2n-1)\pi/2$ for $n \in Z$,

$$D_x \tan u = D_x \left(\frac{\sin u}{\cos u} \right) = \frac{\cos u \, D_x \sin u - \sin u \, D_x \cos u}{\cos^2 u}$$

$$= \frac{\cos^2 u \, D_x u + \sin^2 u \, D_x u}{\cos^2 u}$$

$$= \frac{(\cos^2 u + \sin^2 u) \, D_x u}{\cos^2 u} = \frac{1}{\cos^2 u} D_x u$$

$$= \sec^2 u \, D_x u.$$

Therefore,

$$D_x \tan u = \sec^2 u \, D_x u, \qquad u \neq \frac{(2n-1)\pi}{2},$$

where n is any integer.

The formulas for the derivatives of the cotangent, secant, and cosecant functions are easily obtained from the first three by writing

$$\cot u = \frac{1}{\tan u}, \qquad \sec u = \frac{1}{\cos u}, \qquad \csc u = \frac{1}{\sin u}.$$

Their proofs are left as exercises. ∎

Since $D_y \sin x = \cos x$ and $\cos x$ exists for all numbers x, *the sine function is everywhere continuous* (by 4.5.2). Similarly, *the cosine function is continuous for all numbers* x. From $D_x \tan x = \sec^2 x$ and the fact that $\sec x$ exists for all numbers x except odd multiples of $\frac{1}{2}\pi$, we see that *the tangent function is continuous for all numbers* x *except* $x = \pm\frac{1}{2}\pi, \pm\frac{3}{2}\pi, \pm\frac{5}{2}\pi$, and so on.

Example 1. $D_x \sin (3x^2 - 4) = \cos (3x^2 - 4) D_x(3x^2 - 4) = 6x \cos (3x^2 - 4).$

Example 2. Find $D_x \cos^3 (\sqrt{x})$.

Solution. Starting with the general power rule, we write

$$D_x \cos^3 (\sqrt{x}) = 3 \cos^2 \sqrt{x} \, D_x \cos \sqrt{x}$$
$$= 3 \cos^2 \sqrt{x} (-\sin \sqrt{x}) \, D_x \sqrt{x}$$
$$= \frac{-3 \cos^2 \sqrt{x} \, \sin \sqrt{x}}{2\sqrt{x}}.$$

Example 3. Find $D_x \tan^2 9x$.

Solution.

$$D_x \tan^2 9x = 2 \tan 9x \cdot D_x \tan 9x$$
$$= 2 \tan 9x \sec^2 9x \, D_x(9x)$$
$$= 18 \tan 9x \sec^2 9x.$$

Example 4. If $y = \sec x/(1 - \cot x)$, find dy/dx.

Solution. We differentiate this quotient as follows.

$$\frac{dy}{dx} = \frac{(1 - \cot x) D_x \sec x - \sec x \, D_x(1 - \cot x)}{(1 - \cot x)^2}$$
$$= \frac{(1 - \cot x) \sec x \tan x - \sec x \, (\csc^2 x)}{(1 - \cot x)^2}$$
$$= \frac{\sec x \, (\tan x - \tan x \cot x - \csc^2 x)}{(1 - \cot x)^2}$$
$$= \frac{\sec x \, (\tan x - 1 - \csc^2 x)}{(1 - \cot x)^2}.$$

Example 5. Analyze and sketch the graph of $y = \sin x$.

Solution. Let $f(x) = \sin x$. We saw earlier that the sine function is continuous on $(-\infty, \infty)$. Moreover, it has the period 2π; so the form of its graph is the same on all successive intervals of length 2π. Thus we may confine our attention to the graph of $f(x) = \sin x$ on the interval $[0, 2\pi]$. Now $f'(x) = \cos x$, and it can be seen from the unit circle in Fig. 10-7 that $\cos x = u$

is positive for $0 \leq x < \pi/2$, zero for $x = \pi/2$, negative for $\pi/2 < x < 3\pi/2$, zero for $x = 3\pi/2$, and positive for $3\pi/2 < x \leq 2\pi$. So $f(x) = \sin x$ is increasing on $[0, \pi/2]$, has its maximum value 1 at $x = \pi/2$, is decreasing on $[\pi/2, 3\pi/2]$, has its minimum value -1 at $x = 3\pi/2$, and increases on $[3\pi/2, 2\pi]$.

Again, $f''(x) = -\sin x$; and from the behavior of $v = \sin x$ on the unit circle (Fig. 10-7), $f''(0) = 0$, $f''(x) < 0$ for $0 < x < \pi$, $f''(\pi) = 0$, $f''(x) > 0$ for $\pi < x < 2\pi$, and $f''(2\pi) = 0$. Thus the graph of $f(x) = \sin x$ is concave downward on $(0, \pi)$, concave upward on $(\pi, 2\pi)$, and has a point of inflection at $x = 0$, at $x = \pi$, and at $x = 2\pi$.

From this information, and from the values of $\sin x$ found in Section 10.5, an accurate graph of $y = \sin x$ for $0 \leq x \leq \pi/2$ can be drawn (Fig. 10-12).

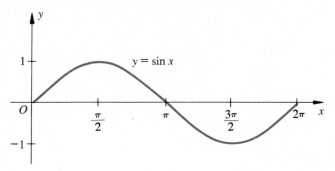

Figure 10-12

It is easy to draw the graph of the cosine function when we know the shape of the graph of $y = \sin x$. From the reduction formula $\cos x = \sin (x + \pi/2)$, it follows that the graph of $y = \cos x$ for $0 \leq x < \pi/2$ is the same as the graph of $y = \sin x$ for $\pi/2 \leq x \leq \pi$. Therefore, to draw the graph of $y = \cos x$, sketch $y = \sin x$ and then move the origin $\pi/2$ units to the right (Fig. 10-10).

Example 6. Analyze and sketch the graph of $y = \tan x$.

Solution. The period of the tangent function is π. Since $\tan x = \sin x/\cos x$ and $\cos (-\pi/2) = \cos \pi/2 = 0$, $\tan x$ does not exist at $\pm\pi/2$. We shall therefore confine our attention to the graph of $y = \tan x$ on the open interval $(-\pi/2, \pi/2)$.

Let $f(x) = \tan x$. Then $f'(x) = \sec^2 x = 1/\cos^2 x$. Since $\cos x > 0$ for $-\pi/2 < x < \pi/2$, $f'(x) > 0$ for $-\pi/2 < x < \pi/2$. Thus $f(x) = \tan x$ is increasing on $(-\pi/2, \pi/2)$ and has no maximum or minimum. In fact,

$$\lim_{x \to \pi/2^-} \tan x = \lim_{x \to \pi/2^-} \frac{\sin x}{\cos x} = \infty,$$

because $\lim_{x \to \pi/2} \sin x = \sin \pi/2 = 1$ and $\lim_{x \to \pi/2} \cos x = \cos \pi/2 = 0$, from the continuity of the sine and cosine. So the line $x = \pi/2$ is a vertical asymptote for the graph of $y = \tan x$. Similarly, the line $x = -\pi/2$ is also a vertical asymptote for the curve $y = \tan x$ (Fig. 10-10).

Again, $f''(x) = 2 \sec x (\sec x \tan x) = 2 \sec^2 x \tan x = 2 \sin x/\cos^3 x$. Since $\sin x < 0$ for $-\pi/2 < x < 0$, $\sin 0 = 0$, $\sin x > 0$ for $0 < x < \pi/2$, and $\cos x > 0$ for $-\pi/2 < x < \pi/2$, we have $f''(x) = 2 \sin x/\cos^3 x < 0$ for $-\pi/2 < x < 0$, $f''(0) = 0$, and $f''(x) > 0$ for $0 < x < \pi/2$. Thus the graph of $f(x) = \tan x$ is concave downward on $(-\pi/2, 0)$, concave upward on $(0, \pi/2)$, and has a point of inflection at the origin. Finally, $f'(0) = 1/\cos^2 0 = 1$; so the slope of the graph at its point of inflection is 1.

With all this information, and the values of tan x found in Section 10.5, it is easy to make an accurate sketch of the curve $y = \tan x$ (Fig. 10-10).

The graphs of the other trigonometric functions can be sketched quickly if we *remember the shapes of the sine and tangent curves.*

We saw, above, that the graph of $y = \cos x$ may be sketched by drawing the graph of $y = \sin x$ and then choosing a new y axis $\pi/2$ units to the right of the old y axis (Fig. 10-10). This curve is easily sketched by recalling that the sine and cosine curves have the same overall shape and period but that $\sin 0 = 0$ whereas $\cos 0 = 1$.

A simple way to draw the graph of $y = \cot x$ is to rewrite the equation as $y = 1/\tan x$. Then sketch lightly from memory the graph of $y = \tan x$, and find the reciprocal of tan x for a sufficient number of values of x. In finding these reciprocals, we estimate the values of tan x from the graph of $y = \tan x$, not from tables. In Fig. 10-13, the graph of $y = \tan x$ is dotted. Starting with this, we choose any point A on the x axis (except $\frac{1}{2}n\pi$, where n is an integer) and draw a vertical line through A, intersecting $y = \tan x$ in some point B. Estimate the directed length $\overline{AB}$. In the diagram this length seems to be about $\frac{3}{5}$. Since $\overline{AB} = \tan x$, then $1/\overline{AB} = \cot x$. We therefore mark C on the vertical line so that $\overline{AC} \doteq \frac{5}{3}$. After finding a number of points in this manner, we draw a smooth curve through them. This is the graph of $y = \cot x$.

Notice that the graphs of $y = \tan x$ and $y = \cot x$ intersect on the horizontal lines $y = 1$ and $y = -1$. This is because the reciprocal of 1 is 1 and the reciprocal of -1 is -1. Since $y = \tan x$ crosses the x axis at the origin, the curve $y = \cot x$ has the y axis as an asymptote.

This method can also be used to sketch $y = \sec x$, since $y = \sec x$ can be rewritten $y = 1/\cos x$ (Fig. 10-14).

The sketching of $y = \csc x$ as $y = 1/\sin x$ is left for the reader.

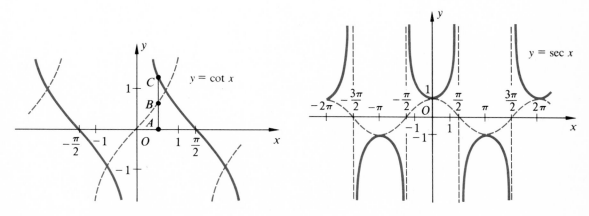

Figure 10-13　　　　　　　　　　　　　　　　Figure 10-14

Example 7. Sketch the graph of $y = 5 \cos \frac{1}{2}x$.

Solution. Since the period of $\cos x$ is 2π, and since $\frac{1}{2}x$ increases by 2π only when x increases by 4π, the period of $\cos \frac{1}{2}x$ is 4π. Similarly, the periods of $\cos kx$ and $\sin kx$ are $2\pi/k$. The graph of $y = 5 \cos \frac{1}{2}x$ crosses the x axis when $\cos \frac{1}{2}x$ is zero—that is, when $x = \pm\pi, \pm3\pi, \pm5\pi$, and so on. Moreover, y takes on its maximum value, 5, when $x = 0, \pm4\pi, \pm8\pi$, etc., and its minimum value, -5, when $x = \pm2\pi, \pm6\pi$, etc. The curve is shown in Fig. 10-15.

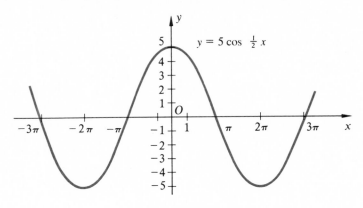

Figure 10-15

The antidifferentiation formulas that are counterparts of the formulas for the derivatives of the trigonometric functions follow.

10.7.2 Corollary.

$$A_x(\sin u \, D_x u) = -\cos u + C.$$

$$A_x(\cos u \, D_x u) = \sin u + C.$$

$$A_x(\sec^2 u \, D_x u) = \tan u + C.$$

$$A_x(\csc^2 u \, D_x u) = -\cot u + C.$$

$$A_x(\sec u \tan u \, D_x u) = \sec u + C.$$

$$A_x(\csc u \cot u \, D_x u) = -\csc u + C.$$

Of course, if $u = x$ in these formulas, $D_x u = 1$ and the factor $D_x u$ disappears. Thus the first formula becomes $A_x \sin x = -\cos x + C$, and similarly for the others.

Example 8. Find the antiderivatives with respect to x of $\dfrac{\cos 5x}{\sin 5x}$.

Solution. Notice that the numerator of the given expression is almost the derivative of the denominator. So we let $u = \sin 5x$ and use 10.1.4. Then $D_x u = 5 \cos 5x$, and we can write

$$A_x \left(\frac{\cos 5x}{\sin 5x} \right) = \frac{1}{5} A_x \left(\frac{5 \cos 5x}{\sin 5x} \right)$$

$$= \frac{1}{5} A_x \left[\frac{1}{\sin 5x} D_x (\sin 5x) \right]$$

$$= \frac{1}{5} \ln \sin 5x + C.$$

Example 9. Find the antiderivatives of $\cot^2 x$.

Solution. We do not yet have a formula for the antiderivatives of $\cot^2 x$, but we do have a formula for $A_x \csc^2 x$. Using the identity $1 + \cot^2 x = \csc^2 x$, we obtain

$$A_x \cot^2 x = A_x(\csc^2 x - 1) = A_x \csc^2 x - A_x 1$$

$$= -\cot x - x + C.$$

Example 10. Evaluate $\displaystyle\int_0^{\pi/12} \sec 3x \tan 3x \, dx$.

Solution. Let $u = 3x$; then $D_x u = 3$. Using the fundamental theorem and 10.7.2, we have

$$\int_0^{\pi/12} \sec 3x \tan 3x \, dx = \frac{1}{3} \int_0^{\pi/12} (\sec 3x \tan 3x)(3) \, dx$$

$$= \frac{1}{3} \int_0^{\pi/12} \sec 3x \tan 3x \, D_x(3x) \, dx$$

$$= \frac{1}{3} \sec 3x \Big]_0^{\pi/12} = \frac{1}{3} \left(\sec \frac{\pi}{4} - \sec 0 \right)$$

$$= \frac{1}{3} \left(\frac{1}{\cos \pi/4} - \frac{1}{\cos 0} \right)$$

$$= \frac{\sqrt{2} - 1}{3}.$$

Example 11. Evaluate $\displaystyle\int_{\pi/6}^{\pi/4} \sin^3 2x \cos 2x \, dx$.

Solution. We use the power rule. Let $u = \sin 2x$; then $D_x u = \cos 2x \, D_x(2x) = 2 \cos 2x$. Thus

$$\int_{\pi/6}^{\pi/4} \sin^3 2x \cos 2x \, dx = \frac{1}{2} \int_{\pi/6}^{\pi/4} (\sin 2x)^3 (2 \cos 2x) \, dx$$

$$= \frac{1}{2} \int_{\pi/6}^{\pi/4} (\sin 2x)^3 \, D_x(\sin 2x) \, dx$$

$$= \frac{1}{2} \left[\frac{(\sin 2x)^4}{4} \right]_{\pi/6}^{\pi/4}$$

$$= \frac{1}{8} \left[\left(\sin \frac{\pi}{2} \right)^4 - \left(\sin \frac{\pi}{3} \right)^4 \right] = \frac{1 - \frac{1}{2}}{12}$$

$$= \frac{1 - \frac{9}{16}}{8} = \frac{7}{128}$$

Exercises

In Exercises 1 through 12, differentiate the indicated functions.

1. $y = \cos^2 (x^2 - 2)$.

2. $y = \sin \sqrt{2 + 7x}$.

3. $y = \cot x \csc x$.

4. $y = \dfrac{x}{1 - \tan x}$.

5. $s = \csc \dfrac{1}{t}$.

6. $s = \tan^2 \sqrt{t}$.

7. $w = e^{\cot z}$.

8. $w = e^z \cos z$.

9. $u = \ln \tan v$.

10. $u = \ln \cos e^{2v}$.

11. $y = 4 \cot (\ln x - 2)$.

12. $y = \ln (\sec x + \tan x)$.

In Exercises 13 to 22, find the antiderivatives of the indicated functions, and verify your answers by differentiation.

13. $y = x \sin x^2$.

14. $y = \sin 2x \cos 2x$.

15. $y = \tan x$. $\left(\textit{Hint: } \text{Write } \tan x = \dfrac{\sin x}{\cos x}.\right)$

16. $y = \cot x$.

17. $y = \tan^2 x$.

18. $y = \dfrac{\sec^2 x}{\tan x}$.

19. $y = \dfrac{\sin x}{\cos^2 x}$.

20. $y = e^{2x} \cos e^{2x}$.

21. $y = \sec x \csc x \ln \cot x$.

22. $y = \dfrac{\sec \sqrt{x} \tan \sqrt{x}}{\sqrt{x}}$.

23. Evaluate $\displaystyle\int_0^{\pi/2} \sin^2 x \cos x \, dx$.

24. Evaluate $\displaystyle\int_0^{\pi/3} \dfrac{\sin t}{\cos^2 t} \, dt$.

25. Find the area of the region bounded by the curve $y = \tan x$, the x axis, and the line $x = \pi/4$. Make a sketch.

26. Find the volume of the solid generated by revolving about the x axis the region of Exercise 25.

In Exercises 27 and 28, find the extrema and the points of inflection of the indicated functions f. Find the intervals on which the graph is concave upward and those on which it is concave downward. Then sketch the graph.

27. $f(x) = \sin x + \cos x$, $-\pi < x < \pi$.

28. $f(x) = \sin x - \tan x$, $-\dfrac{\pi}{2} < x < \dfrac{\pi}{2}$.

Evaluate the limits in Exercises 29 through 36.

29. $\displaystyle\lim_{x \to 0} \left(\dfrac{\sin 2x}{x}\right)$.

30. $\displaystyle\lim_{x \to 0} \left(\dfrac{\sin \frac{1}{2}x}{3x}\right)$.

31. $\displaystyle\lim_{x \to 0} \left(\dfrac{\sin x}{\tan x}\right)$.

32. $\displaystyle\lim_{x \to 0} \left(\dfrac{1 - \cos x}{\sin x}\right)$.

33. $\displaystyle\lim_{x \to \pi/2} \left(\dfrac{2 \tan x}{\sec x}\right)$.

34. $\displaystyle\lim_{x \to 0} \left(\dfrac{\cot x}{\csc x - 1}\right)$.

35. $\displaystyle\lim_{x \to \pi/2} \left(\dfrac{\cos x}{\pi/2 - x}\right)$.

36. $\displaystyle\lim_{x \to 0} \left(\dfrac{\tan x - \sin x}{x \cos x}\right)$.

37. A steel cable to guy an electric pole is to pass over an 8-foot wall without touching it. If the wall is $24\sqrt{3}$ feet from the pole and we neglect the thickness of the wall, what is the greatest lower bound for the length of the cable? (*Hint:* Let y be the length of the cable, x the distance from the foot of the wall to the point where the cable enters the ground, and

ϕ the angle the cable makes with the ground. Express y in terms of ϕ and minimize y.)

38.   A square plaque, 6 feet high, is fastened to a wall with its lower edge horizontal and 2 feet above the eye level of a certain viewer. How far from the wall should he stand in order to obtain the best view? (Here "best view" means maximum angle from eye to plaque.)

10.8 INVERSE TRIGONOMETRIC FUNCTIONS

Recall that if a function has an inverse, then to each number of its range there corresponds *exactly one* number of its domain.

The sine function does not have a unique inverse because it is periodic; $\sin x = \sin(x + 2n\pi)$, where n is any integer. Thus to each number y in its range $[-1, 1]$, there correspond many numbers x such that $y = \sin x$. For instance, if $y = \frac{1}{2}$, then $x = \frac{1}{6}\pi, \frac{5}{6}\pi, \frac{13}{6}\pi, -\frac{7}{6}\pi$, etc. will all satisfy $\frac{1}{2} = \sin x$. Geometrically, any horizontal line between $y = -1$ and $y = 1$ will cut the graph of $y = \sin x$ indefinitely many times (Fig. 10-10).

It is customary to define a new function, Sine (spelled with a capital S), sometimes called the **principal part** of the sine function, whose domain is $[-\frac{1}{2}\pi, \frac{1}{2}\pi]$, and whose value is defined by

(1) $\qquad\qquad \text{Sin } x = \sin x \qquad \text{for} \quad -\tfrac{1}{2}\pi \le x \le \tfrac{1}{2}\pi.$

Since the sine function is continuous and increasing throughout the closed interval $[-\frac{1}{2}\pi, \frac{1}{2}\pi]$, Sine is continuous and increasing throughout its domain [Fig. 10-16(a)]. Therefore Sine has an inverse, called the **inverse Sine function** and indicated by Sin^{-1} or Arcsin, that is continuous and increasing on $[-1, 1]$ (by the inverse function theorem). Thus

(2) $\qquad\qquad y = \text{Sin}^{-1} x \qquad \text{if and only if} \qquad x = \text{Sin } y.$

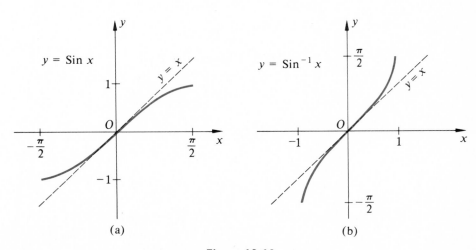

(a)   (b)

Figure 10-16

The domain of the inverse Sine function is $[-1, 1]$, and its range is $[-\frac{1}{2}\pi, \frac{1}{2}\pi]$.

The graph of Sin^{-1}, the inverse Sine function, is shown in Fig. 10-16(b); it may be obtained by reflecting the graph of $y = \text{Sin } x$ about the line $y = x$.

To find the derivative of $y = \text{Sin}^{-1} x$, we start with the equivalent equation [see (2)]

$$x = \text{Sin } y.$$

Using (1), we can differentiate both members of this equation and obtain

$$\frac{dx}{dy} = \frac{d}{dy} \text{Sin } y = \frac{d}{dy} (\sin y) = \cos y = \sqrt{1 - \sin^2 y}$$

$$= \sqrt{1 - \text{Sin}^2 y} = \sqrt{1 - x^2} \quad \text{for} \quad -1 < x < 1.$$

Since $D_x y = 1/D_y x$,

$$(3) \qquad \frac{dy}{dx} = \frac{1}{\sqrt{1 - x^2}}, \qquad -1 < x < 1.$$

If $y = \text{Sin}^{-1} u$ and u is a differentiable function of x, then (3) and the chain rule yield the formula

10.8.1
$$D_x \text{Sin}^{-1} u = \frac{1}{\sqrt{1 - u^2}} D_x u, \qquad -1 < u < 1.$$

Example 1. Find $D_x \text{Sin}^{-1} (3x - 5)$.

Solution. Using 10.8.1 with $u = 3x - 5$, we find

$$D_x \text{Sin}^{-1} (3x - 5) = \frac{1}{\sqrt{1 - (3x - 5)^2}} D_x(3x - 5) = \frac{3}{\sqrt{-9x^2 + 30x - 24}}.$$

Of course, 10.8.1 leads to a formula for antiderivatives.

10.8.2
$$A_x \left[\frac{D_x u}{\sqrt{1 - u^2}} \right] = \text{Sin}^{-1} u + C, \qquad -1 < u < 1.$$

Example 2. Evaluate $\displaystyle\int_0^{1/2} \frac{dx}{\sqrt{1 - x^2}}$.

Solution. From 10.8.2 and the fundamental theorem,

$$\int_0^{1/2} \frac{dx}{\sqrt{1 - x^2}} = \text{Sin}^{-1} \frac{1}{2} - \text{Sin}^{-1} 0 = \frac{\pi}{6} - 0 = \frac{\pi}{6}.$$

The cosine function is treated similarly. Since the cosine function does not possess an inverse, we confine our attention to that part of the cosine function whose domain is $[0, \pi]$ [Fig. 10-17(a)]. This is called the **principal part** of the cosine function and is symbolized by Cosine. The Cosine function is defined by

$$\text{Cos } x = \cos x \quad \text{for} \quad 0 \le x \le \pi.$$

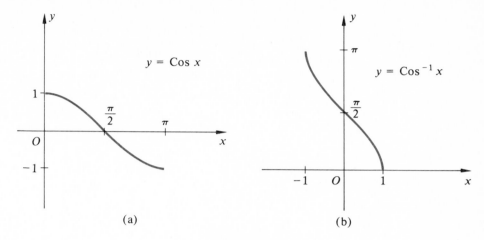

Figure 10-17

Since Cosine is continuous and decreasing throughout its domain $[0, \pi]$ it has an inverse, Cosine^{-1} or Arccosine, called the **inverse cosine function** [Fig. 10-17(b)]. It is defined by

$$y = \text{Cos}^{-1} x \qquad \text{if and only if} \qquad x = \text{Cos } y.$$

Its domain is $[-1, 1]$, and its range is $[0, \pi]$.

The formula for the derivative of $y = \text{Cos}^{-1} x$ is found by starting with the equivalent equation $x = \text{Cos } y$ and differentiating both sides with respect to y. We get

$$\frac{dx}{dy} = \frac{d}{dy} \text{Cos } y = \frac{d}{dy} \cos y = -\sin y$$

$$= -\sqrt{1 - \cos^2 y} = -\sqrt{1 - \text{Cos}^2 y} = -\sqrt{1 - x^2}.$$

Hence

(4) $$\frac{dy}{dx} = \frac{-1}{\sqrt{1 - x^2}}, \qquad -1 < x < 1.$$

If $y = \text{Cos}^{-1} u$ and u is a differentiable function of x, then from (4) and the chain rule we obtain

10.8.3 $$D_x \text{Cos}^{-1} u = \frac{-1}{\sqrt{1 - u^2}} D_x u, \qquad -1 < u < 1.$$

The following formula for antiderivatives is a counterpart of 10.8.3.

10.8.4 $$A_x \left[\frac{-D_x u}{\sqrt{1 - u^2}} \right] = \text{Cos}^{-1} u + C, \qquad -1 < u < 1.$$

The Tangent function (capital T) is that part of the tangent function whose domain is $-\frac{1}{2}\pi < x < \frac{1}{2}\pi$. It is defined by

$$\text{Tan } x = \tan x \quad \text{if} \quad -\frac{1}{2}\pi < x < \frac{1}{2}\pi.$$

The Tangent function is continuous and increasing throughout its domain, which is the open interval $(-\frac{1}{2}\pi, \frac{1}{2}\pi)$ [Fig. 10-18(a)]. Therefore Tangent has an inverse, Tangent^{-1} or Arctangent, called the **inverse Tangent function** and defined by

$$y = \text{Tan}^{-1} x \quad \text{if and only if} \quad x = \text{Tan } y.$$

The graph of $y = \text{Tan}^{-1} x$ is shown in Fig. 10-18(b). Its domain is the set of all real numbers and its range is the open interval $(-\frac{1}{2}\pi, \frac{1}{2}\pi)$.

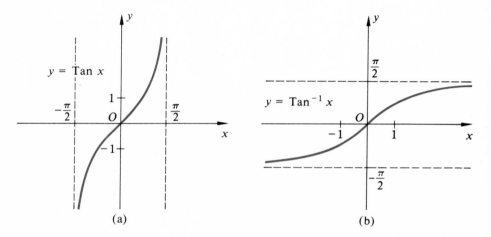

Figure 10-18

To find the derivative of $y = \text{Tan}^{-1} x$, we differentiate both members of $x = \text{Tan } y$ with respect to y, obtaining

$$\frac{dx}{dy} = \frac{d}{dy}\text{Tan } y = \frac{d}{dy}\tan y = \sec^2 y = 1 + \tan^2 y$$

$$= 1 + \text{Tan}^2 y = 1 + x^2 \quad \text{for} \quad -\frac{1}{2}\pi < x < \frac{1}{2}\pi.$$

Thus

(5) $$\frac{dy}{dx} = \frac{1}{1 + x^2}.$$

Using the chain rule, we get the more general formula

10.8.5 $$D_x \text{Tan}^{-1} u = \frac{1}{1 + u^2} D_x u.$$

From it we have the formula for antiderivatives

10.8.6
$$A_x\left(\frac{1}{1+u^2}\,D_xu\right) = \mathrm{Tan}^{-1}\,u + C.$$

The three remaining trigonometric functions are treated similarly. The graphs of Cotangent^{-1}, Secant^{-1}, and Cosecant^{-1} are shown in Fig. 10-19. The Cotangent function (capital C) is that part of the cotangent function whose domain is $(0, \pi)$; thus the range of its inverse, Cotangent^{-1}, is also $(0, \pi)$. We define the Secant function to be that part of the secant function whose domain is $[0, \pi/2) \cup (\pi/2, \pi]$, and this is the range of its inverse, Secant^{-1}. (It should be mentioned here that there is no universal agreement on what part of the secant function should be chosen for its principal part.) The range of Cosecant^{-1} is $[-\pi/2, 0) \cup (0, \pi/2]$.

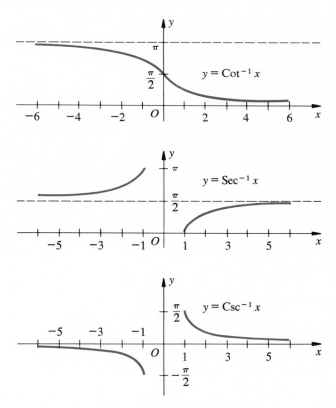

Figure 10-19

Formulas for the derivatives of Cotangent^{-1}, Secant^{-1}, and Cosecant^{-1} follow; their proofs are left as exercises.

10.8.7
$$D_x\,\mathrm{Cot}^{-1}\,u = \frac{-1}{1+u^2}\,D_xu.$$

10.8.8
$$D_x \operatorname{Sec}^{-1} u = \frac{1}{|u|\sqrt{u^2 - 1}} D_x u, \qquad |u| > 1.$$

10.8.9
$$D_x \operatorname{Csc}^{-1} u = \frac{-1}{|u|\sqrt{u^2 - 1}} D_x u, \qquad |u| > 1.$$

Since the secant function is continuous and increasing on $[0, \pi/2) \cup (\pi/2, \pi]$, $y = \operatorname{Sec}^{-1} u$ is increasing for $|u| > 1$. Thus its derivative must be positive, and we write $|u|\sqrt{u^2 - 1}$ in 10.8.8 instead of $u\sqrt{u^2 - 1}$.

Example 3. A man at the top of a vertical cliff, 200 feet above a lake, watches a motorboat move directly away from the foot of the cliff at the rate of 25 feet per second. How fast is the angle of depression of his line of sight changing when the boat is 150 feet from the foot of the cliff?

Solution. Let x be the distance in feet from the foot of the cliff to the boat, and denote by θ the angle of depression of the man's line of sight. Then $\operatorname{Cot} \theta = x/200$, and

(6)
$$\theta = \operatorname{Cot}^{-1}\left(\frac{x}{200}\right).$$

We are given that $dx/dt = 25$, and we seek the value of $d\theta/dt$ when $x = 150$.

If we differentiate both members of (6) with respect to t, we obtain

$$\frac{d\theta}{dt} = \frac{d}{dt}\left[\operatorname{Cot}^{-1}\left(\frac{x}{200}\right)\right] = \frac{-1}{1 + (x/200)^2} \cdot \frac{d}{dt}\left(\frac{x}{200}\right),$$

or

$$\frac{d\theta}{dt} = \frac{-200}{40,000 + x^2}\frac{dx}{dt}.$$

By substituting $dx/dt = 25$ and $x = 150$ in the latter equation, we find

$$\frac{d\theta}{dt} = -0.08 \text{ radian per second.}$$

Exercises

Differentiate the indicated functions in Exercises 1 through 10.

1. $y = \operatorname{Sin}^{-1} x^2.$

2. $y = \frac{1}{2} \operatorname{Sin}^{-1} e^{2x}.$

3. $y = 7 \operatorname{Cos}^{-1} \sqrt{2x}.$

4. $y = (3x - 1) \operatorname{Cos}^{-1} x^2.$

5. $y = \operatorname{Tan}^{-1} e^x.$

6. $y = e^x \operatorname{Sin}^{-1} x.$

7. $s = t \operatorname{Cot}^{-1} t.$

8. $s = (\operatorname{Tan}^{-1} t)^2.$

9. $w = \sin(\operatorname{Tan}^{-1} t).$

10. $w = \tan(\operatorname{Sin}^{-1} t).$

11. Prove: $D_x \operatorname{Cot}^{-1} u = \dfrac{-1}{1 + u^2} D_x u.$

12. Prove: $D_x \operatorname{Sec}^{-1} u = \dfrac{1}{|u|\sqrt{u^2 - 1}} D_x u.$

In Exercises 13 to 20, find the antiderivatives of the indicated functions.

13. $y = \dfrac{1}{1 + 4x^2}.$

14. $y = \dfrac{e^x}{1 + e^{2x}}.$

15. $y = \dfrac{1}{x^2 + 4x + 5}.$

16. $y = \dfrac{1}{\sqrt{2x - x^2}}$. (*Hint:* Complete the square under the radical sign.)

17. $s = \dfrac{-2 \sin 2t}{1 + \cos^2 2t}$.

18. $s = \dfrac{1}{(t + 2)\sqrt{t^2 + 4t + 3}}$.

19. $w = \dfrac{1}{v\sqrt{v^4 - 1}}$.

20. $w = \dfrac{1 + \tan^2 v}{\sqrt{1 - \tan^2 v}}$.

21. Find the area of the region bounded by the curve $x^2 y + y - 4 = 0$, the coordinate axes, and the line $x = 1$. Make a sketch.

22. Find the area of the region in the first quadrant bounded by the curve $x^2 y^2 - 9y^2 + 36 = 0$, the coordinate axes, and the line $x = \frac{3}{2}$. Make a sketch.

23. Find the volume of the solid generated by revolving about the x axis the region bounded by the curves $y = 5(x^2 + 1)^{-1/2}$, $y = 0$, $x = 0$, and $x = 4$. Make a sketch.

24. Find the volume of the solid generated by revolving about the x axis the region in the first quadrant bounded by the curves $x^4 y^4 - x^2 y^4 - 81 = 0$, $y = 0$, $x = \sqrt{2}$, and $x = 2$. Make a sketch. (*Hint:* See 10.8.8.)

25. The structural steel work of a new office building is finished. Across the street, 60 feet from the ground floor of the freight elevator shaft in the building, a spectator is standing and watching the freight elevator ascend at a constant rate of 15 feet per second. How fast is the angle of elevation of the spectator's line of sight to the elevator increasing 6 seconds after his line of sight passes the horizontal?

26. An airplane is flying at a constant altitude of 2 miles and a constant speed of 600 miles per hour, on a straight course that will take it directly over an observer. How fast is the angle of elevation of the observer's line of sight increasing when the distance from him to the plane is 3 miles? Give your result in radians per minute.

27. A revolving beacon light is located on an island, 2 miles away from the nearest point P of the straight shoreline of the mainland. The beacon throws a spot of light that moves along the shoreline as the beacon revolves. If the speed of the spot of light on the shoreline is 50π miles per minute when the spot is 1 mile from P, how fast is the beacon revolving?

28. A man on a dock is pulling in a rope attached to a rowboat at a rate of 5 feet per second. If the man's hands are 8 feet higher than the point where the rope is attached to the boat, how fast is the angle of depression of the rope changing when there are still 17 feet of rope out?

10.9 HYPERBOLIC FUNCTIONS

In applied mathematics certain combinations of e^x and e^{-x} occur so frequently that special names are given to the functions that these combinations define. The two principal ones are called the *hyperbolic sine* function and the *hyperbolic cosine* function.

10.9.1 Definition. The **hyperbolic sine** function is defined by

$$\sinh x = \frac{e^x - e^{-x}}{2};$$

both its domain and its range are the set of all real numbers (Fig. 10-20). The **hyperbolic cosine** function is defined by

$$\cosh x = \frac{e^x + e^{-x}}{2};$$

its domain is the set of all real numbers and its range is the set of all numbers greater than or equal to 1 (Fig. 10-20).

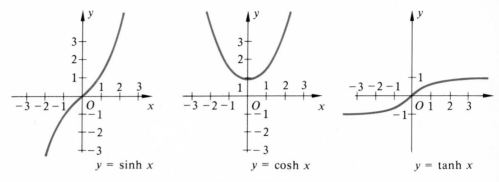

Figure 10-20

The graphs of $y = \sinh x$ and $y = \cosh x$ are shown in Fig. 10-20. They are readily constructed from the familiar graphs of $y = e^x$ and $y = e^{-x}$ by a method known as **addition of ordinates**. For instance, to draw the graph of $y = \cosh x = \frac{1}{2}(e^x + e^{-x})$, sketch lightly on the same pair of axes the graphs of $y = e^x$ and $y = e^{-x}$. For each value of x the corresponding ordinate (or y coordinate) of $y = \cosh x$ is half the sum of the ordinates of $y = e^x$ and $y = e^{-x}$. This addition of ordinates should be done graphically by using a pair of dividers or the edge of a sheet of paper; no numbers need to be computed.

The graph of $y = \cosh x$ is called a *catenary*; it is the shape assumed by a perfectly flexible, uniform cable hanging between two supports.

The definitions of the remaining hyperbolic functions may be stated as combinations of the first two; they are easy to remember, since they are analogous to well-known relations between trigonometric functions.

10.9.2 Definition.

hyperbolic tangent:	$\tanh x = \dfrac{\sinh x}{\cosh x} = \dfrac{e^x - e^{-x}}{e^x + e^{-x}}.$
hyperbolic cotangent:	$\coth x = \dfrac{\cosh x}{\sinh x} = \dfrac{e^x + e^{-x}}{e^x - e^{-x}}.$
hyperbolic secant:	$\operatorname{sech} x = \dfrac{1}{\cosh x} = \dfrac{2}{e^x + e^{-x}}.$
hyperbolic cosecant:	$\operatorname{csch} x = \dfrac{1}{\sinh x} = \dfrac{2}{e^x - e^{-x}}.$

The graph of the hyperbolic tangent function is shown in Fig. 10-20.

There are identical relations among the hyperbolic functions that are strikingly reminiscent of the familiar trigonometric identities. Three basic ones are

10.9.3
$$-\sinh^2 x + \cosh^2 x = 1.$$

10.9.4
$$1 - \tanh^2 x = \operatorname{sech}^2 x.$$

10.9.5
$$1 - \coth^2 x = -\operatorname{csch}^2 x.$$

They are easily proved from the definitions of the hyperbolic functions (10.9.1 and 10.9.2). As an illustration,

$$-\sinh^2 x + \cosh^2 x = -\left(\frac{e^x - e^{-x}}{2}\right)^2 + \left(\frac{e^x + e^{-x}}{2}\right)^2$$

$$= \frac{-(e^{2x} - 2e^0 + e^{-2x}) + (e^{2x} + 2e^0 + e^{-2x})}{4} = 1,$$

which proves the first. The second may be proved as follows.

$$1 - \tanh^2 x = 1 - \frac{\sinh^2 x}{\cosh^2 x} = \frac{\cosh^2 x - \sinh^2 x}{\cosh^2 x} = \frac{1}{\cosh^2 x} = \operatorname{sech}^2 x.$$

There are other identities among the hyperbolic functions similar to the other well-known trigonometric identities.

Example 1. Prove that $\cosh(x + y) = \cosh x \cosh y + \sinh x \sinh y.$

Solution.

$$\cosh x \cosh y + \sinh x \sinh y = \left(\frac{e^x + e^{-x}}{2}\right)\left(\frac{e^y + e^{-y}}{2}\right) + \left(\frac{e^x - e^{-x}}{2}\right)\left(\frac{e^y - e^{-y}}{2}\right)$$

$$= \frac{2e^x e^y + 2e^{-x}e^{-y}}{4} = \frac{e^{x+y} + e^{-(x+y)}}{2} = \cosh(x + y).$$

Formulas for differentiating the hyperbolic functions are

10.9.6
$$D_x \sinh x = \cosh x.$$

10.9.7
$$D_x \cosh x = \sinh x.$$

10.9.8
$$D_x \tanh x = \operatorname{sech}^2 x.$$

10.9.9
$$D_x \coth x = -\operatorname{csch}^2 x.$$

10.9.10
$$D_x \operatorname{sech} x = -\operatorname{sech} x \tanh x.$$

10.9.11
$$D_x \operatorname{csch} x = -\operatorname{csch} x \coth x.$$

These formulas are easily proved from the definitions of the hyperbolic functions (10.9.1 and 10.9.2) and the basic identities (10.9.3, etc.). For instance,

$$D_x \sinh x = D_x \left(\frac{e^x + e^{-x}}{2}\right) = \frac{e^x - e^{-x}}{2} = \cosh x;$$

and

$$D_x \tanh x = D_x \left(\frac{\sinh x}{\cosh x}\right) = \frac{\cosh^2 x - \sinh^2 x}{\cosh^2 x} = \frac{1}{\cosh^2 x} = \operatorname{sech}^2 x.$$

Example 2. Differentiate $y = \cosh^2 (3x - 1)$.

Solution. From the power rule, from 10.9.7, and from the chain rule, we obtain

$$\frac{dy}{dx} = 2 \cosh (3x - 1) \cdot \frac{d}{dx} \cosh (3x - 1)$$

$$= 2 \cosh (3x - 1) \sinh (3x - 1) \cdot \frac{d}{dx} (3x - 1)$$

$$= 6 \sinh (3x - 1) \cosh (3x - 1).$$

Example 3. Find $D_x \tanh (\sin x)$.

Solution. By using 10.9.8 and the chain rule, we get

$$D_x \tanh (\sin x) = \operatorname{sech}^2 (\sin x) \cdot D_x (\sin x)$$

$$= \cos x \operatorname{sech}^2 (\sin x).$$

Example 4. Find the antiderivatives of $\coth x$.

Solution.

$$A_x \coth x = A_x \left(\frac{\cosh x}{\sinh x}\right) = A_x \left[\frac{D_x (\sinh x)}{\sinh x}\right]$$

$$= \ln (\sinh x) + C.$$

Example 5. Find the antiderivatives of $\tanh^2 x$.

Solution. From 10.9.4 and 10.9.8, we find

$$A_x \tanh^2 x = A_x (1 - \operatorname{sech}^2 x)$$

$$= A_x (1) - A_x \operatorname{sech}^2 x$$

$$= x - \tanh x + C.$$

It is because of the close similarity between the identities and differentiation formulas of the trigonometric functions and the hyperbolic functions that the latter are called hyperbolic *sine*, hyperbolic *cosine*, hyperbolic *tangent*, and so on.

The word *hyperbolic* is used to indicate the geometric relations that these functions have to a hyperbola, analogous to the relations of the trigonometric functions to a unit circle.

Figure 10-21 shows a *unit circle*, $x^2 + y^2 = 1$, and a *unit hyperbola*, $x^2 - y^2 = 1$ (Section 12.6). Three geometric analogies are listed below; in each, (a) refers to the unit circle shown in Fig. 10-21(a), and (b) refers to the unit equilateral hyperbola in Fig. 10-21(b).

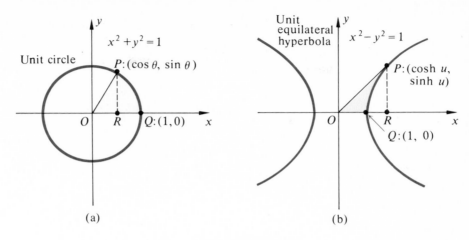

Figure 10-21

1. (a) For each number θ, $0 \leq \theta \leq \frac{1}{2}\pi$, there is a point P:$(\cos\theta, \sin\theta)$ in the first quadrant that is on the unit circle $x^2 + y^2 = 1$ because $\cos^2\theta + \sin^2\theta = 1$;

    (b) For each number u, $0 \leq u$, there is a point P:$(\cosh u, \sinh u)$ in the first quadrant that is on the unit hyperbola $x^2 - y^2 = 1$ because $\cosh^2 u - \sinh^2 u = 1$.

2. (a) $\overline{RP} = \sin\theta$, $\overline{OR} = \cos\theta$;

    (b) $\overline{RP} = \sinh u$, $\overline{OR} = \cosh u$.

3. (a) Area of shaded region $OQP = \frac{1}{2}\theta$;

    (b) Area of shaded region $OQP = \frac{1}{2}u$, as will be shown in Exercises 27 and 28 or Section 11.9.

Incidentally, this also shows why the trigonometric functions are often called **circular functions.**

Exercises

Prove the given identities in Exercises 1 to 8.

1. (a) $\sinh(-x) = -\sinh x$;
   (b) $\cosh(-x) = \cosh x$;
   (c) $\tanh(-x) = -\tanh x$.

2. $\sinh(x + y) = \sinh x \cosh y + \cosh x \sinh y$.

3. $\sinh(x - y) = \sinh x \cosh y - \cosh x \sinh y$.

4. $\tanh(x + y) = \dfrac{\tanh x + \tanh y}{1 + \tanh x \tanh y}$.

5. $\sinh 2x = 2 \sinh x \cosh x$.

6. $\sinh \dfrac{x}{2} = \pm\sqrt{\dfrac{\cosh x - 1}{2}}$.

7. $e^x = \cosh x + \sinh x$.

8. $e^{-x} = \cosh x - \sinh x$.

In Exercises 9 to 12, prove the given differentiation formulas.

9. $D_x \cosh x = \sinh x$.

10. $D_x \coth x = -\operatorname{csch}^2 x$.

11. $D_x \operatorname{sech} x = -\operatorname{sech} x \tanh x$.

12. $D_x \operatorname{csch} x = -\operatorname{csch} x \coth x$.

In Exercises 13 to 24, differentiate the indicated functions.

13. $y = \sinh^2 x$.

14. $y = 5 \sinh^3 x$.

15. $y = \cosh (x^2 - 1)$.

16. $y = \ln (\coth x)$.

17. $y = x^2 \operatorname{sech} x$.

18. $y = e^x \operatorname{sech} x$.

19. $y = \sinh 4x \cosh 2x$.

20. $y = \sinh 3x \cosh 5x$.

21. $y = \cosh e^{2x}$.

22. $y = e^x \tanh 2x$.

23. $y = \dfrac{\cosh x}{x}$.

24. $y = \sinh (\cos x)$.

In Exercises 25 to 30, find the antiderivatives of the indicated functions.

25. $y = x \cosh (x^2 + 3)$.

26. $y = \dfrac{\sinh \sqrt{x}}{\sqrt{x}}$.

27. $y = \tanh x$.

28. $y = \coth^2 x$.

29. $y = \tanh x \ln (\cosh x)$. (*Hint:* Rearrange to the form $u \cdot D_k u$.)

30. $y = \coth x \ln (\sinh x)$.

31. Find the area of the region bounded by the curves $y = \cosh x$, $y = 0$, $x = 0$, and $x = 1$.

32. Find the area of the region bounded by the curves $y = \sinh x$, $y = 0$, and $x = 2$. Make a sketch.

33. Find the area of the region bounded by the curves $y = \tanh x$, $y = 0$, and $x = 3$. Make a sketch.

34. What is the length of the segment of the catenary $y = \cosh x$ between the y axis and the line $x = 1$?

35. Find the asymptotes of the curve $y = \coth x$ and sketch the curve.

36. Find the extrema and the points of inflection of $y = \operatorname{sech} x$ and the intervals on which its graph is concave upward and concave downward. Then sketch the graph.

10.10 INVERSE HYPERBOLIC FUNCTIONS

Since

$$D_x \sinh x = D_x \frac{e^x - e^{-x}}{2} = \frac{e^x + e^{-x}}{2} = \frac{e^{2x} + 1}{2e^x}$$

and $e^x > 0$ for all $x \in R$, then $D_x \sinh x > 0$ for all real numbers x. Thus the hyperbolic sine function is everywhere continuous and increasing (Fig. 10-20), and therefore the hyperbolic sine has an inverse that is continuous and increasing, by the inverse function theorem (10.2.1).

The **inverse hyperbolic sine** function is defined by

$$y = \sinh^{-1} x \qquad \text{if and only if} \qquad x = \sinh y.$$

On the other hand, the graph of $y = \cosh x$ (Fig. 10-20) cuts any horizontal line $y = k$, where $k > 1$, in two points; that is, to each value of y greater than 1 there correspond two numbers x, related by $y = \cosh x$. So the hyperbolic cosine does not have a unique inverse. However, if we limit the domain by excluding negative values of x, the function defined by $y = \cosh x$ for $x \geq 0$ does have a well-defined inverse.

The **inverse hyperbolic cosine** function is defined by

$$y = \cosh^{-1} x \qquad \text{if and only if} \qquad x = \cosh y \quad \text{and} \quad y \geq 0$$

or by

$$\{(x, y) \mid x = \cosh y, y \geq 0\}.$$

Similarly, tanh, coth, and csch have inverses, denoted by $\tanh^{-1}$, $\coth^{-1}$, and csch^{-1}, but sech does not have a unique inverse. Accordingly, we define the **inverse hyperbolic secant** by

$$y = \text{sech}^{-1} x \qquad \text{if and only if} \qquad x = \text{sech } y \quad \text{and} \quad y \geq 0$$

or

$$\{(x, y) \mid x = \text{sech } y, y \geq 0\}.$$

Since the natural logarithmic function is the inverse of the exponential function, it is not surprising that the inverse hyperbolic functions can be expressed in terms of natural logarithms. Let $y = \cosh^{-1} x$, where $x \geq 1$. Then $x = \cosh y = \frac{1}{2}(e^y + e^{-y})$ for $y \geq 0$. Multiplying both members of $x = \frac{1}{2}(e^y + e^{-y})$ by $2e^y$, we get $2xe^y = e^{2y} + 1$ or

$$(e^y)^2 - 2x(e^y) + 1 = 0.$$

If we solve this quadratic equation in e^y, we obtain $e^y = x \pm \sqrt{x^2 - 1}$, or

$$y = \ln (x \pm \sqrt{x^2 - 1}).$$

Since $\cosh^{-1} x$ is the larger of these two values of y, we have

$$\cosh^{-1} x = \ln (x + \sqrt{x^2 - 1}).$$

The other inverse hyperbolic functions may be treated similarly. Thus

10.10.1
$$\sinh^{-1} x = \ln (x + \sqrt{x^2 + 1}) \qquad \text{(any } x\text{)}.$$

10.10.2
$$\cosh^{-1} x = \ln (x + \sqrt{x^2 - 1}), \qquad x \geq 1.$$

10.10.3
$$\tanh^{-1} x = \frac{1}{2} \ln \frac{1 + x}{1 - x}, \qquad -1 < x < 1.$$

To find the derivative of the inverse hyperbolic sine, let $y = \sinh^{-1} u$, where u is a differentiable function of x. Then $u = \sinh y$ and $du/dy = \cosh y$. Thus

$$\frac{dy}{du} = \frac{1}{du/dy} = \frac{1}{\cosh y} = \frac{1}{\sqrt{1 + \sinh^2 y}} = \frac{1}{\sqrt{u^2 + 1}},$$

and

$$\frac{dy}{dx} = \frac{dy}{du} \cdot \frac{du}{dx} = \frac{1}{\sqrt{u^2 + 1}} \cdot \frac{du}{dx}.$$

We may treat the other inverse hyperbolic functions similarly. These derivatives may also be obtained by differentiating the formulas 10.10.1 to 10.10.3. Thus

10.10.4
$$D_x(\sinh^{-1} u) = \frac{1}{\sqrt{u^2 + 1}} \cdot D_x u, \qquad x \in R.$$

10.10.5
$$D_x(\cosh^{-1} u) = \frac{1}{\sqrt{u^2 - 1}} \cdot D_x u, \quad u > 1.$$

10.10.6
$$D_x(\tanh^{-1} u) = \frac{1}{1 - u^2} \cdot D_x u, \quad -1 < u < 1.$$

10.10.7
$$D_x(\coth^{-1} u) = \frac{1}{1 - u^2} \cdot D_x u, \quad u^2 > 1.$$

Exercises

In Exercises 1 to 10, differentiate the indicated functions.

1. $y = \cosh^{-1} x^3$.

2. $y = \tanh^{-1}(2x^5 - 1)$.

3. $y = x \sinh^{-1} 2x$.

4. $y = \sqrt{\coth^{-1} x}$.

5. $y = \tanh^{-1} e^x$.

6. $y = \ln \sinh^{-1} x$.

7. $y = \cosh^{-1} \dfrac{3}{x^2}$.

8. $y = \dfrac{\tanh^{-1} x}{x}$.

9. $y = \coth^{-1}(\ln x)$.

10. $y = \sinh^{-1}(\sin x)$.

11. Prove: $\sinh^{-1} x = \ln(x + \sqrt{x^2 - 1})$, $x \in R$.

12. Prove: $\tanh^{-1} x = \dfrac{1}{2} \ln\left(\dfrac{1 + x}{1 - x}\right)$.

13. Prove: $D_x \cosh^{-1} u = \dfrac{1}{\sqrt{u^2 - 1}} D_x u$, $u > 1$.

14. Prove: $D_x \tanh^{-1} u = \dfrac{1}{1 - u^2} D_x u$, $-1 < u < 1$.

In Exercises 15 to 20, find the antiderivatives of the indicated functions. (*Hint:* See 10.10.4 to 10.10.7.)

15. $y = \dfrac{1}{\sqrt{4x^2 + 1}}$.

16. $y = \dfrac{1}{x\sqrt{(\ln x)^2 - 1}}$.

17. $y = \csc x$. $\left(\text{*Hint:* } \csc x = \dfrac{\sin x}{1 - \cos^2 x}\right)$.

18. $y = \dfrac{e^x}{1 - e^{2x}}$.

19. $y = \dfrac{x^2}{\sqrt{25x^6 + 1}}$.

20. $y = \dfrac{-1}{x^2 + 4x + 3}$.

$\left[\text{*Hint:* } \dfrac{-1}{x^2 + 4x + 3} = \dfrac{1}{1 - [1/(x + 2)]^2} \cdot \dfrac{-1}{(x + 2)^2}\right]$.

21. Without using 10.10.2, show that $D_x(\cosh^{-1} x) = 1/\sqrt{x^2 - 1}$, $x > 1$.

22. Without using 10.10.3, show that $D_x(\tanh^{-1} x) = 1/(1 - x^2)$, $-1 < x < 1$. $\frac{1}{2} \ln[(1 + x)/(1 - x)]$.

23. Sketch the graph of $y = \tanh^{-1} x$. (*Hint:* Sketch $x = \tanh y$.)

24. Sketch the graph of $y = \operatorname{csch}^{-1} x$.

10.11 REVIEW EXERCISES

In Exercises 1 through 22, differentiate the indicated functions.

1. $\ln \dfrac{x}{2}$.

2. $\sin^2(x^3)$.

3. $e^{x^2 - 4x}$.

4. $\log_{10}(x^5 - 1)$.

5. $\tan (\ln e^x)$.

6. $e^{\ln \cot x}$

7. $2 \tanh \sqrt{x}$.

8. $\tanh^{-1} (\sin x)$.

9. $\sinh^{-1} (\tan x)$.

10. $2 \operatorname{Sin}^{-1} \sqrt{3x}$.

11. $\operatorname{Sec}^{-1} e^x$.

12. $2 \ln \sin \dfrac{x}{2}$.

13. $3 \ln (e^{5x} + 1)$.

14. $\ln (2x^3 - 4x + 5)$.

15. $\cos e^{\sqrt{x}}$.

16. $\ln (\tanh x)$.

17. $2 \cosh^{-1} \sqrt{x}$.

18. $e^{\sec 2x}$

19. $2 \csc e^{\ln \sqrt{x}}$.

20. $(\log_{10} 2x)^{2/3}$.

21. $4 \tan 5x \sec 5x$.

22. $x \operatorname{Tan}^{-1} \dfrac{x^2}{2}$.

In Exercises 23 to 32, find the antiderivatives of the indicated functions and verify your results by differentiation.

23. e^{3x-1}.

24. $6 \cot 3x$.

25. $e^x \sin e^x$.

26. $\dfrac{6x + 3}{x^2 + x - 5}$.

27. $\dfrac{e^{x+2}}{e^{x+3} + 1}$.

28. $4x \cos x^2$.

29. $\dfrac{3}{\sqrt{1 - x^2}}$

30. $\dfrac{\cos x}{1 + \sin^2 x}$.

31. $\dfrac{-1}{x + x(\ln x)^2}$.

32. $\operatorname{sech}^2 (x - 3)$.

In Exercises 33 and 34, find the intervals on which f is increasing and the intervals on which f is decreasing. Find where the graph of f is concave upward and where it is concave downward. Find any extreme values and points of inflection. Then sketch the graph of f.

33. $f(x) = \sin x + \cos x$, $-\pi \le x \le \pi$.

34. $f(x) = \dfrac{x^2}{e^x}$, $x \in R$.

35. Evaluate $\displaystyle\int_1^3 \dfrac{2^x + 2^{-x}}{2^x - 2^{-x}} \, dx$.

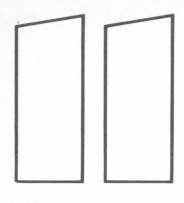

techniques of integration

11.1 INTRODUCTION

The definite integral

$$(1) \qquad \int_a^b f(x)\, dx$$

was defined (8.3.4) to be a limit of Riemann sums of the form $\sum_{i=1}^{n} f(w_i)\, \Delta x_i$. This limit always exists when f is continuous on the closed interval $[a, b]$.

According to the fundamental theorem of calculus (8.4.3), the value of the definite integral (1) can be calculated exactly if we can find an antiderivative of its integrand $f(x)$. By an antiderivative of $f(x)$ we mean any function F whose derivative is $f(x)$.

There are, of course, some definite integrals that exist but whose integrand has no antiderivative that can be expressed in terms of elementary functions (the functions that we have been studying). But the great majority of definite integrals encountered in college mathematics *can* be evaluated by the fundamental theorem. Our purpose here is to systematize and enlarge our technique for finding antiderivatives of given integrands.

Differentiation is a direct process; the formulas for writing down the derivative of any differentiable function are well known, and we apply them without hesitation. But antidifferentiation is an *inverse* process, the inverse of differentiation. A function f is given and we are asked to discover which function was differentiated to give f. Although this search for a function F whose derivative is the given integrand $f(x)$ is somewhat of a "fishing expedition," we can systematize the search and improve our techniques, as will be shown in this chapter.

Before proceeding, we shall adopt some standard terminology. Instead of the symbol $A_x f(x)$ for the most general antiderivative of f with respect to x, it is customary to write $\int f(x)\, dx$. We can read the symbol $\int f(x)\, dx$ just as we did our $A_x f(x)$, "the antiderivatives of f with respect to x," but it is usually read "the indefinite integral of $f(x)$" or simply "the integral of $f(x)$."

11.1.1 Definition. The **indefinite integral** of a function f, written

$$\int f(x)\, dx,$$

is the most general antiderivative of f; that is,

$$\int f(x)\, dx = F(x) + C$$

if and only if $F'(x) = f(x)$.

As an illustration, $\int x^3\, dx = \frac{1}{4}x^4 + C$ because $D_x(\frac{1}{4}x^4 + C) = x^3$; and $\int \cos u\, du = \sin u + C$, since $D_u(\sin u + C) = \cos u$.

When we speak of **integrating** $f(x)$, we shall mean finding the most general antiderivative of $f(x)$. The expression **technique of integration** will mean a systematic procedure for finding antiderivatives, and a **formula for integration** will mean a formula for antidifferentiation.

We delayed introduction of this terminology so that the reader would make a clear distinction between definite integrals and antiderivatives (now to be called indefinite integrals or integrals). Despite the similarity of the symbols for the definite and indefinite integrals (antiderivatives), their meanings are very different and should not be confused. We repeat that the definite integral $\int_a^b f(x)\, dx$ is a limit of Riemann sums, and the indefinite integral $\int f(x)\, dx$ is the most general antiderivative of $f(x)$. The only connection between them is provided by the fundamental theorem of calculus—namely, that the value of the definite integral can be computed by using the indefinite integral (antiderivative) of its integrand.

The notation $\int f(x)\, dx$ is due to Leibniz. An advantage of this symbol is that it can be thought of in two ways: (a) as consisting of two parts, $\int$ and dx, with $f(x)$ sandwiched between them, meaning the most general antiderivative of f; or (b) as one part, $\int$, followed by the differential $f(x)\, dx$, meaning the most general function whose *differential* is $f(x)\, dx$.

11.2 THE BASIC INTEGRATION FORMULAS

Since an antiderivative of $f(x)$ is a function F such that $F'(x) = f(x)$, each formula for differentiation gives rise to a formula for antidifferentiation. In Chapters 7 and 10 we learned and used many formulas for antidifferentiation. They will now be listed in the Leibniz notation as indefinite integrals, and each can be verified by differentiating the right-hand member. Since these integration formulas are inverses of our familiar differentiation formulas, it should be easy to remember them.

11.2.1
$$\int u^n \, du = \frac{u^{n+1}}{n+1} + C, \qquad n \neq -1.$$

11.2.2
$$\int \frac{du}{u} = \ln|u| + C \qquad \text{if} \qquad u \neq 0.$$

11.2.3
$$\int e^u \, du = e^u + C.$$

11.2.4
$$\int a^u \, du = \frac{a^u}{\ln a} + C.$$

11.2.5
$$\int \sin u \, du = -\cos u + C.$$

11.2.6
$$\int \cos u \, du = \sin u + C.$$

11.2.7
$$\int \sec^2 u \, du = \tan u + C.$$

11.2.8
$$\int \csc^2 u \, du = -\cot u + C.$$

11.2.9
$$\int \sec u \tan u \, du = \sec u + C.$$

11.2.10
$$\int \csc u \cot u \, du = -\csc u + C.$$

11.2.11
$$\int \frac{du}{\sqrt{1-u^2}} = \text{Sin}^{-1} u + C, \qquad |u| < 1.$$

11.2.12
$$\int \frac{du}{1+u^2} = \text{Tan}^{-1} u + C.$$

11.2.13
$$\int \frac{du}{|u|\sqrt{u^2-1}} = \text{Sec}^{-1} u + C, \qquad |u| > 1.$$

Four other basic integration formulas are listed here for reference. They are not simply reversals of standard differentiation formulas but can be derived easily from 11.2.2, as will be shown below.

11.2.14
$$\int \tan u \, du = -\ln |\cos u| + C.$$

11.2.15
$$\int \cot u \, du = \ln |\sin u| + C.$$

11.2.16
$$\int \sec u \, du = \ln |\sec u + \tan u| + C.$$

11.2.17
$$\int \csc u \, du = \ln |\csc u - \cot u| + C.$$

Theorem 7.2.3. is restated in the following two general formulas for integration.

11.2.18
$$\int [f(u) + g(u)] \, du = \int f(u) \, du + \int g(u) \, du.$$

11.2.19
$$\int kf(u) \, du = k \int f(u) \, du \qquad \text{if } k \text{ is a constant.}$$

The first says that the indefinite integral of a sum of functions is equal to the sum of the integrals of the functions; and the second says that a *constant* factor may be moved across the integral sign.

In order to derive the integration formulas 11.2.14 to 11.2.17, we shall change each of the integrals to the form

$$\int \frac{dw}{w}$$

and then use 11.2.2.

For 11.2.14, we write

$$\int \tan u \, du = \int \frac{\sin u}{\cos u} \, du = -\int \frac{(-\sin u \, du)}{\cos u}$$

Let $w = \cos u$. Then $dw = -\sin u \, du$, and

$$-\int \frac{(-\sin u \, du)}{\cos u} = -\int \frac{dw}{w} = -\ln |w| + C,$$

by 11.2.2. Since $-\ln |w| = -\ln |\cos u|$,

$$\int \tan u \, du = -\ln |\cos u| + C,$$

which is 11.2.14.

Formula 11.2.15, $\int \cot u \, du = \ln |\sin u| + C$, is derived similarly by writing $\cot u = \cos u/\sin u$.

Example 1. Find $\int (x - 2)^2 \, dx$.

Solution. We use integration formula 11.2.1, letting $u = x - 2$. Then $du = dx$, and

$$\int (x - 2)^2 \, dx = \int u^2 \, du = \frac{u^3}{3} + C = \frac{1}{3}(x - 2)^3 + C.$$

Example 2. Perform the integration $\int \frac{1}{3x + 5} \, dx$.

Solution. Since $d(3x + 5) = 3 \, dx$, we use 11.2.2 with $u = 3x + 5$. Then

$$\int \frac{1}{3x + 5} \, dx = \frac{1}{3} \int \frac{3 \, dx}{3x + 5} = \frac{1}{3} \int \frac{d(3x + 5)}{3x + 5}$$

$$= \frac{1}{3} \int \frac{du}{u} = \frac{1}{3} \ln |u| + C$$

$$= \frac{1}{3} \ln |3x + 5| + C.$$

As to 11.2.16,

$$\int \sec u \, du = \int \sec u \left(\frac{\sec u + \tan u}{\sec u + \tan u} \right) du$$

$$= \int \frac{(\sec u \tan u + \sec^2 u) \, du}{\sec u + \tan u}$$

$$= \int \frac{d(\sec u + \tan u)}{\sec u + \tan u}$$

$$= \ln |\sec u + \tan u| + C,$$

by 11.2.2.

The formula $\int \csc u \, du = \ln |\csc u - \cot u| + C$ is obtained similarly, after multiplying the integrand by $(\csc u - \cot u)/(\csc u - \cot u)$.

Notice that in deriving 11.2.14, we replace $\tan u$ by $\sin u/\cos u$, in 11.2.15 we replace $\cot u$ by $\cos u/\sin u$, in 11.2.16 we insert the factor $(\sec u + \tan u)/(\sec u + \tan u)$, and in deriving 11.2.17 we insert the factor $(\csc u - \cot u)/(\csc u - \cot u)$. In each case, we reduce the given integral to the form $\int (1/w) \, dw$. Some people prefer to remember these devices and apply them to each individual problem, instead of memorizing formulas 11.2.14 to 11.2.17 (see Example 6 below).

Example 3. Find $\int 6e^{2x-7} \, dx$.

Solution. We let $u = 2x - 7$ in 11.2.3. Then $du = 2 \, dx$, and

$$\int 6e^{2x-7} \, dx = 3 \int e^{2x-7}(2 \, dx)$$

$$= 3 \int e^{2x-7} \, d(2x - 7) = 3 \int e^u \, du$$

$$= 3e^u + C = 3e^{2x-7} + C.$$

Example 4. Find $\displaystyle\int \frac{x \, dx}{\cos^2 (x^2)}$.

Solution. To bring the given integral into one of the basic forms, we use $\sec u = 1/\cos u$. Thus

$$\int \frac{x \, dx}{\cos^2 (x^2)} = \int [\sec^2 (x^2)] x \, dx.$$

Let $u = x^2$ in 11.2.7. Then $du = 2x \, dx$, and

$$\int [\sec^2 (x^2)] x \, dx = \frac{1}{2} \int [\sec^2 (x^2)] 2x \, dx$$

$$= \frac{1}{2} \int [\sec^2 (x^2)] \, d(x^2) = \frac{1}{2} \int \sec^2 u \, du$$

$$= \frac{1}{2} \tan u + C = \frac{1}{2} \tan x^2 + C.$$

Example 5. Find $\displaystyle\int \frac{3 \, dx}{\sqrt{1 - 4x^2}}$.

Solution. Let $u = 2x$ in 11.2.11. Then $du = 2 \, dx$, and

$$\int \frac{3 \, dx}{\sqrt{1 - 4x^2}} = \frac{3}{2} \int \frac{2 \, dx}{\sqrt{1 - (2x)^2}} = \frac{3}{2} \int \frac{d(2x)}{\sqrt{1 - (2x)^2}}$$

$$= \frac{3}{2} \int \frac{du}{\sqrt{1 - u^2}} = \frac{3}{2} \operatorname{Sin}^{-1} u + C$$

$$= \frac{3}{2} \operatorname{Sin}^{-1} 2x + C.$$

Example 6. Find $\displaystyle\int \cot (3x + 5) \, dx$.

Solution. Since $\cot u = \cos u / \sin u$, we can express the given integral in the form $\int du/u$ and then use 11.2.2. Thus

$$\int \cot (3x + 5) \, dx = \int \frac{\cos (3x + 5)}{\sin (3x + 5)} \, dx.$$

Let $u = \sin (3x + 5)$ in 11.2.2. Then $du = \cos (3x + 5) \, d(3x + 5) = \cos (3x + 5)(3 \, dx)$, and

$$\int \frac{\cos (3x + 5)}{\sin (3x + 5)} \, dx = \frac{1}{3} \int \frac{\cos (3x + 5)(3 \, dx)}{\sin (3x + 5)}$$

$$= \frac{1}{3} \int \frac{d[\sin (3x + 5)]}{\sin (3x + 5)}$$

$$= \frac{1}{3} \ln |\sin (3x + 5)| + C.$$

Of course, if we remember formula 11.2.15, we can write

$$\int \cot(3x + 5)\, dx = \frac{1}{3} \int [\cot(3x + 5)](3\, dx)$$

$$= \frac{1}{3} \int \cot(3x + 5)\, d(3x + 5)$$

$$= \frac{1}{3} \ln|\sin(3x + 5)| + C,$$

where $u = 3x + 5$ and $du = 3\, dx$.

A set of easy exercises follows. It is advisable to do all of them as a review of the basic formulas for antidifferentiation and to become accustomed to the Leibniz notation for indefinite integrals.

Exercises

Perform the indicated integrations.

1. $\int (x - 1)^4\, dx.$

2. $\int \sqrt{2x}\, dx.$

3. $\int \frac{dx}{x + 1}.$

4. $\int \frac{e^x}{1 + 2e^x}\, dx.$

5. $\int 3t\sqrt{2 + t^2}\, dt.$

6. $\int \frac{dt}{\sqrt{1 + t}}.$

7. $\int \sec^2(s + 3)\, ds.$

8. $\int x \csc^2(x^2)\, dx.$

9. $\int \frac{\tan z}{\cos z}\, dz. \left(Hint: \sec z = \frac{1}{\cos z}.\right)$

10. $\int 5 \sec 5x \tan 5x\, dx.$

11. $\int \frac{\cos \sqrt{x}}{\sqrt{x}}\, dx.$

12. $\int e^{2x+1}\, dx.$

13. $\int \frac{e^{\sin x}}{\sec x}\, dx. \left(Hint: \cos x = \frac{1}{\sec x}.\right)$

14. $\int \frac{dt}{\csc t}.$ (*Hint:* Use 11.2.5.)

15. $\int \frac{x\, dx}{\sqrt{1 - x^4}}.$ (*Hint:* Let $u = x^2$.)

16. $\int \frac{\cos x}{1 + \sin^2 x}\, dx.$ (*Hint:* Let $u = \sin x$.)

17. $\int x \csc x^2 \cot x^2\, dx.$ (*Hint:* Let $u = x^2$.)

18. $\int \frac{\csc t}{\tan t}\, dt. \left(Hint: \cot t = \frac{1}{\tan t}.\right)$

19. $\int \tan(5x - 1)\, dx.$

20. $\int e^x (\cot e^x)\, dx.$ (*Hint:* Let $u = e^x$.)

11.3 INTEGRATION BY SUBSTITUTION

The basic integration formulas can be extended to a surprisingly large variety of integrands by a technique known as **substitution.**

Notice that if we substitute $u = g(x)$ throughout the integration formula

$$\int f(u)\, du = F(u) + C,$$

we get

(1) $$\int f(g(x))\, d(g(x)) = F(g(x)) + C;$$

and if we replace $d(g(x))$ by $g'(x)\, dx$ as suggested by the differential $du = g'(x)\, dx$, (1) becomes

$$\int f(g(x)) \cdot g'(x)\, dx = F(g(x)) + C.$$

11.3.1 **Method of Substitution.** Let $u = g(x)$ define a differentiable function g whose range is an interval I, and let f be a function defined on I. If we make the formal substitution $u = g(x)$ throughout an integration formula

$$\int f(u)\, du = F(u) + C,$$

then the result

$$\int f(g(x))\, d(g(x)) = F(g(x)) + C$$

is valid.

The mathematical justification for this important method of integration by substitution is given at the end of this section.

Example 1. Find $\displaystyle\int \frac{\sin \sqrt{x}}{\sqrt{x}}\, dx.$

Solution. As it stands, the given integral does not come under 11.2.5. But if we substitute $u = g(x) = \sqrt{x}$, then $du = g'(x)\, dx = [1/(2\sqrt{x})]\, dx$ and, by 11.3.1 and 11.2.5,

$$\int \frac{\sin \sqrt{x}}{\sqrt{x}}\, dx = 2 \int \sin \sqrt{x} \left[\frac{1}{2\sqrt{x}} \right] dx$$

$$= 2 \int \sin \sqrt{x}\, d(\sqrt{x})$$

$$= 2 \int \sin u\, du = -2 \cos u + C$$

$$= -2 \cos \sqrt{x} + C.$$

Example 2. Find $\displaystyle\int \sin^2 x \cos x\, dx.$

Solution. Let $u = \sin x$ and $du = \cos x\, dx$ in 11.2.1. Then

$$\int \sin^2 x\, (\cos x\, dx) = \int \sin^2 x\, d(\sin x)$$

$$= \int u^2\, du = \frac{u^3}{3} + C = \frac{\sin^3 x}{3} + C.$$

Example 3. Find $\int x\sqrt{x^2 + 11}\, dx$.

Solution. Let $u = x^2 + 11$ in 11.2.1. Then $du = 2x\, dx$, and

$$\int x\sqrt{x^2 + 11}\, dx = \frac{1}{2} \int (x^2 + 11)^{1/2}(2x\, dx)$$

$$= \frac{1}{2} \int (x^2 + 11)^{1/2}\, d(x^2 + 11) = \frac{1}{2} \int u^{1/2}\, du$$

$$= \frac{1}{2} \frac{u^{3/2}}{\frac{3}{2}} + C = \frac{u^{3/2}}{3} + C$$

$$= \frac{1}{3} (x^2 + 11)^{3/2} + C.$$

Example 4. Find $\int \frac{6e^{1/x}}{x^2}\, dx$.

Solution. Let $u = 1/x$. Then $du = (-1/x^2)\, dx$ and, by 11.2.3,

$$\int \frac{6e^{1/x}}{x^2}\, dx = -6 \int e^{1/x}\left(-\frac{1}{x^2}\, dx\right)$$

$$= -6 \int e^{1/x}\, d\left(\frac{1}{x}\right) = -6 \int e^u\, du$$

$$= -6e^u + C = -6e^{1/x} + C.$$

After a little practice, the reader will probably omit writing the steps that involve u explicitly. But he should continue to write down what he chooses for u and what du is.

Example 5. Find $\int \frac{\sec^2 (\sin x)}{\sec x}\, dx$.

Solution. Let $u = \sin x$ in 11.2.7. Then $du = \cos u\, du$. Since $\cos u = 1/\sec u$, we can write

$$\int \frac{\sec^2 (\sin x)}{\sec x}\, dx = \int [\sec^2 (\sin x)] \cos x\, dx$$

$$= \int \sec^2 (\sin x)\, d(\sin x)$$

$$= \tan (\sin x) + C.$$

Example 6. Find $\int \frac{4e^x}{\sqrt{1 - e^{2x}}}\, dx$.

Solution. We use 11.2.1, with $u = e^x$. Then $du = e^x\, dx$, and

$$\int \frac{4e^x}{\sqrt{1 - e^{2x}}}\, dx = 4 \int \frac{e^x\, dx}{\sqrt{1 - (e^x)^2}} = 4 \int \frac{d(e^x)}{\sqrt{1 - (e^x)^2}}$$

$$= 4 \operatorname{Sin}^{-1} e^x + C.$$

Example 7. Integrate $\displaystyle\int \frac{x^2 - x}{x + 1}\, dx$.

Solution. When the integrand is the quotient of two polynomials and the numerator is of equal or greater degree than the denominator, *always divide the denominator into the numerator first.* Using long division, we obtain

$$\frac{x^2 - x}{x + 1} = x - 2 + \frac{2}{x + 1}.$$

Hence

$$\int \frac{x^2 - x}{x + 1}\, dx = \int (x - 2)\, dx + 2 \int \frac{dx}{x + 1}$$

$$= \int (x - 2)\, d(x - 2) + 2 \int \frac{d(x + 1)}{x + 1}$$

$$= \frac{(x - 2)^2}{2} + 2 \ln |x + 1| + C.$$

Example 8. Find $\displaystyle\int \frac{a^{\tan t}\, dt}{\cos^2 t}$.

Solution. We use 11.2.4, with $u = \tan t$. Then $du = \sec^2 t\, dt$, and

$$\int \frac{a^{\tan t}\, dt}{\cos^2 t} = \int a^{\tan t}\, (\sec^2 t\, dt)$$

$$= \int a^{\tan t}\, d(\tan t) = \frac{a^{\tan t}}{\ln a} + C.$$

Success in *integration by substitution* depends on a judicious choice of part of the integrand to be equated to a new variable u. Skill in this is acquired by practice.

The mathematical justification for the method of substitution is established in the theorem that follows.

11.3.2 Theorem. If $u = g(x)$ defines a differentiable function g whose range is an interval I, and if f is a function defined on I such that $\int f(u)\, du = F(u) + C$, *then*

$$\int f(g(x)) \cdot g'(x)\, dx = F(g(x)) + C.$$

Proof. It follows from $\int f(u)\, du = F(u) + C$ and the definition of an indefinite integral that

(2) $$\frac{d}{du} F(u) = f(u);$$

and since $u = g(x)$, we have

(3) $$\frac{du}{dx} = g'(x).$$

From the chain rule, and from (2) and (3),

$$\frac{d}{dx}F(g(x)) = \frac{d}{dx}F(u) = F'(u) \cdot \frac{du}{dx}$$

$$= f(u) \cdot \frac{du}{dx} = f(g(x)) \cdot g'(x).$$

Thus

(4) $$\frac{d}{dx}F(g(x)) = f(g(x)) \cdot g'(x).$$

But (4) and the definition of an indefinite integral imply

$$\int f(g(x)) \cdot g'(x)\, dx = F(g(x)) + C. \quad \blacksquare$$

Exercises

In each of the following exercises, perform the indicated integration.

1. $\int (7x - 1)^{12}\, dx.$

2. $\int (4x^3 + 3x - 1)^4(4x^2 + 1)\, dx.$

3. $\int (12x^5 - x)(8x^6 - 2x^2 + 19)^{10}\, dx.$

4. $\int (\sin 6x)^3 \cos 6x\, dx.$

5. $\int \cos \frac{x}{3} \sin \frac{x}{3}\, dx.$

6. $\int x^2\sqrt{7x^3 + 5}\, dx.$

7. $\int x \sin 2x^2\, dx.$

8. $\int \frac{\cos \sqrt{1 - x}}{\sqrt{1 - x}}\, dx.$

9. $\int e^{x^2+2x-1}(x + 1)\, dx.$

10. $\int \frac{dt}{e^{3t}}.$

11. $\int \frac{e^{\sqrt{2x+1}}}{\sqrt{2x + 1}}\, dx.$

12. $\int \frac{e^{\csc t} \cot t}{\sin t}\, dt.$

13. $\int \frac{(\ln t^2)^9}{t}\, dt.$

14. $\int \frac{(\text{Cos}^{-1} 2w)^7}{\sqrt{1 - 4w^2}}\, dw.$

15. $\int \frac{\sin x}{\cos^5 x}\, dx.$

16. $\int (\cos^2 x - \sin^2 x)^3(\sin x \cos x)\, dx.$ (*Hint:* Let $u = \cos^2 x - \sin^2 x.$)

17. $\int \frac{x\, dx}{6x^2 - 19}.$

18. $\int \frac{\sin t}{\cos t}\, dt.$

19. $\int \frac{2x^2 + x}{x + 1}\, dx.$

20. $\int \frac{3x^4 - 15x^3 + 2x - 5}{x^2 - 5x}\, dx.$

21. $\int \frac{4x^4 + x^3 + 20x^2 + 2x}{x^2 + 5}\, dx.$

22. $\int (e^t - e^{-t})^2(e^t + e^{-t})\, dt.$

23. $\int \sin^2 y \cos y\sqrt{\sin^3 y + 4}\, dy.$ (*Hint:* Let $u = \sin^3 y + 4.$)

24. $\int \dfrac{\sqrt{\tan x}}{1 - \sin^2 x}\, dx.$ $\left(Hint\!: \sec^2 x = \dfrac{1}{\cos^2 x}.\right)$

25. $\int x5^{x^2-1}\, dx.$ (*Hint:* Use 11.2.4.)

26. $\int 6^{\cos x} \sin x\, dx.$

27. $\int e^{\sin \theta \cos \theta} \cos 2\theta\, d\theta.$
    (*Hint:* Let $u = \sin \theta \cos \theta$.)

28. $\int \dfrac{e^{2 \ln (2x-1)}}{2x - 1}\, dx.$ [*Hint:* See equation (3), Section 10.3.]

11.4 SUBSTITUTION IN THE REMAINING BASIC FORMULAS FOR INTEGRATION

In the preceding section we used the method of substitution to transform certain integrals so that the basic integration formulas could be used. Three purposes were served: the method of substitution was introduced, the reader became familiar with the Leibniz notation for indefinite integrals (antiderivatives), and the first four basic integration formulas were reviewed.

We now apply the method of substitution to integrals that can be transformed into the remaining basic integration formulas, 11.2.5 to 11.2.17. As noted, success with the method of substitution depends on the choice of an appropriate part of the given integrand to be represented by a new variable u. Skill in this, and in all other techniques for integration, can only be acquired by practice.

Example 1. Find $\int \cos (8x + 3)\, dx.$

Solution. If we let $u = 8x + 3$ in 11.2.6, then $du = 8\, dx$ and

$$\int \cos (8x + 3)\, dx = \frac{1}{8} \int \cos (8x + 3)(8\, dx)$$

$$= \frac{1}{8} \int \cos (8x + 3)\, d(8x + 3)$$

$$= \frac{1}{8} \int \cos u\, du = \frac{1}{8} \sin u + C$$

$$= \frac{1}{8} \sin (8x + 3) + C.$$

Example 2. Find $\int e^{2x} \csc^2 e^{2x}\, dx.$

Solution. Let $u = e^{2x}$ in 11.2.8; then $du = e^{2x} 2\, dx$. So

$$\int e^{2x} \csc^2 e^{2x}\, dx = \frac{1}{2} \int \csc^2 e^{2x}(e^{2x}2\, dx)$$

$$= \frac{1}{2} \int \csc^2 e^{2x}\, d(e^{2x})$$

$$= \frac{1}{2} \int \csc^2 u\, du = -\frac{1}{2} \cot u + C$$

$$= -\frac{1}{2} \cot e^{2x} + C.$$

Example 3. Find $\int \tan^2 y \, dy$.

Solution. None of the basic integration formulas has $\tan^2 y$ for its integrand. But 11.2.7 has $\sec^2 y$, and we know that $1 + \tan^2 y = \sec^2 y$. Thus

$$\int \tan^2 y \, dy = \int (\sec^2 y - 1) \, dy = \int \sec^2 y \, dy - \int dy$$

$$= \tan y - y + C.$$

Example 4. Find $\int \dfrac{dx}{\sqrt{1 - 16x^2}}$.

Solution. We let $u = 4x$ in 11.2.11. Then $du = 4 \, dx$, and

$$\int \frac{dx}{\sqrt{1 - 16x^2}} = \frac{1}{4} \int \frac{4 \, dx}{\sqrt{1 - (4x)^2}} = \frac{1}{4} \int \frac{d(4x)}{\sqrt{1 - (4x)^2}}$$

$$= \frac{1}{4} \operatorname{Sin}^{-1} 4x + C.$$

Example 5. Find $\int \dfrac{dx}{4 + 9x^2}$.

Solution. To put the given integral into the form of 11.2.12, we first factor 4 from the denominator.

$$\int \frac{dx}{4 + 9x^2} = \frac{1}{4} \int \frac{dx}{1 + 9x^2/4} = \frac{1}{4} \int \frac{dx}{1 + (3x/2)^2}$$

$$= \frac{1}{2 \cdot 3} \int \frac{(3/2) \, dx}{1 + (3x/2)^2}$$

$$= \frac{1}{6} \int \frac{d(3x/2)}{1 + (3x/2)^2}$$

$$= \frac{1}{6} \operatorname{Tan}^{-1} \frac{3x}{2} + C,$$

by 11.2.12 with $u = 3x/2$.

Since integrals depending on formulas 11.2.11 to 11.2.13 for their solution seldom have a one in the denominator (see Example 5, above), it may be worthwhile to memorize the following three formulas, in place of 11.2.11 to 11.2.13.

11.4.1
$$\int \frac{du}{\sqrt{a^2 - u^2}} = \operatorname{Sin}^{-1} \frac{u}{a} + C, \qquad \left| \frac{u}{a} \right| < 1.$$

11.4.2
$$\int \frac{du}{a^2 + u^2} = \frac{1}{a} \operatorname{Tan}^{-1} \frac{u}{a} + C.$$

11.4.3
$$\int \frac{du}{|u|\sqrt{u^2 - a^2}} = \frac{1}{a} \operatorname{Sec}^{-1} \frac{u}{a} + C, \qquad \left| \frac{u}{a} \right| > 1.$$

Each may be proved by showing that the derivative of the right member is the same as the integrand of the corresponding left member. For instance, to prove 11.4.1:

$$D_u \left[\text{Sin}^{-1} \frac{u}{a} + C \right] = \frac{1}{\sqrt{1 - (u/a)^2}} \, D_u \left(\frac{u}{a} \right)$$

$$= \frac{1}{a\sqrt{1 - (u/a)^2}} = \frac{1}{\sqrt{a^2 - u^2}}.$$

Therefore

$$\int \frac{du}{\sqrt{a^2 - u^2}} = \text{Sin}^{-1} \frac{u}{a} + C.$$

Example 6. Find $\displaystyle\int \frac{dy}{\sqrt{25 - 9y^2}}$.

Solution. By letting $a = 5$ and $u = 3y$ in 11.4.1, and we obtain

$$\int \frac{dy}{\sqrt{25 - 9y^2}} = \frac{1}{3} \int \frac{dy}{\sqrt{5^2 - (3y)^2}} = \frac{1}{3} \, \text{Sin}^{-1} \frac{3y}{5} + C.$$

Example 7. Find $\displaystyle\int \frac{7 \, dx}{x^2 - 6x + 25}$.

Solution. Many integrals with quadratic expressions in their denominators can be reduced to basic forms by **completing the square** in the quadratic expression. This procedure is often useful and should be tried when a quadratic expression appears. Thus

$$\int \frac{7 \, dx}{x^2 - 6x + 25} = \int \frac{7 \, dx}{(x^2 - 6x + 9) + (25 - 9)}$$

$$= \int \frac{7 \, dx}{(x - 3)^2 + 16}$$

$$= 7 \int \frac{d(x - 3)}{4^2 + (x - 3)^2}$$

$$= \frac{7}{4} \, \text{Tan}^{-1} \frac{x - 3}{4} + C,$$

by 11.4.2 with $a = 4$ and $u = x - 3$.

Notice that in the first step we "completed the square" in the quadratic expression $x^2 - 6x$ by adding and subtracting 9.

Example 8. Find $\displaystyle\int \frac{\text{Tan}^{-1} x}{1 + x^2} \, dx$.

Solution. In order to integrate, we use the very first basic integration formula, 11.2.1, with $u = \text{Tan}^{-1} x$. Then $du = [1/(1 + x^2)] \, dx$ and

$$\int \frac{\text{Tan}^{-1} x}{1 + x^2} \, dx = \int \text{Tan}^{-1} x \left(\frac{dx}{1 + x^2} \right)$$

$$= \int \text{Tan}^{-1} x \, d(\text{Tan}^{-1} x)$$

$$= \frac{(\text{Tan}^{-1} x)^2}{2} + C.$$

Exercises

Perform the indicated integration in Exercises 1 through 40.

1. $\int \sin (13x - 11) \, dx.$

2. $\int x \sin (x^2 - 1) \, dx.$

3. $\int \cos (\tfrac{1}{2}\pi x) \, dx.$

4. $\int \dfrac{\cos \ln 4x^2}{x} \, dx.$

5. $\int \csc^2 \pi x \, dx.$

6. $\int 3x \sec^2 2x^2 \, dx.$

7. $\int e^{3t} \sin e^{3t-2} \, dt.$

8. $\int \dfrac{\cos \ln t^2}{t} \, dt.$

9. $\int e^{2y} \cot^2 e^{2y} \, dy.$

10. $\int \dfrac{\tan^2 \ln 5y^2}{y} \, dy.$

11. $\int w \sec (3w^2 + 1) \tan (3w^2 + 1) \, dw.$

12. $\int \dfrac{\csc e^{-w} \cot e^{-w}}{e^w} \, dw.$

13. $\int \dfrac{\sin x - \cos x}{\sin x} \, dx.$

14. $\int x^2 \sec x^3 \, dx.$

15. $\int \dfrac{\tan 3\theta}{\cos 3\theta} \, d\theta.$

16. $\int \dfrac{\sec^2 \sqrt{2\theta - 3}}{\sqrt{2\theta - 3}} \, d\theta.$

17. $\int x \sec^2 (5x^2 - 1) \, dx.$

18. $\int e^x \sec^2 e^x \, dx.$

19. $\int \tan t \, dt.$

20. $\int \cot 3x \, dx.$

21. $\int \tan (5u - 2) \, du.$

22. $\int 3u \cot (5u^2 + 1) \, du.$

23. $\int \dfrac{\csc^2 (\ln x)}{x} \, dx.$

24. $\int \dfrac{\cot^2 (\ln 3y^2)}{y} \, dy.$

25. $\int \tan^2 3y \, dy.$

26. $\int \dfrac{\sqrt{x} \, dx}{\csc (2x^{3/2} + 1)}.$

27. $\int \dfrac{5e^x}{\sqrt{1 - e^{2x}}} \, dx.$

28. $\int \dfrac{5e^{2x}}{\sqrt{1 - e^{2x}}} \, dx.$

29. $\int \dfrac{\sec 4x}{\cot 4x} \, dx.$

30. $\int \dfrac{x}{x^4 + 1} \, dx.$

31. $\int \dfrac{\csc (\ln x^2) \cot (\ln x^2)}{x} \, dx.$

32. $\int \dfrac{3x \cos x^2}{\sin x^2} \, dx.$

33. $\int \dfrac{\sin (4t - 1)}{1 - \sin^2 (4t - 1)} \, dt.$

34. $\int \dfrac{z + 2}{\cot (z^2 + 4z - 3)} \, dz.$

35. $\int \dfrac{z \csc \sqrt{3z^2 - 5}}{\sqrt{3z^2 - 5}} \, dz.$

36. $\int e^{2x} \sec e^{2x} \, dx.$

37. $\displaystyle\int \frac{\sec^3 x + e^{\sin x}}{\sec x}\, dx.$

38. $\displaystyle\int \frac{(6t - 1) \sin \sqrt{3t^2 - t - 1}}{\sqrt{3t^2 - t - 1}}\, dt.$

39. $\displaystyle\int \frac{t^2 \cos (t^3 - 2)}{\sin^2 (t^3 - 2)}\, dt.$

40. $\displaystyle\int \frac{\sec x \tan x}{1 + \sec^2 x}\, dx.$ (*Hint:* Use 11.2.12.)

In Exercises 41 through 50, use formulas 11.4.1 to 11.4.3 (or 11.2.11 to 11.2.13) to perform the indicated integrations.

41. $\displaystyle\int \frac{5\, dx}{\sqrt{9 - 4x^2}}.$

42. $\displaystyle\int \frac{\sin x}{16 + \cos^2 x}\, dx.$

43. $\displaystyle\int \frac{dt}{|2t|\sqrt{4t^2 - 1}}.$

44. $\displaystyle\int \frac{e^{3t}\, dt}{\sqrt{9 - e^{6t}}}.$

45. $\displaystyle\int \frac{d\theta}{(1 + \theta)\sqrt{\theta}}.$

46. $\displaystyle\int \frac{d\theta}{\theta^2 - 6\theta + 10}.$

47. $\displaystyle\int \frac{dx}{\sqrt{3x - x^2}}.$

48. $\displaystyle\int \frac{dx}{\sqrt{3x + 1}\sqrt{3x}}.$

49. $\displaystyle\int \frac{y\, dy}{\sqrt{16 - 9y^4}}.$

50. $\displaystyle\int \frac{\sec^2 2y}{9 + \tan^2 2y}\, dy.$

51. Find the volume of the solid generated by rotating about the x axis the region bounded by the x axis, the curve $y = \sec x$, and the lines $x = 0$ and $x = \pi/4$.

52. Find the volume of the solid generated by revolving the region

$$\{P{:}(x, y)\,|\,{-3} \leq x \leq 0,\, 0 \leq y \leq 2e^x \tan e^{2x}\}$$

about the x axis. Make a sketch.

53. Find the length of the curve $y = \ln \cos x$, from $x = 0$ to $x = \pi/4$.

11.5 INTEGRATION BY PARTS

One of the most powerful methods of integration is called integration by parts. It is based on the inversion of the formula for the derivative of the product of two functions.

Let $u = f(x)$ and $v = g(x)$ define smooth functions (that is, functions having continuous derivatives). We know that

$$D_x[f(x)g(x)] = f(x)g'(x) + g(x)f'(x);$$

and by integrating both members of this equation, we obtain

$$f(x)g(x) = \int f(x)g'(x)\, dx + \int g(x)f'(x)\, dx$$

or

$$\int f(x)g'(x)\, dx = f(x)g(x) - \int g(x)f'(x)\, dx.$$

Since $dv = g'(x)\, dx$ and $du = f'(x)\, dx$, the preceding equation is usually written symbolically as

11.5.1
$$\int u \, dv = uv - \int v \, du$$

and is known as the formula for **integration by parts.** It enables us to shift the problem from integrating one form, $\int u \, dv$, to that of integrating another, $\int v \, du$, which may be easier to handle. Success with this method usually depends on our choice of u and dv. It is often (but not always) best to include in our selection for dv the most complicated part of the given integrand that we are able to integrate. Certainly we should choose u as a function that is simplified by differentiation but that leaves a dv that we can integrate.

Example 1. Find $\int x \cos x \, dx$.

Solution. Our first task is to decide which part of $x \cos x \, dx$ is to be the u of formula 11.5.1, and which part dv. Obviously, dx must be included in the dv. We can integrate either $x \, dx$ or $\cos x \, dx$; but since x is simplified by differentiation and $\int \cos x \, dx$ is known, we let $dv = \cos x \, dx$ and let $u = x$. Then $v = \int \cos x \, dx = \sin x$ and $du = dx$. Therefore (by 11.5.1)

$$\int x \cos x \, dx = x \sin x - \int \sin x \, dx = x \sin x + \cos x + C.$$

Example 2. Find $\int \ln x \, dx$.

Solution. Let $u = \ln x$ and $dv = dx$. Then $du = (1/x) \, dx$ and $v = \int dx = x$. By the formula for integration by parts (11.5.1),

$$\int \ln x \, dx = x \ln x - \int x \left(\frac{1}{x} \right) dx$$

$$= x \ln x - \int dx = x \ln x - x + C.$$

Example 3. Integrate $\int \text{Cot}^{-1} x \, dx$.

Solution. Let $u = \text{Cot}^{-1} x$ and let $dv = dx$; then $du = [-1/(1 + x^2)] \, dx$ and $v = x$. Applying the formula for integration by parts, we have

$$\int \text{Cot}^{-1} x \, dx = x \, \text{Cot}^{-1} x + \int \frac{x}{1 + x^2} \, dx = x \, \text{Cot}^{-1} x + \frac{1}{2} \int \frac{2x \, dx}{1 + x^2}$$

$$= x \, \text{Cot}^{-1} x + \frac{1}{2} \int \frac{d(1 + x^2)}{1 + x^2} = x \, \text{Cot}^{-1} x + \frac{1}{2} \ln |1 + x^2| + C.$$

Sometimes it is necessary to apply integration by parts several times.

Example 4. Find $\int x^2 \cos x \, dx$.

Solution. Using 11.5.1, we first let $u = x^2$ and $dv = \cos x \, dx$. Then $du = 2x \, dx$ and $v = \int \cos x \, dx = \sin x$. Therefore

(1)
$$\int x^2 \cos x \, dx = x^2 \sin x - \int \sin x \cdot 2x \, dx$$
$$= x^2 \sin x - 2 \int x \sin x \, dx.$$

We will again apply integration by parts, this time to $\int x \sin x \, dx$. Let $u = x$ and $dv = \sin x$ dx. Then $du = dx$ and $v = \int \sin x \, dx = -\cos x$. Thus

$$\int x \sin x \, dx = -x \cos x + \int \cos x \, dx$$
$$= -x \cos x + \sin x + C.$$

By substituting this result in (1), we get

$$\int x^2 \cos x \, dx = x^2 \sin x + 2x \cos x - 2 \sin x + C.$$

The next example of integration by parts has an interesting and useful variation.

Example 5. Integrate $\int e^x \sin x \, dx$.

Solution. Take $u = e^x$ and $dv = \sin x \, dx$. Then $du = e^x \, dx$ and $v = -\cos x$. Thus (11.5.1)

(2)
$$\int e^x \sin x \, dx = -e^x \cos x + \int e^x \cos x \, dx.$$

We now apply integration by parts to $\int e^x \cos x \, dx$. Let $u = e^x$ and $dv = \cos x \, dx$; then $du = e^x \, dx$, $v = \int \cos x \, dx = \sin x$, and

(3)
$$\int e^x \cos x \, dx = e^x \sin x - \int e^x \sin x \, dx.$$

Substituting (3) in (2) gives

(4)
$$\int e^x \sin x \, dx = -e^x \cos x + e^x \sin x - \int e^x \sin x \, dx.$$

By transposing the last term of (4) to the left side and combining terms, we obtain

$$2 \int e^x \sin x \, dx = e^x(\sin x - \cos x) + C_1,$$

from which

$$\int e^x \sin x \, dx = \tfrac{1}{2}e^x(\sin x - \cos x) + C.$$

Example 6. Find $\int \sec^3 \theta \, d\theta$.

Solution. Let $u = \sec \theta$ and $dv = \sec^2 \theta \, d\theta$. Then $du = \sec \theta \tan \theta \, d\theta$, $v = \tan \theta$, and

$$\int \sec^3 \theta \, d\theta = \sec \theta \tan \theta - \int \sec \theta \tan^2 \theta \, d\theta$$
$$= \sec \theta \tan \theta - \int \sec \theta \, (\sec^2 \theta - 1) \, d\theta$$
$$= \sec \theta \tan \theta - \int \sec^3 \theta \, d\theta + \int \sec \theta \, d\theta.$$

Consequently,

$$2 \int \sec^3 \theta \, d\theta = \sec \theta \tan \theta + \ln |\sec \theta + \tan \theta| + C_1$$

from which

$$\int \sec^3 \theta \, d\theta = \tfrac{1}{2} \sec \theta \tan \theta + \tfrac{1}{2} \ln |\sec \theta + \tan \theta| + C.$$

Exercises

In Exercises 1 through 18, use integration by parts to perform the indicated integrations.

1. $\int xe^x \, dx.$

2. $\int ye^{3y} \, dy.$

3. $\int x \sin 3x \, dx.$

4. $\int \ln 3w \, dw.$

5. $\int \mathrm{Tan}^{-1} x \, dx.$

6. $\int x\sqrt{x+1} \, dx.$

7. $\int t \sec^2 5t \, dt.$

8. $\int \mathrm{Cot}^{-1} y \, dy.$

9. $\int \sqrt{x} \ln x \, dx.$

10. $\int z^3 \ln z \, dz.$

11. $\int x \, \mathrm{Tan}^{-1} x \, dx.$

12. $\int t \cos 4t \, dt.$

13. $\int w \ln w \, dw.$

14. $\int x \cos^2 x \sin x \, dx.$

15. $\int \sin^2 x \, dx.$

16. $\int \csc^3 x \, dx.$

17. $\int xa^x \, dx.$

18. $\int x \sin^3 x \, dx.$ [*Hint:* $\sin^3 x = (1 - \cos^2 x) \sin x$. Use Exercise 14.]

In Exercises 19 to 24, use the formula for integration by parts *twice* to perform the indicated integrations (see Examples 4 and 5).

19. $\int x^2 e^x \, dx.$

20. $\int \ln^2 x \, dx.$

21. $\int e^t \cos t \, dt.$

22. $\int x^2 \sin x \, dx.$

23. $\int \sin (\ln x) \, dx.$

24. $\int (\ln x)^3 \, dx.$ (*Hint:* Use Exercise 20.)

25. Find the area of the region bounded by the x axis and the arch of the sine curve from $x = 0$ to $x = \pi$. Make a sketch.

26. Find the volume of the solid generated by revolving about the x axis the region of Exercise 25.

27. Find the centroid of the plane region in Exercise 25.

28. Find the area of the region bounded by the curves $y = 3xe^{-x/3}$, $y = 0$, and $x = 9$. Make a sketch.

29. Find the volume of the solid generated by revolving about the x axis the region described in Exercise 28.

30. Find the centroid of the plane region in Exercise 28.

11.6 SOME TRIGONOMETRIC INTEGRALS

We have seen that an integral containing trigonometric functions can often be transformed by a suitable trigonometric identity (10.5.2) into one or more of the basic integral forms. We shall now apply this method to some commonly encountered types of integrals with integrands that are powers of trigonometric functions.

I. $$\int \sin^2 x \, dx \quad \text{and} \quad \int \cos^2 x \, dx.$$

II. $$\int \sin^n x \, dx \quad \text{and} \quad \int \cos^n x \, dx, \quad n \in N.$$

III. $$\int \tan^n x \, dx, \quad n \in N.$$

IV. $$\int \sin^m x \cos^n x \, dx.$$

V. $$\int \tan^m x \sec^n x \, dx.$$

VI. $$\int \sin mx \cos nx \, dx, \quad \int \sin mx \sin nx \, dx,$$

and $$\int \cos mx \cos nx \, dx.$$

Type I. $$\int \sin^2 x \, dx \quad \text{and} \quad \int \cos^2 x \, dx.$$

These integrals can be integrated by parts (Exercise 15, Section 11.5), but it is easier to integrate them after transforming their integrands by the half-angle formulas

$$\sin^2 x = \frac{1 - \cos 2x}{2} \quad \text{and} \quad \cos^2 x = \frac{1 + \cos 2x}{2}.$$

Example 1.

$$\int \sin^2 x \, dx = \int \frac{1 - \cos 2x}{2} \, dx$$

$$= \frac{1}{2} \int dx - \frac{1}{4} \int \cos 2x \, (2 \, dx)$$

$$= \frac{1}{2} \int dx - \frac{1}{4} \int \cos 2x \, d(2x)$$

$$= \frac{x}{2} - \frac{\sin 2x}{4} + C.$$

Type II. $$\int \sin^n x \, dx \quad \text{and} \quad \int \cos^n x \, dx, \quad n \in N.$$

Two cases will be considered, depending on whether (1) n is an even positive integer or (2) n is an odd positive integer.

CASE 1. *n is an even positive integer.*

The half-angle formulas,

$$\sin^2 x = \frac{1 - \cos 2x}{2} \quad \text{and} \quad \cos^2 x = \frac{1 + \cos 2x}{2},$$

will reduce the degree of the integrand, and a sufficient number of applications of these formulas will reduce the problem to Type I.

Example 2.

$$\int \sin^4 x \, dx = \int \left(\frac{1 - \cos 2x}{2}\right)^2 dx$$

$$= \frac{1}{4} \int (1 - 2 \cos 2x + \cos^2 2x) \, dx$$

$$= \frac{1}{4} \int dx - \frac{1}{4} \int \cos 2x (2 \, dx) + \frac{1}{4} \int \left(\frac{1 + \cos 4x}{2}\right) dx$$

$$= \frac{3}{8} \int dx - \frac{1}{4} \int \cos 2x \, d(2x) + \frac{1}{32} \int \cos 4x \, (4 \, dx)$$

$$= \frac{3x}{8} - \frac{\sin 2x}{4} + \frac{1}{32} \int \cos 4x \, d(4x)$$

$$= \frac{3x}{8} - \frac{\sin 2x}{4} + \frac{\sin 4x}{32} + C.$$

CASE 2. *n is an odd positive integer.*

After taking out the factor $\sin x \, dx$ or $\cos x \, dx$, use the identity $\sin^2 x + \cos^2 x = 1$, as in the next example.

Example 3.

$$\int \sin^7 x \, dx = \int (\sin^2 x)^3 \sin x \, dx$$

$$= \int (1 - \cos^2 x)^3 \sin x \, dx$$

$$= \int (1 - 3 \cos^2 x + 3 \cos^4 x - \cos^6 x) \sin x \, dx$$

$$= \int \sin x \, dx + 3 \int \cos^2 x \, (-\sin x \, dx)$$

$$- 3 \int \cos^4 x \, (-\sin x \, dx) + \int \cos^6 x \, (-\sin x \, dx)$$

$$= \int \sin x \, dx + 3 \int \cos^2 x \, d(\cos x)$$

$$- 3 \int \cos^4 x \, d(\cos x) + \int \cos^6 x \, d(\cos x)$$

$$= -\cos x + \cos^3 x - \frac{3 \cos^5 x}{5} + \frac{\cos^7 x}{7} + C.$$

Type III.
$$\int \tan^n x \, dx, \qquad n \in N.$$

Again we consider two cases: (1) n is an even positive integer and (2) n is an odd positive integer. In both cases, the trigonometeric identity $1 + \tan^2 x = \sec^2 x$ is used.

CASE 1. *n is an even positive integer.*

Example 4.

$$\int \tan^4 x \, dx = \int \tan^2 x \, (\sec^2 x - 1) \, dx$$

$$= \int \tan^2 x \, (\sec^2 x \, dx) - \int \tan^2 x \, dx$$

$$= \int \tan^2 x \, d(\tan x) - \int (\sec^2 x - 1) \, dx$$

$$= \int \tan^2 x \, d(\tan x) - \int \sec^2 x \, dx + \int dx$$

$$= \frac{\tan^3 x}{3} - \tan x + x + C.$$

CASE 2. *n is an odd positive integer.*

Example 5.

$$\int \tan^5 x \, dx = \int \tan^3 x \, (\sec^2 x - 1) \, dx$$

$$= \int \tan^3 x \, (\sec^2 x \, dx) - \int \tan^3 x \, dx$$

$$= \int \tan^3 x \, d(\tan x) - \int \tan x \, (\sec^2 x - 1) \, dx$$

$$= \int \tan^3 x \, d(\tan x) - \int \tan x \, (\sec^2 x \, dx) + \int \tan x \, dx$$

$$= \int \tan^3 x \, d(\tan x) - \int \tan x \, d(\tan x) - \int \frac{-\sin x \, dx}{\cos x}$$

$$= \int \tan^3 x \, d(\tan x) - \int \tan x \, d(\tan x) - \int \frac{d(\cos x)}{\cos x}$$

$$= \frac{\tan^4 x}{4} - \frac{\tan^2 x}{2} - \ln|\cos x| + C.$$

Type IV.
$$\int \sin^m x \cos^n x \, dx.$$

CASE 1.   *Either m or n is an odd positive integer; the other may be* **any number.**

Let m be odd, and write $\sin^m x = (\sin^{m-1} x) \sin x$. Since $m - 1$ is even, we use the identity $\sin^2 x = 1 - \cos^2 x$ to transform $\sin^{m-1} x$ into powers of $\cos x$.

Example 6.

$$\int \sin^3 x \cos^{-4} x \, dx = -\int (1 - \cos^2 x) \cos^{-4} x \, (-\sin x \, dx)$$

$$= -\int \cos^{-4} x \, d(\cos x) + \int \cos^{-2} x \, d(\cos x)$$

$$= \frac{\cos^{-3} x}{3} - \cos^{-1} x + C$$

$$= \frac{\sec^3 x}{3} - \sec x + C.$$

CASE 2.   *Both m and n are even positive integers.*

The half-angle formulas,

$$\sin^2 x = \frac{1 - \cos 2x}{2} \qquad \text{and} \qquad \cos^2 x = \frac{1 + \cos 2x}{2},$$

can be used to reduce the degree of the integrand.

Example 7.

$$\int \sin^2 x \cos^4 x \, dx$$

$$= \int \left(\frac{1 - \cos 2x}{2} \right) \left(\frac{1 + \cos 2x}{2} \right)^2 dx$$

$$= \frac{1}{8} \int (1 + \cos 2x - \cos^2 2x - \cos^3 2x) \, dx$$

$$= \frac{1}{8} \int dx + \frac{1}{16} \int \cos 2x (2 \, dx) - \frac{1}{8} \int \frac{1 + \cos 4x}{2} dx$$

$$- \frac{1}{8} \int (1 - \sin^2 2x) \cos 2x \, dx$$

$$= \frac{1}{8} \int dx + \frac{1}{16} \int \cos 2x \, d(2x) - \frac{1}{16} \int dx$$

$$- \frac{1}{64} \int \cos 4x (4 \, dx) - \frac{1}{16} \int \cos 2x \, (2 \, dx)$$

$$- \frac{1}{16} \int \sin^2 x \, [\cos 2x \, (2 \, dx)]$$

$$= \frac{1}{16} \int dx - \frac{1}{64} \int \cos 4x \, d(4x) - \frac{1}{16} \int \sin^2 2x \, d(\sin 2x)$$

$$= \frac{x}{16} - \frac{\sin 4x}{64} - \frac{\sin^3 2x}{48} + C.$$

Occasionally it is better to use only one of the half-angle formulas along with

$$\sin x \cos x = \frac{\sin 2x}{2},$$

which comes from the double-angle formula.

Type V. $$\int \tan^m x \sec^n x \, dx \qquad \text{and} \qquad \int \cot^m x \csc^n x \, dx.$$

CASE 1. *n is an even positive integer and m is any number.*

In the first integral, keep $\sec^2 x \, dx$ and transform the remaining $\sec^{n-2} x$ to powers of $\tan x$ by the identity $1 + \tan^2 x = \sec^2 x$. In the second integral, keep $\csc^2 x \, dx$ and use $1 + \cot^2 x = \csc^2 x$.

Example 8.

$$\int \tan^{-3/2} x \sec^4 x \, dx = \int \tan^{-3/2} x (1 + \tan^2 x) \sec^2 x \, dx$$

$$= \int \tan^{-3/2} x \, (\sec^2 x \, dx) + \int \tan^{1/2} x \, (\sec^2 x \, dx)$$

$$= \int \tan^{-3/2} x \, d(\tan x) + \int \tan^{1/2} x \, d(\tan x)$$

$$= -2 \tan^{-1/2} x + \frac{2 \tan^{3/2} x}{3} + C.$$

CASE 2. *m is an odd positive integer, n is any number.*

Since m is odd, $m - 1$ is even. Keep $\sec x \tan x \, dx$ and transform $\tan^{m-1} x$ to powers of $\sec x$ by means of the identity $\tan^2 x = \sec^2 x - 1$.

Example 9.

$$\int \tan^3 x \sec^{-1/2} x \, dx = \int \tan^2 x \sec^{-3/2} x \, (\sec x \tan x \, dx)$$

$$= \int (\sec^2 x - 1) \sec^{-3/2} x \, d(\sec x)$$

$$= \int \sec^{1/2} x \, d(\sec x) - \int \sec^{-3/2} x \, d(\sec x)$$

$$= \frac{2 \sec^{3/2} x}{3} + 2 \sec^{-1/2} x + C.$$

Type VI. $\int \sin mx \cos nx\, dx,$ $\int \sin mx \sin nx\, dx,$ and $\int \cos mx \cos nx\, dx,$

m and n any numbers.

Integrals of this type occur in alternating current theory and heat transfer problems where Fourier series are used. Each is integrated easily after its integrand has been transformed by the appropriate one of the following trigonometric identities.

$$\sin mx \cos nx = \tfrac{1}{2}[\sin (m + n)x + \sin (m - n)x],$$
$$\sin mx \sin nx = -\tfrac{1}{2}[\cos (m + n)x - \cos (m - n)x],$$
$$\cos mx \cos nx = \tfrac{1}{2}[\cos (m + n)x + \cos (m - n)x].$$

These equations follow easily from the addition identities (10.5.2). Thus to derive the first, we add $\sin (u + v) = \sin u \cos v + \cos u \sin v$ and $\sin (u - v) = \sin u \cos v - \cos u \sin v$, member by member, obtaining

$$2 \sin u \cos u = \sin (u + v) + \sin (u - v);$$

and by letting $u = mx$ and $v = nx$, this equation becomes

$$\sin mx \cos nx = \tfrac{1}{2}[\sin (m + n)x + \sin (m - n)x].$$

Example 10.

$$\int \sin 2x \cos 3x\, dx = \int \frac{\sin 5x + \sin (-x)}{2}\, dx = \frac{1}{10} \int \sin 5x \, d(5x) - \frac{1}{2} \int \sin x \, dx$$
$$= -\frac{\cos 5x}{10} + \frac{\cos x}{2} + C.$$

Exercises

In each of the following exercises, perform the indicated integration.

1. $\int \cos^2 x\, dx.$

2. $\int \sin^4 5x\, dx.$

3. $\int \cos^3 x\, dx.$

4. $\int \sin^3 x\, dx.$

5. $\int \sin^5 t\, dt.$

6. $\int \cos^6 t\, dt.$

7. $\int \tan^3 y\, dy.$

8. $\int \tan^6 2x\, dx.$

9. $\int \sin^7 3x \cos^2 3x\, dx.$

10. $\int \sin^3 t \sqrt[5]{\cos t}\, dt.$

11. $\int \cos^3 \theta \sin^{-2} \theta\, d\theta.$

12. $\int \sin^{1/2} \theta \cos^3 \theta\, d\theta.$

13. $\int \sin^4 2t \cos^4 2t\, dt.$

14. $\int \sin^6 t \cos^2 t\, dt.$

15. $\int \tan^3 3y \sec^3 3y\, dy.$

16. $\int \cos^4 \frac{w}{2} \sin^2 \frac{w}{2} \, dw.$

17. $\int \cot x \csc^3 x \, dx.$

18. $\int \cot x \sec^2 x \, dx.$

19. $\int \tan^{-3} t \sec^2 t \, dt.$

20. $\int \tan^5 t \sec^{-3/2} t \, dt.$

21. $\int \sin 4y \cos 5y \, dy.$

22. $\int \cos y \cos 4y \, dy.$

23. $\int \cot^4 2x \, dx.$ [*Hint:* $\cot^4 u = \cot^2 u \, (\csc^2 u$

     $- 1).$]

24. $\int \csc^4 3y \, dy.$

25. $\int \sec^4 7x \, dx.$

26. $\int \cot^6 4w \, dw.$

27. $\int \cot^3 x \, dx.$

28. $\int \frac{d\theta}{\cos 5\theta}.$

29. $\int (\tan x + \cot x)^2 \, dx.$

30. $\int \cos mx \cos nx \, dx, \, m \neq n.$

11.7 RATIONALIZING SUBSTITUTIONS

The expression $\sqrt[n]{ax + b}$ is called a *linear irrationality* because $ax + b$ is linear in the variable x and it is under a radical sign (or has a fractional exponent).

If a linear irrationality $\sqrt[n]{ax + b}$ appears in an integral, the substitution

$$z^n = ax + b$$

will eliminate its radical sign.

Example 1.   Find $\int x\sqrt[3]{x - 4} \, dx.$

Solution.   Let $x - 4 = z^3$. Then $x = z^3 + 4$ and $dx = 3z^2 \, dz$. Thus

$$\int x\sqrt[3]{x - 4} \, dx = \int (z^3 + 4)\sqrt[3]{z^3}(3z^2 \, dz)$$

$$= 3 \int (z^6 + 4z^3) \, dz$$

$$= \frac{3z^7}{7} + 3z^4 + C.$$

To express this answer in terms of x, we use the relation $x - 4 = z^3$ to obtain $z = (x - 4)^{1/3}$. Then

$$\frac{3z^7}{7} + 3z^4 + C = \frac{3(x - 4)^{7/3}}{7} + 3(x - 4)^{4/3} + C,$$

which is the final answer.

Example 2.   Integrate $\int x\sqrt[5]{(x + 1)^2} \, dx.$

Solution.   The given integral can be expressed as

$$\int x(x+1)^{2/5} \, dx.$$

Let $x + 1 = z^5$. Then $x = z^5 - 1$ and $dx = 5z^4 \, dz$. We can now write

$$\int x(x+1)^{2/5} \, dx = \int (z^5 - 1)(z^5)^{2/5}(5z^4 \, dz)$$

$$= 5 \int (z^{11} - z^6) \, dz$$

$$= \frac{5z^{12}}{12} - \frac{5z^7}{7} + C.$$

Since $z = (x+1)^{1/5}$, this can be written

$$\frac{5(x+1)^{12/5}}{12} - \frac{5(x+1)^{7/5}}{7} + C.$$

Example 3.   Find $\int \sqrt[7]{ax + b} \, dx.$

Solution.   This integral comes under one of the basic integration formulas, and nothing is to be gained by using the substitution $z^7 = ax + b$. Instead, we use formula 11.2.1 and write immediately

$$\int \sqrt[7]{ax + b} \, dx = \frac{1}{a} \int (ax + b)^{1/7}(a \, dx)$$

$$= \frac{1}{a} \int (ax + b)^{1/7} \, d(ax + b)$$

$$= \frac{7(ax + b)^{8/7}}{8a} + C.$$

When an integral contains several different fractional powers of the variable x, the substitution $z^n = x$, where n is the least common multiple of the denominators of the exponents, is often effective.

Example 4.   Integrate $\int \dfrac{dx}{\sqrt{x}\,(1 + \sqrt[3]{x}\,)}.$

Solution.   The given integral can be written

$$\int \frac{dx}{x^{1/2}(1 + x^{1/3})}.$$

Let $z^6 = x$. Then $dx = 6z^5 \, dz$, and

$$\int \frac{dx}{x^{1/2}(1 + x^{1/3})} = \int \frac{6z^5 \, dz}{(z^6)^{1/2}[1 + (z^6)^{1/3}]}$$

$$= 6 \int \frac{z^5}{z^3(1 + z^2)} \, dz = 6 \int \frac{z^2}{1 + z^2} \, dz.$$

When the integrand is a quotient of polynomials and the numerator is of equal or greater degree than the denominator, *always divide the denominator into the numerator* (by long division). Here

$$\frac{z^2}{z^2 + 1} = 1 - \frac{1}{z^2 + 1}.$$

Thus

$$6 \int \frac{z^2}{1 + z^2} \, dz = 6 \int dz - 6 \int \frac{dz}{1 + z^2}$$

$$= 6z - 6 \text{ Tan}^{-1} z + C.$$

From $z^6 = x$, we have $z = x^{1/6}$, and the answer in terms of x is

$$6x^{1/6} - 6 \text{ Tan}^{-1} (x^{1/6}) + C.$$

Expressions like $\sqrt{a^2 - u^2}$, $\sqrt{a^2 + u^2}$, and $\sqrt{u^2 - a^2}$ are called *quadratic irrationalities* because the expressions under the radical sign are quadratic in the variable u. Quadratic irrationalities in the integrand can often be removed by a *trignometric substitution* based on the Pythagorean identities, $\sin^2 \theta + \cos^2 \theta = 1$ and $1 + \tan^2 \theta = \sec^2 \theta$.

After multiplying both identities through by a^2, where a is a positive constant, they may be written

$$a^2 - a^2 \sin^2 \theta = a^2 \cos^2 \theta,$$

$$a^2 + a^2 \tan^2 \theta = a^2 \sec^2 \theta,$$

$$a^2 \sec^2 \theta - a^2 = a^2 \tan^2 \theta.$$

It follows that if we substitute $u = a \sin \theta$ in $\sqrt{a^2 - u^2}$, we obtain $\sqrt{a^2 - a^2 \sin^2 \theta} = \sqrt{a^2 \cos^2 \theta} = a \cos \theta$, thus removing the irrationality. Similarly, if we let $u = a \tan \theta$ in $\sqrt{a^2 + u^2}$, we have $\sqrt{a^2 + a^2 \tan^2 \theta} = \sqrt{a^2 \sec^2 \theta} = a \sec \theta$; and if we substitute $u = a \sec \theta$ in $\sqrt{u^2 - a^2}$, the result is $\sqrt{a^2 \sec^2 \theta - a^2} = a \tan \theta$.

To recapitulate,

when $\sqrt{a^2 - u^2}$ occurs, let $u = a \text{ Sin } \theta,$ $\quad \frac{-\pi}{2} \leq \theta \leq \frac{\pi}{2};$

when $\sqrt{a^2 + u^2}$ occurs, let $u = a \text{ Tan } \theta,$ $\quad \frac{-\pi}{2} < \theta < \frac{\pi}{2};$

when $\sqrt{u^2 - a^2}$ occurs, let $u = a \text{ Sec } \theta,$ $\quad \theta \in \left[0, \frac{\pi}{2}\right) \cup \left(\frac{\pi}{2}, \pi\right].$

See Fig. 11-1.

Example 5. Find $\int \frac{\sqrt{4 - x^2}}{x^2} \, dx,$ $\quad 0 < \left|\frac{x}{2}\right| \leq 1.$

Solution. Let $x = 2 \text{ Sin } \theta$. Then $dx = 2 \cos \theta \, d\theta$ and

$$\int \frac{\sqrt{4 - x^2}}{x^2} \, dx = \int \frac{\sqrt{4 - 4 \text{ Sin}^2 \theta} \cdot 2 \cos \theta}{4 \text{ Sin}^2 \theta} \, d\theta$$

$$= \int \frac{\sqrt{1 - \text{ Sin}^2 \theta} \cdot \cos \theta}{\text{ Sin}^2 \theta} \, d\theta$$

$$= \int \frac{\cos^2 \theta}{\text{Sin}^2 \theta} \, d\theta = \int \cot^2 \theta \, d\theta$$

$$= \int (\csc^2 \theta - 1) \, d\theta = \int \csc^2 \theta \, d\theta - \int d\theta$$

$$= -\cot \theta - \theta + C.$$

From a sketch (Fig. 11-2) illustrating $\text{Sin } \theta = x/2$, we read $\cot \theta = \sqrt{4 - x^2}/x$. Also, $\theta = \text{Sin}^{-1}(x/2)$. Therefore

$$\int \frac{\sqrt{4 - x^2}}{x^2} \, dx = -\cot \theta - \theta + C$$

$$= -\frac{\sqrt{4 - x^2}}{x} - \text{Sin}^{-1} \frac{x}{2} + C.$$

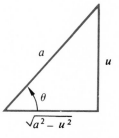

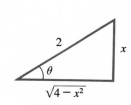

Figure 11-1   Figure 11-2

Example 6.   Integrate $\displaystyle\int \frac{dx}{\sqrt{9 + x^2}}$.

Solution.   Let $x = 3 \text{ Tan } \theta$. Then $dx = 3 \sec^2 \theta \, d\theta$ and

$$\int \frac{dx}{\sqrt{9 + x^2}} = \int \frac{3 \sec^2 \theta}{\sqrt{9 + 9 \text{ Tan}^2 \theta}} \, d\theta = \int \frac{\sec^2 \theta}{\sqrt{1 + \text{Tan}^2 \theta}} \, d\theta$$

$$= \int \sec \theta \, d\theta = \ln |\sec \theta + \tan \theta| + C_1.$$

But $\tan \theta = x/3$, and from a sketch (Fig. 11-3) we read $\sec \theta = \sqrt{9 + x^2}/3$. Thus

$$\ln |\sec \theta + \tan \theta| + C_1 = \ln \left| \frac{\sqrt{9 + x^2} + x}{3} \right| + C_1$$

$$= \ln |x + \sqrt{9 + x^2}| + C,$$

where $C = C_1 - \ln 3$.

Example 7.   Find $\displaystyle\int \frac{\sqrt{x^2 - 4}}{x} \, dx$, $|x| \geq 2$.

Solution.   Let $x = 2 \text{ Sec } \theta$. Then $dx = 2 \sec \theta \tan \theta \, d\theta$. Thus

$$\int \frac{\sqrt{x^2 - 4}}{x} \, dx = \int \frac{\sqrt{4 \text{ Sec}^2 \theta - 4} \, (2 \sec \theta \tan \theta)}{2 \text{ Sec } \theta} \, d\theta$$

$$= 2 \int \sqrt{\text{Sec}^2 \, \theta - 1} \, (\tan \theta) \, d\theta$$

$$= 2 \int \tan^2 \theta \, d\theta = 2 \int (\sec^2 \theta - 1) \, d\theta$$

$$= 2 \int \sec^2 \theta \, d\theta - 2 \int d\theta$$

$$= 2 \tan \theta - 2\theta + C.$$

But $\theta = \text{Sec}^{-1}(x/2)$ and (Fig. 11-4) $\tan \theta = \sqrt{x^2 - 4}/2$. Therefore

$$\int \frac{\sqrt{x^2 - 4}}{x} \, dx = 2 \tan \theta - 2\theta + C$$

$$= \sqrt{x^2 - 4} - 2 \, \text{Sec}^{-1} \frac{x}{2} + C.$$

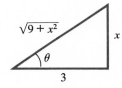

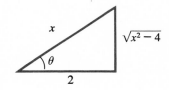

Figure 11-3   Figure 11-4

Example 8. Find $\int \dfrac{dx}{x\sqrt{x^4 - 4}}$.

Solution. Although the expression under the radical sign is not quadratic, the present method is well suited to this integration.

Let $x^2 = 2 \, \text{Sec} \, \theta$; then $x = \sqrt{2 \, \text{Sec} \, \theta}$, and $dx = [(\sec \theta \tan \theta)/\sqrt{2 \, \text{Sec} \, \theta}] \, d\theta$. By substituting this in the given integral, we get

$$\int \frac{dx}{x\sqrt{x^4 - 4}} = \int \frac{(\sec \theta \tan \theta)/\sqrt{2 \, \text{Sec} \, \theta}}{\sqrt{2 \, \text{Sec} \, \theta}\sqrt{4 \, \text{Sec}^2 \, \theta - 4}} \, d\theta$$

$$= \frac{1}{4} \int d\theta = \frac{\theta}{4} + C.$$

But $\theta = \text{Sec}^{-1}(x^2/2)$ (Fig. 11-5). Therefore

$$\int \frac{dx}{x\sqrt{x^4 - 4}} = \frac{1}{4} \, \text{Sec}^{-1} \frac{x^2}{2} + C.$$

Example 9. Integrate $\int \dfrac{dz}{(z^2 + 2z + 2)^{3/2}}$.

Solution. By completing the square in the denominator, we can write

$$\int \frac{dz}{(z^2 + 2z + 2)^{3/2}} = \int \frac{dz}{[(z^2 + 2z + 1) + 1]^{3/2}} = \int \frac{dz}{[(z + 1)^2 + 1]^{3/2}}.$$

Let $z + 1 = \text{Tan} \, \theta$; then $dz = \sec^2 \theta \, d\theta$ and

$$\int \frac{dz}{[(z + 1)^2 + 1]^{3/2}} = \int \frac{\sec^2 \theta \, d\theta}{[\text{Tan}^2 \theta + 1]^{3/2}} = \int \frac{d\theta}{\sec \theta} = \int \cos \theta \, d\theta = \sin \theta + C.$$

Since Tan $\theta = z + 1$, then $\sin \theta = (z + 1)/\sqrt{z^2 + 2z + 2}$ (Fig. 11-6). Therefore

$$\int \frac{dz}{(z^2 + 2z + 2)^{3/2}} = \frac{z + 1}{\sqrt{z^2 + 2z + 2}} + C.$$

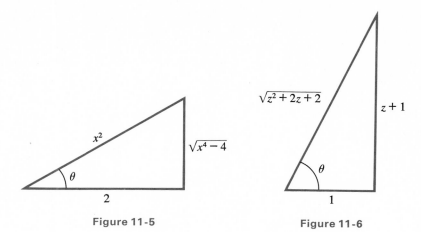

Figure 11-5 Figure 11-6

Exercises

In each of the following exercises, perform the indicated integration.

1. $\int x\sqrt{x + 3}\, dx.$

2. $\int x\sqrt[3]{x + 4}\, dx.$

3. $\int \frac{t\, dt}{\sqrt{2t + 7}}.$

4. $\int \frac{x^2 + 2x}{(x + 1)^{1/2}}\, dx.$

5. $\int \frac{dx}{\sqrt{x} + 2}.$

6. $\int \frac{\sqrt{y}}{y + 1}\, dy.$

7. $\int 7t(2t + 1)^{3/2}\, dt.$

8. $\int x(x + 4)^{1/3}\, dx.$

9. $\int \cos \sqrt{2y}\, dy.$

10. $\int \sin \sqrt{x + 1}\, dx.$

11. $\int \frac{dx}{x^{1/3} + x^{2/3}}.$

12. $\int \frac{x^{1/3}\, dx}{x + x^{5/6}}.$

13. $\int \frac{dx}{x^{1/2} + x^{1/3}}.$

14. $\int \frac{t^2\, dt}{\sqrt{t + 3}}.$

15. $\int \frac{\sqrt{1 - x^2}}{x}\, dx.$

16. $\int \frac{x^2\, dx}{\sqrt{9 - x^2}}.$

17. $\int \frac{dx}{x\sqrt{x^2 + 9}}.$

18. $\int \frac{dx}{(x^2 + 9)^{3/2}}.$

19. $\int \frac{dx}{x^2\sqrt{x^2 - 16}}.$

20. $\displaystyle\int \frac{3x - 2}{\sqrt{4 - x^2}}\, dx.$

21. $\displaystyle\int \frac{dt}{(t - 1)\sqrt{t^2 - 1}}.$

22. $\displaystyle\int \frac{\sqrt{t^2 - 16}}{t^3}\, dt.$

23. $\displaystyle\int \frac{dx}{\sqrt{12 + 4x - x^2}}.$ (*Hint:* Complete the square under the radical sign.)

24. $\displaystyle\int \frac{x^2 - 2x + 5}{\sqrt{9 - x^2}}\, dx.$

25. $\displaystyle\int \frac{y^3\, dy}{(y^2 + 4)^{3/2}}.$

26. $\displaystyle\int \frac{y^2\, dy}{(9 - y^2)^{5/2}}.$

27. $\displaystyle\int \frac{t\, dt}{\sqrt{16 - t^2}}.$

28. $\displaystyle\int \frac{t\, dt}{(t^2 + 1)^{3/2}}.$

29. $\displaystyle\int \frac{u\, du}{u^4 \sqrt{u^4 - 1}}.$

30. $\displaystyle\int \frac{dy}{(25 - y^2)^{3/2}}.$

11.8 INTEGRANDS INVOLVING $Ax^2 + Bx + C$

When a quadratic polynomial, $Ax^2 + Bx + C$, appears in an integrand, it is usually worthwhile to "complete the square," and this procedure should be tried.

First factor A from the polynomial:

$$Ax^2 + Bx + C = A\left(x^2 + \frac{B}{A}x + \frac{C}{A}\right)$$

$$= A(x^2 + Dx + E),$$

where $D = B/A$ and $E = C/A$. To complete the square in $x^2 + Dx + E$, we add and subtract $(D/2)^2$, obtaining

$$x^2 + Dx + E = \left(x^2 + Dx + \frac{D^2}{4}\right) + \left(E - \frac{D^2}{4}\right)$$

$$= \left(x + \frac{D}{2}\right)^2 + \frac{4E - D^2}{4}.$$

The latter expression can be put into one of three forms, $u^2 + a^2$, $u^2 - a^2$, or $a^2 - u^2$, as shown in the first three examples.

Example 1.

$$x^2 + 4x + 13 = (x^2 + 4x) + 13 = (x^2 + 4x + 4) + 13 - 4$$

$$= (x + 2)^2 + 9 = u^2 + a^2,$$

where $u = x + 2$ and $a = 3$.

Example 2.

$$x^2 - 6x - 7 = (x^2 - 6x) - 7 = (x^2 - 6x + 9) - 7 - 9$$

$$= (x - 3)^2 - 16 = u^2 - a^2,$$

in which $u = x - 3$ and $a = 4$.

Example 3.

$$11 + 10x - x^2 = 11 - (x^2 - 10x) = 11 + 25 - (x^2 - 10x + 25)$$
$$= 36 - (x - 5)^2 = a^2 - u^2,$$

where $a = 6$ and $u = x - 5$.

When the quadratic expression is under a radical sign, completing the square will prepare it for one of the trigonometric substitutions discussed in Section 11.7.

Example 4. Integrate $\displaystyle\int \frac{dx}{\sqrt{x^2 + 2x + 26}}$.

Solution. By completing the square,

$$x^2 + 2x + 26 = (x^2 + 2x + 1) + 25 = (x + 1)^2 + 25.$$

Let $u = x + 1$ and $a = 5$. Then $dx = du$, and

$$\int \frac{dx}{\sqrt{x^2 + 2x + 26}} = \int \frac{dx}{\sqrt{(x + 1)^2 + 25}} = \int \frac{du}{\sqrt{u^2 + a^2}}.$$

As in Section 11.7, let $u = a \, \mathrm{Tan} \, \theta$. Then $du = a \sec^2 \theta \, d\theta$, and

$$\int \frac{du}{\sqrt{u^2 + a^2}} = \int \frac{a \sec^2 \theta \, d\theta}{\sqrt{a^2 \, \mathrm{Tan}^2 \, \theta + a^2}} = \int \frac{\sec^2 \theta \, d\theta}{\sec \theta}$$
$$= \int \sec \theta \, d\theta = \ln |\sec \theta + \tan \theta| + C_1.$$

From a sketch of a right triangle illustrating $\mathrm{Tan} \, \theta = u/a$, we read $\sec \theta = \sqrt{u^2 + a^2}/a$. Thus

$$\ln |\sec \theta + \tan \theta| + C_1 = \ln \left| \frac{\sqrt{u^2 + a^2} + u}{a} \right| + C_1$$
$$= \ln |\sqrt{u^2 + a^2} + u| + C,$$

where $C = C_1 - \ln a$. But $u = x + 1$ and $a = 5$. Therefore

$$\ln |\sqrt{u^2 + a^2} + u| + C = \ln |\sqrt{x^2 + 2x + 26} + x + 1| + C,$$

which is the answer in terms of the original variable x.

Example 5. Find $\int \sqrt{5 - 4x - x^2} \, dx$.

Solution. By completing the square under the radical sign, we can write

$$\int \sqrt{5 - 4x - x^2} \, dx = \int \sqrt{9 - (x + 2)^2} \, dx,$$

as in Example 3, above. The integrand is now in the form $\sqrt{a^2 - u^2}$ and calls for the trigonometric substitution $u = a \, \mathrm{Sin} \, \theta$. After a little practice, one can omit the step where u and a are introduced and let $x + 2 = 3 \, \mathrm{Sin} \, \theta$ directly. Then $dx = 3 \cos \theta \, d\theta$, and

$$\int \sqrt{9 - (x + 2)^2} \, dx = \int \sqrt{9 - 9 \, \mathrm{Sin}^2 \, \theta} \cdot 3 \cos \theta \, d\theta$$

$$= 9 \int \cos^2 \theta \, d\theta = \frac{9}{2} \int (1 + \cos 2\theta) \, d\theta$$

$$= \frac{9}{2} \int d\theta + \frac{9}{4} \int \cos 2\theta \, d(2\theta)$$

$$= \frac{9\theta}{2} + \frac{9 \sin 2\theta}{2} + C$$

$$= \frac{9\theta}{2} + 9 \sin \theta \cos \theta + C.$$

But

$$\sin \theta = \frac{x+2}{3}, \qquad \cos \theta = \frac{\sqrt{5 - 4x - x^2}}{3},$$

and

$$\theta = \sin^{-1} \frac{x+2}{3}.$$

Therefore

$$\frac{9\theta}{2} + 9 \sin \theta \cos \theta + C$$

$$= \frac{9}{2} \sin^{-1} \frac{x+2}{3} + (x+2)\sqrt{5 - 4x - x^2} + C.$$

Even when the quadratic expression is not under a radical sign, it is often useful to complete the square.

Example 6. Integrate $\int \dfrac{dx}{9x^2 + 18x + 10}$.

Solution. $9x^2 + 18x + 10 = 9(x^2 + 2x) + 10 = 9(x^2 + 2x + 1) + 10 - 9 = 9(x + 1)^2 + 1$. Thus

$$\int \frac{dx}{9x^2 + 18x + 10} = \frac{1}{9} \int \frac{dx}{(x+1)^2 + \frac{1}{9}}.$$

Let $u = x + 1$ and $a = \frac{1}{3}$. Then $du = dx$, and

$$\frac{1}{9} \int \frac{dx}{(x+1)^2 + \frac{1}{9}} = \frac{1}{9} \int \frac{du}{u^2 + a^2} = \frac{1}{9} \left(\frac{1}{a} \tan^{-1} \frac{u}{a} \right) + C$$

$$= \frac{1}{3} \tan^{-1} (3x + 3) + C.$$

Example 7. Find $\int \dfrac{4x + 7}{2x^2 + x + 5} \, dx$.

Solution.

$$\int \frac{4x + 7}{2x^2 + x + 5} \, dx = \int \frac{4x + 1}{2x^2 + x + 5} \, dx + 6 \int \frac{dx}{2x^2 + x + 5}$$

$$= \int \frac{d(2x^2 + x + 5)}{2x^2 + x + 5} + 6 \int \frac{dx}{2(x^2 + \frac{1}{2}x) + 5}$$

$$= \ln |2x^2 + x + 5| + 6 \int \frac{dx}{2(x^2 + \frac{1}{2}x + \frac{1}{4}) + (5 - \frac{2}{4})}$$

$$= \ln |2x^2 + x + 5| + 6 \int \frac{dx}{2(x + \frac{1}{2})^2 + \frac{18}{4}}$$

$$= \ln |2x^2 + x + 5| + 3 \int \frac{dx}{(x + \frac{1}{2})^2 + \frac{9}{4}}.$$

In the second term, let $u = x + \frac{1}{2}$ and $a = \frac{3}{2}$. Then $dx = du$, and the preceding expression becomes

$$\ln |2x^2 + x + 5| + 3 \int \frac{du}{u^2 + a^2} = \ln |2x^2 + x + 5| + \frac{3}{a} \operatorname{Tan}^{-1} \frac{u}{a} + C$$

$$= \ln |2x^2 + x + 5| + 2 \operatorname{Tan}^{-1} \frac{x + \frac{1}{2}}{\frac{3}{2}} + C$$

$$= \ln |2x^2 + x + 5| + 2 \operatorname{Tan}^{-1} \frac{2x + 1}{3} + C.$$

Example 8. Integrate $\displaystyle\int \frac{x}{(x^2 - 2x + 10)^2} \, dx$.

Solution.

$$\int \frac{x}{(x^2 - 2x + 10)^2} \, dx = \frac{1}{2} \int \frac{2x - 2}{(x^2 - 2x + 10^2)} \, dx$$

$$+ \int \frac{dx}{(x^2 - 2x + 10)^2}$$

$$= \frac{1}{2} \int (x^2 - 2x + 10)^{-2} \, d(x^2 - 2x + 10)$$

$$+ \int \frac{dx}{[(x^2 - 2x + 1) + 9]^2}$$

$$= \frac{-1}{2(x^2 - 2x + 10)} + \int \frac{dx}{[(x - 1)^2 + 3^2]^2}$$

$$= \frac{-1}{2(x^2 - 2x + 10)} + \int \frac{du}{(u^2 + a^2)^2}$$

where $u = x - 1$, $du = dx$, and $a = 3$.

To integrate the remaining integral, we let $u = a \operatorname{Tan} \theta$. Then $du = a \sec^2 \theta \, d\theta$, and

$$\int \frac{du}{(u^2 + a^2)^2} = \int \frac{a \sec^2 \theta \, d\theta}{(a^2 \operatorname{Tan}^2 \theta + a^2)^2}$$

$$= \int \frac{a \sec^2 \theta \, d\theta}{a^4 \sec^4 \theta} = \frac{1}{a^3} \int \frac{d\theta}{\sec^2 \theta}$$

$$= \frac{1}{a^3} \int \cos^2 \theta \, d\theta = \frac{1}{a^3} \int \frac{1 + \cos 2\theta}{2} \, d\theta$$

$$= \frac{1}{2a^3} \int d\theta + \frac{1}{4a^3} \int \cos 2\theta \, d(2\theta)$$

$$= \frac{\theta}{2a^3} + \frac{\sin 2\theta}{4a^3} + C = \frac{\theta}{2a^3} + \frac{2 \sin \theta \cos \theta}{4a^3} + C.$$

But $\theta = \text{Tan}^{-1}(u/a)$, $\sin\theta = (u/\sqrt{u^2 + a^2})$, and $\cos\theta = (a/\sqrt{u^2 + a^2})$. So

$$\frac{\theta}{2a^3} + \frac{2\sin\theta\cos\theta}{4a^3} + C = \frac{1}{2a^3}\text{Tan}^{-1}\frac{u}{a} + \frac{2au}{4a^3(u^2 + a^2)} + C$$

$$= \frac{1}{2(3)^3}\text{Tan}^{-1}\frac{x-1}{3} + \frac{x-1}{2(3)^2[(x-1)^2 + 3^2]} + C$$

$$= \frac{1}{54}\text{Tan}^{-1}\frac{x-1}{3} + \frac{x-1}{18(x^2 - 2x + 10)} + C.$$

Thus

$$\int\frac{dx}{[(x-1)^2 + 3^2]^2} = \frac{1}{54}\text{Tan}^{-1}\frac{x-1}{3} + \frac{x-1}{18(x^2 - 2x + 10)} + C.$$

Therefore

$$\int\frac{x}{(x^2 - 2x + 10)^2}\,dx = \frac{-1}{2(x^2 - 2x + 10)}$$

$$+ \frac{1}{54}\text{Tan}^{-1}\frac{x-1}{3} + \frac{x-1}{18(x^2 - 2x + 10)} + C.$$

Exercises

Integrate.

1. $\int\dfrac{dx}{x^2 - 2x - 3}.$

2. $\int\dfrac{2x - 1}{x^2 + 2x + 2}\,dx.$

3. $\int\dfrac{2x + 1}{x^2 - 6x + 18}\,dx.$

4. $\int\dfrac{x - 5}{3 - 2x - x^2}\,dx.$

5. $\int\dfrac{dx}{\sqrt{16 + 6x - x^2}}.$

6. $\int\dfrac{dx}{4x^2 + 4x + 5}.$

7. $\int\dfrac{dx}{\sqrt{4x - x^2}}.$

8. $\int\dfrac{dx}{\sqrt{x^2 - 4x - 5}}.$

9. $\int\dfrac{x\,dx}{\sqrt{x^2 - 4x + 5}}.$

10. $\int\dfrac{x\,dx}{\sqrt{x^2 + 2x + 5}}.$

11. $\int\dfrac{dx}{(x^2 + 1)^2}.$

12. $\int\dfrac{du}{(a^2 - u^2)^2}.$ (*Hint:* Let $u = a\,\text{Sin}\,\theta$, and use Example 6 of Section 11.5.)

13. $\int\dfrac{2x + 3}{(x^2 - 4x + 5)^2}\,dx.$

14. $\int\dfrac{3x + 4}{(x^2 - 8x + 17)^2}\,dx.$

11.9 DEFINITE INTEGRALS. CHANGE OF LIMITS

The value of $\int_a^b f(x)\,dx$ can be calculated if we can integrate the indefinite integral $\int f(x)\,dx$ (that is, find the antiderivatives of f). For, in that event, the fundamental theorem of calculus gives

$$\int_a^b f(x)\,dx = \int f(x)\,dx\bigg]_a^b.$$

Example 1. Find the value of the definite integral $\int_4^9 \dfrac{dx}{x - \sqrt{x}}$.

Solution. To integrate the indefinite integral

$$\int \frac{dx}{x - \sqrt{x}},$$

we use the substitution $t^2 = x$. Then $dx = 2t\,dt$, and

$$\int \frac{dx}{x - \sqrt{x}} = \int \frac{2t\,dt}{t^2 - t} = 2 \int \frac{dt}{t - 1} = 2 \ln |t - 1| + C.$$

Changing back to the original variable, we have

$$\int \frac{dx}{x - \sqrt{x}} = 2 \ln |\sqrt{x} - 1| + C,$$

and therefore

$$\int_4^9 \frac{dx}{x - \sqrt{x}} = \left[2 \ln |\sqrt{x} - 1| \right]_4^9 = 2(\ln 2 - \ln 1)$$

$$= \ln 2^2 - 0 = \ln 4.$$

Nevertheless, there is a better way of evaluating the definite integral in Example 1. The same result will be obtained with less labor if, instead of changing back from the variable t to the original variable x, we *change the limits* of the given definite integral to correspond to the change of variable

From the relation between x and t in the substitution $t^2 = x$, we see that when $x = 4$, $t = 2$, and when $x = 9$, $t = 3$. Thus

$$\int_4^9 \frac{dx}{x - \sqrt{x}} = 2 \int_2^3 \frac{dt}{t - 1} = 2 \left[\ln |t - 1| \right]_2^3$$

$$= 2(\ln 2 - \ln 1) = \ln 2^2 - 2(0) = \ln 4.$$

This process of substitution throughout a definite integral is justified in the following theorem.

11.9.1 Theorem. Let g be a function that is smooth on a closed interval $[c, d]$, with $g(c) = a$ and $g(d) = b$. If f is a function that is continuous on the closed interval $[a, b]$, then

$$\int_a^b f(x)\,dx = \int_c^d f(g(t)) \cdot g'(t)\,dt.$$

Proof. Since f is continuous on $[a, b]$, then f has an antiderivative F such that

$$F'(x) = f(x), \qquad a \leq x \leq b.$$

Thus, by the fundamental theorem of calculus,

$$(1) \qquad \int_a^b f(x)\,dx = F(b) - F(a).$$

If we let $x = g(t)$, where t is in $[c, d]$, and use the chain rule, we can write

$$\frac{d}{dt} F(g(t)) = \frac{dF}{dg} \frac{dg}{dt} = F'(g(t)) \cdot g'(t).$$

Therefore (again by the fundamental theorem)

(2) $$\int_c^d f(g(t)) \cdot g'(t)\, dt = F(g(t)) \Big]_c^d$$

$$= F(g(d)) - F(g(c)) = F(b) - F(a).$$

But (1) and (2) imply that

$$\int_a^b f(x)\, dx = \int_c^d f(g(t)) \cdot g'(t)\, dt. \quad \blacksquare$$

Example 2.   Find the value of $\int_{-2}^4 \sqrt{16 - x^2}\, dx$.

Solution.   Let $x = 4 \sin \theta$. Then $dx = 4 \cos \theta\, d\theta$ and $\theta = \sin^{-1}(x/4)$. When $x = -2$, $\theta = \sin^{-1}(-1/2) = -\pi/6$; when $x = 4$, $\theta = \sin^{-1} 1 = \pi/2$. So

$$\int_{-2}^4 \sqrt{16 - x^2}\, dx = \int_{-\pi/6}^{\pi/2} \sqrt{16 - 16 \sin^2 \theta}\, (4 \cos \theta\, d\theta)$$

$$= 16 \int_{-\pi/6}^{\pi/2} \cos^2 \theta\, d\theta = 16 \int_{-\pi/6}^{\pi/2} \frac{1 + \cos 2\theta}{2}\, d\theta$$

$$= 8 \int_{-\pi/6}^{\pi/2} d\theta + 4 \int_{-\pi/6}^{\pi/2} \cos 2\theta\, d(2\theta)$$

$$= \Big[8\theta + 4 \sin 2\theta \Big]_{-\pi/6}^{\pi/2} = \frac{16\pi}{3} + 2\sqrt{3}.$$

Exercises

Evaluate the definite integrals in Exercises 1 through 16.

1. $\displaystyle\int_{3/2}^{12} \frac{x\, dx}{\sqrt{2x + 1}}$.

2. $\displaystyle\int_0^2 x^2\sqrt{x^3 + 1}\, dx$.

3. $\displaystyle\int_{-1}^0 x(x + 1)^{5/3}\, dx$.

4. $\displaystyle\int_{-1}^7 y(y + 1)^{1/3}\, dy$.

5. $\displaystyle\int_0^{\pi/3} \sin^5 t\, dt$.

6. $\displaystyle\int_0^{\pi/4} \sec^6 t\, dt$.

7. $\displaystyle\int_{-1}^{\sqrt{2}} \sqrt{4 - x^2}\, dx$.

8. $\displaystyle\int_{-\sqrt{3}}^{3\sqrt{3}} \frac{dx}{\sqrt{9 + x^2}}$.

9. $\displaystyle\int_1^6 t\sqrt{3t - 2}\, dt$.

10. $\displaystyle\int_0^5 y\sqrt{y + 4}\, dy$.

11. $\displaystyle\int_0^{\pi/12} \tan^2 3x\, dx$.

12. $\displaystyle\int_{-5}^3 2x\, (x + 5)^{4/3}\, dx$.

13. $\displaystyle\int_3^4 x^3\sqrt{x - 3}\, dx$.

14. $\int_{\pi/9}^{\pi/6} \sin^3 3x \cos^2 3x \, dx.$

15. $\int_{10/\sqrt{3}}^{5\sqrt{2}} \frac{dt}{\sqrt{t^2 - 25}}.$

16. $\int_{2}^{4} \frac{\sqrt{16 - x^2}}{x^2} \, dx.$

17. Find the area of the region bounded by the curve $y = x/\sqrt{x + 1}$, and the lines $y = 0$ and $x = 3$. Make a sketch.

18. Find the area of the region bounded by the curve $y = 1/(1 + \sqrt{x})$ and the lines $y = 0$, $x = 0$, and $x = 4$. Make a sketch.

19. Find the area of the region bounded by the curve $y = x/(x + 1)^{2/3}$ and the lines $y = 0$ and $x = 7$. Make a sketch.

20. Find the volume of the solid generated by revolving about the x axis the region described in Exercise 19.

21. Find the centroid of the region in Exercise 19.

22. Find the length of the segment of the semicircle $y = \sqrt{16 - x^2}$ from $x = 2$ to $x = 2\sqrt{3}$.

23. Find the area of the region enclosed by the curve $y = 4/(x\sqrt{x^2 + 4})$ and the lines $y = 0$, $x = 1$, and $x = 2$. Make a sketch.

24. Find the volume of the solid generated by revolving about the y axis the region described in Exercise 19.

25. Find the area of the region bounded by the upper half of the ellipse $b^2x^2 + a^2y^2 = a^2b^2$ and the x axis, $(a > 0, b > 0)$. (*Hint:* The region is symmetric with respect to the y axis.)

26. Find the centroid of the region in Exercise 19.

27. Show that the area of the region in the first quadrant [Fig. 10-21(b)] bounded by the curve $y = \sqrt{x^2 - 1}$, the x axis, and the vertical line $x = \cosh u$ is $\frac{1}{4} \sinh 2u - \frac{1}{2}u$. (*Hint:* In the definite integral, make the substitution $x = \cosh t$, and then use the identities $-\sinh^2 t + \cosh^2 t = 1$ and $\cosh 2t = 2 \sinh^2 t - 1$.)

28. In Fig. 10-21(b), find the area of the shaded region OQP.
(*Hint:* Area of triangle ORP is $\frac{1}{2}|OR|\cdot|RP| = \frac{1}{2}\cosh u \sinh u = \frac{1}{4}\sinh 2u$. Now use Exercise 27.)

11.10 INTEGRATION OF RATIONAL FUNCTIONS BY PARTIAL FRACTIONS

A **rational function** is one that can be expressed as the quotient of two polynomial functions.

An expression in the form $f(x)/g(x)$, where $f(x)$ and $g(x)$ are polynomials, is called a **rational fraction.** If the degree of $f(x)$ is less than the degree of $g(x)$, the rational fraction $f(x)/g(x)$ is said to be **proper**; otherwise it is **improper.**

When integrating an improper rational fraction, the first step is to divide the denominator into the numerator by long division until a remainder is reached that is of lower degree than the denominator. For instance,

$$\frac{x^5 + 2x^3 - x + 1}{x^3 + 5x} = x^2 - 3 + \frac{14x + 1}{x^3 + 5x}.$$

We assume throughout this section that the rational integrands being considered are proper.

It is a well-known theorem of algebra that a polynomial with real coefficients is factorable into linear and quadratic factors with real coefficients. Of course, in practice, it may be difficult to find these factors, but we know that they exist.

It is proved in advanced algebra that a proper rational fraction $f(x)/g(x)$ can be decomposed into a sum of *partial fractions*, each of whose denominators is a power of a single linear or quadratic factor of $g(x)$. As an illustration,

$$\frac{5x^3 - 15x^2 + 14x - 29}{(x^2 + x + 1)(x - 3)^2} = \frac{x - 2}{x^2 + x + 1} + \frac{1}{(x - 3)^2} + \frac{4}{x - 3},$$

which the reader should verify by adding the fractions in the right member.

To integrate a proper rational fraction that does not come under any of the basic forms, we decompose the rational expression into a sum of partial fractions and integrate each partial fraction separately by methods previously explained. This is called the **method of partial fractions**, and it turns out that the results can always be expressed in terms of algebraic, logarithmic, and inverse trigonometric functions.

Let $f(x)$ and $g(x)$ be polynomials with $f(x)$ of lower degree than $g(x)$. In integrating

$$\int \frac{f(x)}{g(x)}\, dx$$

by partial fractions, we distinguish two cases, depending on whether the irreducible real factors of the denominator $g(x)$ are all linear or include at least one quadratic factor.

CASE 1.   The irreducible real factors of the denominator are all linear.

It is proved in advanced algebra that to each factor $(ax + b)^r$ of $g(x)$, where r is a positive integer, there correspond r terms in the partial fraction expansion of $f(x)/g(x)$ of the form

$$\frac{A}{(ax + b)^r} + \frac{B}{(ax + b)^{r-1}} + \frac{C}{(ax + b)^{r-2}} + \cdots + \frac{K}{ax + b},$$

where $A, B, C, \cdots, K$ are constants. Two common methods for determining the values of these constants are shown in the following examples.

Example 1.   Integrate $\displaystyle\int \frac{1}{x^2 - 4}\, dx$.

Solution.   Since $x^2 - 4 = (x - 2)(x + 2)$, the denominator of the integrand is the product of two distinct linear factors. Thus the integrand can be written

(1) $$\frac{1}{x^2 - 4} = \frac{A}{x - 2} + \frac{B}{x + 2},$$

where A and B are (unknown) constants. To determine the values of A and B, we clear (1) of fractions and obtain

$$1 = A(x + 2) + B(x - 2).$$

Collecting terms in x, we have

(2) $$1 = (A + B)x + 2(A - B).$$

Since (2) is valid for all values of x, the coefficients of like powers of x on both sides of this identity must be equal. Consequently,

$$A + B = 0,$$
$$2(A - B) = 1.$$

By solving these simultaneous equations for A and B, we find that $A = \frac{1}{4}$ and $B = -\frac{1}{4}$. Therefore

$$\frac{1}{x^2 - 4} = \frac{\frac{1}{4}}{x - 2} + \frac{-\frac{1}{4}}{x + 2},$$

and

$$\int \frac{1}{x^2 - 4}\, dx = \frac{1}{4} \int \frac{dx}{x - 2} - \frac{1}{4} \int \frac{dx}{x + 2}$$

$$= \frac{1}{4} \ln|x - 2| - \frac{1}{4} \ln|x + 2| + C$$

$$= \frac{1}{4} \ln\left|\frac{x - 2}{x + 2}\right| + C.$$

Example 2.   Integrate $\displaystyle\int \frac{5x + 3}{x^3 - 2x^2 - 3x}\, dx.$

Solution.   Since the denominator is $x^3 - 2x^2 - 3x = x(x + 1)(x - 3)$, the integrand can be reduced to partial fractions in the form

$$\frac{5x + 3}{x(x + 1)(x - 3)} = \frac{A}{x} + \frac{B}{x + 1} + \frac{C}{x - 3},$$

where A, B, and C are constants, yet to be determined. Clearing of fractions, we get

$$(3) \qquad 5x + 3 = A(x + 1)(x - 3) + Bx(x - 3) + Cx(x + 1).$$

Then, collecting terms,

$$5x + 3 = (A + B + C)x^2 + (-2A - 3B + C)x - 3A,$$

which is an identity that is valid for all values of x. Therefore the coefficients of like powers of x must be equal:

$$A + B + C = 0,$$

$$-2A - 3B + C = 5,$$

$$-3A = 3.$$

Solving these three simultaneous equations, we find $A = -1$, $B = -\frac{1}{2}$ and $C = \frac{3}{2}$.

There is a shorter way of finding the values of A, B, and C in this example. Although the given integrand is not defined when $x = 0$, $x = -1$, or $x = 3$, equation (3) is an identity that is true for all values of x, including 0, -1, and 3. If we substitute $x = 0$ in (3), we get $3 = -3A$ or $A = -1$. Substituting $x = -1$ in (3), we get $-5 + 3 = B(-1)(-4)$, or $B = -\frac{1}{2}$. When we substitute $x = 3$ in (3), we find $C = \frac{3}{2}$.

By either method,

$$(4) \qquad \frac{5x + 3}{x^2 - 2x^2 - 3x} = -\frac{1}{x} - \frac{\frac{1}{2}}{x + 1} + \frac{\frac{3}{2}}{x - 3}.$$

Therefore

$$\int \frac{(5x + 3)\, dx}{x^3 - 2x^2 - 3x} = -\int \frac{dx}{x} - \frac{1}{2} \int \frac{dx}{x + 1} + \frac{3}{2} \int \frac{dx}{x - 3}$$

$$= -\ln|x| - \tfrac{1}{2} \ln|x + 1| + \tfrac{3}{2} \ln|x - 3| + C.$$

Example 3.   Find $\displaystyle\int \frac{x}{(x - 3)^2}\, dx.$

Solution. The denominator of the integrand is $(x - 3)^2$, which is a linear factor *repeated*. Thus

(5)
$$\frac{x}{(x - 3)^2} = \frac{A}{(x - 3)^2} + \frac{B}{x - 3},$$

in which we must determine the values of the constants A and B. Multiplying both members of (5) by $(x - 3)^2$ gives

$$x = A + B(x - 3),$$

or

$$x = Bx + (A - 3B),$$

which is an identity in x. Equating coefficients of like powers of x in both sides of this identity, we obtain the simultaneous equations

$$B = 1,$$
$$A - 3B = 0.$$

From them, $A = 3$ and $B = 1$. Therefore

$$\frac{x}{(x - 3)^2} = \frac{3}{(x - 3)^2} + \frac{1}{x - 3},$$

and

$$\int \frac{x}{(x - 3)^2}\, dx = 3 \int (x - 3)^{-2}\, dx + \int \frac{dx}{x - 3}$$

$$= \frac{-3}{x - 3} + \ln|x - 3| + C.$$

Example 4. Integrate $\displaystyle\int \frac{3x^2 - 8x + 13}{x^3 + x^2 - 5x + 3}\, dx.$

Solution. The denominator of the integrand factors into $(x + 3)(x - 1)^2$; so we write

(6)
$$\frac{3x^2 - 8x + 13}{x^3 + x^2 - 5x + 3} = \frac{A}{x + 3} + \frac{B}{x - 1} + \frac{C}{(x - 1)^2},$$

where A, B, and C are constants that are still to be determined. Clearing of fractions, we change this to

(7)
$$3x^2 - 8x + 13 = A(x - 1)^2 + B(x + 3)(x - 1) + C(x + 3),$$

or, collecting terms,

(8)
$$3x^2 - 8x + 13 = (A + B)x^2 + (-2A + 2B + C)x + (A - 3B + 3C).$$

If we equate the coefficients of corresponding powers of x in (8), we obtain

$$A + B = 3,$$
$$-2A + 2B + C = -8,$$
$$A - 3B + 3C = 13.$$

Solving these equations simultaneously for A, B, and C, we find $A = 4$, $B = -1$, $C = 2$. Then the substitution of these values for A, B, and C in (6) gives

(9)
$$\frac{3x^2 - 8x + 13}{x^3 + x^2 - 5x + 3} = \frac{4}{x + 3} - \frac{1}{x - 1} + \frac{2}{(x - 1)^2}.$$

An easier way to find the values of A, B, and C is to let $x = 1$ in (7), obtaining $4C = 8$ or $C = 2$; then let $x = -3$ in (7), getting $64 = 16A$ or $A = 4$. To find B, we put $A = 4$

and $C = 2$ in (7) and then let x be any number *except* 1 and -3, say $x = 0$. This gives $B = -1$.

Whichever method for finding A, B, and C is used, we now integrate (9) term by term:

$$\int \frac{3x^2 - 8x + 13}{x^3 + x^2 - 5x + 3}\, dx = 4 \int \frac{dx}{x + 3} - \int \frac{dx}{x - 1} + 2 \int (x - 1)^{-2}\, dx$$

$$= 4 \ln|x + 3| - \ln|x - 1| - \frac{2}{x - 1} + C.$$

CASE 2. At least one of the irreducible real factors of the denominator is quadratic.

It is shown in algebra that, in the decomposition of $f(x)/g(x)$ into partial fractions, to each irreducible factor $(ax^2 + bx + c)^r$ of $g(x)$ there correspond r terms of the form

$$\frac{Ax + B}{(ax^2 + bx + c)^r} + \frac{Cx + D}{(ax^2 + bx + c)^{r-1}} + \cdots + \frac{Kx + L}{ax^2 + bx + c},$$

where A, B, C, $\cdots$, L are constants yet to be determined.

Example 5. Integrate $\displaystyle \int \frac{6x^2 - 3x + 1}{(4x + 1)(x^2 + 1)}\, dx.$

Solution. Since the denominator of the integrand is the product of one linear factor and one irreducible quadratic factor, we write

(10)
$$\frac{6x^2 - 3x + 1}{(4x + 1)(x^2 + 1)} = \frac{A}{4x + 1} + \frac{Bx + C}{x^2 + 1}.$$

To determine the values of A, B, and C, we multiply both members of this equation by $(4x + 1)(x^2 + 1)$ and obtain

$$6x^2 - 3x + 1 = A(x^2 + 1) + (Bx + C)(4x + 1)$$

or, after collecting terms,

$$6x^2 - 3x + 1 = (A + 4B)x^2 + (B + 4C)x + (A + C).$$

Because this is an identity in x, the coefficients of like powers of x are equal. Thus

$$A + 4B = 6,$$
$$B + 4C = -3,$$
$$A + C = 1.$$

By solving these simultaneous equations, we find $A = 2$, $B = 1$, and $C = -1$. Substituting these values in (10), we obtain

$$\frac{6x^2 - 3x + 1}{(4x + 1)(x^2 + 1)} = \frac{2}{4x + 1} + \frac{x - 1}{x^2 + 1}.$$

Thus

$$\int \frac{6x^2 - 3x + 1}{(4x + 1)(x^2 + 1)}\, dx = \int \frac{2\, dx}{4x + 1} + \int \frac{x - 1}{x^2 + 1}\, dx$$

$$= \frac{1}{2} \int \frac{4\, dx}{4x + 1} + \frac{1}{2} \int \frac{2x\, dx}{x^2 + 1} - \int \frac{dx}{x^2 + 1}$$

$$= \frac{1}{2} \int \frac{d(4x + 1)}{4x + 1} + \frac{1}{2} \int \frac{d(x^2 + 1)}{x^2 + 1} - \int \frac{dx}{x^2 + 1}$$

$$= \frac{1}{2} \ln |4x + 1| + \frac{1}{2} \ln (x^2 + 1) - \text{Tan}^{-1} x + C$$

$$= \frac{1}{2} \ln |(4x + 1)(x^2 + 1)| - \text{Tan}^{-1} x + C.$$

Example 6.   Integrate $\int \frac{6x^2 - 15x + 22}{(x + 3)(x^2 + 2)^2} dx$.

Solution.   Let

$$\frac{6x^2 - 15x + 22}{(x + 3)(x^2 + 2)^2} = \frac{A}{x + 3} + \frac{Bx + C}{x^2 + 2} + \frac{Dx + E}{(x^2 + 2)^2}.$$

Clearing of fractions, we have

(11) $6x^2 - 15x + 22 = A(x^2 + 2)^2 + (Bx + C)(x + 3)(x^2 + 2) + (Dx + E)(x + 3).$

By letting $x = -3$ in (11), we find $A = 1$. Collecting terms in (11), we have

$$6x^2 - 15x + 22 = (A + B)x^4 + (3B + C)x^3 + (4A + 2B + 3C + D)x^2$$
$$+ (6B + 2C + 3D + E)x + (4A + 6C + 3E).$$

If we equate the corresponding coefficients of x^4, x^3, x^2, and the constant terms, we obtain

$$A + B = 0,$$

$$3B + C = 0,$$

$$4A + 2B + 3C + D = 6,$$

$$4A + 6C + 3E = 22.$$

With $A = 1$, we solve these four equations simultaneously for B, C, D, and E, getting $A = 1$, $B = -1$, $C = 3$, $D = -5$, $E = 0$. Thus the integrand can be written

$$\frac{6x^2 - 15x + 22}{(x + 3)(x^2 + 2)^2} = \frac{1}{x + 3} + \frac{-x + 3}{x^2 + 2} + \frac{-5x}{(x^2 + 2)^2}.$$

Therefore

$$\int \frac{6x^2 - 15x + 22}{(x + 3)(x^2 + 2)^2} dx = \int \frac{dx}{x + 3} - \int \frac{x - 3}{x^2 + 2} dx - 5 \int \frac{x}{(x^2 + 2)^2} dx$$

$$= \int \frac{dx}{x + 3} - \frac{1}{2} \int \frac{2x \, dx}{x^2 + 2} + 3 \int \frac{dx}{x^2 + 2} - \frac{5}{2} \int (x^2 + 2)^{-2}(2x \, dx)$$

$$= \int \frac{dx}{x + 3} - \frac{1}{2} \int \frac{d(x^2 + 2)}{x^2 + 2} + 3 \int \frac{dx}{x^2 + 2}$$

$$- \frac{5}{2} \int (x^2 + 2)^{-2} d(x^2 + 2)$$

$$= \ln |x + 3| - \frac{1}{2} \ln (x^2 + 2) + \frac{3}{\sqrt{2}} \text{Tan}^{-1} \frac{x}{\sqrt{2}}$$

$$+ \frac{5}{2(x^2 + 2)} + C.$$

Exercises

Integrate.

1. $\int \dfrac{2}{x^2 + 2x} \, dx.$

2. $\int \dfrac{2}{x^2 - 1} \, dx.$

3. $\int \dfrac{5x + 3}{x^2 - 9} \, dx.$

4. $\int \dfrac{x - 6}{x^2 - 2x} \, dx.$

5. $\int \dfrac{x - 11}{x^2 + 3x - 4} \, dx.$

6. $\int \dfrac{3x - 13}{x^2 + 3x - 10} \, dx.$

7. $\int \dfrac{2x + 21}{2x^2 + 9x - 5} \, dx.$

8. $\int \dfrac{17x - 3}{3x^2 + x - 2} \, dx.$

9. $\int \dfrac{2x^2 + x - 4}{x^3 - x^2 - 2x} \, dx.$

10. $\int \dfrac{6x^2 + 22x - 23}{(2x - 1)(x^2 + x - 6)} \, dx.$

11. $\int \dfrac{3x^3}{x^2 + x - 2} \, dx.$

12. $\int \dfrac{x^4 + 8x^2 + 8}{x^3 - 4x} \, dx.$

13. $\int \dfrac{x + 1}{(x - 3)^2} \, dx.$

14. $\int \dfrac{5x + 7}{x^2 + 4x + 4} \, dx.$

15. $\int \dfrac{3x^2 - 21x + 32}{x^3 - 8x^2 + 16x} \, dx.$

16. $\int \dfrac{x^2 + 19x + 10}{2x^4 + 5x^3} \, dx.$

17. $\int \dfrac{2x^2 + x - 8}{x^3 + 4x} \, dx.$

18. $\int \dfrac{2x^2 - 3x - 36}{(2x - 1)(x^2 + 9)} \, dx.$

19. $\int \dfrac{x^3 - 8x^2 - 1}{(x + 3)(x - 2)(x^2 + 1)} \, dx.$

20. $\int \dfrac{20x - 11}{(3x + 2)(x^2 - 4x + 5)} \, dx.$

21. $\int \dfrac{x^3 - 4x}{(x^2 + 1)^2} \, dx.$

22. $\int \dfrac{2x^3 + 5x^2 + 16x}{x^5 + 8x^3 + 16x} \, dx.$ [*Hint:* To integrate $\int (x^2 + 4)^{-2} \, dx$, let $x = 2 \operatorname{Tan} \theta$.]

23. Evaluate $\displaystyle\int_4^6 \dfrac{x - 17}{x^2 + x - 12} \, dx.$

24. Find the value of $\displaystyle\int_1^5 \dfrac{3x + 13}{x^2 + 4x + 3} \, dx.$

11.11   RATIONAL FUNCTIONS OF sin x
AND cos x^*

An integrand that is a rational function of sin x or cos x or both, can be changed to a rational function of z by the substitution

(1) $$z = \tan \tfrac{1}{2} x;$$

and the resulting rational function of z can be integrated by the method of partial fractions.

*Optional.

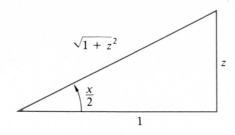

Figure 11-7

From Fig. 11-7 we see that

$$\sin \frac{1}{2} x = \frac{z}{\sqrt{1 + z^2}} \qquad \text{and} \qquad \cos \frac{1}{2} x = \frac{1}{\sqrt{1 + z^2}}.$$

If these results are substituted in the double-angle formulas, we obtain

$$(2) \qquad \sin x = 2 \sin \frac{1}{2} x \cos \frac{1}{2} x = \frac{2z}{1 + z^2},$$

and

$$(3) \qquad \cos x = \cos^2 \frac{1}{2} x - \sin^2 \frac{1}{2} x = \frac{1 - z^2}{1 + z^2};$$

and since $x = 2 \, \text{Tan}^{-1} z$ [from (1)], we also have

$$(4) \qquad dx = \frac{2 \, dz}{1 + z^2}.$$

For reference, we collect the preceding formulas.

11.11.1 If $z = \tan \frac{1}{2} x$, then

$$\sin x = \frac{2z}{1 + z^2}, \qquad \cos x = \frac{1 - z^2}{1 + z^2}, \qquad dx = \frac{2 \, dz}{1 + z^2}.$$

Example. Integrate $\displaystyle\int \frac{dx}{3 \sin x + 4 \cos x}.$

Solution. Substituting from 11.11.1, we get

$$\int \frac{dx}{3 \sin x + 4 \cos x} = -\int \frac{dz}{2z^2 - 3z - 2} = -\int \frac{dz}{(2z + 1)(z - 2)}$$

$$= \frac{1}{5} \int \frac{2 \, dz}{2z + 1} - \frac{1}{5} \int \frac{dz}{z - 2} = \frac{1}{5} \ln \left| \frac{2z + 1}{z - 2} \right| + C$$

$$= \frac{1}{5} \ln \left| \frac{2 \tan \frac{1}{2} x + 1}{\tan \frac{1}{2} x - 2} \right| + C.$$

Exercises

Perform the indicated integrations.

1. $\int \dfrac{dx}{1 - \sin x}$.

2. $\int \dfrac{dx}{3 + \cos x}$.

3. $\int \dfrac{dx}{4 \sin x + 3 \cos x}$.

4. $\int \dfrac{\sin x}{2 + \cos x} \, dx$.

5. $\int \dfrac{\cot x}{1 + \cos x} \, dx$.

6. $\int \dfrac{\cot x}{1 + \sin x} \, dx$.

7. $\int \dfrac{dx}{\cos x - \sin x + 1}$.

8. $\int \dfrac{\sin x}{2 - \sin x} \, dx$.

9. $\int \dfrac{dx}{4 + \sin x}$.

10. $\int \dfrac{2 \, dx}{\sin x + \tan x}$.

11.12 REVIEW EXERCISES FOR INTEGRATION

The purpose of this section is to give the reader practice in choosing the proper technique for a specific integration.

In Exercises 1 through 50, perform the indicated integrations by any methods.

1. $\int \dfrac{dx}{x^{1/3} + x^{2/3}}$.

2. $\int \dfrac{\tan x}{\ln |\cos x|} \, dx$.

3. $\int \dfrac{3 \, dt}{t^3 - 1}$.

4. $\int \sinh x \, dx$.

5. $\int \dfrac{(\ln y)^5}{y} \, dy$.

6. $\int x \cot^2 x \, dx$.

7. $\int \dfrac{\sin \sqrt{x}}{\sqrt{x}} \, dx$.

8. $\int e^{\cot x} \csc^2 x \, dx$.

9. $\int \dfrac{\ln t^2}{t} \, dt$.

10. $\int x^2 \operatorname{Cos}^{-1} x \, dx$.

11. $\int \ln (y^2 + 9) \, dy$.

12. $\int \operatorname{Sin}^{-1} \sqrt{\dfrac{t}{t + 1}} \, dt$.

13. $\int \operatorname{sech} x \, dx$.

14. $\int e^{t/3} \sin 3t \, dt$.

15. $\int \dfrac{t + 9}{t^3 + 9t} \, dt$.

16. $\int \cosh x \, dx$.

17. $\int \sin \dfrac{3x}{2} \cos \dfrac{x}{2} \, dx$.

18. $\int \dfrac{\sin 2x - 1}{\sin 2x + 1} \, dx$.

19. $\int \sinh^2 x \, dx$.

20. $\int \cosh^2 x \, dx$.

21. $\int t^2 e^{4t}\, dt.$

22. $\int \ln \sqrt{2w - 1}\, dw.$

23. $\int \dfrac{144}{x(x^2 - 1)(x^2 - 9)}\, dx.$

24. $\int \tanh y\, dy.$

25. $\int \theta \sin 2\theta \cos 2\theta\, d\theta.$

26. $\int \dfrac{\sqrt{x}}{1 + \sqrt{x}}\, dx.$

27. $3 \int \dfrac{x^3 + 1}{x^3 - 1}\, dx.$

28. $\int \tan^{3/2} x \sec^4 x\, dx.$

29. $\int \dfrac{dt}{t(t^{1/6} + 1)}.$

30. $\int \coth x\, dx.$

31. $\int \dfrac{e^{2y}\, dy}{\sqrt{9 - e^{2y}}}.$

32. $\int \cos^5 x \sqrt{\sin x}\, dx.$

33. $\int e^{\ln |3 \cos x|}\, dx.$

34. $\int t^n \ln t\, dt,\ n \neq -1.$

35. $\int \dfrac{\sqrt{7 - y^2}}{y}\, dy.$

36. $\int \dfrac{e^{4x}}{1 + e^{8x}}\, dx.$

37. $\int \dfrac{\operatorname{Tan}^{-1} x}{x^3}\, dx.$

38. $\int \dfrac{\sqrt{x^2 + a^2}}{x^3}\, dx.$

39. $\int \dfrac{w}{\sqrt{w + 5}}\, dw.$

40. $\int (\ln 2x)^2\, dx.$

41. $\int \dfrac{\sin t\, dt}{\sqrt{1 + \cos t}}.$

42. $\int \dfrac{\sin y \cos y}{9 + \cos^4 y}\, dy.$

43. $\int \operatorname{csch} x\, dx.$

44. $\int \dfrac{\sec^2 x\, dx}{\tan x(\ln \tan x)}.$

45. $\int \dfrac{e^y\, dy}{(a^2 - e^{2y})^{3/2}}.$

46. $\int \dfrac{dx}{\sqrt{1 - 6x - x^2}}.$

47. $\int \dfrac{4x^2 + 3x + 6}{x^2(x^2 + 3)}\, dx.$

48. $\int \dfrac{e^{\tan x}}{1 - \sin^2 x}\, dx.$

49. $\int \dfrac{dx}{(16 + x^2)^{3/2}}.$

50. $\int \sinh^{-1} x\, dx.$

51. Find the area of the region bounded by the x axis, the curve $y = 18/(x^2\sqrt{x^2 + 9})$, and the lines $x = \sqrt{3}$ and $x = 3\sqrt{3}$. Make a sketch.

52. Find the area of the region bounded by the curve $y = x^3 \cos 2x$ and the x axis, between $x = -\pi/4$ and $x = \pi/4$. Make a sketch.

53. Find the area of the region bounded by the curve $s = t/(t - 1)^2$, $s = 0$, $t = -6$, and $t = 0$. Make a sketch.

54. Find the volume of the solid generated by revolving the region

$$\{P{:}(x, y) \mid 0 \leq x \leq 2,\ 0 \leq y \leq xe^x\}$$

about the x axis. Make a sketch.

55. Find the volume of the solid generated by revolving the region

$$\left\{ P{:}(x, y) \mid -3 \leq x \leq -1,\ \dfrac{6}{x\sqrt{x + 4}} \leq y \leq 0 \right\}$$

about the x axis. Make a sketch.

56. Show that the volume of the solid generated by revolving the region

$$\left\{P{:}(x, y)\,|\,0 \le x \le 2\cos^3 y, \; -\frac{\pi}{2} \le y \le \frac{\pi}{2}\right\}$$

about the y axis is equal to the volume of the right circular cylinder with height π and diameter of base $\sqrt{5}$.

57. Find the centroid of the region in the first quadrant bounded by the ellipse $4x^2 + 25y^2 = 100$ and the coordinate axes.

58. Find the centroid of the region

$$\{P{:}(x, y)\,|\,0 \le x \le 3, \, 0 \le y \le xe^{-x}\}.$$

59. Find the centroid of the region

$$\{P{:}(x, y)\,|\,-2 \le x \le 2, \, 0 \le y \le \cosh x\}.$$

60. Find the centroid of the region

$$\{P{:}(x, y)\,|\,0 \le x \le 2, \, 0 \le y \le xe^{x}\}.$$

Use Pappus' theorem (9.8.9) to show that the volume of the solid in Exercise 54 can be obtained by multiplying the area of the region in the present exercise by $2\pi \bar{y}$.

61. Find the length of the segment of the curve $y = \ln(\sin x)$ from $x = \pi/6$ to $x = \pi/3$. Make a sketch.

62. Find the length of the segment of the curve $y = e^x$ from $x = 0$ to $x = \ln(4\sqrt{3})$. Make a sketch.

63. Let f be a function with a continuous, nonzero derivative on $[a, b]$. Prove that

$$\int_a^b 2xf(x)\,dx = b^2 d - a^2 c - \int_c^d [f^{-1}(y)]^2\,dy,$$

where $c = f(a)$ and $d = f(b)$. [*Hint:* $D_x[x^2 f(x)] = x^2 f'(x) + 2xf(x)$.]

64. Let f be the function in Exercise 63 but with $a \ge 0$ and $f(a) \ge 0$. Denote by R the region bounded by $y = f(x)$, $x = a$, $x = b$, and the x axis, and indicate by S the solid of revolution generated by revolving R about the y axis. Prove that the "cylindrical shell method" (9.3) gives the same volume for S as the "washer method" (9.2). (*Hint:* Use Exercise 63.)

11.13 TABLES OF INTEGRALS

The techniques and devices learned in this chapter will enable the reader to reduce many indefinite integrals to types that are treated in the basic integration formulas (11.2).

Much longer lists of integration formulas are found in *tables of integrals*. Engineers and scientists usually resort to a table of integrals when integrating a complicated integrand. A brief table of integrals appears in Appendix A.7.

But the reader is advised to become familiar with the methods of this chapter before attempting to use a table of integrals. Often, complicated integrands do not appear in the tables in just the form wanted, but some of the techniques explained in this chapter may enable us to transform the given integral into a form that does appear in the table of integrals.

11.14 SEPARABLE DIFFERENTIAL EQUATIONS. APPLICATIONS

Differential equations have been studied since the beginning of calculus. Many problems encountered by the engineer and scientist lead to differential equations when expressed in mathematical symbols. In fact, most of the differential equations studied by mathematicians arose from physical problems. Although some differential equations are difficult to solve, the present section is chiefly concerned with a class of simple differential equations that have remarkable applications.

An equation involving derivatives or differentials is called a **differential equation**. Examples of differential equations are

(a) $y\dfrac{dy}{dx} = x$;

(b) $\dfrac{d^2y}{dx^2} = -32$;

(c) $\dfrac{d^3y}{dx^3} + \sin x\,\dfrac{d^2y}{dx^2} + y = \cos x$;

(d) $x\,dy + y\,dx = 0$.

The **order** of a differential equation is the order of the derivative of highest order in the equation. In the examples shown, (c) is of order 3, (b) is of order 2, (a) is of order 1, and (d) is of order 1 (since it can be written $dy/dx + y/x = 0$.

A function f, defined by $y = f(x)$, is a **solution** of a differential equation if the equation becomes an identity when y and its derivatives are replaced by $f(x)$ and its corresponding derivatives.

For example, $y = x \ln x - x$ is a solution of the differential equation

$$\frac{dy}{dx} = \frac{x + y}{x}$$

because substitution of $y = x \ln x - x$ throughout this equation gives

$$\frac{d}{dx}(x \ln x - x) = \frac{x + (x \ln x - x)}{x}$$

or

$$x \cdot \frac{1}{x} + \ln x - 1 = \frac{x \ln x}{x},$$

which simplifies to

$$\ln x = \ln x,$$

an identity for all $x > 0$.

Consider the first-order differential equation

(1) $$\frac{dy}{dx} = \sin x.$$

If we integrate both members of this equation we obtain

$$y = -\cos x + C.$$

This is the **general solution** of (1) because it satisfies (1) identically and contains one arbitrary constant, C. All particular solutions of this differential equation are obtained by assigning particular values to C. Thus $y = -\cos x + 5$ is a particular solution of (1) and $y = -\cos x - 7\pi/2$ is another.

The second-order differential equation

(2) $$\frac{d^2y}{dx^2} = -32$$

can be written

$$\frac{d}{dx}\left(\frac{dy}{dx}\right) = -32.$$

By integrating both members of this equation once, with respect to x, we get

$$\frac{dy}{dx} = -32x + C_1,$$

where C_1 is an arbitrary constant of integration. And if we integrate this latter equation, we obtain

$$(3) \qquad\qquad y = -16x^2 + C_1 x + C_2,$$

where C_2 is another arbitrary constant, independent of C_1. We can assign any particular value to either C_1 or C_2 and have complete freedom in choosing a particular value for the other. Equation (3) is the *general* solution of the second-order differential equation (2) because it contains two independent constants.

Similarly, the **general solution** of an nth-order differential equation is a solution that contains n independent arbitrary constants. *Particular solutions* are obtained from it by assigning particular values to the arbitrary constants.

We shall now confine our attention to a class of first-order differential equations called **separable**.

Let M be a continuous function of x and let N be a continuous function of y. The differential equation

$$\frac{dy}{dx} = \frac{M(x)}{N(y)}$$

is of the first order. Because it can be written in the form

$$(4) \qquad\qquad M(x)\,dx - N(y)\,dy = 0,$$

its variables are said to be *separable*. The general solution of (4) is

$$\int M(x)\,dx - \int N(y)\,dy = C,$$

where C is an arbitrary constant.

Example 1. Find the general solution of the differential equation $xy^2\,dx - (x+5)\,dy = 0$.

Solution. The variables are separable, since the given differential equation can be written

$$\frac{x}{x+5}\,dx - \frac{1}{y^2}\,dy = 0, \qquad x \neq -5, y \neq 0.$$

Its general solution, which must contain one arbitrary constant because the equation is of the first order, is

$$\int \frac{x}{x+5}\,dx - \int y^{-2}\,dy = C,$$

or

$$x - 5\ln|x+5| + \frac{1}{y} = C,$$

where C is an arbitrary constant. This can be expressed as an explicit function of x,

$$y = \frac{1}{C - x + \ln|x + 5|^5}.$$

The reader should always verify that the solution satisfies the given differential equation identically by substituting it and its derivatives in the given equation.

In many problems of physics, chemistry, biology, and economics, the time rate of change of a varying quantity is proportional to the magnitude of the quantity at any instant. For instance, under favorable conditions a *population* (the number of people in a country, the number of bacteria in a colony, and so on.) increases at a rate that is proportional to the number of individuals present at any instant. A radioactive substance disintegrates at a rate proportional to the amount present. Often in a chemical reaction a substance is used up at a rate that is proportional to the amount of the substance present at any time.

If y is a positive-valued function of t and k is a constant, the differential equation

$$\frac{dy}{dt} = ky$$

states that the time rate of change of y is proportional to the magnitude of y at any time t.

This is a separable differential equation and it can be written

$$\frac{dy}{y} = k \, dt.$$

By integrating both members, we get

$$\ln y = kt + C,$$

from which (by 10.3.2)

$$y = e^{kt+C} = e^{kt} \cdot e^C$$

or

(5) $$y = Ae^{kt},$$

where $A = e^C$ is an arbitrary constant.

The value of the constant k is (5) depends on the nature of the problem. If k is positive, (5) is called the law of **exponential growth**. If k is negative, (5) is the law of **exponential decay**.

Example 2. The number of bacteria in a culture was 10,000 at a certain time and 2 hours later became 40,000. Under ideal conditions for growth, what would the bacteria population be at the end of 5 hours?

Solution. Under favorable conditions the rate of growth of the number of bacteria in a culture is proportional to the number of bacteria present. If y is the number of bacteria in the culture at time t, the rate of growth is given by

$$\frac{dy}{dt} = ky,$$

where k is a factor of proportionality.

After separating the variables in this differential equation by rewriting it

$$\frac{dy}{y} = k \, dt,$$

we integrate both members to obtain the general solution

$$\ln y = kt + C.$$

Since 10,000 bacteria were present at the beginning, we substitute $y = 10,000$ and $t = 0$ in this equation to get $\ln 10,000 = k(0) + C$, or $C = \ln 10,000$. The solution becomes

(6) $$\ln y = kt + \ln 10,000.$$

We now evaluate k from the **initial conditions** (the given conditions). At the end of 2 hours there were 40,000 bacteria; so we substitute $y = 40,000$ and $t = 2$ in this equation, obtaining $\ln 40,000 = 2k + \ln 10,000$, or

$$k = \frac{1}{2} (\ln 40,000 - \ln 10,000)$$

$$= \frac{1}{2} \ln \frac{40,000}{10,000} = \ln \sqrt{4} = \ln 2.$$

Thus (6) becomes

$$\ln y = (\ln 2)t + \ln 10,000.$$

At the end of 5 hours,

$$\ln y = (\ln 2)5 + \ln 10,000$$

$$= \ln 2^5 + \ln 10,000$$

$$= \ln 320,000.$$

Finally,

$$y = 320,000.$$

There are 320,000 bacteria present at the end of 5 hours.

Example 3. Radium decomposes as it emits alpha rays. Assuming that the rate of decomposition is proportional to the amount present at any time t, and that 25% of the original amount is lost in 664 years, what is the half-life of radium? (The **half-life** of a radioactive substance is the time it takes any given amount of the substance to lose half its mass by disintegration.)

Solution. Let y be the amount of radium present at any time t, where t is measured in years. Since the rate of decomposition is proportional to the amount present,

(7) $$\frac{dy}{dt} = ky,$$

in which k is a constant of proportionality. To get the general solution of this first-order differential equation, we separate the variables,

$$\frac{dy}{y} = k \, dt,$$

and integrate both members, obtaining

$$\ln y = kt + C.$$

If we denote the original amount of radium by R_0, then $y = R_0$ when $t = 0$ and we have $\ln R_0 = k(0) + C$, or $C = \ln R_0$. The solution now is

$$\ln y = kt + \ln R_0.$$

Since $y = 0.75R_0$ when $t = 664$, we determine k by substituting these initial conditions in the preceding equation, getting $\ln 0.75R_0 = 664k + \ln R_0$, from which $k = (\ln 0.75)/664 \doteq -0.000433$. Thus the particular solution of the differential equation (7) for this problem is

$$\ln y \doteq -0.000433t + \ln R_0.$$

In order to find t when y disintegrates to half its original mass R_0, we substitute $y = \frac{1}{2}R_0$ in the preceding equation and solve for t. This gives

$$t \doteq \frac{\ln R_0 - \ln (\tfrac{1}{2}R_0)}{0.000433} = \frac{\ln 2}{0.000433} \doteq 1600 \text{ years.}$$

Thus the half-life of radium is approximately 1600 years.

Example 4. *Temperature Change.* Newton's law of cooling states that the rate of change of the temperature of a body at any time t is proportional to the difference in the temperatures of the body and the surrounding medium at time t.

Let B be the temperature of the body at any time t and let M be the temperature of the surrounding medium at time t. Then Newton's law of cooling is expressed by the first-order differential equation,

$$(8) \qquad \frac{dB}{dt} = k(B - M),$$

where k is a factor of proportionality.

If the centigrade temperature of the surrounding air remains $40°$ and the temperature of an object drops from $170°$ to $105°$ in 45 minutes, what will the temperature of the body be at the end of 2 hours, 15 minutes?

Since $M = 40$, a constant, the variables in (8) are separable for this problem, and (8) can be written

$$\frac{dB}{B - 40} = k \, dt.$$

Integrating, we get

$$\int \frac{dB}{B - 40} = k \int dt,$$

or

$$\ln (B - 40) = kt + C.$$

To determine C, we substitute the initial condition that $B = 170$ when $t = 0$. This gives $C = \ln 130$, and the solution is

$$\ln (B - 40) = kt + \ln 130.$$

We shall use a minute as the unit of time. When $t = 45$, $B = 105$, and substitution of this result in the preceding equation yields $\ln (105 - 40) = k(45) + \ln 130$, from which

$$k = \frac{-(\ln 130 - \ln 65)}{45} = \frac{-\ln 2}{45} = -\ln 2^{1/45}.$$

The solution now becomes

$$\ln (B - 40) = (-\ln 2^{1/45})t + \ln 130 = \ln \frac{130}{2^{t/45}},$$

and so

$$B = 40 + \frac{130}{2^{t/45}}.$$

To find the temperature of the body at the end of 2 hours, 15 minutes, we substitute $t = 135$ in the preceding equation and find

$$B = 40 + \frac{130}{2^3} = 56\tfrac{1}{4}.$$

Thus the temperature of the body at the end of 2 hours and 15 minutes is 56.25°C.

Example 5.   *Falling Bodies.*   Suppose that an object falls from a height above the earth, and that the only forces acting on it are the attraction due to gravity and air resistance, which we will assume is proportional to the velocity.

Let m represent the mass of the falling body, s the number of feet it falls in t seconds, v its velocity in feet per second at time t, and g ($\doteq 32$) the constant of acceleration due to gravity. Because force = mass × acceleration, the force of gravity on the object is mg, directed downward, and the force of air resistance is directed upward and is equal to $-kv$, where k is a positive constant of proportionality. The resultant of these two forces acting on the object is $mg - kv$. Since $a = dv/dt$, the differential equation expressing this situation is

$$m\frac{dv}{dt} = mg - kv.$$

This is a first-order separable differential equation, which can be written

$$\frac{dv}{mg - kv} = \frac{1}{m}\, dt.$$

To find its general solution, we integrate both members, obtaining

$$-\frac{1}{k} \int \frac{k\, dv}{kv - mg} = \frac{1}{m} \int dt,$$

or

$$\ln(kv - mg) = \frac{-kt}{m} + C.$$

Thus

$$kv - mg = e^{-kt/m + C},$$

from which

$$v = \frac{(e^{-kt/m})e^C + mg}{k}$$

or

(9)
$$v = \frac{mg}{k} + Ae^{-kt/m},$$

where $A = e^C/k$. This is the general solution.

Assuming that the body started falling from rest when $t = 0$, we determine the value of A for this problem by substituting the initial conditions, $v = 0$ and $t = 0$, in the general solution (9). This gives $A = -mg/k$, and the solution now becomes

(10)
$$v = \frac{mg}{k}(1 - e^{-kt/m}).$$

Notice that as $t \longrightarrow \infty$, $e^{-kt/m} \longrightarrow 0$. Hence the speed at which the body falls approaches a *terminal speed* of mg/k feet per second.

Because $v = ds/dt$, equation (10) can be written as a first-order differential equation in the variables s and t:

$$\frac{ds}{dt} = \frac{mg}{k}(1 - e^{-kt/m}).$$

Again the variables can be separated, giving

$$ds = \frac{mg}{k} \, dt - \frac{mg}{k} e^{-kt/m} \, dt,$$

whose general solution is

$$s = \frac{mg}{k} t + \frac{m^2 g}{k^2} e^{-kt/m} + C.$$

To determine C, we substitute the initial condition, $s = 0$ when $t = 0$, in this equation, and find $C = -m^2 g/k^2$. Thus

$$s = \frac{mg}{k} t + \frac{m^2 g}{k^2} e^{-kt/m} - \frac{m^2 g}{k^2}$$

gives the number of feet that the body falls in t seconds.

Exercises

In each of Exercises 1 to 6, verify that the given solution satisfies the given differential equation identically.

1. $x\dfrac{dy}{dx} + \sqrt{1 - \left(\dfrac{dy}{dx}\right)^2} - y = 0.$
Solution: $y = Cx + \sqrt{1 - C^2}.$

2. $x\left(\dfrac{dy}{dx}\right)^2 - 2y\dfrac{dy}{dx} - x = 0.$
Solution: $y = \dfrac{x^2 - C^2}{2C}.$

3. $x\dfrac{d^2y}{dx^2} = \dfrac{dy}{dx}.$
Solution: $y = C_1 x^2 + C_2.$

4. $\dfrac{d^2y}{dx^2} + K^2 y = 0.$
Solution: $y = C_1 \cos Kx + C_2 \sin Kx.$

5. $\dfrac{d^3y}{dx^3} = -\dfrac{3}{x}\dfrac{d^2y}{dx^2}.$
Solution: $y = C_1 x + \dfrac{C_2}{x} + C_3.$

6. $x\dfrac{dy}{dx} = y - x\sqrt{x^2 - y^2}.$
Solution: $y = x \sin (C - x).$

In each of Exercises 7 through 16, find the general solution of the given differential equation.

7. $x \, dy + y \, dx = 0.$

8. $x^2 \, dy + y^2 \, dx = 0.$

9. $dy + y \tan x \, dx = 0.$ (*Hint:* $\ln C$ is an arbitrary constant.)

10. $y \sec x \, dy + \sqrt{1 - y^2} \, dx = 0.$

11. $y \cos x \, dy + \sqrt{1 + y^2} \, dx = 0.$

12. $\dfrac{dy}{dx} = \dfrac{1 + y}{1 + x}.$

13. $x^2 y \, dy + (x^2 + 1) \, dx = 0.$

14. $xy \, dy + (x^2 + 1) \, dx = 0.$

15. $(1 - x^2) \, dy + (1 + y^2) \, dx = 0.$

16. $\dfrac{dy}{dx} = \dfrac{x - xy^2}{x^2 y - y}.$

17. A bacterial population increases at a rate proportional to the population. If the population triples in one hour, in how many hours will it be 100 times its original size?

18. Radioactive carbon has a half-life of approximately 5600 years. In how many years will it decay to 20% of its original amount? To 10%?

19. If a certain radioactive substance loses 15% of its radioactivity in 3 days, what is its half-life?

20. In the preceding problem, how long will it take for the substance to be 95% dissipated?

21. To what rate of interest, payable annually, is 6% interest, compounded *continuously*, equivalent?

22. A spherical drop of liquid evaporates at a rate proportional to its surface area. If a given drop evaporates to one-eighth its original volume in 5 minutes, in how many minutes will it evaporate completely?

23. Since a thermometer is itself a physical body, it is subject to Newton's law of cooling. A thermometer is taken from inside a house, where the temperature is 80°, to the outdoors, where the temperature is 40°. Five minutes later it reads 50°. What will it read 15 minutes after it is taken outdoors?

24. The acceleration of a body outside the earth, due to the gravitational attraction of the earth alone, is inversely proportional to the square of the distance from the center of the earth and is directed toward the center of the earth. Show that this acceleration is given by $a = g(R/s)^2$, where R is the radius of the earth (approximately 4000 miles) and s is the distance from the center of the earth.

25. Since acceleration is given by

$$a = \frac{dv}{dt},$$

we can write

$$a = \frac{dv}{dt} = \frac{dv}{ds}\frac{ds}{dt} = v\frac{dv}{ds}.$$

Use this relationship to find the velocity with which

a body, initially at rest 10,000 miles from the *center* of the earth, strikes the surface of the earth if it falls freely, neglecting air resistance. (*Hint:* Use the expression for acceleration in Exercise 24. Take R, the radius of the earth, as 4000 miles.)

26. Show that the velocity with which a body, initially at rest, falling through any distance s_0 from the center of the earth, strikes the earth's surface is given by

$$v = \sqrt{2gR^2\left(\frac{1}{R} - \frac{1}{s_0}\right)}.$$

Hence conclude that a body falling to earth from a great distance strikes the earth with a velocity of approximately $\sqrt{2gR} = 7$ miles per second, approximately.

11.15 SIMPSON'S RULE

If the definite integral $\int_a^b f(x)\,dx$ exists, and if we can find an antiderivative $F(x)$ of $f(x)$, the value of the definite integral is $F(b) - F(a)$

But sometimes it is difficult or impossible to find an antiderivative of $f(x)$ that can be expressed in terms of elementary functions (those we have been studying). In such event we resort to methods that enable us to approximate the value of the given definite integral as accurately as wanted. One such method, the trapezoidal rule, was discussed in 8.7.6. Another is **Simpson's rule.**

These approximation methods were suggested by the fact that if $f(x)$ is nonnegative between $x = a$ and $x = b$, the value of the definite integral $\int_a^b f(x)\,dx$ is the same as the area of the region bounded by $y = f(x)$, the x axis, $x = a$, and $x = b$.

Partition the interval $[a, b]$ into n subintervals, and erect ordinates to the graph from each of the partitioning points. In the trapezoidal rule the points in which successive ordinates met the graph were connected by straight-line segments; in Simpson's rule the points are connected by segments of parabolas. For a given partition of $[a, b]$, Simpson's rule usually gives the better approximation to the value of the definite integral, although it is a little more trouble to apply.

Let x_1 be the midpoint of the interval $[x_0, x_2]$ whose length is $2h$; then $x_0 = x_1 - h$ and $x_2 = x_1 + h$ (Fig. 11-8).

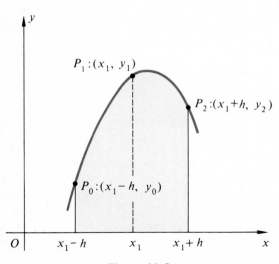

Figure 11-8

Consider three noncollinear points P_0:$(x_1 - h, y_0)$, P_1:(x_1, y_1), and P_2:$(x_1 + h, y_2)$, where it is assumed that y_0, y_1, and y_2 are nonnegative (Fig. 11-8). Any parabola whose principal axis is vertical has an equation of the form $y = Ax^2 + Bx + C$. Such a parabola goes through P_0, P_1, and P_2 if and only if

$$y_0 = A(x_1 - h)^2 + B(x_1 - h) + C,$$
$$y_1 = Ax_1^2 + Bx_1 + C,$$
$$y_2 = A(x_1 + h)^2 + B(x_1 + h) + C.$$

Thus

(1) $$y_0 + 4y_1 + y_2 = A(6x_1^2 + 2h^2) + B(6x_1) + 6C.$$

The area of the region bounded by the parabola, the x axis, $x = x_1 - h$, and $x = x_1 + h$ (Fig. 11-8) is given by

$$\text{Area} = \int_{x_1-h}^{x_1+h} (Ax^2 + Bx + C)\, dx = \left[\frac{A}{3}x^3 + \frac{B}{2}x^2 + Cx\right]_{x_1-h}^{x_1+h}$$

$$= \frac{A}{3}[(x_1 + h)^3 - (x_1 - h)^3] + \frac{B}{2}[(x_1 + h)^2 - (x_1 - h)^2]$$

$$+ C[(x_1 + h) - (x_1 - h)]$$

$$= \frac{h}{3}[A(6x_1^2 + 2h^2) + B(6x_1) + 6C]$$

$$= \frac{h}{3}(y_0 + 4y_1 + y_2) \qquad \text{[by (1)]},$$

which proves the following lemma.

11.15.1 Lemma. If the parabola with vertical axis, $y = Ax^2 + Bx + C$, passes through three noncollinear points P_0:$(x_1 - h, y_0)$, P_1:(x_1, y_1), and P_2:$(x_1 + h, y_2)$, where y_0, y_1, and y_2 are assumed to be nonnegative, then the area of the region bounded by the parabola, the x axis, and the vertical lines $x = x_1 - h$ and $x = x_1 + h$ is given by

$$\text{Area} = \frac{h}{3}(y_0 + 4y_1 + y_2).$$

Let f be a function that is continuous on $[a, b]$, and partition the interval $[a, b]$ into an *even* number n of subintervals, each of length $(b - a)/n$, by means of the points $a = x_0, x_1, x_2, \cdots, x_{n-1}, x_n = b$. Denote by P_0:(x_0, y_0), P_1:(x_1, y_1), $\cdots$, P_n:(x_n, y_n) the points on the curve $y = f(x)$ whose abscissas are the partitioning points $x_0, x_1, x_2, \cdots, x_n$ (Fig. 11-9). The segment of the curve $y = f(x)$ from P_0 to P_2 is approximated by the segment of the parabola, with vertical axis, that goes through P_0, P_1, and P_2; and the area of the region R_1 bounded by the parabola, the x axis, $x = x_0$, and $x = x_2$ is

$$\frac{b - a}{3n}(y_0 + 4y_1 + y_2),$$

by the lemma (11.15.1).

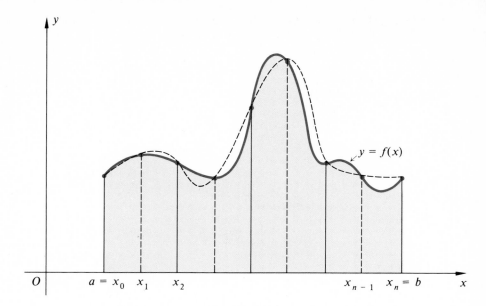

Figure 11-9

Next, the segment of the curve $y = f(x)$ from P_2 to P_4 is approximated by a parabolic segment through P_2, P_3, and P_4, the area of the region under which is

$$\frac{b - a}{3n}(y_2 + 4y_3 + y_4)$$

(by 11.15.1). This process is continued until $P_n:(x_n, y_n)$ is reached. The sum of the area of the regions R_1, R_2, $\cdots$, $R_{n/2}$, under the parabolic segments, approximates the area of the region bounded by the curve $y = f(x)$, the x axis, $x = a$, and $x = b$, and therefore it approximates the value of the definite integral $\int_a^b f(x)\, dx$. This is Simpson's rule.

11.15.2 Simpson's Rule.

$$\int_a^b f(x)\, dx \doteq \frac{h}{3}(y_0 + 4y_1 + 2y_2 + 4y_3 + 2y_4 + \cdots + 4y_{n-1} + y_n),$$

where $h = (b - a)/n$.

Of course, the larger we take the value of n, the better the approximation becomes.

Example 1. Find an approximate value of $\int_0^{\pi/2} \sqrt{\sin x}\, dx$ by Simpson's rule, taking $n = 6$. The integrand cannot be integrated in terms of elementary functions.

Solution. Here $n = 6$ and $b - a = \pi/2$; so $h = \pi/12$. Simpson's rule is

$$\int_0^{\pi/2} \sqrt{\sin x} \, dx \doteq \frac{\pi}{36}(y_0 + 4y_1 + 2y_2 + 4y_3 + 2y_4 + 4y_5 + y_6).$$

The calculation of the value of the expression in parentheses may be arranged in a table as follows. Neatness and orderliness in such calculations make for accuracy.

i	x_i	y_i	c_i	$c_i y_i$
0	0	0	1	0
1	$\frac{1}{12}\pi$	0.50 875	4	2.03 500
2	$\frac{1}{6}\pi$	0.70 710	2	1.41 420
3	$\frac{1}{4}\pi$	0.84 090	4	3.36 360
4	$\frac{1}{3}\pi$	0.93 060	2	1.86 120
5	$\frac{5}{12}\pi$	0.98 280	4	3.93 120
6	$\frac{1}{2}\pi$	1	1	1

Sum = 13.60 520

Thus

$$\int_0^{\pi/2} \sqrt{\sin x} \, dx \doteq \frac{\pi}{36}(13.60\ 520) \doteq 1.18\ 726.$$

Example 2. Use Simpson's rule, with $n = 6$, to approximate the value of $\int_0^\pi \sin x \, dx$ and compare the result with the exact value found by integration.

Solution. $h = \pi/6$. By Simpson's rule,

$$\int_0^\pi \sin x \, dx \doteq \frac{\pi}{18}(y_0 + 4y_1 + 2y_2 + 4y_3 + 2y_4 + 4y_5 + y_6).$$

As in Example 1, the computation of the value of the expression in parentheses looks like this:

i	x_i	y_i	c_i	$c_i y_i$
0	0	0	1	0
1	$\frac{1}{6}\pi$	0.50 000	4	2
2	$\frac{1}{3}\pi$	0.86 603	2	1.73 206
3	$\frac{1}{2}\pi$	1	4	4
4	$\frac{2}{3}\pi$	0.86 603	2	1.73 206
5	$\frac{5}{6}\pi$	0.50 000	4	2
6	π	0	1	0

Sum = 11.46 412

$$\int_0^\pi \sin x \, dx \doteq \frac{\pi}{18}(11.46\ 412) \doteq 2.00\ 087.$$

The correct result, by integration, is 2.

As with any approximation process, we want to know how good the approximation is. A bound for the error, E_n, in using a Simpson's approximation in place of the true value of a definite integral is provided by the following theorem.*

11.15.3 Theorem. Let f be a function whose fourth derivative, $f^{(4)}$, exists on $[a, b]$, If

$$E_n = \int_a^b f(x)\, dx - S_n,$$

where S_n is the approximate value of the integral, as given by Simpson's rule using n subdivisions, then there is a number z in (a, b) such that

$$E_n = -\frac{h^4(b-a)}{180} f^{(4)}(z),$$

where $h = (b-a)/n$.

Example 3. Use Simpson's rule, with $n = 8$, to approximate the value of $\int_1^2 \frac{dx}{1+x}$, and find bounds for the error.

Solution. Since $n = 8$ and $b - a = 1$, $h = \frac{1}{8}$. Simpson's rule, for this problem, is

$$\int_1^2 \frac{dx}{1+x} = \frac{1}{24}(y_0 + 4y_1 + 2y_2 + 4y_3 + 2y_4 + 4y_5 + 2y_6 + 4y_7 + y_8).$$

The computation of the value of the expression in parentheses is shown in the following table.

i	x_i	y_i	c_i	$c_i y_i$
0	1	0.50 000	1	0.50 000
1	1.125	0.47 059	4	1.88 236
2	1.25	0.44 444	2	0.88 888
3	1.375	0.42 105	4	1.68 420
4	1.5	0.40 000	2	0.80 000
5	1.625	0.38 095	4	1.52 380
6	1.75	0.36 364	2	0.72 728
7	1.875	0.34 783	4	1.39 132
8	2	0.33 333	1	0.33 333

Sum $= 9.73\ 117$

Thus

$$\int_1^2 \frac{dx}{1+x} \doteq \frac{1}{24}(9.73\ 117) \doteq 0.40\ 547 = S_8.$$

To find a bound for E_8, the error in the Simpson's approximation S_8, we use 11.15.3. Since $f^{(4)}(x) = 24/(1+x)^5$,

$$E_8 = -\frac{h^4(b-a)}{180} f^{(4)}(z) = -\frac{(0.125)^4}{180} \cdot \frac{24}{(1+z)^5},$$

where z is some number in $(1, 2)$.

*For a proof of this theorem see J. M. H. Olmsted, *Intermediate Analysis*, New York, Appleton-Century-Crofts, 1956, p. 146.

Because $f^{(5)}(x) = -120/(1 + x)^6 < 0$ for $1 \le x \le 2$, $f^{(4)}(x)$ is decreasing on $[1, 2]$. Thus $f^{(4)}(1)$ is the maximum value of $f^{(4)}$ on $[1, 2]$ and $f^{(4)}(2)$ is the minimum. Therefore

$$-\frac{(0.125)^4}{180} \cdot \frac{24}{(1 + 1)^5} \le E_8 \le -\frac{(0.125)^4}{180} \cdot \frac{24}{(1 + 2)^5},$$

or

$$-0.00\,00\,0102 \le E_8 \le -0.00\,00\,00\,134.$$

Since $S_8 \doteq 0.40\,547$,

$$0.40\,547 - 0.00\,00\,01\,02 \le S_8 + E_8 \le 0.40\,547 - 0.00\,00\,00\,13.$$

But

$$S_8 + E_8 = \int_1^2 \frac{dx}{1 + x}.$$

Therefore

$$0.40\,5469 \le \int_1^2 \frac{dx}{1 + x} \le 0.40\,5470.$$

Exercises

For each of the elementary integrals in Exercises 1 to 6, find (a) an approximate value by Simpson's rule and (b) its value, to five decimal places, by integration.

1. $\int_0^1 x^4 \, dx; n = 4.$

2. $\int_0^{1/2} \sqrt{1 - x^2} \, dx; n = 4.$

3. $\int_0^1 \frac{dx}{1 + x^2}; n = 4.$

4. $\int_0^1 \frac{dx}{\sqrt{1 + x^2}}; n = 4.$

5. $\int_0^{\pi/2} \frac{dx}{1 + \sin x}; n = 4.$

6. $\int_{-1/2}^{1/2} \frac{dx}{\sqrt{1 - x^2}}; n = 4.$

In each of Exercises 7 and 8, approximate the value of the given integral by Simpson's rule and find bounds for the error.

7. $\int_0^{\pi/2} \cos x \, dx; n = 6.$

8. $\int_1^3 \frac{1}{x} \, dx = \ln 3; n = 8.$

The integrands in Exercises 9 to 14 cannot be integrated in terms of elementary functions. Use Simpson's rule to find approximate values of the given definite integrals.

9. $\int_0^1 e^{-x^2} \, dx; n = 10.$

10. $\int_0^3 \sqrt{1 + x^4} \, dx; n = 6.$

11. $\int_0^1 \sqrt{1 + x^3} \, dx; n = 6.$

12. $\int_{1/2}^2 \frac{e^x}{x} \, dx; n = 6.$

13. $\int_0^2 \frac{dx}{\sqrt{1 + x^3}}; n = 8.$

14. $\int_{\pi/4}^{3\pi/4} \frac{\sin x}{x} \, dx; n = 4.$

15. Show that Simpson's rule gives the exact value of $\int_a^b f(x) \, dx$ if $f(x)$ is a polynomial of degree 3 or less.

16. Use Simpson's rule, with $n = 4$, to find the volume of the solid generated by revolving about the x axis the region bounded by $y = \ln x$, $y = 0$, $x = 1$, and $x = 3$.

17. Use Simpson's rule, with $n = 6$, to find an approximation to the length of one arch of the sine curve, $y = \sin x$.

12

conics. polar coordinates

12.1 CONIC SECTIONS

In about 200 B.C. the Greek mathematician Apollonius of Perga studied the curves of intersection of a plane and a right circular cone. Although the methods at his disposal were elementary, he discovered many properties of these *conic sections* and was long known as "the great geometer."

The cone was thought of as having two nappes that extended indefinitely far in both directions. Figure 12-1 shows a portion of a right circular cone. A line lying entirely on the cone is called a **generator** of the cone, and all generators of a cone pass through its **vertex**.

A **conic section**, as studied by Apollonius, is the curve in which a plane cuts a right circular cone. There are three types of conic sections, depending on whether the cutting plane is parallel to no generator, to one and only one generator, or to two generators.

If the cutting plane is parallel to no generator, it of course cuts all generators. The curve of intersection is called an **ellipse** (Fig. 12-2). A circle is a limiting case of the ellipse. The ellipse reduces to a single point when the cutting plane goes through the vertex of the cone and contains no generator.

If the cutting plane is parallel to one and only one generator, the curve of intersection is a **parabola** (Fig. 12-2). Should the cutting plane pass through the vertex of the cone and contain one and only one generator, the parabola degenerates into a straight line. If the vertex of the cone moves far away, the cone approaches the form of a right circular cylinder and the parabola has as a limiting case two parallel lines.

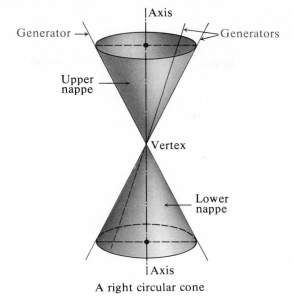

A right circular cone

Figure 12-1

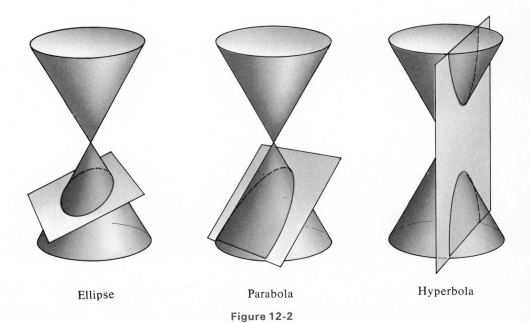

Ellipse   Parabola   Hyperbola

Figure 12-2

If the cutting plane is parallel to two generators, the curve of intersection is a **hyperbola** (Fig. 12-2). The hyperbola degenerates into a pair of intersecting lines when the cutting plane goes through the vertex of the cone and contains two generators.

These curves of intersection of a plane and a right circular cone are called conic sections, or **conics** for short.

The purpose of the foregoing historical introduction is to show how the curves known as conic sections got their name. It is not intended as a working definition of conics, for it involves the three-dimensional concept of cone. Since, for the present, we are confining our attention to two dimensions, we need a definition of conics that uses only two-dimensional ideas. Such a definition, quite independent of the notion of a cone, will be given in 12.2.1. It can be shown, however, that the two definitions are consistent and that plane sections of a cone are, in fact, truly "conics" as we shall define them.

12.2 DEFINITION OF A CONIC

We are working in two-dimensional space, and so it is desirable to have a definition of conics that avoids the three-dimensional concept of cone.

12.2.1 Definition. A **conic** is a set of points whose undirected distances from a fixed point are in constant ratio to their undirected distances from a fixed line (not through the fixed point). The fixed point F is called a **focus** of the conic and the fixed line d is the corresponding **directrix**. The constant ratio e is the **eccentricity** of the conic. If P is a point and Q is the projection of P on d (Fig. 12-3), then P is on the conic if and only if

$$|FP| = e\,|QP|.$$

In Fig. 12-4, the undirected distance of each point, $P_1, P_2, \cdots, P_6$, from F is one-half of its undirected distance from d. That is, $|FP_1| = \frac{1}{2}|Q_1P_1|$, $|FP_2| = \frac{1}{2}|Q_2P_2|$, $\cdots$, $|FP_6| = \frac{1}{2}|Q_6P_6|$. The set of all such points is a conic whose eccentricity is one-half.

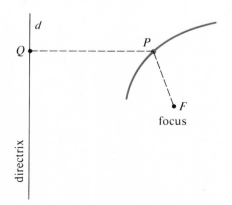

Figure 12-3

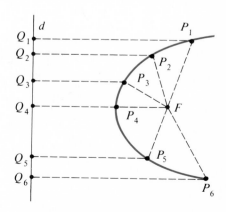

Figure 12-4

The line through a focus perpendicular to its directrix is called the **principal axis** of the conic. The points of intersection of the conic and its principal axis are the **vertices** of the conic. In Fig. 12-5, F is a focus, d is the corresponding directrix, and A is a vertex.

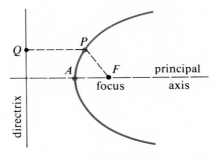

Figure 12-5

The eccentricity e, being a ratio of undirected distances, is a positive number.

12.2.2 Definition.   When $e = 1$, the conic is a **parabola**. When $e < 1$, the conic is an **ellipse**. When $e > 1$, the conic is a **hyperbola** (Fig. 12-6).

It appears from Fig. 12-6 that a conic is symmetric with respect to its principal axis, that a parabola has one and only one vertex, but that an ellipse or a hyperbola has two vertices. This we shall now prove.

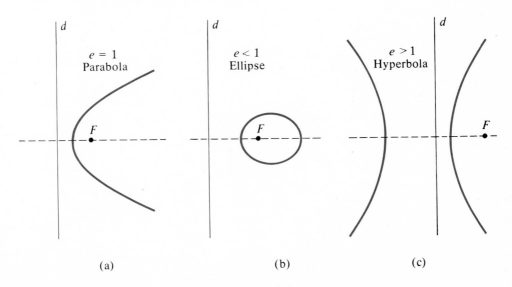

(a)   (b)   (c)

Figure 12-6

12.2.3 Theorem. A conic is symmetric with respect to its principal axis. If the conic is a parabola, it has one and only one vertex: if it is an ellipse or a hyperbola, it has two vertices.

Proof. Place the y axis along the directrix of the conic, and the x axis on the focus. Then the equation of the directrix is $x = 0$ and the focus is $F:(k, 0)$, where $k \neq 0$ is the directed distance from the origin to the focus (Fig. 12-7).

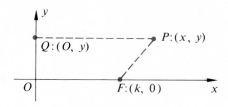

Figure 12-7

Let $P:(x, y)$ be any point on the conic; its projection on the y axis is $Q:(0, y)$. By 12.2.1,

$$(1) \qquad\qquad |FP| = e|QP|,$$

where $e > 0$ is the eccentricity of the conic. But

$$|FP| = \sqrt{(x - k)^2 + y^2} \qquad \text{and} \qquad |QP| = |x|.$$

Therefore

$$\sqrt{(x - k)^2 + y^2} = e|x|,$$

or

$$(2) \qquad\qquad (1 - e^2)x^2 + y^2 - 2kx + k^2 = 0,$$

which is an equation that is satisfied by the coordinates of all points $P:(x, y)$ on the conic.

Since the x axis goes through the focus and is perpendicular to the directrix, the x axis is the principal axis of the conic. Because y appears only to even degree in (2), *the conic is symmetric with respect to its principal axis.*

The vertices of a conic are its points of intersection with its principal axis. To find the x coordinates of the vertices of our conic, we eliminate y between the equation of the conic (2) and the equation of the directrix, $y = 0$, obtaining

$$(3) \qquad\qquad (1 - e^2)x^2 - 2kx + k^2 = 0.$$

If $e = 1$, the conic is a parabola and (3) becomes $-2kx + k^2 = 0$, which has exactly one root. Thus *a parabola has one and only one vertex.*

If $e \neq 1$, the conic is an ellipse or a hyperbola (12.2.2), and (3) is a quadratic equation in x whose discriminant is $4k^2 - 4(1 - e^2)k^2$, or

$$(4) \qquad\qquad 4e^2k^2.$$

Since $e > 0$ and $k \neq 0$, (4) is a positive number. Thus (3) has two distinct real roots, which are the x coordinates of the vertices of the conic. Therefore, *an ellipse or a hyperbola has two vertices.* ∎

Exercise

Turn a large sheet of paper so that its long edges are horizontal. Draw a vertical directrix d down the center of the sheet. About 2 inches to the right of d and about halfway down the paper, mark a focus F.

(a) Locate about 10 points on the paper, each of which is equidistant from F and d. (By distance from a line we mean, of course, perpendicular distance.) Draw a smooth curve through the 10 points and name the resulting conic. What is the value of its eccentricity?

(b) Locate about 10 points on the paper, each of which is half as far from F as from d. Draw a smooth curve through these points and name the conic. What is its eccentricity?

(c) Locate about 16 points on the paper, each of which is twice as far from F as from d. About half these points should be to the left of d. Draw smooth curves through these points and name the conic. State its eccentricity.

12.3 THE PARABOLA (e = 1)

When $e = 1$, the conic is a parabola and 12.2.1 becomes

12.3.1 Definition. The set of points equidistant from the focus and the directrix is a **parabola** (Fig. 12-8).

We wish to derive an equation of the parabola from its definition, and we want the equation to be as simple as possible. The position of the coordinate axes has no effect on the curve but does affect the simplicity of its equation. Since all conics are symmetric with respect to their principle axis, it is natural to place one of the coordinate axes, say the x axis, along the principal axis of the conic.

From 12.3.1, the vertex of the parabola is midway between the focus and the directrix. Choose the vertex of the parabola to be the origin of coordinates (Fig. 12-9).

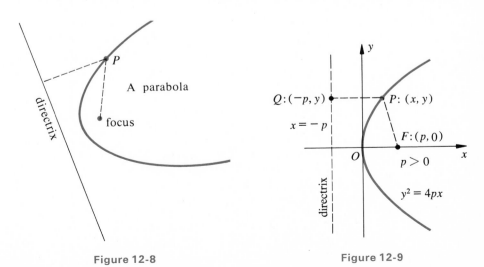

Figure 12-8   Figure 12-9

Let the focus be the point F:$(0, p)$ on the x axis; then the directrix is the vertical line $x = -p$. Any point P:(x, y) is on the parabola if and only if

$$|FP| = |QP|,$$

where Q:$(-p, y)$ is the projection of P on the directrix. But

$$|FP| = \sqrt{(x - p)^2 + y^2} \qquad \text{and} \qquad |QP| = |x - (-p)| = |x + p|.$$

Therefore P is on the parabola if and only if

$$\sqrt{(x - p)^2 + y^2} = |x + p|.$$

By squaring both members of this equation and simplifying, we obtain

$$y^2 = 4px.$$

This proves the following theorem.

12.3.2 **Theorem.** The equation of the parabola whose focus is $(p, 0)$ and whose directrix is $x = -p$ is

$$y^2 = 4px.$$

Figure 12-9 shows the parabola when $p > 0$, and Fig. 12-10 shows the parabola when $p < 0$.

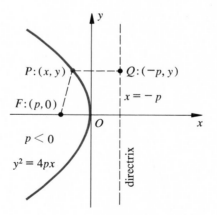

Figure 12-10

A parabola is said to be in **standard position** when its vertex is the origin and its principal axis is one of the coordinate axes. The equation of a parabola in standard position is called a **standard equation**. Thus the equation in 12.3.2 is a standard equation of a parabola.

Example 1. Find the coordinates of the focus and the equation of the directrix of the parabola $y^2 = 12x$. Sketch the parabola, its focus, and its directrix.

Solution.   Comparing the given equation with 12.3.2, we see that $p = 3$. Thus its focus is $(3, 0)$, its directrix is $x = -3$, and its vertex is at the origin. This parabola opens to the right, as in Fig. 12-9, because $p > 0$.

An equally simple standard equation of a parabola is obtained by interchanging the x and y axes in the preceding derivation. That is, we start by placing the y axis along the principal axis of the parabola, with the origin at the vertex.

12.3.3   Theorem.   The standard equation of the parabola with focus $(0, p)$ and directrix $y = -p$ is

$$x^2 = 4py.$$

If $p > 0$, the parabola $x^2 = 4py$ opens upward as in Fig. 12-11(a). If $p < 0$, the parabola opens downward [Fig. 12-11(b)].

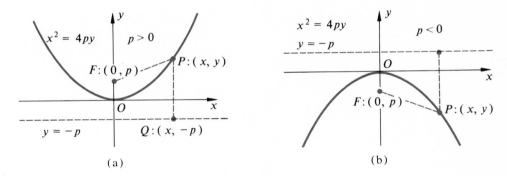

Figure 12-11

Example 2.   Find the equation of the parabola in standard position that goes through the points $(4, -2)$ and $(-4, -2)$. Make a sketch.

Solution.   From the positions of the given points, the parabola must open downward and thus its standard equation will be in the form $x^2 = 4py$, where p is a negative number yet to be determined. Since $(4, -2)$ is on the parabola, $(4)^2 = 4p(-2)$, from which $p = -2$. The desired equation is, then, $x^2 = -8y$ (Fig. 12-12).

A simple geometric property of parabolas is the basis for some important applications. If F is the focus of a parabola, the tangent to the parabola at any point P_1 makes equal angles with the line FP_1 and the line through P_1 that is parallel to the principal axis (Fig. 12-13). A law of physics says that when a light ray strikes a reflecting surface, the angle of incidence is equal to the angle of reflection.

It follows that if a parabola is revolved about its principal axis to form a hollow reflecting shell, all light rays from the focus that strike the inside of the shell are

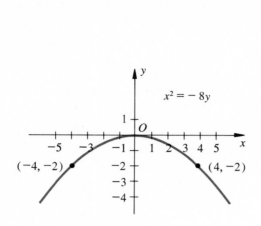

Figure 12-12

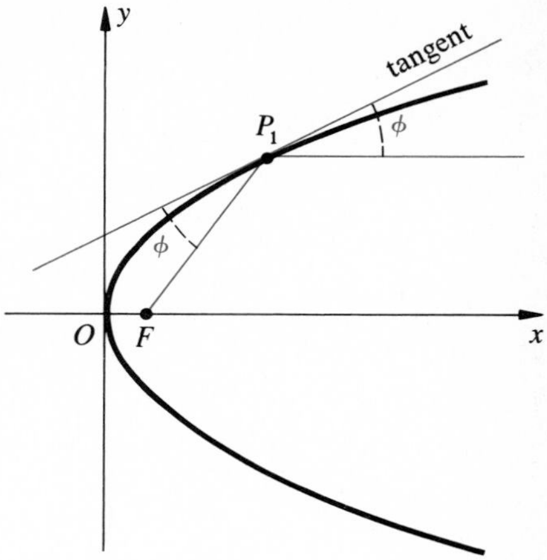

Figure 12-13

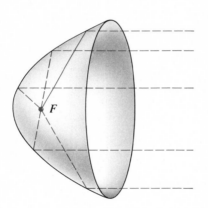

Figure 12-14

reflected outward parallel to the principal axis (Fig. 12-14). This property of parabolas is used in designing automobile headlights and searchlights, the light source being placed at the focus. Of course, incoming rays that are parallel to the principal axis will be reflected to the focus, and, since the light rays from a distant star are parallel for all practical purposes, parabolic mirrors are sometimes used in telescopes.

Sound obeys the same laws of reflection as light, and parabolic microphones are used to pick up and concentrate sounds from, for example, a distant part of a football stadium. Radar and radio telescopes are also based on these same principles.

A few of the many other applications of parabolas are as follows. The path of a projectile is a parabola if air resistance and similar factors are neglected. The cable of an evenly loaded suspension bridge takes the form of a parabola. Arches are often parabolic. The paths of a few comets are parabolic.

Exercises

In Exercises 1 to 8, find the coordinates of the focus and the equation of the directrix for each parabola. Make a sketch, showing the parabola, its focus, and its directrix.

1. $y^2 = 16x$.

2. $y^2 = -28x$.

3. $x^2 = -16y$.

4. $x^2 = 12y$.

5. $y^2 = 2x$.

6. $y^2 - 7x = 0$.

7. $x^2 - 6y = 0$.

8. $2x^2 - 7y = 0$.

In Exercises 9 to 14, find the standard equation of each parabola from the given information. Make a sketch.

9. Focus is $(3, 0)$.

10. Directrix is $x = 2$.

11. Directrix is $y + 4 = 0$.

12. Focus is $(0, -\frac{1}{3})$.

13. Focus is $(-5, 0)$.

14. Directrix is $y = \frac{5}{2}$.

15. Find the equation of the parabola with vertex at the origin and principal axis along the x axis, if the parabola passes through the point $(3, -1)$. Make a sketch.

16. Find the equation of the parabola through the point $(-2, 4)$ if its vertex is the origin and its principal axis is the x axis. Make a sketch.

17. Find the equation of the parabola through the point $(6, -5)$ if its vertex is at the origin and its principal axis is along the y axis. Make a sketch.

18. Find the equation of the parabola whose vertex is the origin and whose principal axis is the y axis, if the parabola passes through the point $(-3, 5)$. Make a sketch.

In Exercises 19 to 26, find the equations of the tangent and the normal to the given parabola at the given point. Sketch the parabola, the tangent, and the normal.

19. $y^2 = 16x$, $(1, -4)$.

20. $x^2 = -10y$, $(2\sqrt{5}, -2)$.

21. $x^2 = 2y$, $(2\sqrt{3}, 6)$.

22. $y^2 = -9x$, $(-1, -3)$.

23. $y^2 = -15x$, $(-3, -3\sqrt{5})$.

24. $x^2 = 4y$, $(4, 4)$.

25. $x^2 = -6y$, $(3\sqrt{2}, -3)$.

26. $y^2 = 20x$, $(2, -2\sqrt{10})$.

27. The slope of the tangent to the parabola $y^2 = 5x$ at a certain point on the parabola is $\sqrt{5}/4$. Find the coordinates of that point. Make a sketch.

28. The slope of the tangent to the parabola $x^2 = -14y$ at a certain point on the parabola is $-2\sqrt{7}/7$. Find the coordinates of that point.

29. Find the equation of the tangent to the parabola $y^2 = -18x$ that is parallel to the line $3x - 2y + 4 = 0$.

30. Any line segment through the focus of a parabola, with endpoints on the parabola, is a **focal chord**. Prove that the tangents to a parabola at the extremities of any focal chord intersect on the directrix.

31. Prove that the tangents to a parabola at the extremities of any focal chord are perpendicular to each other (see Exercise 30).

12.4 CENTRAL CONICS ($e \neq 1$)

A conic for which $e \neq 1$ is an ellipse or a hyperbola (12.2.2), and we saw that every ellipse or hyperbola has two vertices. The point midway between the vertices is called the **center** of the conic.

Since a parabola has only one vertex, it cannot have a center. Ellipses and hyperbolas are called **central conics**.

A central conic is said to be in **standard position** when its center is the origin and its principal axis is one of the coordinate axes. An equation of a conic in standard position is called a **standard equation** of the conic.

Place the origin of coordinates at the center of a central conic and the x axis along its principal axis (Figs. 12-15 and 12-16). Call the vertices A' and A, with A' to

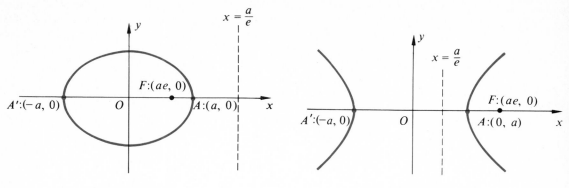

Figure 12-15   Figure 12-16

the left of A. Let the directed distance $\overline{OA} = a$ be a positive number. Then the coordinates of A are $(a, 0)$ and the coordinates of A' are $(-a, 0)$.

12.4.1 Lemma. If the coordinates of the vertices of a central conic with eccentricity e are $(\pm a, 0)$ where a is a positive number, the coordinates of one focus are $(ae, 0)$ and the equation of the corresponding directrix is $x = a/e$.

Proof of this lemma is easy and is left as an exercise.

We shall now derive a standard equation of a central conic.

12.4.2 Theorem. The (standard) equation of a central conic with focus at $(ae, 0)$ and directrix $x = a/e$ is

$$\frac{x^2}{a^2} + \frac{y^2}{a^2(1 - e^2)} = 1,$$

where $a > 0$.

Proof. Let $P:(x, y)$ be any point on the conic (Fig. 12-17). Denote by Q the projection of P on the directrix. The coordinates of Q are $(a/e, y)$.

By 12.2.1, P is on the conic if and only if

$$|FP| = e|QP|.$$

But $|FP| = \sqrt{(x - ae)^2 + y^2}$ and $|QP| = |x - a/e|$. Therefore

$$\sqrt{(x - ae)^2 + y^2} = e\left|x - \frac{a}{e}\right|.$$

Squaring both members of this equation and collecting terms give

$$(1 - e^2)x^2 + y^2 = a^2(1 - e^2),$$

or

$$\frac{x^2}{a^2} + \frac{y^2}{a^2(1 - e^2)} = 1. \quad\blacksquare$$

Since the equation in 12.4.2 contains x and y only to even degree, this central conic is symmetric with respect to both coordinate axes. Because of this symmetry, it is clear that the ellipse or hyperbola must have a second focus, $F:(-ae, 0)$, and a second directrix, $x = -a/e$ (Fig. 12-18). Either focus, along with its corresponding directrix, can be used to construct the whole conic.

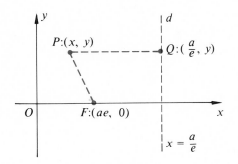

Figure 12-17   Figure 12-18

12.4.3 Corollary. The central conic

$$\frac{x^2}{a^2} + \frac{y^2}{a^2(1 - e^2)} = 1$$

has two foci, $F:(ae, 0)$ and $F':(-ae, 0)$, whose corresponding directrices are $x = a/e$ and $x = -a/e$, respectively.

12.5 THE ELLIPSE (e < 1)

For the ellipse, $e < 1$, so $a^2(1 - e^2)$ is positive and $\sqrt{a^2(1 - e^2)}$ is real. For simplicity of notation, let

$$b = a\sqrt{1 - e^2}.$$

On substituting this in the equation of 12.4.2, we obtain the theorem

12.5.1 Theorem. The standard equation of the ellipse with foci at $F:(ae, 0)$ and $F':(-ae, 0)$ and corresponding directrices $x = a/e$ and $x = -a/e$ is

$$\frac{x^2}{a^2} + \frac{y^2}{b^2} = 1,$$

where a and b are positive numbers and $b^2 = a^2(1 - e^2)$.

The segment AA' of the principal axis of the ellipse lying between the vertices $(\pm a, 0)$ is called the **major axis** of the ellipse (Fig. 12-18). Its length is $2a$.

The segment BB' of the y axis intercepted by the ellipse is the **minor axis** of the ellipse (Fig. 12-18). Its length is $2b$.

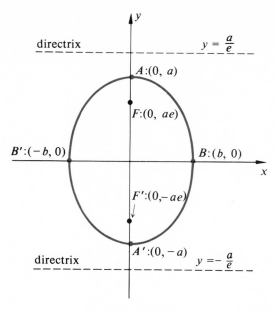

Figure 12-19

If, in deriving a standard equation of the ellipse, the y axis had been placed along the principal axis of the ellipse (Fig. 12-19), we would have obtained the theorem

12.5.2 Theorem. If the foci of an ellipse are at $(0, \pm ae)$ and the directrices are $y = \pm a/e$, then the standard equation of the ellipse is

$$\frac{x^2}{b^2} + \frac{y^2}{a^2} = 1,$$

where a and b are positive numbers and $b^2 = a^2(1 - e^2)$.

We know that, for the ellipse, $0 < e < 1$ and so $1 - e^2$ is a positive number less than 1. From this and from $b^2 = a^2(1 - e^2)$ we see that *always $a^2 > b^2$ in the standard equation of the ellipse*. This makes it easy for us to decide in a given standard equation of an ellipse whether the foci are on the x axis or on the y axis. For examples, in the ellipse

$$\frac{x^2}{9} + \frac{y^2}{4} = 1$$

we have $9 > 4$, and therefore 9 must be a^2 and 4 must be b^2. Thus 12.5.1 applies and the foci are on the x axis. On the other hand, in

$$\frac{x^2}{6} + \frac{y^2}{16} = 1$$

we have $b^2 = 6$ and $a^2 = 16$, so that 12.5.2 applies and the foci are on the y axis.

Example. Find the vertices, foci, directrices, eccentricity, and the lengths of the major and minor axes of the ellipse

$$\frac{x^2}{25} + \frac{y^2}{9} = 1.$$

Solution. Since $a^2 > b^2$, we have $a^2 = 25$ and $b^2 = 9$. Thus 12.5.1 applies. The vertices are $(\pm 5, 0)$, and the lengths of the major and minor axes are 10 and 6, respectively. Now, $b^2 = a^2(1 - e^2)$. Substituting the values of b^2 and a^2, we have $9 = 25(1 - e^2)$. Solving for e, we find $e = 4/5$. Therefore $ae = 4$ and $a/e = 25/4$, and so the coordinates of the foci are $(\pm 4, 0)$ and the equations of the directrices are $x = \pm 25/4$ (Fig. 12-20).

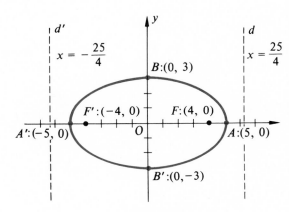

Figure 12-20

Exercises

Make a sketch for each exercise.

Find the coordinates of the vertices, the eccentricity, the coordinates of the foci, the equations of the directrices, and the lengths of the major and minor axes of the ellipses in Exercises 1 to 6.

1. $\dfrac{x^2}{9} + \dfrac{y^2}{4} = 1.$

2. $\dfrac{x^2}{16} + \dfrac{y^2}{4} = 1.$

3. $\dfrac{x^2}{9} + \dfrac{y^2}{16} = 1.$

4. $\dfrac{x^2}{10} + \dfrac{y^2}{5} = 1.$

5. $4x^2 + y^2 = 36.$

6. $25x^2 + 4y^2 = 100.$

In each of Exercises 7 to 14, find the equation of the ellipse in standard position from the given data.

7. A focus at $(-3, 0)$ and a vertex at $(6, 0)$.

8. A focus at $(6, 0)$ and eccentricity $\frac{2}{3}$.

9. A focus at $(0, -5)$ and eccentricity $\frac{1}{3}$.

10. A focus at $(0, 3)$ and length of minor axis 8.

11. A focus at $(0, -2)$ and directrix $y = -6$.

12. A vertex at $(5, 0)$ and passing through the point $(2, 3)$.

13. It goes through the points $(-5, 1)$ and $(-4, -2)$.

14. Through the points $(1, 4)$ and $(-3, 2)$.

In Exercises 15 to 18, find the equations of the tangent and the normal to the given ellipse at the given point.

15. $4x^2 + 9y^2 - 72 = 0$, $(3, 2)$.

16. $x^2 + 3y^2 - 12 = 0$, $(-3, -1)$.

17. $8x^2 + 5y^2 - 133 = 0$, $(1, -5)$.

18. $25x^2 + 16y^2 - 356 = 0$, $(-2, 4)$.

19. Show that the area of the ellipse $x^2/a^2 + y^2/b^2 = 1$ is πab.

20. The slope of the tangent to the ellipse $5x^2 + 3y^2 - 17 = 0$ at a certain point on the ellipse is $\frac{5}{6}$. What are the coordinates of the point of tangency? (Two solutions.)

21. Find the equations of the tangents to the ellipse $x^2 + 2y^2 - 2 = 0$ that are parallel to the line $3x - 3\sqrt{2}\,y - 7 = 0$.

22. Find the equations of the tangents to the ellipse $4x^2 + y^2 - 20 = 0$ whose y intercepts are 5.

23. A dam with semielliptical cross section has a maximum depth of 40 feet and a width of 100 feet at the top. When it is full of water, what is the total force (in tons) exerted on the face of the dam by the weight of the water?

24. A coffer dam of the same shape, size, and elevation as that of the preceding exercise is built 100 feet upstream from the dam and the space between them is filled with water to a maximum depth of 30 feet. How much work is done in pumping all of the water between the two dams over the top of the downstream dam?

25. Find the dimensions of the rectangle having greatest possible area that can be inscribed in the ellipse $b^2x^2 + a^2y^2 = a^2b^2$.

12.6 THE HYPERBOLA ($e > 1$)

In the case of the hyperbola, $e > 1$, so that $a^2(e^2 - 1)$ is positive and therefore $\sqrt{a^2(e^2 - 1)}$ is real. For simplicity of notation, let

$$b = a\sqrt{e^2 - 1}.$$

On substituting this in the equation of 12.4.2, we obtain

12.6.1 Theorem.   The standard equation of the hyperbola with foci $F{:}(ae, 0)$ and $F'{:}(-ae, 0)$ and corresponding directrices $x = a/e$ and $x = -a/e$ is

$$\frac{x^2}{a^2} - \frac{y^2}{b^2} = 1,$$

where a and b are positive numbers and $b^2 = a^2(e^2 - 1)$.

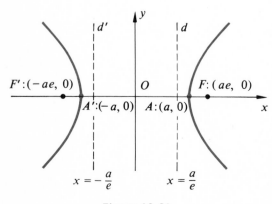

Figure 12-21

The segment AA' of the principal axis lying between the vertices $(\pm a, 0)$ is called the **transverse axis** of the hyperbola (Fig. 12-21). Its length is $2a$.

Since the standard equation of the hyperbola involves x and y only to even powers, the hyperbola is symmetric with respect to both of the coordinate axes.

The hyperbola does not intersect the y axis; for if we substitute $x = 0$ in the equation of the hyperbola (12.6.1), we get $y = \pm b\sqrt{-1}$, which is imaginary. However, the line segment terminated by the points $(0, \pm b)$ is called the **conjugate axis** of the hyperbola. Its length is $2b$.

Solving the equation of the hyperbola for y in

terms of x, we obtain

(1) $$y = \pm \frac{b}{a} \sqrt{x^2 - a^2}.$$

If $x^2 < a^2$, the expression $\sqrt{x^2 - a^2}$ is imaginary. Therefore y is imaginary in equation (1) if $|x| < a$. Thus the hyperbola fails to exist in the vertical band lying between the lines $x = a$ and $x = -a$.

Moreover, y is real in equation (1) if $|x| \geq a$. This means that the hyperbola is not a closed curve but consists of two branches, one passing through the vertex $A{:}(a, 0)$ and extending indefinitely far to the right of A, and the other going through the other vertex, $A'{:}(-a, 0)$, and extending indefinitely far to the left of A'. The shape of a hyperbola is shown in Fig. 12-21.

An equally simple standard equation of a hyperbola is obtained by choosing the y axis to be its principal axis.

12.6.2   Theorem. The standard equation of the hyperbola with foci at $(0, \pm ae)$ and directrices $y = \pm a/e$ is

$$\frac{y^2}{a^2} - \frac{x^2}{b^2} = 1,$$

where a and b are positive numbers and $b^2 = a^2(e^2 - 1)$.

Example 1.   Find the *standard* equation of the conic with a focus at $(-4, 0)$ and corresponding directrix $x = -1$.

Solution.   The x axis is the principal axis of the conic, since it goes through the given focus and is perpendicular to the given directrix (Fig. 12-22). Then $(4, 0)$ and $x = 1$ are also a focus and a directrix. Using 12.6.1, we have $ae = 4$ and $a/e = 1$. Hence $a^2 = 4$, $a = 2$, and $e = 2$.

Substituting these results in $b^2 = a^2(e^2 - 1)$, we obtain $b^2 = 12$. Therefore the desired equation (12.6.1) is

$$\frac{x^2}{4} - \frac{y^2}{12} = 1.$$

Unlike the parabola and the ellipse, hyperbolas have asymptotes.

12.6.3   Theorem.   The two lines

$$\frac{x}{a} - \frac{y}{b} = 0 \qquad \text{and} \qquad \frac{x}{a} + \frac{y}{b} = 0$$

are asymptotes of the hyperbola

$$\frac{x^2}{a^2} - \frac{y^2}{b^2} = 1.$$

Proof.   (See Fig. 12-23). The equation of the upper half of the hyperbola is

$$y = \frac{b}{a} \sqrt{x^2 - a^2},$$

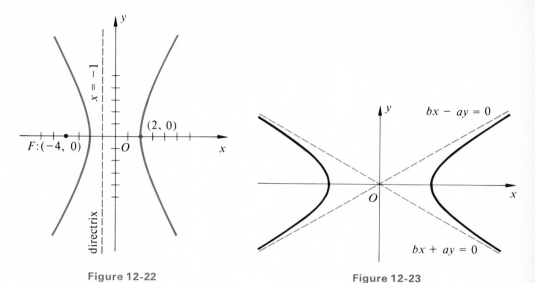

Figure 12-22   Figure 12-23

and the equation of the line $(x/a) - (y/b) = 0$ can be written

$$y = \frac{bx}{a}.$$

Since

$$\lim_{x \to \infty} \left[\frac{b}{a} \sqrt{x^2 - a^2} - \frac{bx}{a} \right] = \frac{b}{a} \lim_{x \to \infty} \frac{(\sqrt{x^2 - a^2} - x)(\sqrt{x^2 - a^2} + x)}{\sqrt{x^2 - a^2} + x}$$

$$= \frac{b}{a} \lim_{x \to \infty} \frac{-a^2}{\sqrt{x^2 - a^2} + x} = 0,$$

$y = bx/a$, or $bx - ay = 0$, is an asymptote of the given hyperbola (by 6.11.7). Similarly (or by symmetry), the other line $bx + ay = 0$ is also an asymptote. ∎

Asymptotes are guiding lines and are of great help in sketching a hyperbola. When the equation is given in a standard form, say 12.6.1, the reader should first draw lightly the **fundamental rectangle**, whose vertices are the points (a, b), $(-a, b)$, $(-a, -b)$, and $(a, -b)$ (see Fig. 12-24). Its diagonals, extended, are the asymptotes, as may be verified. The vertices of the hyperbola are the points of intersection of the x axis and the fundamental rectangle. A good sketch of the hyperbola can be made quickly without plotting more than a single point by simply drawing the hyperbola through its vertices, tangent to the vertical sides of the fundamental rectangle, through the plotted point, and approaching closer and closer to the asymptotes as it moves farther out.

A hyperbola whose asymptotes are perpendicular to each other is called an **equilateral hyperbola**. Its fundamental rectangle is a square, and its transverse and conjugate axes are equal.

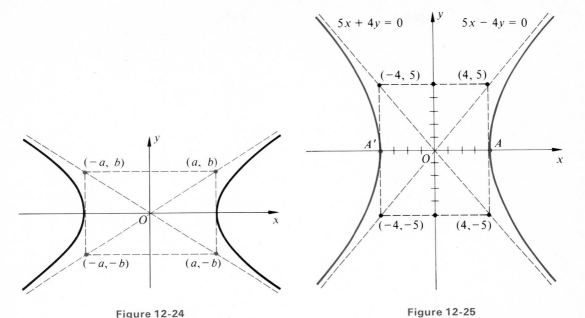

Figure 12-24   Figure 12-25

Example 2.   Find the asymptotes of the hyperbola $25x^2 - 16y^2 = 400$. Sketch the curve (Fig. 12-25).

Solution.   This equation can be written

$$\frac{x^2}{16} - \frac{y^2}{25} = 1.$$

By 12.6.3, the asymptotes are $x/4 - y/5 = 0$ and $x/4 + y/5 = 0$. To draw the hyperbola, we first construct the fundamental rectangle whose vertices are $(\pm 4, \pm 5)$. The asymptotes of the hyperbola are along the diagonals of this rectangle. By substitution, we find that the curve goes through the points $(\pm 4\sqrt{2}, \pm 5) = (\pm 5.66, \pm 5)$. We draw the hyperbola through these points, tangent to the vertical sides of the fundamental rectangle at A and A' and approaching the asymptotes as it goes out to the right or left.

Exercises

Make a sketch for each exercise.

In each of Exercises 1 to 6, the equation of a hyperbola is given. Find its eccentricity, the coordinates of its vertices and foci, the equations of its directrices, the lengths of the transverse and conjugate axes, and the equations of its asymptotes. Then construct its fundamental rectangle, draw its asymptotes, and sketch the hyperbola.

1. $\dfrac{x^2}{25} - \dfrac{y^2}{16} = 1.$

2. $\dfrac{x^2}{9} - \dfrac{y^2}{49} = 1.$

3. $4x^2 - 9y^2 = 36.$

4. $4y^2 - 25x^2 = 100.$

5. $4y^2 - 9x^2 = 36.$

6. $4x^2 - y^2 = 4.$

In Exercises 7 to 12, the hyperbolas are in standard position. Find their equations from the given data. Make a sketch.

7. A focus is $(5, 0)$ and a vertex is $(4, 0)$.

8. A focus is $(-5, 0)$ and the corresponding directrix is $x = -16/5$.

9. The y axis is its principal axis, its eccentricity is $\sqrt{6}/2$, and it passes through the point $(2, 4)$.

10. A vertex is $(0, -3)$ and its eccentricity is $3/2$.

11. A focus is $(2\sqrt{5}, 0)$ and the asymptotes are $2x \pm 4y = 0$.

12. The hyperbola goes through the point $(2, -3\sqrt{5})$ and its asymptotes are $y \pm 3x = 0$.

13. Show that all hyperbolas of the family

$$\frac{x^2}{9} - \frac{y^2}{16} = k,$$

where k is any nonzero constant, have the same asymptotes. Draw the asymptotes and on the same set of coordinate axes sketch the four members of this family for which $k = 1, 3, -1$, and -2.

14. Show that all members of the family of hyperbolas in Exercise 13 for which k has the same sign have the same eccentricity. State the eccentricity when $k > 0$ and $k < 0$.

15. Find the equation of the hyperbola whose asymptotes are $x \pm 2y = 0$ and whose vertices are $(\pm 2\sqrt{3}, 0)$. Make a sketch.

16. Find the equation of the hyperbola whose asymptotes are $x \pm 2y = 0$ and that goes through the point $(4, 3)$. Make a sketch.

17. Find the equations of the tangent and the normal to the hyperbola $16x^2 - 9y^2 - 31 = 0$ at the point $(4, 5)$.

18. Find the equations of the tangent and the normal to the hyperbola $4x^2 - 25y^2 - 96 = 0$ at the point $(-7, 2)$.

19. The slope of the tangent to the hyperbola $2x^2 - 7y^2 - 35 = 0$ at a certain point on the hyperbola is $-\frac{2}{3}$. What are the coordinates of the point of tangency? (Two solutions.)

20. Find the equations of the tangents to the hyperbola $2x^2 - 3y^2 - 6 = 0$ whose x intercepts are 1. Make a sketch.

21. Find the equation of the tangent to the hyperbola $3x^2 - 4y^2 - 12 = 0$ that is parallel to the line $7x - 7y - 13 = 0$. Make a sketch.

22. Prove that the point of contact of any tangent to a hyperbola is midway between the points in which the tangent intersects the asymptotes.

12.7 OTHER DEFINITIONS OF CENTRAL CONICS*

Ellipses and hyperbolas have other characteristic properties that might have been used for definitions.

Consider any point $P_1:(x_1, y_1)$ on the ellipse

$$\frac{x^2}{a^2} + \frac{y^2}{b^2} = 1.$$

The undirected distances between P_1 and the foci, namely $|FP_1|$ and $|F'P_1|$, are called the **focal radii** of P_1.

Draw a horizontal line through P_1, intersecting the directrices d' and d in Q' and Q, respectively (Fig. 12-26).

From the definition of a conic

$$|F'P_1| = e|Q'P_1| \qquad \text{and} \qquad |FP_1| = e|QP_1|;$$

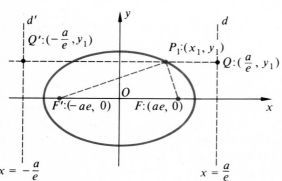

Figure 12-26

*Optional.

then

(1) $$|F'P_1| + |FP_1| = e[|Q'P_1| + |QP_1|].$$

The ellipse lies between its directrices, to the right of d' and to the left of d. Therefore $\overline{Q'P_1}$ and $\overline{P_1Q}$ are positive and

$$|Q'P_1| + |QP_1| = \overline{Q'P_1} + \overline{P_1Q} = \overline{Q'Q}.$$

But

$$\overline{Q'Q} = \frac{a}{e} - \left(-\frac{a}{e}\right) = \frac{2a}{e}.$$

Therefore

$$|Q'P_1| + |QP_1| = \frac{2a}{e}.$$

Substituting this result in (1), we have

$$|F'P_1| + |FP_1| = 2a.$$

That is, the sum of the focal radii of any point on an ellipse is constant and is equal to the length of the major axis.

This property is characteristic of ellipses. In fact, an **ellipse** is often defined as the set of points, the sum of whose undirected distances from two fixed points is constant.

It is easy to show, in analogous fashion, that for any point on the hyperbola, the absolute value of the difference of its focal radii is $2a$. This property is characteristic of the **hyperbola**, which is often defined as the set of points, the absolute value of the difference of whose undirected distances from two fixed points is constant.

The former property suggests a simple method for drawing an ellipse mechanically. Cover a board with a sheet of paper and push thumb tacks halfway in at F and F'. Tie the ends of a piece of string firmly together so as to form a loop of constant length. Place the loop over the tacks and pull taut with a pencil (Fig. 12-27). The string forms a triangle, PFF'. Move the pencil around the paper, keeping the string taut. The pencil will trace an ellipse with foci at F' and F, for $|FP| + |PF'| + |F'F|$ is constant for all positions of the pencil point, P, since this is the total length of the string. But $|F'F|$ is constant in length. Therefore $|FP| + |PF'|$ is constant for all positions of P, and the curve is an ellipse.

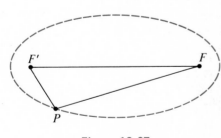

Figure 12-27

Note. In the introduction to this chapter we mentioned that ellipses, parabolas, and hyperbolas are obtained as the intersections of right circular cones and planes. A circle is also the intersection of a right circular cone and a plane perpendicular to the axis of the cone. But our definition of a conic (12.2.1) does not yield a circle for any value of e.

However, if we let the eccentricity approach zero while a remains constant in the standard equation of the ellipse,

(2) $$\frac{x^2}{a^2} + \frac{y^2}{b^2} = 1,$$

the abscissas of the foci, $(\pm ae, 0)$, approach zero. Moreover, since $b^2 = a^2(1 - e^2)$, b^2 approaches a^2 as e approaches zero. Thus equation (2) has the limiting form

$$\frac{x^2}{a^2} + \frac{y^2}{a^2} = 1 \quad \text{or} \quad x^2 + y^2 = a^2,$$

which is the equation of a circle with center at the orign and radius a. *A circle is a limiting form of the ellipse.*

This might have been surmised from the mechanical construction of the ellipse which we discussed above. When F and F' are coincident, the pencil traces a circle whose center is F and whose radius is equal to half the length of the loop of string.

Exercises

Make a sketch for each exercise.

1. Find the equation of the set of points, the sum of whose undirected distances from the two points $(\pm 3, 0)$ is equal to 10. Name the graph.

2. Find the equation of the set of points, the sum of whose undirected distances from the two points $(\pm 4, 0)$ is equal to 16. Name the graph.

3. Find the equation of the set of points, the sum of whose undirected distances from $(0, \pm 2)$ is $4\sqrt{2}$.

4. Find the equation of the set of points, the sum of whose undirected distances from $(0, \pm\sqrt{7})$ is $4\sqrt{3}$.

5. Find the equation of the set of points, the absolute value of the difference of whose undirected distances from the points $(\pm 5, 0)$ is equal to 4. Name their graph.

6. Find the equation of the set of points, the absolute value of the differences of whose undirected distances from the points $(\pm 6, 0)$ is equal to 4.

7. Find the equation of the set of points, the sum of whose undirected distances from the two points $(\pm ae, 0)$ is equal to $2a$.

8. Find the equation of the set of points, the absolute value of the difference of whose undirected distances from the points $(\pm ae, 0)$ is equal to $2a$.

9. Taking F' at $(-\sqrt{a^2 - b^2}, 0)$ and F at $(\sqrt{a^2 - b^2}, 0)$, find the equation of the curve

$$\{P : (x, y) \,|\, |F'P| + |FP| = 2a.\}$$

10. Find the equation of the curve

$$\{P : (x, y) \,|\, |F'P| - |FP| = 2a\},$$

where F' and F are the points whose coordinates are $(-\sqrt{a^2 + b^2}, 0)$ and $(\sqrt{a^2 + b^2}, 0)$, respectively.

12.8 TRANSLATION OF COORDINATE AXES

When a circle of radius 5, say, has its center at $(2, 3)$, its equation is $(x - 2)^2 + (y - 3)^2 = 25$, or

$$x^2 + y^2 - 4x - 6y - 12 = 0.$$

The same circle, if its center were at the origin (Fig. 12-28), would have the much simpler equation

$$x^2 + y^2 - 25 = 0.$$

The position of the coordinate axes has nothing whatever to do with the size or shape of the graph, but it does affect its algebraic representation. It is often desirable to be able to choose new coordinate axes, without changing the appearance of a graph, in order to simplify the algebra involved in a problem.

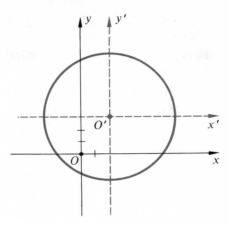

Figure 12-28

If new axes are chosen in the plane, every point will have two sets of coordinates, the old ones (x, y) relative to the old axes and the new ones (x', y') relative to the new axes. The original coordinates are said to undergo a **transformation**. If the new axes are respectively parallel to the original axes and have the same directions (Fig. 12-28), the transformation is called a **translation**. We need formulas connecting the two sets of coordinates in a translation.

12.8.1   **Theorem.**   If new Cartesian axes are chosen in the plane, respectively parallel to the old axes (and having the same directions), so that the new origin has coordinates (h, k) relative to the original axes, then the old coordinates (x, y) and the new coordinates (x', y') of any point in the plane are connected by the **equations of translation**

$$x = x' + h, \qquad y = y' + k.$$

Proof.   Let P be any point in the plane (Fig. 12-29). A line through P, parallel to the x axis, intersects the old y axis in a point A and the new y' axis in A'. A line through P, parallel to the y axis, intersects the old x axis in B and the new x' axis in B'.

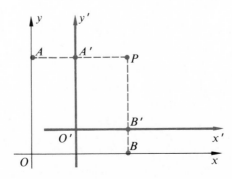

Figure 12-29

In the original coordinate system, the coordinates of P are (x, y), the coordinates of A' are (h, y), and those of A are $(0, y)$. Therefore

$$\overline{A'P} = x - h.$$

But $\overline{A'P} = \overline{O'B'} = x'$. So $x' = x - h$, or

$$x = x' + h.$$

Similarly,

$$y = y' + k. \quad \blacksquare$$

Example 1.   Find the new coordinates of the point P:$(-6, 5)$ after a translation to a new origin at $(2, -4)$.

Solution.   The original coordinates of P are $x = -6$ and $y = 5$, and the coordinates of the new origin are given as $h = 2$ and $k = -4$ (Fig. 12-30). Substituting in the equations of 12.8.1, we get $-6 = x' + 2$ and $5 = y' + (-4)$, or $x' = -8$ and $y' = 9$. Thus the new coordinates of P are $(-8, 9)$.

Example 2.   Given the equation $x^2 + y^2 - 4x + 6y - 3 = 0$, find the new equation of its graph after a translation with new origin at $(2, -3)$.

Solution.   By 12.8.1, every point in the plane has old coordinates (x, y) and new coordinates (x', y'), connected by $x = x' + 2$ and $y = y' - 3$. Substituting these expressions for x and y in the given equation, we have

$$(x' + 2)^2 + (y' - 3)^2 - 4(x' + 2) + 6(y' - 3) - 3 = 0.$$

This result simplifies to

$$x'^2 + y'^2 - 16 = 0,$$

the graph of which is a circle with center at the new origin and radius 4 (Fig. 12-31).

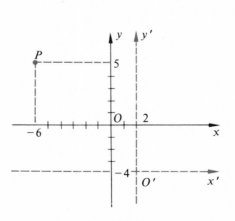

Figure 12-30

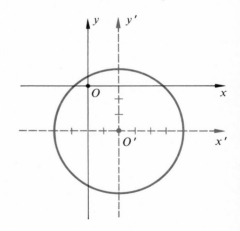

Figure 12-31

By the **degree** of a term in a polynomial with two variables is meant the sum of the exponents of the variables in that term. For example, in

$$4x^2y^3 + 3xy + 5x^2 - 7y + 11,$$

the degree of the first term, $4x^2y^3$, is $2 + 3 = 5$; the degree of $3xy$ is $1 + 1 = 2$; the degree of $5x^2$ is 2; the degree of $-7y$ is 1; and the degree of 11 is zero.

An equation

$$Ax^2 + Bxy + Cy^2 + Dx + Ey + F = 0,$$

in which not all of A, B and C are zero, is said to be of the **second degree** in the variables x and y. If $B = 0$, this second-degree equation will contain no term in xy and can often be simplified by a translation. Two methods will be explained in the examples.

Example 3. Transform the coordinates so that the new equation of the graph of

$$4x^2 + 9y^2 + 8x - 90y + 193 = 0$$

will contain no first-degree terms.

Solution. The given second-degree equation contains no xy term but does contain terms in x^2 and y^2. That is, the equation is in the form

$$Ax^2 + Cy^2 + Dx + Ey + F = 0,$$

where $A \neq 0$ and $C \neq 0$. It is not difficult to show that such an equation can always be reduced to the form

$$Ax'^2 + Cy'^2 + F' = 0$$

by an appropriate translation (see Exercise 19 below). From the latter equation it is apparent that the graph is symmetric with respect to the new x' axis and also with respect to the new y' axis.

FIRST METHOD. The given equation can be rewritten

$$4(x^2 + 2x) + 9(y^2 - 10y) = -193.$$

Completing the squares in the parentheses by adding $4(1) = 4$ and $9(25) = 225$ to both sides of this equation, we have

$$4(x^2 + 2x + 1) + 9(y^2 - 10y + 25) = 36,$$

or

(1) $$4(x + 1)^2 + 9(y - 5)^2 = 36.$$

If we substitute x' for $x + 1$ and y' for $y - 5$ in (1), we obtain

(2) $$4x'^2 + 9y'^2 = 36,$$

which is the desired equation.

The substitution $x' = x + 1$, $y' = y - 5$ is a translation with equations $x = x' - 1$ and $y = y' + 5$; the new origin is at $(-1, 5)$ (by 12.8.1). Drawing both sets of axes (Fig. 12-32), we graph (2) relative to the new axes after noting that its graph will be symmetric with respect to the new axes.

SECOND METHOD. Substituting the equations of translation, $x = x' + h$ and $y = y' + k$, in the given equation, we have

$$4(x' + h)^2 + 9(y' + k)^2 + 8(x' + h) - 90(y' + k) + 193 = 0.$$

If we expand the parentheses and collect terms, this becomes

(3) $$4x'^2 + 9y'^2 + (8h + 8)x' + (18k - 90)y' + (4h^2 + 9k^2 + 8h - 90k + 193) = 0.$$

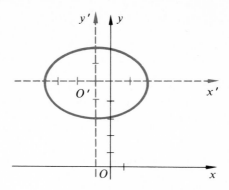

Figure 12-32

We seek values of h and k so that equation (3) will contain no first-degree terms in either x' or y'. To determine the proper values of h and k, we set the coefficients of x' and y' equal to zero. We get

$$8h + 8 = 0, \qquad 18k - 90 = 0,$$

from which $h = -1$ and $k = 5$. This is equivalent to a translation with the new origin at $(-1, 5)$.

Substituting $h = -1$ and $k = 5$ in (3), we obtain

$$4x'^2 + 9y'^2 = 36.$$

Example 4. Simplify the equation $y^2 - 4x - 12y + 28 = 0$ by a translation.

Solution. The given second-degree equation contains no terms in either x^2 or xy. However, it does contain a term in y^2 and also a term in x. It is in the form

$$Cy^2 + Dx + Ey + F = 0,$$

where $C \neq 0$ and $D \neq 0$. It is easy to show that such an equation can always be reduced to the form

$$Cy'^2 + Dx' = 0$$

by an appropriate translation (see Exercise 20). The graph will be symmetric with respect to the new x' axis and will pass through the new origin.

FIRST METHOD.   The given equation can be rewritten

$$y^2 - 12y = 4x - 28.$$

Completing the square in the left-hand member by adding 36 to both sides of the equation, we obtain

$$y^2 - 12y + 36 = 4x + 8,$$

or

$$(y - 6)^2 = 4(x + 2).$$

If we let $x' = x + 2$ and $y' = y - 6$, and thus make a translation with the new origin at $(-2, 6)$, the equation becomes

$$y'^2 = 4x'.$$

Figure 12-33 shows the graph of this equation, sketched relative to the new axes.

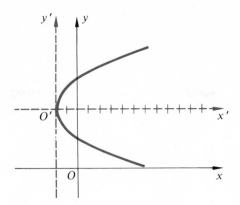

Figure 12-33

SECOND METHOD. Substituting the equations of translation in the given equation, we have

$$(y' + k)^2 - 4(x' + h) - 12(y' + k) + 28 = 0.$$

Expanding, and collecting terms, we obtain

(4) $$y'^2 - 4x' + (2k - 12)y' + (k^2 - 4h - 12k + 28) = 0.$$

We seek values of h and k so that the resulting equation will contain no first-degree term in y' and no constant term. Accordingly, let

$$2k - 12 = 0 \quad \text{and} \quad k^2 - 4h - 12k + 28 = 0,$$

from which $h = -2$ and $k = 6$. Substituting in (4), we have

$$y'^2 - 4x' = 0.$$

Exercises

In each of Exercises 1 through 10, use both methods of translation to simplify the given equation.

1. $x^2 + y^2 - 2x + 4y + 4 = 0.$

2. $x^2 + y^2 - 6x + 2y + 6 = 0.$

3. $9x^2 - 16y^2 + 90x + 192y - 495 = 0.$

4. $4x^2 + 9y^2 - 16x + 72y + 124 = 0.$

5. $y^2 - 10x - 8y - 14 = 0.$

6. $x^2 + 2x + 6y - 23 = 0.$

7. $16x^2 + 5y^2 - 128x + 30y + 221 = 0.$

8. $25x^2 + 4y^2 + 150x - 8y + 129 = 0.$

9. $x^2 + 6x - 16y - 71 = 0.$

10. $y^2 + 7x - 6y + 44 = 0.$

The graphs of the equations in Exercises 11 to 14 are central conics that are not in standard position. By suitable translations (either method) reduce the given equations to standard equations. Then draw both sets of axes and sketch the graph. Name the conic.

11. $x^2 + 4y^2 - 2x + 16y + 1 = 0.$

12. $25x^2 + 9y^2 + 150x - 18y + 9 = 0.$

13. $9x^2 - 16y^2 + 54x + 64y - 127 = 0.$

14. $x^2 - 4y^2 - 14x - 32y - 11 = 0.$

15. Use the second method of translation to reduce

$$xy + 2x - y - 10 = 0$$

to an equation with no first-degree terms. Then draw both sets of axes and sketch the graph.

16. By means of the second method of translation, reduce the equation

$$xy + 8x - 7y - 59 = 0$$

to an equation with no first-degree terms. Then draw both sets of axes and sketch the graph.

17. Use the second method of translation to reduce

$$8x^3 - 24x^2 + 24x - y + 1 = 0$$

to an equation containing no second-degree term and no constant term. Then draw both sets of axes and sketch the graph.

18. By means of the second method of translation, reduce the equation

$$x^2y - 8x^2 + 8xy - 64x + 16y - 129 = 0$$

to an equation containing no second-degree terms

and no first-degree terms. Then draw both sets of axes and sketch the graph.

19. Find the equations of translation of axes that reduce the equation

$$Ax^2 + Cy^2 + Dx + Ey + F = 0, \qquad AC \neq 0,$$

to the form

$$Ax'^2 + Cy'^2 + F' = 0,$$

where

$$F' = (4ACF - CD^2 - AE^2)/4AC.$$

(*Hint:* Use the first method, as in Example 3.)

20. Find the equations of translation of axes that reduce the equation

$$Ax^2 + Dx + Ey + F = 0, \qquad AE \neq 0,$$

to the form

$$Ax'^2 + Ey' = 0.$$

(*Hint:* Use the first method, as in Example 4.)

12.9 THE ROTATION TRANSFORMATION

Consider a pair of rectangular Cartesian coordinate axes, Ox and Oy (Fig. 12-34). If two new mutually perpendicular lines, Ox' and Oy', through the old origin are chosen for new coordinate axes, then every point P in the plane (except O) will have two sets of coordinates, (x, y) and (x', y'). The original coordinates (x, y) are the projections of P on the original axes, Ox and Oy, and the new coordinates (x', y') are the projections of P on the new axes, Ox' and Oy'. We seek formulas connecting the old and the new coordinates of any point P.

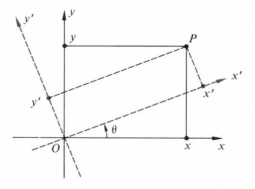

Figure 12-34

12.9.1 Theorem. If new rectangular coordinate axes having the same origin as the original axes are chosen, then the coordinates (x, y) of any point P, relative to the original axes, and its coordinates (x', y'), relative to the new axes, are connected by the equations

$$x = x' \cos \theta - y' \sin \theta,$$

$$y = x' \sin \theta + y' \cos \theta,$$

where θ is the angle from Ox to Ox'.

Proof. Let P be any point other than the origin (Fig. 12-35). Denote the angle from the original x axis to the new x' axis by θ. Draw OP and indicate the angle from the

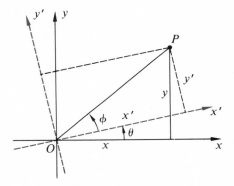

Figure 12-35

new x' axis to OP by ϕ. Then

$$\frac{x}{|OP|} = \cos (\phi + \theta),$$

or

$$x = |OP| \cos (\phi + \theta) = |OP| (\cos \phi \cos \theta - \sin \phi \sin \theta).$$

But

(1) $$\sin \phi = \frac{y'}{|OP|} \qquad \text{and} \qquad \cos \phi = \frac{x'}{|OP|}.$$

Substituting above, we obtain

(2) $$x = x' \cos \theta - y' \sin \theta.$$

Similarly,

$$y = |OP| \sin (\phi + \theta) = |OP| (\sin \phi \cos \theta + \cos \phi \sin \theta).$$

By means of (1), this becomes

(3) $$y = x' \sin \theta + y' \cos \theta. \quad \blacksquare$$

Example 1. Find the new coordinates of the point P:(2, 6) after a rotation transformation in which $\theta = \pi/6$ (Fig. 12-36).

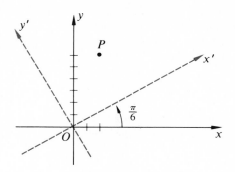

Figure 12-36

Solution. By substituting $x = 2$, $y = 6$, $\sin \theta = \sin \pi/6 = 1/2$, and $\cos \theta = \cos \pi/6 = (1/2)\sqrt{3}$ in the equations of rotation (12.9.1), we obtain

$$2 = \frac{1}{2}\sqrt{3}\,x' - \frac{1}{2}y',$$

$$6 = \frac{1}{2}x' + \frac{1}{2}\sqrt{3}\,y'.$$

If we solve these equations simultaneously for x' and y', we find $x' = \sqrt{3} + 3$ and $y' = -1 + 3\sqrt{3}$. Thus the coordinates of P relative to the new axes are

$$(\sqrt{3} + 3, -1 + 3\sqrt{3}).$$

Example 2. Find the new equation of the graph of $xy = 1$ after a rotation in which $\theta = \pi/4$.

Solution. Since $\sin \frac{1}{4}\pi = \frac{1}{2}\sqrt{2}$ and $\cos \frac{1}{4}\pi = \frac{1}{2}\sqrt{2}$, the equations of rotation (12.9.1) become

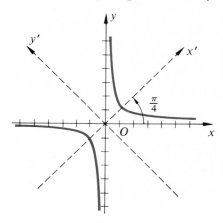

Figure 12-37

$$x = \frac{1}{2}\sqrt{2}\,(x' - y'),$$

$$y = \frac{1}{2}\sqrt{2}\,(x' + y').$$

Substituting these expressions in the given equation, we obtain

$$\frac{1}{2}(x' - y')(x' + y') = 1,$$

or

$$\frac{x'^2}{2} - \frac{y'^2}{2} = 1,$$

which represents an equilateral hyperbola in standard position relative to the new axes, Ox' and Oy' (Fig. 12-37).

The most general equation of the second degree in x and y is

(4) $$Ax^2 + Bxy + Cy^2 + Dx + Ey + F = 0,$$

where not all of A, B, and C are zero. As usual, we assume the coefficients to be real numbers.

We shall now show that if equation (4) contains a term in xy (that is, if $B \neq 0$), then it is always possible to find a rotation transformation that transforms (4) into a second-degree equation in x' and y' that has no $x'y'$ term.

Substituting the equations of rotation,

$$x = x' \cos \theta - y' \sin \theta, \qquad y = x' \sin \theta + y' \cos \theta,$$

in (4), we obtain

$$A(x'^2 \cos^2 \theta - 2x'y' \cos \theta \sin \theta + y'^2 \sin^2 \theta)$$
$$+ B(x'^2 \cos \theta \sin \theta + x'y' \cos^2 \theta - x'y' \sin^2 \theta - y'^2 \sin \theta \cos \theta)$$
$$+ C(x'^2 \sin^2 \theta + 2x'y' \sin \theta \cos \theta + y'^2 \cos^2 \theta)$$
$$+ D(x' \cos \theta - y' \sin \theta) + E(x' \sin \theta + y' \cos \theta) + F = 0.$$

If we collect terms, this becomes

(5) $$A'x'^2 + B'x'y' + C'y'^2 + D'x' + E'y' + F' = 0,$$

where

(6)
$$A' \equiv A \cos^2 \theta + B \sin \theta \cos \theta + C \sin^2 \theta,$$
$$B' \equiv B(\cos^2 \theta - \sin^2 \theta) - 2(A - C) \sin \theta \cos \theta,$$
$$C' \equiv A \sin^2 \theta - B \sin \theta \cos \theta + C \cos^2 \theta,$$
$$D' \equiv D \cos \theta + E \sin \theta,$$
$$E' \equiv -D \sin \theta + E \cos \theta,$$
$$F' \equiv F.$$

Since $\cos^2 \theta - \sin^2 \theta = \cos 2\theta$ and $2 \sin \theta \cos \theta = \sin 2\theta$, the expression for B' in (6) can be rewritten

(7) $$B' = B \cos 2\theta - (A - C) \sin 2\theta.$$

Thus B' will be zero if

$$B \cos 2\theta - (A - C) \sin 2\theta = 0,$$

or

$$\cot 2\theta = \frac{A - C}{B}.$$

12.9.2 Theorem. The equation

$$Ax^2 + Bxy + Cy^2 + Dx + Ey + F = 0,$$

where $B \neq 0$, is transformed into

$$A'x'^2 + C'y'^2 + D'x' + E'y' + F' = 0$$

by a rotation through an angle θ for which

$$\cot 2\theta = \frac{A - C}{B}.$$

Example 3. By means of a rotation, find a new equation of the graph of $41x^2 - 24xy + 34y^2 - 90x + 5y + 25 = 0$ that contains no $x'y'$ term.

Solution. We choose θ, the angle of rotation, so that

$$\cot 2\theta = \frac{A - C}{B} = -\frac{7}{24}.$$

From $\cot 2\theta = -\frac{7}{24}$, we must find $\sin \theta$ and $\cos \theta$. From the sketch (Fig. 12-38) we read $\cos 2\theta = -\frac{7}{25}$. Using the half-angle formulas from trigonometry, we obtain

$$\cos \theta = \sqrt{\frac{1 - \frac{7}{25}}{2}} = \frac{3}{5}, \qquad \sin \theta = \frac{4}{5}.$$

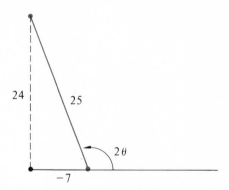

Figure 12-38

Thus the equations of rotation are

$$x = \frac{1}{5}(3x' - 4y'), \qquad y = \frac{1}{5}(4x' + 3y').$$

Substituting in the given equation, we obtain

$$\frac{1}{25}[41(9x'^2 - 24x'y' + 16y'^2) - 24(12x'^2 - 7x'y' - 12y'^2) + 34(16x'^2 + 24x'y' + 9y'^2)]$$

$$+ \frac{1}{5}[-90(3x' - 4y') + 5(4x' + 3y')] + 25 = 0.$$

If we collect terms, this equation becomes

$$\frac{1}{25}[(369 - 288 + 544)x'^2 + 24(-41 + 7 + 34)x'y' + (656 + 288 + 306)y'^2]$$

$$+ \frac{1}{5}[(-270 + 20)x' + (360 + 15)y'] + 25 = 0,$$

or

$$\frac{1}{25}(625x'^2 + 1250y'^2) + \frac{1}{5}(-250x' + 375y') + 25 = 0,$$

or

$$x'^2 + 2y'^2 - 2x' + 3y' + 1 = 0,$$

which is the desired equation.

A plane **algebraic curve** is the graph of a polynomial in two variables, say x and y, equated to zero. Straight lines and conics are simple examples of algebraic curves. Since the equations of rotation,

$$x = x' \cos \theta - y' \sin \theta, \qquad y = x' \sin \theta + y' \cos \theta,$$

are of the first degree in the variables x, y, x', and y', their substitution in the equation of an algebraic curve cannot result in a new equation of higher degree. That is, rotation cannot raise the degree of the equation of a curve.

Nor can rotation result in a new equation of lower degree for the algebraic curve. For if it did, the reverse rotation, with angle $-\theta$, that restores the original coordinate system would transform the new, lower-degree equation of the curve back into the original, higher-degree equation of the curve, which is impossible.

Therefore the degree of the equation of a plane algebraic curve is unchanged by a rotation.

The same reasoning applies to the equations of translation (12.8.1).

This discussion is summarized in

12.9.3 Theorem. The degree of the equation of a plane algebraic curve is invariant under a rotation or a translation.

Exercises

In each of Exercises 1 to 6, make a sketch and show both pairs of axes.

1. Find the new coordinates of the points $(3, -1)$, $(-2, 2)$, $(0, \sqrt{2})$, and $(2\sqrt{2}, -\sqrt{2})$ after a rotation transformation in which $\theta = \pi/4$.

2. Find the new coordinates of the points $(-4, 6)$, $(1, \sqrt{3})$, and $(-2\sqrt{3}, 2)$ after a rotation transformation in which $\theta = \pi/3$.

3. Find the new equation of the line $x - \sqrt{3}\,y - 4 = 0$ after a rotation transformation in which $\theta = \pi/6$.

4. Find the new equation of the line $4x + 3y - 5 = 0$ after a rotation in which $\theta = \text{Tan}^{-1}(3/4)$.

5. Find the new equation of the circle $x^2 + y^2 = 36$ after a rotation in which $\theta = \pi/3$.

6. Find the new equation of the curve

$$5x^2 - 6xy + 5y^2 - 36 = 0$$

after a rotation in which $\theta = \pi/4$.

In Exercises 7 to 16, transform the coordinates by an appropriate rotation so that the new equation will contain no $x'y'$ term.

7. $8x^2 + 5xy - 4y^2 + 2 = 0$.

8. $3x^2 - 3xy + 7y^2 - 5 = 0$.

9. $2x^2 + 9xy + 14y^2 + 5 = 0$.

10. $12x^2 + 7xy - 12y^2 - 5 = 0$.

11. $x^2 + 12xy + 6y^2 - 2x + 5y + 1 = 0$.

12. $9x^2 + 24xy + 2y^2 - 126 = 0$.

13. $x^2 + 14xy + 49y^2 + \sqrt{2}\,y + 1 = 0$.

14. $2x^2 + xy + 2y^2 + 3\sqrt{2}\,x + 7 = 0$.

15. $27x^2 + 5xy + 15y^2 - 2 = 0$.

16. $5x^2 + 12xy + 10y^2 - 5 = 0$.

12.10 THE GRAPH OF ANY SECOND-DEGREE EQUATION*

In the preceding section we showed that, by a suitable rotation of the coordinate axes, any second-degree equation in x and y can be reduced to the form

$$(1) \qquad Ax^2 + Cy^2 + Dx + Ey + F = 0,$$

which has no term in xy.

Since the degree of the equation of a plane algebraic curve is unchanged by a rotation, equation (1) is also of the second degree. Thus the coefficients A and C in (1) cannot both be zero.

Two cases will be considered, depending on whether both A and C are different from zero or only one of A and C is different from zero.

CASE 1.   If $A \neq 0$ and $C \neq 0$, by the method of completing squares we can write (1) in the form

$$(2) \qquad A\left(x + \frac{D}{2A}\right)^2 + C\left(y + \frac{E}{2C}\right)^2 = \frac{D^2}{4A} + \frac{E^2}{4C} - F.$$

If we substitute

$$x' = x + \frac{D}{2A}, \qquad y' = y + \frac{E}{2C},$$

which is a translation with new origin at

$$O' : \left(-\frac{D}{2A}, -\frac{E}{2C}\right) \qquad \text{(by 12.8.1),}$$

we have

$$(3) \qquad Ax'^2 + Cy'^2 = F',$$

where

$$F' = \frac{D^2}{4A} + \frac{E^2}{4C} - F.$$

If A, C, and F' have the same sign, (3) can be rewritten in the form 12.5.1 or 12.5.2, and its graph is an *ellipse* or a *circle*, which is a limiting form of an ellipse. If A and C have the same sign and F' has the opposite sign, (3) has *no graph*. If A and C have the same sign and $F' = 0$, the graph reduces to a single point. This is a limiting form of an ellipse and is sometimes called a *point-ellipse*.

If A and C have opposite signs in (3) and $F' \neq 0$, equation (3) can be written in the form 12.6.1 or 12.6.2, and the graph is a *hyperbola*. If A and C have opposite signs and $F' = 0$, equation (3) can be factored into two linear factors and the graph is two intersecting lines, a *degenerate hyperbola*.

CASE 2.   If either A or C in (1) is zero, say $A = 0$ and $C \neq 0$, then (1) can be rewritten

$$(4) \qquad C\left(y^2 + \frac{E}{C}y\right) + Dx + F = 0.$$

*Optional.

Completing the square in the parentheses, we obtain

(5) $$C\left(y + \frac{E}{2C}\right)^2 + Dx + F - \frac{E^2}{4C} = 0.$$

If $D \neq 0$, (5) can be rewritten

(6) $$C\left(y + \frac{E}{2C}\right)^2 + D\left(x + \frac{F}{D} - \frac{E^2}{4CD}\right) = 0.$$

Substituting

$$y' = y + \frac{E}{2C}, \qquad x' = x + \frac{4CF - E^2}{4CD}$$

in (6), which is equivalent to a translation with the new origin O' at the point

$$\left(-\frac{4CF - E^2}{4CD}, -\frac{E}{2C}\right),$$

we get

(7) $$Cy'^2 + Dx' = 0.$$

The graph of (7) is a *parabola* (12.3.2).

If $D = 0$, (5) can be written

(8) $$C\left(y + \frac{E}{2C}\right)^2 + F - \frac{F^2}{4C} = 0.$$

If we substitute

$$y' = y + \frac{E}{2C}$$

in (8), which is equivalent to a translation in which the new origin is at $(0, -E/2C)$, then (8) becomes

(9) $$Cy'^2 + K = 0,$$

where $K = F - E^2/4C$. If C and K have the same sign, there is *no graph*. If C and K have opposite signs, equation (9) can be factored into two linear factors whose graph is two parallel lines, a *degenerate parabola*. If $K = 0$, the graph of (9) is two coincident lines, a *degenerate parabola*.

Similar results are obtained if $A \neq 0$ and $C = 0$.

Now, the translation and rotation transformations of the coordinates have no effect on the graph. Its equation is simplified, but the graph remains unchanged. We conclude, then, that if the general second-degree equation in x and y,

(10) $$Ax^2 + Bxy + Cy^2 + Dx + Ey + F = 0$$

has a graph, it is always a conic, or a limiting form of a conic, or a degenerate conic, as stated in the following theorem.

12.10.1 **Theorem.** The graph, if any, of a second-degree equation in plane Cartesian coordinates is always a conic, or a limiting form of a conic, or a degenerate conic.

Since a translation is sufficient to reduce equation (1) to a standard form, the principal axis of the conic represented by (1) is parallel to, or coincident with, one of the coordinate axes. Moreover, since a properly chosen rotation, in which the angle is acute, will remove the xy term in equation (10), it follows that the principal axis of the conic represented by a second-degree equation containing a term in xy is never parallel to a coordinate axis.

12.10.2 Corollary. The xy term is present in a second-degree equation if and only if the principal axis of the conic it represents is not parallel to either coordinate axis.

Example 1. Reduce to standard form the equation $5x^2 - 3xy + y^2 + 65x - 25y + 203 = 0$. Then sketch the graph, showing all three sets of axes.

Solution. We choose the angle of rotation θ so that

$$\cot 2\theta = \frac{A - C}{B} = \frac{5 - 1}{-3} = -\frac{4}{3}.$$

Then $\cos 2\theta = -\frac{4}{5}$ and

$$\sin \theta = \sqrt{\frac{1 - \cos 2\theta}{2}} = \sqrt{\frac{1 + \frac{4}{5}}{2}} = \frac{3}{\sqrt{10}},$$

$$\cos \theta = \frac{1}{\sqrt{10}}.$$

Thus the equations of rotation are

$$x = \frac{1}{10}\sqrt{10}(x' - 3y'), \qquad y = \frac{1}{10}\sqrt{10}(3x' + y').$$

Substituting the equations of rotation in the given equation, we obtain

$$\frac{1}{10}[5(x'^2 - 6x'y' + 9y'^2) - 3(3x'^2 - 8x'y' - 3y'^2) + (9x'^2 + 6x'y' + y'^2)]$$

$$+ \frac{1}{10}\sqrt{10}\,[65(x' - 3y') - 25(3x' + y')] + 203 = 0,$$

or, if we collect terms,

$$x'^2 + 11y'^2 - 2\sqrt{10}\,x' - 44\sqrt{10}\,y' + 406 = 0.$$

By the method of completing the squares, this equation can be rewritten

(11) $(x' - \sqrt{10})^2 + 11(y' - 2\sqrt{10})^2 = 44.$

If we substitute

(12) $x'' = x' - \sqrt{10}, \qquad y'' = y' - 2\sqrt{10}$

in equation (11), we find

(13) $\dfrac{x''^2}{44} + \dfrac{y''^2}{4} = 1.$

The substitution (12) is a translation with new origin at $(\sqrt{10}, 2\sqrt{10})$.

The graph of (13) is an ellipse in standard position relative to the x'' and y'' axes. To sketch it, we draw the original x and y axes (Fig. 12-39). Then we draw the x' and y' axes through the same origin but so that $\theta = \text{Tan}^{-1} 3$. We now locate the point $(\sqrt{10}, 2\sqrt{10})$ relative to the x' and y' axes and draw the x'' and y'' axes through that point parallel to the x' and y' axes, respectively. Finally, we sketch the ellipse in standard position on the x'' and y'' axes.

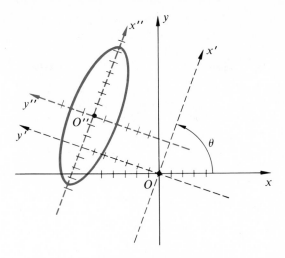

Figure 12-39

Example 2. Reduce the equation

$$x^2 - 4xy + y^2 + 10\sqrt{2}\,x - 2\sqrt{2}\,y - 4 = 0$$

to standard form. Then sketch the graph, showing all three sets of axes.

Solution. To remove the xy term, we rotate through an angle θ such that

$$\text{Cot } 2\theta = \frac{A - C}{B} = \frac{1 - 1}{-4} = 0.$$

Thus $2\theta = \pi/2$, $\theta = \pi/4$, $\sin\theta = 1/\sqrt{2}$, and $\cos\theta = 1/\sqrt{2}$. The equations of rotation are

$$x = \frac{x' - y'}{\sqrt{2}}, \qquad y = \frac{x' + y'}{\sqrt{2}}.$$

By substituting these expressions in the given equation and collecting terms, we get

$$x'^2 - 3y'^2 - 8x' + 12y' + 4 = 0.$$

This equation can be written $(x'^2 - 8x' + 16) - 3(y'^2 - 4y' + 4) + 4 - 16 + 12 = 0$
or

$$(14) \qquad\qquad (x' - 4)^2 - 3(y' - 2)^2 = 0.$$

Thus the substitutions $x'' = x' - 4$ and $y'' = y' - 2$, which are equivalent to the translation

$$x' = x'' + 4, \qquad y' = y'' + 2,$$

reduce equation (14) to the standard form

$$(15) \qquad\qquad x''^2 - 3y''^2 = 0.$$

Since this is equivalent to

$$(x'' - \sqrt{3}\,y'')(x'' + \sqrt{3}\,y'') = 0,$$

the graph of (15) is the graph of the two intersecting lines $x'' - \sqrt{3}\,y'' = 0$ and $x'' + \sqrt{3}\,y'' = 0$ (Fig. 12-40), a degenerate hyperbola.

Therefore the graph of the given equation is a *degenerate hyperbola*.

The standard equations of the conics are all of the second degree. But in deriving those equations, the coordinate axes were placed in a special position relative to the conic.

Consider any conic not in standard position (Fig. 12-41). We have seen that by means of an appropriate rotation and translation we can always select new x'' and

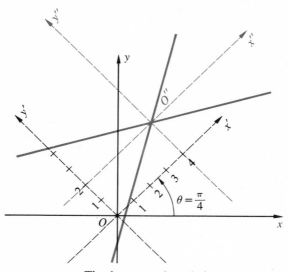

The degenerate hyperbola
$$x^2 - 4xy + y^2 + 10\sqrt{2}\,x - 2\sqrt{2}\,y - 4 = 0$$

Figure 12-40

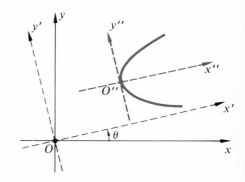

Figure 12-41

y'' axes, with respect to which the conic *is* in standard position. Since the standard equation of the conic is of the second degree, and since the degree of an equation is unchanged by a rotation or a translation, the original equation of the conic must also be of the second degree.

This proves the theorem

12.10.3 Theorem. The Cartesian equation of a conic is always of the second degree.

Exercises

By a rotation transformation and a translation, reduce the following equations to standard form. Then sketch the curve, showing all three sets of axes.

1. $62x^2 + 168xy + 13y^2 + 380x - 90y - 575 = 0.$

2. $6x^2 + 3xy + 2y^2 + 81x + 17y + 236 = 0.$

3. $16x^2 - 24xy + 9y^2 + 85x + 30y + 175 = 0.$

4. $3x^2 + 2xy + 3y^2 - 12\sqrt{2}\,x + 4\sqrt{2}\,y + 24 = 0.$

5. $5x^2 - 6xy - 3y^2 + 168x + 24y + 612 = 0.$

6. $6x^2 - 5xy - 6y^2 + 78x + 52y + 26 = 0.$

7. $x^2 + 2xy + y^2 + 12\sqrt{2}\,x - 6 = 0.$

8. $16x^2 - 24xy + 9y^2 + 3x + 4y + 5 = 0.$

9. $1396x^2 - 600xy + 801y^2 + 1768x - 3666y$
$- 1859 = 0.$

10. $7x^2 + 12xy + 2y^2 + 140x + 76y + 548 = 0.$

11. $12x^2 - 12xy + 7y^2 + 60x - 118y + 511 = 0.$

12. $670x^2 + 840xy - 163y^2 + 2340x + 3172y$
$+ 2028 = 0.$

12.11 POLAR COORDINATES

The reader will recall that Descartes' invention of the rectangular coordinate system, which now bears his name, was the basis for his development of analytic geometry. It established a correspondence between the points of the plane and ordered pairs of real numbers.

There are, however, many other ways of establishing such a correspondence. Although each is useful in simplifying some particular class of problems, the only other two-dimensional coordinates sufficiently important for us to consider here are **polar coordinates.**

We start with a fixed point O, called the **pole** or **origin**, and a fixed half-line Ox, called the **polar axis** or **initial ray.** The polar axis starts at the pole and extends indefinitely far in one direction, which is customarily taken to be horizontally to the right (Fig. 12-42).

Let P be any point in the plane, and draw the half-line that starts at O and goes through P. Let θ be a measure of the angle xOP having the polar axis for its initial side and the half-line from O through P as its terminal side (Fig. 12-43). Denote the undirected distance between O and P by r; that is, let $|OP| = r$. Then (r, θ), always written in that order, is one pair of polar coordinates of P.

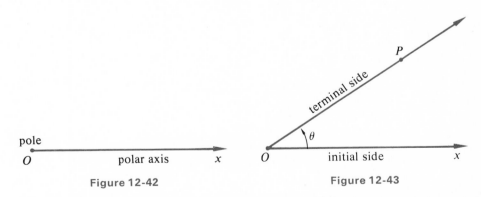

Figure 12-42   Figure 12-43

For instance, to locate the point P whose polar coordinates are $(5, \frac{2}{3}\pi)$, we construct the positive angle whose measure is $\frac{2}{3}\pi$ with the polar axis for its initial side and the pole for vertex [Fig. 12-44(a)]. The point P is on the terminal side of this angle, five units from the origin.

But this same point, P, has many other sets of polar coordinates, among them $(5, \frac{8}{3}\pi)$ and $(5, -\frac{4}{3}\pi)$ [Fig. 12-44(b) and (c)].

So while a given set of polar coordinates determines just one point, a given point has indefinitely many sets of polar coordinates. Unlike Cartesian coordinates, the

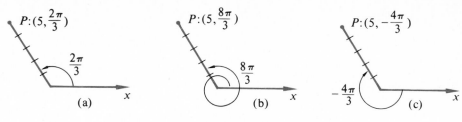

Figure 12-44

polar coordinate system fails to establish a one-to-one correspondence between the points of the plane and ordered pairs of real numbers. This failure introduces certain difficulties not encountered in Cartesian coordinates, but the advantages of polar coordinates for some types of problems far outweigh this disadvantage.

In what we have said so far, r has been positive or zero. Other polar coordinates in which r is negative are defined as follows.

The terminal side of θ was a half-line issuing from the pole. By the **extension of the terminal side** of θ, we mean the half-line issuing from the pole in the opposite direction to the terminal side of θ (Fig. 12-45). In our diagrams the extension of the terminal side of θ is indicated by a dotted line.

Let P be any point in the plane. Construct an angle θ so that P is on the extension of its terminal side (Fig. 12-46).* Let $r = -|OP|$. Then (r, θ) is defined to be a set of polar coordinates of P.

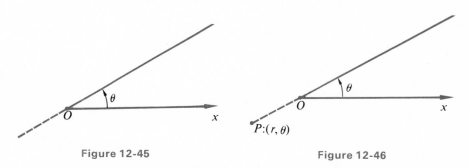

Figure 12-45   Figure 12-46

For instance, to locate the point $P:(-3, \frac{3}{2}\pi)$, we construct the positive angle $\frac{3}{2}\pi$ that has Ox for its initial side (Fig. 12-47). Because r is negative, we locate P on the extension of the terminal side of θ, three units from the pole.

The point $(-5, 5\pi/3)$ in Fig. 12-48 also has polar coordinates $(5, 2\pi/3)$.

Returning to Fig. 12-47, the reader should observe that if the horizontal and vertical lines were Cartesian axes, the y coordinate of P would be positive 3 because the direction from O to P is upward. But in polar coordinates, upward and downward do not *in themselves* determine the sign of r; whenever P is on the terminal side of the

*For brevity, we shall sometimes follow the custom of saying "angle θ" when what is meant is "angle whose measure is θ."

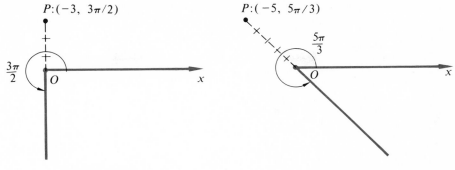

Figure 12-47   Figure 12-48

given polar angle, r is positive; and whenever P is on the extension of the terminal side of the polar angle, r is negative.

Polar coordinate paper has equally spaced circles with centers at the pole, and half-lines radiating from the pole. It is useful for plotting points and graphing equations in polar coordinates.

In numerical examples, when no degree, minute, or second symbol appears in the second polar coordinate, it is understood that the angle is measured in radians. Thus the polar coordinates $(-2, 3)$ denote the point for which $r = -2$ and $\theta = 3$ *radians*.

The preceding discussion can be summarized as follows. *To plot the point* $P:(r, \theta)$, *first construct the angle* θ *with the pole as its vertex and the polar axis as its initial side. If r is positive, locate P on the terminal side of* θ *so that* $|OP| = r$. *If r is negative, locate P on the extension of the terminal side of* θ *so that* $|OP| = -r$.

Example 1. Plot the point whose polar coordinates are $(-4, -\frac{1}{4}\pi)$ and find three other sets of polar coordinates for the same point.

Solution. Draw the angle $-\frac{1}{4}\pi$ in the clockwise direction from its initial line, the polar axis [Fig. 12-49(a)]. Locate P on the extension of the terminal side of θ, four units from the pole. Other sets of polar coordinates of this same point are $(4, \frac{3}{4}\pi)$, $(4, -\frac{5}{4}\pi)$, and $(-4, \frac{7}{4}\pi)$ [Fig. 12-49(b), (c), and (d)].

When a new coordinate system is introduced, it is desirable to have formulas for finding the coordinates of a point in either system when its coordinates in the other system are known.

Let the pole of a polar coordinate system be the origin of a rectangular Cartesian system, and let the polar axis coincide with the positive x axis [Fig. 12-50(a)]. Then any point P in the plane has polar coordinates (r, θ) and also Cartesian coordinates (x, y).

If P is on the terminal side of θ [Fig. 12-50(a)], r is positive and $r = |OP|$. Then

$$\cos \theta = \frac{x}{|OP|} = \frac{x}{r}, \qquad \sin \theta = \frac{y}{|OP|} = \frac{y}{r},$$

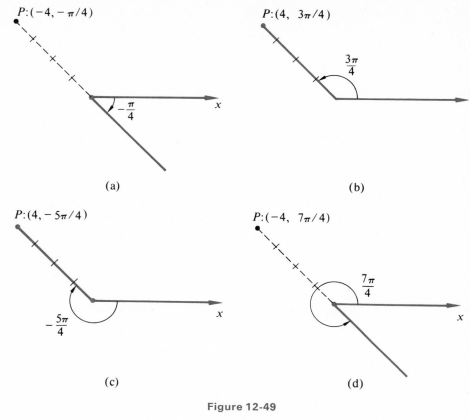

Figure 12-49

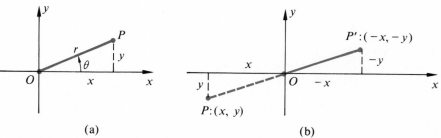

Figure 12-50

or

(1) $x = r \cos \theta, \qquad y = r \sin \theta.$

If P is on the extension of the terminal side of θ, r is negative and $r = -|OP|$. Let P' be the point that is symmetric to P with respect to the origin [Fig. 12-50(b)]. Since the Cartesian coordinates of P are (x, y), those of P' are $(-x, -y)$ and $|OP| =$

$|OP'|$. Now

$$\cos \theta = \frac{-x}{|OP'|} = \frac{-x}{|OP|} = \frac{-x}{-r} = \frac{x}{r},$$

$$\sin \theta = \frac{-y}{|OP'|} = \frac{-y}{|OP|} = \frac{-y}{-r} = \frac{y}{r},$$

or

$$x = r \cos \theta, \qquad y = r \sin \theta,$$

which are the same as equations (1). Thus formulas (1) are valid in every case.

These formulas enable us to find the Cartesian coordinates of a point whose polar coordinates are known. For example, the Cartesian coordinates of $P:(10, \frac{1}{6}\pi)$ are $x = 10 \cos \frac{1}{6}\pi = 5\sqrt{3}$ and $y = 10 \sin \frac{1}{6}\pi = 5$. That is, the Cartesian coordinates of P are $(5\sqrt{3}, 5)$.

If we square equations (1) and add them, member by member, we obtain $x^2 + y^2 = r^2(\sin^2 \theta + \cos^2 \theta)$ or

$$(2) \qquad\qquad r = \pm\sqrt{x^2 + y^2}.$$

If we divide the second formula of (1) by the first, member by member, we get

$$\tan \theta = \frac{y}{x},$$

where θ can be any one of the numbers for which $\tan \theta = y/x$.

Finally, by substituting (2) in (1), we obtain

$$\sin \theta = \frac{y}{\pm\sqrt{x^2 + y^2}}, \qquad \cos \theta = \frac{x}{\pm\sqrt{x^2 + y^2}}.$$

12.11.1 Theorem. If the origin and the positive x axis of a rectangular Cartesian coordinate system are the pole and the polar axis, respectively, of a polar coordinate system, then the Cartesian and polar coordinates of any point P are related by the formulas

$$x = r \cos \theta, \qquad y = r \sin \theta,$$

$$r = \pm\sqrt{x^2 + y^2}, \qquad \tan \theta = \frac{y}{x},$$

and

$$\sin \theta = \frac{y}{\pm\sqrt{x^2 + y^2}}, \qquad \cos \theta = \frac{x}{\pm\sqrt{x^2 + y^2}},$$

where the sign before the radical is positive if P is on the terminal side of θ and negative if P is on the extension of the terminal side of θ.

Example 2. Find polar coordinates of the point P whose rectangular Cartesian coordinates are $(-1, -1)$.

Solution. By 12.11.1, $r = \pm\sqrt{(-1)^2 + (-1)^2} = \pm\sqrt{2}$ and $\tan \theta = 1$. One such value of θ is $\frac{1}{4}\pi$. From Fig. 12-51 it is clear that if $\theta = \frac{1}{4}\pi$, P is on the extension of the terminal side of θ and $r = -\sqrt{2}$. Therefore, polar coordinates of P are $(-\sqrt{2}, \frac{1}{4}\pi)$. Another pair of polar coordinates of the same point is $(\sqrt{2}, \frac{5}{4}\pi)$.

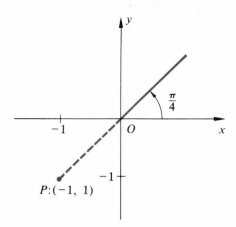

Figure 12-51

Example 3. Find a polar equation of an ellipse in standard position.

Solution. The Cartesian equation of an ellipse in standard position is

$$\frac{x^2}{a^2} + \frac{y^2}{b^2} = 1.$$

Using the substitutions $x = r \cos \theta$, $y = r \sin \theta$ (from 12.11.1), we obtain

$$\frac{r^2 \cos^2 \theta}{a^2} + \frac{r^2 \sin^2 \theta}{b^2} = 1,$$

or

$$r^2(b^2 \cos^2 \theta + a^2 \sin^2 \theta) = a^2 b^2.$$

Example 4. Transform the general form of the equation of a straight line in Cartesian coordinates into a polar equation.

Solution. The general equation of a straight line in Cartesian coordinates is

$$Ax + By + C = 0.$$

Substituting $x = r \cos \theta$, $y = r \sin \theta$ in this equation gives

$$Ar \cos \theta + Br \sin \theta + C = 0,$$

or

$$r(A \cos \theta + B \sin \theta) + C = 0.$$

Example 5. Find the Cartesian equation of the graph of $r = 4/(1 - \cos \theta)$.

Solution. The given equation can be rewritten $r - r \cos \theta = 4$. By means of the substitutions $r = \pm\sqrt{x^2 + y^2}$ and $x = r \cos \theta$ (from 12.11.1), our equation becomes

$$\pm\sqrt{x^2 + y^2} - x = 4.$$

Transposing the term in x and squaring both members, we have

$$x^2 + y^2 = x^2 + 8x + 16,$$

or

$$y^2 = 8x + 16.$$

Its graph is a parabola that is symmetric with respect to the x axis. (It is not in standard position.)

Exercises

1. Plot the points whose polar coordinates are $(4, \frac{1}{3}\pi)$, $(2, \frac{1}{2}\pi)$, $(5, \frac{1}{6}\pi)$, $(6, \frac{2}{3}\pi)$, $(0, \frac{11}{7}\pi)$, $(3, \frac{3}{2}\pi)$, $(7, \frac{4}{3}\pi)$, $(4, 0)$.

2. Plot the points whose polar coordinates are $(3, \pi)$, $(7, 2\pi)$, $(2, \frac{11}{6}\pi)$, $(4, -\frac{1}{3}\pi)$, $(5, -\frac{3}{4}\pi)$, $(0, 0)$, $(0, \frac{2}{3}\pi)$, $(3, -\frac{3}{2}\pi)$.

3. Plot the points whose polar coordinates are $(-5, \frac{1}{4}\pi)$, $(5, -\frac{3}{4}\pi)$, $(2, -\frac{1}{3}\pi)$, $(-2, \frac{2}{3}\pi)$, $(-6, 0)$, $(3, -\pi)$, $(-4, -\frac{2}{3}\pi)$, $(-3, \pi)$.

4. Plot the points whose polar coordinates are $(2, \frac{5}{8}\pi)$, $(-4, \frac{17}{6}\pi)$, $(5, -\frac{7}{3}\pi)$, $(7, 1)$, $(-3, -2)$, $(3, 0)$, $(-3, -\pi)$, $(0, -\frac{5}{4}\pi)$.

5. Plot the points whose polar coordinates follow. Then give four other pairs of polar coordinates, two with positive r and two with negative r, for each point.

(a) $(4, \frac{1}{3}\pi)$; (b) $(-3, \frac{5}{4}\pi)$;
(c) $(-5, \frac{1}{6}\pi)$; (d) $(7, -\frac{2}{3}\pi)$.

6. Plot the points whose polar coordinates follow. Then give four other pairs of polar coordinates, two with positive r and two with negative r.

(a) $(-2, \pi)$; (b) $(5, -\frac{1}{12}\pi)$;
(c) $(4, \frac{19}{4}\pi)$; (d) $(-7, -\frac{3}{2}\pi)$.

7. Find the Cartesian coordinates of the points in Exercise 5.

8. Find the Cartesian coordinates of the points in Exercise 6.

9. Find polar coordinates of the points whose

Cartesian coordinates are
(a) $(-2\sqrt{3}, -2)$; (b) $(1, \sqrt{3})$;
(c) $(\sqrt{2}, -\sqrt{2})$; (d) $(0, 0)$.

10. Find polar coordinates of the points whose Cartesian coordinates are
(a) $(1 + \sqrt{2}/2, \sqrt{2}/2)$; (b) $(-7\sqrt{3}/2, 7/2)$;
(c) $(0, -4)$; (d) $(5, -12)$.

In each of Exercises 11 to 16, sketch the graph of the given Cartesian equation and then find a polar equation for it.

11. $x - 4y + 2 = 0$.

12. $x = 0$.

13. $y = -5$.

14. $x + y = 0$.

15. $x^2 + y^2 = 16$.

16. $y^2 = 4px$.

In Exercises 17 to 22, find the Cartesian equations of the graphs of the given polar equations.

17. $\theta = \frac{1}{3}\pi$.

18. $r = 2$.

19. $r \cos \theta + 6 = 0$.

20. $r - 6 \cos \theta = 0$.

21. $r \sin \theta - 4 = 0$.

22. $r^2 - 8r \cos \theta - 4r \sin \theta + 11 = 0$.

12.12 THE GRAPH OF A POLAR EQUATION

It is sometimes easy to sketch the graph of an equation in polar coordinates by simply assigning a series of values to θ and computing the corresponding values of r. We then plot the points and draw a smooth curve through them.

Example 1.   Sketch the graph of the polar equation $r = 8 \cos \theta$.

Solution.   If we substitute 0 for θ in this equation, we obtain $r = 8$. Thus the point whose polar coordinates are (8, 0) is on the graph of $r = 8 \cos \theta$. Again, if we substitute $\frac{1}{6}\pi$ for θ in the given equation, we obtain $r = 4\sqrt{3}$, and so the point $(4\sqrt{3}, \frac{1}{6}\pi)$ is on the graph. Continuing this process, our results are shown in the accompanying table of approximate values. After plotting all of these points, we draw a smooth curve through them, proceeding

θ	0	$\frac{1}{6}\pi$	$\frac{1}{4}\pi$	$\frac{1}{3}\pi$	$\frac{1}{2}\pi$	$\frac{2}{3}\pi$	$\frac{3}{4}\pi$	$\frac{5}{6}\pi$	π
r	8	6.8	5.6	4	0	-4	-5.6	-6.8	-8

from one point to the next in the order of their appearance as θ increases from 0 to π (Fig. 12-52).

The graph seems to be a circle passing through the pole and symmetric with respect to the polar axis. This will be proved in Section 12.13.

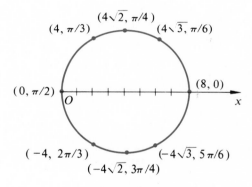

Figure 12-52

Example 2.   Sketch the graph of the polar equation

(1) $$r = \frac{2}{1 - \cos \theta}.$$

Solution.   As in Example 1, we construct a table of approximate values by substituting a series of increasing values for θ in (1) and computing the corresponding values of r. It will be proved

θ	0	$\frac{1}{4}\pi$	$\frac{1}{2}\pi$	$\frac{3}{4}\pi$	π	$\frac{5}{4}\pi$	$\frac{3}{2}\pi$	$\frac{7}{4}\pi$
r	—	6.8	2	1.2	1	1.2	2	6.8

in 12.13 that the graph of (1) is a parabola with its focus at the pole (Fig. 12-53).

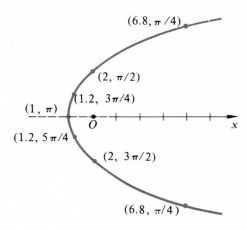

Figure 12-53

Ths is reminiscent of curve sketching in Cartesian coordinates, but there is one essential difference. Notice that the coordinates $(-2, \frac{3}{2}\pi)$ do not satisfy equation (1). But the point $P:(-2, \frac{3}{2}\pi)$ must be on the curve because another set of polar coordinates of the same point, $(2, \frac{1}{2}\pi)$, does satisfy equation (1). Thus, the point $P:(-2, \frac{3}{2}\pi)$ is on the graph of (1) even though its coordinates $(-2, \frac{3}{2}\pi)$ fail to satisfy the equation. We must conclude that *in polar coordinates, failure of a particular set of coordinates to satisfy a given equation is no guarantee that the point is not on the graph of that equation.*

12.12.1 **Definition.** The **graph of a polar equation** is the set of points, each of which has at least one pair of polar coordinates that satisfy the equation.

In Cartesian coordinates it was usually helpful to make a preliminary analysis of an equation before attempting to sketch its graph. Such is often the case with polar coordinates, too. There is, however, one difficulty. Each point has indefinitely many pairs of polar coordinates, and even though one pair may not satisfy a given polar equation, another pair may. Nevertheless, the following preliminary tests are worthwhile and should be tried.

1. Transform to Cartesian Coordinates.

If a polar equation is transformed into a Cartesian equation by means of the substitutions 12.11.1, the graph of the Cartesian equation may be familiar to us. Since the graph is the same in both coordinate systems, this is sometimes an easy way to sketch the graph of a polar equation. As an illustration, the polar equation $r^2 \sin 2\theta = 2$ can be written $r^2 \sin \theta \cos \theta = 1$. Using the formulas of 12.11.1, the latter becomes $xy = 1$. Thus the graph of $r^2 \sin 2\theta = 2$ is the familiar equilateral hyperbola (Example 2, Section 12.9).

Again, the polar equation $r = 8 \cos \theta$ in Example 1, above, is transformed by 12.11.1 into the Cartesian equation $\sqrt{x^2 + y^2} = 8x/\sqrt{x^2 + y^2}$, or $(x - 4)^2 + y^2 = 16$. Its graph is the circle with center at $(4, 0)$ and passing through the pole (Fig. 12-52).

When the corresponding Cartesian equation turns out to be complicated or unfamiliar, as is often the case, we return to the original polar equation and make the following tests.

2. Some Sufficient Conditions for Symmetry.

The points $P{:}(r, \theta)$ and $P'{:}(r, -\theta)$ are symmetric with respect to the polar axis (Fig. 12-54). Also, the points $P{:}(r, \theta)$ and $P{:}(-r, \pi - \theta)$ are symmetric with respect to the polar axis (Fig. 12-55). This leads to the following rule.

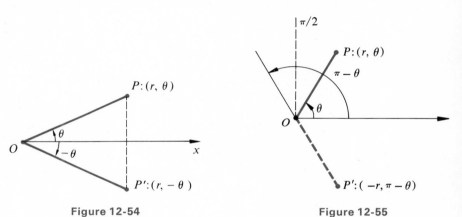

Figure 12-54   Figure 12-55

(i) *If a polar equation is unchanged when (r, θ) is replaced throughout by either $(r, -\theta)$ or by $(-r, \pi - \theta)$, its graph is symmetric with respect to the* **polar axis** *(and its extension).*

The points $P{:}(r, \theta)$ and $P'{:}(r, \pi - \theta)$ are symmetric with respect to the $\frac{1}{2}\pi$ axis (Fig. 12-56); and the points $P{:}(r, \theta)$ and $P'{:}(-r, -\theta)$ are symmetric with respect to the $\frac{1}{2}\pi$ axis (Fig. 12-57).

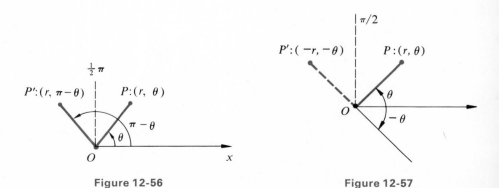

Figure 12-56   Figure 12-57

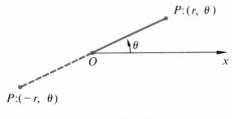

Figure 12-58

(ii) *If a polar equation is unchanged when* (r, θ) *is replaced throughout by either* $(r, \pi - \theta)$ *or by* $(-r, -\theta)$, *its graph is symmetric with respect to the* $\frac{1}{2}\pi$ **axis** *(and its extension).*

The points $P{:}(r, \theta)$ and $P'{:}(-r, \theta)$ are symmetric with respect to the pole (Fig. 12-58).

(iii) *If a polar equation is unchanged when* r *is replaced by* $-r$ *throughout, its graph is symmetric with respect to the* **pole.**

For example, the parabola $r = 2/(1 - \cos \theta)$ is symmetric with respect to the polar axis (Fig. 12-53) because replacement of (r, θ) by $(r, -\theta)$ yields $r = 2/[1 - \cos(-\theta)]$; and since $\cos(-\theta) = \cos \theta$, this can be written $r = 2/(1 - \cos \theta)$, which is the original equation.

The reader is warned, however, that because each point has many pairs of polar coordinates, an equation may fail any of the preceding tests even though its graph has the symmetry that is being tested for. In other words, these tests are *sufficient* conditions for symmetries but not *necessary* conditions.

As an illustration, our test for symmetry with respect to the pole fails in the equation $r = 4 \sin 2\theta$, for replacement of r by $-r$ gives $-r = 4 \sin 2\theta$, which essentially changes the original equation. Yet the graph of $r = 4 \sin 2\theta$ *is* symmetric with respect to the origin, as shown in Fig. 12-63.

Although a long list of other tests for symmetry could be compiled, it would not be worthwhile remembering them.

3. Domain. Periodicity.

In $r = f(\theta)$ the domain is the set of values of θ for which r is real. When trigonometric functions of θ appear in polar equations, however, we use the periods of the trigonometric functions involved to find a conveniently small part of the domain on which the whole curve is traced once.

The equation $r = 8 \cos \theta$ in Example 1 has the set of all real numbers for its domain. But because the cosine function is periodic, the whole circle is traced once as θ takes on the values in the interval $[0, \pi)$. The same circle is traced again on any other interval of length π.

The interval for θ can often be further reduced when the graph is symmetric. Since $\cos(-\theta) = \cos \theta$, the equation $r = 8 \cos \theta$ remains unchanged when (r, θ) is replaced by $(r, -\theta)$. Thus its graph is symmetric with respect to the polar axis. It follows that we could have confined θ to the interval $[0, \frac{1}{2}\pi]$ because the upper half of the circle is generated there, and then obtain the lower half by symmetry.

These ideas are developed further in the examples below.

4. Tangents at the Pole.

The polar equation of any line through the pole can be written

$$\theta = k,$$

where k is some constant; the line goes through the pole because $(0, k)$ is one pair of

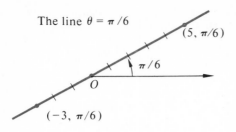

The line $\theta = \pi/6$

$(5, \pi/6)$

$\pi/6$

O

$(-3, \pi/6)$

Figure 12-59

polar coordinates of the pole and $(0, k)$ satisfies the equation $\theta = k$. As an illustration, $\theta = \frac{1}{6}\pi$ is an equation of the line through the pole whose polar angle is $\frac{1}{6}\pi$ (Fig. 12-59). Some points on this line are $(5, \frac{1}{6}\pi)$, $(-3, \frac{1}{6}\pi)$, and the origin $(0, \frac{1}{6}\pi)$.

It will be shown in Section 12-14 that *when a polar curve goes through the pole, the polar equations of the tangents to the curve at the pole may be found by substituting $r = 0$ in the polar equation of the curve; if the real roots of the resulting equation in θ are $\theta_1, \theta_2, \theta_3, \cdots$, then $\theta = \theta_1, \theta = \theta_2, \theta = \theta_3, \cdots$ are equations of the tangents to the curve at the pole.*

For instance, substitution of zero for r in the equation of the *lemniscate*, $r^2 = 8 \cos 2\theta$, gives $\cos 2\theta = 0$; whence $2\theta = \frac{1}{2}\pi$ or $\frac{3}{2}\pi$ and $\theta = \frac{1}{4}\pi$ or $\frac{3}{4}\pi$. The coordinates $(0, \frac{1}{4}\pi)$ and $(0, \frac{3}{4}\pi)$ satisfy the given equation and the lemniscate goes through the pole twice (Fig. 12-62). The equations of the tangent lines at the pole are $\theta = \frac{1}{4}\pi$ and $\theta = \frac{3}{4}\pi$.

Example 3. Analyze the curve $r = 2 \cos \theta + 2$ and then sketch its graph.

Solution. By substituting $r = \pm\sqrt{x^2 + y^2}$ and $\cos \theta = x/\pm\sqrt{x^2 + y^2}$ in the given equation and simplifying, we find the corresponding Cartesian equation to be $(x^2 + y^2 - 2x)^2 - 4(x^2 + y^2) = 0$. This equation is far too complicated to be helpful, and so we return to the given polar equation.

Since $\cos(-\theta) = \cos \theta$, the equation $r = 2 \cos \theta + 2$ is unchanged when (r, θ) is replaced by $(r, -\theta)$. Thus the graph is symmetric with respect to the polar axis. Our other tests for symmetry fail.

The period of the cosine function is 2π, and the whole curve is generated as θ takes on the values in $[0, 2\pi)$. But because the graph is symmetric with respect to the polar axis, we can confine θ to $[0, \pi]$ and sketch the upper half of the curve. The lower half can then be quickly sketched from symmetry.

By substituting $r = 0$ in the given equation, we obtain $2 \cos \theta + 2 = 0$, or $\cos \theta = -1$. The only solution to this equation when $0 \leq \theta \leq 2\pi$ is $\theta = \pi$. Thus the graph is tangent to the extension of the polar axis at the pole.

After constructing a table of approximate values, we plot the points with these polar coordinates. We then sketch the upper half of the graph through these points, remembering

θ	0	$\frac{1}{6}\pi$	$\frac{1}{4}\pi$	$\frac{1}{3}\pi$	$\frac{1}{2}\pi$	$\frac{2}{3}\pi$	$\frac{3}{4}\pi$	$\frac{5}{6}\pi$	π
r	4	3.7	3.4	3	2	1	0.6	0.3	0

that the curve is tangent to the extension of the polar axis at the pole. The lower half of the curve is sketched from symmetry (Fig. 12-60). Because of its heart shape, this curve is called a **cardioid**.

Example 4. Analyze the equation $r = 4 \cos \theta + 2$ and then sketch its graph.

Solution. The Cartesian equation of this curve is too complicated to be useful here; so we proceed to analyze the given polar equation.

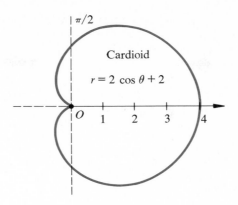

Figure 12-60

Substitution of $(r, -\theta)$ for (r, θ) in the given equation yields $r = 4 \cos(-\theta) + 2 = 4 \cos \theta + 2$, which is the original equation. Thus its graph is symmetric with respect to the polar axis. Our other tests for symmetry fail, so we do not know whether the curve is symmetric with respect to the axis or the origin.

Since the period of the cosine function is 2π, the whole graph is obtained when θ takes on the values in $[0, 2\pi)$. But because the curve is symmetric with respect to the polar axis, we shall confine θ to $[0, \pi]$ and draw the upper half of the curve.

If 0 is substituted for r, the given equation becomes $4 \cos \theta + 2 = 0$, the real roots of which are $\theta = \frac{2}{3}\pi$ and $\theta = \frac{4}{3}\pi$ when $0 \le \theta < 2\pi$. So there are two distinct tangents to the curve at the pole, $\theta = \frac{2}{3}\pi$ and $\theta = \frac{4}{3}\pi$.

After drawing the tangent lines at the pole, we compute a table of approximate values of r for $0 \le \theta \le \pi$ and then draw the upper half of the curve (Fig. 12-61). The rest of this

θ	0	$\frac{1}{6}\pi$	$\frac{1}{3}\pi$	$\frac{1}{2}\pi$	$\frac{7}{12}\pi$	$\frac{2}{3}\pi$	$\frac{3}{4}\pi$	$\frac{5}{6}\pi$	$\frac{11}{12}\pi$	π
r	6	5.5	4	2	1	0	-0.8	-1.5	-1.9	-2

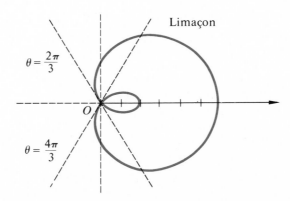

Figure 12-61

graph is easily sketched because of its symmetry with respect to the polar axis. The curve is called a **limaçon**.

Example 5.   Analyze the equation $r^2 = 8 \cos 2\theta$ and sketch its graph.

Solution.   The corresponding Cartesian equation is complicated, so we stay with the polar equation.

    If we replace (r, θ) by $(r, -\theta)$ in the given equation, we obtain $r^2 = 8 \cos(-2\theta) = 8 \cos 2\theta$, the original equation; so its graph is symmetric with respect to the polar axis.

    Since $8 \cos 2(\pi - \theta) = 8 \cos(2\pi - 2\theta) = 8(\cos 2\pi \cos 2\theta + \sin 2\pi \sin 2\theta) = 8 \cos 2\theta$, the equation $r^2 = 8 \cos 2\theta$ is unchanged when (r, θ) is replaced by $(r, \pi - \theta)$. Thus the curve is symmetric with respect to the $\tfrac{1}{2}\pi$ axis.

    The equation is also unchanged if (r, θ) is replaced by $(-r, \theta)$; so the graph is symmetric with respect to the origin.

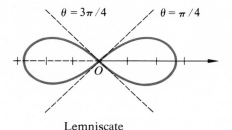

Lemniscate

Figure 12-62

    By substituting $r = 0$ in the given equation, we obtain $\cos 2\theta = 0$. Since $2\theta = \tfrac{1}{2}\pi$ and $2\theta = \tfrac{3}{2}\pi$ are solutions to this equation, $\theta = \tfrac{1}{4}\pi$ and $\theta = \tfrac{3}{4}\pi$ are tangents to the curve at the pole.

    The given equation can be rewritten $r = \pm 2\sqrt{2 \cos 2\theta}$. Because of the symmetry of the graph with respect to the polar axis and the $\tfrac{1}{2}\pi$ axis, we can concentrate on the first quadrant. In choosing a domain for θ from $[0, \tfrac{1}{2}\pi]$, we must exclude numbers that make $\cos 2\theta$ negative. Thus θ must not be in $(\tfrac{1}{4}\pi, \tfrac{1}{2}\pi)$, and we shall confine θ to the domain $[0, \tfrac{1}{4}\pi]$. The rest of the curve can then be sketched from its symmetries.

    A table of approximate values when $0 \le \theta \le \tfrac{1}{4}\pi$ follows. When drawing a smooth curve through these points, remember that it must be tangent to $\theta = \tfrac{1}{4}\pi$ at the pole.

θ	0	$\tfrac{1}{12}\pi$	$\tfrac{1}{6}\pi$	$\tfrac{1}{4}\pi$
r	± 2.8	± 2.6	± 2	0

The curve is shown in Fig. 12-62. It is called a **lemniscate**.

Example 6.   Analyze the equation $r = 4 \sin 2\theta$ and sketch its graph.

Solution.   The corresponding Cartesian equation of this curve turns out to be a sixth-degree polynomial in x and y, so we shall stay with the given polar equation.

    Since $\sin 2(\pi - \theta) = \sin(2\pi - 2\theta) = \sin 2\pi \cos 2\theta - \cos 2\pi \sin 2\theta = -\sin 2\theta$,

replacement of (r, θ) by $(-r, \pi - \theta)$ yields $-r = 4 \sin 2(\pi - \theta) = -4 \sin 2\theta$ or $r =$

$4 \sin 2\theta$, which is the original equation. Consequently, its graph is symmetric with respect to the polar axis.

Because $\sin 2(-\theta) = \sin(-2\theta) = -\sin 2\theta$, replacement of (r, θ) by $(-r, -\theta)$ in $r = 4 \sin 2\theta$ leaves the given equation unchanged. Therefore its graph is also symmetric with respect to the $\frac{1}{2}\pi$ axis.

Although our test for symmetry with respect to the pole fails, the graph is symmetric with respect to the pole because it is symmetric with respect to both the polar axis and the $\frac{1}{2}\pi$ axis.

To find the tangents (if any) to the curve at the pole, we put $r = 0$ in the given equation and obtain $\sin 2\theta = 0$. Therefore the lines $\theta = 0$ and $\theta = \frac{1}{2}\pi$ are tangents to the graph at the pole.

Four-leaved rose

$r = 4 \sin 2\theta$

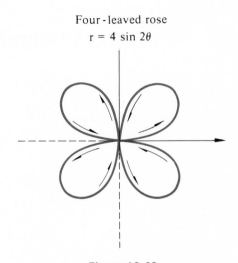

Figure 12-63

Because of its symmetries, the whole curve is easily obtained from the part in the first quadrant. Thus we need only consider values of θ in $[0, \frac{1}{2}\pi]$. Some corresponding values of r and θ are shown in the table. After plotting these points in the first quadrant, we draw a

θ	0	$\frac{1}{12}\pi$	$\frac{1}{8}\pi$	$\frac{1}{6}\pi$	$\frac{1}{4}\pi$	$\frac{1}{3}\pi$	$\frac{3}{8}\pi$	$\frac{5}{12}\pi$	$\frac{1}{2}\pi$
r	0	2	2.8	3.5	4	3.5	2.8	2	0

smooth curve through them. It starts at the pole, tangent to the polar axis, and returns to the pole, tangent to the $\frac{1}{2}\pi$ axis, as indicated by the arrows (Fig. 12-63).

This curve is called a **four-leaved rose**.

Example 7.   Draw the graph of $r = \theta$.

Solution.   All our tests for symmetry fail.

The curve crosses the polar axis (or its extension) when $\theta = 0$, $\theta = \pm\pi$, $\theta = \pm 2\pi$, $\theta = \pm 3\pi$, $\cdots$. In fact, the curve crosses the polar axis or its extension whenever $\theta = n\pi$ (n any integer). It crosses the $\frac{1}{2}\pi$ axis or its extension when $\theta = \frac{1}{2}n\pi$ (n any odd integer).

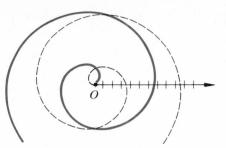

Spiral of Archimedes

Figure 12-64

If $r = 0$, then $\theta = 0$, and thus the tangent to the curve at the pole is the polar axis.

To plot points on the curve, we see that when $\theta = 0$, $r = 0$; when $\theta = 1$ (radian), $r = 1$; when $\theta = \frac{1}{2}\pi$, $r = \frac{1}{2}\pi \doteq 1.6$; and when $\theta = -\pi$, $r = -\pi \doteq -3.1$, and so forth.

This curve is called a **spiral of Archimedes** (Fig. 12-64). The dotted part of the curve corresponds to negative values of θ.

The curves in Examples 3 to 7, above, are representatives of the general classes of polar curves listed here for reference. Let a and b be positive constants. The graph of a polar equation having the form

$$r = a \pm b \cos\theta \qquad \text{or} \qquad r = a \pm b \sin\theta$$

is a **limaçon**. If $a < b$, the limaçon has a loop (Fig. 12-61). If $a \geq b$, the limaçon has no loop (Fig. 12-62). When $a = b$, the limaçon is heart-shaped (Fig. 12-60) and has the special name **cardioid.**

The graph of an equation of the form

$$r^2 = a^2 \cos 2\theta \qquad \text{or} \qquad r^2 = a^2 \sin 2\theta$$

is a **lemniscate**. It is shaped like a figure eight (Fig. 12-62).

The graph of an equation of the form

$$r = a \sin n\theta \qquad \text{or} \qquad r = a \cos n\theta,$$

where n is a positive integer, is a **rose**. It has n leaves if n is odd, and $2n$ leaves if n is even (Fig. 12-63). If $n = 1$, its single petal is a circle (Fig. 12-52).

The graph of an equation having the form

$$r = a\theta$$

is a **spiral of Archimedes** (Fig. 12-64).

The graph of

$$r = ae^{b\theta}$$

is called a **logarithmic spiral** (Fig. 12-76).

Exercises

In Exercises 1 to 6, sketch the graph of the given polar equation.

1. (a) $\theta = \frac{1}{3}\pi$;   (b) $\theta = -\frac{3}{4}\pi$;
   (c) $\theta = 2$;   (d) $r = 5$.

2. (a) $r = -3$;
   (b) $\theta(\theta - 1) = 0$;

   (c) $\theta^2 - 1 = 0$;
   (d) $(r - 2)(\theta + \frac{1}{4}\pi) = 0$.

3. $r \cos\theta = 4$. (*Hint:* Change to Cartesian coordinates.)

4. $r \sin\theta + 6 = 0$.

5. $r^2 \sin 2\theta = 0$. (*Hint:* $\sin 2\theta = 2 \sin\theta \cos\theta$.)

6. $r = 6 \sin \theta$.

In each of Exercises 7 through 32, analyze the equation and then sketch its graph.

7. $r = \dfrac{4}{1 - \cos \theta}$.

8. $r = \dfrac{2}{1 + \sin \theta}$.

9. $r = 5 - 5 \sin \theta$ (cardioid).

10. $r = 4 - 4 \cos \theta$ (cardioid).

11. $r = 3 - 3 \cos \theta$ (cardioid).

12. $r = 4 + 4 \sin \theta$ (cardioid).

13. $r = 2 - 4 \cos \theta$ (limaçon).

14. $r = 4 - 2 \cos \theta$ (limaçon).

15. $r = 4 - 3 \sin \theta$ (limaçon).

16. $r = 3 - 4 \sin \theta$ (limaçon).

17. $r^2 = 9 \sin 2\theta$ (lemniscate).

18. $r^2 = 4 \cos 2\theta$ (lemniscate).

19. $r^2 = -16 \cos 2\theta$ (lemniscate).

20. $r^2 = -4 \sin 2\theta$ (lemniscate).

21. $r = 5 \cos 3\theta$ (three-leaved rose).

22. $r = 3 \sin 3\theta$ (three-leaved rose).

23. $r = 6 \sin 2\theta$ (four-leaved rose).

24. $r = 4 \cos 2\theta$ (four-leaved rose).

25. $r = 7 \cos 5\theta$ (five-leaved rose).

26. $r = 3 \sin 5\theta$ (five-leaved rose).

27. $r = \frac{1}{2}\theta$ (spiral of Archimedes).

28. $r = 2\theta$ (spiral of Archimedes).

29. $r = e^{\theta}$ (logarithmic spiral).

30. $r = e^{\theta/2}$ (logarithmic spiral).

31. $r = \dfrac{2}{\theta}$ (reciprocal spiral).

32. $r = -\dfrac{1}{\theta}$ (reciprocal spiral).

12.13 LINES, CIRCLES, AND CONICS

The polar equation of a straight line or circle in general position is usually not as convenient as the rectangular equation and is seldom used. When the line or circle is in one of certain special positions, however, the polar equation is simple and very useful.

Consider a line through the point $A{:}(a, 0)$, perpendicular to the polar axis (or its extension). Its Cartesian equation is $x = a$. Therefore (by 12.11.1)

$$r \cos \theta = a$$

is the polar equation of any line perpendicular to the polar axis. When a is positive, the line is to the right of the pole (Fig. 12-65); and when a is negative, the line is to the left of the pole.

For instance, the graph of $r \cos \theta = -5$ is a straight line perpendicular to the polar axis and five units to the left of the pole. Solving the equation for r, we obtain

$$r = \frac{-5}{\cos \theta}.$$

When $\theta = 0$, $\cos \theta = 1$, and $r = -5$. Thus the point $(-5, 0)$ is on the line. Similarly, the points whose polar coordinates are $(-10, \frac{1}{3}\pi)$ and $(10, \frac{2}{3}\pi)$ are also on the line (Fig. 12-66).

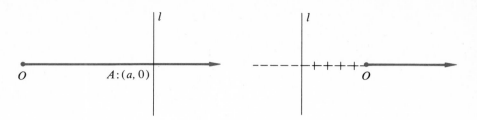

Figure 12-65   Figure 12-66

Next, consider a line l through the point A:$(a, \frac{1}{2}\pi)$ and parallel to the polar axis. Its Cartesian equation is $y = a$. Therefore (by 12.11.1) its polar equation is

$$r \sin \theta = a.$$

This is *the polar equation of any line parallel to the polar axis*. If a is positive, the line is above the polar axis; and if a is negative, the line is below the polar axis.

Any *line through the pole* has the very simple polar equation

$$\theta = k,$$

where the constant k is the measure of an angle that the line makes with the polar axis, for a point lies on such a line if and only if one value of its second polar coordinate is k.

For example, the polar equation of the line through the pole, making the angle $\frac{1}{4}\pi$ with the polar axis is $\theta = \frac{1}{4}\pi$. But notice that this line has other polar equations, among them $\theta = \frac{9}{4}\pi$ and $\theta = -\frac{3}{4}\pi$. This is because each point has indefinitely many sets of polar coordinates; and if any one of these sets satisfies a given equation, the point lies on the graph of that equation.

As in the case of the straight line, the general equation of the circle in polar coordinates is so complicated that it is seldom used (see Exercise 18). Nevertheless, when the circle is in certain special positions, its polar equation is simple and useful.

A *circle with its center at the pole* and radius k has the polar equations

$$r = \pm k.$$

We turn next to a circle with center at the point A:$(a, 0)$ and radius equal to $|a|$. It passes through the pole. Denote its second intersection with the polar axis by Q (Fig. 12-67). Let P:(r, θ) be any point on the circle different from O and Q. Then OPQ is a right angle, since it is inscribed in a semicircle. Thus $\cos \theta = r/(2a)$. Therefore

$$r = 2a \cos \theta$$

is *the polar equation of the circle with radius* $|a|$, *having its center on the polar axis or its extension, and passing through the pole.*

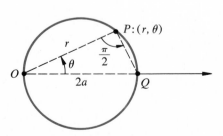

Figure 12-67

When a is positive in the above equation, the circle is to the right of the pole. When a is negative, the circle is to the left of the pole. The reader should verify that the coordinates of the points $Q:(0, \frac{1}{2}\pi)$ and $Q:(2a, 0)$ satisfy the equation.

Similarly,

$$r = 2a \sin \theta$$

is *the polar equation of the circle with radius* $|a|$ *and center on the* $\frac{1}{2}\pi$ *axis or its extension, and passing through the pole.*

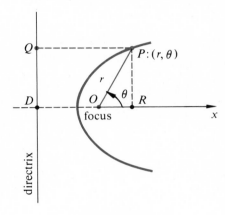

Figure 12-68

From the definition of a conic (12.2.1), we recall that the parabola, ellipse, and hyperbola have the property that the ratio of the undirected distance of any point of the curve from a focus to its undirected distance from the corresponding directrix is always equal to the eccentricity. Very simple polar equations of conics are obtained by placing a focus at the pole and making the corresponding directrix perpendicular to the polar axis.

Let the directrix be to the left of the focus, and denote its intersection with the extension of the polar axis by D (Fig. 12-68). Call the undirected distance between the focus and the directrix d; that is, $|DO| = d$.

Let $P:(r, \theta)$ be any point on the conic, and denote its projection on the directrix by Q and its projection on the polar axis (or its extension) by R. From 12.2.1 P is on the conic if and only if

(1) $$|OP| = e|QP|,$$

where e is the eccentricity.

Assume for the moment that P is to the right of the directrix and on the terminal side of θ. Then $r > 0$, and

(2) $$|OP| = r.$$

Since $\overline{OR} = r \cos \theta$, $|DR| = |DO| + r \cos \theta = d + r \cos \theta$. But $|QP| = |DR|$. Therefore

(3) $$|QP| = d + r \cos \theta.$$

By substituting (2) and (3) in (1), we get

$$r = e(d + r \cos \theta),$$

or, after solving for r,

(4) $$r = \frac{ed}{1 - e \cos \theta}.$$

The graph of (4) is a parabola if $e = 1$ and an ellipse if $0 < e < 1$. Both curves lie entirely to the right of the given directrix. But under our assumption that P is to the right of the directrix and that r is positive, the graph of (4) for $e > 1$ is only one branch of a hyperbola. It can be shown, however, that by allowing r to take on negative as well as positive values, the graph of (4) for $e > 1$ is the whole hyperbola (see the Example, below).

Similarly, if the directrix is to the right of the corresponding focus (Fig. 12-69), a polar equation of the conic is

$$r = \frac{ed}{1 + e \cos \theta}.$$

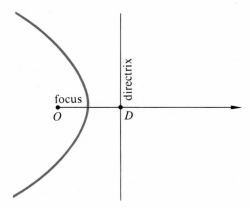

Figure 12-69

All this is summarized in:

12.13.1 Theorem. A **polar equation of a conic** with a focus at the pole and the corresponding directrix perpendicular to the polar axis is

$$r = \frac{ed}{1 \pm e \cos \theta},$$

where e is the eccentricity and d is the undirected distance between the focus and the corresponding directrix. The positive sign applies when the given directrix is to the right of the corresponding focus, and the negative sign when the given directrix is to the left of the focus.

Equally simple polar equations of conics are obtained if a focus is at the pole and the directrix is parallel to the polar axis.

12.13.2 Theorem. A **polar equation of a conic** having a focus at the pole and the corresponding directrix parallel to the polar axis is

$$r = \frac{ed}{1 \pm e \sin \theta},$$

where e is the eccentricity and d is the undirected distance between the focus and the corresponding directrix. The positive sign is used when the given directrix is above the polar axis, and the negative sign when the directrix is below the polar axis.

Example.   Name the conic that has the equation

$$r = \frac{30}{5 - 10 \cos \theta}$$

and find its vertices, center, foci, and directrices. Then plot this data and sketch the conic.

Solution.   In order to compare the given equation with 12.13.1, we must make the first term in the denominator of the right member equal to unity. Accordingly, we divide both numerator and denominator of the right member by 5 and obtain

$$r = \frac{6}{1 - 2 \cos \theta}.$$

Comparing it with 12.13.1, we see that $e = 2$ and $d = 3$. Therefore the graph is a hyperbola with a focus at the pole and the corresponding directrix three units to the left of the pole (Fig. 12-70).

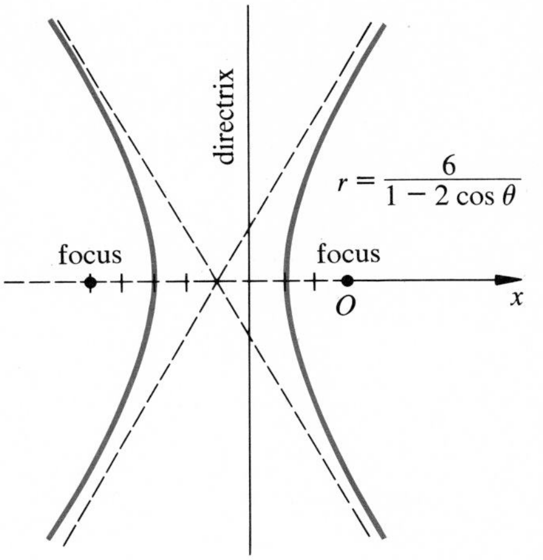

Figure 12-70

The vertices of the hyperbola are its points of intersection with its principal axis, which is the polar axis (and its extension). By letting $\theta = 0$ in equation (1), we get $r = -6$. Thus $(-6, 0)$ is a vertex. The other vertex is $(2, \pi)$, found by substituting $\theta = \pi$ in (1). Since the center is midway between the vertices, a set of polar coordinates of the center is $(0, -4)$.

One focus is the pole. The distance between a focus and the corresponding directrix is $d = 3$. Thus the vertical line $r \cos \theta = -3$ is a directrix. By symmetry with respect to the center, $(0, -4)$, the other focus is $(0, -8)$ and the corresponding directrix is $r \cos \theta = -5$.

From the table of approximate values it is clear that as θ increases from 0 and approaches $\frac{1}{3}\pi$, the point $P:(r, \theta)$ starts from $(-6, 0)$, the left vertex of the hyperbola, and moves indefinitely far away along the lower half of the left branch. For $\theta = \frac{1}{3}\pi$ there is no value of r because the denominator, $1 - 2 \cos \frac{1}{3}\pi$, would be zero. The line $\theta = \frac{1}{3}\pi$ is parallel to an asymptote. As θ increases, taking on values between $\frac{1}{3}\pi$ and $\frac{5}{3}\pi$, the point traces the entire right branch of the hyperbola from top to bottom. For $\theta = \frac{5}{3}\pi$ there is no corresponding r. As θ continues to increase, taking on values between $\frac{5}{3}\pi$ and 2π, the point moves in, along the upper half of the left branch, completing the curve.

θ	0	$\frac{1}{6}\pi$	$\frac{1}{4}\pi$	$\frac{1}{3}\pi$	$\frac{1}{2}\pi$	$\frac{2}{3}\pi$	$\frac{3}{4}\pi$	$\frac{5}{6}\pi$	π	$\frac{7}{6}\pi$	$\frac{5}{4}\pi$	$\frac{4}{3}\pi$	$\frac{3}{2}\pi$	$\frac{5}{3}\pi$	$\frac{7}{4}\pi$	$\frac{11}{6}\pi$
r	-6	-8.1	-15	—	6	3	2.5	2.2	2	2.2	2.5	3	6	—	-15	-8.1

The center is four units to the left of the pole and the asymptotes go through the center, making angles of $\pm\frac{1}{3}\pi$ with the axis.

Exercises

Make a sketch for each exercise.

1. Without using Cartesian coordinates, write a polar equation of the line that is

(a) perpendicular to the polar axis and eight units to the right of the pole.

(b) parallel to the polar axis and three units below it.

(c) through the pole with slope $\sqrt{3}$.

(d) the polar axis and its extension.

2. Without using Cartesian coordinates, write a polar equation of the line that is

(a) parallel to the polar axis and ten units above it.

(b) through the pole with slope 1.

(c) perpendicular to the polar axis and two units to the left of the pole.

(d) the $\frac{1}{2}\pi$ axis and its extension.

3. Without using Cartesian coordinates, find the polar equation of the line through the point $(4, \frac{1}{3}\pi)$ and perpendicular to the polar axis.

4. Without using Cartesian coordinates, find the polar equation of the line through the point $(8, \frac{5}{4}\pi)$ and parallel to the polar axis.

5. Find the polar equation of a circle

(a) with center at the pole and radius 8.

(b) with center $(5, 0)$ and radius 5.

(c) with center $(6, \frac{1}{2}\pi)$ and radius 6.

(d) with center $(4, \pi)$ and radius 4.

(e) with radius 3 and tangent to the polar axis at the pole.

6. Find the polar equation of a circle

(a) with center $(-7, 0)$ and radius 7.

(b) with center $(-3, \pi)$ and radius 3.

(c) with center $(2, -\frac{1}{2}\pi)$ and radius 2.

(d) with center $(-6, \frac{3}{2}\pi)$ and radius 6.

(e) with radius 5 and tangent to the $\frac{1}{2}\pi$ axis at the pole.

In each of Exercises 7 to 16, name the conic whose equation is given. If it is a parabola, find its vertex and its directrix; if an ellipse or hyperbola, find its vertices, center, foci, and directrices. Then plot the data found and draw the conic.

7. $r = \dfrac{2}{1 - \cos\theta}$.

8. $r = \dfrac{8}{3 + 3\sin\theta}$.

9. $r = \dfrac{4}{2 - \cos\theta}$.

10. $r = \dfrac{9}{3 - \cos\theta}$.

11. $r = \dfrac{15}{2 + 3\cos\theta}$.

12. $r = \dfrac{4}{2 - 4\sin\theta}$.

13. $r = \dfrac{10}{7 - 2\sin\theta}$.

14. $r = \dfrac{12}{2 + 2\cos\theta}$.

15. $r = \dfrac{7}{3 + 4\sin\theta}$.

16. $r = \dfrac{2}{2 + \cos\theta}$.

17. Show that the undirected distance between any two points, $P_1:(r_1, \theta_1)$ and $P_2:(r_2, \theta_2)$, is given by

$$|P_1 P_2| = \sqrt{r_1^2 + r_2^2 - 2r_1 r_2 \cos(\theta_1 - \theta_2)}.$$

(*Hint:* Use the law of cosines in the triangle OP_1P_2.)

18. Without using Cartesian coordinates, show that a polar equation of the circle with center (c, α) and radius a is

$$r^2 + c^2 - 2rc\cos(\theta - \alpha) = a^2.$$

(*Hint:* Use Exercise 17.)

12.14 ANGLE FROM THE RADIUS VECTOR TO THE TANGENT LINE

Let l_1 and l_2 be any two intersecting lines. **The angle from l_1 to l_2** is the smallest positive angle having l_1 for its initial side and l_2 for its terminal side (Fig. 12-71).

Consider a line, l, and let P_1 be an arbitrarily chosen point on l. Denote by h the horizontal half-line starting at P_1 and extending indefinitely to the right (Fig. 12-72).

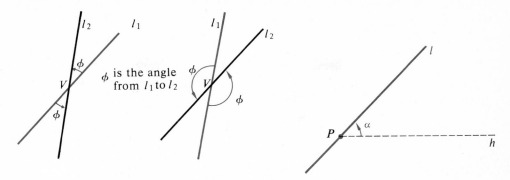

Figure 12-71

Figure 12-72

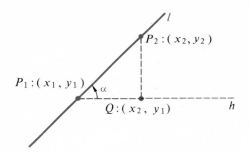

Figure 12-73

ϕ is the angle from l_1 to l_2

The angle, α, from h to l is called the **angle of inclination** of l or, briefly, the **inclination** of l. Notice that the inclination of a line is independent of any coordinate system.

Now let l be a nonvertical line in a Cartesian plane and let $P_1:(x_1, y_1)$ and $P_2:(x_2, y_2)$ be arbitrarily chosen distinct points on l (Fig. 12-73). The slope of l is

$$m = \frac{y_2 - y_1}{x_2 - x_1} = \tan \alpha.$$

Since $m = \tan \alpha$, *the slope of line is equal to the tangent of its inclination*, regardless of the coordinate system. The slope of the tangent to a plane curve assumed early importance in Cartesian coordinates, partly because it provides a simple geometric interpretation of the derivative. It also gives the direction of the graph of a Cartesian equation at any point on it, which is very useful in curve sketching and other applications.

But in polar coordinates $dr/d\theta$ does *not* give the slope of the tangent to the graph of a differentiable equation $r = f(\theta)$. It turns out that ψ (psi), *the angle from the radius vector OP to the tangent* at P (Fig. 12-74), is much easier to work with than the inclination, α, of the tangent line; and if the direction of the tangent at P is wanted, its slope, $\tan \alpha$, is easily found from $\tan \psi$ because of the relation

(1) $$\alpha = (\theta + \psi) - n\pi.$$

Consider a polar curve whose equation is

$$r = f(\theta),$$

where f is a differentiable function of θ. We seek a formula for $\tan \psi$ in terms of θ.

If a rectangular Cartesian coordinate system is superposed on the polar system so that the origin is the pole and the positive x axis is the polar axis, then (12.11.1) the two sets of coordinates of any point on the curve $r = f(\theta)$ are connected by the equations

Angle ψ from radius
vector to tangent line

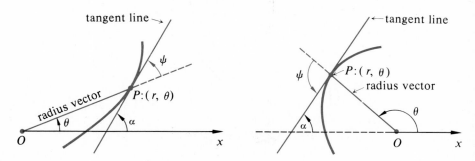

Figure 12-74

$$
\begin{aligned}
x &= r \cos \theta = f(\theta) \cos \theta, \\
y &= r \sin \theta = f(\theta) \sin \theta.
\end{aligned}
$$
(2)

Here x and y are functions of θ; and by taking differentials, we have

$$
\begin{aligned}
dx &= [-f(\theta) \sin \theta + f'(\theta) \cos \theta] \, d\theta, \\
dy &= [f(\theta) \cos \theta + f'(\theta) \sin \theta] \, d\theta.
\end{aligned}
$$
(3)

From (1), $\psi = (\alpha - \theta) + n\pi$; and $\tan \psi = \tan [(\alpha - \theta) + n\pi] = \tan (\alpha - \theta)$.
Thus

$$
\tan \psi = \frac{\tan \alpha - \tan \theta}{1 + \tan \alpha \tan \theta}.
$$
(4)

But

$$
\tan \alpha = \frac{dy}{dx} \quad \text{and} \quad \tan \theta = \frac{\sin \theta}{\cos \theta}.
$$
(5)

From (4) and (5),

$$
\tan \psi = \frac{\dfrac{dy}{dx} - \dfrac{\sin \theta}{\cos \theta}}{1 + \dfrac{\sin \theta}{\cos \theta} \dfrac{dy}{dx}},
$$

or

$$
\tan \psi = \frac{\cos \theta \, dy - \sin \theta \, dx}{\cos \theta \, dx + \sin \theta \, dy}.
$$
(6)

From (3),

$$
\begin{aligned}
\cos \theta \, dy - \sin \theta \, dx &= [f(\theta) \cos^2 \theta + f'(\theta) \sin \theta \cos \theta \\
&\quad + f(\theta) \sin^2 \theta - f'(\theta) \sin \theta \cos \theta] \, d\theta \\
&= f(\theta) (\sin^2 \theta + \cos^2 \theta) \, d\theta = f(\theta) \, d\theta.
\end{aligned}
$$

So

$$
\cos \theta \, dy - \sin \theta \, dx = f(\theta) \, d\theta.
$$
(7)

Similarly,

(8) $$\cos \theta \, dx + \sin \theta \, dy = f'(\theta) \, d\theta.$$

From (6), (7), and (8), we have

$$\tan \psi = \frac{f(\theta)}{f'(\theta)},$$

which proves the following theorem.

12.14.1 Theorem. Let $P:(r, \theta)$, $r \neq 0$, be a point on the curve whose polar equation is $r = f(\theta)$, and let ψ be the angle from the radius vector OP to the tangent to the curve at P. If $f'(\theta)$ exists and is not zero, then

$$\tan \psi = \frac{f(\theta)}{f'(\theta)}.$$

Example 1. Find the slope of the tangent to the cardioid $r = 1 + \sin \theta$ (Fig. 12-75) at the point $(1 + \sqrt{2}/2, 3\pi/4)$.

Solution. We seek $\tan \alpha$, the slope of the tangent line to the curve at P. Here $\alpha = (\theta + \psi) - \pi$, and $\tan[(\theta + \psi) - \pi] = \tan(\theta + \psi)$. Thus

$$\tan \alpha = \tan(\theta + \psi) = \frac{\tan \theta + \tan \psi}{1 - \tan \theta \tan \psi}.$$

From 12.14.1, $\tan \psi = f(\theta)/f'(\theta) = (1 + \sin \theta)/\cos \theta$. Since $\sin \frac{3}{4}\pi = \frac{1}{2}\sqrt{2}$ and $\cos \frac{3}{4}\pi = -\frac{1}{2}\sqrt{2}$,

$$\tan \psi = \frac{1 + \frac{1}{2}\sqrt{2}}{-\frac{1}{2}\sqrt{2}} = -(1 + \sqrt{2}).$$

Moreover, $\tan \theta = \tan \frac{3}{4}\pi = -1$. Therefore

$$\tan \alpha = \frac{-1 - (1 + \sqrt{2})}{1 - (1 + \sqrt{2})} = 1 + \sqrt{2}.$$

Example 2. Show that ψ is constant at all points on the logarithmic spiral $r = e^{\theta/2}$ (Fig. 12-76).

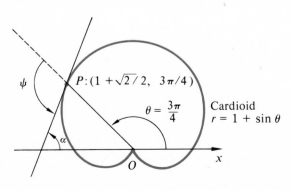

Figure 12-75

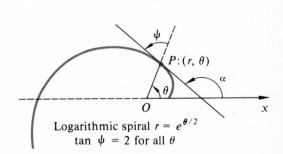

Figure 12-76

Solution.   For all values of θ,

$$\tan \psi = \frac{f(\theta)}{f'(\theta)} = \frac{e^{\theta/2}}{\frac{1}{2}e^{\theta/2}} = 2.$$

In sketching a polar curve that passes one or more times through the pole, it is helpful to know the equations of the tangents to the curve at the pole. A working rule for finding the equations of such tangents was stated, without proof, in Section 12.12. Its justification is the following theorem.

12.14.2   Theorem.   Let $r = f(\theta)$, where f is a differentiable function of θ, be the equation of a polar curve that passes one or more times through the pole. If $\theta_1, \theta_2, \theta_3, \cdots$ are the real values of θ that satisfy $r = f(\theta)$ when $r = 0$, then $\tan \theta_1, \tan \theta_2, \cdots$ (when they exist) are the slopes of the nonvertical tangents to the curve at the pole. If $(0, n\pi/2)$ satisfies the equation of the curve for some odd integer n, the curve also has a vertical tangent at the pole.

Proof.   Denote by α the inclination of a tangent line to the curve $r = f(\theta)$ at the pole. From (3) and (5),

$$(9) \qquad\qquad \tan \alpha = \frac{dy}{dx} = \frac{f(\theta) \cos \theta + f'(\theta) \sin \theta}{-f(\theta) \sin \theta + f'(\theta) \cos \theta}.$$

When $r = 0$, $f(\theta) = 0$, and (9) becomes

$$\tan \alpha = \frac{\sin \theta}{\cos \theta} = \tan \theta,$$

provided that $f'(\theta) \cos \theta \neq 0$. This proves the first part of the theorem.

The proof that the curve has a vertical tangent at the pole when $(0, n\pi/2)$ satisfies its equation is similar and is left as an exercise.   ∎

Exercise

Let $r = f(\theta)$, where f is a differentiable function of θ, be the polar equation of a curve. Prove that if $(0, n\pi/2)$ satisfies this equation for some odd integer n, the curve has a vertical tangent at the pole. [*Hint:* If $\sin \theta \neq 0$,

$$\cot \alpha = \frac{-f(\theta) \sin \theta + f'(\theta) \cos \theta}{f(\theta) \cos \theta + f'(\theta) \sin \theta}.$$

Show that if $(0, n\pi/2)$ satisfies $r = f(\theta)$, then $\cot \alpha = \cot \theta = 0$ and the curve has a vertical tangent at the pole.]

12.15   INTERSECTION OF CURVES IN POLAR COORDINATES

In Cartesian coordinates, *all* the points of intersection of two curves were found when we solved the equations of the curves simultaneously. But in polar coordinates such is often not the case.

The explanation is that a point P has many pairs of polar coordinates, and one pair may satisfy the polar equation of one curve and a different pair may satisfy the

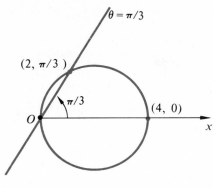

Figure 12-77

polar equation of the other curve. Even though P may be a point of intersection of the two curves, it is possible that no single pair of polar coordinates of P satisfies both polar equations simultaneously.

For instance (Fig. 12-77), the circle $r = 4 \cos \theta$ intersects the line $\theta = \frac{1}{3}\pi$ in *two* points, the pole and $(2, \frac{1}{3}\pi)$, and yet only the latter is a common solution of the two given equations. This is because the coordinates of the pole that satisfy the equation of the line are $(0, \frac{1}{3}\pi)$ and the coordinates of the pole that satisfy the equation of the circle are $(0, \frac{1}{2}\pi)$.

Thus, in order *to find all the intersections of two curves* whose polar equations are given, solve the equations simultaneously and then graph the two curves carefully so as to discover other points of intersection whose polar coordinates did not appear in the common solution of the two given equations. Since the pole presents particular difficulties, set $r = 0$ in both equations to determine whether both curves go through the pole.

Example. Find the points of intersection of the two cardioids, $r = 1 + \cos \theta$ and $r = 1 - \sin \theta$ (Fig. 12-78).

Solution. If we eliminate r between the two given equations, we get $\cos \theta = -\sin \theta$, or $\tan \theta = -1$. Thus $\theta = \frac{3}{4}\pi$ or $\frac{7}{4}\pi$. By substituting $\theta = \frac{3}{4}\pi$ in the first equation, we have $r = \frac{1}{2}(2 - \sqrt{2})$; thus the point $(\frac{1}{2}(2 - \sqrt{2}), \frac{3}{4}\pi)$ is on the first cardioid. When $\theta = \frac{7}{4}\pi$ is substituted in the first equation, we obtain $r = \frac{1}{2}(2 + \sqrt{2})$; so the point $(\frac{1}{2}(2 + \sqrt{2}), \frac{7}{4}\pi)$ is also on the first cardioid.

It is easy to verify that both pairs of coordinates also satisfy the second equation. There-

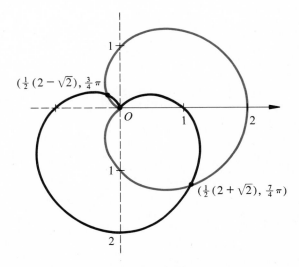

Figure 12-78

fore $(\frac{1}{2}(2 - \sqrt{2}), \frac{3}{4}\pi)$ and $(\frac{1}{2}(2 + \sqrt{2}), \frac{7}{4}\pi)$ are two points of intersection of the given cardioids (Fig. 12-78).

The graphs, however, show that the pole is also a point of intersection of the two cardioids. The reason why this does not appear in the analytic solutions above is that $r = 0$ in $r = 1 + \cos\theta$ when $\theta = \pi$, but $r = 0$ in $r = 1 - \sin\theta$ when $\theta = \frac{1}{2}\pi$.

Exercises

In each of Exercises 1 through 10, find $\tan\psi$ and $\tan\alpha$ for the given curve at the given point. Sketch the curve, showing ψ and α.

1. $r = 6\sin 2\theta;\ \left(3\sqrt{3}, \frac{\pi}{3}\right)$.

2. $r = \dfrac{6}{1 - \cos\theta};\ \left(12, \frac{\pi}{3}\right)$.

3. $r = \dfrac{12}{2 - \cos\theta};\ \left(\frac{24}{5}, \frac{2\pi}{3}\right)$.

4. $r = 3 - 3\sin\theta;\ \left(3 + \frac{3\sqrt{2}}{2}, \frac{-\pi}{4}\right)$.

5. $r = 2 + 4\sin\theta;\ \left(2 - 2\sqrt{3}, \frac{4\pi}{3}\right)$.

6. $r^2 = 16\sin 2\theta;\ \left(2\sqrt{2}, \frac{\pi}{12}\right)$.

7. $r = 4 + 4\cos\theta;\ \left(4 - 2\sqrt{3}, \frac{5\pi}{6}\right)$.

8. $r = 4\cos 3\theta;\ \left(2\sqrt{2}, \frac{3\pi}{4}\right)$.

9. $r = \frac{1}{2}\theta;\ \left(\frac{5\pi}{8}, \frac{5\pi}{4}\right)$.

10. $r = \dfrac{1}{\theta};\ \left(\frac{1}{2}, 2\right)$

11. Show that $\psi = \theta$ at every point on the circle $r = 2c\sin\theta$, $c > 0$. Make a sketch.

12. Show that $\psi = \theta + \pi/2$ at every point on the circle $r = 2c\cos\theta$, $c > 0$. Make a sketch.

In Exercises 13 to 18, find equations of the tangents to the given curves at the pole.

13. $r^2 = 9\sin 2\theta$.

14. $r = 6 - 6\cos\theta$.

15. $r = 2 - 2\sin\theta$.

16. $r = 2\cos 3\theta$.

17. $r = 5\sin 3\theta$.

18. $r = 9\cos 4\theta$.

In each of Exercises 19 to 22, sketch the pair of curves whose equations are given, and find coordinates of their points of intersection.

19. $r = 3\sqrt{3}\cos\theta$, $r = 3\sin\theta$.

20. $r = 5$, $r = \dfrac{5}{1 - 2\cos\theta}$.

21. $r = 6\sin\theta$, $r = \dfrac{6}{1 + 2\sin\theta}$.

22. $r^2 = 4\cos 2\theta$, $r = 2\sqrt{2}\sin\theta$.

23. If the graphs of $r = f_1(\theta)$ and $r = f_2(\theta)$ intersect at a point P, and if ϕ denotes the angle from the first curve to the second curve at P, show that

$$\tan\phi = \frac{\tan\psi_2 - \tan\psi_1}{1 + \tan\psi_1\tan\psi_2}, \qquad \phi \neq \frac{\pi}{2}.$$

24. Use Exercise 23 to find the tangent of the angle ϕ from the curve $r = 6\sin\theta$ to the curve $r = 6/(1 + 2\sin\theta)$ at the point $(3, \pi/6)$.

25. Use Exercise 23 to find the tangent of the angle ϕ from the curve $r^2 = 4\cos 2\theta$ to the curve $r = 2\sqrt{2}\sin\theta$ at the point $(\sqrt{2}, 5\pi/6)$.

26. State a condition for two curves to intersect orthogonally. (*Hint:* See Exercise 23.)

27. Show that for every value of θ the directions of the two curves, $r = \sqrt{3}\cos\theta$ and $r = 3\sin\theta$, are orthogonal. Make a sketch.

28. Find the points on the cardioid $r = 2\cos\theta + 2$ at which the tangent is horizontal. Make a sketch.

29. Find a formula for the slope of the tangent to the parabola

$$r = \frac{d}{1 - \cos\theta}$$

at any point on the parabola except the vertex. Simplify your answer.

30. Use the preceding exercise to prove the reflection property (Section 12.3) of the parabola

$$r = \frac{d}{1 - \cos \theta}.$$

12.16 PLANE AREAS IN POLAR COORDINATES

Areas of certain regions in the polar plane can be found in much the same way as we found areas of regions in the Cartesian plane. The element of area in Cartesian coordinates was a rectangle, but in polar coordinates it is a sector of a circle.

Let f be a function that is continuous and nonnegative on a closed interval $[a, b]$, $b - a \le 2\pi$. We seek the area, A, of the region, R, that is bounded by the half-lines $\theta = a$ and $\theta = b$, and the curve $r = f(\theta)$ (Fig. 12-79).

Partition $[a, b]$ into n subintervals by $n - 1$ numbers, $\theta_1, \theta_2, \cdots, \theta_{n-1}$, so that $a = \theta_0 < \theta_1 < \theta_2 < \cdots < \theta_{n-1} < \theta_n = b$, and let $\Delta \theta_i = \theta_i - \theta_{i-1}$, $i = 1, 2, 3, \cdots,$ n (Fig. 12-80).

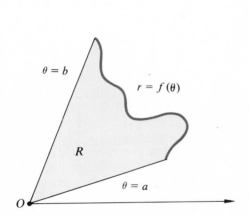

Figure 12-79

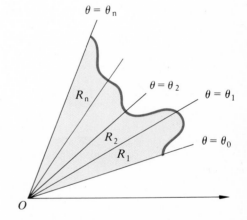

Figure 12-80

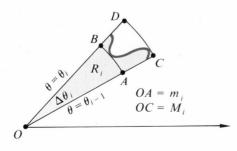

Figure 12-81

Consider any particular subregion, R_i, bounded by the polar rays $\theta = \theta_{i-1}$ and $\theta = \theta_i$, and the curve $r = f(\theta)$ (Fig. 12-81). Denote its area by A_i. Since f is continuous on the closed subinterval $[\theta_{i-1}, \theta_i]$, there are numbers u_i and v_i in that subinterval such that $f(u_i)$ is the minimum of f on $[\theta_{i-1}, \theta_i]$ and $f(v_i)$ is the maximum (by the maximum-minimum theorem). Thus

(1) $$f(u_i) \le f(\theta) \le f(v_i)$$

for all $\theta \in [\theta_{i-1}, \theta_i]$.

Let OAB be the circular sector with radius $\overline{OA} = f(u_i)$, and let OCD be the circular sector having radius $\overline{OC} = f(v_i)$ (Fig. 12-81). Because of (1),

(2) $$\text{sector } OAB \subseteq R_i \subseteq \text{sector } OCD.$$

We recall that the area of a sector of a circle is $\frac{1}{2}r^2\theta$, where r is the radius of the circle and θ is the radian measure of the central angle. Then the area of the sector $OAB = \frac{1}{2}[f(u_i)]^2 \Delta\theta_i$, and the area of the sector $OCD = \frac{1}{2}[f(v_i)]^2 \Delta\theta_i$. It follows from (2) that

(3) $$\frac{1}{2}[f(u_i)]^2 \Delta\theta_i \leq A_i \leq \frac{1}{2}[f(v_i)]^2 \Delta\theta_i.$$

By summing each member of (3) from 1 to n, we get

(4) $$\sum_{i=1}^{n} \frac{1}{2}[f(u_i)]^2 \Delta\theta_i \leq A \leq \sum_{i=1}^{n} \frac{1}{2}[f(v_i)]^2 \Delta\theta_i,$$

in which $A = \sum_{i=1}^{n} A_i$. The first and third members of (4) are Riemann sums for the function f^2 on $[a, b]$; and f^2 is continuous on $[a, b]$ because f is. Thus

(5) $$\lim_{|P| \to 0} \sum_{i=1}^{n} \frac{1}{2}[f(u_i)]^2 \Delta\theta_i = \int_a^b \frac{1}{2}[f(\theta)]^2 \, d\theta$$

and

(6) $$\lim_{|P| \to 0} \sum_{i=1}^{n} \frac{1}{2}[f(v_i)]^2 \Delta\theta_i = \int_a^b \frac{1}{2}[f(\theta)]^2 \, d\theta.$$

It follows from (4), (5), and (6) that

$$A = \frac{1}{2} \int_a^b [f(\theta)]^2 \, d\theta.$$

Therefore if f is continuous and nonnegative on $[a, b]$, the **area** of the region bounded by the curve $r = f(\theta)$ and the half-lines $\theta = a$ and $\theta = b$ is given by

12.16.1 $$A = \frac{1}{2} \int_a^b r^2 \, d\theta.$$

Example 1. Find the area of the region bounded by the limaçon

$$r = 2 + \cos\theta$$

(Fig. 12-82).

Solution. The complete curve is traced once as θ takes on all values in the interval $[0, 2\pi)$. So we could find the desired area by using 12.16.1 and integrating from 0 to 2π. The curve is symmetric with respect to the polar axis, however, and it will suffice to find the area of the region above the polar axis by integrating from 0 to π, and then multiplying this result by 2. Thus

$$\frac{A}{2} = \frac{1}{2} \int_0^\pi r^2 \, d\theta = \frac{1}{2} \int_0^\pi (2 + \cos\theta)^2 \, d\theta.$$

From this,

$$A = \int_0^\pi (2 + \cos\theta)^2 \, d\theta = \int_0^\pi (4 + 4\cos\theta + \cos^2\theta) \, d\theta$$

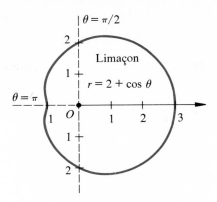

Figure 12-82

$$= 4 \int_0^\pi d\theta + 4 \int_0^\pi \cos\theta \, d\theta + \int_0^\pi \frac{1 + \cos 2\theta}{2} \, d\theta$$

$$= 4 \int_0^\pi d\theta + 4 \int_0^\pi \cos\theta \, d\theta + \frac{1}{2} \int_0^\pi d\theta + \frac{1}{4} \int_0^\pi \cos 2\theta \, d(2\theta)$$

$$= 4\theta \Big]_0^\pi + 4 \sin\theta \Big]_0^\pi + \frac{\theta}{2} \Big]_0^\pi + \frac{1}{4} \sin 2\theta \Big]_0^\pi$$

$$= 4\pi + \frac{\pi}{2} = \frac{9\pi}{2}.$$

The area of the region enclosed by the given limaçon is $\frac{9}{2}\pi$ square units.

Example 2. Find the area of the region enclosed by one leaf of the four-leaved rose

$$r = 4 \sin 2\theta$$

(Fig. 12-63).

Solution. Since the leaf in the first quadrant is traced once as θ assumes all values from 0 to $\frac{1}{2}\pi$, the area of the region enclosed by the leaf is

$$A = \frac{1}{2} \int_0^{\pi/2} (4 \sin 2\theta)^2 \, d\theta = 8 \int_0^{\pi/2} \sin^2 2\theta \, d\theta$$

$$= 8 \int_0^{\pi/2} \frac{1 - \cos 2(2\theta)}{2} \, d\theta$$

$$= 4 \int_0^{\pi/2} d\theta - \int_0^{\pi/2} \cos 4\theta \, d(4\theta)$$

$$= 4\theta \Big]_0^{\pi/2} - \sin 4\theta \Big]_0^{\pi/2} = 2\pi.$$

Example 3. Find the area of the region enclosed by the lemniscate $r^2 = 8 \cos 2\theta$ (Fig. 12-83).

Solution. The given equation can be written $r = \pm 2\sqrt{2 \cos 2\theta}$, which defines two functions of θ. Each function has the whole lemniscate for its graph, so we may confine our attention to the function defined by

(7) $r = 2\sqrt{2 \cos 2\theta}.$

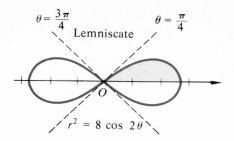

$\theta = \dfrac{3\pi}{4}$ ... $\theta = \dfrac{\pi}{4}$

Lemniscate

O

$r^2 = 8 \cos 2\theta$

Figure 12-83

Because of the radical sign, all values of θ that make $\cos 2\theta$ negative must be excluded. Consequently, we exclude $\frac{1}{4}\pi < \theta < \frac{3}{4}\pi$ and $\frac{5}{4}\pi < \theta < \frac{7}{4}\pi$, and choose $[-\frac{1}{4}\pi, \frac{1}{4}\pi] \cup [\frac{3}{4}\pi, \frac{5}{4}\pi]$ for the domain of (7). On $[-\frac{1}{4}\pi, \frac{1}{4}\pi]$ the right-hand loop is traced by (7), and on $[\frac{3}{4}\pi, \frac{5}{4}\pi]$ the left-hand loop is traced.

Since the lemniscate is symmetric with respect to the polar axis and the $\frac{1}{2}\pi$ axis, it will be sufficient to find the area in the first quadrant and multiply it by 4. The graph of (7) for $0 \le \theta \le \frac{1}{4}\pi$ is the part of the lemniscate in the first quadrant, so one-fourth of the area is given by

$$\frac{A}{4} = \frac{1}{2} \int_0^{\pi/4} [2\sqrt{2 \cos 2\theta}]^2 \, d\theta$$

$$= 4 \int_0^{\pi/4} \cos 2\theta \, d\theta = 2 \int_0^{\pi/4} \cos 2\theta \, d(2\theta)$$

$$= 2 \sin 2\theta \Big]_0^{\pi/4} = 2(1 - 0) = 2.$$

Therefore the area of the region enclosed by the whole lemniscate is eight square units.

We can find the area of a region bounded by two polar curves and, possibly, by one or two radius vectors, by subtracting areas, just as we did in Cartesian coordinates.

Example 4. Find the area of the region that is inside the circle $r = \sin \theta$ and outside the cardioid $r = 1 + \cos \theta$ (Fig. 12-84).

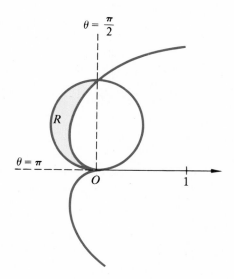

$\theta = \dfrac{\pi}{2}$

$\theta = \pi$

R

O ... 1

Figure 12-84

Solution. To find coordinates of the points of intersection of the two curves, we solve their equations simultaneously. By eliminating r between the equations, we get $\sin \theta = 1 + \cos \theta$. If we square both members of this equation and then replace $\sin^2 \theta$ by $1 - \cos^2 \theta$, we obtain $1 - \cos^2 \theta = 1 + 2 \cos \theta + \cos^2 \theta$, or $\cos \theta (1 + \cos \theta) = 0$. Thus $\cos \theta = 0$ or $\cos \theta = -1$. When $\cos \theta = 0$, $\theta = \frac{1}{2}\pi$ or $\theta = \frac{3}{2}\pi$; when $\cos \theta = -1$, $\theta = \pi$. By substituting these values of θ in both the given equations, we find that the points of intersection of the curves are $(1, \frac{1}{2}\pi)$ and $(0, \pi)$.

As θ increases from $\frac{1}{2}\pi$ to π, the semicircle that is the outer boundary of region R is traced, and also the part of the cardioid that forms the inner boundary of R is traced. Thus the desired area A is the area of the region

$$R = \{P{:}(r, \theta) \mid \tfrac{1}{2}\pi \le \theta \le \pi, 1 + \cos \theta \le r \le \sin \theta\}.$$

Therefore,

$$A = \frac{1}{2} \int_{\pi/2}^{\pi} \sin^2 \theta \, d\theta - \frac{1}{2} \int_{\pi/2}^{\pi} (1 + \cos \theta)^2 \, d\theta$$

$$= \frac{1}{2} \int_{\pi/2}^{\pi} [\sin^2 \theta - (1 + \cos \theta)^2] \, d\theta$$

$$= \frac{1}{2} \int_{\pi/2}^{\pi} [-(\cos^2 \theta - \sin^2 \theta) - 1 - 2 \cos \theta] \, d\theta$$

$$= \frac{1}{2} \int_{\pi/2}^{\pi} -(\cos 2\theta + 1 + 2 \cos \theta) \, d\theta$$

$$= -\frac{1}{4} \int_{\pi/2}^{\pi} \cos 2\theta \, d(2\theta) - \frac{1}{2} \int_{\pi/2}^{\pi} (1 + 2 \cos \theta) \, d\theta$$

$$= -\frac{1}{4} \sin 2\theta \Big]_{\pi/2}^{\pi} - \frac{1}{2}(\theta + 2 \sin \theta) \Big]_{\pi/2}^{\pi}$$

$$= 1 - \frac{1}{4}\pi \text{ square unit.}$$

Another way to find the definite integral for such an area is to take as the element of area that part of a circular sector that remains when a smaller sector with the same central angle is subtracted from it, as illustrated in Fig. 12-85. If we let $r_2 = \sin \theta$ and $r_1 = 1 + \cos \theta$, the "truncated sector" element of area is

$$\frac{1}{2} r_2^2 \, \Delta\theta - \frac{1}{2} r_1^2 \, \Delta\theta = \frac{1}{2}(r_2^2 - r_1^2) \, \Delta\theta.$$

Then the definite integral for A is

$$A = \frac{1}{2} \int_{\pi/2}^{\pi} (r_2^2 - r_1^2) \, d\theta$$

$$= \frac{1}{2} \int_{\pi/2}^{\pi} [\sin^2 \theta - (1 + \cos \theta)^2] \, d\theta$$

$$= 1 - \frac{1}{4}\pi \text{ square unit,}$$

as we saw above.

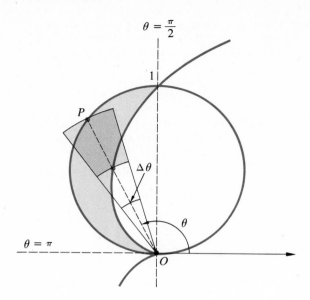

$\theta = \dfrac{\pi}{2}$

$\theta = \pi$

Figure 12-85

Exercises

In each of Exercises 1 to 10, sketch the graph of the given equation and find the area of the region bounded by it.

1. $r = a$, $a > 0$.

2. $r = 2a \sin \theta$, $a > 0$.

3. $r = 3 + \cos \theta$.

4. $r = 4 + 3 \cos \theta$.

5. $r = 4 - 4 \cos \theta$.

6. $r = 7 - 7 \sin \theta$.

7. $r = a(1 + \sin \theta)$, $a > 0$.

8. $r^2 = 5 \cos 2\theta$.

9. $r^2 = 4 \sin 2\theta$.

10. $r^2 = a \sin 2\theta$, $a > 0$.

11. Sketch the limaçon $r = 2 - 4 \cos \theta$, and find the area of the region inside its small loop.

12. Sketch the limaçon $r = 3 - 6 \sin \theta$, and find the area of the region inside its small loop.

13. Sketch the limaçon $r = 2 - 4 \sin \theta$, and find the area of the region inside its large loop.

14. Sketch one leaf of the four-leaved rose $r = 3 \cos 2\theta$ and find the area of the region enclosed by it.

15. Sketch the three-leaved rose $r = 4 \cos 3\theta$ and find the area of the total region enclosed by it.

16. Sketch the three-leaved rose $r = 2 \sin 3\theta$ and find the area of the region bounded by it.

17. Find the area of the region between the two concentric circles $r = 7$ and $r = 10$.

18. Sketch the region that is inside the circle $r = 3 \sin \theta$ and outside the cardioid $r = 1 + \sin \theta$, and find its area.

19. Sketch the region that is outside the circle $r = 2$ and inside the lemniscate $r^2 = 8 \cos 2\theta$, and find its area.

20. Sketch the limaçon $r = 3 - 6 \sin \theta$ and find

the area of the region that is inside its large loop and outside its small loop.

21. Sketch the region in the first quadrant that is inside the cardioid $r = 3 + 3 \cos \theta$ and outside the cardioid $r = 3 + 3 \sin \theta$, and find its area.

22. Sketch the region in the second quadrant that is inside the cardioid $r = 2 + 2 \sin \theta$ and outside the cardioid $r = 2 + 2 \cos \theta$, and find its area.

12.17 REVIEW EXERCISES

In each of Exercises 1 to 12, name the conic that has the given equation. Find its vertices and foci, and sketch its graph.

1. $y^2 - 6x = 0$.

2. $9x^2 + 4y^2 - 36 = 0$.

3. $r = \dfrac{6}{1 + 2 \cos \theta}$.

4. $25x^2 - 36y^2 + 900 = 0$.

5. $x^2 + 9y = 0$.

6. $x^2 - 4y^2 - 16 = 0$.

7. $r = \dfrac{4}{1 - \cos \theta}$.

8. $r(3 - \sin \theta) = 4$.

9. $9x^2 + 25y^2 - 225 = 0$.

10. $r = \dfrac{5}{2 + 2 \sin \theta}$.

11. $r(2 + \cos \theta) = 3$.

12. $r(2 - 3 \sin \theta) = 6$.

In each of Exercises 13 to 18, find the Cartesian equation of the conic having the given properties. Make a sketch.

13. Vertices $(\pm 4, 0)$ and eccentricity $\frac{1}{2}$.

14. Eccentricity 1, focus $(0, -3)$, and vertex $(0, 0)$.

15. Eccentricity 1, vertex $(0, 0)$, symmetric with respect to the x axis, and passing through the point $(-1, 3)$.

16. Eccentricity $\frac{5}{3}$ and vertices $(0, \pm 3)$.

17. Vertices $(\pm 2, 0)$ and asymptotes $x \pm 2y = 0$.

18. Vertices $(0, \pm 5)$ and passing through the point $(2, 5\sqrt{5}/3)$.

In Exercises 19 through 32, analyze the given equation and sketch its graph.

19. $r = 6 \cos \theta$.

20. $r = \dfrac{5}{\sin \theta}$.

21. $r = 4 \cos 2\theta$.

22. $r = \dfrac{3}{\cos \theta}$.

23. $r = 4$.

24. $r = 5 - 5 \cos \theta$.

25. $r = 4 - 3 \cos \theta$.

26. $r = 2 - 3 \cos \theta$.

27. $\theta = \frac{2}{3}\pi$.

28. $r = 4 \sin 3\theta$.

29. $r^2 = 16 \sin 2\theta$.

30. $r = -4 \sin \theta$.

31. $r = -\theta, \theta \geq 0$.

32. $r = e^{2\theta}, \theta \geq 0$.

In Exercises 33 to 36, reduce the given equation to a standard form by a translation. Then sketch the conic, showing both pairs of axes.

33. $4x^2 + 9y^2 - 24x - 36y + 36 = 0$.

34. $x^2 + 8x + 6y + 28 = 0$.

35. $3x^2 - 10y^2 + 36x - 20y + 68 = 0$.

36. $x^2 - 4y^2 + 6x + 32y - 55 = 0$.

37. Find a Cartesian equation of the graph of
$$r^2 - 6r(\cos \theta + \sin \theta) + 9 = 0,$$
and then sketch the graph.

38. Find a Cartesian equation of the graph of
$$r^2 \cos 2\theta = 9,$$
and then sketch the graph.

39. Find the slope of the tangent to the graph of
$$r = 3 + 3 \cos \theta$$
at the point on the graph where $\theta = \frac{1}{6}\pi$. Make a sketch.

40. Sketch the graphs of

$$r = 5 \sin \theta \quad \text{and} \quad r = 2 + \sin \theta,$$
and find their points of intersection.

41. Find the area of the region bounded by the graph of
$$r = 2 - \cos \theta.$$
Make a sketch.

42. Find the area of the region that is outside the limaçon $r = 2 + \sin \theta$ and inside the circle $r = 5 \sin \theta$. (See Exercise 40.)

13

indeterminate forms. improper integrals

When limits were first discussed, it was shown (3.3.1) that the limit of a quotient of functions at a number c is equal to the quotient of their limits at c, provided that the limit of the denominator function is not zero. That is, if $\lim\limits_{x \to c} f(x)$ and $\lim\limits_{x \to c} g(x)$ exist, and if $\lim\limits_{x \to c} g(x) \neq 0$, then

(1)
$$\lim_{x \to c} \frac{f(x)}{g(x)} = \frac{\lim\limits_{x \to c} f(x)}{\lim\limits_{x \to c} g(x)}.$$

We shall soon see that even if $\lim\limits_{x \to c} g(x) = 0$ and $\lim\limits_{x \to c} f(x) = 0$,

$$\lim_{x \to c} \frac{f(x)}{g(x)}$$

sometimes exists. Although equation (1) no longer applies, there is a powerful method for finding such limits known as *l'Hôpital's rule*. It will be explained and used in the early part of this chapter.

In the remainder of the chapter the concept of a definite integral will be extended. Our definition of $\int_a^b f(x)\, dx$ assumed that f was defined and bounded on a closed interval $[a, b]$. Under certain circumstances a meaning and value can be given to a definite integral when the interval of integration is infinite.

Another extension of definite integrals is to integrands having one or more infinite discontinuities in $[a, b]$.

This chapter provides essential preparation for the very important chapter on infinite series that follows it.

13.1 CAUCHY'S MEAN VALUE THEOREM

The mean value theorem (6.2.1), which dealt with one function f, was extended to two functions f and g by the French mathematician A. L. Cauchy (1789–1857). Cauchy's mean value theorem is one of those basic theorems in calculus that enable us to prove other, more immediately "practical" theorems.

13.1.1 Cauchy's Mean Value Theorem. If f and g are functions that are continuous on a closed interval $[a, b]$ and differentiable on the open interval (a, b), and if $g'(x) \neq 0$ for all x in (a, b), then there exists a number z between a and b such that

$$\frac{f(b) - f(a)}{g(b) - g(a)} = \frac{f'(z)}{g'(z)}.$$

Proof. Let H be the function defined by

$$H(x) = [f(b) - f(a)]g(x) - [g(b) - g(a)]f(x).$$

Then

$$H(a) = f(b)g(a) - f(a)g(b) = H(b).$$

Since f and g are continuous on $[a, b]$, H is continuous on $[a, b]$. Moreover,

(1)
$$H'(x) = [f(b) - f(a)]g'(x) - [g(b) - g(a)]f'(x)$$

exists on (a, b) because $f'(x)$ and $g'(x)$ exist there.

Thus the function H satisfies the conditions for Rolle's theorem (6.1.4), and so there exists a number z between a and b such that $H'(z) = 0$. By substituting this in (1), we obtain

(2)
$$[f(b) - f(a)]g'(z) = [g(b) - g(a)]f'(z).$$

Now $g'(x) \neq 0$ for all x in (a, b) by hypothesis and thus $g(a) \neq g(b)$, since otherwise Rolle's theorem would ensure the existence of a number z in (a, b) such that $g'(z) = 0$. Therefore (2) can be rewritten

(3)
$$\frac{f(b) - f(a)}{g(b) - g(a)} = \frac{f'(z)}{g'(z)},$$

where z is some number such that $a < z < b$. ∎

Notice that if we let $g(x) = x$, Cauchy's extended mean value theorem becomes our former mean value theorem (6.2.1). Thus 6.2.1 is a special case of the present 13.1.1.

A very simple geometric interpretation of Cauchy's mean value theorem will be given in Section 15.7 with the aid of vectors. It is strikingly similar to that given in Section 6.2 for the original mean value theorem.

13.2 INDETERMINATE FORMS

The function F defined by

(1)
$$F(x) = \frac{x^2 - 3x + 2}{x^2 + 3x - 10}$$

is defined for all numbers x, except $x = 2$ and $x = -5$ where the denominator is zero. But at $x = 2$ the numerator also is zero, and $F(x)$ is said to have the *indeterminate form* $0/0$ at $x = 2$. We cannot "find the value" of this indeterminate form at $x = 2$ because F has no value there.

But $F(x)$ may have a limit as $x \longrightarrow 2$. Some functions that have the indeterminate form $0/0$ at a particular number do have a limit there, whereas others do not. The function F defined in (1) has a limit at 2 that is easy to find. Thus

$$\lim_{x \to 2} F(x) = \lim_{x \to 2} \frac{x^2 - 3x + 2}{x^2 + 3x - 10} = \lim_{x \to 2} \frac{(x - 2)(x - 1)}{(x - 2)(x + 5)}.$$

Now

$$F(x) = \frac{(x - 2)(x - 1)}{(x - 2)(x + 5)} = \frac{x - 1}{x + 5} \qquad \text{if} \qquad x \neq 2 \quad \text{and} \quad x \neq -5;$$

and since $\lim_{x \to 2} F(x)$ has nothing to do with the value of F at 2, we can write

$$\lim_{x \to 2} F(x) = \lim_{x \to 2} \frac{(x - 2)(x - 1)}{(x - 2)(x + 5)} = \lim_{x \to 2} \frac{x - 1}{x + 5} = \frac{1}{7}.$$

In the last step we used the limit of a quotient theorem (3.3.1). Recall that this theorem requires that the limit of the denominator be a number different from zero and thus could not have been used to find the limit of (1) directly.

On the other hand, the function G defined by

(2)
$$G(x) = \frac{x^2 + 3x - 10}{x^2 - 4x + 4},$$

which also has the indeterminate form $0/0$ at 2, fails to have a limit as $x \longrightarrow 2$. For

$$\lim_{x \to 2} G(x) = \lim_{x \to 2} \frac{x^2 + 3x - 10}{x^2 - 4x + 4} = \lim_{x \to 2} \frac{x + 5}{x - 2}$$

$$= \lim_{x \to 2} \left(1 + \frac{7}{x - 2} \right) = 1 + \lim_{x \to 2} \frac{7}{x - 2} = \infty.$$

Finding the limit of an indeterminate form is really not new to us. In Section 10.6 we showed that

$$\lim_{x \to 0} \frac{\sin x}{x} = 1 \qquad \text{and} \qquad \lim_{x \to 0} \frac{1 - \cos x}{x} = 0.$$

Moreover, every time we established a formula for differentiation we found the limit of $\Delta y / \Delta x$ as $\Delta x \longrightarrow 0$, and a necessary condition for the limit to exist is that Δy approach zero as $\Delta x \longrightarrow 0$.

13.2.1 **Definition.** If f and g are functions such that $\lim_{x \to c} f(x) = \lim_{x \to c} g(x) = 0$, the function defined by $f(x)/g(x)$ is said to have the **indeterminate form** 0/0 at c.

There is another type of indeterminate form that is symbolized by ∞/∞. As an illustration, $\lim_{x \to 0} (-\ln |x|) = \infty$ and $\lim_{x \to 0} \cot |x| = \infty$; for this reason, the function G defined by

$$G(x) = \frac{-\ln |x|}{\cot |x|}$$

is said to have the indeterminate form ∞/∞ at $x = 0$. Although $G(x)$ does not exist at $x = 0$, it so happens that

$$\lim_{x \to 0} G(x) = \lim_{x \to 0} \frac{-\ln |x|}{\cot |x|} = 0,$$

but we cannot find this by the methods already familiar to us.

13.2.2 **Definition.** If f and g are functions such that $\lim_{x \to c} f(x) = \pm \infty$ and $\lim_{x \to c} g(x) = \pm \infty$, then the function defined by $f(x)/g(x)$ is said to have the **indeterminate form** $\pm\infty/\infty$ at c.

The next two sections will be devoted to a powerful method, called **l'Hôpital's rule**, for finding the limits of functions at points where they have an indeterminate form.

13.3 L'HÔPITAL'S RULES

The French mathematician G. F. A. de l'Hôpital (1661–1704) wrote the first calculus textbook. In it he published a method for finding the limit, if any, of a quotient of functions when both the numerator and the denominator approach zero. It came to him from his teacher, Johann Bernoulli.

13.3.1 **Theorem** (*l'Hôpital's Rule*). Let f and g be functions that are differentiable in an open interval I containing the point c, except possibly at c itself, and let $g'(x) \neq 0$ for all $x \neq c$ in I. If $\lim_{x \to c} f(x) = \lim_{x \to c} g(x) = 0$, and if

$$\lim_{x \to c} \frac{f'(x)}{g'(x)} = L,$$

then

$$\lim_{x \to c} \frac{f(x)}{g(x)} = L.$$

Proof. We did not assume that the functions f and g were defined at c; so we now define new functions F and G as follows.

(1)
$$F(x) = f(x) \quad \text{for} \quad x \neq c, \quad \text{and} \quad F(c) = 0;$$
$$G(x) = g(x) \quad \text{for} \quad x \neq c, \quad \text{and} \quad G(c) = 0.$$

Then F and G satisfy the hypotheses of Cauchy's mean value theorem (13.1.1).

Let x be a point in I different from c.

CASE 1. $c < x$. If we let $a = c$ and $b = x$ in Cauchy's mean value theorem, we have

$$\frac{F(x) - F(c)}{G(x) - G(c)} = \frac{F'(z)}{G'(z)},$$

or, using (1),

$$\frac{f(x)}{g(x)} = \frac{f'(z)}{g'(z)},$$

where z is some number in (c, x). Therefore

$$\lim_{x \to c^+} \frac{f(x)}{g(x)} = \lim_{x \to c^+} \frac{f'(z)}{g'(z)}.$$

Since $\lim_{x \to c} [f'(x)/g'(x)] = L$ by hypothesis, then $\lim_{x \to c^+} [f'(x)/g'(x)] = L$; and since z is always between c and x, it follows that $\lim_{x \to c^+} [f'(z)/g'(z)] = L$. Consequently,

(2)
$$\lim_{x \to c^+} \frac{f(x)}{g(x)} = \lim_{x \to c^+} \frac{f'(x)}{g'(x)} = L.$$

CASE 2. $x < c, \; x \in I$. If $b = c$ and $a = x$, the equation in Cauchy's mean value theorem becomes

$$\frac{F(c) - F(x)}{G(c) - G(x)} = \frac{F'(z)}{G'(z)}$$

or

$$\frac{F(x) - F(c)}{G(x) - G(c)} = \frac{F'(z)}{G'(z)},$$

where z is some number in (x, c). By means of (1), this can be written

$$\frac{f(x)}{g(x)} = \frac{f'(x)}{g'(x)}.$$

By hypothesis, $\lim_{x \to c} [f'(x)/g'(x)] = L$, and since c is always between x and c, this implies that $\lim_{x \to c} [f'(z)/g'(z)] = L$. Thus

(3)
$$\lim_{x \to c^-} \frac{f(x)}{g(x)} = \lim_{x \to c^-} \frac{f'(x)}{g'(x)} = L.$$

Therefore, from (2) and (3),

$$\lim_{x \to c} \frac{f'(x)}{g'(x)} = L \implies \lim_{x \to c} \frac{f(x)}{g(x)} = L. \quad \blacksquare$$

Example 1.   Find $\displaystyle \lim_{x \to 0} \frac{\tan 2x}{\ln (1 + x)}$.

Solution. $\lim\limits_{x\to 0} \tan 2x = \lim\limits_{x\to 0} \ln(1+x) = 0$. If we apply l'Hôpital's rule, we get

$$\lim_{x\to 0} \frac{\tan 2x}{\ln(1+x)} = \lim_{x\to 0} \frac{2\sec^2 2x}{1/(1+x)} = \lim_{x\to 0} [2(1+x)\sec^2 2x] = 2.$$

Sometimes when we apply l'Hôpital's rule to $f(x)/g(x)$ at a point c, we find that $f'(x)/g'(x)$ also has the indeterminate form $0/0$ at c. In this case, we apply l'Hôpital's rule again, this time to $f'(x)/g'(x)$.

Example 2. Find $\lim\limits_{x\to 0} \dfrac{\sin x - x}{x^3}$.

Solution. By l'Hôpital's rule,

$$\lim_{x\to 0} \frac{\sin x - x}{x^3} = \lim_{x\to 0} \frac{\cos x - 1}{3x^2}.$$

But $(\cos x - 1)/3x^2$ is also indeterminate, of the form $0/0$, at $x = 0$. If we again apply l'Hôpital's rule, we get

$$\lim_{x\to 0} \frac{\cos x - 1}{3x^2} = \lim_{x\to 0} \frac{-\sin x}{6x},$$

and $(-\sin x)/6x$ is also indeterminate at $x = 0$. But a third application of l'Hôpital's rule yields

$$\lim_{x\to 0} \frac{-\sin x}{6x} = \lim_{x\to 0} \frac{-\cos x}{6} = -\frac{1}{6}.$$

Therefore

$$\lim_{x\to 0} \frac{\sin x - x}{x^3} = -\frac{1}{6}.$$

Several variations on l'Hôpital's rule hold true. In 13.3.1 all limits can be right-hand limits or all limits can be left-hand limits and the theorem remains valid. Also, c or L can be replaced by $\pm\infty$ in 13.3.1 without affecting the validity of the theorem. The proofs of all these variations on l'Hôpital's rule are analogous to that of 13.3.1.

A second rule of l'Hôpital, which applies to ∞/∞ indeterminate forms, is stated below. Its proof is more difficult and is left for a later course.

13.3.2 **Theorem** (*l'Hôpital's Second Rule*). Let f and g be functions that are differentiable in an open interval I containing the point c, except at c itself, and let $g'(x) \neq 0$ for all $x \neq c$ in I. If $\lim\limits_{x\to c} |f(x)| = \lim\limits_{x\to c} |g(x)| = \infty$, and if

$$\lim_{x\to c} \frac{f'(x)}{g'(x)} = L,$$

then

$$\lim_{x\to c} \frac{f(x)}{g(x)} = L.$$

Again, all the limits in 13.3.2 can be replaced by right-hand limits or by left-hand limits without affecting the validity of the theorem. Also, c or L can be $+\infty$ or $-\infty$ and both rules remain true.

Example 3. Find $\lim\limits_{x \to 0} \dfrac{-\ln |x|}{\cot |x|}$.

Solution. Since $\lim\limits_{x \to 0} (-\ln |x|) = \lim\limits_{x \to 0} \cot |x| = \infty$, the given quotient has the indeterminate form ∞/∞ at 0. By differentiating numerator and denominator and applying l'Hôpital's second rule, we find

$$\lim_{x \to 0} \frac{-\ln |x|}{|\cot x|} = \lim_{x \to 0} \frac{-1/|x|}{\csc^2 |x|} = \lim_{x \to 0} \left[-\sin |x| \left(\frac{\sin |x|}{|x|} \right) \right] = 0(1) = 0.$$

Example 4. Find $\lim\limits_{x \to \infty} \dfrac{2^x}{x^2}$ if it exists.

Solution. Since $\lim\limits_{x \to \infty} 2^x = \lim\limits_{x \to \infty} x^2 = \infty$, we apply l'Hôpital's second rule (twice):

$$\lim_{x \to \infty} \frac{2^x}{x^2} = \lim_{x \to \infty} \frac{2^x \ln 2}{2x} = \lim_{x \to \infty} \frac{2^x (\ln 2)^2}{2} = \infty.$$

Therefore $\lim\limits_{x \to \infty} \dfrac{2^x}{x^2}$ does not exist.

Exercises

Find the following limits

1. $\lim\limits_{x \to 0} \dfrac{\sin x - 2x}{x}$.

2. $\lim\limits_{x \to \pi/2} \dfrac{\cos x}{x - \frac{1}{2}\pi}$.

3. $\lim\limits_{x \to \infty} \dfrac{\ln x^{100}}{x}$.

4. $\lim\limits_{x \to 0^+} \dfrac{\ln x}{1/x}$.

5. $\lim\limits_{x \to \infty} \dfrac{\ln x}{x^{10}}$.

6. $\lim\limits_{x \to 0} \dfrac{x - 2\sin x}{\tan x}$.

7. $\lim\limits_{x \to 3} \dfrac{2x^2 - x - 15}{3x^2 - 8x - 3}$.

8. $\lim\limits_{x \to \infty} \dfrac{x \ln x}{x^2 + 1}$.

9. $\lim\limits_{x \to 0} \dfrac{\operatorname{Sin}^{-1} x}{3 \operatorname{Tan}^{-1} x}$.

10. $\lim\limits_{x \to 0} \dfrac{\tan x}{3x}$.

11. $\lim\limits_{x \to 0^+} \dfrac{2\sin x}{\sqrt{x}}$.

12. $\lim\limits_{x \to \infty} \dfrac{\ln x}{x}$.

13. $\lim\limits_{x \to \infty} \dfrac{\ln x}{a^x}$.

14. $\lim\limits_{x \to \pi/2} \dfrac{\ln \sin x}{\frac{1}{2}\pi - x}$.

15. $\lim\limits_{x \to \infty} \dfrac{10^x}{x^{10}}$.

16. $\lim\limits_{x \to 0} \dfrac{e^x - e^{-x}}{4 \sin x}$.

17. $\lim\limits_{x \to 0} \dfrac{\ln \cos 3x}{2x^2}$.

18. $\lim\limits_{x \to \infty} \dfrac{x^{250}}{e^x}$.

19. $\lim\limits_{x \to 0} \dfrac{3 - \csc x}{7 + \cot x}$.

20. $\lim\limits_{x \to 0} \dfrac{\sinh 3x}{2 \sin x}$.

21. $\lim\limits_{t \to 1} \dfrac{\sqrt{t} - t}{\ln t}$.

22. $\lim\limits_{x \to 0^+} \dfrac{8^{\sqrt{x}} - 1}{3^{\sqrt{x}} - 1}$.

23. $\lim\limits_{x\to\infty} \dfrac{2x}{\ln(3x+e^x)}$.

24. $\lim\limits_{x\to0} \dfrac{x\cos x}{\text{Sin}^{-1} x}$.

25. $\lim\limits_{x\to0} \dfrac{\tan x - x}{\sin x - x}$.

26. $\lim\limits_{x\to\pi/2} \dfrac{\sin 10x}{\sin 4x}$.

27. $\lim\limits_{x\to4^+} \dfrac{\sqrt{x}-2}{\sqrt{x-4}}$.

28. $\lim\limits_{x\to1} \dfrac{\text{Cos}^{-1} x}{1-x}$.

29. $\lim\limits_{x\to0^+} \dfrac{\cot x}{\ln x}$.

30. $\lim\limits_{\theta\to\pi^-} \dfrac{2\sec\frac{1}{2}\theta}{\tan\frac{1}{2}\theta}$.

31. $\lim\limits_{x\to0} \dfrac{x - \ln(1+x)}{\cos x - 1}$.

32. $\lim\limits_{x\to\pi/4} \dfrac{1-\tan x}{\frac{1}{4}\pi - x}$.

33. $\lim\limits_{x\to0} \dfrac{\text{Tan}^{-1} x - x}{8x^3}$.

34. $\lim\limits_{x\to0} \dfrac{x^2}{\ln\cos x}$.

35. $\lim\limits_{t\to0} \dfrac{t-\tan t}{\sin t - t}$.

36. $\lim\limits_{x\to0^+} \dfrac{\ln\csc x}{\ln\cot x}$.

13.4 OTHER INDETERMINATE FORMS

If $\lim\limits_{x\to c} f(x) = 0$ and $\lim\limits_{x\to c} g(x) = \infty$, then the function defined by $f(x)g(x)$ is said to have the **indeterminate form $0\cdot\infty$** at c.

If $\lim\limits_{x\to c} f(x) = \lim\limits_{x\to c} g(x) = \infty$, then the function defined by $f(x) - g(x)$ has the **indeterminate form $\infty - \infty$** at c.

When $\lim\limits_{x\to c} f(x) = \lim\limits_{x\to c} g(x) = 0$, the function defined by $f(x)^{g(x)}$ has the **indeterminate form 0^0** at c.

Similar definitions apply to the **indeterminate forms ∞^0 and 1^∞**.

All these indeterminate forms can be reduced to the indeterminate forms $0/0$ and ∞/∞ by algebraic manipulation so that l'Hôpital's rules can be applied.

Example 1. Find $\lim\limits_{x\to\pi/2} (\tan x \cdot \ln\sin x)$.

Solution. Since $\lim\limits_{x\to\pi/2} \tan x = \infty$ and $\lim\limits_{x\to\pi/2} \ln\sin x = 0$, this is an $\infty\cdot 0$ indeterminate form. Rewriting the given expression as the quotient $(\ln\sin x)/\cot x$ and applying l'Hôpital's rule, we obtain

$$\lim_{x\to\pi/2} (\tan x \cdot \ln\sin x) = \lim_{x\to\pi/2} \frac{\ln\sin x}{\cot x} = \lim_{x\to\pi/2} \frac{\frac{1}{\sin x}\cdot\cos x}{-\csc^2 x}$$

$$= \lim_{x\to\pi/2} (-\cos x\cdot\sin x) = 0.$$

Example 2. Find $\lim\limits_{x\to1} \left(\dfrac{x}{x-1} - \dfrac{1}{\ln x}\right)$.

Solution. This is an $\infty - \infty$ indeterminate form. By combining the two fractions, we have

$$\lim_{x\to1} \left(\frac{x}{x-1} - \frac{1}{\ln x}\right) = \lim_{x\to1} \frac{x\ln x - x + 1}{(x-1)\ln x},$$

which has the indeterminate form 0/0 at 1. Applying l'Hôpital's rule twice gives

$$\lim_{x \to 1} \frac{x \ln x - x + 1}{(x - 1) \ln x} = \lim_{x \to 1} \frac{x \ln x}{x - 1 + x \ln x} = \lim_{x \to 1} \frac{1 + \ln x}{2 + \ln x} = \frac{1}{2}.$$

Example 3. Find $\lim_{x \to 0} (1 + x)^{1/x}$.

Solution. Since $\lim_{x \to 0} (1 + x) = 1$ and $\lim_{x \to 0} (1/x) = \infty$, the given function has the indeterminate form 1^∞ at 0.

Let $y = (1 + x)^{1/x}$ and take the natural logarithm of both members; the result is

$$\ln y = \frac{\ln (1 + x)}{x}.$$

Since the right-hand member of this latter equation has the indeterminate form 0/0 at $x = 0$, we apply l'Hôpital's rule and obtain

$$\lim_{x \to 0} \ln y = \lim_{x \to 0} \frac{\ln (1 + x)}{x} = \lim_{x \to 0} \frac{1}{1 + x} = 1.$$

That is, $\lim_{x \to 0} \ln y = 1$. Since $\ln y$ and its inverse are continuous functions, $\ln (\lim_{x \to 0} y) = \lim_{x \to 0} \ln y$ $= 1 = \ln e$. Therefore $\lim_{x \to 0} y = e$ and

$$\lim_{x \to 0} (1 + x)^{1/x} = e,$$

which is an important result that is sometimes taken as the definition of the number e (see Section 10.1).

Exercises

Evaluate the following limits.

1. $\lim_{x \to 0^+} x^x$.

2. $\lim_{x \to 0} (\csc x - \cot x)$.

3. $\lim_{x \to \pi/2} (\sec x - \tan x)$.

4. $\lim_{x \to 0} (2x \ln x^2)$.

5. $\lim_{x \to 0} (2x)^{x^2}$.

6. $\lim_{x \to 0} (\cos x)^{\cot x}$.

7. $\lim_{x \to 0} (x + e^{x/2})^{2/x}$.

8. $\lim_{x \to 0} \left(\csc^2 x - \frac{1}{x^2} \right)$.

9. $\lim_{x \to \pi/2} (\tan x)^{\cos x}$.

10. $\lim_{x \to \infty} x^{1/x}$.

11. $\lim_{x \to \pi/2} [(x - \tfrac{1}{2}\pi) \sec x]$.

12. $\lim_{x \to 0} (x^2 \csc x)$.

13. $\lim_{x \to 0^+} x \ln x$.

14. $\lim_{x \to 0} (\cos x)^{1/x^2}$.

15. $\lim_{x \to 0} (\cot x)^x$.

16. $\lim_{x \to 0^+} (\sin x)^x$.

17. $\lim_{x \to 0} (\cos x - \sin x)^{1/x}$.

18. $\lim_{x \to \pi/2} (\cos x)^{x - \pi/2}$.

19. $\lim_{x \to 0} (x^2 + 3x + 1)^{2/(3x)}$.

20. $\lim_{x \to 0} \left(\csc x - \frac{1}{x} \right)$.

21. $\lim_{x \to 0} (\sin x)^{\sin x}$.

22. $\lim_{x \to \infty} \left(1 + \dfrac{1}{x}\right)^x$.

23. $\lim_{x \to 0} (x^{3/2} \ln x)$.

24. $\lim_{x \to 0} (\cos x)^{1/x}$.

25. $\lim_{x \to 1} \left(\dfrac{1}{x-1} - \dfrac{x}{\ln x}\right)$.

26. $\lim_{x \to \infty} x^{1/x^2}$.

27. $\lim_{x \to 0} (\sinh x)^{\tan x}$.

28. $\lim_{x \to 0^+} [(\tan x) \ln \sin x]$.

29. $\lim_{x \to 0} \left(\dfrac{1}{x} - \dfrac{1}{\tan x}\right)$.

30. $\lim_{x \to \infty} (1 + 2e^x)^{1/x}$.

13.5 INFINITE LIMITS OF INTEGRATION*

In the definition of the definite integral $\int_a^b f(x)\, dx$ the interval $[a, b]$ was assumed to be finite. Is it possible to extend the definition of the definite integral so that the interval of integration is infinite, and, if so, how is the value of such an "integral" to be found?

Consider the definite integral

(1) $$\int_1^b \frac{dx}{x^2} \qquad \text{where} \quad b > 1.$$

We have at once

$$\int_1^b \frac{1}{x^2}\, dx = -\frac{1}{x}\Big]_1^b = 1 - \frac{1}{b}.$$

Clearly, for each value of b greater than 1 the definite integral (1) exists, and we can make the value of (1) as close as we please to 1 by taking b sufficiently large. This is expressed by writing

$$\int_1^\infty \frac{dx}{x^2} = 1.$$

13.5.1 Definition. If f is continuous on the infinite interval $[a, \infty)$, then

$$\int_a^\infty f(x)\, dx = \lim_{t \to \infty} \int_a^t f(x)\, dx,$$

provided that this limit exists. If this limit does exist, the integral is said to be **convergent**; otherwise it is **divergent**.

Definite integrals with infinite limits of integration are included in what are called **improper integrals**.

*The reader is advised to review Section 3.4 and the first part of Section 6.11 before proceeding.

Example 1. Determine whether the improper integral $\int_5^\infty \frac{dx}{x}$ converges or diverges.

Solution. This improper integral diverges because (by 13.5.1)

$$\int_5^\infty \frac{dx}{x} = \lim_{t\to\infty} \int_5^t \frac{dx}{x} = \lim_{t\to\infty} \left[\ln x \right]_5^t = \lim_{t\to\infty} (\ln t - \ln 5) = \lim_{t\to\infty} \ln t - \ln 5 = \infty.$$

Example 2. If possible, find the value of the improper integral

$$\int_1^\infty e^{-x}\, dx.$$

Solution.

$$\int_1^\infty e^{-x}\, dx = \lim_{t\to\infty} \int_1^t e^{-x}\, dx = \lim_{t\to\infty} \int_1^t -e^{-x}\, d(-x)$$

$$= -\lim_{t\to\infty} \int_1^t e^{-x}\, d(-x) = -\lim_{t\to\infty} \left[e^{-x} \right]_1^t$$

$$= -\lim_{t\to\infty} (e^{-t} - e^{-1}) = -(0 - e^{-1}) = \frac{1}{e}.$$

Example 3. If possible, evaluate the improper integral

$$\int_0^\infty xe^{-x^2}\, dx.$$

Solution.

$$\int_0^\infty xe^{-x^2}\, dx = \lim_{t\to\infty} \int_0^t xe^{-x^2}\, dx = \lim_{t\to\infty} -\frac{1}{2} \int_0^\infty e^{-x^2}(-2x)\, dx$$

$$= -\frac{1}{2} \lim_{t\to\infty} \int_0^t e^{-x^2}\, d(-x^2)$$

$$= -\frac{1}{2} \lim_{t\to\infty} (e^{-t^2} - 1) = \frac{1}{2}.$$

Similar definitions apply when the lower limit of integration is infinite and when both limits of integration are infinite.

13.5.2 Definition. If f is continuous on $(-\infty, b]$, then

$$\int_{-\infty}^b f(x)\, dx = \lim_{t\to-\infty} \int_t^b f(x)\, dx,$$

provided that this limit exists. If this limit exists, the improper integral is **convergent;** otherwise it is **divergent**.

13.5.3 Definition. If f is everywhere continuous, then

$$\int_{-\infty}^\infty f(x)\, dx = \int_{-\infty}^c f(x)\, dx + \int_c^\infty f(x)\, dx,$$

where c is an arbitrarily chosen number, provided that both of the improper integrals in the right-hand member are convergent.

Example 4. If the improper integral $\displaystyle\int_{-\infty}^{\infty} \frac{dx}{(x^2 + 1)^2}$ is convergent, evaluate it.

Solution. By means of the trigonometric substitution $x = \tan\theta$, we find

$$\int \frac{dx}{(x^2 + 1)^2} = \frac{x}{2(x^2 + 1)} + \frac{1}{2}\,\mathrm{Tan}^{-1}\,x + C.$$

If we set $c = 0$ in definition 13.5.3, we obtain

$$\int_{-\infty}^{\infty} \frac{dx}{(x^2 + 1)^2} = \int_{-\infty}^{0} \frac{dx}{(x^2 + 1)^2} + \int_{0}^{\infty} \frac{dx}{(x^2 + 1)^2}$$

$$= \lim_{t \to -\infty} \int_{t}^{0} \frac{dx}{(x^2 + 1)^2} + \lim_{s \to \infty} \int_{0}^{s} \frac{dx}{(x^2 + 1)^2}$$

$$= \lim_{t \to -\infty} \left[\frac{x}{2(x^2 + 1)} + \frac{1}{2}\,\mathrm{Tan}^{-1}\,x\right]_{t}^{0} + \lim_{s \to \infty} \left[\frac{x}{2(x^2 + 1)} + \frac{1}{2}\,\mathrm{Tan}^{-1}\,x\right]_{0}^{s}$$

$$= \frac{\pi}{4} + \frac{\pi}{4} = \frac{\pi}{2}.$$

It is instructive to interpret an improper integral as an area. To illustrate, consider two integrals discussed above,

$$\int_{1}^{\infty} \frac{dx}{x} \qquad \text{and} \qquad \int_{1}^{\infty} \frac{dx}{x^2}.$$

The graphs of their integrands are shown in Fig. 13-1. Both graphs have the positive x axis as an asymptote; yet the shaded area in (a) becomes greater than any preassigned number as it extends indefinitely far to the right, while the shaded area in (b) is always less than 1 no matter how far to the right it goes. Neither graph ever intersects the

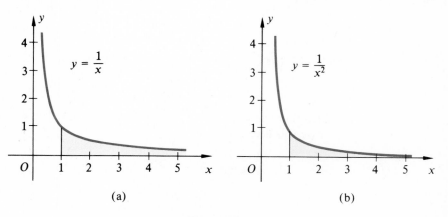

(a)   (b)

Figure 13-1

x axis; but as x increases, the graph of the second approaches the x axis more rapidly than the graph of the first.

13.5.4 Definition.   Let f be continuous on the infinite interval $[a, \infty)$. The **area** A of the region between the curve $y = f(x)$ and the x axis, and to the right of $x = a$, is

$$A = \int_0^\infty |f(x)| \, dx$$

if this improper integral converges.

Similar interpretations hold for other convergent integrals with infinite limits of integration.

Example 5.   Find the area of the region under the curve $y = 1/(x^2 + 1)$ and above the x axis.

Solution.   The region whose area is wanted extends to left and right indefinitely (Fig. 13-2). Thus the area A is given by

$$A = \int_{-\infty}^\infty \frac{dx}{x^2 + 1}.$$

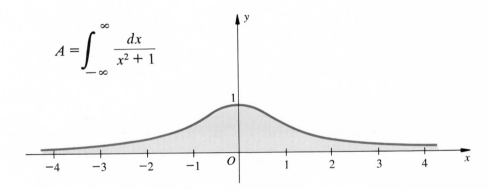

Figure 13-2

We evaluate this improper integral as follows.

$$\int_{-\infty}^\infty \frac{dx}{x^2 + 1} = \int_{-\infty}^0 \frac{dx}{x^2 + 1} + \int_0^\infty \frac{dx}{x^2 + 1}$$

$$= \lim_{t \to -\infty} \int_t^0 \frac{dx}{x^2 + 1} + \lim_{t \to \infty} \int_0^t \frac{dx}{x^2 + 1}$$

$$= \lim_{t \to -\infty} (-\mathrm{Tan}^{-1} t) + \lim_{t \to \infty} \mathrm{Tan}^{-1} t$$

$$= \frac{\pi}{2} + \frac{\pi}{2} = \pi.$$

Hence $A = \pi$ square units.

Exercises

In each of Exercises 1 through 28, evaluate the given improper integral or show that it is divergent.

1. $\displaystyle\int_1^\infty e^x \, dx.$

2. $\displaystyle\int_{-\infty}^\infty \frac{dx}{e^x + e^{-x}}.$

3. $\displaystyle\int_{-\infty}^{-2} \frac{dx}{x^5}.$

4. $\displaystyle\int_4^\infty x e^{-x^2} \, dx.$

5. $\displaystyle\int_{-\infty}^0 \frac{dx}{1 + x^2}.$

6. $\displaystyle\int_0^\infty e^{-x} \cos x \, dx.$

7. $\displaystyle\int_{-\infty}^0 e^{3x} \, dx.$

8. $\displaystyle\int_3^\infty \frac{x \, dx}{\sqrt{9 + x^2}}.$

9. $\displaystyle\int_3^\infty \frac{dx}{x^2}.$

10. $\displaystyle\int_1^\infty \frac{dx}{x^{7/5}}.$

11. $\displaystyle\int_1^\infty \frac{dx}{x^{1.01}}.$

12. $\displaystyle\int_1^\infty \frac{dx}{x^{0.99}}.$

13. $\displaystyle\int_1^\infty \frac{dx}{\sqrt{3x}}.$

14. $\displaystyle\int_{-\infty}^\infty \frac{dx}{x^2 + 2x + 4}.$

15. $\displaystyle\int_2^\infty \frac{dx}{(1 - x)^{2/3}}.$

16. $\displaystyle\int_1^\infty \frac{x \, dx}{(2 + x^2)^{3/2}}.$

17. $\displaystyle\int_2^\infty e^{-x} \sin x \, dx.$

18. $\displaystyle\int_1^\infty \frac{dx}{e^x - e^{-x}}.$

19. $\displaystyle\int_1^\infty \frac{\ln x}{x} \, dx.$

20. $\displaystyle\int_3^\infty \frac{dx}{x(\ln x)^2}.$

21. $\displaystyle\int_0^\infty \frac{\sin x \, dx}{e^x}.$

22. $\displaystyle\int_{-\infty}^{-1} \frac{dx}{x^4}.$

23. $\displaystyle\int_{-4}^\infty e^{-x} \, dx.$

24. $\displaystyle\int_0^\infty x e^{-x} \, dx.$

25. $\displaystyle\int_{-\infty}^\infty \frac{x^2 \, dx}{e^{|x|}}.$

26. $\displaystyle\int_1^\infty \frac{x^2 \, dx}{(2 + x^2)^{3/2}}.$

27. $\displaystyle\int_{-\infty}^\infty \frac{dx}{2x^2 + 2x + 1}.$

28. $\displaystyle\int_0^\infty \frac{dx}{x^{3/2} + 4x^{1/2}}.$

29. Find the area of the region to the right of the line $x = 3$ and between the curve $y = 8/(4x^2 - 1)$ and the x axis. Make a sketch.

30. Find the area of the region between the positive x axis and the curve $y = x/(x^3 + 1)$. Make a sketch.

31. Find the area of the region in the first quadrant and below the curve $y = e^{-x}$.

32. Find the area of the region between the negative x axis and the curve $y = 2xe^x$. Make a sketch.

33. Show that if $a > 0$, then $\displaystyle\int_a^\infty \frac{dx}{x^n}$ converges for all numbers $n > 1$.

34. Extend the definition of volume of a solid of revolution to find the volume of the solid generated by revolving about the x axis the region to the right of the line $x = 1$ and between the curve $y = x^{-3/2}$ and the x axis. Make a sketch.

13.6 BOUNDED SETS. THE COMPLETENESS AXIOM

We say that a set S of real numbers has an **upper bound** U, or is **bounded above**, if there exists a real number U such that $x \leq U$ for all $x \in S$. Thus 4 is an upper bound for the set of negative numbers and so is $\frac{7}{3}$; in fact, any nonnegative number is an upper bound for the set of negative numbers.

13.6.1 Definition.   Let S be a set of real numbers. Then L is the **least upper bound** (abbreviated "l.u.b") of S if
  (i) L is an upper bound of S and
  (ii) $L \leq U$ for every upper bound U of S.

For example, 0 is the least upper bound of the set of negative numbers, and 2 is the least upper bound of the set $\{x \mid x \in R, x^2 < 4\}$.

As an exercise, the reader should write definitions of **lower bound**, **bounded below**, and **greatest lower bound** (g.l.b) parallel to what has been given here.

We now come to an essential property of real numbers.

13.6.2 Axiom of Completeness.   Every nonempty subset of real numbers that is bounded above has a least upper bound that is a real number; and every nonempty subset of real numbers that is bounded below has a greatest lower bound that is a real number.

The importance of this axiom cannot be overemphasized. It will be used later in proving some of our most basic theorems. Right now, we illustrate its use by proving a simple property of positive integers that is familiar to everyone.

Example.   Show that the set N of natural numbers has no upper bound.

Proof.   Assume the contrary—namely, that N is bounded above. Since N is nonempty, N has a least upper bound L (by the completeness axiom). Now $L - 1 < L$; so $L - 1$ cannot be an upper bound for N (because L is the *least* upper bound of N). Thus there exists a natural number n such that $n > L - 1$, which implies that $n + 1 > L$. But $(n + 1) \in N$ because $n \in N$, and so there exists a natural number, $n + 1$, that is greater than the upper bound L of N, a contradiction.

Therefore our assumption that N is bounded above was false, and the theorem is proved.

Exercises

State the least upper bound for each of the following sets and whether the l.u.b is a member of the set. (R is the set of all real numbers and Q is the set of rational numbers.)

(a) $\{x \mid x \in R, x^2 \leq 9\}$.

(b) $\{x \mid x \in R, x^2 \leq 5\}$.

(c) $\{x \mid x \in Q, x^2 \leq 3\}$.

(d) $\{x \mid x \in Q, x^2 \leq 5\}$.

(e) $\left\{x \mid x = \dfrac{2^n - 1}{2^n}, n \in N\right\}$.

(f) $\left\{x \mid x = -2 + \dfrac{8}{n}, n \in N\right\}$.

13.7 COMPARISON TEST FOR CONVERGENCE

In Section 13.5 our evaluation of improper integrals having infinite intervals of integration involved finding an indefinite integral. But some important improper integrals have integrands for which no antiderivative exists that can be expressed in terms of elementary functions (the functions we study in a first course). For instance, the improper integral

$$\int_0^\infty e^{-x^2}\,dx$$

is basic in the study of statistics and probability and yet the indefinite integral of e^{-x^2} cannot be found in terms of elementary functions.

In order to evaluate such integrals by approximation methods, we must be sure that the improper integral converges. Fortunately, there is a test for convergence that can be applied to many such improper integrals. It is called the **comparison test**. In order to prove it we shall first prove another theorem, on bounded monotonic functions.

13.7.1 **Theorem.** Let F be a function that is defined and nondecreasing on the infinite interval $[a, \infty)$. If F is bounded above by U, then $\lim_{x\to\infty} F(x)$ exists and is not greater than U.

Proof. The set of real numbers

$$S = \{F(x) \mid x \geq a\}$$

has an upper bound U by hypothesis. Therefore, by the completeness axiom (13.6.2), S has a least upper bound L.

Let ϵ be an arbitrarily chosen positive number. Then $L - \epsilon < L$, and there exists a number $F(x_1)$ in S that is greater than $L - \epsilon$, since otherwise $L - \epsilon$ would be an upper bound for S that is less than the least upper bound L for S. Therefore

$$L - \epsilon < F(x) \leq L \qquad \text{for all} \quad x > x_1.$$

But this result implies

$$-\epsilon < F(x) - L \leq 0 < \epsilon,$$

or

$$|F(x) - L| < \epsilon$$

for all $x > x_1$. Therefore (by 6.11.1)

$$\lim_{x\to\infty} F(x) = L \leq U. \quad \blacksquare$$

13.7.2 **Definition.** A function g is said to **dominate** a function f on an interval I, finite or infinite, if both functions are defined on I and if $|f(x)| \leq g(x)$ for all x in I.

Example 1. Let $f(x) = e^{-x^2}$ and $g(x) = e^{-x}$, for $x \geq 1$. Show that the function g dominates the function f on the infinite interval $[1, \infty)$ (Fig. 13-3).

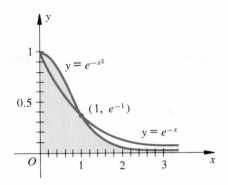

Figure 13-3

Solution. For $x = 1$, $e^{-x^2} = e^{-x} = 1/e^x > 0$. Thus

(1) $$|f(x)| = g(x) \qquad \text{for} \quad x = 1.$$

Let $x > 1$. Then $1 < x \implies 0 < x < x^2 \implies e^x < e^{x^2} \implies e^{-x^2} < e^{-x} \implies |e^{-x^2}| < e^{-x}$. Thus

(2) $$|f(x)| < g(x) \qquad \text{for} \quad x > 1.$$

From (1), (2), and 13.7.2, g dominates f on $[1, \infty)$.

13.7.3 Comparison Test. Let g be a function that dominates a nonnegative-valued function f on the infinite interval $[a, \infty)$ and let f and g be integrable on $[a, b]$ for every $b > a$.

   (i)   If the improper integral $\int_a^\infty g(x)\, dx$ converges, then $\int_a^\infty f(x)\, dx$ converges also.

   (ii)   If $\int_a^\infty f(x)\, dx$ diverges, then $\int_a^\infty g(x)\, dx$ diverges.

Proof. (i) Since f is nonnegative valued and integrable on every interval $[a, b]$, $b > a$, the function F defined by

$$F(x) = \int_a^x f(t)\, dt, \qquad x \geq a,$$

is a nondecreasing function of x. Because $\int_a^x g(t)\, dt \leq \int_a^\infty g(t)\, dt = L$, and g dominates f on $[a, \infty)$,

$$0 \leq \int_a^x f(t)\, dt \leq \int_a^x g(t)\, dt \leq L$$

for all $x \geq a$. Thus

$$0 \leq F(x) \leq L \qquad \text{for all} \quad x \geq a.$$

Therefore $\lim_{x \to \infty} F(x)$ exists (by 13.7.1). That is, $\int_a^\infty f(x)\, dx$ converges.

(ii) From (i), if $\int_a^\infty g(x)\,dx$ converges, then $\int_a^\infty f(x)\,dx$ must also converge. But $\int_a^\infty f(x)\,dx$ diverges. Therefore $\int_a^\infty g(x)\,dx$ diverges. ∎

Example 2. Prove that the improper integral $\int_0^\infty e^{-x^2}\,dx$ converges.

Solution.

$$\text{(3)}\qquad \int_0^\infty e^{-x^2}\,dx = \int_0^1 e^{-x^2}\,dx + \int_1^\infty e^{-x^2}\,dx,$$

provided that both integrals in the right-hand member exist.

Let $f(x) = e^{-x^2}$ and $g(x) = e^{-x}$. Both f and g are continuous and positive valued for all $x \in R$. Since f is continuous on the closed interval $[0, 1]$,

$$\text{(4)}\qquad \int_0^1 e^{-x^2}\,dx = L_1$$

exists.

Because f and g are continuous everywhere, both are integrable on every closed interval $[1, b]$, $b > 1$. Moreover, g dominates f on $[1, \infty)$ as shown in Example 1 above. Since all the conditions for the comparison test are fulfilled and $\int_1^\infty g(x)\,dx$ converges (see Example 2 of Section 13.5), then $\int_1^\infty e^{-x^2}\,dx$ converges also. That is,

$$\text{(5)}\qquad \int_1^\infty e^{-x^2}\,dx = L_2$$

for some number L_2.

From (3), (4), and (5)

$$\int_0^\infty e^{-x^2}\,dx = L_1 + L_2,$$

so $\int_0^\infty e^{-x^2}\,dx$ converges.

Example 3. Show that $\displaystyle\int_1^\infty \frac{dx}{\sqrt{x^3 + x}}$ converges.

Solution. Let $f(x) = 1/\sqrt{x^3 + x}$ and $g(x) = 1/\sqrt{x^3}$, $x \geq 1$. We shall first show that the function g dominates the nonnegative-valued function f on the infinite interval $[1, \infty)$.

Since $x^3 + x > x^3 > 0$ for all $x \geq 1$, then $\sqrt{x^3 + x} > \sqrt{x^3}$. Thus

$$\frac{1}{\sqrt{x^3 + x}} < \frac{1}{\sqrt{x^3}}, \qquad x \geq 1,$$

so g dominates f on $[1, \infty)$.

Because f and g are continuous on $[1, \infty)$, f and g are integrable on $[1, b]$ for every $b > 1$. Consequently, the conditions for using the comparison test are met.

Now

$$\int_1^\infty g(x)\,dx = \int_1^\infty \frac{dx}{\sqrt{x^3}} = \int_1^\infty x^{-3/2}\,dx$$

$$= \lim_{t \to \infty} \int_1^t x^{-3/2} \, dx = \lim_{t \to \infty} \left[\frac{x^{-1/2}}{-\frac{1}{2}} \right]_1^t$$

$$= \lim_{t \to \infty} \left[\frac{-2}{\sqrt{x}} \right]_1^t = \lim_{t \to \infty} \left(\frac{-2}{\sqrt{t}} + 2 \right) = 2.$$

Thus $\int_1^\infty g(x) \, dx$ converges.

Therefore (by the comparison test)

$$\int_1^\infty f(x) \, dx = \int_1^\infty \frac{dx}{\sqrt{x^3 + x}}$$

converges.

Example 4. For what numbers n does the improper integral

$$\int_1^\infty x^n \, dx$$

converge? For what numbers n does it diverge?

Solution. The function F, defined by $F(x) = x^n$, n an arbitrarily chosen real number, is positive valued and continuous on the infinite interval $[1, \infty)$. Thus F is integrable on $[1, b]$ for all $b > 1$.

For $x \geq 1$, $n > -1$ implies that $x^n \geq x^{-1}$. But we saw in Example 1 of Section 13.5 that $\int_1^\infty x^{-1} \, dx$ diverges. Therefore [by part (ii) of the comparison test] the given improper integral diverges for $n \geq -1$.

Next, let $n < -1$.

$$\int_1^\infty x^n \, dx = \lim_{t \to \infty} \int_1^t x^n \, dx = \lim_{t \to \infty} \left[\frac{x^{n+1}}{n+1} \right]_1^t$$

$$= \lim_{t \to \infty} \left(\frac{t^{n+1}}{n+1} - \frac{1}{n+1} \right)$$

$$= \lim_{t \to \infty} \frac{t^{n+1}}{n+1} - \frac{1}{n+1}.$$

Since $n < -1$, $n + 1 < 0$, and we can write $n + 1 = -k$, where $k > 0$. Then

$$\lim_{t \to \infty} \frac{t^{n+1}}{n+1} - \frac{1}{n+1} = \frac{1}{n+1} \lim_{t \to \infty} t^{-k} - \frac{1}{n+1}$$

$$= \frac{1}{n+1} \lim_{t \to \infty} \frac{1}{t^k} - \frac{1}{n+1}$$

$$= \frac{1}{n+1}(0) - \frac{1}{n+1} = -\frac{1}{n+1}.$$

Therefore the given integral converges to $-1/(n+1)$ for all $n < -1$.

To summarize, *the improper integral* $\int_1^\infty x^n \, dx$ *converges when $n < -1$ and diverges when $n \geq -1$.*

Exercises

1. Three functions, f, g, and h, are defined by $f(x) = x^{-3}$, $g(x) = 2/(2x^3 - x^2)$, and $h(x) = 1/(x^3 + 4x^2 + 10)$. Which of these functions dominates which others on the infinite interval $[1, \infty)$?

2. If $f(x) = 1/(e^x + k)$, $k > 0$, and $g(x) = 1/e^x$, show that g dominates f on $[0, \infty)$.

3. Show that g dominates f on $[1, \infty)$ if $f(x) = (x^3 + 1)^{-1/2}$ and $g(x) = x^{-3/2}$.

4. Let $f(x) = x^{-2} \ln x$ and $g(x) = x^{-3/2}$. Show that g dominates f on $[1, \infty)$.

In each of Exercises 5 to 16, show that the given improper integral converges or show that it diverges.

5. $\displaystyle\int_1^\infty \frac{dx}{\sqrt{x^3 + 1}}.$

6. $\displaystyle\int_1^\infty \frac{|\sin x|}{x^2}\, dx.$

7. $\displaystyle\int_0^\infty \frac{dx}{2 + e^x}.$

8. $\displaystyle\int_{-\infty}^\infty e^{-x^4}\, dx.$ (*Hint:* See Exercise 14 of Section 8.5.)

9. $\displaystyle\int_1^\infty \frac{dx}{\sqrt{x}}.$

10. $\displaystyle\int_1^\infty \frac{dx}{\sqrt{x^6 + x^4 + 1}}.$

11. $\displaystyle\int_1^\infty \frac{dx}{\sqrt{x^4 + 2x}}.$

12. $\displaystyle\int_1^\infty x^2 e^{-x^2}\, dx.$

13. $\displaystyle\int_1^\infty \frac{dx}{\sqrt{2x - 1}}.$

14. $\displaystyle\int_0^\infty x e^{-x}\, dx.$

15. $\displaystyle\int_1^\infty \frac{x^2 e^{-x}}{x + 3}\, dx.$

16. $\displaystyle\int_1^\infty \frac{\ln x}{x^2}\, dx.$

13.8 INFINITE INTEGRANDS

Another type of improper integral has finite limits of integration, but its integrand becomes infinite at one or more points in the interval of integration.

13.8.1 Definition. Let f be continuous on the half-open interval $[a, b)$ and let $\lim\limits_{x \to b^-} f(x) = \pm\infty$. Then

$$\int_a^b f(x)\, dx = \lim_{t \to b^-} \int_a^t f(x)\, dx,$$

provided that this limit exists.

Example 1. Evaluate, if possible, the improper integral $\displaystyle\int_0^2 \frac{dx}{\sqrt{4 - x^2}}.$

Solution. The integrand is continuous on $[0, 2)$, but $\lim\limits_{x \to 2^-} [1/\sqrt{4 - x^2}] = \infty$. By 13.8.1,

$$\int_0^2 \frac{dx}{\sqrt{4 - x^2}} = \lim_{t \to 2^-} \int_0^t \frac{dx}{\sqrt{4 - x^2}} = \lim_{t \to 2^-} \left[\operatorname{Sin}^{-1} \frac{x}{2} \right]_0^t$$

$$= \lim_{t \to 2^-} \left(\operatorname{Sin}^{-1} \frac{t}{2} - \operatorname{Sin}^{-1} 0 \right) = \frac{\pi}{2}.$$

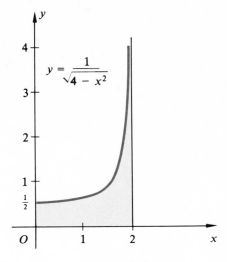

$$y = \frac{1}{\sqrt{4 - x^2}}$$

Figure 13-4

This improper integral may be interpreted as the area of the region bounded by the curve $y = 1/\sqrt{4 - x^2}$, the coordinate axes, and the line $x = 2$ (Fig. 13-4).

A similar definition applies when the integrand becomes infinite at the lower limit of integration.

13.8.2 Definition. If f is continuous on $(a, b]$ and if $\lim_{x \to a^+} f(x) = \pm\infty$, then

$$\int_a^b f(x)\, dx = \lim_{t \to a^+} \int_t^b f(x)\, dx,$$

provided that this limit exists.

If the integrand, $f(x)$, is continuous throughout its interval of integration except at an interior point c where $\lim_{x \to c}|f(x)| = \infty$, the following definition applies. Notice that the symbol $\lim_{x \to c}|f(x)| = \infty$ includes the four cases shown in Fig. 13-5(a), (b), (c), and (d).

13.8.3 Definition. If f is continuous on $[a, b]$ except at a number c, where $a < c < b$, and if $\lim_{x \to c}|f(x)| = \infty$, then

$$\int_a^b f(x)\, dx = \int_a^c f(x)\, dx + \int_c^b f(x)\, dx,$$

provided that both integrals in the right member exist.

This definition (13.8.3) is readily modified to apply to infinite intervals of integration by replacing a by $-\infty$, b by ∞, or both (see Exercise 16).

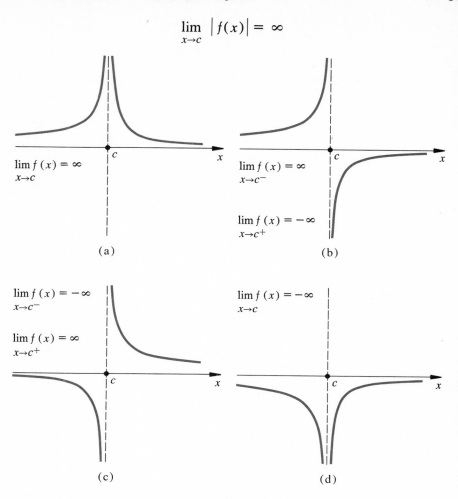

$$\lim_{x \to c} |f(x)| = \infty$$

$\lim_{x \to c} f(x) = \infty$

(a)

$\lim_{x \to c^-} f(x) = \infty$

$\lim_{x \to c^+} f(x) = -\infty$

(b)

$\lim_{x \to c^-} f(x) = -\infty$

$\lim_{x \to c^+} f(x) = \infty$

(c)

$\lim_{x \to c} f(x) = -\infty$

(d)

Figure 13-5

Example 2. Evaluate, if possible, the improper integral $\displaystyle\int_0^3 \frac{dx}{(x-1)^{2/3}}$.

Solution. The integrand is continuous for all values of x in $[0, 3]$ except $x = 1$ where $\lim_{x \to 1} [1/(x-1)^{2/3}] = \infty$. Applying 13.8.3, we have

$$\int_0^3 \frac{dx}{(x-1)^{2/3}} = \int_0^1 \frac{dx}{(x-1)^{2/3}} + \int_1^3 \frac{dx}{(x-1)^{2/3}}$$

$$= \lim_{t \to 1^-} \int_0^t \frac{dx}{(x-1)^{2/3}} + \lim_{s \to 1^+} \int_s^3 \frac{dx}{(x-1)^{2/3}}$$

$$= \lim_{t \to 1^-} \left[3(x-1)^{1/3} \right]_0^t + \lim_{s \to 1^+} \left[3(x-1)^{1/3} \right]_s^3$$

$$= 3 \lim_{t \to 1^-} [(t-1)^{1/3} + 1] + 3 \lim_{s \to 1^+} [2^{1/3} - (s-1)^{1/3}]$$

$$= 3 + 3(2^{1/3}) = 3(1 + \sqrt[3]{2}) \doteq 6.78.$$

This result may be interpreted as the area of the shaded region shown in Fig. 13-6.

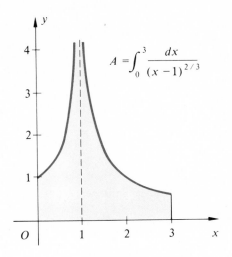

$$A = \int_0^3 \frac{dx}{(x-1)^{2/3}}$$

Figure 13-6

Example 3. Evaluate, if possible, $\displaystyle\int_0^4 \frac{dx}{(x-2)^2}$.

Solution. If we failed to notice that the integrand is discontinuous at $x = 2$, we might be tempted to say that, since an antiderivative of the integrand is $-1/(x-2)$, the value of this integral is

$$\frac{-1}{x-2}\bigg]_0^4 = -\frac{1}{2} - \frac{1}{2} = -1.$$

Yet a glance at Fig. 13-7 tells us that it is impossible for this integral to have a negative value, since the curve is everywhere above the x axis.

But if we proceed correctly, using 13.8.3, we find that

$$\int_0^4 \frac{dx}{(x-2)^2} = \int_0^2 \frac{dx}{(x-2)^2} + \int_2^4 \frac{dx}{(x-2)^2}$$

$$= \lim_{t \to 2^-} \left[\frac{-1}{x-2}\right]_0^t + \lim_{s \to 2^+} \left[\frac{-1}{x-2}\right]_s^4$$

$$= \lim_{t \to 2^-} \frac{-1}{t-2} - \frac{1}{2} - \frac{1}{2} - \lim_{s \to 2^+} \frac{-1}{s-2}$$

$$= \infty - \frac{1}{2} - \frac{1}{2} + \infty.$$

Therefore the given integral diverges and has no value.

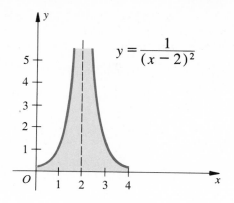

Figure 13-7

Exercises

In Exercises 1 to 12, evaluate the given improper integral or show that it diverges.

1. $\displaystyle\int_1^2 \frac{dx}{(x-1)^{1/3}}$.

2. $\displaystyle\int_2^4 \frac{dx}{(3-x)^{2/3}}$.

3. $\displaystyle\int_0^2 \frac{3\,dx}{x^2+x-2}$.

4. $\displaystyle\int_3^7 \frac{dx}{\sqrt{x-3}}$.

5. $\displaystyle\int_{-3}^3 \frac{dx}{\sqrt{9-x^2}}$.

6. $\displaystyle\int_{-3}^0 \frac{x\,dx}{(x^2-4)^{2/3}}$.

7. $\displaystyle\int_0^3 \frac{x\,dx}{\sqrt{9-x^2}}$.

8. $\displaystyle\int_1^3 \frac{dx}{\sqrt{-x^2+4x-3}}$.

9. $\displaystyle\int_{-2}^{-1} \frac{dx}{(x+1)^{4/3}}$.

10. $\displaystyle\int_{-3}^2 \frac{dx}{x^2}$.

11. $\displaystyle\int_{-1}^{27} x^{-2/3}\,dx$.

12. $\displaystyle\int_0^1 \ln x\,dx$.

13. Find the area of the region bounded by the curve $y=(x-8)^{-2/3}$, the x axis, and the lines $x=0$ and $x=8$. Make a sketch.

14. Find the area of the region between the curves $y=1/x$ and $y=1/(x^3+x)$, from $x=0$ to $x=1$. Make a sketch.

15. Show that $\displaystyle\int_0^1 (1/x^n)\,dx$ converges for all numbers $n<1$ and diverges for $n\ge 1$.

16. Let f be a function that is continuous on the infinite interval $[a,\infty)$ except at one interior point c $(a<c)$ and let $\lim\limits_{x\to c}|f(x)|=\infty$. Combine the ideas in 13.5.1 and 13.8.3 to formulate a definition of $\displaystyle\int_a^\infty f(x)\,dx$.

17. Let $f(x)=x^{-2/3}$, $0<x\le 1$. Denote by R the region in the first quadrant that is to the left of the line $x=1$ and below the curve $y=x^{-2/3}$.
(a) Show that the area of R is finite by finding its value.
(b) Show that the volume of the solid generated by revolving the region R about the x axis is infinite.

Exercises 18 to 23 are more challenging. They combine ideas from throughout Chapter 13 and provide an introduction to the **gamma function.**

18. Prove that for each positive number n there is a positive number M such that

$$0 < \frac{x^{n-1}}{e^x} < \frac{1}{x^2} \quad \text{for all} \quad x \geq M.$$

19. Prove that for each arbitrarily chosen positive number n, the improper integral

$$\int_1^\infty x^{n-1}e^{-x}\, dx$$

converges.

20. Prove that the integral

$$\int_0^1 x^{n-1}e^{-x}\, dx$$

exists for every positive number n.

The **gamma function** of Leonhard Euler (1707–1783) is of considerable importance in mathematics and the physical sciences. It is defined by

$$\Gamma(n) = \int_0^\infty x^{n-1}e^{-x}\, dx, \qquad n > 0.$$

For this definition to be meaningful, the improper integral must converge.

21. Prove that for every positive number n, the improper integral $\int_0^\infty x^{n-1}e^{-x}\, dx$ converges. (*Hint:* Use Exercises 19 and 20.)

22. Prove that

$$\Gamma(n + 1) = n\Gamma(n)$$

for every positive number n.

23. Show that if n is a positive *integer,*

$$\Gamma(n + 1) = n!.$$

13.9 REVIEW EXERCISES

Find the limits in Exercises 1 through 14.

1. $\lim\limits_{x \to 0} \dfrac{4x}{\tan x}.$

2. $\lim\limits_{x \to 0} \dfrac{\tan 2x}{\sin 3x}.$

3. $\lim\limits_{x \to 0} \dfrac{\sin x - x}{\frac{1}{3}x^3}.$

4. $\lim\limits_{x \to 0} 2x \cot x.$

5. $\lim\limits_{x \to \infty} \dfrac{\ln t}{t^2}.$

6. $\lim\limits_{x \to \infty} \dfrac{2x^3}{\ln x}.$

7. $\lim\limits_{s \to 0^+} s \ln s.$

8. $\lim\limits_{x \to 0^+} x^x.$

9. $\lim\limits_{x \to 0} (1 + \sin x)^{2/x}.$

10. $\lim\limits_{x \to 0^+} x^2 \ln x.$

11. $\lim\limits_{t \to \infty} t^{1/t}.$

12. $\lim\limits_{x \to 0^+} \left(\dfrac{1}{\sin x} - \dfrac{1}{x} \right).$

13. $\lim\limits_{x \to \pi/2} \dfrac{\tan 3x}{\tan x}.$

14. $\lim\limits_{x \to \pi/2} (\sin x)^{\tan x}.$

In Exercises 15 to 30, evaluate the given improper integral or show that it diverges.

15. $\displaystyle\int_0^\infty \dfrac{dx}{(x + 1)^2}.$

16. $\displaystyle\int_0^\infty \dfrac{dx}{1 + x^2}.$

17. $\displaystyle\int_{-\infty}^1 e^{2x}\, dx.$

18. $\displaystyle\int_{-1}^1 \dfrac{dx}{1 - x}.$

19. $\displaystyle\int_0^\infty \dfrac{dx}{x + 1}.$

20. $\displaystyle\int_{1/2}^2 \dfrac{dx}{x(\ln x)^{1/5}}.$

21. $\displaystyle\int_0^\infty \frac{dx}{e^x + e^{-x}}.$

22. $\displaystyle\int_{-\infty}^1 \frac{dx}{(2-x)^2}.$

23. $\displaystyle\int_{-2}^0 \frac{dx}{2x+3}.$

24. $\displaystyle\int_1^4 \frac{dx}{\sqrt{x-1}}.$

25. $\displaystyle\int_2^\infty \frac{dx}{x(\ln x)^2}.$

26. $\displaystyle\int_0^\infty \frac{dx}{e^{x/2}}.$

27. $\displaystyle\int_3^5 \frac{dx}{(4-x)^{2/3}}.$

28. $\displaystyle\int_2^\infty xe^{-x^2}\,dx.$

29. $\displaystyle\int_0^2 \frac{x\,dx}{(x^2-1)^{2/3}}.$

30. $\displaystyle\int_0^2 \frac{dx}{\sqrt{2x-x^2}}.$

In each of Exercises 31 to 36, show that the given improper integral converges or that it diverges.

31. $\displaystyle\int_1^\infty \frac{dx}{\sqrt{x^6+x}}.$

32. $\displaystyle\int_0^\infty e^{-x^2}\,dx.$

33. $\displaystyle\int_1^\infty \frac{\ln x}{e^x}\,dx.$

34. $\displaystyle\int_1^\infty \frac{\ln x}{x}\,dx.$

35. $\displaystyle\int_1^\infty \frac{\ln x}{x^3}\,dx.$

36. $\displaystyle\int_1^\infty \frac{dx}{xe^x}.$

14

infinite series

Twenty-four hundred years ago the Greek philosopher Zeno of Elea said, in one of his famous paradoxes, that a runner cannot finish any race because he must first cover half the distance, then half of the remaining distance, then half of the still remaining distance, and so on, forever. Since the runner's time is finite, he cannot traverse the infinite number of segments of the course.

This amounts to saying that if the course is one mile, say, he must run $\frac{1}{2}$ mile, plus $\frac{1}{4}$ mile, plus $\frac{1}{8}$ mile, and so on, which may be written

(1) $$\frac{1}{2} + \frac{1}{4} + \frac{1}{8} + \frac{1}{16} + \frac{1}{32} + \cdots$$

Yet we all know that many runners finish a mile race. The catch is in the three dots in (1), indicating that the operation of addition is to be repeated "indefinitely."

Up to now, the word "sum" has been defined only for a finite number of quantities. The indicated "infinite sum" in (1) has, as yet, no meaning for us. It is an example of an *infinite series*.

By computing the partial sums

$$s_1 = \frac{1}{2},$$

$$s_2 = \frac{1}{2} + \frac{1}{4} = \frac{3}{4},$$

$$s_3 = \frac{1}{2} + \frac{1}{4} + \frac{1}{8} = \frac{7}{8},$$

$$\cdots \cdots \cdots \cdots \cdots$$

$$s_n = \frac{1}{2} + \frac{1}{4} + \frac{1}{8} + \cdots + \frac{1}{2^n} = \frac{2^n - 1}{2^n} = 1 - \frac{1}{2^n},$$

we can write the *infinite sequence* of partial sums, $s_1, s_2, s_3, \cdots, s_n, \cdots$, as

(2)
$$\frac{1}{2}, \frac{3}{4}, \frac{7}{8}, \frac{15}{16}, \frac{31}{32}, \cdots, 1 - \frac{1}{2^n}, \cdots,$$

from which it is apparent that the sum, s_n, of the first n terms of the infinite series (1) can be made as close as we please to 1 by taking n sufficiently large. We shall say that the limit of the sequence (2) is

$$\lim_{n \to \infty} s_n = \lim_{n \to \infty} \left(1 - \frac{1}{2^n} \right) = 1 - \lim_{n \to \infty} \frac{1}{2^n} = 1,$$

and that the *sum* of the *infinite series* (1) is 1.

14.1 SEQUENCES

Recall that a function (of one variable) is a correspondence between the elements of two sets, its domain and its range, such that to each element of the domain there corresponds one and only one element of the range. Up to now, the domains of the functions of one variable usually consisted of *all* real numbers (with possibly some exceptions) in one or more intervals, open or closed, finite or infinite.

If the domain of a function is restricted to integers, the function is called a *sequence*.

14.1.1 Definition. An infinite **sequence** is a function whose domain is the set of all integers greater than or equal to some specified integer.

As an illustration, the sequence defined by $f(n) = 1/n$, where n is any positive integer, is $\{(n, 1/n) \mid n$ is a positive integer$\}$; its range is $\{f(1) = 1, f(2) = 1/2, f(3) = 1/3, \cdots, f(n) = 1/n, \cdots\}$. Unless stated to the contrary, *the domain of a sequence will be assumed to be the set of all positive integers*. This enables us to write $\{1/n\}$ for the above sequence instead of $\{(n, 1/n) \mid n$ is a positive integer$\}$.

Another example of an infinite sequence is $\{n/2^n\}$. It is defined by $f(n) = n/2^n$ and its range is $\{1/2, 2/4, 3/8, \cdots, n/2^n, \cdots\}$.

The numbers in the range of a sequence f, namely $f(1), f(2), f(3), \cdots$, are called the **elements** of the sequence; and $f(n)$ is called the nth *element* or **general element** of the sequence. It is customary to use subscript notation in writing the elements of a sequence. Thus we write $u_1, u_2, u_3, \cdots, u_n, \cdots$ instead of $f(1), f(2), f(3), \cdots, f(n), \cdots$ for the elements of a sequence, and the symbol $\{u_n\}$ will mean the infinite sequence whose elements are $u_1, u_2, u_3, \cdots, u_n, \cdots$. For instance, the elements of the sequence $\{n!/10^n\}$ are $u_1 = 1/10, u_2 = 2/100, u_3 = 6/1000, \cdots, u_n = n!/10^n, \cdots$.

In the case of a sequence, the definition (6.11.1) of the limit of a function as the independent variable approaches infinity takes on the following form.

14.1.2 Definition.   A sequence $\{u_n\}$ has the **limit** L, written

$$\lim_{n \to \infty} u_n = L,$$

if to each positive number ϵ there corresponds a positive integer N such that $|u_n - L|$ $< \epsilon$ whenever $n \geq N$.

That is, a sequence has a limit if, corresponding to each positive number ϵ, we can specify an element of the sequence such that this element and every following element will differ from the limit by less than ϵ. The reader should compare this with 6.11.1.

A sequence that has a limit is said to be **convergent**, and to **converge** to that limit. A sequence that fails to have a limit is **divergent**.

The limit theorems for sums, products, and quotients of functions also apply when the functions are sequences.

14.1.3 Theorem.   If $\{u_n\}$ and $\{v_n\}$ are convergent sequences and if k is a number, then

(i) $\lim\limits_{n \to \infty} k u_n = k \lim\limits_{n \to \infty} u_n;$

(ii) $\lim\limits_{n \to \infty} (u_n + v_n) = \lim\limits_{n \to \infty} u_n + \lim\limits_{n \to \infty} v_n;$

(iii) $\lim\limits_{n \to \infty} u_n v_n = \lim\limits_{n \to \infty} u_n \lim\limits_{n \to \infty} v_n;$

and, provided that $\lim v_n \neq 0$,

(iv) $\lim\limits_{n \to \infty} \dfrac{u_n}{v_n} = \dfrac{\lim\limits_{n \to \infty} u_n}{\lim\limits_{n \to \infty} v_n}.$

A proof of this theorem is analogous to the proof of 3.3.1.

Example 1.   Find the limit of the sequence defined by $u_n = \dfrac{3n^2}{7n^2 + 1}$, $n \in N$.

Solution.   Since $n \neq 0$,

$$u_n = \frac{3n^2}{7n^2 + 1} = \frac{3}{7 + 1/n^2}.$$

Thus

$$\lim_{n \to \infty} u_n = \lim_{n \to \infty} \frac{3}{7 + 1/n^2} = \frac{3}{7 + \lim\limits_{n \to \infty} 1/n^2} = \frac{3}{7}.$$

The proof that $\lim\limits_{x \to \infty} 1/x^p = 0$, $p \in N$, given in Example 2 of Section 6.11, assumed that $1/x^p$ was defined for all positive numbers x. But this implies that

$$\lim_{x \to \infty} \frac{1}{n^p} = 0, \qquad p \in N,$$

when n is restricted to positive integers.

As a direct consequence of the earlier definition (6.11.1) of $\lim_{x \to \infty} f(x) = L$, we have the following important specialization. Its usefulness will be shown in Examples 2 and 3, below.

14.1.4 If $\lim_{x \to \infty} f(x) = L$ when f is defined for all real numbers greater than some number K, then

$$\lim_{n \to \infty} f(n) = L$$

when the domain of f is restricted to **positive integers** greater than K.

Example 2. Does the sequence $\left\{ \dfrac{\ln n}{e^n} \right\}$ converge?

Solution. Since $\ln x$ and e^x exist for all positive numbers x, and $\lim_{x \to \infty} \ln x = \lim_{x \to \infty} e^x = \infty$, we can use l'Hôpital's rule to find

$$\lim_{x \to \infty} \frac{\ln x}{e^x} = \lim_{x \to \infty} \frac{1/x}{e^x} = 0.$$

Therefore (by 14.1.4)

$$\lim_{n \to \infty} \frac{\ln n}{e^n} = 0,$$

when n is restricted to the positive integers, and thus the sequence $\{\ln n/e^n\}$ converges to the limit 0.

Example 3. Show that the sequence $\{r^n\}$ converges to the limit zero if $0 \le r < 1$ and diverges if $r > 1$.

Solution. We are asked to show that $\lim_{n \to \infty} r^n = 0$ if $0 \le r < 1$ and that $\lim_{n \to \infty} r^n$ fails to exist if $r > 1$, $n \in N$.

This will be accomplished (14.1.4) by proving that, for $x \in R$, $\lim_{x \to \infty} r^x = 0$ when $0 \le r < 1$ and $\lim_{x \to \infty} r^x = \infty$ when $r > 1$.

The proof is trivial if $r = 0$; so it is assumed in what follows that $r > 0$.

Let $0 < r < 1$ and let ϵ be any positive number less than 1. Then (Section 10.2) $\ln r < 0$ and $\ln \epsilon < 0$; consequently, $(\ln \epsilon)/(\ln r) > 0$. Thus

$$x > \frac{\ln \epsilon}{\ln r} \iff x \ln r < \ln \epsilon$$

$$\iff \ln r^x < \ln \epsilon$$

$$\iff r^x < \epsilon \iff |r^x| < \epsilon.$$

That is, $|r^x - 0| < \epsilon$ for all numbers x such that $x > (\ln \epsilon)/(\ln r)$. Therefore (by 6.11.1) $\lim_{x \to \infty} r^x = 0$ when $0 < r < 1$.

Now assume that $r > 1$, and let K be an arbitrarily chosen number greater than 1. Since $\ln r > 0$ and $\ln K > 0$,

$$x > \frac{\ln K}{\ln r} \iff x \ln r > \ln K$$

$$\Longleftrightarrow \ln r^x > \ln K$$

$$\Longleftrightarrow r^x > K.$$

Therefore (by 6.11.4) $\lim\limits_{x \to \infty} r^x = \infty$ if $r > 1$.

The student should compare the following definition with 6.3.1.

14.1.5 Definition.   A sequence $\{u_n\}$ is said to be

   (i)   **increasing** if and only if $u_n < u_{n+1}$ for every positive integer n,
   (ii) **nondecreasing** if and only if $u_n \leq u_{n+1}$ for every positive integer n,
   (iii) **nonincreasing** if and only if $u_n \geq u_{n+1}$ for every positive integer n,
   (iv) **decreasing** if and only if $u_n > u_{n+1}$ for every positive integer n.
   (v)   If any one of the above holds, the sequence $\{u_n\}$ is **monotonic**.

As an illustration, the sequence $\{(2^n - 1)/2^n\}$ is increasing, and the sequence $\{1/n\}$ is decreasing.

The number A is said to be an **upper bound** for the sequence $\{u_n\}$ if $u_n \leq A$ for $n = 1, 2, 3, \cdots$, and such a sequence is said to be **bounded above**. If there is a number B such that $B \leq u_n$ for $n = 1, 2, 3, \cdots$, then B is a **lower bound** for the sequence $\{u_n\}$, and the sequence is **bounded below**. The sequence $\{u_n\}$ is said to be **bounded** if there is a number C such that $|u_n| \leq C$ for all positive integers n.

A necessary and sufficient condition for a monotonic sequence to be convergent is given in the following theorem.

14.1.6 Theorem.   A monotonic sequence is convergent if and only if it is bounded.

Proof.   Assume that the sequence $\{u_n\}$ is nondecreasing and that it has an upper bound A. Since the set of real numbers $S = \{u_1, u_2, u_3, \cdots\}$ has an upper bound, it has a least upper bound L, by the completeness axiom (13.6.2).

Let ϵ be a positive number. Then $L - \epsilon < L$, and $L - \epsilon$ cannot be an upper bound for S, since L is the least upper bound for S. Thus there exists an element u_k of the sequence that is greater than $L - \epsilon$, and $L - \epsilon < u_n \leq L$ for all $n > k$. But this situation implies that $-\epsilon < u_n - L \leq 0 < \epsilon$, or $|u_n - L| < \epsilon$, for all $n > k$. Therefore $\lim\limits_{n \to \infty} u_n = L$, and $\{u_n\}$ is convergent. We have proved that if $\{u_n\}$ is nondecreasing and is bounded, it is convergent.

We will now assume that $\{u_n\}$ is nondecreasing and convergent, and we will prove that it is bounded. Of course, any nondecreasing sequence is bounded below by its first term; so our task is to find an upper bound. Since $\lim\limits_{n \to \infty} u_n = L$ (say,) corresponding to each $\epsilon > 0$ there exists an integer N such that $|u_n - L| < \epsilon$ whenever $n \geq N$; that is, $-\epsilon < u_n - L < \epsilon$, or $L - \epsilon < u_n < L + \epsilon$, for all $n \geq N$. Since the sequence is nondecreasing, the implication is that $u_n < L + \epsilon$ for all positive integers n. Therefore $\{u_n\}$ is bounded above by $L + \epsilon$.

The proof of the theorem when $\{u_n\}$ is nonincreasing is immediate, since $\{-u_n\}$ is nondecreasing. ■

14.1.7   Corollary.   If U is an upper bound of a nondecreasing sequence, the sequence converges to a limit that is less than or equal to U; if V is a lower bound of a nonincreasing sequence, the sequence converges to a limit that is greater than or equal to V.

It should be noted that since $\lim_{n \to \infty} u_n$ depends only on what happens to u_n after n exceeds a certain number, any finite number of elements of the sequence can be discarded without affecting the existence of a limit or the value of the limit. It follows that 14.1.6 and 14.1.7 apply to sequences that are not monotonic, provided that they become monotonic for all n greater than some integer k. *If there is a number k such that for $n > k$ the sequence $\{u_n\}$ is monotonic and bounded, then $\{u_n\}$ is convergent.*

Example 4.   The sequence $\{10^n/2^{n^2}\}$ is not monotonic, since $u_1 < u_2 > u_3 > u_4 > \cdots$. But if we discard the first element and consider the sequence $u_n = 10^n/2^{n^2}$ for $n = 2, 3, 4, \cdots$, this new sequence *is* decreasing. Moreover, it is bounded below by 0, since $10^n/2^{n^2}$ cannot be negative for any positive integer n. Therefore it converges (by 14.1.6).

Exercises

Write out the first five elements of the sequences in Exercises 1 through 10. Determine whether the sequences converge and, if so, find the limit.

1. $\left\{\dfrac{n}{2n-1}\right\}$.

2. $\left\{\dfrac{2n+1}{n+2}\right\}$.

3. $\left\{\dfrac{n^2+1}{n^2-2n+3}\right\}$.

4. $\left\{\dfrac{3n^2+2}{n+4}\right\}$.

5. $\left\{\dfrac{n}{\ln(n+1)}\right\}$.

6. $\left\{\dfrac{1}{n}\cos n\right\}$.

7. $\left\{\dfrac{\ln n}{n^2}\right\}$.

8. $\left\{\dfrac{e^n}{2^n}\right\}$.

9. $\left\{\dfrac{e^n}{n^2}\right\}$.

10. $\{e^{-n}\sin n\}$.

Find the simplest possible expression for the nth element of the sequences indicated in Exercises 11 to 20. Tell whether the sequence converges and, if so, to what limit.

11. $\{\tfrac{1}{2}, \tfrac{2}{3}, \tfrac{3}{4}, \tfrac{4}{5}, \cdots\}$.

12. $\{2, 1, 2, 1\tfrac{1}{2}, 2, 1\tfrac{3}{4}, 2, 1\tfrac{7}{8}, \cdots\}$.

13. $\left\{1, \dfrac{1}{1-\tfrac{1}{2}}, \dfrac{1}{1-\tfrac{2}{3}}, \dfrac{1}{1-\tfrac{3}{4}}, \cdots\right\}$.

14. $\left\{1, \dfrac{1}{4}, 1, \dfrac{3}{16}, 1, \dfrac{5}{64}, 1, \dfrac{7}{256}, 1, \cdots\right\}$.

15. $\left\{0, \dfrac{\ln 2}{2}, \dfrac{\ln 3}{3}, \dfrac{\ln 4}{4}, \cdots\right\}$.

16. $\left\{0, \dfrac{1}{2^2}, \dfrac{2}{3^2}, \dfrac{3}{4^2}, \cdots\right\}$.

17. $\left\{1, \dfrac{2}{2^2-1^2}, \dfrac{3}{3^2-2^2}, \dfrac{4}{4^2-3^2}, \cdots\right\}$.

18. $\left\{\dfrac{1}{2-\tfrac{1}{2}}, \dfrac{2}{3-\tfrac{1}{3}}, \dfrac{3}{4-\tfrac{1}{4}}, \cdots\right\}$.

19. $\left\{2, 1, \dfrac{2^3}{3^2}, \dfrac{2^4}{4^2}, \dfrac{2^5}{5^2}, \cdots\right\}$.

20. $\left\{\sin 1°, \dfrac{\sin 2°}{2}, \dfrac{\sin 3°}{3}, \cdots\right\}$.

21. Prove part (ii) of 14.1.3.

22. Prove that if $\lim_{x \to \infty} f(x) = L$ when f is defined for all real numbers greater than some number A, then $\lim_{x \to \infty} f(n) = L$ when the domain of f is restricted to *positive integers* greater than A.

23. The proof in Example 3, above, depends in part on the validity of the statement that when a and b are positive numbers,

$$a < b \iff \ln a < \ln b.$$

Prove this equivalence.

14.2 INFINITE SERIES

Associated with any infinite sequence

$$\{a_n\} = a_1, a_2, a_3, \cdots, a_n, \cdots$$

is another sequence

$$\{s_n\} = s_1, s_2, s_3, \cdots, s_n, \cdots$$

defined by

$$s_1 = a_1,$$
$$s_2 = a_1 + a_2,$$
$$s_3 = a_1 + a_2 + a_3,$$
$$\cdot \quad \cdot \quad \cdot \quad \cdot \quad \cdot$$
$$s_n = a_1 + a_2 + a_3 + \cdots + a_n,$$
$$\cdot \quad \cdot \quad \cdot \quad \cdot \quad \cdot \quad \cdot \quad \cdot$$

The sequence $\{s_n\}$ is called an **infinite series** and is commonly symbolized by

$$a_1 + a_2 + a_3 + \cdots + a_n + \cdots \qquad \text{or by} \qquad \sum_{i=1}^{\infty} a_i.$$

To illustrate, if the given sequence is

$$\left\{\frac{1}{2^n}\right\} = \frac{1}{2}, \frac{1}{4}, \frac{1}{8}, \cdots, \frac{1}{2^n}, \cdots,$$

then the infinite series associated with it is $\{s_n\}$, where

$$s_1 = \frac{1}{2},$$

$$s_2 = \frac{1}{2} + \frac{1}{4} = \frac{3}{4},$$

$$s_3 = \frac{1}{2} + \frac{1}{4} + \frac{1}{8} = \frac{7}{8},$$

$$\cdot \quad \cdot \quad \cdot \quad \cdot \quad \cdot \quad \cdot$$

$$s_n = \frac{1}{2} + \frac{1}{4} + \frac{1}{8} + \cdots + \frac{1}{2^n} = \frac{2^n - 1}{2^n},$$

$$\cdot \quad \cdot \quad \cdot \quad \cdot \quad \cdot \quad \cdot \quad \cdot \quad \cdot \quad \cdot$$

This infinite series is designated by

$$\frac{1}{2} + \frac{1}{4} + \frac{1}{8} + \cdots + \frac{1}{2^n} + \cdots \qquad \text{or by} \qquad \sum_{n=1}^{\infty} \frac{1}{2^n}.$$

14.2.1 Definition. If $\{a_n\} = a_1, a_2, a_3, \cdots, a_n, \cdots$ is an infinite sequence, the sequence $\{s_n\}$, defined by

$$s_n = a_1 + a_2 + a_3 + \cdots + a_n$$

for $n = 1, 2, 3, \cdots$, is called an **infinite series** and is usually symbolized by

$$a_1 + a_2 + a_3 + \cdots + a_n + \cdots \qquad \text{or by} \qquad \sum_{i=1}^{\infty} a_i.$$

The numbers $a_1, a_2, a_3, \cdots$ are called the **terms** of the infinite series $a_1 + a_2 + a_3 + \cdots + a_n + \cdots$, and the numbers $s_1, s_2, s_3, \cdots$ are the **partial sums** of the infinite series.

We wish to assign a number to this infinite series and call that number the **sum** of the series. Of course, the word "sum" has been defined only for a finite number of terms. In the following definition its meaning is extended to certain infinite series.

14.2.2 Definition. Let

$$\sum_{i=1}^{\infty} a_i = a_1 + a_2 + a_3 + \cdots + a_n + \cdots$$

be an infinite series and let $\{s_n\}$, where $s_n = a_1 + a_2 + a_3 + \cdots + a_n$ for $n = 1, 2, 3, \cdots$, be the sequence of partial sums of the infinite series. If

$$\lim_{n \to \infty} s_n = S$$

exists, the series is said to be **convergent** (and to converge to the value S) and S is called the **sum** of the infinite series $\sum_{i=1}^{\infty} a_i$. If $\lim_{n \to \infty} s_n$ fails to exist, the series is **divergent** and has no sum.

If the series $\sum_{i=1}^{\infty} a_i$ has a sum S, we write

(1) $$\sum_{i=1}^{\infty} a_i = S.$$

This does *not* mean that we have added the infinite number of terms of the series. *Equation (1) simply means that the sequence of partial sums, $\{s_n\}$, of the infinite series* $\sum_{i=1}^{\infty} a_i$ *converges to the limit S.*

Example 1. Show that the infinite series

$$1 + \frac{1}{2} + \frac{1}{4} + \frac{1}{8} + \cdots + \frac{1}{2^{n-1}} + \cdots$$

converges and find its sum.

Solution.

$$1 + \frac{1}{2} + \frac{1}{4} + \cdots + \frac{1}{2^{n-1}}$$

is a finite geometric series of n terms. Its sum is

$$s_n = 1 + \frac{1}{2} + \frac{1}{4} + \cdots + \frac{1}{2^{n-1}} = 2 - \frac{1}{2^{n-1}}.$$

Since

$$\lim_{n \to \infty} s_n = \lim_{n \to \infty} \left[2 - \frac{1}{2^{n-1}} \right] = 2 - \lim_{n \to \infty} \frac{1}{2^{n-1}} = 2,$$

the given infinite series converges and its sum is 2 (by 14.2.2).

Sometimes we wish to start the series with the term a_0 or a_2 or some other term. If k is an arbitrarily chosen nonnegative integer, we write

$$\sum_{n=k}^{\infty} a_n = a_k + a_{k+1} + a_{k+2} + \cdots.$$

When it is not important which index we assign to the first term, or when it is clear from the context, we often write $\sum a_n$ for an infinite series.

Since $\lim_{n \to \infty} (s_n - C)$, where C is a constant, exists if and only if $\lim_{n \to \infty} s_n$ exists, it follows that we can omit a finite number of terms at the beginning of an infinite series without affecting its convergence or divergence. Of course, the *value* of the sum, if any, will be affected.

14.2.3 Theorem.   The two infinite series

$$\sum_{i=1}^{\infty} a_i \qquad \text{and} \qquad \sum_{i=k}^{\infty} a_i,$$

where k is an arbitrarily chosen positive integer, both converge or both diverge.

A test for divergence of an infinite series, one that is often easy to apply, is given in the next theorem. It should be tried first when testing a series for convergence, for if the series is divergent, no further tests are needed.

14.2.4 Theorem.   A **necessary** condition for the series

$$\sum a_n = a_1 + a_2 + a_3 + \cdots + a_n + \cdots$$

to converge is $\lim_{n \to \infty} a_n = 0$.

Proof.   Assume that $\sum a_n$ converges; that is, assume that $\lim_{n \to \infty} s_n = S$, where S is a number.

It follows from the definition of the limit of a sequence (14.1.2) that $\lim_{n \to \infty} s_n = S$ implies $\lim_{n \to \infty} s_{n-1} = S$. But $a_n = s_n - s_{n-1}$. Therefore

$$\lim_{n \to \infty} a_n = \lim_{n \to \infty} (s_n - s_{n-1}) = \lim_{n \to \infty} s_n - \lim_{n \to \infty} s_{n-1} = S - S = 0.$$

Thus if $\sum a_n$ converges, then $\lim_{n \to \infty} a_n = 0$.   ∎

14.2.5 Corollary. If $\lim\limits_{n\to\infty} a_n \neq 0$, then the infinite series $a_1 + a_2 + a_3 + \cdots + a_n + \cdots$ diverges.

It is important to notice that 14.2.4 gives a *necessary* condition for the convergence of an infinite series, not a *sufficient* one. That is, a series might not converge even if its nth term approaches zero. But if $\lim\limits_{n\to\infty} a_n \neq 0$, the series diverges.

Example 2. The series

$$\sum_{i=1}^{\infty} ar^{i-1} = a + ar + ar^2 + \cdots + ar^{n-1} + \cdots,$$

where $a \neq 0$ and $r \neq 0$, is called a **geometric series**. Show that *the geometric series converges if $|r| < 1$ and diverges if $|r| \geq 1$.*

Solution. Write $s_n = a + ar + ar^2 + \cdots + ar^{n-1}$. Then $(1 - r)s_n = a - ar^n$ and, if $r \neq 1$,

$$s_n = \frac{a - ar^n}{1 - r}.$$

Since $\lim\limits_{n\to\infty} r^n = 0$ if $|r| < 1$ (by Example 2 of Section 14.1), we have

$$\lim_{n\to\infty} s_n = \frac{a}{1 - r} \quad \text{if} \quad |r| < 1.$$

Thus the geometric series converges if $|r| < 1$.

If $|r| > 1$, then

$$\lim_{n\to\infty} |r|^n = \infty$$

(by Example 2 of Section 14.1), and

$$\lim_{n\to\infty} a_n = \lim_{n\to\infty} (ar^{n-1}) = a \lim_{n\to\infty} r^{n-1} \neq 0.$$

Therefore (by 14.2.5) the geometric series diverges when $|r| > 1$.

If $r = 1$, then $s_n = na$ and $\lim\limits_{n\to\infty} s_n = \pm\infty$. If $r = -1$, the series is $a - a + a - \cdots + (-1)^{n+1}a + \cdots$ and $\lim\limits_{n\to\infty} a_n \neq 0$. So (by 14.2.2 and 14.2.5) the geometric series diverges when $|r| = 1$.

Therefore the geometric series converges if $|r| < 1$ and diverges if $|r| \geq 1$.

Example 3. Consider the **harmonic series**

$$\sum \frac{1}{n} = 1 + \frac{1}{2} + \frac{1}{3} + \cdots + \frac{1}{n} + \cdots.$$

Its nth term is $1/n$ and $\lim\limits_{n\to\infty} 1/n = 0$. But the series diverges. For if $k = 2^{n-1}$,

$$s_k = 1 + \frac{1}{2} + \frac{1}{3} + \cdots + \frac{1}{2^{n-1}}$$

$$= 1 + \frac{1}{2} + \left(\frac{1}{3} + \frac{1}{4}\right) + \left(\frac{1}{5} + \frac{1}{6} + \frac{1}{7} + \frac{1}{8}\right) + \left(\frac{1}{9} + \cdots + \frac{1}{16}\right) + \cdots$$

$$+ \left(\frac{1}{2^{n-2} + 1} + \frac{1}{2^{n-2} + 2} + \cdots + \frac{1}{2^{n-1}}\right)$$

$$\geq 1 + \frac{1}{2} + \left(\frac{1}{4} + \frac{1}{4}\right) + \left(\frac{1}{8} + \frac{1}{8} + \frac{1}{8} + \frac{1}{8}\right) + \left(\frac{1}{16} + \frac{1}{16} + \cdots + \frac{1}{16}\right) + \cdots$$

$$+ \left(\frac{1}{2^{n-1}} + \frac{1}{2^{n-1}} + \cdots + \frac{1}{2^{n-1}}\right)$$

$$= 1 + \frac{1}{2} + \frac{1}{2} + \frac{1}{2} + \cdots + \frac{1}{2} = 1 + (n-1)\frac{1}{2} = \frac{n+1}{2}.$$

Thus $s_k \geq \frac{1}{2}(n+1)$; and since $\lim_{n\to\infty} \frac{1}{2}(n+1) = \infty$, $\lim_{n\to\infty} s_k = \infty$. Therefore the monotonic sequence $\{s_n\}$ is unbounded and **the harmonic series diverges** (14.1.5).

An infinite series may be multiplied, term by term, by a nonzero constant without affecting its convergence or divergence.

14.2.6 Theorem. If c is a constant and the series $\sum a_n$ converges, then so does $\sum ca_n$, and its limit is $c\sum a_n$. If $\sum a_n$ diverges, then $\sum ca_n$ diverges, provided that $c \neq 0$.

Proof. Assume that $\sum a_n$ converges, so that $\lim_{n\to\infty} s_n = S$, where $s_n = a_1 + a_2 + \cdots + a_n$. The sum of the first n terms of the series $\sum ca_n$ is cs_n. But $\lim_{n\to\infty} cs_n = c \lim_{n\to\infty} s_n = cS$. Therefore the series $\sum ca_n$ converges to the limit $cS = c\sum a_n$.

If $\sum a_n$ diverges and $c \neq 0$, then $\sum ca_n$ diverges, for otherwise, by the first part of the theorem, the convergence of $\sum ca_n$ would imply the convergence of $\sum (1/c)(ca_n)$ $= \sum a_n$. ∎

An infinite series is not an ordinary sum, and so it is not surprising that some of the laws of operation on finite sums fail to carry over to infinite series. For instance, the associative law for addition permits us to group the terms of a finite sum in any way we please by the insertion of parentheses, without affecting the sum. But consider the infinite series

$$1 - 1 + 1 - 1 + \cdots + (-1)^{n+1} + \cdots.$$

Its nth term does not approach zero as $n \longrightarrow \infty$, so the series is divergent and has no sum. If we group the terms of the series so that it becomes

$$(1-1) + (1-1) + (1-1) + \cdots = 0 + 0 + 0 + \cdots,$$

it converges with sum zero; and if we regroup the terms another way,

$$1 - (1-1) - (1-1) - (1-1) - \cdots = 1 - 0 - 0 - 0 - \cdots,$$

its sum is 1.

However, *if a series is convergent*, it can be treated in many ways like an ordinary finite sum, as shown in the following theorems.

14.2.7 Theorem. The terms of a convergent series may be grouped in any way (without changing the order of the terms) and the new series will converge to the same limit as the original series.

Proof. Let $\sum a_n$ be a convergent series whose sequence of partial sums is $\{s_n\}$ with limit S. If $\sum b_m$ is a series formed by grouping the terms of $\sum a_n$, and if $\{t_m\}$ is the sequence of partial sums of $\sum b_n$, then for each m, $t_m \in \{s_n\}$, and as m increases indefinitely, so does the corresponding n. Therefore the limit of $\{t_m\}$ exists and is equal to S. ∎

Two convergent series may be added term by term and the resulting series will converge to the sum of the limits of the original series.

14.2.8 Theorem. If $\sum a_n$ and $\sum b_n$ are convergent infinite series, then $\sum (a_n + b_n)$ is convergent and

$$\sum_{n=1}^{\infty} (a_n + b_n) = \sum_{n=1}^{\infty} a_n + \sum_{n=1}^{\infty} b_n.$$

Proof. Let $s_n = a_1 + a_2 + a_3 + \cdots + a_n$, $t_n = b_1 + b_2 + b_3 + \cdots + b_n$, and $w_n = (a_1 + b_1) + (a_2 + b_2) + \cdots + (a_n + b_n)$. Then

$$w_n = s_n + t_n.$$

Assume that $\sum a_n$ and $\sum b_n$ converge. Then $\lim_{n \to \infty} s_n$ and $\lim_{n \to \infty} t_n$ exist and

$$\lim_{n \to \infty} w_n = \lim_{n \to \infty} (s_n + t_n) = \lim_{n \to \infty} s_n + \lim_{n \to \infty} t_n.$$

Therefore

$$\sum_{n=1}^{\infty} (a_n + b_n) = \sum_{n=1}^{\infty} a_n + \sum_{n=1}^{\infty} b_n. \quad ∎$$

Note. In Examples 1 and 2, above, we found a formula for the partial sum $s_n = a_1 + a_2 + a_3 + \cdots + a_n$ and then used that formula to find $\lim_{n \to \infty} s_n$. This not only proved that the given infinite series converged (14.2.2) but also gave its sum, $S = \lim_{n \to \infty} s_n$.

However, it is not always easy to find such a formula for s_n; and because of this we shall soon develop other methods for testing an infinite series for convergence. Curious as it may seem right now, in working with infinite series it is often sufficient to know that the series converges, even though we cannot find its sum.

Exercises

In each of Exercises 1 to 4, (a) write the first five terms of the indicated infinite series; (b) find a formula for s_n; (c) tell whether the series converges or diverges; (d) if the series converges, find its sum.

1. $\displaystyle\sum_{i=1}^{\infty} \left(\frac{1}{5}\right)^i$.

2. $\displaystyle\sum_{i=1}^{\infty} 2\left(-\frac{1}{3}\right)^{i-1}$.

3. $\displaystyle\sum_{i=1}^{\infty} \frac{1}{i(i+1)}$. [*Hint:* In (b), compute s_1, $s_2, \cdots, s_5$.]

4. $\displaystyle\sum_{i=1}^{\infty}\left[3\left(\frac{2}{3}\right)^i + 2\left(\frac{1}{2}\right)^i\right].$

8. $\displaystyle\sum_{i=1}^{\infty}\frac{2^i}{i^2}.$

In Exercises 5 to 8, show that the given series diverge.

5. $\displaystyle\sum_{i=1}^{\infty}\frac{i^2}{i+1}.$

6. $\displaystyle\sum_{i=1}^{\infty}\frac{i!}{10^i}.$

7. $\displaystyle\sum_{i=1}^{\infty}\left(\frac{4}{3}\right)^i.$

In Exercises 9 to 12, find the rational numbers represented by the given repeating decimals by writing the decimals as infinite series and obtaining their sums.

9. $0.2\,2\,2\,2\,2\cdots$.

10. $0.21\ 21\ 21\ 21\cdots$.

11. $3.013\ 013\ 013\cdots$.

12. $0.40\ 125\ 125\ 125\cdots$.

14.3 TESTS FOR CONVERGENCE OF SERIES OF POSITIVE TERMS

The terms of the infinite series considered so far were any real numbers, positive, negative, or zero. The definitions and theorems in the preceding section apply equally to all infinite series, regardless of the signs of the terms.

In the present section we shall confine our attention to series all of whose terms are *positive numbers* (or, more generally, *nonnegative numbers*).

Notice that if all the terms of a series are nonnegative, its sequence of partial sums is nondecreasing. Thus 14.1.6 has the following corollary.

14.3.1 **Theorem.** A necessary and sufficient condition for an infinite series of nonnegative terms to converge is that its sequence of partial sums have an upper bound.

Testing a given series for convergence would often be difficult if we depended on the definition of convergence alone. The problem is that it is usually not easy to find an expression for s_n, the nth partial sum, that is simple enough for us to decide whether or not it possesses a limit as $n \longrightarrow \infty$. In the tests for convergence given below, it is not necessary to find an expression for the nth partial sum.

14.3.2 **Definition.** The series $\sum A_n = A_1 + A_2 + A_3 + \cdots + A_n + \cdots$ is said to **dominate** the series $\sum a_n = a_1 + a_2 + a_3 + \cdots + a_n + \cdots$ if $|a_n| \le A_n$ for all positive integers n.

Clearly, the terms of a dominating series must all be nonnegative, although this is not necessarily true of the series that is dominated.

14.3.3 **Comparison Test.** Let $\sum a_n$ and $\sum b_n$ be series of nonnegative terms and let $\sum b_n$ dominate $\sum a_n$ so that $a_n \le b_n$ for all positive integers n.

(i) If the dominating series $\sum b_n$ is convergent, then $\sum a_n$ is convergent.

(ii) If $\sum a_n$ is divergent, then the dominating series $\sum b_n$ is divergent.

Proof. Assume that the dominating series $\sum b_n$ is convergent and let

$$s_n = a_1 + a_2 + a_3 + \cdots + a_n \quad \text{and} \quad t_n = b_1 + b_2 + b_3 + \cdots + b_n.$$

Since the nonnegative series $\sum b_n$ is convergent, its sequence of partial sums $\{t_n\}$ has an upper bound (by 14.3.1). But $0 \le s_n \le t_n$ for all positive integers n because $\sum b_n$ dominates $\sum a_n$. Hence $\{s_n\}$ is bounded above and $\sum a_n$ converges.

Part (ii) follows immediately, for if $\sum a_n$ is divergent, it is not convergent, and therefore $\sum b_n$ cannot be convergent [by part (i)]. ∎

Example 1. Show that the series $\displaystyle\sum \frac{n}{2^n(n+1)}$ converges.

Solution. Since $0 < n/(n+1) < 1$ for all positive integers n,

$$\frac{n}{2^n(n+1)} < \frac{1}{2^n}.$$

But $\sum 1/2^n$ is a geometric series with $r = 1/2$, and so it is convergent. Therefore the given series is convergent by the comparison test.

Example 2. Test for convergence $\displaystyle\sum \frac{n}{5n^2 - 4}$.

Solution.

$$\frac{n}{5n^2 - 4} > \frac{n}{5n^2} = \frac{1}{5n} = \frac{1}{5}\left(\frac{1}{n}\right).$$

Now $\sum 1/n$ is the harmonic series, which is divergent (Example, 3, Section 14.2). Therefore $\sum 1/5n = \sum n/5n^2$ is divergent (by 14.2.6), and it follows that the given series is divergent by the comparison test.

The next test follows from the comparison test.

14.3.4 **Theorem.** If $\sum a_n$ and $\sum b_n$ are series of positive terms and if

$$\lim_{n \to \infty} \frac{a_n}{b_n} = L > 0,$$

then either both series converge or both series diverge.

Proof. Since $\lim_{n \to \infty} a_n/b_n = L > 0$, there is a positive integer N such that

$$\left| \frac{a_n}{b_n} - L \right| < \frac{L}{2}$$

for all $n \ge N$. This is equivalent to

$$-\frac{L}{2} < \frac{a_n}{b_n} - L < \frac{L}{2},$$

or

$$\frac{L}{2} < \frac{a_n}{b_n} < \frac{3L}{2}$$

for $n \geq N$. Hence

$$\frac{L}{2} b_n < a_n \qquad \text{and} \qquad \frac{2a_n}{3L} < b_n$$

for all $n \geq N$. Thus the series $\sum a_n$ dominates the series $(L/2) \sum b_n$, and the series $\sum b_n$ dominates the series $(2/3L) \sum a_n$. It follows from the comparison test that either both the series $\sum a_n$ and $\sum b_n$ converge or both diverge. ∎

14.3.5 Theorem. If $\sum a_n$ and $\sum b_n$ are series of positive terms and if $\lim_{n \to \infty} (a_n/b_n) = 0$ and $\sum b_n$ converges, then $\sum a_n$ also converges.

For, $\lim_{n \to \infty} (a_n/b_n) = 0$ implies that there exists a positive integer N such that

$$\left| \frac{a_n}{b_n} \right| < 1$$

for all $n > N$. That is, $|a_n| < |b_n|$ for all $n > N$. But $a_n > 0$ and $b_n > 0$. Therefore $0 < a_n < b_n$ for all $n > N$. Thus $\sum b_n$ dominates $\sum a_n$ and, since $\sum b_n$ converges, so also does $\sum a_n$ (by 14.3.3).

Example 3. Show that the series $\sum \dfrac{1}{\sqrt{4n^2 - 7}}$ is divergent.

Solution. The limit of the ratio of the general term of the harmonic series $\sum 1/n$ to the general term of the given series, as $n \longrightarrow \infty$, is

$$\lim_{n \to \infty} \frac{1/n}{1/\sqrt{4n^2 - 7}} = \lim_{n \to \infty} \frac{\sqrt{4n^2 - 7}}{n} = \lim_{n \to \infty} \sqrt{\frac{4n^2 - 7}{n^2}} = \lim_{n \to \infty} \sqrt{4 - \frac{7}{n^2}} = 2.$$

Since the harmonic series is divergent, the given series is divergent (by 14.3.4).

To use the two preceding tests effectively, we need some series, whose convergence or divergence we know, as bases for comparison. The next test, known as the **Maclaurin–Cauchy integral test**, will provide such series; and it is interesting and worthwhile in its own right.

14.3.6 Integral Test. Let f be a continuous, positive, nonincreasing function defined for all real numbers $x \geq 1$ and let $a_i = f(i)$ for all positive integers i. Then the infinite series

$$\sum_{i=1}^{\infty} a_i$$

converges if and only if the improper integral

$$\int_1^{\infty} f(x) \, dx$$

converges.

Proof. By the mean value theorem for integrals,

$$\text{(1)} \qquad \int_i^{i+1} f(x)\,dx = 1 \cdot f(\xi),$$

where $i < \xi < i + 1$. Since f is nonincreasing,

$$a_i = f(i) \geq f(\xi) \geq f(i+1) = a_{i+1}.$$

Therefore

$$\text{(2)} \qquad a_i \geq \int_i^{i+1} f(x)\,dx \geq a_{i+1}.$$

It follows from (2) that for every n

$$\text{(3)} \qquad \sum_{i=1}^n a_i \geq \sum_{i=1}^n \int_i^{i+1} f(x)\,dx = \int_1^{n+1} f(x)\,dx \geq \sum_{i=1}^n a_{i+1} = \left(\sum_{i=1}^{n+1} a_i \right) - a_1$$

(see Fig. 14–1). Therefore all three of the expressions

$$\sum_{i=1}^\infty a_i, \qquad \int_1^\infty f(x)\,dx, \qquad \left(\sum_{i=1}^\infty a_i \right) - a_1$$

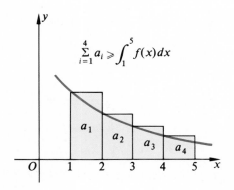

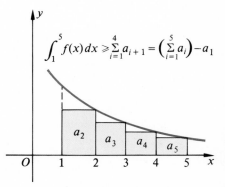

Figure 14-1

converge or diverge together. Also, in case of convergence, if $\sum_{i=1}^\infty a_i = S$, then

$$S \geq \int_1^\infty f(x)\,dx \geq S - a_1. \quad \blacksquare$$

Example 4. The series

$$\sum_{n=1}^\infty \frac{1}{n^p} = 1 + \frac{1}{2^p} + \frac{1}{3^p} + \cdots + \frac{1}{n^p} + \cdots,$$

where p is a constant, is called the **p series**. Prove that
   (a)   the p series converges if $p > 1$.
   (b)   the p series diverges if $p \leq 1$.

Solution. The function defined by $1/x^p$, where p is a *nonnegative constant*, is continuous, positive, and nonincreasing for $x \geq 1$. Hence, by the integral test, for $p \geq 0$, the p series converges if and only if the improper integral $\int_1^\infty x^{-p}\,dx = \lim_{t \to \infty} \int_1^t x^{-p}\,dx$ exists.

Now

$$\int_1^t x^{-p}\,dx = \frac{x^{1-p}}{1-p}\bigg|_1^t = \frac{t^{1-p}-1}{1-p} \qquad \text{when} \quad p \neq 1,$$

and

$$\int_1^t x^{-p}\,dx = \ln x \bigg|_1^t = \ln t \qquad \text{when} \quad p = 1.$$

Since

$$\lim_{t \to \infty} t^{1-p} = 0 \quad \text{if} \quad p > 1 \qquad \text{and} \qquad \lim_{t \to \infty} t^{1-p} = \infty \quad \text{if} \quad p < 1,$$

and since

$$\lim_{t \to \infty} \ln t = \infty,$$

we have proved that the p series converges if $p > 1$ and diverges if $0 \leq p \leq 1$.

When $p = 1$, the p series is the harmonic series $1 + \frac{1}{2} + \frac{1}{3} + \frac{1}{4} + \cdots$. We have just proved that the harmonic series is divergent. (In Example 3 of Section 14.2 we proved that the harmonic series is divergent without using the integral test.)

Now consider the p series when $p < 0$, say $p = -q$, where q is some positive number. Then

$$\sum \frac{1}{n^p} = \sum n^q, \qquad \text{where} \quad q > 0.$$

Since $\lim_{n \to \infty} n^q = \infty$ for $q > 0$, the series $\sum n^q = \sum 1/n^p$ is divergent if $p < 0$.

The p series is a useful comparison series.

Example 5. Test for convergence or divergence the series $\sum \dfrac{\ln n}{n^3}$.

Solution. Since $\ln n < n$ for any positive integer n, we have

$$\frac{\ln n}{n^3} < \frac{1}{n^2},$$

and the convergent p series $\sum 1/n^2$ dominates the given series of positive terms. Hence the given series converges.

Example 6. By means of an improper integral, find an upper bound for the error in using the sum of the first five terms of the convergent series

$$\sum_{n=1}^\infty \frac{n}{e^{n^2}}$$

to approximate the sum of the series.

Solution. The error is

$$\sum_{n=6}^\infty \frac{n}{e^{n^2}};$$

and

$$\sum_{n=6}^{\infty} \frac{n}{e^{n^2}} < \int_5^{\infty} x e^{-x^2}\, dx = -\frac{1}{2} \int_5^{\infty} e^{-x^2}\, d(-x^2)$$

$$= -\frac{1}{2} \lim_{t \to \infty} \int_5^t e^{-x^2}\, d(-x^2) = -\frac{1}{2} \lim_{t \to \infty} \left[e^{-x^2} \right]_5^t$$

$$= -\frac{1}{2} \lim_{t \to \infty} \left(\frac{1}{e^{t^2}} - \frac{1}{e^{25}} \right) = \frac{1}{2e^{25}} < \frac{1}{2(2.7)^{25}} < 8 \times 10^{-12}.$$

Thus the error is less than 8×10^{-12}.

14.3.7   Ratio Test.   Let $a_1 + a_2 + a_3 + \cdots + a_n + \cdots$ be a series of positive terms and let

$$\lim_{n \to \infty} \frac{a_{n+1}}{a_n} = \rho.$$

(i)   If $\rho < 1$, the series converges.
(ii)   If $\rho > 1$, the series diverges.
(iii)   If $\rho = 1$, the test is inconclusive.

Proof of (i).   ($\rho < 1$.) Since $a_n > 0$ for all positive integers n, $a_{n+1}/a_n > 0$ for all n and ρ cannot be negative.

Figure 14-2

Choose a number r between ρ and 1 so that $0 \le \rho < r < 1$ (Fig. 14–2); then $r - \rho > 0$. Since $\lim_{n \to \infty} (a_{n+1}/a_n) = \rho$, there exists a positive integer N such that for all $n \ge N$,

$$\left| \frac{a_{n+1}}{a_n} - \rho \right| < r - \rho,$$

or

$$\rho - (r - \rho) < \frac{a_{n+1}}{a_n} < \rho + (r - \rho).$$

Therefore $a_{n+1} < a_n r$ for all $n \ge N$. Thus

$$a_{N+1} < a_N r,$$

$$a_{N+2} < a_{N+1} r < a_N r^2,$$

$$a_{N+3} < a_{N+2} r < a_N r^3,$$

$$\cdot \quad \cdot \quad \cdot \quad \cdot \quad \cdot \quad \cdot$$

Hence the geometric series $a_N r + a_N r^2 + a_N r^3 + \cdots$ dominates the series $a_{N+1} + a_{N+2} + a_{N+3} + \cdots$. Since $r = |r| < 1$, the geometric series converges, and so the series $\sum_{n=N+1}^{\infty} a_n$ converges. Therefore the series $\sum_{n=1}^{\infty} a_n$ converges.

Proof of (ii).   ($\rho > 1$.) As above, there is a positive integer N such that $a_{n+1}/a_n > 1$ for all $n \ge N$. That is, $a_{n+1} > a_n$ for all $n \ge N$. But $a_N > 0$, since this is a series of positive terms. Hence

$$a_n > a_N > 0$$

for all $n > N$, and $\lim\limits_{n \to \infty} a_n$ cannot be zero. It follows that $\sum a_n$ diverges.

Proof of (iii).   ($p = 1$.) For the p series,

$$1 + \frac{1}{2^p} + \frac{1}{3^p} + \cdots + \frac{1}{n^p} + \cdots,$$

the ratio $a_{n+1}/a_n = n^p/(n + 1)^p = [n/(n + 1)]^p$; and

$$\lim_{n \to \infty} \left(\frac{n}{n + 1}\right)^p = \left(\lim_{n \to \infty} \frac{n}{n + 1}\right)^p = 1^p = 1.$$

But if $0 < p < 1$, the series diverges; and if $p > 1$, the series converges (Example 4 above). Hence when $p = 1$, the given series $\sum a_n$ may converge or may diverge.   ∎

Example 7. Test for convergence the infinite series $\sum \dfrac{n^{20}}{2^n}$.

Solution.

$$\frac{a_{n+1}}{a_n} = \frac{(n + 1)^{20}}{2^{n+1}} \div \frac{n^{20}}{2^n} = \frac{1}{2} \left(\frac{n + 1}{n}\right)^{20} = \frac{1}{2} \left(1 + \frac{1}{n}\right)^{20}$$

and

$$\lim_{n \to \infty} \frac{a_{n+1}}{a_n} = \lim_{n \to \infty} \frac{1}{2} \left(\frac{n + 1}{n}\right)^{20} = \frac{1}{2} \lim_{n \to \infty} \left(1 + \frac{1}{n}\right)^{20} = \frac{1}{2}.$$

Therefore, by the ratio test, the given series is convergent.

Exercises

In Exercises 1 to 6, test the given series for convergence or divergence by means of the comparison test or by means of theorem 14.3.4. Give the reason for your choice of test.

1. $1 + \dfrac{1}{1!} + \dfrac{1}{2!} + \dfrac{1}{3!} + \dfrac{1}{4!} + \cdots.$

2. $\dfrac{1}{1 \cdot 2} + \dfrac{1}{2 \cdot 3} + \dfrac{1}{3 \cdot 4} + \dfrac{1}{4 \cdot 5} + \dfrac{1}{5 \cdot 6} + \cdots.$

3. $\dfrac{1}{2^2} + \dfrac{2}{3^2} + \dfrac{3}{4^2} + \dfrac{4}{5^2} + \dfrac{5}{6^2} + \cdots.$

4. $\dfrac{1}{2^3} + \dfrac{2}{3^3} + \dfrac{3}{4^3} + \dfrac{4}{5^3} + \dfrac{5}{6^3} + \cdots.$

5. $\dfrac{2}{1 \cdot 3} + \dfrac{3}{2 \cdot 4} + \dfrac{4}{3 \cdot 5} + \dfrac{5}{4 \cdot 6} + \dfrac{6}{5 \cdot 7} + \cdots.$

6. $\dfrac{2}{1 \cdot 3 \cdot 4} + \dfrac{3}{2 \cdot 4 \cdot 5} + \dfrac{4}{3 \cdot 5 \cdot 6} + \dfrac{5}{4 \cdot 6 \cdot 7}$
$+ \dfrac{6}{5 \cdot 7 \cdot 8} + \cdots.$

In Exercises 7 to 11, test for convergence by means of the integral test.

7. $\dfrac{1}{1 \cdot 2} + \dfrac{1}{2 \cdot 3} + \dfrac{1}{3 \cdot 4} + \dfrac{1}{4 \cdot 5} + \dfrac{1}{5 \cdot 6} + \cdots.$

8. $\dfrac{1}{1^2 + 1} + \dfrac{2}{2^2 + 1} + \dfrac{3}{3^2 + 1} + \dfrac{4}{4^2 + 1}$
$+ \dfrac{5}{5^2 + 1} + \cdots.$

9. $\dfrac{1}{2 \ln 2} + \dfrac{1}{3 \ln 3} + \dfrac{1}{4 \ln 4} + \dfrac{1}{5 \ln 5}$
$+ \dfrac{1}{6 \ln 6} + \cdots.$

10. $\dfrac{1}{\sqrt{1^2 + 1}} + \dfrac{1}{\sqrt{2^2 + 1}} + \dfrac{1}{\sqrt{3^2 + 1}} + \dfrac{1}{\sqrt{4^2 + 1}}$
$\dfrac{1}{\sqrt{5^2 + 1}} + \cdots.$

11. $\dfrac{1}{\sqrt{1^3 + 1}} + \dfrac{1}{\sqrt{2^3 + 1}} + \dfrac{1}{\sqrt{3^3 + 1}} + \dfrac{1}{\sqrt{4^3 + 1}}$
$\dfrac{1}{\sqrt{5^3 + 1}} + \cdots.$

Note. The integral involved in Exercise 11 cannot be evaluated directly. Can you show, however, that it is dominated by an integral that can be evaluated?

In Exercises 12 to 16, estimate the error involved in taking the first five terms as the sum of the series.

12. $\dfrac{1}{1\cdot 2} + \dfrac{1}{2\cdot 3} + \dfrac{1}{3\cdot 4} + \dfrac{1}{4\cdot 5} + \dfrac{1}{5\cdot 6} + \cdots.$

13. $\dfrac{1}{2^2 - 1} + \dfrac{1}{3^2 - 1} + \dfrac{1}{4^2 - 1} + \dfrac{1}{5^2 - 1}$
$+ \dfrac{1}{6^2 - 1} + \cdots.$

14. $\dfrac{1}{e} + \dfrac{2}{e^2} + \dfrac{3}{e^3} + \dfrac{4}{e^4} + \dfrac{5}{e^5} + \cdots.$

15. $\dfrac{1}{e} + \dfrac{2^2}{e^2} + \dfrac{3^2}{e^3} + \dfrac{4^2}{e^4} + \dfrac{5^2}{e^5} + \cdots.$

16. $1 + \dfrac{1}{2\sqrt{2}} + \dfrac{1}{3\sqrt{3}} + \dfrac{1}{4\sqrt{4}} + \dfrac{1}{5\sqrt{5}} + \cdots.$

In Exercises 17 to 21, test for convergence by means of the ratio test.

17. $1 + \dfrac{1}{2!} + \dfrac{1}{3!} + \dfrac{1}{4!} + \dfrac{1}{5!} + \cdots.$

18. $\dfrac{1}{2} + \dfrac{2}{2^2} + \dfrac{3}{2^3} + \dfrac{4}{2^4} + \dfrac{5}{2^5} + \cdots.$

19. $1 + \dfrac{1}{2^{3/2}} + \dfrac{1}{3^{3/2}} + \dfrac{1}{4^{3/2}} + \dfrac{1}{5^{3/2}} + \cdots.$

20. $3 + \dfrac{3^2}{2!} + \dfrac{3^3}{3!} + \dfrac{3^4}{4!} + \dfrac{3^5}{5!} + \cdots.$

21. $\dfrac{\ln 2}{2^2} + \dfrac{\ln 3}{2^3} + \dfrac{\ln 4}{2^4} + \dfrac{\ln 5}{2^5} + \dfrac{\ln 6}{2^6} + \cdots.$

22. In the solution to Example 4 above, it is stated that the function f defined by $f(x) = 1/x^p$, where p is a nonnegative constant, is continuous, positive, and nonincreasing for $x \geq 1$. Prove this statement.

23. Prove that

$$\lim_{n\to\infty} \frac{n!}{n^n} = 0,$$

where n denotes a positive integer.

14.4 ALTERNATING SERIES. ABSOLUTE CONVERGENCE

An infinite series of the form

$$\sum_{i=1}^{\infty} (-1)^{i+1} a_i = a_1 - a_2 + a_3 - \cdots + (-1)^{n+1} a_n + \cdots,$$

where $a_i > 0$ for $i = 1, 2, 3, \cdots$, is called an **alternating series.**

14.4.1 Theorem. The alternating series

$$\sum_{n=1}^{\infty} (-1)^{n+1} a_n = a_1 - a_2 + a_3 - \cdots$$

converges if $0 < a_{n+1} < a_n$ for all positive integers n and if $\lim\limits_{n\to\infty} a_n = 0$.

Proof. In the two partial sums

(1) $$s_{2n} = (a_1 - a_2) + (a_3 - a_4) + \cdots + (a_{2n-1} - a_{2n})$$

and

(2) $$s_{2n+1} = a_1 - (a_2 - a_3) - (a_4 - a_5) - \cdots - (a_{2n} - a_{2n+1}),$$

the quantities in parentheses are all positive numbers because, by hypothesis, $a_n > a_{n+1}$ for all positive integers n. Hence the sequence $\{s_{2n}\}$ is increasing and the sequence $\{s_{2n+1}\}$ is decreasing.

The fact that all the quantities in parentheses in (1) and (2) are positive numbers also implies that $s_{2n} > 0$ and $s_{2n+1} < a_1$, for all positive integers n. Since $s_{2n+1} = s_{2n} + a_{2n+1}$, then $s_{2n} < s_{2n+1}$. Combining these results, we have

$$(3) \qquad\qquad 0 < s_{2n} < s_{2n+1} < a_1$$

for all positive integers n. Therefore the monotonic sequences $\{s_{2n}\}$ and $\{s_{2n+1}\}$ are bounded below by 0 and above by a_1, and they are therefore convergent.

Since $\lim\limits_{n\to\infty} a_{2n+1} = 0$ by hypothesis and $s_{2n+1} = s_{2n} + a_{2n+1}$,

$$\lim_{n\to\infty} s_{2n+1} = \lim_{n\to\infty} (s_{2n} + a_{2n+1}) = \lim_{n\to\infty} s_{2n} + \lim_{n\to\infty} a_{2n+1} = \lim_{n\to\infty} s_{2n}.$$

That is, both the sequences $\{s_{2n}\}$ and $\{s_{2n+1}\}$ have the same limit, which we call S.

Therefore $\lim\limits_{m\to\infty} s_m = S$ whether m takes on odd or even values or both, and

$$\sum_{n=1}^{\infty} (-1)^{n+1} a_n$$

converges to the value S. ∎

Example 1. The alternating series

$$1 - \frac{1}{2} + \frac{1}{3} - \cdots + (-1)^{n+1} \frac{1}{n} + \cdots$$

converges, since $a_{n+1} = 1/(n+1) < 1/n = a_n$ for all positive integers n, and

$$\lim_{n\to\infty} a_n = \lim_{n\to\infty} \frac{1}{n} = 0.$$

14.4.2 Theorem. If

$$\sum (-1)^{n+1} a_n = a_1 - a_2 + a_3 - \cdots + (-1)^{n+1} a_n + \cdots$$

is an alternating series such that $a_n > a_{n+1} > 0$ and $\lim\limits_{n\to\infty} a_n = 0$, the error made by using the sum of the first k terms as an approximation for the sum of the series is less than a_{n+1}, the absolute value of the first neglected term.

Proof. Write

$$\sum_{i=1}^{\infty} (-1)^{i+1} a_i = a_1 - a_2 + a_3 - \cdots + (-1)^{k+1} a_k + R_k,$$

where $R_k = \sum\limits_{i=k+1}^{\infty} (-1)^{i+1} a_i$ is the remainder of the given series after the first k terms. The error in approximating the sum of the given series by the sum of its first k terms is the sum of the series of terms neglected—that is, R_k. But $R_k = \sum\limits_{i=k+1}^{\infty} (-1)^{i+1} a_i$ is an alternating series with the characteristics assumed in 14.4.1; so the absolute value of each of the partial sums of the series $R_k = (-1)^{k+2}a_{k+1} + (-1)^{k+3}a_{k+2} + (-1)^{k+4}a_{k+3} + \cdots$ is less than the term a_{k+1}, as shown in the inequalities (3) in the proof of 14.4.1. ∎

This characteristic of *convergent alternating series* of the type we have been discussing makes them very valuable for computation. For example, we shall soon see that

$$\ln 1.2 = 0.2 - \frac{(0.2)^2}{2} + \frac{(0.2)^3}{3} - \frac{(0.2)^4}{4} + \frac{(0.2)^5}{5} - \cdots .$$

By 14.4.2, the error in using the sum of the first four terms of this series to approximate the value of $\ln 1.2$ is less than $a_5 = (0.2)^5/5 = 0.000064$.

14.4.3 Definition. The infinite series

$$\sum a_n = a_1 + a_2 + a_3 + \cdots + a_n + \cdots$$

is said to **converge absolutely** if

$$\sum |a_n| = |a_1| + |a_2| + |a_3| + \cdots + |a_n| + \cdots$$

converges.

Example 2. The series

$$1 - \frac{1}{3} + \frac{1}{9} - \frac{1}{27} + \cdots + \frac{(-1)^{n+1}}{3^{n-1}} + \cdots$$

converges absolutely because the series

$$|1| + \left|-\frac{1}{3}\right| + \left|\frac{1}{9}\right| + \left|-\frac{1}{27}\right| + \cdots + \left|\frac{(-1)^{n+1}}{3^{n-1}}\right| + \cdots = 1 + \frac{1}{3} + \frac{1}{9} + \frac{1}{27} + \cdots + \frac{1}{3^{n-1}} + \cdots .$$

converges since it is a geometric series with $r = \frac{1}{3} < 1$.

But the convergent alternating series

$$1 - \frac{1}{2} + \frac{1}{3} - \frac{1}{4} + \cdots + \frac{(-1)^{n+1}}{n} + \cdots$$

does not converge absolutely, since

$$1 + \frac{1}{2} + \frac{1}{3} + \frac{1}{4} + \cdots + \frac{1}{n} + \cdots$$

is the harmonic series, which we found to be divergent.

14.4.4 Definition. If a series is convergent but not absolutely convergent, it is said to be **conditionally convergent**.

Thus the alternating series $\sum (-1)^{n+1}/n$ in Example 1 is conditionally convergent.

14.4.5 Theorem. If a series converges absolutely, it converges.

Proof. Assume that the series $\sum |a_n|$ converges and let $b_n = a_n + |a_n|$. Notice that for each positive integer n, either $b_n = 0$ or $b_n = 2|a_n|$. Thus

$$0 \leq b_n \leq 2|a_n|;$$

and if we let

$$A = \sum_{i=1}^{\infty} |a_i|, \qquad A_n = \sum_{i=1}^{n} |a_i|, \qquad B_n = \sum_{i=1}^{n} b_i,$$

then $0 \le B_n \le 2A_n \le 2A$. Therefore (by 14.3.1) $\lim_{n \to \infty} B_n$ exists and $\sum_{i=1}^{\infty} b_n$ converges.

Since $\sum |a_n|$ and $\sum b_n$ both converge, then (by 14.2.8)

$$\sum_{i=1}^{\infty} (b_i - |a_i|) = \sum_{i=1}^{\infty} a_i$$

converges. ∎

Of course, all convergent series of nonnegative terms are absolutely convergent. Our tests for convergence of series of positive terms (14.3) may be used to determine whether a series containing an infinite number of positive terms and an infinite number of negative terms is absolutely convergent. It is often easier to establish the absolute convergence of such a series and then infer its convergence by 14.4.5 than to prove convergence directly.

Example 3. Determine whether the series

$$\sum \frac{10 \sin \frac{1}{6} n\pi}{n^{1.1}}$$

converges or diverges.

Solution. The first five terms of this series are positive, the sixth is zero, the next five are negative, the twelfth is zero, and so forth. Moreover, the series of absolute values of its terms is not monotonic, since sometimes a later term is greater than an earlier term and sometimes it is less. Nevertheless, the absolute convergence of the series is easily established.

Since $-1 \le \sin \frac{1}{6} n\pi \le 1$ for all positive integers n, $|(10 \sin \frac{1}{6} n\pi)/n^{1.1}| \le 10/n^{1.1}$. Thus the series

$$\sum \left| \frac{10 \sin \frac{1}{6} n\pi}{n^{1.1}} \right|$$

is dominated by the series $\sum 10/n^{1.1} = 10 \sum 1/n^{1.1}$. Since $\sum 1/n^{1.1}$ is a convergent p series (with $p = 1.1 > 1$), the given series converges absolutely. Therefore the given series converges.

14.4.6 Theorem. For any series of nonzero terms, $\sum u_n$, let

$$\lim_{n \to \infty} \left| \frac{u_{n+1}}{u_n} \right| = \rho.$$

(i) If $\rho < 1$, the series converges absolutely.
(ii) If $\rho > 1$, the series diverges.

Proof. Part (i) follows immediately from the ratio test (14.3.7) and the definition of absolute convergence. As to (ii), we showed in the proof of part (ii) of the ratio test that $\lim_{n \to \infty} |u_n|$ cannot be zero. Consequently, $\lim_{n \to \infty} u_n$ cannot be zero and the series $\sum u_n$ diverges. ∎

We saw in Section 14.2 that a convergent series behaves in some ways like a finite sum. In particular, we showed in 14.2.6 and 14.2.7 that convergent series obey laws resembling the associative and distributive laws for finite sums. If a series is *absolutely* convergent, the commutative law of addition also holds; that is, any re-arrangement of the terms of an absolutely convergent series will result in a series that is also absolutely convergent to the same sum. We state this, without proof, in the following theorem.

14.4.7 Theorem. If a series is absolutely convergent, its terms may be rearranged without affecting the absolute convergence of the series or its sum.

It was proved in 14.2.8 that if any two series are convergent (conditionally or absolutely), they can be added term by term and the resulting series also will be con-vergent, with a sum equal to the sum of the sums of the original series.

If two series are *absolutely convergent*, the series formed by multiplying the terms of one series by the terms of the other series, as indicated in the next theorem, will converge absolutely and its sum will be the product of the sums of the original series.

14.4.8 Theorem. If the series

$$\sum a_n = a_1 + a_2 + a_3 + \cdots \qquad \text{and} \qquad \sum b_n = b_1 + b_2 + b_3 + \cdots$$

are absolutely convergent with sums A and B, then the series

$$a_1 b_1 + a_1 b_2 + a_2 b_1 + a_1 b_3 + a_2 b_2 + a_3 b_1 + \cdots$$

converges absolutely and its sum is AB.

Notice that the terms of the product series consist of all possible products of the form $a_i b_j$ taken in the order for which first $i + j = 2$, then $i + j = 3$, then $i + j = 4$, and so on.

A proof of this theorem may be found in books on advanced calculus.

Exercises

In each of Exercises 1 to 10, determine whether the given series converges absolutely, converges con-ditionally, or diverges.

1. $\dfrac{1}{1 \cdot 2} - \dfrac{1}{2 \cdot 3} + \dfrac{1}{3 \cdot 4} - \dfrac{1}{4 \cdot 5} + \dfrac{1}{5 \cdot 6} - \cdots.$

2. $\dfrac{1}{2 \ln 2} - \dfrac{1}{3 \ln 3} + \dfrac{1}{4 \ln 4} - \dfrac{1}{5 \ln 5} + \dfrac{1}{6 \ln 6} - \cdots.$

3. $\dfrac{1}{2\sqrt{2}} - \dfrac{1}{3\sqrt{3}} + \dfrac{1}{4\sqrt{4}} - \dfrac{1}{5\sqrt{5}} + \dfrac{1}{6\sqrt{6}} - \cdots.$

4. $\dfrac{1}{\sqrt{2^2 - 1}} - \dfrac{1}{\sqrt{3^2 - 1}} + \dfrac{1}{\sqrt{4^2 - 1}} - \dfrac{1}{\sqrt{5^2 - 1}}$
   $+ \dfrac{1}{\sqrt{6^2 - 1}} - \cdots.$

5. $1 - \dfrac{2}{2^2} + \dfrac{3}{2^3} - \dfrac{4}{2^4} + \dfrac{5}{2^5} - \cdots.$

6. $1 - \dfrac{1}{2!} + \dfrac{1}{3!} - \dfrac{1}{4!} + \dfrac{1}{5!} - \cdots.$

7. $\dfrac{1}{1^2 + 1} - \dfrac{2}{2^2 + 1} + \dfrac{3}{3^2 + 1} - \dfrac{4}{4^2 + 1}$

$+ \dfrac{5}{5^2 + 1} - \cdots$.

8. $\dfrac{1}{2!} - \dfrac{2}{3!} + \dfrac{3}{4!} - \dfrac{4}{5!} + \dfrac{5}{6!} - \cdots$.

9. $\dfrac{1}{e} - \dfrac{2^2}{e^2} + \dfrac{2^3}{e^3} - \dfrac{2^4}{e^4} + \dfrac{2^5}{e^5} - \cdots$.

10. $1 - \dfrac{1}{2} + \dfrac{2}{3} - \dfrac{3}{4} + \dfrac{4}{5} - \cdots$.

11. Prove that a convergent infinite series, all of whose terms are negative, converges absolutely.

12. Prove that if an infinite series converges conditionally, its positive terms form a divergent series and its negative terms form another divergent series.

14.5 POWER SERIES

If $\{a_n\}$ is a sequence of constants,

$$(1) \qquad \sum_{i=0}^{\infty} a_i x^i = a_0 + a_1 x + a_2 x^2 + a_3 x^3 + \cdots + a_n x^n + \cdots$$

is called a **power series** in x (where $x^0 = 1$ for all values of x, including $x = 0$).

When a particular number is substituted for x, the power series (1) becomes an infinite series whose terms are numbers, such as we have been studying. Clearly, the series (1) must converge if $x = 0$. But are there any other numbers x for which the power series converges?

Example 1. Find all numbers x for which the following power series converges:

$$\sum_{n=0}^{\infty} \frac{x^n}{(n + 1)2^n} = 1 + \frac{1}{2 \cdot 2} x + \frac{1}{3 \cdot 4} x^2 + \frac{1}{4 \cdot 8} x^3 + \cdots + \frac{1}{(n + 1)2^n} x^n + \cdots.$$

Solution. Applying the ratio test (14.4.6), we find

$$\rho = \lim_{n \to \infty} \left| \frac{x^{n+1}}{(n + 2)2^{n+1}} \div \frac{x^n}{(n + 1)2^n} \right| = \frac{|x|}{2} \lim_{n \to \infty} \left(\frac{n + 1}{n + 2} \right) = \frac{|x|}{2}.$$

Thus the power series converges absolutely when x is a number that makes $\rho = |x|/2 < 1$, and it diverges when $\rho = |x|/2 > 1$. The series is absolutely convergent when $|x| < 2$ and diverges when $|x| > 2$.

If $x = 2$ or $x = -2$, the ratio test fails. But when $x = 2$, the power series becomes the harmonic series $1 + \frac{1}{2} + \frac{1}{3} + \cdots$, which diverges; and when $x = -2$, the power series is the convergent alternating series $1 - \frac{1}{2} + \frac{1}{3} - \frac{1}{4} + \cdots$.

Hence the set of numbers for which this power series converges is the half-open interval $[-2, 2)$. It is absolutely convergent for all numbers x in the open interval $(-2, 2)$ and converges conditionally for $x = -2$; it diverges for all other numbers.

The set of numbers for which a power series converges is called the **convergence set** for the power series.

The foregoing example suggests that any power series in x to which we can apply the ratio test will have for its convergence set one of the following:

1. The origin alone.

2. All points in an open interval $(-r, r)$ whose center is the origin and possibly one or both endpoints.

3. All real numbers.

In 14.5.2 and 14.5.3 we shall prove that this statement is true for *all* power series in x, even for those to which we cannot apply the ratio test.

Power series in x

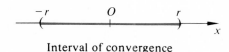

Interval of convergence

Figure 14-3

In case 2 the series converges for $|x| < r$ and diverges for $|x| > r$; and we call r the **radius of convergence** and the open interval $(-r, r)$ the **interval of convergence** (Fig. 14-3).

In case 1, where the series diverges for $|x| > 0$, we take the radius of convergence to be $r = 0$. In case 3, the interval of convergence is $(-\infty, \infty)$.

Example 2. Find the convergence set for the power series

$$\sum n!\, x^n = 1 + x + 2x^2 + 6x^3 + \cdots + n!\, x^n + \cdots.$$

Solution. Using the ratio test, we have

$$\rho = \lim_{n \to \infty} \left| \frac{(n+1)!\, x^{n+1}}{n!\, x^n} \right| = \lim_{n \to \infty} |(n+1)x|.$$

For $x = 0$, $\rho = 0$. But for any number $x \neq 0$, $\rho = \infty$. Thus this power series in x converges only for $x = 0$. For any other number, it diverges.

Example 3. Show that the power series

$$\sum \frac{x^n}{n!} = 1 + x + \frac{x^2}{2} + \frac{x^3}{6} + \cdots + \frac{x^n}{n!} + \cdots$$

converges for all real numbers x. (Define $0! = 1$.)

Solution. The limit of the absolute value of the ratio of the $(n+1)$st term to the nth term, as $n \longrightarrow \infty$, is

$$\rho = \lim_{n \to \infty} \left| \frac{x^{n+1}}{(n+1)!} \div \frac{x^n}{n!} \right| = |x| \lim_{n \to \infty} \frac{1}{n+1} = |x| \cdot 0 = 0$$

for all numbers x. Since $\rho < 1$, the series converges (by the ratio test). Hence the given power series converges for all real numbers x.

A useful result follows from Example 3. Since $\sum x^n/n!$ converges for all real numbers, the limit of its nth term as $n \longrightarrow \infty$ must be zero. Therefore

14.5.1
$$\lim_{n \to \infty} \frac{x^n}{n!} = 0$$

for all real numbers x.

Example 4. Prove that if k is an arbitrary real number and $-1 < x < 1$, then

$$\lim_{n \to \infty} \frac{k(k-1)(k-2) \cdots (k-n)}{n!} x^n = 0.$$

Solution. Applying the ratio test to the series

$$\sum_{n=1}^{\infty} \frac{k(k-1)(k-2) \cdots (k-n)}{n!} x^n,$$

we find

$$\rho = \lim_{n \to \infty} \left| \frac{k(k-1) \cdots (k-n)(k-n-1)x^{n+1}}{(n+1)!} \cdot \frac{n!}{k(k-1) \cdots (k-n)x^n} \right|$$

$$= \lim_{n \to \infty} \left| \frac{k-n-1}{n+1} \right| |x| = |x|.$$

When $|x| < 1$, we have $\rho < 1$ and the series converges. Thus the limit of its nth term as $n \longrightarrow \infty$ is zero; that is,

$$\lim_{n \to \infty} \frac{k(k-1)(k-2) \cdots (k-n)}{n!} x^n = 0$$

if $-1 < x < 1$.

14.5.2 **Theorem.** If the power series $\sum a_n x^n$ converges for a number $x_1 \neq 0$, then it converges absolutely for all numbers x such that $|x| < |x_1|$.

Proof. Assume that $\sum a_n x_1^n$ converges. Then $\lim_{n \to \infty} a_n x_1^n = 0$. Thus there is a positive integer N such that $|a_n x_1^n| < 1$ for all $n \geq N$. For any number x such that $|x| < |x_1|$,

$$|a_n x^n| = |a_n x_1^n| \left| \frac{x}{x_1} \right|^n < \left| \frac{x}{x_1} \right|^n$$

when $n \geq N$. Thus the series $\sum_{n=N}^{\infty} |a_n x^n|$ is dominated by the series

$$\sum_{n=N}^{\infty} \left| \frac{x}{x_1} \right|^n.$$

But this latter series is a convergent geometric series because $|x| < |x_1|$ and $|x/x_1| < 1$. Therefore the series $\sum_{n=N}^{\infty} |a_n x^n|$ converges, and so the given power series converges absolutely for all numbers x such that $|x| < |x_1|$. ∎

14.5.3 **Corollary.** If the power series $\sum a_n x^n$ diverges for a number x_2, then it diverges for all numbers x such that $|x| > |x_2|$.

Proof. If the series converges for some number x such that $|x| > |x_2|$, it must converge for x_2 (by 14.5.2), which is contrary to hypothesis. Therefore the series diverges for all x such that $|x| > |x_2|$. ∎

From these theorems it follows that if a power series converges at some point other than the origin, either the series converges absolutely everywhere or there is a positive number r such that the series converges absolutely throughout the open interval $(-r, r)$ and diverges everywhere outside the closed interval $[-r, r]$. These theorems say nothing about the convergence or divergence of the power series at the endpoints of the interval of convergence.

More generally, if $\{a_n\}$ is a sequence, the expression

$$\sum a_n(x - b)^n = a_0 + a_1(x - b) + a_2(x - b)^2 + \cdots + a_n(x - b)^n + \cdots$$

is called a **power series in** $(x - b)$.

Since this power series in $(x - b)$ can be obtained from a power series in x by the translation $x = x' - b$, all that we have said about power series in x applies equally to power series in $(x - b)$. The interval of convergence is now $(b - r, b + r)$, with center at the point b (Fig. 14-4). If the power series in $(x - b)$ converges only for $x = b$, we say that $r = 0$. If it converges for all real numbers, we say that $r = \infty$.

Power series in $(x - b)$

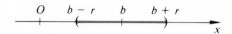

Interval of convergence

Figure 14-4

Example 5. Find the interval of convergence and the convergence set of the power series in $(x - 2)$:

$$\sum_{n=0}^{\infty} (-1)^n \frac{(x - 2)^n}{n + 1} = 1 - \frac{x - 2}{2} + \frac{(x - 2)^2}{3} - \cdots + \frac{(-1)^n(x - 2)^n}{n + 1} + \cdots.$$

Solution. Again we use the ratio test.

$$\rho = \lim_{n\to\infty} \left| \frac{(x - 2)^{n+1}}{n + 2} \cdot \frac{n + 1}{(x - 2)^n} \right| = |x - 2| \lim_{n\to\infty} \left(\frac{n + 1}{n + 2} \right) = |x - 2|.$$

Now $\rho < 1 \iff |x - 2| < 1 \iff -1 < x - 2 < 1 \iff 1 < x < 3$. Hence the interval of convergence of this power series in $(x - 2)$ is $(1, 3)$.

When $x = 1$, the series is $\sum 1/(n + 1)$, which is the (divergent) harmonic series.

When $x = 3$, the series is $\sum (-1)^n 1/(n + 1) = 1 - \frac{1}{2} + \frac{1}{3} - \cdots$, which is a convergent alternating series.

Therefore the convergence set of this power series in $(x - 2)$ is $(1, 3]$.

Exercises

Find the convergence set of the given power series.

1. $\dfrac{x}{1 \cdot 2} - \dfrac{x^2}{2 \cdot 3} + \dfrac{x^3}{3 \cdot 4} - \dfrac{x^4}{4 \cdot 5} + \dfrac{x^5}{5 \cdot 6} - \cdots.$

2. $1 + x + \dfrac{x^2}{2!} + \dfrac{x^3}{3!} + \dfrac{x^4}{4!} + \cdots.$

3. $x - \dfrac{x^3}{3!} + \dfrac{x^5}{5!} - \dfrac{x^7}{7!} + \dfrac{x^9}{9!} - \cdots.$

4. $1 - \dfrac{x^2}{2!} + \dfrac{x^4}{4!} - \dfrac{x^6}{6!} + \dfrac{x^8}{8!} - \dfrac{x^{10}}{10!} + \cdots.$

5. $x + 2x^2 + 3x^3 + 4x^4 + \cdots.$

6. $x + 2^2x^2 + 3^2x^3 + 4^2x^4 + \cdots.$

7. $1 - x + \dfrac{x^2}{2} - \dfrac{x^3}{3} + \dfrac{x^4}{4} - \cdots.$

8. $1 + x + \dfrac{x^2}{\sqrt{2}} + \dfrac{x^3}{\sqrt{3}} + \dfrac{x^4}{\sqrt{4}} + \dfrac{x^5}{\sqrt{5}} + \cdots.$

9. $1 - \dfrac{x}{1 \cdot 3} + \dfrac{x^2}{2 \cdot 4} - \dfrac{x^3}{3 \cdot 5} + \dfrac{x^4}{4 \cdot 6} - \cdots.$

10. $\dfrac{x}{2^2 - 1} + \dfrac{x^2}{3^2 - 1} + \dfrac{x^3}{4^2 - 1} + \dfrac{x^4}{5^2 - 1} + \cdots.$

11. $1 - \dfrac{x}{2} + \dfrac{x^2}{2^2} - \dfrac{x^3}{2^3} + \dfrac{x^4}{2^4} - \cdots.$

12. $1 + 2x + 2^2 x^2 + 2^3 x^3 + 2^4 x^4 + \cdots.$

13. $1 + 2x + \dfrac{2^2 x^2}{2!} + \dfrac{2^3 x^3}{3!} + \dfrac{2^4 x^4}{4!} + \cdots.$

14. $\dfrac{x}{2} + \dfrac{2x^2}{3} + \dfrac{3x^3}{4} + \dfrac{4x^4}{5} + \dfrac{5x^5}{6} + \cdots.$

15. $\dfrac{(x - 1)}{1} + \dfrac{(x - 1)^2}{2} + \dfrac{(x - 1)^3}{3}$
$+ \dfrac{(x - 1)^4}{4} + \cdots.$

16. $1 + (x + 2) + \dfrac{(x + 2)^2}{2!} + \dfrac{(x + 2)^3}{3!}$
$+ \dfrac{(x + 2)^4}{4!} + \cdots.$

17. $1 + \dfrac{(x + 1)}{2} + \dfrac{(x + 1)^2}{2^2} + \dfrac{(x + 1)^3}{2^3}$
$+ \dfrac{(x + 1)^4}{2^4} + \cdots.$

18. $\dfrac{(x - 2)}{1^2} + \dfrac{(x - 2)^2}{2^2} + \dfrac{(x - 2)^3}{3^2}$
$+ \dfrac{(x - 2)^4}{4^2} + \cdots.$

19. $\dfrac{(x + 5)}{1 \cdot 2} + \dfrac{(x + 5)^2}{2 \cdot 3} + \dfrac{(x + 5)^3}{3 \cdot 4}$
$+ \dfrac{(x + 5)^4}{4 \cdot 5} + \cdots.$

20. $(x + 3) - 2(x + 3)^2 + 3(x + 3)^3$
$- 4(x + 3)^4 + \cdots.$

14.6 FUNCTIONS DEFINED BY POWER SERIES

Since the power series

$$\sum a_n (x - b)^n = a_0 + a_1(x - b) + a_2(x - b)^2 + \cdots + a_n(x - b)^n + \cdots$$

has a unique sum at each point in its convergence set, a function f is defined by

$$f(x) = a_0 + a_1(x - b) + a_2(x - b)^2 + \cdots + a_n(x - b)^n + \cdots.$$

The domain of f is the convergence set of the power series, and the value of f at any point x in this domain is the sum of the power series at that point.

We state, without proof, the following important theorem about such functions. From parts (ii) and (iii) of this theorem we see that a power series can be differentiated or integrated term by term and that the resulting series represents the derivative and integral, respectively, of the function defined by the original power series at every number x inside the interval of convergence of the original power series.

14.6.1 **Theorem.** Let $\sum\limits_{n=0}^{\infty} a_n(x-b)^n$ be a power series in $(x-b)$ having a nonzero radius of convergence r, and let f be the function defined by

$$f(x) = a_0 + a_1(x-b) + a_2(x-b)^2 + \cdots$$

for $|x-b| < r$. Then at every point x in the interval of convergence of the given series,

    (i) f is continuous.

    (ii) $f'(x) = \sum\limits_{n=0}^{\infty} D_x[a_n(x-b)^n] = \sum\limits_{n=1}^{\infty} na_x(x-b)^{n-1}$

$$= a_1 + 2a_2(x-b) + 3a_3(x-b)^2 + \cdots + na_n(x-b)^{n-1} + \cdots.$$

    (iii) $\int_b^x f(t)\,dt = \sum\limits_{n=0}^{\infty} \int_b^x a_n(t-b)^n\,dt = \sum\limits_{n=0}^{\infty} \dfrac{a_n(x-b)^{n+1}}{n+1}$

$$= a_0(x-b) + a_1\frac{(x-b)^2}{2} + \frac{a_2(x-b)^3}{3} + \cdots$$

$$+ \frac{a_n(x-b)^{n+1}}{n+1} + \cdots.$$

Example 1. We know that the sum of the convergent geometric series $1 + x + x^2 + \cdots + x^n + \cdots$, where $|x| < 1$, is $1/(1-x)$. That is,

$$\frac{1}{1-x} = 1 + x + x^2 + \cdots + x^n + \cdots \qquad \text{for} \quad |x| < 1.$$

According to the preceding theorem, we can integrate both members of this equation and obtain

$$\int_0^x \frac{1}{1-t}\,dt = x + \frac{x^2}{2} + \frac{x^3}{3} + \cdots + \frac{x^{n+1}}{n+1} + \cdots.$$

But

$$\int_0^x \frac{1}{1-t}\,dt = -\ln(1-t)\Big|_0^x = -\ln(1-x)$$

for $|x| < 1$. Hence

$$-\ln(1-x) = x + \frac{x^2}{2} + \frac{x^3}{3} + \cdots + \frac{x^{n+1}}{n+1} + \cdots$$

is valid for all numbers x such that $-1 < x < 1$.

    If we let $x = \frac{1}{3}$, say, we obtain

$$-\ln\left(1 - \frac{1}{3}\right) = -\ln\left(\frac{2}{3}\right) = \ln 1 - \ln\left(\frac{2}{3}\right) = \ln\left(1 \div \frac{2}{3}\right) = \ln 1.5$$

$$= \frac{1}{3} + \frac{1}{2}\cdot\frac{1}{9} + \frac{1}{3}\cdot\frac{1}{27} + \cdots.$$

The sum of the first three terms of this latter series is 0.40 to two decimal places, and the value of $\ln 1.5$ (from a table) is 0.41 to two decimal places. By using a sufficient number of terms of this series, we can find $\ln 1.5$ correct to any desired number of decimal places.

If we start multiplying the power series $\sum a_n x^n$ by the power series $\sum b_n x^n$ term by term, as we would two polynomials, we get

$$a_0 + a_1 x + a_2 x^2 + a_3 x^3 + a_4 x^4 + \cdots$$
$$b_0 + b_1 x + b_2 x^2 + b_3 x^3 + b_4 x^4 + \cdots$$

$$a_0 b_0 + a_1 b_0 x + a_2 b_0 x^2 + a_3 b_0 x^3 + a_4 b_0 x^4 + \cdots$$
$$a_0 b_1 x + a_1 b_1 x^2 + a_2 b_1 x^3 + a_3 b_1 x^4 + \cdots$$
$$a_0 b_2 x^2 + a_1 b_2 x^3 + a_2 b_2 x^4 + \cdots$$
$$a_0 b_3 x^3 + a_1 b_3 x^4 + \cdots$$
$$a_0 b_4 x^4 + \cdots$$

$$a_0 b_0 + (a_1 b_0 + a_0 b_1)x + (a_2 b_0 + a_1 b_1 + a_0 b_2)x^2$$
$$+ (a_3 b_0 + a_2 b_1 + a_1 b_2 + a_0 b_3)x^3$$
$$+ (a_4 b_0 + a_3 b_1 + a_2 b_2 + a_1 b_3 + a_0 b_4)x^4 + \cdots$$

Notice the pattern in the way that the resulting product series has been arranged. The coefficient of x^k, $k = 1, 2, 3, \cdots$, is the sum of all products $a_i b_j$ such that $i + j = k$.

14.6.2 Definition.   The **Cauchy product** of the power series $\sum a_n(x - b)^n$ and $\sum b_n(x - b)^n$ is the power series

$$c_0 + c_1(x - b) + c_2(x - b)^2 + \cdots + c_n(x - b)^n + \cdots,$$

where $c_n = a_n b_0 + a_{n-1} b_1 + \cdots + a_1 b_{n-1} + a_0 b_n$.

Again, if we start dividing the power series $\sum a_n x^n$ by the power series $\sum b_n x^n$, by long division just as we would for two polynomials, and in the quotient collect terms having like powers of x, we obtain a power series

$$c_0 + c_1 x + c_2 x^2 + c_3 x^3 + \cdots.$$

That these formal manipulations lead to useful results under certain circumstances is ensured by the next theorem, which is easy to understand and use but not easy to prove.

14.6.3 Theorem.   If $f(x) = \sum a_n(x - b)^n$ for $|x - b| < r_1$ and $g(x) = \sum b_n(x - b)^n$ for $|x - b| < r_2$, and if r is the minimum of r_1 and r_2, then for all x such that $|x - b| < r$,
(i)   the Cauchy product of the two series is $f(x) \cdot g(x)$.
(ii)   the quotient of the first series by the second is $f(x)/g(x)$, provided that $b_0 \neq 0$.

Now consider any power series in $x - b$,

$$\sum a_n(x - b)^n = a_0 + a_1(x - b) + a_2(x - b)^2 + \cdots + a_n(x - b)^n + \cdots$$

and denote its interval of convergence by $(b - r, b + r)$, where $0 \leq r \leq \infty$, and let f be the function defined by

(1) $f(x) = a_0 + a_1(x - b) + a_2(x - b)^2 + \cdots + a_n(x - b)^n + \cdots$

for all x in $(b - r, b + r)$. By 14.6.1(ii)

$$f'(x) = a_1 + 2a_2(x - b) + 3a_3(x - b)^2 + \cdots$$
$$f''(x) = 2a_2 + 2 \cdot 3a_3(x - b) + 3 \cdot 4a_4(x - b)^2 + \cdots$$
$$f'''(x) = 2 \cdot 3a_3 + 2 \cdot 3 \cdot 4a_4(x - b) + 3 \cdot 4 \cdot 5a_5(x - b)^2 + \cdots$$

$\cdot \quad \cdot \quad \cdot \quad \cdot \quad \cdot \quad \cdot \quad \cdot \quad \cdot \quad \cdot \quad \cdot \quad \cdot \quad \cdot \quad \cdot \quad \cdot \quad \cdot$

$$f^{(n)}(x) = n!\, a_n + (n + 1)!\, a_{n+1}(x - b) + \frac{(n + 2)!}{2!} a_{n+2}(x - b)^2 + \cdots$$

$\cdot \quad \cdot \quad \cdot \quad \cdot \quad \cdot \quad \cdot \quad \cdot \quad \cdot \quad \cdot \quad \cdot \quad \cdot \quad \cdot \quad \cdot \quad \cdot \quad \cdot$

If we let $x = b$ in these equations, we get $f(b) = a_0, f'(b) = a_1, f''(b) = 2!\, a_2, f'''(b) = 3!\, a_3, \cdots, f^{(n)}(b) = n!\, a_n, \cdots$; and if we solve these latter equations for the coefficients $a_0, a_1, a_2, \cdots, a_n, \cdots$ and substitute in (1), we obtain

$$f(x) = f(b) + f'(b)(x - b) + \frac{f''(b)}{2!}(x - b)^2 + \cdots + \frac{f^{(n)}(b)}{n!}(x - b)^n + \cdots.$$

So we see that if a function f can be represented by a power series in $(x - b)$ on an interval $(b - r, b + r)$, the series must have the precise form

$$\sum_{n=0}^{\infty} \frac{f^{(n)}(b)(x - b)^n}{n!} = f(b) + f'(b)(x - b) + \frac{f''(b)(x - b)^2}{2!} + \cdots$$
$$+ \frac{f^{(n)}(b)(x - b)^n}{n!} + \cdots.$$

This is called the **Taylor's series** in $(x - b)$ of the function f.

14.6.4 Thoerem. Consider any power series in $(x - b)$,

(2) $\qquad a_0 + a_1(x - b) + a_2(x - b)^2 + \cdots + a_n(x - b)^n + \cdots,$

and denote its interval of convergence by $(b - r, b + r)$, where $0 \leq r \leq \infty$. Denote by f the function defined by

(3) $\qquad f(x) = a_0 + a_1(x - b) + a_2(x - b)^2 + \cdots + a_n(x - b)^n + \cdots,$

where x is any number in $(b - r, b + r)$. Then the power series (2) is the **Taylor's series** in $(x - b)$ of the function f, namely

$$f(x) = f(b) + f'(b)(x - b) + \frac{f''(b)(x - b)^2}{2!} + \cdots + \frac{f^{(n)}(b)(x - b)^n}{n!} + \cdots$$

when $b - r < x < b + r$.

Thus there is only one power series in $(x - b)$ that defines a function f, and that is the Taylor's series in $(x - b)$ of the function f.

14.6.5 Definition. When $b = 0$, the Taylor's series in x,

$$f(x) = f(0) + f'(0)x + \frac{f''(0)x^2}{2!} + \cdots + \frac{f^{(n)}(0)x^n}{n!} + \cdots,$$

is called a **Maclaurin's series**.

All the derivatives that appear in the coefficients of a Maclaurin's series are evaluated at the origin, and the interval of convergence has the origin for center.

We sometimes say that the Taylor's series in $(x - b)$ of a function f is the **Taylor's expansion** of f about the point b, and the Maclaurin's series of a function f is the **Maclaurin's expansion** of f about the origin.

Exercises

1. By algebraic division, find the power series representation for $1/(1 + x)$. What is its interval of convergence?

2. Integrate the power series of Exercise 1 term by term to secure the power series for $\ln (1 + x)$. What will the interval of convergence of this series be?

3. Add the series of Example 1 (14.6) and Exercise 2, above, to secure the series for $\ln [(1 + x)/(1 - x)]$. What will the interval of convergence of this series be?

4. Show that any positive number M can be represented by $(1 + x)/(1 - x)$, where x lies within the interval of convergence of the series of Exercise 3. Hence conclude that the natural logarithm of any positive number can be found by means of this series.

5. By algebraic division, find the series expansion for $2/(1 - x^2)$, and obtain the same result as that of Exercise 3 by integrating this series term by term.

6. By algebraic division, find the series expansion for $1/(1 + x^2)$. What is the interval of convergence of this series?

7. Integrate the series of Exercise 6 term by term to find the series expansion of $\text{Tan}^{-1} x$. What is the interval of convergence of this series?

8. Use the series of Exercise 7 to secure a famous series expression for $\frac{1}{4}\pi$. From theorem 14.4.2, how many terms of this series would be necessary to compute π accurately to six decimal places?

9. Integrate the series for $\ln (1 + x)$ term by term and then show by manipulating the series for $\ln (x + 1)$ that

$$\int \ln (1 + x) \, dx = (1 + x) \ln (1 + x) - x.$$

10. Show that if $f(x) = P_n(x)$, an nth-degree polynomial in x, then the Maclaurin series terminates after $n + 1$ terms and is the polynomial itself.

14.7 TAYLOR'S THEOREM

We stated that a function f defined by a power series $\sum a_n(x - b)^n$ [so that the sum of the power series at any number x in its convergence set is the value $f(x)$ of the function there] has derivatives of every order, and that the value of the nth derivative at b is $f^{(n)}(b) = n! \, a_n$.

Now consider a function F that is defined in some way other than by a power series [for example, by a closed expression, such as $F(x) = e^{x-1}$ or $F(x) = \sin x$]. If such a function possesses derivatives of all orders at some particular point b, then we can certainly *write down* a power series in the form of a Taylor's series,

$$f(x) = F(b) + F'(b)(x - b) + \frac{F''(b)(x - b)^2}{2!} + \cdots + \frac{F^{(n)}(b)(x - b)^n}{n!} + \cdots.$$

This series, the **Taylor's series** (or Maclaurin's series if $b = 0$) generated by the function F, may or may not converge for numbers other than b. Even if it does converge for all points in some interval about b, say $(b - r, b + r)$, and thus defines some function f there, *is the function f the same as the function F that we started with?*

Certainly, from what we have seen, both f and F have equal derivatives of all orders *at b*, since

$$F^{(n)}(b) = n!\, a_n = f^{(n)}(b)$$

for $n = 0, 1, 2, 3, \cdots$. [$F^{(0)}(b) = F(b)$ by convention.] But the answer to the preceding question can be yes or no, depending on the function F.

For each particular function, the question as to whether its Taylor's expansion (the Taylor's series generated by the function) represents the function may be answered by the use of a remarkable formula known as **Taylor's formula with remainder.**

14.7.1 **Taylor's Theorem.** Let f be a function that is defined in some open interval $(b - r, b + r)$, $0 \le r < \infty$, and whose first $n + 1$ derivatives exist and are continuous throughout the interval. Then for all numbers x in $(b - r, b + r)$,

$$f(x) = \frac{f(b)}{0!} + \frac{f'(b)(x - b)}{1!} + \frac{f''(b)(x - b)^2}{2!} + \cdots + \frac{f^{(n)}(b)(x - b)^n}{n!} + R_{n+1}(x),$$

where

$$R_{n+1}(x) = \left(\frac{1}{n!}\right) \int_b^x (x - t)^n f^{(n+1)}(t)\, dt.$$

Proof. For any particular number x in $(b - r, b + r)$,

$$\int_b^x f'(t)\, dt = f(x) - f(b);$$

and this may be written

(1) $$f(x) = f(b) + \int_b^x f'(t)\, dt.$$

Let us apply the method of integration by parts to the integral on the right in (1), with $u = f'(t)$, $du = f''(t)\, dt$, $dv = dt$, and $v = t - x$, where x is a constant with respect to the variable of integration t.

$$\int_b^x f'(t)\, dt = \left[f'(t)(t - x) - \int (t - x)f''(t)\, dt \right]\Big|_b^x = f'(b)(x - b) - \int_b^x (t - x)f''(t)\, dt.$$

Substituting this result in (1) gives

(2) $$f(x) = f(b) + f'(b)(x - b) + \int_b^x (x - t)f''(t)\, dt.$$

Let us again apply integration by parts, this time to the integral in (2). If $u = f''(t)$, $du = f'''(t)\, dt$, $dv = (x - t)\, dt$, and $v = -\frac{1}{2}(x - t)^2$, we obtain

$$\int_b^x (x - t)f''(t)\, dt = \left[f''(t)\left(-\frac{(x - t)^2}{2} \right) + \int \frac{(x - t)^2}{2} f'''(t)\, dt \right]\Big|_b^x$$

$$= f''(b)\frac{(x - b)^2}{2} + \int_b^x \frac{(x - t)^2}{2} f'''(t)\, dt.$$

Substitution of this equation in (2) gives

(3) $$f(x) = f(b) + f'(b)(x - b) + \frac{f''(b)(x - b)^2}{2!} + \int_b^x \frac{(x - t)^2}{2} f'''(t)\, dt.$$

By mathematical induction we can show that by applying this process n times, we obtain

$$f(x) = f(b) + f'(b)(x - b) + \frac{f''(b)(x - b)^2}{2!} + \cdots + \frac{f^{(n)}(b)(x - b)^n}{n!} + R_{n+1}(x),$$

where

$$R_{n+1}(x) = \left(\frac{1}{n!}\right) \int_b^x (x - t)^n f^{(n+1)}(t) \, dt. \quad \blacksquare$$

The expression

$$R_{n+1}(x) = \frac{1}{n!} \int_b^x (x - t)^n f^{(n+1)}(t) \, dt$$

is called the **integral form** of the remainder after $(n + 1)$ terms in Taylor's formula.
Since

$$f(x) = \sum_{i=0}^n \frac{f^{(i)}(b)(x - b)^i}{i!} + R_{n+1}(x),$$

then

$$f(x) - \sum_{i=0}^n \frac{f^{(i)}(b)(x - b)^i}{i!} = R_{n+1}(x).$$

Thus the remainder term in Taylor's formula measures the difference between the value of the function f at x and the sum of the first $(n + 1)$ terms of the Taylor's series in $(x - b)$ generated by f. Therefore, a necessary and sufficient condition for the Taylor's series in $(x - b)$, generated by a function f, to represent f at all points in some interval $(b - r, b + r)$ is that $\lim_{n \to \infty} R_{n+1}(x) = 0$ whenever $|x - b| < r$.

14.7.2 Theorem. Let f be a function having derivatives of all orders in some interval $(b - r, b + r)$. A necessary and sufficient condition that the Taylor's series

$$\sum_{n=0}^\infty \frac{f^{(n)}(b)(x - b)^n}{n!}$$

represent the function f at all points in the interval is that whenever $|x - b| < r$,

$$\lim_{n \to \infty} R_{n+1}(x) = 0,$$

where $R_{n+1}(x)$ is the remainder after $(n + 1)$ terms in Taylor's formula (14.7.1).

Sometimes, it is not easy in practice to apply this theorem because of the difficulty in proving that $\lim_{n \to \infty} R_{n+1}(x) = 0$. Two other forms of $R_{n+1}(x)$ will be given in the next section. For some functions one of the three forms is the most convenient to use, while for other functions another form is preferable. We usually try to find bounds for $R_{n+1}(x)$ and then prove that the bound approaches zero as $n \longrightarrow \infty$.

Example 1. Prove that the Taylor's series in x, generated by e^x, converges for all real numbers x and represents the function defined by e^x at all real numbers x.

Solution. Since the nth derivative of e^x is e^x for all real numbers x and $e^0 = 1$, the Taylor's series about 0 generated by e^x is

$$(4) \qquad 1 + x + \frac{x^2}{2!} + \frac{x^3}{3!} + \cdots + \frac{x^n}{n!} + \cdots.$$

From Taylor's formula, we have

$$e^x = 1 + x + \frac{x^2}{2!} + \cdots + \frac{x^n}{n!} + R_{n+1}(x),$$

where

$$R_{n+1}(x) = \frac{1}{n!} \int_0^x (x - t)^n e^t \, dt.$$

Choose an arbitrary number x, different from zero, and hold it fixed. Then, by the mean value theorem for integrals (8.6.1), there exists a number ξ_n between 0 and x such that

$$R_{n+1}(x) = \frac{1}{n!} \int_0^x (x - t)^n e^t \, dt = \frac{x}{n!} (x - \xi_n)^n e^{\xi_n}.$$

Thus

$$|R_{n+1}(x)| = \left| \frac{x}{n!} \right| \cdot |(x - \xi_n)^n| \cdot |e^{\xi_n}|.$$

But $|x/n!| = |x|/n!$; and since $0 < |x - \xi_n| < |x|$, then $|(x - \xi_n)|^n = |x - \xi_n|^n < |x|^n$. Also, $0 < e^{\xi_n} < e^{|x|}$ and thus $|e^{\xi_n}| = e^{\xi_n} < e^{|x|}$. Consequently,

$$|R_{n+1}(x)| < \frac{|x|}{n!} \cdot |x|^n e^{|x|} = C \frac{|x|^n}{n!},$$

where C is a constant. So (using 14.5.1)

$$\lim_{n \to \infty} |R_{n+1}(x)| \le C \lim_{n \to \infty} \frac{|x|^n}{n!} = C \cdot 0 = 0,$$

which implies that $\lim_{n \to \infty} R_{n+1}(x) = 0$ for every real number $x \ne 0$.

Therefore the Taylor's series (4) converges and represents e^x for all real numbers x.

Example 2. Compute the value of e correct to five decimal places.

Solution. Letting $x = 1$ in Example 1, we have

$$e = 1 + 1 + \frac{1}{2!} + \frac{1}{3!} + \cdots + \frac{1}{n!} + R_{n+1}(1).$$

We showed in Example 1 that

$$R_{n+1}(1) < \frac{e}{n!}.$$

In 8.7 we found $\ln 2 \doteq 0.7$ by the trapezoidal rule. By again using the trapezoidal rule, we could find that $\ln 3 \doteq 1.1$. Since $\ln x$ was shown to be continuous for $x > 0$, and e was defined to be that number for which $\ln e = 1$, it follows from the intermediate value theorem that $2 < e < 3$. Therefore

$$R_{n+1}(1) < \frac{e}{n!} < \frac{3}{n!}.$$

We wish to take enough terms of the series so that the error in neglecting the remaining ones will be less than 0.000 005. Thus we seek n so that

$$R_{n+1}(1) < \frac{3}{n!} < 0.000\ 005,$$

or

$$n! > \frac{3}{0.000\ 005} = 600,000.$$

Hence $n = 10$ will do, and the value of e correct to five decimal places may be obtained from

$$e \doteq 1 + 1 + \frac{1}{2!} + \frac{1}{3!} + \cdots + \frac{1}{10!}.$$

In performing this computation, we use the fact that $1/n!$ can be found by dividing $1/(n-1)!$ by n. The work may be conveniently arranged as follows.

$$
\begin{array}{lll}
1.000\ 000 \le\ \ 1\ \ \le 1.000\ 000 & \\
1.000\ 000 \le\ \ 1\ \ \le 1.000\ 000 & \text{(divide by 2)} \\
0.500\ 000 \le 1/2! \le 0.500\ 000 & \text{(divide by 3)} \\
0.166\ 666 \le 1/3! \le 0.166\ 667 & \text{(divide by 4)} \\
0.041\ 666 \le 1/4! \le 0.041\ 667 & \text{(divide by 5)} \\
0.008\ 333 \le 1/5! \le 0.008\ 334 & \\
0.001\ 388 \le 1/6! \le 0.001\ 389 & \\
0.000\ 198 \le 1/7! \le 0.000\ 199 & \\
0.000\ 024 \le 1/8! \le 0.000\ 025 & \text{(divide by 9)} \\
0.000\ 002 \le 1/9! \le 0.000\ 003 & \\
\hline
2.718\ 277 \le \text{sum} \le 2.718\ 284 &
\end{array}
$$

Thus the value of e, correct to 5 decimal places, is 2.718 28.

Exercises

1. Derive the formula

$$D_x^n \ln (1 + x) = (n - 1)! \cdot (1 + x)^{-n}(-1)^{n+1}$$

and use it to find the Maclaurin expansion for $\ln (1 + x)$, with the integral form of the remainder.

2. As in Exercise 1, find the Maclaurin expansion for $\ln (1 - x)$.

3. If an attempt is made to secure a general formula for the nth derivative of $\text{Tan}^{-1} x$, the result is unworkably complicated. Show, however, that if the first derivative of $\text{Tan}^{-1} x$ is expanded in a power series by algebraic division, a simple formula can be found for $f^n(0)$ by differentiating this series term by term successively. Give this formula.

4. Use the result of Exercise 3 to find the Maclaurin expansion for $\text{Tan}^{-1} x$ to five nonzero terms. Do not attempt to find a form for the remainder.

5. Find the Maclaurin expansion for e^{-x}, with the integral form of the remainder.

6. Find the Maclaurin expansion for 2^x, with integral form of the remainder.

In Exercises 7 to 10, find the first four non-vanishing terms of the Maclaurin expansion of the given function, but do not attempt to find the remainder.

7. $\sec x$.

8. $\tan x$.

9. $\text{Sin}^{-1} x$.

10. $\sqrt{1 + x}$.

It will be shown in the next section that the Maclaurin's expansions of $\sin x$ and $\cos x$,

$$\sin x = x - \frac{x^3}{3!} + \frac{x^5}{5!} - \frac{x^7}{7!} + \cdots$$
$$+ \frac{(-1)^{n+1}x^{2n-1}}{(2n-1)!} + \cdots$$

and

$$\cos x = 1 - \frac{x^2}{2!} + \frac{x^4}{4!} - \frac{x^6}{6!} + \cdots$$
$$+ \frac{(-1)^{n+1}x^{2n-2}}{(2n-2)!} + \cdots,$$

represent $\sin x$ and $\cos x$ for all real numbers x.

In Exercises 11 to 15, find the first four non-vanishing terms of the power series for the given functions by performing the indicated algebraic divisions.

11. Find the series for e^{-x} by dividing 1 by the series for e^x.

12. Find the series for $\sec x$ by dividing 1 by the series for $\cos x$.

13. Find the series for $\tan x$ by dividing the series for $\sin x$ by the series for $\cos x$.

14. Find the series for $1/\sqrt{1 + x}$ by dividing 1 by the series for $\sqrt{1 + x}$.

15. Find the series for $(\sin x)/x$ by dividing the series for $\sin x$ by x.

In Exercises 16 to 20, find the first four terms of the Taylor series for the given function about the given point. In Exercises 16 to 18, show the integral form of the remainder.

16. e^x, about $x = 1$.

17. $\sin x$, about $x = \frac{1}{6}\pi$.

18. $\cos x$, about $x = \frac{1}{3}\pi$.

19. $\tan x$, about $x = \frac{1}{4}\pi$.

20. $\sec x$, about $x = \frac{1}{4}\pi$.

14.8 OTHER FORMS OF THE REMAINDER IN TAYLOR'S THEOREM

In the preceding section we proved Taylor's theorem with the remainder in the integral form. Two other frequently used forms of the remainder will now be derived. With some functions one form of the remainder is more convenient; with other functions another form is preferable.

The integral form of the remainder is

$$(1) \qquad R_{n+1}(x) = \frac{1}{n!} \int_b^x (x - t)^n f^{(n+1)}(t) \, dt,$$

where $f^{(n+1)}$ is assumed to be continuous in $(b - r, b + r)$. Let x be an arbitrary number in $(b - r, b + r)$. By the mean value theorem for integrals (8.6.1), there is a number ξ_n between b and x such that

$$R_{n+1}(x) = \frac{(x - b)(x - \xi_n)^n}{n!} f^{(n+1)}(\xi_n).$$

This is called the **Cauchy form** of the remainder in Taylor's theorem.

14.8.1 Cauchy's Form of the Remainder. If $f^{(n+1)}$ is continuous in the open interval $(b - r, b + r)$, where $0 \le r < \infty$, and if x is a number in this interval, then the Cauchy form of the remainder after $n + 1$ terms in Taylor's theorem (14.7.1) is

$$R_{n+1}(x) = \frac{(x - b)(x - \xi_n)^n}{n!} \cdot f^{(n+1)}(\xi_n),$$

where ξ_n is some number between b and x.

To derive the third form of the remainder, we go back to the integral form (1). Let x be any particular number in $(b - r, b + r)$. Since $f^{(n+1)}$ is continuous in the closed interval I whose endpoints are b and x, $f^{(n+1)}$ has minimum and maximum values on I; denote them by m and M, respectively. Then

$$m \le f^{(n+1)}(t) \le M$$

for all t in I.

If $b < x$, then $(x - t)^n$ is nonnegative for all numbers t in $I = [b, x]$, and

$$m(x - t)^n \le f^{(n+1)}(t)(x - t)^n \le M(x - t)^n.$$

Therefore

$$m \int_b^x (x - t)^n \, dt \le \int_b^x f^{(n+1)}(t)(x - t)^n \, dt \le M \int_b^x (x - t)^n \, dt,$$

or, since $\int_b^x f^{(n+1)}(t)(x - t)^n \, dt = n! \, R_{n+1}(t)$ [by (1)],

$$m \le \frac{n! \, R_{n+1}(t)}{\int_b^x (x - t)^n \, dt} \le M.$$

If $x < b$, the preceding inequalities are reversed. In either case, let

(2)
$$H = \frac{n! \, R_{n+1}(t)}{\int_b^x (x - t)^n \, dt}.$$

Then H is a number between the minimum and the maximum values of the continuous function $f^{(n+1)}$ on I and, by the intermediate value theorem (3.5.8), there exists a number μ_n between b and x such that $f^{(n+1)}(\mu_n) = H$. Substituting this in (2), we get

$$R_{n+1}(x) = \frac{f^{(n+1)}(\mu_n)}{n!} \int_b^x (x - t)^n \, dt,$$

or, since $\int_b^x (x - t)^n \, dt = (x - b)^{n+1}/(n + 1)$,

$$R_{n+1}(x) = \frac{(x - b)^{n+1}}{(n + 1)!} \cdot f^{(n+1)}(\mu_n).$$

This is called the **Lagrange form** of the remainder in Taylor's theorem.

14.8.2 Lagrange's Form of the Remainder. If $f^{(n+1)}$ is continuous in $(b - r, b + r)$, where $0 \le r < \infty$, and x is in this interval, then the Lagrange form of the remainder after $n + 1$ terms in Taylor's theorem (14.7.1) is

$$R_{n+1}(x) = \frac{(x - b)^{n+1}}{(n + 1)!} \cdot f^{(n+1)}(\mu_n),$$

where μ_n is some number between b and x.

The Lagrange form of the remainder in Taylor's theorem can be derived directly without using the integral form, and with the less restrictive condition that $f^{(n+1)}$ *exist* in $(b - r, b + r)$ and not necessarily be continuous there. When $n = 0$, Taylor's theorem with the remainder in the Lagrange form is simply the mean value theorem (6.2.1):

$$f(x) = f(b) + f'(\xi)(x - b),$$

where ξ is a number between b and x. For this reason, Taylor's theorem, with the remainder in Lagrange's form, is sometimes called the **extended theorem of mean value**.

Example 1. Find the Maclaurin's series generated by $\sin x$ and show that the series represents the function for all real numbers x.

Solution.

$$\begin{aligned} f(x) &= \sin x, & f(0) &= 0, \\ f'(x) &= \cos x, & f'(0) &= 1, \\ f''(x) &= -\sin x, & f''(0) &= 0, \\ f'''(x) &= -\cos x, & f'''(0) &= -1, \\ f^{(4)}(x) &= \sin x, & f^{(4)}(0) &= 0. \end{aligned}$$

Hence, by Taylor's theorem,

$$\sin x = x - \frac{x^3}{3!} + \frac{x^5}{5!} - \frac{x^7}{7!} + \cdots + (-1)^{n+1}\frac{x^{2n-1}}{(2n-1)!} + R_{n+1}(x).$$

Using the Cauchy form of the remainder, we have

$$R_{n+1}(x) = \frac{x(x - \xi)^n}{n!} \cdot f^{(n+1)}(\xi),$$

where ξ is some number between 0 and x.

Now $f^{(n+1)}(\xi)$ is $\pm\sin \xi$ or $\pm\cos \xi$, depending on n. Thus, for all n, $|f^{(n+1)}(\xi)| \le 1$. Hence

$$|R_{n+1}(x)| = \left|\frac{x(x - \xi)^n}{n!}\right| \cdot |f^{(n+1)}(\xi)| \le \left|\frac{x(x - \xi)^n}{n!}\right|,$$

and

$$\lim_{n\to\infty} |R_{n+1}(x)| = |x|\lim_{n\to\infty} \frac{|(x - \xi)^n|}{n!} = 0$$

for all real numbers x (by 14.5.1). Therefore $\lim_{n\to\infty} R_{n+1}(x) = 0$, and

$$\sin x = x - \frac{x^3}{3!} + \frac{x^5}{5!} - \frac{x^7}{7!} + \cdots + (-1)^{n+1}\frac{x^{2n-1}}{(2n-1)!} + \cdots$$

for all real numbers x.

Example 2. Find the Maclaurin's series for $\cos x$ and show that it represents the function for all real numbers x.

Solution. Since

$$\sin x = x - \frac{x^3}{3!} + \frac{x^5}{5!} - \frac{x^7}{7!} + \cdots + (-1)^{n+1}\frac{x^{2n-1}}{(2n-1)!} + \cdots$$

for all numbers x, we can differentiate both members, term by term, and get

$$\cos x = 1 - \frac{x^2}{2!} + \frac{x^4}{4!} - \frac{x^6}{6!} + \cdots + \frac{(-1)^{n+1}x^{2n-2}}{(2n-2)!} + \cdots,$$

which is true for all numbers x (by 14.6.1).

Example 3.   Show that the binomial theorem

$$(1 + x)^k = 1 + kx + \frac{k(k-1)}{2!}x^2 + \frac{k(k-1)(k-2)}{3!}x^3 + \cdots$$

$$+ \frac{k(k-1)(k-2)\cdots(k-n+1)}{n!}x^n + \cdots$$

is valid for all real numbers k if $-1 < x < 1$.

Solution.   By substitution, we see that the formula is valid for $x = 0$. We therefore assume in what follows that $0 < |x| < 1$.

For the function f defined by $f(x) = (1 + x)^k$, we have

$$
\begin{aligned}
f(x) &= (1 + x)^k, & f(0) &= 1, \\
f'(x) &= k(1 + x)^{k-1}, & f'(0) &= k, \\
f''(x) &= k(k-1)(1 + x)^{k-2}, & f''(0) &= k(k-1), \\
f'''(x) &= k(k-1)(k-2)(1 + x)^{k-3}, & f'''(0) &= k(k-1)(k-2), \\
& \cdot\ \cdot\ \cdot\ \cdot\ \cdot\ \cdot\ \cdot\ \cdot & & \cdot\ \cdot\ \cdot\ \cdot\ \cdot\ \cdot\ \cdot\ \cdot \\
f^{(n)}(x) &= k(k-1)(k-2)\cdots(k-n+1)(1 + x)^{k-n}, \\
f^{(n)}(0) &= k(k-1)(k-2)\cdots(k-n+1), \\
f^{(n+1)}(x) &= k(k-1)\cdots(k-n)(1 + x)^{k-n-1}.
\end{aligned}
$$

Thus, by Taylor's theorem,

$$(1 + x)^k = 1 + kx + \frac{k(k-1)}{2!}x^2 + \frac{k(k-1)(k-2)}{3!}x^3 + \cdots$$

$$+ \frac{k(k-1)(k-2)\cdots(k-n+1)}{n!}x^n + R_{n+1}(x).$$

Of course, if k is a positive integer or zero, the Taylor's series has but a finite number of nonzero terms, and $R_{n+1}(x) = 0$ for some n; this is the binomial theorem of elementary algebra. We will consider the situation where k is any real number that is not a positive integer or zero. In doing so, we will use Cauchy's form of the remainder,

$$R_{n+1}(x) = \frac{x(x-\xi)^n}{n!} \cdot k(k-1)(k-2)\cdots(k-n)(1 + \xi)^{k-n-1},$$

where ξ is a number between 0 and x.

This remainder can be rewritten

$$R_{n+1}(x) = \frac{k(k-1)(k-2)\cdots(k-n)}{n!}\left[x(1 + \xi)^{k-1}\right]\left(\frac{x-\xi}{1+\xi}\right)^n.$$

We will first show that for $-1 < x < 1$,

$$\left|\frac{x-\xi}{1+\xi}\right| \le |x|.$$

If $0 < \xi < 1$,

$$\left|\frac{x-\xi}{1+\xi}\right| = \left|\frac{x(1-\xi/x)}{1+\xi}\right| < |x|.$$

If $-1 < x < \xi < 0$,

$$\frac{x-\xi}{1+\xi} = \frac{-[-x+\xi]}{1+\xi} = \frac{-[|x|-|\xi|]}{1-|\xi|} = \frac{-|x|(1-|\xi/x|)}{1-|\xi|}.$$

Since $|\xi| < |x| < 1$,

$$1 < \left|\frac{x}{\xi}\right| < \frac{1}{|\xi|} \qquad \text{and} \qquad 1 > \left|\frac{\xi}{x}\right| > |\xi|.$$

Therefore

$$0 > \left|\frac{\xi}{x}\right| - 1 > |\xi| - 1, \qquad 1 - |\xi| > 1 - \left|\frac{\xi}{x}\right| > 0.$$

Hence

$$\left|\frac{-|x|(1 - \xi/x)}{1 - |\xi|}\right| < |x|$$

and so

$$\left|\frac{x - \xi}{1 + \xi}\right| < |x| \qquad \text{if} \qquad 0 < |\xi| < |x|.$$

Thus

$$|R_{n+1}(x)| = |x(1 + \xi)^{k-1}| \cdot \left|\frac{k(k - 1)(k - 2) \cdots (k - n)}{n!}\left(\frac{x - \xi}{1 + \xi}\right)^n\right|$$

$$\leq |x(1 + \xi)^{k-1}| \cdot \left|\frac{k(k - 1)(k - 2) \cdots (k - n)}{n!}\right| |x|^n$$

for $-1 < x < 1$. In this latter expression, $|x(1 + \xi)^{k-1}|$ does not depend on n, and we have previously shown (in Example 4, Section 14.5) that

$$\lim_{n \to \infty} \left|\frac{k(k - 1)(k - 2) \cdots (k - n)}{n!}\right| |x|^n = 0$$

when $-1 < x < 1$. Therefore

$$\lim_{n \to \infty} R_{n+1}(x) = 0 \qquad \text{if} \qquad -1 < x < 1.$$

It follows from 14.7.2 that

$$(1 + x)^k = 1 + kx + \frac{k(k - 1)}{2!}x^2 + \cdots + \frac{k(k - 1)(k - 2) \cdots (k - n + 1)}{n!}x^n + \cdots$$

for any real number k, if $-1 < x < 1$.

Example 4. Find the Maclaurin's series for the function f defined by $f(x) = \text{Sin}^{-1} x$ and prove that it represents the function for $-1 < x < 1$.

Solution. $D_x \text{Sin}^{-1} x = (1 - x^2)^{-1/2}$. By the binomial theorem,

$$(1 - x^2)^{-1/2} = 1 + \frac{1}{2}x^2 + \frac{1 \cdot 3}{2 \cdot 2}\frac{x^4}{2!} + \frac{1 \cdot 3 \cdot 5}{2 \cdot 2 \cdot 2}\frac{x^6}{3!} + \cdots + \frac{1 \cdot 3 \cdot 5 \cdot \ldots \cdot (2n - 1)}{2^n}\frac{x^{2n}}{n!} + \cdots,$$

or

$$(1 - x^2)^{-1/2} = 1 + \frac{1}{2}x^2 + \frac{1 \cdot 3}{2 \cdot 4}x^4 + \frac{1 \cdot 3 \cdot 5}{2 \cdot 4 \cdot 6}x^6 + \cdots + \frac{1 \cdot 3 \cdot 5 \cdot \ldots \cdot (2n - 1)}{2 \cdot 4 \cdot 6 \cdot \ldots \cdot 2n}x^{2n} + \cdots,$$

which is valid for $-1 < x < 1$. Integrating this term by term, we get

$$\text{Sin}^{-1} x = x + \frac{1}{2}\frac{x^3}{3} + \frac{1 \cdot 3}{2 \cdot 4}\frac{x^5}{5} + \frac{1 \cdot 3 \cdot 5}{2 \cdot 4 \cdot 6}\frac{x^7}{7} + \cdots + \frac{1 \cdot 3 \cdot 5 \cdot \ldots \cdot (2n - 1)}{2 \cdot 4 \cdot 6 \cdot \ldots \cdot 2n}\frac{x^{2n+1}}{2n + 1} + \cdots,$$

which is valid for $-1 < x < 1$. The constant of integration is zero, since $\text{Sin}^{-1} 0 = 0$.

Example 5. To how many decimal places of accuracy can $\sin 10°$ be computed by using the first two nonvanishing terms of the Maclaurin expansion for $\sin x$?

Solution. The answer to this and to many similar computational problems is best found by using the Lagrange form of the remainder. Since the angles 10° and 0.17453 radian are approximately equal, the series in which we are interested is

$$\sin x = 0 + x + 0 \cdot \frac{x^2}{2} - \frac{1}{3!}x^3 + 0 \cdot x^4 + \frac{1}{5!}x^5 + \cdots,$$

where the real number x is approximately equal to 0.17453. In using the Lagrange form of the remainder after the first two nonvanishing terms of the series, we have the option of using either a fourth-degree remainder of a fifth-degree remainder. Since the latter will, in general, yield a better estimate of the accuracy than the former, we choose the fifth-degree form,

$$R_5(x) = \frac{\cos \xi}{5!}x^5, \qquad 0 < \xi < x,$$

where $x \doteq 0.17453$.

Since $0 < \cos \xi < 1$ when $0 < \xi < x \doteq 0.17453$, $R_5(x) < x^5/5!$, or

$$R_5(0.17453)_{\text{max}} = \frac{1}{5!}(0.17453)^5 = \frac{1.619 \times 10^{-4}}{120} = 1.35 \times 10^{-6}.$$

Hence the error is less than 2×10^{-6}, and the computation of $\sin 10°$, using two nonvanishing terms of the Maclaurin series for $\sin x$, is good to five decimal places.

Since the series for $\sin x$ is an alternating series, we could also have used theorem 14.4.2 to achieve the same result.

Had we elected to use the fourth-degree form of the remainder term, the remainder would have been given by

$$R_4(x) = \frac{\sin \xi}{4!}x^4, \qquad 0 < \xi < x.$$

The best that could have been done in bounding $\sin \xi$ in the interval $0 < \xi < x$, without introducing needlessly complicated computational difficulties, would have been to say that $\sin 10°$ was less than $\sin 30°$ and so $\sin \xi$ was less than $\frac{1}{2}$.

Then

$$R_4(0.17453) < \frac{1}{2} \frac{(0.17453)^4}{4!}$$

or

$$R_4(0.17453) < \frac{9.279 \times 10^{-4}}{48} = 1.933 \times 10^{-4}.$$

In other words, this remainder term would only ensure that the first two nonvanishing terms of the series for $\sin x$ would yield three-decimal-place accuracy, whereas, as was seen above, the accuracy is actually good to five decimal places.

Exercises

1. To what accuracy can $\sin 1°$ be computed by using the approximation $\sin x \doteq x$?

2. To what accuracy can $\sin 1°$ be computed by using the first two nonvanishing terms of the Maclaurin expansion for $\sin x$?

3. How many nonvanishing terms of the Maclaurin expansion for $\sin x$ must be used to secure six-decimal-place accuracy in computing $\sin 28°$?

4. If we expand $\sin x$ about $\frac{1}{6}\pi$, as in Exercise 17, Section 14.7, how many terms of this series must be used to secure six-decimal-place accuracy in the computation of sin 28°?

5. How many terms of the series

$$e = 1 + 1 + \frac{1}{2!} + \frac{1}{3!} + \frac{1}{4!} + \cdots$$

suffice to compute e accurately to four decimal places?

6. How many terms of the Maclaurin series for e^x must be used to secure six-decimal-place accuracy in the computation of $e^{1.1}$?

7. Expand e^x about $x = 1$. How many terms of this series must be used to secure six-decimal-place accuracy in the computation of $e^{1.1}$?

8. How accurate is the computation of $e^{1.1}$ if we use only the first two terms of the series of Exercise 7?

9. From the Maclaurin series for e^x, state an approximation formula for e^x when $|x|$ is small.

10. From the series of Exercise 7, state an approximation formula for e^{1+x} when $|x|$ is small.

11. Give an approximation formula for $\sin(\frac{1}{6}\pi + x)$ when $|x|$ is small.

12. From Exercise 18, Section 14.7, deduce an approximation formula for $\cos(\frac{1}{3}\pi + x)$ when $|x|$ is small.

13. Show that the approximation formulas of Exercises 9 to 12 can be deduced simply by using differentials.

14. From the Maclaurin expansion for $\ln(1 + x)$, compute $\ln(1.1)$ to three-decimal-place accuracy.

15. For what values of x does the approximation $\ln(1 + x) \doteq x$ give two-decimal-place accuracy?

16. Use the series for $\ln[(1 + x)/(1 - x)]$ to compute $\ln 2$ to three decimal places.

17. Use the series for e^x and the relation $e^{\ln 2} = 2$ to check your answer to Exercise 16.

18. Use the series for $\operatorname{Sin}^{-1} x$ to compute $\operatorname{Sin}^{-1} 0.2$ to three decimal places. (Even though an expression for the remainder term is not given, can you tell from the series itself how many terms must be taken?)

19. Put the answer to Exercise 18 back in the series for $\sin x$ to check the work.

20. From the series for $\operatorname{Sin}^{-1} x$, derive an approximation for $\operatorname{Sin}^{-1} x$ good to two-decimal-place accuracy when $-\frac{1}{2} \le x \le \frac{1}{2}$.

21. It is stated in the text that the Lagrange form of the remainder in Taylor's theorem can be derived directly without using the integral form, and with the less restrictive condition thet $f^{(n+1)}$ *exist* in $(b - r, b + r)$ and not necessarily be continuous there. Prove this statement by proving the following extension of the mean value theorem, without using Section 14.7 or Section 14.8.

Theorem. If a function f and its first n derivatives are continuous on a closed interval $[a, b]$, and if $f^{(n+1)}(x)$ exists on the open interval (a, b), then there is a number z in (a, b) such that

$$f(b) = f(a) + f'(a)\frac{(b - a)}{1!} + f''(a)\frac{(b - a)^2}{2!} + \cdots + f^{(n)}(a)\frac{(b - a)^n}{n!} + f^{(n+1)}(z)\frac{(b - a)^{n+1}}{(n + 1)!}.$$

[*Hint:* Define a constant K by

$$f(b) = f(a) + f'(a)\frac{(b - a)}{1!} + f''(a)\frac{(b - a)^2}{2!} + \cdots + f^{(n)}(a)\frac{(b - a)^n}{n!} + K\frac{(b - a)^{n+1}}{(n + 1)!}$$

and define a function F by

$$F(x) = f(b) - f(x) - f'(x)\frac{(b - x)}{1!} - f''(x)\frac{(b - x)^2}{2!} - \cdots - f^{(n)}(x)\frac{(b - x)^n}{n!} - K\frac{(b - x)^{n+1}}{(n + 1)!}.$$

Show that the function F satisfies the conditions of Rolle's theorem on $[a, b]$ and therefore that $F'(z) = 0$. Then show that this implies that $K = f^{(n+1)}(z)$.]

22. The second derivative test (6.6.1) for the local extrema of a function f of one variable fails at a critical number c if $f''(c) = 0$. Prove the following more useful generalization.

Theorem. Let f be a function that is defined in some open interval containing c, and whose first n derivatives $(n > 1)$ exist and are continuous throughout the interval. If $f'(c) = f''(c) = \cdots = f^{(n-1)}(c) = 0$, and if $f^{(n)}(c) \neq 0$, then

    (i)   if n is even and $f^{(n)}(c) > 0$, f has a local minimum value at c.

    (ii) if n is even and $f^{(n)}(c) < 0$, f has a local maximum value at c.

    (iii) if n is odd, f has a point of inflection at c.

[*Hint:* Use Taylor's theorem with the remainder after n terms in the Lagrange form

$$R_n = \frac{(x-c)^n}{n!} f^{(n)}(\mu),$$

where μ is some number between c and x. Also recall 3.3.7 and 3.3.8.]

23. The integral

$$\int \frac{\sin x}{x}\, dx$$

cannot be expressed in terms of a finite number of elementary functions (see Exercise 14, Section 11.15). Express the above integral as a Maclaurin's series and state its interval of convergence.

24. Sketch the **normal probability curve** $y = e^{-x^2}$. In statistics it is important to be able to find area under this curve but, as stated in Exercise 9, Section 11.15, the indefinite integral $\int e^{-x^2}\, dx$ cannot be found in terms of a finite number of elementary functions. Write the Maclaurin's series for

$$\int_0^1 e^{-x^2}\, dx$$

and tell how many terms are needed to compute the value of this integral correct to four decimal places.

25. The indefinite integral $\int \sqrt{1 + x^3}\, dx$ cannot be found in terms of a finite number of elementary functions (Exercise 11, Section 11.15). Write a Maclaurin's series for

$$\int \sqrt{1 + x^3}\, dx$$

and find its interval of convergence.

14.9 REVIEW EXERCISES

In Exercises 1 to 4, find the limit of the given sequence or show that it is divergent.

1. $\left\{ \dfrac{9n}{\sqrt{9n^2 + 1}} \right\}.$

2. $\left\{ \dfrac{\ln n}{n} \right\}.$

3. $\left\{ \dfrac{50n}{n^{4/3} + 1} \right\}.$

4. $\left\{ \dfrac{n!}{3^n} \right\}.$

In Exercises 5 to 12, tell whether the given series converges absolutely, converges conditionally, or diverges.

5. $\displaystyle\sum n^2 \left(\frac{2}{3} \right)^n.$

6. $\displaystyle\sum \frac{(-1)^n}{1 + \ln n}.$

7. $\displaystyle\sum (-1)^n 2^{1/n}.$

8. $\displaystyle\sum \frac{1}{(n + 3)(n + 4)}.$

9. $\displaystyle\sum \frac{(-1)^n}{3^n + 2}.$

10. $\displaystyle\sum \frac{1}{\sqrt[4]{n^3 + 2}}.$

11. $\displaystyle\sum \frac{(-1)^n n}{n^2 + 1}.$

12. $\displaystyle\sum \frac{1}{3n - 1}.$

In Exercises 13 to 20, find the convergence set of the given power series.

13. $\displaystyle\sum \frac{x^n}{n^3 + 1}.$

14. $\displaystyle\sum \frac{(-1)^{n+1} x^n}{n}.$

15. $\displaystyle\sum \frac{(-1)^n (x - 4)^n}{n + 1}.$

16. $\displaystyle\sum \frac{3^n x^{3n}}{(3n)!}.$

17. $\displaystyle\sum \frac{(x - 3)^n}{2n + 1}.$

18. $\displaystyle\sum n!(x + 1)^n.$

19. $\displaystyle\sum \frac{(x - 6)^n}{2^n \ln (n + 1)}.$

20. $\displaystyle\sum \frac{(2x)^n}{5^{n+1}}.$

21. By differentiating the geometric series,

$$\frac{1}{1 - x} = 1 + x + x^2 + x^3 + \cdots + x^n + \cdots, |x| < 1,$$

find a power series that represents $1/(1 - x)^2$. What is its interval of convergence?

22. Find a power series that represents $1/(1 - x)^3$ on the interval $(-1, 1)$.

23. Find the Maclaurin's series for $\sin^2 x$. For what values of x does the series represent the function?

24. Find the first four terms of the Taylor's series for e^x about the point $x = 2$, and show the integral form of the remainder.

25. Write the Maclaurin expansion of $f(x) = \sin x + \cos x$. For what values of x does it represent f?

26. Write the Maclaurin's series for

$$\int_0^1 \cos x^2 \, dx.$$

How many terms of the series are needed to compute the value of this integral correct to four decimal places?

27. Write the Maclaurin's series for

$$\int_0^1 \frac{e^x - 1}{x} \, dx.$$

28. In a table of natural logarithms we find that $\ln 5 > 1.6$.
(a) Verify that for $x = 5$, $\ln x > 1 + 1/x$.
(b) Show that for $x \in R$ and $x \geq 5$, $\ln x > 1 + 1/x$, which is equivalent to $(x + 1) - x \ln x < 0$.
(c) Let $f(x) = \ln x/(x + 1)$. Find $f'(x)$ and use (b) to show that $f'(x) < 0$ for $x \geq 5$, and thus that f is decreasing on $[5, \infty)$.
(d) From (c), show that the sequence

$$\left\{ \frac{\ln n}{n + 1} \right\}, \quad n \in N,$$

is decreasing for $n > 5$.

29. Find $\displaystyle\lim_{n \to \infty} \frac{\ln n}{n + 1}.$

30. Use Exercise 29 and part (d) of Exercise 28 to show that the alternating series

$$\sum \frac{(-1)^n \ln n}{n + 1}$$

is convergent.

31. Show that the alternating series of Exercise 30 does not converge absolutely.

32. Without further tests, use Exercises 30 and 31 to find the convergence set of the power series

$$\sum \frac{(-1)^n \ln n}{n + 1} x^n.$$

15

parametric equations and vectors in the plane

The plane curves discussed so far have been graphs of a single equation, and most of them have been graphs of functions. But the concept of a *plane curve* is much broader and will soon be defined precisely by means of *parametric equations*. This leads naturally to *vectors*, which are essential tools in engineering and science. Incidentally, vectors also enable us to express many physical and geometric facts in a surprisingly concise and elegant manner.

15.1 PLANE CURVES

Consider the equations

(1) $$x = t^2, \qquad y = t - 3, \qquad -\infty < t < \infty.$$

If we assign a particular real number to t, the equations (1) determine the coordinates (x, y) of a point in the Cartesian plane; and as t varies, the corresponding point $P{:}(x, y)$ traces out a graph. As an illustration, if $t = 0$, the corresponding point on the graph has Cartesian coordinates $(0, -3)$; if $t = 1$, the point is $(1, -2)$; if $t = 2$, the point is $(4, -1)$, and so on. A table of values of t and the corresponding Cartesian coordinates (x, y) of P follows.

t	-4	-3	-2	-1	0	1	2	3	4	5
x	16	9	4	1	0	1	4	9	16	25
y	-7	-6	-5	-4	-3	-2	-1	0	1	2

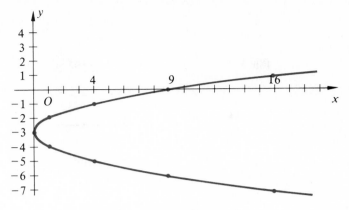

Figure 15-1

If we plot the points whose coordinates are the pairs (x, y) in this table, they are seen to lie on a parabola (Fig. 15-1).

The equations (1) are called **parametric equations** of the parabola and t is the **parameter.**

The (nonparametric) *Cartesian equation*, that is, the x, y equation, of this parabola may be found by eliminating the parameter t from equations (1). From the second of the two equations, $t = y + 3$; and if we substitute this result in the first equation, we obtain $(y + 3)^2 = x$, which is the Cartesian equation of a parabola, tangent to the y axis at $(0, -3)$, with principal axis $y = -3$, and opening to the right (Fig. 15-1).

If f and g are functions having a common domain $\mathfrak{D}$, then

$$x = f(t), \qquad y = g(t), \quad t \in \mathfrak{D},$$

are **parametric equations** in which t is the **parameter**. The graph of these parametric equations is the set of points

$$\{P{:}(x, y) \,|\, x = f(t), y = g(t), t \in \mathfrak{D}\}.$$

Example 1. Draw the graph of the parametric equations

$$x = 4 \cos t, \qquad y = 3 \sin t, \quad 0 \le t \le 2\pi.$$

Solution. When $t = 0$, $x = 4$ and $y = 0$; so the point $P{:}(4, 0)$ corresponds to $t = 0$. As t increases from 0 to π, the point $P{:}(x, y)$ traces out the part of the graph that lies in the first and second quadrants, as indicated in the following table of approximate values.

t	0	$\frac{1}{6}\pi$	$\frac{1}{4}\pi$	$\frac{1}{3}\pi$	$\frac{1}{2}\pi$	$\frac{2}{3}\pi$	$\frac{3}{4}\pi$	$\frac{5}{6}\pi$	π
x	4	3.5	2.8	2	0	-2	-2.8	-3.5	-4
y	0	1.5	2.1	2.6	3	2.6	2.1	1.5	0

Because of the properties of the cosine function, the graph is symmetric with respect to the x axis. Thus we need only reflect the upper half of the graph about the x axis to complete the graph of the given parametric equations (Fig. 15-2).

To find the Cartesian equation of this graph, we obtain $\cos t = x/4$ and $\sin t = y/3$ from the given parametric equations and substitute in the trigonometric identity $\sin^2 t + \cos^2 t = 1$. This gives $y^2/9 + x^2/16 = 1$, or

$$\frac{x^2}{16} + \frac{y^2}{9} = 1,$$

which is a standard equation of an ellipse (Fig. 15-2).

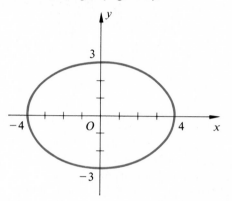

Figure 15-2

Example 2. Draw the graph of the parametric equations

$$x = 2 - 2t, \qquad y = t, \quad 0 \le t \le 1.$$

Solution. When $t = 0$, the corresponding point on the graph is $P{:}(2, 0)$; when $t = \frac{1}{2}, P$ is $(1, \frac{1}{2})$; and when $t = 1$, P is $(0, 1)$. By eliminating the parameter t between the given equations, we obtain the Cartesian equation $x = 2 - 2y$, or

$$x + 2y - 2 = 0.$$

The graph of this Cartesian equation is the straight line passing through the points $(2, 0)$ and $(0, 1)$. But the graph of the given parametric equation is only the line segment whose endpoints are $(2, 0)$ and $(0, 1)$, because the parameter is restricted to values in the closed interval $[0, 1]$ on the t axis (Fig. 15-3).

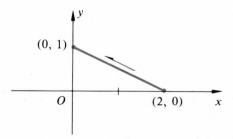

Figure 15-3

Example 3.   Find parametric equations for the graph of the Cartesian equation

$$x^2 + y^2 - 4y = 0.$$

Solution.   The graph is a circle with center (0, 2) and radius 2.

By "completing the square," we can write the given equation in the form $x^2 + (y - 2)^2 = 4$, or

$$\left(\frac{x}{2}\right)^2 + \left(\frac{y - 2}{2}\right)^2 = 1.$$

Comparison of this equation with the trigonometric identity $\sin^2 \theta + \cos^2 \theta = 1$ suggests that we let $x/2 = \sin \theta$ and $(y - 2)/2 = \cos \theta$. Then parametric equations of the graph are

$$x = 2 \sin \theta, \qquad y = 2 \cos \theta + 2, \quad 0 \leq \theta \leq 2\pi.$$

We are now prepared for a definition of what is meant by a plane curve.

15.1.1   Definition.   A **plane curve** is the graph of a pair of parametric equations

(2) $$x = f(t), \qquad y = g(t), \quad t \in I,$$

in which the functions f and g are continuous on an interval I (finite or infinite, open, half-open, or closed). The equations (2) are a **parametric representation** of the curve. If the interval I is closed, the curve is said to be an **arc**.

The parabola, the ellipse, the line segment, and the circle, discussed above, are curves. The ellipse and the circle are also called arcs because both have a closed interval of definition, $I = [0, 2\pi]$. The line segment in Example 2 is an arc because $I = [0, 1]$ is closed. But the parabola (1) is not an arc because its interval $I = (-\infty, \infty)$ is not closed.

An arc, then, is a curve that has endpoints. If the endpoints coincide, the curve is a **closed arc**. Thus the ellipse in Example 1 is a closed arc because its endpoints, $(4 \cos 0, 3 \sin 0) = (4, 0)$ and $(4 \cos 2\pi, 3 \sin 2\pi) = (4, 0)$, are identical.

A curve such that distinct values of its parameter always correspond to distinct points on the curve is called a **simple curve**. A simple curve cannot "cross itself." The parabola in Fig. 15-1 is a simple curve, and the line segment in Fig. 15-3 is both a simple curve and a simple arc. However, the curve in Fig. 15-4 is not a simple curve because two distinct values of the parameter give the same point O on the curve, as will be shown in Example 4.

Example 4.   Sketch the curve having the parametric representation

$$x = t^2 - 2, \qquad y = t(t^2 - 2), \qquad I = [-2, 2].$$

Is it an arc? Is it a simple curve?

Solution.   A table of values of t and the Cartesian coordinates (x, y) of the corresponding points on the curve is easily constructed from the given parametric equations.

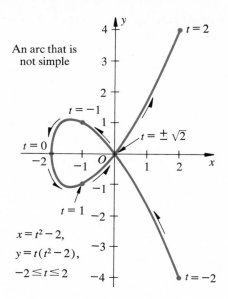

An arc that is not simple

$x = t^2 - 2,$
$y = t(t^2 - 2),$
$-2 \le t \le 2$

Figure 15-4

t	-2	$-\sqrt{2}$	-1	0	1	$\sqrt{2}$	2
x	2	0	-1	-2	-1	0	2
y	-4	0	1	0	-1	0	4

As t increases from -2 to 2, the curve starts at the point $(2, -4)$ in the fourth quadrant, moves upward through the origin into the second quadrant, and cuts the x axis at $(-2, 0)$ as it moves down into the third quadrant; it then goes through the origin *for the second time* and terminates in the first quadrant at the point $(2, 4)$ (see Fig. 15-4).

This curve is an arc because the t interval, $I = [-2, 2]$, is closed, and thus the curve has endpoints. But the curve is *not* a simple curve because it crosses itself; to the distinct values $t = \sqrt{2}$ and $t = -\sqrt{2}$ of the parameter corresponds the same point $(0, 0)$ on the curve, as may be seen in the table of values.

The requirement in our definition of a curve (15.1.1) that f and g be continuous on I prevents the curve from breaking up, as would be the case if either f or g had a jump discontinuity. Actually, the restriction of continuity is not nearly strong enough for most purposes. With only that restriction a curve might have no tangents and an arc might not have a finite length. Even more bizarre curves are included in our definition. Therefore we shall usually confine our attention to *smooth* curves.

15.1.2 Definition. The curve with parametric representation

$$x = f(t), \qquad y = g(t), \quad t \in I,$$

is a **smooth curve** if f and g are smooth functions—that is, if f and g have continuous first derivatives on I.

Every curve has indefinitely *many* parametric representations. For instance, the arc C, defined by

$$x = 5 \cos \theta, \qquad y = 4 \sin \theta, \quad -\tfrac{1}{2}\pi \le \theta \le \tfrac{1}{2}\pi,$$

is the right-hand half of the ellipse $16x^2 + 25y^2 = 400$ (Fig. 15-5). Some other parametric representations of the same curve C are

$$x = 5 \cos (2\phi + 1), \qquad y = 4 \sin (2\phi + 1), \quad -\tfrac{1}{4}(\pi + 2) \le \phi \le \tfrac{1}{4}(\pi - 2),$$

and

$$x = \frac{5(1 - t^2)}{1 + t^2}, \qquad y = \frac{8t}{1 + t^2}, \quad -1 \le t \le 1,$$

as the reader should verify.

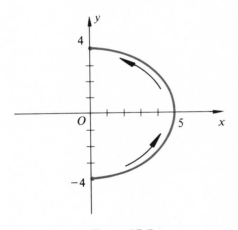

Figure 15-5

Parametric equations of a curve constitute a continuous mapping of the interval I onto the curve. In this mapping, the curve is the continuous image of the t interval.

Since the numbers in any interval I are ordered, the points of a simple arc may be ordered by agreeing that a point $P_1{:}(f(t_1)), g(t_1))$ precedes a point $P_2{:}(f(t_2), g(t_2))$ if and only if $t_1 < t_2$. We define the **positive direction** on a simple arc to be with increasing t. Notice that the positive direction on C (Fig. 15-5) in all three of its parametric representations discussed above is from $(0, -4)$ to $(5, 0)$ to $(0, 4)$, as indicated by the arrow.

When a continuous *function* is defined on an interval I by a Cartesian equation $y = f(x)$, it is particularly easy to write a parametric representation of its graph. By simply letting $x = t$, parametric equations of the graph are

$$x = t, \qquad y = f(t), \quad t \in I.$$

Similarly, in polar coordinates, if $r = f(\theta)$, in which f is continuous on a θ interval I, parametric equations of its graph, in which θ is taken for the parameter, are

$$x = r \cos \theta = f(\theta) \cos \theta, \qquad y = r \sin \theta = f(\theta) \sin \theta$$

for $\theta \in I$.

Sometimes the Cartesian equation of a curve C can be found by eliminating the parameter between a pair of parametric equations that represent C. The graph of the resulting Cartesian equation always includes all the points of C, but *the reader should make sure that it does not include points not on C.* As an illustration, consider the curve

$$C = \{P{:}(x, y) \,|\, x = 3 \sin \theta, y = 2 \cos \theta, \theta \in [0, \pi]\}.$$

If we eliminate the parameter θ between these equations by substituting $\sin \theta = x/3$ and $\cos \theta = y/2$ in the trigonometric identity $\sin^2 \theta + \cos^2 \theta = 1$, we get the Cartesian equation $4x^2 + 9y^2 = 36$, whose graph is an ellipse. But the curve C is only the right-hand half of this ellipse [including the points $(0, 2)$ and $(0, -2)$] because of the restriction $0 \leq \theta \leq \pi$.

Parametric equations are particularly useful in physics and engineering when dealing with problems of motion. If we let the parameter t designate time in a parametric representation of a curve C, we not only know the path traced by a particle P on C but also exactly where P is at any time t in the t interval. When the parameter t represents time, the curve is often called a **trajectory**.

By way of illustration, if air resistance is neglected, the position of a projectile at the end of t seconds, fired with an initial velocity of v_0 feet per second and at an angle θ from the horizontal (Fig. 15-6), is given by the parametric equations

(3) $$x = (v_0 \cos \theta)t, \qquad y = (v_0 \sin \theta)t - 16t^2, \quad 0 \leq t \leq k,$$

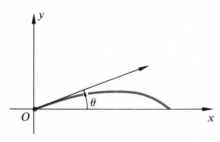

where the unit of distance is a foot. Here x, y, and t are variables; all other symbols represent constants. The Cartesian equation of the trajectory of the projectile can be found by eliminating the parameter t between the parametric equations (3). From the first of these equations we get $t = x/(v_0 \cos \theta)$; and by substituting this result in the second equation and simplifying, we obtain

$$y = (\tan \theta)x - \left[\frac{4}{v_0 \cos \theta}\right]^2 x^2,$$

Figure 15-6

whose graph is a parabola. Thus the trajectory is the arc of this parabola from the origin to the point of impact.

A **cycloid** is the curve traced by a point on a circle as the circle rolls on a straight line without slipping (Fig. 15-7). The Cartesian equation of a cycloid is so complicated that it is seldom used. But simple parametric equations are readily found, as shown in the next example.

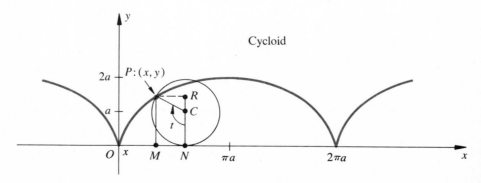

Figure 15-7

Example 5.   Find parametric equations of a cycloid.

Solution.   Let the line on which the circle rolls be the x axis and place the origin of coordinates at one of the places where the tracing point P comes into contact with the x axis (Fig. 15-7).

Denote the center of the rolling circle by C and the radius by a. Let $P:(x, y)$ be any position of the tracing point and choose for parameter the radian measure t of the angle through which the line segment CP has turned from its vertical position when P was at the origin. We assume that the circle is rolling to the right on top of the x axis.

Since $\overline{ON} = \text{arc } \widehat{PN} = at$,

$$x = \overline{OM} = \overline{ON} - \overline{MN} = at - a \sin t = a(t - \sin t)$$

and

$$y = \overline{MP} = \overline{NR} = \overline{NC} + \overline{CR} = a - a \cos t = a(1 - \cos t).$$

Thus parametric equations of the cycloid are

$$x = a(t - \sin t), \qquad y = a(1 - \cos t).$$

The cycloid has a number of interesting applications, especially in mechanics. It is the "curve of fastest descent." If a particle, acted on only by gravity, is allowed to slide down some curve from a point A to a lower point B, not on the same vertical line, it completes its journey in the shortest time when the curve is an inverted cycloid (Fig. 15-8). Of course, the shortest distance is along the straight line segment AB, but the least time is used when the path is along a cycloid; this is because the acceleration when it is released depends on the steepness of descent, and along a cycloid it builds up velocity much more quickly than along a straight line.

Another interesting property is this: if L is the lowest point on an arch of an inverted cycloid (Fig. 15-9), the time it takes a particle P to slide down the cycloid to L is the same no matter where P starts from on the inverted arch; thus if several particles, P_1, P_2, and P_3, in different positions on the cycloid (Fig. 15-9), start to slide at the same instant, all will reach the low point L at the same time.

As a consequence of this latter property, the period of a cycloidal pendulum is independent of the amplitude of the oscillation.

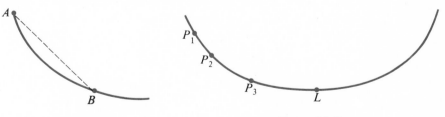

Figure 15-8   Figure 15-9

15.2 FUNCTIONS DEFINED BY PARAMETRIC EQUATIONS

Let f and g be smooth functions with a common domain $[\alpha, \beta]$ and let

(1) $x = f(t) \quad \text{and} \quad y = g(t)$.

If $f'(t) \neq 0$ for $\alpha \leq t \leq \beta$, then f is strictly monotonic on $[\alpha, \beta]$. Therefore f has a continuous inverse f^{-1} such that $t = f^{-1}(x)$ for all x on the closed interval whose endpoints are $f(\alpha)$ and $f(\beta)$ (by 10.2.4). Thus

(2) $$y = g(t) = g(f^{-1}(x)) = F(x),$$

where $F = g(f^{-1})$ is a continuous function whose domain is the closed interval with endpoints $f(\alpha)$ and $f(\beta)$.

Hence the parametric equations (1) define y as a continuous function of x, whose rule of correspondence is given in (2).

By substituting $x = f(t)$ into $g(t) = F(x)$ from (2), we obtain the identity in t:

(3) $$g(t) = F(f(t)).$$

If we differentiate both members of (3) with respect to t by the chain rule, we obtain

$$g'(t) = F'(f(t)) \cdot f'(t),$$

which in view of equations (1) can be written

$$\frac{dy}{dt} = \frac{dF}{dx} \cdot \frac{dx}{dt}.$$

Therefore, since $dx/dt = f'(t) \neq 0$,

$$\frac{dF}{dx} = \frac{dy}{dt} \div \frac{dx}{dt}.$$

This proves the following theorem.

15.2.1 Theorem. Let f and g be smooth functions with a common domain $[\alpha, \beta]$. If $f'(t) \neq 0$ for $\alpha \leq t \leq \beta$, then the parametric equations

$$x = f(t) \qquad \text{and} \qquad y = g(t)$$

define y as a differentiable function of x, and

$$\frac{dy}{dx} = \frac{dy}{dt} \div \frac{dx}{dt}.$$

Example 1. If $x = t^2$, $y = 1/(2t)$, find $D_x y$.

Solution.

$$\frac{dy}{dx} = \frac{dy}{dt} \div \frac{dx}{dt} = -\frac{1}{2t^2} \div 2t = -\frac{1}{4t^3}.$$

Example 2. Find the first two derivatives of the function defined by the parametric equations

$$x = t^2 - 1, \qquad y = \frac{2}{t}.$$

Solution. $dy/dt = -2/t^2$ and $dx/dt = 2t$. From 15.2.1

$$\frac{dy}{dx} = \frac{dy}{dt} \div \frac{dx}{dt} = \frac{-2}{t^2} \div \frac{2t}{1} = \frac{-2}{t^2} \cdot \frac{1}{2t} = \frac{-1}{t^3}.$$

Thus $y' = -1/t^3$, in terms of t. Since $dy'/dt = 3/t^4$, we again use 15.2.1 to write

$$\frac{d^2y}{dx^2} = \frac{dy'}{dx} = \frac{dy'}{dt} \div \frac{dx}{dt} = \frac{3}{t^4} \div \frac{2t}{1} = \frac{3}{t^4} \cdot \frac{1}{2t} = \frac{3}{2t^5}.$$

Sometimes a definite integral involves two variables, say x and y, in the integrand and differential, and y may be defined as a function of x by equations that give x and y in terms of a parameter such as t. In such cases, it is often convenient to evaluate the definite integral by expressing the integrand and the differential in terms of t and dt, and adjusting the limits of integration before integrating with respect to t.

Example 3.   Evaluate (a) $\int_1^3 y \, dx$ and (b) $\int_1^3 xy^2 \, dx$, using $x = 2t - 1$ and $y = t^2 + 2$.

Solution.   From $x = 2t - 1$ we have $dx = 2 \, dt$; when $x = 1$, $t = 1$; and when $x = 3$, $t = 2$. Thus

(a) $\int_1^3 y \, dx = \int_1^2 (t^2 + 2)2 \, dt = 2\left[\dfrac{t^3}{3} + 2t\right]_1^2 = \dfrac{26}{3}$.

(b) $\int_1^3 xy^2 \, dx = \int_1^2 (2t - 1)(t^2 + 2)^2 2 \, dt$

$\qquad = 2 \int_1^2 (2t^5 - t^4 + 8t^3 - 4t^2 + 8t - 4) \, dt = 86\tfrac{14}{15}$.

Exercises

In each of Exercises 1 to 14, a parametric representation of a curve is given.
(a)   Sketch the curve by assigning values to the parameter.
(b)   Which of the following apply to the curve: simple, smooth, closed, arc?
(c)   Obtain the Cartesian equation of the curve by eliminating the parameter.

1. $x = 2t$, $y = 3t$; $t \in R$.

2. $x = 4t - 1$, $y = 2t$; $0 \le t \le 3$.

3. $x = t - 4$, $y = \sqrt{t}$; $0 \le t \le 4$.

4. $x = t$, $y = \dfrac{1}{t}$; $0 < t$.

5. $x = t^2$, $y = t^3$; $-1 \le t \le 2$.

6. $x = t^2 - 1$, $y = t^3 - t$; $-3 \le t \le 3$.

7. $x = t^3 - 4t$, $y = t^2 - 4$; $-3 \le t \le 3$.

8. $x = 3\sqrt{t - 3}$, $y = 2\sqrt{4 - t}$; $3 \le t \le 4$.

9. $x = 3 \sin t$, $y = 5 \cos t$; $0 \le t \le 2\pi$.

10. $x = 3 \sin \theta - 1$, $y = 2 \cos \theta + 2$; $0 < \theta < \pi$.

11. $x = 4 \sec \theta$, $y = 3 \tan \theta$; $-\tfrac{1}{2}\pi < \theta < \tfrac{1}{2}\pi$.
(*Hint:* $1 + \tan^2 \theta = \sec^2 \theta$.)

12. $x = 4 \sec t - 4$, $y = 5 \tan t + 5$;
$-\tfrac{1}{4}\pi \le t \le \tfrac{1}{4}\pi$.

13. $x = 4 \sin^4 t$, $y = 4 \cos^4 t$; $0 \le t \le \tfrac{1}{2}\pi$.

14. $x = 2 \cos \theta$, $y = 2 \cos \tfrac{1}{2}\theta$; $\theta \in R$.

In Exercises 15 to 20, find dy/dx without eliminating the parameter.

15. $x = 3t^2$, $y = 2t^3$; $t \ne 0$.

16. $x = 6t^2$, $y = t^3$; $t \ne 0$.

17. $x = 2t - \dfrac{3}{t}$, $y = 2t + \dfrac{3}{t}$; $t \ne 0$.

18. $x = 1 - \cos t$, $y = 2 + 3 \sin t$; $t \ne n\pi$, $n \in Z$.

19. $x = 3 \tan t - 1$, $y = 5 \sec t + 2$; $t \ne \dfrac{(2n + 1)\pi}{2}$, $n \in Z$.

20. $x = \dfrac{2}{1 + t^2}$, $y = \dfrac{2}{t(1 + t^2)}$; $t \ne 0$.

In Exercises 21 to 26, find the equations of the tangent and normal to the given curve at the given point without eliminating the parameter. Make a sketch.

21. $x = t^2$, $y = t^3$; $t = 2$.

22. $x = 3t$, $t = 8t^3$; $t = -\tfrac{1}{2}$.

23. $x = 3 \sin t$, $y = 3 \cos t$; $t = \tfrac{1}{6}\pi$.

24. $x = 4 \cos \theta - 3$, $y = 4 \sin \theta + 1$; $\theta = \dfrac{2\pi}{3}$.

25. $x = 2 \sec t$, $y = 2 \tan t$; $t = -\dfrac{\pi}{6}$.

26. $x = 2e^t$, $y = \frac{1}{3}e^{-t}$; $t = 0$.

In Exercises 27 to 30, integrate and express the result in terms of x.

27. $\displaystyle\int (x^2 - 4y)\, dx$, where $x = t + 1$, $y = t^3 + 4$.

28. $\displaystyle\int \frac{3y}{x^2 - 1}\, dx$, where $x = \cos t$, $y = \sin t$.

29. $\displaystyle\int xy\, dy$, where $x = \sec t$, $y = \tan t$.

30. $\displaystyle\int (x^2y - y^3)\, dy$, where $x = 3\cos 2t$, $y = 3\sin 2t$.

31. Find the area of the region between the curve $x = e^{2t}$, $y = e^{-t}$ and the x axis from $t = 0$ to $t = \ln 5$. Make a sketch.

32. Find the area of the region bounded by the curve

$$x = t + \frac{1}{t}, \qquad y = t - \frac{1}{t},$$

and the line $3x - 10 = 0$, without eliminating the parameter. Make a sketch.

33. Find the area of the region between the arc

$$x = \sin t, \qquad y = e^t, \qquad -\frac{\pi}{2} \le t \le \frac{\pi}{2},$$

and the x axis. Make a sketch.

34. Sketch carefully the graph of one arch of the cycloid

$$x = 4(t - \sin t), \qquad y = 4(1 - \cos t).$$

35. Find the area under one arch of the cycloid

$$x = a(t - \sin t), \qquad y = a(1 - \cos t).$$

36. Each value of the parameter t in the equations of a cycloid,

$$x = a(t - \sin t), \qquad y = a(1 - \cos t),$$

determines a unique point $P{:}(x, y)$ on the cycloid and also determines a unique position of the rolling circle that generates the cycloid. Let $P_1{:}(x_1, y_1)$ be the point on the cycloid determined by the value t_1 of the parameter t. Prove that the tangent to the cycloid at P_1 passes through the highest point on the rolling circle when the circle is in the position deter-

mined by t_1. Through what point on the circle does the normal pass?

37. Let a circle of radius b roll, without slipping, inside a fixed circle of radius a, $a > b$. A point P on the circumference of the rolling circle traces out a curve called a **hypocycloid**. Find parametric equations of the hypocycloid. [*Hint:* Place the origin O of Cartesian coordinates at the center of the fixed, larger circle and let the point $A{:}(a, 0)$ be one position of the tracing point P. Denote by B the moving point of tangency of the two circles and let t, the radian measure of the angle AOB, be the parameter.]

38. Show that if $b = a/4$ in Exercise 3, the parametric equations of the hypocycloid may be simplified to

$$x = a\cos^3 t, \qquad y = a\sin^3 t.$$

This is called a **hypocycloid of four cusps.** Sketch it carefully and show that its Cartesian equation is $x^{2/3} + y^{2/3} = a^{2/3}$.

39. The curve traced out by a point on the circumference of a circle of radius b as it rolls without slipping on the outside of a fixed circle of radius a is called an **epicycloid.** Show that it has parametric equations

$$x = (a + b)\cos t - b\cos\frac{a + b}{b} t,$$

$$y = (a + b)\sin t - b\sin\frac{a + b}{b} t.$$

(See Hint in Exercise 37.)

40. If $b = a$, the equations in Exercise 39 are

$$(1) \qquad x = 2a\cos t - a\cos 2t,$$
$$y = 2a\sin t - a\sin 2t.$$

Show that this special epicycloid is the cardioid $r = 2a(1 - \cos\theta)$, where the pole of the polar coordinate system is the point $(a, 0)$ in the Cartesian system and the polar axis has the direction of the positive x axis. [*Hint:* Find a Cartesian equation of the epicycloid by eliminating the parameter t between equations (1). Then show that the equations connecting the Cartesian and polar systems are

$$x = r\cos\theta + a, \qquad y = r\sin\theta,$$

and use these equations to transform the Cartesian equation into $r = 2a(1 - \cos\theta)$.]

15.3 LENGTH OF A PLANE ARC

In Section 9.6 we developed a formula for the length of the arc of a curve $y = F(x)$ from one point on the curve to another. But the graph of a function F is a specialized curve that cannot be cut by a vertical line in more than one point.

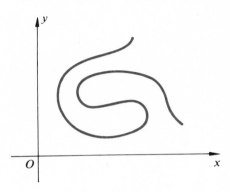

Figure 15-10

The lengths of many other arcs cannot be found by our earlier formula (9.6.1). For example, a curve such as that shown in Fig. 15-10 is not the graph of a function and cannot be rectified by our previous formula, except by separating the curve into several arcs and finding their lengths. Yet if the curve has parametric equations $x = f(t)$ and $y = g(t)$, in which f and g are smooth functions on a closed interval $[a, b]$, then the length of the arc from $t = a$ to $t = b$ can be found in a manner about to explained.

Let f and g be functions that are smooth on the closed interval $[a, b]$, that is, let f' and g' be continuous on $[a, b]$, and consider the smooth arc having parametric equations

$$(1) \qquad x = f(t), \qquad y = g(t), \qquad t \in [a, b].$$

Partition the interval $[a, b]$ into n subintervals by the points $a = t_0 < t_1 < t_2 < \cdots t_{i-1} < t_i < \cdots < t_n = b$ in the usual manner; call this partition P and denote its norm by $|P|$. If $\Delta t_i = t_i - t_{i-1}$, then $\Delta t_i \leq |P|$ for $i = 1, 2, 3, \cdots, n$.

Corresponding to the numbers t_i are n points $P_i:(f(t_i), g(t_i))$ on the arc (1); denote by P_0 the point $(f(a), g(a))$ (Fig. 15-11). Form the sum $\sum_{i=1}^{n} |P_{i-1}P_i|$ of the lengths of the chords of the curve that join the points P_{i-1} and P_i $(i = 1, 2, 3, \cdots, n)$. Then the length L of the arc (1) from the point $(f(a), g(a))$ to the point $(f(b), g(b))$ is

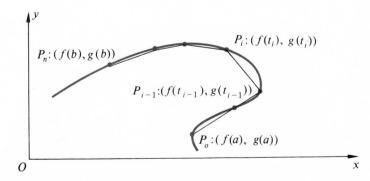

Figure 15-11

defined to be

$$(2) \qquad L = \lim_{|P| \to 0} \sum_{i=1}^{n} |P_{i-1}P_i|,$$

provided that this limit exists.

The notation in (2) means that corresponding to each positive number ϵ there is a $\delta > 0$ such that

$$\left| \sum_{i=1}^{n} |P_{i-1}P_i| - L \right| < \epsilon$$

for *all* partitions P whose norms satisfy $|P| < \delta$.

Using the distance formula, we can write

$$(3) \qquad \sum_{i=1}^{n} |P_{i-1}P_i| = \sum_{i=1}^{n} \sqrt{[f(t_i) - f(t_{i-1})]^2 + [g(t_i) - g(t_{i-1})]^2}.$$

Since f' and g' are continuous on the subinterval $[t_{i-1}, t_i]$, we know by the mean value theorem that there exist numbers z_i and w_i between t_{i-1} and t_i such that

$$f(t_i) - f(t_{i-1}) = f'(z_i)\, \Delta t_i,$$
$$g(t_i) - g(t_{i-1}) = g'(w_i)\, \Delta t_i.$$

Thus (3) can be rewritten

$$(4) \qquad \sum_{i=1}^{n} \sqrt{[f'(z_i)]^2 + [g'(w_i)]^2}\, \Delta t_i,$$

where $t_{i-1} < z_i < t_i$ and $t_{i-1} < w_i < t_i$. From (2), (3), and (4) the length of the arc (1) from $(f(a), g(a))$ to $(f(b), g(b))$ is

$$(5) \qquad \lim_{|P| \to 0} \sum_{i=1}^{n} \sqrt{[f'(z_i)]^2 + [g'(w_i)]^2}\, \Delta t_i.$$

Now (4) is not a Riemann sum because z_i and w_i are not necessarily the same point in $[t_{i-1}, t_i]$. But it can be shown[*] that the value of (5) is just what it would be if z_i and w_i *were* the same point in the ith subinterval (for $i = 1, 2, 3, \cdots, n$), namely

$$\int_a^b \sqrt{[f'(t)]^2 + [g'(t)]^2}\, dt.$$

This is summarized in the following important theorem.

15.3.1   **Theorem.**   Let f and g be functions that are smooth on a closed interval $[a, b]$. Then the length of the smooth arc

$$x = f(t), \qquad y = g(t), \qquad a \le t \le b,$$

from the point $(f(a), g(a))$ to the point $(f(b), g(b))$ is given by

$$L = \int_a^b \sqrt{[f'(t)]^2 + [g'(t)]^2}\, dt.$$

---

[*]See, for example, J. M. H. Olmsted, *Advanced Calculus*, New York, Appleton-Century-Crofts, 1961, p. 238.

Let $s(t)$ be the length of the arc from the fixed point $P(a) = (f(a), g(a))$ to the variable point $P(t) = (f(t), g(t))$, $a \leq t \leq b$. Then

$$s(t) = \int_a^t \sqrt{[f'(w)]^2 + [g'(w)]^2} \, dw,$$

where w is a dummy variable. Then

$$\frac{ds}{dt} = \sqrt{[f'(t)]^2 + [g'(t)]^2} = \sqrt{\left(\frac{dx}{dt}\right)^2 + \left(\frac{dy}{dt}\right)^2},$$

from which

(6) $$ds^2 = dx^2 + dy^2.$$

Example 1. Find the length of the arc of the curve with parametric equations $x = 2t^2$ and $y = 2t^3$, (a) from $t = 0$ to $t = 2$, and (b) from $t = -2$ to $t = 0$.

Solution. The curve is shown in Fig. 15-12. Observe that as t increases from -2 to 0, the corresponding point (x, y) moves along the curve in the fourth quadrant from the point $(8, -16)$ to the

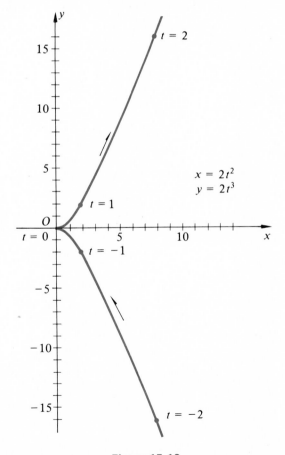

Figure 15-12

origin, and that as t continues to increase, from 0 to 2, the corresponding point moves away from the origin, along the curve in the first quadrant to the point $(8, 16)$. The positive direction on the curve is indicated by arrows.

(a) $D_t x = 4t$ and $D_t y = 6t^2$. By 15.3.1, the length of the arc from $t = 0$ to $t = 2$ is equal to

$$\int_0^2 \sqrt{16t^2 + 36t^4} \, dt = \int_0^2 2t\sqrt{4 + 9t^2} \, dt$$

$$= \frac{1}{9} \int_0^2 (4 + 9t^2)^{1/2} \, d(4 + 9t^2) = \frac{2}{27}(4 + 9t^2)^{3/2} \Big]_0^2$$

$$= \frac{16}{27}(10^{3/2} - 1) \doteq 18.1.$$

(b) The length of the arc from $t = -2$ to $t = 0$ is

$$\int_{-2}^0 \sqrt{16t^2 + 36t^4} \, dt.$$

Notice that for negative values of t, $\sqrt{t^2} = -t$, since the symbol $\sqrt{t^2}$ means the nonnegative square root of t^2. Thus

$$\int_{-2}^0 \sqrt{16t^2 + 36t^4} \, dt = \int_{-2}^0 2\sqrt{t^2}\sqrt{4 + 9t^2} \, dt$$

$$= \int_{-2}^0 2(-t)\sqrt{4 + 9t^2} \, dt = -\frac{1}{9} \int_{-2}^0 (4 + 9t^2)^{1/2} \, d(4 + 9t^2)$$

$$= -\frac{2}{27}(4 + 9t^2)^{3/2} \Big]_{-2}^0 = -\frac{16}{27}(1 - 10^{3/2}) \doteq 18.1.$$

It is easy to show that when the curve in 15.3.1 is the graph of a function F having a continuous derivative on $[a, b]$, our present formula for arc length becomes the earlier formula in 9.6.1. For if $y = F(x)$, parametric equations of its graph, in which x is the parameter, are

$$x = x, \qquad y = F(x), \quad a \le x \le b.$$

Then the formula for arc length in 15.3.1 becomes

$$L = \int_a^b \sqrt{1 + [F'(x)]^2} \, dx,$$

which is 9.6.1.

The formula for arc length given in 15.3.1 can be written

(7) $$L = \int_a^b \sqrt{\left(\frac{dx}{dt}\right)^2 + \left(\frac{dy}{dt}\right)^2} \, dt.$$

The equations $x = r \cos \theta$ and $y = r \sin \theta$ express the relation between the Cartesian and polar coordinates of any point in the plane when the Cartesian and polar frames of reference are superposed; by differentiating them with respect to t, we get

$$\frac{dx}{dt} = \cos \theta \frac{dr}{dt} - r \sin \theta \frac{d\theta}{dt},$$

$$\frac{dy}{dt} = \sin \theta \frac{dr}{dt} + r \cos \theta \frac{d\theta}{dt}.$$

If we substitute these results in (7) and simplify, we obtain

(8)
$$L = \int_a^b \sqrt{\left(\frac{dr}{dt}\right)^2 + \left(r\frac{d\theta}{dt}\right)^2}\, dt.$$

When a curve is given by a polar equation $r = \phi(\theta)$, we can think of θ instead of t as the parameter. In this case, $dr/dt = dr/d\theta$ and $d\theta/dt = dt/dt = 1$. By substituting this in formula (8), we obtain the following theorem.

15.3.2 **Theorem.** If ϕ is a function such that ϕ' is continuous on the closed interval $[\alpha, \beta]$, then the length of the arc of the polar curve $r = \phi(\theta)$ from $\theta = \alpha$ to $\theta = \beta$ is

$$L = \int_\alpha^\beta \sqrt{r^2 + \left(\frac{dr}{d\theta}\right)^2}\, d\theta.$$

Example 2. Find the perimeter of the cardioid $r = a(1 + \cos \theta)$ (see Fig. 15-13).

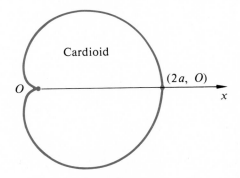

Figure 15-13

Solution. In applying 15.3.2, we find that
$$\sqrt{r^2 + r'^2} = a\sqrt{2}\sqrt{1 + \cos \theta} = 2a \cos \tfrac{1}{2}\theta.$$

The entire cardioid is swept out when θ varies from 0 to 2π, and we may be tempted to take 0 and 2π for the lower and upper limits of our definite integral. But $\cos \tfrac{1}{2}\theta < 0$ when $\pi < \theta < 2\pi$ and $\sqrt{r^2 + r'^2} \geq 0$ (by definition of the radical sign). Thus $\sqrt{r^2 + r'^2} \neq 2a \cos \tfrac{1}{2}\theta$ when $\pi < \theta < 2\pi$. This difficulty may be avoided by finding half the length of the cardioid by integrating from 0 to π (see Fig. 15-13) or by getting the entire length by integrating from $-\pi$ to π. Using the latter method, we find the complete length of the cardioid to be

$$\int_{-\pi}^{\pi} \sqrt{r^2 + r'^2}\, d\theta = 4a \int_{-\pi}^{\pi} \cos \tfrac{1}{2}\theta \, (\tfrac{1}{2}\, d\theta) = 8a.$$

Exercises

In each of Exercises 1 to 10, sketch the given arc and find its length.

1. $x = t^3, y = t^2; 0 \leq t \leq 4.$

2. $x = 3t^2 + 2, y = 2t^3 - 1; 1 \leq t \leq 3.$

3. $x = 3 \sin t, y = 3 \cos t - 3; 0 \leq t \leq 2\pi.$

4. $x = 4 \cos t + 5, y = 4 \sin t - 1; 0 \leq t \leq 2\pi.$

5. $x = e^t \cos t$, $y = e^t \sin t$; $0 \le t \le \dfrac{\pi}{2}$.

6. $x = e^{-t} \sin t$, $y = e^{-t} \cos t$; $0 \le t \le \pi$.

7. $x = 2 \tan t$, $y = \tan^2 t$; $0 \le t \le \frac{1}{3}\pi$. (*Hint:* See Example 6 of Section 11.5)

8. $x = 6t$, $y = 4t^{3/2}$; $3 \le t \le 8$.

9. $x = t^2$, $y = 2t - 1$; $-2 \le t \le 2$.

10. $x = t^2 + 4$, $y = t - 3$; $0 \le t \le \dfrac{\sqrt{3}}{2}$.

11. Find the entire length of the four-cusped hypocycloid

$$x = a \sin^3 t, \qquad y = a \cos^3 t.$$

Sketch the curve.

12. Show that the length of one arch of the cycloid

$$x = a(t - \sin t), \qquad y = a(1 - \cos t)$$

is equal to eight times the radius of the rolling circle that generated it.

13. Find the perimeter of the cardioid $r = a \cos^2 \frac{1}{2}\theta$. Make a sketch.

14. Sketch the curve whose polar equation is $r = a \sin^3 \frac{1}{3}\theta$ and find its entire length.

15. Find the length of the logarithmic spiral $r = e^{\theta/2}$ from $\theta = 0$ to $\theta = 2\pi$. Make a sketch.

16. Find the length of the spiral of Archimedes $r = a\theta$ from $\theta = 0$ to $\theta = \pi$. Make a sketch. (*Hint:* See Example 6, Section 11.5.)

15.4 VECTORS IN THE PLANE

Many concepts in physics and engineering involve both *magnitude and direction*. Familiar examples are velocity, acceleration, and force. Vectors are ideal for the study of such physical concepts, and vectors often enable us to state and prove geometric theorems in a concise and elegant manner. Moreover, vector analysis is an interesting and important branch of pure mathematics.

Vector analysis was suggested by the work of the Irish mathematician William Rowan Hamilton (1805–1865) on quaternions, and developed to its present form by the American mathematician Josiah Willard Gibbs (1839–1903) and the English engineer Oliver Heaviside (1850–1925).

15.4.1 Definition. A two-dimensional **vector** is an ordered pair of real numbers $[x, y]$; and the numbers x and y are the **components** of the vector.

For example, $[2, 3]$, $[-1, 5]$, $[\pi, \sqrt{2}]$, and $[0.03, -19]$ are vectors. Square brackets are used here in writing vectors to avoid confusion with points. Thus $[4, -1]$ is a vector and $(4, -1)$ is a point.

Symbols for vectors are printed in boldface type in this book; thus $\mathbf{a} = [a_1, a_2]$, where a_1 and a_2 are numbers, is a vector. When writing by hand, it is convenient to indicate a vector $\mathbf{a}$ by putting a small arrow over the a; thus $\mathbf{a}$ is often handwritten $\vec{a}$.

The vector $[0, 0]$, both of whose components are zero, is called the **zero vector** and is given the special symbol $\mathbf{0}$; that is,

$$\mathbf{0} = [0, 0] \text{ is the \textbf{zero vector}.}$$

15.4.2 Definition. Two vectors, $\mathbf{a} = [a_1, a_2]$ and $\mathbf{b} = [b_1, b_2]$ are said to be **equal** if and only if $a_1 = b_1$ and $a_2 = b_2$; that is,

$$\mathbf{a} = \mathbf{b} \iff a_1 = b_1 \text{ and } a_2 = b_2.$$

The **negative** of a vector $\mathbf{a} = [a_1, a_2]$ is the vector $-\mathbf{a} = [-a_1, -a_2]$. Thus the negative of the vector $\mathbf{b} = [3, -11]$ is $-\mathbf{b} = [-3, 11]$.

Vectors are added according to the following definition.

15.4.3 Definition. The **sum** of two vectors $\mathbf{a} = [a_1, a_2]$ and $\mathbf{b} = [b_1, b_2]$ is the vector

$$\mathbf{a} + \mathbf{b} = [a_1 + b_1, a_2 + b_2].$$

As an illustration, $[2, -5] + [-3, 4] = [-1, -1]$.

The following rules or "laws" for the addition of vectors are easily derived from the preceding definition and the rules for operating with real numbers.

15.4.4 Theorem. Let $\mathbf{a}, \mathbf{b}$, and $\mathbf{c}$ be any three vectors; then

$$\mathbf{a} + \mathbf{b} = \mathbf{b} + \mathbf{a} \qquad \text{(commutative law for addition)},$$
$$\mathbf{a} + (\mathbf{b} + \mathbf{c}) = (\mathbf{a} + \mathbf{b}) + \mathbf{c} \qquad \text{(associative law for addition)},$$
$$\mathbf{a} + \mathbf{0} = \mathbf{a},$$
$$\mathbf{a} + (-\mathbf{a}) = \mathbf{0}.$$

In proving the first, we have $\mathbf{a} + \mathbf{b} = [a_1 + b_1, a_2 + b_2]$ (by 15.4.3). But $a_1 + b_1 = b_1 + a_1$ and $a_2 + b_2 = b_2 + a_2$ by the commutative law for the addition of numbers. Therefore

$$\mathbf{a} + \mathbf{b} = [a_1 + b_1, a_2 + b_2] = [b_1 + a_1, b_2 + a_2] = \mathbf{b} + \mathbf{a}.$$

The proofs of the other three rules for addition of vectors are left as exercises.

Engineers and scientists commonly think of vectors as directed line segments. The reason for this will soon be apparent.

The two-dimensional vector $\mathbf{a} = [a_1, a_2]$ may be interpreted as a **directed line segment** whose initial point is any point (x, y) in the Cartesian plane and whose terminal point is the point whose coordinates are $(x + a_1, y + a_2)$. The zero vector $\mathbf{0}$ may be represented by any point, and its direction is unspecified.

For example, the vector $[4, 2]$ is represented in Fig. 15-14 by the directed line segment $\overrightarrow{QP}$ whose initial point is $Q:(1, 2)$ and whose terminal point P has coordinates $(1 + 4, 2 + 2) = (5, 4)$; it is also represented by $\overrightarrow{RS}$, where R is the point $(-2, -3)$ and S is the point $(2, -1)$, and by any other directed line segment having the same length as $\overrightarrow{QP}$ and $\overrightarrow{RS}$ and the same direction.

It is easy to prove that if $\overrightarrow{QP}$ and $\overrightarrow{RS}$ are two directed line segments representing the same vector, and if $\overrightarrow{QP}$ and $\overrightarrow{RS}$ are not on the same line, then $QPSR$ is a parallelogram (Fig. 15-15).

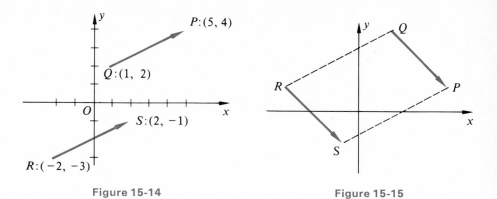

Figure 15-14   Figure 15-15

By the slope formula, the slope of the line segment from the point (x, y) to the point $(x + a_1, y + a_2)$ is a_2/a_1 if $a_1 \neq 0$ (Fig. 15-16). Hence the **slope** of the vector $\mathbf{a} = [a_1, a_2]$ is a_2/a_1 if $a_1 \neq 0$. For example, the slope of the vector $[4, 2]$ is $\frac{1}{2}$ (see Fig. 15-14).

Our definition of the sum of two vectors (15.4.3) may be interpreted geometrically as follows (Fig. 15-17). To find the vector $\mathbf{a} + \mathbf{b}$ that is the sum of the vectors $\mathbf{a}$ and $\mathbf{b}$, let Q be an arbitrary point in the plane and represent $\mathbf{a}$ and $\mathbf{b}$ by directed line segments $\overrightarrow{QP}$ and $\overrightarrow{PR}$; then the directed line segment $\overrightarrow{QR}$ represents $\mathbf{a} + \mathbf{b}$ (Fig. 15-17).

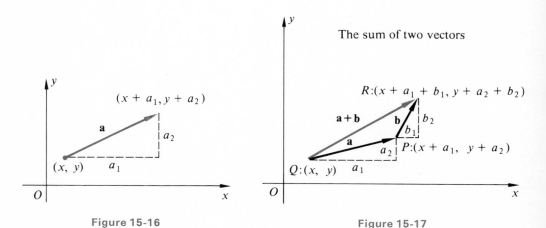

Figure 15-16   Figure 15-17

Since the vectors $\mathbf{a}$ and $\mathbf{b}$ in this construction are represented by adjacent sides of a parallelogram and $\mathbf{a} + \mathbf{b}$ is a diagonal, this is called the **parallelogram law** for the addition of vectors. The importance of vectors in engineering and the physical sciences is partly due to the fact that many physical concepts involving magnitude and direction, such as velocities, accelerations, and forces, combine in accordance with the parallelogram law, just as direct line segments do.

It is easy to verify geometrically, by means of the parallelogram law, the rules for addition of vectors given in 15.4.4. For example, the commutative law, $\mathbf{a} + \mathbf{b} = \mathbf{b} + \mathbf{a}$, is verified by Fig. 15-18, since the lower triangle shows the construction for finding $\mathbf{a} + \mathbf{b}$ and the upper triangle for $\mathbf{b} + \mathbf{a}$; in both cases, the result is the same diagonal.

If the vector $\mathbf{a}$ is represented by the directed line segment $\overrightarrow{QP}$ (Fig. 15-19), then $-\mathbf{a}$, the negative of $\mathbf{a}$, is represented by $\overrightarrow{PQ}$; $-\mathbf{a}$ is also represented by any directed

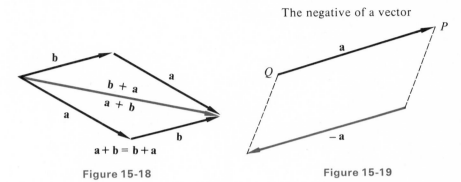

Figure 15-18   Figure 15-19

line segment that is parallel to $\overrightarrow{QP}$, has the same length as $\overrightarrow{QP}$, and points in the opposite direction.

Subtraction of vectors is defined as follows.

15.4.5 Definition.   The **difference $\mathbf{a} - \mathbf{b}$** is the vector obtained by adding $\mathbf{a}$ and the negative of $\mathbf{b}$:

$$\mathbf{a} - \mathbf{b} = \mathbf{a} + (-\mathbf{b}) = [a_1 - b_1, a_2 - b_2].$$

For example, $[6, -1] - [2, -8] = [6, -1] + [-2, 8] = [4, 7]$.

There is a simple geometric interpretation for subtraction of vectors. Since $\mathbf{b} + (\mathbf{a} - \mathbf{b}) = (\mathbf{a} - \mathbf{b}) + \mathbf{b} = [\mathbf{a} + (-\mathbf{b})] + \mathbf{b} = \mathbf{a} + [(-\mathbf{b}) + \mathbf{b}] = \mathbf{a} + [\mathbf{b} + (-\mathbf{b})]$ $= \mathbf{a} + \mathbf{0} = \mathbf{a}$, we have

$$\mathbf{b} + (\mathbf{a} - \mathbf{b}) = \mathbf{a}.$$

Hence if we represent the vectors $\mathbf{a}$ and $\mathbf{b}$ by directed line segments having the same initial point (Fig. 15-20), *the directed line segment from the terminal point of $\mathbf{b}$ to the terminal point of $\mathbf{a}$ represents $\mathbf{a} - \mathbf{b}$*, since this is the parallelogram construction for $\mathbf{b} + (\mathbf{a} - \mathbf{b}) = \mathbf{a}$.

15.4.6 Definition.   By the **length** or **magnitude**, $|\mathbf{a}|$, of the vector $\mathbf{a} = [a_1, a_2]$ is meant the number (not vector)

$$|\mathbf{a}| = \sqrt{a_1^2 + a_2^2}.$$

Thus the length of the vector $[4, 2]$ is $\sqrt{4^2 + 2^2} = 2\sqrt{5}$; it is, of course, the same as the length of any line segment representing it (see Fig. 15-14). Clearly, the length of the zero vector is zero, and the lengths of all other vectors are positive numbers.

Subtraction of vectors

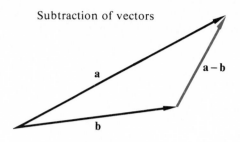

Figure 15-20

Exercises

1. Represent the following vectors by line, segments in the plane: $[3, 1]$, $[5, -3]$, $[0, 4]$, $[-3, 6]$, $[-5, -3]$, $[-4, 0]$, $[3, -\sqrt{2}]$, $[\pi, e]$, $[4, -\frac{3}{2}\pi]$, $[2, -\sqrt{3}]$.

2. Find the slope of each of the vectors in Exercise 1 or state that it has no slope.

3. Find the negatives of the vectors given in Exercise 1 and represent each as a directed line segment on your diagrams for Exercise 1.

4. Find the sum of each of the following pairs of vectors algebraically and illustrate by diagrams like Fig. 15-17.
(a) $[2, 5]$, $[-4, 3]$;
(b) $[5, -2]$, $[0, 7]$;
(c) $[-3, 2]$, $[0, 0]$;
(d) $[-4, -6]$, $[\sqrt{3}, 0]$.

5. In each pair of vectors given in Exercise 4, subtract the second from the first. Illustrate by diagrams like Fig. 15-20.

6. Find the length of each of the vectors in Exercise 1.

7. In each part of this exercise, the coordinates of a pair of points, A and B, are given. Find the vector (ordered pair of numbers) represented by the directed line segment $\overrightarrow{AB}$.
(a) $A:(-4, 1)$, $B:(2, 6)$;
(b) $A:(3, 0)$, $B:(-5, -5)$;
(c) $A:(5, -7)$, $B:(0, 0)$;
(d) $A:(\sqrt{3}, 4)$, $B:(4\sqrt{3}, -2)$.

8. The vertices of a triangle are as follows, $A:(-3, 1)$, $B:(5, -2)$ and, $C:(4, 4)$. Find the sum of the vectors represented by the directed line segments $\overrightarrow{AB}$, $\overrightarrow{BC}$, and $\overrightarrow{CA}$.

9. The vertices of a triangle are the three points $A:(a_1, a_2)$, $B:(b_1, b_2)$, and $C:(c_1, c_2)$. Prove that the sum of the vectors represented by the directed line segments $\overrightarrow{AB}$, $\overrightarrow{BC}$, and $\overrightarrow{CA}$ is the zero vector.

10. Prove that if A, B, C, and D are the vertices of a quadrilateral, the sum of the vectors represented by the directed line segments $\overrightarrow{AB}$, $\overrightarrow{BC}$, $\overrightarrow{CD}$, and $\overrightarrow{DA}$ is the zero vector.

11. Prove the associative law for addition of vectors: $\mathbf{a} + (\mathbf{b} + \mathbf{c}) = (\mathbf{a} + \mathbf{b}) + \mathbf{c}$.

12. Verify by a labeled diagram the associative law for addition of vectors.

13. An airplane flies directly east for 400 miles and then southeast for 280 miles. Represent these displacements as vectors and find the resultant displacement (the sum of the two vectors) both graphically and algebraically.

14. A ship sails due north for 10 miles and then northwest for 8 miles. Represent these displacements as vectors and find the resultant displacement (the sum of the two vectors) both graphically and algebraically.

15. The pilot of a light plane wishes to fly southeast. If the air speed of the plane is 120 miles per hour, and a wind from the south blows 40 miles per hour, along what vector should he set his course?

16. The pilot of a light airplane wishes to fly southwest. If the air speed of the airplane is 150 miles per hour and a wind from the west blows 50 miles per hour, along what vector should he set his course?

17. At a stretch of a river where it flows due south between parallel banks, a man on the west bank wants to run his motorboat straight across the river. If the boat can go 10 miles per hour in still water and the speed of the current is 4 miles per hour, on what vector should he head his boat?

15.5 SCALARS, DOT PRODUCT, AND BASIS VECTORS

In physics, quantities that can be measured with a scale are called scalars. For example, we say that the temperature outdoors is 85° Fahrenheit. In vector analysis, **scalar** means real number.

On the other hand, a velocity or a force is a vector, not a scalar, because each has both magnitude and direction.

A scalar and a vector can be multiplied according to the following definition.

15.5.1 Definition. If k is a scalar (a real number) and $\mathbf{a} = [a_1, a_2]$ is a vector, then the **scalar product** of k and $\mathbf{a}$ is the vector

$$k\mathbf{a} = k[a_1, a_2] = [ka_1, ka_2].$$

Geometrically, if $k > 0$, the product $k\mathbf{a}$ is a vector that is k times as long as $\mathbf{a}$ and has the same direction as $\mathbf{a}$ (Fig. 15-21); if $k < 0$, $k\mathbf{a}$ is a vector that is $|k|$ times as long as $\mathbf{a}$ and points in the opposite direction (Fig. 15-22).

Figure 15-21   Figure 15-22

It is easy to deduce the following rules for multiplying scalars and vectors.

15.5.2 Theorem. Let $\mathbf{a}$ and $\mathbf{b}$ be vectors and let k and m be scalars. Then

$$k(\mathbf{a} + \mathbf{b}) = k\mathbf{a} + k\mathbf{b},$$
$$(k + m)\mathbf{a} = k\mathbf{a} + m\mathbf{a},$$
$$(km)\mathbf{a} = k(m\mathbf{a}), \qquad (-k)\mathbf{a} = -(k\mathbf{a}),$$
$$|k\mathbf{a}| = |k||\mathbf{a}|,$$
$$0\mathbf{a} = \mathbf{0}, \qquad 1\mathbf{a} = \mathbf{a}, \qquad k\mathbf{0} = \mathbf{0}.$$

We can prove that $(k + m)\mathbf{a} = k\mathbf{a} + m\mathbf{a}$ as follows.

$$(k + m)\mathbf{a} = [(k + m)a_1, (k + m)a_2] = [ka_1 + ma_1, ka_2 + ma_2]$$
$$= [ka_1, ka_2] + [ma_1, ma_2] = k\mathbf{a} + m\mathbf{a}.$$

The proofs of the other rules are left as exercises.

Example 1. Let $\mathbf{a} = [3, -1]$ and $\mathbf{b} = [-2, 5]$. Then

$$4(\mathbf{a} + \mathbf{b}) = 4[3, -1] + 4[-2, 5] = [12, -4] + [-8, 20] = [4, 16].$$

Again,

$$(2 + \sqrt{7})\mathbf{a} = (2 + \sqrt{7})[3, -1] = 2[3, -1] + \sqrt{7}[3, -1]$$
$$= [6, -2] + [3\sqrt{7}, -\sqrt{7}] = [6 + 3\sqrt{7}, -2 - \sqrt{7}].$$

Also, $|-9\mathbf{a}| = |-9||\mathbf{a}| = |-9|\sqrt{3^2 + (-1)^2} = 9\sqrt{10}$.

In two dimensions, the only product of two vectors is the dot product.

15.5.3 **Definition.** Let $\mathbf{a} = [a_1, a_2]$ and $\mathbf{b} = [b_1, b_2]$ be vectors. The **dot product** of $\mathbf{a}$ and $\mathbf{b}$, symbolized by $\mathbf{a} \cdot \mathbf{b}$, is the number

$$\mathbf{a} \cdot \mathbf{b} = [a_1, a_2] \cdot [b_1, b_2] = a_1b_1 + a_2b_2.$$

For instance, if $\mathbf{a} = [-4, 1]$ and $\mathbf{b} = [2, \frac{1}{3}]$, then

$$\mathbf{a} \cdot \mathbf{b} = -4(2) + 1(\tfrac{1}{3}) = -7\tfrac{2}{3}.$$

It is to be emphasized that the dot product of two vectors is a *scalar* (number), not a vector. For this reason, the dot product is often called the **scalar product** of two vectors.

Rules for dot multiplication of vectors are given as follows.

15.5.4 **Theorem.** If $\mathbf{a}$, $\mathbf{b}$, and $\mathbf{c}$ are vectors and k is a scalar, then

$$\mathbf{a} \cdot \mathbf{b} = \mathbf{b} \cdot \mathbf{a} \qquad \text{(commutative law)},$$
$$\mathbf{a} \cdot (\mathbf{b} + \mathbf{c}) = \mathbf{a} \cdot \mathbf{b} + \mathbf{a} \cdot \mathbf{c} \qquad \text{(distributive law)},$$
$$k(\mathbf{a} \cdot \mathbf{b}) = (k\mathbf{a}) \cdot \mathbf{b} = \mathbf{a} \cdot (k\mathbf{b}),$$
$$\mathbf{0} \cdot \mathbf{a} = 0,$$
$$\mathbf{a} \cdot \mathbf{a} = |\mathbf{a}|^2.$$

To prove the commutative law, we write

$$\mathbf{a} \cdot \mathbf{b} = [a_1, a_2] \cdot [b_1, b_2] = a_1b_1 + a_2b_2 = b_1a_1 + b_2a_2 = [b_1, b_2] \cdot [a_1, a_2] = \mathbf{b} \cdot \mathbf{a}.$$

The remaining proofs are left as exercises.

15.5.5 **Definition.** Let two nonzero vectors, $\mathbf{a}$ and $\mathbf{b}$, be represented by directed line segments $\overrightarrow{QA}$ and $\overrightarrow{QB}$ having the same initial point Q (Fig. 15-23). By the **angle between a**

The angle θ between **a** and **b**

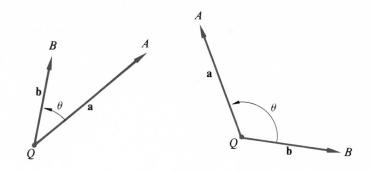

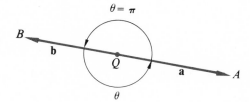
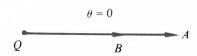

Figure 15-23

and **b** is meant the smallest nonnegative angle θ formed by $\overrightarrow{QA}$ and $\overrightarrow{QB}$. Always

$$0 \leq \theta \leq \pi.$$

A formula for the angle between two vectors can be written in terms of their dot product.

15.5.6 Theorem.   If $\mathbf{a} = [a_1, a_2]$ and $\mathbf{b} = [b_1, b_2]$ are nonzero vectors, the angle θ between them is given by

$$\cos \theta = \frac{\mathbf{a} \cdot \mathbf{b}}{|\mathbf{a}||\mathbf{b}|}.$$

Proof.   Let **a** and **b** be represented by directed line segments having a common initial point (Fig. 15-24). From the law of cosines

(1) $$|\mathbf{a} - \mathbf{b}|^2 = |\mathbf{a}|^2 + |\mathbf{b}|^2 - 2|\mathbf{a}||\mathbf{b}| \cos \theta.$$

But (by 15.5.4)

$$|\mathbf{a} - \mathbf{b}|^2 = (\mathbf{a} - \mathbf{b}) \cdot (\mathbf{a} - \mathbf{b}) = \mathbf{a} \cdot (\mathbf{a} - \mathbf{b}) - \mathbf{b} \cdot (\mathbf{a} - \mathbf{b})$$
$$= \mathbf{a} \cdot \mathbf{a} + \mathbf{b} \cdot \mathbf{b} - 2\mathbf{a} \cdot \mathbf{b} = |\mathbf{a}|^2 + |\mathbf{b}|^2 - 2\mathbf{a} \cdot \mathbf{b},$$

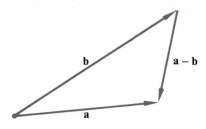

Figure 15-24

or

(2) $$|\mathbf{a} - \mathbf{b}|^2 = |\mathbf{a}|^2 + |\mathbf{b}|^2 - 2\mathbf{a} \cdot \mathbf{b}.$$

From (1) and (2) we get $\mathbf{a} \cdot \mathbf{b} = |\mathbf{a}||\mathbf{b}| \cos \theta$, or

$$\cos \theta = \frac{\mathbf{a} \cdot \mathbf{b}}{|\mathbf{a}||\mathbf{b}|}. \quad \blacksquare$$

Since $0 \leq \theta \leq \pi$, $\cos \theta = 0$ if and only if $\theta = \tfrac{1}{2}\pi$.

15.5.7 Corollary.   Two nonzero vectors are **perpendicular** if and only if their dot product is zero.

We recall that the length of a vector $\mathbf{a} = [a_1, a_2]$ was defined to be $|\mathbf{a}| = \sqrt{a_1^2 + a_2^2}$. A vector of length 1 is called a **unit vector**. Examples of unit vectors are $[\tfrac{3}{5}, \tfrac{4}{5}]$ and $[\tfrac{5}{13}, \tfrac{12}{13}]$.

Proof of the following theorem is easy and is left as an exercise.

15.5.8 Theorem.   Let $\mathbf{a}$ be any nonzero vector. Then

$$\frac{\mathbf{a}}{|\mathbf{a}|}$$

is the **unit vector in the direction of a**.

Two particular unit vectors play an important role and are given special symbols. They are the **basis vectors**

$$\mathbf{i} = [1, 0] \quad \text{and} \quad \mathbf{j} = [0, 1].$$

Geometrically, both $\mathbf{i}$ and $\mathbf{j}$ may be represented with their initial points at the origin, $\mathbf{i}$ lying along the positive x axis and $\mathbf{j}$ along the positive y axis (Fig. 15-25).

Their importance lies in the fact that every two-dimensional vector can be expressed in one and only one way as a linear combination of $\mathbf{i}$ and $\mathbf{j}$.

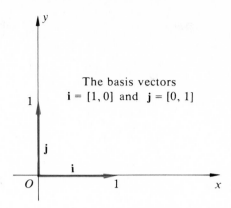

The basis vectors
$\mathbf{i} = [1, 0]$ and $\mathbf{j} = [0, 1]$

Figure 15-25

15.5.9 Theorem. If $\mathbf{a} = [a_1, a_2]$ is an arbitrary vector in the plane, then

(3) $$\mathbf{a} = a_1\mathbf{i} + a_2\mathbf{j}$$

uniquely.

Proof. By 15.5.1 and 15.4.3,

$$\mathbf{a} = [a_1, a_2] = [a_1, 0] + [0, a_2] = a_1[1, 0] + a_2[0, 1] = a_1\mathbf{i} + a_2\mathbf{j}.$$

Moreover, if $b_1\mathbf{i} + b_2\mathbf{j}$ is any linear combination of $\mathbf{i}$ and $\mathbf{j}$ that is equal to $\mathbf{a}$, then

$$\mathbf{a} = b_1\mathbf{i} + b_2\mathbf{j} = b_1[1, 0] + b_2[0, 1] = [b_1, 0] + [0, b_2] = [b_1, b_2].$$

Since $\mathbf{a} = [a_1, a_2]$ and $\mathbf{a} = [b_1, b_2]$, then $a_1 = b_1$ and $a_2 = b_2$ by 15.4.2. Therefore (3) is unique. ∎

As illustrations, $[5, -1] = 5\mathbf{i} - \mathbf{j}$; $[-2, 10] = -2\mathbf{i} + 10\mathbf{j}$; and $[0, 6] = 0\mathbf{i} + 6\mathbf{j} = \mathbf{0} + 6\mathbf{j} = 6\mathbf{j}$.

It is easy to verify the useful properties

$$\mathbf{i} \cdot \mathbf{i} = \mathbf{j} \cdot \mathbf{j} = 1 \qquad \text{and} \qquad \mathbf{i} \cdot \mathbf{j} = \mathbf{j} \cdot \mathbf{i} = 0.$$

Sometimes it is more convenient to express the vector $\mathbf{a}$ in the concise form $[a_1, a_2]$, but in other circumstances the form $\mathbf{a} = a_1\mathbf{i} + a_2\mathbf{j}$ has advantages, as we shall see.

Whereas a_1 and a_2 were defined to be the *components* of $\mathbf{a}$, $a_1\mathbf{i}$ and $a_2\mathbf{j}$ are called **vector components** of $\mathbf{a}$, in particular, **orthonormal** vector components of $\mathbf{a}$.

Example 2. If $\mathbf{a}$ is the vector from the point $(2, -1)$ to the point $(-3, 7)$, write $\mathbf{a}$ in the form $\mathbf{a} = a_1\mathbf{i} + a_2\mathbf{j}$.

Solution. $\mathbf{a} = [-3 - 2, 7 - (-1)] = [-5, 8]$. Therefore, by 15.5.9, $\mathbf{a} = -5\mathbf{i} + 8\mathbf{j}$.

We conclude this section with a remark. The set of all vectors in the plane, with addition and multiplication by a scalar, as defined in 15.4.3 and 15.5.1, is an example of a **vector space**. There are many other important vector spaces.

Exercises

1. Let $\mathbf{a} = -3\mathbf{i} + 4\mathbf{j}$, $\mathbf{b} = 2\mathbf{i} - 3\mathbf{j}$, and $\mathbf{c} = -5\mathbf{j}$. Find
(a) $2\mathbf{a} - 4\mathbf{b}$;
(b) $\mathbf{a} \cdot \mathbf{b}$;
(c) $\mathbf{a} \cdot (\mathbf{b} + \mathbf{c})$;
(d) $(-2\mathbf{a} + 3\mathbf{b}) \cdot 5\mathbf{c}$;
(e) $|\mathbf{a}| \mathbf{c} \cdot \mathbf{a}$;
(f) $\mathbf{b} \cdot \mathbf{b} - |\mathbf{b}|$.

2. Let $\mathbf{a} = 4\mathbf{i} - \mathbf{j}$, $\mathbf{b} = \mathbf{i} - \mathbf{j}$, and $\mathbf{c} = 6\mathbf{j}$. Find
(a) $-4\mathbf{a} + 3\mathbf{b}$;
(b) $\mathbf{b} \cdot \mathbf{c}$;
(c) $(\mathbf{a} + \mathbf{b}) \cdot \mathbf{c}$;
(d) $2\mathbf{c} \cdot (3\mathbf{a} + 4\mathbf{b})$;
(e) $|\mathbf{b}| \mathbf{b} \cdot \mathbf{a}$;
(f) $|\mathbf{c}|^2 - \mathbf{c} \cdot \mathbf{c}$.

3. Find the cosine of the angle between $\mathbf{a}$ and $\mathbf{b}$ and make a sketch.
(a) $\mathbf{a} = 2\mathbf{i} - 3\mathbf{j}$, $\mathbf{b} = -\mathbf{i} + 4\mathbf{j}$;
(b) $\mathbf{a} = -5\mathbf{i} - 2\mathbf{j}$, $\mathbf{b} = 6\mathbf{i}$;
(c) $\mathbf{a} = -3\mathbf{i} - \mathbf{j}$, $\mathbf{b} = -2\mathbf{i} - 4\mathbf{j}$;
(d) $\mathbf{a} = 4\mathbf{i} - 5\mathbf{j}$, $\mathbf{b} = -8\mathbf{i} + 10\mathbf{j}$.

4. Find the angle between $\mathbf{a}$ and $\mathbf{b}$ and make a sketch.
(a) $\mathbf{a} = 12\mathbf{i}$, $\mathbf{b} = -7\mathbf{i}$;
(b) $\mathbf{a} = 4\mathbf{i} + 3\mathbf{j}$, $\mathbf{b} = 12\mathbf{i} + 9\mathbf{j}$;
(c) $\mathbf{a} = -\mathbf{i} + 3\mathbf{j}$, $\mathbf{b} = 3\mathbf{i} - 9\mathbf{j}$;
(d) $\mathbf{a} = \sqrt{3}\mathbf{i} + \mathbf{j}$, $\mathbf{b} = -3\mathbf{i} + \sqrt{3}\mathbf{j}$.

5. Write in the form $\mathbf{a} = a_1\mathbf{i} + a_2\mathbf{j}$ the vector represented by $\overrightarrow{AB}$.
(a) $A:(2, -1)$, $B:(-3, 4)$;
(b) $A:(0, 5)$, $B:(-6, 0)$;
(c) $A:(\sqrt{2}, -e)$, $B:(0, 0)$;
(d) $A:(-7, 2)$, $B:(-4, \frac{1}{3})$.

6. Find a unit vector $\mathbf{u}$ in the direction of $\mathbf{a}$ and express it in the form $\mathbf{u} = u_1\mathbf{i} + u_2\mathbf{j}$.
(a) $\mathbf{a} = [-3, 4]$;
(b) $\mathbf{a} = [1, -7]$;
(c) $\mathbf{a} = [0, -2]$;
(d) $\mathbf{a} = [-5, -12]$.

7. If $\mathbf{a}$ and $\mathbf{b}$ are vectors and k and m are scalars, prove that $k(\mathbf{a} + \mathbf{b}) = k\mathbf{a} + k\mathbf{b}$, $km(\mathbf{a}) = k(m\mathbf{a})$, and $(-k)\mathbf{a} = -(k\mathbf{a})$.

8. Show that $\mathbf{i} \cdot \mathbf{i} = \mathbf{j} \cdot \mathbf{j} = 1$ and $\mathbf{i} \cdot \mathbf{j} = \mathbf{j} \cdot \mathbf{i} = 0$.

9. If $\mathbf{a}$ and $\mathbf{b}$ are vectors and k is a scalar, prove that $|k\mathbf{a}| = |k||\mathbf{a}|$, $0\mathbf{a} = \mathbf{0}$, $1\mathbf{a} = \mathbf{a}$, and $k\mathbf{0} = \mathbf{0}$.

10. Prove that if $\mathbf{a}$ and $\mathbf{b}$ are vectors and k is a

scalar, then $k(\mathbf{a} \cdot \mathbf{b}) = (k\mathbf{a}) \cdot \mathbf{b} = \mathbf{a} \cdot (k\mathbf{b})$, $\mathbf{0} \cdot \mathbf{a} = 0$, and $\mathbf{a} \cdot \mathbf{a} = |\mathbf{a}|^2$.

11. Prove that if $\mathbf{a}$, $\mathbf{b}$, and $\mathbf{c}$ are vectors, then $\mathbf{a} \cdot (\mathbf{b} + \mathbf{c}) = \mathbf{a} \cdot \mathbf{b} + \mathbf{a} \cdot \mathbf{c}$.

12. Show that $(\mathbf{a} + \mathbf{b}) \cdot (\mathbf{a} + \mathbf{b}) = \mathbf{a} \cdot \mathbf{a} + 2\mathbf{a} \cdot \mathbf{b} + \mathbf{b} \cdot \mathbf{b}$.

13. Show that the vectors $6\mathbf{i} + 3\mathbf{j}$ and $-\mathbf{i} + 2\mathbf{j}$ are orthogonal (mutually perpendicular).

14. Show that the vectors $-5\mathbf{i} + \sqrt{3}\mathbf{j}$ and $\sqrt{27}\mathbf{i} + 15\mathbf{j}$ are orthogonal.

15. Given $\mathbf{a} = 3\mathbf{i} - 2\mathbf{j}$ and $\mathbf{b} = -\mathbf{i} + 4\mathbf{j}$, two noncollinear vectors (that is, vectors such that the angle θ between them satisfies $0 < \theta < \pi$), and another vector $\mathbf{r} = 7\mathbf{i} - 8\mathbf{j}$. Find scalars, k and m, such that $\mathbf{r} = k\mathbf{a} + m\mathbf{b}$.

16. Given $\mathbf{a} = -4\mathbf{i} + 3\mathbf{j}$ and $\mathbf{b} = 3\mathbf{i} - \mathbf{j}$, two noncollinear vectors, and another vector $\mathbf{r} = 6\mathbf{i} - 7\mathbf{j}$. Find scalars, k and m, such that $\mathbf{r} = k\mathbf{a} + m\mathbf{b}$.

17. Let $\mathbf{a} = a_1\mathbf{i} + a_2\mathbf{j}$ and $\mathbf{b} = b_1\mathbf{i} + b_2\mathbf{j}$ be noncollinear vectors. If $\mathbf{r} = r_1\mathbf{i} + r_2\mathbf{j}$ is an arbitrarily chosen vector in the plane of $\mathbf{a}$ and $\mathbf{b}$, find scalars, k and m, such that $\mathbf{r} = k\mathbf{a} + m\mathbf{b}$.

18. Let $\mathbf{a}$ and $\mathbf{b}$ be vectors with terminal points A and B, and the same initial point 0. Show that the vector from 0 to the midpoint of the line segment AB is $\frac{1}{2}(\mathbf{a} + \mathbf{b})$.

19. Given three vectors, $\mathbf{a}$, $\mathbf{b}$, and $\mathbf{c}$, with the initial point of $\mathbf{b}$ at the terminal point of $\mathbf{a}$, and the initial point of $\mathbf{c}$ at the terminal point of $\mathbf{b}$. What is the condition for $\mathbf{a}$, $\mathbf{b}$, and $\mathbf{c}$ to be the sides of a triangle?

20. Use vector methods to prove that the line segment joining the midpoints of two sides of a triangle is parallel to the third side and equal to half the length of the third side.

21. By vector methods, prove that the diagonals of a parallelogram bisect each other.

22. By vectors, prove that the line segments joining in succession the midpoints of the sides of any quadrilateral form a parallelogram.

15.6 VECTOR FUNCTIONS

If the vector $\mathbf{a} = [a_1, a_2]$ is represented by a directed line segment *whose initial point is the origin*, then $\mathbf{a}$ is called a **position vector**. In this section we shall be chiefly interested in position vectors.

Notice that when $\mathbf{a} = [a_1, a_2]$ is a position vector (Fig. 15-26), its terminal point has the coordinates (a_1, a_2).

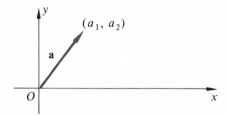

Figure 15-26

15.6.1 Definition. Let f and g be functions with a common domain $\mathfrak{D}$. Then $[f(t), g(t)]$ defines a vector for each number t in $\mathfrak{D}$. We shall say that a **vector function r** with domain $\mathfrak{D}$ is defined and has the (vector) **value** $\mathbf{r}(t)$, where

$$\mathbf{r}(t) = [f(t), g(t)].$$

If $\mathbf{r}(t)$ is represented by a position vector for each number t in $\mathfrak{D}$, then as t takes on all values in $\mathfrak{D}$, the terminal point $(f(t), g(t))$ of the position vector traces out the plane curve (Fig. 15-27) whose parametric representation is

$$x = f(t), \qquad y = g(t), \qquad t \in \mathfrak{D}.$$

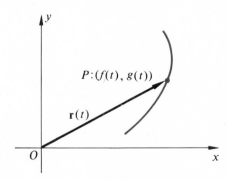

Figure 15-27

15.6.2 Definition. Let $\mathbf{r}$ be a vector function defined by $\mathbf{r}(t) = [f(t), g(t)]$. Then the **limit** of $\mathbf{r}(t)$ as t approaches t_0 is

$$\lim_{t \to t_0} \mathbf{r}(t) = [\lim_{t \to t_0} f(t), \lim_{t \to t_0} g(t)],$$

provided that the two latter limits exist.

As an illustration, let $\mathbf{r}(t) = [2t, e^{t-1}]$; then

$$\lim_{t \to 1} \mathbf{r}(t) = [\lim_{t \to 1} 2t, \lim_{t \to 1} e^{t-1}] = [2, 1].$$

Thus the limit is the vector $[2, 1]$.

15.6.3 **Definition.** The vector function $\mathbf{r}$ is **continuous** at t_0 if
  (i) $\mathbf{r}$ is defined at t_0,
  (ii) $\lim_{t \to t_0} \mathbf{r}(t)$ exists, and
  (iii) $\lim_{t \to t_0} \mathbf{r}(t) = \mathbf{r}(t_0)$.

Clearly, a necessary and sufficient condition for $\mathbf{r} = [f(t), g(t)]$ to be continuous at t_0 is that both f and g be continuous there.

As t varies, both the length and direction of $\mathbf{r}(t)$ vary, in general. The derivative of a vector function is defined as follows.

15.6.4 **Definition.** Let $\mathbf{r}$ be a vector function. The **derivative** of $\mathbf{r}$ is another vector function $\mathbf{r}'$ defined by

$$\mathbf{r}'(t) = \lim_{\Delta t \to 0} \frac{\mathbf{r}(t + \Delta t) - \mathbf{r}(t)}{\Delta t},$$

provided that this limit exists.

It follows from 15.6.2 that the derivative of the vector function $\mathbf{r}$ can be expressed in terms of the derivatives of f and g:

$$\mathbf{r}'(t) = \lim_{\Delta t \to 0} \frac{\mathbf{r}(t + \Delta t) - \mathbf{r}(t)}{\Delta t} = \lim_{\Delta t \to 0} \frac{[f(t + \Delta t), g(t + \Delta t)] - [f(t), g(t)]}{\Delta t}$$

$$= \lim_{\Delta t \to 0} \left[\frac{f(t + \Delta t) - f(t)}{\Delta t}, \frac{g(t + \Delta t) - g(t)}{\Delta t} \right]$$

$$= \left[\lim_{\Delta t \to 0} \frac{f(t + \Delta t) - f(t)}{\Delta t}, \lim_{\Delta t \to 0} \frac{g(t + \Delta t) - g(t)}{\Delta t} \right] = [f'(t), g'(t)].$$

15.6.5 **Theorem.** If $\mathbf{r}$ is a vector function defined by $\mathbf{r}(t) = [f(t), g(t)]$, then the derivative of $\mathbf{r}$ at t is given by

$$\mathbf{r}'(t) = [f'(t), g'(t)].$$

The **second derivative $\mathbf{r}''$** is defined as the derivative of the first derivative. It is the vector function defined by

$$\mathbf{r}''(t) = [f''(t), g''(t)].$$

Higher derivatives are defined similarly.

The rule for differentiating the dot product of two vector functions is *formally* the same as the familiar rule for differentiating a product of (scalar-valued) functions.

15.6.6 Theorem. If **r** and **u** are differentiable vector functions of t, then

$$\frac{d}{dt}(\mathbf{r} \cdot \mathbf{u}) = \frac{d\mathbf{r}}{dt} \cdot \mathbf{u} + \mathbf{r} \cdot \frac{d\mathbf{u}}{dt}.$$

Proof. Let $\mathbf{r} = [f(t), g(t)]$ and $\mathbf{u} = [F(t), G(t)]$, where f, g, F, and G are differentiable. By the definition of a dot product,

$$\mathbf{r} \cdot \mathbf{u} = f(t)F(t) + g(t)G(t).$$

Thus

(1) $$\frac{d}{dt}(\mathbf{r} \cdot \mathbf{u}) = f(t)F'(t) + f'(t)F(t) + g(t)G'(t) + g'(t)G(t).$$

Also,

$$\frac{d\mathbf{r}}{dt} \cdot \mathbf{u} + \mathbf{r} \cdot \frac{d\mathbf{u}}{dt} = [f'(t), g'(t)] \cdot [F(t), G(t)] + [f(t), g(t)] \cdot [F'(t), G'(t)]$$

$$= f'(t)F(t) + g'(t)G(t) + f(t)F'(t) + g(t)G'(t).$$

That is,

(2) $$\frac{d\mathbf{r}}{dt} \cdot \mathbf{u} + \mathbf{r} \cdot \frac{d\mathbf{u}}{dt} = f(t)F'(t) + f'(t)F(t) + g(t)G'(t) + g'(t)G(t).$$

From (1) and (2) we get

$$\frac{d}{dt}(\mathbf{r} \cdot \mathbf{u}) = \frac{d\mathbf{r}}{dt} \cdot \mathbf{u} + \mathbf{r} \cdot \frac{d\mathbf{u}}{dt}. \quad \blacksquare$$

Proofs of the two following formulas, one for differentiating the product of a scalar-valued function and a vector-valued function and the other a chain rule, are easy and are left as exercises.

15.6.7 Theorem. If a vector-valued function **r** and a scalar-valued function h are differentiable on an interval I, then for all $t \in I$

$$D_t\{h(t)\mathbf{r}(t)\} = h(t)\mathbf{r}'(t) + h'(t)\mathbf{r}(t).$$

15.6.8 Theorem. Let **f** be a vector-valued function, let g be a scalar-valued function with $\mathfrak{R}_g \subseteq \mathfrak{D}_\mathbf{f}$, and denote by **F** the vector-valued function defined by $\mathbf{F}(t) = \mathbf{f}(g(t))$. If **f** is differentiable on $\mathfrak{D}_\mathbf{f}$ and g is differentiable on $\mathfrak{D}_g$, then for every $t \in \mathfrak{D}_g$

$$\mathbf{F}'(t) = \mathbf{f}'(u)g'(t),$$

where $u = g(t)$.

Exercises

1. When no domain is given in the definition of a vector-valued function, it is to be understood that the domain is the set of all (real) scalars to which correspond real vectors (that is, vectors with real components). Find the domain of each of the following vector-valued functions.

(a) $\mathbf{r}(t) = \left(\dfrac{1}{t-2}\right)\mathbf{i} + \sqrt{4+t}\,\mathbf{j}$;

(b) $\mathbf{r}(t) = [\![t]\!]\mathbf{i} - \sqrt{t^2+1}\,\mathbf{j}$;

(c) $\mathbf{r}(t) = \ln(1-t)\mathbf{i} + (\text{Cos}^{-1}\,t)\mathbf{j}$;

(d) $\mathbf{r}(t) = -\ln(2-t)\mathbf{i} + \sqrt{t-2}\,\mathbf{j}$.

2. State the domain of each of the following vector-valued functions.

(a) $\mathbf{r}(t) = \ln(t^2+1)\mathbf{i} + (\text{Tan}^{-1}\,t)\mathbf{j}$;

(b) $\mathbf{r}(t) = \ln(2t^{-1})\mathbf{i} - (-t)^{1/2}\mathbf{j}$;

(c) $\mathbf{r}(t) = (\cosh^{-1}\,t)\mathbf{i} + [\![2t-1]\!]\mathbf{j}$;

(d) $\mathbf{r}(t) = (5-t)^{-1/2}\mathbf{i} + (t-2)^{3/2}\mathbf{j}$.

3. For what values of t is each of the functions in Exercise 1 continuous?

4. For what values of t is each of the functions in Exercise 2 continuous?

5. Find $D_t\mathbf{r}(t)$ and $D_t^2\mathbf{r}(t)$ for each of the following.

(a) $\mathbf{r}(t) = (2t+3)^2\mathbf{i} - e^{2t}\mathbf{j}$;

(b) $\mathbf{r}(t) = \cos 2t\mathbf{i} - \sin^3 t\mathbf{j}$;

(c) $\mathbf{r}(t) = e^{-t}\mathbf{i} + \ln^2(t^3)\mathbf{j}$;

(d) $\mathbf{r}(t) = \sqrt{t+1}\,\mathbf{i} + \text{Cos}^{-1}\,t\mathbf{j}$.

6. Find $\mathbf{r}'(t)$ and $\mathbf{r}''(t)$ for each of the following .

(a) $\mathbf{r}(t) = (e^t + e^{-t})\mathbf{i} - e^{t^2}\mathbf{j}$;

(b) $\mathbf{r}(t) = \tan t\mathbf{i} - 2t^{5/3}\mathbf{j}$;

(c) $\mathbf{r}(t) = \text{Tan}^{-1}\,t\mathbf{i} + \sinh(t-1)\mathbf{j}$;

(d) $\mathbf{r}(t) = \ln^3(t^2-1)\mathbf{i} + e^{-1/t}\mathbf{j}$.

7. If $\mathbf{r}(t) = \sqrt{t^2+3}\,\mathbf{i} - t^3\mathbf{j}$, find $D_t\{\mathbf{r}'(t)\cdot\mathbf{r}(t)\}$.

8. If $\mathbf{r}(t) = e^{2t}\mathbf{i} + \ln(t^3)\mathbf{j}$, find $D_t\{\mathbf{r}(t)\cdot\mathbf{r}'(t)\}$.

9. If $\mathbf{r}(t) = \sin 3t\mathbf{i} - \cos 2t\mathbf{j}$, find $D_t\{\mathbf{r}'(t)\cdot\mathbf{r}''(t)\}$.

10. If $\mathbf{r}(t) = t^{-2}\mathbf{i} + \text{Sin}^{-1}\sqrt{t}\,\mathbf{j}$, find $D_t\{\mathbf{r}'(t)\cdot\mathbf{r}''(t)\}$.

11. If $\mathbf{r}(t) = \sqrt{t-1}\,\mathbf{i} + \ln(2t^2)\mathbf{j}$ and $h(t) = e^{-3t}$, find $D_t\{h(t)\mathbf{r}(t)\}$.

12. If $\mathbf{r}(t) = \sin 2t\mathbf{i} + \cosh t\mathbf{j}$ and $h(t) = \ln(3t-2)$, find $D_t\{h(t)\mathbf{r}(t)\}$.

13. Let $\mathbf{F}(t) = \mathbf{f}(g(t))$, where $\mathbf{f}(u) = \cos^2 u\mathbf{i} - e^{1-3u}\mathbf{j}$ and $g(t) = (3t^2-4)^{1/2}$. Find $\mathbf{F}'(t)$ in terms of t.

14. Let $\mathbf{F}(t) = \mathbf{f}(g(t))$, where $\mathbf{f}(u) = \cosh u\mathbf{i} - 2e^{-u}\mathbf{j}$ and $g(t) = \ln(t^2+4)$. Find $\mathbf{F}'(t)$ in terms of t.

15. Prove 15.6.7.

16. Prove 15.6.8.

15.7 CURVILINEAR MOTION

In our earlier discussion of motion we defined velocity and acceleration for straight-line motion only. For the more general case of motion along any trajectory (curve) in the plane, new definitions are needed.

Let $r(t) = f(t)\mathbf{i} + g(t)\mathbf{j}$ be a vector function that is differentiable on some interval I. If $\mathbf{r}(t)$ is a *position vector*, represented by the directed line segment $\overrightarrow{OP}$ (Fig. 15-28), then as t varies, the terminal point $P{:}(f(t), g(t))$ traces the curve having parametric

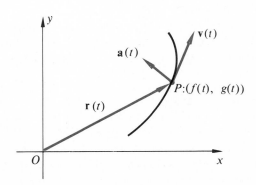

Figure 15-28

equations

(1) $$x = f(t), \qquad y = g(t), \qquad t \in I.$$

In this connection, $\mathbf{r}(t) = f(t)\mathbf{i} + g(t)\mathbf{j}$ is called a **vector equation** of the curve (1) and $\mathbf{r}'(t)$ is called the **velocity vector** of the moving point P, often designated by the special symbol $\mathbf{v}(t)$. Thus

$$\mathbf{v}(t) = \mathbf{r}'(t).$$

15.7.1 Definition. The **instantaneous velocity** at time t of the point $P:(x, y)$, moving on the trajectory $x = f(t)$, $y = g(t)$ at t varies, is defined to be

$$\mathbf{v}(t) = f'(t)\mathbf{i} + g'(t)\mathbf{j},$$

provided that these derivatives exist.

At any time t, the velocity vector $\mathbf{v}(t)$ gives the magnitude and direction of the instantaneous velocity of P.

The slope of the velocity vector $\mathbf{v}(t) = f'(t)\mathbf{i} + g'(t)\mathbf{j}$ is $g'(t)/f'(t)$, which is equal to dy/dx, the slope of the tangent to the trajectory (1) traced out by P. Thus the **direction** of the velocity vector is along the tangent to the trajectory traced out by P.

The **magnitude** or **length** of the velocity vector

(2) $$|\mathbf{v}(t)| = \sqrt{[f'(t)]^2 + [g'(t)]^2}$$

is called the **speed** of the moving point P at the instant t.

15.7.2 Definition. The derivative of the velocity vector is called the **acceleration vector** and is denoted by

$$\mathbf{a}(t) = \mathbf{v}'(t) = \mathbf{r}''(t).$$

It gives the instantaneous rate of change of the velocity vector at time t.

The acceleration vector does not, in general, have the same direction as the tangent line to the trajectory.

If the components $f(t)$ and $g(t)$ of the position vector $\mathbf{r}(t) = f(t)\mathbf{i} + g(t)\mathbf{j}$ are expressed in feet and the parameter t is in seconds, the speed is found in terms of feet per second, and the magnitude $|\mathbf{a}(t)|$ of the acceleration will be in feet per second per second.

Example. The parametric equations of the trajectory of a point moving in the plane are $x = 3 \cos t$ and $y = 2 \sin t$, where t represents time. Find and sketch the velocity and acceleration vectors when $t = \frac{1}{3}\pi$. What is the speed of P when $t = \frac{1}{3}\pi$?

Solution. Since $x/3 = \cos t$ and $y/2 = \sin t$, the trajectory of the moving point is the ellipse $x^2/9 + y^2/4 = 1$ (Fig. 15-29). The position vector is $\mathbf{r}(t) = 3 \cos t\mathbf{i} + 2 \sin t\mathbf{j}$. Thus the velocity vector is

$$\mathbf{v}(t) = \mathbf{r}'(t) = -3 \sin t\mathbf{i} + 2 \cos t\mathbf{j}$$

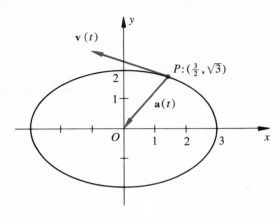

Figure 15-29

and the acceleration vector is

$$a(t) = v'(t) = -3 \cos t \, \mathbf{i} - 2 \sin t \, \mathbf{j}.$$

For $t = \frac{1}{3}\pi$,

$$v(\tfrac{1}{3}\pi) = -\tfrac{3}{2}\sqrt{3}\,\mathbf{i} + \mathbf{j}$$

and

$$a(\tfrac{1}{3}\pi) = -\tfrac{3}{2}\mathbf{i} - \sqrt{3}\,\mathbf{j}.$$

The speed of the moving point at $t = \frac{1}{3}\pi$ is

$$|v(t)| = 9 \sin^2 t + 4 \cos^2 t = \sqrt{\tfrac{27}{4} + 1} = \tfrac{1}{2}\sqrt{31}.$$

When $t = \frac{1}{3}\pi$, the position of the moving point P on the ellipse is $(\frac{3}{2}, \sqrt{3})$. It is easy to draw the velocity vector $v(\frac{1}{3}\pi)$ with its initial point at $P:(\frac{3}{2}, \sqrt{3})$ when we know its terminal point. Since the components of $v(\frac{1}{3}\pi)$ are $-\frac{3}{2}\sqrt{3}$ and 1, its terminal point is $(\frac{3}{2} - \frac{3}{2}\sqrt{3}, \sqrt{3} + 1) \doteq (-1.1, 2.7)$.

Similarly, the acceleration vector $a(\frac{1}{3}\pi) = -\frac{3}{2}\mathbf{i} - \sqrt{3}\,\mathbf{j}$ is drawn with its initial point $P:(\frac{3}{2}, \sqrt{3})$ on the ellipse and its terminal point $(\frac{3}{2} - \frac{3}{2}, \sqrt{3} - \sqrt{3}) = (0, 0)$. See Fig. 15-29.

Since $|r'(t)| = |v(t)| = \sqrt{[f'(t)]^2 + [g'(t)]^2}$, the expression for arc length $L = \int_a^b \sqrt{[f'(t)]^2 + [g'(t)]^2} \, dt$ given in 15.3.1 can be rewritten $L = \int_a^b |r'(t)| \, dt$. Hence 15.3.1 can be stated in terms of vectors as follows.

15.7.3 Theorem. If the functions f and g are smooth on the closed interval $[a, b]$, then the length of the arc traced by the terminal point of the position vector $r(t) = f(t)\mathbf{i} + g(t)\mathbf{j}$ as t increases from $t = a$ to $t = b$ is

$$L = \int_a^b |r'(t)| \, dt.$$

Notice that the integrand, $|r'(t)|$, of the above integral is the *speed* of the point $P:(f(t), g(t))$ as it traces the trajectory.

We conclude this section with a simple geometric interpretation of Cauchy's mean value theorem (13.1.1), suggested by vectors. Recall that Cauchy's theorem stated that if f and g are functions that are differentiable on $[a, b]$ and if $g'(t) \neq 0$ for $a < t < b$, then there exists a number z between a and b such that

(3) $$\frac{f(b) - f(a)}{g(b) - g(a)} = \frac{f'(z)}{g'(z)}.$$

Consider the differentiable position vector function

$$r(t) = [f(t), g(t)], \qquad a \le t \le b.$$

As t increases from a to b, the terminal point P of $r(t)$ traces out an arc of a plane

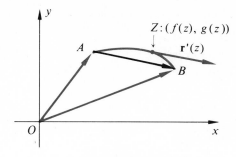

Figure 15-30

curve whose endpoints are $A:(f(a), g(a))$ and $B:(f(b), g(b))$ (Fig. 15-30). The chord vector $\overrightarrow{AB}$ is $\mathbf{r}(b) - \mathbf{r}(a)$ or

$$(4) \qquad \overrightarrow{AB} = [f(b) - f(a), g(b) - g(a)].$$

The derivative $\mathbf{r}'(t)$ for any number t between a and b is a tangent vector to the curve at $P:(f(t), g(t))$. In particular,

$$(5) \qquad \mathbf{r}'(z) = [f'(z), g'(z)]$$

is a tangent vector to the curve at some point $Z:(f(z), g(z))$ on the arc between A and B.

Because of (3), there is a constant (of proportionality) c such that

$$(6) \qquad f(b) - f(a) = cf'(z), \qquad g(b) - g(a) = cg'(z).$$

If follows from (4), (5), and (6) that

$$\overrightarrow{AB} = c\mathbf{r}'(z), \qquad c \neq 0,$$

and thus the vectors $\overrightarrow{AB}$ and $\mathbf{r}'(z)$ are parallel. This gives our geometric interpretation of Cauchy's mean value theorem: *there is at least one point $Z:(f(z), g(z))$ on the arc $\overparen{AB}$ at which the tangent is parallel to the chord AB* (Fig. 15-30).

Exercises

In each of Exercises 1 to 12, the position of a moving particle at time t is given by $\mathbf{r}(t)$. Find the velocity and acceleration vectors, $\mathbf{v}(t)$ and $\mathbf{a}(t)$, and their values at the given time $t = t_1$, and the speed of the particle then. Sketch a portion of the graph of $\mathbf{r}(t)$ containing the position P of the particle when $t = t_1$ and draw $\mathbf{v}(t_1)$ and $\mathbf{a}(t_1)$ with their initial points at P.

1. $\mathbf{r}(t) = e^{-t}\mathbf{i} + e^{t}\mathbf{j}; t_1 = 1.$

2. $\mathbf{r}(t) = (3t^2 - 1)\mathbf{i} + t\mathbf{j}; t_1 = \frac{1}{2}.$

3. $\mathbf{r}(t) = 2\cos t\mathbf{i} - 3\sin^2 t\mathbf{j}; t_1 = \frac{\pi}{3}.$

4. $\mathbf{r}(t) = \tan t\mathbf{i} + \sin t\mathbf{j}; t_1 = \frac{\pi}{6}.$

5. $\mathbf{r}(t) = 3t^2\mathbf{i} + t^3\mathbf{j}; t_1 = 2.$

6. $\mathbf{r}(t) = a\sin t\mathbf{i} + b\cos t\mathbf{j}; t_1 = \frac{\pi}{4}.$

7. $\mathbf{r}(t) = \cos t\mathbf{i} - 2\tan t\mathbf{j}; t_1 = -\frac{\pi}{4}.$

8. $\mathbf{r}(t) = e^{t/2}\mathbf{i} + e^{-t}\mathbf{j}; t_1 = 2.$

9. $\mathbf{r}(t) = 3(t - \sin t)\mathbf{i} + 3(1 - \cos t)\mathbf{j}; t_1 = \frac{\pi}{3}.$

10. $\mathbf{r}(t) = 3t^{-1}\mathbf{i} - \frac{1}{3}t\mathbf{j}; t_1 = 2.$

11. $\mathbf{r}(t) = a\sinh bt\mathbf{i} + a\cosh bt\mathbf{j}; t_1 = 0.$

12. $\mathbf{r}(t) = \sin t\mathbf{i} + \cos 2t\mathbf{j}; t_1 = \frac{\pi}{4}.$

In Exercises 13 to 16, a position vector function $\mathbf{r}(t)$ and two numbers, t_1 and t_2, are given. Use 15.7.3 to find the length of the arc traced by $\mathbf{r}(t)$ from $t = t_1$ to $t = t_2$.

13. $\mathbf{r}(t) = 9t\mathbf{i} + 6t^{3/2}\mathbf{j}; t_1 = 3, t_2 = 8.$

14. $\mathbf{r}(t) = e^{-t}\sin t\mathbf{i} + e^{-t}\cos t\mathbf{j}; t_1 = 0, t_2 = \pi.$

15. $\mathbf{r}(t) = (\sin t - t\cos t)\mathbf{i} + (\cos t + t\sin t)\mathbf{j};$ $t_1 = \frac{\pi}{2}, t_2 = 2\pi.$

16. $\mathbf{r}(t) = t^3\mathbf{i} + 2t^2\mathbf{j}; t_1 = -1, t_2 = 2.$

15.8 CURVATURE

Let $\mathbf{r}(t) = f(t)\mathbf{i} + g(t)\mathbf{j}$ be a position vector whose derivative is continuous for $a \leq t \leq b$. As t increases, the terminal point P of $\mathbf{r}(t)$ traces out an arc C having a parametric representation

$$x = f(t), \qquad y = g(t), \qquad a \leq t \leq b.$$

The velocity vector at $P(t) = (f(t), g(t))$ is $\mathbf{v}(t) = \mathbf{r}'(t)$.

The vector

$$\text{(1)} \qquad \mathbf{T}(t) = \frac{\mathbf{v}(t)}{|\mathbf{v}(t)|}$$

is called the **unit tangent vector** at $P(t)$. Its length is 1, and its direction is the same as the direction of $\mathbf{v}(t)$ (by 15.5.8), which is along the tangent to the curve at P.

Let $s(t)$ be the length of the arc from $P(a)$ to $P(t)$. From 15.7.3

$$s(t) = \int_a^t |\mathbf{r}'(w)| \, dw, \qquad a \leq t \leq b.$$

Then the **speed** of the moving point at time t is

$$\text{(2)} \qquad \frac{ds}{dt} = |\mathbf{r}'(t)| = |\mathbf{v}(t)|.$$

We assume that s increases as t increases.

The instantaneous rate of change of the unit tangent vector $\mathbf{T}(t)$ with respect to s is called the **curvature vector** at P; that is, the curvature vector at P is

$$\frac{d\mathbf{T}(t)}{ds}.$$

Since the length of the unit tangent vector $\mathbf{T}(t)$ is always 1, the change in $\mathbf{T}(t)$ as its initial point $P(t)$ moves along the curve is a change in direction only. Thus the curvature vector measures the rate of change of the direction of the tangent to the curve with respect to the distance along the curve from some fixed point.

To express the curvature vector in terms of derivatives with respect to time, we use the chain rule to obtain

$$\frac{d\mathbf{T}}{ds} = \frac{d\mathbf{T}}{dt} \frac{dt}{ds} = \frac{\mathbf{T}'(t)}{|\mathbf{v}(t)|},$$

since $ds/dt = |\mathbf{v}(t)|$ by (2).

15.8.1 Definition. The **curvature** $K(t)$ of the arc $x = f(t)$, $y = g(t)$, $a \leq t \leq b$, at a point $P(t)$ is the magnitude of the curvature vector at P. That is,

$$K(t) = \left| \frac{d\mathbf{T}(t)}{ds} \right| = \left| \frac{\mathbf{T}'(t)}{|\mathbf{v}(t)|} \right|.$$

Example 1. Find the curvature of the four-cusped hypocycloid

$$x = a \cos^3 t, \qquad y = a \sin^3 t,$$

at the point where $t = \pi/4$.

Solution.   Since
$$\mathbf{v}(t) = -3a\cos^2 t \sin t\mathbf{i} + 3a\sin^2 t \cos t\mathbf{j}$$
and
$$|\mathbf{v}(t)| = 3a|\sin t \cos t|,$$
the unit tangent vector, when $0 < t < \pi/2$, is
$$\mathbf{T}(t) = -\cos t\mathbf{i} + \sin t\mathbf{j}.$$
Thus $\mathbf{T}'(t) = \sin t\mathbf{i} + \cos t\mathbf{j}$, and the curvature is
$$K(t) = \left|\frac{\sin t\mathbf{i} + \cos t\mathbf{j}}{3a|\sin t \cos t|}\right| = \frac{\sqrt{\sin^2 t + \cos^2 t}}{3a|\sin t \cos t|} = \frac{1}{3a|\sin t \cos t|}.$$
Therefore $K(\pi/4) = 2/(3a)$.

The reciprocal of the curvature of a curve at any point P on it is called the **radius of curvature** of the curve at P. If we denote the radius of curvature by R, then

$$R = \frac{1}{K}, \qquad K \neq 0.$$

Example 2.   Show that the radius of curvature of any circle is constant at all points of the circle and is equal to the radius of the circle.

Solution.   Let the radius of the circle be any positive number a. We shall show that the radius of curvature of the circle at any point P on it is equal to a.

A vector equation of the circle is

$$\mathbf{r}(t) = a\cos t\mathbf{i} + a\sin t\mathbf{j},$$

and may be verified by eliminating t between its parametric equations $x = a\cos t$, $y = a\sin t$.

Then $\mathbf{v}(t) = -a\sin t\mathbf{i} + a\cos t\mathbf{j}$, $|\mathbf{v}(t)| = (a^2\sin^2 t + a^2\cos^2 t)^{1/2} = a$, and the unit tangent vector at $P(t)$ is

$$\mathbf{T}(t) = \frac{\mathbf{v}(t)}{|\mathbf{v}(t)|} = -\sin t\mathbf{i} + \cos t\mathbf{j}.$$

From this,
$$\mathbf{T}'(t) = -\cos t\mathbf{i} - \sin t\mathbf{j}.$$
Thus the curvature vector is
$$\frac{\mathbf{T}'(t)}{|\mathbf{v}(t)|} = -\frac{\cos t}{a}\mathbf{i} - \frac{\sin t}{a}\mathbf{j},$$
and the curvature of the circle at any point $P(t)$ on it is
$$K(t) = \left|\frac{\mathbf{T}'(t)}{|\mathbf{v}(t)|}\right| = \left(\frac{\cos^2 t + \sin^2 t}{a^2}\right)^{1/2} = \frac{1}{a},$$
a constant. Therefore the radius of curvature of the circle at any point P on it is
$$R = \frac{1}{K} = a,$$
which is also the radius of the circle.

In general, the curvature and the radius of curvature change as P moves along a curve. The only plane curves that have constant nonzero curvature (and thus a constant radius of curvature) are circles.

The **center of curvature** of a curve at P is the point Q on the normal to the curve at P that is on the concave side of the curve and at a distance R from P. The circle with center Q and radius R is called the **circle of curvature** of the curve at P (Fig. 15-31). It is tangent to the curve at P and is the circle of "closest fit."

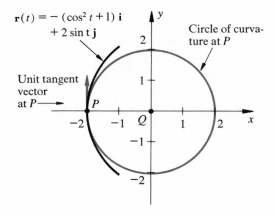

Figure 15-31

Example 3. Find the unit tangent vector and the curvature of the curve

$$\mathbf{r}(t) = -(\cos^2 t + 1)\mathbf{i} + 2 \sin t\mathbf{j}$$

at the point $P(t)$ on the curve where $t = 0$. Then draw carefully a part of the curve that includes $P(0)$, draw the unit tangent vector at P, and draw the circle of curvature at P (Fig. 15-31).

Solution. From the given vector equation of the curve we obtain

$$\mathbf{v}(t) = 2 \sin t \cos t\mathbf{i} + 2 \cos t\mathbf{j}$$

and

$$|\mathbf{v}(t)| = 2 \cos t(\sin^2 t + 1)^{1/2}.$$

Then the unit tangent vector is

$$\mathbf{T}(t) = \frac{\mathbf{v}(t)}{|\mathbf{v}(t)|} = \frac{\sin t}{(\sin^2 t + 1)^{1/2}}\mathbf{i} + \frac{1}{(\sin^2 t + 1)^{1/2}}\mathbf{j},$$

from which

$$\mathbf{T}'(t) = \frac{\cos t}{(\sin^2 t + 1)^{3/2}}\mathbf{i} - \frac{\sin t \cos t}{(\sin^2 t + 1)^{3/2}}\mathbf{j}.$$

The curvature is

$$K(t) = \left|\frac{\mathbf{T}'(t)}{|\mathbf{v}(t)|}\right| = \left|\frac{\mathbf{i} - \sin t\mathbf{j}}{2(\sin^2 t + 1)^2}\right|$$

$$= \frac{(1 + \sin^2 t)^{1/2}}{2(\sin^2 t + 1)^2} = \frac{1}{2(\sin^2 t + 1)^{3/2}}.$$

Thus the unit tangent vector at the point $P(t)$ on the curve where $t = 0$ is $T(0) = \mathbf{j}$, and the curvature at $P(0)$ is $K(0) = \frac{1}{2}$. The radius of curvature at $P(0)$ is $R = 1/K(0) = 2$, and the normal to the curve at $P(0) = (-2, 0)$ is the x axis.

To draw the circle of curvature of the curve at $P(0) = (-2, 0)$, we locate its center Q two units to the right of P on the x axis; thus Q is at the origin. The circle of curvature, then, of the given curve at $P(0) = (-2, 0)$ is the circle with center Q:$(0, 0)$ and radius $R = 2$ (Fig. 15-31).

It is useful to have a formula in rectangular Cartesian coordinates for the curvature of the graph of a *function*.

The curvature of a curve at a point P on the curve has just been defined to be the length of the curvature vector at P. As shown above, the curvature vector measures the rate of change of the direction of the tangent to the curve with respect to the distance s along the curve from some fixed point. Since the direction of the tangent line is given by its inclination α, we seek a formula in rectangular coordinates for the magnitude of the rate of change of α with respect to s, that is, $|d\alpha/ds|$.

Consider the curve $y = f(x)$, where f is a function such that f'' exists on an interval $[a, b]$, and let c be some number in $[a, b]$. If s is the length of the arc of the curve $y = f(x)$ from some fixed point $(c, f(c))$ to an arbitrary point P:(x, y) on the curve and α is the inclination of the tangent to the curve at P, then $|d\alpha/ds|$ gives the curvature $K(x)$ of the curve at P. For conciseness, we shall denote $f'(x)$ by y'.

Since $\tan \alpha = f'(x) = y'$ at P:(x, y) on the curve, $\alpha = \text{Tan}^{-1} y'$ and, by the chain rule,

$$K(x) = \left|\frac{d\alpha}{ds}\right| = \left|\frac{d\alpha}{dx} \cdot \frac{dx}{ds}\right| = \left|\frac{d}{dx} \text{Tan}^{-1} y'\right| \left|\frac{dx}{ds}\right| = \left|\frac{y''}{1 + y'^2}\right| \left|\frac{dx}{ds}\right|.$$

But $ds = \sqrt{1 + y'^2}\, dx$ (by 9.6.3); thus

$$\frac{ds}{dx} = \sqrt{1 + y'^2}.$$

By substituting this result in the preceding equation, we obtain

$$K(x) = \frac{|y''|}{(1 + y'^2)}.$$

15.8.2 Theorem. If f is a function that is twice differentiable on a closed interval $[a, b]$, the **curvature** of the graph of $y = f(x)$ at any point P:(x, y) on the graph is given by

$$K(x) = \frac{|y''|}{(1 + y'^2)^{3/2}}, \qquad a \le x \le b.$$

It is left as an exercise for the reader to show that the coordinates (q_1, q_2) of the center of curvature Q of the curve $y = f(x)$, at the point (x, y) on the curve, are given by

$$q_1 = x - \frac{y'[1 + (y')^2]}{y''}, \qquad q_2 = y + \frac{1 + (y')^2}{y''}.$$

Exercises

In Exercises 1 to 6, a curve $\mathbf{r}(t)$ and a number t_1 are given.

(a) Find the unit tangent vector $\mathbf{T}(t_1)$ at the point $P(t_1)$ on the curve where $t = t_1$.

(b) Find $K(t_1)$, the curvature of the curve at $P(t_1)$.

(c) Draw a part of the curve containing the point $P(t_1)$, and draw the unit tangent vector and the circle of curvature for the curve at $P(t_1)$.

1. $\mathbf{r}(t) = 4t^2\mathbf{i} + 4t\mathbf{j}$; $t_1 = \frac{1}{2}$.

2. $\mathbf{r}(t) = \frac{1}{2}t^2\mathbf{i} + \frac{1}{3}t^3\mathbf{j}$; $t_1 = 1$.

3. $\mathbf{r}(t) = 4\cos t\mathbf{i} + 3\sin t\mathbf{j}$; $t_1 = \dfrac{\pi}{4}$.

4. $\mathbf{r}(t) = e^t\sin t\mathbf{i} + e^t\cos t\mathbf{j}$; $t_1 = \dfrac{\pi}{2}$.

5. $\mathbf{r}(t) = \sqrt{3}\cos^3 t\mathbf{i} + \sqrt{3}\sin^3 t\mathbf{j}$; $t_1 = \dfrac{\pi}{3}$.

6. $\mathbf{r}(t) = 4t\mathbf{i} + 2(1 - t^2)\mathbf{j}$; $t_1 = 1$.

In Exercises 7 to 16, use 15.8.2 to find the curvature of the given curve at the given point. Find the center of curvature for the curve at that point. Sketch the curve and its tangent line at the given point; then draw the circle of curvature there.

7. $y = x^2$, $(1, 1)$.

8. $y = x(x - 2)^2$, $(2, 0)$.

9. $y^2 = x + 4$, $(-3, -1)$.

10. $y = \ln x$, $(1, 0)$.

11. $y = e^x - x$, $(0, 1)$.

12. $y = \cos 2x$, $(\frac{1}{6}\pi, \frac{1}{2})$.

13. $y = \cosh \frac{1}{2}x$, $(0, 1)$.

14. $y = \ln \sin x$, $(\frac{1}{4}\pi, -\ln\sqrt{2})$.

15. $y = \sin x$, $\left(\dfrac{\pi}{4}, \dfrac{\sqrt{2}}{2}\right)$.

16. $y = \tan x$, $\left(\dfrac{\pi}{4}, 1\right)$.

In Exercises 17 to 20, find the point on the given curve at which the curvature is a maximum.

17. $y = \ln x$.

18. $y = e^x$.

19. $y = \sin x$; $-\pi \le x \le \pi$.

20. $y = \cosh x$.

15.9 TANGENTIAL AND NORMAL COMPONENTS OF ACCELERATION*

Let us return to the unit tangent vector $\mathbf{T}(t)$ discussed in the preceding section. If α is the angle of inclination of the tangent line to the curve at $P(t)$, then

$$\mathbf{T}(t) = \cos \alpha\mathbf{i} + \sin \alpha\mathbf{j},$$

for its length is $\sqrt{\cos^2 \alpha + \sin^2 \alpha} = 1$ and its direction is α (Fig. 15-32).

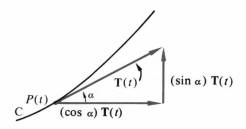

Figure 15-32

*Optional.

The derivative of $\mathbf{T}(t)$ with respect to t is the vector

$$\mathbf{T}'(t) = (-\sin \alpha \mathbf{i} + \cos \alpha \mathbf{j}) \frac{d\alpha}{dt}.$$

Since the dot product of $\mathbf{T}(t)$ and $\mathbf{T}'(t)$ is

$$\mathbf{T}(t) \cdot \mathbf{T}'(t) = (-\cos \alpha \sin \alpha + \sin \alpha \cos \alpha) \frac{d\alpha}{dt} = 0,$$

the vector $\mathbf{T}'(t)$ is perpendicular to the unit tangent vector $\mathbf{T}(t)$ at every point $P(t)$ on the curve (by 15.5.7).

If $|\mathbf{T}'(t)| \neq 0$, the vector

$$\text{(1)} \qquad \mathbf{N}(t) = \frac{\mathbf{T}'(t)}{|\mathbf{T}'(t)|}$$

is called the **unit normal vector** to the curve C at $P(t)$. It has length 1 and its direction is that of $\mathbf{T}'(t)$, namely perpendicular to the unit tangent vector at P (15.5.8).

We know that if $\mathbf{r}(t) = f(t)\mathbf{i} + g(t)\mathbf{j}$ is twice differentiable for $a \le t \le b$, the acceleration vector at any point $P(t)$ on the curve traced out by $\mathbf{r}(t)$ can be written

$$\mathbf{a}(t) = f''(t)\mathbf{i} + g''(t)\mathbf{j}.$$

In this equation the vector components of $\mathbf{a}(t)$ are $f''(t)\mathbf{i}$ and $g''(t)\mathbf{j}$; they are parallel to the coordinate axes and are scalar multiples of the orthogonal unit vectors $\mathbf{i}$ and $\mathbf{j}$.

We shall now show how $\mathbf{a}(t)$ can be expressed as the sum of a tangential vector component and a normal vector component, both of which are scalar multiples of the unit tangent vector and the unit normal vector at $P(t)$.

From equations (1) and (2) of Section 15.8, the velocity vector at any point $P(t)$ on the curve can be written $\mathbf{v}(t) = |\mathbf{v}(t)|\mathbf{T}(t)$, or

$$\mathbf{v}(t) = \frac{ds}{dt}\mathbf{T}(t).$$

If we differentiate both members of this equation with respect to t, we obtain

$$\mathbf{a}(t) = \frac{d^2s}{dt^2}\mathbf{T}(t) + \frac{ds}{dt}\mathbf{T}'(t),$$

or, using equation (1),

15.9.1 $$\mathbf{a}(t) = \frac{d^2s}{dt^2}\mathbf{T}(t) + \frac{ds}{dt}|\mathbf{T}'(t)|\mathbf{N}(t).$$

The equation 15.9.1 expressed the acceleration vector $\mathbf{a}(t)$ at $P(t)$ on the curve $\mathbf{r}(t)$ as the sum of its **tangential vector component**

$$\frac{d^2s}{dt^2}\mathbf{T}(t)$$

and its **normal vector component**

$$\frac{ds}{dt}|\mathbf{T}'(t)|\mathbf{N}(t).$$

Example. Find the tangential and normal vector components of the acceleration vector $\mathbf{a}(t)$ at the point $P:(4, \frac{8}{3})$ on the curve $\mathbf{r}(t) = t^2\mathbf{i} + \frac{1}{3}t^3\mathbf{j}$.

Solution. Since $\mathbf{v}(t) = 2t\mathbf{i} + t^2\mathbf{j}$ and $|\mathbf{v}(t)| = \sqrt{4t^2 + t^4} = t\sqrt{4 + t^2}$, the unit tangent vector is

$$\mathbf{T}(t) = \frac{\mathbf{v}(t)}{|\mathbf{v}(t)|} = \frac{2\mathbf{i} + t\mathbf{j}}{\sqrt{4 + t^2}}.$$

From

$$\mathbf{T}'(t) = \frac{-2t\mathbf{i} + 4\mathbf{j}}{(4 + t^2)^{3/2}}$$

and

$$|\mathbf{T}'(t)| = \frac{2}{4 + t^2},$$

we obtain the unit normal vector

$$\mathbf{N}(t) = \frac{\mathbf{T}'(t)}{|\mathbf{T}'(t)|} = \frac{-t\mathbf{i} + 2\mathbf{j}}{\sqrt{4 + t^2}}.$$

Since $t = 2$ at $P:(4, \frac{8}{3})$, we find

$$\mathbf{T}(2) = \frac{\mathbf{i} + \mathbf{j}}{\sqrt{2}}, \qquad |T'(2)| = \frac{1}{4}, \qquad \mathbf{N}(2) = \frac{-\mathbf{i} + \mathbf{j}}{\sqrt{2}}.$$

Now

$$\frac{ds}{dt} = |\mathbf{v}(t)| = t\sqrt{4 + t^2}$$

and

$$\frac{d^2s}{dt^2} = \frac{4 + 2t^2}{\sqrt{4 + t^2}};$$

so at $P:(4, \frac{8}{3})$, where $t = 2$,

$$\frac{ds}{dt} = 4\sqrt{2} \qquad \text{and} \qquad \frac{d^2s}{dt^2} = 3\sqrt{2}.$$

By substituting these results in

$$\mathbf{a}(t) = \frac{d^2s}{dt^2}\mathbf{T}(t) + \frac{ds}{dt}|\mathbf{T}'(t)|\mathbf{N}(t),$$

we get

$$\mathbf{a}(2) = 3\sqrt{2}\,\mathbf{T}(2) + \sqrt{2}\,\mathbf{N}(2).$$

Therefore (Fig. 15-33) the tangential vector component of $\mathbf{a}(t)$ at $P:(4, \frac{8}{3})$ is

$$3\sqrt{2}\,\mathbf{T}(2) = 3\mathbf{i} + 3\mathbf{j},$$

and the normal vector component of $\mathbf{a}(2)$ is

$$\sqrt{2}\,\mathbf{N}(2) = -\mathbf{i} + \mathbf{j}.$$

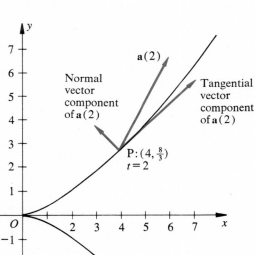

Normal vector component of $\mathbf{a}(2)$

Tangential vector component of $\mathbf{a}(2)$

$\mathbf{a}(2)$

$P:(4, \frac{8}{3})$
$t = 2$

Figure 15-33

Exercises

In each of Exercises 1 to 6, a curve $\mathbf{r}(t)$ and a number t_1 are given.
(a) Find the tangential and normal vector compo-

nents of the acceleration vector $\mathbf{a}(t)$ in terms of t.
(b) Find those vector components when $t = t_1$.

(c) Sketch a part of the curve containing the point $P(t_1)$ and draw the tangential and normal vector components of $\mathbf{a}(t_1)$ at $P(t_1)$.

1. $\mathbf{r}(t) = t^2\mathbf{i} + t\mathbf{j}; t_1 = 1.$

2. $\mathbf{r}(t) = (2t + 1)\mathbf{i} + (t^2 - 2)\mathbf{j}; t_1 = -1.$

3. $\mathbf{r}(t) = a \cos t\mathbf{i} + a \sin t\mathbf{j}; t_1 = \dfrac{\pi}{6}.$

4. $\mathbf{r}(t) = 4 \cos^3 t\mathbf{i} + 4 \sin^3 t\mathbf{j}; t_1 = \dfrac{\pi}{3}.$

5. $\mathbf{r}(t) = (\tfrac{1}{3}t^3 - t)\mathbf{i} + (t^2 - 1)\mathbf{j}; t_1 = 2.$

6. $\mathbf{r}(t) = e^t\mathbf{i} + e^{-t}\mathbf{j}; t_1 = \ln 2.$

7. Show that if a particle moves on a trajectory $\mathbf{r}(t)$ with constant speed, its acceleration vector is always normal to the path.

8.* In polar coordinates it is sometimes useful to express the velocity vector $\mathbf{v}(t)$ or the acceleration vector $\mathbf{a}(t)$ as the sum of a **radial vector component** in the direction of $\mathbf{r}(t)$ and a **transverse vector component** at right angles to it.

Superpose a polar coordinate system on the Cartesian system so that the pole is at the origin, the polar axis is the positive x axis, and the units of length are the same. Then the curve C traced out by

$$\mathbf{r}(t) = f(t)\mathbf{i} + g(t)\mathbf{j},$$

which has the parametric equations $x = f(t)$, $y = g(t)$ in Cartesian coordinates, has polar parametric equations $r = F(t)$, $\theta = G(t)$, where the Cartesian and polar coordinates of any point P are connected by $x = r \cos \theta$ and $y = r \sin \theta$.

Denote by $\mathbf{u}_r$ the unit vector in the direction of $\mathbf{r}(t)$ and let $\mathbf{u}_\theta$ be the unit vector in the direction that is $\pi/2$ counterclockwise from $\mathbf{u}_r$ (Fig. 15-34).

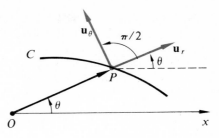

Figure 15-34

(a) Show that
$$\mathbf{u}_r = \cos \theta\mathbf{i} + \sin \theta\mathbf{j},$$
$$\mathbf{u}_\theta = -\sin \theta\mathbf{i} + \cos \theta\mathbf{j},$$

and
$$\frac{d\mathbf{u}_r}{d\theta} = \mathbf{u}_\theta, \qquad \frac{d\mathbf{u}_\theta}{dr} = -\mathbf{u}_r.$$

(b) Although $\mathbf{i}$ and $\mathbf{j}$ were constant unit vectors, the unit vectors $\mathbf{u}_r$ and $\mathbf{u}_\theta$ are functions of θ and hence of t. This is indicated by writing $\mathbf{u}_r(t)$ and $\mathbf{u}_\theta(t)$. Show that
$$\frac{d\mathbf{u}_r}{dt} = \frac{d\mathbf{u}_r}{d\theta}\frac{d\theta}{dt} = \mathbf{u}_\theta\frac{d\theta}{dt},$$
$$\frac{d\mathbf{u}_\theta}{dt} = \frac{d\mathbf{u}_\theta}{d\theta}\frac{d\theta}{dt} = -\mathbf{u}_r\frac{d\theta}{dt}.$$

(c) For conciseness, denote $|\mathbf{r}(t)|$ by $r(t)$; then $\mathbf{r}(t) = r(t)\mathbf{u}_r(t)$. Show that the velocity vector of the terminal point P of $\mathbf{r}(t)$ is
$$\mathbf{v}(t) = \frac{d\mathbf{r}(t)}{dt} = \frac{dr(t)}{dt}\mathbf{u}_r(t) + r(t)\frac{d\mathbf{u}_r(t)}{dt},$$
or
$$\mathbf{v}(t) = \frac{dr}{dt}\mathbf{u}_r + r\frac{d\theta}{dt}\mathbf{u}_\theta.$$

Here $\mathbf{v}(t)$ is expressed as the sum of a radial vector component $(dr/dt)\mathbf{u}_r$ and a transverse vector component $(r\,d\theta/dt)\mathbf{u}_\theta$.

15.10 REVIEW EXERCISES

In each of Exercises 1 to 6, a parametric representation of a curve is given. State which of the following terms apply to the given curve: simple, closed, arc. By eliminating the parameter, write the Cartesian equation of the curve. Sketch the curve.

1. $x = 6t + 2,\ y = 2t; t \in R.$

2. $x = 4t^2,\ y = 4t; -1 \le t \le 2.$

3. $x = 2t,\ y = t^3; -1 \le t \le 2.$

4. $x = t^2 - 2,\ y = t^3 - 2t; -2 \le t \le 2.$

*Used in optional Section 17.4.

5. $x = 4 \sin t - 2, \ y = 3 \cos t - 2; \ 0 \leq t \leq 2\pi.$

6. $x = 2 \sec t, \ y = \tan t; \ -\frac{1}{2}\pi < t < \frac{1}{2}\pi.$

In Exercises 7 and 8, find the equations of the tangent line and the normal line to the given curve at the given point without eliminating the parameter. Make a sketch.

7. $x = 3 \sin t - 4, \ y = 5 + 2 \cos t; \ t_1 = \frac{5}{4}\pi.$

8. $x = 3e^{-t}, \ y = \frac{1}{2}e^t; \ t_1 = 2.$

9. Find the length of the arc

$$x = \cos t + t \sin t$$

$$y = \sin t - t \cos t$$

from $t = 0$ to $t = 2\pi$. Make a sketch.

10. Find the length of the cardioid $r = 2(1 - \cos\theta)$. Sketch the curve.

In each of Exercises 11 and 12, a pair of vectors is given.
(a) Represent them by directed line segments.
(b) Find their slopes.
(c) Add them algebraically and graphically.
(d) Subtract the second from the first, algebraically and graphically.
(e) Find their lengths.

11. $[-3, 2]$ and $[5, 6]$.

12. $[0, 4]$ and $[-6, 1]$.

13. Let $a = 2i - 5j, \ b = i + j,$ and $c = -6i$. Find
(a) $3a - 2b;$ (b) $a \cdot b;$
(c) $a \cdot (b + c);$ (d) $(4a + 5b) \cdot 3c;$
(e) $|c| c \cdot b;$ (f) $c \cdot c - |c|.$

14. Find the cosine of the angle between a and b and make a sketch.
(a) $a = 3i + 2j, \ b = -i + 4j;$
(b) $a = -5i - 3j, \ b = 2i - j;$

(c) $a = 4i, \ b = -3j;$
(d) $a = 7i, \ b = 5i + j.$

15. Write in the form $a = a_1 i + a_2 j$ the vector represented by $\overrightarrow{AB}$.
(a) $A:(-7, 2), \ B:(3, 4);$
(b) $A:(0, 5), \ B:(-6, -1);$
(c) $A:(\frac{1}{2}, \sqrt{2}), \ B:(-7, 5\sqrt{2});$
(d) $A:(0, 0), \ B:(3, -\pi).$

16. Given $a = -2i$ and $b = 3i - 2j$, two non-collinear vectors, and another vector $r = 5i - 4j$. Find scalars, k and m, such that $r = ka + mb$.

17. Find $r'(t)$ and $r''(t)$ for each of the following.
(a) $r(t) = (\ln t)i - 3t^2 j;$
(b) $r(t) = \sin ti + \cos 2tj;$
(c) $r(t) = \tan ti - t^4 j;$
(d) $r(t) = (e^t - e^{-2t})j.$

18. Find the length of the arc $r(t) = 4t^{3/2}i + 3tj$ from $t = 2$ to $t = 5$. Make a sketch.

In Exercises 19, 20, and 21, the position of a moving particle at time t is given by $r(t)$. Find the velocity and acceleration vectors, $v(t)$ and $a(t)$, and their values at the given time $t = t_1$. Also find the speed of the particle for $t = t_1$. Sketch a part of the graph of $r(t)$ containing the position P of the particle when $t = t_1$, and draw $v(t_1)$ and $a(t_1)$ with their initial points at P.

19. $r(t) = 2t^2 i + (4t + 2)j; \ t_1 = -1.$

20. $r(t) = \frac{1}{2}t^3 i + t^2 j; \ t_1 = 2.$

21. $r(t) = 4(1 - \sin t)i + 4(t + \cos t)j; \ t_1 = \frac{2}{3}\pi.$

22. Find the unit tangent vector $T(t)$ for the curve $r(t) = ti + \frac{1}{3}t^3 j$. At the point $P(1)$ on the curve where $t = 1$, find $T(1)$. Find the curvature $K(1)$ of the curve at $P(1)$. Sketch the curve and draw the unit tangent vector $T(1)$ with its initial point at $P(1)$.

16

three-dimensional spaces

Cartesian coordinates and vectors extend easily and naturally to three-dimensional space. Most of the definitions and theorems will be simple restatements, with one more variable, of those we are familiar with in the plane. Because the proofs are usually analogous, less time need be given them.

16.1 CARTESIAN COORDINATES IN THREE-SPACE

A Cartesian coordinate system for three-dimensional space can be established in much the same way as was done in the plane.

We start with *three* mutually perpendicular coordinate lines with their origins at a common point O and having the same unit of length. Although these lines can be oriented in any way one pleases, we shall follow the usual custom of thinking of Oy and Oz as lying in the plane of the paper with their positive directions to the right and upward, respectively, and Ox as perpendicular to the paper with its positive direction toward us (Fig. 16-1). This is called a **right-hand system** of rectangular Cartesian coordinates because if a screw with a right-hand thread points upward, it will advance in the positive direction of the z axis when turned in the direction from Ox to Oy.

The plane of the x and y axes is called the xy plane, the plane of the y and z axes (the plane of the paper) is the yz plane, and the plane of the z and x axes is the zx plane (Fig. 16-2). They are often called the **coordinate planes**.

Let P be any point in space. The coordinates of the projections of P on the x axis, the y axis, and the z axis are its Cartesian coordinates x, y, and z, respectively (Fig.

A right-hand system of rectangular Cartesian coordinate axes

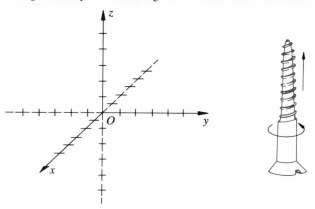

Figure 16-1

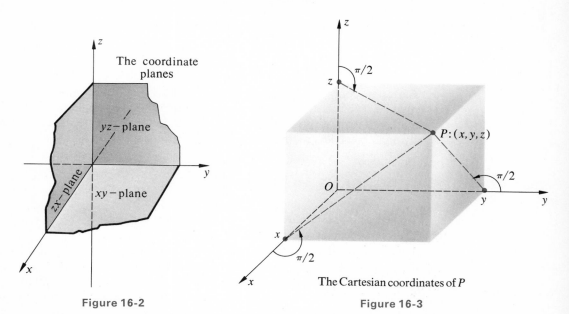

The coordinate planes

Figure 16-2

The Cartesian coordinates of P

Figure 16-3

16-3). Thus to each point P in space there corresponds an ordered triple of numbers (x, y, z), its Cartesian coordinates relative to the given axes. Conversely, by reversing this process, to each ordered triple of real numbers (x, y, z) there corresponds one and only one point P. For example, to **plot** the point whose coordinates are $(2, -3, 4)$, we start from the origin and count 2 units toward us along the positive x axis, then 3 units to left, and 4 units upward (Fig. 16-4). Again, the point $(-3, 2, -5)$ is located

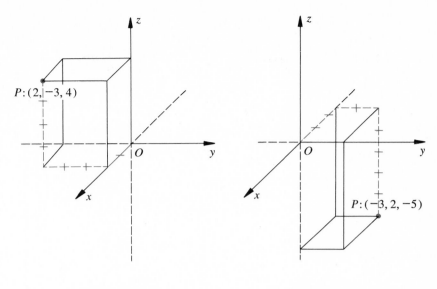

Figure 16-4   Figure 16-5

by going 3 units from O along the negative x axis, 2 units to the right, and 5 units downward (Fig. 16-5).

The coordinate planes separate Cartesian three-space into eight octants. In the **first octant** the coordinates of all points are positive. There is no general agreement on how the other octants should be numbered.

In our two-dimensional diagrams of three-dimensional figures, the y and z axes are drawn perpendicular to each other, since we think of them as lying in the plane of the paper. To make the x axis seem to come toward us, we draw it so that angle yOx is about $-135°$ (Fig. 16-1).

Although the units on the three coordinate axes are actually all the same length, we draw the units on the x axis about three-quarters as long as those on the y and z axes so as to aid the illusion of three-dimensional space. Of course, this is not true perspective drawing, since lines parallel to the x axis do not appear to converge; but it is simple to execute and gives a surpringly good illusion of three-dimensional space.

The reader is advised to "complete the box" at first when plotting a point, as we did in Figs. 16-4 and 16-5. This will help in making the transition from two-dimensional to three-dimensional visualization. The box is easy to draw, since all its edges are parallel to, or segments of, the coordinate axes.

A line is parallel to the x axis if all points on it have the same y coordinate and the same z coordinate; and similarly for lines parallel to either of the other coordinate axes. We wish opposite sides of any rectangle to have equal lengths, so we extend 1.4.1 (ii) as follows.

16.1.1 Definition.

    (i) If P_1:(x_1, y_1, z_1) and P_2:(x_2, y_2, z_2) are two points on a line parallel to the x axis, then the (undirected) **distance** between P_1 and P_2 is

$$|P_1P_2| = |x_2 - x_1|.$$

    (ii) If P_1 and P_2 are on a line parallel to the y axis, then

$$|P_1P_2| = |y_2 - y_1|.$$

    (iii) If P_1 and P_2 are on a line parallel to the z axis, then

$$|P_1P_2| = |z_2 - z_1|.$$

    Consider two arbitrarily selected but fixed points P_1:(x_1, y_1, z_1) and P_2:(x_2, y_2, z_2) in three-space. Draw the rectangular parallelepiped having P_1 and P_2 as opposite vertices, with its edges parallel to the coordinate axes (Fig. 16-6). The triangles P_1RQ

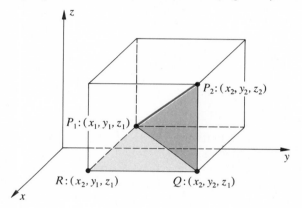

Figure 16-6

and P_1QP_2 are right triangles, and, by the Pythagorean theorem,

$$|P_1P_2|^2 = |P_1Q|^2 + |QP_2|^2$$

and

$$|P_1Q|^2 = |P_1R|^2 + |RQ|^2.$$

Thus

(1) $$|P_1P_2|^2 = |P_1R|^2 + |RQ|^2 + |QP_2|^2.$$

But $|P_1R| = |x_2 - x_1|$, $|RQ| = |y_2 - y_1|$, and $|QP_2| = |z_2 - z_1|$ (by 16.1.1). By substituting these results in (1), we obtain

$$|P_1P_2|^2 = |x_2 - x_1|^2 + |y_2 - y_1|^2 + |z_2 - z_1|^2.$$

This suggests the following definition.

16.1.2 Definition. The (undirected) **distance** between any two points P_1:(x_1, y_1, z_1) and P_2:(x_2, y_2, z_2), or the **length** of the line segment P_1P_2, is

$$|P_1P_2| = \sqrt{(x_2 - x_1)^2 + (y_2 - y_1)^2 + (z_2 - z_1)^2}.$$

The set of all ordered triples of real numbers (x, y, z), together with the definition of distance given in 16.1.2, constitutes a **three-dimensional Cartesian space**.

A **sphere** is the set of points at a constant distance from a fixed point called the center. The constant distance is the radius of the sphere.

16.1.3 Theorem. An equation of the **sphere** whose center is (h, k, l) and whose radius is r is

(2) $$(x - h)^2 + (y - k)^2 + (z - l)^2 = r^2.$$

This follows immediately from the distance formula (16.1.2) and the definition of a sphere.

The equation (2) is particularly simple if the center is the origin. The equation of a sphere with radius r and center at the origin is

$$x^2 + y^2 + z^2 = r^2.$$

By expanding the terms in parentheses, equation (2) can be rewritten

$$x^2 + y^2 + z^2 - 2hx - 2ky - 2lz + (h^2 + k^2 + l^2 - r^2) = 0,$$

which is in the form

(3) $$x^2 + y^2 + z^2 + Gx + Hy + Iz + J = 0,$$

where G, H, I, and J are constants. Thus every sphere has an equation of the form (3).

Conversely, (3) can be written

$$(x^2 + Gx) + (y^2 + Hy) + (z^2 + Iz) = -J.$$

If we complete the squares in the parentheses by adding $G^2/4$, $H^2/4$, and $I^2/4$ to both members of this equation, we have

(4) $$\left(x + \frac{G}{2}\right)^2 + \left(y + \frac{H}{2}\right)^2 + \left(z + \frac{I}{2}\right)^2 = \frac{G^2 + H^2 + I^2 - 4J}{4}.$$

If the right-hand member of (4) is positive, the graph of (4) is a sphere; if the right-hand member of (4) is zero, the graph is a single point; if the right-hand member of (4) is negative, there is no graph.

16.1.4 Theorem. The graph, if any, of

$$x^2 + y^2 + z^2 + Gx + Hy + Iz + J = 0$$

is a sphere or a single point, and every sphere has an equation of this form.

Example. Sketch the graph of $x^2 + y^2 + z^2 - 10x - 8y - 12z + 68 = 0$.

Solution. If this equation has a graph, it is a sphere or a single point. We rewrite the given equation as

$$(x^2 - 10x) + (y^2 - 8y) + (z^2 - 12z) = -68$$

and complete the squares in the parentheses by adding 25, 16, and 36 to both members, obtaining

$$(x^2 - 10x + 25) + (y^2 - 8y + 16) + (z^2 - 12z + 36) = 9,$$

or

$$(x - 5)^2 + (y - 4)^2 + (z - 6)^2 = 9.$$

Therefore (16.1.3) the graph is a sphere with center $(5, 4, 6)$ and radius 3 (Fig. 16-7).

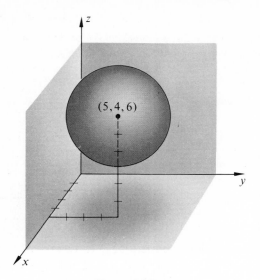

Figure 16-7   **Figure 16-8**

Since the line segment P_1P_2 in Fig. 16-8 is the hypotenuse of three right triangles, P_1RP_2, P_1SP_2, and P_1TP_2, each having an adjacent side parallel to a coordinate axis, it is easy to deduce the following formula from 16.1.1 and 1.4.2.

16.1.5 Midpoint Formula. The coordinates of the **midpoint** of the line segment terminated by P_1:(x_1, y_1, z_1) and P_2:(x_2, y_2, z_2) are

$$x = \frac{x_1 + x_2}{2}, \qquad y = \frac{y_1 + y_2}{2}, \qquad z = \frac{z_1 + z_2}{2}.$$

Exercises

1. Plot the points whose coordinates are $(3, 2, 4)$, $(2, 0, 3)$, $(-3, 4, 5)$, $(0, -5, -3)$, $(-4, 0, 0)$, $(-2, -6, -1)$, $(0, 0, 0)$, $(0, 4, 0)$. In each instance, "complete the box."

2. Plot the points whose coordinates are as follows, $(\sqrt{2}, -2, 2)$, $(0, \pi, -7)$, $(-3, \frac{1}{2}, \sqrt{3})$, $(1, 1, 1)$, $(\frac{7}{3}, -\frac{8}{3}, 3)$, $(0, 0, e)$, $(-4, -3, \sqrt{5})$, $(2, 0, -5)$. In each case, "complete the box."

3. What is peculiar to the coordinates of all points in the zx plane? In the xy plane?

4. What is peculiar to the coordinates of all points on the x axis? On the negative z axis?

5. Name each of the following sets of points.
(a) $\{P{:}(x, y, z) \,|\, x = 0\}$;

(b) $\{P{:}(x, y, z) \,|\, x = 0, z = 0\}$;
(c) $\{P{:}(x, y, z) \,|\, x < 0, y = 0, z = 0\}$.

6. Is the line through the points P:$(2, 4, 3)$ and Q:$(2, -5, 3)$ parallel to one of the coordinate axes? Find $|PQ|$.

7. Is the line through the points P:$(-1, -2, 4)$ and Q:$(-6, -2, 4)$ parallel to a coordinate axis? Find $|PQ|$.

8. Define, in terms of the coordinates of the points on it, what is meant by saying that a line is parallel to (a) the x axis; (b) the y axis; (c) the z axis.

9. Find the distance between each of the following pairs of points.
(a) $(-1, 3, 2)$ and $(4, 0, -5)$;

(b) $(6, -2, 0)$ and $(2, 3, 7)$;

(c) $(0, 0, 2)$ and $(-1, -3, 0)$;

(d) $(\sqrt{2}, 0, \sqrt{3})$ and $(0, 2, 0)$.

10.   Express by an equation the statement that the distance between the point P:(x, y, z) and the fixed point C:$(3, -2, 4)$ is always equal to 5. Name the graph of the set of all such points P.

11.   Write the equation of the sphere whose center and radius are given.

(a) $(2, 4, 1)$, 3;   (b) $(-4, 6, 0)$, 7;

(c) $(-6, 2, -3)$, 1;   (d) $(8, 0, -1)$, 2.

12.   Write the equation of the sphere whose center and radius are:

(a) $(3, 0, 4)$, 5;   (b) $(-2, -4, 3)$, 3;

(c) $(1, 5, 1)$, 2;   (d) $(3, -5, 2)$, 7.

In Exercises 13 to 16, find the center and radius of the sphere whose equation is given.

13. $x^2 + y^2 + z^2 - 12x + 14y - 8z + 1 = 0$.

14. $x^2 + y^2 + z^2 + 2x - 6y - 10z + 34 = 0$.

15. $36x^2 + 36y^2 + 36z^2 - 48x + 36y - 360z - 1379 = 0$.

16. $x^2 + y^2 + z^2 + 8x - 4y - 22z + 77 = 0$.

17.   Find the equation of the sphere whose center is $(2, 4, 5)$ and that is tangent to the xy plane. Make a sketch.

18.   Find the equation of the sphere whose center is $(-3, 0, -2)$ and that is tangent to the yz plane.

19.   Find the equation of the sphere that is tangent to all the coordinate planes, if its radius is 6 and its center is in the first octant. Make a sketch.

20.   Find the equation of the sphere that is tangent to the zx plane at the origin, if its radius is 4 and its center is on the negative y axis. Make a sketch.

21.   Find the equations of the tangent spheres whose centers are $(-3, 1, 2)$ and $(5, -3, 6)$, and whose radii are equal.

16.2   THREE-DIMENSIONAL VECTORS

The definitions and theorems for two dimensional vectors are extended in an obvious manner to three-space.

16.2.1   Definition.   Three-dimensional vectors are ordered triples of numbers, $[x, y, z]$, that obey the following laws.

If $\mathbf{a} = [a_1, a_2, a_3]$ and $\mathbf{b} = [b_1, b_2, b_3]$ are vectors and k is a scalar, then

(i) $\mathbf{a} = \mathbf{b}$ if and only if $a_1 = b_1$, $a_2 = b_2$, and $a_3 = b_3$.

(ii) $k\mathbf{a} = [ka_1, ka_2, ka_3]$.

(iii) $\mathbf{a} + \mathbf{b} = [a_1 + b_1, a_2 + b_2, a_3 + b_3]$.

(iv) $\mathbf{a} - \mathbf{b} = [a_1 - b_1, a_2 - b_2, a_3 - b_3]$.

The numbers x, y, and z that appear in the vector $[x, y, z]$ are the **components** of the vector.

The set of all ordered triples of numbers and the operations of addition and multiplication by a scalar, as defined above, form a **vector space** of three dimensions.

Three-dimensional vectors are commonly represented by directed line segments. Specifically, the vector $\mathbf{a} = [a_1, a_2, a_3]$ can be represented by the directed line segment $\overrightarrow{QP}$ whose initial point Q:(x, y, z) is any selected point and whose terminal point is

$$P:(x + a_1, y + a_2, z + a_3)$$

(Fig. 16-9).

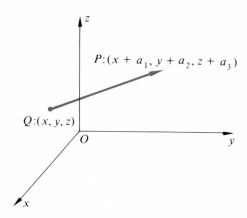

Figure 16-9

The **direction** of a vector is the direction of any directed line segment that represents the vector.

The **length** (or **magnitude**) of a vector $\mathbf{a} = [a_1, a_2, a_3]$ is the number

$$|\mathbf{a}| = \sqrt{a_1^2 + a_2^2 + a_3^2}.$$

It is equal to the length of any directed line segment that represents the vector.

The vector $\mathbf{0} = [0, 0, 0]$ is called the **zero vector** and may be represented by any point; its direction is unspecified.

The **negative** of a vector $\mathbf{a} = [a_1, a_2, a_3]$ is the vector

$$-\mathbf{a} = [-a_1, -a_2, -a_3].$$

It is represented by any directed line segment that represents $\mathbf{a}$, *but with its direction reversed.* As an illustration, the negative of the vector $\mathbf{b} = [-1, -3, 4]$ is $-\mathbf{b} = [1, 3, -4]$ (Fig. 16-10).

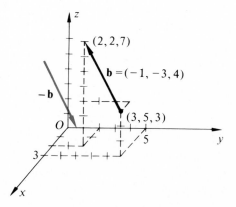

Figure 16-10

Geometrically, if $P:(x, y, z)$ is a point in three-dimensional space and the vector $\mathbf{a} = [a_1, a_2, a_3]$ has its initial point at P and its terminal point at

$$Q:(x + a_1, y + a_2, z + a_3),$$

and if the vector $\mathbf{b} = [b_1, b_2, b_3]$ has its initial point at Q and its terminal point at $R:(x + a_1 + b_1, y + a_2 + b_2, z + a_3 + b_3)$, then the vector

$$\mathbf{a} + \mathbf{b} = [a_1 + b_1, a_2 + b_2, a_3 + b_3]$$

is represented by the directed line segment $\overrightarrow{PR}$ [Fig. 16-11(a)].

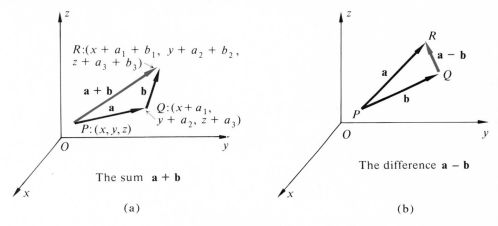

Figure 16-11

Figure 16-11(b) shows a geometric interpretation of $\mathbf{a} - \mathbf{b}$. If $\mathbf{a}$ and $\mathbf{b}$ are represented by the directed line segments $\overrightarrow{PQ}$ and $\overrightarrow{PR}$ having the same initial point P, then $\mathbf{a} - \mathbf{b}$ is represented by the directed line segment $\overrightarrow{QR}$ because $\mathbf{b} + (\mathbf{a} - \mathbf{b}) = \mathbf{a}$ by the previous construction for the sum of two vectors.

This geometric interpretation of $\mathbf{a} - \mathbf{b}$ has a useful corollary.

16.2.2 Corollary. If $P_1:(x_1, y_1, z_1)$ and $P_2:(x_2, y_2, z_2)$ are distinct points in three-space, the directed line segment $\overrightarrow{P_1P_2}$ represents the vector

$$[x_2 - x_1, y_2 - y_1, z_2 - z_1]$$

(see Fig. 16-12).

For if $\mathbf{a} = [x_2, y_2, z_2]$ and $\mathbf{b} = [x_1, y_1, z_1]$ are *position* vectors having the terminal points P_1 and P_2, respectively, then $\mathbf{a} - \mathbf{b} = [x_2 - x_1, y_2 - y_1, z_2 - z_1]$ has the geometric interpretation $\overrightarrow{P_1P_2}$ (Fig. 16-12).

If k is a positive number (a positive scalar), the product $k\mathbf{a}$ is a vector k times as long as $\mathbf{a}$ and having the same direction as $\mathbf{a}$ (see Fig. 15-21); if $k < 0$, $k\mathbf{a}$ is a vector $|k|$ times as long as $\mathbf{a}$ and pointing in the opposite direction (see Fig. 15-22).

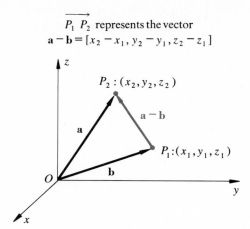

$\overrightarrow{P_1\,P_2}$ represents the vector
$$\mathbf{a} - \mathbf{b} = [x_2 - x_1, y_2 - y_1, z_2 - z_1]$$

Figure 16-12

Three-dimensional vectors obey the same rules for addition and subtraction as vectors in the plane.

16.2.3 Theorem. If $\mathbf{a}, \mathbf{b}, \mathbf{c}$, and $\mathbf{0}$ are vectors and $k, m, 1$, and 0 are scalars (numbers), then

$$\mathbf{a} + \mathbf{b} = \mathbf{b} + \mathbf{a} \qquad \text{(commutative law),}$$

$$\mathbf{a} + (\mathbf{b} + \mathbf{c}) = (\mathbf{a} + \mathbf{b}) + \mathbf{c} \qquad \text{(associative law),}$$

$$\mathbf{a} + \mathbf{0} = \mathbf{a}, \qquad \mathbf{a} + (-\mathbf{a}) = \mathbf{0},$$

$$k(\mathbf{a} + \mathbf{b}) = k\mathbf{a} + k\mathbf{b}, \qquad (k + m)\mathbf{a} = k\mathbf{a} + m\mathbf{a},$$

$$(km)\mathbf{a} = k(m\mathbf{a}), \qquad (-k)\mathbf{a} = -(k\mathbf{a}), \qquad |k\mathbf{a}| = |k||\mathbf{a}|,$$

$$0\mathbf{a} = \mathbf{0}, \qquad 1\mathbf{a} = \mathbf{a}, \qquad k\mathbf{0} = \mathbf{0}.$$

A proof of 16.2.3 is similar to the proofs of 15.4.4 and 15.5.2.

The definition of the *dot product* of two vectors in the plane is readily extended to higher spaces.

16.2.4 Definition. If $\mathbf{a} = [a_1, a_2, a_3]$ and $\mathbf{b} = [b_1, b_2, b_3]$ are vectors, then their **dot product** (or **inner product**) is the number

$$\mathbf{a} \cdot \mathbf{b} = a_1 b_1 + a_2 b_2 + a_3 b_3.$$

Proof of the following properties of dot multiplication of vectors is similar to that of 15.5.4 and is left as an exercise.

16.2.5 Properties of the Dot Product. If **a**, **b**, and **c** are vectors and k is a scalar, then

$$\mathbf{a} \cdot \mathbf{b} = \mathbf{b} \cdot \mathbf{a},$$

$$\mathbf{a} \cdot (\mathbf{b} + \mathbf{c}) = \mathbf{a} \cdot \mathbf{b} + \mathbf{a} \cdot \mathbf{c},$$

$$k(\mathbf{a} \cdot \mathbf{b}) = (k\mathbf{a}) \cdot \mathbf{b} = \mathbf{a} \cdot (k\mathbf{b}),$$

$$\mathbf{0} \cdot \mathbf{a} = 0, \qquad \mathbf{a} \cdot \mathbf{a} = |\mathbf{a}|^2.$$

Represent two nonzero vectors, **a** and **b**, by directed line segments $\overrightarrow{QA}$ and $\overrightarrow{QB}$ (see Fig. 15-23). The **angle between a** and **b** is the smallest nonnegative angle θ formed by $\overrightarrow{QA}$ and $\overrightarrow{QB}$. Always

$$0 \le \theta \le \pi.$$

A proof of the next theorem is exactly the same as the proof of 15.5.6.

16.2.6 Theorem. If **a** and **b** are nonzero vectors, the angle θ between them is given by

$$\cos \theta = \frac{\mathbf{a} \cdot \mathbf{b}}{|\mathbf{a}||\mathbf{b}|}, \qquad 0 \le \theta \le \pi.$$

16.2.7 Corollary. Two nonzero vectors are **perpendicular** (or **orthogonal**) if and only if their dot product is zero.

Example 1. Find the angle between the vectors $\mathbf{a} = [4, 5, -2]$ and $\mathbf{b} = [-3, 6, 0]$.

Solution. From 16.2.6 the cosine of the angle θ between these two vectors is

$$\cos \theta = \frac{\mathbf{a} \cdot \mathbf{b}}{|\mathbf{a}||\mathbf{b}|} = \frac{4(-3) + 5(6) - 2(0)}{\sqrt{16 + 25 + 4}\sqrt{9 + 36 + 0}} = \frac{18}{45} = \frac{2}{5}.$$

Thus $\theta \doteq 66°25'$.

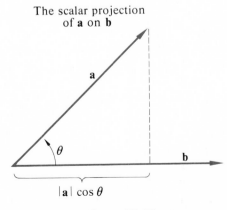

The scalar projection of **a** on **b**

Figure 16-13

A **unit vector** is a vector whose length is 1. Thus $\mathbf{a} = [\frac{3}{7}, \frac{2}{7}, \frac{6}{7}]$ is a unit vector because $|\mathbf{a}| = \sqrt{\frac{9}{49} + \frac{4}{49} + \frac{36}{49}} = 1$.

It is easy to verify that if $\mathbf{a} = [a_1, a_2, a_3]$ is a nonzero vector, then

$$\frac{\mathbf{a}}{|\mathbf{a}|}$$

is the **unit vector in the direction of a**.

The **scalar projection** of a vector **a** on a vector **b** means the number

$$|\mathbf{a}| \cos \theta,$$

where θ is the angle between **a** and **b** (Fig. 16-13). This number is positive, zero, or negative, depending on whether θ is acute, right, or obtuse.

The **projection** (or **vector projection**) of a vector **a** on a vector **b** is the vector

$$|\mathbf{a}| \cos \theta \, \frac{\mathbf{b}}{|\mathbf{b}|},$$

where $|\mathbf{a}| \cos \theta$ is the scalar projection of **a** on **b** and $\mathbf{b}/|\mathbf{b}|$ is the unit vector in the direction of **b**. The projection of **a** on **b** is also called the **vector component** of **a** along **b**.

The unit vectors $\mathbf{i} = [1, 0, 0], \mathbf{j} = [0, 1, 0]$, and $\mathbf{k} = [0, 0, 1]$ are called **basis vectors** because every nonzero, three-dimensional vector can be expressed in one and only one way as a linear combination of **i**, **j**, and **k**. Thus if $\mathbf{a} = [a_1, a_2, a_3]$, then

$$\mathbf{a} = a_1\mathbf{i} + a_2\mathbf{j} + a_3\mathbf{k}$$

uniquely.

It is easy to verify that

(1) $$\mathbf{i} \cdot \mathbf{i} = \mathbf{j} \cdot \mathbf{j} = \mathbf{k} \cdot \mathbf{k} = 1$$

and

(2) $$\mathbf{i} \cdot \mathbf{j} = \mathbf{j} \cdot \mathbf{k} = \mathbf{k} \cdot \mathbf{i} = 0.$$

Example 2. Show that the dot product of two vectors, $\mathbf{a} = a_1\mathbf{i} + a_2\mathbf{j} + a_3\mathbf{k}$ and $\mathbf{b} = b_1\mathbf{i} + b_2\mathbf{j} + b_3\mathbf{k}$, can be found by multiplying the right members as though they were ordinary algebraic polynomials.

Solution.

$$(a_1\mathbf{i} + a_2\mathbf{j} + a_3\mathbf{k}) \cdot (b_1\mathbf{i} + b_2\mathbf{j} + b_3\mathbf{k}) = a_1b_1\mathbf{i} \cdot \mathbf{i} + a_1b_2\mathbf{i} \cdot \mathbf{j} + a_1b_3\mathbf{i} \cdot \mathbf{k} + a_2b_1\mathbf{j} \cdot \mathbf{i}$$
$$+ a_2b_2\mathbf{j} \cdot \mathbf{j} + a_2b_3\mathbf{j} \cdot \mathbf{k} + a_3b_1\mathbf{k} \cdot \mathbf{i} + a_3b_2\mathbf{k} \cdot \mathbf{j}$$
$$+ a_3b_3\mathbf{k} \cdot \mathbf{k}$$
$$= a_1b_1 + a_2b_2 + a_3b_3 \qquad \text{[by (1) and (2)].}$$

This result is verified by the definition of $\mathbf{a} \cdot \mathbf{b}$.

We may think of the unit vectors $\mathbf{i} = [1, 0, 0], \mathbf{j} = [0, 1, 0]$, and $\mathbf{k} = [0, 0, 1]$ as having their initial points at the origin and pointing in the directions of the positive x, y, and z axes, respectively (Fig. 16-14).

The angles between a nonzero vector **a** and the unit vectors **i**, **j**, and **k** are called the **direction angles** of the vector **a**; they are designated by α, β, and γ, respectively (Fig. 16-15).

Generally it is more convenient to work with the cosines of these angles than with the direction angles themselves. The numbers $\cos \alpha$, $\cos \beta$, and $\cos \gamma$ are called the **direction cosines** of the vector **a**.

If $\mathbf{a} = [a_1, a_2, a_3]$, then

$$\cos \alpha = \frac{\mathbf{a} \cdot \mathbf{i}}{|\mathbf{a}||\mathbf{i}|} = \frac{a_1}{|\mathbf{a}|}$$

(by 16.2.5), and similarly for $\cos \beta$ and $\cos \gamma$.

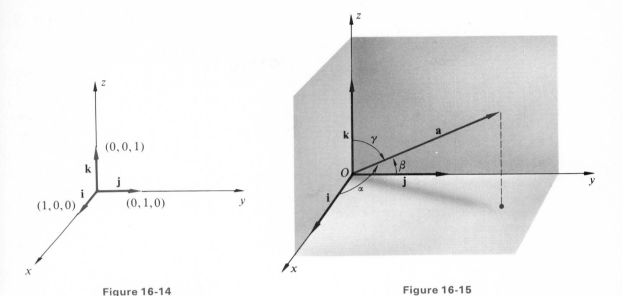

Figure 16-14   **Figure 16-15**

16.2.8 Theorem. If $\mathbf{a} = [a_1, a_2, a_3]$ is a nonzero vector, then its **direction cosines** are

$$\cos \alpha = \frac{a_1}{|\mathbf{a}|}, \qquad \cos \beta = \frac{a_2}{|\mathbf{a}|}, \qquad \cos \gamma = \frac{a_3}{|\mathbf{a}|},$$

where $|\mathbf{a}| = \sqrt{a_1^2 + a_2^2 + a_3^2}$ is the length of $\mathbf{a}$.

Thus *the components of a vector are proportional to its direction cosines.* It follows from 16.2.8 that if we know the length of a vector and its direction cosines, we can find the components of the vector and hence the vector itself. It is also apparent from 16.2.8 that the direction cosines of a vector are not independent of each other but have the following useful relation.

If $\cos \alpha$, $\cos \beta$, and $\cos \gamma$ are the direction cosines of a nonzero vector, then

16.2.9 $$\cos^2 \alpha + \cos^2 \beta + \cos^2 \gamma = 1,$$

and vice versa.

Example 3. Find the direction cosines of the vector represented by the directed line segment whose initial point is Q:$(-1, 3, 2)$ and whose terminal point is P:$(3, -2, 5)$.

Solution. The vector represented by $\overrightarrow{QP}$ is $[4, -5, 3]$. Its length is $\sqrt{(4)^2 + (-5)^2 + (3)^2} = 5\sqrt{2}$. Thus (by 16.2.8) its direction cosines are

$$\cos \alpha = \frac{2\sqrt{2}}{5}, \qquad \cos \beta = \frac{-\sqrt{2}}{2}, \qquad \cos \gamma = \frac{3\sqrt{2}}{10}.$$

16.2.10 Definition. Two nonzero vectors **a** and **b** are said to be **parallel** if and only if there is a scalar k such that

$$\mathbf{a} = k\mathbf{b}.$$

Thus the components of parallel vectors are proportional. It follows from 16.2.8 and 16.2.10 that if two parallel vectors point in the same direction, their direction cosines are equal; and if they point in opposite directions, the direction cosines of one are the negatives of the direction cosines of the other.

Exercises

1. For each pair of points, P_1 and P_2, given below, sketch the directed line segment $\overrightarrow{P_1P_2}$ and find the vector represented by it.
(a) P_1:(1, 2, 3) and P_2:(4, 5, 1).
(b) P_1:(−2, 1, 1) and P_2:(4, 3, 5).
(c) P_1:(−2, 2, −1) and P_2:(3, 0, 4).
(d) P_1:(0, 0, 0) and P_2:(−3, 1, 6).

2. Express the following vectors in the form $a_1\mathbf{i} + a_2\mathbf{j} + a_3\mathbf{k}$ and draw a directed line segment representing each.
(a) $[2, -1, 3]$;   (b) $[-5, 4, 0]$;
(c) $[3, 7, -4]$;   (d) $[1, -1, -1]$;
(e) $[0, 0, 9]$;   (f) $[-4, 0, -6]$.

3. Find the length of each of the following vectors.
(a) $4\mathbf{i} + \mathbf{j} + 2\mathbf{k}$;   (b) $-3\mathbf{j} - 4\mathbf{j} - 5\mathbf{k}$;
(c) $4\mathbf{i} - 2\mathbf{j} - 4\mathbf{k}$;   (d) $-\mathbf{i} - 2\mathbf{j} + 2\mathbf{k}$.

4. Find the direction cosines of each of the vectors in Exercise 2.

5. Find a unit vector having the same direction as
(a) $3\mathbf{i} + \mathbf{j} - 7\mathbf{k}$;   (b) $-2\mathbf{i} + 5\mathbf{j} - 3\mathbf{k}$.

6. Find a unit vector having the same direction cosines as
(a) $[3, -2\sqrt{3}, -2]$;   (b) $[-6, 2, 3]$.

7. Find the cosine of the angle between the vectors (c) and (d) in Exercise 3.

8. Find the angle between the vectors $4\mathbf{i} - 3\mathbf{j} - 2\mathbf{k}$ and $-\mathbf{i} + 2\mathbf{j} - 5\mathbf{k}$.

9. Find the angle between the vectors $3\mathbf{i} - \mathbf{j} - 2\mathbf{k}$ and $-6\mathbf{i} + 2\mathbf{j} + 4\mathbf{k}$.

10. Find the angle between the vectors $\frac{1}{3}\mathbf{i} + \frac{3}{5}\mathbf{j} - \frac{2}{3}\mathbf{k}$ and $5\mathbf{i} + 9\mathbf{j} - 10\mathbf{k}$.

11. Find all vectors that are perpendicular to both of the vectors $[1, -3, -2]$ and $[-3, 6, 5]$.

12. Find two vectors of length 21, each of which is perpendicular to both of the vectors $4\mathbf{i} + 3\mathbf{j} + 6\mathbf{k}$ and $-2\mathbf{i} - 3\mathbf{j} - 2\mathbf{k}$.

13. Find two perpendicular vectors **a** and **b** such that each is also perpendicular to the vector $\mathbf{c} = [-4, 2, 5]$.

14. Find two perpendicular vectors **a** and **b** such that each is also perpendicular to the vector $\mathbf{i} + 3\mathbf{j} - 2\mathbf{k}$.

15. Find the scalar projection of the vector $-4\mathbf{i} + \mathbf{j} - 2\mathbf{k}$ on the vector $\mathbf{i} + 3\mathbf{j} - 3\mathbf{k}$.

16. Find the scalar projection of the vector $2\mathbf{i} - 5\mathbf{j} + \mathbf{k}$ on the vector $-3\mathbf{i} + \mathbf{j} + 7\mathbf{k}$.

17. Find the (vector) projection of $-3\mathbf{i} + \mathbf{j} - 2\mathbf{k}$ on $2\mathbf{i} + 4\mathbf{j} - 5\mathbf{k}$.

18. Find the vector component of $\mathbf{i} + 3\mathbf{j} - 2\mathbf{k}$ along $2\mathbf{i} - 5\mathbf{j} - 2\mathbf{k}$.

19. If **a** and **b** are three-dimensional position vectors with terminal points A and B, write the position vector whose terminal point is the midpoint of the line segment AB.

20. Let **a**, **b**, and **c** be nonzero three-dimensional position vectors. Prove that **c** is coplanar with **a** and **b** if and only if there are numbers s and t such that

$$\mathbf{c} = s\mathbf{a} + t\mathbf{b}.$$

(*Hint:* Show that a position vector that is perpendicular to **a** and **b** is also perpendicular to **c**.)

21. Let **a**, **b**, **c**, and **d** be three-dimensional position vectors whose terminal points are A, B, C, and D,

respectively. Express in vector notation a necessary and sufficient condition for the figure $ABCD$ to be a parallelogram.

22. Prove the commutative law and the associative law for addition of three-dimensional vectors.

23. Find a vector $\mathbf{c}$ that bisects the angle between the nonzero three-dimensional vectors $\mathbf{a}$ and $\mathbf{b}$. [*Hint:* The angle between $\mathbf{a}$ and $\mathbf{c}$ must equal the angle between $\mathbf{b}$ and $\mathbf{c}$; and $\mathbf{a}$, $\mathbf{b}$, and $\mathbf{c}$ must be coplanar (see Exercise 20).]

24. Prove that $\mathbf{a} \cdot \mathbf{b} = \mathbf{b} \cdot \mathbf{a}$, where $\mathbf{a}$ and $\mathbf{b}$ are three-dimensional vectors.

25. If $\mathbf{a}$ and $\mathbf{b}$ are three-dimensional vectors and k is a scalar, prove that $k(\mathbf{a} + \mathbf{b}) = k\mathbf{a} + k\mathbf{b}$.

26. If $\mathbf{a}$, $\mathbf{b}$, and $\mathbf{c}$ are three-dimensional vectors, prove that $\mathbf{a} \cdot (\mathbf{b} + \mathbf{c}) = \mathbf{a} \cdot \mathbf{b} + \mathbf{a} \cdot \mathbf{c}$.

16.3 THE CROSS PRODUCT

For three-dimensional vectors (only), there is another product of two vectors called the **cross product**. Whereas the dot product of two vectors is a scalar, the cross product of two vectors is a vector. Consequently, a cross product is sometimes called a **vector product**.

16.3.1 Definition. The **cross product** (or **vector product**) of two vectors, $\mathbf{a} = [a_1, a_2, a_3]$ and $\mathbf{b} = [b_1, b_2, b_3]$, written $\mathbf{a} \times \mathbf{b}$ and read "$\mathbf{a}$ cross $\mathbf{b}$", is the vector

$$\mathbf{a} \times \mathbf{b} = [a_1, a_2, a_3] \times [b_1, b_2, b_3] = [a_2b_3 - a_3b_2, a_3b_1 - a_1b_3, a_1b_2 - a_2b_1].$$

In particular, the various cross products of the unit vectors $\mathbf{i}, \mathbf{j}$, and $\mathbf{k}$ are

$$\mathbf{i} \times \mathbf{i} = \mathbf{j} \times \mathbf{j} = \mathbf{k} \times \mathbf{k} = \mathbf{0},$$
$$\mathbf{i} \times \mathbf{j} = \mathbf{k}, \quad \mathbf{j} \times \mathbf{k} = \mathbf{i}, \quad \mathbf{k} \times \mathbf{i} = \mathbf{j},$$
$$\mathbf{j} \times \mathbf{i} = -\mathbf{k}, \quad \mathbf{k} \times \mathbf{j} = -\mathbf{i}, \quad \mathbf{i} \times \mathbf{k} = -\mathbf{j},$$

as can easily be verified.

If we express the vectors $\mathbf{a}$ and $\mathbf{b}$ in terms of $\mathbf{i}, \mathbf{j}$, and $\mathbf{k}$, so that $\mathbf{a} = a_1\mathbf{i} + a_2\mathbf{j} + a_3\mathbf{k}$ and $\mathbf{b} = b_1\mathbf{i} + b_2\mathbf{j} + b_3\mathbf{k}$, the cross product

$$\mathbf{a} \times \mathbf{b} = (a_1\mathbf{i} + a_2\mathbf{j} + a_3\mathbf{k}) \times (b_1\mathbf{i} + b_2\mathbf{j} + b_3\mathbf{k})$$

can be obtained by multiplying the two expressions in parentheses just as polynomials are multiplied:

$$\begin{aligned}
\mathbf{a} \times \mathbf{b} &= (a_1\mathbf{i} + a_2\mathbf{j} + a_3\mathbf{k}) \times (b_1\mathbf{i} + b_2\mathbf{j} + b_3\mathbf{k}) \\
&= a_1b_1(\mathbf{i} \times \mathbf{i}) + a_1b_2(\mathbf{i} \times \mathbf{j}) + a_1b_3(\mathbf{i} \times \mathbf{k}) \\
&\quad + a_2b_1(\mathbf{j} \times \mathbf{i}) + a_2b_2(\mathbf{j} \times \mathbf{j}) + a_2b_3(\mathbf{j} \times \mathbf{k}) \\
&\quad + a_3b_1(\mathbf{k} \times \mathbf{i}) + a_3b_2(\mathbf{k} \times \mathbf{j}) + a_3b_3(\mathbf{k} \times \mathbf{k}) \\
&= (a_2b_3 - a_3b_2)\mathbf{i} + (a_3b_1 - a_1b_3)\mathbf{j} + (a_1b_2 - a_2b_1)\mathbf{k}.
\end{aligned}$$

Notice that this can be written*

*Elementary properties of determinants are reviewed in Appendix A.5.

$$(1) \qquad \mathbf{a} \times \mathbf{b} = \begin{vmatrix} a_2 & a_3 \\ b_2 & b_3 \end{vmatrix} \mathbf{i} + \begin{vmatrix} a_3 & a_1 \\ b_3 & b_1 \end{vmatrix} \mathbf{j} + \begin{vmatrix} a_1 & a_2 \\ b_1 & b_2 \end{vmatrix} \mathbf{k},$$

which is often expressed as

$$\mathbf{a} \times \mathbf{b} = \begin{vmatrix} \mathbf{i} & \mathbf{j} & \mathbf{k} \\ a_1 & a_2 & a_3 \\ b_1 & b_2 & b_3 \end{vmatrix}.$$

It must be remembered that this last expression is not actually a determinant because the elements in the first row are vectors. But if it is "expanded" according to the elements of the first row, we obtain the expression (1).

Example 1. The cross product of the vectors $[2, -1, 6]$ and $[0, 4, -2]$ is the vector

$$\begin{vmatrix} \mathbf{i} & \mathbf{j} & \mathbf{k} \\ 2 & -1 & 6 \\ 0 & 4 & -2 \end{vmatrix} = \begin{vmatrix} -1 & 6 \\ 4 & -2 \end{vmatrix} \mathbf{i} + \begin{vmatrix} 6 & 2 \\ -2 & 0 \end{vmatrix} \mathbf{j} + \begin{vmatrix} 2 & -1 \\ 0 & 4 \end{vmatrix} \mathbf{k}$$

$$= -22\mathbf{i} + 4\mathbf{j} + 8\mathbf{k} = [-22, 4, 8].$$

The cross product obeys the following laws, which the student should verify.

16.3.2 Theorem. If $\mathbf{a}$, $\mathbf{b}$, and $\mathbf{c}$ are vectors and m is a scalar, then

$$\mathbf{a} \times \mathbf{b} = -(\mathbf{b} \times \mathbf{a}) \qquad \text{(anticommutative law)},$$

$$\mathbf{a} \times (\mathbf{b} + \mathbf{c}) = \mathbf{a} \times \mathbf{b} + \mathbf{a} \times \mathbf{c} \qquad \text{(distributive law)},$$

$$\mathbf{a} \times (m\mathbf{b}) = (m\mathbf{a}) \times \mathbf{b} = m(\mathbf{a} \times \mathbf{b}),$$

$$\mathbf{a} \times \mathbf{a} = 0, \qquad 0 \times \mathbf{a} = \mathbf{a} \times 0 = 0,$$

$$(\mathbf{a} \times \mathbf{b}) \cdot \mathbf{c} = \mathbf{a} \cdot (\mathbf{b} \times \mathbf{c}), \qquad \mathbf{a} \times (\mathbf{b} \times \mathbf{c}) = (\mathbf{a} \cdot \mathbf{c})\mathbf{b} - (\mathbf{a} \cdot \mathbf{b})\mathbf{c}.$$

Example 2. Show that $|\mathbf{a} \times \mathbf{b}|^2 = |\mathbf{a}|^2 |\mathbf{b}|^2 - (\mathbf{a} \cdot \mathbf{b})^2$.

Solution.

$$|\mathbf{a} \times \mathbf{b}|^2 = (a_2 b_3 - a_3 b_2)^2 + (a_3 b_1 - a_1 b_3)^2 + (a_1 b_2 - a_2 b_1)^2$$

$$= a_2^2 b_3^2 - 2a_2 a_3 b_2 b_3 + a_3^2 b_2^2 + a_3^2 b_1^2 - 2a_1 a_3 b_1 b_3$$

$$+ a_1^2 b_3^2 + a_1^2 b_2^2 - 2a_1 a_2 b_1 b_2 + a_2^2 b_1^2.$$

Also,

$$|\mathbf{a}|^2 |\mathbf{b}|^2 - (\mathbf{a} \cdot \mathbf{b})^2 = (a_1^2 + a_2^2 + a_3^2)(b_1^2 + b_2^2 + b_3^2) - (a_1 b_1 + a_2 b_2 + a_3 b_3)^2$$

$$= a_1^2 b_2^2 + a_1^2 b_3^2 + a_2^2 b_1^2 + a_2^2 b_3^2 + a_3^2 b_1^2 + a_3^2 b_2^2$$

$$- 2a_1 a_3 b_1 b_3 - 2a_2 a_3 b_2 b_3 - 2a_1 a_2 b_1 b_2.$$

Therefore

$$|\mathbf{a} \times \mathbf{b}|^2 = |\mathbf{a}|^2 |\mathbf{b}|^2 - (\mathbf{a} \cdot \mathbf{b})^2.$$

16.3.3 Theorem. If $\mathbf{a}$ and $\mathbf{b}$ are nonzero vectors and θ is the angle between them, then

$$|\mathbf{a} \times \mathbf{b}| = |\mathbf{a}||\mathbf{b}| \sin \theta.$$

Proof.   We showed in Example 2 that

(2) $$|\mathbf{a} \times \mathbf{b}|^2 = |\mathbf{a}|^2|\mathbf{b}|^2 - (\mathbf{a} \cdot \mathbf{b})^2.$$

But (from 16.2.5) $\mathbf{a} \cdot \mathbf{b} = |\mathbf{a}||\mathbf{b}| \cos \theta$; and by substituting this result in (2), we obtain

$$|\mathbf{a} \times \mathbf{b}|^2 = |\mathbf{a}|^2|\mathbf{b}|^2 - |\mathbf{a}|^2|\mathbf{b}|^2 \cos^2 \theta = |\mathbf{a}|^2|\mathbf{b}|^2(1 - \cos^2 \theta) = |\mathbf{a}|^2|\mathbf{b}|^2 \sin^2 \theta$$

or

$$|\mathbf{a} \times \mathbf{b}|^2 = |\mathbf{a}|^2|\mathbf{b}|^2 \sin^2 \theta.$$

Since $0 \leq \theta \leq \pi$, $\sin \theta \geq 0$. Therefore

$$|\mathbf{a} \times \mathbf{b}| = |\mathbf{a}||\mathbf{b}| \sin \theta. \quad \blacksquare$$

Since $\mathbf{a}$ and $\mathbf{b}$ are nonzero vectors in 16.3.3, $|\mathbf{a} \times \mathbf{b}| = 0$ if and only if $\sin \theta = 0$, —that is, if and only if $\theta = 0$ or $\theta = \pi$.

16.3.4   Corollary.   In three-space, two nonzero vectors $\mathbf{a}$ and $\mathbf{b}$ are **parallel** if and only if

$$\mathbf{a} \times \mathbf{b} = \mathbf{0}.$$

There is an interesting geometric interpretation of the number $|\mathbf{a} \times \mathbf{b}|$. When $\mathbf{a}$ and $\mathbf{b}$ have the same initial point, $|\mathbf{a} \times \mathbf{b}|$ *is equal to the area of the parallelogram having $\mathbf{a}$ and $\mathbf{b}$ as adjacent sides* (Fig. 16-16) because an altitude of the parallelogram is $|\mathbf{b}| \sin \theta$ and the length of the base is $|\mathbf{a}|$.

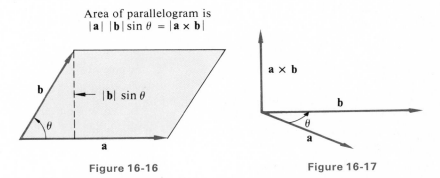

Area of parallelogram is
$|\mathbf{a}| \, |\mathbf{b}| \sin \theta = |\mathbf{a} \times \mathbf{b}|$

Figure 16-16   Figure 16-17

For a geometric interpretation of the vector $\mathbf{a} \times \mathbf{b}$, let $\mathbf{a}, \mathbf{b},$ and $\mathbf{a} \times \mathbf{b}$ have the same initial point (Fig. 16-17). Then

$$\mathbf{a} \cdot (\mathbf{a} \times \mathbf{b}) = (\mathbf{a} \times \mathbf{a}) \cdot \mathbf{b} = \mathbf{0} \cdot \mathbf{b} = 0$$

(by 16.3.2), and hence the vector $\mathbf{a}$ is perpendicular to the vector $\mathbf{a} \times \mathbf{b}$ (by 16.2.7). Similarly, $\mathbf{b}$ is perpendicular to $\mathbf{a} \times \mathbf{b}$. Thus $\mathbf{a} \times \mathbf{b}$ is a vector that is perpendicular to the plane of the vectors $\mathbf{a}$ and $\mathbf{b}$. Moreover, it can be shown that $\mathbf{a} \times \mathbf{b}$ points in the direction that a right-hand screw would move if rotated through an angle θ from $\mathbf{a}$ to $\mathbf{b}$.

16.3.5   Theorem.   If $\mathbf{a}$ and $\mathbf{b}$ are three-dimensional vectors, their cross product $\mathbf{a} \times \mathbf{b}$ is a vector that is perpendicular to both $\mathbf{a}$ and $\mathbf{b}$.

A nonzero vector **a** is said to be **normalized** when it is divided by its length. Thus the normalized vector

$$\frac{\mathbf{a}}{|\mathbf{a}|}$$

is always a unit vector having the same direction as **a** (see Section 16.2).

Example 3. Find a unit vector perpendicular to the vectors $\mathbf{a} = [2, -1, 6]$ and $\mathbf{b} = [0, 4, -2]$ given in Example 1.

Solution. The vector

$$\mathbf{a} \times \mathbf{b} = \begin{vmatrix} \mathbf{i} & \mathbf{j} & \mathbf{k} \\ 2 & -1 & 6 \\ 0 & 4 & -2 \end{vmatrix} = -22\mathbf{i} + 4\mathbf{j} + 8\mathbf{k}$$

is perpendicular to both **a** and **b**, and

$$\frac{\mathbf{a} \times \mathbf{b}}{|\mathbf{a} \times \mathbf{b}|} = \frac{-22\mathbf{i} + 4\mathbf{j} + 8\mathbf{k}}{\sqrt{(-22)^2 + 4^2 + 8^2}} = \frac{-11}{\sqrt{141}}\mathbf{i} + \frac{2}{\sqrt{141}}\mathbf{j} + \frac{4}{\sqrt{141}}\mathbf{k}$$

is a unit vector perpendicular to the plane of **a** and **b** and therefore to the vectors **a** and **b**.

Exercises

1. Let $\mathbf{a} = -3\mathbf{i} + 2\mathbf{j} - 2\mathbf{k}$, $\mathbf{b} = \mathbf{i} - 4\mathbf{j} + 3\mathbf{k}$, and $\mathbf{c} = 2\mathbf{i} - \mathbf{j} + 4\mathbf{k}$. Find
(a) $\mathbf{a} \times \mathbf{b}$; (b) $\mathbf{a} \times (\mathbf{b} + \mathbf{c})$;
(c) $\mathbf{a} \cdot (\mathbf{b} \times \mathbf{c})$; (d) $\mathbf{a} \times (\mathbf{b} \times \mathbf{c})$.

2. If $\mathbf{a} = [4, -2, 1]$, $\mathbf{b} = [-2, 3, -2]$, and $\mathbf{c} = [-1, 4, 3]$, find
(a) $\mathbf{a} \times \mathbf{b}$; (b) $\mathbf{a} \times (\mathbf{b} + \mathbf{c})$;
(c) $\mathbf{a} \cdot (\mathbf{b} \times \mathbf{c})$; (d) $\mathbf{a} \times (\mathbf{b} \times \mathbf{c})$.

3. Find all vectors that are perpendicular to both of the vectors $-2\mathbf{i} + \mathbf{j} - 4\mathbf{k}$ and $3\mathbf{i} - 4\mathbf{j} + 5\mathbf{k}$.

4. Find all vectors that are perpendicular to both of the vectors $\mathbf{i} - 4\mathbf{j} - 3\mathbf{k}$ and $4\mathbf{i} + \mathbf{j} - \mathbf{k}$.

5. Find the unit vectors that are perpendicular to the plane determined by the three points $(-1, 2, 0)$, $(5, -1, 3)$, and $(4, 0, -2)$.

6. Find the unit vectors that are perpendicular to the plane determined by the three points $(0, -1, 5)$, $(2, 2, 2)$, and $(-3, -4, 1)$.

7. Find the area of the parallelogram having $-2\mathbf{i} + \mathbf{j} + 4\mathbf{k}$ and $4\mathbf{i} - 2\mathbf{j} - 5\mathbf{k}$ as adjacent sides.

8. Find the area of the parallelogram having $2\mathbf{i} - 5\mathbf{j} + 2\mathbf{k}$ and $3\mathbf{i} - 3\mathbf{j} + 6\mathbf{k}$ as adjacent sides.

9. Show that the area of the triangle whose vertices are the terminal points of the position vectors **a**, **b**, and **c** is $\frac{1}{2}|(\mathbf{b} - \mathbf{a}) \times (\mathbf{c} - \mathbf{a})|$.

10. Find the area of the triangle whose vertices are the points $(-2, 1, 5)$, $(4, 0, 6)$, and $(3, -3, 2)$.

11. Show that $\mathbf{a} \times \mathbf{a} = \mathbf{0}$ and $\mathbf{0} \times \mathbf{a} = \mathbf{a} \times \mathbf{0} = \mathbf{0}$.

12. Prove the distributive law for cross products: $\mathbf{a} \times (\mathbf{b} + \mathbf{c}) = \mathbf{a} \times \mathbf{b} + \mathbf{a} \times \mathbf{c}$.

13. Prove the anticommutative law for the cross product of two vectors, namely $\mathbf{a} \times \mathbf{b} = -(\mathbf{b} \times \mathbf{a})$

14. Prove that $\mathbf{a} \times (\mathbf{b} \times \mathbf{c}) = (\mathbf{a} \cdot \mathbf{c})\mathbf{b} - (\mathbf{a} \cdot \mathbf{b})\mathbf{c}$.

15. Show that if **a**, **b**, and **c** are three-dimensional vectors having the same initial point, then $(\mathbf{b} - \mathbf{a}) \times (\mathbf{c} - \mathbf{a}) = (\mathbf{a} \times \mathbf{b}) + (\mathbf{b} \times \mathbf{c}) + (\mathbf{c} \times \mathbf{a})$.

16. Show that if **a**, **b**, and **c** are three-dimensional vectors having the same initial point, and with terminal points A, B, and C, then the vector $(\mathbf{a} \times \mathbf{b}) + (\mathbf{b} \times \mathbf{c}) + (\mathbf{c} \times \mathbf{a})$ is perpendicular to the plane determined by A, B, and C.

17. Prove that $\mathbf{a} \times (\mathbf{b} \times \mathbf{c}) + \mathbf{b} \times (\mathbf{c} \times \mathbf{a}) + \mathbf{c} \times (\mathbf{a} \times \mathbf{b}) = \mathbf{0}$.

18. The number $\mathbf{a} \cdot (\mathbf{b} \times \mathbf{c})$ is called the **scalar triple product** of the vectors $\mathbf{a}$, $\mathbf{b}$, and $\mathbf{c}$. Show that if $\mathbf{a}$, $\mathbf{b}$, and $\mathbf{c}$ are vectors having the same initial point, then the magnitude of the scalar triple product $\mathbf{a} \cdot (\mathbf{b} \times \mathbf{c})$ gives the volume of the parallelepiped having $\mathbf{a}$, $\mathbf{b}$, and $\mathbf{c}$ as edges. (*Hint:* Call the parallelogram having $\mathbf{b}$ and $\mathbf{c}$ for adjacent edges the base of the parallelepiped; its area is $|\mathbf{b} \times \mathbf{c}|$ (see Fig. 16-16). Use 16.2.6, 16.3.5, and Fig. 16-13 to show that the altitude of the parallelepiped is $\mathbf{a} \cdot (\mathbf{b} \times \mathbf{c})/|\mathbf{b} \times \mathbf{c}|$.)

19. If $\mathbf{a}$ and $\mathbf{b}$ are nonzero three-dimensional vectors, show that $\mathbf{a} \cdot (\mathbf{a} \times \mathbf{b}) = \mathbf{0}$.

20. Show that
$$\mathbf{a} \cdot (\mathbf{b} \times \mathbf{c}) = \begin{vmatrix} a_1 & a_2 & a_3 \\ b_1 & b_2 & b_3 \\ c_1 & c_2 & c_3 \end{vmatrix}.$$

21. Find the volume of the parallelepiped having $-2\mathbf{i} + 3\mathbf{j} + \mathbf{k}$, $4\mathbf{i} - \mathbf{j} + 3\mathbf{k}$, and $3\mathbf{i} + 2\mathbf{j} - \mathbf{k}$ as adjacent edges. (*Hint:* Use Exercise 20.)

22. Find the volume of the parallelepiped having $\mathbf{i} - \mathbf{j} + 6\mathbf{k}$, $2\mathbf{i} + 4\mathbf{j} - \mathbf{k}$, and $5\mathbf{i} - 3\mathbf{j} + 2\mathbf{k}$ as adjacent edges.

23. Show that three nonzero vectors $\mathbf{a}$, $\mathbf{b}$, and $\mathbf{c}$, having the same initial point, are coplanar if and only if
$$\begin{vmatrix} a_1 & a_2 & a_3 \\ b_1 & b_2 & b_3 \\ c_1 & c_2 & c_3 \end{vmatrix} = 0.$$

Note. This condition is equivalent to the three vectors being linearly dependent, that is, that each can be expressed as a linear combination of the other two. Compare with Exercise 20, Section 16.2.

24. Find all position vectors coplanar with $-4\mathbf{i} + 3\mathbf{j} - 6\mathbf{k}$ and $7\mathbf{i} - 5\mathbf{j} + 2\mathbf{k}$.

25. Show that
$$(\mathbf{a} \times \mathbf{b}) \cdot (\mathbf{c} \times \mathbf{d}) = \begin{vmatrix} \mathbf{a} \cdot \mathbf{c} & \mathbf{b} \cdot \mathbf{c} \\ \mathbf{a} \cdot \mathbf{d} & \mathbf{b} \cdot \mathbf{d} \end{vmatrix}.$$

16.4 PLANES IN THREE-DIMENSIONAL SPACE

Euclid did not defined a plane. However, he deduced many properties of planes, some of which are sufficient to characterize the concept completely.

Let P_1 be a point in three-space and $\mathbf{n}$ a nonzero vector. The set of points P such that the vector $\overrightarrow{P_1 P}$ is perpendicular to $\mathbf{n}$ is the **plane** containing P_1 and perpendicular to $\mathbf{n}$ (Fig. 16-18). Since every plane contains a point and is perpendicular to some vector, every plane can be characterized in this way.

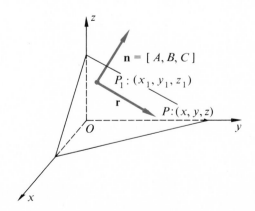

Figure 16-18

16.4.1 Theorem. The **vector equation of the plane** that passes through the point $P_1:(x_1, y_1, z_1)$ and is perpendicular to the nonzero vector $\mathbf{n} = [A, B, C]$ is

$$\mathbf{n} \cdot \mathbf{r} = 0,$$

where $\mathbf{r} = [x - x_1, y - y_1, z - z_1]$ is the vector represented by the directed line segment from P_1 to an arbitrary point $P:(x, y, z)$ in the plane.

Proof. Consider the plane that passes through the fixed point $P_1:(x_1, y_1, z_1)$ and is perpendicular to the vector $\mathbf{n} = [A, B, C]$ (Fig. 16-18). Then a point $P:(x, y, z)$ is on the plane if and only if the vector $\mathbf{r} = [x - x_1, y - y_1, z - z_1]$, whose initial point is P_1 and whose terminal point is P, is perpendicular to the vector $\mathbf{n}$; that is, P is on the plane if and only if

$$\mathbf{n} \cdot \mathbf{r} = 0.$$

If $P = P_1$, the vector $\mathbf{r}$ becomes the zero vector, which, having no assigned direction, can be included in the vectors that are perpendicular to $\mathbf{n}$. ▌

The foregoing vector equation $\mathbf{n} \cdot \mathbf{r} = 0$ can be written

$$[A, B, C] \cdot [x - x_1, y - y_1, z - z_1] = 0,$$

which is

$$A(x - x_1) + B(y - y_1) + C(z - z_1) = 0,$$

a Cartesian equation of the same plane.

16.4.2 Corollary. A Cartesian equation of the plane that passes through the point $P_1:(x_1, y_1, z_1)$ and is perpendicular to the nonzero vector $\mathbf{n} = [A, B, C]$ is

$$A(x - x_1) + B(y - y_1) + C(z - z_1) = 0,$$

or

$$Ax + By + Cz + D = 0,$$

where $D = -(Ax_1 + By_1 + Cz_1)$ is a constant.

As an illustration, the plane through the point $(4, -3, 1)$ and perpendicular to the vector $[2, 5, -6]$ has the Cartesian equation $2(x - 4) + 5(y + 3) - 6(z - 1) = 0$, or $2x + 5y - 6z + 13 = 0$.

The equation

$$(1) \qquad\qquad Ax + By + Cz + D = 0$$

is called the **general form** of a Cartesian equation of a plane. It is important to notice that *the coefficients A, B, and C in (1) are the components of a vector $\mathbf{n} = [A, B, C]$ that is normal* (perpendicular) *to the plane.*

Example 1. The line through the origin that is perpendicular to a certain plane intersects the plane in the point $Q:(3, -1, 5)$. Find a Cartesian equation of the plane.

Solution. From 16.2.2, the vector represented by $\overrightarrow{OQ}$ is $[3 - 0, -1 - 0, 5 - 0] = [3, -1, 5]$. Therefore (16.4.2) a Cartesian equation of the plane through the point $Q:(3, -1, 5)$ and perpendicular to the vector $[3, -1, 5]$ is

$$3(x - 3) - (y + 1) + 5(z - 5) = 0$$

or

$$3x - y + 5z - 35 = 0.$$

16.4.3 Definition. A point is on the graph of a Cartesian equation (or a set of simultaneous Cartesian equations) if and only if its Cartesian coordinates satisfy the equation (or equations).

Earlier we saw that every plane has a Cartesian equation of the first degree in in x, y, and z. We shall now show that, conversely, in three-space the Cartesian graph of every first-degree equation is a plane.

The most general equation of the first degree in three-dimensional Cartesian coordinates is

(2) $$Ax + By + Cz + D = 0,$$

where A, B, and C are not all zero. Let $P_1:(x_1, y_1, z_1)$ be a point whose coordinates satisfy (2); that is, let

(3) $$Ax_1 + By_1 + Cz_1 + D = 0.$$

If we subtract equation (3) from equation (2), member by member, we obtain

(4) $$A(x - x_1) + B(y - y_1) + C(z - z_1) = 0.$$

But (by 16.4.2) the Cartesian graph of (4) is the plane containing P_1 and perpendicular to the vector $[A, B, C]$. Since the preceding steps are reversible, the graph of (2) is the same as the graph of (4).

16.4.4 Theorem. In three-space, every plane has a Cartesian equation of the first degree, and the graph of every first-degree equation in Cartesian coordinates is a plane.

Notice that the yz plane has the equation $x = 0$. For the yz plane contains the origin $O:(0, 0, 0)$ and is perpendicular to the basis vector $\mathbf{i} = [1, 0, 0]$. Therefore, by 16.4.1, the vector equation of the yz plane is $[1, 0, 0] \cdot [x, y, z] = 0$, from which the Cartesian equation is $x = 0$. Similarly for the other coordinate planes.

16.4.5 Corollary. The Cartesian equation of the yz plane is $x = 0$, the equation of the zx plane is $y = 0$, and the equation of the xy plane is $z = 0$.

It follows that every point on the x axis has zero for its y coordinate and zero for its z coordinate, and vice versa. Thus $(-2, 0, 0)$ and $(5, 0, 0)$ are points on the x axis. Analogous statements hold for points on the other coordinate axes.

To illustrate, if we substitute zero for y and zero for z in the equation $3x + 4y + 2z - 12 = 0$, we get $x = 4$; thus the point of intersection of this plane with the x

axis is (4, 0, 0). Similarly, its point of intersection with the y axis is (0, 3, 0), and its intersection with the z axis is (0, 0, 6) (Fig. 16-19).

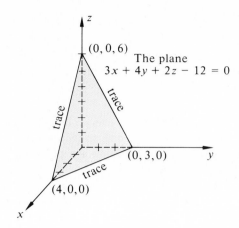

Figure 16-19

The lines of intersection of a plane with the coordinate planes are called its **traces**. Usually the easiest way to sketch a plane from its Cartesian equation is to draw its traces. To draw the traces of a plane, not through the origin and not parallel to a coordinate axis, simply find the points of intersection of the plane with the coordinate axes and connect them by straight lines. This is shown in Fig. 16-19 for the plane $3x + 4y + 2z - 12 = 0$.

Two planes are **parallel** if and only if a normal vector to one is parallel to a normal vector of the other (Fig. 16-20); and two vectors are parallel if and only if their respective components are proportional. Since the coefficients A, B, and C in the Cartesian equation $Ax + By + Cz + D = 0$ of a plane are the components of the vector $[A, B, C]$ that is normal to the plane, we have the following theorem.

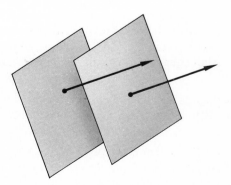

Figure 16-20

16.4.6 Theorem. Two planes

$$A_1x + B_1y + C_1z + D_1 = 0$$

and

$$A_2x + B_2y + C_2z + D_2 = 0$$

are **parallel** if and only if A_1, B_1, C_1 are proportional to A_2, B_2, C_2.

For example, the planes $2x - 3y + z - 1 = 0$ and $4x - 6y + 2z + 4 = 0$ are parallel. But the equations $2x - 3y + z - 1 = 0$ and $4x - 6y + 2z - 2 = 0$ represent the same plane, not two parallel planes, since the second equation can be obtained by multiplying both members of the first by 2.

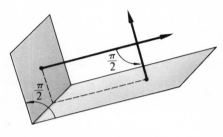

Figure 16-21

Two planes are **perpendicular** if each normal vector to the first plane is perpendicular to every normal vector to the second plane, and vice versa (Fig. 16-21). Moreover, we saw that the coefficients A, B, and C in the general form of the Cartesian equation of a plane, $Ax + By + Cz + D = 0$, are the components of a vector $\mathbf{n} = [A, B, C]$ that is normal (perpendicular) to the plane. From this, and from 16.2.7, we have the following theorem.

16.4.7 Theorem. Two planes

$$A_1x + B_1y + C_1z + D_1 = 0$$

and

$$A_2x + B_2y + C_2z + D_2 = 0$$

are **perpendicular** if and only if

$$A_1A_2 + B_1B_2 + C_1C_2 = 0.$$

Thus the planes $4x - 2y + z + 7 = 0$ and $x + 3y + 2z - 1 = 0$ are perpendicular because $4(1) - 2(3) + 1(2) = 0$.

Since the equation of the xy plane is $z = 0$, the plane $Ax + By + Cz + D = 0$ is perpendicular to the xy plane if and only if $A(0) + B(0) + C(1) = 0$—that is, if and only if $C = 0$. Similar statements are true for planes perpendicular to the yz or zx plane.

16.4.8 Corollary. A plane is perpendicular to the xy plane if and only if its Cartesian equation contains no term in z; it is perpendicular to the yz plane if and only if its equation contains no term in x; it is perpendicular to the zx plane if and only if its equation contains no term in y.

As illustrations, $2x + 3y - 7 = 0$ is perpendicular to the xy plane, and the plane $y - 2z + 1 = 0$ is perpendicular to the yz plane. The plane $y = 7$ is perpendicular to the yz plane because it contains no term in x, and it is also perpendicular to the xy plane because it contains no term in z. Thus it is parallel to the zx plane.

Example 2. Find the direction cosines of all vectors that are normal to the plane

$$4x + 3y - 5z + 6 = 0.$$

Solution. One vector that is normal to this plane is $\mathbf{n} = [4, 3, -5]$. Its direction cosines are

$$\cos \alpha = \frac{4}{\pm 5\sqrt{2}}, \qquad \cos \beta = \frac{3}{\pm 5\sqrt{2}}, \qquad \cos \gamma = \frac{-1}{\pm \sqrt{2}}.$$

All vectors normal to the plane (and thus parallel to $\mathbf{n}$) have these direction cosines. The positive sign before the radical applies to those that have the same direction as $\mathbf{n}$; the negative sign applies to those that point in the opposite direction.

Example 3. Find a vector equation of the plane whose Cartesian equation is $4x + 2y - 7z + 10 = 0$.

Solution. The coefficients of x, y. and z in the given Cartesian equation are the components of a vector $\mathbf{n} = [4, 2, -7]$ that is normal to the plane. To use 16.4.1, we must find the coordinates of some point P_1 on the plane. An easy point to find is the point in which the given plane intersects a coordinate axis because two of its coordinates will be zero. The x and z coordinates of any point on the y axis are zero; so if we substitute zero for x and zero for z in the given equation, $4x + 2y - 7z + 10 = 0$, we obtain $y = -5$. Thus P_1:$(0, -5, 0)$ is a point on the plane.

From 16.4.1 the vector equation of the plane that passes through the point P_1:$(0, -5, 0)$ and is perpendicular to the vector $\mathbf{n} = [4, 2, -7]$ is

$$\mathbf{n} \cdot \mathbf{r} = 0,$$

where $\mathbf{r} = [x - 0, y - (-5), z - 0] = [x, y + 5, z]$ is the vector represented by the directed line segment from P_1 to any point P:(x, y, z) in the plane.

Example 4. A plane through the point P_1:$(2, 3, 2)$ is perpendicular to the line segment joining points Q:$(0, 5, -4)$ and R:$(6, 9, 8)$.
    (a)   Find a vector equation of the plane.
    (b)   Write a Cartesian equation of the plane.
    (c)   Sketch the plane by drawing its traces, and sketch the given line segment.

Solution. The vector represented by the directed line segment $\overrightarrow{QR}$, from the point Q:$(0, 5, -4)$ to the point R:$(6, 9, 8)$, is $\mathbf{n} = [6 - 0, 9 - 5, 8 - (-4)] = [6, 4, 12]$ (by 16.2.2).

From 16.4.1, the vector equation of the plane containing the point P_1:$(2, 3, 2)$ and perpendicular to $\mathbf{n} = [6, 4, 12]$ is

$$\mathbf{n} \cdot \mathbf{r} = 0,$$

where $\mathbf{r} = [x - 2, y - 3, z - 2]$ is the vector represented by the directed line segment from P_1 to any point P:(x, y, z) in the plane.

Since $\mathbf{n} \cdot \mathbf{r} = [6, 4, 12] \cdot [x - 2, y - 3, z - 2] = 6(x - 2) + 4(y - 3) + 12(z - 2) = 0$, a Cartesian equation of the plane is

$$3(x - 2) + 2(y - 3) + 6(z - 2) = 0,$$

or

$$3x + 2y + 6z - 24 = 0.$$

To draw the traces of this plane, we find its points of intersection with the coordinate axes and connect them by lines. Every point on the x axis has zero for its y coordinate and zero for its z coordinate. To find the x coordinate of the point of intersection of the plane and the x axis, we substitute 0 for y and 0 for z in the equation $3x + 2y + 6z - 24 = 0$ and obtain $x = 8$. Thus the plane intersects the x axis at $(8, 0, 0)$. Similarly, the plane intersects the y axis at $(0, 12, 0)$ and the z axis at $(0, 0, 4)$. The plane and its traces are shown in Fig. 16-22.

To draw the given line segment $\overrightarrow{QR}$, perpendicular to the plane, we plot the points Q:$(0, 5, -4)$ and R:$(6, 9, 8)$ and connect them by a line segment (Fig. 16-22).

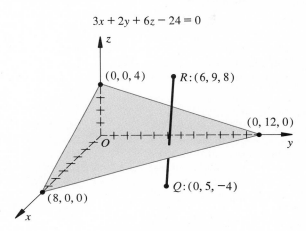

$$3x + 2y + 6z - 24 = 0$$

Figure 16-22

Example 5. Find the Cartesian equation of the plane through the three points P:$(2, -1, 4)$, Q:$(1, 3, -2)$, and R:$(-3, 1, 2)$.

Solution. The directed line segments $\overrightarrow{PQ}$ and $\overrightarrow{PR}$ lie in the plane and (by 16.2.2) represent the vectors

$$\mathbf{a} = [-1, 4, -6] \quad \text{and} \quad \mathbf{b} = [-5, 2, -2].$$

From 16.3.5, the cross product $\mathbf{a} \times \mathbf{b}$ is a vector that is perpendicular to both $\mathbf{a}$ and $\mathbf{b}$ and thus to the plane of $\overrightarrow{PQ}$ and $\overrightarrow{PR}$. Since

$$\mathbf{a} \times \mathbf{b} = \begin{vmatrix} 4 & -6 \\ 2 & -2 \end{vmatrix} \mathbf{i} + \begin{vmatrix} -6 & -1 \\ -2 & -5 \end{vmatrix} \mathbf{j} + \begin{vmatrix} -1 & 4 \\ -5 & 2 \end{vmatrix} \mathbf{k}$$

$$= 4\mathbf{i} + 28\mathbf{j} + 18\mathbf{k},$$

the vector

$$\mathbf{n} = [4, 28, 18]$$

is normal to the plane of P, Q, and R.

Therefore (16.4.2) a Cartesian equation of the plane through the point P:$(2, -1, 4)$ and perpendicular to the vector $\mathbf{n} = [4, 28, 18]$ is

$$4(x - 2) + 28(y + 1) + 18(z - 4) = 0,$$

or

$$2x + 14y + 9z - 26 = 0.$$

The reader should verify that the coordinates of each of the three given points satisfy this equation.

Exercises

1. The line through the origin, perpendicular to a plane, intersects the plane in the point $(3, -1, 2)$.
(a) Find a vector equation of the plane.
(b) Write a Cartesian equation of the plane.
(c) Sketch the plane by drawing its traces.

2. The line through the origin, perpendicular to a plane, intersects the plane in the point $(-5, 3, -4)$.
(a) Find a vector equation of the plane.
(b) Write a Cartesian equation of the plane.
(c) Sketch the plane by drawing its traces.

3. A plane through the point $(4, 2, 5)$ is perpendicular to the line segment that joins the points $(-2, 3, 6)$ and $(7, -4, 1)$.
(a) Find a vector equation of the plane.
(b) Write a Cartesian equation of the plane.
(c) Sketch the plane by drawing its traces, and sketch the given line segment.

4. A plane through the point $(-1, 1, 6)$ is perpendicular to the line segment which joins the points $(3, 2, -4)$ and $(-2, 5, 1)$.
(a) Write a vector equation of the plane.
(b) Find a Cartesian equation of the plane.
(c) Sketch the plane by drawing its traces and sketch the given line.

5. A line through the point $(2, -3, 7)$, perpendicular to a certain plane, intersects that plane in the point $(-3, 2, 0)$.
(a) Find a vector equation of the plane.
(b) Find a Cartesian equation of the plane.
(c) Sketch the plane by drawing its traces and sketch the given line.

6. A line through the point $(-7, 5, 2)$, perpendicular to a certain plane, intersects that plane in the point $(-4, -2, 6)$.
(a) Find a vector equation of the plane.
(b) Write a Cartesian equation of the plane.
(c) Sketch the plane by drawing its traces, and sketch the given line.

7. Find a vector equation of the plane whose Cartesian equation is $2x - 6y + z + 3 = 0$.

8. Find a vector equation of the plane whose Cartesian equation is $5x + 2y - 2z - 6 = 0$.

9. Which of the following planes are identical, which are parallel to each other, and which are perpendicular to each other?
(a) $x - 2y + z - 4 = 0$;
(b) $4x + 3y + 2z + 1 = 0$;
(c) $-8x - 6y - 4z + 1 = 0$;
(d) $3x - 6y + 3z - 12 = 0$.

10. Which of the following planes are identical, which are parallel, and which are perpendicular?
(a) $2x + y + 2z - 6 = 0$;
(b) $2x + 2y - 3z + 1 = 0$;
(c) $-6x - 6y + 9z - 3 = 0$;
(d) $-6x - 3y - 6z + 7 = 0$.

11. Write the equation of the plane through the point $(4, -3, 6)$ that is
(a) parallel to the yz plane;
(b) perpendicular to the y axis;
(c) parallel to both the x and y axes.

12. Write the equation of the plane through the point $(-2, 5, -4)$ that is
(a) perpendicular to the x axis;
(b) parallel to the zx plane;
(c) perpendicular to the z axis.

13. Find the equation of the plane through the origin that is parallel to the plane $5x - 7y + 12z - 14 = 0$. Make a sketch.

14. Find the equation of the plane through the origin that is parallel to the plane $x + 2y - 6z - 12 = 0$. Make a sketch.

15. Find the equation of the plane through the point $(-2, 1, 5)$ and parallel to the plane $2x - 3y + z + 6 = 0$. Sketch both planes.

16. Find the equation of the plane through the point $(0, -4, -6)$ and parallel to the plane $3x + 7y - 6z + 20 = 0$. Sketch both planes.

17. Find the value of B if the plane $x + By - 4z + 1 = 0$ is perpendicular to the plane $7x + y + 3z + 6 = 0$.

18. Find the value of C if the plane $3x - y + Cz + 10 = 0$ is perpendicular to the plane $5x + 6y + 3z - 5 = 0$.

19. Find the equation of the plane through the three points $(5, 2, 3)$, $(-7, -3, 7)$, and $(2, 1, 5)$.

20. Find the equation of the plane through the three points $(-7, -2, 1)$, $(5, 4, -1)$, and $(3, 1, 4)$.

21. Find the equation of the plane that is perpendicular to both of the planes $x - 3y + 2z - 7 = 0$ and $2x - 2y - z + 3 = 0$, and that passes through the point $(-1, -2, 3)$.

22. Find the equation of the plane through the point $(6, 2, -1)$ and perpendicular to the line of intersection of the planes $4x - 3y + 2z + 5 = 0$ and $3x + 2y - z + 11 = 0$.

23. Find the equation of the plane through the point $(-3, 1, 4)$ and perpendicular to the trace of the plane $x - 3y + 7z - 3 = 0$ in the xy plane.

24. Find the equation of the plane containing the points $(2, -3, 1)$ and $(4, 1, 5)$, and parallel to the line through the points $(1, -2, 2)$ end $(3, 0, 1)$.

25. Show that the plane through the terminal point of the vector $a_1\mathbf{i} + a_2\mathbf{j} + a_3\mathbf{k}$ and parallel to the plane containing the vectors $b_1\mathbf{i} + b_2\mathbf{j} + b_3\mathbf{k}$ and $c_1\mathbf{i} + c_2\mathbf{j} + c_3\mathbf{k}$ has the vector equation

$$(\mathbf{r} - \mathbf{a}) \cdot (\mathbf{b} \times \mathbf{c}) = 0,$$

where $\mathbf{r} = x\mathbf{i} + y\mathbf{j} + z\mathbf{k}$. (*Hint:* Use Exercise 21 of Section 16.3.)

26. Use Exercise 25 to find the Cartesian equation of the plane through the point $(2, -3, 2)$ and parallel to the plane of the vectors $4\mathbf{i} + 3\mathbf{j} - \mathbf{k}$ and $2\mathbf{i} - 5\mathbf{j} + 6\mathbf{k}$. Make a sketch.

27. Let $\mathbf{a}$ and $\mathbf{b}$ be nonzero vectors having the same initial point and let $\mathbf{c}$ be any nonzero vector. Show that $(\mathbf{a} \times \mathbf{b}) \times \mathbf{c}$ is a vector parallel to the plane of $\mathbf{a}$ and $\mathbf{b}$. How is the vector $(\mathbf{a} \times \mathbf{b}) \times \mathbf{c}$ related to $\mathbf{c}$?

28. Find a vector that is perpendicular to the vector $2\mathbf{i} - 5\mathbf{j} + \mathbf{k}$ and parallel to the plane of the position vectors $-3\mathbf{i} + \mathbf{j} - 2\mathbf{k}$ and $4\mathbf{i} - 2\mathbf{j} + 3\mathbf{k}$.

29. Let $\mathbf{a}$ and $\mathbf{b}$ be vectors in a plane π_1 and let $\mathbf{c}$ and $\mathbf{d}$ be vectors in another plane π_2. Show that $(\mathbf{a} \times \mathbf{b}) \times (\mathbf{c} \times \mathbf{d})$ is a vector that is perpendicular to the line of intersection of π_1 and π_2.

30. Let the vectors $2\mathbf{i} - \mathbf{j} + \mathbf{k}$ and $-3\mathbf{i} + \mathbf{j} + 2\mathbf{k}$ lie in a plane π_1 and let the vectors $4\mathbf{i} + 2\mathbf{j}$ and $-\mathbf{i} - 3\mathbf{j} + 5\mathbf{k}$ lie in a plane π_2. Find a vector that is perpendicular to the line of intersection of π_1 and π_2.

16.5 LINES IN THREE-SPACE

16.5.1 Definition. Let $\mathbf{a}$ be a nonzero vector and let P_1 be an arbitrarily chosen, but fixed, point in three-space. The set of points P, such that the directed line segment $\overrightarrow{P_1P}$ represents the vector $t\mathbf{a}$ for some number t, is called the **line** through P_1 that is parallel to $\mathbf{a}$ (Fig. 16-23).

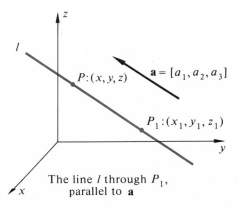

The line l through P_1, parallel to **a**

Figure 16-23

We shall denote this line by l.

Let $\mathbf{a} = [a_1, a_2, a_3]$. The directed line segment whose fixed initial point is $P_1:(x_1, y_1, z_1)$ and whose terminal point is $P:(x, y, z)$ represents the vector $[x - x_1, y - y_1, z - z_1]$. This vector is equal to $t\mathbf{a} = [ta_1, ta_2, ta_3]$ if and only if $x - x_1 = ta_1$, $y - y_1 = ta_2$, and $z - z_1 = ta_3$, or

$$x = ta_1 + x_1, \qquad y = ta_2 + y_1, \qquad z = ta_3 + z_1, \quad t \in R.$$

These equations are called **parametric equations** of the line l, and t is the parameter.

16.5.2 Theorem. Parametric equations of the line through the point $P_1:(x_1, y_1, z_1)$ and parallel to the nonzero vector $\mathbf{a} = [a_1, a_2, a_3]$ are

(1) $$x = ta_1 + x_1, \qquad y = ta_2 + y_1, \qquad z = ta_3 + z_1, \quad t \in R.$$

To draw a line whose parametric equations are known, find two points on it by assigning any two convenient numbers to the parameter, and then draw the line through the two points. For example, if the line has parametric equations

$$x = t + 3, \qquad y = 4t - 2, \qquad z = -3t + 4,$$

$t = 0$ gives the point $(3, -2, 4)$, and $t = 2$ gives the point $(5, 6, -2)$. We plot these points and draw the straight line through them (Fig. 16-24).

If none of a_1, a_2, a_3 is zero, we can solve each of the equations in (1) for t:

$$t = \frac{x - x_1}{a_1}, \qquad t = \frac{y - y_1}{a_2}, \qquad t = \frac{z - z_1}{a_3}.$$

By equating the right members of these equations in pairs, we obtain

(2) $$\frac{x - x_1}{a_1} = \frac{y - y_1}{a_2} \quad \text{and} \quad \frac{y - y_1}{a_2} = \frac{z - z_1}{a_3},$$

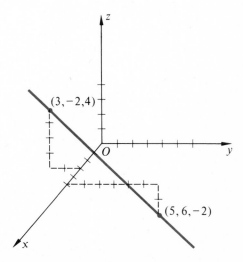

Figure 16-24

which are customarily written

$$\frac{x - x_1}{a_1} = \frac{y - y_1}{a_2} = \frac{z - z_1}{a_3}.$$

16.5.3 Theorem. Symmetric equations of the line through the point $P_1{:}(x_1, y_1, z_1)$ and parallel to the vector $\mathbf{a} = [a_1, a_2, a_3]$ are

(3) $$\frac{x - x_1}{a_1} = \frac{y - y_1}{a_2} = \frac{z - z_1}{a_3},$$

provided that a_1, a_2, and a_3 are all different from zero.

If one of a_1, a_2, a_3 is zero, say $a_3 = 0$, the symmetric equations of l are

$$\frac{x - x_1}{a_1} = \frac{y - y_1}{a_2}, \qquad z = z_3;$$

here the line lies in the horizontal plane $z = z_3$.

A vector is **parallel** to a line if some directed segment of the line represents the vector (Fig. 16-25).

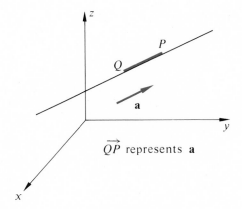

$\overrightarrow{QP}$ represents $\mathbf{a}$

Figure 16-25

The components of any nonzero vector that is parallel to a given line are **direction numbers of the line.** Thus a line has indefinitely many sets of direction numbers, but each set is proportional to every other. In particular, *the line l, with symmetric equations* (3), *has direction numbers* a_1, a_2, a_3.

16.5.4 Corollary. Distinct lines having proportional direction numbers are parallel, and conversely.

By 16.2.7, two nonzero vectors $[a_1, a_2, a_3]$ and $[b_1, b_2, b_3]$ are perpendicular if and only if $a_1b_1 + a_2b_2 + a_3b_3 = 0$. Since the components of a nonzero vector are direction numbers of any line parallel to it, we have the following corollary.

16.5.5 Corollary. Two lines having direction numbers a_1, a_2, a_3 and b_1, b_2, b_3, respectively, are perpendicular if and only if

$$a_1b_1 + a_2b_2 + a_3b_3 = 0.$$

A line is considered undirected unless a positive direction is assigned to it. By the **direction cosines** of a *directed* line is meant the direction cosines of any vector having the same direction as the line. Of course, a directed line has only one set of direction cosines.

Example 1. Find symmetric equations and parametric equations of the line that is parallel to the vector $[4, -3, 2]$ and that goes through the point $(2, 5, 1)$. If the line is assigned a positive direction that is the same as the direction of the given vector, find the direction cosines of the line.

Solution. Direction numbers of the line are $4, -3, 2$. By substituting these numbers for a_1, a_2, a_3 in the symmetric equations (3) and substituting $2, 5, 1$ for x_1, y_1, z_1, we find symmetric equations of the line to be

$$\frac{x - 2}{4} = \frac{y - 5}{-3} = \frac{z - 1}{2}.$$

If we set each member of these equations equal to t and then solve for x, y, and z, we obtain the parametric equations

$$x = 4t + 2, \qquad y = -3t + 5, \qquad z = 2t + 1, \quad t \in R.$$

The direction cosines of the line are

$$\frac{4}{\sqrt{29}}, \qquad \frac{-3}{\sqrt{29}}, \qquad \frac{2}{\sqrt{29}}.$$

They are found by dividing the direction numbers $4, -3, 2$ by $\sqrt{(4)^2 + (-3)^2 + (2)^2} = \sqrt{29}$ (see 16.2.8).

There are, of course, infinitely many planes through a line, and any two of them will determine the line. However, those planes through a line that are perpendicular to a coordinate plane have the simplest equations because each lacks a variable (16.4.8).

The planes through a line that are perpendicular to a coordinate plane are called the **projecting planes** of the line.

In general, a line has three projecting planes (Fig. 16-26). But if the line is parallel to a coordinate plane, two of the projecting planes coincide.

Three equations can be formed by equating the three members of the symmetric equations (3) in pairs, but they are equivalent to two independent equations, such as those in (2). Since the equation

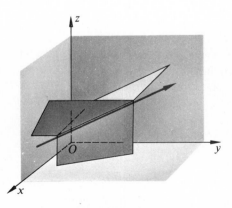

Figure 16-26

(4)
$$\frac{x - x_1}{a_1} = \frac{y - y_1}{a_2}$$

is of the first degree in the variables x, y, z but

contains no term in z, it *separately* represents a plane perpendicular to the xy plane. Moreover, it contains l. Therefore (4) is a projecting plane of l. Similarly, the other symmetric equations *separately* give the other projecting planes of l.

The equations of the projecting planes of the line through the point $P_1:(x_1, y_1, z_1)$, with nonzero direction numbers a_1, a_2, a_3, are

$$\frac{x - x_1}{a_1} = \frac{y - y_1}{a_2}, \qquad \frac{y - y_1}{a_2} = \frac{z - z_1}{a_3}, \qquad \frac{z - z_1}{a_3} = \frac{x - x_1}{a_1}.$$

The first projecting plane is perpendicular to the xy plane, the second is perpendicular to the yz plane, and the third is perpendicular to the zx plane.

Any two nonparallel planes intersect in a line.

16.5.6 Definition. The set of points whose coordinates satisfy the equations

$$A_1 x + B_1 y + C_1 z + D_1 = 0 \qquad \text{and} \qquad A_2 x + B_2 y + C_2 z + D_2 = 0$$

simultaneously, is the **line of intersection** of the planes $A_1 x + B_1 y + C_1 z + D_1 = 0$ and $A_2 x + B_2 y + C_2 z + D_2 = 0$. It is assumed that A_1, B_1, C_1 are not proportional to A_2, B_2, C_2.

The simultaneous equations in 16.5.6 are called **general equations** of the line.

The **piercing points** of a line are the points in which it intersects the coordinate planes. To draw a line whose general equations are given, find the coordinates of two piercing points and connect them by a straight line. Any other points on the line would have served equally well, but piercing points are easier to find.

Example 2. Draw the line having general equations

$$2x - y - 5z + 14 = 0 \qquad \text{and} \qquad 4x + 5y + 4z - 28 = 0.$$

Then find parametric equations of the line, symmetric equations of the line, and its projecting planes.

Solution. The piercing point of the given line in the yz plane is $(0, 4, 2)$, found by solving simultaneously with the given equations of the line the equation $x = 0$ of the yz plane. Similarly, the piercing point in the zx plane is $(3, 0, 4)$. The line l joining these two points is the desired line (Fig. 16-27).

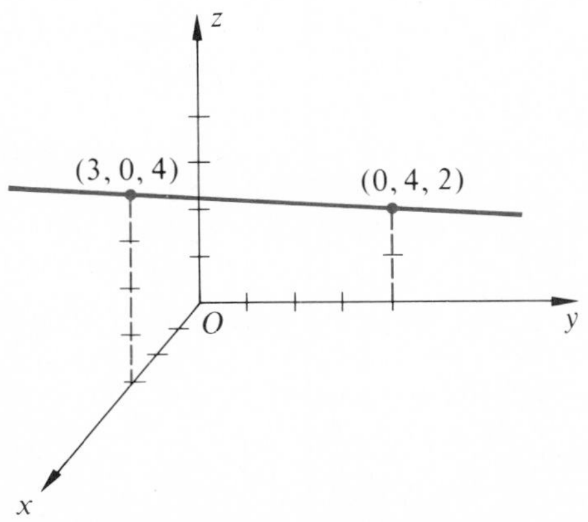

Since the points $(0, 4, 2)$ and $(3, 0, 4)$ are on the line, a vector parallel to the line is $[3 - 0, 0 - 4, 4 - 2]$, or $[3, -4, 2]$. Therefore $3, -4, 2$ are direction numbers of the line. If we substitute these direction numbers for a_1, a_2, a_3 in equations (1) and (3), and substitute $3, 0, 4$ for x_1, y_1, z_1 in the same equations, we obtain parametric equations of the line,

$$x = 3t + 3, \qquad y = -4t, \qquad z = 2t + 4, \quad t \in R,$$

and symmetric equations of the line,

$$\frac{x - 3}{3} = \frac{y}{-4} = \frac{z - 4}{2}.$$

Figure 16-27

From the symmetric equations we find the projecting

planes of the line,

$$\frac{x-3}{3} = \frac{y}{-4}, \qquad \frac{y}{-4} = \frac{z-4}{2}, \qquad \frac{z-4}{2} = \frac{x-3}{3}.$$

Exercises

In each of Exercises 1 to 8, a point and a vector are given. Sketch the line through the given point and parallel to the given vector. Then
(a) find parametric equations of the line;
(b) find symmetric equations of the line;
(c) find the projecting planes of the line.

1. $(4, -6, 3)$, $[-2, 1, 5]$.

2. $(-1, 3, 2)$, $[4, 2, -1]$.

3. $(2, 5, -4)$, $[-3, 4, 2]$.

4. $(-3, 0, -3)$, $[7, -6, 3]$.

5. $(4, 2, 7)$, $[-2, 4, 6]$.

6. $(-5, -4, -6)$, $[0, 2, -5]$.

7. $(6, -2, 5)$, $[3, 0, 0]$.

8. $(-1, 2, 0)$, $[5, -7, 2]$.

In Exercises 9 to 16, general equations of a line are given.
(a) Find the piercing points of the line in the two indicated coordinate planes.
(b) Draw the line.
(c) Find parametric equations of the line.
(d) Find symmetric equations of the line.

9. $5x + 2y - 5z - 5 = 0$, $10x + 6y - 5z - 25 = 0$; piercing points in the zx and yz planes.

10. $x + y - z - 1 = 0$, $3x - 3y + 7z - 9 = 0$; piercing points in the xy and yz planes.

11. $x + 4y + 2z - 13 = 0$, $2x - y - 2z + 10 = 0$; piercing points in the xy and zx planes.

12. $x - 2y + z - 1 = 0$, $6x - 5y + 4z - 10 = 0$; piercing points in the yz and zx planes.

13. $4x - 6y - z + 22 = 0$, $3x - 12y - 7z + 54 = 0$; piercing points in the zx and xy planes.

14. $3x + 5y - 9z + 26 = 0$, $9x + 11y - 18z + 50 = 0$; piercing points in the xy and yz planes.

15. $10x + 7y - 7z - 42 = 0$, $5x - 7y - 14z + 21 = 0$; piercing points in the yz and zx planes.

16. $x - y + z - 1 = 0$, $8x - 12y + 9z - 3 = 0$; piercing points in the zx and yz planes.

17. Find symmetric equations of the line through the point $(4, 0, 6)$ and perpendicular to the plane $x - 5y + 2z + 10 = 0$.

18. Find symmetric equations of the line through the point $(-5, 7, -2)$ and perpendicular to the plane $3x + 4y - 6z + 11 = 0$.

19. Find symmetric equations of the line through the points $(7, 4, 6)$ and $(-1, 3, -5)$. Make a sketch.

20. Find symmetric equations of the line through the points $(-4, 0, -3)$ and $(2, 6, 1)$. Make a sketch.

21. Sketch the line through the point $(5, -3, 4)$, that intersects the z axis at a right angle. Then find its symmetric equations.

22. Find symmetric equations of the line through the point $(2, -4, 5)$, that is parallel to the plane $3x + y - 2z + 5 = 0$ and perpendicular to the line

$$\frac{x+8}{2} = \frac{y-5}{3} = \frac{z-1}{-1}.$$

23. Find parametric equations of the line through the origin that is perpendicular to the plane $6x + 5y - 8z + 19 = 0$.

24. Find parametric equations of the line through the point $(-2, 7, -4)$ and perpendicular to the plane $4x - 7y + z + 23 = 0$.

25. Find parametric equations of the line through the point $(2, -1, 5)$ and parallel to the line whose general equations are $x - 2y + 3z + 1 = 0$, $2x + y + z + 7 = 0$. Then sketch the line.

26. Find parametric equations of the line through the point $(-4, 3, 2)$ and parallel to the line whose general equations are $3x + 4y - z + 5 = 0$, $3x + y - 2z - 8 = 0$.

27. Let $A_1x + B_1y + C_1z + D_1 = 0$, $A_2x + B_2y + C_2z + D_2 = 0$ be general equations of a line l. Prove that for each real value of λ, the graph of

$\backsim_1 y + C_1 z + D_1)$
$$+ \lambda(A_2 x + B_2 y + C_2 z + D_2) = 0$$

is a plane through l. The set of all such planes is called the **pencil of planes** on l.

28. Find the value of λ that makes the graph of

$$(3x - y + 2z + 4) + \lambda(2x + 5y - z + 6) = 0$$

be the projecting plane, perpendicular to the yz plane, of the line

$$3x - y + 2z + 4 = 0, \qquad 2x + 5y - z + 6 = 0.$$

Write the equation of this projecting plane. [*Hint:* Rewrite the given equation as

$$(3 + 2\lambda)x + (-1 + 5\lambda)y + (2 - \lambda)z$$
$$+ (4 + 6\lambda) = 0.$$

Let $3 + 2\lambda = 0$, and solve for $\lambda = -\frac{3}{2}$. Then substitute this value of λ in the given equation, obtaining $17y - 7z + 10 = 0$.]

In each of Exercises 29 to 32, use the method of Exercise 28 to find the three projecting planes of the line whose general equations are given.

29. $x - 2y + z + 4 = 0, 4x + y - 2z + 5 = 0.$

30. $3x + 5y - 2z + 1 = 0,$
$\quad 4x - 2y + z + 6 = 0.$

31. $4x + 4y - z - 5 = 0, 3x + y + 7z + 1 = 0.$

32. $6x + y + 2z + 3 = 0, 2x - y + z - 4 = 0.$

16.6 SURFACES

The set of points

$$(1) \qquad \{P{:}(x, y, z) \mid x \in R, y \in R, z \in R\},$$

with no other restriction on the coordinates, fills three-dimensional Cartesian space. Such a set is said to have *three degrees of freedom*, for we can assign any value whatever to each of the three coordinates quite independently of our choice of values for the other two.

If a restriction is placed on the set of points (1) by requiring that the coordinates of P satisfy an equation in x, y, and z, the freedom of the set is reduced. Thus the set

$$(2) \qquad \{P{:}(x, y, z) \mid x^2 + y^2 + z^2 = 1\}$$

has two degrees of freedom. For we can choose any value for z such that $-1 \le z \le 1$ and quite *independently* any y, $-1 \le y \le 1$; but, having done so, (2) becomes a quadratic equation in the variable x alone, which is only satisfied by its two roots.

The set (2) is a sphere. Had the equation in (2) been linear in x, y, and z, the set would have been a plane. Planes and spheres are simple examples of surfaces.

The set of points $P{:}(x, y, z)$ whose coordinates satisfy one equation in x, y, and z is usually a **surface**.

As illustrations, the set $\{P{:}(x, y, z) \mid x^2/a^2 + y^2/b^2 - z^2/c^2 = 1\}$ is a surface called a hyperboloid (see Fig. 16-40), and the set $\{P{:}(x, y, z) \mid x^2 + y^2 + z^2 = 9\}$ is a sphere whose radius is 3 and whose center is at the origin.

While *in general* the set of points $P{:}(x, y, z)$ whose coordinates satisfy one equation is a surface, under certain circumstances this may not be so. For example, the set $\{P{:}(x, y, z) \mid x^2 + y^2 + z^2 = 0\}$ is a single point, the origin; it may be thought of as the limiting case of the sphere $\{P{:}(x, y, z) \mid x^2 + y^2 + z^2 = r^2\}$ when $r = 0$. Again, the set $\{P{:}(x, y, z) \mid x^2 + y^2 + z^2 = -1\}$ is empty.

If two restrictions are placed on the points (1) by requiring that the coordinates of P satisfy two independent equations in x, y, and z simultaneously, the freedom of the set is further reduced. Thus the set

(3) $$\{P:(x, y, z) \,|\, x^2 + y^2 + z^2 = 9, \; 2x + y - 3z = 0\}$$

is the curve of intersection of two surfaces, the sphere $x^2 + y^2 + z^2 = 9$ and the plane $2x + y - 3z = 0$; the curve (3) is a circle that lies in the plane $2x + y - 3z = 0$ and whose center is the origin. It has *one degree of freedom*, since any value can be assigned to one of the variables, say z, but after this has been done, there are two independent simultaneous equations in x and y whose solutions are determined.

A *curve* in three-dimensional space will be defined in Section 17.1. Not every curve in three-space can be represented as the complete intersection of two surfaces.

16.7 CYLINDERS. SURFACES OF REVOLUTION

The student is familiar with right circular cylinders from high school geometry. In analytic geometry the word *cylinder* denotes a much more extensive class of surfaces.

Let c be a plane curve, and let l be a line that is neither in, nor parallel to, the plane of c. The set of all points on lines that are parallel to l and that intersect c is called a **cylinder**. The lines that are parallel to l and intersect c are called **generators** of the cylinder, and the plane curve c is a **directrix** of the cylinder.

For the cylinders considered in this book, the directrix is always a curve in a coordinate plane and the generators are straight lines perpendicular to that plane.

Figure 16-28 shows a circular cylinder whose directrix is a circle in the xy plane with its center at the origin and radius 4. Its generators are perpendicular to the xy

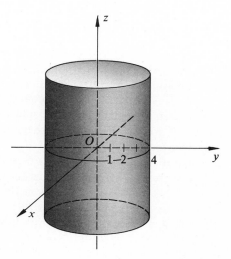

Figure 16-28

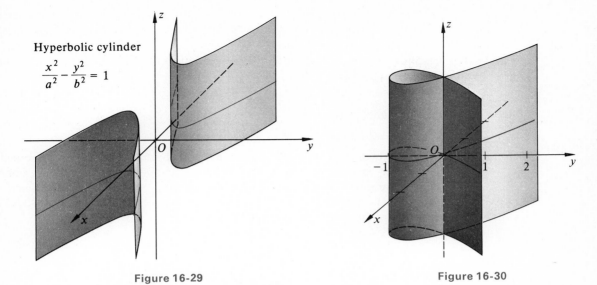

Hyperbolic cylinder

$$\frac{x^2}{a^2} - \frac{y^2}{b^2} = 1$$

Figure 16-29

Figure 16-30

plane. Figures 16-29 and 16-30 show a hyperbolic cylinder and a cubic cylinder. They are named after their directrix curves.

Let us find an equation of the circular cylinder shown in Fig. 16-28. Since its directrix curve is the circle in the xy plane,

(1) $$x^2 + y^2 = 16, \qquad z = 0,$$

the equation of the cylinder in three-dimensional Cartesian coordinates is

(2) $$x^2 + y^2 = 16.$$

For, if $(x_1, y_1, 0)$ is a point on the circle (1), then $x_1^2 + y_1^2 = 16$. Thus the coordinates of the point (x_1, y_1, z) satisfy the equation (2) *for every real number z*. As z runs through all real values, the point (x_1, y_1, z) traces out a straight line parallel to the z axis and intersecting the directrix circle in the point $(x_1, y_1, 0)$.

Similarly, the equation, in three-dimensional Cartesian coordinates, of the hyperbolic cylinder (Fig. 16-29) is

$$\frac{x^2}{a^2} - \frac{y^2}{b^2} = 1;$$

and the equation of the cubic cylinder (Fig. 16-30) is

$$x^2 = (y + 1)y^2.$$

This discussion is generalized in the following theorem.

16.7.1 Theorem. In three-dimensional Cartesian space the graph, if any, of an equation in only two of the three variables, x, y, z, is a cylinder whose generators are parallel to the axis of the missing variable, and the two-dimensional equation of whose directrix curve is the same as the given equation.

For example,

$$\frac{y^2}{b^2} + \frac{z^2}{c^2} = 1$$

is an equation of the elliptic cylinder whose generators are parallel to the x axis and whose directrix curve is the ellipse in the yz plane

$$c^2y^2 + b^2z^2 = b^2c^2, \qquad x = 0.$$

The simultaneous equations

$$y^2 = 4px, \qquad z = 0,$$

represent a parabola. The first equation alone represents a parabolic cylinder perpendicular to the xy plane, and the two equations simultaneously represent the intersection of the parabolic cylinder and the xy plane. That is, they represent the directrix of the parabolic cylinder.

At this point the reader should observe that although the equations

(3) $$\begin{cases} x^2 + y^2 + z^2 = 9, \\ y = 2, \end{cases}$$

represent the circle of intersection of the sphere $x^2 + y^2 + z^2 = 9$ and the plane $y = 2$, simpler and more useful equations of the same circle are readily available. Eliminating y between equations (3) by substituting $y = 2$ from the second equation into the first equation, we have

(4) $$x^2 + z^2 = 5,$$

which represents a circular cylinder perpendicular to the zx plane. Clearly, every set of coordinates that satisfy the simultaneous equations (3) also satisfy (4). That is, the circle (3) is on the cylinder (4). But the circle is also on the plane $y = 2$. Therefore new equations of the circle (3) are

(5) $$x^2 + z^2 = 5, \qquad y = 2.$$

It is obvious from equations (5) [but not from equations (3)] that the radius of the circle is $\sqrt{5}$ and that its center is on the y axis. Equation (4) represents a **projecting cylinder** of the circle (3).

This method is general, and the reader should use it to obtain the simplest possible equations when sketching the curve of intersection of two surfaces.

Example. Find simpler equations of the curve

$$45x^2 - 180y^2 - 4z^2 = 36, \qquad z = 6,$$

and draw it.

Solution. The projecting cylinder of this curve perpendicular to the xy plane is

$$\frac{x^2}{4} - \frac{y^2}{1} = 1,$$

found by substituting $z = 6$ from the second equation into the first and simplifying. Therefore simpler equations of the given curve are

(6) $$\frac{x^2}{4} - \frac{y^2}{1} = 1, \qquad z = 6.$$

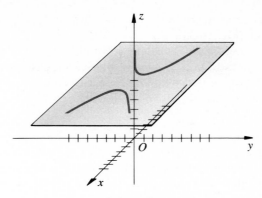

Figure 16-31

It is a hyperbola in standard position in the plane $z = 6$ (Fig. 16-31). The length of its transverse axis is 4 and the length of its conjugate axis is 2.

An extensive class of surfaces, used in mathematical applications, are **surfaces of revolution**. Such a surface can be thought of as swept out by a plane curve revolving about a fixed line lying in the plane of the curve. The fixed line is called the **axis** of the surface of revolution.

For instance, a sphere can be generated by revolving a circle about one of its diameters. A surface of revolution generated by revolving a parabola about its principal axis is shown in Fig. 16-32. Clearly, all plane sections of a surface of revolution that are perpendicular to its axis are circles.

Let $y = f(z)$, $x = 0$ be the equations of a plane curve that either lies in the right-hand half of the yz plane or else is symmetric with respect to the z axis. We seek an

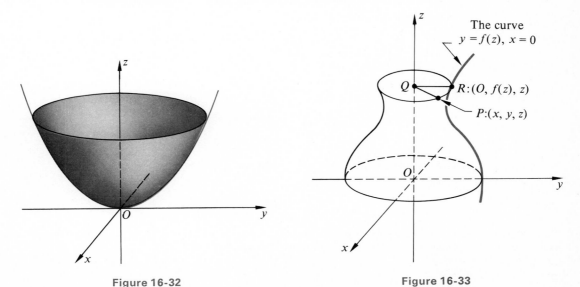

Figure 16-32

Figure 16-33

equation of the surface of revolution swept out by revolving this curve about the z axis (Fig. 16-33).

Let $P:(x, y, z)$ be any point on the surface and pass a plane through P perpendicular to the z axis. The plane intersects the surface in a circle whose center is the point $Q:(0, 0, z)$; it also intersects the curve $y = f(z)$ in the point $R:(0, f(z), z)$. Since P is on the surface, $|QP| = |QR|$. But $|QP| = \sqrt{x^2 + y^2}$ and $|QR| = f(z)$. Therefore $\sqrt{x^2 + y^2} = f(z)$, or

$$x^2 + y^2 = [f(z)]^2.$$

This is an equation of the surface of revolution.

An analogous discussion holds for surfaces of revolution symmetric with respect to either of the other two coordinate axes.

16.7.2 Theorem.

(i) If $f(z) \geq 0$, an equation of the surface of revolution swept out by revolving the plane curve $y = f(z)$, $x = 0$ about the z axis is

$$x^2 + y^2 = [f(z)]^2.$$

(ii) If $f(x) \geq 0$, an equation of the surface of revolution swept out by revolving the plane curve $z = f(x)$, $y = 0$ about the x axis is

$$y^2 + z^2 = [f(x)]^2.$$

(iii) If $f(y) \geq 0$, an equation of the surface of revolution swept out by revolving the plane curve $x = f(y)$, $z = 0$ about the y axis is

$$z^2 + x^2 = [f(y)]^2.$$

For example, the equation of the surface of revolution generated by revolving one branch of the hyperbola

$$\frac{x^2}{a^2} - \frac{y^2}{b^2} = 1, \qquad z = 0,$$

about the y axis is

$$z^2 + x^2 = \frac{a^2(y^2 + b^2)}{b^2},$$

or

$$\frac{x^2 + z^2}{a^2} - \frac{y^2}{b^2} = 1.$$

The reader should sketch this surface.

Exercises

1. Name and sketch the graph in Cartesian three-space of each of the following equations.
(a) $25x^2 + 16y^2 = 400$;
(b) $y^2 + z^2 = 9$;
(c) $2x + 5z - 12 = 0$;
(d) $z^2 = 6y$;
(e) $x^2 + y^2 - 8x + 2y + 13 = 0$;
(f) $z^2 = y^3$.

2. Name and sketch the graph in Cartesian three-space of each of the following equations.
(a) $x^2 + z^2 = 25$;
(b) $2y - 3z - 6 = 0$;
(c) $y^2 = 8x$;
(d) $y^2 - z^2 = 16$;
(e) $4x^2 + 9y^2 - 36 = 0$;
(f) $4x^2 - 9z^2 = 0$.

3. Write the equation of a parabolic cylinder whose generators are parallel to the x axis.

4. Write the equation of the circular cylinder whose generators are parallel to the z axis, if its directrix curve has its center at $(3, -2, 0)$ and its radius is 4. Make a sketch.

5. Write the equation of the circular cylinder whose generators are parallel to the x axis, if its directrix curve has its center at $(0, 4, 3)$ and its radius is 3. Make a sketch.

6. Write the equation of a parabolic cylinder whose generators are parallel to the z axis.

7. In three-space, write Cartesian equations of the parabola in standard position in the yz plane if its focus is $(0, -3, 0)$. Make a sketch.

8. Write three-space Cartesian equations of the ellipse in the yz plane whose vertices are $(0, 0, \pm3)$ and the length of whose minor axis is 2.

9. Name and sketch the graph of $\{P:(x, y, z) \mid x^2 + y^2 + z^2 = 36, z = 3\}$.

10. Name and sketch the graph of $\{P:(x, y, z) \mid x^2 + y^2 + z^2 = 9, y = -1\}$.

11. Find a projecting cylinder of the curve in Exercise 9.

12. Find a projecting cylinder of the circle in Exercise 10.

In Exercises 13 to 22, find the equation of the surface of revolution generated by revolving the given plane curve about the indicated axis. Sketch each surface.

13. $y = 2\sqrt{z}$, $x = 0$, about the z axis.

14. The same curve as in Exercise 13, about the y axis. [*Hint:* The given curve can be written $z = y^2/4$, $x = 0$, for nonnegative values of y. It sweeps out the same surface of revolution about the y axis as the curve $x = y^2/4$, $z = 0$ $(y \geq 0)$. Apply 16.7.2(iii) to the latter.]

15. $z = 5\sqrt{16 - x^2}/4$, $y = 0$, about the x axis.

16. The same curve as in Exercise 15, about the z axis. (*Hint:* See Exercise 14.)

17. $x = y^{3/2}$, $z = 0$, about the y axis.

18. $z = x^3$, $y = 0$ $(x \geq 0)$, about the x axis.

19. $z = 1/x$, $y = 0$ $(x > 0)$, about the x axis.

20. $y = \ln z$, $x = 0$ $(z \geq 1)$, about the z axis.

21. $y = e^z$, $x = 0$, about the z axis.

22. $x = \cos y$, $z = 0$ $(0 \leq y \leq \pi/2)$, about the y axis.

23. Find the axis and the generating curve for each of the following surfaces of revolution.
(a) $x^2 + y^2 - z^6 = 0$, $z \geq 0$;
(b) $x^2y^2 + y^2z^2 = 1$, $y > 0$;
(c) $9x^2 + 9y^2 - 4z^2 + 24z - 36 = 0$, $z \geq 3$.

24. Which of the following equations has for its graph a surface of revolution? Find the axis and the generating curve for each such surface of revolution.
(a) $9x^2 - 36y^2 + 4z^2 = 36$;
(b) $9x^2 + 25y^2 + 9z^2 = 225$;
(c) $9x^2 + 9y^2 - 16z = 0$.

16.8 SYMMETRY, TRACES, AND PLANE SECTIONS OF A SURFACE

It is desirable to be able to sketch a surface when its equation is known. Several of the ideas used in analyzing plane curves can be readily extended to surfaces in three-space.

Two points P and P' are **symmetric** with respect to a plane π if π is a perpendicular bisector of the line segment PP' (Fig. 16-34). Each of the points P and P' is said to be the **reflection** of the other about the plane π. A surface is symmetric with respect to a plane π if the reflection about π of every point on the surface is also on the surface.

Clearly, two points are symmetric with respect to the xy plane if and only if their x coordinates are the same and their y coordinates are the same, but the z coordinate of one is the negative of the z coordinate of the other (Fig. 16-35).

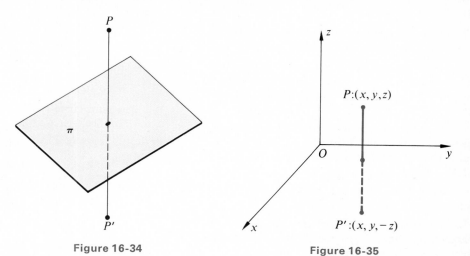

Figure 16-34   Figure 16-35

Similar statements hold for pairs of points symmetric with respect to either of the other coordinate planes.

A surface is **symmetric with respect to the xy plane** if its equation is unchanged by replacing z by $-z$ throughout; it is **symmetric with respect to the yz plane** if its equation is unchanged by replacing x by $-x$ throughout; and a surface is **symmetric with respect to the zx plane** if its equation is unchanged when y is replaced by $-y$ throughout the equation.

As an illustration, the surface $x^2 + y^2 = 4z$ is symmetric with respect to the yz plane, since the equation is unchanged when we replace x by $-x$. It is also symmetric with respect to the zx plane, since its equation is unaffected when we replace y by $-y$ (Fig. 16-41).

Two points P and P' are **symmetric with respect to a line l** if l bisects perpendicularly the line segment PP'. Thus the two points $P{:}(x, y, z)$ and $P{:}(x, -y, -z)$ are symmetric with respect to the x axis (Fig. 16-36).

A surface is **symmetric with respect to the x axis** if its equation is unchanged when y and z are replaced by $-y$ and $-z$ throughout.

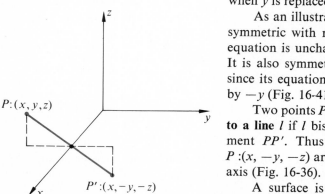

Figure 16-36

Analogous statements hold for symmetry of a surface with respect to the y or z axis. Thus the surface $x^2 + y^2 = 4z$, referred to above, is symmetric with respect to the z axis (Fig. 16-41).

The curve of intersection of a surface with a coordinate plane is called a **trace** of the surface. The three traces of a surface are of great help in sketching it. To find the trace of a surface in the xy plane, set $z = 0$ simultaneously with the given equation of the surface. We proceed similarly in finding the other traces.

For example, the trace of the surface $x^2 + y^2 - z^2 = 1$ in the xy plane is the plane curve

$$x^2 + y^2 - z^2 = 1, \qquad z = 0.$$

As in 16.7, we obtain simpler equations of this curve by finding its projecting cylinder. Eliminating z between the two equations, we obtain $x^2 + y^2 = 1$. Thus equivalent but simpler equations of the trace of the given surface in the xy plane are

$$x^2 + y^2 = 1, \qquad z = 0,$$

which represent a circle with center at the origin and radius 1 (Fig. 16-37).

Similarly, the trace of the surface $x^2 + y^2 - z^2 = 1$ in the zx plane is

$$x^2 + y^2 - z^2 = 1, \qquad y = 0,$$

or

$$x^2 - z^2 = 1, \qquad y = 0,$$

which is an equilateral hyperbola in standard position in the zx plane.

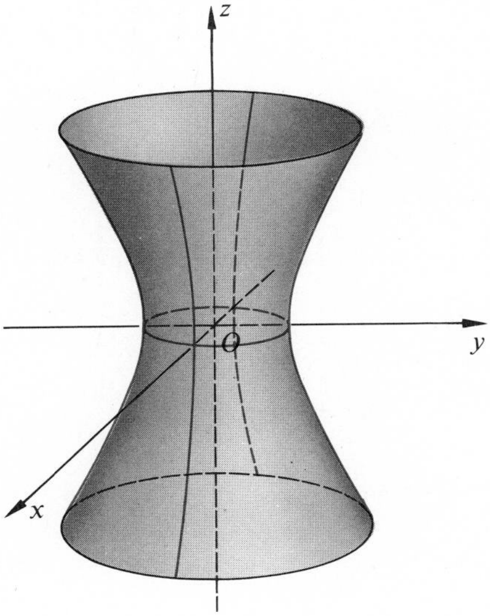

Figure 16-37

The trace of the surface in the yz plane is the equilateral hyperbola

$$y^2 - z^2 = 1, \qquad x = 0.$$

If further information is needed to sketch the surface, we resort to a series of plane sections of the surface parallel to a coordinate plane. For example, the plane $z = 1$ cuts the surface (Fig. 16-37) in the curve

$$x^2 + y^2 - z^2 = 1, \qquad z = 1,$$

or

$$x^2 + y^2 = 2, \qquad z = 1,$$

which is a circle with radius $\sqrt{2}$, lying in the plane $z = 1$ and with its center on the z axis. The plane $z = -5$ cuts the surface in the circle

$$x^2 + y^2 - z^2 = 1, \qquad z = -5,$$

or

$$x^2 + y^2 = 26, \qquad z = -5.$$

Similarly, all plane sections parallel to the zx plane are hyperbolas. For example, the plane $y = 3$ cuts the surface in the hyperbola

$$x^2 + y^2 - z^2 = 1, \qquad y = 3,$$

or

$$\frac{z^2}{8} - \frac{x^2}{8} = 1, \qquad y = 3.$$

Example. Sketch the surface whose equation is $2x^2 - y^2 + 2z^2 - 2y + 1 = 0$.

Solution. Our tests show that the surface is symmetric with respect to the yz plane and also to the xy plane. Therefore the surface is symmetric with respect to the y axis.

The trace in the xy plane is

$$2x^2 - y^2 - 2y + 1 = 0, \qquad z = 0,$$

or

$$\frac{(y + 1)^2}{2} - \frac{x^2}{1} = 1, \qquad z = 0,$$

which is a hyperbola in the xy plane, symmetric with respect to the y axis and with center at $(0, -1, 0)$.

Similarly, the trace in the zx plane is

$$2x^2 + 2z^2 + 1 = 0, \qquad y = 0,$$

which is imaginary. So the surface does not intersect the zx plane.

The trace in the yz plane is

$$y^2 - 2z^2 + 2y - 1 = 0, \qquad x = 0,$$

or

$$\frac{(y + 1)^2}{2} - \frac{z^2}{1} = 1, \qquad x = 0,$$

which is a hyperbola that is symmetric with respect to the y axis and has its center at $(0, -1, 0)$.

The intersection of the surface with planes perpendicular to the y axis and to the right of $y = -1 + \sqrt{2}$ (or $y \doteq 0.4$) are circles that get larger and larger as we move to the right. Similarly, the sections perpendicular to the y axis and to the left of $y = -1 - \sqrt{2}$ (or $y \doteq -2.4$) are circles that increase in size as we move to the left. Thus the surface is a surface of revolution about the y axis. It is now easy to sketch it (Fig. 16-38).

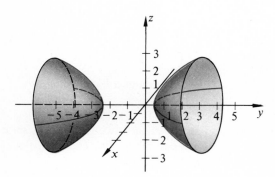

Figure 16-38

Exercises

Sketch the surfaces whose equations are given below.

1. $9x^2 + 36y^2 + 16z^2 - 144 = 0$.

2. $4x^2 - y^2 + 4z^2 - 4 = 0$.

3. $-16x^2 - 100y^2 + 25z^2 - 400 = 0$.

4. $4x^2 + 9y^2 + 36z^2 - 108 = 0$.

5. $4x^2 + 9y^2 - 12z = 0$.

6. $x^2 - y^2 + z^2 = 0$.

7. $400x^2 - 144y^2 + 225z^2 - 3{,}600 = 0$.

8. $x^2 + y^2 + 4z = 0$.

9. $9x^2 - 36y^2 + 4z^2 + 36 = 0$.

10. $4x^2 + 4y^2 - 25z^2 + 100 = 0$.

11. $x^2 - z^2 = 0$.

12. $y^2 + z^2 - 12y = 0$.

13. $x^2 - y^2 + z^2 + 16 = 0$.

14. $y = e^{2z}$.

15. $y = \sin x$.

16. $9x^2 + 4z^2 + 36y = 0$.

16.9 QUADRIC SURFACES

If a surface is the graph of an equation of second degree in three-dimensional Cartesian coordinates, it is called a **quadric surface**. All plane sections of a quadric surface are conics.

It can be shown that any equation of second degree in x, y, and z can be reduced, by rotation and translation of the coordinate axes, to one of the two forms

(1) $$Ax^2 + By^2 + Cz^2 + J = 0$$

and

(2) $$Ax^2 + By^2 + Iz = 0.$$

The quadric surfaces represented by equation (1) are symmetric with respect to all three coordinate planes (by 16.8) and hence are symmetric with respect to the origin. For this reason, the graphs of (1) are called **central quadrics**, and the origin is their center. If the coefficients in (1) are all different from zero, it can be rewritten

(3) $$\pm \frac{x^2}{a^2} \pm \frac{y^2}{b^2} \pm \frac{z^2}{c^2} = 1,$$

where a, b, and c are positive constants. The equations (3) represent the following three types of central quadrics.

1. Ellipsoid (Fig. 16-39).

The standard equation of an ellipsoid is

(4) $$\frac{x^2}{a^2} + \frac{y^2}{b^2} + \frac{z^2}{c^2} = 1.$$

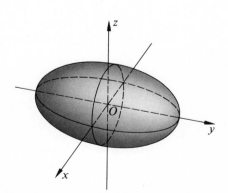

The ellipsoid intersects the x axis in the points $(\pm a, 0, 0)$, found by substituting $y = 0$ and $z = 0$ in (4). Similarly, it intersects the y axis in $(0, \pm b, 0)$ and z axis in $(0, 0, \pm c)$.

The trace of the ellipsoid in the xy plane is the ellipse

$$\frac{x^2}{a^2} + \frac{y^2}{b^2} = 1, \qquad z = 0.$$

Figure 16-39

All plane sections of the surface parallel to the xy plane are ellipses if $a \neq b$ and are circles if $a = b$. They become smaller as the cutting plane recedes from the origin. For, the equations of the curve of intersection of the ellipsoid and the plane $z = k$ are

$$\frac{x^2}{a^2} + \frac{y^2}{b^2} + \frac{k^2}{c^2} = 1, \qquad z = k,$$

or

$$\frac{x^2}{a^2} + \frac{y^2}{b^2} = 1 - \frac{k^2}{c^2}, \qquad z = k,$$

which may be rewritten

(5)
$$\frac{x^2}{a^2\left(1 - \frac{k^2}{c^2}\right)} + \frac{y^2}{b^2\left(1 - \frac{k^2}{c^2}\right)} = 1, \qquad z = k.$$

When $|k| < c$, equations (5) represent an ellipse, the lengths of whose semiaxes are $a\sqrt{1 - k^2/c^2}$ and $b\sqrt{1 - k^2/c^2}$. The ellipse is largest when $k = 0$—that is, when it is the trace of the ellipsoid in the plane $z = 0$. As $|k|$ increases, the ellipses grow smaller and then reduce to a point when $|k| = c$. For $|k| > c$, the ellipse is imaginary.

Similarly, the traces of the ellipsoid in the yz and zx planes are also ellipses, and every real plane section of the surface, parallel to either of those planes, is an ellipse.

If two of the three quantities a, b, c are equal, the ellipsoid is a surface of revolution; if all three are equal, the ellipsoid is a sphere.

If $a = c$ and $b > a$ in the equation of the ellipsoid, the graph is a **prolate spheroid**; it is shaped like a football.

If $a = b$ and $c < a$, the ellipsoid is an **oblate spheroid**. The planet Earth is an oblate spheroid, somewhat flattened at the poles.

2. Hyperboloid of One Sheet (Fig. 16-40).

The standard equation of this surface is

(6)
$$\frac{x^2}{a^2} + \frac{y^2}{b^2} - \frac{z^2}{c^2} = 1.$$

Its trace in the xy plane is an ellipse, and all plane sections parallel to the xy plane are ellipses or circles that become larger as the cutting plane recedes from the origin. The trace of this hyperboloid in the yz or zx plane is a hyperbola, and all plane sections parallel to either of those coordinate planes are hyperbolas. If $a = b$, the hyperboloid of one sheet is a surface of revolution. If the negative sign in (6) had been before the first or second term, instead of the third, the surface would still have been a hyperboloid of one sheet, but about one of the other axes.

3. Hyperboloid of Two Sheets (Fig. 16-41).

The standard equation of this surface is

(7)
$$\frac{x^2}{a^2} - \frac{y^2}{b^2} - \frac{z^2}{c^2} = 1.$$

The traces of this surface in the xy and zx planes are hyperbolas, as are all plane sections parallel to either of them. The transverse axis of these hyperbolas gets longer as the cutting plane recedes from the origin. This hyperboloid of two sheets does not intersect the yz plane, and thus there is no trace in that plane. Planes parallel to the yz plane, between the vertices $(\pm a, 0, 0)$, fail to intersect the surface. All other planes

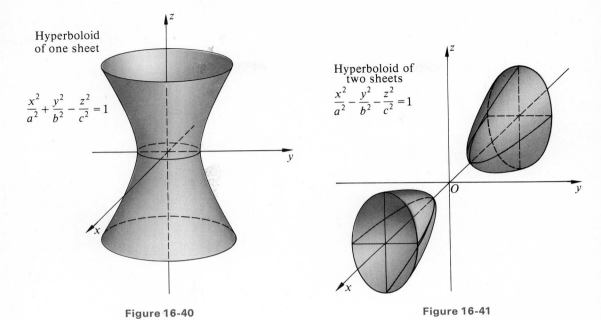

$$\frac{x^2}{a^2} + \frac{y^2}{b^2} - \frac{z^2}{c^2} = 1$$

Hyperboloid of one sheet

Hyperboloid of two sheets

$$\frac{x^2}{a^2} - \frac{y^2}{b^2} - \frac{z^2}{c^2} = 1$$

Figure 16-40   Figure 16-41

that are parallel to the yz plane intersect this hyperboloid in ellipses (or circles if $b = c$) that get larger as their plane recedes from the origin. If the negative signs in equation (7) had been before the first and second terms or the first and third terms, instead of the second and third terms, the hyperboloid of two sheets would have been about one of the other axes.

We turn next to equation (2), which represents the **noncentral quadrics**. As equation (2) is written, the surfaces that it represents are symmetric with respect to the yz and zx planes but not the xy plane. Thus they are symmetric with respect to the z axis. If all the coefficients in (2) are different from zero, it can be rewritten

$$(8) \qquad \pm\frac{x^2}{a^2} \pm \frac{y^2}{b^2} = z.$$

This represents the following two distinct types of noncentral quadric surfaces. (For simplicity of language, we shall assume that the coordinate axes are oriented as in our diagrams.)

4. Elliptic Paraboloid (Fig. 16-42).

The standard equation of this paraboloid is

$$(9) \qquad \frac{x^2}{a^2} + \frac{y^2}{b^2} = z.$$

Its trace in the xy plane is a single point—the origin. Planes parallel to and above the xy plane cut the surface in ellipses that are larger when the cutting plane is higher. The surface does not extend below the xy plane. The traces in the yz and zx planes are parabolas and so are all plane sections of the surface parallel to either of those coordinate planes. If there had been a negative sign before z in (9), the elliptic paraboloid would have opened downward. Of course, the equation $x^2/a^2 + z^2/b^2 = y$ also represents an elliptic paraboloid—one that opens to the right along the y axis.

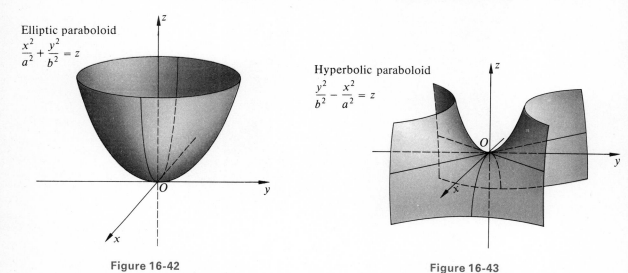

Elliptic paraboloid
$$\frac{x^2}{a^2} + \frac{y^2}{b^2} = z$$

Figure 16-42

Hyperbolic paraboloid
$$\frac{y^2}{b^2} - \frac{x^2}{a^2} = z$$

Figure 16-43

5. Hyperbolic Paraboloid (Fig. 16-43).

The standard equation of this paraboloid is

(10) $$\frac{y^2}{b^2} - \frac{x^2}{a^2} = z.$$

The trace of this surface in the xy plane is

$$\frac{y^2}{b^2} - \frac{x^2}{a^2} = 0, \qquad z = 0,$$

which consists of two straight lines through the origin, since the first of the two simultaneous equations can be factored. All plane sections parallel to the xy plane and above it are hyperbolas whose transverse axes are parallel to the y axis. All plane sections parallel to and below the xy plane are hyperbolas whose transverse axes are parallel to the x axis. The trace of the hyperbolic paraboloid in the yz plane is a parabola opening upward, as are all plane sections parallel to the yz plane. The trace in the zx plane is a parabola opening downward, and so are all plane sections of the surface parallel to the zx plane.

Singular or degenerate quadric surfaces may occur when one or more of the coefficients in equations (1) or (2) are zero. The most important are the **quadric cone** and the **quadric cylinders**.

6. Quadric Cone (Fig. 16-44).

The equation of this cone is

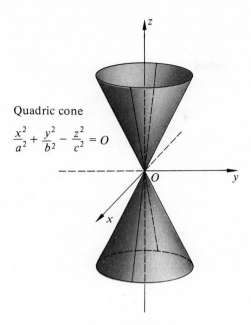

Quadric cone

$$\frac{x^2}{a^2} + \frac{y^2}{b^2} - \frac{z^2}{c^2} = O$$

Figure 16-44

(11)
$$\frac{x^2}{a^2} + \frac{y^2}{b^2} - \frac{z^2}{c^2} = 0.$$

Its trace in the xy plane is a single point—the origin. All plane sections parallel to the xy plane are ellipses (or circles) that get larger as they recede from the origin. The trace in the yz plane is a pair of intersecting lines through the origin. All plane sections parallel to the yz plane are hyperbolas. The trace in the zx plane is also a pair of intersecting lines through the origin, and plane sections parallel to the zx plane are hyperbolas.

7. Quadric Cylinders.
The equations

$$\frac{x^2}{a^2} \pm \frac{y^2}{b^2} = 1$$

represent an elliptic cylinder and a hyperbolic cylinder (Fig. 16-29) perpendicular to the xy plane (by 16.7.1). The equation $y^2 = kx$ has for its graph a parabolic cylinder perpendicular to the xy plane.

16.10 PROCEDURE FOR SKETCHING A SURFACE

It may be helpful to outline a systematic procedure for sketching a surface when its equation is known. In this discussion we shall assume that the equation does have a real graph that does not degenerate by a limiting process or by factorization of the equation into simpler graphs.

1.   If the equation is of the first degree, the surface is a plane. When the plane does not go through the origin, that is, when its equation has a nonzero constant term, find its points of intersection with the coordinate axes and connect them by straight lines (its traces). When the plane does go through the origin, as shown by the absence of any constant term in the equation, find two traces of the plane and draw a triangular portion of the plane.

2.   If the equation is of the second degree, the graph is a quadric surface. If the second-degree equation is similar to one of those discussed in Section 16.9, the quadric is in standard position and is readily sketched from its traces. Otherwise proceed as follows.

3.   If the equation lacks one of the variables, its graph is a cylinder perpendicular to a coordinate plane (16.7).

4.   Test the given equation for symmetry with respect to the coordinate planes (16.8).

5.   Find and draw its traces (16.8).

6.   If further information is necessary, draw a series of plane sections of the surface parallel to a coordinate plane (16.8).

Example 1.   By inspection, name the graph of each of the following equations.

(a) $4x^2 + 4y^2 - 25z^2 + 100 = 0$.
(b) $y^2 + z^2 - 12y = 0$.
(c) $x^2 - z^2 = 0$.
(d) $9x^2 + 4z^2 - 36y = 0$.

Solution.

(a)   After dividing both members of the given equation by -100, it can be written

$$-\frac{x^2}{25} - \frac{y^2}{25} + \frac{z^2}{4} = 1.$$

Since this is in the form of equation (7) of Section 16.9, it is a *hyperboloid of two sheets*. Its vertices are at $(0, 0, \pm 2)$.

(b)   The variable x does not appear in the given equation; so its graph (if any) is a cylinder with generators parallel to the x axis. Since this equation can be written in the form $(y - 6)^2 + z^2 = 36$, its graph is a *circular cylinder* whose directrix is the circle in the yz plane having center $(0, 6, 0)$ and radius 6.

(c)   Because y does not appear in this equation, its graph is a cylinder. Moreover, the given equation can be written $(x - z)(x + z) = 0$; so its graph consists of the *two planes* $x - y = 0$ and $x + y = 0$. This is a degenerate quadric cylinder.

(d)   Since this equation can be written

$$\frac{x^2}{4} + \frac{z^2}{9} = y,$$

its graph is an *elliptic paraboloid* with its vertex at the origin, and symmetric with respect to the positive y axis.

Example 2.   Sketch the surface whose equation is $x^2 + z^2 - \cos^2 y = 0$, following the procedure outlined above.

Solution. This equation is neither of the first nor second degree; so parts 1 and 2 of the foregoing procedure do not apply. In fact, the expression "degree of an equation" is only pertinent to *polynomial* equations (or equations that can be reduced to polynomial form).

The given equation contains all the variables, x, y, and z; so we also omit part 3.

Part 4. Since the equation is unchanged when $-x$ is substituted for x, its graph is symmetric with respect to the yz plane; and because the equation is unchanged when $-z$ is substituted for z, the surface is also symmetric with respect to the xy plane. Thus the surface is symmetric with respect to the y axis.

Part 5. The trace of the surface in the yz plane, found by substituting zero for x in the given equation, is the pair of curves $z = \pm\cos y$, $x = 0$.

The trace in the xy plane is again a cosine curve and its reflection, $x = \pm\cos y$, $z = 0$.

The trace of the surface in the zx plane is the unit circle $x^2 + z^2 = 1$, $y = 0$.

Part 6. Since all plane sections perpendicular to the y axis are circles,

$$x^2 + z^2 = \cos^2 k, \qquad k \in R,$$

the graph of the given equation is a surface of revolution about the y axis. Notice that the radii of the circular sections oscillate between 0 and 1 as $y = k$ increases.

Having all this information, it is easy to sketch the surface (Fig. 16-45).

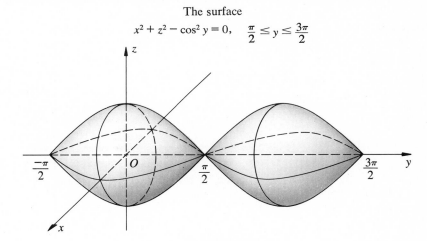

The surface

$$x^2 + z^2 - \cos^2 y = 0, \qquad \frac{\pi}{2} \le y \le \frac{3\pi}{2}$$

Figure 16-45

Exercises

1. By inspection of the equations in Exercises 1, 2, 3, 5, and 6 of Section 16.8, name their graphs.

2. By inspection of the equations in Exercises 4, 7, 8, 9, and 14 of Section 16.8, name their graphs.

In Exercises 3 to 14, name and sketch the quadric surfaces whose equations are given.

3. $100x^2 + 225y^2 - 36z^2 = 0.$

4. $36x^2 + 9y^2 + 16z^2 - 144 = 0.$

5. $16x^2 - 25y^2 + 400z = 0.$

6. $9x^2 - 4y^2 + 9z^2 - 36 = 0.$

7. $x^2 - z^2 + y = 0.$

8. $x^2 + y^2 - 4z^2 + 4 = 0.$

9. $400x^2 + 25y^2 + 16z^2 - 400 = 0.$

10. $9x^2 + 4z^2 - 36y = 0.$

11. $x^2 + 4z^2 - 8y = 0.$

12. $4x^2 - y^2 + 4z^2 = 0.$

13. $225x^2 - 100y^2 + 144z^2 = 0.$

14. $9x^2 - 25y^2 + 9z^2 - 225 = 0.$

In Exercises 15 to 22, sketch the surfaces whose equations are given.

15. $5x + 8y - 2z = 0.$

16. $y = e^{-x}.$

17. $z - y^4 = 0.$

18. $x^2 + y^2 = \sin z.$

19. $x^2 + y^2 = 4|(y - 1)y(y + 1)|.$

20. $y = 2 \cos x.$

21. $x = \tan 2z.$

22. $4x^2 + 4y^2 + 4z^2 - 40x + 48y - 32z + 283 = 0.$

16.11 REVIEW EXERCISES

1. Sketch the two position vectors $\mathbf{a} = 2\mathbf{i} - \mathbf{j} + 2\mathbf{k}$ and $\mathbf{b} = 5\mathbf{i} + \mathbf{j} - 3\mathbf{k}$. Then find
(a) their lengths;
(b) their direction cosines;
(c) the unit vector having the same direction as $\mathbf{a}$;
(d) the angle θ between $\mathbf{a}$ and $\mathbf{b}$.

2. Let $\mathbf{a} = 2\mathbf{i} - \mathbf{j} + \mathbf{k}$, $\mathbf{b} = -\mathbf{i} + 3\mathbf{j} + 2\mathbf{k}$, and $\mathbf{c} = \mathbf{i} + 2\mathbf{j} - \mathbf{k}$. Find
(a) $\mathbf{a} \times \mathbf{b}$.   (b) $\mathbf{a} \times (\mathbf{b} + \mathbf{c})$.
(c) $\mathbf{a} \cdot (\mathbf{b} \times \mathbf{c})$.   (d) $\mathbf{a} \times (\mathbf{b} \times \mathbf{c})$.

3. Find all vectors that are perpendicular to both of the vectors $3\mathbf{i} + 3\mathbf{j} - \mathbf{k}$ and $-\mathbf{i} - 2\mathbf{j} + 4\mathbf{k}$.

4. Find the unit vectors that are perpendicular to the plane determined by the three points $(3, -6, 4)$, $(2, 1, 1)$, and $(5, 0, -2)$.

5. Write the equation of the plane through the point $(-5, 7, -2)$ that is
(a) parallel to the zx plane;
(b) perpendicular to the x axis;
(c) parallel to both the x and y axes.
Sketch each plane.

6. The line through the origin and perpendicular to a certain plane intersects the plane in the point $(1, -1, -2)$.
(a) Write a vector equation of the plane.
(b) Find a Cartesian equation of the plane.
(c) Sketch the plane by drawing its traces, and sketch the given line.

7. A plane through the point $(2, -4, -5)$ is perpendicular to the line joining the points $(-1, 5, -7)$ and $(4, 1, 1)$.
(a) Write a vector equation of the plane.
(b) Find a Cartesian equation of the plane.
(c) Sketch the plane by drawing its traces, and sketch the given line.

8. A plane bisects perpendicularly the line segment having endpoints $(-3, 1, 2)$ and $(5, 7, 4)$.
(a) Write a vector equation of the plane.
(b) Find a Cartesian equation of the plane.
(c) Sketch the plane by drawing its traces, and sketch the given line segment.

9. A plane passes through the point $(1, -2, 4)$ and is parallel to the plane $6x + 3y - z + 10 = 0$.
(a) Write a vector equation of the plane.
(b) Write a Cartesian equation of the plane.

10. Find the value of B if the plane $2x + By - z + 8 = 0$ is perpendicular to the plane $3x - 2y + 10z + 1 = 0$.

11. Find the value of C if the plane $x + 5y + Cz + 6 = 0$ is perpendicular to the plane $4x - y + z - 17 = 0$.

12. Find a Cartesian equation of the plane through the three points $(2, 3, -1)$, $(-1, 5, 2)$, and $(-4, -2, 2)$.

13. Find a Cartesian equation of the plane through the three points $(0, 5, -2)$, $(3, 4, 4)$, and $(2, -6, -1)$

In Exercises 14, 15, and 16, general equations of a line are given.
(a) Find the piercing points of the line in the two indicated coordinate planes.
(b) Draw the line.
(c) Find parametric equations of the line.
(d) Find symmetric equations of the line.

14. $x - 2y + 4z - 14 = 0$, $x + 20y - 18z + 30 = 0$; piercing points in the yz and zx planes.

15. $2x + 5y + 4z - 4 = 0$, $28x - 4y + 19z + 92 = 0$; piercing points in the yz and xy planes.

16. $4x - 3y + 16z - 36 = 0$, $6x + 3y + 9z - 39 = 0$; piercing points in the yz and zx planes.

17. Find symmetric equations of the line through the point $(4, 5, 8)$ and perpendicular to the plane $3x + 5y + 2z - 30 = 0$. Sketch the plane and the line.

18. Find a vector equation of the line through the points $(2, -2, 1)$ and $(-3, 2, 4)$. Find parametric equations of the line. Sketch the line.

19. Name and sketch the graph (if any) of each of the following equations.
(a) $x^2 + y^2 = 81$.
(b) $x^2 + y^2 + z^2 = 81$.

(c) $z^2 = 4y$.
(d) $3y - 6z + 3 = 0$.
(e) $9x^2 - 4z^2 = 36$.
(f) $4y^2 + 9z^2 + 36 = 0$.

20. Write the equation of the circular cylinder whose generators are parallel to the x axis, if its directrix curve has its center at $(0, -2, 5)$ and its radius is 3. Sketch the cylinder.

21. Find the equation of the surface of revolution swept out by revolving the plane curve $z^2 = 6y$, $x = 0$ about
(a) the z axis;   (b) the y axis.
Sketch both surfaces.

In Exercises 22 to 28, sketch the surface whose equation is given.

22. $x^2 + y^2 - z^2 - 1 = 0$.

23. $3x^2 + 4y^2 + 9z^2 - 24 = 0$.

24. $x^2 + z^2 + y = 0$.

25. $x^2 + y^2 - z^2 = 0$.

26. $9x^2 + 9y^2 - 4z^2 + 36 = 0$.

27. $x^2 + y^2 - 4x = 0$.

28. $x^2 + z^2 - y = 0$.

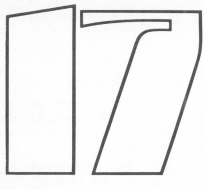

vector functions in three-dimensional space

17.1 VECTOR FUNCTIONS

The definitions and theorems for vector functions in the plane are extended in a natural manner to three-dimensional space. They are collected here for easy reference.

17.1.1 Let f, g, and h be functions defined on an interval I.

    (i) A **vector function r** is defined on I by
$$\mathbf{r}(t) = [f(t), g(t), h(t)], \qquad t \in I.$$

    (ii) If $t_0 \in I$, the **limit** of the vector function $\mathbf{r}$ at t_0 is the vector
$$\lim_{t \to t_0} \mathbf{r}(t) = [\lim_{t \to t_0} f(t), \lim_{t \to t_0} g(t), \lim_{t \to t_0} h(t)],$$
provided that the last three limits exist.

    (iii) The vector function $\mathbf{r}$, defined by $\mathbf{r}(t) = [f(t), g(t), h(t)]$, is **continuous** at t_0 if and only if f, g, and h are continuous at t_0.

    (iv) If f', g', and h' exist on I, the **derivative** of the vector function $\mathbf{r}$ is another vector function $\mathbf{r}'$, defined by
$$\mathbf{r}'(t) = [f'(t), g'(t), h'(t)], \qquad t \in I.$$

    (v) The **second derivative, $\mathbf{r}''$,** is the derivative of the first derivative, $\mathbf{r}'$; it is defined by
$$\mathbf{r}''(t) = [f''(t), g''(t), h''(t)].$$
Similar definitions hold for higher derivatives.

Example 1. A vector function **r** is defined by

$$\mathbf{r}(t) = e^t \cos t\mathbf{i} + e^t \sin t\mathbf{j} + e^t\mathbf{k}.$$

Because the exponential, sine, and cosine functions exist for all real numbers, the domain of **r** is R.

The first and second derivatives, **r**′ and **r**″, are given by

$$\mathbf{r}'(t) = (-e^t \sin t + e^t \cos t)\mathbf{i}$$
$$+ (e^t \cos t + e^t \sin t)\mathbf{j} + e^t\mathbf{k}$$

and

$$\mathbf{r}''(t) = (-e^t \cos t - e^t \sin t - e^t \sin t + e^t \cos t)\mathbf{i}$$
$$+ (-e^t \sin t + e^t \cos t + e^t \cos t + e^t \sin t)\mathbf{j} + e^t\mathbf{k}$$
$$= -2e^t \sin t\mathbf{i} + 2e^t \cos t\mathbf{j} + e^t\mathbf{k}.$$

Since **r**′ and **r**″ exist for all $t \in R$, the vector functions **r** and **r**′ are differentiable on $(-\infty, \infty)$ and thus are everywhere continuous.

When a vector is represented by a directed line segment whose initial point is the origin, it is called a **position vector**. Observe that any position vector $\mathbf{a} = [a_1, a_2, a_3]$ extends from the origin to the point (a_1, a_2, a_3) (Fig. 17-1).

17.1.2 Definition. Let a vector function **r** be continuous on an interval I. If

(1) $$\mathbf{r}(t) = [f(t), g(t), h(t)]$$

is a position vector for every $t \in I$, then the set of its terminal points, $P{:}(f(t), g(t), h(t))$, as t assumes all values in I, is a **space curve**. Equation (1) is a **vector equation** of the space curve, and equations

(2) $$x = f(t), \qquad y = g(t), \qquad z = h(t), \qquad t \in I,$$

are **parametric equations** of the space curve (Fig. 17-2).

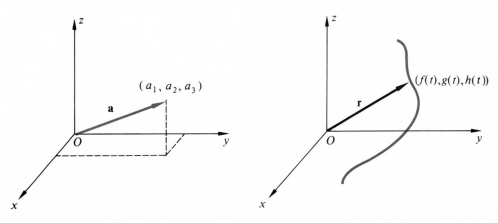

Figure 17-1   Figure 17-2

If the interval I in the preceding definition is *closed*, say $I = [a, b]$, the space curve is said to be an **arc**; and if to distinct values of t in open (a, b) there correspond distinct points on the arc (that is, if the arc does not cross itself), it is a **simple arc** (Fig. 17-3).

An arc   A simple arc

(a)   (b)

Figure 17-3

Let P_1 and P_2 be points on a simple arc that correspond to the values t_1 and t_2 of the parameter. We shall say that P_1 **precedes** P_2 on the simple arc if and only if $t_1 < t_2$. The **positive direction** on a simple arc is defined to be the direction in which the value of the parameter increases.

Let f, g, and h be differentiable functions on an interval I and let $P:(f(t), g(t), h(t))$ be an arbitrary point on the curve

(3) $$x = f(t), \qquad y = g(t), \qquad z = h(t), \quad t \in I.$$

If $Q:(f(t + \Delta t), g(t + \Delta t), h(t + \Delta t))$, $\Delta t \neq 0$, is another point on the curve, then the components of the vector

$$\overrightarrow{PQ} = [f(t + \Delta t) - f(t), g(t + \Delta t) - g(t), h(t + \Delta t) - h(t)]$$

are direction numbers of the secant line, PQ (Fig. 17-4). Since any three numbers

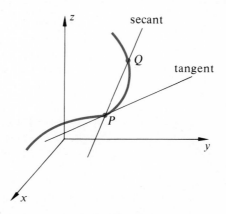

Figure 17-4

proportional to these components are also direction numbers of PQ,

$$(4) \qquad \frac{f(t + \Delta t) - f(t)}{\Delta t}, \qquad \frac{g(t + \Delta t) - g(t)}{\Delta t}, \qquad \frac{h(t + \Delta t) - h(t)}{\Delta t}$$

is another set of direction numbers of the secant line PQ.

The limits of the direction numbers (4) as $\Delta t \longrightarrow 0$,

$$\lim_{\Delta t \to 0} \frac{f(t + \Delta t) - f(t)}{\Delta t} = f'(t),$$

$$\lim_{\Delta t \to 0} \frac{g(t + \Delta t) - g(t)}{\Delta t} = g'(t),$$

$$\lim_{\Delta t \to 0} \frac{h(t + \Delta t) - h(t)}{\Delta t} = h'(t),$$

are defined to be direction numbers of the **tangent line** to the curve (3) at the point P.

17.1.3 Definition. Let f, g, and h be differentiable functions on an interval I. Then a set of **direction numbers of the tangent line** to the curve $x = f(t)$, $y = g(t)$, $z = h(t)$, $t \in I$, at the point $P:(f(t), g(t), h(t))$ is

$$f'(t), \quad g'(t), \quad h'(t).$$

Example 2. To sketch the arc whose vector equation is

$$\mathbf{r}(t) = t\mathbf{i} + \tfrac{1}{2}t^2\mathbf{j} + \tfrac{1}{3}t^3\mathbf{k}, \qquad -2 \leq t \leq 3,$$

we compute the following table of values, plot the points, and draw a smooth curve through them (Fig. 17-5).

t	-2	-1	0	1	2	3
x	-2	-1	0	1	2	3
y	2	$\frac{1}{2}$	0	$\frac{1}{2}$	2	$\frac{9}{2}$
z	$-\frac{8}{3}$	$-\frac{1}{3}$	0	$\frac{1}{3}$	$\frac{8}{3}$	9

A vector in the direction of the tangent line to the arc at any point $P(t)$ on the arc is

$$\mathbf{r}'(t) = \mathbf{i} + t\mathbf{j} + t^2\mathbf{k}.$$

Thus direction numbers of the tangent line to the arc at $P(2) = (2, 2, \frac{8}{3})$ are 1, 2, 4; and symmetric equations of this tangent line are

$$\frac{x - 2}{1} = \frac{y - 2}{2} = \frac{z - \frac{8}{3}}{4}.$$

The curve and its tangent line at $P(2)$ are shown in Fig. 17-5.

It is easy to find formulas for differentiating products involving vector functions.

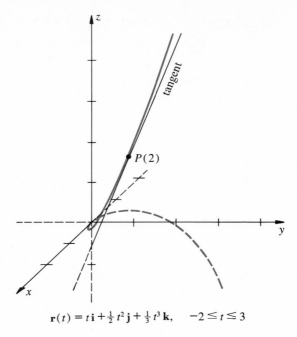

$$\mathbf{r}(t) = t\,\mathbf{i} + \tfrac{1}{2}\,t^2\,\mathbf{j} + \tfrac{1}{3}\,t^3\,\mathbf{k}, \quad -2 \le t \le 3$$

Figure 17-5

17.1.4 Theorem. If $\mathbf{r}$ and $\mathbf{s}$ are differentiable vector functions and k is a scalar constant, then

(i) $\quad \dfrac{d}{dt} k\mathbf{r} = k\,\dfrac{d\mathbf{r}}{dt}.$

(ii) $\quad \dfrac{d}{dt}(\mathbf{r} \cdot \mathbf{s}) = \dfrac{d\mathbf{r}}{dt} \cdot \mathbf{s} + \mathbf{r} \cdot \dfrac{d\mathbf{s}}{dt}.$

(iii) $\quad \dfrac{d}{dt}(\mathbf{r} \times \mathbf{s}) = \dfrac{d\mathbf{r}}{dt} \times \mathbf{s} + \mathbf{r} \times \dfrac{d\mathbf{s}}{dt}.$

It is important in (iii) that the order of $\mathbf{r}$ and $\mathbf{s}$ be maintained.

Proof. The proof of (i) is left for the reader. To prove (ii) and (iii), let $\mathbf{r}(t) = [f_1(t),\, g_1(t),\, h_1(t)]$ and $\mathbf{s}(t) = [f_2(t),\, g_2(t),\, h_2(t)]$. Since $\mathbf{r} \cdot \mathbf{s} = f_1 f_2 + g_1 g_2 + h_1 h_2$, then

$$\frac{d}{dt}(\mathbf{r} \cdot \mathbf{s}) = \left(f_1 \frac{df_2}{dt} + \frac{df_1}{dt} f_2 \right) + \left(g_1 \frac{dg_2}{dt} + \frac{dg_1}{dt} g_2 \right) + \left(h_1 \frac{dh_2}{dt} + \frac{dh_1}{dt} h_2 \right) = \frac{d\mathbf{r}}{dt} \cdot \mathbf{s} + \mathbf{r} \cdot \frac{d\mathbf{s}}{dt}.$$

Again, $\mathbf{r} \times \mathbf{s} = [g_1 h_2 - h_1 g_2,\ h_1 f_2 - h_2 f_1,\ f_1 g_2 - g_1 f_2]$. Therefore

$$\frac{d}{dt}(\mathbf{r} \times \mathbf{s})$$

$$= [g_1 h_2' + g_1' h_2 - h_1 g_2' - h_1' g_2,\ h_1 f_2' + h_1' f_2 - h_2 f_1' - h_2' f_1,\ f_1 g_2' + f_1' g_2 - g_1 f_2' - g_1' f_2]$$

$$= [g_1' h_2 - h_1' g_2,\ h_1' f_2 - h_2 f_1',\ f_1' g_2 - g_1' f_2] + [g_1 h_2' - h_1 g_2',\ h_1 f_2' - h_2' f_1,\ f_1 g_2' - g_1 f_2']$$

$$= [f'_1, g'_1, h'_1] \times [f_2, g_2, h_2] + [f_1, g_1, h_1] \times [f'_2, g'_2, h'_2]$$

$$= \frac{d\mathbf{r}}{dt} \times \mathbf{s} + \mathbf{r} \times \frac{d\mathbf{s}}{dt}. \quad \blacksquare$$

Exercises

1. Sketch the arc whose vector equation is

$$\mathbf{r}(t) = 2 \cos t\mathbf{i} + 6 \sin t\mathbf{j} + t\mathbf{k}, \qquad 0 \le t \le 2\pi.$$

2. Sketch the arc having parametric equations

$$x = 2t^2, \qquad y = 4t, \qquad z = t^3, \quad -2 \le t \le 2.$$

3. Sketch the graph of

$$x = 3t, \qquad y = 2t^2, \qquad z = t^5, \quad -2 \le t \le 2.$$

4. Sketch the arc whose vector equation is

$$\mathbf{r}(t) = t \sin t\mathbf{i} + 3t\mathbf{j} + 2t \cos t\mathbf{k}, \qquad -\pi \le t \le \pi.$$

5. Find symmetric equations of the tangent line to the arc in Exercise 1 at the point where $t = \pi/3$. Sketch it on the curve.

6. Find parametric equations of the tangent line to the curve in Exercise 2 at the point where $t = 1$. Sketch it on the curve.

7. Find symmetric equations of the tangent line to the curve in Exercise 3 at the point where $t = -1$. Sketch it on the curve.

8. Find symmetric equations of the tangent line to the curve in Exercise 4 at the point where $t = \pi/2$. Sketch it on the curve.

9. Prove that if $\mathbf{r}$ and $\mathbf{s}$ are differentiable vector functions of t,

$$\frac{d}{dt}(\mathbf{r} + \mathbf{s}) = \frac{d\mathbf{r}}{dt} + \frac{d\mathbf{s}}{dt}.$$

10. Prove that if ϕ is a differentiable scalar function of t and $\mathbf{r}$ is a differentiable vector function of t,

then

$$\frac{d}{dt}(\phi\mathbf{r}) = \phi\frac{d\mathbf{r}}{dt} + \frac{d\phi}{dt}\mathbf{r}.$$

11. If $\mathbf{r}(t) = \cos t\mathbf{i} - \sin t\mathbf{j} + 5t\mathbf{k}$ and $\mathbf{s}(t) = 2t\mathbf{i} - t^3\mathbf{j} + 4t^2\mathbf{k}$, find (a) $D_t(\mathbf{r} \cdot \mathbf{s})$ and (b) $D_t(\mathbf{r} \times \mathbf{s})$.

12. If $\mathbf{r}(t) = e^t\mathbf{i} + e^{-t}\mathbf{j} - 3t^2\mathbf{k}$ and $\mathbf{s}(t) = \ln t\mathbf{i} - 4t^3\mathbf{j} + e^{3t}\mathbf{k}$, find (a) $D_t(\mathbf{r} \cdot \mathbf{s})$ and (b) $D_t(\mathbf{r} \times \mathbf{s})$.

13. Let $\mathbf{r}$ be a differentiable vector function of t. Show that

$$D_t|\mathbf{r}| = \frac{\mathbf{r} \cdot \mathbf{r}'}{|\mathbf{r}|}.$$

14. If $\mathbf{r}(t) = te^t\mathbf{i} - \sin 2t\mathbf{j} + 3t^2\mathbf{k}$, find $D_t|\mathbf{r}(t)|$.

15. If $\mathbf{r}, \mathbf{s},$ and $\mathbf{w}$ are differentiable vector functions of t, show that

$$D_t[\mathbf{r} \cdot (\mathbf{s} \times \mathbf{w})] = \mathbf{r}' \cdot (\mathbf{s} \times \mathbf{w}) + \mathbf{r} \cdot (\mathbf{s}' \times \mathbf{w})$$
$$+ \mathbf{r} \cdot (\mathbf{s} \times \mathbf{w}').$$

16. If $\mathbf{r}(t) = t\mathbf{i} + t^2\mathbf{j} + t^3\mathbf{k}$, $\mathbf{s}(t) = \sin t\mathbf{i} - \cos t\mathbf{k}$, and $\mathbf{w}(t) = e^t\mathbf{j} + e^{-t}\mathbf{k}$, find

$$D_t[\mathbf{r} \cdot (\mathbf{s} \times \mathbf{w})].$$

17. If $\mathbf{r}, \mathbf{s},$ and $\mathbf{w}$ are differentiable vector functions of t, show that

$$D_t[\mathbf{r} \times (\mathbf{s} \times \mathbf{w})] = \mathbf{r}' \times (\mathbf{s} \times \mathbf{w})$$
$$+ \mathbf{r} \times (\mathbf{s}' \times \mathbf{w}) + \mathbf{r} \times (\mathbf{s} \times \mathbf{w}').$$

18. If $\mathbf{r}(t) = t^2\mathbf{i} - t\mathbf{k}$, $\mathbf{s}(t) = 3t\mathbf{j} + 2t^3\mathbf{k}$, and $\mathbf{w}(t) = \cos t\mathbf{i} + \sin t\mathbf{j}$, find

$$D_t[\mathbf{r} \times (\mathbf{s} \times \mathbf{w})].$$

17.2 VELOCITY, ACCELERATION, AND ARC LENGTH

Consider the motion of a particle on the space curve

$$x = f(t), \qquad y = g(t), \qquad z = h(t), \qquad t \in I,$$

where t represents time and $P:(f(t), g(t), h(t))$ is the position of the particle at time t. In this connection, the path of the particle (the space curve) is called its **trajectory**. We shall now define what is meant by the instantaneous velocity and acceleration of the particle at time t.

17.2.1 **Definition.** Let $\mathbf{r}$ be a differentiable vector function on an interval I and let

(1) $$\mathbf{r}(t) = [f(t), g(t), h(t)]$$

be a position vector for all $t \in I$. If t represents the number of units of time that have elapsed since $t = t_0$ and if $P:(f(t), g(t), h(t))$ is the position of a particle moving on the trajectory (1), then the **instantaneous velocity** of the particle at time t is the vector

$$\mathbf{v}(t) = \mathbf{r}'(t).$$

The **acceleration** of the particle at time t is defined to be the vector

$$\mathbf{a}(t) = \mathbf{v}'(t) = \mathbf{r}''(t).$$

The velocity vector

(2) $$\mathbf{v}(t) = [f'(t), g'(t), h'(t)]$$

gives the magnitude and the direction of the instantaneous velocity of the particle at time t. Since the components of the velocity vector (2) are a set of direction numbers for the tangent line to the curve at P (see 17.1.3), *the direction of the velocity of the particle at time t is along the tangent to the trajectory at P* (Fig. 17-6).

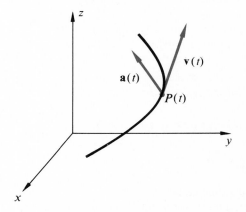

Figure 17-6

The magnitude of the velocity is given by

$$|\mathbf{v}(t)| = \sqrt{[f'(t)]^2 + [g'(t)]^2 + [h'(t)]^2},$$

which is the length of the velocity vector. It is called the **speed** of the moving particle at time t.

The acceleration vector, $\mathbf{a}(t) = \mathbf{v}'(t)$, is the derivative of the velocity vector and gives the instantaneous rate of change of the velocity vector at time t. In general, the direction of the acceleration vector is *not* along the tangent to the curve.

Example 1. Let a particle move according to the vector equation

$$\overrightarrow{OP} = a \cos t\mathbf{i} + b \sin t\mathbf{j} + t\mathbf{k},$$

where O is the origin of Cartesian coordinates and P is the position of the particle at time t. Describe the trajectory of the particle, and find its velocity and acceleration vectors at any time t.

Solution. Parametric equations of the path are

$$x = a \cos t, \qquad y = b \sin t, \qquad z = t.$$

By plotting the coordinates (x, y, z) of P for various values of t, we see that the trajectory from $t = 0$ to $t = 2\pi$ is as shown in Fig. 17-7. It is called an (elliptic) **helix**. This helix lies

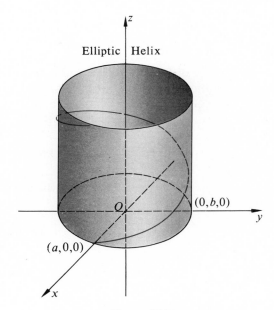

Figure 17-7

entirely on the elliptic cylinder

$$\frac{x^2}{a^2} + \frac{y^2}{b^2} = 1.$$

The velocity vector of the moving particle at time t is

$$\mathbf{v}(t) = -a \sin t\mathbf{i} + b \cos t\mathbf{j} + \mathbf{k},$$

and the acceleration vector is

$$\mathbf{a}(t) = -a \cos t\mathbf{i} - b \sin t\mathbf{j}.$$

The definition of the length of an arc in three-space is similar to that given in Section 15.3 for an arc in the plane, and a formula for it can be derived analogously.

17.2.2 Theorem. Let $f, g,$ and h be functions having continuous derivatives on a closed interval $[a, b]$ and let $\mathbf{r}(t) = [f(t), g(t), h(t)]$ be a position vector for $a \le t \le b$. Then the length of the arc

$$x = f(t), \qquad y = g(t), \qquad z = h(t)$$

from the point $(f(a), g(a), h(a))$ to the point $(f(b), g(b), h(b))$ is given by

$$L = \int_a^b |\mathbf{v}(t)|\, dt = \int_a^b \sqrt{[f'(t)]^2 + [g'(t)]^2 + [h'(t)]^2}\, dt.$$

Notice that if t represents time, $|\mathbf{v}(t)|$ in the above integral is the speed at which a particle moves along the curve; for if we write

17.2.3
$$s(t) = \int_a^t |\mathbf{v}(w)|\, dw,$$

then $s(t)$ is the distance along the curve from the fixed point given by $t = a$ to the variable position $t = t$, and the **speed** of the moving particle is

$$\frac{ds}{dt} = |\mathbf{v}(t)|.$$

Example 2. Find the length of the arc of the circular helix whose parametric equations are $x = \cos t$, $y = \sin t$, $z = t$, from $t = 0$ to $t = 2\pi$.

Solution. The position vector of a point P on the helix is $\mathbf{r}(t) = \cos t\mathbf{i} + \sin t\mathbf{j} + t\mathbf{k}$, and the length of the arc from $t = 0$ to $t = 2\pi$ is

$$s = \int_0^{2\pi} |\mathbf{r}'(t)|\, dt = \int_0^{2\pi} \sqrt{(-\sin t)^2 + (\cos t)^2 + 1^2}\, dt = \int_0^{2\pi} \sqrt{2}\, dt = 2\sqrt{2}\,\pi.$$

This can be interpreted as saying that a particle that travels for 2π seconds at the constant speed of $\sqrt{2}$ feet per second will travel a distance equal to $2\sqrt{2}\,\pi$ feet.

Exercises

In each of Exercises 1 to 8, a particle moves along the curve whose parametric equations are given. Find the velocity and acceleration vectors at any time t, and give the speed of the moving particle at the indicated time t_1. Sketch a portion of the curve that includes the point at which $t = t_1$, and carefully draw the velocity and acceleration vectors at that point.

1. $x = 3t, y = 4(t^2 - 1), z = 2t^3; t_1 = 1.$

2. $x = 4t^2, y = 9t, z = t^4; t_1 = 2.$

3. $x = (2t - t^2), y = 3t, z = t^3 + 1; t_1 = 1.$

4. $x = 2 \sin t, y = 4 \cos t, z = e^{2t}; t_1 = \dfrac{\pi}{6}.$

5. $x = \cos 2t, y = 2e^{-t}, z = 3 \sin t; t_1 = 0.$

6. $x = t \sin t, y = e^t, z = \cos t; t_1 = 0.$

7. $x = 2 \cos \tfrac{1}{2}t, y = 2 \sin \tfrac{1}{2}t, z = 3 \ln (\cos \tfrac{1}{2}t); t_1 = \dfrac{\pi}{3}.$

8. $x = 4 \sin \pi t, y = 4 \cos \pi t, z = 4 \sin \pi t; t_1 = \tfrac{1}{4}.$

9. Show that if the speed of a moving particle is constant, its acceleration vector is always perpendicular to its velocity vector.

10. The curve with parametric equations $x = e^t \cos t$, $y = e^t \sin t$, $z = e^t$ lies on a right circular cone. Find a Cartesian equation of the cone and show that the tangent vector at any point on the curve makes a constant angle with the generator of the cone passing through that point.

In each of Exercises 11 to 16, find the length of the arc whose vector equation is given.

11. $\mathbf{r}(t) = t \sin t\,\mathbf{i} + t \cos t\,\mathbf{j} + \sqrt{8}\,t\mathbf{k}; \ 0 \le t \le 4.$

12. $\mathbf{r}(t) = \sqrt{6}\,t^2\mathbf{i} + \tfrac{2}{3}t^3\mathbf{j} + 6t\mathbf{k}; \ 3 \le t \le 6.$

13. $\mathbf{r}(t) = 2t^{3/2}\mathbf{i} + (t - 3)\mathbf{j} + 3t\mathbf{k}; \ 6 \le t \le 10.$

14. $\mathbf{r}(t) = e^{2t} \cos t\,\mathbf{i} + e^{2t} \sin t\,\mathbf{j} + e^{2t}\mathbf{k}; 0 \le t \le 5.$

15. $\mathbf{r}(t) = \sinh 2t\mathbf{i} + \cosh 2t\mathbf{j} + 2t\mathbf{k}; \ -\tfrac{1}{2} \le t \le 1.$

16. $\mathbf{r}(t) = \tfrac{1}{3}t^3\mathbf{i} + \tfrac{2}{9}(t^3 + 4)^{3/2}\mathbf{j} + \tfrac{2}{3}t^3\mathbf{k}; \ 0 \le t \le 3.$

17.3 CURVATURE. VECTOR COMPONENTS

The discussion in Sections 15.8 and 15.9 of curvature and of tangential and normal vector components in two-dimensional Cartesian coordinates carries over, word for word, to three-dimensional Cartesian space. We need, then, only restate the principal results here, with one more variable.

Let

$$\mathbf{r}(t) = f(t)\mathbf{i} + g(t)\mathbf{j} + h(t)\mathbf{k}$$

define a position vector function having a continuous second derivative for $a \le t \le b$. Its set of terminal points $P{:}(f(t), g(t), h(t))$ is a smooth arc C in three-space with parametric equations

$$x = f(t), \qquad y = g(t), \qquad z = h(t), \qquad a \le t \le b.$$

The vector

$$\text{(1)} \qquad\qquad \mathbf{T}(t) = \frac{\mathbf{v}(t)}{|\mathbf{v}(t)|},$$

where $\mathbf{v}(t) = \mathbf{r}'(t)$, is the **unit tangent vector** at $P(t)$. Its length is 1, and its direction is along the tangent vector, $\mathbf{v}(t)$, to the curve C at P.

Now

$$s(t) = \int_a^t |\mathbf{v}(w)|\,dw, \qquad a \le t \le b$$

is the length of the arc C from $P(a)$ to $P(t)$; and

$$\text{(2)} \qquad\qquad \frac{ds}{dt} = |\mathbf{v}(t)|$$

is the **speed** of a particle moving on C, at time t. It is assumed that s increases with t.

The vector

$$\frac{d\mathbf{T}(t)}{ds}$$

is the **curvature vector** at P on C. It is the instantaneous rate of change of the unit tangent vector $\mathbf{T}(t)$ with respect to s.

Since the length of the unit tangent vector does not change, the curvature vector measures the rate of change of the *direction* of the tangent to the curve with respect to the distance along the curve from some fixed point.

By the chain rule,

$$\frac{d\mathbf{T}}{ds} = \frac{d\mathbf{T}}{dt}\frac{dt}{ds} = \frac{\mathbf{T}'(t)}{|\mathbf{v}(t)|},$$

since $ds/dt = |\mathbf{v}(t)|$ by (2).

17.3.1 Definition. The **curvature** $K(t)$ of the arc

$$x = f(t), \qquad y = g(t), \qquad z = h(t), \qquad a \le t \le b$$

at a point $P(t)$ is the magnitude of the curvature vector at P:

$$K(t) = \left|\frac{d\mathbf{T}(t)}{ds}\right| = \left|\frac{\mathbf{T}'(t)}{|\mathbf{v}(t)|}\right|.$$

Example 1.   Find the curvature of the circular helix

$$x = \cos t, \qquad y = \sin t, \qquad z = t$$

at any time t.

Solution.   The unit tangent vector to the helix at time t is

$$\mathbf{T}(t) = \frac{\mathbf{v}(t)}{|\mathbf{v}(t)|} = \frac{-\sin t}{\sqrt{2}}\mathbf{i} + \frac{\cos t}{\sqrt{2}}\mathbf{j} + \frac{1}{\sqrt{2}}\mathbf{k}.$$

The curvature of the circular helix at time t is

$$K(t) = \left|\frac{\mathbf{T}'(t)}{|\mathbf{v}(t)|}\right| = \left|\frac{-\cos t}{\sqrt{2}}\mathbf{i} - \frac{\sin t}{\sqrt{2}}\mathbf{j}\right| = \frac{1}{2}.$$

Notice that the curvature of the *circular* helix is constant.

The reciprocal of the curvature of a curve at a point P on it is the **radius of curvature** at P.

It is often easier to find the curvature of a space curve by using the following formula instead of applying definition 17.3.1 directly.

17.3.2 Theorem. If $\mathbf{r}(t)$ is a differentiable position vector having velocity and acceleration vectors $\mathbf{v}(t)$ and $\mathbf{a}(t)$, the **curvature** of the space curve traced out by the terminal point of $\mathbf{r}(t)$ is given by

$$K(t) = \frac{|\mathbf{a}(t) \times \mathbf{v}(t)|}{|\mathbf{v}(t)|^3}.$$

A proof of this theorem will be given at the end of this section.

The next theorem will be useful as we continue our discussion of vector functions.

17.3.3 Theorem. If a differentiable vector function $\mathbf{r}$ has constant length on an interval I, then $\mathbf{r}'(t)$ is perpendicular to $\mathbf{r}(t)$ for all $t \in I$.

Proof.   Denote $|\mathbf{r}(t)|^2$ by $F(t)$. Since $|\mathbf{r}(t)|$ was assumed constant, $F'(t) = 0$ for all $t \in I$. But (by 16.2.5)

$$F(t) = |\mathbf{r}(t)|^2 = \mathbf{r}(t) \cdot \mathbf{r}(t),$$

and (by 17.1.4)

$$F'(t) = \mathbf{r}'(t) \cdot \mathbf{r}(t) + \mathbf{r}(t) \cdot \mathbf{r}'(t) = 2\mathbf{r}'(t) \cdot \mathbf{r}(t).$$

Therefore $\mathbf{r}'(t) \cdot \mathbf{r}(t) = 0$, and (by 16.2.7) $\mathbf{r}'(t)$ is perpendicular to $\mathbf{r}(t)$ for all $t \in I$. ∎

It follows from 17.3.3 that the vector $\mathbf{T}'(t)$ is perpendicular to the unit tangent vector $\mathbf{T}(t)$ at every point on the curve.

If $|\mathbf{T}'(t)| \neq 0$, the unit vector

(3) $$\mathbf{N}(t) = \frac{\mathbf{T}'(t)}{|\mathbf{T}'(t)|}$$

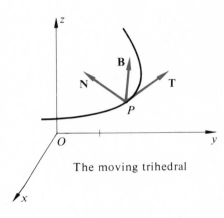

The moving trihedral

Figure 17-8

is called the **principal normal** vector to the curve C at $P(t)$. Its length is 1, and its direction is that of $\mathbf{T}'(t)$, namely perpendicular to the unit tangent vector at P (Fig. 17-8).

The vector

$$\mathbf{B}(t) = \mathbf{T}(t) \times \mathbf{N}(t)$$

is called the **binormal**. It, too, is a unit vector, and it is perpendicular to both $\mathbf{T}(t)$ and $\mathbf{N}(t)$ (Section 16.3).

If the unit tangent vector $\mathbf{T}(t)$, the principal normal $\mathbf{N}(t)$, and the binormal $\mathbf{B}(t)$ to a curve at a given point P on the curve have their initial points at P, they form a mutually perpendicular triple of unit vectors known as the **trihedral** at P (Fig. 17-8). This **moving trihedral** plays an important part in differential geometry. The plane of $\mathbf{T}(t)$ and $\mathbf{N}(t)$ is called the **osculating plane** at P.

Example 2. Find the unit tangent vector $\mathbf{T}(t)$, the principal normal $\mathbf{N}(t)$, and the binormal $\mathbf{B}(t)$ to the elliptic helix

$$x = 2 \cos t, \qquad y = 3 \sin t, \qquad z = t$$

at the point P on the helix where $t = \pi/2$. Sketch the curve from $t = 0$ to $t = 2\pi$, and draw the trihedral at P (Fig. 17-9).

Solution. Since $\mathbf{v}(t) = [-2 \sin t, 3 \cos t, 1]$ and $|\mathbf{v}(t)| = \sqrt{5 + 5 \cos^2 t}$,

$$\mathbf{T}(t) = \frac{-2 \sin t}{\sqrt{5 + 5 \cos^2 t}}\mathbf{i} + \frac{3 \cos t}{\sqrt{5 + 5 \cos^2 t}}\mathbf{j} + \frac{1}{\sqrt{5 + 5 \cos^2 t}}\mathbf{k}$$

and

$$\mathbf{T}'(t) = \frac{-20 \cos t}{(5 + 5 \cos^2 t)^{3/2}}\mathbf{i} - \frac{15 \sin t}{(5 + 5 \cos^2 t)^{3/2}}\mathbf{j} + \frac{5 \sin t \cos t}{(5 + 5 \cos^2 t)^{3/2}}\mathbf{k}.$$

Thus

$$\mathbf{T}\left(\frac{\pi}{2}\right) = \left[\frac{-2}{\sqrt{5}}, 0, \frac{1}{\sqrt{5}}\right], \qquad \mathbf{T}'\left(\frac{\pi}{2}\right) = \left[0, \frac{-3}{\sqrt{5}}, 0\right], \qquad \mathbf{N}\left(\frac{\pi}{2}\right) = [0, -1, 0].$$

Therefore

$$\mathbf{B}\left(\frac{\pi}{2}\right) = \mathbf{T}\left(\frac{\pi}{2}\right) \times \mathbf{N}\left(\frac{\pi}{2}\right) = \left[\frac{1}{\sqrt{5}}, 0, 2\sqrt{5}\right].$$

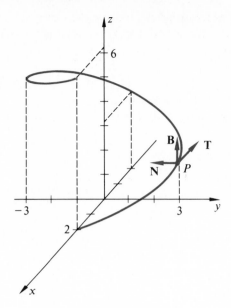

Figure 17-9

The acceleration vector at $P(t)$ on the curve C can be written

$$\mathbf{a}(t) = f''(t)\mathbf{i} + g''(t)\mathbf{j} + h''(t)\mathbf{k}.$$

Here the vector components of $\mathbf{a}(t)$ are the three terms in the right-hand member. They are parallel to the coordinate axes.

To write $\mathbf{a}(t)$ as the sum of a tangential vector component and a normal vector component, we start with $\mathbf{v}(t) = |\mathbf{v}(t)|\mathbf{T}(t)$ from (1), or

$$\mathbf{v}(t) = \frac{ds}{dt}\mathbf{T}(t).$$

Differentiating both members of this equation with respect to t gives

$$\mathbf{a}(t) = \frac{d^2s}{dt^2}\mathbf{T}(t) + \frac{ds}{dt}\frac{d}{dt}\mathbf{T}(t)$$

or, from equation (3),

(4) $$\mathbf{a}(t) = \frac{d^2s}{dt^2}\mathbf{T}(t) + |\mathbf{v}(t)||\mathbf{T}'(t)|\mathbf{N}(t).$$

This equation expresses the acceleration vector $\mathbf{a}(t)$ at any point $P(t)$ on the curve C as the sum of its **tangential vector component**

$$\frac{d^2s}{dt^2}\mathbf{T}(t)$$

and its **normal vector component**

$$|\mathbf{v}(t)||\mathbf{T}'(t)|\mathbf{N}(t).$$

It follows from the parallelogram law that if the acceleration vector at a point P on the curve is drawn with its initial point at P (as is customary), *the acceleration vector will lie in the osculating plane at* P.

Proof of 17.3.2. Equation (1) can be written

(5)
$$\mathbf{v}(t) = |\mathbf{v}(t)|\mathbf{T}(t).$$

Since the unit vectors $\mathbf{N}(t)$ and $\mathbf{T}(t)$ are perpendicular, it follows from 16.3.3 that

(6)
$$|\mathbf{N}(t) \times \mathbf{T}(t)| = |\mathbf{N}(t)||\mathbf{T}(t)| = 1.$$

From (4), (5), and 17.3.1 [and remembering that $\mathbf{T}(t) \times \mathbf{T}(t) = \mathbf{0}$],

$$\mathbf{a}(t) \times \mathbf{v}(t) = \left\{ \frac{d^2s}{dt^2}\mathbf{T}(t) + |\mathbf{v}(t)||\mathbf{T}'(t)|\mathbf{N}(t) \right\} \times |\mathbf{v}(t)|\mathbf{T}(t)$$

$$= |\mathbf{v}(t)|^2|\mathbf{T}'(t)|\mathbf{N}(t) \times \mathbf{T}(t)$$

$$= |\mathbf{v}(t)|^3 K(t)\mathbf{N}(t) \times \mathbf{T}(t).$$

Thus

$$|\mathbf{a}(t) \times \mathbf{v}(t)| = |\mathbf{v}(t)|^3 K(t)|\mathbf{N}(t) \times \mathbf{T}(t)| = |\mathbf{v}(t)|^3 K(t),$$

by (6). Therefore

$$K(t) = \frac{|\mathbf{a}(t) \times \mathbf{v}(t)|}{|\mathbf{v}(t)|^3}. \quad \blacksquare$$

Exercises

In each of Exercises 1 to 8, find the curvature, the unit tangent vector, the principal normal, and the binormal for the given curve at the given value, t_1, of t. Then sketch a part of the curve containing the indicated point $P_1(t_1)$ and carefully draw the trihedral there.

1. $\mathbf{r}(t) = (t^2 - 1)\mathbf{i} + (2t + 3)\mathbf{j} + (t^2 - 4t)\mathbf{k}$; $t_1 = 2$.

2. $\mathbf{r}(t) = \frac{1}{2}t^2\mathbf{i} + t\mathbf{j} + \frac{1}{3}t^3\mathbf{k}$; $t_1 = 2$.

3. $x = \sin 3t$, $y = \cos 3t$, $z = t$; $t_1 = \frac{\pi}{9}$.

4. $x = 8 \sin t$, $y = 8 \cos t$, $z = 4t$; $t_1 = \frac{\pi}{6}$.

5. $\mathbf{r}(t) = e^t \sin t\mathbf{i} + e^t \cos t\mathbf{j} + e^t\mathbf{k}$; $t_1 = \frac{\pi}{2}$.

6. $\mathbf{r}(t) = e^{-2t}\mathbf{i} + e^{2t}\mathbf{j} + 2\sqrt{2}\,t\mathbf{k}$; $t_1 = 0$.

7. $x = 2e^{-t}$, $y = \sin t$, $z = -2 \cos t$; $t_1 = 0$.

8. $x = \cos t$, $y = e^t$, $z = t^2 + 2$; $t_1 = 0$.

9. Let $\mathbf{r}$ be a twice differentiable vector function of t. Show that the normal vector component of the acceleration vector $\mathbf{a}(t)$ can be written $|\mathbf{v}(t)||\mathbf{T}'(t)|$.

In Exercises 10 to 13, find the tangential and normal vector components of the acceleration vector for the given curve at any point on it.

10. $x = t$, $y = t^2$, $z = t^3$.

11. $x = e^{-t}$, $y = 2t$, $z = e^t$.

12. $\mathbf{r}(t) = t\mathbf{i} + \frac{1}{3}t^3\mathbf{j} + t^{-1}\mathbf{k}$.

13. $\mathbf{r}(t) = \ln \sin t\mathbf{i} + \ln \cos t\mathbf{j} + t\mathbf{k}$.

14. Show that if the binormal to the arc

$$x = f(t), \qquad y = g(t), \qquad z = h(t), \qquad a \le t \le b$$

at the point $P_1(t_1)$ is

$$\mathbf{B}(t_1) = b_1(t_1)\mathbf{i} + b_2(t_1)\mathbf{j} + b_3(t_1)\mathbf{k},$$

then a Cartesian equation of the *osculating plane* at P_1 is

$$(x - x_1)b_1(t_1) + (y - y_1)b_2(t_1)$$
$$+ (z - z_1)b_3(t_1) = 0.$$

17.4 THE LAWS OF PLANETARY MOTION*

From a large mass of empirical data, due principally to the astronomical observations of Tycho Brahe, Johannes Kepler (1571–1630) after years of struggle announced the three laws of planetary motion that bear his name. About a half century later Isaac Newton (1642–1727) used his newly invented calculus to deduce Kepler's laws from Newton's own laws of gravitation and motion. Both Kepler's and Newton's work, in quite different ways, were tremendous achievements.

We shall now prove Kepler's laws. Comprehension of these proofs can hardly fail to impress the student with the great power of the calculus he has already mastered.

Kepler's three **laws of planetary motion** are:

1. Each planet moves in an elliptical orbit, with the sun at one focus of the ellipse.

2. The radius vector from the sun to the moving planet sweeps out area at a constant rate.

3. The square of the time required for each planet to travel once around its elliptical orbit is proportional to the cube of the length of the semimajor axis of the ellipse.

According to Newton's law of gravitation, the force of attraction between the sun and any one of its planets is directly proportional to the product of the masses of the sun and the planet, and inversely proportional to the square of the distance between them. Since the mass of the sun is so very much greater than the mass of the planet, we shall think of the sun as fixed in position and that the force of attraction is acting on the planet as it moves about the sun. In the discussion that follows, all other forces are neglected.

Let $\mathbf{r}(t)$ be a position vector whose initial point O is the center of the stationary sun and whose terminal point P is the center of the moving planet. Denote by $\mathbf{u}_r(t)$ a unit vector in the direction of $\mathbf{r}(t)$. Then

$$\mathbf{r}(t) = r(t)\mathbf{u}_r(t),$$

where $r(t) = |\mathbf{r}(t)|$. By Newton's law of gravitation, the force of attraction acting on the moving planet is

(1)
$$\mathbf{F}(t) = -\frac{GmM}{r^2(t)}\mathbf{u}_r(t),$$

where m is the mass of the planet, M is the mass of the sun, G is the universal gravitational constant, and $r^2(t) = |\mathbf{r}(t)|^2$ is the square of the distance between the sun and the planet at any time t; the negative sign indicates that the direction of the force is opposite to the direction of $\mathbf{r}(t)$.

But from Newton's second law of motion

(2)
$$\mathbf{F}(t) = m\mathbf{a}(t),$$

*Optional.

where $\mathbf{a}(t)$ is the acceleration vector of the moving planet. By equating the right-hand members of (1) and (2), we find

(3) $$\mathbf{a}(t) = -\frac{GM}{r^2(t)}\mathbf{u}_r(t).$$

We shall first prove that the path of the planet (the curve traced out by P as t increases) is a *plane* curve.

It is evident from (3) that $\mathbf{u}_r(t)$ and $\mathbf{a}(t)$ are parallel; and since $\mathbf{r}(t)$ has the same direction as $\mathbf{u}_r(t)$, $\mathbf{r}(t)$ and $\mathbf{a}(t)$ are parallel for all values of t. Thus $\mathbf{r}(t) \times \mathbf{a}(t) = \mathbf{0}$ (by 16.3.4). Moreover, $\mathbf{v}(t) \times \mathbf{v}(t) = \mathbf{0}$. Therefore (17.1.4)

$$D_t(\mathbf{r}(t) \times \mathbf{v}(t)) = \mathbf{r}' \times \mathbf{v} + \mathbf{r} \times \mathbf{v}' = \mathbf{v} \times \mathbf{v} + \mathbf{r} \times \mathbf{a} = \mathbf{0}.$$

Since the derivative of the vector function $\mathbf{r}(t) \times \mathbf{v}(t)$ is $\mathbf{0}$,

(4) $$\mathbf{r}(t) \times \mathbf{v}(t) = \mathbf{h},$$

where $\mathbf{h}$ is some constant vector. Thus the vectors $\mathbf{r}(t)$, for all values of t, are perpendicular to the constant vector $\mathbf{h}$; and since all $\mathbf{r}(t)$ have O for their initial point, P traces out a plane curve as t increases. Therefore *the path of the planet about the sun is a plane curve.*

We shall now prove Kepler's first law, that the path of the planet is an ellipse having one focus at the sun.

Proof of Kepler's First Law. Assume a polar coordinate system in the plane of the planet's path, with the pole at O, the center of the sun. We saw in Exercise 8 of Section 15.9 that

(5) $$\mathbf{v}(t) = \frac{dr}{dt}\mathbf{u}_r + r\frac{d\theta}{dt}\mathbf{u}_\theta.$$

From (4) and (5) we obtain

(6) $$\mathbf{h} = r\mathbf{u}_r \times \left(\frac{dr}{dt}\mathbf{u}_r + r\frac{d\theta}{dt}\mathbf{u}_\theta\right) = r^2\frac{d\theta}{dt}(\mathbf{u}_r \times \mathbf{u}_\theta),$$

since $\mathbf{u}_r \times \mathbf{u}_r = \mathbf{0}$.

From (3), (6), 16.2.7 and 16.3.2, we get

$$\mathbf{a}(t) \times \mathbf{h} = -\frac{GM}{r^2}\mathbf{u}_r \times r^2\frac{d\theta}{dt}(\mathbf{u}_r \times \mathbf{u}_\theta)$$

$$= -GM\frac{d\theta}{dt}\mathbf{u}_r \times (\mathbf{u}_r \times \mathbf{u}_\theta)$$

$$= -GM\frac{d\theta}{dt}[(\mathbf{u}_r \cdot \mathbf{u}_\theta)\mathbf{u}_r - (\mathbf{u}_r \cdot \mathbf{u}_r)\mathbf{u}_\theta]$$

$$= GM\frac{d\theta}{dt}\mathbf{u}_\theta.$$

Since $\mathbf{a}(t) = d\mathbf{v}(t)/dt$ and $\mathbf{u}_\theta = d\mathbf{u}_r/d\theta$ (see Exercise 8 of Section 15.9), the preceding equation can be rewritten

(7) $$\frac{d}{dt}(\mathbf{v}(t) \times \mathbf{h}) = \frac{d}{dt}(GM\mathbf{u}_r).$$

By integrating both members of (7), we obtain

$$\mathbf{v}(t) \times \mathbf{h} = GM\mathbf{u}_r + \mathbf{c},$$

in which $\mathbf{c}$ is a constant vector. Let $\mathbf{c} = GM\mathbf{e}$, where $\mathbf{e}$ is another constant vector. Then the preceding equation becomes

(8)
$$\mathbf{v}(t) \times \mathbf{h} = GM(\mathbf{u}_r(t) + \mathbf{e}).$$

We now proceed to find an equation for r (the length of $\mathbf{r}$) from (4) and (8). By using (4), we can write

$$\mathbf{h} \cdot \mathbf{h} = (\mathbf{r}(t) \times \mathbf{v}(t)) \cdot \mathbf{h} = \mathbf{r}(t) \cdot (\mathbf{v}(t) \times \mathbf{h}).$$

Substituting from (8) in this result gives

$$\mathbf{h} \cdot \mathbf{h} = \mathbf{r}(t) \cdot GM(\mathbf{u}_r(t) + \mathbf{e}) = GM(\mathbf{r} \cdot \mathbf{u}_r + \mathbf{r} \cdot \mathbf{e}),$$

or

$$|\mathbf{h}|^2 = GM(|\mathbf{r}|\mathbf{u}_r \cdot \mathbf{u}_r + \mathbf{r} \cdot \mathbf{e}).$$

Denote $|\mathbf{h}|$ by h and $|\mathbf{r}(t)|$ by r. Then this equation can be written

(9)
$$h^2 = GM(r + \mathbf{r}(t) \cdot \mathbf{e}).$$

Let the initial point of the constant vector $\mathbf{e}$ be at O. We see from (8) that the vector $\mathbf{u}_r(t) + \mathbf{e}$ is perpendicular to $\mathbf{h}$, and thus $\mathbf{u}_r(t) + \mathbf{e}$ is in the plane of the path of the planet. Since $\mathbf{u}_r(t)$ is also in that plane, $\mathbf{e}$ must be in the plane of the planet's path. Denote the angle from $\mathbf{e}$ to $\mathbf{r}(t)$ by θ. Then $\mathbf{r}(t) \cdot \mathbf{e} = re \cos \theta$, where $e = |\mathbf{e}|$, and (9) becomes

$$h^2 = rGM(1 + e \cos \theta).$$

Let $d = h^2/(GMe)$; then $h^2 = dGMe$, and the preceding equation can be written $dGMe = rGM(1 + e \cos \theta)$, or

(10)
$$r = \frac{ed}{1 + e \cos \theta}.$$

This is the polar equation of a conic with eccentricity e, and having one focus at the sun's center. Since the orbit of a planet is a closed curve, the conic is an ellipse; thus $0 < e < 1$. This completes the proof of **Kepler's first law.** ∎

Proof of Kepler's Second Law. From equation (6)

$$\mathbf{h} = r^2 \frac{d\theta}{dt}(\mathbf{u}_r \times \mathbf{u}_\theta).$$

Since $\mathbf{u}_r$ and $\mathbf{u}_\theta$ are unit vectors in the plane of the planet's orbit, $\mathbf{u}_r \times \mathbf{u}_\theta$ is a unit vector perpendicular to the plane of the planet's orbit. Let θ increase as t increases; then

$$|\mathbf{h}| = h = \left| r^2 \frac{d\theta}{dt} \right| = r^2 \frac{d\theta}{dt},$$

or

(11)
$$[r(t)]^2 \frac{d\theta}{dt} = h.$$

The polar coordinates of P, on the planet's orbit, are functions of t; so $r = r(t)$ and $\theta = \theta(t)$.

Let $\theta_1 = \theta(t_1)$ and $\theta_2 = \theta(t_1 + t)$, $t \geq 0$. The area swept out by $\mathbf{r}(t)$ from time t_1 to time $t_1 + t$ is

$$A = \int_{\theta_1}^{\theta_2} r^2(t) \, d\theta,$$

by 12.16.1. By changing to the variable t and using (11), we can write

$$A = \frac{1}{2} \int_{t_1}^{t_1+t} r^2(t) \frac{d\theta}{dt} \, dt = \frac{1}{2} \int_{t_1}^{t_1+t} h \, dt = \frac{ht}{2}.$$

Notice that this result is independent of t_1 and thus of the initial position of P. The area swept out by $\mathbf{r}(t)$ in t units of time is

$$A(t) = \frac{ht}{2}.$$

Therefore *the rate at which $\mathbf{r}(t)$ sweeps out area as the planet moves in its orbit is $\frac{1}{2}h$, a constant.* This proves **Kepler's second law.** ∎

Proof of Kepler's Third Law. According to Kepler's second law, the constant rate at which $\mathbf{r}(t)$ sweeps out area as the planet moves in its orbit is $\frac{1}{2}h$ per unit of time. If T is the total time that it takes the planet to make one complete journey about its orbit, $\mathbf{r}(t)$ sweeps out $\frac{1}{2}hT$ square units of area in T units of time. But this is the area of the entire ellipse, πab (Exercise 19, Section 12.5); here, as usual, a and b denote the lengths of the semimajor and semiminor axes of the ellipse. Thus $\frac{1}{2}hT = \pi ab$; and since $b = a\sqrt{1 - e^2}$ (see 12.5.1), this may be written $T = (2\pi a^2 \sqrt{1 - e^2})/h$ or

(12) $$T^2 = \frac{4\pi^2 a^4 (1 - e^2)}{h^2}.$$

When $\theta = 0$ in the polar equation of the ellipse (10), $r = ed/(1 + e)$; and when $\theta = \pi$, $r = ed/(1 - e)$. The sum of these two values of r is the length of the major axis of the ellipse, $2a$. That is,

$$\frac{ed}{1 + e} + \frac{ed}{1 - e} = 2a,$$

or

(13) $$a = \frac{ed}{1 - e^2}.$$

Since d was defined by $d = h^2/(GMe)$, equation (13) can be rewritten as $a = h^2/GM\,(1 - e^2)$, from which

(14) $$h^2 = GMa(1 - e^2).$$

By combining (14) and (12), and simplifying, we obtain

$$T^2 = \left(\frac{4\pi^2}{GM}\right) a^3.$$

That is, *the square of the time it takes a planet to go once around its elliptical orbit is proportional to the cube of the length of its semimajor axis.* This completes the proof of **Kepler's third law.** ∎

17.5 REVIEW EXERCISES

1. Sketch the arc whose vector equation is
$$\mathbf{r}(t) = t\mathbf{i} + \tfrac{1}{2}t^2\mathbf{j} + \tfrac{1}{3}t^3\mathbf{k}, \qquad -2 \le t \le 3.$$

2. Find a vector in the direction of the tangent to the curve in Exercise 1 at the point where $t = 2$.

3. Sketch the arc whose vector equation is
$$\mathbf{r}(t) = 2\cos t\mathbf{i} + 3\sin t\mathbf{j} + t\mathbf{k}, \qquad 0 \le t \le 2\pi.$$

4. Find a tangent vector to the curve in Exercise 3 at the point where $t = \tfrac{1}{6}\pi$.

5. Sketch the arc whose vector equation is
$$\mathbf{r}(t) = e^t \cos t\mathbf{i} + e^t \sin t\mathbf{j} + e^t\mathbf{k}, \qquad -\pi \le t \le 2\pi.$$

6. Find the first and second derivatives of the vector function $\mathbf{r}$ defined in Exercise 5.

7. Sketch the arc having the vector equation
$$\mathbf{r}(t) = t\cos t\mathbf{i} + t\sin t\mathbf{j} + 2t\mathbf{k}, \qquad 0 \le t \le 2\pi.$$

8. Find a unit tangent vector to the arc in Exercise 7 at the point where $t = \tfrac{1}{2}\pi$.

9. Prove the theorem: If $\mathbf{r}(t)$ is a differentiable vector function having constant magnitude, then its derivative, $\mathbf{r}'(t)$, is perpendicular to $\mathbf{r}(t)$.

10. Find symmetric equations of the tangent line to the arc in Exercise 7 at the point where $t = \tfrac{1}{2}\pi$. Sketch it on the curve.

In Exercises 11 and 12, a particle moves along the curve whose parametric equations are given. Find the velocity and acceleration vectors at any time t, and give the speed of the moving particle at the indicated time t_1. Sketch a portion of the curve that includes the point $P_1(t_1)$ at which $t = t_1$, and carefully draw the velocity and acceleration vectors at that point.

11. $x = 2t, y = t^3, z = t^2 - 1; t_1 = 2$.

12. $x = \cos t, y = \sin t, z = e^t; t_1 = \tfrac{1}{2}\pi$.

13. Show that a particle moving on the elliptic helix
$$x = 2\cos t, \qquad y = 3\sin t, \qquad z = t,$$
where t represents time, rises on a cylinder at a constant rate. Sketch the helix for $0 \le t \le 2\pi$, and sketch the part of the cylinder on which it lies.

14. Find the length of the arc
$$\mathbf{r}(t) = e^t \sin t\mathbf{i} + e^t \cos t\mathbf{j} + e^t\mathbf{k}, \qquad 1 \le t \le 5.$$

In Exercises 15 and 16, find the curvature $K(t_1)$, the unit tangent vector $\mathbf{T}(t_1)$, and the principal normal $\mathbf{N}(t_1)$ for the given curve at the given value t_1 of t. Then sketch a part of the curve containing the indicated point $P_1(t_1)$ and draw the two unit vectors with their initial points at $P_1(t_1)$.

15. $x = t, y = t^2, z = t^3; t_1 = 1$.

16. $x = e^t, y = e^{-t}, z = 2t; t_1 = \ln 2$.

18

functions of two or more variables

Calculus of functions of more than one independent variable will be discussed in this chapter. Because most of the interesting and useful features of the calculus of several variables appear when we extend our definitions and theorems from functions of one variable to functions of two independent variables, much of our attention will be given to the two-variable cases. From the pattern thus established, it is easy to go on to functions of three or more variables.

We first extend the concepts of limit, continuity, and differentiation to functions of two or more variables, and then study directional derivatives, the gradient vector, tangent planes and normal lines to a surface, and maximum and minimum values of functions of two variables.

The chapter concludes with a discussion of line integrals; they are an extension of the familiar Riemann integral to functions of two variables in which the interval of integration is an arc of a plane curve instead of a segment of the x axis.

18.1 FUNCTIONS OF MORE THAN ONE INDEPENDENT VARIABLE

We defined a *function* to be a correspondence between two sets of numbers, the domain and the range, such that to each number in the domain there corresponds one and only one number in the range and each number of the range is the correspondent of at least one number of the domain (2.8.1).

Because of the concept about to be introduced—namely, a function of two variables—it will sometimes be convenient to refer to the function defined above as a **function of one variable**. Thus $y = 3x^2 - 1$ defines a function of one (independent) variable x because to each number x in the domain of real numbers there corresponds exactly one number y such that $y = 3x^2 - 1$.

The domain of a function of two variables is not a set of numbers but a set of *ordered pairs* of numbers, and a function of two variables associates with each ordered pair of numbers in its domain one and only one number of the range.

For instance, $z = x^2 + 3y^2$ defines a function of two (independent) variables x and y, because to each ordered pair of numbers (x, y) there corresponds one and only one number z such that $z = x^2 + 3y^2$. If the ordered pair is $(x, y) = (2, 3)$, then the corresponding number z in the range is $z = 2^2 + 3(3^2) = 31$; if the ordered pair is $(x, y) = (-1, 0)$, the corresponding number z is $z = (-1)^2 + 3(0^2) = 1$.

18.1.1   **Definition.**   Let $\mathfrak{D}$ be a set of ordered pairs of real numbers and let $\mathfrak{R}$ be a set of real numbers. A rule of correspondence that assigns to each ordered pair of $\mathfrak{D}$ one and only one number of $\mathfrak{R}$, and that assigns each number of $\mathfrak{R}$ to at least one ordered pair of $\mathfrak{D}$, is called a **function of two variables**. The set $\mathfrak{D}$ is the **domain** of the function and the set $\mathfrak{R}$ is its **range**.

A function of two variables, then, involves three things: its *domain* $\mathfrak{D}$, which is a set of ordered pairs of real numbers, its *range* $\mathfrak{R}$, which is a set of real numbers, and a *rule of correspondence*, that enables us to determine which unique number of the range corresponds to each ordered pair of the domain.

It is possible, of course, for several ordered pairs of $\mathfrak{D}$ to have the same corresponding number in $\mathfrak{R}$, but it is *not* possible for more than one number of $\mathfrak{R}$ to correspond to any particular ordered pair of $\mathfrak{D}$.

Let f be a function of two variables. Then $f(x, y)$, read "f of x, y," is the unique number in the range that corresponds to the ordered pair (x, y) in the domain; $f(x, y)$ is called the **value** of the function f at (x, y). The variables x and y are spoken of as the **independent variables** of the function; and if $z = f(x, y)$, then z is the **dependent variable**.

When the domain is not explicitly stated in the definition of a function of two variables, we will understand that it consists of all ordered pairs of *real* numbers (x, y) that make $f(x, y)$ a unique *real* number. As an illustration, the domain of the function ϕ defined by $\phi(s, t) = 2/\sqrt{s^2 - 4t^2}$ is the set of all ordered pairs of real numbers (s, t) such that $|s| > 2|t|$, and the range is the set of all positive numbers.

A function f of two variables x and y gives rise to a set of ordered triples of numbers $(x, y, f(x, y))$, or (x, y, z) where $z = f(x, y)$. If we think of these ordered triples of numbers (x, y, z) as the Cartesian coordinates of a point in three-dimensional space, the set of points $\{P{:}(x, y, z) \,|\, (x, y) \in \mathfrak{D} \text{ and } z = f(x, y)\}$ is the Cartesian **graph** of the function f. The domain $\mathfrak{D}$ is represented by the set of points in the xy plane whose coordinates are the ordered pairs belonging to $\mathfrak{D}$. Since to each point (x, y) in $\mathfrak{D}$ there corresponds one and only one number z, *no line perpendicular to the xy plane can interesect the graph of f in more than one point.*

It will often be convenient, in what follows, to use the notation $f(x, y)$ for a function f whose values are given by $f(x, y)$. This notation shows immediately the number of independent variables and the letters used to represent them.

Example 1. The graph of the function f defined by $f(x, y) = \frac{1}{3}\sqrt{36 - 9x^2 - 4y^2}$ is the upper half of the ellipsoid (Fig. 18-1). Its domain is $\mathfrak{D} = \{(x, y) \mid 36 - 9x^2 - 4y^2 \geq 0\}$ and its range is $\mathfrak{R} = \{z \mid 0 \leq z \leq 2\}$.

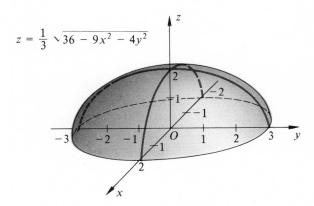

$$z = \frac{1}{3}\sqrt{36 - 9x^2 - 4y^2}$$

Figure 18-1

As we have just seen, the graph of a function f of two variables is usually a surface in three-dimensional space. But a surface, whose equation $z = f(x, y)$ is given, is often difficult to sketch. Fortunately, there is another, easier way to represent a function of two variables graphically.

If a plane $z = k$, parallel to the xy plane, intersects the surface $z = f(x, y)$, and the curve of intersection is projected onto the xy plane, each point on the projected curve will correspond to the unique point on the surface that is k units above (or below) it (Fig. 18-2). If a set of n such planes $z = k_i$ ($i = 1, 2, 3, \cdots, n$), all parallel to the xy plane, intersects a surface $z = f(x, y)$ and all the curves of intersection are plotted *in the xy plane,* this set of curves is a mapping onto the xy plane of the intersection curves. Each projected curve in the xy plane is called a **level curve** because the points on the surface that correspond to it are all at the same height, k. Level curves for the function f in Example 1 are shown in Fig. 18-3(a).

This device is used in topographical maps. A person in a mountainous region who follows the path of a level curve on a topographical map stays at the same elevation; he neither ascends nor descends (Fig. 18-2).

If the temperature t of a point (x, y) on a metal plate is given by $t = f(x, y)$, the curve $f(x, y) = k$, where k is a constant, is called an **isothermal**, because all points on it have the same temperature k.

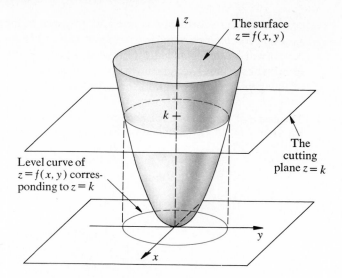

Figure 18-2

Level curves for
$$z = \tfrac{1}{3}\sqrt{36 - 9x^2 - 4y^2} \quad \text{and} \quad z = \sqrt{x^2 - y^2 - 1}$$

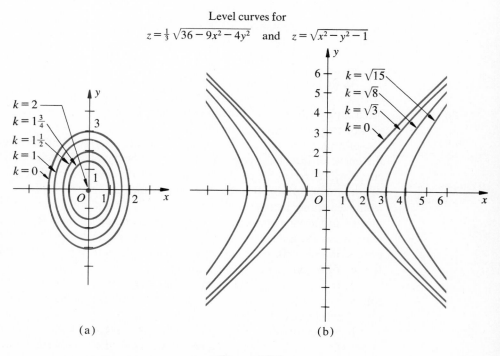

(a)   (b)

Figure 18-3

Again, the electrostatic potential of a point in the xy plane may be given as a function f of the coordinates (x, y) of the point; and the curves $f(x, y) = k_i (i = 1, 2, 3, \cdots, n)$ are called **equipotential curves**, since all points on such a curve have the same electrostatic potential.

Example 2. Find level curves for the function defined by $z = \sqrt{x^2 - y^2 - 1}$.

Solution. The graph of the this function is the upper half of the hyperboloid of two sheets shown in Fig. 16-41 (with $a = b = c = 1$).

We seek curves in the xy plane that are projections of the curve of intersection of the surface $z = \sqrt{x^2 - y^2 - 1}$ and the plane $z = k$, for various nonnegative values of k. By substituting k for z in the given equation and simplifying, we find the equation of the level curves in the xy plane to be

$$\frac{x^2}{k^2 + 1} - \frac{y^2}{k^2 + 1} = 1.$$

For each value assigned to k, this is the equation of a hyperbola in standard position. Thus for $k = \sqrt{3}$, the hyperbola is $x^2/4 - y^2/4 = 1$. The computation is simplified by choosing values of k that make $k^2 + 1$ a square. The level curves for $k = 0$, $k = \sqrt{3}$, $k = \sqrt{8}$, and $k = \sqrt{15}$ are shown in Fig. 18-3(b).

There is no difficulty in extending our definitions of a function of one variable and a function of two variables to functions of three or more variables.

Let $\mathfrak{D}$ be a set of ordered triples of real numbers and let $\mathfrak{R}$ be a set of real numbers. A rule of correspondence that assigns to each ordered triple of $\mathfrak{D}$ one and only one number of $\mathfrak{R}$, and that assigns each number of $\mathfrak{R}$ to at least one ordered triple of $\mathfrak{D}$, is called a **function of three variables**. The set $\mathfrak{D}$ is the **domain** of the function and the set $\mathfrak{R}$ is its **range**.

Exercises

Find and sketch the Cartesian graph of the *domain* of each of the functions of two variables defined in Exercises 1 to 10, showing clearly any points on the boundaries of the domain that belong to the domain.

1. $z = \dfrac{1}{x - y}$.

2. $z = \dfrac{1}{x^2 + y^2}$.

3. $z = 4x^2 + 9y^2$.

4. $z = \sqrt{x^2 + y^2 + z^2}$.

5. $z = \sqrt{16 - 4x^2 - y^2}$.

6. $z = \ln\left(x - \dfrac{1}{y}\right)$.

7. $z = \sqrt{36 - 4x^2 - 9y^2}$.

8. $z = \dfrac{\sqrt{9 - x^2 - y^2}}{x}$.

9. $z = -\sqrt{49x^2 + 36y^2 - 1764}$.

10. $z = \sqrt{-4x + 3y + 12}$.

11. Find the range of the functions in Exercises 2 and 7.

12. Find the ranges of the functions in Exercises 8 and 10.

In Exercises 13 to 18, sketch a convenient part of the graph of the indicated function when x, y, and z are nonnegative.

13. $z = 2 - x - y^2$.

14. $z = 3 - x^2 - y^2$.

15. $z = 3 - xy$.

16. $z = x^2 \ln y$.

17. $z = e^{-x} \sin y$.

18. $z = 2 - [\![x]\!] + [\![y]\!]$.

19. Sketch the level curves of $z = \frac{1}{2}(x^2 + y^2)$ for $z = 0, 2, 4, 6$, and 8.

20. Sketch the level curves of $z = e^{(x^2-y)}$ for $z = 1, 3, 5$, and 8.

21. Sketch the level curves of

$$z = \frac{x^2 + y}{x + y^2}$$

for $z = 0, 1, 2$, and 4. Draw the curve in the xy plane whose points are *not* in the domain of this function.

22. If the temperature at any given point in the xy plane is given by $t = x^2 - y^2$, draw the isothermals for $t = -4, -2, 0, 2, 4$.

23. The temperature at any point in the xy plane is given by $t = y - \sin x$. Draw the isothermals for $t = -2, -1, 0, 1, 2$.

24. If the electrostatic potential at any point (except one) in the xy plane is given by

$$p = \frac{4}{\sqrt{(x-2)^2 + (y+3)^2}},$$

draw the equipotential curves for $p = \frac{1}{2}, 1, 2$, and 4.

In Exercises 25 to 28, describe geometrically the domain of each of the indicated functions of three variables.

25. $f(x, y, z) = \sqrt{x^2 + y^2 + z^2 - 16}$.

26. $f(x, y, z) = \sqrt{x^2 + y^2 - z^2 - 9}$.

27. $f(x, y, z) = \sqrt{144 - 16x^2 - 9y^2 - 144z^2}$.

28. $f(x, y, z) = \dfrac{(144 - 16x^2 - 16y^2 + 9z^2)^{3/2}}{xyz}$.

A **level surface** for a function f of three variables is the graph of the set of points in three-dimensional space whose coordinates satisfy an equation $f(x, y, z) = k$, where k is a constant; that is,

$$\{P{:}(x, y, z) \mid f(x, y, z) = k, k \text{ is a constant}\}.$$

Describe geometrically the level surfaces for the functions defined in Exercises 29 to 33.

29. $f(x, y, z) = x^2 + y^2 + z^2; k > 0$.

30. $f(x, y, z) = 100x^2 + 16y^2 + 25z^2; k > 0$.

31. $f(x, y, z) = 16x^2 + 16y^2 - 9z^2; k \in R$.

32. $f(x, y, z) = 9x^2 - 4y^2 - z^2; k \in R$.

33. $f(x, y, z) = 4x^2 - 9y^2; k \in R$.

18.2 PARTIAL DERIVATIVES

If in $f(x, y)$ the value of y is held constant, f becomes a function of one independent variable, x, and may be differentiated with respect to x, with y being treated as a constant. Such a derivative is called the *partial derivative* of f with respect to x and is commonly denoted by $\partial f/\partial x$ or f_x.

For example, if $f(x, y) = x^2 - 3xy + \ln(x^2 + y^2)$, by differentiating with respect to x and treating y as a constant we obtain the value at (x, y) of the partial derivative of f with respect to x,

$$\frac{\partial}{\partial x} f(x, y) = f_x(x, y) = 2x - 3y + \frac{1}{x^2 + y^2}(2x).$$

Similarly, the partial derivative of f with respect to y is found by treating x as a constant and differentiating with respect to y. It is indicated by $\partial f/\partial y$ or f_y. As an illustration, if $f(x, y) = x^2 - 3xy + \ln(x^2 + y^2)$, then the value of $\partial f/\partial y$ at (x, y) is

$$\frac{\partial}{\partial y} f(x, y) = -3x + \frac{1}{x^2 + y^2}(2y).$$

18.2.1   Definition.   Let f be a function of two variables x and y. The **partial derivative of f with respect to** x is the function f_x (or $\partial f/\partial x$) whose value at any point (x, y) in the domain of f is

$$f_x(x, y) = \frac{\partial}{\partial x} f(x, y) = \lim_{\Delta x \to 0} \frac{f(x + \Delta x, y) - f(x, y)}{\Delta x},$$

provided that this limit exists; and the **partial derivative of f with respect to** y is the function f_y whose value at any point (x, y) in the domain of f is

$$f_y(x, y) = \frac{\partial}{\partial y} f(x, y) = \lim_{\Delta y \to 0} \frac{f(x, y + \Delta y) - f(x, y)}{\Delta y},$$

provided that this limit exists.

Partial derivatives have simple geometric interpretations. Consider the surface whose equation is $z = f(x, y)$. The plane $y = c$ intersects this surface in the plane curve QPR [Fig. 18-4(a)], and the value of the partial derivative of f with respect to x, as y keeps the constant value c, is the slope of the tangent to the plane curve QPR at the point $P{:}(x, c, f(x, c))$. Similarly, the plane $x = k$ cuts the surface $z = f(x, y)$ in the plane curve LPM [Fig. 18-4(b)], and $f_y(k, y)$ is the slope of the tangent to this curve at the point $P{:}(k, y, f(k, y))$.

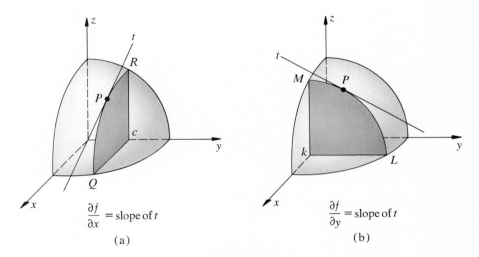

$$\frac{\partial f}{\partial x} = \text{slope of } t \qquad\qquad \frac{\partial f}{\partial y} = \text{slope of } t$$

(a)   (b)

Figure 18-4

For functions of one variable, the derivative gives the rate of change of the function with respect to the independent variable. In finding the partial derivative of a function of two variables, $f(x, y)$, with respect to x, we hold y constant, say $y = k$, and differentiate $f(x, k)$ with respect to x; so we are again differentiating a function of one variable, x. Thus f_x gives the rate of change of the function with respect to x only. Similarly, f_y gives the rate of change of f with respect to y.

Example 1. The volume V of a certain gas is related to its temperature T and its pressure P by the gas law

$$PV = 10T,$$

where V is measured in cubic inches, T in degrees, and P in pounds per square inch.

If the volume of the gas is kept constant at 200 cubic inches, what is the instantaneous rate of change of its pressure with respect to the temperature?

Solution. The gas equation can be written $P = 10T/V$. The rate of change of P with respect to T, as V is held constant, is given by the partial derivative

$$\frac{\partial P}{\partial T} = \frac{10}{V}.$$

Since $V = 200$, the rate of change of the pressure with respect to the temperature is $10/200 = 0.05$.

Since a partial derivative of a function of two variables is, in general, another function of the same two variables, it may be differentiated partially with respect to either x or y to obtain four **second partial derivatives** of f:

$$\frac{\partial}{\partial x}\left(\frac{\partial f}{\partial x}\right) = \frac{\partial^2 f}{\partial x^2} = f_{xx}, \qquad \frac{\partial}{\partial y}\left(\frac{\partial f}{\partial y}\right) = \frac{\partial^2 f}{\partial y^2} = f_{yy},$$

$$\frac{\partial}{\partial x}\left(\frac{\partial f}{\partial y}\right) = \frac{\partial^2 f}{\partial x\,\partial y} = f_{yx}, \qquad \frac{\partial}{\partial y}\left(\frac{\partial f}{\partial x}\right) = \frac{\partial^2 f}{\partial y\,\partial x} = f_{xy}.$$

For instance, if $f(x, y) = xe^y - \sin(x/y) + x^3y^2$, then

$$f_x(x, y) = e^y - \frac{1}{y}\cos\left(\frac{x}{y}\right) + 3x^2y^2,$$

$$f_y(x, y) = xe^y + \frac{x}{y^2}\cos\left(\frac{x}{y}\right) + 2x^3y,$$

$$f_{xx}(x, y) = \frac{1}{y^2}\sin\left(\frac{x}{y}\right) + 6xy^2,$$

$$f_{yy}(x, y) = xe^y + \frac{x^2}{y^4}\sin\left(\frac{x}{y}\right) - \frac{2x}{y^3}\cos\left(\frac{x}{y}\right) + 2x^3,$$

$$f_{xy}(x, y) = e^y - \frac{x}{y^3}\sin\left(\frac{x}{y}\right) + \frac{1}{y^2}\cos\left(\frac{x}{y}\right) + 6x^2y,$$

$$f_{yx}(x, y) = e^y - \frac{x}{y^3}\sin\left(\frac{x}{y}\right) + \frac{1}{y^2}\cos\left(\frac{x}{y}\right) + 6x^2y.$$

Notice that in the above example $f_{xy} = f_{yx}$, which is usually the case for the functions of two variables encountered in a first course. A criterion for this will be given in 18.3.4.

Partial derivatives of the third, and higher, orders are defined analogously, and the notation for them is similar. Thus if f is a function of the two variables x and y, the third partial derivative of f that is obtained by differentiating f partially, first with respect to x and then twice with respect to y, will be indicated by

$$\frac{\partial}{\partial y}\left(\frac{\partial^2 f}{\partial y\,\partial x}\right) = \frac{\partial^3 f}{\partial y^2\,\partial x} = f_{xyy}.$$

It is easy to extend the concept of partial derivatives to functions of three or more variables.

Let f be a function of three variables, x, y, and z. The **partial derivative of f with respect to** x is the function f_x (or $\partial f/\partial x$) whose value at any point (x, y, z) in the domain of f is

$$f_x(x, y, z) = \frac{\partial}{\partial x} f(x, y, z)$$

$$= \lim_{\Delta x \to 0} \frac{f(x + \Delta x, y, z) - f(x, y, z)}{\Delta x},$$

provided that this limit exists. The **partial derivative of f with respect to** y is the function f_y (or $\partial f/\partial y$) whose value at any point (x, y, z) in the domain of f is

$$f_y(x, y, z) = \frac{\partial}{\partial y} f(x, y, z)$$

$$= \lim_{\Delta y \to 0} \frac{f(x, y + \Delta y, z) - f(x, y, z)}{\Delta y},$$

when this limit exists. The **partial derivative of f with respect to** z is the function f_z (or $\partial f/\partial z$) whose value at any point (x, y, z) in the domain of f is

$$f_z(x, y, z) = \frac{\partial}{\partial z} f(x, y, z)$$

$$= \lim_{\Delta z \to 0} \frac{f(x, y, z + \Delta z) - f(x, y, z)}{\Delta z},$$

provided that this limit exists.

Example 2. If $f(x, y, z) = xy + 2yz + 3zx$, find f_x, f_y, and f_z.

Solution. To get f_x, we think of y and z as constants and differentiate with respect to the variable x. Thus

$$f_x(x, y, z) = y + 3z.$$

To find f_y, we treat x and z as constants and differentiate with respect to y:

$$f_y(x, y, z) = x + 2z.$$

Similarly,

$$f_z(x, y, z) = 2y + 3x.$$

Example 3. If $f(x, y, z) = x \cos(y - z)$, find $\partial f/\partial x$, $\partial f/\partial y$, and $\partial f/\partial z$.

Solution.

$$\frac{\partial}{\partial x}[x \cos(y - z)] = \cos(y - z),$$

$$\frac{\partial}{\partial y}[x \cos(y - z)] = -x \sin(y - z),$$

$$\frac{\partial}{\partial z}[x \cos(y - z)] = x \sin(y - z).$$

Exercises

In Exercises 1 to 16, find the first partial derivatives of the given functions with respect to each independent variable.

1. $f(x, y) = (2x - 4)^4$.

2. $f(x, y) = (4x - y^2)^{3/2}$.

3. $f(x, y) = \dfrac{x^2 - y^2}{xy}$.

4. $f(x, y) = e^x \cos y$.

5. $f(x, y) = e^y \sin x$.

6. $f(x, y) = (3x^2 + y^2)^{-1/3}$.

7. $f(x, y) = \sqrt{x^2 - y^2}$.

8. $f(u, v) = e^{uv}$.

9. $g(x, y) = e^{-xy}$.

10. $f(s, t) = \ln (s^2 - t^2)$.

11. $f(x, y) = \text{Tan}^{-1} (4x - 7y)$.

12. $F(w, z) = w \, \text{Sin}^{-1} \left(\dfrac{w}{z} \right)$.

13. $f(x, y) = y \cos (x^2 + y^2)$.

14. $f(s, t) = e^{t^2 - s^2}$.

15. $F(x, y) = 2 \sin x \cos y$.

16. $f(r, \theta) = 3r^3 \cos 2\theta$.

In Exercises 17 to 22, verify that

$$\frac{\partial^2 f}{\partial y \, \partial x} = \frac{\partial^2 f}{\partial x \, \partial y}.$$

17. $f(x, y) = 2x^2 y^3 - x^3 y^5$.

18. $f(x, y) = (x^3 + y^2)^5$.

19. $f(x, y) = 3e^{2x} \cos y$.

20. $f(x, y) = \text{Tan}^{-1} xy$.

21. $f(x, y) = \ln \left(\dfrac{x - 1}{y - 1} \right)$.

22. $f(x, y) = \ln \left(\dfrac{x^2 - 2}{y^4 + 3} \right)$.

23. If $F(x, y) = \dfrac{2x - y}{xy}$, find $F_x(3, -2)$ and $F_y(3, -2)$.

24. If $F(x, y) = \ln (x^2 + xy + y^2)$, find $F_x(-1, 4)$ and $F_y(-1, 4)$.

25. If $f(x, y) = \text{Tan}^{-1} (y^2/x)$, find $f_x(\sqrt{5}, -2)$ and $f_y(\sqrt{5}, -2)$.

26. If $f(x, y) = e^y \cosh x$, find $f_x(-1, 1)$ and $f_y(-1, 1)$.

27. Find the slope of the tangent to the curve of intersection of the surface $36z = 4x^2 + 9y^2$ and the plane $x = 3$, at the point $(3, 2, 2)$. Make a sketch.

28. Find the slope of the tangent to the curve of intersection of the surface $3z = \sqrt{36 - 9x^2 - 4y^2}$ and the plane $x = 1$, at the point $(1, -2, \sqrt{11}/3)$. Make a sketch.

29. Find the slope of the tangent to the curve of intersection of the surface $2z = \sqrt{9x^2 + 9y^2 - 36}$ and the plane $y = 1$, at the point $(2, 1, \frac{3}{2})$. Make a sketch.

30. Find the slope of the tangent to the curve of intersection of the cylinder $4z = 5\sqrt{16 - x^2}$ and the plane $y = 3$ at the point $(2, 3, 5\sqrt{3}/2)$. Make a sketch.

A function of two variables that satisfies **Laplace's equation**,

$$\frac{\partial^2 f}{\partial x^2} + \frac{\partial^2 f}{\partial y^2} = 0,$$

is said to be **harmonic**. Show that the functions defined in Exercises 31 to 34 are harmonic functions.

31. $f(x, y) = x^3 y - xy^3$.

32. $f(x, y) = \ln (4x^2 + 4y^2)$.

33. $f(x, y) = e^{-y} \cos x$.

34. $f(x, y) = \text{Tan}^{-1} \left(\dfrac{x}{y} \right)$.

35. If $F(x, y) = 3x^4 y^5 - 2x^2 y^3$, find $\partial^3 F(x, y)/\partial y^3$.

36. If $f(x, y) = \cos (2x^2 - y^2)$, find $\partial^3 f(x, y)/\partial y \, \partial x^2$.

37. If f is the function of three variables defined by $f(x, y, z) = 3x^2 y - xyz + y^2 z^2$, find $f_x(x, y, z)$, $f_y(x, y, z)$, and $f_z(x, y, z)$.

38. If f is the function of three variables defined by

$$f(x, y, z) = (x^3 + y^2 + z)^4, \quad \text{find} \quad \frac{\partial}{\partial x} f(x, y, z),$$

$\frac{\partial}{\partial y} f(x, y, z)$, and $\frac{\partial}{\partial z} f(x, y, z)$.

39. If $f(x, y, z) = e^{-xyz} - \ln(xy - z^2)$, find $f_x(x, y, z), f_y(x, y, z)$, and $f_z(x, y, z)$.

40. If $F(x, y, z) = (xy/z)^{1/2}$, find $F_x(-2, -1, 8)$, $F_y(-2, -1, 8)$, and $F_z(-2, -1, 8)$.

18.3 LIMITS AND CONTINUITY

The set of points in the xy plane that are inside a rectangle, with sides parallel to the coordinate axes, is called a two-dimensional **open rectangular region** [Fig. 18-5(a)]. The **boundary** of this region is the set of points on the rectangle itself. No boundary points are included in an *open* rectangular region.

The union of an open rectangular region and its boundary is called a **closed rectangular region** [Fig. 18-5(b)]. It consists of all points in the xy plane that are inside, or on, a rectangle whose sides are parallel to the coordinate axes.

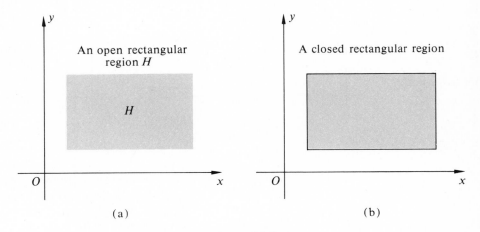

Figure 18-5

An **open disk** is the set of points that are inside a circle [Fig. 18-6(a)]. The **boundary** of the disk is the set of points on the circle. No boundary points are included in an *open* disk.

A **closed disk** is the set of points that are inside, or on, a circle [Fig. 18-6(b)]. It is the union of an open disk and its boundary.

A function f of two variables, x and y, has a limit L at a point C of the xy plane if the value $f(x, y)$ of the function can be made as close as we please to the number L by taking (x, y) sufficiently close to C.

When reading the following definition, recall that $\sqrt{(x - a)^2 + (y - b)^2}$ is the distance between the points (x, y) and (a, b) in the xy plane.

18.3.1 Definition. Let $C{:}(a, b)$ be a point in some open rectangular region H in the xy plane, and let f be a function of two independent variables x and y that is defined at every point $P{:}(x, y)$ of open H except possibly at $C{:}(a, b)$. Then the **limit of f** at $C{:}(a, b)$ is L, written

$$\lim_{(x, y)\to(a, b)} f(x, y) = L \quad \text{or} \quad \lim_{P\to C} f(x, y) = L,$$

if for each positive number ϵ (no matter how small) there exists some positive number δ such that whenever $P{:}(x, y)$ is in H and

$$0 < \sqrt{(x - a)^2 + (y - b)^2} < \delta,$$

then

$$|f(x, y) - L| < \epsilon.$$

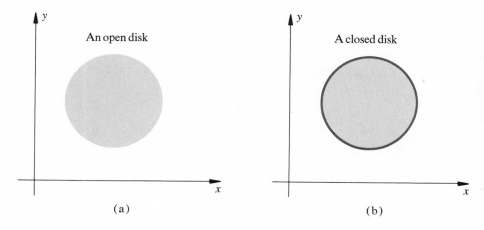

(a)   (b)

Figure 18-6

This definition may be interpreted geometrically to mean that corresponding to each positive number ϵ (no matter how small) there is always some open disk inside H with radius δ and center $C{:}(a, b)$, small enough so that whenever $P{:}(x, y)$ is in the disk but distinct from $C{:}(a, b)$, then the value of the function at $P{:}(x, y)$ will differ from L by less than ϵ [Fig. 18-7(a)]. In other words, we can make the value $f(x, y)$ as close to the number L as we please by making the disk small enough. It is sometimes convenient to refer to such a disk as a δ disk at C.

Since $f(x, y)$ must differ from L by less than ϵ at all points $P{:}(x, y)$ in some open δ disk at $C{:}(a, b)$, $f(x, y)$ will differ from L by less than ϵ when the points P are confined to a line segment or curve segment in the disk that passes through C [Fig. 18-7(b)]. Such a line or curve through C is called a *path of approach* to C. Clearly, then, the limit of f as P approaches C must be the same along all paths of approach when f has a limit at C.

It follows that *if the limit of $f(x, y)$ is not the same when P approaches C along two different paths, then* $\lim_{P\to C} f(x, y)$ *fails to exist.* The converse, however, is not true;

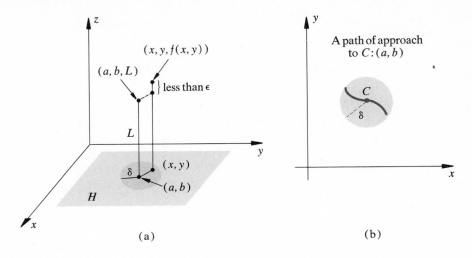

Figure 18-7

$f(x, y)$ can have the same limit as P approaches C along two different paths, and yet $\lim\limits_{P \to C} f(x, y)$ may or may not exist.

Example 1. Show that the function f defined by

$$f(x, y) = \frac{x^2 - y^2}{x^2 + y^2}$$

has no limit at the origin.

Solution. The function f is defined everywhere in the xy plane except at the origin.
At all points on the x axis, different from the origin, the value of f is

$$f(x, 0) = \frac{x^2 - 0}{x^2 + 0} = 1;$$

thus the limit of $f(x, y)$ as (x, y) approaches $(0, 0)$ along the x axis is

$$\lim_{(x,\, 0) \to (0,\, 0)} f(x, 0) = \lim_{(x,\, 0) \to (0,\, 0)} \frac{x^2 - 0}{x^2 + 0} = +1.$$

Similarly, the limit of $f(x, y)$ as (x, y) approaches $(0, 0)$ along the y axis is

$$\lim_{(0,\, y) \to (0,\, 0)} f(0, y) = \lim_{(0,\, y) \to (0,\, 0)} \frac{0 - y^2}{0 + y^2} = -1.$$

Since the limit of $f(x, y)$ is not the same when $P \longrightarrow 0$ along these two different paths, $\lim\limits_{P \to 0} f(x, y)$ fails to exist.

Most limit theorems for functions of one variable are true for functions of two variables, and the proofs are similar.

When a function of two variables is defined at a point $C:(a, b)$ and has a limit there, then if this limit is the same as the value of the function at $C:(a, b)$, the function is said to be *continuous* at (a, b).

18.3.2 **Definition.** A function f of two variables is **continuous** at (a, b) if

(i) f is defined at (a, b),

(ii) $\displaystyle\lim_{(x, y) \to (a, b)} f(x, y)$ exists, and

(iii) $\displaystyle\lim_{(x, y) \to (a, b)} f(x, y) = f(a, b)$.

A function is **continuous in a region** if it is continuous at every point in the region. The graph of a continuous function is a surface that has no breaks or sudden jumps.

If f and g are functions of two variables, x and y, that are defined on a common domain $\mathfrak{D}$, then their **sum**, $f + g$, is the function H defined on $\mathfrak{D}$ by $H(x, y) = f(x, y) + g(x, y)$. Similar definitions hold for the **product** $f \cdot g$ and the **quotient** f/g of the functions f and g.

As in functions of one variable, the sum or product of continuous functions of several variables are continuous, and the quotient of two continuous functions is continuous if the denominator function is not zero at any point in the region.

It follows that *polynomial functions* of two variables are continuous everywhere, since they are sums and products of the continuous functions ax, by, and c, where a, b, and c are constants. For example, the function $f(x, y) = 5x^4y^2 - 2xy^3 + 4$ is continuous at all points in the xy plane.

Rational functions of two variables are quotients of polynomial functions and thus are continuous wherever the denominator is not zero. To illustrate, $f(x, y) = (2x + 3y)/(y^2 - 4x)$ is continuous everywhere in the xy plane except at points on the parabola $y^2 = 4x$.

As with functions of one variable (3.5.4), a continuous function of a continuous function is continuous.

18.3.3 **Theorem.** If a function g of two variables is continuous at (a, b) and a function f of one variable is continuous at $g(a, b)$, then the composite function $f \circ g$, defined by $(f \circ g)(x, y) = f(g(x, y))$, is continuous at (a, b).

Proof of this theorem is similar to the proof of (3.5.4).

Example 2. Since $\cos t$ is continuous for all $t \in R$ and the polynomial function $x^3 - 4xy + y^2$ is continuous everywhere, the composite function $\cos(x^3 - 4xy + y^2)$ is continuous for all $x, y \in R$.

The square root function $\sqrt{t}$ is continuous for $t \geq 0$ and the rational function $(2x^2 + 1)/(x^2 + y^2)$ is positive valued and continuous at all points in the xy plane except the origin. Thus the composite function $\sqrt{(2x^2 + 1)/(x^2 + y^2)}$ is continuous everywhere except at the origin.

We said in Section 18.2 that for most functions of two variables studied in a first course, $f_{xy} = f_{yx}$; that is, the order of differentiation in mixed partial derivatives is immaterial. Now that continuity is understood, conditions for this to be true can be simply stated.

18.3.4 Theorem. Let f be a function of two variables, x and y, such that f_{xy} and f_{yx} are continuous in an open rectangular region H of the xy plane; then

$$f_{xy} = f_{yx}$$

at any point in H.

A proof of this theorem is given in books on advanced calculus.

The definitions and properties of limits and continuity of functions of two variables discussed above are easily extended to functions of three or more independent variables.

By a *three-dimensional* **open rectangular region** is meant the set of points inside a rectangular parallelepiped whose faces are parallel to the coordinate planes [Fig. 18-8(a)]. Its **boundary** is the set of points on the faces of the parallelepiped. The union of a three-dimensional open rectangular region and its boundary is a three-dimensional **closed rectangular region**.

An **open ball** is the set of points inside a sphere [Fig. 18-8(b)]. Its **boundary** is the set of points that form the surface of the sphere. The union of an open ball and its boundary is a **closed ball**.

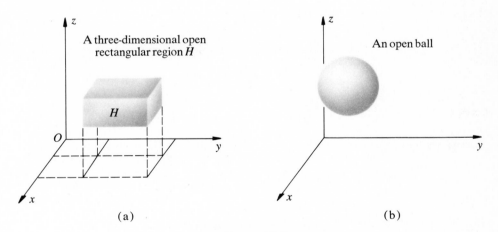

(a)   (b)

Figure 18-8

18.3.5 Definition. Let $C:(a, b, c)$ be a point in some three-dimensional open rectangular region H and let f be a function of three variables, $x, y,$ and z, that is defined at every point $P:(x, y, z)$ of open H except possibly at C. Then the **limit of f** at $C:(a, b, c)$ is L, written

$$\lim_{P \to C} f(x, y, z) = L,$$

if for each positive number ϵ (no matter how small) there exists some positive number δ such that whenever $P:(x, y, z)$ is in H and

$$0 < \sqrt{(x - a)^2 + (y - b)^2 + (z - c)^2} < \delta,$$

then

$$|f(x, y, z) - L| < \epsilon.$$

Notice the slight change in notation here. It would have been awkward and needlessly cumbersome to have written

$$\lim_{(x, y, z) \to (a, b, c)} f(x, y, z) = L;$$

so we have replaced it with the simpler

$$\lim_{P \to C} f(x, y, z) = L,$$

in which P and C represent their coordinates, (x, y, z) and (a, b, c). The rest of the definition remains unchanged from 18.3.1 except for the presence of one more variable. The convenience of this new notation increases with the number of independent variables in the function; but it can also be used for functions of one or two variables if desired.

A function f of three variables, x, y, and z, is **continuous** at a point $C:(a, b, c)$ if

1. f is defined at C,
2. $\lim_{P \to C} f(x, y, z)$ exists, and
3. $\lim_{P \to C} f(x, y, z) = f(a, b, c)$.

A function of three variables is *continuous in a three-dimensional region* if it is continuous at every point of the region.

Exercises

1. At what points is the function

$$f(x, y) = \frac{x^3 + xy - 5}{x^2 + y^2 + 1}$$

continuous?

2. In what region of the xy plane is $F(x, y) = \sqrt{x - y + 1}$ continuous?

3. Find the limit at $(1, -2)$ of the function f defined in Exercise 1.

4. What is the limit at $(4, 1)$ of the function F in Exercise 2?

5. Let $f(x, y) = \ln(x - 2y + 1)$. At what points in the xy plane is f continuous?

6. Let $f(x, y) = \sqrt{4 - x^2 - y^2}$. At what points in the xy plane is f continuous?

7. Let $F(x, y) = x^2 y^2 / (x^3 - 8y^3)$. Is F continuous

in the open rectangular region $R = \{P:(x, y) \mid 0 < x < 5, 0 < y < 3\}$? If your answer is no, explain.

8. For what points is the function defined by $\phi(x, y) = x^3 / (x^2 + 4y^2 - 4)$ continuous?

9. Does

$$\lim_{(x, y) \to (0, 0)} \frac{xy}{x^2 + y^2}$$

exist? Explain. [*Hint:* Compare the limit found by letting $(x, y) \longrightarrow (0, 0)$ along the x axis with the limit obtained when $(x, y) \longrightarrow (0, 0)$ along the line $x - y = 0$.]

10. Does $\lim_{(x, y) \to (0, 0)} y^2 / (x + y^2)$ exist? Explain.

11. Show that

$$\lim_{(x, y) \to (2, -2)} \ln \left(\frac{4 + x^2}{y + 2} \right)$$

does not exist.

12. Does

$$\lim_{(x,y)\to(0,0)} \frac{x^2 y^2}{x^4 + 3y^4}$$

exist? Explain.

13. Prove the theorem: Let f and g be functions of two variables. If the limits of f and g exist at a point

(a, b), then

$$\lim_{(x,y)\to(a,b)} [f(x, y) + g(x, y)]$$
$$= \lim_{(x,y)\to(a,b)} f(x, y) + \lim_{(x,y)\to(a,b)} g(x, y).$$

(*Hint:* See the proof of 3.3.1.)

18.4 INCREMENTS AND DIFFERENTIALS

18.4.1 Definition. Let f be a function of two variables and let $z = f(x, y)$. If Δx and Δy are increments of the independent variables, x and y, then

$$\Delta z = \Delta f(x, y) = f(x + \Delta x, y + \Delta y) - f(x, y)$$

is called the **increment of the dependent variable** z (or the **increment of the function**) that corresponds to the increments Δx and Δy of the independent variables at the point (x, y).

Here we have four independent variables, $x, y, \Delta x$, and Δy; and, in general, Δz depends on all of them. The coordinates, x and y, and the increments, Δx and Δy, can be any real numbers such that (x, y) and $(x + \Delta x, y + \Delta y)$ are points in the domain of f.

The next theorem enables us to express the increment of a function of two variables in a useful form.

18.4.2 Theorem. Let f be a function of two variables, x and y, with f_x and f_y continuous in an open rectangular region H, and let $z = f(x, y)$. If (x, y) is a point in H and Δx and Δy are any two numbers such that $(x + \Delta x, y + \Delta y)$ is also a point in H, then

$$\Delta z = \Delta f(x, y) = f_x(x, y)\, \Delta x + f_y(x, y)\, \Delta y + \theta_1\, \Delta x + \theta_2\, \Delta y,$$

where θ_1 and θ_2 are functions of both Δx and Δy such that $\theta_1 \longrightarrow 0$ and $\theta_2 \longrightarrow 0$ as $\Delta x \longrightarrow 0$ and $\Delta y \longrightarrow 0$.

Proof. From 18.4.1

(1) $$\Delta z = f(x + \Delta x, y + \Delta y) - f(x, y).$$

Since

$$f(x + \Delta x, y + \Delta y) - f(x, y)$$
$$= [f(x + \Delta x, y + \Delta y) - f(x, y + \Delta y)] + [f(x, y + \Delta y) - f(x, y)],$$

equation (1) can be rewritten

(2) $$\Delta z = [f(x + \Delta x, y + \Delta y) - f(x, y + \Delta y)] + [f(x, y + \Delta y) - f(x, y)].$$

If we keep $y + \Delta y$ fixed, then $f(x, y + \Delta y)$ defines a function of x alone. By applying the alternate mean value theorem for functions of one variable (6.2.2) to it, we obtain

$$(3) \qquad f(x + \Delta x, y + \Delta y) - f(x, y + \Delta y) = f_x(x + \lambda_1 \Delta x, y + \Delta y)\, \Delta x,$$

where $0 < \lambda_1 < 1$. In like manner, if the value of x is fixed, $f(x, y)$ defines a function of y alone. So, by the alternate mean value theorem,

$$(4) \qquad f(x, y + \Delta y) - f(x, y) = f_y(x, y + \lambda_2 \Delta y)\, \Delta y,$$

where $0 < \lambda_2 < 1$. From (2), (3), and (4), we can write

$$(5) \qquad \Delta z = f_x(x + \lambda_1 \Delta x, y + \Delta y)\, \Delta x + f_y(x, y + \lambda_2 \Delta y)\, \Delta y$$

for some numbers λ_1 and λ_2 between 0 and 1.

Let

$$(6) \qquad \theta_1(\Delta x, \Delta y) = f_x(x + \lambda_1 \Delta x, y + \Delta y) - f_x(x, y),$$
$$\theta_2(\Delta x, \Delta y) = f_y(x, y + \lambda_2 \Delta y) - f_y(x, y).$$

Then

$$\lim_{(\Delta x,\, \Delta y) \to (0,\, 0)} \theta_1(\Delta x, \Delta y) = 0 \qquad \text{and} \qquad \lim_{(\Delta x,\, \Delta y) \to (0,\, 0)} \theta_2(\Delta x, \Delta y) = 0,$$

since f_x and f_y were assumed continuous in R. From (6)

$$(7) \qquad f_x(x + \lambda_1 \Delta x, y + \Delta y) = f_x(x, y) + \theta_1(\Delta x, \Delta y),$$
$$f_y(x, y + \lambda_2 \Delta y) = f_y(x, y) + \theta_2(\Delta x, \Delta y).$$

Substituting (7) in (5) gives

$$\Delta z = f_x(x, y)\, \Delta x + f_y(x, y)\, \Delta y + \theta_1 \Delta x + \theta_2 \Delta y,$$

where $\displaystyle \lim_{(\Delta x,\, \Delta y) \to (0,\, 0)} \theta_1 = 0$ and $\displaystyle \lim_{(\Delta x,\, \Delta y) \to (0,\, 0)} \theta_2 = 0.$ ∎

18.4.3 Definition. Let f be a function of two independent variables, x and y, with continuous first partial derivatives, f_x and f_y, in an open rectangular region H, and let $z = f(x, y)$. If (x, y) is a point in H and Δx and Δy are any numbers such that $(x + \Delta x, y + \Delta y)$ is also a point in H, then

(i) the **differentials of the independent variables**, dx and dy, are defined by

$$dx = \Delta x \qquad \text{and} \qquad dy = \Delta y.$$

(ii) the **differential of the dependent variable**, dz, or the **total differential of the function**, $df(x, y)$, is defined by

$$dz = df(x, y) = f_x(x, y)\, dx + f_y(x, y)\, dy.$$

Comparing this definition of dz with the expression for Δz in 18.4.2, we see that the differential dz and the increment Δz differ by $\theta_1 \Delta x + \theta_2 \Delta y$. Since $\theta_1 \longrightarrow 0$ and $\theta_2 \longrightarrow 0$ as $\Delta x \longrightarrow 0$ and $\Delta y \longrightarrow 0$, dz is usually a good approximation to Δz when the increments Δx and Δy are small.

Example. The formula $P = k(T/V)$, where k is a constant, gives the pressure P of a confined gas of volume V and temperature T. Find, approximately, the maximum percentage error in P introduced by an error of $\pm 0.4\%$ in measuring the temperature and an error of $\pm 0.9\%$ in measuring the volume.

Solution. The error in P is ΔP, which we will approximate by dP. Thus

$$|\Delta P| \doteq |dP| = \left| \frac{\partial P}{\partial T} \Delta T + \frac{\partial P}{\partial V} \Delta V \right|$$

$$\leq \left| \frac{k}{V} (\pm 0.004T) \right| + \left| -\frac{kT}{V^2} (\pm 0.009 V) \right|$$

$$= \frac{kT}{V} (0.004 + 0.009) = 0.013 \frac{kT}{V} = 0.013P.$$

The maximum relative error, $\Delta P/P$, is approximately 0.013, and the maximum percentage error is approximately 1.3%.

This discussion is easily extended to functions of three or more variables. If $z = f(x, y, z)$ defines a function f of three independent variables, then under conditions analogous to those stated above, the **increment** Δz is given by

$$\Delta z = f_x(x, y, z)\, \Delta x + f_y(x, y, z)\, \Delta y + f_z(x, y, z)\, \Delta z + \theta_1 \Delta x + \theta_2 \Delta y + \theta_3 \Delta z,$$

where $\theta_1 \longrightarrow 0, \theta_2 \longrightarrow 0, \theta_3 \longrightarrow 0$ as $\Delta x \longrightarrow 0, \Delta y \longrightarrow 0$, and $\Delta z \longrightarrow 0$; and the **differential** dz is defined to be

$$dz = f_x(x, y, z)\, dx + f_y(x, y, z)\, dy + f_z(x, y, z)\, dz.$$

Exercises

In Exercises 1 to 6, find dz.

1. $z = x^3 - xy^2 + 3y$.

2. $z = \dfrac{x^2 y}{y^3 + 1}$.

3. $z = \text{Tan}^{-1}(xy)$.

4. $z = x^2 \sin 3y$.

5. $z = e^x e^{-y}$.

6. $z = \ln(x^2 - y^2)$.

In Exercises 7 to 10, find dw.

7. $w = r \sin \theta + r^2$.

8. $w = \ln(rst)$.

9. $w = e^{-rst}$.

10. $w = \sin(x - y) + \cos(y - z)$.

In Exercises 11 to 16, use 18.4.1 and 18.4.3 to compute the increment and the differential of the indicated function with the given data.

11. $z = 2x^2 y^3$ at $(1, 1)$, with $\Delta x = -0.01$ and $\Delta y = 0.02$.

12. $z = x^2 - 5xy + y$ at $(2, 3)$, with $\Delta x = 0.03$ and $\Delta y = -0.02$.

13. $w = 3st^{-2}$ at $(-1, 2)$, with $\Delta s = 0.04$ and $\Delta t = 0.02$.

14. $w = \sqrt{s + t^2}$ at $(3, -1)$, with $\Delta s = 0.01$ and $\Delta t = -0.05$.

15. $z = \ln(x^2 y)$ at $(-2, 4)$, with $\Delta x = -0.02$ and $\Delta y = 0.08$.

16. $z = y \, \text{Tan}^{-1} xy$ at $(-2, -0.5)$, with $\Delta x = -0.03$ and $\Delta y = -0.01$.

17. If errors are made in measuring the length x and the width y of a room, find the error in the computed area A of the floor. (*Hint:* Denote by Δx and Δy the errors in the length and width measurements. Find ΔA.)

18. Assume that the errors in measurement in Exercise 17 are relatively small. Approximate the

error in the computed area by finding dA, the differential of A.

19. In determining the specific gravity of an object, its weight in air is found to be $A = 36$ pounds and its weight in water is $W = 20$ pounds, with a possible error in each measurement of 0.02 pound. Find, approximately, the maximum possible error

in calculating its specific gravity S, where $S = A/(A - W)$.

20. Use differentials to find the approximate amount of copper in the four sides and bottom of a rectangular copper tank that is 6 feet long, 4 feet wide, and 3 feet deep *inside*, if the sheet copper is $\frac{1}{4}$ inch thick. (*Hint:* Make a sketch.)

18.5 CHAIN RULE

The chain rule for differentiating functions of one variable can be extended to partial differentiation of functions of two or more variables.

18.5.1 Theorem (*Chain Rule*). Let u and v be functions defined by $u = u(x, y)$ and $v = v(x, y)$, such that u and v are continuous and possess first partial derivatives at (x, y). Let F be a function of u and v having continuous first partial derivatives in some open rectangular region that contains the point $(u, v) = (u(x, y), v(x, y))$. Then

$$\frac{\partial F}{\partial x} = \frac{\partial F}{\partial u}\frac{\partial u}{\partial x} + \frac{\partial F}{\partial v}\frac{\partial v}{\partial x}$$

and

$$\frac{\partial F}{\partial y} = \frac{\partial F}{\partial u}\frac{\partial u}{\partial y} + \frac{\partial F}{\partial v}\frac{\partial v}{\partial y}.$$

Proof. From the definition of partial derivatives,

$$\frac{\partial}{\partial x} F(u(x, y), v(x, y)) = \lim_{\Delta x \to 0} \frac{F(u(x + \Delta x, y), v(x + \Delta x, y)) - F(u(x, y), v(x, y))}{\Delta x}$$

$$= \lim_{\Delta x \to 0} \frac{F(u + \Delta u, v + \Delta v) - F(u, v)}{\Delta x},$$

where $\Delta u = u(x + \Delta x, y) - u(x, y)$ and $\Delta v = v(x + \Delta x, y) - v(x, y)$. Using 18.4.1 and 18.4.2, we can write

$$\frac{\Delta F(u, v)}{\Delta x} = \frac{F(u + \Delta u, v + \Delta v) - F(u, v)}{\Delta x}$$

$$= F_u(u, v)\frac{\Delta u}{\Delta x} + F_v(u, v)\frac{\Delta v}{\Delta x} + \theta_1\frac{\Delta u}{\Delta x} + \theta_2\frac{\Delta v}{\Delta x},$$

where θ_1 and θ_2 are functions of both Δu and Δv such that $\theta_1 \longrightarrow 0$ and $\theta_2 \longrightarrow 0$ as $\Delta u \longrightarrow 0$ and $\Delta v \longrightarrow 0$. Since $\Delta u \longrightarrow 0$ and $\Delta v \longrightarrow 0$ as $\Delta x \longrightarrow 0$, then $\theta_1 \longrightarrow 0$ and $\theta_2 \longrightarrow 0$ as $\Delta x \longrightarrow 0$. Taking the limiting form of this equation as $\Delta x \longrightarrow 0$, we have

$$\frac{\partial}{\partial x} F(u(x, y), v(x, y)) = \frac{\partial F}{\partial u}\frac{\partial u}{\partial x} + \frac{\partial F}{\partial v}\frac{\partial v}{\partial x}.$$

The proof of the second part is similar. ∎

Analogously, for functions of three variables, we have

18.5.2 Theorem. If F is a function of three variables, u, v, and w, if u, v, and w are continuous functions of two variables defined by $u = u(x, y)$, $v = v(x, y)$, and $w = w(x, y)$, and if all these functions have first partial derivatives, with those of F continuous, then

$$\frac{\partial F}{\partial x} = \frac{\partial F}{\partial u}\frac{\partial u}{\partial x} + \frac{\partial F}{\partial v}\frac{\partial v}{\partial x} + \frac{\partial F}{\partial w}\frac{\partial w}{\partial x}$$

and

$$\frac{\partial F}{\partial y} = \frac{\partial F}{\partial u}\frac{\partial u}{\partial y} + \frac{\partial F}{\partial v}\frac{\partial v}{\partial y} + \frac{\partial F}{\partial w}\frac{\partial w}{\partial y}.$$

This theorem is readily extended to functions of any number of dependent and independent variables.

The pattern of all these chain rules is easy to remember, particularly in the notation that we have used. The reader is warned, however, that the symbol $\partial f/\partial x$ for a partial derivative is not a quotient; *neither ∂f nor ∂x has any meaning by itself.*

Example 1. Let $F(u, v) = 3u^2 - v^2$, $u = 2x + 7y$, and $v = 5xy$. Find $\partial F/\partial x$ and $\partial F/\partial y$ in terms of x and y.

Solution. By 18.5.1,

$$\frac{\partial F}{\partial x} = \frac{\partial F}{\partial u}\frac{\partial u}{\partial x} + \frac{\partial F}{\partial v}\frac{\partial v}{\partial x}.$$

Since

$$\frac{\partial F}{\partial u} = 6u, \qquad \frac{\partial F}{\partial v} = -2v, \qquad \frac{\partial u}{\partial x} = 2, \qquad \frac{\partial v}{\partial x} = 5y,$$

we have

$$\frac{\partial F}{\partial x} = (6u)(2) + (-2v)(5y) = 12u - 10yv$$

$$= 12(2x + 7y) - 10y(5xy)$$

$$= 24x + 84y - 50xy^2.$$

Again,

$$\frac{\partial F}{\partial y} = \frac{\partial F}{\partial u}\frac{\partial u}{\partial y} + \frac{\partial F}{\partial v}\frac{\partial v}{\partial y}.$$

But

$$\frac{\partial u}{\partial y} = 7 \qquad \text{and} \qquad \frac{\partial v}{\partial y} = 5x.$$

Therefore

$$\frac{\partial F}{\partial y} = (6u)(7) + (-2v)(5x) = 42u - 10xv$$

$$= 42(2x + 7y) - 10x(5xy)$$

$$= 84x + 294y - 50x^2y.$$

Example 2. Let $f(x, y) = e^{xy}$, $x = r\cos\theta$, and $y = r\sin\theta$. Find f_r and f_θ in terms of r and θ.

Solution. By 18.5.1,

$$\frac{\partial f}{\partial r} = \frac{\partial f}{\partial x}\frac{\partial x}{\partial r} + \frac{\partial f}{\partial y}\frac{\partial y}{\partial r} = (ye^{xy})\cos\theta + (xe^{xy})\sin\theta$$

$$= (r\sin\theta\, e^{r^2\sin\theta\cos\theta})\cos\theta + (r\cos\theta\, e^{r^2\sin\theta\cos\theta})\sin\theta = re^{r^2\sin\theta\cos\theta}\sin 2\theta.$$

Again,

$$\frac{\partial f}{\partial\theta} = \frac{\partial f}{\partial x}\frac{\partial x}{\partial\theta} + \frac{\partial f}{\partial y}\frac{\partial y}{\partial\theta} = (ye^{xy})(-r\sin\theta) + (xe^{xy})r\cos\theta$$

$$= (r\sin\theta\, e^{r^2\sin\theta\cos\theta})(-r\sin\theta) + r\cos\theta\, e^{r^2\sin\theta\cos\theta}(r\cos\theta) = r^2 e^{r^2\sin\theta\cos\theta}\cos 2\theta.$$

Example 3. If $\phi(u, v, w) = u^3 + 2uvw + vw^2$, $u = xy$, $v = x - y$, and $w = x/y$, find $\partial\phi/\partial y$
(a) in terms of u, v, w, x and y;
(b) in terms of x and y, only.

Solution. (a) From 18.5.2,

$$\frac{\partial\phi}{\partial y} = \frac{\partial\phi}{\partial u}\frac{\partial u}{\partial y} + \frac{\partial\phi}{\partial v}\frac{\partial v}{\partial y} + \frac{\partial\phi}{\partial w}\frac{\partial w}{\partial y} = (3u^2 + 2vw)x + (2uw + w^2)(-1) + (2uv + 2vw)\left(-\frac{x}{y^2}\right)$$

$$= (3u^2 + 2vw)x = (2uw + w^2) - (2uv + 2vw)\frac{x}{y^2}.$$

(b) By substituting the given values of u, v and w in terms of x and y in the preceding equation and simplifying, we obtain

$$\frac{\partial\phi}{\partial y} = \frac{x^2}{y^3}(3xy^5 - 2y^3 - 2x + y).$$

Let $z = F(x, y)$ and let $y = g(x)$. Then $z = F(x, g(x))$ defines z as a function of one variable, x, and (by 18.5.1)

$$\frac{dz}{dx} = \frac{\partial F}{\partial x}\frac{\partial x}{\partial x} + \frac{\partial F}{\partial y}\frac{\partial y}{\partial x},$$

or

(1)
$$\frac{dz}{dx} = \frac{\partial F}{\partial x} + \frac{\partial F}{\partial y}\frac{dy}{dx}.$$

In Section 5.6 we learned how to find dy/dx when $F(x, y) = 0$ defined y *implicitly* as a differentiable function of x. The preceding formula provides another way of accomplishing this. If $F(x, y) = 0$, then $z = 0$ identically in (1) and $dz/dx = 0$. Here (1) becomes

$$0 = \frac{\partial F}{\partial x} + \frac{\partial F}{\partial y}\frac{dy}{dx},$$

or

(2)
$$\frac{dy}{dx} = \frac{-\dfrac{\partial F}{\partial x}}{\dfrac{\partial F}{\partial y}},$$

provided that $\partial F/\partial y \neq 0$.

Equation (2) shows another method of finding dy/dx when $F(x, y) = 0$ defines y implicitly as a function of x. Compare it with our earlier method in Section 5.6.

The following more general theorem is proved similarly.

18.5.3 Theorem. If z is an implicit function of x and y defined by the equation $F(x, y, z) = 0$, then

$$\frac{\partial z}{\partial x} = -\frac{\dfrac{\partial F}{\partial x}}{\dfrac{\partial F}{\partial z}}, \qquad \frac{\partial z}{\partial y} = -\frac{\dfrac{\partial F}{\partial y}}{\dfrac{\partial F}{\partial z}},$$

provided that $\partial F/\partial z \neq 0$.

Example 4. If $f(x, y, z) = x^3 e^{y+z} - y \sin(x - z) = 0$ defines z implicitly as a function of x and y in some rectangular domain, find $\partial z/\partial x$ and $\partial z/\partial y$.

Solution.

$$\frac{\partial f}{\partial x} = 3x^2 e^{y+z} - y \cos(x - z), \qquad \frac{\partial f}{\partial y} = x^3 e^{y+z} - \sin(x - z),$$

$$\frac{\partial f}{\partial z} = x^3 e^{y+z} + y \cos(x - z).$$

By 18.5.3,

$$\frac{\partial z}{\partial x} = -\frac{\dfrac{\partial f}{\partial x}}{\dfrac{\partial f}{\partial z}} \quad \text{and} \quad \frac{\partial z}{\partial y} = -\frac{\dfrac{\partial f}{\partial y}}{\dfrac{\partial f}{\partial z}}.$$

Therefore

$$\frac{\partial z}{\partial x} = -\frac{[3x^2 e^{y+z} - y \cos(x - z)]}{[x^3 e^{y+z} + y \cos(x - z)]}$$

and

$$\frac{\partial z}{\partial y} = -\frac{[x^3 e^{y+z} - \sin(x - z)]}{[x^3 e^{y+z} + y \cos(x - z)]}.$$

Exercises

In Exercises 1 to 6, find $\dfrac{\partial F}{\partial x}$ and $\dfrac{\partial F}{\partial y}$ in terms of u, v, x, and y, as in Example 3(a).

1. $F(u, v) = u^2 - 3uv + v^2$, $u = 2x^2 - y$, $v = x + 3xy$.

2. $F(u, v) = ue^v - ve^u$, $u = xy$, $v = x^2 - y^2$.

3. $F(u, v) = 2u^3 + uv - 3v^3$, $u = x^2 + y$, $v = 5xy - y^2$.

4. $F(u, v) = e^{u^2 + v^2}$, $u = x \sin y$, $v = y \sin x$.

5. $F(u, v) = v \sin u^3$, $u = \cos(x - y)$, $v = \ln(x + y)$.

6. $F(u, v) = \ln(u + v) - \ln(u - v)$, $u = ye^x$, $v = y + x^{-1}$.

In Exercises 7 to 10, find f_x, f_y, and f_z in terms of u, v, x, y, z.

7. $f(u, v) = uv$, $u = \ln(x + y + z)$, $v = e^{x-y-z}$.

8. $f(u, v) = u \ln v$, $u = x \sin y + y \sin z$, $v = x \cos y + y \cos z$.

9. $f(u, v) = 2u^2 v$, $u = e^{x+y+z}$, $v = \ln(x + y + z)$.

10. $f(u, v) = u \sin v + v \sin u$, $u = x \ln(y + z)$, $v = y \ln(x + z)$.

In Exercises 11 to 14, find $\frac{\partial \phi}{\partial r}$ and $\frac{\partial \phi}{\partial \theta}$ in terms of x, y, r, θ.

11. $\phi(x, y) = 3x^2 - 5y^2$, $x = r \sin 2\theta$, $y = \cos(\theta - r)$.

12. $\phi(w, z) = \text{Sin}^{-1}\frac{w}{z}$, $w = r^2 - \theta^2$, $z = \cos(r + \theta)$.

13. $\phi(x, y) = 4x^2 + 7y^2$, $x = \sin(r - \theta)$, $y = r \cos 3\theta$.

14. $\phi(x, y) = \sin x \cos 3y$, $x = e^{r+\theta}$, $y = \ln(r - \theta)$.

In 18.5.2, if $u = f(t)$ and $v = g(t)$ define functions of one independent variable t, then F is a function of t alone and the formula of 18.5.2 becomes

$$\frac{dF}{dt} = \frac{\partial F}{\partial u}\frac{du}{dt} + \frac{\partial F}{\partial v}\frac{dv}{dt} + \frac{\partial F}{\partial w}\frac{dw}{dt}.$$

Use this result in Exercises 15 to 18 to find dF/dt in terms of x, y, z, t.

15. $F(x, y) = 3x^2 + 5xy + y^2$, $x = \sin 2t$, $y = \cos 2t$.

16. $F(x, y) = \ln\left(\frac{x}{y}\right)$, $x = \tan t$, $y = \sec^2 t$.

17. $F(x, y, z) = \sin(xyz^2)$, $x = t^3$, $y = t^2$, $z = t$.

18. $F(x, y, z) = \text{Tan}^{-1}\left(\frac{xy}{z}\right)$, $x = \ln t$, $y = e^{-t}$, $z = t^2 - 1$.

In Exercises 19 to 22, use equation (2) to find dy/dx.

19. $x^3 + 2x^2y - y^3 = 0$.

20. $ye^{-x} + 5x - 17 = 0$.

21. $x \sin y + y \cos x = 0$.

22. $x^2 \cos y - y^2 \sin x = 0$.

23. If $xy - z^2 + 2xyz = 0$, find $\partial z/\partial x$ and $\partial z/\partial y$.

24. If $3x^2z + y^3 - xyz^3 = 0$, find $\partial z/\partial x$ and $\partial z/\partial y$.

25. The altitude of a right circular cone is increasing at the rate of 3 inches per second and the radius of its base is increasing at the rate of 2 inches per second. If the altitude is 60 inches when the radius of the base is 25 inches, how fast is the volume of the cone increasing at that instant?

26. A triangle has vertices A, B, and C. The length of the side $b = AC$ is increasing at the rate of 6 inches per second, the side $c = AB$ is decreasing at 2 inches per second, and the included angle α is increasing at 0.2 radian per second. If at the same instant, $b = 20$ inches, $c = 30$ inches, and $\alpha = \pi/4$, how fast is the area changing then?

27. Let $w = f(x, y)$. If the coordinate axes in the xy plane are rotated by the substitutions $x = x' \cos \theta - y' \sin \theta$ and $y = x' \sin \theta + y' \cos \theta$, find $\partial w/\partial x'$ and $\partial w/\partial y'$.

18.6 THE DIRECTIONAL DERIVATIVE AND THE GRADIENT

Let f be a function of two variables. Since partial derivatives are ordinary derivatives of functions of one independent variable, $f_x(x, y)$ gives the rate of change of $f(x, y)$ with respect to distance *in the x direction* and $f_y(x, y)$ gives the rate of change of $f(x, y)$ *in the y direction* (Fig. 18-4).

A generalization of partial derivatives is the **directional derivative**, which gives the rate of change of $f(x, y)$ with respect to distance in *any chosen direction* in the xy plane.

Consider a function f of two variables, x and y, and fixed point $P_1:(x_1, y_1)$ in the domain of f. Let $\mathbf{u} = \cos\theta\mathbf{i} + \sin\theta\mathbf{j}$ be a unit vector with its initial point at P_1 and making an angle θ with the x axis [Fig. 18-9(a)]; and let l be a directed line through P_1 in the direction $\mathbf{u}$ [Fig. 18-9(b)].

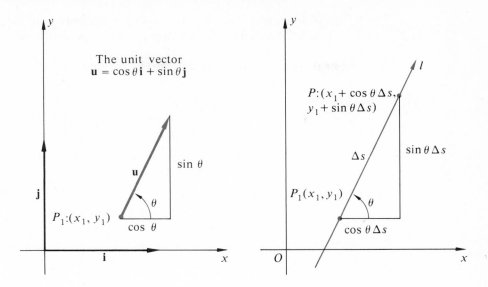

Figure 18-9

If P is any point on l, distinct from P_1, such that the line segment P_1P is the domain of f, and if we denote the distance from P_1 to P by Δs, then the coordinates of P are $(x_1 + \cos \theta \, \Delta s, y_1 + \sin \theta \, \Delta s)$ [Fig. 18-9(b)].

The average rate of change of $f(x, y)$ from P_1 to P, with respect to the distance Δs, is

$$\frac{f(x_1 + \cos \theta \, \Delta s, y_1 + \sin \theta \, \Delta s) - f(x_1, y_1)}{\Delta s},$$

in which x_1, y_1, and θ are constants and Δs is variable. The instantaneous rate of change of $f(x, y)$ at P_1, with respect to distance in the direction $\mathbf{u}$, is

$$\lim_{\Delta s \to 0} \frac{f(x_1 + \cos \theta \, \Delta s, y_1 + \sin \theta \, \Delta s) - f(x_1, y_1)}{\Delta s},$$

provided that this limit exists. It is the value at (x_1, y_1) of what is known as the directional derivative of f in the direction $\mathbf{u}$ and is denoted by $D_{\mathbf{u}} f(x_1, y_1)$.

18.6.1 Definition. Let f be a function of two variables. The **directional derivative** $D_{\mathbf{u}} f$ in the direction $\mathbf{u} = \cos \theta \mathbf{i} + \sin \theta \mathbf{j}$ is defined by

$$D_{\mathbf{u}} f(x_1, y_1) = \lim_{\Delta s \to 0} \frac{f(x_1 + \cos \theta \, \Delta s, y_1 + \sin \theta \, \Delta s) - f(x_1, y_1)}{\Delta s}$$

at any point (x_1, y_1) in the domain of f such that this limit exists.

Notice that when $\theta = 0$, the direction $\mathbf{u}$ is the direction of the x axis, $\cos \theta = 1$, $\sin \theta = 0$, and the directional derivative is simply the partial derivative with respect to x. Again, when $\theta = \pi/2$, the direction $\mathbf{u}$ is the direction of the y axis, $\cos \theta = 0$, $\sin \theta = 1$, and the directional derivative becomes the partial derivative with respect to y.

In working with directional derivatives, the following theorem is very useful.

18.6.2   Theorem.   Let f be a function of two variables, x and y, having continuous first partial derivatives. Then the value of the directional derivative $D_{\mathbf{u}}f$ at (x_1, y_1) in the direction $\mathbf{u} = \cos \theta \mathbf{i} + \sin \theta \mathbf{j}$ is given by

$$D_{\mathbf{u}}f(x_1, y_1) = f_x(x_1, y_1) \cos \theta + f_y(x_1, y_1) \sin \theta.$$

Proof.   By definition (18.6.1),

$$(1) \qquad D_{\mathbf{u}}f(x_1, y_1) = \lim_{\Delta s \to 0} \frac{f(x_1 + \cos \theta \, \Delta s, y_1 + \sin \theta \, \Delta s) - f(x_1, y_1)}{\Delta s}.$$

Since the differences, Δx and Δy, in the coordinates of the points $P_1:(x_1, y_1)$ and $P:(x_1 + \cos \theta \, \Delta s, y_1 + \sin \theta \, \Delta s)$ are

$$(2) \qquad \Delta x = \cos \theta \, \Delta s, \qquad \Delta y = \sin \theta \, \Delta s$$

(Fig. 18-10), the point P can be expressed as $P:(x_1 + \Delta x, y_1 + \Delta y)$. In this notation, equation (1) can be written

$$(3) \qquad D_{\mathbf{u}}f(x_1, y_1) = \lim_{\Delta s \to 0} \frac{f(x_1 + \Delta x, y_1 + \Delta y) - f(x_1, y_1)}{\Delta s},$$

and (by 18.4.1) its right-hand member is equal to

$$\lim_{\Delta s \to 0} \frac{\Delta f(x_1, y_1)}{\Delta s}.$$

But from 18.4.2

$$\frac{\Delta f(x_1, y_1)}{\Delta s} = f_x(x_1, y_1)\frac{\Delta x}{\Delta s} + f_y(x_1, y_1)\frac{\Delta y}{\Delta s} + \phi_1 \frac{\Delta x}{\Delta s} + \phi_2 \frac{\Delta y}{\Delta s},$$

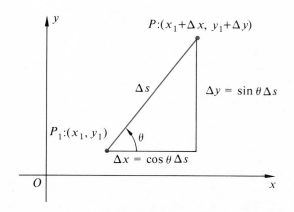

Figure 18-10

where $\phi_1 \longrightarrow 0$ and $\phi_2 \longrightarrow 0$ as $\Delta x \longrightarrow 0$ and $\Delta y \longrightarrow 0$. Therefore

(4)
$$D_{\mathbf{u}}f(x_1, y_1) = \lim_{\Delta s \to 0} \frac{\Delta f(x_1, y_1)}{\Delta s} = f_x(x_1, y_1) \lim_{\Delta s \to 0} \frac{\Delta x}{\Delta s}$$
$$+ f_y(x_1, y_1) \lim_{\Delta s \to 0} \frac{\Delta y}{\Delta s} + \left(\lim_{\Delta s \to 0} \phi_1\right)\left(\lim_{\Delta s \to 0} \frac{\Delta x}{\Delta s}\right) + \left(\lim_{\Delta s \to 0} \phi_2\right)\left(\lim_{\Delta s \to 0} \frac{\Delta y}{\Delta s}\right)$$

From (2), $\Delta x = \cos \theta \, \Delta s$ and $\Delta y = \sin \theta \, \Delta s$, in which θ is constant. Thus

(5)
$$\Delta s \longrightarrow 0 \implies \Delta x \longrightarrow 0 \quad \text{and} \quad \Delta y \longrightarrow 0.$$

Moreover, $\Delta x/\Delta s = \cos \theta$ and $\Delta y/\Delta s = \sin \theta$, both of which are constants; and so

(6)
$$\lim_{\Delta s \to 0} \frac{\Delta x}{\Delta s} = \cos \theta, \qquad \lim_{\Delta s \to 0} \frac{\Delta y}{\Delta s} = \sin \theta.$$

From (4), (5), and (6)

$$D_{\mathbf{u}}f(x_1, y_1) = f_x(x_1, y_1) \cos \theta + f_y(x_1, y_1) \sin \theta + (0) \cos \theta + (0) \sin \theta. \quad \blacksquare$$

Example 1. If $f(x, y) = 4x^2 - xy + 3y^2$, find the directional derivative of f at $P{:}(2, -1)$ in the direction of the vector $\mathbf{a} = 4\mathbf{i} + 3\mathbf{j}$.

Solution. The unit vector $\mathbf{u}$ in the direction of $\mathbf{a}$ is $\mathbf{a}/|\mathbf{a}| = (4/5)\mathbf{i} + (3/5)\mathbf{j}$, from which $\cos \theta = 4/5$ and $\sin \theta = 3/5$.

Now $f_x(x, y) = 8x - y$ and $f_y(x, y) = -x + 6y$; thus $f_x(2, -1) = 17$ and $f_y(2, -1) = -8$. Consequently, by 18.6.2, the directional derivative of f at P in the direction of $\mathbf{a} = 4\mathbf{i} + 3\mathbf{j}$ is

$$f_x(2, -1) \cos \theta + f_y(2, -1) \sin \theta = (17)\tfrac{4}{5} + (-8)\tfrac{3}{5} = \tfrac{44}{5}.$$

Example 2. Find the direction θ in the xy plane for which the value of the directional derivative of $f(x, y) = x^2 - y^2$ at $(\sqrt{3}, 1)$ is a maximum. What is this maximum value?

Solution. From 18.6.2, the value of the directional derivative is given by

$$f_x(x, y) \cos \theta + f_y(x, y) \sin \theta = 2x \cos \theta - 2y \sin \theta.$$

At $(\sqrt{3}, 1)$, this is $2\sqrt{3} \cos \theta - 2 \sin \theta$. To find the direction θ that makes this function of θ a maximum, we differentiate it with respect to θ, obtaining $-2\sqrt{3} \sin \theta - 2 \cos \theta$, and then set this result equal to zero. The roots of $\sqrt{3} \sin \theta + \cos \theta = 0$, or $\tan \theta = -1/\sqrt{3}$, are the critical numbers $\theta = \tfrac{5}{6}\pi$ and $\theta = \tfrac{11}{6}\pi$. Testing these critical numbers by the second derivative test, we find that when $\theta = \tfrac{5}{6}\pi$, the value of the directional derivative of f at $(\sqrt{3}, 1)$ is a minimum; and when $\theta = \tfrac{11}{6}\pi$, the value of the directional derivative is a maximum.

The maximum value of this directional derivative at $(\sqrt{3}, 1)$ is 4.

As with partial derivatives, the directional derivative has a simple geometric interpretation. Let $P_1{:}(x_1, y_1)$ be a point in the domain of a function f of two variables, x and y, and let $\mathbf{u} = \cos \theta \mathbf{i} + \sin \theta \mathbf{j}$ be a unit vector in the xy plane with its initial point at P_1. Denote by l the line through P_1 in the direction $\mathbf{u}$ (Fig. 18-11). Then the plane through l that is perpendicular to the xy plane intersects the surface $z = f(x, y)$ in a plane curve C, and the directional derivative of f at (x_1, y_1) gives the slope of the tangent line to C at the point $(x_1, y_1, f(x_1, y_1))$.

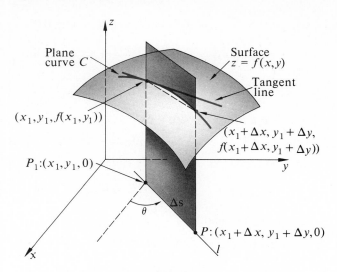

Figure 18-11

The value of the directional derivative at (x, y) in the direction $\mathbf{u} = \cos\theta\mathbf{i} + \sin\theta\mathbf{j}$,

$$D_{\mathbf{u}}f(x, y) = f_x(x, y)\cos\theta + f_y(x, y)\sin\theta,$$

can be expressed as the dot product of two vectors,

(7) $$D_{\mathbf{u}}f(x, y) = (\cos\theta\mathbf{i} + \sin\theta\mathbf{j}) \cdot (f_x(x, y)\mathbf{i} + f_y(x, y)\mathbf{j}),$$

as is easily verified. The first of the two vectors in this dot product is the unit vector $\mathbf{u} = \cos\theta\mathbf{i} + \sin\theta\mathbf{j}$ [Fig. 18-9(a)], and the second vector is an extremely important one called the gradient.

18.6.3 Definition. If f is a function of two variables having first partial derivatives f_x and f_y, then the **gradient** vector of f, denoted by ∇f (read "del f"), is defined by

$$\nabla f(x, y) = f_x(x, y)\mathbf{i} + f_y(x, y)\mathbf{j}.$$

18.6.4 Corollary. If f is a function of two variables having continuous first partial derivatives, its directional derivative in the xy direction $\mathbf{u}$ may be written

$$D_{\mathbf{u}}f(x, y) = \mathbf{u} \cdot \nabla f(x, y),$$

in which $\mathbf{u} = \cos\theta\mathbf{i} + \sin\theta\mathbf{j}$ is a unit vector and $\nabla f(x, y)$ is the gradient vector of f.

In this expression for the directional derivative of f, the unit vector $\mathbf{u}$ depends only on the angle θ, and the gradient vector $\nabla f(x, y)$ depends only on the function f and the point (x, y) in the domain of f.

The gradient vector $\nabla f(x, y) = f_x(x, y)\mathbf{i} + f_y(x, y)\mathbf{j}$ is always in the domain of f.

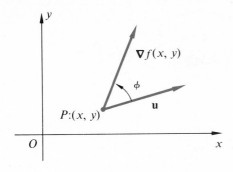

Figure 18-12

Let ϕ be the angle between the unit vector **u** and the gradient vector $\nabla f(x, y)$ (Fig. 18-12). By 15.5.6,

$$\mathbf{u} \cdot \nabla f(x, y) = |\mathbf{u}||\nabla f(x, y)| \cos \phi,$$

in which $|\mathbf{u}| = 1$. Thus 18.6.4 can be written

$$(8) \qquad D_{\mathbf{u}} f(x, y) = |\nabla f(x, y)| \cos \phi,$$

which is a maximum when $\phi = 0$ (so that $\cos \phi = 1$). Therefore the directional derivative is a maximum when its direction **u** is the same as the direction of the gradient.

18.6.5 Theorem. The directional derivative of $f(x, y)$ at a point $P:(x, y)$ in its domain is a **maximum** in the direction of the gradient of f at P, and that maximum value is $|\nabla f(x, y)|$.

Example 3. Use 18.6.5 to find the direction θ in which the directional derivative of $f(x, y) = x^2 - y^2$ at $(\sqrt{3}, 1)$ is a maximum and find that maximum value.

Solution. The gradient vector of $f(x, y) = x^2 - y^2$ at the point $(\sqrt{3}, 1)$ is

$$\nabla f(\sqrt{3}, 1) = f_x(\sqrt{3}, 1)\mathbf{i} + f_y(\sqrt{3}, 1)\mathbf{j}.$$

Since $f_x(x, y) = 2x, f_x(\sqrt{3}, 1) = 2\sqrt{3}, f_y(x, y) = -2y$, and $f_y(\sqrt{3}, 1) = -2$,

$$\nabla f(\sqrt{3}, 1) = 2\sqrt{3}\,\mathbf{i} - 2\mathbf{j}.$$

The slope of this gradient is $\tan \theta = -2/2\sqrt{3} = -1/\sqrt{3}$, from which $\theta = 11\pi/6$. The magnitude of the gradient is $|\nabla f(\sqrt{3}, 1)| = 4$. Therefore, by 18.6.5, the directional derivative $D_{\mathbf{u}} f(\sqrt{3}, 1)$ has its maximum value 4 in the direction $\theta = 11\pi/6$.

This is the same problem as Example 2. The usefulness of 18.6.5 will be evident if the reader compares the two solutions.

Since the directional derivative $D_{\mathbf{u}} f(x, y)$ is the rate of change of $f(x, y)$ in the direction **u**, the greatest increase in the value of f at any point $P:(x, y)$ in its domain is in the direction of its gradient at P.

Recall from Section 18.1 that the *level curves* of a surface $z = f(x, y)$ are the projections onto the xy plane of the curves of intersection of the surface with planes $z = k$ that are parallel to the xy plane. The value of the function at all points on the same level curve is constant (Fig. 18-13).

Denote by L_1 the level curve of $f(x, y)$ that passes through an arbitrarily chosen point $P_1(x_1, y_1)$ in the domain of f and let the unit vector **u** be *tangent* to L_1 at P_1. Since the value of f is the same at all points on the level curve L_1, its directional derivative $D_{\mathbf{u}} f(x_1, y_1)$, which is the rate of change of $f(x, y)$ in the direction **u**, is zero when **u** is tangent to L_1. But from equation (8) the direction derivative is zero if and only if $\phi = \pi/2$. Therefore the *gradient vector of f at any point P_1 in its domain is normal to the level curve of f that goes through P_1* (Fig. 18-13).

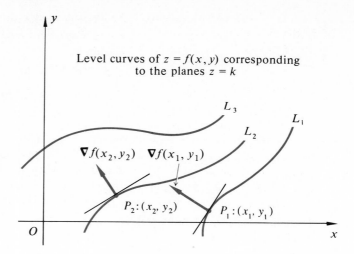

Figure 18-13

It follows from this, and from 18.6.5, that the direction of greatest increase in $f(x, y)$ at any point in its domain is perpendicular to its level curve that goes through the point.

Example 4. For the paraboloid $z = x^2/4 + y^2$, find the equation of its level curve that passes through the point P_1:(2, 1) and sketch it. Find the gradient vector of the paraboloid at P_1 and draw the gradient with its initial point at P_1 (Fig. 18-14).

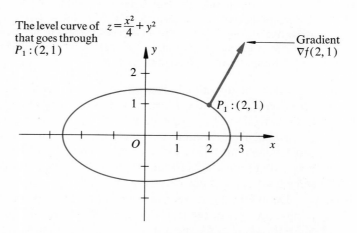

Figure 18-14

Solution. The level curve of the paraboloid that corresponds to the cutting plane $z = k$, $k > 0$, has the equation

$$\frac{x^2}{4} + y^2 = k.$$

To find the value of k belonging to the level curve through P_1, we substitute $(2,1)$ for (x, y) in the above equation and obtain $k = 2$. Thus the equation of the level curve that goes through P_1 is

$$\frac{x^2}{8} + \frac{y^2}{2} = 1.$$

It is shown in Fig. 18-14.

Let $f(x, y) = x^2/4 + y^2$. Since $f_x(x, y) = x/2$ and $f_y(x, y) = 2y$, the gradient of the paraboloid at $P_1:(2, 1)$ is

$$\nabla f(2, 1) = f_x(2, 1)\mathbf{i} + f_y(2, 1)\mathbf{j} = \mathbf{i} + 2\mathbf{j}.$$

The definitions and theorems for directional derivatives and gradients of functions of two variables extend in a natural way to functions of three (or more) variables, and the proofs of the theorems are similar. They are stated informally below.

There is one minor change, however. The three-dimensional unit vector $\mathbf{u}$, which gives the direction of the directional derivative at P in the domain of the function, will be written

$$\mathbf{u} = \cos\alpha\,\mathbf{i} + \cos\beta\,\mathbf{j} + \cos\gamma\,\mathbf{k};$$

its components are the direction cosines of $\mathbf{u}$ because

$$|\mathbf{u}| = (\cos^2\alpha + \cos^2\beta + \cos^2\gamma)^{1/2} = 1$$

(Fig. 18-15). For functions of more than three variables, the unit vector is written analogously.

If f is a function of three variables, its **directional derivative** at the point $P:(x, y, z)$ in the direction $\mathbf{u} = \cos\alpha\,\mathbf{i} + \cos\beta\,\mathbf{j} + \cos\gamma\,\mathbf{k}$ is defined by

$$D_{\mathbf{u}}f(x, y, z) = \lim_{\Delta s \to 0} \frac{f(x + \cos\alpha\,\Delta s, y + \cos\beta\,\Delta s, z + \cos\gamma\,\Delta s) - f(x, y, z)}{\Delta s}$$

if this limit exists. It gives the rate of change of $f(x, y, z)$ in the direction $\mathbf{u}$. The value of this directional derivative is given by

$$D_{\mathbf{u}}f(x, y, z) = f_x(x, y, z)\cos\alpha + f_y(x, y, z)\cos\beta \\ + f_z(x, y, z)\cos\gamma.$$

The unit vector at P,
$\mathbf{u} = \cos\alpha\,\mathbf{i} + \cos\beta\,\mathbf{j} + \cos\gamma\,\mathbf{k}$

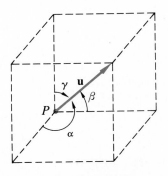

Figure 18-15

The **gradient** vector of $f(x, y, z)$ is

$$\nabla f(x, y, z) = f_x(x, y, z)\mathbf{i} + f_y(x, y, z)\mathbf{j} + f_z(x, y, z)\mathbf{k}.$$

As with functions of two variables, the directional derivative and the gradient of $f(x, y, z)$ are related by

$$D_{\mathbf{u}}f(x, y, z) = \mathbf{u} \cdot \nabla f(x, y, z) = |\nabla f(x, y, z)| \cos \phi,$$

where ϕ is the angle between the unit vector $\mathbf{u}$ and the gradient vector ∇f.

Moreover, the directional derivative of a function of three variables at a point $P:(x, y, z)$ in its domain is a *maximum* in the direction of the gradient of the function at P, and that maximum value is $|\nabla f(x, y, z)|$.

The concept of level curves for functions of two variables generalizes to *level surfaces* for functions of three variables. If f is a function of three variables, the surface $f(x, y, z) = k$, where k is a constant, is called a **level surface** for f. At all points on a level surface the value of the function is the same, and the gradient vector of $f(x, y, z)$ at a point $P:(x, y, z)$ in its domain is normal to the level surface of f that goes through P.

In problems of heat conduction in a homogeneous body where $w = f(x, y, z)$ gives the temperature at the point (x, y, z), the level surface $f(x, y, z) = k$ is called an **isothermal surface** because all points on it have the same temperature k. At any given point of the body, heat flows in the direction opposite to the gradient (that is, in the direction of the greatest decrease in temperature) and therefore perpendicular to the isothermal surface through the point. Again, if $w = f(x, y, z)$ gives the electrostatic potential (voltage) at any point in an electric potential field, the level surfaces of the function are called **equipotential surfaces**. All points on an equipotential surface have the same electrostatic potential, and the direction of flow of electricity is along the negative gradient—that is, in the direction of greatest drop in potential.

Example 5. If the temperature at any point in a homogeneous body is given by $T = e^{xy} - xy^2 - x^2yz$, what is the direction of the greatest drop in temperature at the point $(1, -1, 2)$?

Solution. The greatest decrease in temperature at $(1, -1, 2)$ is in the direction of the negative gradient at that point.

Since $\nabla T = (ye^{xy} - y^2 - 2xyz)\mathbf{i} + (xe^{xy} - 2xy - x^2z)\mathbf{j} + (-x^2y)\mathbf{k}$, $\quad -\nabla T \quad$ at $(1, -1, 2)$ is

$$(e^{-1} - 3)\mathbf{i} - e^{-1}\mathbf{j} - \mathbf{k}.$$

Exercises

In Exercises 1 to 6, find the directional derivative of the indicated function. What is its value at the given point P in the given direction?

1. $f(x, y) = x^2 - 3xy + 2y^2$; $P = (-1, 2)$, in the direction $\theta = \pi/6$.

2. $f(x, y) = 2x^2 + xy - y^2$; $P = (3, -2)$, in the direction $\theta = -\pi/4$.

3. $f(x, y) = x^3 + 4x^2y - y^4$; $P = (1, -3)$, in the direction $\mathbf{u} = (1/\sqrt{2})\mathbf{i} + (1/\sqrt{2})\mathbf{j}$.

4. $f(x, y) = e^x \sin y$; $P = (0, \pi/4)$, in the direction $\mathbf{u} = (1/2)\mathbf{i} + (\sqrt{3}/2)\mathbf{j}$.

5. $f(x, y) = e^{-xy}$; $P = (1, -1)$, in the direction of the vector $\mathbf{a} = -\mathbf{i} + \sqrt{3}\,\mathbf{j}$.

6. $f(x, y) = \ln(x^2 - y^2)$; $P = (3, -2)$, in the direction of the vector $\mathbf{a} = 3\mathbf{i} + \sqrt{3}\mathbf{j}$.

In Exercises 7 to 10, find the directional derivative of the indicated function of three variables at the given point P in the direction $\mathbf{u} = \cos\alpha\mathbf{i} + \cos\beta\mathbf{j} + \cos\gamma\mathbf{k}$.

7. $f(x, y, z) = x^2 + y^2 + z^2$, $P = (1, -1, 2)$.

8. $f(x, y, z) = x^3y - y^2z^2$, $P = (-2, 1, 3)$.

9. $f(x, y, z) = \ln(y^2 - xz) + e^{xyz}$, $P = (1, -2, 2)$.

10. $f(x, y, z) = (xy^{-2}z)^{1/2}$, $P = (4, -1, 1)$.

In Exercises 11 to 16, find the gradient vector of the given function at the given point P.

11. $f(x, y) = x^2y - xy^2$, $P = (-2, 3)$.

12. $f(x, y) = e^x \sin y + e^y \sin x$, $P = \left(\dfrac{\pi}{6}, \dfrac{\pi}{3}\right)$.

13. $f(x, y) = \ln(x^3 - xy^2 + 4y^3)$, $P = (-3, 3)$.

14. $f(x, y) = e^{-y}\ln x - e^{-x}\ln y$, $P = (5, 7)$.

15. $f(x, y, z) = 3x^2 - 5y^2 + z^2$, $P = (1, 1, -7)$.

16. $f(x, y, z) = e^{xy} - 2e^{yz} + 3e^{xz}$, $P = (1, -2, -1)$.

17. In what direction is $f(x, y) = 2x^2 + y^2 - y$ increasing most rapidly at $(2, 3)$?

18. Find the direction of greatest increase in $f(x, y) = x^2y + 2xy^2$ at $(2, 1)$.

In Exercises 19 to 22, find the direction θ in which the directional derivative of the indicated function at the given point P is a maximum. What is that maximum value?

19. $f(x, y) = x^3 - y^3$, $P = (2, -1)$.

20. $f(x, y) = e^y \sin x$, $P = \left(\dfrac{5\pi}{6}, 3\right)$.

21. $f(x, y) = x^4 - xy^2$, $P = (1, -3)$.

22. $f(x, y) = 2\cos(x + y)$, $P = \left(\dfrac{\pi}{2}, \dfrac{\pi}{4}\right)$.

23. For the elliptic paraboloid $z = x^2/9 + y^2/4$,
(a) find the equation of its level curve that goes through the point $(6, 2)$;
(b) find the gradient vector ∇z at $(6, 2)$;
(c) draw the level curve, and also draw the gradient vector with its initial point at $(6, 2)$.

18.7 TANGENT PLANES TO A SURFACE. EXTREMA OF A FUNCTION OF TWO VARIABLES

The Cartesian graph of an equation

$$F(x, y, z) = 0$$

is usually a surface. When it is possible to solve this equation for z in terms of x and y, the equation of the surface can be written in the form $z = \phi(x, y)$, but this is not assumed in the discussion that follows.

Let $P:(a, b, c)$ be a point on a surface $F(x, y, z) = 0$, and let

(1) $$\mathbf{r}(t) = f(t)\mathbf{i} + g(t)\mathbf{j} + h(t)\mathbf{k}$$

be a position vector that traces out a curve C, which is on the surface and passes through $P:(a, b, c)$, with $a = f(t_0)$, $b = g(t_0)$, and $c = h(t_0)$. Then

(2) $$F(f(t), g(t), h(t)) = 0$$

for all numbers t in the domain of $\mathbf{r}$.

If F_x, F_y, and F_z are continuous and not all zero at P, and if $f'(t_0)$, $g'(t_0)$, and $h'(t_0)$ exist, then (by 18.5.2)

(3) $$F_t(a, b, c) = F_x(a, b, c)f'(t_0) + F_y(a, b, c)g'(t_0) + F_z(a, b, c)h'(t_0) = 0.$$

It is easy to verify that equation (3) can be written in the vector form

$$\mathbf{\nabla}F(a, b, c) \cdot \mathbf{r}'(t_0) = 0,$$

where $\mathbf{\nabla}F(a, b, c)$ is the vector value of the gradient vector of F at the point $P{:}(a, b, c)$ in the domain of F. Since $\mathbf{r}'(t_0)$ is the velocity vector of (1) at $P{:}(a, b, c)$, then $\mathbf{r}'(t_0)$ is tangent to the curve C at P; and it follows (from 16.2.7) that the gradient vector of F at $P{:}(a, b, c)$ is perpendicular to the tangents at P of all curves, such as C, that lie on the surface and go through P (Fig. 18-16). Thus *all these tangent lines at P lie in a plane that is perpendicular to the gradient of F at P.*

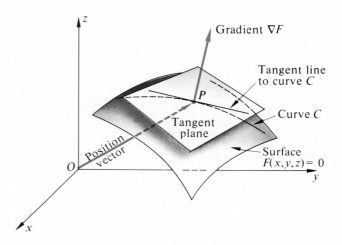

Figure 18-16

18.7.1 **Definition.** If a surface $F(x, y, z) = 0$ has a nonzero gradient vector at a point $P{:}(a, b, c)$ on it, the plane through P that is perpendicular to the gradient $\mathbf{\nabla}F{:}(a, b, c)$ is called the **tangent plane** to the surface at P.

The components of the gradient $\mathbf{\nabla}F(a, b, c)$ are $F_x(a, b, c)$, $F_y(a, b, c)$, and $F_z(a, b, c)$. Therefore (16.4.1) a **vector equation of the tangent plane** to the surface $F(x, y, z) = 0$ at a point $P{:}(a, b, c)$ on the surface is

18.7.2
$$\mathbf{\nabla}F(a, b, c) \cdot [(x - a)\mathbf{i} + (y - b)\mathbf{j} + (z - c)\mathbf{k}] = 0;$$

and by performing the indicated dot multiplication, we obtain a **Cartesian equation of the tangent plane** to the surface $F(x, y, z) = 0$ at $P{:}(a, b, c)$,

18.7.3
$$(x - z)F_x(a, b, c) + (y - b)F_y(a, b, c) + (z - c)F_z(a, b, c) = 0,$$

provided that F_x, F_y, and F_z are continuous and not all zero at P.

The line through $P:(a, b, c)$ having direction numbers proportional to $F_x(a, b, c)$, $F_y(a, b, c)$, and $F_z(a, b, c)$ is called the **normal** to the surface $F(x, y, z) = 0$ at P.

If the equation of the surface is given in the form $z = f(x, y)$, we can write $F(x, y, z) = f(x, y) - z = 0$. Here $F_x = f_x$, $F_y = f_y$, and $F_z = -1$.

Thus a Cartesian equation of the **tangent plane** to the surface $z = f(x, y)$ at a point $P:(a, b, c)$ on the surface is

$$(x - a)f_x(a, b) + (y - b)f_y(a, b) - (z - c) = 0,$$

if f_x and f_y are continuous at (a, b).

Example 1. Find a vector equation and a Cartesian equation of the tangent plane to the hyperboloid $4x^2 - 9y^2 - 9z^2 - 36 = 0$ (Fig. 16-41) at the point $P:(3\sqrt{3}, 2, 2)$, and write symmetric equations of the normal to the surface at P.

Solution. Let $F(x, y, z) = 4x^2 - 9y^2 - 9z^2 - 36 = 0$. Then $F_x(x, y, z) = 8x$, $F_y(x, y, z) = -18y$, and $F_z(x, y, z) = -18z$. Thus $F_x(3\sqrt{3}, 2, 2) = 24\sqrt{3}$, $F_y(3\sqrt{3}, 2, 2) = -36$, and $F_z(3\sqrt{3}, 2, 2) = -36$; and the gradient vector of F at P is

$$\nabla F(3\sqrt{3}, 2, 2) = 24\sqrt{3}\,\mathbf{i} - 36\mathbf{j} - 36\mathbf{k}.$$

From 18.7.2 and 18.7.3, a vector equation of the tangent plane is

$$\nabla F(3\sqrt{3}, 2, 2) \cdot [(x - 3\sqrt{3})\mathbf{i} + (y - 2)\mathbf{j} + (z - 2)\mathbf{k}] = 0,$$

and its Cartesian equation is

$$24\sqrt{3}(x - 3\sqrt{3}) - 36(y - 2) - 36(z - 2) = 0,$$

or

$$2\sqrt{3}\,x - 3y - 3z - 6 = 0.$$

The normal line to the hyperboloid at $P:(3\sqrt{3}, 2, 2)$ has direction numbers $2\sqrt{3}$, -3, -3; thus it has symmetric equations

$$\frac{x - 3\sqrt{3}}{2\sqrt{3}} = \frac{y - 2}{-3} = \frac{z - 2}{-3}.$$

The definitions of the local extreme values of a function of two variables are simple extensions of those for a function of one variable (6.5.1).

We say that $f(a, b)$ is a **local maximum** value of f if there is an open rectangular region H with (a, b) as an interior point such that $f(x, y) \leq f(a, b)$ for all points (x, y) in H; and $f(a, b)$ is a **local minimum** of f if $f(x, y) \geq f(a, b)$ for all points (x, y) in H.

The local maximum values and the local minimum values of a function are its **extrema**.

Let f be a function of two variables that is continuous in an open rectangular region H of the xy plane and let (a, b) be a point of H. If $f(a, b)$ is a local maximum value of f, then the plane curve of intersection of the surface $z = f(x, y)$ and any plane through $(a, b, 0)$ that is perpendicular to the xy plane will have a local maximum point there (Fig. 18-17).

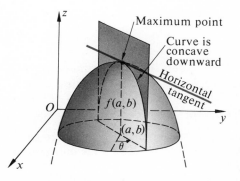

Figure 18-17

In particular, the curves of intersection of $z = f(x, y)$ and the planes $x = a$ and $y = b$ will have local maximum points there. If $f_x(a, b)$ and $f_y(a, b)$ exist, then necessarily $f_x(a, b) = 0$ and $f_y(a, b) = 0$ (by 6.5.2). So the tangent plane to $z = f(x, y)$ at a local maximum point is horizontal. Notice that $f_x(a, b) = 0$ and $f_y(a, b) = 0$ can be expressed compactly by $\nabla f(a, b) = \mathbf{0}$, where $\mathbf{0} = [0, 0]$ is the zero vector (by 15.4.2).

A similar discussion applies to a local minimum point on $z = f(x, y)$.

18.7.4 **A Necessary Condition for Extrema.** Let f be a function of two variables that is continuous in a rectangular region H in the xy plane. If (a, b) is an interior point of H and if $f_x(a, b)$ and $f_y(a, b)$ exist, then a **necessary** condition for $f(a, b)$ to be an extremum of f is $f_x(a, b) = f_y(a, b) = 0$, or

$$\nabla f(a, b) = \mathbf{0}.$$

That the preceding conditions are not sufficient to guarantee an extremum is seen in the following example.

Example 2. For the hyperbolic paraboloid $z = x^2/a^2 - y^2/b^2$, we have $f_x(x, y) = 2x/a^2$ and $f_y(x, y) = -2y/b^2$; thus $f_x(0, 0) = f_y(0, 0) = 0$. But it is clear from Fig. 18-18 that $(0, 0, 0)$ is neither a local maximum point nor a local minimum point of this surface.

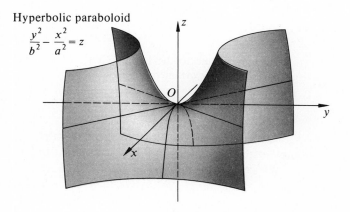

Hyperbolic paraboloid
$$\frac{y^2}{b^2} - \frac{x^2}{a^2} = z$$

Figure 18-18

Any point (a, b) in the interior of the domain of a continuous function f of two variables at which $f_x(a, b) = 0$ and $f_y(a, b) = 0$ is called a **critical point** for f.

Example 3. Find the local maximum or minimum values of the function f defined by $f(x, y) = x^2 + y^2/4$.

Solution. The function f is continuous throughout its domain, the xy plane. Since $f_x(x, y) = 2x$ and $f_y(x, y) = y/2$ are zero only when $x = 0$ and $y = 0$, $(0, 0)$ is the only critical point for f. Thus, by 18.7.4, if f has any relative maximum or minimum value, it must be $f(0, 0) = 0$. But as shown in Example 2, the value of a function at a critical point does not have to be an extremum.

For the given function it is easy to prove that $f(0, 0) = 0$ is a local minimum value of f. Let (x, y) be any point in the xy plane other than $(0, 0)$. Then $f(x, y) = x^2 + y^2/4 > 0$, since it is the sum of the squares of two real numbers, not both of which are zero. Therefore $f(0, 0) = 0$ is a local minimum value of f. The reader will find it instructive to sketch the graph of f.

When a function of two variables is fairly simple, we can sometimes determine whether its value at a known critical point is an extremum or not by using only the definitions of extrema and a little ingenuity. Such was the situation in Example 3. However, there is a powerful criterion that is often helpful when simpler methods fail. It constitutes a set of *sufficient* conditions for a function of two variables to have a local maximum value or a local minimum value at a critical point. We now state this test, which involves second partial derivatives, without proof.

18.7.5 Sufficient Conditions for Extrema. Let f be a function of two variables that is continuous and has continuous first and second partial derivatives in some open rectangular region H of the xy plane. Let (a, b) be a point of H at which

$$\nabla f(a, b) = 0$$

and let $\Delta = f_{xx}(a, b) f_{yy}(a, b) - f_{xy}^2(a, b)$.

Then
(i)   if $\Delta > 0$ and $f_{xx}(a, b) < 0$, $f(a, b)$ is a local maximum value of f.
(ii) if $\Delta > 0$ and $f_{xx}(a, b) > 0$, $f(a, b)$ is a local minimum value of f.
(iii) if $\Delta < 0$, $f(a, b)$ is not an extremum of f.
(iv) if $\Delta = 0$, the test is inconclusive.

Example 4.   Find the extrema, if any, of the function F defined by $F(x, y) = 3x^3 + y^2 - 9x + 4y$.

Solution.   Since $F_x(x, y) = 9x^2 - 9$ and $F_y(x, y) = 2y + 4$, the critical points, obtained by solving the simultaneous equation $F_x(x, y) = F_y(x, y) = 0$, are $(1, -2)$ and $(-1, -2)$.

Now $F_{xx}(x, y) = 18x$, $F_{yy}(x, y) = 2$ and $F_{xy} = 0$. Thus at the critical point $(1, -2)$, $\Delta = F_{xx}(1, -2) F_{yy}(1, -2) - F_{xy}^2(1, -2) = 18(2) - 0 = 36 > 0$. Furthermore, $F_{xx}(1, -2) = 18 > 0$, and therefore, by 18.7.5(ii), $F(1, -2) = -10$ is a local minimum value of F.

In testing the given function at the other critical point, $(-1, -2)$, we find $F_{xx}(-1, -2) = -18$, $F_{yy}(-1, -2) = 2$, and $F_{xy}(-1, -2) = 0$, which make $\Delta = -36 < 0$. Thus, by (iii) of 18.7.5, $F(-1, -2)$ is not an extremum.

Example 5.   Find the minimum distance between the origin and the surface $z^2 = x^2 y + 4$.

Solution.   Let P:(x, y, z) be any point on the surface. The square of the distance between the origin and P is $d^2 = x^2 + y^2 + z^2$. We seek the coordinates of P that make d^2 (and hence d) a minimum.

Since P is on the surface, its coordinates satisfy the equation of the surface. Eliminating z^2 between $z^2 = x^2 y + 4$ and $d^2 = x^2 + y^2 + z^2$, we obtain d^2 as a function of two variables x and y:

(4) $$d^2 = f(x, y) = x^2 + y^2 + x^2 y + 4.$$

To find the critical points, we set $f_x(x, y) = 0$ and $f_y(x, y) = 0$, obtaining

(5) $$2x + 2xy = 0 \qquad \text{and} \qquad 2y + x^2 = 0.$$

By eliminating y between these equtions, we get

$$2x - x^3 = 0.$$

Thus $x = 0$ and $x = \pm\sqrt{2}$. Substituting these values in the second of the equations (5), we obtain $y = 0$ and $y = -1$. Therefore the critical points are $(0, 0)$, $(\sqrt{2}, -1)$, and $(-\sqrt{2}, -1)$.

To test each of these by 18.7.5, we need $f_{xx}(x, y) = 2 + 2y, f_{yy}(x, y) = 2, f_{xy}(x, y) = 2x$, and $\Delta(x, y) = f_{xx}f_{yy} - f_{xy}^2 = 4 + 4y - 4x^2$. Since $\Delta(\pm\sqrt{2}, -1) = -8 < 0$, neither $(\sqrt{2}, -1)$ nor $(-\sqrt{2}, -1)$ yields an extremum. However, $\Delta(0, 0) = 4 > 0$ and $f_{xx}(0, 0) = 2 > 0$; and so $(0, 0)$ yields the minimum distance. Substituting $x = 0$ and $y = 0$ in (4), we find $d^2 = 4$.

The minimum distance between the origin and the given surface is 2.

The results of our discussion of the extrema of functions of two variables may be used to establish the important **method of least squares**.

Consider a set of n points in the plane, $P_1:(x_1, y_1)$, $P_2:(x_2, y_2)$, $\cdots$, $P_n:(x_n, y_n)$, where not all the x_i are equal. They may represent observed data for some project in the physical or biological sciences, in engineering, in economics, or in sociology.

Assume that there is an idealized relation between the x and y values that is *linear*. We seek the equation $y = mx + b$ of the straight line that "best approximates" or "best fits" the observed data so that we can compute an approximate value of y for each value of x between the successive observed values x_i, $i = 1, 2, 3, \cdots, n$.

After finding such a line, we will usually have two values of y for each of the observed x_i, $i = 1, 2, 3, \cdots, n$; one y value is the observed y_i, and the other is the computed value of y, $y = mx_i + b$. Their difference, $y_i - (mx_i + b)$, is called a *deviation*. Some of the deviations may be positive and some negative, but the squares of the deviations will always be nonnegative.

In the method of least squares, the line of "best fit" to the observed data is defined to be the line for which the sum of the squares of the n deviations is a minimum. Such a line is called the *line of regression* of y on x.

The sum of the squares of the deviations can be written

$$f(m, b) = \sum_{i=1}^{n} (y_i - mx_i - b)^2.$$

In Exercises 23 and 24 the reader is asked to find values of m and b that minimize $f(m, b)$.

Exercises

In Exercises 1 to 6, find the gradient vector of the given surface at the point P. Write the equation of the tangent plane to the surface at P in both the vector form and the Cartesian form. Sketch the part of the surface that is in the first octant and draw the tangent plane and the gradient vector at P.

1. $8x^2 + y^2 + 8z^2 = 16$, $P = \left(1, 2, \frac{\sqrt{2}}{2}\right)$.

2. $x^2 + y^2 + z^2 = 16$, $P = (2, 3, \sqrt{3})$.

3. $x^2 - y^2 + z^2 + 1 = 0$, $P = (1, 3, \sqrt{7})$.

4. $x^2 + y^2 - z^2 = 4$, $P = (2, 1, 1)$.

5. $x^2 - y^2 + z^2 = -9$, $P = (3, 6, 3\sqrt{2})$.

6. $z = \frac{x^2}{4} + \frac{y^2}{4}$, $P = (2, 2, 2)$.

Let C be a curve of intersection of two surfaces, $F(x, y, z) = 0$ and $G(x, y, z) = 0$, and let $P_1:(x_1, y_1, z_1)$ be a point on C. Since the curve C lies on both surfaces, its tangent line, l, at P is in the tangent plane to each surface at P_1, and thus is the line of intersection of the two tangent planes. It follows that l is perpendicular to both $\nabla F(x_1, y_1, z_1)$ and $\nabla G(x_1, y_1, z_1)$, the gradients of F and G at P_1 (see 17.7.1). But the vector $\nabla F(x_1, y_1, z_1) \times \nabla G(x_1, y_1, z_1)$ is perpendicular to each gradient (Section 16.3). Therefore *direction numbers of the vector $\nabla F(x_1, y_1, z_1) \times \nabla G(x_1, y_1, z_1)$ are also direction numbers of the tangent line l to the curve C at $P_1:(x_1, y_1, z_1)$.*

These direction numbers and the coordinates of P enable us to write symmetric equations of the tangent line to C at P.

7. Find direction numbers of the line l that is tangent to the curve of intersection of the surfaces

$$9x^2 + 4y^2 + 4z^2 - 41 = 0$$

and

$$2x^2 - y^2 + 3z^2 - 10 = 0$$

at the point $(1, 2, 2)$. Write symmetric equations of l.

8. The graph of $\mathbf{r}(t) = t^2\mathbf{i} + t^3\mathbf{j} + t\mathbf{k}$ is a curve in three-space known as the *twisted cubic curve*. Verify that it lies on both of the cylinders $x = z^2$ and $y = z^3$ for all $t \in R$. Thus it is contained in the intersection of the two cylinders. Draw the cubic curve carefully for $0 \le t \le 2$. Find the equation of the tangent plane to each cylinder at $(1, 1, 1)$ and draw them; their line of intersection is the tangent line to the twisted cubic curve at the point $(1, 1, 1)$. Use the method outlined above to find symmetric equations of this tangent line and sketch it. Verify that its direction numbers are the components of the tangent vector $\mathbf{r}'(t)$ when $t = 1$.

9. Show that the surfaces $x^2 + z^2 + 4y = 0$ and $x^2 + y^2 + z^2 - 6z + 7 = 0$ are tangent to each other at the point $(0, -1, 2)$. Make a sketch. [*Hint:* Show that the two surfaces have the same tangent plane at $(0, -1, 2)$.]

In Exercises 10 to 16, find the extrema of the indicated functions.

10. $f(x, y) = 2x^4 - x^2 + 3y^2$.

11. $f(x, y) = xy^2 - 6x^2 - 3y^2$.

12. $f(x, y) = x^2 - xy + 3y^2 + 11x - 6$.

13. $f(x, y) = x^3 + y^3 - 6xy$.

14. $f(x, y) = x^4 + y^4 + 4x - 32y$.

15. $f(x, y) = xy$.

16. $f(x, y) = \cos x + \cos y + \cos(x + y)$.

17. A rectangular metal tank with open top is to hold 256 cubic feet of liquid. What are the dimensions of the tank that require the least material to build?

18. A rectangular box, whose edges are parallel to the coordinate axes, is inscribed in the ellipsoid $96x^2 + 4y^2 + 4z^2 = 36$. What is the greatest possible volume for such as box?

19. Find the three-dimensional vector having length 9, the sum of whose components is a maximum.

20. Find the minimum distance between the point $(1, 2, 0)$ and the quadric cone $z^2 = x^2 + y^2$.

21. An open gutter with cross section in the form of a trapezoid having equal base angles is to be made from a long piece of metal, 12 inches wide, by bending up equal strips along both sides. Find the base angles and the width of the sides for maximum carrying capacity.

22. Find the minimum distance between the lines having parametric equations $x = t - 1$, $y = 2t$, $z = t + 3$, and $x = 3s$, $y = s + 2$, $z = 2s - 1$.

23. In the *method of least squares*, the sum of the squares of the deviations can be written

$$f(m, b) = \sum_{i=1}^{n} (y_i - mx_i - b)^2.$$

In seeking values of m and b that minimize $f(m, b)$, show that the solution of the two simultaneous linear equations in m and b

(6)
$$\begin{aligned} m \sum x_i^2 + b \sum x_i &= \sum x_i y_i, \\ m \sum x_i + nb &= \sum y_i, \end{aligned}$$

where each summand is from $i = 1$ to $i = n$, is the only critical pair (m, b) for $f(m, b)$.

24. Prove that the unique values of m and b found by solving equations (6) always make $f(m, b)$ a minimum and therefore make $y = mx + b$ the line of regression of y on x for the given set of n points, $P_i:(x_i, y_i)$.

25. Prove that the line of regression of y on x in the method of least squares always passes through the centroid (Section 9.8) of the given set of n points.

26. Show that when the line of regression of y on x is found by the method of least squares, the sum of the deviations is zero.

27. Use the method of least squares to find the line of regression of y on x for the set of six coplanar

points, (1.0, 3.6), (2.3, 3.0), (3.7, 3.2), (4.2, 5.1), (6.1, 5.3), and (7.0, 6.8).

28. On large-scale rectangular coordinate paper, plot the six points given in the preceding exercise; the result is called a **scatter diagram**. Carefully draw the graph of the line of regression and use a scale to measure each of the six deviations. Their sum should be approximately zero.

18.8 CONSTRAINED MAXIMA AND MINIMA. LAGRANGE MULTIPLIERS

The definition of local extreme values of a function of two variables and the necessary condition for their occurrence, which were discussed in the preceding section, are easily extended to functions of three or more variables.

If f is a function of three variables, x, y, and z, having first partial derivatives with respect to each variable, then a *necessary* condition for $f(a, b, c)$ to be an extremum of f is that $f_x(a, b, c) = 0$, $f_y(a, b, c) = 0$, and $f_z(a, b, c) = 0$. This condition can be compactly written as

$$\nabla f(a, b, c) = \mathbf{0}.$$

The points (a, b, c) that satisfy this equation are *critical points* for f. All points at which f has a local maximum or a local minimum are included in its critical points. However, as we have seen, not all critical points yield extrema.

Let us return to Examples 4 and 5 in the preceding section. In Example 4 we found the extrema of the function of two variables defined by $F(x, y) = 3x^3 + y^2 - 9x + 4y$. In Example 5 we were asked to find the minimum distance of a point $P:(x, y, z)$ from the origin, with the restriction that P must be on the surface $z^2 = x^2y + 4$; that is, we found the minimum of a function of three variables, $f(x, y, z) = x^2 + y^2 + z^2$, subject to the **constraint** (or **side condition**) $g(x, y, z) = x^2y - z^2 + 4 = 0$. Example 4 is a problem in free extrema, and Example 5 is a problem in **constrained extrema**, the subject of discussion in the present section.

Example 5, in the preceding section, was solved by eliminating z between the two equations and then finding the critical points of the resulting function of two variables. Since it is not always feasible to eliminate a variable between two equations, and because finding partial derivatives of the resulting equation may be complicated, we present another method for handling problems in constrained extrema, known as the method of **Lagrange multipliers**. We start with an example of how it is used.

Example 1. What is the greatest area that a rectangle can have if the length of its diagonal is 2?

Solution. Place the rectangle in the first quadrant with two of its sides along the coordinate axes; then the vertex opposite the origin has coordinates (x, y), where neither x nor y can be negative. The length of its diagonal is $\sqrt{x^2 + y^2} = 2$, and its area is given by $f(x, y) = xy$.

Thus we seek the maximum value of $f(x, y) = xy$, subject to the constraint $g(x, y) = x^2 + y^2 - 4 = 0$. The gradients of f and g are the vectors

$$\nabla f(x, y) = f_x(x, y)\mathbf{i} + f_y(x, y)\mathbf{j} = y\mathbf{i} + x\mathbf{j},$$
$$\nabla g(x, y) = g_x(x, y)\mathbf{i} + g_y(x, y)\mathbf{j} = 2x\mathbf{i} + 2y\mathbf{j}.$$

Lagrange saw that the critical points of f can be found from the solutions of the simultaneous equations

$$\nabla f(x, y) = \lambda \nabla g(x, y) \qquad \text{and} \qquad g(x, y) = 0,$$

in which λ is a number yet to be determined. The number λ is called a **Lagrange multiplier**.

These equations are equivalent to three simultaneous equations in the variables x, y, and λ,

$$f_x(x, y) = \lambda g_x(x, y), \qquad f_y(x, y) = \lambda g_y(x, y), \qquad g(x, y) = 0,$$

which in the present problem are

(1) $$y = \lambda(2x),$$
(2) $$x = \lambda(2y),$$
(3) $$x^2 + y^2 = 4.$$

If we multiply the first equation by y and the second equation by x, we get

$$y^2 = 2xy\lambda \qquad \text{and} \qquad x^2 = 2xy\lambda,$$

from which

(4) $$y^2 = x^2.$$

From equations (3) and (4) we find $x = \sqrt{2}$ and $y = \sqrt{2}$; and by substituting these values in the equation (1), we obtain $\lambda = \frac{1}{2}$. Thus the solution of the simultaneous equations (1), (2), and (3) is $x = \sqrt{2}$, $y = \sqrt{2}$, $\lambda = \frac{1}{2}$.

When these values of x and y are substituted in $f(x, y) = xy$, we have $f(\sqrt{2}, \sqrt{2}) = 2$, the maximum value of f under the constraint $g(x, y) = 0$.

Therefore the greatest area that a rectangle with diagonal 2 can have is 2. The rectangle is a square.

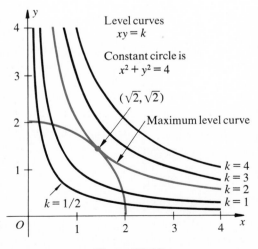

Level curves
$xy = k$

Constant circle is
$x^2 + y^2 = 4$

$(\sqrt{2}, \sqrt{2})$

Maximum level curve

$k = 4$
$k = 3$
$k = 2$
$k = 1$

$k = 1/2$

Figure 18-19

A geometric interpretation of Lagrange's method will show why it works.

In Example 1, above, we found the maximum value of the function $f(x, y) = xy$ when (x, y) satisfies the constraint equation $g(x, y) = x^2 + y^2 - 4 = 0$. The level curves of f are the curves $xy = k$, where k is an arbitrary constant. They are shown as black hyperbolas in Fig. 18-19, for several values of k. At all points on a particular level curve $xy = k$, the function f has the (same) value k.

The graph of the constraint $g(x, y) = x^2 + y^2 - 4 = 0$ is the colored circle shown in Fig. 18-19. When a level curve $xy = k$ intersects the constraint circle, each point of intersection has coordinates (x, y) that make $f(x, y) = xy$ have the value k, yet also satisfy the constraint $g(x, y) = x^2 + y^2 - 4 = 0$. We seek the level curve with greatest possible k that intersects the constraint circle. It is apparent in Fig. 18-19 that such a level curve is tangent to the constraint circle; its equation is $xy = 2$. Thus the maximum value of $f(x, y) = xy$, subject to the constraint $x + y - 4 = 0$, is 2. Notice that the critical point for f is the point of tangency, $(\sqrt{2}, \sqrt{2})$.

It was easy to solve Example 1 geometrically because the equations involved are so simple. But when the function f and the constraint g are more complicated, an analytic method is needed. How can we find the point of tangency of a level curve of f and the constraint curve $g(x, y) = 0$ without graphing them?

At their point of tangency the two curves have a common tangent and thus have the same normal line. But the gradient vector of f at any point on a level curve of f is normal to the curve, and the same is true for the gradient vector of g at any point on the constraint curve $g(x, y) = 0$. At a point of tangency of the two curves, their gradient vectors are on the common normal line, and so

$$\nabla f(x, y) = \lambda \, \nabla g(x, y)$$

for some nonzero number λ.

This suggests the following formulation of **Lagrange's method of multipliers**.

To find the maximum or minimum of $f(x, y)$, subject to the constraint $g(x, y) = 0$, solve the simultaneous equations

$$\nabla f(x, y) = \lambda \, \nabla g(x, y) \qquad \text{and} \qquad g(x, y) = 0$$

for (x, y, λ). Each resulting point (x, y) is a critical point for f. The number λ is called a **Lagrange multiplier**.

This involves solving for (x, y, λ) the three simultaneous equations

$$f_x(x, y) = \lambda g_x(x, y),$$
$$f_y(x, y) = \lambda g_y(x, y),$$
$$g(x, y) = 0.$$

Lagrange's method is similar for functions of three variables, $f(x, y, z)$, with the constraint $g(x, y, z) = 0$.

Example 2.   Find the minimum of $f(x, y, z) = 3x + 2y + z + 5$, subject to the constraint $g(x, y, z) = 9x^2 + 4y^2 - z = 0$.

Solution.   The gradients of f and g are $\nabla f(x, y, z) = 3\mathbf{i} + 2\mathbf{j} + \mathbf{k}$ and $\nabla g(x, y, z) = 18x\mathbf{i} + 8y\mathbf{j} - \mathbf{k}$. To find the critical points of f, we solve the equations

$$\nabla f(x, y, z) = \lambda \, \nabla g(x, y, z) \qquad \text{and} \qquad g(x, y, z) = 0$$

for (x, y, z, λ), in which λ is a Lagrange multiplier. This is equivalent, in the present problem, to solving the following system of four simultaneous equations in the four variables, x, y, z, λ,

(1) $\qquad\qquad\qquad\qquad\qquad\qquad 3 = 18x\lambda,$

(2) $\qquad\qquad\qquad\qquad\qquad\qquad 2 = 8y\lambda,$

(3) $\qquad\qquad\qquad\qquad\qquad\qquad 1 = -\lambda,$

(4) $\qquad\qquad\qquad\qquad 9x^2 + 4y^2 - z = 0.$

From (3), $\lambda = -1$. Substituting this result in equations (1) and (2), we get $x = -\frac{1}{6}$ and $y = -\frac{1}{4}$. By putting these values for x and y in the equation (4), we obtain $z = \frac{1}{2}$. Thus the solution of the foregoing system of four simultaneous equations is $(-\frac{1}{6}, -\frac{1}{4}, \frac{1}{2}, -1)$, and the only critical point for f is $(-\frac{1}{6}, -\frac{1}{4}, \frac{1}{2})$. Therefore the minimum of $f(x, y, z)$, subject to the constraint $g(x, y, z) = 0$, is $f(-\frac{1}{6}, -\frac{1}{4}, \frac{1}{2}) = 4\frac{1}{2}$.

Example 3. Find the least distance between the origin and the plane $3x + 2y - z = 14$.

Solution. The square of the distance between the origin and a point $P{:}(x, y, z)$ is $f(x, y, z) = x^2 + y^2 + z^2$. We seek the minimum of $f(x, y, z)$, subject to the constraint $g(x, y, z) = 3x + 2y - z - 14 = 0$.

The critical points for f will be obtained from the solutions of the simultaneous equations

$$\nabla f(x, y, z) = \lambda \nabla g(x, y, z), \qquad g(x, y, z) = 0,$$

where λ is a Lagrange multiplier. Since $\nabla f(x, y, z) = 2x\mathbf{i} + 2y\mathbf{j} + 2z\mathbf{k}$ and $\nabla g(x, y, z) = 3\mathbf{i} + 2\mathbf{j} - \mathbf{k}$ here, we must find the solutions to the following system of four simultaneous equations in the four variables x, y, z, λ.

(1) $\qquad\qquad\qquad\qquad\qquad 2x = 3\lambda,$

(2) $\qquad\qquad\qquad\qquad\qquad 2y = 2\lambda,$

(3) $\qquad\qquad\qquad\qquad\qquad 2z = -\lambda,$

(4) $\qquad\qquad\qquad 3x + 2y - z - 14 = 0.$

From (3), $\lambda = -2z$; and by substituting this in equations (1) and (2), we find $x = -3z$ and $y = -2z$. When these relations are substituted for x and y in equation (4), we get $z = -1$. Thus $x = 3$, $y = 2$, $z = -1$, and $\lambda = 2$ are the solutions of the system of four simultaneous equations. From this, the critical point for f is $(3, 2, -1)$. Thus the minimum of f, subject to the constraint $g(x, y, z) = 0$, is $f(3, 2, -1) = (3)^2 + (2)^2 + (-1)^2 = 14$.

Therefore the least distance between the origin and the given plane is $\sqrt{14}$.

When more than one constraint is imposed on the variables of a function that is to be maximized or minimized, additional Lagrange multipliers are used (one for each constraint). For example, if we seek the extrema of a function f of three variables, subject to the two constraints $g(x, y, z) = 0$ and $h(x, y, z) = 0$, we solve for x, y, z, λ, and μ the equations

$$\nabla f(x, y, z) = \lambda \nabla g(x, y, z) + \mu \nabla h(x, y, z),$$
$$g(x, y, z) = 0 \qquad h(x, y, z) = 0,$$

where λ and μ are Lagrange multipliers. This is equivalent to finding the solutions of the system of five simultaneous equations in the variables $x, y, z, \lambda, \mu,$

(1) $\qquad\qquad f_x(x, y, z) = \lambda g_x(x, y, z) + \mu h_x(x, y, z),$

(2) $\qquad\qquad f_y(x, y, z) = \lambda g_y(x, y, z) + \mu h_y(x, y, z),$

(3) $\qquad\qquad f_z(x, y, z) = \lambda g_z(x, y, z) + \mu h_z(x, y, z),$

(4) $\qquad\qquad g(x, y, z) = 0,$

(5) $\qquad\qquad h(x, y, z) = 0.$

From the solutions of this system we obtain the critical points (x, y, z) for f.

Exercises

1. Find the minimum of $f(x, y) = x^2 + y^2$, subject to the constraint $g(x, y) = xy - 3$.

2. Find the maximum of $f(x, y) = xy$, subject to the constraint $g(x, y) = 4x^2 + 9y^2 - 36 = 0$.

3. Find the maximum of $f(x, y) = 4x^2 - 4xy + y^2$, subject to the constraint $x^2 + y^2 = 1$.

4. Find the minimum of $f(x, y) = x^2 + 4xy + y^2$, subject to the constraint $x - y - 6 = 0$.

5. Find the minimum of $f(x, y, z) = x^2 + y^2 + z^2$, subject to the constraint $x + 3y - 2z = 12$.

6. Find the minimum of $f(x, y, z) = 4x - 2y + 3z$, subject to the constraint $2x^2 + y^2 - 3z = 0$.

7. What are the dimensions of the rectangular box, open at the top, which has maximum volume when the surface area is 48?

8. Find the least distance between the origin and the plane $x + 3y - 2z = 4$.

9. The material for the bottom of a rectangular box costs three times as much per square foot as the material for the sides and top. Find the greatest capacity such a box can have if the total amount of money available for material is $12.

10. Use the method of Lagrange multipliers to find the least distance between the origin and the surface $x^2y - z^2 + 9 = 0$.

18.9   EXACT DIFFERENTIALS

In our discussion of line integrals (Sections 18.10 and 18.11), we shall be working with expressions having the form

(1) $$P(x, y)\, dx + Q(x, y)\, dy.$$

Interestingly, such expressions occur in many applications of mathematics.

We saw in 18.4.3 that if f is a function of two variables, its differential is defined by

$$df(x, y) = \frac{\partial}{\partial x} f(x, y)\, dx + \frac{\partial}{\partial y} f(x, y)\, dy.$$

It will be important for us to be able to tell whether an expression like (1) is the differential of some function and, if so, to know how to find the function.

18.9.1   Definition.   The expression

$$P(x, y)\, dx + Q(x, y)\, dy$$

is said to be an **exact differential** if and only if there is a function f such that

$$df(x, y) = P(x, y)\, dx + Q(x, y)\, dy.$$

It is easy to verify that $(9x^2y + ye^{xy})\, dx + (3x^3 + xe^{xy})\, dy$ is the exact differential of $3x^3y + e^{xy}$, and it can be proved that $(\ln xy)\, dx + (xe^{-y})\, dy$ is not the differential of any function.

The next theorem gives a necessary and sufficient condition for the expression (1) to be an exact differential.

18.9.2 Theorem. Let P and Q be functions of two variables such that P, Q, $\partial P/\partial y$, and $\partial Q/\partial x$ are continuous in an open rectangular region H. Then a necessary and sufficient condition for

$$(2) \qquad P(x, y)\, dx + Q(x, y)\, dy$$

to be an exact differential at all points (x, y) in H is

$$(3) \qquad \frac{\partial}{\partial y} P(x, y) = \frac{\partial}{\partial x} Q(x, y).$$

It is easy to prove that (3) is a necessary condition for (2) to be an exact differential, but the proof of its sufficiency is more subtle and will be left for a later course.

Proof that (3) *is a necessary condition for* (2) *to be an exact differential.* Assume that $P(x, y)\, dx + Q(x, y)\, dy$ is an exact differential. Then (by (18.9.1)) there is a function f such that

$$(4) \qquad df(x, y) = P(x, y)\, dx + Q(x, y)\, dy.$$

By 18.4.3,

$$(5) \qquad df(x, y) = \frac{\partial}{\partial x} f(x, y)\, dx + \frac{\partial}{\partial y} f(x, y)\, dy.$$

From (4) and (5)

$$\frac{\partial f}{\partial x} = P \qquad \text{and} \qquad \frac{\partial f}{\partial y} = Q.$$

By differentiating the first of these two equations partially with respect to y, and the second with respect to x, we get

$$\frac{\partial^2 f}{\partial y\, \partial x} = \frac{\partial P}{\partial y} \qquad \text{and} \qquad \frac{\partial^2 f}{\partial x\, \partial y} = \frac{\partial Q}{\partial x}.$$

Since $\partial P/\partial y$ and $\partial Q/\partial x$ are continuous in H by hypothesis, the order of differentiation in the mixed second partial derivatives of F is immaterial (by 18.3.4). Therefore

$$\frac{\partial P}{\partial y} = \frac{\partial Q}{\partial x},$$

which proves that if $P(x, y)\, dx + Q(x, y)\, dy$ is an exact differential, then $\partial P/\partial y = \partial Q/\partial x$. ∎

Example 1. Determine whether

$$(4x^3 + 9x^2 y^2)\, dx + (6x^3 y + 6y^5)\, dy$$

is an exact differential and, if so, find the function f whose differential it is.

Solution. Let $P(x, y) = 4x^3 + 9x^2 y^2$ and $Q(x, y) = 6x^3 y + 6y^5$. Since

$$\frac{\partial}{\partial y} P(x, y) = 18x^2 y = \frac{\partial}{\partial x} Q(x, y),$$

the given expression is an exact differential of some function f (by 18.9.2).

Now

$$df(x, y) = \frac{\partial}{\partial x} f(x, y)\, dx + \frac{\partial}{\partial y} f(x, y)\, dy,$$

and in the present problem we have

(6) $$\frac{\partial}{\partial x} f(x, y) = 4x^3 + 9x^2 y^2$$

and

(7) $$\frac{\partial}{\partial y} f(x, y) = 6x^3 y + 6y^5.$$

By integrating both members of (6) with respect to x, we obtain

(8) $$f(x, y) = x^4 + 3x^3 y^2 + C_1(y),$$

in which the "constant" of integration (with respect to x) is a function of y. If we now differentiate (8) partially with respect to y, we get

(9) $$\frac{\partial}{\partial y} f(x, y) = 6x^3 y + C_1'(y).$$

From (7) and (9)

$$6x^3 y + 6y^5 = 6x^3 y + C_1'(y).$$

Thus $C_1'(y) = 6y^5$ and $C_1(y) = y^6 + C$. By substituting this result in (8), we have

$$f(x, y) = x^4 + 3x^3 y^2 + y^6 + C.$$

Example 2. Show that

$$(e^x \ln y - e^y x^{-1})\, dx + (e^x y^{-1} - e^y \ln x)\, dy$$

is the exact differential of some function f of two variables, and find f.

Solution. Let $P(x, y) = e^x \ln y - e^y x^{-1}$ and $Q(x, y) = e^x y^{-1} - e^y \ln x$. Since

$$\frac{\partial}{\partial y} P(x, y) = e^x y^{-1} - e^y x^{-1} = \frac{\partial}{\partial x} Q(x, y),$$

the given expression is the exact differential of some function f of two variables.
Here

(10) $$f_x(x, y) = e^x \ln y - e^y x^{-1}$$

and

(11) $$f_y(x, y) = e^x y^{-1} - e^y \ln x.$$

By integrating (10) with respect to x, we get

(12) $$f(x, y) = e^x \ln y - e^y \ln x + C_1(y);$$

and differentiating (12) with respect to y gives

(13) $$f_y(x, y) = e^x y^{-1} - e^y \ln x + C_1'(y).$$

From (11) and (13)

$$e^x y^{-1} - e^y \ln x = e^x y^{-1} - e^y \ln x + C_1'(y);$$

thus $C_1'(y) = 0$, from which $C_1(y) = C$, a constant.
By substituting this value of $C_1(y)$ in (12), we obtain

$$f(x, y) = e^x \ln y - e^y \ln x + C.$$

The extension of 18.9.1 and 18.9.2 to functions of three variables follows.

18.9.3   Definition.   Let P, Q, and R be functions of three variables. The expression
$$P(x, y, z)\, dx + Q(x, y, z)\, dy + R(x, y, z)\, dz$$
is an **exact differential** if and only if there is a function f of three variables such that
$$df(x, y, z) = P(x, y, z)\, dx + Q(x, y, z)\, dy + R(x, y, z)\, dz.$$

18.9.4   Theorem.   Let P, Q, and R be functions of three variables such that P, Q, R, P_y, P_z, Q_z, Q_x, R_x, and R_y are continuous in some open three-dimensional rectangular region H. Then
$$P(x, y, z)\, dx + Q(x, y, z)\, dy + R(x, y, z)\, dz$$
defines an **exact differential** in H if and only if
$$\frac{\partial Q}{\partial x} = \frac{\partial P}{\partial y}, \qquad \frac{\partial R}{\partial y} = \frac{\partial Q}{\partial z}, \qquad \frac{\partial P}{\partial z} = \frac{\partial R}{\partial x}.$$

Exercises

In each of the following exercises, determine whether the given expression is an exact differential. If so, find the function f whose differential it is; if not, state that it is not.

1. $(10x - 7y)\, dx - (7x - 2y)\, dy.$

2. $(12x^2 + 3y^2 + 5y)\, dx + (6xy - 3y^2 + 5x)\, dy.$

3. $(45x^4 y^2 - 6y^6 + 3)\, dx + (18x^5 y - 12xy^5 + 7)\, dy.$

4. $(35x^4 - 3x^2 y^4 + y^9)\, dx - (4x^3 y^3 - 9xy^8)\, dy.$

5. $\left(\dfrac{6x^2}{5y^2}\right) dx - \left(\dfrac{4x^3}{5y^3}\right) dy.$

6. $\dfrac{21x^2}{y^4 + 1}\, dx - \dfrac{4y^3(7x^3 - 2)}{(y^4 + 1)^2}\, dy.$

7. $(2e^y - ye^x)\, dx + (2xe^y - e^y)\, dy.$

8. $4y^2 \cos(xy^2)\, dx + 8x \cos(xy^2)\, dy.$

9. $3x^2 \cos(y^2)\, dx - 2x^3 y \sin(y^2)\, dy.$

10. $-e^{-x} \ln y\, dx + e^{-x} y^{-1}\, dy.$

11. $3x^2\, dx + 6y^2\, dy + 9z^2\, dz.$

12. $-e^{-y} \sin x \sin z\, dx - e^{-y} \cos x \sin z\, dy + e^{-y} \cos x \cos z\, dz.$

18.10   LINE INTEGRALS

The familiar concept of the Riemann integral, $\displaystyle\int_a^b f(x)\, dx$, can be extended in several ways. One of these is the *line integral* (more descriptively called the *curve integral*).

18.10.1   Definition.   Let P and Q be functions of two variables whose first partial derivatives are continuous in an open rectangular region H of the xy plane. Consider an arc C in H whose parametric equations
$$x = f(t), \qquad y = g(t), \qquad a \le t \le b$$
are such that as t increases from a to b, the corresponding point $(f(t), g(t))$ traces the arc C from the point $A:(f(a), g(a))$ to the point $B:(f(b), g(b))$, entirely in H (Fig. 18-20). Let f' and g' be continuous for $a \le t \le b$. Then
$$\int_C P(x, y)\, dx + Q(x, y)\, dy = \int_a^b [P(f(t), g(t))f'(t) + Q(f(t), g(t))g'(t)]\, dt$$
is called the **line integral** (or **curve integral**) of $P(x, y)\, dx + Q(x, y)\, dy$ along C from A to B.

Notice that the right-hand member of the preceding equation expresses the line integral as a Riemann integral whose integrand defines a continuous function of t.

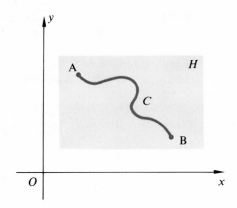

Figure 18-20

Example 1.   Evaluate the line integral

$$\int_C (x^2 - y^2)\, dx + 2xy\, dy$$

along the curve C whose parametric equations are $x = t^2$, $y = t^3$, from $t = 0$ to $t = \frac{3}{2}$.

Solution. $dx = 2t\, dt$ and $dy = 3t^2\, dt$. Thus

$$\int_C (x^2 - y^2)\, dx + 2xy\, dy = \int_0^{3/2} [(t^4 - t^6)2t + 2t^5(3t^2)]\, dt$$

$$= \int_0^{3/2} (2t^5 + 4t^7)\, dt = \frac{8505}{512}.$$

Since the vector equation of the arc C is $\mathbf{r}(t) = f(t)\mathbf{i} + g(t)\mathbf{j}$, $a \le t \le b$,

$$\mathbf{r}'(t) = f'(t)\mathbf{i} + g'(t)\mathbf{j}.$$

If we write $\mathbf{F}(x, y) = P(x, y)\mathbf{i} + Q(x, y)\mathbf{j}$ so that

$$\mathbf{F}(f(t), g(t)) = P(f(t), g(t))\mathbf{i} + Q(f(t), g(t))\mathbf{j}, \quad a \le t \le b,$$

then

$$\mathbf{F}(f(t), g(t)) \cdot \mathbf{r}'(t) = P(f(t), g(t))f'(t) + Q(f(t), g(t))g'(t).$$

18.10.2   Corollary.   The line integral defined in 18.10.1 can be written

$$\int_C P(x, y)\, dx + Q(x, y)\, dy = \int_b^a \mathbf{F}(f(t), g(t)) \cdot \mathbf{r}'(t)\, dt,$$

where $\mathbf{r}(t) = f(t)\mathbf{i} + g(t)\mathbf{j}$, $a \leq t \leq b$, is the vector equation of the arc C, and

$$\mathbf{F}(x, y) = P(x, y)\mathbf{i} + Q(x, y)\mathbf{j}.$$

Usually the value of a line integral, as defined in 18.10.1, depends on the integrand, on the endpoints A and B, and on the arc C from A to B, along which the integration is performed. However, there are certain line integrals, of great importance in applications, whose values depend only on the integrand and the endpoints A and B, and not at all on the arc C from A to B. Such line integrals are said to be **independent of the path.**

18.10.3 Theorem. Let $P(x, y)\,dx + Q(x, y)\,dy$ be an exact differential of some function G in an open rectangular region H. If C is an arc, lying entirely in H, with parametric equations

$$x = f(t), \qquad y = g(t), \qquad t_1 \leq t \leq t_2,$$

in which f' and g' are continuous, then

$$\int_C P(x, y)\,dx + Q(x, y)\,dy = G(x_2, y_2) - G(x_1, y_1),$$

where $A:(x_1, y_1) = (f(t_1), g(t_1))$ and $B:(x_2, y_2) = (f(t_2), g(t_2))$ are the endpoints of C.

Proof. By hypothesis, there is a function G such that

$$dG(x, y) = P(x, y)\,dx + Q(x, y)\,dy$$

for all (x, y) in H. If we restrict G to points on the arc C, then

(1) $$dG(f(t), g(t)) = [P(f(t), g(t))f'(t) + Q(f(t), g(t))g'(t)]\,dt,$$

for $t_1 \leq t \leq t_2$.

Let Φ be the function of one variable defined by

$$\Phi(t) = G(f(t), g(t)), \qquad t_1 \leq t \leq t_2.$$

Then

(2) $$d\Phi(t) = \Phi'(t)\,dt = dG(f(t), g(t)).$$

From (1) and (2)

$$\Phi'(t)\,dt = [P(f(t), g(t))f'(t) + Q(f(t), g(t))g'(t)]\,dt.$$

Thus

(3) $$\int_{t_1}^{t_2} \Phi'(t)\,dt = \int_{t_1}^{t_2} [P(f(t), g(t))f'(t) + Q(f(t), g(t))g'(t)]\,dt.$$

It is left as an exercise for the reader to show that $\Phi'(t)$ is continuous for $t_1 \leq t \leq t_2$. By the fundamental theorem, the left-hand member of (3) is equal to

$$\Phi(t_2) - \Phi(t_1) = G(f(t_2), g(t_2)) - G(f(t_1), g(t_1)) = G(x_2, y_2) - G(x_1, y_1).$$

By 18.10.1, the right-hand member of (3) is equal to $\int_C P(x, y)\, dx + Q(x, y)\, dy$. Therefore

$$\int_C P(x, y)\, dx + Q(x, y)\, dy = G(x_2, y_2) - G(x_1, y_1). \quad \blacksquare$$

Notice that the value of the line integral in 18.10.3 depends only on the function G and the endpoints $A{:}(x_1, y_1)$ and $B{:}(x_2, y_2)$. Any arc from A to B will give the same result, provided that it lies entirely in H and that the functions in its parametric representation have continuous derivatives.

18.10.4 Corollary. If $P(x, y)\, dx + Q(x, y)\, dy$ is an exact differential in an open rectangular region H, and if A and B are points in H, then the value of the line integral

$$\int_C P(x, y)\, dx + Q(x, y)\, dy$$

from A to B is independent of the path.

Example 2. Find the value of the line integral

$$\int_C (3x^2 - 6xy)\, dx + (-3x^2 + 4y + 1)\, dy$$

from the point $(-1, 2)$ to the point $(4, 3)$.

Solution. This is in the form

$$\int_C P(x, y)\, dx + Q(x, y)\, dy.$$

Since $P_y(x, y) = -6x = Q_x(x, y)$, the integrand is the exact differential of some function G such that

$$G_x(x, y) = 3x^2 - 6xy, \qquad G_y(x, y) = -3x^2 + 4y + 1.$$

By inspection, we see that

$$G(x, y) = x^3 - 3x^2 y + 2y^2 + y.$$

Therefore (by 18.10.2), the value of the given line integral is $G(4, 3) - G(-1, 2) = -62$.

Example 3. Find the value of the line integral

$$\int_C (e^x \ln y - e^y x^{-1})\, dx + (e^x y^{-1} - e^y \ln x)\, dy$$

from the point $(1, 1)$ to the point $(5, 2)$.

Solution. Since

$$P_y(x, y) = e^x y^{-1} - e^y x^{-1} = Q_x(x, y),$$

the given integrand is the exact differential of some function G for $x > 0$ and $y > 0$ and may be evaluated by using 18.10.2.

If we proceed as in the solution of Example 2, Section 18.9, we find

$$G(x, y) = e^x \ln y - e^y \ln x, \qquad x > 0 \quad \text{and} \quad y > 0.$$

Therefore the value of the given line integral is

$$G(5, 2) - G(1, 1) = e^5 \ln 2 - e^2 \ln 5.$$

An arc, like C in the definition (18.10.1) of a line integral, such that the functions f and g in its parametric representation *have continuous derivatives*, f' and g', for $a \le t \le b$, is called a **smooth arc**.

In evaluating path-independent line integrals, it is often convenient to choose for the path a finite number of smooth arcs, $C_1, C_2, \cdots, C_n$, joined sequentially so that the terminal point of C_1 is the initial point of C_2, the terminal point of C_2 is the initial point of C_3, and so on. (Fig. 18-21). Such a path is said to be a **sectionally smooth arc**.

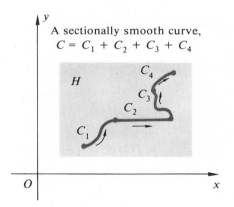

Figure 18-21

18.10.5 **Definition.**   Let $C = C_1 + C_2 + \cdots + C_n$ be a sectionally smooth arc composed of the smooth arcs $C_1, C_2, \cdots, C_n$. The line integral over C,

$$\int_C P(x, y)\, dx + Q(x, y)\, dy,$$

is defined to be the sum of the line integrals

$$\int_{C_i} P(x, y)\, dx + Q(x, y)\, dy, \qquad i = 1, 2, \cdots, n,$$

over $C_1, C_2, \cdots, C_n$.

It is easy to prove that 18.10.3 holds when C is a sectionally smooth curve. This is left for the student.

Second Solution for Example 2.   Since the given line integral is independent of the path, we can first integrate along the horizontal line segment from $(-1, 2)$ to $(4, 2)$, during which $y = 2$ and $dy = 0$, and then integrate along the vertical line segment from $(4, 2)$ to $(4, 3)$, with $x = 4$ and $dx = 0$. We obtain

$$\int_C P(x, y)\, dx + Q(x, y)\, dy = \int_{-1}^4 P(x, 2)\, dx + \int_2^3 Q(4, y)\, dy$$

$$= \int_{-1}^4 (3x^2 - 12x)\, dx + \int_2^3 (4y - 47)\, dy = -62.$$

Second Solution for Example 3. Because this line integral is independent of the path, we can first integrate along the horizontal line segment from $(1, 1)$ to $(5, 1)$ and then along the vertical line segment from $(5, 1)$ to $(5, 2)$.

$$\int_C (e^x \ln y - e^y x^{-1})\, dx + (e^x y^{-1} - e^y \ln x)\, dy$$

$$= \int_1^5 (e^x \ln 1 - ex^{-1})\, dx + \int_1^2 (e^5 y^{-1} - e^y \ln 5)\, dy$$

$$= e^5 \ln 2 - e^2 \ln 5.$$

We saw (18.6.3) that the gradient vector of a function G of two variables, having continuous first partial derivatives G_x and G_y, was defined by

(4) $$\nabla G(x, y) = G_x(x, y)\mathbf{i} + G_y(x, y)\mathbf{j}.$$

Let us indicate the vector $dx\, \mathbf{i} + dy\, \mathbf{j}$ by $d\mathbf{r}$; that is, let

(5) $$d\mathbf{r} = dx\, \mathbf{i} + dy\, \mathbf{j}.$$

From (4) and (5)

$$\nabla G(x, y) \cdot d\mathbf{r} = G_x(x, y)\, dx + G_y(x, y)\, dy.$$

Thus when $P(x, y)\, dx + Q(x, y)\, dy$ is the exact differential of a function G,

$$\int_C P(x, y)\, dx + Q(x, y)\, dy = \int_C \nabla G(x, y) \cdot d\mathbf{r}.$$

18.10.6 Corollary. If $P(x, y)\, dx + Q(x, y)\, dy$ is the exact differential of a function G in an open rectangular region H, and if C is any path in H whose endpoints are $A:(x_1, y_1)$ and $B:(x_2, y_2)$, then

$$\int_C P(x, y)\, dx + Q(x, y)\, dy = \int_C \nabla G(x, y) \cdot d\mathbf{r}$$

$$= G(x_2, y_2) - G(x_1, y_1),$$

where ∇G is the gradient vector of G in H and the symbol $d\mathbf{r}$ denotes the vector $dx\, \mathbf{i} + dy\, \mathbf{j}$.

The next theorem shows that if $A = B$ in 18.10.4, so that C is a *closed* smooth arc, the value of the line integral taken once around C is zero.

18.10.7 Theorem. Let C be a closed smooth arc lying entirely in an open rectangular region H of the xy plane. If $P(x, y)\, dx + Q(x, y)\, dy$ is an exact differential in H, then

$$\int_C P(x, y)\, dx + Q(x, y)\, dy = 0.$$

Proof. Denote by C_1 the segment of C from A to K on which t increases, and by C_2 the other segment of C from A to K (Fig. 18-22). By 18.10.4, the line integral from from A to K along C_1 is equal to the line integral *from A to K* along C_2. But the line integral along C_2 *from K to A* is the negative of the line integral along C_2 from A to K.

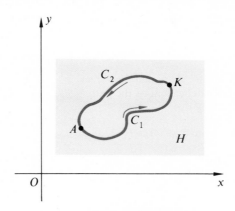

Figure 18-22

Thus the value of the line integral around C in the positive direction (with increasing t) is

$$\int_{C_1} P(x, y)\, dx + Q(x, y)\, dy - \int_{C_2} P(x, y)\, dx + Q(x, y)\, dy = 0. \quad \blacksquare$$

Line integrals are extended to arcs in three-dimensional space in the exercises.

Exercises

1. Evaluate the line integral $\int_C y\, dx + x^2\, dy$
   (a) along the curve $x = 2t$, $y = t^2 - 1$ from $t = 0$ to $t = 2$;
   (b) from the point A:(0, −1) to the point B:(4, 3) along the line segments AC and CB, where C is the point (4, −1). (*Hint:* Along AC, $dy = 0$; along CB, $dx = 0$.)

2. Find the value of $\int_C y^2\, dx + (x - y)\, dy$ from the point A:(0, −2) to the point B:(28, 6)
   (a) along the arc $x = t^3 + 1$, $y = 2t$, $-1 \leq t \leq 3$;
   (b) along the straight-line segment AB.

3. Evaluate $\int_C y^3\, dx + x^3\, dy$ from the point

   A:(−4, 1) to the point B:(2, −2),
   (a) along the arc $x = 2t$, $y = t^2 - 3$;
   (b) along the line segments AC and CB, where C is the point (−4, −2).

   In Exercises 4 to 8, find the value of the line integral along the arc whose parametric equations are given.

4. $\int_C y^2\, dx + 2xy\, dy$; $x = t^3 + 3$, $y = 2t^2$, from $t = -2$ to $t = 1$.

5. $\int_C (2x - xy)\, dx + x^2\, dy$; $x = t^{3/2}$, $y = t^{1/2} - 1$, $1 \leq t \leq 4$.

6. $\int_C \sqrt{y} \, dx + (2x - y) \, dy; \quad x = 5t^3, \quad y = t^2,$ from $t = 0$ to $t = 2$.

7. $\int_C \frac{dx}{x + y^2} + ye^x \, dy; \, x = 4t^2, \, y = t, \, 1 \leq t \leq 3.$

8. $\int_C (x - 3y) \, dx + (x^3 + 2)^{1/2} \, dy; \, x = 2t, \, y = 8t^3,$ $0 \leq t \leq 1$.

In each of Exercises 9 to 16, show that the value of the given line integral is independent of the path, and evaluate it.

9. $\int_C xy^2 \, dx + x^2y \, dy, \quad \text{from } (1, -4) \text{ to } (-2, 3).$

10. $\int_C (4xy - y^2) \, dx + 2(x^2 - xy) \, dy, \, \text{from } (-1, 2)$ to $(3, 3)$.

11. $\int_C (3x - y) \, dx + (y^2 - x) \, dy, \quad \text{from} \quad (0, 0) \quad \text{to}$ $(1, 2)$

12. $\int_C 8xy^3 \, dx + 12x^2 \, y^2 \, dy, \, \text{from } (2, -1) \text{ to } (4, 2).$

13. $\int_C \ln y^2 \, dx + 2xy^{-1} \, dy, \, \text{from } (-3, 1) \text{ to } (4, 4).$

14. $\int_C (x^3 + 3x^2y + y) \, dx + (x^3 + x) \, dy, \quad \text{from}$ $(-1, -1) \text{ to } (2, 5).$

15.
$\int_C (6x^2y^2 + y^4 - 3y) \, dx + (4x^3y + 4xy^3 - 3x) \, dy,$
from $(0, 1)$ to $(-2, -3)$.

16. $\int_C -e^{-x} \sin y \, dx + e^{-x} \cos y \, dy, \quad \text{from } (-1,$ $\pi/6)$ to $(0, \pi/2)$.

17. Show that Φ' in equation (3) is continuous for

$t_1 \leq t \leq t_2$. [*Hint:* See 18.9.2 and equations (1) and (2) of the present section.]

18. Let P, Q, and R be functions of three variables such that they, and their first partial derivatives, are continuous in an open three-dimensional rectangular region H. Assume that

$$x = f(t), \qquad y = g(t), \qquad z = h(t), \qquad a \leq t \leq b,$$

are parametric equations of an arc C in H with endpoints $A:(f(a), g(a), h(a))$ and $B:(f(b), g(b), h(b))$, and that f', g', and h' are continuous for $a \leq t \leq b$.
(a) Define the line integral

$$\int_C P(x, y, z) \, dx + Q(x, y, z) \, dy + R(x, y, z) \, dz$$

in a manner analogous to 18.10.1.
(b) Show that this line integral can be written in the form

$$\int_a^b \mathbf{F}(f(t), g(t), h(t)) \cdot \mathbf{r}'(t) \, dt,$$

where $\mathbf{r}(t) = f(t)\mathbf{i} + g(t)\mathbf{j} + h(t)\mathbf{k}, \, a \leq t \leq b,$ is the (position) vector equation of the arc C, and

$$\mathbf{F}(x, y, z) = P(x, y, z)\mathbf{i} + Q(x, y, z)\mathbf{j}$$
$$+ R(x, y, z)\mathbf{k}.$$

19. Show that the line integral

$$\int_C (2xy + 4yz) \, dx + (x^2 + 4xz - 2z^2) \, dy$$
$$+ (4xy - 4yz) \, dz$$

from $(-1, 1, 2)$ to $(3, -2, 1)$ is independent of the path and find its value. (*Hint:* Use 18.9.3.)

20. Show that the line integral

$$\int_C ye^{xy} \, dx + [xe^{xy} - z \sin (yz)] \, dy - y \sin (yz) \, dz$$

from $(0, 1, \pi/4)$ to $(2, 1/2, 2\pi/3)$ is independent of the path, and find its value.

18.11 WORK

Recall (Section 9.4) that when a constant force acts on a particle that is moving along a straight line, the work done by this force equals the component of the force along the line multiplied by the magnitude of the displacement of the particle.

Thus if $\mathbf{r}_1 = x_1\mathbf{i} + y_1\mathbf{j}$ and $\mathbf{r}_2 = x_2\mathbf{i} + y_2\mathbf{j}$ are position vectors with terminal

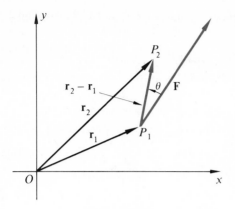

Figure 18-23

points P_1:(x_1, y_1) and P_2:(x_2, y_2) (Fig. 18-23) and if **F** is a constant force vector making an angle θ with the vector $\mathbf{r}_2 - \mathbf{r}_1$, the work done by **F** on a particle moving along the straight-line segment from P_1 to P_2 is

$$|\mathbf{F}| \cos \theta \, |\mathbf{r}_2 - \mathbf{r}_1|,$$

in which $|\mathbf{F}| \cos \theta$ is the scalar projection (Section 16.2) of the vector **F** on the vector $\mathbf{r}_2 - \mathbf{r}_1$, and $|\mathbf{r}_2 - \mathbf{r}_1|$ is the magnitude of the displacement of the particle. By 16.2.6, this can be written

(1) $$W = \mathbf{F} \cdot (\mathbf{r}_2 - \mathbf{r}_1).$$

Now consider a force vector

(2) $$\mathbf{F}(x, y) = P(x, y)\mathbf{i} + Q(x, y)\mathbf{j},$$

where P and Q are functions of two variables having continuous first partial derivatives at every point (x, y) in an open rectangular region H, and let the initial point of each such force vector be the point (x, y) in H. This establishes a **force field** in H (Fig. 18-24).

Let C be an arc lying entirely in H and having the (position) vector equation

(3) $$\mathbf{r}(t) = f(t)\mathbf{i} + g(t)\mathbf{j}, \qquad a \le t \le b.$$

The purpose of the following discussion is to motivate the definition (to be given below) of the work done by the force (2) on a particle traversing the arc C.

Let $p = \{[t_0, t_1], [t_1, t_2], \cdots, [t_{n-1}, t_n]\}$ be a partition of the time interval $[a, b]$, in which $t_0 = a$ and $t_n = b$, and denote by P_i the point $(x_i, y_i) = (f(t_i), g(t_i))$ on C. Then the vector $\overrightarrow{P_{i-1}P_i} = \mathbf{r}(t_i) - \mathbf{r}(t_{i-1})$ (Fig. 18-25) can be written

$$\overrightarrow{P_{i-1}P_i} = (f(t_i) - f(t_{i-1}))\mathbf{i} + (g(t_i) - g(t_{i-1}))\mathbf{j};$$

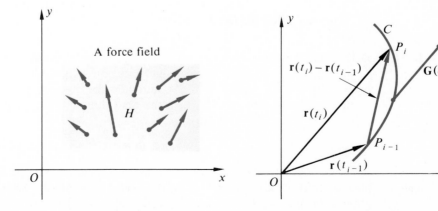

Figure 18-24   **Figure 18-25**

and by the mean value theorem, there are numbers σ_i and τ_i between t_{i-1} and t_i such that

$$\overrightarrow{P_{i-1}P_i} = f'(\sigma_i)(t_i - t_{i-1})\mathbf{i} + g'(\tau_i)(t_i - t_{i-1})\mathbf{j}$$
$$= (f'(\sigma_i)\mathbf{i} + g'(\tau_i)\mathbf{j}) \, \Delta t_i,$$

where $\Delta t_i = t_i - t_{i-1}$.

If the force (2) acting on the particle at time t is written

(4) $$\mathbf{G}(t) = \mathbf{F}(f(t), g(t)) = P(f(t), g(t))\mathbf{i} + Q(f(t), g(t))\mathbf{j},$$

then the work done by this force in the time interval $[t_{i-1}, t_i]$ is approximately

(5) $$W_i = \mathbf{G}(\sigma_i) \cdot (f'(\sigma_i)\mathbf{i} + g'(\tau_i)\mathbf{j}) \, \Delta t_i;$$

and the total work W done by $\mathbf{F}$ on the particle as it moves on the arc C from $A{:}(f(a), g(a))$ to $B{:}(f(b), g(b))$ is approximated by

(6) $$W \doteq \sum_{i=1}^{n} W_i = \sum_{i=1}^{n} \mathbf{G}(\sigma_i) \cdot (f'(\sigma_i)\mathbf{i} + g'(\tau_i)\mathbf{j}) \, \Delta t_i.$$

If σ_i and τ_i were the same number, say λ_i, in $[t_{i-1}, t_i]$, the right-hand member of (6) could be written

$$\sum_{i=1}^{n} \mathbf{G}(\lambda_i) \cdot \mathbf{r}'(\lambda_i) \, \Delta t_i,$$

which is a Riemann sum, and

$$\lim_{|p| \to 0} \sum_{i=1}^{n} \mathbf{G}(\lambda_i) \cdot \mathbf{r}'(\lambda_i) \, \Delta t_i = \int_{a}^{b} \mathbf{G}(t) \cdot \mathbf{r}'(t) \, dt.$$

It can be shown that the limit of (6) as $|p| \longrightarrow 0$ is also

$$\int_{a}^{b} \mathbf{G}(t) \cdot \mathbf{r}'(t) \, dt = \int_{a}^{b} \mathbf{F}(f(t), g(t)) \cdot \mathbf{r}'(t) \, dt,$$

which (by 18.10.2) is the line integral

$$\int_{C} P(x, y) \, dx + Q(x, y) \, dy.$$

This suggested the following definition.

18.11.1 Definition. Let $\mathbf{F}(x, y) = P(x, y)\mathbf{i} + Q(x, y)\mathbf{j}$, where P and Q have continuous first partial derivatives in an open rectangular region H, establish a force field in H. If $\mathbf{r}(t) = f(t)\mathbf{i} + g(t)\mathbf{j}$, with f' and g' continuous for $a \leq t \leq b$, is the vector equation of an arc C lying entirely in H, then the **work** done by $\mathbf{F}$ on a particle moving along C from $A{:}(f(a), g(a))$ to $B{:}(f(b), g(b))$ is

$$W = \int_{C} P(x, y) \, dx + Q(x, y) \, dy = \int_{a}^{b} \mathbf{F}(f(t), g(t)) \cdot \mathbf{r}'(t) \, dt.$$

Example 1. Find the work done by the force $\mathbf{F}(x, y) = xy\mathbf{i} + (x + y)\mathbf{j}$ on a particle moving from the origin to the point $(4, 6)$

(a) along the curve $\mathbf{r}(t) = t^2\mathbf{i} + 3t\mathbf{j}$ (Fig. 18-26);

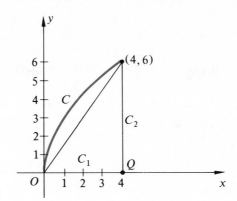

Figure 18-26

(b)   along the straight line joining the origin to (4, 6);

(c)   along the broken line consisting of the line segment C_1 from O to Q:(4, 0) and the line segment C_2 from Q to R:(4, 6).

Solution.   (a) Since $\mathbf{F}(t^2, 3t) = 3t^3\mathbf{i} + (t^2 + 3t)\mathbf{j}$ and $\mathbf{r}'(t) = 2t\mathbf{i} + 3\mathbf{j}$, the work done by $\mathbf{F}$ on the particle as it moves from (0, 0) to (4, 6) along the curve C is

$$W = \int_0^2 \mathbf{F}(t^2, 3t) \cdot \mathbf{r}'(t) \, dt$$

$$= \int_0^2 (3t^3\mathbf{i} + (t^2 + 3t)\mathbf{j}) \cdot (2t\mathbf{i} + 3\mathbf{j}) \, dt$$

$$= 3 \int_0^2 (2t^4 + t^2 + 3t) \, dt = \frac{322}{5}.$$

(b) The line joining (0, 0) to (4, 6) has the vector equation $\mathbf{r}(t) = 2t\mathbf{i} + 3t\mathbf{j}$. Here $\mathbf{F}(2t, 3t) = 6t^2\mathbf{i} + 5t\mathbf{j}$ and $\mathbf{r}'(t) = 2\mathbf{i} + 3\mathbf{j}$. Thus

$$W = \int_0^2 (6t^2\mathbf{i} + 5t\mathbf{j}) \cdot (2\mathbf{i} + 3\mathbf{j}) \, dt$$

$$= \int_0^2 (12t^2 + 15t) \, dt = 62.$$

(c) Write the force vector in the form $\mathbf{F}(x, y) = P(x, y)\mathbf{i} + Q(x, y)\mathbf{j} = xy\mathbf{i} + (x + y)\mathbf{j}$. Then

$$W = \int_{C_1} P(x, y) \, dx + Q(x, y) \, dy + \int_{C_2} P(x, y) \, dx + Q(x, y) \, dy$$

$$= \int_{C_1} xy \, dx + (x + y) \, dy + \int_{C_2} xy \, dx + (x + y) \, dy.$$

Along C_1, which is a segment of the x axis, $y = 0$ and $dy = 0$; and along C_2, which is parallel to the y axis, $x = 4$ and $dx = 0$. Thus the preceding line integrals become

$$W = \int_0^6 (4 + y) \, dy = 42.$$

Example 2. Let $\mathbf{r}(t)$ be the position vector of a particle of mass m moving on a plane arc C, from $t = a$ to $t = b$, in a force field established by $\mathbf{F}$. Prove the **principle of work and energy**—namely, that *the work done is equal to the change in kinetic energy.*

Solution. In mechanics, the kinetic energy at time t of a particle of mass m moving with a velocity $\mathbf{v}$ is defined to be

$$\text{KE} = \frac{km}{2}|\mathbf{v}|^2,$$

where k is a constant.

The work done by $\mathbf{F}$ in displacing the particle from $t = a$ to $t = b$ along the curve C is

$$(7) \qquad\qquad \int_a^b \mathbf{F}(f(t), g(t)) \cdot \mathbf{r}'(t)\, dt.$$

By Newton's second law of motion, $\mathbf{F} = km\mathbf{a}$; so $\mathbf{F}(f(t), g(t)) = km(d\mathbf{v}/dt)$, and (7) can be rewritten

$$\int_a^b \mathbf{F}(f(t), g(t)) \cdot \mathbf{r}'(t)\, dt = \int_a^b \left(km\frac{d\mathbf{v}}{dt} \cdot \mathbf{v}\right) dt = km \int_a^b [f''(t)f'(t) + g''(t)g'(t)]\, dt$$

$$= \frac{km}{2} \int_a^b [2f'(t)f''(t) + 2g'(t)g''(t)]\, dt$$

$$= \frac{km}{2} \int_a^b \frac{d}{dt}(\mathbf{v} \cdot \mathbf{v})\, dt = \frac{km}{2} \int_a^b \frac{d}{dt}|\mathbf{v}|^2\, dt$$

$$= \frac{km}{2}|v|^2 \Big]_{t=a}^{t=b}.$$

Thus the work done in displacing a particle of mass m from $t = a$ to $t = b$ along the curve is equal to the change in kinetic energy.

If $P_y = Q_x$ in H, the force vector (2) becomes the gradient vector

$$(8) \qquad\qquad \nabla G(x, y) = G_x(x, y)\mathbf{i} + G_y(x, y)\mathbf{j}$$

for some function G in H. Such a force field is said to be **conservative** because the physical law of conservation of energy holds. It follows from 18.10.4, 18.10.5, and 18.11.1 that the work done by the force (8) on a particle moving from a point A to a point B in the conservative force field is independent of the path traversed by the particle, and is given by

$$(9) \qquad\qquad W = \int_C G_x(x, y)\, dx + G_y(x, y)\, dy = \int_C \nabla G(x, y) \cdot d\mathbf{r},$$

in which the symbol $d\mathbf{r}$ denotes the vector $dx\,\mathbf{i} + dy\,\mathbf{j}$.

Example 3. Prove the principle of **conservation of energy**, namely that *in a conservative force field the sum of the potential and kinetic energies of a particle is constant.*

Solution. Let $\mathbf{F}(x, y) = P(x, y)\mathbf{i} + Q(x, y)\mathbf{j}$ determine a conservative force field. Then $P_y = Q_x$, and there exists a function G of x and y such that $G_x(x, y) = P(x, y)$ and $G_y(x, y) = Q(x, y)$. Moreover,

$$(10) \qquad \nabla G(x, y) = G_x(x, y)\mathbf{i} + G_y(x, y)\mathbf{j} = P(x, y)\mathbf{i} + Q(x, y)\mathbf{j} = \mathbf{F}(x, y).$$

Consider a particle of mass m moving along the arc traced out by the position vector $\mathbf{r}(t) = f(t)\mathbf{i} + g(t)\mathbf{j}$ in our conservative force field. In physics, the **potential energy** of the particle at (x, y) is defined to be $-G(x, y)$ and its kinetic energy, as we saw in Example 2, is $\frac{1}{2}km|\mathbf{r}'(t)|^2$.

Let $S(t)$ represent the sum of the potential and kinetic energies of the particle at time t; that is, let

$$S(t) = -G(f(t), g(t)) + \frac{km}{2}|\mathbf{r}'(t)|^2.$$

We shall prove that $S(t)$ is constant by showing that $S'(t) = 0$ for all values of t.

$$\begin{aligned}
S'(t) &= -[G_x(f(t), g(t))f'(t) + G_y(f(t), g(t))g'(t)] + \frac{km}{2}[2f'(t)f''(t) + 2g'(t)g''(t)] \\
&= -\nabla G(f(t), g(t)) \cdot \mathbf{r}'(t) + km\mathbf{r}'(t) \cdot \mathbf{r}''(t) \\
&= [-\nabla G(f(t), g(t)) + km\mathbf{r}''(t)] \cdot \mathbf{r}'(t) \\
&= [-\mathbf{F}(f(t), g(t)) + km\mathbf{r}''(t)] \cdot \mathbf{r}'(t) \qquad \text{[by (10)]}.
\end{aligned}$$

But, by Newton's second law of motion, $\mathbf{F}(f(t), g(t)) = km\mathbf{r}''(t)$. Therefore

$$S'(t) = [-\mathbf{F}(f(t), g(t)) + \mathbf{F}(f(t), g(t))] \cdot \mathbf{r}'(t) = 0,$$

and $S(t)$ is a constant.

Exercises

In Exercises 1 to 7, find the work done by the force field $\mathbf{F}(x, y)$ as a particle is displaced along the given arc.

1. $\mathbf{F}(x, y) = (x^3 - y^3)\mathbf{i} + xy^2\mathbf{j}$, along the arc traced by the terminal point of the position vector $\mathbf{r}(t) = t^2\mathbf{i} + t^3\mathbf{j}$ from $t = -1$ to $t = 0$.

2. $\mathbf{F}(x, y) = x^2y^2\mathbf{i} + (x^2 - y^2)\mathbf{j}$, along the arc $\mathbf{r}(t) = t^3\mathbf{i} - 2t\mathbf{j}$, $0 \leq t \leq 3$.

3. $\mathbf{F}(x, y) = e^x\mathbf{i} - e^{-y}\mathbf{j}$, along the arc $\mathbf{r}(t) = 3\ln t\mathbf{i} + \ln 2t\mathbf{j}$, $1 \leq t \leq 5$.

4. $\mathbf{F}(x, y) = xy^2\mathbf{i} - x^2y\mathbf{j}$, along the quarter-circle $\mathbf{r}(t) = \cos t\mathbf{i} + \sin t\mathbf{j}$, $0 \leq t \leq \pi/2$.

5. $\mathbf{F}(x, y) = ye^{xy}\mathbf{i} + xy\mathbf{j}$, along the arc $\mathbf{r}(t) = 2t^2\mathbf{i} + t^{-1}\mathbf{j}$, $1 \leq t \leq 3$.

6. $\mathbf{F}(x, y) = 2y\mathbf{i} - 2x\mathbf{j}$, along the quarter-ellipse $\mathbf{r}(t) = a\sin t\mathbf{i} + b\cos t\mathbf{j}$, $0 \leq t \leq \pi/2$.

7. $\mathbf{F}(x, y) = (x + y)\mathbf{i} + (x - y)\mathbf{j}$, along the quarter-ellipse $\mathbf{r}(t) = a\cos t\mathbf{i} + b\sin t\mathbf{j}$, $0 \leq t \leq \pi/2$.

8. A particle at any point in the xy plane is attracted to the origin by a force that is inversely proportional to the distance of the particle from the origin. Write the vector $\mathbf{F}(x, y)$ that defines this force field and show that the field is conservative inside any circle that does not contain the origin.

9. The ellipse $144(x - 5)^2 + 169y^2 = 24,336$ has its left focus at the origin. A particle in the force field of Exercise 8 moves along this ellipse from the point $(5, 12)$ to the point $(-8, 0)$. How much work does the force field do on the particle? (*Hint:* Find parametric equations of the ellipse.)

18.12 DIVERGENCE AND CURL*

Let $\mathbf{F}(x, y, z) = f(x, y, z)\mathbf{i} + g(x, y, z)\mathbf{j} + h(x, y, z)\mathbf{k}$ define a vector-valued function $\mathbf{F}$ of three independent scalar variables x, y, and z.

The **divergence** of $\mathbf{F}$, written div $\mathbf{F}$, is the scalar function defined by

$$\text{div } \mathbf{F} = \frac{\partial f}{\partial x} + \frac{\partial g}{\partial y} + \frac{\partial h}{\partial z}.$$

*Optional.

The **curl** of **F**, written curl **F**, is the vector function defined by

$$\text{curl } \mathbf{F} = \left(\frac{\partial h}{\partial y} - \frac{\partial g}{\partial z}\right)\mathbf{i} - \left(\frac{\partial f}{\partial z} - \frac{\partial h}{\partial x}\right)\mathbf{j} + \left(\frac{\partial g}{\partial x} - \frac{\partial f}{\partial y}\right)\mathbf{k}.$$

Motivation for the names "divergence" *and* "curl."

If a fluid (liquid or gas) is flowing, and $\mathbf{v}(x, y, z)$ is the velocity vector attached to each particle (x, y, z) of the fluid and $\rho(x, y, z)$ is the density of the fluid at (x, y, z), then

$$\mathbf{F}(x, y, z) = \rho(x, y, z)\mathbf{v}(x, y, z)$$

gives the rate of flow of mass of fluid at (x, y, z) in the direction of $\mathbf{v}(x, y, z)$. Assume that it is a **steady state** flow, which means that $\mathbf{v}$ depends only on the position (x, y, z) of the particle, and that at any particular position $\mathbf{v}$ does not change with time.

It can be shown that the divergence of $\mathbf{F}$ at (x, y, z) gives the time rate of change of mass of fluid per unit of volume at the position (x, y, z). If div $\mathbf{F} > 0$ at (x, y, z), there is a "source" of fluid at (x, y, z) or the density is diminishing there. For example, air moves out through a hole in a container if the container is heated, and the remaining air is less dense. If div $\mathbf{F} < 0$ at (x, y, z), there is a "sink" at (x, y, z) or the density is increasing there because the fluid is being compressed. If the fluid is incompressible and there are no sources or sinks, div $\mathbf{F} = 0$ for all (x, y, z). If instead of a fluid we are considering an electric or magnetic field, $\mathbf{F}(x, y, z)$ is called the **flux density**, and analogous interpretations apply.

In fluid flow, curl $\mathbf{F}(x, y, z)$ measures the angular velocity of the fluid *in the neighborhood* of the point (x, y, z).

Exercises

In Exercises 1 through 7, **A** and **B** are three-dimensional vector-valued functions defined by

$$\mathbf{A}(x, y, z) = a_1(x, y, z)\mathbf{i} + a_2(x, y, z)\mathbf{j} + a_3(x, y, z)\mathbf{k},$$

$$\mathbf{B}(x, y, z) = b_1(x, y, z)\mathbf{i} + b_2(x, y, z)\mathbf{j} + b_3(x, y, z)\mathbf{k},$$

whose components a_i and b_i $(i = 1, 2, 3)$ have continuous first and second partial derivatives; and $G(x, y, z)$ defines a scalar function of three variables having continuous first partial derivatives.

We defined the *gradient* of G to be the vector-valued function ∇G, read "del G," where

$$\nabla G = \frac{\partial G}{\partial x}\mathbf{i} + \frac{\partial G}{\partial y}\mathbf{j} + \frac{\partial G}{\partial z}\mathbf{k}.$$

It is useful to think of "del,"

$$\nabla = \frac{\partial}{\partial x}\mathbf{i} + \frac{\partial}{\partial y}\mathbf{j} + \frac{\partial}{\partial z}\mathbf{k},$$

by itself as an operator that produces vector functions when applied to scalar functions. Of course, ∇ is not a vector; but when used symbolically as a vector, it can simplify the writing of many formulas.

1. Show that div $\mathbf{A} = \nabla \cdot \mathbf{A}$ and curl $\mathbf{A} = \nabla \times \mathbf{A}$.

2. Show that div (curl $\mathbf{A}$) = 0.

3. Show that
$$\text{div } (\mathbf{A} + \mathbf{B}) = \text{div } \mathbf{A} + \text{div } \mathbf{B};$$
$$\text{div } (G\mathbf{A}) = G \text{ div } \mathbf{A} + \nabla G \cdot \mathbf{A}.$$

4. Show that
$$\text{curl } (\mathbf{A} + \mathbf{B}) = \text{curl } \mathbf{A} + \text{curl } \mathbf{B};$$
$$\text{curl } (G\mathbf{A}) = G \text{ curl } \mathbf{A} + \nabla G \cdot \mathbf{A}.$$

5. Show that
$$\text{curl } \nabla G = \mathbf{0}.$$

6. Show that
$$\text{div } (\mathbf{A} \times \mathbf{B}) = \mathbf{B} \cdot \text{curl } \mathbf{A} - \mathbf{A} \cdot \text{curl } \mathbf{B}.$$

7. The equation

$$\frac{\partial^2 G}{\partial x^2} + \frac{\partial^2 G}{\partial y^2} + \frac{\partial^2 G}{\partial z^2} = 0$$

is called *Laplace's equation*, and its left member is the Laplacian of $G(x, y, z)$. Show that Laplace's equation can be written

$$\text{div } \nabla G = 0,$$

or

$$\nabla^2 G = 0,$$

where $\nabla^2 G$ means $\nabla \cdot \nabla G$.

Angular Velocity. Let P be a particle moving around a circle whose center is C. If we denote by θ the radian measure of the angle from CP_1 to CP, where P_1 is an arbitrarily selected position of P, the **angular speed** of P is $\omega = |d\theta/dt|$.

The **angular velocity vector** ω of P is defined to be the vector along the axis of rotation in the direction of progress of a right-hand screw that is turned in the way that the point P is rotating, and whose magnitude is equal to the angular speed of P; that is, $|\omega| = \omega$.

8. Let the axis of rotation of the particle P (described above) be the z axis, with the circle not necessarily in the xy plane. If $\mathbf{r}(t)$ is the position vector of P, $\mathbf{v}(t)$ its velocity vector, and $\omega(t)$ its angular velocity vector, prove that

$$\mathbf{v} = \omega \times \mathbf{r}.$$

Note. This relation is valid no matter how the axis of rotation is oriented with respect to the coordinate axes.

9. A homogeneous rigid body is rotating about the z axis. Let $P:(x, y, z)$ be an arbitrarily chosen point in the body and denote its velocity vector by $\mathbf{v}$. If the angular velocity vector ω of the rotating body is constant and in the direction of the positive z axis, show that curl $\mathbf{v} = 2\omega$. (*Hint:* The position vector of P can be written

$$\mathbf{r}(t) = c \cos \omega t \mathbf{i} + c \sin \omega t \mathbf{j} + z\mathbf{k},$$

where c is the radius of the circle of rotation of P, $\omega = |\omega|$ is the angular speed of P, and z is constant.)

18.13 REVIEW EXERCISES

1. Find and sketch the domain of each indicated function of two variables, showing clearly any points on the boundary of the domain that belong to the domain.
(a) $z = \sqrt{x^2 + 4y^2 - 100}$.
(b) $z = -\sqrt{2x - y - 1}$.

2. Describe geometrically the domain of each of the following indicated functions of three variables.
(a) $f(x, y, z) = \sqrt{400 - 100x^2 - 16y^2 - 15z^2}$.
(b) $f(x, y, z) = \dfrac{(144 - 16x^2 - 16y^2 + 9z^2)^{3/2}}{xyz}$.

3. Sketch the level curves of $f(x, y) = x^2 + y^2/9$ for $k = 0, \frac{1}{2}, 1, 4$.

4. Sketch the level curves of $f(x, y) = (x + y^2)/(x^2 + y)$ for $k = 0, 1, 2, 4$. Draw the curve in the xy plane whose points are *not* in the domain of the indicated function.

In Exercises 5 to 8, find all first and second partial derivatives.

5. $f(x, y) = 3x^4y^2 + 7x^2y^7$.

6. $f(x, y) = \cos^2 x - \sin^2 y$.

7. $f(x, y) = e^{-y} \tan x$.

8. $f(x, y) = e^{-x} \sin y$.

9. If $F(x, y) = 5x^3y^6 - xy^7$, find $\partial^3 F(x, y)/\partial x \, \partial y^2$.

10. If f is the function of three variables defined by $f(x, y, z) = xy^3 - 5x^2yz^4$, find $f_x(x, y, z)$, $f_y(x, y, z)$, and $f_z(x, y, z)$.

11. Find the slope of the tangent to the curve of intersection of the surface $z = x^2 + y^2/4$ and the plane $x = 2$ at the point $(2, 2, 5)$. Make a sketch.

12. For what points is the function defined by $f(x, y) = xy/(x^2 - y)$ continuous?

13. Does

$$\lim_{(x, y) \to (0, 0)} \frac{x - y}{x + y}$$

exist? Explain.

14. If an error of 1% is made in measuring the radius of a cylinder and an error of 2% is made in measuring its altitude, use differentials to find approximately the greatest possible error in calculating the volume.

15. If $F(u, v) = \text{Tan}^{-1}(uv)$, $u = \sqrt{xy}$, $v = \sqrt{x} - \sqrt{y}$, find $\partial F/\partial x$ and $\partial F/\partial y$ in terms of u, v, x, and y.

16. If $f(u, v) = u/v$, $u = \sqrt{x^2 - 3y + 4z}$, $v = xyz$, find f_x, f_y, and f_z in terms of u, v, x, y, and z.

17. If $f(u, v) = ve^{-u}$, $u = \text{Tan}^{-1}(xyz)$, $v = \ln(xy + yz)$, find f_x, f_y, and f_z in terms of u, v, x, y, and z.

18. If $F(x, y) = x^3 - xy^2 - y^4$, $x = 2 \cos 3t$, $y = 3 \sin t$, find dF/dt in terms of x, y, and t.

19. If $F(x, y, z) = \text{Tan}^{-1}(5x^2 y/z^3)$, $x = t^{3/2} + 2$, $y = \ln 4t$, $z = e^{3t}$, find dF/dt in terms of x, y, z, and t.

20. A triangle has vertices A, B, and C. The length of the side $c = AB$ is increasing at the rate of 3 inches per second, the side $b = AC$ is decreasing at 1 inch per second, and the included angle α is increasing at 0.1 radian per second. If $c = 10$ inches and $b = 8$ inches when $\alpha = \pi/6$, how fast is the area changing then?

21. Find the directional derivative of $f(x, y) = \text{Tan}^{-1}(3xy)$. What is its value at the point $(4, 2)$ in the direction $\mathbf{u} = (\sqrt{3}/2)\mathbf{i} - (1/2)\mathbf{j}$?

22. Find the slope of the tangent line to the curve of intersection of the vertical plane $x - \sqrt{3}y + 2\sqrt{3} - 1 = 0$ and the surface $z = x^2 + y^2$ at the point $(1, 2, 5)$. Sketch the curve and the tangent.

23. In what direction is $f(x, y) = 9x^4 + 4y^2$ increasing most rapidly at $(1, 2)$?

24. For $f(x, y) = x^2/2 + y^2$,
(a) find the equation of its level curve that goes through the point $(4, 1)$ in its domain;

(b) find the gradient vector ∇z at $(4, 1)$;
(c) draw the level curve and draw the gradient vector with its initial point at $(4, 1)$.

25. Find the gradient vector of the ellipsoid $9x^2 + 4y^2 + 9z^2 = 36$ at the point $P:(1, 2, \sqrt{11}/3)$. Write the equation of the tangent plane to this surface in the vector form and in the Cartesian form. Sketch the part of the ellipsoid that is in the first octant and draw the tangent plane and the gradient vector of the surface at P.

26. Find the extrema of $f(x, y) = \cos(x + y) + \sin x + \cos y$.

27. A rectangular box whose edges are parallel to the coordinate axes is inscribed in the ellipsoid $36x^2 + 4y^2 + 9z^2 = 36$. What is the greatest possible volume for such a box?

28. Use Lagrange multipliers to find the maximum and the minimum of $f(x, y) = xy$, subject to the constraint $x^2 + y^2 = 1$.

29. Use Lagrange multipliers to find the dimensions of the right circular cylinder with maximum volume if its surface area is 24π.

30. Find the value of the line integral $\int_C xy^2 \, dx + 3xy \, dy$ along the arc that has parametric equations $x = t^2 - 1$, $y = 2t^3$; $-1 \leq t \leq 2$.

31. Show that the value of the line integral $\int_C y \sec^2 x \, dx + \tan x \, dy$ from $(0, -1)$ to $(\pi/4, 9)$ is independent of the path, and evaluate it.

32. Find the work done by the force field $\mathbf{F}(x, y) = xy\mathbf{i} + (x^2 + y^2)\mathbf{j}$ as a particle is displaced along the arc $\mathbf{r}(t) = t\mathbf{i} + 4t^2\mathbf{j}$ from $t = 0$ to $t = 3$.

33. Find the work done by the force field $\mathbf{F}(x, y) = (x - y)\mathbf{i} + (x + y)\mathbf{j}$ as a particle is displaced along the quarter-circle $\mathbf{r}(t) = \sin t\mathbf{i} + \cos t\mathbf{j}$, $0 \leq t \leq \pi/2$.

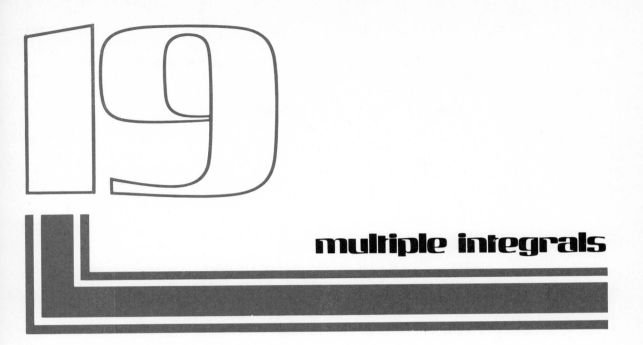

19

multiple integrals

19.1 DOUBLE INTEGRALS

Another way in which the concept of the Riemann integral $\int_a^b f(x)\,dx$ can be extended is to **multiple integrals**—that is, to double integrals, triple integrals, and the like.

Whereas the integrand of our earlier integral was a function of one variable on a closed interval of the x axis, the integrands of the **double integrals** that we are about to discuss are functions of two variables on a closed region of the xy plane whose boundary is what is known as a closed, **piecewise smooth curve**.

We recall from 15.1.1 that a simple arc in the xy plane is the graph of parametric equations

$$x = f(t), \qquad y = g(t), \qquad a \leq t \leq b,$$

in which f and g are continuous, and such that to distinct values of t between a and b correspond distinct points on the graph. Thus a simple arc does not cross itself.

If, in addition, the first derivatives f' and g' are continuous for $a \leq t \leq b$, the simple arc is said to be **smooth** [Fig. 19-1(a)]. Its tangent line turns continuously as t increases from a to b.

A **piecewise smooth curve** is either a simple smooth arc or a succession of simple smooth arcs joined end to end such that to distinct values of t in $[a, b]$ (except possibly a and b) correspond distinct points on the graph [Fig. 19-1(b)]. If the points corresponding to $t = a$ and $t = b$ are identical, the piecewise smooth curve is **closed**.

A simple smooth arc

A piecewise smooth arc

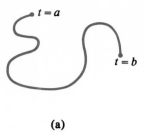

(a)

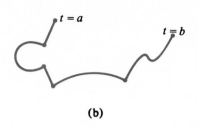

(b)

Figure 19-1

Let f be a function of two variables that is defined in an open rectangular region H and let C be a closed, piecewise smooth curve lying entirely inside H (Fig. 19-2).

Consider the closed region R consisting of all points inside or on the piecewise smooth curve C. By means of lines parallel to the coordinate axes, cover the region R with a grid consisting of a finite number of rectangles (Fig. 19-3). Those rectangles lying inside or on the boundary of R form a **partition** p of R; number them from 1 to n in any convenient manner. The partitioning rectangles are shaded in Fig. 19-3. Denote the lengths of the sides of the ith partitioning rectangle by Δx_i and Δy_i and its area by $\Delta A_i = \Delta x_i \, \Delta y_i$ $(i = 1, 2, \cdots, n)$. By the **norm** $|p|$ of a partition p of R is meant the longest diagonal of any of the rectangles belonging to the partition p.

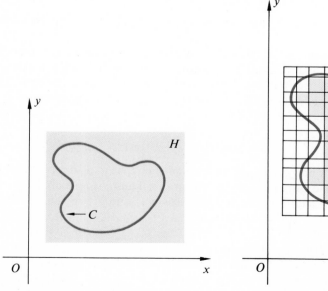

Figure 19-2

Figure 19-3

Since f is defined in H, it is defined inside and on the boundary of R. Let (x_i, y_i) be an arbitrarily chosen point in the ith partitioning rectangle, $i = 1, 2, \cdots, n$. Then the sum

$$(1) \qquad \sum_{i=1}^{n} f(x_i, y_i) \, \Delta A_i$$

is called a **Riemann sum** for $f(x, y)$ on R.

It seems evident that the unshaded white space inside the region R in Fig. 19-3 can be made as small as desired by using enough lines in forming the grid, so that the norm $|p|$ of the partition will be sufficiently small.

Let L be a number. If for each $\epsilon > 0$ there is a $\delta > 0$ such that

$$\left| \sum_{i=1}^{n} f(x_i, y_i) \, \Delta A_i - L \right| < \epsilon$$

for every partition p of R whose norm $|p| < \delta$, then L is said to be the **limit** of the Riemann sum (1) as $|p| \longrightarrow 0$. This is written

$$\lim_{|p| \to 0} \sum_{i=1}^{n} f(x_i, y_i) \, \Delta A_i = L.$$

19.1.1 **Definition of Double Integral.**   Let R be a closed region in the xy plane bounded by a piecewise smooth curve. If f is a function of two variables that is defined on the region R, then the **double integral** of f on R, written

$$\iint_R f(x, y) \, dA \qquad \text{or} \qquad \iint_R f(x, y) \, dx \, dy,$$

is defined by

$$\iint_R f(x, y) \, dA = \lim_{|p| \to 0} \sum_{i=1}^{n} f(x_i, y_i) \, \Delta A_i,$$

provided that this limit exists. When this limit does exist, the function f is said to be **integrable** on the region R.

It can be proved* that if f is *continuous* in the open rectangular region H, and thus inside and on the boundary of the closed region R (Fig. 19-2), then this double integral exists.

19.1.2 **Existence of Double Integral.**   If f is a function of two variables that is continuous on a closed region R whose boundary is a closed, piecewise smooth curve in the xy plane, then the double integral of f on R

$$\iint_R f(x, y) \, dA$$

exists.

*See, for example, J. M. H. Olmsted, *Real Variables*, New York, Appleton-Century-Crofts, 1959, Sections 1303 and 1306.

This theorem gives *sufficient* conditions for the double integral to exist, but they are not *necessary* conditions. Less stringent conditions are discussed in more advanced works.

Double integrals have basic properties that are analogous to those for integrals of functions of one variable.

19.1.3 Properties of Double Integrals. Let R be a closed region in the xy plane.

  (i) If c is a constant and if f is a function of two variables that is integrable on R, then cf is integrable on R and

$$\iint\limits_{R} cf(x, y)\, dA = c \iint\limits_{R} f(x, y)\, dA.$$

  (ii) If the functions f and g are integrable on R, then

$$\iint\limits_{R} [f(x, y) + g(x, y)]\, dA = \iint\limits_{R} f(x, y)\, dA + \iint\limits_{R} g(x, y)\, dA.$$

  (iii) If the functions f and g are integrable on R and $f(x, y) \le g(x, y)$ for all (x, y) in R, then

$$\iint\limits_{R} f(x, y)\, dA \le \iint\limits_{R} g(x, y)\, dA.$$

  (iv) If f is continuous on R, and R is decomposed into regions R_1 and R_2, then

$$\iint\limits_{R} f(x, y)\, dA = \iint\limits_{R_1} f(x, y)\, dA + \iint\limits_{R_2} f(x, y)\, dA.$$

Proofs of the four properties in this theorem are similar to those of the analogous properties of integrals of functions of one variable.

Just as the definite integral of a function of one variable can be interpreted as a plane area, a double integral can be interpreted as a *volume*.

Assume that $f(x, y) \ge 0$ for all (x, y) in a closed region R. Then $f(x_i, y_i)\, \Delta A_i$, the ith element in the Riemann sum (1), is the volume of a rectangular parallelepiped whose base has area ΔA_i and whose altitude is $f(x_i, y_i)$ (Fig. 19-4). The sum (1) is the volume of a sum of such parallelepipeds. By taking the norm $|p|$ of the partition sufficiently small, we can make the sum of the parallelepipeds approximate as closely as we please the solid under the surface $z = f(x, y)$, whose base is R and whose sides are cylinders that are perpendicular to the xy plane (Fig. 19-4).

19.1.4 Definition of Volume. Let f be a function of two variables, continuous on a closed region R that is bounded by a closed, piecewise smooth curve in the xy plane. If $f(x, y) \ge 0$ for all (x, y) in R, the **volume** V of the solid under the surface $z = f(x, y)$ and above the region R is defined to be the value of the double integral of $f(x, y)$ on R:

$$V = \iint\limits_{R} f(x, y)\, dA.$$

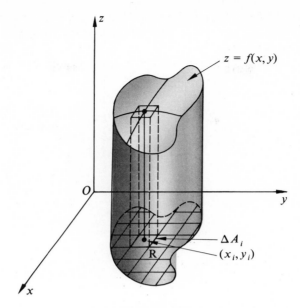

Figure 19-4

Other interpretations of the double integral will be discussed in Section 19.3.

Example. Find an approximation to the volume of the solid in the first octant, bounded by the coordinate planes, the planes $x = 4$ and $y = 8$, and the surface

$$z = f(x, y) = \frac{64 - 8x + y^2}{16}.$$

Solution. The solid is shown in Fig. 19-5. For simplicity of computation, we partition its rectangular base, R, into 8 equal squares and choose for the points (x_i, y_i), $i = 1, 2, \cdots, 8$, the midpoints of the squares. Thus

$$(x_1, y_1) = (1, 1), \qquad f(x_1, y_1) = 57/16.$$
$$(x_2, y_2) = (1, 3), \qquad f(x_2, y_2) = 65/16.$$
$$(x_3, y_3) = (1, 5), \qquad f(x_3, y_3) = 81/16.$$
$$(x_4, y_4) = (1, 7), \qquad f(x_4, y_4) = 105/16.$$
$$(x_5, y_5) = (3, 1), \qquad f(x_5, y_5) = 41/16.$$
$$(x_6, y_6) = (3, 3), \qquad f(x_6, y_6) = 49/16.$$
$$(x_7, y_7) = (3, 5), \qquad f(x_7, y_7) = 65/16.$$
$$(x_8, y_8) = (3, 7), \qquad f(x_8, y_8) = 89/16.$$

Then

$$V = \iint\limits_{R} f(x, y)\, dA \doteq \sum_{i=1}^{8} f(x_i, y_i)\, \Delta A_i.$$

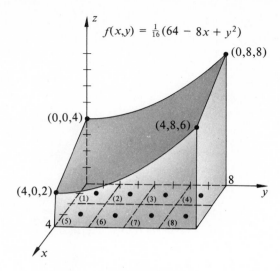

Figure 19-5

Since the areas of the equal squares are $\Delta A_i = 4$, $i = 1, 2, \cdots, 8$, we can write

$$\sum_{i=1}^{8} f(x_i, y_i) \, \Delta A_i = 4 \sum_{i=1}^{8} f(x_i, y_i) = \frac{4(57 + 65 + 81 + 105 + 41 + 49 + 65 + 89)}{16} = 138.$$

So an approximation to the volume of the given solid is

$$V = \iint_R f(x, y) \, dA \doteq 138 \text{ cubic units.}$$

In Section 19.2 we shall learn how to find the *exact* volume of this solid. It is $138\frac{2}{3}$ cubic units.

Exercises

In Exercises 1 to 4, R is the rectangular region in the xy plane bounded by the lines $x = 0$, $x = 6$, $y = 0$, and $y = 4$, and p is the partition of R into 6 equal squares by the lines $x = 2$, $x = 4$, and $y = 2$. The (x_i, y_i) are the midpoints of the 6 squares. Find the approximate value of the double integral $\iint_R f(x, y) \, dA$ for this partition of R and this choice of the (x_i, y_i), $i = 1, 2, \cdots, 6$. Make a sketch.

1. $f(x, y) = \frac{1}{6}(48 - 4x - 3y)$.

2. $f(x, y) = \frac{1}{9}(72 - y^2)$.

3. $f(x, y) = \frac{1}{6}(36 - x^2 + 3y)$.

4. $f(x, y) = \frac{1}{12}(2x^2 + 3y^2 + 24)$.

In Exercises 5 to 8, R and p are the same as in Exercises 1 to 4, but in each square the (x_i, y_i) is the vertex of the square that is closest to the origin, $i = 1, 2, \cdots, 6$. Find the approximate value of the double integral $\iint_R f(x, y) \, dA$ for this partition of R and this choice of the (x_i, y_i).

5. $f(x, y) = \frac{1}{6}(48 - 4x - 3y)$.

6. $f(x, y) = \frac{1}{9}(72 - y^2)$.

7. $f(x, y) = \frac{1}{6}(36 - x^2 + 3y)$.

8. $f(x, y) = \frac{1}{12}(2x^2 + 3y^2 + 24)$.

9. Find an approximate volume of the solid bounded by the planes $z = 0$, $x = 0$, $x = 2$, $y = -1$, $y = 2$, and the surface $z = \frac{1}{2}(12 - 2x + y^2)$, if the rectangular base R is partitioned into 6 equal squares and the point (x_i, y_i) in each square is that vertex of the square that is closest to the origin. Make a sketch.

10. Find an approximate volume of the solid bounded by the planes $z = 0$, $x = 3$, $x = -1$, $y = 2$, $y = -2$, and the surface $z = \frac{1}{2}(20 - x^2 - y^2)$, if its base R is partitioned into 3 rectangles by the lines $y = 0$ and $y = 1$, and the (x_i, y_i) are the midpoints of the 3 rectangles. Make a sketch.

19.2 EVALUATION OF DOUBLE INTEGRALS BY ITERATED INTEGRALS

In the preceding section we defined the double integral over a rather general closed region R. The region R was bounded by any closed, piecewise smooth curve in the xy plane. However, the plane regions encountered in a first course in calculus are of the two simple types shown in Fig. 19-6, or they may be decomposed into a finite number of regions of these types.

A **type I region** in the xy plane is the closed region R bounded by the lines $x = a$ and $x = b$, and the smooth curves $y = \phi_1(x)$ and $y = \phi_2(x)$, where $\phi_1(x) \le \phi_2(x)$ for all x in $[a, b]$ (Fig. 19-6). That is,

$$(1) \qquad R = \{(x, y) \mid a \le x \le b, \phi_1(x) \le y \le \phi_2(x)\}.$$

A **type II region** in the xy plane is a closed region R that is bounded by the lines $y = c$ and $y = d$, with $c < d$, and by smooth curves $x = \psi_1(y)$ and $x = \psi_2(y)$, where $\psi_1(y) \le \psi_2(y)$ for all y in $[c, d]$ (Fig. 19-6). When of type II,

$$(2) \qquad R = \{(x, y) \mid \psi_1(y) \le x \le \psi_2(y), c \le y \le d\}.$$

In the rest of this chapter *the symbol R will always mean a region of one of the above two types.* Notice that the boundaries of such regions are simple examples of closed, piecewise smooth curves.

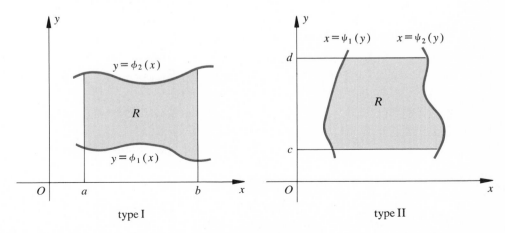

type I   type II

Figure 19-6

As with functions of one variable, it is usually impossible to find the value of a double integral from its definition alone. In functions of one variable, the fundamental theorem of calculus enables us to evaluate a definite integral by means of an antiderivative of the integrand. A double integral can be evaluated by what is known as an *iterated integral*.

When forming mixed second partial derivatives of a function of two variables, we first differentiate with respect to one variable while holding the other constant, and then differentiate the result with respect to the other variable. An iterated integral involves a similar procedure with the inverse process, integration; a function of two variables is first integrated with respect to one variable while holding the other constant, and then the resulting function is integrated with respect to the other variable.

The symbol

$$\int_a^b \int_{\phi_1(x)}^{\phi_2(x)} f(x, y) \, dy \, dx$$

represents an **iterated integral**. It is a shorter way of writing

$$\int_a^b \left(\int_{\phi_1(x)}^{\phi_2(x)} f(x, y) \, dy \right) dx,$$

which means that the inner definite integral $\int_{\phi_1(x)}^{\phi_2(x)} f(x, y) \, dy$ is evaluated first, thinking of x as a constant during this integration with respect to y. After substitution of the limits $\phi_2(x)$ and $\phi_1(x)$ for y, the result is a function of x alone, say $F(x)$, which now becomes the integrand of the outer definite integral $\int_a^b F(x) \, dx$.

Example 1. Evaluate the iterated integral

$$\int_3^5 \int_{-x}^{x^2} (4x + 10y) \, dy \, dx.$$

Solution.

$$\int_3^5 \int_{-x}^{x^2} (4x + 10y) \, dy \, dx = \int_3^5 \left(\int_{-x}^{x^2} (4x + 10y) \, dy \right) dx.$$

We first perform the inner integration with respect to y, temporarily thinking of x as constant, and obtain

$$\int_3^5 \left(\int_{-x}^{x^2} (4x + 10y) \, dy \right) dx = \int_3^5 \left[4xy + 5y^2 \right]_{-x}^{x^2} dx$$

$$= \int_3^5 [(4x^3 + 5x^4) - (-4x^2 + 5x^2)] \, dx$$

$$= \int_3^5 (5x^4 + 4x^3 - x^2) \, dx = \left[x^5 + x^4 - \frac{x^3}{3} \right]_3^5$$

$$= 3393\tfrac{1}{3}.$$

Notice in iterated integrals that *the outer integral has constant limits*; thus the last integration has constant limits.

Example 2.   Evaluate the iterated integral

$$\int_0^1 \int_0^{y^2} 2ye^x \, dx \, dy.$$

Solution.

$$\int_0^1 \int_0^{y^2} 2ye^x \, dx \, dy = \int_0^1 \left(\int_0^{y^2} 2ye^x \, dx \right) dy$$

$$= \int_0^1 \left[2ye^x \right]_0^{y^2} dy = \int_0^1 (2ye^{y^2} - 2ye^0) \, dy$$

$$= \int_0^1 e^{y^2} \, d(y^2) - 2 \int_0^1 y \, dy$$

$$= \left[e^{y^2} \right]_0^1 - 2\left[\frac{y^2}{2} \right]_0^1 = e - 1 - 2\left(\frac{1}{2} \right) = e - 2.$$

Our chief interest in iterated integrals stems from the fact that under suitable conditions they can be used to evaluate double integrals. We state, without proof, a basic theorem.

19.2.1   Evaluation of Double Integrals.   Let f be a function of two variables that is continuous inside and on the boundary of a region R in the xy plane.

If R is a type I region (Fig. 19-6), then

$$\iint_R f(x, y) \, dA = \int_a^b \int_{\phi_1(x)}^{\phi_2(x)} f(x, y) \, dy \, dx.$$

If R is of type II, then

$$\iint_R f(x, y) \, dA = \int_c^d \int_{\psi_1(y)}^{\psi_2(y)} f(x, y) \, dx \, dy.$$

It is interesting and instructive to interpret an iterated integral geometrically. Let R be a type I region and let $f(x, y)$ define a function that is continuous and non-negative on R.

The intersection of the surface $z = f(x, y)$ and a plane $x = \bar{x}$, where $a \leq \bar{x} \leq b$, is a plane curve K, a segment of which is over the region R (Fig. 19-7). The area of the plane region under this curve segment and above the xy plane is given by

$$A(\bar{x}) = \int_{\phi_1(\bar{x})}^{\phi_2(\bar{x})} f(\bar{x}, y) \, dy.$$

Thus the iterated integral

$$\int_a^b \int_{\phi_1(x)}^{\phi_2(x)} f(x, y) \, dy \, dx = \int_a^b \left(\int_{\phi_1(x)}^{\phi_2(x)} f(x, y) \, dy \right) dx$$

$$= \int_a^b A(x) \, dx = \lim_{|p| \to 0} \sum_{i=1}^n A(x_i) \, \Delta x_i$$

by 8.3.4.

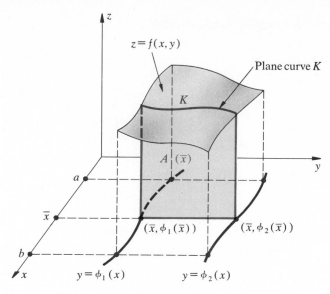

Figure 19-7

But $A(x_i)\,\Delta x_i$ is the volume of a solid (a "slab") with parallel faces of area $A(x_i)$ $= \int_{\phi_1(x_i)}^{\phi_2(x_i)} f(x_i, y)\,dy$ and of thickness Δx_i (Fig. 19-8); and the sum of the volumes of these solids ($i = 1, 2, \cdots, n$) approximates the volume of the solid that is under the surface $z = f(x, y)$, above the xy plane, between the planes $x = a$ and $x = b$, and

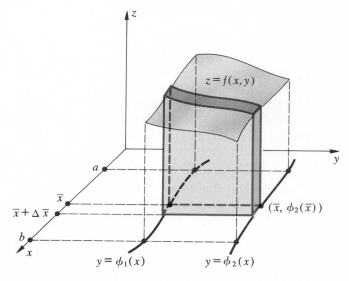

Figure 19-8

between the cylindrical surfaces $y = \phi_1(x)$ and $y = \phi_2(x)$. Furthermore, this approximation can be made as good as desired by taking $|p|$ sufficiently small. It is hardly surprising, then, that the double integral, which is equal to the volume of this solid (by 19.1.4), can be evaluated by this iterated integral.

The reader should notice that this is a generalization of the procedure used in finding the volume of a solid of revolution by the washer method (Section 9.2). The first integration in the iterated integral, above, gave $A(x)$, the area of a typical cross section of the solid. This step was not necessary for a solid of revolution because the typical cross section was a circle whose area we knew to be $\pi[f(x)]^2$. The second integration in the iterated integral, $\int_a^b A(x)\, dx$, corresponds to the only integration in 9.2.1, $\int_a^b \pi[f(x)]^2\, dx$.

Example 3. Use double integration to find the volume of the tetrahedron bounded by the coordinate planes and the plane $3x + 6y + 4z - 12 = 0$.

Solution. Denote by R the triangular region in the xy plane that forms the base of the tetrahedron (Fig. 19-9). We seek the volume of the solid under the surface $z = \frac{3}{4}(4 - x - 2y)$ and above the region R.

The given plane intersects the xy plane in the line $x + 2y - 4 = 0$, a segment of which belongs to the boundary of R. Since this equation can be written $y = 2 - x/2$ and $x = 4 - 2y$, R can be thought of as the type I region

(3) $$R = \left\{(x, y)\,|\,0 \leq x \leq 4,\ 0 \leq y \leq 2 - \frac{x}{2}\right\}$$

or as the type II region

$$R = \{(x, y)\,|\,0 \leq x \leq 4 - 2y,\ 0 \leq y \leq 2\}.$$

We shall treat R as a type I region; the result would be the same either way, as the reader should verify.

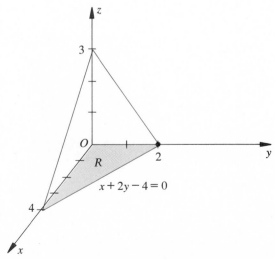

Figure 19-9

From 19.1.4, the volume V of the solid is

$$V = \iint_R \frac{3}{4}(4 - x - 2y) \, dA.$$

In setting up an iterated integral to evaluate this double integral, it will be easy to determine the limits of integration from (3) if we remember that the outer integral has constant limits and that this integration is performed last. Thus

$$V = \int_0^4 \int_0^{2-x/2} \frac{3}{4}(4 - x - 2y) \, dy \, dx$$

$$= \int_0^4 \left(\frac{3}{4} \int_0^{2-x/2} (4 - x - 2y) \, dy \right) dx$$

$$= \int_0^4 \frac{3}{4} \Big[4y - xy - y^2 \Big]_0^{2-x/2} dx$$

$$= \frac{3}{16} \int_0^4 (16 - 8x + x^2) \, dx$$

$$= \frac{3}{16} \Big[16x - 4x^2 + \frac{x^3}{3} \Big]_0^4 = 4.$$

Example 4. Find the volume of the solid in the first octant ($x \geq 0, y \geq 0, z \geq 0$) bounded by the circular paraboloid $z = x^2 + y^2$, the cylinder $x^2 + y^2 = 4$, and the coordinate planes (Fig. 19-10).

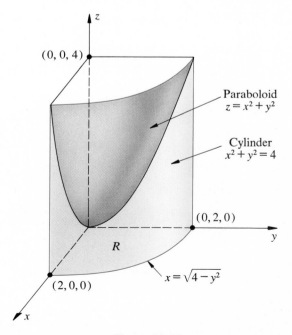

Figure 19-10

Solution. The region R in the first quadrant of the xy plane is bounded by a quarter of the circle $x^2 + y^2 = 4$ and the lines $x = 0$ and $y = 0$. Although R can be thought of as either a type I or a type II region, we shall treat R as a type II region and write its boundary curves as $x = \sqrt{4 - y^2}$, $x = 0$, and $y = 0$. Thus

$$(4) \qquad R = \{(x, y) \mid 0 \le x \le \sqrt{4 - y^2}, 0 \le y \le 2\}.$$

From 19.1.5, the volume is given by

$$V = \iint_R (x^2 + y^2) \, dA.$$

We evaluate this double integral by an iterated integral (19.2.1). Keeping in mind that the outer integral in an iterated integral has constant limits, it is easy to read the limits from (4) and write

$$V = \iint_R (x^2 + y^2) \, dA = \int_0^2 \int_0^{\sqrt{4 - x^2}} (x^2 + y^2) \, dx \, dy$$

$$= \int_0^2 \left[\frac{1}{3} (4 - y^2)^{3/2} + y^2 \sqrt{4 - y^2} \right] dy.$$

By the trigonometric substitution $y = 2 \sin \theta$, the latter integral can be rewritten

$$\int_0^{\pi/2} \left[\frac{8}{3} \cos^3 \theta + 8 \sin^2 \theta \cos \theta \right] 2 \cos \theta \, d\theta$$

$$= \int_0^{\pi/2} \left[\frac{16}{3} \cos^4 \theta + 16 \sin^2 \theta \cos^2 \theta \right] d\theta$$

$$= \frac{16}{3} \int_0^{\pi/2} \cos^2 \theta (1 - \sin^2 \theta + 3 \sin^2 \theta) \, d\theta$$

$$= \frac{16}{3} \int_0^{\pi/2} (\cos^2 \theta + 2 \sin^2 \theta \cos^2 \theta) \, d\theta$$

$$= \frac{16}{3} \int_0^{\pi/2} \left(\cos^2 \theta + \frac{1}{2} \sin^2 2\theta \right) d\theta = 2\pi.$$

Exercises

Evaluate the iterated integrals in Exercises 1 through 12.

1. $\int_0^{-1} \int_0^3 x^4 \, dy \, dx.$

2. $\int_1^2 \int_0^{x-1} y \, dy \, dx.$

3. $\int_{-1}^3 \int_0^{3y} (x^2 + y^2) \, dx \, dy.$

4. $\int_{-3}^1 \int_0^{4x} (x^3 - y^3) \, dy \, dx.$

5. $\int_1^3 \int_{-y}^{2y} x e^{y^3} \, dx \, dy.$

6. $\int_1^5 \int_0^x \frac{3}{x^2 + y^2} \, dy \, dx.$

7. $\int_{1/2}^1 \int_0^{2x} \cos (\pi x^2) \, dy \, dx.$

8. $\int_0^{\pi/4} \int_{\sqrt{2}}^{\sqrt{2} \cos \theta} r \, dr \, d\theta.$

9. $\int_{\pi/4}^{\pi/2} \int_{\pi/2}^r \csc^2 \theta \, d\theta \, dr.$

10. $\displaystyle\int_0^{\pi/9}\int_{\pi/4}^{3r}\sec^2\theta\,d\theta\,dr.$

11. $\displaystyle\int_0^{\pi/2}\int_0^{2y} y\sin\left(\tfrac{1}{2}x\right)dx\,dy.$

12. $\displaystyle\int_{\pi/6}^{\pi/2}\int_0^{\sin\theta} 6r\cos\theta\,dr\,d\theta.$

In Exercises 13 to 26, sketch the indicated solid. Express the region R as shown in equation (1) or (2) above, and from this determine the limits of integration.

13. By double integration, find the volume of the tetrahedron bounded by the coordinate planes and the plane $20x + 12y + 15z - 60 = 0$.

14. By double integration, find the volume of the tetrahedron bounded by the coordinate planes and the plane $3x + 4y + 2z - 12 = 0$.

15. By double integration, find the volume of the wedge bounded by the coordinate planes and the planes $x = 5$ and $y + 2z - 4 = 0$.

16. Find the volume of the solid in the first octant bounded by the coordinate planes and the planes $2x + y - 4 = 0$ and $8x + y - 4z = 0$.

17. Find the volume of the solid in the first octant bounded by the surface $9x^2 + 4y^2 = 36$ and the plane $9x + 4y - 6z = 0$.

18. Find the volume of the solid in the first octant

bounded by the surface $z = 9 - x^2 - y^2$ and the coordinate planes.

19. Find the volume of the region bounded by the parabolic cylinder $x^2 = 4y$ and the planes $z = 0$ and $5y + 9z - 45 = 0$.

20. Find the volume of the region in the first octant bounded by the cylinder $y = x^2$ and the planes $x = 0$, $z = 0$, and $y + z = 1$.

21. Find the volume of the region in the first octant bounded by the surface $16x^2 + 9y^2 - 144 = 0$, the plane $16x + 9y - 12z = 0$, and the coordinate planes.

22. Find the volume of the region in the first octant bounded by the cylinder $z = \tan x^2$ and the planes $x = y$, $x = 1$, and $y = 0$.

23. Find the volume of the region in the first octant bounded by the surface $z = e^{x-y}$, the plane $x + y = 1$, and the coordinate planes.

24. Find the volume of the region in the first octant bounded by the surface $9z = 36 - 9x^2 - 4y^2$ and the coordinate planes.

25. Find the volume of the region in the first octant bounded by the circular cylinders $x^2 + z^2 = 16$ and $y^2 + z^2 = 16$, and the coordinate planes.

26. Find the volume of the region in the first octant bounded by the surface $z = e^{x+y}$ and the planes $x = 2y - 1$ and $y = 1$.

19.3 OTHER APPLICATIONS OF DOUBLE INTEGRALS

Throughout this section f is assumed to be a function of two variables that is continuous on a closed region R in the xy plane.

If f is the constant function whose value is 1, so that $f(x, y) = 1$ for all (x, y) in R, the double integral (19.1.1) becomes $\iint_R dA$. When evaluated by an iterated integral, it becomes either

$$\iint_R dA = \int_a^b\int_{\phi_1(x)}^{\phi_2(x)} dy\,dx = \int_a^b (\phi_2(x) - \phi_1(x))\,dx$$

or

$$\iint_R dA = \int_c^d\int_{\psi_1(y)}^{\psi_2(y)} dx\,dy = \int_c^d (\psi_2(y) - \psi_1(y))\,dy.$$

In both cases, the right-hand integrals give the area of R (by 9.1.2).

Thus the **area** of R is

19.3.1
$$A = \iint_R dA.$$

Notice that when $f(x, y) = 1$ for all (x, y) in R, $\iint_R 1 \, dA$ can also be interpreted as the volume of a right cylinder with base R and altitude 1 (by 19.1.4). The number of cubic units in this volume is the same as the number of square units in the area of the base R.

Example 1. Find the area of the region R in the xy plane, between the y axis and the line $x = \pi/4$, below the curve $y = \cos x$, and above $y = \sin x$.

Solution. Since
$$R = \left\{ (x, y) \,|\, 0 \le x \le \frac{\pi}{4}, \, \sin x \le y \le \cos x \right\},$$
we can write
$$A = \iint_R dA = \int_0^{\pi/4} \int_{\sin x}^{\cos x} dy \, dx$$
$$= \int_0^{\pi/4} \left[y \right]_{\sin x}^{\cos x} dx = \int_0^{\pi/4} (\cos x - \sin x) \, dx$$
$$= \left[\sin x + \cos x \right]_0^{\pi/4} = \sqrt{2} - 1.$$

Consider a *lamina L* (a thin plate of uniform thickness) whose base is a region R in the xy plane. If the lamina is homogeneous, its **density** is defined to be its mass per unit of area of R.

In Section 9.8 we discussed the moments of a *homogeneous* lamina about the x and y axes, and its center of mass. We wish to define the moments about the coordinate axes, and also the center of mass, of a *nonhomogeneous* lamina, one whose density varies from point to point in R. First, however, we must see what is meant by the density at a point of a nonhomogeneous lamina.

Let $P:(x, y)$ be an arbitrarily chosen (but fixed) point in a region R, and let $Q:(x + \Delta x, y + \Delta y)$ be any other point of R such that the closed rectangular region, having P and Q as opposite vertices, lies entirely in R (Fig. 19-11). Indicate, the length $|PQ|$ of a diagonal of the rectangle by d, the area of the rectangle by ΔA, and the mass of the rectangular lamina based on the rectangle by Δm. Then
$$\frac{\Delta m}{\Delta A}$$
is called the average density per unit of area of the rectangular lamina. If

(1)
$$\lim_{d \to 0} \frac{\Delta m}{\Delta A}$$
exists, this limit is called the **density** of the lamina L at the point $P:(x, y)$.

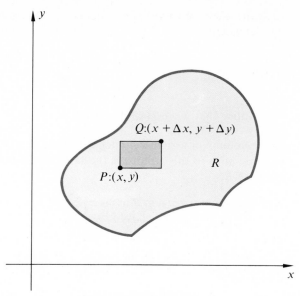

Figure 19-11

A function ρ of two variables is called a **density function** for a plane region R if

1. ρ is continuous on R;
2. $\rho(x, y) \geq 0$ for all (x, y) in R;
3. when ρ is a constant function whose value is k, the mass Δm of any rectangular lamina in R, as described above, is $k \, \Delta A$.

The total **mass** of the lamina with base R is defined to be

$$M = \iint_R \rho(x, y) \, dA,$$

where $\rho(x, y)$ is its density at the point (x, y) of R.

We shall now define the moments about the coordinate axes, and also the center of mass, of a nonhomogeneous lamina.

19.3.2 Definition. The **moment** (or **first moment**) of the mass M of the lamina with respect to the x axis is

$$M_x = \iint_R y \, \rho(x, y) \, dA;$$

the **moment** (or **first moment**) of M with respect to the y axis is

$$M_y = \iint_R x \, \rho(x, y) \, dA;$$

and the **center of mass** of the lamina, or the **centroid** of the region R, is the point $(\bar{x}, \bar{y})$, where

$$\bar{x} = \frac{M_y}{M} = \frac{\iint\limits_R x\,\rho(x, y)\,dA}{\iint\limits_R \rho(x, y)\,dA},$$

$$\bar{y} = \frac{M_x}{M} = \frac{\iint\limits_R y\,\rho(x, y)\,dA}{\iint\limits_R \rho(x, y)\,dA}.$$

Thus $M_y = M\bar{x}$ and $M_x = M\bar{y}$, which may be interpreted to mean that the first moment of the lamina about a coordinate axis is equal to the mass of the lamina multiplied by the length of the "lever arm" which is the perpendicular distance from that axis to the centroid. It follows that in discussing the first moments of a lamina, the mass of the lamina may conveniently be thought of as concentrated at the centroid.

Example 2. A lamina, whose variable density is given by $\rho(x, y) = xy$, is bounded by the x axis, the line $x = 8$, and the curve $y = x^{2/3}$ (Fig. 19-12). Find its mass M, its first moments about the coordinate axes, M_y and M_x, and its center of mass $(\bar{x}, \bar{y})$.

Solution. The region R is

$$R = \{(x, y)\,|\,0 \le x \le 8, 0 \le y \le x^{2/3}\}.$$

So

$$M = \iint\limits_R \rho(x, y)\,dA = \int_0^8 \int_0^{x^{2/3}} xy\,dy\,dx = \frac{1}{2}\int_0^8 x^{7/3}\,dx = \frac{768}{5}.$$

$$M_x = \iint\limits_R y\,\rho(x, y)\,dA = \int_0^8 \int_0^{x^{2/3}} xy^2\,dy\,dx = \frac{1}{3}\int_0^8 x^3\,dx = \frac{1024}{3}.$$

$$M_y = \iint\limits_R x\,\rho(x, y)\,dA = \int_0^8 \int_0^{x^{2/3}} x^2 y\,dy\,dx = \frac{1}{2}\int_0^8 x^{10/3}\,dx = \frac{12{,}288}{13}.$$

$$\bar{x} = \frac{M_y}{M} = 6\tfrac{2}{13}, \qquad \bar{y} = \frac{M_x}{M} = 2\tfrac{2}{9}.$$

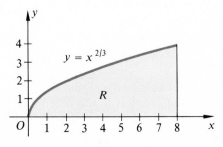

Figure 19-12

In Chapter 9 we defined the first moments of a *homogeneous* lamina about the coordinate axes. Those definitions are easily seen to be specializations of the more general definitions of the present section. Since the lamina was homogeneous, $\rho(x, y) = k$, where k is a constant, for all (x, y) in R. Thus 19.3.1 would give $M_x = \iint\limits_{R} ky \, dA$. If we express R as a type II region, then

$$M_x = \iint\limits_{R} ky \, dA = \int_c^d \int_{\psi_1(y)}^{\psi_2(y)} ky \, dx \, dy = k \int_c^d y[\psi_2(y) - \psi_1(y)] \, dy;$$

and if we express R as a type I region, then

$$M_y = \iint\limits_{R} kx \, dA = \int_a^b \int_{\phi_1(x)}^{\phi_2(x)} kx \, dy \, dx = k \int_a^b x[\phi_2(x) - \phi_1(x)] \, dx.$$

This integral for M_y is identical to that in 9.8.4(i); and the integral for M_x, in which y is the independent variable on $[c, d]$, can be obtained from 9.8.4(i) by interchanging x and y and replacing a and b by c and d, respectively.

We have seen that the kinetic energy of a particle of mass m and velocity v, moving on a *straight line*, is

(2) $$\text{KE} = \tfrac{1}{2}kmv^2,$$

where k is a constant.

If, instead of moving on a straight line, the particle rotates about an axis with an angular velocity of ω radians per second, its linear velocity is $v = r\omega$, where r is the radius of its circular path. By substituting this in (2), we obtain

$$\text{KE} = \tfrac{1}{2}km(r\omega)^2 = \tfrac{1}{2}k(r^2m)\omega^2.$$

The expression r^2m in the right-hand member is called the *moment of inertia* of the particle and is indicated by I; that is,

$$I = r^2m.$$

Thus for the rotating particle,

(3) $$\text{KE} = \tfrac{1}{2}kI\omega^2.$$

It is clear from (2) and (3) that the moment of inertia of a body in circular motion plays a part similar to that of the mass of the body in rectilinear motion.

For a system of n particles in a plane, having masses $m_1, m_2, \cdots, m_n$ and at distances $r_1, r_2, \cdots, r_n$ from a line l in their plane, the moment of inertia of the system about l is defined to be

$$I = m_1r_1^2 + m_2r_2^2 + \cdots + m_nr_n^2 = \sum_{i=1}^{n} m_ir_i^2.$$

If L is a lamina on a region R in the xy plane, it is natural to define its moments of intertia with respect to the coordinate axes as follows.

19.3.3 Definition. Let L be a lamina on a region R of the xy plane whose density function is ρ. The **moments of inertia** (or **second moments**) of L about the x and y axes are, respectively,

$$I_x = \iint\limits_R y^2\, \rho(x, y)\, dA \qquad \text{and} \qquad I_y = \iint\limits_R x^2\, \rho(x, y)\, dA.$$

Its **moment of inertia** about the z axis (or its **polar moment of inertia** about the origin O) is

$$I_0 = I_x + I_y = \iint\limits_R (x^2 + y^2)\, \rho(x, y)\, dA.$$

Example 3. Find the moments of inertia about the x and y axes of the lamina in Example 2, and also find its polar moment of inertia.

Solution. From 19.3.3

$$I_x = \iint\limits_R xy^3\, dA = \int_0^8 \int_0^{x^{2/3}} xy^3\, dy\, dx = \frac{1}{4}\int_0^8 x^{11/3}\, dx = \frac{6144}{7}.$$

$$I_y = \iint\limits_R x^3 y\, dA = \int_0^8 \int_0^{x^{2/3}} x^3 y\, dy\, dx = \frac{1}{2}\int_0^8 x^{13/3}\, dx = 6144.$$

$$I_0 = I_x + I_y = \frac{49{,}152}{7}.$$

If we compare the definitions of moments (of mass) (19.3.1) with the definitions of moments of inertia (19.3.2), we see that the former use the *first* power of the distance from the axis whereas the latter use the *second* power. For this reason, a moment (of mass) is often called a *first moment* and a moment of inertia is called a *second moment*.

Although we introduced the concepts of first and second moments by their use in mechanics, they are also important in other fields, notably statistics and probability.

The **radius of gyration** of a lamina L about an axis is the number r such that

19.3.4
$$r^2 = \frac{I}{M},$$

where I is the moment of inertia of L about the axis, and M is the mass of L.

It is often convenient when discussing the moment of inertia of a lamina to think of the entire mass of the lamina to be concentrated at a point of the lamina whose distance from the axis is equal to r, the radius of gyration.

Exercises

In each of Exercises 1 to 9, sketch the region R and find its area by double integration. (*Hint:* In an iterated integral, the outer integral has constant limits.)

1. $R = \{P{:}(x, y)\,|\,0 \leq x \leq 1,\ \frac{1}{4}x^3 \leq y \leq 2 - x^4\}$.

2. $R = \{P{:}(x, y)\,|\,-1 \leq x \leq 0,\ e^x \leq y \leq 2e^{-x}\}$.

3. $R = \{P{:}(x, y)\,|\,0 \leq x \leq 1,\ \tan x \leq y \leq 2\sqrt{x}\}$.

4. $R = \{P{:}(x, y) \mid 1 \leq x \leq 2, \ln x \leq y \leq e^x\}$.

5. $R = \left\{P{:}(x, y) \mid 0 \leq x \leq \dfrac{\pi}{3}, \sin x \leq y \leq \sec^2 x\right\}$.

6. $R = \left\{P{:}(x, y) \mid 0 \leq x \leq \dfrac{\pi}{4}, \right.$

$\left. x^2 \leq y \leq x + \cos 2x\right\}$.

7. $R = \left\{P{:}(x, y) \mid \tan y \leq x \leq 2 \cos y, \right.$

$\left. 0 \leq y \leq \dfrac{\pi}{4}\right\}$.

8. $R = \{P{:}(x, y) \mid 0 \leq x \leq 1, \sinh x \leq y \leq e^x\}$.

9. $R = \{P{:}(x, y) \mid \tfrac{1}{2}y^{1/4} \leq x \leq (1 + y^2)^{-1},$
$0 \leq y \leq \tfrac{1}{3}\sqrt{3}\,\}$.

In each of Exercises 10 to 14, a face R of a lamina and its variable density $\rho(x, y)$ are given. Find (a) the center of mass $(\bar{x}, \bar{y})$ of the lamina; (b) its moments of inertia with respect to the x and y axes, I_x and I_y; and (c) its polar moment of inertia, I_0.

10. $R = \{P{:}(x, y) \mid 0 \leq x \leq 2, 0 \leq y \leq \sqrt{4 - x^2}\}$;
$\rho(x, y) = y$.

11. $R = \{P{:}(x, y) \mid 0 \leq x \leq 4, 0 \leq y \leq 3\}$;
$\rho(x, y) = y + 1$.

12. Same R as in Exercise 10; $\rho(x, y) = x + y$.

13. $R = \{P{:}(x, y) \mid 0 \leq x \leq 1, 0 \leq y \leq e^x\}$;
$\rho(x, y) = 2 - x + y$.

14. Same R as in Exercise 13; $\rho(x, y) = xy$.

15. There is a mean value theorem for double integrals that is analogous to 10.7.2:

Mean Value Theorem for Double Integrals. If a function of two variables is continuous throughout a plane region R, then there exists a point (λ, μ) in the interior of R such that

$$\iint\limits_{R} f(x, y)\, dA = f(\lambda, \mu)\, A,$$

where A is the area of R.

Assuming the validity of this theorem, show that the value $\rho(x, y)$ of the density function ρ at each point $P{:}(x, y)$ of R is consistent with (1); that is, show that

$$\lim_{d \to 0} \frac{\Delta m}{\Delta A} = \rho(x, y).$$

19.4 GREEN'S THEOREM*

There is a remarkable theorem that identifies a double integral over a region R with a line integral around its boundary curve, enabling us to express each in terms of the other. It is named for George Green (1793–1841), a self-taught English mathematician who wrote on the mathematical theory of electricity and magnetism, fluid flow, and light and sound.

19.4.1 Green's Theorem. Let P and Q be functions of two variables that are continuous and have continuous first partial derivatives in some rectangular region H in the xy plane. If C is a simple, closed, piecewise smooth curve lying entirely in H, and if R is the bounded region enclosed by C, then

$$\oint_C P(x, y)\, dx + Q(x, y)\, dy = \iint\limits_{R} \left(\frac{\partial Q}{\partial x} - \frac{\partial P}{\partial y}\right) dA.$$

The symbol $\oint_C$ means that the line integral is to be taken once around the closed path C *in the counterclockwise direction* as indicated by the arrow.

*Optional.

Green's theorem is easy to understand and use. But a proof of the theorem for *every* bounded region having a simple, closed, piecewise smooth curve for its boundary is too difficult for a first course in calculus.

It is comparatively easy, however, to prove Green's theorem for regions R that are of both type I and type II (Fig. 19-13), as well as for regions that can be decomposed by appropriate line segments or curve segments into a finite number of subregions, each of which is of both type I and type II (Fig. 19-15).

A proof of Green's theorem for such regions follows.

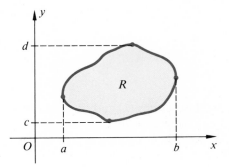

Figure 19-13

Proof. Assume that R is a region in the xy plane that is both type I and type II. Treating it as type I region [Fig. 19-14(a)], we have (from 18.10.1 and 19.2.1)

$$\oint_C P(x, y)\, dx = \int_a^b P(x, \phi_1(x))\, dx - \int_a^b P(x, \phi_2(x))\, dx$$

$$= - \int_a^b \int_{\phi_1(x)}^{\phi_2(x)} \frac{\partial}{\partial y} P(x, y)\, dy\, dx = - \iint_R \frac{\partial}{\partial y} P(x, y)\, dA,$$

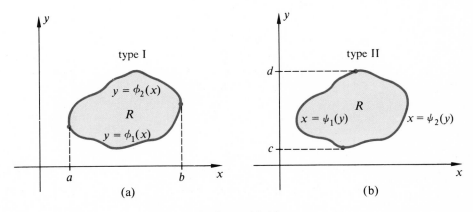

Figure 19-14

in which x is the parameter of C; that is,

$$(1) \qquad \oint_C P(x, y)\, dx = -\iint_R \frac{\partial}{\partial y} P(x, y)\, dA.$$

Similarly, by treating R as a type II region [Fig. 19-14(b)] and taking y as the parameter of C, we can write

$$\oint_C Q(x, y)\, dy = \int_c^d Q(\psi_2(y), y)\, dy - \int_c^d Q(\psi_1(y), y)\, dy$$

$$= \int_c^d \int_{\psi_1(y)}^{\psi_2(y)} \frac{\partial}{\partial x} Q(x, y)\, dx\, dy = \iint_R \frac{\partial}{\partial x} Q(x, y)\, dA,$$

or

$$(2) \qquad \oint_C Q(x, y)\, dy = \iint_R \frac{\partial}{\partial x} Q(x, y)\, dA.$$

If we add equations (1) and (2), member by member, we obtain Green's theorem for the simplified region R.

When R is a region that can be decomposed by appropriate line or curve segments into a finite number of subregions, each of which is of both type I and type II (Fig. 19-15), the simplified version of Green's theorem, which we have just proved, can be applied to each of the subregions. Since the line integral is taken around the boundary of each subregion in the counterclockwise direction, it is taken along each dotted line twice, once in one direction and once in the opposite direction, thereby canceling each other. Thus the sum of the line integrals around the boundaries of the subregions is simply the value of the line integral around C. ▮

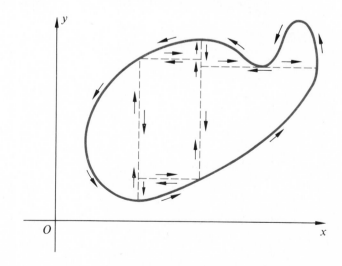

Figure 19-15

Notice that 18.10.7 is a simple corollary of Green's theorem, for if $P(x, y)\, dx + Q(x, y)\, dy$ is an exact differential, $\partial Q/\partial x = \partial P/\partial y$ and the right-hand member of the equation in Green's theorem is zero. Therefore

$$\oint_C P(x, y)\, dx + Q(x, y)\, dy = 0.$$

The following consequence of Green's theorem is often useful.

19.4.2   Theorem.   If R is a type I or type II region, or a combination of such types, then the area of R is given by

$$A = \frac{1}{2} \oint_C x\, dy - y\, dx,$$

where C is the boundary of R.

Proof.   If we let $P(x, y) = -y/2$ and $Q(x, y) = x/2$, then $\partial P/\partial y = -\frac{1}{2}$ and $\partial Q/\partial x = \frac{1}{2}$. By substituting these results in Green's theorem, we obtain

$$\oint_C \left(-\frac{y}{2}\, dx + \frac{x}{2}\, dy \right) = \iint_R \left(\frac{1}{2} + \frac{1}{2} \right) dA = \iint_R dA.$$

But $\iint_R dA$ is the area A of R. Therefore

$$\frac{1}{2} \oint_C x\, dy - y\, dx = A. \quad \blacksquare$$

Example 1.   Use 19.4.2 to find the area of the region enclosed by the ellipse $b^2x^2 + a^2y^2 = a^2b^2$.

Solution.   From 19.4.2 the area of the region enclosed by the given ellipse is

$$A = \frac{1}{2} \oint_C x\, dy - y\, dx,$$

where C is the ellipse.

To evaluate this line integral, we use parametric equations of the elipse, $x = a \cos t$, $y = b \sin t$, $0 \le t \le 2\pi$. From them, $dx = -a \sin t\, dt$ and $dy = b \cos t\, dt$. Then

$$A = \frac{1}{2} \oint_C x\, dy - y\, dx,$$

$$= \frac{1}{2} \int_0^{2\pi} a \cos t\, (b \cos t\, dt) - b \sin t\, (-a \sin t\, dt)$$

$$= \frac{1}{2} \int_0^{2\pi} ab\, (\cos^2 t + \sin^2 t)\, dt$$

$$= \frac{1}{2} ab \int_0^{2\pi} dt = \frac{1}{2} ab \left[\, t\, \right]_0^{2\pi} = \frac{1}{2} ab(2\pi) = \pi ab.$$

Example 2.   Find the area of the region R bounded by the curves $y = x^3$ and $y = x^{1/2}$ (Fig. 19-16).

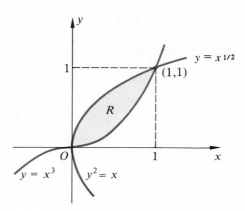

Figure 19-16

Solution. Let C_1 be the segment of the curve $y = x^3$ from $(0, 0)$ to $(1, 1)$, and let C_2 be the segment of the curve $y = x^{1/2}$ from $(1, 1)$ to $(0, 0)$. Then $C = C_1 + C_2$. By taking x for the parameter of C and using 19.4.2, we can write

$$A = \frac{1}{2} \int_{C_1} (x \, dy - y \, dx) + \frac{1}{2} \int_{C_2} (x \, dy - y \, dx)$$

$$= \frac{1}{2} \int_0^1 (3x^3 \, dx - x^3 \, dx) + \frac{1}{2} \int_1^0 \left(\frac{x}{2x^{1/2}} \, dx - x^{1/2} \, dx \right)$$

$$= \int_0^1 x^3 \, dx - \frac{1}{4} \int_1^0 x^{1/2} \, dx = \frac{5}{12} \text{ square unit.}$$

Example 3. Use Green's theorem to evaluate the line integral

$$\oint_C (x^3 + 2y) \, dx + (4x - 3y^2) \, dy,$$

where C is the ellipse $b^2x^2 + a^2y^2 = a^2b^2$.

Solution. Let $P(x, y) = x^3 + 2y$ and $Q(x, y) = 4x - 3y^2$. Then $\partial Q/\partial x = 4$ and $\partial P/\partial y = 2$. Applying Green's theorem, we have

$$\oint_C (x^3 + 2y) \, dx + (4x - 3y^2) \, dy = \iint_R (4 - 2) \, dA = 2 \iint_R dA.$$

But (by 19.3.1) $\iint_R dA$ is the area of the ellipse, which is πab, as we saw in Example 1 above. Therefore

$$\oint_C (x^3 + 2y) \, dx + (4x - 3y^2) \, dy = 2\pi ab.$$

Exercises

In Exercises 1 to 6, use Green's theorem to evaluate the given line integrals. Sketch the region R.

1. $\oint_C 2xy \, dx + y^2 \, dy$, where C is the closed curve formed by $y = x/2$ and $y = \sqrt{x}$ between $(0, 0)$ and $(4, 2)$.

2. $\oint_C \sqrt{y} \, dx + \sqrt{x} \, dy$, where C is the closed curve formed by $y = 0$, $x = 2$, and $y = x^2/2$.

3. $\oint_C (2x + y^2) \, dx + (x^2 + 2y) \, dy$, where C is the closed curve formed by $y = 0$, $x = 2$, and $y = x^3/4$.

4. $\oint_C xy \, dx + (x + y) \, dy$, where C is the triangle with vertices $(0, 0)$, $(2, 0)$, and $(0, 1)$.

5. $\oint_C (x^2 + 4xy) \, dx + (2x^2 + 3y) \, dy$, where C is the ellipse $9x^2 + 16y^2 = 144$.

6. $\oint_C (e^{3x} + 2y) \, dx + (x^2 + \sin y) \, dy$, where C is the rectangle with vertices $(2, 1)$, $(6, 1)$, $(6, 4)$, and $(2, 4)$.

7. Use 19.4.2 to find the area of the region R that is enclosed by the curve C, if C has parametric equations $x = \cos t$, $y = \sin t$, $0 \le t \le 2\pi$. Sketch the region R.

8. Use 19.4.2 to find the area of the region R enclosed by the curve $C = C_1 + C_2$, where C_1 is the segment of the curve $x = 2t$, $y = t^2 - 1$, from $t = -1$ to $t = 2$, and C_2 is the segment of the line $x = 2t - 2$, $y = t$, from $t = 3$ to $t = 0$. Sketch the region R.

9. Use 19.4.2 to find the area of the closed region R that is bounded by the curves $y = 4x$ and $y = 2x^2$. Make a sketch. (*Hint:* $C = C_1 + C_2$, where C_1 is $y = 2x^2$ from $x = 0$ to $x = 2$ and C_2 is $y = 4x$ from $x = 2$ to $x = 0$).

10. Use 19.4.2 to find the area of the closed region bounded by the curves $y = \frac{1}{2}x^3$ and $y = x^2$. Make a sketch.

11. Let P and Q be functions of two variables that are continuous and have continuous first partial derivatives in some open rectangular region H in the xy plane, and let A and B be two arbitrarily selected points in H. Assuming that C is a piecewise smooth curve having A and B as endpoints and lying entirely in H, prove (without using 18.10.3 or 18.10.4) that a necessary and sufficient condition for the value of the line integral from A to B,

$$\int_C P(x, y) \, dx + Q(x, y) \, dy,$$

to be independent of the choice of the path C is that

$$\frac{\partial P}{\partial y} = \frac{\partial Q}{\partial x}$$

at every point (x, y) in H. (*Hint:* Denote by C_1 and C_2 any two arbitrarily selected curves that fulfill the restrictions on C, and use Green's theorem.)

12. Find the work done by the force field

$$\mathbf{F}(x, y) = (2x - y)\mathbf{i} + (3x + 7y)\mathbf{j}$$

on a particle that moves once around the ellipse $4x^2 + 9y^2 = 36$ in the counterclockwise direction. (*Hint:* Use Green's theorem and 19.3.1.)

13. Show that the equation in Green's theorem can be written in the form

$$\oint_C \mathbf{F} \cdot d\mathbf{r} = \iint_R (\text{curl } \mathbf{F}) \cdot \mathbf{k} \, dA,$$

where $\mathbf{F}(x, y, z) = P(x, y)\mathbf{i} + Q(x, y)\mathbf{j} + 0\mathbf{k}$, $\mathbf{r}(t) = f(t)\mathbf{i} + g(t)\mathbf{j} + 0\mathbf{k}$ is the position vector of C and $d\mathbf{r} = dx \, \mathbf{i} + dy \, \mathbf{j} + 0\mathbf{k}$. (*Hint:* See Section 18.12.)

14. Show that if α is the angle of inclination of the tangent line to C at an arbitrary point of C, then

$$\int_C P(x, y) \, dx + Q(x, y) \, dy$$

$$= \int [P(x, y) \cos \alpha + Q(x, y) \sin \alpha] \, ds,$$

where s is the arc length along C. Also show that if $\mathbf{V}(x, y) = Q(x, y)\mathbf{i} - P(x, y)\mathbf{j}$ defines a vector field and $\mathbf{N}(x, y) = \cos (\alpha - \frac{1}{2}\pi)\mathbf{i} + \sin (\alpha - \frac{1}{2}\pi)\mathbf{j}$ is the outer unit normal vector at an arbitrary point on C,

then

$$\int_C P(x, y)\, dx + Q(x, y)\, dy = \int V(x, y) \cdot N(x, y)\, ds.$$

15. Show that the equation in Green's theorem can be expressed in the form

$$\oint_C V(x, y) \cdot N(x, y)\, ds = \iint_C \operatorname{div} V(x, y)\, dA,$$

where $V(x, y) = Q(x, y)\mathbf{i} - P(x, y)\mathbf{j}$ defines a vector field and $N(x, y)$ is the outer unit normal vector at any point on C. (*Hint:* See Exercise 14.)

19.5 DOUBLE INTEGRALS IN POLAR COORDINATES

By a **polar rectangle** is meant a region in the polar coordinate plane bounded by two concentric circles about the pole and two rays emanating from the pole (Fig. 19-17). If the radii of the two circles are r and $r + \Delta r$, and if the angle between the two rays is $\Delta\theta$, with $\Delta r > 0$ and $0 < \Delta\theta \le 2\pi$, then the area of the polar rectangle is equal to the area of the sector OBC minus the area of the sector OAD (Fig. 19-17). Since the area of a sector of a circle is one-half the product of the square of the radius and the central angle, the area of this polar rectangle is

$$\tfrac{1}{2}(r + \Delta r)^2 \Delta\theta - \tfrac{1}{2}r^2 \Delta\theta = (r + \tfrac{1}{2}\Delta r)\, \Delta r\, \Delta\theta = \bar{r}\, \Delta r\, \Delta\theta,$$

where $\bar{r} = r + \tfrac{1}{2}\Delta r$ is the number midway between r and $r + \Delta r$. Thus

(1)   Area of polar rectangle $= \bar{r}\, \Delta r\, \Delta\theta.$

A polar rectangle

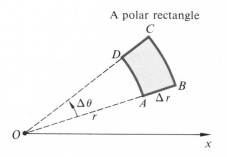

Figure 19-17

Denote by T_I a **polar region** bounded by rays $\theta = \alpha$ and $\theta = \beta$, with $\alpha < \beta$, and by segments of plane curves $r = \Phi_1(\theta)$ and $r = \Phi_2(\theta)$, where Φ_1 and Φ_2 are functions that are continuous and have continuous first partial derivatives on $[\alpha, \beta]$, with $\Phi_1(\theta) \le \Phi_2(\theta)$ for $\alpha \le \theta \le \beta$ (Fig. 19-18). Let F be a function of r and θ that is continuous inside and on the boundaries of T_I.

Cover T_I with a **polar net** consisting of arcs of circles with centers at the pole and rays emanating from the pole (Fig. 19-18). This polar net forms a set of polar rectangles. Those polar rectangles that lie entirely inside or on the boundary of T_I form a **polar partition** of T_I; number them in any systematic way from 1 to n. The **norm**, μ,

A polar partition

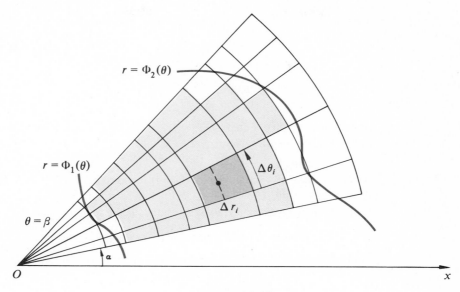

Figure 19-18

of a polar partition is the length of the longest diagonal of any polar rectangle in the partition.

The area of the ith polar rectangle in the partition is $\Delta A_i = \bar{r}_i \, \Delta r_i \, \Delta \theta_i$, where $\bar{r}_i = r_i + \frac{1}{2}\Delta r_i$ [by (1)]. Form the sum

$$(2) \qquad \sum_{i=1}^{n} F(\bar{r}_i, \bar{\theta}_i)\, \bar{r}_i \, \Delta r_i \, \Delta \theta_i$$

in which $\bar{r}_i = r_i + \frac{1}{2}\Delta r_i$ and $\bar{\theta}_i = \theta_i + \frac{1}{2}\Delta \theta_i$ are the polar coordinates of the "center" of the ith polar rectangle.

It can be shown (although we do not do so here) that the limit of the sum (2) exists, as the number of polar rectangles increases indefinitely and the norm, μ, of the polar partition of T_{I} approaches zero. This limit is written

$$\lim_{\mu \to 0} \sum_{i=1}^{n} F(\bar{r}_i, \bar{\theta}_i)\, \bar{r}_i \, \Delta r_i \, \Delta \theta_i.$$

19.5.1 Definition. The **double integral** of F on the polar region T_{I} is

$$\iint\limits_{T_{\mathrm{I}}} F(r, \theta)\, r \, dr \, d\theta = \lim_{\mu \to 0} \sum_{i=1}^{n} F(\bar{r}_i, \bar{\theta}_i)\, \bar{r}_i \, \Delta r_i \, \Delta \theta_i.$$

It can also be shown that this double integral can be evaluated by an iterated integral.

19.5.2   Theorem.   The value of the double integral defined in 19.5.1 is given by

$$\iint\limits_{T_I} F(r, \theta)\, r\, dr\, d\theta = \int_\alpha^\beta \int_{\Phi_1(\theta)}^{\Phi_2(\theta)} F(r, \theta)\, r\, dr\, d\theta,$$

where

$$\int_\alpha^\beta \int_{\Phi_1(\theta)}^{\Phi_2(\theta)} F(r, \theta)\, r\, dr\, d\theta = \int_\alpha^\beta \left(\int_{\Phi_1(\theta)}^{\Phi_2(\theta)} F(r, \theta)\, r\, dr \right) d\theta$$

means that we first integrate $F(r, \theta)\, r$ with respect to r while holding θ fixed and then integrate the result with respect to θ.

Example 1.   Find the mass of a lamina whose face is the cardioid $r = a(1 + \cos \theta)$ and whose density is proportional to the distance from the pole (Fig. 19-19).

Solution.   The density is given by $\rho(r, \theta) = kr$ and R_* is the region bounded by the cardioid. Thus

$$\text{Mass} = \iint\limits_{R_*} \rho(r, \theta)\, dA_* = \iint\limits_{R_*} kr^2\, dr\, d\theta = 2k \int_0^\pi \int_0^{a(1+\cos\theta)} r^2\, dr\, d\theta$$

$$= \frac{2ka^3}{3} \int_0^\pi (1 + \cos \theta)^3\, d\theta = \frac{5\pi ka^3}{3}.$$

Although the region of integration in the polar plane need not be as restricted as above, it is assumed here that the region is either of the type T_I just discussed or of one other type that will now be described.

The **second type** of polar region, T_{II}, is bounded by the arcs of two circles with their centers at the pole and of radii γ and δ, with $\gamma < \delta$, and by segments of two curves, $\theta = \Psi_1(r)$ and $\theta = \Psi_2(r)$, where Ψ_1 and Ψ_2 are functions that are continuous and have continuous derivatives in $[\gamma, \delta]$, with $\Psi_1(r) \leq \Psi_2(r)$ for $\gamma \leq r \leq \delta$ (Fig. 19-20).

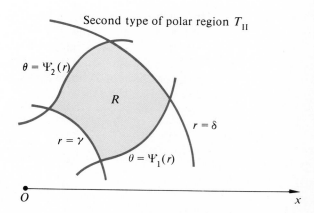

Second type of polar region T_{II}

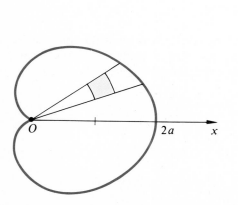

Figure 19-19   Figure 19-20

We state, without proof, the following theorem.

19.5.3 Theorem.   If F is a function of r and θ that is continuous inside and on the boundaries of this second type of polar region T_{II}, the double integral

$$\iint\limits_{T_{\text{II}}} F(r, \theta)\, r\, d\theta\, dr$$

exists and may be evaluated by an iterated integral:

$$\iint\limits_{T_{\text{II}}} F(r, \theta)\, r\, d\theta\, dr = \int_{\gamma}^{\delta} \int_{\Psi_1(r)}^{\Psi_2(r)} F(r, \theta)\, r\, d\theta\, dr,$$

where

$$\int_{\gamma}^{\delta} \int_{\Psi_1(r)}^{\Psi_2(r)} F(r, \theta)\, r\, d\theta\, dr = \int_{\gamma}^{\delta} \left(\int_{\Psi_1(r)}^{\Psi_2(r)} F(r, \theta)\, r\, d\theta \right) dr$$

means that $F(r, \theta)\, r$ is to be integrated first with respect to θ, holding r constant, and then this result is to be integrated with respect to r.

Notice that the first type of polar region T_{I}, (Fig. 19-18) is *formally* the same as a type I region in the xy plane (Fig. 19-6). This is easy to see if the polar region [Fig. 19-21(a)] is sketched on an $r\theta$ rectangular system of coordinates [Fig. 19-21(b)].

Similarly, the second type of polar region T_{II} is formally the same as a type II region in the xy plane (Fig. 19-6).

When a double integral is given in one system of coordinates, it is sometimes easier to evaluate it by changing to the other system of coordinates. The equations of transformation are

$$x = r \cos \theta, \qquad y = r \sin \theta, \qquad r > 0, \qquad -\frac{\pi}{2} < \theta < \frac{\pi}{2}.$$

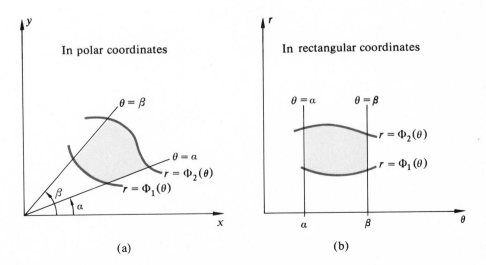

(a)   (b)

Figure 19-21

This procedure is justified by the following theorem, which is stated without proof.

19.5.4   Theorem.   Let R be a type I or a type II region in the xy plane and let T be the polar region that is the image of R in the transformation

$$x = r \cos \theta, \qquad y = r \sin \theta, \qquad r > 0, \qquad -\frac{\pi}{2} < \theta < \frac{\pi}{2}.$$

If f is a function of two variables that is continuous on R, then

$$\iint_R f(x, y)\, dA = \iint_T f(r \cos \theta, r \sin \theta)\, r\, dr\, d\theta$$

$$= \iint_T F(r, \theta)\, r\, dr\, d\theta,$$

in which F is continuous on T.

Example 2.   Evaluate the double integral

$$\iint_R \frac{1}{x^2 + y^2}\, dA,$$

where R is the region in the first quadrant bounded by the circles $x^2 + y^2 = 1$ and $x^2 + y^2 = 4$ (Fig. 19-22).

Solution.   This double integral is easier to evaluate in polar coordinates where the region T_{I} is bounded by the circles $r = 1$ and $r = 2$ and the rays $\theta = 0$ and $\theta = \frac{1}{2}\pi$. Since $x^2 + y^2 = r^2$, we can write

$$\iint_R \frac{1}{x^2 + y^2}\, dA = \iint_{T_{\mathrm{I}}} \frac{1}{r^2}\, r\, dr\, d\theta = \int_0^{\pi/2} \int_1^2 \frac{1}{r^2}\, r\, dr\, d\theta$$

$$= \int_0^{\pi/2} \int_1^2 \frac{1}{r}\, dr\, d\theta = \ln 2 \int_0^{\pi/2} d\theta = \frac{1}{2}\pi \ln 2 = \pi \ln \sqrt{2}.$$

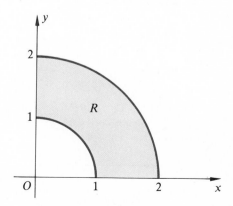

Figure 19-22

Exercises

Use iterated integration in each of the following exercises. Be sure to make a sketch first.

1. Find the area of the region inside the circle $r = 4 \cos \theta$ and outside the circle $r = 2$.

2. Find the area of the smaller region bounded by $\theta = \pi/6$ and $r = 4 \sin \theta$.

3. Find the area of one leaf of the four-leaved rose $r = a \sin 2\theta$.

4. Find the area of the region bounded by the cardioid $r = 6 - 6 \sin \theta$.

5. Find the area of the region between the concentric circles $r = a$ and $r = b$ $(0 < a < b)$.

6. Find the area of the region inside the circle $r = \sin \theta$ and outside the cardioid $r = 1 + \cos \theta$. (Compare your solution, using an iterated integral, with Example 2, Section 12.16.)

7. Find the area of the region inside the four-leaved rose $r = a \cos 2\theta$.

8. Find the area of the region enclosed by one leaf of the three-leaved rose $r = \sin 3\theta$.

9. Find the area of the region that is inside the larger loop of the limaçon $r = a - b \sin \theta$, where $0 < a < b$.

10. Find the area of the region that is outside the circle $r = b$ and inside the lemniscate $r^2 = a^2 \cos 2\theta$, $(0 < b < a)$.

In Exercises 11 and 12, change to polar coordinates to evaluate the given double integrals.

11. $\iint\limits_{R} (1/e^{x^2+y^2}) \, dA$, where R is the region in the first quadrant bounded by $y = \sqrt{9 - x^2}$ and the coordinate axes.

12. $\iint\limits_{R} (x^2 + y^2)^{-1/2} \, dA$, where R is the region bounded by the lines $y = \sqrt{3}\, x$, $x = 3$, and $y = 0$.

If rectangular coordinate axes are superposed on a polar system so that their origin is at the pole and the positive x axis is along the polar axis, it follows from 19.5.4 that the formulas 19.3.2 and 19.3.3 for the first and second moments of a lamina with respect to the x and y axes become

$$M_x = \iint\limits_{T} \sigma(r, \theta) \, r^2 \sin \theta \, dr \, d\theta,$$

$$M_y = \iint\limits_{T} \sigma(r, \theta) \, r^2 \cos \theta \, dr \, d\theta,$$

$$I_x = \iint\limits_{T} \sigma(r, \theta) \, r^3 \sin^2 \theta \, dr \, d\theta,$$

$$I_y = \iint\limits_{T} \sigma(r, \theta) \, r^3 \cos^2 \theta \, dr \, d\theta,$$

in polar coordinates, where $\sigma(r, \theta) = \rho(r \cos \theta, r \sin \theta)$ is the density of the lamina at any point (r, θ) of it.

In Exercises 13 to 16, find the center of mass of the lamina whose face is the indicated region, if the density $\sigma(r, \theta)$ of the lamina at any point on it is proportional to the distance of that point from the pole.

13. The region inside the circle $r = 2 \sin \theta$.

14. The region in the first quadrant bounded by the curve $r = a \sin 2\theta$.

15. The sector of the circle $r = 5$ from $\theta = -\pi/6$ to $\theta = \pi/6$.

16. The region inside the cardioid $r = a(1 + \cos \theta)$.

In Exercises 17 to 20, find the moments of inertia, I_x and I_y, of the lamina whose face is the given region, if the density is $\sigma(r, \theta) = kr$, where k is a factor of proportionality.

17. The region bounded by the cardioid $r = a(1 + \cos \theta)$.

18. The region given in Exercise 14.

19. The region between the concentric circles $r = a$ and $r = b$, $(0 < a < b)$.

20. The region given in Exercise 6.

19.6 SURFACE AREA

Let F be a function of three variables having continuous first partial derivatives F_x, F_y, and F_z, with $F_z \neq 0$. Consider the surface $F(x, y, z) = 0$, and denote by G a part of this surface that projects onto a closed region R in the xy plane (Fig. 19-23). We wish to define what is meant by the *area* of G and find a formula for calculating it.

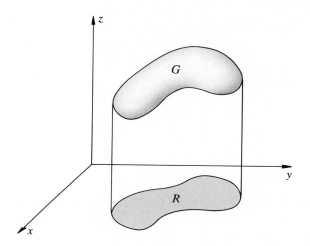

Figure 19-23

Form a partition p of R into n rectangles R_i, $i = 1, 2, 3, \cdots, n$, by lines parallel to the x and y axes, as shown in Fig. 19-3. A typical rectangle R_i of the partition p has vertices $(x_i, y_i, 0)$, $(x_i + \Delta x_i, y_i, 0)$, $(x_i, y_i + \Delta y_i, 0)$ and $(x_i + \Delta x_i, y_i + \Delta y_i, 0)$, $\Delta x_i > 0$ and $\Delta y_i > 0$. Its area is $\Delta A_i = \Delta x_i \, \Delta y_i$.

Let G_i be the part of the surface G that projects onto R_i and indicate by $P_i{:}(x_i, y_i, z_i)$ the point of G_i that is directly over the vertex $(x_i, y_i, 0)$ of R_i [Fig. 19-24(a)].

The surface has a tangent plane at P_i. Denote by T_i the parallelogram in this tangent plane that projects onto R_i [Fig. 19-24(b)] and designate the area of T_i by ΔT_i.

If the norm $|p|$ of the partition of R is small, the set of tangent parallelograms T_i will approximately conform to the surface G, and the smaller $|p|$ is taken, the better the conformation. This suggests the following definition of the area of G.

The **area** of the surface G is defined to be

(1)
$$\lim_{|p| \to 0} \sum_{i=1}^{n} \Delta T_i.$$

We shall now find a formula for the area of G. The gradient vector of G at P_i is $\mathbf{V}F(P_i)$; it is normal to the surface at P_i and thus normal to the tangent parallelogram T_i. Denote by γ_i the angle between this gradient vector and the vertical unit

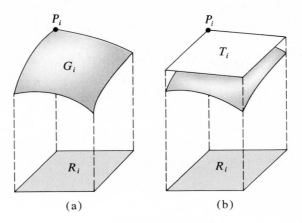

(a)   (b)

Figure 19-24

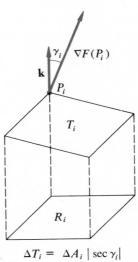

$$\Delta T_i = \Delta A_i \, |\sec \gamma_i|$$

Figure 19-25

vector $\mathbf{k}$ (Fig. 19-25). It is left for the reader (Exercise 10) to show that the area, ΔT_i, of T_i is

$$\Delta T_i = \Delta A_i |\sec \gamma_i|.$$

By substituting this in equation (1), we obtain

(2) $$\text{Area of } G = \lim_{|p| \to 0} \sum_{i=1}^{n} |\sec \gamma_i| \, \Delta A_i.$$

By 16.2.6,

$$|\cos \gamma_i| = \frac{|\boldsymbol{\nabla} F(P_i) \cdot \mathbf{k}|}{|\boldsymbol{\nabla} F(P_i)|},$$

from which

$$|\sec \gamma_i| = \frac{|\nabla F(P_i)|}{|\nabla F(P_i) \cdot \mathbf{k}|}.$$

Since $|\nabla F(P_i) \cdot \mathbf{k}| = |[F_x(P_i)\mathbf{i} + F_y(P_i)\mathbf{j} + F_z(P_i)\mathbf{k}] \cdot \mathbf{k}| = |F_z(P_i)|$, we have

$$|\sec \gamma_i| = \frac{\sqrt{[F_x(P_i)]^2 + [F_y(P_i)]^2 + [F_z(P_i)]^2}}{|F_z(P_i)|}.$$

From this result and from (2),

$$\text{Area of } G = \lim_{|p| \to 0} \sum_{i=1}^{n} \frac{\sqrt{[F_x(P_i)]^2 + [F_y(P_i)]^2 + [F_z(P_i)]^2}}{|F_z(P_i)|} \Delta A_i.$$

Therefore (by 19.1.1)

19.6.1
$$\text{Area of } G = \iint_R \frac{\sqrt{F_x^2 + F_y^2 + F_z^2}}{|F_z|} \, dA.$$

This double integral exists because F_x, F_y, and F_z are continuous and $F_z \neq 0$.

When the equation of the surface is in the form $z = f(x, y)$, where f_x and f_y are continuous, let $F(x, y, z) = f(x, y) - z$. Then $F_x = f_x$, $F_y = f_y$, and $F_z = -1$. By substituting this in 19.6.1, we get

19.6.2
$$\text{Area of } G = \iint_R \sqrt{f_x^2 + f_y^2 + 1} \, dA.$$

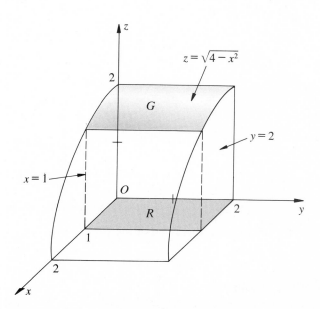

Figure 19-26

Example 1. If R is the rectangular region in the xy plane that is bounded by the lines $x = 0$, $x = 1$, $y = 0$, and $y = 2$, find the area of the part of the semicylindrical surface $z = \sqrt{4 - x^2}$ that projects onto R (Fig. 19-26).

Solution. Let $f(x, y) = \sqrt{4 - x^2}$. Then $f_x = -x/\sqrt{4 - x^2}$ and $f_y = 0$. From 19.6.2, the area is

$$\iint_R \sqrt{f_x^2 + f_y^2 + 1} \, dA = \iint_R \sqrt{\frac{x^2}{4 - x^2} + 1} \, dA$$

$$= \iint_R \frac{2}{(4 - x^2)^{1/2}} \, dA = \int_0^1 \int_0^2 \frac{2}{(4 - x^2)^{1/2}} \, dx \, dy$$

$$= \int_0^1 2 \left[\text{Sin}^{-1} \frac{x}{2} \right]_0^2 \, dy = \int_0^1 2 \left(\frac{\pi}{2} \right) \, dy = \pi.$$

Example 2. Find the area of the part of the surface $z = x^2 + y^2$ in the first octant that is cut off by the plane $z = 9$.

Solution. The designated part of the surface projects onto the quarter disk R in the xy plane, $0 \leq x \leq \sqrt{9 - y^2}$, $0 \leq y \leq 3$.

Let $f(x, y) = x^2 + y^2$. Then $f_x = 2x$ and $f_y = 2y$; and it follows from 19.6.2 that the wanted area is

$$\iint_R \sqrt{4x^2 + 4y^2 + 1} \, dA = \int_0^3 \int_0^{\sqrt{9 - y^2}} \sqrt{4(x^2 + y^2) + 1} \, dx \, dy.$$

Because of the presence of $x^2 + y^2$ in the integrand, the integration will be easier if we change to polar coordinates. Then $x^2 + y^2 = r^2$ and $dx \, dy = r \, dr \, d\theta$. The limits for r are 0 and 3, and the limits of θ are 0 to $\pi/2$. So

$$\int_0^3 \int_0^{\sqrt{9 - y^2}} \sqrt{4(x^2 + y^2) + 1} \, dx \, dy$$

$$= \int_0^{\pi/2} \int_0^3 \sqrt{4r^2 + 1} \, r \, dr \, d\theta$$

$$= \int_0^{\pi/2} \left(\frac{1}{8} \int_0^3 (4r^2 + 1)^{1/2} \, d(4r^2 + 1) \right) d\theta$$

$$= \int_0^{\pi/2} \frac{1}{8} \left[\frac{2}{3} (4r^2 + 1)^{3/2} \right]_0^3 \, d\theta$$

$$= \frac{1}{12} (37^{3/2} - 1) \int_0^{\pi/2} d\theta = \frac{(37\sqrt{37} - 1)\pi}{24}.$$

Exercises

Make a sketch for each exercise.

1. Find the area of the part of the plane $3x + 4y + 6z = 12$ that is directly above the rectangle whose vertices are $(0, 0, 0)$, $(2, 0, 0)$, $(2, 1, 0)$, and $(0, 1, 0)$.

2. Find the area of the part of the plane $3x - 2y + 6z - 12 = 0$ that is intercepted by the planes $x = 0$, $y = 0$, and $3x + 2y - 12 = 0$.

3. Find the area of the part of the surface $z =$

$\sqrt{4 - y^2}$ that is directly above the square having vertices $(1, 0, 0)$, $(2, 0, 0)$, $(2, 1, 0)$, and $(1, 1, 0)$.

4. Find the area of the part of the surface $z = \sqrt{4 - y^2}$ in the first octant that is directly above the circle $x^2 + y^2 = 4$, $z = 0$.

5. Find the area of the part of the cylinder $x^2 + z^2 = 9$ that is directly over the rectangle whose vertices are $(0, 0, 0)$, $(2, 0, 0)$, $(2, 3, 0)$, and $(0, 3, 0)$.

6. Find area of the part of the paraboloid $z = x^2 + y^2$ that is cut off by the plane $z = 4$.

7. Find the area of the part of the conical surface $x^2 + y^2 = z^2$ that is directly over the triangle whose vertices are $(0, 0, 0)$, $(4, 0, 0)$, and $(0, 4, 0)$.

8. Find the area of the surface that is cut from the cylinder $z = x^2/4 + 4$ by the planes $x = 0$, $x = 1$, $y = 0$, and $y = 2$.

9. Find the area of the part of the sphere $x^2 + y^2 + z^2 = 25$ in the first octant that is inside the cylinder $x^2 + y^2 = 16$.

10. Show that the area, ΔT_i, of T_i is given by

$$\Delta T_i = \Delta A_i |\sec \gamma_i|,$$

where ΔA_i is the area of R_i and γ_i is the angle between normals to T_i and R_i. [*Hint:* Denote by **a** and **b** vectors that have their initial points at P_i and coincide with two adjacent sides of the parallelogram T_i; and let **c** and **d**, with initial point $(x_i, y_i, 0)$, coincide with two adjacent sides of the rectangle R_i. Then $|\mathbf{a} \times \mathbf{b}| = \Delta T_i$ and $|\mathbf{c} \times \mathbf{d}| = \Delta A_i$. Since the projection of T_i onto the xy plane is R_i, the area of the projection of T_i onto the xy plane is ΔA_i. But the scalar projection of $\mathbf{a} \times \mathbf{b}$ onto $\mathbf{c} \times \mathbf{d}$ is $|\mathbf{a} \times \mathbf{b}| \cos \gamma_i = \Delta T_i \cos \gamma_i$. Therefore $\Delta T_i = \Delta A_i |\sec \gamma_i|.$]

19.7 TRIPLE INTEGRALS

The extension of the Riemann double integral to triple integrals, quadruple integrals, and the like, should be fairly clear by analogy.

Let R be a type I region of the xy plane and let F_1 and F_2 be functions that are defined and continuous on R, with $F_1(x, y) \leq F_2(x, y)$ there. Let S be the closed three-dimensional region bounded by the planes $x = a$ and $x = b$, the cylinders $y = \phi_1(x)$ and $y = \phi_2(x)$, and the surfaces $z = F_1(x, y)$ and $z = F_2(x, y)$ (Fig. 19-27).

By means of a number of planes parallel to the coordinate planes, construct a set of rectangular parallelepipeds so as to contain the region S completely. Those

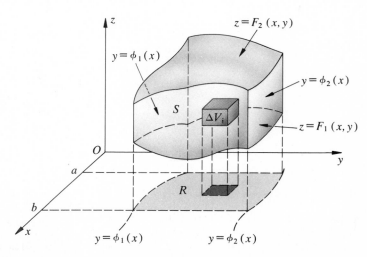

Figure 19-27

parallelepipeds that are entirely inside S or on the boundary of S form a **partition** p of S; number them from 1 to n according to some system.

The *norm* $|p|$ of this partition of S is the length of the longest diagonal of any parallelepiped belonging to this partition.

The volume of the ith partitioning parallelepiped is ΔV_i (Fig. 19-27), $i = 1$, $2, \cdots, n$.

Let f be a function of three variables that is continuous inside and on the boundaries of S, and denote by (x_i, y_i, z_i) an arbitrary point in the ith partitioning parallelpiped. Form the sum

$$(1) \qquad \sum_{i=1}^{n} f(x_i, y_i, z_i)\, \Delta V_i.$$

It can be proved that the sum (1) has a limit as the number of partitioning parallelepipeds increases indefinitely and $|p| \longrightarrow 0$, and that this limit is independent of the choice of the partitioning planes and the choice of the arbitrary point (x_i, y_i, z_i) in each parallelepiped. This limit is called the **triple integral** of f on S and is denoted by

$$\iiint_S f(x, y, z)\, dV.$$

19.7.1 **Definition of Triple Integral.** Let R be a type I region in the xy plane and let F_1 and F_2 be functions of two variables that are continuous on R, with $F_1(x, y) \le F_2(x, y)$. Denote by S the closed three-dimensional region bounded by the planes $x = a$ and $x = b$, the cylinders $y = \phi_1(x)$ and $y = \phi_2(x)$, and the surfaces $z = F_1(x, y)$ and $z = F_2(x, y)$. If f is a function of three variables that is continuous inside and on the boundaries of S, then the **triple integral**

$$\iiint_S f(x, y, z)\, dV$$

is defined by

$$\iiint_S f(x, y, z)\, dV = \lim_{|p| \to 0} \sum_{i=1}^{n} f(x_i, y_i, z_i)\, \Delta V_i.$$

It can also be shown that this triple integral can be evaluated by an iterated integral.

19.7.2 **Evaluation of Triple Integrals.** The triple integral, defined above can be evaluated by

$$\iiint_S f(x, y, z)\, dV = \int_a^b \int_{\phi_1(x)}^{\phi_2(x)} \int_{F_1(x, y)}^{F_2(x, y)} f(x, y, z)\, dz\, dy\, dx,$$

where the iterated integral

$$\int_a^b \int_{\phi_1(x)}^{\phi_2(x)} \int_{F_1(x, y)}^{F_2(x, y)} f(x, y, z)\, dz\, dy\, zx = \int_a^b \left[\int_{\phi_1(z)}^{\phi_2(z)} \left\{ \int_{F_1(x, y)}^{F_2(x, y)} f(x, y, z)\, dz \right\} dy \right] dx$$

means that $f(x, y, z)$ is first integrated with respect to z while x and y remain constant, then this result is integrated with respect to y while x is constant, and, finally, the latter result is integrated with respect to x. Of course, *after each integration the indicated limits are substituted before the next integration.*

Example 1. Evaluate the iterated integral

$$\int_{-2}^{5}\int_{0}^{3x}\int_{y}^{x+2} 4\, dz\, dy\, dx.$$

Solution.

$$\int_{-2}^{5}\int_{0}^{3x}\int_{y}^{x+2} 4\, dz\, dy\, dx = \int_{-2}^{5}\int_{0}^{3x}\left(\int_{y}^{x+2} 4\, dz\right) dy\, dx$$

$$= \int_{-2}^{5}\int_{0}^{3x}\left[4z\right]_{y}^{x+2} dy\, dx$$

$$= \int_{-2}^{5}\int_{0}^{3x}(4x - 4y + 8)\, dy\, dx$$

$$= \int_{-2}^{5}\left[4xy - 2y^2 + 8y\right]_{0}^{3x} dx$$

$$= \int_{-2}^{5}(-6x^2 + 24x)\, dx = -14.$$

In setting up a triple iterated integral, remember that the innermost integral has limits that are functions of the two remaining variables and this integration is done first; the middle integral has limits that are functions of the one remaining variable and this integration is performed next; and, finally, the outermost integral has constant limits and this integration is done last.

Notice that the limits in the middle and outer integrals of the (triple) iterated integral of 19.7.2 are the same as the limits in the iterated integral used to evaluate the double integral over a type I region R in 19.2.1.

Also notice that this plane region R is the projection of the solid region S onto the xy plane (Fig. 19-27). It follows that in setting up a triple iterated integral, *the limits of the last two integrations* (*the middle and outer integrals*) *may be found from the boundaries of R.*

Example 2. Evaluate the triple integral of $f(x, y, z) = 2xyz$ over the solid region S that is bounded by the parabolic cylinder $z = 2 - \frac{1}{2}x^2$ and the planes $z = 0$, $y = x$, and $y = 0$.

Solution. The solid region S is shown in Fig. 19-28. The triple integral

$$\iiint\limits_{S} 2xyz\, dA$$

can be evaluated by an iterated integral.

To set up an appropriate iterated integral, we observe from the figure that the projection of S onto the xy plane is the triangular region R that is bounded by the lines $x = 2$, $y = 0$, and $y = x$. It can be expressed as the type I region

$$R = \{(x, y)\,|\,0 \le x \le 2, 0 \le y \le x\}.$$

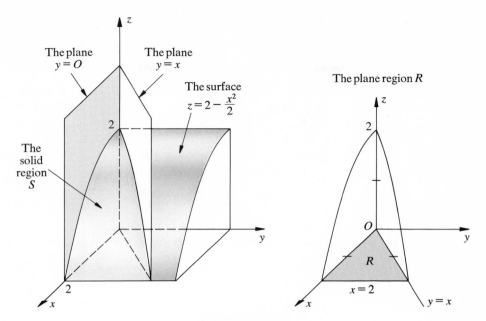

Figure 19-28

This indicates that in the outer integral of the iterated integral x is to be integrated between the constant limits 0 and 2, and in the middle integral y will be integrated from 0 to x.

Only the inner integral and the variable z are left. From the figure it is clear that the z coordinate of points of S varies from $z = 0$ in the xy plane to $z = 2 - \frac{1}{2}x^2$ on the cylinder. Thus an iterated integral for evaluating the triple integral is

$$\int_0^2 \int_0^x \int_0^{2-(1/2)x^2} 2xyz \, dz \, dy \, dx.$$

Therefore

$$\iiint_S xyz \, dV = \int_0^2 \int_0^x \int_0^{2-x^2/2} 2xyz \, dz \, dy \, dx$$

$$= \int_0^2 \int_0^x \left[xyz^2 \right]_0^{2-x^2/2} dy \, dx$$

$$= \int_0^2 \int_0^x (4xy - 2x^3y + \tfrac{1}{4}x^5y) \, dy \, dx$$

$$= \int_0^2 (2x^3 - x^5 + \tfrac{1}{8}x^7) \, dx = 1\tfrac{1}{3}.$$

Although a triple integral can exist when the solid region S is less restricted than ours, it is assumed in this book that S is of the type described above, or that it is one of the other five types that can be defined similarly by permuting the variables. For example, the region S can be bounded by two parallel planes $y = a$ and $y = b$, the two cylinders $z = \phi_1(y)$ and $z = \phi_2(y)$, and two surfaces $x = F_1(y, z)$ and $x =$

$F_2(y, z)$. Here S is the set of points (x, y, z) such that

$$F_1(y, z) \leq x \leq F_2(y, z), \qquad a \leq y \leq b, \qquad \phi_1(y) \leq z \leq \phi_2(y).$$

Thus the projection of S onto the yz plane is the plane region

$$R = \{(y, z) \,|\, a \leq y \leq b, \phi_1(y) \leq z \leq \phi_2(y)\},$$

which suggests that in the outer integral of the iterated integral we integrate y between the constant limits a and b, and in the middle integral we integrate z from $\phi_1(y)$ to $\phi_2(y)$.

We are left with x to be integrated from $F_1(y, z)$ to $F_2(y, z)$ in the inner integral (which is evaluated first). Therefore

$$\iiint\limits_{S} f(x, y, z) \, dV = \int_a^b \int_{\phi_1(y)}^{\phi_2(y)} \int_{F_1(y, z)}^{F_2(y, z)} f(x, y, z) \, dx \, dz \, dy.$$

Example 3. Write an iterated integral that would evaluate the triple integral

$$\iiint\limits_{S} f(x, y, z) \, dV,$$

if S is the solid region in the first octant bounded by the cylinder $z = \sqrt{y}$ and the planes $x + y = 4$, $x = 0$, and $z = 0$ (Fig. 19-29).

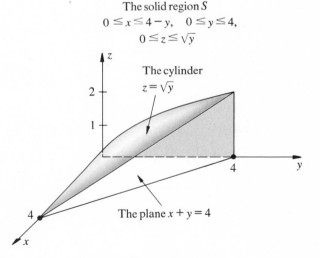

The solid region S
$0 \leq x \leq 4 - y, \quad 0 \leq y \leq 4,$
$0 \leq z \leq \sqrt{y}$

The cylinder
$z = \sqrt{y}$

The plane $x + y = 4$

Figure 19-29

Solution. The solid region S is shown in Fig. 19-29. It is apparent in the figure that the projection of S onto the yz plane is the plane region

$$R = \{(y, z) \,|\, 0 \leq y \leq 4, 0 \leq z \leq \sqrt{y}\}.$$

This suggests that in the outer integral we integrate y from the constant limit 0 to the constant limit 4, and in the middle integral we integrate z from 0 to $\sqrt{y}$.

Again from the figure, x can vary from 0 to $4 - y$. So in the inner integral we integrate x from 0 to $4 - y$. Therefore an iterated integral that would evaluate the given triple integral is

$$\int_0^4 \int_0^{\sqrt{y}} \int_0^{4-y} f(x, y, z) \, dx \, dz \, dy.$$

Exercises

Evaluate the iterated integrals in Exercises 1 to 9.

1. $\int_{-3}^{7} \int_0^{2x} \int_y^{x-1} dz \, dy \, dx.$

2. $\int_0^2 \int_{-1}^4 \int_0^{3y+x} dz \, dy \, dx.$

3. $\int_1^4 \int_{z-1}^{2z} \int_0^{y+2z} dx \, dy \, dz.$

4. $\int_0^5 \int_{-2}^4 \int_1^2 6xy^2z^3 \, dx \, dy \, dz.$

5. $\int_0^2 \int_1^z \int_0^{\sqrt{x/z}} 2xyz \, dy \, dx \, dz.$

6. $\int_0^{\pi/2} \int_0^z \int_0^y \sin(x + y + z) \, dx \, dy \, dz.$

7. $\int_{-2}^4 \int_{x-1}^{x+1} \int_0^{\sqrt{2y/x}} 3xyz \, dz \, dy \, dx.$

8. $\int_0^{\pi/2} \int_{\sin 2z}^0 \int_0^{2yz} \sin\left(\frac{x}{y}\right) dx \, dy \, dz.$

9. $\int_{\pi/3}^{\pi} \int_{\cos y}^1 \int_0^{xy} \cos\left(\frac{z}{x}\right) dz \, dx \, dy.$

In each of Exercises 10 to 19, sketch the given solid region S and write an iterated integral that is equal to the given triple integral over S. If the instructor requests it, evaluate your iterated integral.

10. $\iiint_S xyz \, dV;$

$S = \{(x, y, z) \mid 0 \le x \le 1, 0 \le y \le 3,$
$$0 \le z \le \tfrac{1}{6}(12 - 3x - 2y)\}.$$

11. $\iiint_S (x + 2y - 3z) \, dV;$

$S = \{(x, y, z) \mid 0 \le x \le \sqrt{4 - y^2}, 0 \le y \le 2,$
$$0 \le z \le 3\}.$$

12. $\iiint_S (2xz - y) \, dV;$

$S = \{(x, y, z) \mid 0 \le x \le \tfrac{1}{2}y, 0 \le y \le 4, 0 \le z \le 2\}.$

13. $\iiint_S (x^2 + y^2 + z^2) \, dV;$

$S = \{(x, y, z) \mid 0 \le x \le \sqrt{y}, 0 \le y \le 4,$
$$0 \le z \le \tfrac{3}{2}x\}.$$

14. $\iiint_S (3x^2 - y^2 + 2z^2) \, dV;$

$S = \{(x, y, z) \mid 0 \le x \le 3z, 0 \le y \le 4 - x - 2z,$
$$0 \le z \le 1\}.$$

15. $\iiint_S (x^{1/2} + y^{1/2} + z^{1/2}) \, dV;$

$S = \{(x, y, z) \mid 0 \le x \le y^2, 0 \le y \le \sqrt{z},$
$$0 \le z \le 1\}.$$

16. $\iiint_S e^{x+y+z} \, dV;$ S is the tetrahedron having vertices $(0, 0, 0)$, $(3, 2, 0)$, $(0, 3, 0)$, and $(0, 0, 2)$.

17. $\iiint_S dV;$ S is the region in the first octant bounded by the surface $z = 9 - x^2 - y^2$ and the coordinate planes.

18. $\iiint_S \frac{y - 2z}{x} \, dV;$ S is the region in the first octant bounded by the cylinder $y^2 + z^2 = 1$ and the planes $x = 1$ and $x = 4$.

19. $\iiint_S (xy - yz + zx) \, dV;$ S is the smaller region bounded by the cylinder $x^2 + y^2 - 2y = 0$ and the planes $x - y = 0$, $z = 0$, and $z = 3$.

19.8 APPLICATIONS OF TRIPLE INTEGRALS

It is assumed in this section that S is a solid region of one of the six types discussed in 19.7 and that f is a function of three variables that is continuous inside and on the boundaries of S.

Just as the double integral $\iint\limits_{R} dA$ can be interpreted as the area of the plane region R, the triple integral $\iiint\limits_{S} dV$ can be interpreted as the **volume** of the three-dimensional region S.

19.8.1
$$\text{Volume of } S = \iiint\limits_{S} dV.$$

Example 1. Find the volume of the region S in the first octant bounded by the elliptic paraboloids $x^2 + 4y^2 = 4z$ and $x^2 + 4y^2 = 48 - 8z$.

Solution. We seek the volume of the solid region S, shown in Fig. 19-30. The volume of S is

$$V = \iiint\limits_{S} dV.$$

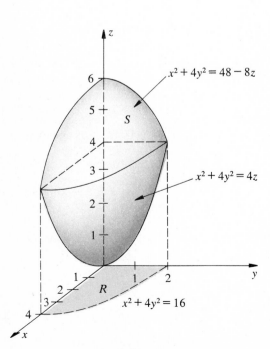

Figure 19-30

This triple integral will be evaluated by an iterated integral, and our first task is to determine the limits for such an iterated integral.

The curve of intersection of the two given paraboloids is found by eliminating z between their equations. This gives $x^2 + 4y^2 = 16$, an ellipse in the horizontal plane $z = 4$. Its projection onto the xy plane is the ellipse $x^2 + 4y^2 = 16$, $z = 0$. The segment of this ellipse that is in the first quadrant of the xy plane is $x = 2\sqrt{4 - y^2}$, $0 \le y \le 2$.

It is clear from Fig. 19-30 that the projection of S onto the xy plane is the region

$$R = \{(x, y) \,|\, 0 \le x \le 2\sqrt{4 - y^2}, 0 \le y \le 2\}.$$

This suggests that in the outer integral of the iterated integral we integrate y between the constant limits 0 and 2, and in the middle integral we integrate x from 0 to $2\sqrt{4 - y^2}$.

Since the z coordinate of points of S varies from $z = (x^2 + 4y^2)/4$ on the lower paraboloid to $z = (48 - x^2 - 4y^2)/8$ on the upper paraboloid (Fig. 19-30), the inner integral must integrate z from $(x^2 + 4y^2)/4$ to $(48 - x^2 - 4y^2)/8$. As always, the inner integration is performed first.

Thus an iterated integral that will evaluate the triple integral is

$$\int_0^2 \int_0^{2\sqrt{4-y^2}} \int_{(x^2+4y^2)/4}^{(48-x^2-4y^2)/8} dz\,dx\,dy.$$

Therefore

$$\iiint_S dV = \int_0^2 \int_0^{2\sqrt{4-y^2}} \int_{(x^2+4y^2)/4}^{(48-x^2-4y^2)/8} dz\,dx\,dy$$

$$= \int_0^2 \int_0^{2\sqrt{4-y^2}} \tfrac{3}{8}(16 - x^2 - 4y^2)\,dx\,dy$$

$$= 2\int_0^2 (4 - y^2)^{3/2}\,dy = 6\pi.$$

If ρ is a density function of three variables such that $\rho(x, y, z)$ gives the density of the solid S at any point (x, y, z) in S, then the total **mass** of S is defined to be

19.8.2
$$\text{Mass of } S = M = \iiint_S \rho(x, y, z)\,dV.$$

The **center of mass** (center of gravity) of S is the point $(\bar{x}, \bar{y}, \bar{z})$, where

19.8.3
$$M\bar{x} = \iiint_S x\,\rho(x, y, z)\,dV,$$

$$M\bar{y} = \iiint_S y\,\rho(x, y, z)\,dV,$$

$$M\bar{z} = \iiint_S z\,\rho(x, y, z)\,dV.$$

The moments of inertia of the mass M about the x, y, and z axes, respectively, are

19.8.4
$$I_x = \iiint_S (y^2 + z^2)\,\rho(x, y, z)\,dV,$$

$$I_y = \iiint_S (z^2 + x^2)\,\rho(x, y, z)\,dV,$$

$$I_z = \iiint_S (x^2 + y^2)\,\rho(x, y, z)\,dV.$$

Example 2. Find the moment of inertia about the y axis of the solid in the first octant bounded by the cylinders $x^2 + z^2 = 1$ and $y^2 + z^2 = 1$, if its density is proportional to the distance from the yz plane.

Solution. The region S is shown in Fig. 19-31.

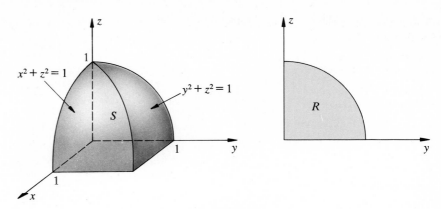

Figure 19-31

The density is $\rho(x, y, z) = kx$, where k is a constant factor of proportionality. By 19.8.4,

$$I_y = \iiint\limits_S (z^2 + x^2)\, \rho(x, y, z)\, dV = \iiint\limits_S kx(z^2 + x^2)\, dx\, dy\, dz.$$

This triple integral will be evaluated by an iterated integral, the limits of which we shall now determine.

The projection of S onto the yz plane is the circular sector R, bounded by the circle $x^2 + y^2 = 1$ and the positive y and z axes (Fig. 19-31). From R we see that the y and z coordinates of the points (x, y, z) of S satisfy the inequalities

$$0 \leq y \leq \sqrt{1 - z^2}, \qquad 0 \leq z \leq 1.$$

So in the outer integral of the iterated integral, z will be integrated between the constant limits 0 and 1; and in the middle integral y will be integrated from 0 to $\sqrt{1 - z^2}$.

This leaves x for the inner integral. From Fig. 19-31 we see that x varies from $x = 0$ in the yz plane to $x = \sqrt{1 - z^2}$ on the cylinder $x^2 + z^2 = 1$. As always, the inner integration is done first.

Therefore

$$I_y = \iiint\limits_S kx(z^2 + x^2)\, dV = k \int_0^1 \int_0^{\sqrt{1-z^2}} \int_0^{\sqrt{1-z^2}} (xz^2 + x^3)\, dx\, dy\, dz$$

$$= \frac{k}{4} \int_0^1 \int_0^{\sqrt{1-z^2}} (1 - z^4)\, dy\, dz = \frac{k}{4} \int_0^1 (1 - z^4)\sqrt{1 - z^2}\, dz = \frac{7k\pi}{128}.$$

Exercises

In each of Exercises 1 to 13, sketch the given solid region S and write an iterated integral for the volume V of S. If the instructor requests it, evaluate V.

1. The region bounded by the elliptic paraboloid $4x^2 + 9y^2 + 36z = 72$ and the plane $z = 0$.

2. The region bounded by the paraboloid $9x^2 + y^2 - 9z = 0$ and the plane $z = 4$.

3. The region in the first octant bounded by the cylinder $y^2 = 4 - x$, the coordinate planes, and the plane $2y + z - 4 = 0$.

4. The first octant region bounded by the cylinders $x^2 + y^2 = 16$ and $y^2 + z^2 = 16$, and the coordinate planes.

5. The region common to the two cylinders $x^2 + y^2 = 9$ and $x^2 + z^2 = 9$.

6. The region bounded by the cylinder $y = x^2 + 2$ and the planes $y = 4$, $z = 0$ and $3y - 4z = 0$.

7. The region bounded by the paraboloid $x^2 + y^2 + 2z - 16 = 0$ and the plane $z = 2$.

8. The region bounded by the circular paraboloids $z = 6 - x^2 - y^2$ and $2z = x^2 + y^2$.

9. The region in the first octant bounded by the surfaces $x^2 + y^2 - 3z = 0$, $x^2 + y^2 + z - 12 = 0$, and the planes $x = 0$ and $y = 0$.

10. The sphere $x^2 + y^2 + z^2 = a^2$.

11. The region bounded by the parabolic cylinders $x^2 = y$ and $z^2 = y$, and the plane $y = 1$.

12. The region in the first octant bounded by the surface $z = 1/(x + 2)$, the cylinder $x^2 + y^2 = 4$, and the coordinate planes.

13. The ellipsoid

$$\frac{x^2}{a^2} + \frac{y^2}{b^2} + \frac{z^2}{c^2} = 1.$$

14. Find the center of mass of the solid bounded by the hemisphere $z = \sqrt{4 - x^2 - y^2}$ and the plane $z = 0$, if its density is proportional to the distance from the xy plane.

15. Find the center of mass of the tetrahedron bounded by the planes $x + y + z = 1$, $x = 0$, $y = 0$, and $z = 0$, if the density at any point is proportional to the sum of the coordinates of the point.

16. Find the center of mass of the solid bounded by the cylinder $x^2 + y^2 = 9$ and the planes $z = 0$ and $z = 4$, if its density is proportional to the distance from the xy plane.

17. Find the center of mass of the solid bounded by the cylinder $x^2 + z^2 - 16 = 0$ and the planes $y = 0$ and $y = 5$, if its density is porportional to the distance from the xz plane.

18. Find the center of mass of the solid in Exercise 16 if its density at any point is proportional to the square of the distance of the point from the origin.

19. Find the center of mass of the solid in Exercise 17 if its density at any point is proportional to the square of the distance of the point from the origin.

20. Find the center of mass of the solid bounded by $y = 4 - x^2$, $y = 0$, $z = 0$, and $z = 4$ if its density is proportional to the distance from the zx plane.

21. Find the center of mass of the solid bounded by $z = 9 - x^2$, $z = 0$, $y = 0$, and $y = 9$ if its density is proportional to the distance from the xy plane.

22. Find the center of mass of the solid in the first octant bounded by the cylinders $x^2 + y^2 = 9$ and $x^2 + z^2 = 9$, and the coordinate planes, if its density is proportional to the distance from the xy plane.

23. Find the center of mass of the solid bounded by the surface $z = \sqrt{9 - x^2 - y^2}$ and the plane $z = 0$, if its density is proportional to the distance from the xy plane.

24. Find I_x, the moment of inertia about the x axis, of the solid region bounded by the cylinder $y^2 + z^2 = 4$ and the planes $x - y = 0$, $x = 0$, $z = 0$, if the density is proportional to the distance from the xy plane.

25. Write an iterated integral for the moment of inertia about the x axis of the tetrahedron of Exercise 15, if its density is proportional to the distance from the yz plane.

26. Find the moment of inertia about the z axis of the solid in the first octant bounded by the cylinders $x^2 + y^2 = 9$ and $x^2 + z^2 = 9$, and the coordinate planes, if its density at any point is proportional to the product of the distances of the point from the coordinate planes.

27. State a **mean value theorem for triple integrals** analogous to 10.7.1 and the one given in Exercise 15, Section 19.3.

19.9 CYLINDRICAL COORDINATES

Another system of coordinates in three-dimensional space is called *cylindrical coordinates*. In place of the x and y coordinates of the Cartesian system, it uses the polar coordinates, r and θ. The z coordinate is the same as in Cartesian coordinates.

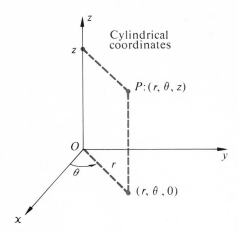

Cylindrical coordinates

Figure 19-32

Like other alternatives to Cartesian coordinates, cylindrical coordinates were devised to facilitate working with a special class of problems, those in which the three-dimensional region S has an axis of symmetry. We saw in Example 2 of Section 19.6 how the integrand was simplified by changing from x, y coordinates in the plane to polar coordinates, r and θ; this was because the $x^2 + y^2$ that appeared in the integrand was replaced by r^2 from the polar system. In that example we were working with a paraboloid $z = x^2 + y^2$, which is symmetric with respect to the z axis.

If (x, y, z) are the Cartesian coordinates of a point P in three-space, the **cylindrical coordinates** of P are (r, θ, z), where (r, θ) are polar coordinates of the projection of P on the xy plane (now the $r\theta$ plane) and z is the Cartesian coordinate of the projection of P onto the z axis (Fig. 19-32).

In a cylindrical coordinate system, the graph of $r = c$, where c is a nonzero constant, is a circular cylinder of radius $|c|$, having the z axis for a line of symmetry (Fig. 19-33). The graph of this equation is the set of points $P{:}(r, \theta, z)$ in which θ and z can have any values while r holds the constant value c. The graph of $r = 0$ is the z axis.

The graph of $\theta = c$ is a plane containing the z axis that makes an angle $\theta = c$ with the xz plane (Fig. 19-34); for, the coordinates r and z in $P{:}(r, \theta, z)$ can have any values, whereas θ keeps the constant value c.

The graph of $z = c$ in cylindrical coordinates is, as formerly, a plane parallel to the xy plane.

Cylindrical and Cartesian coordinates are related by the equations

(1) $$x = r \cos \theta, \qquad y = r \sin \theta, \qquad z = z,$$

and

(2) $$r^2 = x^2 + y^2, \qquad \tan \theta = \frac{y}{x} \quad \text{if } x \neq 0, \qquad z = z.$$

Example 1. Find the equation in cylindrical coordinates of the plane whose Cartesian equation is $Ax + By + Cz + D = 0$.

Solution. Substituting from equations (1) into the given equation, we obtain $Ar \cos \theta + Br \sin \theta + Cz + D = 0$, or
$$r(A \cos \theta + B \sin \theta) + Cz + D = 0.$$

Example 2. Find the equations in cylindrical coordinates of the paraboloid and cylinder whose Cartesian equations are $x^2 + y^2 = 4 - z$ and $x^2 + y^2 = 2x$.

Solution. Since $x^2 + y^2 = r^2$, the equation in cylindrical coordinates of the paraboloid $x^2 + y^2 = 4 - z$ is $z = 4 - r^2$; and by using $x = r \cos \theta$, we find the equation of the cylinder in cylindrical coordinates to be $r^2 = 2r \cos \theta$, or $r = 2 \cos \theta$.

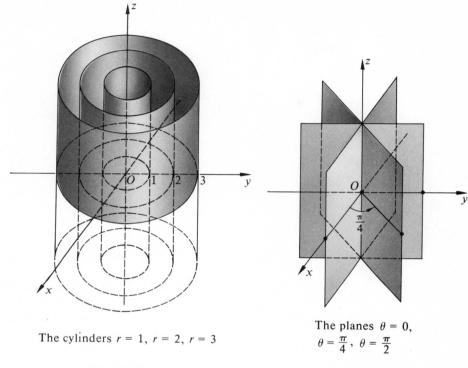

The cylinders $r = 1$, $r = 2$, $r = 3$

The planes $\theta = 0$,
$\theta = \dfrac{\pi}{4}$, $\theta = \dfrac{\pi}{2}$

Figure 19-33   **Figure 19-34**

Example 3. Find Cartesian equations of the surfaces whose equations in cylindrical coordinates are $r^2 + 4z^2 = 16$ and $r^2 \cos 2\theta = z$.

Solution. Since $r^2 = x^2 + y^2$, the surface $r^2 + 4z^2 = 16$ has the Cartesian equation $x^2 + y^2 + 4z^2 = 16$, or $x^2/16 + y^2/16 + z^2/4 = 1$. Its graph is an ellipsoid.

From $\cos 2\theta = \cos^2 \theta - \sin^2 \theta$, the second equation can be written $r^2 \cos^2 \theta - r^2 \sin^2 \theta = z$. In Cartesian coordinates it becomes $x^2 - y^2 = z$, the graph of which is a hyperbolic paraboloid.

When a solid region S in three-space has an axis of symmetry, the evaluation of the triple integrals over S is often facilitated by using cylindrical coordinates.

Partition the solid region S by planes containing the z axis, planes perpendicular to the z axis, and circular cylinders about the z axis. Typical of the resulting subregions is the element of volume shown in Fig. 19-35. The set of these elements that lie entirely inside or on the boundary of S forms a **cylindrical partition** p of S; number them from 1 to n. The **norm** $|p|$ of this partition is the length of the longest "diagonal" of any of the subregions of the partition.

The volume of the ith element (Fig. 19-35) is equal to the area of its base times its altitude. But the area of the base is $\bar{r}_i \, \Delta r_i \, \Delta \theta_i$, where $\bar{r}_i = r_i + \frac{1}{2}\Delta r_i$ (Section 19.5). Thus if Δz_i is the altitude of the ith partitioning element, the volume of the ith element is $\bar{r}_i \, \Delta r_i \, \Delta \theta_i \, \Delta z_i$.

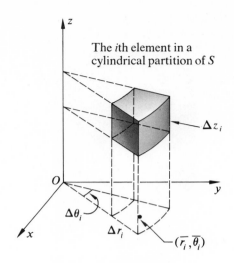

The ith element in a cylindrical partition of S

Δz_i

$\Delta \theta_i$

Δr_i

$(\bar{r}_i, \bar{\theta}_i)$

Figure 19-35

Let f be a function of r, θ, and z that is continuous inside and on the boundary of S, and let $(\bar{r}_i, \bar{\theta}_i, \bar{z}_i)$ be the coordinates of a point of the ith partitioning element, such that $\theta_i \leq \bar{\theta}_i \leq \theta_i + \Delta \theta_i$ and $z_i \leq \bar{z}_i \leq z_i + \Delta z_i$. Form the sum

(3)
$$\sum_{i=1}^{n} f(\bar{r}_i, \bar{\theta}_i, \bar{z}_i) \, \bar{r}_i \, \Delta r_i \, \Delta \theta_i \, \Delta z_i.$$

Then it can be shown that when S is a solid region of the type described below, the limit as $|p| \longrightarrow 0$ of the sum (3) exists and is independent of the choice of the point $(\bar{r}_i, \bar{\theta}_i, \bar{z}_i)$ in the ith element. This limit is called the **triple integral in cylindrical coordinates** of the function f over the solid region S, and it is denoted by $\iiint\limits_{S} f(r, \theta, z) \, dV$.

Thus

$$\iiint\limits_{S} f(r, \theta, z) \, dV = \lim_{|p| \to 0} \sum_{i=1}^{n} f(\bar{r}_i, \bar{\theta}_i, \bar{z}_i) \, r_i \, \Delta r_i \, \Delta \theta_i \, \Delta z_i.$$

This triple integral is evaluated by an iterated integral. Let the solid region S be bounded by the planes $\theta = \alpha$ and $\theta = \beta$, with $\alpha < \beta$, the cylinders $r = \phi_1(\theta)$ and $r = \phi_2(\theta)$, where ϕ_1 and ϕ_2 are functions that are continuous on $[\alpha, \beta]$ with $\phi_1(\theta) \leq \phi_2(\theta)$ for $\alpha \leq \theta \leq \beta$, and by surfaces $z = F_1(r, \theta)$ and $z = F_2(r, \theta)$, where F_1 and F_2 are functions of two variables that are continuous throughout the region

$$R = \{(r, \theta) \mid \phi_1(\theta) \leq r \leq \phi_2(\theta), \, \alpha \leq \theta \leq \beta\},$$

Let $F_1(r, \theta) \leq F_2(r, \theta)$ for every point (r, θ) in R. Then

19.9.1
$$\iiint\limits_{S} f(r, \theta, z) \, dV = \int_{\alpha}^{\beta} \int_{\phi_1(\theta)}^{\phi_2(\theta)} \int_{F_1(r, \theta)}^{F_2(r, \theta)} r \, f(r, \theta, z) \, dz \, dr \, d\theta.$$

Five other regions S can be similarly defined (by permuting the variables) for which statements analogous to 19.9.1 are true. As before, these are *sufficient* restrictions on S, not necessary ones.

Example 4. Find the volume of the solid region S that is bounded by the paraboloid $z = 6 - x^2 - y^2$ and the plane $z = 0$ (Fig. 19-36).

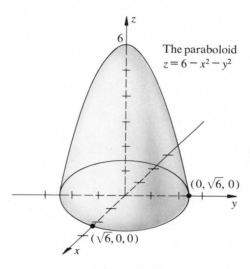

The paraboloid
$z = 6 - x^2 - y^2$

$(0, \sqrt{6}, 0)$

$(\sqrt{6}, 0, 0)$

Figure 19-36

Solution. Since S is symmetric with respect to the z axis, we shall use cylindrical coordinates. The equation of the paraboloid in cylindrical coordinates is $z = 6 - r^2$.

The volume of S is given by

$$V = \iiint_S dV,$$

where $dV = dx\,dy\,dz = r\,dr\,d\theta\,dz$. To evaluate this triple integral by an iterated integral, we observe that the projection of S on the plane $z = 0$ is the circular disk

$$0 \le r \le \sqrt{6}, \qquad 0 \le \theta \le 2\pi.$$

Moreover, since S is bounded above by the paraboloid $z = 6 - r^2$ and below by the plane $z = 0$, the limits for z are from 0 to $6 - r^2$. Consequently,

$$V = \iiint_S dV = \int_0^{2\pi} \int_0^{\sqrt{6}} \int_0^{6-r^2} r\,dz\,dr\,d\theta$$

$$= \int_0^{2\pi} \int_0^{\sqrt{6}} \left[rz \right]_0^{6-r^2} dr\,d\theta$$

$$= \int_0^{2\pi} \int_0^{\sqrt{6}} (6r - r^3)\,dr\,d\theta$$

$$= \int_0^{2\pi} \left[3r^2 - \frac{r^4}{4} \right]_0^{\sqrt{6}} d\theta = \int_0^{2\pi} 9\,d\theta = 8\pi.$$

Example 5. Find the mass of the solid in the first octant that is bounded above by the paraboloid $x^2 + y^2 = 4 - z$, below by $z = 0$, and laterally by $y = 0$ and the cylinder $x^2 + y^2 = 2x$, if its density is proportional to the distance from the xy plane (Fig. 19-37).

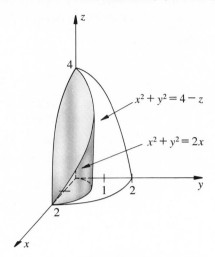

Figure 19-37

Solution. In cylindrical coordinates, the paraboloid is $z = 4 - r^2$ and the cylinder is $r = 2\cos\theta$, as we saw in Example 2. The solid S is bounded above by $z = 4 - r^2$ and below by $z = 0$. The projection of S on the xy plane is the plane region R bounded by that half of the circle $r = 2\cos\theta$ that is in the first quadrant, and the line $\theta = 0$ (Fig. 19-37). Since the density of the solid at any of its points (r, θ, z) is $f(r, \theta, z) = kz$, where k is a factor of proportionality, the mass M of the solid is

$$M = \iiint\limits_{S} f(r, \theta, z)\, dV = k \int_0^{\pi/2} \int_0^{2\cos\theta} \int_0^{4-r^2} zr\, dz\, dr\, d\theta$$

$$= \frac{k}{2} \int_0^{\pi/2} \int_0^{2\cos\theta} (16r - 8r^3 + r^5)\, dr\, d\theta$$

$$= 16k \int_0^{\pi/2} \left(\cos^2\theta - \cos^4\theta + \frac{1}{3}\cos^6\theta \right) d\theta = \frac{11\pi k}{6}.$$

Exercises

1. Find cylindrical coordinates of the points whose Cartesian coordinates are

(a) $(2, 2, 3)$; (b) $(1, \sqrt{3}, -5)$;

(c) $(4\sqrt{3}, -4, 6)$; (d) $(-5, 0, -2)$.

2. Find Cartesian coordinates of the points whose cylindrical coordinates are

(a) $(6, \pi/6, -2)$; (b) $(-5, 0, 7)$;

(c) $(3\sqrt{2}, -\pi/4, 4)$; (d) $(-4, 4\pi/3, -8)$.

In Exercises 3 to 8, find an equation in cylindrical coordinates of the graph of the given Cartesian equation. Make a sketch.

3. $x^2 + y^2 = 9$.

4. $x^2 + 4y^2 = 16$.

5. $x^2 - y^2 = 25$.

6. $x^2 + y^2 = 9z$.

7. $x^2 + y^2 + 4z^2 = 10$.

8. $x^2 + y^2 + 2z^2 - 12z + 14 = 0$.

In Exercises 9 to 12, an equation in cylindrical coordinates is given. Find a Cartesian equation of its graph and sketch the graph.

9. $r^2 + z^2 = 9$.

10. $r^2 \cos^2 \theta + z^2 = 4$.

11. $2r \cos \theta - 3r \sin \theta + z = 6$.

12. $r^2 \cos 2\theta + z^2 = 1$.

13. Use cylindrical coordinates to find the volume of the solid bounded by the circular paraboloid $x^2 + y^2 + z - 4 = 0$ and the plane $z = 0$.

14. Use cylindrical coordinates to find the moment of inertia about the z axis of the solid in Exercise 13, if its density is constant.

15. Use cylindrical coordinates to find the moment of inertia about its axis of a right circular cylinder of altitude h and base radius a, if its density at any point is proportional to the distance from its base.

16. Use cylindrical coordinates to find the mass of the solid in Exercise 15 if its density at any point is proportional to the distance from its axis.

17. The density of the cone bounded by the conical surface $x^2 + y^2 - z^2 = 0$ and the plane $z = 4$ is proportional to the distance from that plane. Use cylindrical coordinates to find its moment of inertia about the z axis.

18. Use cylindrical coordinates to find the center of mass of the cone in Exercise 17.

19. The axis of a right circular cylinder of radius 2 goes through the center of a sphere of radius 3. Use cylindrical coordinates to find the volume of the region inside the sphere and outside the cylinder.

19.10 SPHERICAL COORDINATES

A third system of coordinates in three-dimensional space is called *spherical coordinates*. This system is advantageous in certain geometric and physical problems involving a center of symmetry.

A point $P{:}(x, y, z)$ has **spherical coordinates** (ρ, θ, ϕ), where ρ is the undirected distance $|OP|$ of P from the origin, θ is the polar coordinate of Q, the projection of P on the xy plane, and ϕ is the angle between the positive z axis and the radial line segment OP (Fig. 19-38). Here

$$\rho \geq 0, \qquad 0 \leq \theta < 2\pi, \qquad 0 \leq \phi \leq \pi.$$

Let c be a nonnegative constant. The graph of $\rho = c$ is a sphere with its center at O and radius c, if $c > 0$. The graph of $\rho = 0$ is a single point, the origin.

The graph of $\theta = c$ is a plane containing the z axis, just as in cylindrical coordinates. The graph of $\phi = c$ is a cone with its vertex at the origin; it has the limiting cases $\phi = \pi/2$, which is the xy plane, $\phi = 0$, which is the nonnegative z axis, and $\phi = \pi$, the nonpositive z axis.

The spherical coordinate ρ is called the *radial distance* of P. In analogy with a terrestrical globe, the z axis is often called the *polar axis*, θ the *longitude* of P, and ϕ the *colatitude* of P.

If we notice in Fig. 19-39 that $x = |OQ| \cos \theta$, $y = |OQ| \sin \theta$, $|OQ| = \rho \sin \phi$, and also $|OQ| = \sqrt{x^2 + y^2}$, it will be easy to read from the figure the following equations connecting the Cartesian coordinates (x, y, z) of a point P and its spherical coordinates (ρ, θ, ϕ):

$$(1) \qquad x = \rho \sin \phi \cos \theta, \qquad y = \rho \sin \phi \sin \theta, \qquad z = \rho \cos \phi;$$

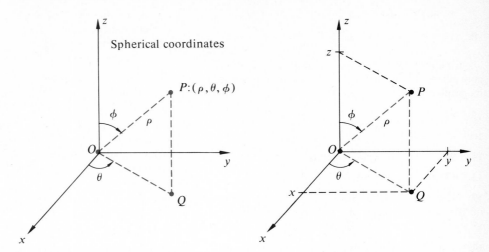

Figure 19-38

Figure 19-39

and

$$\rho^2 = x^2 + y^2 + z^2,$$

(2)
$$\cos \theta = \frac{x}{\sqrt{x^2 + y^2}}, \qquad \sin \theta = \frac{y}{\sqrt{x^2 + y^2}},$$

$$\cos \phi = \frac{z}{\sqrt{x^2 + y^2 + z^2}}, \qquad \sin \phi = \frac{\sqrt{x^2 + y^2}}{\sqrt{x^2 + y^2 + z^2}}.$$

When a solid region S is symmetric with respect to a point, spherical coordinates, with the origin at the center of symmetry, often simplify the task of evaluating a triple integral over S.

To define a triple integral over S in spherical coordinates, we partition S by spheres with centers at the origin, planes containing the z axis, and circular cones about the z axis that have their vertices at the origin. A typical subregion of the partition is shown in Fig. 19-40. Its volume is approximated by

$$\Delta V_i = \rho_i^2 \sin \phi_i \, \Delta \rho_i \, \Delta \theta_i \, \Delta \phi_i,$$

where $(\rho_i, \theta_i, \phi_i)$ is some point in the subregion.

Then the triple integral in spherical coordinates of a function f of ρ, θ, and ϕ, over the solid region S, is defined by

$$\iiint_S f(\rho, \theta, \phi) \, dV = \lim_{|p| \to 0} \sum_{i=1}^{n} f(\rho_i, \theta_i, \phi_i) \, \Delta V_i.$$

This triple integral is evaluated by the iterated integral

$$\iiint_S f(\rho, \theta, \phi) \, dV$$

$$= \int_\alpha^\beta \int_{\theta_1(\phi)}^{\theta_2(\phi)} \int_{\rho_1(\theta, \phi)}^{\rho_2(\theta, \phi)} f(\rho, \theta, \phi) \, \rho^2 \sin \phi \, d\rho \, d\theta \, d\phi.$$

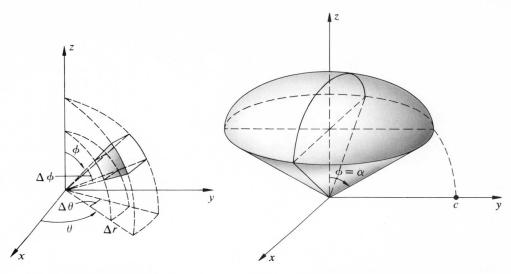

Figure 19-40   **Figure 19-41**

Example. Find the volume and the center of gravity of the homogeneous solid that is bounded above by the sphere $\rho = c$ and below by the cone $\phi = \alpha$ (Fig. 19-41), where $0 < \phi < \pi/2$.

Solution. The volume V of the solid is

$$V = \int_0^\alpha \int_0^{2\pi} \int_0^c \rho^2 \sin \phi \, d\rho \, d\theta \, d\phi$$

$$= \int_0^\alpha \int_0^{2\pi} \left(\frac{c^3}{3}\right) \sin \phi \, d\theta \, d\phi$$

$$= \frac{2\pi c^3}{3} \int_0^\alpha \sin \phi \, d\phi = \frac{2\pi c^3}{3}(1 - \cos \alpha).$$

The mass M of the solid is

$$M = kV = \frac{2}{3}\pi c^3 k(1 - \cos \alpha),$$

where k is its constant density.

From symmetry, the center of gravity of this homogeneous solid is on the z axis. Its z coordinate is given by

$$M\bar{z} = \iiint_S kz \, dV = \iiint_S k\rho \cos \phi \, dV.$$

Moreover,

$$\iiint_S k\rho \cos \phi \, dV$$

$$= \int_0^\alpha \int_0^{2\pi} \int_0^c k\rho^3 \sin \phi \cos \phi \, d\rho \, d\theta \, d\phi$$

$$= \int_0^\alpha \int_0^{2\pi} \frac{1}{4} kc^4 \sin \phi \cos \phi \, d\theta \, d\phi$$

$$= \int_0^\alpha \frac{1}{2}\pi kc^4 \sin \phi \cos \phi \, d\phi = \frac{1}{4}\pi c^4 k \sin^2 \alpha.$$

Therefore

$$\bar{z} = \frac{\frac{1}{4}\pi c^4 k \sin^2 \alpha}{\frac{2}{3}\pi c^3 k (1 - \cos \alpha)} = \frac{3c \sin^3 \alpha}{8(1 - \cos \alpha)}$$

$$= \frac{3}{8} c (1 + \cos \alpha).$$

Exercises

1. Find spherical coordinates of the points whose Cartesian coordinates are:
(a) $(2, -2\sqrt{3}, 4)$;
(b) $(3\sqrt{6}, 3\sqrt{2}, 6\sqrt{2})$;
(c) $(3, -3\sqrt{3}, 0)$;
(d) $(-\sqrt{2}, \sqrt{2}, 2\sqrt{3})$.

2. Without changing to Cartesian coordinates, plot the points whose spherical coordinates are given below.

(a) $\left(8, \frac{\pi}{4}, \frac{\pi}{6}\right)$;   (b) $\left(2\sqrt{2}, \frac{5\pi}{6}, \frac{\pi}{4}\right)$;

(c) $\left(13, 0, \frac{\pi}{2}\right)$;   (d) $\left(4, \frac{\pi}{3}, \frac{3\pi}{4}\right)$.

3. Find Cartesian coordinates of the points in Exercise 2.

In Exercises 4 to 9, find an equation in spherical coordinates of the graph of the given Cartesian equation.

4. $x^2 + y^2 + z^2 = 4$.

5. $2x^2 + 2y^2 - 4z^2 = 0$.

6. $36x^2 + 9y^2 + 16z^2 = 144$.

7. $4x^2 + 9y^2 - 36z^2 = 36$.

8. $x^2 - y^2 - z^2 = 1$.

9. $x^2 + y^2 = z$.

In Exercises 10 to 15, an equation in spherical coordinates is given. Find the Cartesian equation of its graph.

10. (a) $\rho = 3$;   (b) $\theta = 0$;   (c) $\phi = \frac{\pi}{4}$.

11. $\rho^2(\sin^2 \phi \cos^2 \theta + \cos^2 \phi) = 1$.

12. $\rho(2 \sin \phi \cos \theta - 3 \sin \phi \sin \theta - \cos \phi) = 6$.

13. $\rho^2 - 5\rho \cos \phi = 0$.

14. $\rho = 3 \tan \theta$.

15. $\rho^2 \cos 2\phi + 1 = 0$.

16. Find the volume of the homogeneous solid that is bounded above by the sphere $\rho = 2$ and below by the cone $\phi = \pi/4$.

17. Find the center of gravity of the solid in Exercise 16.

18. Find the volume of the solid bounded by the cylinder $\rho^2 \sin^2 \phi = 19$, and cone $\phi = \pi/4$, and the plane $\phi = \pi/2$.

19. Find the moment of inertia of a spherical solid with respect to a diameter if its density is proportional to the square of the distance from the center.

20. Find the center of mass of a hemispherical solid if its density is proportional to the distance from its base.

21. Find the moment of inertia of a right circular cone, of altitude h, about its axis if its density is proportional to the square of the distance from its vertex.

20

differential equations

20.1 INTRODUCTION

Differential equations provide an essential tool for the engineer and scientist. Although some differential equations are easy to solve, others may be difficult or impossible. Many present-day mathematicians are doing original work in the subject and much remains to be known.

In Section 11.14 we discussed a simple class of differential equations with interesting applications. The present chapter is an introduction to some additional types and their applications. We start by recapitulating what we learned in Section 11.14.

An equation involving derivatives or differentials is called a **differential equation**. Some examples of differential equations are

(a)
$$\frac{d^3y}{dx^3} + \sin x \frac{d^2y}{dx^2} + y = \cos x,$$

(b)
$$\frac{d^2s}{dt^2} = -32,$$

(c)
$$\frac{\partial^2 u}{\partial x^2} + \frac{\partial^2 u}{\partial y^2} = 0,$$

(d)
$$\left(\frac{dy}{dx}\right)^4 = \sqrt{3 - \left(\frac{d^3y}{dx^3}\right)^2},$$

(e)
$$D_t y + y = \sin^2 t,$$

(f)
$$(x^2 + 1)\frac{d^2y}{dx^2} + x\frac{dy}{dx} + xy = 0.$$

When the dependent variable in a differential equation is a function of a single independent variable, the derivatives that occur are "ordinary" and the equation is an **ordinary differential equation**. If the dependent variable is a function of two or more independent variables, the derivatives are partial and the equation is called a **partial differential equation**. All the differential equations in the preceding examples are ordinary except (c), which is a partial differential equation. In this book we are concerned only with ordinary differential equations, and from now on differential equation will mean "ordinary differential equation."

The **order** of a differential equation is the order of the highest derivative in the equation. In the examples shown, (a) and (d) are of order 3, (b), (c), and (f) are of order 2, and (e) is of order 1.

When a differential equation can be written in the form of a polynomial equation of degree k *in the highest-order derivative*, the differential equation is said to be of degree k. All the ordinary differential equations above are of the first degree except (d), which is of degree 2, since after the radical is removed, the equation becomes

$$\left(\frac{d^3y}{dx^3}\right)^2 = 3 - \left(\frac{dy}{dx}\right)^8.$$

A function f, defined by $y = f(x)$, is a **solution** of a differential equation if the equation becomes an identity when y and its derivatives are replaced by $f(x)$ and its corresponding derivatives.

For instance, $y = x \ln x - x$ is a solution of the differential equation

$$\frac{dy}{dx} = \frac{x + y}{x}$$

because substitution of $y = x \ln x - x$ throughout the equation gives

$$\frac{d}{dx}(x \ln x - x) = \frac{x + (x \ln x - x)}{x}$$

or

$$\ln x = \ln x,$$

which is an identity in x for all $x > 0$.

If we rewrite the first-order differential equation

(1) $$\frac{dy}{dx} = e^x$$

as $dy = e^x \, dx$ and then integrate both members, we get

$$y = e^x + C,$$

where C is an arbitrary constant (of integration). This result is called the general solution of (1). Particular solutions are obtained by assigning values to C.

By integrating the second-order equation (b), above, twice with respect to the independent variable t, we find its general solution

$$s = -16t^2 + C_1t + C_2,$$

where C_1 and C_2 are arbitrary constants of integration.

If we perform indefinite integration n times, we introduce n independent constants of integration. A **general solution** of an nth-order differential equation is a solution containing n independent arbitrary constants.*

A **particular solution** of a differential equation is any solution that can be obtained by assigning particular values to the arbitrary constants in the general solution. Certain differential equations, which are outside the scope of this brief introduction, have **singular solutions** that are not included in the general solution.

The differential equation

$$(2) \qquad \frac{dy}{dx} + F(x, y) = 0$$

is of the first order and first degree. If it can be written

$$(3) \qquad A(x)\, dx + B(y)\, dy = 0,$$

where A is a function of x alone and B is a function of y, the variables in the differential equation (2) are said to be **separable**. If the functions A and B are continuous, the general solution of (3) is

$$\int A(x)\, dx + \int B(y)\, dy = C,$$

where C is an arbitrary constant.

Example. Find the general solution of the first-order differential equation

$$\frac{dy}{dx} + \frac{x^2(y^2 + 3)}{y(x^3 + 1)} = 0.$$

Solution. The variables are separable, since the given differential equation can be written

$$\frac{y}{y^2 + 3}\, dy + \frac{x^2}{x^3 + 1}\, dx = 0.$$

Its general solution, which must contain one arbitrary constant because the equation is of the first order, is

$$\frac{1}{2} \int \frac{(2y\, dy)}{y^2 + 3} + \frac{1}{3} \int \frac{(3x^2\, dx)}{x^3 + 1} = C,$$

or

$$\frac{1}{2} \ln (y^2 + 3) + \frac{1}{3} \ln (x^3 + 1) = C,$$

where C is an arbitrary constant.

The reader should verify that his or her solution satisfies the given differential equation identically by substituting it and its derivatives in the given equation.

---

*For theoretical justifications, see E. L. Ince, *Ordinary Differential Equations*, 4th ed., New York, Dover, 1953.

Exercises

In each of Exercises 1 to 10, state whether the given differential equation is ordinary or partial and, if ordinary, give the order and degree.

1. $y + x\dfrac{dy}{dx} + \dfrac{dx}{dy} = 0.$

2. $y\dfrac{\partial z}{\partial x} + x\dfrac{\partial z}{\partial y} - z = 0.$

3. $x\dfrac{\partial^2 z}{\partial x^2} + z\dfrac{\partial^2 z}{\partial x\,\partial y} + \dfrac{\partial z}{\partial y} = xyz.$

4. $x^2\left(\dfrac{d^2 y}{dx^2}\right)^2 + \left(\dfrac{dy}{dx}\right)^3 + y = 0.$

5. $\dfrac{d^2 y}{dx^2} = 2x.$

6. $p^2\,dq + q\,dp = 0.$

7. $\dfrac{d^2 y}{dx^2} + 5\dfrac{dy}{dx} + 6y = e^x.$

8. $\cos\theta\,\dfrac{dr}{d\theta} = \theta.$

9. $y\dfrac{dy}{dx} = x.$

10. $\dfrac{\partial^2 z}{\partial y\,\partial x} = e^z.$

In Exercises 11 to 14, verify that the given solution satisfies the given differential equation identically.

11. $y^2\left(\dfrac{dy}{dx}\right)^2 + y^2 = r^2.$
Solution: $(x - C)^2 + y^2 = r^2.$

12. $\dfrac{d^3 y}{dx^3} = 0.$
Solution: $y = C_1 x^2 + C_2 x + C_3.$

13. $y = 2x\dfrac{dy}{dx}.$
Solution: $y^2 - Cx = 0.$

14. $\dfrac{d^2 y}{dx^2} + K^2 y = 0.$
Solution: $y = C_1 \cos Kx + C_2 \sin Kx.$

In Exercises 15 to 20, find the general solution of the given differential equation.

15. $x\,dy - y\,dx = 0.$

16. $\sec x\,dy + \sec y\,dx = 0.$

17. $\dfrac{dy}{dx} = \dfrac{1 + y^2}{1 + x^2}.$

18. $\dfrac{dy}{dx} = \sqrt{\dfrac{1 - y^2}{1 - x^2}}.$

19. $ye^x\,dx + \dfrac{y^2 + 1}{x}\,dy = 0.$

20. $\dfrac{dy}{dx} + \dfrac{\cos^2 y}{e^x \sin y} = 0.$

20.2 HOMOGENEOUS EQUATIONS OF THE FIRST ORDER AND FIRST DEGREE

The differential equation

$$\frac{dy}{dx} + F(x, y) = 0$$

is of the first order and first degree. It can be expressed in the form

$$M(x, y)\,dx + N(x, y)\,dy = 0,$$

where M and N are functions of x, or of y, or of both x and y, or else M or N is constant.

Here, and in the following two sections, we continue our discussion of methods for solving various types of first-order differential equations that are of the first degree.

We say that $f(x, y)$ defines a **homogeneous function** of degree n in x and y if

$$f(kx, ky) = k^n f(x, y)$$

for all $k > 0$ and for all x and y in the domain of f. To illustrate, the expressions $x^2 y - 4y^3$, $y^3 \tan (x/y)$, and $(2x - 5y)^3$ all define homogeneous functions of degree 3 in x and y; but the function defined by $7x^2 - xy + y^2 + 10x - 1$ is not homogeneous.

A differential equation of the first order and first degree,

(1) $$M \, dx + N \, dy = 0,$$

is said to be **homogeneous** if M and N are homogeneous of the same degree in x and y.

A homogeneous equation of the form (1) can be transformed into an equation in v and x in which the variables are separable by the substitution of

(2) $$y = vx, \qquad dy = x \, dv + v \, dx.$$

Example. Find the general solution of $(xy - y^2) \, dx - x^2 \, dy = 0$.

Solution. The coefficients of the equation are homogeneous of degree 2. We substitute $y = vx$ and obtain

$$(x \cdot vx - v^2 x^2) \, dx - x^2 (v \, dx + x \, dv) = 0.$$

After division by the factor x^2, we have

$$(v - v^2) \, dx - (v \, dx + x \, dv) = 0,$$

or

$$\frac{dv}{v^2} + \frac{dx}{x} = 0.$$

Here the variables x and v are separated, and the solution is

$$-\frac{1}{v} + \ln x = C, \qquad \text{or} \qquad \ln x - \frac{x}{y} = C.$$

The general solution can be written

$$y = \frac{x}{\ln cx},$$

where we put $\ln c$ for $-C$. Another form of the general solution is

$$x = c_1 e^{x/y}.$$

The substitution

(3) $$x = vy, \qquad dx = v \, dy + y \, dv,$$

instead of (2), also leads to an equation in which the variables y and v are separable and which may be easier to solve than the equation obtained by the substitution (2).

Exercises

Find the general solution of each of the following differential equations.

1. $y \, dx - x \, dy = 0$.

2. $(2x + y) \, dx + x \, dy = 0$.

3. $(x^2 + y^2) \, dx = 2xy \, dy$.

4. $y^3 \, dx + x^3 \, dy = 0$.

5. $(xy - y^2) \, dx - (xy + x^2) \, dy = 0$.

6. $(x - 3y) \, dx + x \, dy = 0$.

7. $(x + y)\,dx + (x - y)\,dy = 0.$

8. $(y^2 - xy)\,dx + x^2\,dy = 0.$

9. $(6x + y)\,dx + (x - 2y)\,dy = 0.$

10. $(y - 2x)\,dx - 2x\,dy = 0.$

11. $y(x^2 + 2y^2)\,dx + x(2x^2 + y^2)\,dy = 0.$

12. $(x - y)e^{y/x}\,dx + (xe^{y/x} + x)\,dy = 0.$

13. $\left[x + x\sin\dfrac{y}{x} - y\cos\dfrac{y}{x}\right]dx$

$$+ x\cos\dfrac{y}{x}\,dy = 0.$$

14. $xy\,dx - (x^2 + 3y^2)\,dy = 0.$

20.3 EXACT EQUATIONS OF THE FIRST ORDER AND FIRST DEGREE

Let M and N be functions of two variables such that M, N, M_y, and N_x are continuous on a rectangular region R. We saw in Section 18.9 that

$$M(x, y)\,dx + N(x, y)\,dy$$

is the exact differential of some function f with the value $z = f(x, y)$ if and only if

$$\frac{\partial M}{\partial y} = \frac{\partial N}{\partial x}.$$

When this condition is satisfied, the differential equation

(1) $$M(x, y)\,dx + N(x, y)\,dy = 0$$

is said to be **exact**. Since its left member is dz, equation (1) can be written

$$dz = 0,$$

and its general solution is

$$f(x, y) = C.$$

To find this function f, we recall that

$$dz = \frac{\partial f}{\partial x}\,dx + \frac{\partial f}{\partial y}\,dy.$$

Thus we seek a function f such that

$$\frac{\partial f}{\partial x} = M(x, y) \qquad \text{and} \qquad \frac{\partial f}{\partial y} = N(x, y).$$

This will enable us to find by inspection the general solution $f(x, y) = C$ of the exact differential equations considered in this section.

Example 1. Find the general solution of

$$(3x^2 - 2y + e^{x+y})\,dx + (e^{x+y} - 2x + 4)\,dy = 0.$$

Solution. This equation is of the first order and first degree, but its variables are not separable and it is not homogeneous. Testing for exactness, we find that

$$\frac{\partial M}{\partial y} = -2 + e^{x+y} = \frac{\partial N}{\partial x}.$$

Hence it is an exact equation.

To find the function f whose exact differential is the left member of the given equation, we have

$$\frac{\partial f}{\partial x} = 3x^2 - 2y + e^{x+y}.$$

By integrating this with respect to x, we get

$$f(x, y) = x^3 - 2xy + e^{x+y} + A(y).$$

Also,

$$\frac{\partial f}{\partial y} = e^{x+y} - 2x + 4,$$

and by integrating this with respect to y, we obtain

$$f(x, y) = e^{x+y} - 2xy + 4y + B(x).$$

Comparing these two expressions for $f(x, y)$, we see that

$$f(x, y) = x^3 - 2xy + 4y + e^{x+y}.$$

Thus the general solution of the given differential equation is

$$x^3 - 2xy + 4y + e^{x+y} = C.$$

Example 2. Solve the differential equation

$$\left(3x^2 - \frac{1}{x - y^2} - \frac{1}{y}\sin\frac{x}{y}\right)dx + \left(\frac{2y}{x - y^2} + \frac{x}{y^2}\sin\frac{x}{y}\right)dy = 0.$$

Solution. Testing for exactness, we find

$$\frac{\partial M}{\partial y} = -\frac{2y}{(x - y^2)^2} + \frac{x}{y^3}\cos\frac{x}{y} + \frac{1}{y^2}\sin\frac{x}{y} = \frac{\partial N}{\partial x}.$$

Thus the given equation is exact. We seek a function f of x and y whose exact differential is the left member of the given equation.

Since

$$\frac{\partial f}{\partial x} = 3x^2 - \frac{1}{x - y^2} - \frac{1}{y}\sin\frac{x}{y},$$

we integrate with respect to x to obtain

$$f(x, y) = x^3 - \ln|x - y^2| + \cos\frac{x}{y} + A(y).$$

From

$$\frac{\partial f}{\partial y} = \frac{2y}{x - y^2} + \frac{x}{y^2}\sin\frac{x}{y},$$

we get

$$f(x, y) = -\int\frac{-2y\,dy}{x - y^2} + \int -\left(\sin\frac{x}{y}\right)\left(-\frac{x}{y^2}\,dy\right) = -\ln|x - y^2| + \cos\frac{x}{y} + B(x).$$

In comparing these two expressions for $f(x, y)$, we see that

$$f(x, y) = x^3 - \ln|x - y^2| + \cos\frac{x}{y},$$

and the general solution of the given differential equation is

$$x^3 - \ln|x - y^2| + \cos\frac{x}{y} = C.$$

Exercises

Find the general solution of each of the following differential equations.

1. $2xy\,dx + x^2\,dy = 0.$

2. $(3x^2 - y)\,dx - x\,dy = 0.$

3. $(x + y)\,dx + x\,dy = 0.$

4. $(e^y + ye^x)\,dx + (xe^y + e^x)\,dy = 0.$

5. $(x + y^2)\,dy + (y - x^2)\,dx = 0.$

6. $\left(-\dfrac{1}{x}\right)dx + \left(\dfrac{1}{y} + \cos y\right)dy = 0.$

7. $(x^2 + \ln y)\,dx + \dfrac{x}{y}\,dy = 0.$

8. $\dfrac{x^2 + 2xy + y}{(x + y)^2}\,dx - \dfrac{x^2 + x}{(x + y)^2}\,dy = 0.$

9. $(1 + y^2)\,dx + (x^2y + y)\,dy = 0.$

10. $e^{xy}(y\cos x - \sin x)\,dx + xe^{xy}\cos x\,dy = 0.$

20.4 LINEAR EQUATIONS OF THE FIRST ORDER

An equation of the first order that is linear in the dependent variable and its first derivative is called a **linear differential equation of first order**. It can be written in the form

$$(1) \qquad \frac{dy}{dx} + P(x) \cdot y = Q(x).$$

Before finding its general solution, we will solve the simpler equation

$$(2) \qquad \frac{dy}{dx} + P(x) \cdot y = 0,$$

formed by equating the left member of (1) to zero. Since the variables in (2) are separable, it can be written

$$P(x)\,dx + \frac{dy}{y} = 0,$$

and its general solution is

$$\int P(x)\,dx + \ln y = \ln C,$$

or

$$(3) \qquad ye^{\int P(x)\,dx} = C.$$

Let us take the differential of both sides of (3); the result is

$$d[ye^{\int P(x)\,dx}] = ye^{\int P(x)\,dx}P(x)\,dx + e^{\int P(x)\,dx}\,dy = 0,$$

or

$$(4) \qquad d[ye^{\int P(x)\,dx}] = e^{\int P(x)\,dx}[P(x)y\,dx + dy] = 0.$$

It follows from (4) that if we multiply the members of equation (2) by $e^{\int P(x)\,dx}$, it becomes exact.

Then, to find the general solution of (1), we multiply both sides by $e^{\int P(x)\,dx}$. This step makes the left member exact, and the right member is always integrable (theoretically) since it involves the variable x alone. Equation (1) becomes

$$e^{\int P(x)\,dx}[P(x)y\,dx + dy] = e^{\int P(x)\,dx}Q(x)\,dx,$$

or in view of (4)

$$d[ye^{\int P(x)\,dx}] = e^{\int P(x)\,dx}Q(x)\,dx;$$

and its complete solution is

$$ye^{\int P(x)\,dx} = \int e^{\int P(x)\,dx}\,Q(x)\,dx + C.$$

The factor $e^{\int P(x)\,dx}$ is said to be an integrating factor of equation (2). More generally, an **integrating factor** of a nonexact equation is an expression such that the equation becomes exact if it is multiplied by that factor. It can be proved that integrating factors exist for a wide range of differential equations of the first order and first degree. But *finding* an integrating factor for a given equation is another matter and can be very difficult.

20.4.1   Theorem.   If we multiply the first-order linear equation

$$\frac{dy}{dx} + P(x)y = Q(x)$$

by the integrating factor $e^{\int P(x)\,dx}$, the resulting equation

$$d[ye^{\int P(x)\,dx}] = e^{\int P(x)\,dx}\,Q(x)\,dx$$

is exact; and the general solution is

$$ye^{\int P(x)\,dx} = \int e^{\int P(x)\,dx}\,Q(x)\,dx + C.$$

Example.   Solve $x^2\,dy - \sin 3x\,dx + 2xy\,dx = 0$.

Solution.   The given equation can be written

$$(5) \qquad\qquad \frac{dy}{dx} + \frac{2}{x}y = \frac{\sin 3x}{x^2},$$

which is in the form

$$\frac{dy}{dx} + P(x)y = Q(x).$$

Therefore (5) is a linear equation of the first order.

To get its general solution (see 20.4.1), we find

$$\int P(x)\,dx = 2\int \frac{dx}{x} = \ln x^2 + C_1.$$

Since all that we require here is any one particular form of $\int P(x)\,dx = \ln x^2 + C_1$, we choose $C_1 = 0$, which in this example gives us the simplest form of the result. So our integrating factor for equation (5) is

$$e^{\int P(x)\,dx} = e^{\ln x^2} = x^2.$$

By multiplying both members of (5) by the integrating factor x^2, we obtain

$$x^2\frac{dy}{dx} + 2xy = \sin 3x,$$

or

(6)
$$x^2 \, dy + 2xy \, dx = \sin 3x \, dx.$$

To find the function f whose exact differential is the left member of (6), we write
$$df(x, y) = 2xy \, dx + x^2 \, dy.$$
Recalling that
$$df(x, y) = \frac{\partial f}{\partial x} dx + \frac{\partial f}{\partial y} dy,$$
we can write
$$\frac{\partial f}{\partial x} = 2xy \quad \text{and} \quad \frac{\partial f}{\partial y} = x^2.$$

Then integrating the first of these equations with respect to x and the second with respect to y gives
$$f(x, y) = x^2 y + A(y) \quad \text{and} \quad f(x, y) = x^2 y + B(x),$$
from which it is clear that $f(x, y) = x^2 y$. Thus
$$d(x^2 y) = 2xy \, dx + x^2 \, dy,$$
and (6) can be written

(7)
$$d(x^2 y) = \sin 3x \, dx.$$

By integrating both members of (7), we find the general solution of the given differential equation to be
$$x^2 y = \int \sin 3x \, dx + C,$$
or
$$3x^2 y + \cos 3x = C.$$

Exercises

Solve the following differential equations.

1. $\dfrac{dy}{dx} + y = e^{-x}.$

2. $(x + 1) \, dy - (x^2 - y - 1) \, dx = 0.$

3. $(1 - x^2)\dfrac{dy}{dx} + xy = ax.$

4. $y' + y \tan x = \sec x.$

5. $\dfrac{dy}{dx} - \dfrac{y}{x} = xe^x$

6. $y' - ay = f(x).$

7. $\dfrac{dy}{dx} + \dfrac{y}{x} = \dfrac{1}{x}.$

8. $y' + \dfrac{2y}{x + 1} = (x + 1)^3.$

9. $\dfrac{dy}{dx} + 2y = x.$

10. $\dfrac{dy}{dx} + y \cot x = \sec x.$

11. $y' + y f(x) = f(x).$

20.5 SECOND-ORDER EQUATIONS SOLVABLE BY FIRST-ORDER METHODS

A differential equation of the second order can be symbolized by

(1)
$$f\left(\frac{d^2 y}{dx^2}, \frac{dy}{dx}, y, x\right) = 0.$$

Certain second-order equations can be reduced to first-order equations by an appropriate substitution.

1. Dependent Variable Missing.

In the event that y is absent from (1), the second-order equation

(2)
$$f\left(\frac{d^2y}{dx^2}, \frac{dy}{dx}, x\right) = 0$$

can be reduced by the substitutions

$$p = \frac{dy}{dx}, \qquad \frac{dp}{dx} = \frac{d^2y}{dx^2}$$

to

(3)
$$f\left(\frac{dp}{dx}, p, x\right) = 0,$$

which is a first-order differential equation with dependent variable p and independent variable x.

If $p = \phi(x, C_1)$ is the general solution of (3) or, equivalently, if $dy = \phi(x, C_1)\, dx$, another integration gives

$$y = \int \phi(x, C_1)\, dx + C_2,$$

the general solution of the original second-order equation (2).

Example 1. Solve $\dfrac{d^2y}{dx^2} - 2\dfrac{dy}{dx} = e^{2x}$.

Solution. Let $p = dy/dx$ and $dp/dx = d^2y/dx^2$. Then the given equation becomes

$$\frac{dp}{dx} - 2p = e^{2x},$$

or

(4)
$$-2p\, dx + dp = e^{2x}\, dx.$$

This is linear in p and dp. To solve it (see 20.4.1), we find the integrating factor $e^{\int -2\, dx} = e^{-2x}$. If we multiply both members of (4) by this integrating factor, we obtain

(5)
$$-2pe^{-2x}\, dx + e^{-2x}\, dp = dx,$$

which is exact. Since $d(pe^{-2x}) = -2pe^{-2x}\, dx + e^{-2x}\, dp$, equation (5) can be written $d(pe^{-2x}) = dx$. Thus the general solution is

$$pe^{-2x} = \int dx + C_1$$

(by 20.4.1), or

$$p = e^{2x}(x + C_1).$$

But the latter equation is

$$dy = xe^{2x}\, dx + C_1 e^{2x}\, dx.$$

Another integration gives

$$y = \int xe^{2x}\, dx + C_1 \int e^{2x}\, dx + C_2,$$

or

$$y = \frac{e^{2x}(2x - 1 + 2C_1)}{4} + C_2,$$

which is the general solution of the given second-order equation.

2. Independent Variable Missing.

If the independent variable x is missing in (1), the second-order equation is

(6)
$$f\left(\frac{d^2y}{dx^2}, \frac{dy}{dx}, y\right) = 0.$$

The substitutions

$$\frac{dy}{dx} = p, \qquad \frac{d^2y}{dx^2} = \frac{dp}{dy}\frac{dy}{dx} = p\frac{dp}{dy}$$

reduce (6) to

(7)
$$f\left(p\frac{dp}{dy}, p, y\right) = 0.$$

This is first-order equation in which p is the dependent variable and y is the independent variable.

If the general solution of (6) is $p = \psi(y, C_1)$, which can be written

$$dx = \frac{dy}{\psi(y, C_1)}$$

when $\psi(y, C_1) \neq 0$, then another integration gives

$$x = \int \frac{dy}{\psi(y, C_1)} + C_2,$$

which is the general solution of (6).

Example 2. Find the complete solution of $y\frac{d^2y}{dx^2} + 2\left(\frac{dy}{dx}\right)^2 = 0$.

Solution. Let

$$\frac{dy}{dx} = p \qquad \text{and} \qquad \frac{d^2y}{dx^2} = p\frac{dp}{dy}.$$

Then the given equation becomes

$$p(y\, dp + 2p\, dy) = 0.$$

First we will solve $y\, dp + 2p\, dy = 0$. The variables are separable, and this equation can be written

$$\frac{dp}{p} + 2\frac{dy}{y} = 0.$$

Its solution is

$$\ln p + \ln y^2 = \ln C_1,$$

or

$$py^2 = C_1.$$

But this is

$$y^2\, dy = C_1\, dx,$$

whose general solution is

(8)
$$y^3 - 3C_1 x - 3C_2 = 0.$$

Now consider $p = 0$ or $dy = 0$. Its solution is $y = C_3$, which is included in (8), since it can be obtained by letting $C_1 = 0$ and $C_2 = \frac{1}{3}C_3^3$ in (8).

Thus (8) is the general solution of the given second-order differential equation.

An Example From Mechanics. A flexible, inextensible cable of uniformly distributed weight, hanging from two supports, assumes the form of a plane curve called the **catenary** (Fig. 20-1). We seek an equation of the catenary.

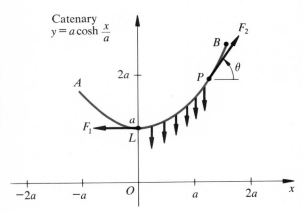

Figure 20-1

Denote by L the lowest point on the cable, and let P be an arbitrarily chosen point on the cable (Fig. 20-1). Draw the y axis vertically through L and ignore, for the present, the exact position of the origin relative to L.

We will confine our attention to the segment LP of the cable. It is acted on by three forces: a force F_1 at L that is tangent to the curve (and therefore horizontal), a force F_2 at P that is tangent to the curve at P, and a downward force, ws, due to the weight of the segment LP, where w is the weight of a linear foot of the cable and s is the length in feet of the segment LP. The forces F_1 and F_2 result from the tension of the cable; F_1 is constant, and F_2 is a function of x since P is an arbitrary point on the cable.

Since the segment LP is in equilibrium, a law of mechanics states that the sum of the horizontal components of these forces must be zero and the sum of their vertical components is zero. So if θ denotes the angle of inclination of F_2, then

(9)
$$F_1 + F_2 \cos \theta = 0$$

and

(10)
$$ws + F_2 \sin \theta = 0.$$

From these two equations we get

$$\tan \theta = \frac{ws}{F_1}.$$

If the desired equation of the catenary is represented by $y = f(x)$, then $\tan \theta = dy/dx$, and the preceding equation becomes

$$\frac{dy}{dx} = \frac{s}{a},$$

where $a = F_1/w$ is constant. By differentiating both sides of this equation with respect to x, we obtain

$$\frac{d^2y}{dx^2} = \frac{1}{a}\frac{ds}{dx}.$$

But

$$\frac{ds}{dx} = \sqrt{1 + \left(\frac{dy}{dx}\right)^2}$$

(by 9.6.3). Therefore

(11)
$$\frac{d^2y}{dx^2} = \frac{1}{a}\sqrt{1 + \left(\frac{dy}{dx}\right)^2}.$$

Although (11) is a differential equation of the second order, the dependent variable y is absent. Thus it can be reduced to a first-order equation, as shown in I above. Let

$$p = \frac{dy}{dx}, \qquad \frac{dp}{dx} = \frac{d^2y}{dx^2}.$$

Then (11) becomes

$$\frac{dp}{dx} = \frac{1}{a}\sqrt{1 + p^2},$$

or

$$\frac{dp}{\sqrt{1 + p^2}} = \frac{dx}{a}.$$

Its solution is

$$\ln\left(p + \sqrt{1 + p^2}\right) = \frac{x}{a} + C.$$

When $x = 0$, the tangent to the catenary is horizontal and thus $dy/dx = p = 0$. By substituting $p = 0$ and $x = 0$ in the preceding equation, we find $C = 0$, and the equation becomes

$$\ln\left(p + \sqrt{1 + p^2}\right) = \frac{x}{a}.$$

Therefore (by 10.10.1)

$$\frac{x}{a} = \sinh^{-1} p = \sinh^{-1}\left(\frac{dy}{dx}\right)$$

and

$$dy = \left(\sinh\frac{x}{a}\right) dx.$$

Integrating, we find

$$y = a\cosh\frac{x}{a} + C_2.$$

If we place the origin a units below the point L, so that $y = a$ when $x = 0$, we have $C_2 = 0$ (since $\cosh 0 = 1$). Therefore the equation of the catenary is

$$y = a\cosh\frac{x}{a}.$$

Exercises

Find the general solution of each of the following differential equations.

1. $\frac{d^2s}{dt^2} = g.$

2. $\frac{d^2y}{dx^2} - \frac{dy}{dx} = x.$

3. $\frac{d^2y}{dx^2} = a^2y.$

4. $\dfrac{d^2y}{dx^2} + \dfrac{1}{x}\dfrac{dy}{dx} = \dfrac{1}{x}.$

8. $y\dfrac{d^2y}{dx^2} - 2\left(\dfrac{dy}{dx}\right)^2 = 0.$

5. $\dfrac{d^2y}{dx^2} - 2\dfrac{dy}{dx} = e^x.$

9. $\dfrac{d^3y}{dx^3} = xe^x.$

6. $\dfrac{d^2y}{dx^2} + a^2y = 0.$

10. $\dfrac{d^2y}{dx^2} + \left(\dfrac{dy}{dx}\right)^2 = 0.$

7. $y'' + yy' = 0.$

20.6 LINEAR EQUATIONS OF ANY ORDER

A differential equation of order n is said to be **linear** if it is in the form of a polynomial of the first degree in the dependent variable and its derivatives. It is of the form

(1) $$\frac{d^ny}{dx^n} + a_1\frac{d^{n-1}y}{dx^{n-1}} + \cdots + a_{n-1}\frac{dy}{dx} + a_ny = f(x),$$

where, in general, the a_i are functions of x.

As an illustration, the equation

$$\frac{d^3y}{dx^3} - 2\frac{d^2y}{dx^2} + (x-6)y = \sin 3x$$

is linear; but the equation

$$\frac{d^2y}{dx^2} + x^3y\frac{dy}{dx} + 7y = 0$$

is not linear, since the degree of the term $x^3y\, dy/dx$ in the variables y and dy/dx is 2 (the sum of the exponents of y and dy/dx).

We shall consider only those linear equations whose coefficients a_i are constants.

When $f(x) = 0$ identically in (1), the linear equation (1) is said to be **homogeneous** (in y, y', y'', $\cdots$, $y^{(n)}$). If $f(x)$ is not identically zero in (1), the linear equation is said to be **nonhomogeneous**.

20.7 HOMOGENEOUS LINEAR EQUATIONS WITH CONSTANT COEFFICIENTS

A homogeneous linear equation with constant coefficients has the form

(1) $$\frac{d^ny}{dx^n} + a_1\frac{d^{n-1}y}{dx^{n-1}} + \cdots + a_{n-1}\frac{dy}{dx} + a_ny = 0,$$

in which the a_i are constants.

If y_1 and y_2 are two particular solutions of (1), then $C_1y_1 + C_2y_2$ is also a solution of (1), where C_1 and C_2 are arbitrary constants.

For

$$\frac{d^ny_1}{dx^n} + a_1\frac{d^{n-1}y_1}{dx^{n-1}} + \cdots + a_{n-1}\frac{dy_1}{dx} + a_ny_1 = 0$$

and

$$\frac{d^n y_2}{dx^n} + a_1 \frac{d^{n-1} y_2}{dx^{n-1}} + \cdots + a_{n-1} \frac{dy_2}{dx} + a_n y_2 = 0$$

identically in x, since y_1 and y_2 are assumed to be solutions of (1). Hence

$$C_1 \left[\frac{d^n y_1}{dx^n} + a_1 \frac{d^{n-1} y_1}{dx^{n-1}} + \cdots + a_n y_1 \right] + C_2 \left[\frac{d^n y_2}{dx^n} + a_1 \frac{d^{n-1} y_2}{dx^{n-1}} + \cdots + a_n y_2 \right] = 0$$

is an identity in x. But this expression can be rewritten

$$\frac{d^n}{dx^n}(C_1 y_1 + C_2 y_2) + a_1 \frac{d^{n-1}}{dx^{n-1}}(C_1 y_1 + C_2 y_2) + \cdots + a_n(C_1 y_1 + C_2 y_2) = 0.$$

Therefore $C_1 y_1 + C_2 y_2$ is a solution of (1).

Let D denote the operation of differentiation with respect to x. Then Dy is the same as dy/dx. Similarly, let $D^2 y = D(Dy) = d^2 y/dx^2$ and, in general, let

$$D^k y = \frac{d^k y}{dx^k},$$

where k is any positive integer.

The "polynomial" in D, $D^n + a_1 D^{n-1} + \cdots + a_{n-1} D + a_n$, called a **linear differential operator**, is defined by

$$(D^n + a_1 D^{n-1} + \cdots + a_{n-1} D + a_n)y = D^n y + a_1 D^{n-1} y + \cdots + a_{n-1} Dy + a_n y$$

$$= \frac{d^n y}{dx^n} + a_1 \frac{d^{n-1} y}{dx^{n-1}} + \cdots + a_{n-1} \frac{dy}{dx} + a_n y.$$

Thus the differential equation

$$\frac{d^2 y}{dx^2} + e^x \frac{dy}{dx} + y \sin x = 0$$

can be written

$$(D^2 + e^x D + \sin x)y = 0.$$

Let $A = D^n + a_1 D^{n-1} + \cdots + a_n$ and $B = D^m + b_1 D^{m-1} + \cdots + b_m$ be two linear operators. We define the **sum** of the linear operators A and B by

$$(A + B)y = Ay + By$$

and the **product** of A and B by

$$(AB)y = A(By).$$

It is not difficult to show that linear operators obey the same algebraic laws for addition, multiplication, and factoring as do ordinary polynomials. As an illustration, $(D^2 - 2D - 3)y = (D - 3)(D + 1)y$. Here $(D - 3)(D + 1)y$ means

$$(D - 3)(D + 1)y = D[(D + 1)y] - 3[(D + 1)y]$$

$$= D(Dy + y) - 3(Dy + y) = D^2 y + Dy - 3Dy - 3y$$

$$= D^2 y - 2Dy - 3y = (D^2 - 2D - 3)y.$$

Consequently, we write

$$D^2 - 2D - 3 = (D - 3)(D + 1)$$

as if the linear differential operator were an ordinary polynomial.

For simplicity, we will confine our attention in the rest of this chapter to linear equations of the second order. However, the methods about to be explained apply to linear equations of any order, provided that they have constant coefficients.

20.8 SOLUTION OF SECOND-ORDER HOMOGENEOUS LINEAR EQUATIONS WITH CONSTANT COEFFICIENTS

Consider any second-order homogeneous linear equation with real constant coefficients,

(1) $$(D^2 + a_1 D + a_2)y = 0.$$

1. Conditions for a General Solution.

We have seen that if y_1 and y_2 are solutions of (1) and C_1 and C_2 are any constants, then

(2) $$y = C_1 y_1 + C_2 y_2$$

is also a solution. However, in order for (2) to be the general solution of (1), it is necessary that y_1 and y_2 be independent solutions. In other words, y_1 is not a mere constant multiple of y_2. For instance, if $y_1 = k y_2$, then

$$y = C_1 y_1 + C_2 y_2 = C_1 k y_2 + C_2 y_2 = C y_2,$$

where $C = C_1 k + C_2$, so that (2) contains only one really essential constant.

If $y = C_1 y_1 + C_2 y_2$ is the general solution of (1), then

(3) $$\begin{vmatrix} y_1 & y_2 \\ y_1' & y_2' \end{vmatrix} \not\equiv 0.$$

For if this determinant (called the *Wronskian* of y_1 and y_2) vanishes identically, then $y_1 y_2' = y_2 y_1'$, or

$$\frac{y_1'}{y_1} = \frac{y_2'}{y_2}.$$

On integration, this becomes $\ln y_1 = \ln y_2 + C$ or $y_1 = k y_2$, which we have seen is not possible when (2) is the general solution. Moreover, if $y_1 = k y_2$ and $y_1' = k y_2'$, the determinant in (3) becomes zero identically, since

$$\begin{vmatrix} y_1 & y_2 \\ y_1' & y_2' \end{vmatrix} = \begin{vmatrix} k y_2 & y_2 \\ k y_2' & y_2' \end{vmatrix} = k \cdot 0 = 0.$$

Therefore (3) is both a necessary and a sufficient condition for y_1 and y_2 to be independent and for $C_1 y_1 + C_2 y_2$ to be the general solution of (1), where C_1 and C_2 are arbitrary constants.

2. Details of Solution.

If we substitute e^{rx} for y in (1), where r is a constant, we obtain

$$(D^2 + a_1 D + a_2)e^{rx} = r^2 e^{rx} + a_1 r e^{rx} + a_2 e^{rx} = 0,$$

or

$$e^{rx}(r^2 + a_1 r + a_2) = 0.$$

Since $e^{rx} \neq 0$, this will be an identity in x if and only if r is a number such that

(4)
$$r^2 + a_1 r + a_2 = 0.$$

Equation (4) is called the **auxiliary equation** of (1).

It follows that $y = e^{r_1 x}$ is a solution of the homogeneous linear differential equation

$$(D^2 + a_1 D + a_2)y = 0$$

if and only if r_1 is a root of the auxiliary equation

$$r^2 + a_1 r + a_2 = 0.$$

Since the auxiliary equation (4) is quadratic, it has two roots, r_1 and r_2. Thus in finding the general solution of (1), three cases must be considered, according as the roots, r_1 and r_2, of (4) are real and distinct, real and equal, or conjugate imaginary.

CASE 1 ($r_1 \neq r_2$, *real*). Since r_1 and r_2 are distinct roots of the auxiliary equation (4), two solutions of the differential equation (1) are $y_1 = e^{r_1 x}$ and $y_2 = e^{r_2 x}$. Since the Wronskian of y_1 and y_2 is not zero, y_1 and y_2 are independent, and

$$y = C_1 e^{r_1 x} + C_2 e^{r_2 x},$$

where C_1 and C_2 are arbitrary constants, is the general solution.

Example 1. Solve $(D^2 + 3D - 10)y = 0$.

Solution. The roots of the auxiliary equation

$$r^2 + 3r - 10 = 0$$

are 2 and -5, and the general solution of the given differential equation is

$$y = C_1 e^{2x} + C_2 e^{-5x}.$$

CASE 2 ($r_1 = r_2$). The auxiliary equation can be written

$$(r - r_1)^2 = r^2 - 2r_1 r + r_1^2 = 0.$$

Therefore the differential equation (1), in this case, is

(5)
$$(D^2 - 2r_1 D + r_1^2)y = 0.$$

Of course, one solution of (5) is $y = e^{r_1 x}$. We will show that another solution is $y = xe^{r_1 x}$.

Now

$$D(xe^{r_1x}) = r_1xe^{r_1x} + e^{r_1x} \quad \text{and} \quad D^2(xe^{r_1x}) = r_1^2xe^{r_1x} + 2r_1e^{r_1x}.$$

So if we substitute xe^{r_1x} for y in the left member of (5), we have

$$(D^2 - 2r_1D + r_1^2)xe^{r_1x} = (r_1^2xe^{r_1x} + 2r_1e^{r_1x}) - 2r_1(r_1xe^{r_1x} + e^{r_1x}) + r_1^2xe^{r_1x}$$

$$= xe^{r_1x}(r_1^2 - r_1^2) = 0$$

identically in x. Therefore $y = xe^{r_1x}$ is a solution of (5).

The general solution of (5) is

$$y = C_1e^{r_1x} + C_2xe^{r_1x},$$

where C_1 and C_2 are arbitrary constants.

Example 2.   Solve $(D^2 - 14D + 49)y = 0$.

Solution.   Both roots of the auxiliary equation $r^2 - 14r + 49 = 0$ are equal to 7. Thus the general solution of the given differential equation is

$$y = C_1e^{7x} + C_2xe^{7x}.$$

CASE 3 (r_1 and r_2 are complex conjugates).

Up to now we have defined e^r only for real exponents r. Let r be the complex number $r = \alpha + i\beta$, where α and β are real and $i = \sqrt{-1}$. We define $e^r = e^{\alpha+i\beta}$ by

(6) $$e^{\alpha+i\beta} = e^{\alpha}(\cos\beta + i\sin\beta).$$

When the distinct roots of the auxiliary equation (4) are the complex conjugate numbers $r_1 = \alpha + i\beta$ and $r_2 = \alpha - i\beta$, the general solution of the given differential equation (1) is

$$y = c_1e^{(\alpha+i\beta)x} + c_2e^{(\alpha-i\beta)x}$$

$$= e^{\alpha x}(c_1e^{i\beta x} + c_2e^{-i\beta x}).$$

By means of (6), this can be written

$$y = e^{\alpha x}[c_1(\cos\beta x + i\sin\beta x) + c_2(\cos\beta x - i\sin\beta x)]$$

$$= e^{\alpha x}[(c_1 + c_2)\cos\beta x + i(c_1 - c_2)\sin\beta x].$$

By letting $c_1 + c_2 = C_1$ and $i(c_1 - c_2) = C_2$, the general solution may be written

$$y = e^{\alpha x}(C_1\cos\beta x + C_2\sin\beta x).$$

It is easy to verify that this relation satisfies the given differential equation identically in x.

The results of this section may be summarized as follows.

20.8.1 Theorem. Let
$$(D^2 + a_1 D + a_2)y = 0$$
be a homogeneous linear differential equation of the second order with real constant coefficients, and let r_1 and r_2 be the roots of the auxiliary equation
$$r^2 + a_1 r + a_2 = 0.$$
  (i) If r_1 and r_2 are real and distinct, the general solution of the given homogeneous linear equation is
$$y = C_1 e^{r_1 x} + C_2 e^{r_2 x}$$
where C_1 and C_2 are arbitrary constants.
  (ii) If $r_1 = r_2$, the general solution is
$$y = C_1 e^{r_1 x} + C_2 x e^{r_1 x}.$$
  (iii) If $r_1 = \alpha + i\beta$ and $r_2 = \alpha - i\beta$, where α and β are real numbers, the general solution of the given differential equation is
$$y = e^{\alpha x}(C_1 \cos \beta x + C_2 \sin \beta x),$$
in which C_1 and C_2 are arbitrary constants.

Example 3. Solve $y'' - 4y' + 13y = 0$.

Solution. The roots of the auxiliary equation $r^2 - 4r + 13 = 0$ are $2 + 3i$ and $2 - 3i$. Hence the general solution is
$$y = e^{2x}(C_1 \cos 3x + C_2 \sin 3x).$$

Exercises

Find the general solution of each of the following differential euqations.

1. $\dfrac{d^2 y}{dx^2} - 5\dfrac{dy}{dx} + 6y = 0.$

2. $\dfrac{d^2 y}{dx^2} + 5\dfrac{dy}{dx} - 6y = 0.$

3. $\dfrac{d^2 y}{dx^2} + 6\dfrac{dy}{dx} - 7y = 0.$

4. $\dfrac{d^2 y}{dx^2} - 6\dfrac{dy}{dx} + 9y = 0.$

5. $6\dfrac{d^2 y}{dx^2} + 7\dfrac{dy}{dx} - 3y = 0.$

6. $\dfrac{d^2 y}{dx^2} - 4y = 0.$

7. $\dfrac{d^2 y}{dx^2} + 9y = 0.$

8. $\dfrac{d^2 y}{dx^2} + 2\dfrac{dy}{dx} + 2y = 0.$

9. $\dfrac{d^2 y}{dx^2} + \dfrac{dy}{dx} + y = 0.$

10. Show that the solution of
$$\frac{d^2 y}{dx^2} - 2b\frac{dy}{dx} + c^2 y = 0,$$
where b and c are real numbers, can be written
$$y = e^{bx}(C_1 \cosh \sqrt{b^2 + c^2}\,x + C_2 \sinh \sqrt{b^2 + c^2}\,x).$$

20.9 NONHOMOGENEOUS LINEAR EQUATIONS OF ORDER TWO

Any nonhomogeneous linear differential equation of the second order, with constant coefficients, can be written

(1) $$(D^2 + a_1 D + a_2)y = f(x),$$

where a_1 and a_2 are constants. We assume that a_1 and a_2 are real.

Consider the associated *homogeneous* equation

(2) $$(D^2 + a_1 D + a_2)y = 0$$

and let its general solution be

(3) $$y_c = C_1 u_1 + C_2 u_2,$$

where u_1 and u_2 are functions of x. Then (3) is called the **complementary function** of the nonhomogeneous linear equation (1).

If we can discover a *particular* solution of (1) in any way whatever, we can write its *general* solution by means of the following theorem.

20.9.1 Theorem. If the differential equation

$$(D^2 + a_1 D + a_2)y = f(x)$$

has a particular solution y_p and the complementary function y_c, then its general solution is

$$y = y_p + y_c.$$

Proof. The equation

$$y_c + y_p = C_1 u_1 + C_2 u_2 + y_p$$

has two arbitrary constants and, when substituted in (1), yields

$$(D^2 + a_1 D + a_2)(y_c + y_p) = f(x),$$

or

$$(D^2 + a_1 D + a_2)y_c + (D^2 + a_1 D + a_2)y_p = f(x).$$

But the first term in the latter equation is zero identically in x, since y_c is a solution of (2); and the remainder of the equation is an identity in x, since y_p is a solution of (1). ∎

Example 1. Find the general solution of the nonhomogeneous linear equation

$$(D^2 - 1)y = -2x.$$

Solution. By trial, we find that $y = 2x$ is a particular solution of the given differential equation. The complementary function is $y = C_1 e^x + C_2 e^{-x}$. Thus the general solution of the given differential equation is

$$y = 2x + C_1 e^x + C_2 e^{-x}.$$

In the above example we "guessed" a particular solution of the nonhomogeneous linear equation. We will now explain a powerful method that enables us to find a particular solution of a nonhomogeneous linear equation with constant coefficients when its complementary function is known. It is called the **method of variation of parameters**.

To find a particular solution of (1), we start by replacing C_1 and C_2 in the complementary function (3) by arbitrary functions of x, v_1 and v_2. We let u_1 and u_2 be independent solutions of (2) and then determine $v_1(x)$ and $v_2(x)$ so that

(4) $$y = v_1(x)\, u_1 + v_2(x)\, u_2$$

is a particular solution of (1).

In order for (4) to be a particular solution of (1), substitution of (4) and its first and second derivatives in (1) must result in an identity in x. Accordingly, we differentiate (4) with respect to x, obtaining

$$(5) \qquad y' = v_1 u_1' + v_2 u_2' + (v_1' u_1 + v_2' u_2).$$

Since v_1 and v_2 are two arbitrary functions of x, we can impose two conditions on v_1 and v_2. In order to simplify (5) before differentiating agian, we let our first condition on v_1 and v_2 be

$$(6) \qquad v_1' u_1 + v_2' u_2 = 0.$$

Then (5) becomes

$$(7) \qquad y' = v_1 u_1' + v_2 u_2'.$$

Differentiating (7) with respect to x, we get

$$(8) \qquad y'' = v_1 u_1'' + v_2 u_2'' + (v_1' u_1' + v_2' u_2').$$

We now substitute (4), (7), and (8) in (1) and, as our second condition on v_1 and v_2, demand that the resulting equation be an identity in x. This substitution gives

$$v_1 u_1'' + v_2 u_2'' + (v_1' u_1' + v_2' u_2') + a_1(v_1 u_1' + v_2 u_2') + a_2(v_1 u_1 + v_2 u_2) = f(x),$$

or

$$v_1(u_1'' + a_1 u_1' + a_2 u_1) + v_2(u_2'' + a_1 u_2' + a_2 u_2) + v_1' u_1' + v_2' u_2' = f(x).$$

Since u_1 and u_2 are solutions of (2), this becomes

$$(9) \qquad v_1' u_1' + v_2' u_2' = f(x).$$

The two conditions imposed on the two arbitrary expressions $v_1(x)$ and $v_2(x)$ are that they satisfy (6) and (9):

$$(10) \qquad v_1' u_1 + v_2' u_2 = 0,$$
$$v_1' u_1' + v_2' u_2' = f(x).$$

These two equations are linear in v_1' and v_2', and it has been shown (Section 20.8) that since u_1 and u_2 are linearly independent,

$$\begin{vmatrix} u_1 & u_2 \\ u_1' & u_2' \end{vmatrix} \neq 0.$$

Thus the system (10) in v_1' and v_2' has a unique solution. After finding this solution for $v_1'(x)$ and $v_2'(x)$, we determine $v_1(x)$ and $v_2(x)$ by integration. This provides a particular solution of (1),

$$y_p = v_1(x)\, u_1 + v_2(x)\, u_2,$$

from (4).

To summarize the **method of variation of parameters**:

A particular solution of the nonhomogeneous linear differential equation with constant coefficients,

$$(D^2 + a_1 D + a_2)y = f(x),$$

is

$$y = v_1(x)\, u_1(x) + v_2(x)\, u_2(x),$$

where u_1 and u_2 are any linearly independent solutions of the corresponding homogeneous equation

$$(D^2 + a_1 D + a_2)y = 0,$$

and v_1 and v_2 are functions of x satisfying the system

$$v_1' u_1 + v_2' u_2 = 0,$$
$$v_1' u_1' + v_2' u_2' = f(x).$$

Example 2. Find the general solution of the nonhomogeneous linear equation

(11) $$(D^2 + 1)y = \sec x.$$

Solution. The general solution of the associated homogeneous equation

$$(D^2 + 1)y = 0$$

is

(12) $$y_c = C_1 \cos x + C_2 \sin x,$$

which is the complementary function of the given equation (11).

To find a particular solution of (11), we replace C_1 and C_2 in (12) by arbitrary expressions in x, $v_1(x)$, and $v_2(x)$, getting

(13) $$y = v_1(x) \cos x + v_2(x) \sin x.$$

We wish to determine v_1 and v_2 so that (13) will be a particular solution of (11). The functions v_1' and v_2' are found by solving the equations

(14) $$v_1' \cos x + v_2' \sin x = 0,$$
$$-v_1' \sin x + v_2' \cos x = \sec x.$$

This gives $v_1'(x) = -\tan x$ and $v_2'(x) = 1$. Thus

$$v_1(x) = \int -\tan x \, dx = \ln \cos x,$$

$$v_2(x) = \int dx = x.$$

Since we are seeking just one particular solution here, not the most general one, we can use the simplest integrals that occur to us, $v_1(x) = \ln \cos x$ (not $\ln \cos x + C$) and $v_2(x) = x$ (not $x + C$). By substituting these results in (13), we find a particular solution of (11) to be

$$y_p = \cos x \ln \cos x + x \sin x.$$

So the general solution of (11) is $y = y_c + y_p$, or

$$y = C_1 \cos x + C_2 \sin x + \cos x \ln \cos x + x \sin x.$$

Exercises

Find the general solution of each of the following equations.

1. $\dfrac{d^2 y}{dx^2} + y = 5.$

2. $\dfrac{d^2 y}{dx^2} - 3\dfrac{dy}{dx} + 2y = 5x + 2.$

3. $\dfrac{d^2 y}{dx^2} + y = \csc x \cot x.$

4. $\dfrac{d^2y}{dx^2} - 4y = e^{2x}$.

5. $\dfrac{d^2y}{dx^2} + y = \tan x$.

6. $\dfrac{d^2y}{dx^2} + y = \cos x$.

7. $\dfrac{d^2y}{dx^2} - 5\dfrac{dy}{dx} + 6y = 2e^x$.

8. $\dfrac{d^2y}{dx^2} + y = \cot x$.

9. $\dfrac{d^2y}{dx^2} - 3\dfrac{dy}{dx} + 2y = \dfrac{e^x}{e^x + 1}$.

10. $\dfrac{d^2y}{dx^2} + y = \sec^2 x$.

20.10 A VIBRATING SPRING

Consider a coiled spring of length l, hanging vertically from a support [Fig. 20-2(a)]. Hooke's law says that the amount s of stretching (or compressing) of the spring, due to a vertical force F, is proportional to $|F|$; that is,

(1) $$|F| = ks,$$

where k is a factor of proportionality. The factor k is unique for a particular spring and depends on the material, thickness, etc., of the spring.

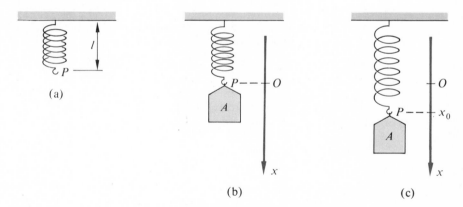

Figure 20-2

Let an object A, of weight w, be attached to the lower end of the spring and let the system come to equilibrium. We assume that there is a vertical coordinate axis whose positive direction is downward and whose origin is on a horizontal line through the lowest point P of the spring [Fig. 20-2(b)].

The object A is now pulled downward a distance x_0 and then released [Fig. 20-2(c)]. We wish to discuss the resulting motion of the lowest point P of the spring.

1. Simple Harmonic Motion.

For the present, we assume that there is no air resistance or other friction.

When the object A is released, there is an upward force on P due to the tension of the spring; this force tends to restore P to the equilibrium position and, by Hooke's law, is equal to $-kx$. But by Newton's second law (definition of force), this force is equal to ma, where $m = w/g$ is the mass of the object A, a is the acceleration of P, and g is the magnitude of the acceleration due to gravity. Thus

$$(2) \qquad \frac{w}{g} \frac{d^2x}{dt^2} = -kx$$

is the differential equation that describes the situation mathematically at any time t after release.

If we let $kg/w = B^2$, equation (2) can be written

$$\frac{d^2x}{dt^2} + B^2x = 0,$$

which is a linear equation with constant coefficients. By 20.8.1, its general solution is

$$(3) \qquad x = C_1 \sin Bt + C_2 \cos Bt,$$

in which C_1 and C_2 are arbitrary constants.

To determine the values of C_1 and C_2 for a particular problem, we differentiate both members of (3) and obtain

$$(4) \qquad \frac{dx}{dt} = C_1 B \cos Bt - C_2 B \sin Bt.$$

At the moment of release, we have $t = 0$, $x = x_0$, and $v = dx/dt = 0$. By substituting these boundary conditions in (3) and (4), we find $C_1 = 0$ and $C_2 = x_0$. Thus the solution of (2) with the boundary conditions $t = 0$, $x = x_0$, and $v = dx/dt = 0$ is

$$(5) \qquad x = x_0 \cos Bt,$$

where $B = \sqrt{kg/w}$. The motion of P is now easily described. As t increases, P oscillates up and down a distance of x_0 units from the origin; x_0 is called the **amplitude** of this periodic motion, and its **period** is $2\pi/B$. The motion described by equation (5) is called **simple harmonic motion** (Fig. 20-3).

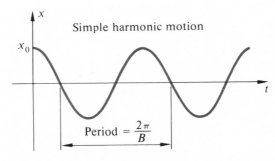

Figure 20-3

Example 1. When an object weighing 5 pounds is attached to the lowest point P of a certain spring that hangs vertically, the spring is extended 6 inches. The 5-pound weight is replaced by a 20-pound weight and the system is allowed to come to equilibrium. If the 20-pound weight is now pulled downward another foot and then released, describe the motion of the lowest point P of the spring. (Assume that there is no air resistance or other friction.)

Solution. Let the acceleration due to gravity be $g = 32$ feet per second per second. To determine the constant k in $|F| = ks$ (Hooke's law), we substitute $F = 5$ and $s = \frac{1}{2}$ and get $k = 10$.

As shown above, the differential equation that expresses the situation is

$$(6) \qquad \frac{w}{g}\frac{d^2x}{dt^2} + 10x = 0.$$

The general solution of (6) is

$$(7) \qquad x = C_1 \sin 4t + C_2 \cos 4t,$$

since $k = 10$, $g = 32$, $w = 20$, and $\sqrt{kg/w} = 4$.

To determine the values of C_1 and C_2 for this particular problem, we substitute the boundary conditions, $t = 0$, $x = 1$, and $v = 0$ in (7) and also in

$$(8) \qquad v = \frac{dx}{dt} = 4C_1 \cos 4t - 4C_2 \sin 4t,$$

obtaining $C_1 = 0$ and $C_2 = 1$. Thus the solution of (6) is

$$x = \cos 4t.$$

The motion of P is simple harmonic motion, with period $\frac{1}{2}\pi$ and amplitude 1 foot. That is, P oscillates up and down from 1 foot below 0 to 1 foot above 0 and then back to 1 foot below 0 every $\frac{1}{2}\pi \doteq 1.57$ seconds.

2. Damped Vibration.

In 1, above, we assumed a simplified situation in which there was no friction. In practice, there is always friction due to air (or other) resistance that causes the motion to be other than simple harmonic. Often the retarding force can be approximately accounted for in the differential equation describing the motion by a term that is proportional to the velocity. A retarding force, such as air resistance, will act in the opposite direction to the motion of the vibrating particle. Thus we replace the Hooke's law equation (1) by

$$F = -kx - qv,$$

where q is a positive constant and v is the velocity of the oscillating particle P. The term $-qv$ in this equation represents the retarding force.

The differential equation that describes this vibration with a retarding force is

$$(9) \qquad \frac{w}{g}\frac{d^2x}{dt^2} = -kx - q\frac{dx}{dt}.$$

By letting $B^2 = kg/w$ and $E = qg/w$, equation (9) can be written

$$(10) \qquad \frac{d^2x}{dt^2} + E\frac{dx}{dt} + B^2x = 0,$$

where E is positive. This is a homogeneous linear equation with constant coefficients whose auxiliary equation is

$$(11) \qquad r^2 + Er + B^2 = 0.$$

Three cases arise, depending on whether $E^2 - 4B^2$ is negative, zero, or positive.

CASE 1 ($E^2 - 4B^2 < 0$). The roots of the auxiliary equation (11) are conjugate imaginary; we will denote them by $-\alpha \pm i\beta$, where α and β are positive numbers.

By 20.8.1, the general solution of (10) is

$$x = e^{-\alpha t}(C_1 \sin \beta t + C_2 \cos \beta t),$$

or

$$(12) \qquad\qquad x = Ce^{-\alpha t} \sin (\beta t + \gamma).$$

The factor $e^{-\alpha t}$ in (12) is called a **damping factor**. Since $\alpha > 0$, $\lim\limits_{t \to \infty} e^{-\alpha t} = 0$. The motion of P described by (12) is **damped harmonic motion**. The amplitude of the vibration is $Ce^{-\alpha t}$, which approaches zero as $t \longrightarrow \infty$ [Fig. 20-4(a)].

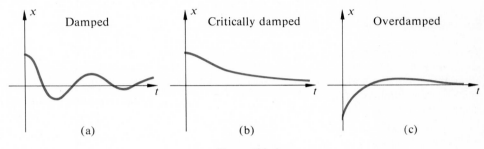

Figure 20-4

CASE 2 $(E^2 - 4B^2 = 0)$.   In this case, the roots of the auxiliary equation are equal. We denote the root by $-\alpha$. Then the general solution of (10) is

$$(13) \qquad\qquad x = C_1 e^{-\alpha t} + C_2 t e^{-\alpha t}.$$

The motion described by this equation is said to be **critically damped**. It is not oscillatory [Fig. 20-4(b)].

CASE 3 $(E^2 - 4B^2 > 0)$.   The roots of the auxiliary equation are real and distinct. If we denote them by $-\alpha_1$ and $-\alpha_2$, the general solution of equation (10) is

$$(14) \qquad\qquad x = C_1 e^{-\alpha_1 t} + C_2 e^{-\alpha_2 t}.$$

The motion described by equation (14) is said to be **overdamped**. It is not oscillatory [Fig. 20-4(c)].

Example 2. If a damping force of magnitude $0.2\,|v|$ is imposed on the system in Example 1, the differential equation that expresses the motion of P becomes

$$(15) \qquad\qquad \frac{d^2x}{dt^2} + 0.32 \frac{dx}{dt} + 16x = 0.$$

The auxiliary equation of this homogeneous linear equation with constant coefficients is

$$r^2 + 0.32r + 16 = 0,$$

having roots $r = -0.16 \pm i\sqrt{15.9744} \doteq -0.16 \pm 4i$. The general solution of (15) is (approximately)

$$(16) \qquad\qquad x = e^{-0.16t}(C_1 \sin 4t + C_2 \cos 4t).$$

The boundary conditions in this problem are $t = 0$, $x = \frac{1}{2}$, and $v = 0$; on substituting them in (16) and in

$$\frac{dx}{dt} = e^{-0.16t}[(4C_1 - 0.16C_2)\cos 4t - (4C_2 + 0.16C_1)\sin 4t],$$

we find $C_1 = 0.02$ and $C_2 = 0.5$. So equation (16) becomes

(17) $$x = e^{-0.16t}(0.02\sin 4t + 0.5\cos 4t).$$

The motion of P, the lowest particle of the spring, is damped harmonic motion and its damping factor is $e^{-0.16t}$. The motion of P is oscillatory with a period of approximately $\frac{1}{2}\pi$.

Exercises

1. A spring with a spring constant of 20 pounds per foot is loaded with a 10-pound weight and allowed to reach equilibrium. It is then raised 1 foot and released. Find the equation of motion, the frequency, and the period. Neglect friction.

2. A spring with a spring constant of 100 pounds per foot is loaded with a 1-pound weight and brought to equilibrium. It is then stretched an additional 1 inch and released. What is the equation of motion, the amplitude, and the period? Neglect friction.

3. In Exercise 1, what is the absolute value of the velocity of the moving weight as it passes through the equilibrium position?

4. A 10-pound weight stretches a certain spring 4 inches. This weight is removed and replaced with a 20-pound weight, which is then allowed to reach equilibrium. The weight is next raised 1 foot and released with an initial velocity of 2 feet per second *downward*. What is the equation of motion? Neglect friction.

5. A spring with a spring constant of 20 pounds per foot is loaded with a 10-pound weight and allowed to reach equilibrium. It is then displaced 1 foot

downward and released. If the weight experiences a retarding force in pounds equal to one-tenth the velocity, find the equation of motion.

6. Determine the motion in Exercise 5 if the retarding force equals 4 times the velocity at every point.

7. In Exercise 5, how long will it take the oscillations to diminish to one-tenth their original amplitude?

8. A spring, assumed weightless, is attached to a weightless damping disk immersed in a viscous fluid. The disk is displaced a distance x_0 and released. If the spring constant is K and if the damping disk is retarded at every point by a force given by $-qv$, what is the equation of motion?

9. In Exercise 5, what will be the equation of motion if the weight is given an upward velocity of 1 foot per second at the moment of release?

10. A weight is hung on a spring, displacing it a distance d. It is then further displaced and released. Neglecting friction, show that the period of the resulting oscillations depends only on the initial displacement d.

20.11 ELECTRIC CIRCUITS

Many problems in electric circuits lead to linear differential equations. A **circuit** is any closed path within an electrical network. Figure 20-5 shows a circuit containing a source of electromotive force E (a battery or a generator), a resistor R, an inductor L, a condenser (or capacitor) C, and a switch s, all in series.

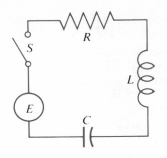

Figure 20-5

The resistor, inductor, and condenser use energy supplied by the source of electromotive force E. A resistor uses energy in resisting the flow of electricity through it; this is analogous to friction resisting the flow of water through a pipe. An inductor tends to stabilize the flow of electricity by opposing any increase or decrease in the current, and in doing so stores and releases energy. A condenser (often called a capacitor) consists of plates separated by insulators; it stores charged particles.

For notation, we use four variables, $q, t, i,$ and E, and three constants, $C, R,$ and L.

q quantity of electricity, measured in *coulombs*, stored or generated in an element of the circuit.

t time in *seconds*.

i current, measured in *amperes*; it is the time rate of change in the quantity of electricity as it flows from one element of the circuit to another, so that

$$i = \frac{dq}{dt}.$$

E electromotive force, measured in *volts*; it is analogous to the pressure that causes water to move through a pipe.

C capacitance, measured in *farads*; it is constant for a particular condenser.

R resistance, measured in *ohms*; it is constant for a particular resistor.

L coefficient of inductance, measured in *henrys*; it is constant for a particular inductor.

It is shown in physics that

1. the voltage drop across a condenser is

$$\frac{1}{C}q,$$

where q is the charge on the condenser at time t;

2. the voltage drop across a resistor is

$$Ri;$$

3. the voltage drop across an inductor is

$$L\frac{di}{dt}.$$

Kirchhoff's second law (voltage law) says that in any closed electric circuit, the sum of the voltage drops at any instant is equal to the impressed electromotive force $E(t)$ at that time.

For the circuit shown in Fig. 20-5, which contains a resistor, an inductor, a condenser, a source of electromotive force $E(t)$, and a switch, Kirchhoff's voltage law is expressed in mathematical terms by the differential equation

$$(1) \qquad L\frac{di}{dt} + Ri + \frac{1}{C}q = E(t).$$

To find the current i at any time t, we substitute dq/dt for i throughout (1), and rewrite it as

$$(2) \qquad \frac{d^2q}{dt^2} + \frac{R}{L}\frac{dq}{dt} + \frac{1}{LC}q = \frac{1}{L}E(t).$$

This is a second-order nonhomogeneous linear equation, and its general solution (see Section 20.9) gives q as a function of t. Differentiation of this solution with respect to t gives $i\,(=dq/dt)$ in terms of t.

We can also get the current i as a function of t by differentiating (1) with respect to t, remembering that $dq/dt = i$:

$$(3) \qquad \frac{d^2i}{dt^2} + \frac{R}{L}\frac{di}{dt} + \frac{1}{LC}i = \frac{1}{L}\frac{d}{dt}E(t).$$

The general solution of this second-order nonhomogeneous linear equation gives i as a function of t.

Example 1. Find the current i as a function of t after the switch is closed in an RCL circuit consisting of a resistor, an inductor, a condenser, a 12-volt battery, and a switch, all in series (Fig. 20-5), if $R = 16$ ohms, $L = 0.02$ henry, and $C = 2 \times 10^{-4}$ farad. Assume that there is no charge on the condenser before the switch is closed, so that $q = 0$ when $t = 0$; also, $i = 0$ when $t = 0$.

Solution. By Kirchhoff's voltage law,

$$\frac{d^2q}{dt^2} + \frac{R}{L}\frac{dq}{dt} + \frac{1}{LC}q = \frac{1}{L}E(t).$$

For the present circuit, this equation becomes

$$(D^2 + 800D + 250{,}000)q = 600.$$

We shall solve this nonhomogeneous linear equation of the second order as in Section 20.9. The equation can be written

$$[D - (-400 + 300i)][D - (-400 - 300i)]q = 600.$$

Its complementary function is

$$q = e^{-400t}(C_1 \cos 300t + C_2 \sin 300t),$$

and a particular solution, found by inspection, is the constant

$$q = 2.4 \times 10^{-3}.$$

Therefore the general solution is

$$q = e^{-400t}(C_1 \cos 300t + C_2 \sin 300t) + 2.4 \times 10^{-3}.$$

By substituting the initial condition, $q = 0$ when $t = 0$, in this equation, we determine that $C_1 = -2.4 \times 10^{-3}$. Hence the equation is

$$q = e^{-400t}(-2.4 \times 10^{-3} \cos 300t + C_2 \sin 300t) + 2.4 \times 10^{-3}.$$

We now differentiate the members of this equation with respect to t, obtaining

$$\frac{dq}{dt} = i = e^{-400t}[-300(-2.4 \times 10^{-3}) \sin 300t + 300C_2 \cos 300t]$$

$$-400e^{-400t}[(-2.4 \times 10^{-3}) \cos 300t + C_2 \sin 300t].$$

Substituting the initial condition, $i = 0$ when $t = 0$, we find $C_2 = -3.2 \times 10^{-3}$.

Thus the required equation for the current i as a function of t, when $t > 0$, is

$$i = 2e^{-400t} \sin 300t.$$

Example 2.   The circuit shown in Fig. 20-6 has a resistance R, an inductance L, and a source of sinusoidal electromotive force $E(t) = K \sin \omega t$. We assume that the switch is closed at $t = 0$ and that $E(0) = 0$.

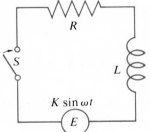

By Kirchhoff's second law,

$$L \frac{di}{dt} + Ri = K \sin \omega t.$$

We seek an expression for the (alternating) current i as a function of t, for $t > 0$.

The above equation can be written

$$\frac{R}{L}i\, dt + di = \frac{K}{L} \sin \omega t\, dt;$$

Figure 20-6

it is linear in i and di, and of the first order. Its general solution (by 20.4.1) is

(4)
$$i = ke^{-Rt/L} + \frac{K}{M^2}(R \sin \omega t - L\omega \cos \omega t),$$

where $M = \sqrt{R^2 + L^2\omega^2}$, the **steady-state impedance** of the circuit, and k is an arbitrary constant.

From the initial condition that $i = 0$ when $t = 0$, we find $k = KL\omega/M^2$. Thus

(5)
$$i = \frac{K}{M^2}(L\omega e^{-Rt/L} + R \sin \omega t - L\omega \cos \omega t)$$

gives the current i at any time $t > 0$.

Notice that as t increases indefinitely, $e^{-Rt/L}$ approaches zero. When t is sufficiently large, the term $KL\omega e^{-Rt/L}/M^2$ in the preceding equation for i becomes negligible. It is customary to separate equation (5) into two parts:

$$i = i_T + i_S,$$

where $i_T = KL\omega e^{-Rt/L}/M^2$ is called the **transient current** and

$$i_S = \frac{K}{M^2}(R \sin \omega t - L\omega \cos \omega t)$$

is the **steady-state current**.

Exercises

1. Using Fig. 20-7, find the charge q on C as a function of time if S is closed at $t = 0$. Assume that C is initially uncharged.

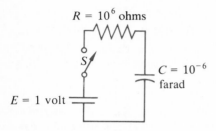

$R = 10^6$ ohms

$C = 10^{-6}$ farad

$E = 1$ volt

Figure 20-7

2. Find the current i as a function of time in Exercise 1 if C is initially charged to 4 volts. (Use Fig. 20-7.)

3. Using Fig. 20-8, find the current as a function of time if S is closed at $t = 0$.

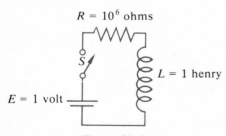

$R = 10^6$ ohms

$L = 1$ henry

$E = 1$ volt

Figure 20-8

4. Using Fig. 20-9, find the current as a function of time if C is initially uncharged and S is closed at $t = 0$. (*Hint:* The current at $t = 0$ will equal 0, since the current through an inductance cannot change instantaneously.)

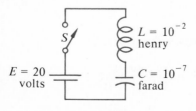

$L = 10^{-2}$ henry

$E = 20$ volts

$C = 10^{-7}$ farad

Figure 20-9

5. Using Fig. 20-10,
(a) Find i as a function of time. Assume C initially uncharged.
(b) Find i as a function of time.

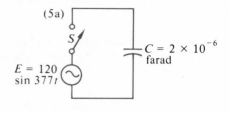

(5a)

S

$C = 2 \times 10^{-6}$ farad

$E = 120$ sin $377t$

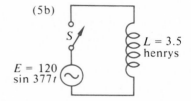

(5b)

S

$L = 3.5$ henrys

$E = 120$ sin $377t$

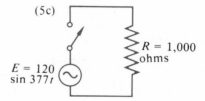

(5c)

$R = 1,000$ ohms

$E = 120$ sin $377t$

Figure 20-10

(c) Find i as a function of time.
(i is the steady-state current in all cases; do not find the transient current.)

6. Using Fig. 20-11, find i as a function of time. Assume C initially uncharged.

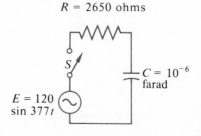

$R = 2650$ ohms

$C = 10^{-6}$ farad

$E = 120$ sin $377t$

Figure 20-11

7. Using Fig. 20-12, find the steady-state current as a function of time.

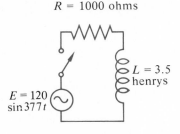

R = 1000 ohms

E = 120 sin 377t

L = 3.5 henrys

Figure 20-12

8. Using Fig. 20-13, find i as a function of time if C has an initial charge of one coulomb at $t = 0$. (See Hint in Exercise 4.)

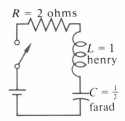

R = 2 ohms

L = 1 henry

$C = \frac{1}{2}$ farad

Figure 20-13

9. Using Fig. 20-14, find the steady-state current as a function of time.

R = 1000 ohms

E = 120 sin 377t

L = 3.5 henrys

$C = 2 \times 10^{-6}$ farads

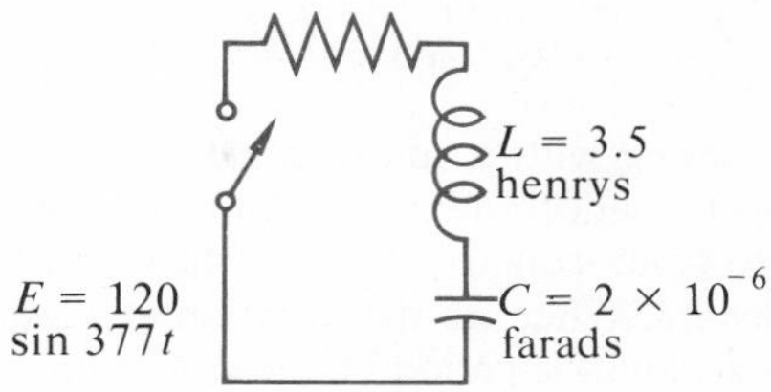

Figure 20-14

10. Using Fig. 20-15, find i as a function of time if C has an initial charge of $\frac{1}{2}$ coulomb. (See Hint in Exercise 4.)

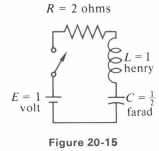

R = 2 ohms

L = 1 henry

E = 1 volt

$C = \frac{1}{2}$ farad

Figure 20-15

20.12 REVIEW EXERCISES

In each of Exercises 1 to 16, find the general solution of the given differential equation.

1. $y\,dx + x\,dy = 0$.

2. $2xy\,dx + (x^2 + 2y)\,dy = 0$.

3. $x\,dy - (x^2 - 2y)\,dx = 0$.

4. $dy + 2x(y - 1)\,dx = 0$.

5. $\dfrac{d^2y}{dx^2} + 3\dfrac{dy}{dx} = e^x$.

6. $\dfrac{d^2y}{dx^2} - y = 0$.

7. $\sin y\,dx + \cos y\,(x + 2\sin y)\,dy = 0$.

8. $\dfrac{dy}{dx} - ay = e^{ax}$.

9. $\dfrac{dy}{dx} - 2y = e^x$.

10. $\left(2x \ln y + \dfrac{1}{xy}\right) dx + \left(\dfrac{x^2}{y} - \dfrac{\ln x}{y^2}\right) dy = 0$.

11. $\dfrac{d^2y}{dx^2} - 3\dfrac{dy}{dx} + 2y = 0$.

12. $4\dfrac{d^2y}{dx^2} + 12\dfrac{dy}{dx} + 9y = 0$.

13. $\dfrac{d^2y}{dx^2} - y = 1$.

14. $\dfrac{d^2y}{dx^2} + 4\dfrac{dy}{dx} + 4y = 3e^x$.

15. $\dfrac{d^2y}{dx^2} + 4y = 0$.

16. $\dfrac{d^2y}{dx^2} + y = \sec x \tan x$.

17. A spring with a spring constant of 5 pounds per foot is loaded with a 10-pound weight and allowed to reach equilibrium. It is then raised 1 foot and released. What are the equation of motion, the amplitude, and the period? Neglect friction.

18. In Exercise 17, what is the absolute value of the velocity of the moving weight as it passes through the equilibrium position?

Exercises 19 and 20 refer to Fig. 20-16.

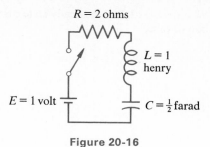

$R = 2$ ohms

$L = 1$ henry

$E = 1$ volt

$C = \frac{1}{2}$ farad

Figure 20-16

19. Find i as a function of time if C is initially uncharged. (*Hint:* The current at $t = 0$ will equal zero, since current through an inductance cannot change instantaneously.)

20. Find i as a function of time if C has an initial charge of $\frac{1}{2}$ coulomb.

appendix

A.1 THEOREMS ON LIMITS

A.1.1 Lemma.   If $\lim_{x \to c} f(x) = L$ and $\lim_{x \to c} g(x) = 0$, then

$$\lim_{x \to c} [f(x)g(x)] = 0.$$

Proof.   Let ϵ be an arbitrarily chosen number. We will show that there is some $\delta > 0$ such that

$$0 < |x - c| < \delta \implies |f(x)g(x) - 0| = |f(x)g(x)| < \epsilon.$$

Since $\lim_{x \to c} f(x) = L$ and 1 is a positive number, there is some $\delta_1 > 0$ such that

$$0 < |x - c| < \delta_1 \implies |f(x) - L| < 1.$$

But, by 1.6.10, $|f(x)| - |L| \leq |f(x) - L|$. Thus $0 < |x - c| < \delta_1 \implies |f(x)| - |L| < 1$, or

(1) $$0 < |x - c| < \delta_1 \implies |f(x)| < 1 + |L|.$$

Since $\lim_{x \to c} g(x) = 0$ and $\epsilon/(1 + |L|) > 0$, there exists some $\delta_2 > 0$ such that

(2) $$0 < |x - c| < \delta_2 \implies |g(x)| < \frac{\epsilon}{1 + |L|}.$$

Let δ be the minimum of the two positive numbers δ_1 and δ_2. Then it follows from (1) and (2) that

$$0 < |x - c| < \delta \implies |f(x)g(x)| < (1 + |L|)\frac{\epsilon}{1 + |L|} = \epsilon.$$

Therefore

$$\lim_{x \to c} [f(x)g(x)] = 0. \quad \blacksquare$$

A.1.2 Theorem. If $\lim_{x \to c} f(x) = L$ and $\lim_{x \to c} g(x) = M$, then

$$\lim_{x \to c} [f(x)g(x)] = LM.$$

Proof. Since $\lim_{x \to c} f(x) = L$ and $\lim_{x \to c} g(x) = M$, then (3.3.6)

(3) $$\lim_{x \to c} (f(x) - L) = 0 \quad \text{and} \quad \lim_{x \to c} (g(x) - M) = 0.$$

Now $f(x)g(x) = (f(x) - L)g(x) + L(g(x) - M) + LM$, Therefore, by 3.3.1(i), A.1.1, and (3),

$$\lim_{x \to c} [f(x)g(x)] = \lim_{x \to c} [(f(x) - L)g(x)] + \lim_{x \to c} [L(g(x) - M)] + \lim_{x \to c} LM = LM. \quad \blacksquare$$

A.1.3 Lemma. If $\lim_{x \to c} g(x) = M \neq 0$, then

$$\lim_{x \to c} \frac{1}{g(x)} = \frac{1}{M}.$$

Proof. Let ϵ be an arbitrarily chosen positive number. We shall prove that there is a $\delta > 0$ such that

$$0 < |x - c| < \delta \implies \left| \frac{1}{g(x)} - \frac{1}{M} \right| < \epsilon,$$

and thus that $\lim_{x \to c} [1/g(x)] = 1/M$.

Since $\lim_{x \to c} g(x) = M$ and $\frac{1}{2}|M| > 0$, there is some $\delta_1 > 0$ such that

$$0 < |x - c| < \delta_1 \implies |M - g(x)| < \frac{1}{2}|M|.$$

But, from 1.6.10, $|M| - |g(x)| \leq |M - g(x)|$. Therefore

$$0 < |x - c| < \delta_1 \implies |M| - |g(x)| < \frac{1}{2}|M|$$

$$\implies |g(x)| > \frac{1}{2}|M| \implies \frac{1}{|g(x)|} < \frac{2}{|M|}.$$

Thus

(4) $$0 < |x - c| < \delta_1 \implies \frac{1}{|Mg(x)|} < \frac{2}{M^2}.$$

Again, corresponding to each $\epsilon > 0$, there is a $\delta_2 > 0$ such that

(5) $$0 < |x - c| < \delta_2 \implies |M - g(x)| < \frac{\epsilon M^2}{2}.$$

Let δ be the minimum of δ_1 and δ_2. Then (4) and (5) hold if $0 < |x - c| < \delta$. Thus

$$0 < |x - c| < \delta \implies \frac{|M - g(x)|}{|Mg(x)|} < \frac{2}{M^2} \cdot \frac{\epsilon M^2}{2} = \epsilon.$$

But

$$\left| \frac{M - g(x)}{Mg(x)} \right| = \left| \frac{1}{g(x)} - \frac{1}{M} \right|.$$

So corresponding to each $\epsilon > 0$ there is a $\delta > 0$ such that

$$0 < |x - c| < \delta, \ x \in \mathfrak{D}_g \implies \left| \frac{1}{g(x)} - \frac{1}{M} \right| < \epsilon.$$

Therefore

$$\lim_{x \to c} \frac{1}{g(x)} = \frac{1}{M}. \quad \blacksquare$$

A.1.4 Theorem. If $\lim\limits_{x \to c} f(x) = L$ and $\lim\limits_{x \to c} g(x) = M \neq 0$, then

$$\lim_{x \to c} \left[\frac{f(x)}{g(x)} \right] = \frac{L}{M}.$$

Proof. By A.1.2 and A.1.3,

$$\lim_{x \to c} \left[\frac{f(x)}{g(x)} \right] = \lim_{x \to c} \left[f(x) \cdot \frac{1}{g(x)} \right] = \lim_{x \to c} f(x) \cdot \lim_{x \to c} \frac{1}{g(x)} = L \left(\frac{1}{M} \right) = \frac{L}{M}. \quad \blacksquare$$

A.1.5 Theorem. If n is a positive integer, then

$$\lim_{x \to c} \sqrt[n]{x} = \sqrt[n]{c},$$

where c may be any real number if n is odd, and $c > 0$ if n is even.

Proof. We assume that n is a positive integer and that if n is odd, c is any number, but if n is even, $c > 0$.

From elementary algebra we know that

$$p^n - q^n = (p - q)(p^{n-1} + p^{n-2}q + p^{n-3}q^2 + \cdots + pq^{n-2} + q^{n-1})$$

for any two numbers p and q. Therefore

(6) $$x - c = (\sqrt[n]{x} - \sqrt[n]{c})[(\sqrt[n]{x})^{n-1} + (\sqrt[n]{x})^{n-2}(\sqrt[n]{c})$$
$$+ \cdots + (\sqrt[n]{x})(\sqrt[n]{c})^{n-2} + (\sqrt[n]{c})^{n-1}].$$

Whether c is positive or negative, there is an open interval containing c in which all numbers x have the same sign as c. Consequently, we can assume without loss of generality that when $c \neq 0$, x has the same sign as c.

Denote the expression in brackets in the right member of (6) by Q; that is, let

$$Q = (\sqrt[n]{x})^{n-1} + (\sqrt[n]{x})^{n-2}(\sqrt[n]{c}) + \cdots + (\sqrt[n]{x})(\sqrt[n]{c})^{n-2} + (\sqrt[n]{c})^{n-1}.$$

If $c > 0$, then $x > 0$ and every term in Q is positive. If $c < 0$, then $x < 0$ and n must be odd; thus the first term of Q is positive because $n - 1$ is even. The second term of Q is also positive, since it is the product of two negative quantities. The third term of Q is the product of two positive quantities; the fourth term of Q is the product of two negative quantities, and so on. Hence all the terms of Q are positive when $c \neq 0$, and it follows that

$$(7) \qquad\qquad 0 < \sqrt[n]{c^{n-1}} < Q,$$

since $\sqrt[n]{c^{n-1}}$ is one term (the last) of Q.

With these preliminary considerations out of the way, let $\epsilon > 0$. We seek a positive number δ so that

$$0 < |x - c| < \delta \implies |\sqrt[n]{x} - \sqrt[n]{c}| < \epsilon$$

for all x in the domain of the principal nth root function. Using (6) and (7), we can write

$$|\sqrt[n]{x} - \sqrt[n]{c}| < \epsilon \iff |\sqrt[n]{x} - \sqrt[n]{c}|\frac{Q}{Q} < \epsilon$$

$$\iff \frac{|(\sqrt[n]{x} - \sqrt[n]{c}|Q}{Q} < \epsilon \iff \frac{|x - c|}{Q} < \epsilon$$

$$\iff |x - c| < \epsilon Q \impliedby 0 < |x - c| < \epsilon\sqrt[n]{c^{n-1}}.$$

Therefore

$$0 < |x - c| < \epsilon\sqrt[n]{c^{n-1}}, \ x \in \mathfrak{D} \implies |\sqrt[n]{x} - \sqrt[n]{c}| < \epsilon.$$

Thus, if corresponding to each $\epsilon > 0$, we choose $\delta = \epsilon\sqrt[n]{c^{n-1}}$, then

$$0 < |x - c| < \delta, \ x \in \mathfrak{D} \implies |\sqrt[n]{x} - \sqrt[n]{c}| < \epsilon.$$

Therefore, if $c \neq 0$,

$$(8) \qquad\qquad \lim_{x \to c} \sqrt[n]{x} = \sqrt[n]{c}.$$

Next, assume that $c = 0$. If n is an odd positive integer and $\epsilon > 0$, then $0 < |x| < \epsilon^n \implies |\sqrt[n]{x}| < \epsilon$. Therefore

$$(9) \qquad\qquad \lim_{x \to 0} \sqrt[n]{x} = 0 \qquad \text{when} \qquad n \text{ is odd.}$$

If n is an even positive integer, then $0 < x < \epsilon^n \implies |\sqrt[n]{x}| < \epsilon$, and thus

$$(10) \qquad\qquad \lim_{x \to 0} \sqrt[n]{x} = 0.$$

It follows from (8), (9), and (10) that the theorem is proved. ∎

A.2   THEOREMS ON CONTINUOUS FUNCTIONS

A.2.1   **Definition.**   Let $[a, b]$ be a closed interval, and let H be a set of open intervals. Then H is said to be an (open) **covering** of $[a, b]$, or to **cover** $[a, b]$, if and only if every point of $[a, b]$ is in at least one of the open intervals of H.

As an illustration, if H is the set of all open intervals of length 1, and if $[a, b] = [1, 7]$, then H covers $[a, b]$.

A.2.2   **Heine-Borel Theorem.**   Let $[a, b]$ be a closed interval. If H is a set of open intervals that covers $[a, b]$, then there exists some finite subset H' of H that also covers $[a, b]$.

Proof.   Denote by A the set of all numbers x of $[a, b]$ such that there is a finite subset H' of H that covers $[a, x]$.

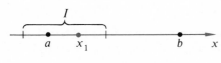

Figure A.2-1

Since H covers $[a, b]$, there is an open interval I of H that contains a. Because I is open, there is a number x_1 in I such that $a < x_1 \leq b$ (Fig. A.2-1). But the set $\{I\}$, whose only element is the open interval I, is a finite subset of H that covers $[a, x_1]$. Therefore $x_1 \in A$ and A *is not empty*.

Moreover, the nonempty set A is bounded above by b. Therefore A has a least upper bound c (by the completeness axiom):

$$c = \text{l.u.b. of } A.$$

Since $a < x_1 \leq c \leq b$, we have

(1) $a < c \leq b.$

We will now show that $c \in A$. Since $c \in [a, b]$, there is an open interval I' of H that contains c (Fig. A.2-2). Because I' is open and $c \in I'$, there is a number x' in I' such that

(2) $x' \in I', \quad x' \in [a, b], \quad$ and $\quad x' < c.$

Then $x' \in A$, for otherwise x' would be an upper bound of A that is less than the least upper bound c, a contradiction. Since $x' \in A$, there is a finite subset G of H that covers $[a, x']$. Then

(3) $H' = G \cup I'$

is a finite subset of H that covers $[a, c]$. Thus $c \in A$.

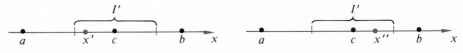

Figure A.2-2   Figure A.2-3

Finally, we will show that $c = b$. If $c < b$, there is a number x'' in I' (Fig. A.2-3) such that

(4) $$x'' \in I', \qquad x'' \in [a, b], \qquad \text{and} \qquad c < x''.$$

So $H' = G \cup I'$ covers $[a, x'']$. But this implies that $x'' \in A$ and $c < x''$, which is impossible because c is the least upper bound of A. Therefore c cannot be less than b, and it follows from (1) that $c = b$.

Since H' is a finite subset of H that covers $[a, b]$, the theorem is proved. ∎

In the illustrative example above, we saw that the (infinite) set H of open intervals having length 1 was an open covering of the closed interval $[1, 7]$. Clearly, the set

$$H' = \{(0.9, 1.9), (1.8, 2.8), (2.7, 3.7), (3.6, 4.6), (4.5, 5.5), (5.4, 6.4), (6.3, 7.3)\}$$

is a finite subset of H that covers $[1, 7]$.

A.2.3 Theorem. A function that is continuous on a closed interval is bounded on that interval.

Proof. Let f be a function that is continuous on a closed interval $[a, b]$ and let c be any point in $[a, b]$. Then f is continuous at c and there is a positive number $\delta(c)$, depending on c, such that

$$c - \delta(c) < x < c + \delta(c) \quad \text{and} \quad x \in [a, b] \implies f(c) - 1 \le f(x) \le f(c) + 1.$$

The infinite set of open intervals $I = (c - \delta(c), c + \delta(c))$, one for each number c in $[a, b]$, is a covering of $[a, b]$. Therefore, by the Heine-Borel theorem (A.2.2), there is a finite subset

$$\{I_1, I_2, \cdots, I_n\}$$

that covers $[a, b]$. But on each interval I_j of this set f is bounded above by the number $f(c_j) + 1$ and bounded below by $f(c_j) - 1$. Thus $f(x)$, for all x in $[a, b]$, has the maximum of the numbers $f(c_j) + 1, j = 1, 2, \cdots, n$, for an upper bound and the minimum of the numbers $f(c_j) - 1$ for a lower bound.

Therefore f is bounded on $[a, b]$. ∎

A.2.4 Theorem. If a function f is continuous on a closed interval $[a, b]$, there is a number x_1 in $[a, b]$ such that $f(x_1)$ is the maximum value of f on $[a, b]$.

Proof. Since f is continuous on the closed interval $[a, b]$, the set

(5) $$S = \{f(x) \,|\, a \le x \le b\}$$

has an upper bound (by A.2.3). So, by the completeness axiom, S has a least upper bound, which will be denoted by U.

Assume that $f(x) \ne U$ for all x in $[a, b]$. Then $f(x) < U$ for all x in $[a, b]$ because U is the (least) upper bound for S. We shall show that this leads to a contradiction and therefore that $f(x_1) = U$ for some number x_1 in $[a, b]$.

Now $f(x) < U$ for all x in $[a, b]$ is equivalent to

(6) $$U - f(x) > 0, \qquad a \leq x \leq b.$$

Since f is continuous on the closed interval $[a, b]$, $U - f(x)$ is continuous on $[a, b]$; and since $U - f(x) \neq 0$ for $a \leq x \leq b$ [by (6)],

(7) $$\frac{1}{U - f(x)}$$

is continuous on $[a, b]$. Moreover, (6) implies that

(8) $$\frac{1}{U - f(x)} > 0 \qquad \text{for} \quad a \leq x \leq b.$$

Therefore the set

$$T = \left\{ \frac{1}{U - f(x)} \,\middle|\, a \leq x \leq b \right\}$$

has 0 for a lower bound and (by A.2.3) some number L for an upper bound. Thus

(9) $$0 < \frac{1}{U - f(x)} \leq L \qquad \text{for} \quad a \leq x \leq b,$$

which implies that

(10) $$U - f(x) \geq \frac{1}{L}, \qquad a \leq x \leq b.$$

But (10) may be written

(11) $$U - \frac{1}{L} \geq f(x), \qquad a \leq x \leq b.$$

Thus $U - 1/L$ is an upper bound for the set S. Since $L > 0$ [from (9)], $1/L > 0$ and

(12) $$U - \frac{1}{L} < U.$$

But (12) is contradicted by the fact that U is the *least* upper bound of S. Therefore our assumption that $f(x) \neq U$ for all x in $[a, b]$ is false, and there exists a number x_1 in $[a, b]$ for which $f(x_1) = U$. But $f(x) \leq U$ for all x in $[a, b]$, since U is an upper bound of the set S. Thus f has the maximum value $f(x_1) = U$, $x_1 \in [a, b]$, on the closed interval $[a, b]$. ∎

A.2.5 Theorem. If f is continuous on the closed interval $[a, b]$, there exists a number x_2 in $[a, b]$ such that $f(x_2)$ is the minimum value of f on $[a, b]$.

The proof of this theorem is similar to the proof of A.2.4, with the inequalities reversed.

A.2.6 Theorem. If a function f is continuous on $[a, b]$ and if $f(a)$ and $f(b)$ have opposite signs, then there is a number z between a and b such that $f(z) = 0$.

Proof. Assume that $f(a) < 0$ and $f(b) > 0$. We shall show that there is a number z between a and b such that $f(z) = 0$.

Let W be the set of numbers

$$W = \{x \mid a \le x \le b, f(x) < 0\}.$$

This set is not empty since it contains the number a. Moreover, it is bounded above by b, and so W has a least upper bound that will be designated by z (by the completeness axiom). By showing that $f(z)$ is neither negative nor positive, we shall prove that $f(z) = 0$.

CASE 1.   Assume that $f(z) < 0$. Since f is continuous on $[a, b]$, there is an interval $(z - \delta, z + \delta)$, $\delta > 0$, such that $f(x) < 0$ for all x in $(z - \delta, z + \delta)$ (by Exercise 27, Section 3.5). Therefore $f(x) < 0$ for some x *greater* than z, in contradiction to the fact that z is an upper bound of the set W.

CASE 2.   Assume that $f(z) > 0$. Since f is continuous on $[a, b]$, there is a subinterval $(z - \delta, z + \delta)$, $\delta > 0$, of $[a, b]$ such that $f(x) > 0$ for all x in $(z - \delta, z + \delta)$. Thus $f(x) > 0$ for all x in $(z - \delta, z)$. But this is impossible since z is the least upper bound of the set W, and so $f(x) < 0$ for all x in $a \le x < z$.

Because $f(z)$ cannot be negative and cannot be positive, yet $f(z)$ exists, it follows that $f(z) = 0$.

We assumed at the beginning of this proof that $f(a) < 0$ and $f(b) > 0$. The proof when $f(a) > 0$ and $f(b) < 0$ is similar.   ∎

A.2.7   **The Intermediate Value Theorem.**   If f is function that is continuous on a closed interval $[a, b]$ and if k is a number between $f(a)$ and $f(b)$, then there is a number z between a and b such that $f(z) = k$.

Proof.   Let k be a number between $f(a)$ and $f(b)$. We define a function ϕ by

$$\phi(x) = f(x) - k.$$

Then ϕ is continuous on $[a, b]$ because f is continuous there, and $\phi(a) = f(a) - k$ and $\phi(b) = f(b) - k$ have opposite signs. Therefore (by A.2.6) there is a number z between a and b such that $\phi(z) = f(z) - k = 0$; that is, there is a number z between a and b such that $f(z) = k$.   ∎

A.3 THE CHAIN RULE

Proof of the chain rule (5.3.1) is made simpler by the following lemma.

A.3.1   **Lemma.**   Let Φ be a function whose domain contains 0, and suppose that Φ is continuous at 0. Let u be a function of x with the property that

$$\lim_{\Delta x \to 0} [u(x + \Delta x) - u(x)] = \lim_{\Delta x \to 0} \Delta u = 0.$$

Then

$$\lim_{\Delta x \to 0} \Phi(\Delta u) = \Phi(0).$$

Proof. Let $\epsilon > 0$. Since Φ is continuous at 0, there exists some positive number δ_1 such that

$$|\Delta u| < \delta_1 \implies |\Phi(\Delta u) - \Phi(0)| < \epsilon.$$

Because $\lim_{\Delta x \to 0} \Delta u = 0$, there is a positive number δ_2 for which

$$0 < |\Delta x| < \delta_2 \implies |\Delta u| < \delta_1.$$

Thus corresponding to each positive number ϵ there is some positive number δ_2 such that

$$0 < |\Delta x| < \delta_2 \implies |\Phi(\Delta u) - \Phi(0)| < \epsilon.$$

Therefore

$$\lim_{\Delta x \to 0} \Phi(\Delta u) = \Phi(0). \quad \blacksquare$$

We are now ready to prove the chain rule.

A.3.2 **Chain Rule.** Let $y = f(u)$ and $u = g(x)$, where f and g are functions. If g is differentiable at x and if f is differentiable at $u = g(x)$, then the composite function defined by $y = f(g(x))$ is differentiable at x and

$$\frac{dy}{dx} = \frac{dy}{du} \cdot \frac{du}{dx}$$

or, equivalently,

$$D_x f(g(x)) = f'(u) \cdot g'(x).$$

Proof. Let $x_1 \in \mathcal{D}_g$ so that $u_1 = g(x_1) \in \mathcal{R}_g \subseteq \mathcal{D}_f$. Throughout this proof it is to be understood that Δx will be a real number such that $x_1 + \Delta x \in \mathcal{D}_g$. It then follows that $g(x_1 + \Delta x) = g(x_1) + g(x_1 + \Delta x) - g(x_1) = u_1 + \Delta u \in \mathcal{R}_g \subseteq \mathcal{D}_f$.

First, g is continuous at x_1 because it is differentiable there. Thus (by 4.5.4 and 3.3.6)

$$(1) \qquad \lim_{\Delta x \to 0} \Delta u = \lim_{\Delta x \to 0} [g(x_1 + \Delta x) - g(x_1)] = 0.$$

By the definition of the derivative of a function f of u, its value at u_1 is $\lim_{\Delta u \to 0} (\Delta f / \Delta u)$ $= f'(u_1)$, and so (by 3.3.6)

$$(2) \qquad \lim_{\Delta u \to 0} \left[\frac{\Delta f}{\Delta u} - f'(u_1) \right] = 0.$$

We now define a function Φ with independent variable Δu as follows:

$$(3) \qquad \Phi(\Delta u) = \begin{cases} \left[\dfrac{\Delta f}{\Delta u} - f'(u_1) \right] & \text{if} \quad \Delta u \neq 0, \\ 0 & \text{if} \quad \Delta u = 0. \end{cases}$$

It follows from (2) that $\lim_{\Delta u \to 0} \Phi(\Delta u) = 0 = \Phi(0)$, so that Φ is continuous at 0. From this and from (1) the conditions of A.3.1 are satisfied, and therefore

$$(4) \qquad \lim_{\Delta x \to 0} \Phi(\Delta u) = \Phi(0) = 0.$$

If we solve the first equation in (3) for Δf, we obtain

$$(5) \qquad \Delta f = f'(u_1)\,\Delta u + \Phi(\Delta u)\,\Delta u \qquad \text{if} \quad \Delta u \neq 0.$$

It is immediate that (5) is also valid if $\Delta u = 0$. Therefore

$$(6) \qquad \frac{\Delta f}{\Delta x} = f'(u_1)\frac{\Delta u}{\Delta x} + \Phi(\Delta u)\frac{\Delta u}{\Delta x}, \qquad \Delta x \neq 0.$$

Taking the limits of both members of (6) as $\Delta x \longrightarrow 0$, and recalling from (4) that $\lim\limits_{\Delta x \to 0} \Phi(\Delta u) = 0$, we get $D_x f(u) = D_x f(g(x)) = f'(u_1)D_x u + 0 \cdot D_x u = f'(u_1)g'(x_1)$. That is,

$$F'(x_1) = f'(u_1)g'(x_1). \quad \blacksquare$$

A.4 PROPERTIES OF DEFINITE INTEGRALS

A.4.1 Theorem. If $\int_a^b f(x)\,dx = \lim\limits_{|P|\to 0} \sum\limits_{i=1}^{n} f(w_i)\,\Delta x_i$ exists, its value is unique.

Proof. Assume that

$$\lim_{|P|\to 0} \sum_{i=1}^{n} f(w_i)\,\Delta x_i = A \qquad \text{and} \qquad \lim_{|P|\to 0} \sum_{i=1}^{n} f(w_i)\,\Delta x_i = B.$$

Then (8.3.3) corresponding to each $\epsilon > 0$ there is a $\delta_1 > 0$ such that

$$(1) \qquad \left| \sum_{i=1}^{n} f(w_i)\,\Delta x_i - A \right| < \frac{\epsilon}{2}$$

for every Riemann sum, $\sum\limits_{i=1}^{n} f(w_i)\,\Delta w_i$, of f on $[a, b]$, whose associated partition P has norm $|P| < \delta_1$; moreover, there is also a $\delta_2 > 0$ such that

$$(2) \qquad \left| B - \sum_{i=1}^{n} f(w_i)\,\Delta x_i \right| < \frac{\epsilon}{2}$$

for all Riemann sums for f on $[a, b]$ with $|P| < \delta_2$.

Denote by δ the minimum of δ_1 and δ_2. Then every Riemann sum, $\sum\limits_{i=1}^{n} f(w_i)\,\Delta x_i$, for f on $[a, b]$ with $|P| < \delta$ satisfies the two inequalities, (1) and (2), simultaneously. Thus

$$\left| \sum_{i=1}^{n} f(w_i)\,\Delta x_i - A \right| + \left| B - \sum_{i=1}^{n} f(w_i)\,\Delta x_i \right| < \epsilon,$$

from which, by the triangle inequality (1.6.8), $|B - A| < \epsilon$. Since $\epsilon > 0$ can be arbitrarily small, $A = B$. $\blacksquare$

A.4.2 Lemma. Let f be integrable on $[a, b]$ and let A be a number. If for each $\epsilon > 0$ and for each $\delta > 0$ there is always some partition P of $[a, b]$ with norm $|P| < \delta$ and at least one associated Riemann sum, $\sum\limits_{i=1}^{n} f(w_i)\,\Delta x_i$, for f on P, such that

$$\left| \sum_{i=1}^{n} f(w_i)\,\Delta x_i - A \right| < \epsilon,$$

then

$$\int_a^b f(x)\,dx = A.$$

Proof. Let $\epsilon > 0$ be an arbitrarily chosen number.

Because $\int_a^b f(x)\,dx$ exists and is unique (A.4.1), it follows from 8.3.3 and 8.3.4 that there is a $\delta > 0$ such that *every* Riemann sum R for f on *every* partition P with norm $|P| < \delta$ satisfies

(3) $$\left| R - \int_a^b f(x)\,dx \right| < \frac{\epsilon}{2}.$$

Assume that there is a number A such that for each δ' $(0 < \delta' \le \delta)$ there is at least one partition P' of $[a, b]$ with norm $|P'| < \delta' \le \delta$, having at least one associated Riemann sum R' for f on P' that satisfies $|R' - A| < \epsilon/2$, which may be written

(4) $$|A - R'| < \frac{\epsilon}{2}.$$

Since R' is a Riemann sum for f on a partition P' with norm $|P'| < \delta' \le \delta$, it follows from (3) that

(5) $$\left| R' - \int_a^b f(x)\,dx \right| < \frac{\epsilon}{2}.$$

If we add the inequalities (4) and (5), member by member, we obtain

$$|A - R'| + \left| R' - \int_a^b f(x)\,dx \right| < \epsilon.$$

By the triangle inequality (1.6.8), this result implies that

$$\left| A - \int_a^b f(x)\,dx \right| < \epsilon.$$

Since $\epsilon > 0$ can be arbitrarily small,

$$A = \int_a^b f(x)\,dx. \quad \blacksquare$$

A.4.3 Interval Additive Property. If f is continuous on an interval containing the three points a, b, and c, then

$$\int_a^b f(x)\,dx + \int_b^c f(x)\,dx = \int_a^c f(x)\,dx,$$

no matter what the order of the points a, b, and c.

Proof. PART 1.   Assume that $a < b < c$.

Let $\epsilon > 0$. Since f is continuous on $[a, b]$ and on $[b, c]$, f is integrable on $[a, b]$ and on $[b, c]$. Thus there is a $\delta_1 > 0$ such that

$$(6) \qquad \left| \sum_{i=1}^{k} f(w_i)\,\Delta x_i - \int_a^b f(x)\,dx \right| < \frac{\epsilon}{2}$$

for every Riemann sum, $\sum_{i=1}^{k} f(w_i)\,\Delta x_i$, for f on $[a, b]$ whose associated partition P_1 of $[a, b]$ has norm $|P_1| < \delta_1$. There is also a $\delta_2 > 0$ so that

$$(7) \qquad \left| \sum_{i=1}^{m} f(w_i)\,\Delta x_i - \int_b^c f(x)\,dx \right| < \frac{\epsilon}{2}$$

for every Riemann sum, $\sum_{i=1}^{m} f(w_i)\,\Delta x_i$, for f on $[b, c]$ whose partition P_2 of $[b, c]$ has norm $|P_2| < \delta_2$.

Let $\delta = \min\{\delta_1, \delta_2\}$. Then (6) and (7) hold with δ replacing δ_1 and δ_2. If we now add (6) and (7), member by member, we have

$$\left| \sum_{i=1}^{k} f(w_i)\,\Delta x_i - \int_a^b f(x)\,dx \right| + \left| \sum_{i=1}^{m} f(w_i)\,\Delta x_i - \int_b^c f(x)\,dx \right| < \frac{\epsilon}{2} + \frac{\epsilon}{2} = \epsilon$$

if $|P_1| < \delta$ and $|P_2| < \delta$. Using the triangle inequality (1.6.8), this result implies that

$$(8) \qquad \left| \sum_{i=1}^{k} f(w_i)\,\Delta x_i - \int_a^b f(x)\,dx + \sum_{i=1}^{m} f(w_i)\,\Delta x_i - \int_b^c f(x)\,dx \right| < \epsilon$$

if $|P_1| < \delta$ and $|P_2| < \delta$.

Denote $P_1 + P_2$ by P^*, a partition of $[a, c]$ having b as a point of division. Then $\sum_{i=1}^{k} f(w_i)\,\Delta x_i + \sum_{i=1}^{m} f(w_i)\,\Delta x_i = \sum_{i=1}^{n} f(w_i)\,\Delta x_i$, a Riemann sum for f on P^*. Since $|P^*| < \delta$ implies that $|P_1| < \delta$ and $|P_2| < \delta$, it follows from (8) that

$$\left| \sum_{i=1}^{n} f(w_i)\,\Delta x_i - \left[\int_a^b f(x)\,dx + \int_b^c f(x)\,dx \right] \right| < \epsilon$$

for every Riemann sum, $\sum_{i=1}^{n} f(w_i)\,\Delta x_i$, for f on $[a, c]$ whose partition P^* of $[a, c]$ has b for a point of division and norm $|P^*| < \delta$. Therefore

$$\int_a^c f(x)\,dx = \int_a^b f(x)\,dx + \int_b^c f(x)\,dx$$

$\left(\text{by A.4.2 with } \int_a^b f(x)\,dx + \int_b^c f(x)\,dx \text{ replacing } A \right).$

PART 2.   There are 6 possible orders for the points a, b, and c:

$$a < b < c, \quad a < c < b, \quad b < a < c$$
$$b < c < a, \quad c < a < b, \quad c < b < a.$$

In Part 1 we showed that the theorem is true for the order $a < b < c$.

Now consider any other order, say $b < c < a$. From Part 1

(9)
$$\int_b^c f(x)\,dx + \int_c^a f(x)\,dx = \int_b^a f(x)\,dx.$$

However,

$$\int_c^a f(x)\,dx = -\int_a^c f(x)\,dx, \qquad \int_b^a f(x)\,dx = -\int_a^b f(x)\,dx.$$

By substituting this in (9), we obtain

$$\int_b^c f(x)\,dx - \int_a^c f(x)\,dx = -\int_a^b f(x)\,dx,$$

or

$$\int_a^b f(x)\,dx + \int_b^c f(x)\,dx = \int_a^c f(x)\,dx,$$

which proves the theorem for the order $b < c < a$.

The proofs for the remaining four orders are analogous. ∎

A.4.4 Theorem. If f is continuous on $[a, b]$ and m and M denote the minimum and maximum of f on $[a, b]$, then

$$m(b - a) \leq \int_a^b f(x)\,dx \leq M(b - a).$$

Proof. To prove that $m(b - a) \leq \int_a^b f(x)\,dx$, assume the contrary, namely that

(10)
$$m(b - a) > \int_a^b f(x)\,dx.$$

We shall show that this assumption leads to a contradiction, and therefore that $m(b - a) \leq \int_a^b f(x)\,dx$, as the theorem asserts.

Since f is continuous on $[a, b]$, f is integrable on $[a, b]$. That is,

$$\lim_{|P| \to 0} \sum_{i=1}^n f(w_i)\,\Delta x_i = \int_a^b f(x)\,dx.$$

By assumption (10), $m(b - a) - \int_a^b f(x)\,dx > 0$. Let

(11)
$$\epsilon = m(b - a) - \int_a^b f(x)\,dx.$$

Then there is a $\delta > 0$ such that for all partitions P of $[a, b]$ with $|P| < \delta$,

$$\left| \sum_{i=1}^n f(w_i)\,\Delta x_i - \int_a^b f(x)\,dx \right| < \epsilon,$$

which is equivalent to

$$\int_a^b f(x)\,dx - \epsilon < \sum_{i=1}^n f(w_i)\,\Delta x_i < \int_a^b f(x)\,dx + \epsilon.$$

By means of (11), the second inequality can be written

$$\sum_{i=1}^n f(w_i)\,\Delta x_i < \int_a^b f(x)\,dx + \epsilon = \int_a^b f(x)\,dx + m(b - a) - \int_a^b f(x)\,dx = m(b - a).$$

Consequently,

$$\sum_{i=1}^{n} f(w_i)\,\Delta x_i < m(b - a) = \sum_{i=1}^{n} m\,\Delta x_i,$$

which implies that for every partition P of $[a, b]$ with $|P| < \delta$,

$$f(w_i) < m$$

for some $w_i \in [a, b]$. But this result is impossible because m is the minimum of f on $[a, b]$.

Therefore the assumption that $m(b - a) > \int_a^b f(x)\,dx$ is false, and it follows that

$$m(b - a) \le \int_a^b f(x)\,dx.$$

The proof that $\int_a^b f(x)\,dx \le M(b - a)$ is similar. ∎

A.5 SOME PROPERTIES OF DETERMINANTS

A square **matrix** of order n is a rectangular array of numbers arranged in n rows and n columns; it is often symbolized by

(1)
$$\begin{bmatrix} a_{11} & a_{12} & a_{13} & \cdots & a_{1n} \\ a_{21} & a_{22} & a_{23} & \cdots & a_{2n} \\ \cdot & \cdot & \cdot & \cdots & \cdot \\ a_{n1} & a_{n2} & a_{n3} & \cdots & a_{nn} \end{bmatrix}.$$

The numbers a_{ij} $(i = 1, 2, 3, \cdots, n; j = 1, 2, 3, \cdots, n)$ are the **elements** of the matrix; the first subscript indicates the row in which the element appears and the second subscript the column. As an illustration, a_{25} is the element in the second row and fifth column.

A matrix does not have a value; it is simply a rectangular array of numbers, like books in a bookcase.

Matrices are important in pure and applied mathematics. They were first used in connection with the solution of systems of linear equations and now have many applications in physics. They are also found in economic theory, genetics, statistics, and psychology.

It is often convenient to indicate a square matrix of order n briefly by

$$[a_{ij}]_n.$$

There is a number associated with each square matrix A called the **determinant** of A and denoted by det A. If

(2)
$$A = \begin{bmatrix} a_{11} & a_{12} & a_{13} & \cdots & a_{1n} \\ a_{21} & a_{22} & a_{23} & \cdots & a_{2n} \\ \cdot & \cdot & \cdot & \cdots & \cdot \\ a_{n1} & a_{n2} & a_{n3} & \cdots & a_{nn} \end{bmatrix},$$

then the determinant of A is symbolized by the same square array but between vertical bars:

$$\det A = \begin{vmatrix} a_{11} & a_{12} & a_{13} & \cdots & a_{1n} \\ a_{21} & a_{22} & a_{23} & \cdots & a_{2n} \\ \cdot & \cdot & \cdot & \cdots & \cdot \\ a_{n1} & a_{n2} & a_{n3} & \cdots & a_{nn} \end{vmatrix}.$$

A square matrix having n rows and n columns is said to be of **order** n. When A is a matrix of order n, we say that det A is a **determinant of order** n. Thus if

$$(3) \qquad A = \begin{bmatrix} -6 & 3 & 1 \\ 0 & 5 & -7 \\ 2 & 0 & 4 \end{bmatrix},$$

then A is a matrix of order 3 and

$$\det A = \begin{vmatrix} -6 & 3 & 1 \\ 0 & 5 & -7 \\ 2 & 0 & 4 \end{vmatrix}$$

is a determinant of order 3.

The **value** of a determinant of order 2 is defined by

$$\begin{vmatrix} a_{11} & a_{12} \\ a_{21} & a_{22} \end{vmatrix} = a_{11}a_{22} - a_{12}a_{21};$$

and the **value** of a determinant of order 3 is

$$\begin{vmatrix} a_{11} & a_{12} & a_{13} \\ a_{21} & a_{22} & a_{23} \\ a_{31} & a_{32} & a_{33} \end{vmatrix} = \begin{aligned} & a_{11}a_{22}a_{33} - a_{11}a_{23}a_{32} \\ & + a_{12}a_{23}a_{31} - a_{12}a_{21}a_{33} \\ & + a_{13}a_{21}a_{32} - a_{13}a_{22}a_{31}. \end{aligned}$$

Notice that the number of terms in an expansion of a determinant of order 2 is 2! and the number of terms in an expansion of a determinant of order 3 is 3!. Also, each term in an expansion is a product of elements, one from each row and column, but no two from the same row or column.

Example 1.

$$\begin{vmatrix} 2 & -1 \\ 4 & 5 \end{vmatrix} = 2(5) - (-1)(4) = 14;$$

$$\begin{vmatrix} 3 & 0 & -4 \\ 5 & 1 & 6 \\ 0 & 2 & -7 \end{vmatrix} = \begin{aligned} & 3(1)(-7) - 3(6)(2) + 0(6)(0) \\ & - 0(5)(-7) + (-4)(5)(2) \\ & - (-4)(1)(0) = -97. \end{aligned}$$

If a row and a column are deleted from a square matrix A, the square array of numbers that remains is called a **submatrix** of A. As an illustration, if we delete the third row and second column from the third-order matrix (3) above, we have its second-order submatrix

$$\begin{bmatrix} -6 & 1 \\ 0 & -7 \end{bmatrix}.$$

In the square matrix (2) of order n, above, the determinant of its submatrix of order $n - 1$ obtained by deleting the row and column in which some particular element a_{ij} appears is called the **minor** of a_{ij} and is denoted by M_{ij}.

The **cofactor** of a_{ij} in (2), indicated by A_{ij}, is defined by

(4)
$$A_{ij} = (-1)^{i+j} M_{ij}.$$

Example 2. In the matrix

$$A = \begin{bmatrix} -2 & 4 & 1 \\ 4 & 5 & 3 \\ -6 & 1 & 0 \end{bmatrix}$$

the element 3 appears in the second row and third column. The submatrix of order 2 obtained by deleting from A its second row and third column is

$$\begin{bmatrix} -2 & 4 \\ -6 & 1 \end{bmatrix};$$

so the minor of the element 7 in A is the determinant

$$M_{23} = \det \begin{bmatrix} -2 & 4 \\ -6 & 1 \end{bmatrix} = \begin{vmatrix} -2 & 4 \\ -6 & 1 \end{vmatrix},$$

and the cofactor of 7 is the signed minor

$$A_{23} = (-1)^{2+3} M_{23} = - \begin{vmatrix} -2 & 4 \\ -6 & 1 \end{vmatrix}.$$

Notice that the definition of the value of a determinant of order 3 above can be written

$$\begin{vmatrix} a_{11} & a_{12} & a_{13} \\ a_{21} & a_{22} & a_{23} \\ a_{31} & a_{32} & a_{33} \end{vmatrix} = \begin{aligned} & a_{11}(a_{22}a_{33} - a_{23}a_{32}) \\ & - a_{12}(a_{21}a_{33} - a_{23}a_{31}) \\ & + a_{13}(a_{21}a_{32} - a_{22}a_{31}) \end{aligned}$$

$$= a_{11}A_{11} + a_{12}A_{12} + a_{13}A_{13}.$$

Thus the value of det A is equal to the sum of the products of each of the elements in the first row by its cofactor. It is easy to verify that the value of det A is also equal to the sum of the products of each of the elements in any selected row (or column) by its cofactor.

We now state a definition of the value of a determinant of any order.

A.5.1 Definition. Let A be a (square) matrix of order n,

$$A = \begin{bmatrix} a_{11} & a_{12} & \cdots & a_{1n} \\ a_{21} & a_{22} & \cdots & a_{2n} \\ \cdot & \cdot & \cdots & \cdot \\ a_{n1} & a_{n2} & \cdots & a_{nn} \end{bmatrix}.$$

If $n = 1$, $A = [a_{11}]$ and det A is defined to be the number a_{11}.

If $n = 2$, det A is the number defined by

$$\begin{vmatrix} a_{11} & a_{12} \\ a_{21} & a_{22} \end{vmatrix} = a_{11}a_{22} - a_{21}a_{12}.$$

If $n > 2$, det A is the number defined by

$$\det A = a_{11}A_{11} + a_{12}A_{12} + \cdots + a_{1n}A_{1n},$$

where A_{1j} is the cofactor of the element a_{1j} in the first row and jth column of A $(j = 1, 2, \cdots, n)$.

Example 3. Find the value of the determinant of the matrix

$$A = \begin{bmatrix} -1 & 4 & 5 \\ 3 & 6 & 2 \\ 4 & -3 & 0 \end{bmatrix}.$$

Solution. Here $n = 3$. From the definition

$$\det A = \begin{vmatrix} -1 & 4 & 5 \\ 3 & 6 & 2 \\ 4 & -3 & 0 \end{vmatrix} = -1A_{11} + 4A_{12} + 5A_{13}.$$

Now

$$A_{11} = (-1)^{1+1} \begin{vmatrix} 6 & 2 \\ -3 & 0 \end{vmatrix} = (-1)^2[6(0) - (-3)(2)] = 6,$$

$$A_{12} = (-1)^{1+2} \begin{vmatrix} 3 & 2 \\ 4 & 0 \end{vmatrix} = (-1)^3[3(0) - 4(2)] = 8,$$

$$A_{13} = (-1)^{1+3} \begin{vmatrix} 3 & 6 \\ 4 & -3 \end{vmatrix} = (-1)^4[3(-3) - 4(6)] = -33.$$

Thus

$$\det A = (-1)(6) + 4(8) + 5(-33) = -139.$$

Example 4. Evaluate the determinant, of order 4,

$$\begin{vmatrix} 2 & 1 & 0 & -1 \\ -5 & 0 & 4 & 2 \\ 1 & -3 & 0 & 4 \\ 0 & 0 & -1 & -2 \end{vmatrix}.$$

Solution. By A.5.1, this determinant is equal to

(5) $$2A_{11} + (1)A_{12} + (0)A_{13} + (-1)A_{14}.$$

Here

$$A_{11} = (-1)^{1+1} \begin{vmatrix} 0 & 4 & 2 \\ -3 & 0 & 4 \\ 0 & -1 & -2 \end{vmatrix}$$

and (as in Example 3)

$$(-1)^2 \begin{vmatrix} 0 & 4 & 2 \\ -3 & 0 & 4 \\ 0 & -1 & -2 \end{vmatrix} = 0(-1)^{1+1} \begin{vmatrix} 0 & 4 \\ -1 & -2 \end{vmatrix} + 4(-1)^{1+2} \begin{vmatrix} -3 & 4 \\ 0 & -2 \end{vmatrix}$$

$$+ 2(-1)^{1+3} \begin{vmatrix} -3 & 0 \\ 0 & -1 \end{vmatrix}$$

$$= 0 - 4[(-3)(-2) - 0(4)] + 2[(-3)(-1) - (0)(0)]$$

$$= -24 + 6 = -18.$$

Thus $A_{11} = -18$. The reader should verify similarly that $A_{12} = 14$ and $A_{14} = 15$. By substituting these results in (5), we find the value of the given fourth-order determinant to be

$$2(-18) + 14 - 15 = -37.$$

In A.5.1 the value of the determinant of the nth-order matrix $A = [a_{ij}]_n$ was defined by the equation

$$\det A = a_{11}A_{11} + a_{12}A_{12} + \cdots + a_{1n}A_{1n}.$$

The right-hand member of this equation is called the expansion of $\det A$ according to the elements of the first row.

It turns out that the (same) value of $\det A$ is also given by its expansion according to the elements of *any* selected row or *any* selected column.

A.5.2 **Theorem.** If A is the nth-order matrix $[a_{ij}]_n$, then

(i) $\det A = a_{i1}A_{i1} + a_{i2}A_{i2} + \cdots + a_{in}A_{in}$, where i is an arbitrarily chosen but fixed integer from the set $\{1, 2, 3, \cdots, n\}$; and

(ii) $\det A = a_{1j}A_{1j} + a_{2j}A_{2j} + \cdots + a_{nj}A_{nj}$, where j is an arbitrarily chosen but fixed integer from the set $\{1, 2, 3, \cdots, n\}$.

It is easy to verify this theorem for $n = 2$ and $n = 3$, but its proof for general n requires more time and space than are warranted here.

The reader should notice that in expanding a determinant according to the elements of some particular row (or column), it is only necessary to use the $(-1)^{i+j}$ in (4) to determine the sign in the *first* term of the expansion, because *the signs alternate from there on*. As an illustration, if we expand $\det [a_{ij}]_3$ according to the elements of the second column, we find $a_{12}A_{12}$ to be $a_{12}(-1)^{2+1}M_{12} = -a_{12}M_{12}$, and the expansion is

$$\det [a_{ij}]_3 = -a_{12}M_{12} + a_{22}M_{22} - a_{32}M_{32},$$

in which M_{i2} is the *minor* of a_{i2}, $i = 1, 2, 3$.

It is clear that a determinant of any particular order, say 17, can be evaluated by repeated application of the process used in Examples 3 and 4, but doing so would be a formidable task. Fortunately, there are properties of determinants that can make their evaluation much easier.

The following useful properties of determinants are stated without proof. They are easily verified for determinants of order 2 or 3.

A.5.3 **Properties of Determinants.**

(i)   If the rows and columns of a square matrix are interchanged—that is, if a_{ij} and a_{ji} are interchanged ($i = 1, 2, \cdots, n; j = 1, 2, \cdots, n$)—the value of the determinant of the matrix is unchanged.

(ii)   If the corresponding elements of two rows (or two columns) of a square matrix are interchanged, the value of its determinant is multiplied by -1.

(iii)   If all the elements in a row (or column) of a square matrix are zeros, the value of its determinant is zero.

(iv)   If the corresponding elements of two rows (or columns) are proportional, the value of its determinant is zero.

(v)   The sum of the products of the elements of any particular row (or column) of a square matrix by the cofactors of the corresponding elements of a different row (or column) is equal to zero.

(vi)   If all the elements of any selected row (or column) of a square matrix are multiplied by a number k, the value of the determinant of the matrix is multiplied by k.

(vii)   If, to each element of a selected row (or column) of a square matrix, k times the corresponding element of another selected row (or column) is added, the value of the determinant of the matrix is unchanged.

Example 5.   Use A.5.3 in evaluating det A, where

$$A = \begin{bmatrix} -1 & 4 & 5 \\ 3 & 6 & 2 \\ 4 & -3 & 0 \end{bmatrix}$$

is the matrix given in Example 3.

Solution.   Since there is already a zero in the third row of A, we shall get another zero element into that row by first multiplying the second column of A by 4 to obtain

$$4 \det A = \det \begin{bmatrix} -1 & 16 & 5 \\ 3 & 24 & 2 \\ 4 & -12 & 0 \end{bmatrix}$$

by property (vi)] and then adding 3 times the first column of the new matrix to its second column [property (vii)]. The result is

$$4 \det A = \det \begin{bmatrix} -1 & 13 & 5 \\ 3 & 33 & 2 \\ 4 & 0 & 0 \end{bmatrix}.$$

We now expand the determinant in the right-hand member according to the elements of the third row, obtaining

$$4 \det A = 4 \begin{vmatrix} 13 & 5 \\ 33 & 2 \end{vmatrix} = 4(-139).$$

Therefore

$$\det A = -139.$$

The reader should compare the amount of work in this evaluation of det A to that shown in Example 3.

Example 6.   Let

$$A = \begin{bmatrix} 3 & -1 & 2 \\ 1 & 6 & -5 \\ -2 & 5 & 4 \end{bmatrix}.$$

Evaluate det A by using A.5.3 to obtain from A a matrix with two zeros in some row (or column), and then expanding according to the elements of that row or column.

Solution. We shall transform A into a 3×3 matrix having two zeros in the third column. In each step the property (from A.5.3) used will be indicated, and the reader should decide by inspection exactly what we have done.

By property (vii),

$$\det A = \det \begin{bmatrix} 3 & -1 & 8 \\ 1 & 6 & -3 \\ -2 & 5 & 0 \end{bmatrix}.$$

By (vi),

$$3 \det A = \det \begin{bmatrix} 9 & -3 & 24 \\ 1 & 6 & -3 \\ -2 & 5 & 0 \end{bmatrix}.$$

By (vii),

$$3 \det A = \det \begin{bmatrix} 17 & 45 & 0 \\ 1 & 6 & -3 \\ -2 & 5 & 0 \end{bmatrix}.$$

We now expand this last determinant according to the elements of the third column and obtain

$$3 \det A = 3 \begin{vmatrix} 17 & 45 \\ -2 & 5 \end{vmatrix} = 3(175).$$

Thus $\det A = 175$.

An immediate application of determinants is *Cramer's rule* for solving systems of n linear equations in n unknowns.

A.5.4   Cramer's Rule.   Let A be the nth order matrix $A = [a_{ij}]_n$ and denote by $A_{(j)}$ the matrix formed by replacing the elements a_{ij} of the jth column of A by the numbers k_i, $i = 1, 2, \cdots, n$. If $\det A \neq 0$, the system of n linear equations in n unknowns

$$a_{11}x_1 + a_{12}x_2 + \cdots + a_{1n}x_n = k_1$$
$$a_{21}x_1 + a_{22}x_2 + \cdots + a_{2n}x_n = k_2$$
$$\begin{matrix} \cdot & \cdot & \cdots & \cdot & \cdot \end{matrix}$$
$$a_{n1}x_1 + a_{n2}x_2 + \cdots + a_{nn}x_n = k_n$$

has the unique solution

$$x_1 = \frac{\det A_{(1)}}{\det A}, \qquad x_2 = \frac{\det A_{(2)}}{\det A}, \ldots, \qquad x_n = \frac{\det A_{(n)}}{\det A}.$$

Proof.   To make the proof easier to understand, it will be written for $n = 3$. The steps in the proof when n is general are exactly the same, and they can be written in less space because the notation is more general.

If (y_1, y_2, y_3) is a solution of the given linear system, then

(6)
$$a_{11}y_1 + a_{12}y_2 + a_{13}y_3 = k_1,$$
$$a_{21}y_1 + a_{22}y_2 + a_{23}y_3 = k_2,$$
$$a_{31}y_1 + a_{32}y_2 + a_{33}y_3 = k_3.$$

If we multiply both members of the first equation by A_{11}, the second by A_{21}, the third by A_{31}, and then add the resulting equations member by member, we obtain

$$(a_{11}A_{11} + a_{21}A_{21} + a_{31}A_{31})y_1 + (a_{12}A_{11} + a_{22}A_{21} + a_{32}A_{31})y_2$$
$$+ (a_{13}A_{11} + a_{23}A_{21} + a_{33}A_{31})y_3 = k_1A_{11} + k_2A_{21} + k_3A_{31}.$$

Since the coefficients of y_2 and y_3 are zero [A.5.3(v)], this equation can be written

$$(7) \qquad\qquad (\det A)y_1 = \det A_{(1)}.$$

In analogous manner, we also deduce from (6) that

$$(8) \qquad\qquad (\det A)y_2 = \det A_{(2)}, \qquad (\det A)y_3 = \det A_{(3)}.$$

From (7) and (8), and because $\det A \neq 0$,

$$(9) \qquad\qquad y_1 = \frac{\det A_{(1)}}{\det A}, \qquad y_2 = \frac{\det A_{(2)}}{\det A}, \qquad y_3 = \frac{\det A_{(3)}}{\det A}.$$

This proves that if (y_1, y_2, y_3) is a solution of the given linear system, then the values of y_1, y_2, and y_3 must be the unique numbers shown in (9).

To complete the proof of Cramer's rule for $n = 3$, we must show that (9) *is* a solution of the given system; that is, we must show that (9) implies (6). But this follows at once from the fact that all the above steps are reversible. ∎

Example 7. Solve the system

$$x + 3y - 2z = 11$$
$$4x - 2y + z = -15$$
$$3x + 4y - z = 3$$

by Cramer's rule.

Solution. Since

$$\det A = \begin{vmatrix} 1 & 3 & -2 \\ 4 & -2 & 1 \\ 3 & 4 & -1 \end{vmatrix} = -25,$$

Cramer's rule gives the solution

$$x = \frac{\begin{vmatrix} 11 & 3 & -2 \\ -15 & -2 & 1 \\ 3 & 4 & -1 \end{vmatrix}}{-25} = \frac{50}{-25} = -2,$$

$$y = \frac{\begin{vmatrix} 1 & 11 & -2 \\ 4 & -15 & 1 \\ 3 & 3 & -1 \end{vmatrix}}{-25} = \frac{-25}{-25} = 1,$$

$$z = \frac{\begin{vmatrix} 1 & 3 & 11 \\ 4 & -2 & -15 \\ 3 & 4 & 3 \end{vmatrix}}{-25} = \frac{125}{-25} = -5.$$

The reader should check the solution $(-2, 1, -5)$ by verifying that it satsfies all three equations of the given system.

A.6 FORMULAS FROM GEOMETRY AND TRIGONOMETRY

A.6.1 Geometry

Let r or R denote radius, h altitude, b length of base, B area of base, s slant height, and θ central angle in radian measure.

1. Circle: circumference $= 2\pi r$, area $= \pi r^2$.
2. Circular arc: length $= r\theta$.
3. Circular sector: area $= \frac{1}{2}r^2\theta$.
4. Circular segment: area $= \frac{1}{2}r^2(\theta - \sin\theta)$.
5. Triangle: area $= \frac{1}{2}bh$.
6. Trapezoid: area $= \frac{1}{2}h(b_1 + b_2)$.
7. Sphere: volume $= \frac{4}{3}\pi r^3$, area of surface $= 4\pi r^2$.
8. Right circular cylinder: volume $= \pi r^2 h$, lateral surface area $= 2\pi rh$.
9. Right circular cone: volume $= \frac{1}{3}\pi r^2 h$, lateral surface area $= \pi rs$.
10. Frustum of cone: volume $= \frac{1}{3}\pi h(R^2 + r^2 + Rr)$, lateral surface area $= \pi s(R + r)$.
11. Prism: volume $= Bh$.
12. Pyramid: volume $= \frac{1}{3}Bh$.

A.6.2 Trigonometry

Law of Cosines.

If A, B, and C are the interior angles of a triangle and a, b, and c are the sides respectively opposite to these angles, then

$$a^2 = b^2 + c^2 - 2bc \cos A.$$

Pythagorean Identities.

$$\sin^2 x + \cos^2 x = 1.$$
$$1 + \tan^2 x = \sec^2 x.$$
$$\cot^2 x + 1 = \csc^2 x.$$

Reduction Formulas.

$$\sin(-x) = -\sin x. \qquad \sin\left(\frac{\pi}{2} + x\right) = \cos x.$$

$$\cos(-x) = \cos x. \qquad \tan\left(\frac{\pi}{2} + x\right) = -\cot x.$$

$$\tan(-x) = -\tan x. \qquad \tan(x - \pi) = \tan x.$$

Addition Formulas.

$$\sin (x \pm y) = \sin x \cos y \pm \cos x \sin y.$$
$$\cos (x \pm y) = \cos x \cos y \mp \sin x \sin y.$$
$$\tan (x \pm y) = \frac{\tan x \pm \tan y}{1 \mp \tan x \tan y}.$$

Double-Angle Formulas.

$$\sin 2x = 2 \sin x \cos x.$$
$$\cos 2x = \cos^2 x - \sin^2 x = 2 \cos^2 x - 1 = 1 - 2 \sin^2 x.$$

Half-Angle Formulas.

$$\sin^2 \frac{x}{2} = \frac{1 - \cos x}{2}. \qquad \cos^2 \frac{x}{2} = \frac{1 + \cos x}{2}.$$

Sums and Products.

$$\sin x \pm \sin y = 2 \sin \left(\frac{x \pm y}{2}\right) \cos \left(\frac{x \mp y}{2}\right).$$

$$\cos x + \cos y = 2 \cos \left(\frac{x + y}{2}\right) \cos \left(\frac{x - y}{2}\right).$$

$$\cos x - \cos y = 2 \sin \left(\frac{x + y}{2}\right) \cos \left(\frac{x - y}{2}\right).$$

A.7 A SHORT TABLE OF INTEGRALS

Some Fundamental Forms.

1. $\int du = u + C.$

2. $\int c \, du = c \int du.$

3. $\int (f + g + \cdots) \, du = \int f \, du + \int g \, du + \cdots$

4. $\int u \, dv = uv - \int v \, du.$

5. $\int u^n \, du = \dfrac{u^{n+1}}{n + 1} + C \qquad (n \neq -1).$

6. $\int \dfrac{du}{u} = \ln u + C.$

Rational Forms Involving $a + bu$.

7. $\int \dfrac{u \, du}{a + bu} = \dfrac{1}{b^2}[a + bu - a \ln (a + bu)] + C.$

8. $\int \dfrac{u^2 \, du}{a + bu} = \dfrac{1}{b^3}\left[\dfrac{1}{2}(a + bu)^2 - 2a(a + bu) + a^2 \ln (a + bu)\right] + C.$

9. $\displaystyle\int \frac{u\,du}{(a+bu)^2} = \frac{1}{b^2}\left[\frac{a}{a+bu} + \ln(a+bu)\right] + C.$

10. $\displaystyle\int \frac{u^2\,du}{(a+bu)^2} = \frac{1}{b^3}\left[a+bu - \frac{a^2}{a+bu} - 2a\ln(a+bu)\right] + C.$

11. $\displaystyle\int \frac{du}{u(a+bu)} = -\frac{1}{a}\ln\frac{a+bu}{u} + C.$

12. $\displaystyle\int \frac{du}{u^2(a+bu)} = -\frac{1}{au} + \frac{b}{a^2}\ln\frac{a+bu}{u} + C.$

13. $\displaystyle\int \frac{du}{u(a+bu)^2} = \frac{1}{a(a+bu)} - \frac{1}{a^2}\ln\frac{a+bu}{u} + C.$

Forms Involving $\sqrt{a+bu}$.

14. $\displaystyle\int u\sqrt{a+bu}\,du = \frac{2(3bu-2a)}{15b^2}(a+bu)^{3/2} + C.$

15. $\displaystyle\int u^2\sqrt{a+bu}\,du = \frac{2(15b^2u^2 - 12abu + 8a^2)}{105b^3}(a+bu)^{3/2} + C.$

16. $\displaystyle\int \frac{u\,du}{\sqrt{a+bu}} = \frac{2(bu-2a)}{3b^2}\sqrt{a+bu} + C.$

17. $\displaystyle\int \frac{u^2\,du}{\sqrt{a+bu}} = \frac{2(3b^2u^2 - 4abu + 8a^2)}{15b^3}\sqrt{a+bu} + C.$

18a. $\displaystyle\int \frac{du}{u\sqrt{a+bu}} = \frac{1}{\sqrt{a}}\ln\frac{\sqrt{a+bu}-\sqrt{a}}{\sqrt{a+bu}+\sqrt{a}} + C \qquad (a>0).$

18b. $\displaystyle\int \frac{du}{u\sqrt{a+bu}} = \frac{2}{\sqrt{-a}}\,\text{Tan}^{-1}\sqrt{\frac{a+bu}{-a}} + C \qquad (a<0).$

19. $\displaystyle\int \frac{du}{u^2\sqrt{a+bu}} = -\frac{\sqrt{a+bu}}{au} - \frac{b}{2a}\int\frac{du}{u\sqrt{a+bu}}.$

20. $\displaystyle\int \frac{\sqrt{a+bu}}{u}\,du = 2\sqrt{a+bu} + a\int\frac{du}{u\sqrt{a+bu}}.$

21. $\displaystyle\int \frac{\sqrt{a+bu}}{u^2}\,du = -\frac{\sqrt{a+bu}}{u} + \frac{b}{2}\int\frac{du}{u\sqrt{a+bu}}.$

Forms Involving $a^2 \pm u^2$ and $u^2 - a^2$.

22. $\displaystyle\int \frac{du}{a^2+u^2} = \frac{1}{a}\,\text{Tan}^{-1}\frac{u}{a} + C \qquad \text{if } a>0.$

23. $\displaystyle\int \frac{du}{a^2-u^2} = \frac{1}{2a}\ln\frac{a+u}{a-u} + C = \frac{1}{a}\tanh^{-1}\frac{u}{a} + C \qquad \text{if } u^2 < a^2.$

24. $\displaystyle\int \frac{du}{u^2-a^2} = \frac{1}{2a}\ln\frac{u-a}{u+a} + C = -\frac{1}{a}\coth^{-1}\frac{u}{a} + C \qquad \text{if } u^2 > a^2.$

Forms Involving $\sqrt{a^2 - u^2}$.

25. $\displaystyle\int \frac{du}{\sqrt{a^2 - u^2}} = \text{Sin}^{-1} \frac{u}{a} + C$ if $u^2 < a^2$, $a > 0$.

26. $\displaystyle\int \sqrt{a^2 - u^2}\, du = \frac{u}{2}\sqrt{a^2 - u^2} + \frac{a^2}{2}\,\text{Sin}^{-1}\frac{u}{a} + C$.

27. $\displaystyle\int u^2\sqrt{a^2 - u^2}\, du = -\frac{u}{4}(a^2 - u^2)^{3/2} + \frac{a^2}{8}u\sqrt{a^2 - u^2} + \frac{a^4}{8}\,\text{Sin}^{-1}\frac{u}{a} + C$.

28. $\displaystyle\int \frac{\sqrt{a^2 - u^2}}{u}\, du = \sqrt{a^2 - u^2} - a\ln\left(\frac{a + \sqrt{a^2 - u^2}}{u}\right) + C$.

29. $\displaystyle\int \frac{\sqrt{a^2 - u^2}}{u^2}\, du = -\frac{\sqrt{a^2 - u^2}}{u} - \text{Sin}^{-1}\frac{u}{a} + C$.

30. $\displaystyle\int \frac{u^2\, du}{\sqrt{a^2 - u^2}} = -\frac{u}{2}\sqrt{a^2 - u^2} + \frac{a^2}{2}\,\text{Sin}^{-1}\frac{u}{a} + C$.

31. $\displaystyle\int \frac{du}{u\sqrt{a^2 - u^2}} = -\frac{1}{a}\ln\left(\frac{a + \sqrt{a^2 - u^2}}{u}\right) + C$.

32. $\displaystyle\int \frac{du}{u^2\sqrt{a^2 - u^2}} = -\frac{\sqrt{a^2 - u^2}}{a^2 u} + C$.

33. $\displaystyle\int (a^2 - u^2)^{3/2}\, du = -\frac{u}{8}(2u^2 - 5a^2)\sqrt{a^2 - u^2} + \frac{3a^4}{8}\,\text{Sin}^{-1}\frac{u}{a} + C$.

34. $\displaystyle\int \frac{du}{(a^2 - u^2)^{3/2}} = \frac{u}{a^2\sqrt{a^2 - u^2}} + C$.

Forms Involving $\sqrt{a^2 + u^2}$.

35. $\displaystyle\int \frac{du}{\sqrt{a^2 + u^2}} = \ln\left(u + \sqrt{a^2 + u^2}\right) + C = \sinh^{-1}\frac{u}{a} + C$.

36. $\displaystyle\int \sqrt{a^2 + u^2}\, du = \frac{u}{2}\sqrt{a^2 + u^2} + \frac{a^2}{2}\ln\left(u + \sqrt{a^2 + u^2}\right) + C$

$\qquad = \frac{u}{2}\sqrt{a^2 + u^2} + \frac{a^2}{2}\sinh^{-1}\frac{u}{a} + C$.

37. $\displaystyle\int u^2\sqrt{a^2 + u^2}\, du = \frac{u}{8}(2u^2 + a^2)\sqrt{a^2 + u^2} - \frac{a^4}{8}\ln\left(u + \sqrt{a^2 + u^2}\right) + C$

$\qquad = \frac{u}{8}(2u^2 + a^2)\sqrt{a^2 + u^2} - \frac{a^4}{8}\sinh^{-1}\frac{u}{a} + C$.

38. $\displaystyle\int \frac{\sqrt{a^2 + u^2}}{u}\, du = \sqrt{a^2 + u^2} - a\ln\left(\frac{a + \sqrt{a^2 + u^2}}{u}\right) + C$.

39. $\displaystyle\int \frac{\sqrt{a^2 + u^2}}{u^2}\, du = -\frac{\sqrt{a^2 + u^2}}{u} + \ln\left(u + \sqrt{a^2 + u^2}\right) + C$

$\qquad = -\frac{\sqrt{a^2 + u^2}}{u} + \sinh^{-1}\frac{u}{a} + C$.

40. $\displaystyle\int \frac{u^2\,du}{\sqrt{a^2+u^2}} = \frac{u}{2}\sqrt{a^2+u^2} - \frac{a^2}{2}\ln\left(u+\sqrt{a^2+u^2}\right) + C$

$\displaystyle\qquad\qquad = \frac{u}{2}\sqrt{a^2+u^2} - \frac{a^2}{2}\sinh^{-1}\frac{u}{a} + C.$

41. $\displaystyle\int \frac{du}{u\sqrt{a^2+u^2}} = -\frac{1}{a}\ln\left(\frac{\sqrt{a^2+u^2}+a}{u}\right) + C.$

42. $\displaystyle\int \frac{du}{u^2\sqrt{a^2+u^2}} = -\frac{\sqrt{a^2+u^2}}{a^2 u} + C.$

43. $\displaystyle\int (a^2+u^2)^{3/2}\,du = \frac{u}{8}(2u^2+5a^2)\sqrt{a^2+u^2} + \frac{3a^4}{8}\ln\left(u+\sqrt{a^2+u^2}\right) + C$

$\displaystyle\qquad\qquad = \frac{u}{8}(2u^2+5a^2)\sqrt{a^2+u^2} + \frac{3a^4}{8}\sinh^{-1}\frac{u}{a} + C.$

44. $\displaystyle\int \frac{du}{(a^2+u^2)^{3/2}} = \frac{u}{a^2\sqrt{a^2+u^2}} + C.$

Forms Involving $\sqrt{u^2-a^2}$.

45. $\displaystyle\int \frac{du}{\sqrt{u^2-a^2}} = \ln\left(u+\sqrt{u^2-a^2}\right) + C = \cosh^{-1}\frac{u}{a} + C.$

46. $\displaystyle\int \sqrt{u^2-a^2}\,du = \frac{u}{2}\sqrt{u^2-a^2} - \frac{a^2}{2}\ln\left(u+\sqrt{u^2-a^2}\right) + C$

$\displaystyle\qquad\qquad = \frac{u}{2}\sqrt{u^2-a^2} - \frac{a^2}{2}\cosh^{-1}\frac{u}{a} + C.$

47. $\displaystyle\int u^2\sqrt{u^2-a^2}\,du = \frac{u}{8}(2u^2-a^2)\sqrt{u^2-a^2} - \frac{a^4}{8}\ln\left(u+\sqrt{u^2-a^2}\right) + C$

$\displaystyle\qquad\qquad = \frac{u}{8}(2u^2-a^2)\sqrt{u^2-a^2} - \frac{a^4}{8}\cosh^{-1}\frac{u}{a} + C.$

48. $\displaystyle\int \frac{\sqrt{u^2-a^2}}{u}\,du = \sqrt{u^2-a^2} - a\,\text{Cos}^{-1}\frac{a}{u} + C$

$\displaystyle\qquad\qquad = \sqrt{u^2-a^2} - a\,\text{Sec}^{-1}\frac{u}{a} + C.$

49. $\displaystyle\int \frac{\sqrt{u^2-a^2}}{u^2}\,du = -\frac{\sqrt{u^2-a^2}}{u} + \ln\left(u+\sqrt{u^2-a^2}\right) + C$

$\displaystyle\qquad\qquad = -\frac{\sqrt{u^2-a^2}}{u} + \cosh^{-1}\frac{u}{a} + C.$

50. $\displaystyle\int \frac{u^2\,du}{\sqrt{u^2-a^2}} = \frac{u}{2}\sqrt{u^2-a^2} + \frac{a^2}{2}\ln\left(u+\sqrt{u^2-a^2}\right) + C$

$\displaystyle\qquad\qquad = \frac{u}{2}\sqrt{u^2-a^2} + \frac{a^2}{2}\cosh^{-1}\frac{u}{a} + C.$

51. $\displaystyle\int \frac{du}{u\sqrt{u^2-a^2}} = \frac{1}{a}\text{Cos}^{-1}\frac{a}{u} + C = \frac{1}{a}\text{Sec}^{-1}\frac{u}{a} + C.$

52. $\displaystyle\int \frac{du}{u^2\sqrt{u^2-a^2}} = \frac{\sqrt{u^2-a^2}}{a^2 u} + C.$

53. $\int (u^2 - a^2)^{3/2} \, du = \frac{u}{8}(2u^2 - 5a^2)\sqrt{u^2 - a^2} + \frac{3a^4}{8} \ln (u + \sqrt{u^2 - a^2}) + C$

$$= \frac{u}{8}(2u^2 - 5a^2)\sqrt{u^2 - a^2} + \frac{3a^4}{8} \cosh^{-1} \frac{u}{a} + C.$$

54. $\int \frac{du}{(u^2 - a^2)^{3/2}} = -\frac{u}{a^2\sqrt{u^2 - a^2}} + C.$

Forms Involving $\sqrt{2au - u^2}$.

55. $\int \sqrt{2au - u^2} \, du = \frac{u - a}{2} \sqrt{2au - u^2} + \frac{a^2}{2} \mathrm{Cos}^{-1} \left(1 - \frac{u}{a}\right) + C.$

56. $\int u\sqrt{2au - u^2} \, du = \frac{2u^2 - au - 3a^2}{6} \sqrt{2au - u^2} + \frac{a^3}{2} \mathrm{Cos}^{-1} \left(1 - \frac{u}{a}\right) + C.$

57. $\int \frac{\sqrt{2au - u^2}}{u} \, du = \sqrt{2au - u^2} + a \, \mathrm{Cos}^{-1} \left(1 - \frac{u}{a}\right) + C.$

58. $\int \frac{\sqrt{2au - u^2}}{u^2} \, du = -\frac{2\sqrt{2au - u^2}}{u} - \mathrm{Cos}^{-1} \left(1 - \frac{u}{a}\right) + C.$

59. $\int \frac{du}{\sqrt{2au - u^2}} = 2 \, \mathrm{Sin}^{-1} \sqrt{\frac{u}{2a}} + C = \mathrm{Cos}^{-1} \left(1 - \frac{u}{a}\right) + C.$

60. $\int \frac{u \, du}{\sqrt{2au - u^2}} = -\sqrt{2au - u^2} + a \, \mathrm{Cos}^{-1} \left(1 - \frac{u}{a}\right) + C.$

61. $\int \frac{u^2 \, du}{\sqrt{2au - u^2}} = -\frac{(u + 3a)}{2}\sqrt{2au - u^2} + \frac{3a^2}{2} \mathrm{Cos}^{-1} \left(1 - \frac{u}{a}\right) + C.$

62. $\int \frac{du}{u\sqrt{2au - u^2}} = -\frac{\sqrt{2au - u^2}}{au} + C.$

63. $\int \frac{du}{(2au - u^2)^{3/2}} = \frac{u - a}{a^2\sqrt{2au - u^2}} + C.$

Trigonometric Forms.

64. $\int \sin u \, du = -\cos u + C.$

65. $\int \cos u \, du = \sin u + C.$

66. $\int \tan u \, du = -\ln \cos u + C = \ln \sec u + C.$

67. $\int \cot u \, du = \ln \sin u + C = -\ln \csc u + C.$

68. $\int \sec u \, du = \ln (\sec u + \tan u) + C = \ln \tan \left(\frac{u}{2} + \frac{\pi}{4}\right) + C.$

69. $\int \csc u \, du = -\ln (\csc u + \cot u) + C = \ln \tan \frac{u}{2} + C.$

70. $\int \sec^2 u \, du = \tan u + C.$

71. $\displaystyle\int \csc^2 u \, du = -\cot u + C.$

72. $\displaystyle\int \sec u \tan u \, du = \sec u + C.$

73. $\displaystyle\int \csc u \cot u \, du = -\csc u + C.$

74. $\displaystyle\int \sin^2 u \, du = \tfrac{1}{2}(u - \sin u \cos u) + C = \tfrac{1}{2}u - \tfrac{1}{4}\sin 2u + C.$

75. $\displaystyle\int \cos^2 u \, du = \tfrac{1}{2}(u + \sin u \cos u) + C = \tfrac{1}{2}u + \tfrac{1}{4}\sin 2u + C.$

76. $\displaystyle\int \tan^2 u \, du = \tan u - u + C.$

77. $\displaystyle\int \sec^3 u \, du = \tfrac{1}{2}\sec u \tan u + \tfrac{1}{2}\ln(\sec u + \tan u) + C.$

78. $\displaystyle\int \sin mu \sin nu \, du = \frac{\sin(m-n)u}{2(m-n)} - \frac{\sin(m+n)u}{2(m+n)} + C.$

79. $\displaystyle\int \sin mu \cos nu \, du = -\frac{\cos(m-n)u}{2(m-n)} - \frac{\cos(m+n)u}{2(m+n)} + C.$

80. $\displaystyle\int \cos mu \cos nu \, du = \frac{\sin(m-n)u}{2(m-n)} + \frac{\sin(m+n)u}{2(m+n)} + C.$

81. $\displaystyle\int u \sin u \, du = \sin u - u \cos u + C.$

82. $\displaystyle\int u \cos u \, du = \cos u + u \sin u + C.$

83. $\displaystyle\int u^2 \sin u \, du = (2 - u^2)\cos u + 2u \sin u + C.$

84. $\displaystyle\int u^2 \cos u \, du = (u^2 - 2)\sin u + 2u \cos u + C.$

85a. $\displaystyle\int \sin^m u \cos^n u \, du = -\frac{\sin^{m-1} u \cos^{n+1} u}{m+n} + \frac{m-1}{m+n}\int \sin^{m-2} u \cos^n u \, du.$

85b. $\displaystyle\int \sin^m u \cos^n u \, du = \frac{\sin^{m+1} u \cos^{n-1} u}{m+n} + \frac{n-1}{m+n}\int \sin^m u \cos^{n-2} u \, du.$

86a. $\displaystyle\int \frac{du}{a + b \cos u} = \frac{2}{\sqrt{a^2 - b^2}}\,\mathrm{Tan}^{-1}\left[\frac{\sqrt{a^2 - b^2}\,\tan(u/2)}{a+b}\right] + C \quad\text{if } a^2 > b^2.$

86b. $\displaystyle\int \frac{du}{a + b \cos u} = \frac{1}{\sqrt{b^2 - a^2}}\ln\left[\frac{a + b + \sqrt{b^2 - a^2}\,\tan(u/2)}{a + b - \sqrt{b^2 - a^2}\,\tan(u/2)}\right] + C$
$$\text{if } b^2 > a^2.$$

87a. $\displaystyle\int \frac{du}{a + b \sin u} = \frac{2}{\sqrt{a^2 - b^2}}\,\mathrm{Tan}^{-1}\left[\frac{a \tan(u/2) + b}{\sqrt{a^2 - b^2}}\right] + C \quad\text{if } a^2 > b^2.$

87b. $\displaystyle\int \frac{du}{a + b \sin u} = \frac{1}{\sqrt{b^2 - a^2}}\ln\left[\frac{a \tan(u/2) + b - \sqrt{b^2 - a^2}}{a \tan(u/2) + b + \sqrt{b^2 - a^2}}\right] + C$
$$\text{if } b^2 > a^2.$$

Inverse Trigonometric Forms.

88. $\displaystyle\int \mathrm{Sin}^{-1} u \, du = u \,\mathrm{Sin}^{-1} u + \sqrt{1 - u^2} + C.$

89. $\int \text{Cos}^{-1} u \, du = u \, \text{Cos}^{-1} u - \sqrt{1 - u^2} + C.$

90. $\int \text{Tan}^{-1} u \, du = u \, \text{Tan}^{-1} u - \frac{1}{2} \ln (1 + u^2) + C.$

Exponential and Logarithmic Forms.

91. $\int e^u \, du = e^u + C.$

92. $\int a^u \, du = \frac{a^u}{\ln a} + C.$

93. $\int u e^u \, du = e^u(u - 1) + C.$

94. $\int u^n e^u \, du = u^n e^u - n \int u^{n-1} e^u \, du.$

95. $\int \frac{e^u}{u^n} \, du = -\frac{e^u}{(n-1)u^{n-1}} + \frac{1}{n-1} \int \frac{e^u \, du}{u^{n-1}}.$

96. $\int \ln u \, du = u \ln u - u + C.$

97. $\int u^n \ln u \, du = u^{n+1} \left[\frac{\ln u}{n+1} - \frac{1}{(n+1)^2} \right] + C.$

98. $\int \frac{du}{u \ln u} = \ln (\ln u) + C.$

99. $\int e^{au} \sin nu \, du = \frac{e^{au}(a \sin nu - n \cos nu)}{a^2 + n^2} + C.$

100. $\int e^{au} \cos nu \, du = \frac{e^{au}(a \cos nu + n \sin nu)}{a^2 + n^2} + C.$

Hyperbolic Forms.

101. $\int \sinh u \, du = \cosh u + C.$

102. $\int \cosh u \, du = \sinh u + C.$

103. $\int \tanh u \, du = \ln \cosh u + C.$

104. $\int \coth u \, du = \ln \sinh u + C.$

105. $\int \text{sech} \, u \, du = \text{Tan}^{-1} (\sinh u) + C.$

106. $\int \text{csch} \, u \, du = \ln \tanh \frac{1}{2} u + C.$

107. $\int \text{sech}^2 u \, du = \tanh u + C.$

108. $\int \text{csch}^2 u \, du = -\coth u + C.$

109. $\int \text{sech} \, u \tanh u \, du = -\text{sech} \, u + C.$

110. $\int \operatorname{csch} u \coth u \, du = -\operatorname{csch} u + C.$

111. $\int \sinh^2 u \, du = \frac{1}{4} \sinh 2u - \frac{1}{2}u + C.$

112. $\int \cosh^2 u \, du = \frac{1}{4} \sinh 2u + \frac{1}{2}u + C.$

113. $\int u \sinh u \, du = u \cosh u - \sinh u + C.$

114. $\int u \cosh u \, du = u \sinh u - \cosh u + C.$

115. $\int e^{au} \sinh nu \, du = \dfrac{e^{au}(a \sinh nu - n \cosh nu)}{a^2 - n^2} + C.$

116. $\int e^{au} \cosh nu \, du = \dfrac{e^{au}(a \cosh nu - n \sinh nu)}{a^2 - n^2} + C.$

Wallis's Formulas

117. $\displaystyle\int_0^{\pi/2} \sin^n u \, du = \int_0^{\pi/2} \cos^n u \, du$

$$= \begin{cases} \dfrac{(n-1)(n-3)\cdots 4\cdot 2}{n(n-2)\cdots 5\cdot 3\cdot 1} & \text{if } n \text{ is an odd integer} > 1; \\[2ex] \dfrac{(n-1)(n-3)\cdots 3\cdot 1}{n(n-2)\cdots 4\cdot 2}\cdot\dfrac{\pi}{2} & \text{if } n \text{ is a positive even integer.} \end{cases}$$

118. $\displaystyle\int_0^{\pi/2} \sin^m u \cos^n u \, du$

$$= \begin{cases} \dfrac{(n-1)(n-3)\cdots 4\cdot 2}{(m+n)(m+n-2)\cdots(m+5)(m+3)(m+1)} \\ \hspace{4cm} \text{if } n \text{ is an odd integer} > 1; \\[2ex] \dfrac{(m-1)(m-3)\cdots 4\cdot 2}{(n+m)(n+m-2)\cdots(n+5)(n+3)(n+1)} \\ \hspace{4cm} \text{if } m \text{ is an odd integer} > 1; \\[2ex] \dfrac{(m-1)(m-3)\cdots 3\cdot 1\cdot(n-1)(n-3)\cdots 3\cdot 1}{(m+n)(m+n-2)\cdots 4\cdot 2}\cdot\dfrac{\pi}{2} \\ \hspace{2cm} \text{if } m \text{ and } n \text{ are both positive even integers.} \end{cases}$$

A.8 NUMERICAL TABLES

A.8.1

powers and roots

n	n^2	$\sqrt{n}$	n^3	$\sqrt[3]{n}$	n	n^2	$\sqrt{n}$	n^3	$\sqrt[3]{n}$
1	1	1.000	1	1.000	51	2601	7.141	132,651	3.708
2	4	1.414	8	1.260	52	2704	7.211	140,608	3,732
3	9	1.732	27	1.442	53	2809	7.280	148,877	3.756
4	16	2.000	64	1.587	54	2916	7.348	157,464	3.780
5	25	2.236	125	1.710	55	3025	7.416	166,376	3.803
6	36	2.449	216	1.817	56	3136	7.483	175,616	3.826
7	49	2.646	343	1.913	57	3249	7.550	185,193	3.848
8	64	2.828	512	2.000	58	3364	7.616	195,112	3.871
9	81	3.000	729	2.080	59	3481	7.681	205,379	3.893
10	100	3.162	1000	2.154	60	3600	7.746	216,000	3.915
11	121	3.317	1331	2.224	61	3721	7.810	226,981	3.936
12	144	3.464	1728	2.289	62	3844	7.874	238,328	3.958
13	169	3.606	2197	2.351	63	3969	7.937	250,047	3.979
14	196	3.742	2744	2.410	64	4096	8.000	262,144	4.000
15	225	3.873	3375	2.466	65	4225	8.062	274,625	4.021
16	256	4.000	4096	2.520	66	4356	8.124	287,496	4.041
17	289	4.123	4913	2.571	67	4489	8.185	300,763	4.062
18	324	4.243	5832	2.621	68	4624	8.246	314,432	4.082
19	361	4.359	6859	2.668	69	4761	8.307	328,509	4.102
20	400	4.472	8000	2.714	70	4900	8.367	343,000	4.121
21	441	4.583	9261	2.759	71	5041	8.426	357,911	4.141
22	484	4.690	10,648	2.802	72	5184	8.485	373,248	4.160
23	529	4.796	12,167	2.844	73	5329	8.544	389,017	4.179
24	576	4.899	13,824	2.884	74	5476	8.602	405,224	4.198
25	625	5.000	15,625	2.924	75	5625	8.660	421,875	4.217
26	676	5.099	17,576	2.962	76	5776	8.718	438,976	4.236
27	729	5.196	19,683	3.000	77	5929	8.775	456,533	4.254
28	784	5.291	21,952	3.037	78	6084	8.832	474,552	4.273
29	841	5.385	24,389	3.072	79	6241	8.888	493.039	4.291
30	900	5.477	27,000	3.107	80	6400	8.944	512,000	4.309
31	961	5.568	29,791	3.141	81	6561	9.000	531,441	4.327
32	1024	5.657	32,768	3.175	82	6724	9.055	551,368	4.344
33	1089	5.745	35,937	3.208	83	6889	9.110	571,787	4.362
34	1156	5.831	39,304	3.240	84	7056	9.165	592,704	4.380
35	1225	5.916	42,875	3.271	85	7225	9.220	614,126	4.397
36	1296	6.000	46,656	3.302	86	7396	9.274	636,056	4.414
37	1369	6.083	50,653	3.332	87	7569	9.327	658,503	4.431
38	1444	6.164	54,872	3.362	88	7744	9.381	681,472	4.448
39	1521	6.245	59,319	3.391	89	7921	9.434	704,969	4.465
40	1600	6.325	64,000	3.420	90	8100	9.487	729,000	4.481
41	1681	6.403	68,921	3.448	91	8281	9.539	753,571	4.498
42	1764	6.481	74,088	3.476	92	8464	9.592	778,688	4.514
43	1849	6.557	79,507	3.503	93	8649	9.643	804,357	4.531
44	1936	6.633	85,184	3.530	94	8836	9.695	830,584	4.547
45	2025	6.708	91,125	3.557	95	9025	9.747	857,375	4.563
46	2116	6.782	97,336	3.583	96	9216	9.798	884,736	4.579
47	2209	6.856	103,823	3.609	97	9409	9.849	912,673	4.595
48	2304	6.928	110,592	3.634	98	9604	9.899	941,192	4.610
49	2401	7.000	117,649	3.659	99	9801	9.950	970,299	4.626
50	2500	7.071	125,000	3.684	100	10,000	10.000	1,000,000	4.642
n	n^2	$\sqrt{n}$	n^3	$\sqrt[3]{n}$	n	n^2	$\sqrt{n}$	n^3	$\sqrt[3]{n}$

A.8.2

trigonometric functions, degree measure

Angle	sin	tan	cot	cos	—
0.0°	0.0000	0.0000	—	1.0000	90.0°
0.5°	0.0087	0.0087	114.59	1.0000	89.5°
1.0°	0.0175	0.0175	57.290	0.9998	89.0°
1.5°	0.0262	0.0262	38.188	0.9997	88.5°
2.0°	0.0349	0.0349	28.636	0.9994	88.0°
2.5°	0.0436	0.0437	22.904	0.9990	87.5°
3.0°	0.0523	0.0524	19.081	0.9986	87.0°
3.5°	0.0610	0.0612	16.350	0.9981	86.5°
4.0°	0.0698	0.0699	14.301	0.9976	86.0°
4.5°	0.0785	0.0787	12.706	0.9969	85.5°
5.0°	0.0872	0.0875	11.430	0.9962	85.0°
5.5°	0.0958	0.0963	10.385	0.9954	84.5°
6.0°	0.1045	0.1051	9.5144	0.9945	84.0°
6.5°	0.1132	0.1139	8.7769	0.9936	83.5°
7.0°	0.1219	0.1228	8.1443	0.9925	83.0°
7.5°	0.1305	0.1317	7.5958	0.9914	82.5°
8.0°	0.1392	0.1405	7.1154	0.9903	82.0°
8.5°	0.1478	0.1495	6.6912	0.9890	81.5°
9.0°	0.1564	0.1584	6.3138	0.9877	81.0°
9.5°	0.1650	0.1673	5.9758	0.9863	80.5°
10.0°	0.1736	0.1763	5.6713	0.9848	80.0°
10.5°	0.1822	0.1853	5.3955	0.9833	79.5°
11.0°	0.1908	0.1944	5.1446	0.9816	79.0°
11.5°	0.1994	0.2035	4.9152	0.9799	78.5°
12.0°	0.2079	0.2126	4.7046	0.9781	78.0°
12.5°	0.2164	0.2217	4.5107	0.9763	77.5°
13.0°	0.2250	0.2309	4.3315	0.9744	77.0°
13.5°	0.2334	0.2401	4.1653	0.9724	76.5°
14.0°	0.2419	0.2493	4.0108	0.9703	76.0°
14.5°	0.2504	0.2586	3.8667	0.9681	75.5°
15.0°	0.2588	0.2679	3.7321	0.9659	75.0°
15.5°	0.2672	0.2773	3.6059	0.9636	74.5°
16.0°	0.2756	0.2867	3.4874	0.9613	74.0°
16.5°	0.2840	0.2962	3.3759	0.9588	73.5°
17.0°	0.2924	0.3057	3.2709	0.9563	73.0°
17.5°	0.3007	0.3153	3.1716	0.9537	72.5°
18.0°	0.3090	0.3249	3.0777	0.9511	72.0°
18.5°	0.3173	0.3346	2.9887	0.9483	71.5°
19.0°	0.3256	0.3443	2.9042	0.9455	71.0°
19.5°	0.3338	0.3541	2.8239	0.9426	70.5°
20.0°	0.3420	0.3640	2.7475	0.9397	70.0°
20.5°	0.3502	0.3739	2.6746	0.9367	69.5°
21.0°	0.3584	0.3839	2.6051	0.9336	69.0°
21.5°	0.3665	0.3939	2.5386	0.9304	68.5°
22.0°	0.3746	0.4040	2.4751	0.9272	68.0°
22.5°	0.3827	0.4142	2.4142	0.9239	67.5°
—	cos	cot	tan	sin	Angle

Angle	sin	tan	cot	cos	—
22.5°	0.3827	0.4142	2.4142	0.9239	67.5°
23.0°	0.3907	0.4245	2.3559	0.9205	67.0°
23.5°	0.3987	0.4348	2.2998	0.9171	66.5°
24.0°	0.4067	0.4452	2.2460	0.9135	66.0°
24.5°	0.4147	0.4557	2.1943	0.9100	65.5°
25.0°	0.4226	0.4663	2.1445	0.9063	65.0°
25.5°	0.4305	0.4770	2.0965	0.9026	64.5°
26.0°	0.4384	0.4877	2.0503	0.8988	64.0°
26.5°	0.4462	0.4986	2.0057	0.8949	63.5°
27.0°	0.4540	0.5095	1.9626	0.8910	63.0°
27.5°	0.4617	0.5206	1.9210	0.8870	62.5°
28.0°	0.4695	0.5317	1.8807	0.8829	62.0°
28.5°	0.4772	0.5430	1.8418	0.8788	61.5°
29.0°	0.4848	0.5543	1.8040	0.8746	61.0°
29.5°	0.4924	0.5658	1.7675	0.8704	60.5°
30.0°	0.5000	0.5774	1.7321	0.8660	60.0°
30.5°	0.5075	0.5890	1.6977	0.8616	59.5°
31.0°	0.5150	0.6009	1.6643	0.8572	59.0°
31.5°	0.5225	0.6128	1.6319	0.8526	58.5°
32.0°	0.5299	0.6249	1.6003	0.8480	58.0°
32.5°	0.5373	0.6371	1.5697	0.8434	57.5°
33.0°	0.5446	0.6494	1.5399	0.8387	57.0°
33.5°	0.5519	0.6619	1.5108	0.8339	56.5°
34.0°	0.5592	0.6745	1.4826	0.8290	56.0°
34.5°	0.5664	0.6873	1.4550	0.8241	55.5°
35.0°	0.5736	0.7002	1.4281	0.8192	55.0°
35.5°	0.5807	0.7133	1.4019	0.8141	54.5°
36.0°	0.5878	0.7265	1.3764	0.8090	54.0°
36.5°	0.5948	0.7400	1.3514	0.8039	53.5°
37.0°	0.6018	0.7536	1.3270	0.7986	53.0°
37.5°	0.6088	0.7673	1.3032	0.7934	52.5°
38.0°	0.6157	0.7813	1.2799	0.7880	52.0°
38.5°	0.6225	0.7954	1.2572	0.7826	51.5°
39.0°	0.6293	0.8098	1.2349	0.7771	51.0°
39.5°	0.6361	0.8243	1.2131	0.7716	50.5°
40.0°	0.6428	0.8391	1.1918	0.7660	50.0°
40.5°	0.6494	0.8541	1.1708	0.7604	49.5°
41.0°	0.6561	0.8693	1.1504	0.7547	49.0°
41.5°	0.6626	0.8847	1.1303	0.7490	48.5°
42.0°	0.6691	0.9004	1.1106	0.7431	48.0°
42.5°	0.6756	0.9163	1.0913	0.7373	47.5°
43.0°	0.6820	0.9325	1.0724	0.7314	47.0°
43.5°	0.6884	0.9490	1.0538	0.7254	46.5°
44.0°	0.6947	0.9657	1.0355	0.7193	46.0°
44.5°	0.7009	0.9827	1.0176	0.7133	45.5°
45.0°	0.7071	1.0000	1.0000	0.7071	45.0°
—	cos	cot	tan	sin	Angle

A.8.3

trigonometric functions, radian measure

Radians	sin	cos	tan	Radians	sin	cos	tan
0.00	0.0000	1.0000	0.0000	0.40	0.3894	0.9211	0.4228
0.01	0.0100	1.0000	0.0100	0.41	0.3986	0.9171	0.4346
0.02	0.0200	0.9998	0.0200	0.42	0.4078	0.9131	0.4466
0.03	0.0300	0.9996	0.0300	0.43	0.4169	0.9090	0.4586
0.04	0.0400	0.9992	0.0400	0.44	0.4259	0.9048	0.4708
0.05	0.0500	0.9988	0.0500	0.45	0.4350	0.9004	0.4831
0.06	0.0600	0.9982	0.0601	0.46	0.4439	0.8961	0.4954
0.07	0.0699	0.9976	0.0701	0.47	0.4529	0.8916	0.5080
0.08	0.0799	0.9968	0.0802	0.48	0.4618	0.8870	0.5206
0.09	0.0899	0.9960	0.0902	0.49	0.4706	0.8823	0.5334
0.10	0.0998	0.9950	0.1003	0.50	0.4794	0.8776	0.5463
0.11	0.1098	0.9940	0.1104	0.51	0.4882	0.8727	0.5594
0.12	0.1197	0.9928	0.1206	0.52	0.4969	0.8678	0.5726
0.13	0.1296	0.9916	0.1307	0.53	0.5055	0.8628	0.5859
0.14	0.1395	0.9902	0.1409	0.54	0.5141	0.8577	0.5994
0.15	0.1494	0.9888	0.1511	0.55	0.5227	0.8525	0.6131
0.16	0.1593	0.9872	0.1614	0.56	0.5312	0.8473	0.6269
0.17	0.1692	0.9856	0.1717	0.57	0.5396	0.8419	0.6410
0.18	0.1790	0.9838	0.1820	0.58	0.5480	0.8365	0.6552
0.19	0.1889	0.9820	0.1923	0.59	0.5564	0.8309	0.6696
0.20	0.1987	0.9801	0.2027	0.60	0.5646	0.8253	0.6841
0.21	0.2085	0.9780	0.2131	0.61	0.5729	0.8196	0.6989
0.22	0.2182	0.9759	0.2236	0.62	0.5810	0.8139	0.7139
0.23	0.2280	0.9737	0.2341	0.63	0.5891	0.8080	0.7291
0.24	0.2377	0.9713	0.2447	0.64	0.5972	0.8021	0.7445
0.25	0.2474	0.9689	0.2553	0.65	0.6052	0.7961	0.7602
0.26	0.2571	0.9664	0.2660	0.66	0.6131	0.7900	0.7761
0.27	0.2667	0.9638	0.2768	0.67	0.6210	0.7838	0.7923
0.28	0.2764	0.9611	0.2876	0.68	0.6288	0.7776	0.8087
0.29	0.2860	0.9582	0.2984	0.69	0.6365	0.7712	0.8253
0.30	0.2955	0.9553	0.3093	0.70	0.6442	0.7648	0.8423
0.31	0.3051	0.9523	0.3203	0.71	0.6518	0.7584	0.8595
0.32	0.3146	0.9492	0.3314	0.72	0.6594	0.7518	0.8771
0.33	0.3240	0.9460	0.3425	0.73	0.6669	0.7452	0.8949
0.34	0.3335	0.9428	0.3537	0.74	0.6743	0.7385	0.9131
0.35	0.3429	0.9394	0.3650	0.75	0.6816	0.7317	0.9316
0.36	0.3523	0.9359	0.3764	0.76	0.6889	0.7248	0.9505
0.37	0.3616	0.9323	0.3879	0.77	0.6961	0.7179	0.9697
0.38	0.3709	0.9287	0.3994	0.78	0.7033	0.7109	0.9893
0.39	0.3802	0.9249	0.4111	0.79	0.7104	0.7038	1.009

A.8.3

trigonometric functions, radian measure (continued)

Radians	sin	cos	tan	Radians	sin	cos	tan
0.80	0.7174	0.6967	1.030	1.20	0.9320	0.3624	2.572
0.81	0.7243	0.6895	1.050	1.21	0.9356	0.3530	2.650
0.82	0.7311	0.6822	1.072	1.22	0.9391	0.3436	2.733
0.83	0.7379	0.6749	1.093	1.23	0.9425	0.3342	2.820
0.84	0.7446	0.6675	1.116	1.24	0.9458	0.3248	2.912
0.85	0.7513	0.6600	1.138	1.25	0.9490	0.3153	3.010
0.86	0.7578	0.6524	1.162	1.26	0.9521	0.3058	3.113
0.87	0.7643	0.6448	1.185	1.27	0.9551	0.2963	3.224
0.88	0.7707	0.6372	1.210	1.28	0.9580	0.2867	3.341
0.89	0.7771	0.6294	1.235	1.29	0.9608	0.2771	3.467
0.90	0.7833	0.6216	1.260	1.30	0.9636	0.2675	3.602
0.91	0.7895	0.6137	1.286	1.31	0.9662	0.2579	3.747
0.92	0.7956	0.6058	1.313	1.32	0.9687	0.2482	3.903
0.93	0.8016	0.5978	1.341	1.33	0.9711	0.2385	4.072
0.94	0.8076	0.5898	1.369	1.34	0.9735	0.2288	4.256
0.95	0.8134	0.5817	1.398	1.35	0.9757	0.2190	4.455
0.96	0.8192	0.5735	1.428	1.36	0.9779	0.2092	4.673
0.97	0.8249	0.5653	1.459	1.37	0.9799	0.1994	4.913
0.98	0.8305	0.5570	1.491	1.38	0.9819	0.1896	5.177
0.99	0.8360	0.5487	1.524	1.39	0.9837	0.1798	5.471
1.00	0.8415	0.5403	1.557	1.40	0.9854	0.1700	5.798
1.01	0.8468	0.5319	1.592	1.41	0.9871	0.1601	6.165
1.02	0.8521	0.5234	1.628	1.42	0.9887	0.1502	6.581
1.03	0.8573	0.5148	1.665	1.43	0.9901	0.1403	7.055
1.04	0.8624	0.5062	1.704	1.44	0.9915	0.1304	7.602
1.05	0.8674	0.4976	1.743	1.45	0.9927	0.1205	8.238
1.06	0.8724	0.4889	1.784	1.46	0.9939	0.1106	8.989
1.07	0.8772	0.4801	1.827	1.47	0.9949	0.1006	9.887
1.08	0.8820	0.4713	1.871	1.48	0.9959	0.0907	10.98
1.09	0.8866	0.4625	1.917	1.49	0.9967	0.0807	12.35
1.10	0.8912	0.4536	1.965	1.50	0.9975	0.0707	14.10
1.11	0.8957	0.4447	2.014	1.51	0.9982	0.0608	16.43
1.12	0.9001	0.4357	2.066	1.52	0.9987	0.0508	19.67
1.13	0.9044	0.4267	2.120	1.53	0.9992	0.0408	24.50
1.14	0.9086	0.4176	2.176	1.54	0.9995	0.0308	32.46
1.15	0.9128	0.4085	2.234	1.55	0.9998	0.0208	48.08
1.16	0.9168	0.3993	2.296	1.56	0.9999	0.0108	92.62
1.17	0.9208	0.3902	2.360	1.57	1.0000	0.0008	1256.
1.18	0.9246	0.3809	2.427				
1.19	0.9284	0.3717	2.498				

A.8.4

common logarithms

n	0	1	2	3	4	5	6	7	8	9
10	0000	0043	0086	0128	0170	0212	0253	0294	0334	0374
11	0414	0453	0492	0531	0569	0607	0645	0682	0719	0755
12	0792	0828	0864	0899	0934	0969	1004	1038	1072	1106
13	1139	1173	1206	1239	1271	1303	1335	1367	1399	1430
14	1461	1492	1523	1553	1584	1614	1644	1673	1703	1732
15	1761	1790	1818	1847	1875	1903	1931	1959	1987	2014
16	2041	2068	2095	2122	2148	2175	2201	2227	2253	2279
17	2304	2330	2355	2380	2405	2430	2455	2480	2504	2529
18	2553	2577	2601	2625	2648	2672	2695	2718	2742	2765
19	2788	2810	2833	2856	2878	2900	2923	2945	2967	2989
20	3010	3032	3054	3075	3096	3118	3139	3160	3181	3201
21	3222	3243	3263	3284	3304	3324	3345	3365	3385	3404
22	3424	3444	3464	3483	3502	3522	3541	3560	3579	3598
23	3617	3636	3655	3674	3692	3711	3729	3747	3766	3784
24	3802	3820	3838	3856	3874	3892	3909	3927	3945	3962
25	3979	3997	4014	4031	4048	4065	4082	4099	4116	4133
26	4150	4166	4183	4200	4216	4232	4249	4265	4281	4298
27	4314	4330	4346	4362	4378	4393	4409	4425	4440	4456
28	4472	4487	4502	4518	4533	4548	4564	4579	4594	4609
29	4624	4639	4654	4669	4683	4698	4713	4728	4742	4757
30	4771	4786	4800	4814	4829	4843	4857	4871	4886	4900
31	4914	4928	4942	4955	4969	4983	4997	5011	5024	5038
32	5051	5065	5079	5092	5105	5119	5132	5145	5159	5172
33	5185	5198	5211	5224	5237	5250	5263	5276	5289	5302
34	5315	5328	5340	5353	5366	5378	5391	5403	5416	5428
35	5441	5453	5465	5478	5490	5502	5514	5527	5539	5551
36	5563	5575	5587	5599	5611	5623	5635	5647	5658	5670
37	5682	5694	5705	5717	5729	5740	5752	5763	5775	5786
38	5798	5809	5821	5832	5843	5855	5866	5877	5888	5899
39	5911	5922	5933	5944	5955	5966	5977	5988	5999	6010
40	6021	6031	6042	6053	6064	6075	6085	6096	6107	6117
41	6128	6138	6149	6160	6170	6180	6191	6201	6212	6222
42	6232	6243	6253	6263	6274	6284	6294	6304	6314	6325
43	6335	6345	6355	6365	6375	6385	6395	6405	6415	6425
44	6435	6444	6454	6464	6474	6484	6493	6503	6513	6522
45	6532	6542	6551	6561	6571	6580	6590	6599	6609	6618
46	6628	6637	6646	6656	6665	6675	6684	6693	6702	6712
47	6721	6730	6739	6749	6758	6767	6776	6785	6794	6803
48	6812	6821	6830	6839	6848	6857	6866	6875	6884	6893
49	6902	6911	6920	6928	6937	6946	6955	6964	6972	6981
50	6990	6998	7007	7016	7024	7033	7042	7050	7059	7067
51	7076	7084	7093	7101	7110	7118	7126	7135	7143	7152
52	7160	7168	7177	7185	7193	7202	7210	7218	7226	7235
53	7243	7251	7259	7267	7275	7284	7292	7300	7308	7316
54	7324	7332	7340	7348	7356	7364	7372	7380	7388	7396
n	0	1	2	3	4	5	6	7	8	9

Proportional Parts

	43	42	41	40
1	4.3	4.2	4.1	4.0
2	8.6	8.4	8.2	8.0
3	12.9	12.6	12.3	12.0
4	17.2	16.8	16.4	16.0
5	21.5	21.0	20.5	20.0
6	25.8	25.2	24.6	24.0
7	30.1	29.4	28.7	28.0
8	34.4	33.6	32.8	32.0
9	38.7	37.8	36.9	36.0

	39	38	37	36
1	3.9	3.8	3.7	3.6
2	7.8	7.6	7.4	7.2
3	11.7	11.4	11.1	10.8
4	15.6	15.2	14.8	14.4
5	19.5	19.0	18.5	18.0
6	23.4	22.8	22.2	21.6
7	27.3	26.6	25.9	25.2
8	31.2	30.4	29.6	28.8
9	35.1	34.2	33.3	32.4

	35	34	33	32
1	3.5	3.4	3.3	3.2
2	7.0	6.8	6.6	6.4
3	10.5	10.2	9.9	9.6
4	14.0	13.6	13.2	12.8
5	17.5	17.0	16.5	16.0
6	21.0	20.4	19.8	19.2
7	24.5	23.8	23.1	22.4
8	28.0	27.2	26.4	25.6
9	31.5	30.6	29.7	28.8

	31	30	29	28
1	3.1	3.0	2.9	2.8
2	6.2	6.0	5.8	5.6
3	9.3	9.0	8.7	8.4
4	12.4	12.0	11.6	11.2
5	15.5	15.0	14.5	14.0
6	18.6	18.0	17.4	16.8
7	21.7	21.0	20.3	19.6
8	24.8	24.0	23.2	22.4
9	27.9	27.0	26.1	25.2

	27	26	25	24
1	2.7	2.6	2.5	2.4
2	5.4	5.2	5.0	4.8
3	8.1	7.8	7.5	7.2
4	10.8	10.4	10.0	9.6
5	13.5	13.0	12.5	12.0
6	16.2	15.6	15.0	14.4
7	18.9	18.2	17.5	16.8
8	21.6	20.8	20.0	19.2
9	24.3	23.4	22.5	21.6

common logarithms (continued)

Proportional Parts				n	0	1	2	3	4	5	6	7	8	9
23	**22**	**21**	**20**	55	7404	7412	7419	7427	7435	7443	7451	7459	7466	7474
1	2.3	2.2	2.1	2.0	56	7482	7490	7497	7505	7513	7520	7528	7536	7543	7551

	23	22	21	20
1	2.3	2.2	2.1	2.0
2	4.6	4.4	4.2	4.0
3	6.9	6.6	6.3	6.0
4	9.2	8.8	8.4	8.0
5	11.5	11.0	10.5	10.0
6	13.8	13.2	12.6	12.0
7	16.1	15.4	14.7	14.0
8	18.4	17.6	16.8	16.0
9	20.7	19.8	18.9	18.0

	19	18	17	16
1	1.9	1.8	1.7	1.6
2	3.8	3.6	3.4	3.2
3	5.7	5.4	5.1	4.8
4	7.6	7.2	6.8	6.4
5	9.5	9.0	8.5	8.0
6	11.4	10.8	10.2	9.6
7	13.3	12.6	11.9	11.2
8	15.2	14.4	13.6	12.8
9	17.1	16.2	15.3	14.4

	15	14	13	12
1	1.5	1.4	1.3	1.2
2	3.0	2.8	2.6	2.4
3	4.5	4.2	3.9	3.6
4	6.0	5.6	5.2	4.8
5	7.5	7.0	6.5	6.0
6	9.0	8.4	7.8	7.2
7	10.5	9.8	9.1	8.4
8	12.0	11.2	10.4	9.6
9	13.5	12.6	11.7	10.8

	11	10	9	8
1	1.1	1.0	0.9	0.8
2	2.2	2.0	1.8	1.6
3	3.3	3.0	2.7	2.4
4	4.4	4.0	3.6	3.2
5	5.5	5.0	4.5	4.0
6	6.6	6.0	5.4	4.8
7	7.7	7.0	6.3	5.6
8	8.8	8.0	7.2	6.4
9	9.9	9.0	8.1	7.2

	7	6	5	4
1	0.7	0.6	0.5	0.4
2	1.4	1.2	1.0	0.8
3	2.1	1.8	1.5	1.2
4	2.8	2.4	2.0	1.6
5	3.5	3.0	2.5	2.0
6	4.2	3.6	3.0	2.4
7	4.9	4.2	3.5	2.8
8	5.6	4.8	4.0	3.2
9	6.3	5.4	4.5	3.6

n	0	1	2	3	4	5	6	7	8	9
55	7404	7412	7419	7427	7435	7443	7451	7459	7466	7474
56	7482	7490	7497	7505	7513	7520	7528	7536	7543	7551
57	7559	7566	7574	7582	7589	7597	7604	7612	7619	7627
58	7634	7642	7649	7657	7664	7672	7679	7686	7694	7701
59	7709	7716	7723	7731	7738	7745	7752	7760	7767	7774
60	7782	7789	7796	7803	7810	7818	7825	7832	7839	7846
61	7853	7860	7868	7875	7882	7889	7896	7903	7910	7917
62	7924	7931	7938	7945	7952	7959	7966	7973	7980	7987
63	7993	8000	8007	8014	8021	8028	8035	8041	8048	8055
64	8062	8069	8075	8082	8089	8096	8102	8109	8116	8122
65	8129	8136	8142	8149	8156	8162	8169	8176	8182	8189
66	8195	8202	8209	8215	8222	8228	8235	8241	8248	8254
67	8261	8267	8274	8280	8287	8293	8299	8306	8312	8319
68	8325	8331	8338	8344	8351	8357	8363	8370	8376	8382
69	8388	8395	8401	8407	8414	8420	8426	8432	8439	7445
70	8451	8457	8463	8470	8476	8482	8488	8494	8500	8506
71	8513	8519	8525	8531	8537	8543	8549	8555	8561	8567
72	8573	8579	8585	8591	8597	8603	8609	8615	8621	8627
73	8633	8639	8645	8651	8657	8663	8669	8675	8681	8686
74	8692	8698	8704	8710	8716	8722	8727	8733	8739	8745
75	8751	8756	8762	8768	8774	8779	8785	8791	8797	8802
76	8808	8814	8820	8825	8831	8837	8842	8848	8854	8859
77	8865	8871	8876	8882	8887	8893	8899	8904	8910	8915
78	8921	8927	8932	8938	8943	8949	8954	8960	8965	8971
79	8976	8982	8987	8993	8998	9004	9009	9015	9020	9025
80	9031	9036	9042	9047	9053	9058	9063	9069	9074	9079
81	9085	9090	9096	9101	9106	9112	9117	9122	9128	9133
82	9138	9143	9149	9154	9159	9165	9170	9175	9180	9186
83	9191	9196	9201	9206	9212	9217	9222	9227	9232	9238
84	9243	9248	9253	9258	9263	9269	9274	9279	9284	9289
85	9294	9299	9304	9309	9315	9320	9325	9330	9335	9340
86	9345	9350	9355	9360	9365	9370	9375	9380	9385	9390
87	9395	9400	9405	9410	9415	9420	9425	9430	9435	9440
88	9445	9450	9455	9460	9465	9469	9474	9479	9484	9489
89	9494	9499	9504	9509	9513	9518	9523	9528	9533	9538
90	9542	9547	9552	9557	9562	9566	9571	9576	9581	9586
91	9590	9595	9600	9605	9609	9614	9619	9624	9628	9633
92	9638	9643	9647	9652	9657	9661	9666	9671	9675	9680
93	9685	9689	9694	9699	9703	9708	9713	9717	9722	9727
94	9731	9736	9741	9745	9750	9754	9759	9763	9768	9773
95	9777	9782	9786	9791	9795	9800	9805	9809	9814	9818
96	9823	9827	9832	9836	9841	9845	9850	9854	9859	9863
97	9868	9872	9877	9881	9886	9890	9894	9899	9903	9908
98	9912	9917	9921	9926	9930	9934	9939	9943	9948	9952
99	9956	9961	9965	9969	9974	9978	9983	9987	9991	9996

Proportional Parts			n	0	1	2	3	4	5	6	7	8	9

A.8.5

natural logarithms

	0.00	0.01	0.02	0.03	0.04	0.05	0.06	0.07	0.08	0.09
1.0	0.0000	0.0100	0.0198	0.0296	0.0392	0.0488	0.0583	0.0677	0.0770	0.0862
1.1	0.0953	0.1044	0.1133	0.1222	0.1310	0.1398	0.1484	0.1570	0.1655	0.1740
1.2	0.1823	0.1906	0.1989	0.2070	0.2151	0.2231	0.2311	0.2390	0.2469	0.2546
1.3	0.2624	0.2700	0.2776	0.2852	0.2927	0.3001	0.3075	0.3148	0.3221	0.3293
1.4	0.3365	0.3436	0.3507	0.3577	0.3646	0.3716	0.3784	0.3853	0.3920	0.3988
1.5	0.4055	0.4121	0.4187	0.4253	0.4318	0.4383	0.4447	0.4511	0.4574	0.4637
1.6	0.4700	0.4762	0.4824	0.4886	0.4947	0.5008	0.5068	0.5128	0.5188	0.5247
1.7	0.5306	0.5365	0.5423	0.5481	0.5539	0.5596	0.5653	0.5710	0.5766	0.5822
1.8	0.5878	0.5933	0.5988	0.6043	0.6098	0.6152	0.6206	0.6259	0.6313	0.6366
1.9	0.6419	0.6471	0.6523	0.6575	0.6627	0.6678	0.6729	0.6780	0.6831	0.6881
2.0	0.6931	0.6981	0.7031	0.7080	0.7130	0.7178	0.7227	0.7275	0.7324	0.7372
2.1	0.7419	0.7467	0.7514	0.7561	0.7608	0.7655	0.7701	0.7747	0.7793	0.7839
2.2	0.7885	0.7930	0.7975	0.8020	0.8065	0.8109	0.8154	0.8198	0.8242	0.8286
2.3	0.8329	0.8372	0.8416	0.8459	0.8502	0.8544	0.8587	0.8629	0.8671	0.8713
2.4	0.8755	0.8796	0.8838	0.8879	0.8920	0.8961	0.9002	0.9042	0.9083	0.9123
2.5	0.9163	0.9203	0.9243	0.9282	0.9322	0.9361	0.9400	0.9439	0.9478	0.9517
2.6	0.9555	0.9594	0.9632	0.9670	0.9708	0.9746	0.9783	0.9821	0.9858	0.9895
2.7	0.9933	0.9969	1.0006	1.0043	1.0080	1.0116	1.0152	1.0188	1.0225	1.0260
2.8	1.0296	1.0332	1.0367	1.0403	1.0438	1.0473	1.0508	1.0543	1.0578	1.0613
2.9	1.0647	1.0682	1.0716	1.0750	1.0784	1.0818	1.0852	1.0886	1.0919	1.0953
3.0	1.0986	1.1019	1.1053	1.1086	1.1119	1.1151	1.1184	1.1217	1.1249	1.1282
3.1	1.1314	1.1346	1.1378	1.1410	1.1442	1.1474	1.1506	1.1537	1.1569	1.1600
3.2	1.1632	1.1663	1.1694	1.1725	1.1756	1.1787	1.1817	1.1848	1.1878	1.1909
3.3	1.1939	1.1970	1.2000	1.2030	1.2060	1.2090	1.2119	1.2149	1.2179	1.2208
3.4	1.2238	1.2267	1.2296	1.2326	1.2355	1.2384	1.2413	1.2442	1.2470	1.2499
3.5	1.2528	1.2556	1.2585	1.2613	1.2641	1.2669	1.2698	1.2726	1.2754	1.2782
3.6	1.2809	1.2837	1.2865	1.2892	1.2920	1.2947	1.2975	1.3002	1.3029	1.3056
3.7	1.3083	1.3110	1.3137	1.3164	1.3191	1.3218	1.3244	1.3271	1.3297	1.3324
3.8	1.3350	1.3376	1.3403	1.3429	1.3455	1.3481	1.3507	1.3533	1.3558	1.3584
3.9	1.3610	1.3635	1.3661	1.3686	1.3712	1.3737	1.3762	1.3788	1.3813	1.3838
4.0	1.3863	1.3888	1.3913	1.3938	1.3962	1.3987	1.4012	1.4036	1.4061	1.4085
4.1	1.4110	1.4134	1.4159	1.4183	1.4207	1.4231	1.4255	1.4279	1.4303	1.4327
4.2	1.4351	1.4375	1.4398	1.4422	1.4446	1.4469	1.4493	1.4516	1.4540	1.4563
4.3	1.4586	1.4609	1.4633	1.4656	1.4679	1.4702	1.4725	1.4748	1.4770	1.4793
4.4	1.4816	1.4839	1.4861	1.4884	1.4907	1.4929	1.4952	1.4974	1.4996	1.5019
4.5	1.5041	1.5063	1.5085	1.5107	1.5129	1.5151	1.5173	1.5195	1.5217	1.5239
4.6	1.5261	1.5282	1.5304	1.5326	1.5347	1.5369	1.5390	1.5412	1.5433	1.5454
4.7	1.5476	1.5497	1.5518	1.5539	1.5560	1.5581	1.5602	1.5623	1.5644	1.5665
4.8	1.5686	1.5707	1.5728	1.5748	1.5769	1.5790	1.5810	1.5831	1.5851	1.5872
4.9	1.5892	1.5913	1.5933	1.5953	1.5974	1.5994	1.6014	1.6034	1.6054	1.6074
5.0	1.6094	1.6114	1.6134	1.6154	1.6174	1.6194	1.6214	1.6233	1.6253	1.6273
5.1	1.6292	1.6312	1.6332	1.6351	1.6371	1.6390	1.6409	1.6429	1.6448	1.6467
5.2	1.6487	1.6506	1.6525	1.6544	1.6563	1.6582	1.6601	1.6620	1.6639	1.6658
5.3	1.6677	1.6696	1.6715	1.6734	1.6752	1.6771	1.6790	1.6808	1.6827	1.6845
5.4	1.6864	1.6882	1.6901	1.6919	1.6938	1.6956	1.6974	1.6993	1.7011	1.7029

$$\ln (N \cdot 10^m) = \ln N + m \ln 10, \qquad \ln 10 = 2.3026$$

natural logarithms (continued)

	0.00	0.01	0.02	0.03	0.04	0.05	0.06	0.07	0.08	0.09
5.5	1.7047	1.7066	1.7084	1.7102	1.7120	1.7138	1.7156	1.7174	1.7192	1.7210
5.6	1.7228	1.7246	1.7263	1.7281	1.7299	1.7317	1.7334	1.7352	1.7370	1.7387
5.7	1.7405	1.7422	1.7440	1.7457	1.7475	1.7492	1.7509	1.7527	1.7544	1.7561
5.8	1.7579	1.7596	1.7613	1.7630	1.7647	1.7664	1.7682	1.7699	1.7716	1.7733
5.9	1.7750	1.7766	1.7783	1.7800	1.7817	1.7834	1.7851	1.7867	1.7884	1.7901
6.0	1.7918	1.7934	1.7951	1.7967	1.7984	1.8001	1.8017	1.8034	1.8050	1.8066
6.1	1.8083	1.8099	1.8116	1.8132	1.8148	1.8165	1.8181	1.8197	1.8213	1.8229
6.2	1.8245	1.8262	1.8278	1.8294	1.8310	1.8326	1.8342	1.8358	1.8374	1.8390
6.3	1.8406	1.8421	1.8437	1.8453	1.8469	1.8485	1.8500	1.8516	1.8532	1.8547
6.4	1.8563	1.8579	1.8594	1.8610	1.8625	1.8641	1.8656	1.8672	1.8687	1.8703
6.5	1.8718	1.8733	1.8749	1.8764	1.8779	1.8795	1.8810	1.8825	1.8840	1.8856
6.6	1.8871	1.8886	1.8901	1.8916	1.8931	1.8946	1.8961	1.8976	1.8991	1.9006
6.7	1.9021	1.9036	1.9051	1.9066	1.9081	1.9095	1.9110	1.9125	1.9140	1.9155
6.8	1.9169	1.9184	1.9199	1.9213	1.9228	1.9242	1.9257	1.9272	1.9286	1.9301
6.9	1.9315	1.9330	1.9344	1.9359	1.9373	1.9387	1.9402	1.9416	1.9430	1.9445
7.0	1.9459	1.9473	1.9488	1.9502	1.9516	1.9530	1.9544	1.9559	1.9573	1.9587
7.1	1.9601	1.9615	1.9629	1.9643	1.9657	1.9671	1.9685	1.9699	1.9713	1.9727
7.2	1.9741	1.9755	1.9769	1.9782	1.9796	1.9810	1.9824	1.9838	1.9851	1.9865
7.3	1.9879	1.9892	1.9906	1.9920	1.9933	1.9947	1.9961	1.9974	1.9988	2.0001
7.4	2.0015	2.0028	2.0042	2.0055	2.0069	2.0082	2.0096	2.0109	2.0122	2.0136
7.5	2.0149	2.0162	2.0176	2.0189	2.0202	2.0215	2.0229	2.0242	2.0255	2.0268
7.6	2.0282	2.0295	2.0308	2.0321	2.0334	2.0347	2.0360	2.0373	2.0386	2.0399
7.7	2.0412	2.0425	2.0438	2.0451	2.0464	2.0477	2.0490	2.0503	2.0516	2.0528
7.8	2.0541	2.0554	2.0567	2.0580	2.0592	2.0605	2.0618	2.0631	2.0643	2.0656
7.9	2.0669	2.0681	2.0694	2.0707	2.0719	2.0732	2.0744	2.0757	2.0769	2.0782
8.0	2.0794	2.0807	2.0819	2.0832	2.0844	2.0857	2.0869	2.0882	2.0894	2.0906
8.1	2.0919	2.0931	2.0943	2.0956	2.0968	2.0980	2.0992	2.1005	2.1017	2.1029
8.2	2.1041	2.1054	2.1066	2.1078	2.1090	2.1102	2.1114	2.1126	2.1138	2.1150
8.3	2.1163	2.1175	2.1187	2.1190	2.1211	2.1223	2.1235	2.1247	2.1258	2.1270
8.4	2.1282	2.1294	2.1306	2.1318	2.1330	2.1342	2.1353	2.1365	2.1377	2.1389
8.5	2.1401	2.1412	2.1424	2.1436	2.1448	2.1459	2.1471	2.1483	2.1494	2.1506
8.6	2.1518	2.1529	2.1541	2.1552	2.1564	2.1576	2.1587	2.1599	2.1610	2.1622
8.7	2.1633	2.1645	2.1656	2.1668	2.1679	2.1691	2.1702	2.1713	2.1725	2.1736
8.8	2.1748	2.1759	2.1770	2.1782	2.1793	2.1804	2.1815	2.1827	2.1838	2.1849
8.9	2.1861	2.1872	2.1883	2.1894	2.1905	2.1917	2.1928	2.1939	2.1950	2.1961
9.0	2.1972	2.1983	2.1994	2.2006	2.2017	2.2028	2.2039	2.2050	2.2061	2.2072
9.1	2.2083	2.2094	2.2105	2.2116	2.2127	2.2138	2.2148	2.2159	2.2170	2.2181
9.2	2.2192	2.2203	2.2214	2.2225	2.2235	2.2246	2.2257	2.2268	2.2279	2.2289
9.3	2.2300	2.2311	2.2322	2.2332	2.2343	2.2354	2.2364	2.2375	2.2386	2.2396
9.4	2.2407	2.2418	2.2428	2.2439	2.2450	2.2460	2.2471	2.2481	2.2492	2.2502
9.5	2.2513	2.2523	2.2534	2.2544	2.2555	2.2565	2.2576	2.2586	2.2597	2.2607
9.6	2.2618	2.2628	2.2638	2.2649	2.2659	2.2670	2.2680	2.2690	2.2701	2.2711
9.7	2.2721	2.2732	2.2742	2.2752	2.2762	2.2773	2.2783	2.2793	2.2803	2.2814
9.8	2.2824	2.2834	2.2844	2.2854	2.2865	2.2875	2.2885	2.2895	2.2905	2.2915
9.9	2.2925	2.2935	2.2946	2.2956	2.2966	2.2976	2.2986	2.2996	2.3006	2.3016

exponential and hyperbolic functions

x	e^x	e^{-x}	$\sinh x$	$\cosh x$	$\tanh x$
0.00	1.0000	1.0000	0.0000	1.0000	0.0000
0.01	1.0101	0.9900	0.0100	1.0001	0.0100
0.02	1.0202	0.9802	0.0200	1.0002	0.0200
0.03	1.0305	0.9704	0.0300	1.0005	0.0300
0.04	1.0408	0.9608	0.0400	1.0008	0.0400
0.05	1.0513	0.9512	0.0500	1.0013	0.0500
0.06	1.0618	0.9418	0.0600	1.0018	0.0599
0.07	1.0725	0.9324	0.0701	1.0025	0.0699
0.08	1.0833	0.9231	0.0801	1.0032	0.0798
0.09	1.0942	0.9139	0.0901	1.0041	0.0898
0.10	1.1052	0.9048	0.1002	1.0050	0.0997
0.11	1.1163	0.8958	0.1102	1.0061	0.1096
0.12	1.1275	0.8869	0.1203	1.0072	0.1194
0.13	1.1388	0.8781	0.1304	1.0085	0.1293
0.14	1.1503	0.9694	0.1405	1.0098	0.1391
0.15	1.1618	0.8607	0.1506	1.0113	0.1489
0.16	1.1735	0.8521	0.1607	1.0128	0.1586
0.17	1.1853	0.8437	0.1708	1.0145	0.1684
0.18	1.1972	0.8353	0.1810	1.0162	0.1781
0.19	1.2092	0.8270	0.1911	1.0181	0.1877
0.20	1.2214	0.8187	0.2013	1.0201	0.1974
0.21	1.2337	0.8106	0.2115	1.0221	0.2070
0.22	1.2461	0.8025	0.2218	1.0243	0.2165
0.23	1.2586	0.7945	0.2320	1.0266	0.2260
0.24	1.2712	0.7866	0.2423	1.0289	0.2355
0.25	1.2840	0.7788	0.2526	1.0314	0.2449
0.26	1.2969	0.7711	0.2629	1.0340	0.2543
0.27	1.3100	0.7634	0.2733	1.0367	0.2636
0.28	1.3231	0.7558	0.2837	1.0395	0.2729
0.29	1.3364	0.7483	0.2941	1.0423	0.2821
0.30	1.3499	0.7408	0.3045	1.0453	0.2913
0.31	1.3634	0.7334	0.3150	1.0484	0.3004
0.32	1.3771	0.7261	0.3255	1.0516	0.3095
0.33	1.3910	0.7189	0.3360	1.0549	0.3185
0.34	1.4049	0.7118	0.3466	1.0584	0.3275
0.35	1.4191	0.7047	0.3572	1.0619	0.3364
0.36	1.4333	0.6977	0.3678	1.0655	0.3452
0.37	1.4477	0.6907	0.3785	1.0692	0.3540
0.38	1.4623	0.6839	0.3892	1.0731	0.3627
0.39	1.4770	0.6771	0.4000	1.0770	0.3714
0.40	1.4918	0.6703	0.4108	1.0811	0.3799
0.41	1.5068	0.6637	0.4216	1.0852	0.3885
0.42	1.5220	0.6570	0.4325	1.0895	0.3969
0.43	1.5373	0.6505	0.4434	1.0939	0.4053
0.44	1.5527	0.6440	0.4543	1.0984	0.4136

exponential and hyperbolic functions (continued)

x	e^x	e^{-x}	$\sinh x$	$\cosh x$	$\tanh x$
0.45	1.5683	0.6376	0.4653	1.1030	0.4219
0.46	1.5841	0.6313	0.4764	1.1077	0.4301
0.47	1.6000	0.6250	0.4875	1.1125	0.4382
0.48	1.6161	0.6188	0.4986	1.1174	0.4462
0.49	1.6323	0.6126	0.5098	1.1225	0.4542
0.50	1.6487	0.6065	0.5211	1.1276	0.4621
0.51	1.6653	0.6005	0.5324	1.1329	0.4699
0.52	1.6820	0.5945	0.5438	1.1383	0.4777
0.53	1.6989	0.5886	0.5552	1.1438	0.4854
0.54	1.7160	0.5827	0.5666	1.1494	0.4930
0.55	1.7333	0.5769	0.5782	1.1551	0.5005
0.56	1.7507	0.5712	0.5897	1.1609	0.5080
0.57	1.7683	0.5655	0.6014	1.1669	0.5154
0.58	1.7860	0.5599	0.6131	1.1730	0.5227
0.59	1.8040	0.5543	0.6248	1.1792	0.5299
0.60	1.8221	0.5488	0.6367	1.1855	0.5370
0.61	1.8044	0.5434	0.6485	1.1919	0.5441
0.62	1.8589	0.5379	0.6605	1.1984	0.5511
0.63	1.8776	0.5326	0.6725	1.2051	0.5581
0.64	1.8965	0.5273	0.6846	1.2119	0.5649
0.65	1.9155	0.5220	0.6967	1.2188	0.5717
0.66	1.9348	0.5169	0.7090	1.2258	0.5784
0.67	1.9542	0.5117	0.7213	1.2330	0.5850
0.68	1.9739	0.5066	0.7336	1.2402	0.5915
0.69	1.9937	0.5016	0.7461	1.2476	0.5980
0.70	2.0138	0.4966	0.7586	1.2552	0.6044
0.71	2.0340	0.4916	0.7712	1.2628	0.6107
0.72	2.0544	0.4868	0.7838	1.2706	0.6169
0.73	2.0751	0.4819	0.7966	1.2785	0.6231
0.74	2.0959	0.4771	0.8094	1.2865	0.6291
0.75	2.1170	0.4724	0.8223	1.2947	0.6351
0.76	2.1383	0.4677	0.8353	1.3030	0.6411
0.77	2.1598	0.4630	0.8484	1.3114	0.6469
0.78	2.1815	0.4584	0.8615	1.3199	0.6527
0.79	2.2034	0.4538	0.8748	1.3286	0.6584
0.80	2.2255	0.4493	0.8881	1.3374	0.6640
0.81	2.2479	0.4449	0.9015	1.3464	0.6696
0.82	2.2705	0.4404	0.9150	1.3555	0.6751
0.83	2.2933	0.4360	0.9286	1.3647	0.6805
0.84	2.3164	0.4317	0.9423	1.3740	0.6858
0.85	2.3396	0.4274	0.9561	1.3835	0.6911
0.86	2.3632	0.4232	0.9700	1.3932	0.6963
0.87	2.3869	0.4190	0.9840	1.4029	0.7014
0.88	2.4109	0.4148	0.9981	1.4128	0.7064
0.89	2.4351	0.4107	1.0122	1.4229	0.7114

exponential and hyperbolic functions (continued)

x	e^x	e^{-x}	sinh x	cosh x	tanh x
0.90	2.4596	0.4066	1.0265	1.4331	0.7163
0.91	2.4843	0.4025	1.0409	1.4434	0.7211
0.92	2.5093	0.3985	1.0554	1.4539	0.7259
0.93	2.5345	0.3946	1.0700	1.4645	0.7306
0.94	2.5600	0.3906	1.0847	1.4753	0.7352
0.95	2.5857	0.3867	1.0995	1.4862	0.7398
0.96	2.6117	0.3829	1.1144	1.4973	0.7443
0.97	2.6379	0.3791	1.1294	1.5085	0.7487
0.98	2.6645	0.3753	1.1446	1.5199	0.7531
0.99	2.6912	0.3716	1.1598	1.5314	0.7574
1.00	2.7183	0.3679	1.1752	1.5431	0.7616
1.05	2.8577	0.3499	1.2539	1.6038	0.7818
1.10	3.0042	0.3329	1.3356	1.6685	0.8005
1.15	3.1582	0.3166	1.4208	1.7374	0.8178
1.20	3.3201	0.3012	1.5085	1.8107	0.8337
1.25	3.4903	0.2865	1.6019	1.8884	0.8483
1.30	3.6693	0.2725	1.6984	1.9709	0.8617
1.35	3.8574	0.2592	1.7991	2.0583	0.8741
1.40	4.0552	0.2466	1.9043	2.1509	0.8854
1.45	4.2631	0.2346	2.0143	2.2488	0.8957
1.50	4.4817	0.2231	2.1293	2.3524	0.9051
1.55	4.7115	0.2122	2.2496	2.4619	0.9138
1.60	4.9530	0.2019	2.3756	2.5775	0.9217
1.65	5.2070	0.1920	2.5075	2.6995	0.9289
1.70	5.4739	0.1827	2.6456	2.8283	0.9354
1.75	5.7546	0.1738	2.7904	2.9642	0.9414
1.80	6.0496	0.1653	2.9422	3.1075	0.9468
1.85	6.3598	0.1572	3.1013	3.2585	0.9517
1.90	6.6859	0.1496	3.2682	3.4177	0.9562
1.95	7.0287	0.1423	3.4432	3.5855	0.9603
2.00	7.3891	0.1353	3.6269	3.7622	0.9640
2.05	7.7679	0.1287	3.8196	3.9483	0.9674
2.10	8.1662	0.1225	4.0219	4.1443	0.9705
2.15	8.5849	0.1165	4.2342	4.3507	0.9732
2.20	9.0250	0.1108	4.4571	4.5679	0.9757
2.25	9.4877	0.1054	4.6912	4.7966	0.9780
2.30	9.9742	0.1003	4.9370	5.0372	0.9801
2.35	10.486	0.0954	5.1951	5.2905	0.9820
2.40	11.023	0.0907	5.4662	5.5569	0.9837
2.45	11.588	0.0863	5.7510	5.8373	0.9852
2.50	12.182	0.0821	6.0502	6.1323	0.9866
2.55	12.807	0.0781	6.3645	6.4426	0.9879
2.60	13.464	0.0743	6.6947	6.7690	0.9890
2.65	14.154	0.0707	7.0417	7.1123	0.9901
2.70	14.880	0.0672	7.4063	7.4735	0.9910

exponential and hyperbolic functions (continued)

x	e^x	e^{-x}	sinh x	cosh x	tanh x
2.75	15.643	0.0639	7.7894	7.8533	0.9919
2.80	16.445	0.0608	8.1919	8.2527	0.9926
2.85	17.288	0.0578	8.6150	8.6728	0.9933
2.90	18.174	0.0550	9.0596	9.1146	0.9940
2.95	19.106	0.0523	9.5268	9.5791	0.9945
3.00	20.086	0.0498	10.018	10.068	0.9951
3.05	21.115	0.0474	10.534	10.581	0.9955
3.10	22.198	0.0450	11.076	11.122	0.9959
3.15	23.336	0.0429	11.647	11.689	0.9963
3.20	24.533	0.0408	12.246	12.287	0.9967
3.25	25.790	0.0388	12.876	12.915	0.9970
3.30	27.113	0.0369	13.538	13.575	0.9973
3.35	28.503	0.0351	14.234	14.269	0.9975
3.40	29.964	0.0334	14.965	14.999	0.9978
3.45	31.500	0.0317	15.734	15.766	0.9980
3.50	33.115	0.0302	16.543	16.573	0.9982
3.55	34.813	0.0287	17.392	17.421	0.9983
3.60	36.598	0.0273	18.286	18.313	0.9985
3.65	38.475	0.0260	19.224	19.250	0.9986
3.70	40.447	0.0247	20.211	20.236	0.9988
3.75	42.521	0.0235	21.249	21.272	0.9989
3.80	44.701	0.0224	22.339	22.362	0.9990
3.85	46.993	0.0213	23.486	23.507	0.9991
3.90	49.402	0.0202	24.691	24.711	0.9992
3.95	51.935	0.0193	25.958	25.977	0.9993
4.00	54.598	0.0183	27.290	27.308	0.9993
4.10	60.340	0.0166	30.162	30.178	0.9995
4.20	66.686	0.0150	33.336	33.351	0.9996
4.30	73.700	0.0136	36.843	36.857	0.9996
4.40	81.451	0.0123	40.719	40.732	0.9997
4.50	90.017	0.0111	45.003	45.014	0.9998
4.60	99.484	0.0101	49.737	49.747	0.9998
4.70	109.95	0.0091	54.969	54.978	0.9998
4.80	121.51	0.0082	60.751	60.759	0.9999
4.90	134.29	0.0074	67.141	67.149	0.9999
5.00	148.41	0.0067	74.203	74.210	0.9999
5.20	181.27	0.0055	90.633	90.639	0.9999
5.40	221.41	0.0045	110.70	110.71	1.0000
5.60	270.43	0.0037	135.21	135.22	1.0000
5.80	330.30	0.0030	165.15	165.15	1.0000
6.00	403.43	0.0025	201.71	201.72	1.0000
7.00	1096.6	0.0009	548.32	548.32	1.0000
8.00	2981.0	0.0003	1490.5	1490.5	1.0000
9.00	8103.1	0.0001	4051.5	4051.5	1.0000
10.00	22026.	0.00005	11013.	11013.	1.0000

1.2   Page 6

1. (a) True; (b) False; (c) False;
   (d) True; (e) False; (f) True;
   (g) True; (h) True; (i) True;
   (j) False; (k) False; (l) True;
   (m) False; (n) True.

5. (a) $\{x \mid x = 2n + 1, n = 1, 2, 3, 4, 5\}$;
   (b) $\{x \mid x = 3n + 1, n = 1, 2, 3, 4, 5\}$;
   (c) $\{x \mid x = 2^n, n \in N\}$;
   (d) $\{x \mid x = 3^n, n \in N\}$;
   (e) $\{x \mid x^2 - 3x - 10 = 0\}$;
   (f) $\{x \mid x^3 - 6x^2 - 7x = 0\}$.

9. (a) $\{a, b, c, d, e, g\}$; (b) $\{a, b, c, d, e, g\}$;
   (c) $\{a, b, c, e, g\}$; (d) $\{a, b, c, e, g\}$.

3. (a) $\{3, 5, 7, 9\}$; (b) $\{8, 64, 125\}$;
   (c) $\{1, 2\}$; (d) $\{-4, 0, 2\}$.

7. (a) $\{a, c, d, e, g\}$; (b) $\{b, c, d, e, f, g\}$;
   (c) $\{d, e, g\}$; (d) $\{d\}$.

11. (a) True;
    (b) False. $\{3\}$ is a set and 3 is not.
    (c) False. Zero is not a set but $\varnothing$ is a set.
    (d) False. The set $\varnothing$ has no elements but the set $\{0\}$ has one element.
    (e) False. The set $\varnothing$ contains no elements.

1.3   Page 12

1. $\sqrt{4}$,   2.718, $(1 + \sqrt{5})(1 - \sqrt{5})$,   7/13,
   $\sqrt{2}/(3\sqrt{2})$, $0.3333\cdots$, $-2.6666\cdots$,
   $1.034\ 034\ 034\cdots$.

3. (a) False; $-20 - 2 = -22$.
   (b) True; $-39 - 1 = -40$.
   (c) True; $5/9 - (-3) = 32/9$.
   (d) True; $-16 - (-4) = -12$.

(e) True; $34/39 - 6/7 = 4/273$.

(f) False; $-44/59 - (-5/7) = -13/413$.

5. (a) 0;   (b) 0;

(c) not defined;   (d) 0;

(e) not defined;   (f) 0.

1.5   Page 20

1. (a)

     $-4 \quad -2 \quad 0 \ 1$

(b)

     $-4 \quad -2 \quad 0 \ 1$

(c)

     $-4 \quad -2 \quad 0 \ 1$

(d)

     $-4 \quad -2 \quad 0 \ 1$

(e)

     1

(f)

     -4

3. $(-\infty, 12)$.

5. $(-5/3, \infty)$.

9. $(-9/2, -2)$.

13. $(1/2, 3)$.

17. $(-\infty, 0) \cup (1/5, \infty)$.

21. $(-3, -21/8)$.

25. $(-1/3, 4)$.

29. $(-\infty, -2/3) \cup (-1/3, 0) \cup (1/3, \infty)$.

7. $(2, \infty)$.

11. $(-2/5, 3/5)$.

15. $\varnothing$.

19. $(-\infty, 2/3) \cup (3/4, \infty)$.

23. $(1/7, 1/2)$.

27. $(-\infty, 2) \cup (3, \infty)$.

1.6   Page 26

1. $(-5, 3)$.

5. $(-12, 24)$.

9. $(-\infty, -3) \cup (2, \infty)$.

13. $|x - 2| < 0.5 \iff -0.5 < x - 2 < 0.5$
$\iff -1.5 < 3x - 6 < 1.5 \iff 7 - 1.5$
$< 3x + 1 < 7 + 1.5$.

17. $|x - 5| < \epsilon/10 \iff -\epsilon/10 < x - 5 < \epsilon/10$
$\iff -\epsilon < 10x - 50 < \epsilon \iff 14 - \epsilon$
$< 10x - 36 < 14 + \epsilon$.

21. $\delta = \epsilon/2$.

25. $(-4/5, 16/3)$.

3. $(-4, 3/4)$.

7. $(-\infty, 2) \cup (5, \infty)$.

11. $(-\infty, -5) \cup (-5/3, 0) \cup (0, \infty)$.

15. $|x - 4| < 0.06 \iff -0.06 < x - 4 < 0.06$
$\iff -0.03 < \frac{1}{2}x - 2 < 0.03 \iff 7 - 0.03$
$< \frac{1}{2}x + 5 < 7 + 0.03$.

19. $\delta = 0.04$.

23. $(-\infty, -23/2) \cup (-19/4, \infty)$.

1.8   Page 29

1. $(-4, \infty)$.

5. $(-6, -3) \cup (-1, 2)$.

9. $(-\infty, -1/3) \cup (7, \infty)$.

13. $(-2, 1]$.

3. $(-\infty, 32/15) \cup (38/13, \infty)$.

7. $(-\infty, -1/2] \cup (1, \infty)$.

11. $(-\infty, 0) \cup (1/5, \infty)$.

15. $(-5, 6)$.

17. $(-1/2, 5)$.

19. $[2/3, 22/9]$.

2.1 Page 34

3. (a) 5;   (b) $\sqrt{17}$;
(c) 4;   (d) $\sqrt{34}$.

7. Length of each side is $3\sqrt{5}$, and the length of each diagonal is $3\sqrt{10}$.

11. The straight line that is the perpendicular bisector of the line segment whose endpoints are $(1, 3)$ and $(5, 5)$.

5. $|P_1P_2|^2 + |P_2P_3|^2 = |P_1P_3|^2$.

9. $3x - 7y - 9 = 0$.

13. $x = \sqrt{(x - 4)^2 + y^2}$.

2.2 Page 38

1. (a) $2/5$;   (b) $-6/7$.

5. $(-11/3, 0)$.

9. No.

13. $x = (2x_1 + x_2)/3, y = (2y_1 + y_2)/3$.

3. (a) $1/2, (1, 3)$;   (b) $-3/2, (4, -1)$;
(c) $0, (1/2, -5)$;
(d) $-1/3, (-7/2, -3/2)$.

7. $-3/10, 4/3$, and -1.

11. $(7, 2)$.

15. $(y - 5)/(x - 2) = 3$, or $3x - y - 1 = 0$,
$x \neq 2$. A straight line.

2.3 Page 41

1. Straight line through $(0, 2)$ and $(4, 0)$.

5. Vertical line through $(2, 0)$.

9. Circle with center at $(0, 0)$ and radius 3.

17. $(4, 3)$.

21. $(1, 1), (-2, 4)$.

3. Line through $(0, -4)$ and $(-2, 0)$.

7. The x axis.

11. Circle with center at $(2, 0)$ and radius 2.

19. $(\sqrt{5}, 2\sqrt{5}), (-\sqrt{5}, -2\sqrt{5})$.

2.4 Page 44

1. $x + y + 4 = 0$.

5. $2x + y - 6 = 0$.

9. $2x - 3y + 6 = 0$.

15. $(-2, 1)$.

3. $y - 5 = 0$.

7. $5x + 7y - 3 = 0$.

13. $2x + y - 3 = 0$.

17. The two straight lines $x - y = 0$ and $x + y = 0$.

2.5 Page 50

1. $x + 3y - 11 = 0$.

5. $7x + 3y = 0$.

9. $(-2, -3)$.

13. $x^2 + y^2 + 6x - 14y = 0$.

17. Center is $(-7, 2)$; radius $= 1$.

21. $x^2 + y^2 - 8x + 10y - 48 = 0$.

3. $2x + y + 6 = 0$.

7. $6x - 8y - 7 = 0$.

11. $(x - 3)^2 + (y + 6)^2 = 36$;
$x^2 + y^2 - 6x + 12y + 9 = 0$.

15. $x^2 + y^2 + 18x + 8y - 72 = 0$.

19. $5x + 2y - 4 = 0$.

2.6 Page 54

1. x intercept: 0; y intercept: 0. Symmetric with respect to the y axis. Exclude $y < 0$.

3. x intercepts: ± 3; y intercepts: ± 2. Symmetric with respect to both coordinate axes. Exclude $|x| > 3$; exclude $|y| > 2$.

5. x intercepts: ± 2. Symmetric with respect to both coordinate axes. Exclude $-2 < x < 2$.

9. x intercepts: ± 2; y intercepts: ± 2. Symmetric with respect to both coordinate axes. Exclude $|x| > 2$; exclude $|y| > 2$.

13. x intercepts: 0, 3, and 5; y intercept: 0.

7. x intercept: 0; y intercept: 0. Symmetric with respect to the origin.

11. y intercept: 2. Symmetric with respect to the y axis. Exclude $y > 2$; exclude $y \le 0$.

2.7 Page 59

7. $\{P{:}(x, y) \mid x > 0, 5x + 4y - 20 > 0\}$.

11. $\{P{:}(x, y) \mid x^2 + y^2 + 4x - 2y + 4 \ge 0,$
$-5x + 6y - 30 \le 0, \quad 3x + 7y + 18 \ge 0,$
$8x + y - 5 \le 0\}$.

2.8 Page 64

1. (a) 0;   (b) 3;   (c) -1;
(d) $k^2 - 1$;   (e) $b^2 - 1$;   (f) $-3/4$;
(g) $4t^2 - 1$;   (h) $9x^2 - 1$;   (i) $1/x^2 - 1$.

5. (c) A function having domain $[-3, 3]$;
(d) A function with domain $\{0\}$.

9. (a) $\mathfrak{D}_F = [-3/2, \infty)$, $\mathfrak{R}_F = [0, \infty)$;
(b) $\mathfrak{D}_g = \{v \mid v \ne 1/4\}$, $\mathfrak{R}_g = \{g(v) \mid g(v) \ne 0\}$;
(c) $\mathfrak{D}_\psi = \{x \mid |x| \ge 3\}$, $\mathfrak{R}_\psi = [0, \infty)$;
(d) $\mathfrak{D}_H = \{y \mid |y| \le 5\}$, $\mathfrak{R}_H = [-25, 0]$.

13. $V(l) = (12 - \pi)\, l^3/12$.

3. (a) -1;   (b) -1000;
(c) 0.01;   (d) $1/(y^2 - 1)$;
(e) $-1/(x + 1)$;   (f) $x^2/(1 - x^2)$.

7. $f(a) = 2a^2 - 1$; $f(a + h) = 2a^2 + 4ah + 2h^2 - 1$; $f(a + h) - f(a) = 4ah + 2h^2$; $[f(a + h) - f(a)]/h = 4a + 2h$.

11. $A(p) = (\sqrt{3}/36)\, p^2$.

15. $V(x) = 4x(5 - x)(8 - x)$, $0 < x < 5$;   maximum $= V(2) = 144$ cubic inches.

2.9 Page 68

1. Constant, polynomial, rational, algebraic. Domain: R; range: $\{-4\}$.

5. Polynomial, rational, algebraic. Domain: R; range: $[-4/3, \infty)$.

9. Irrational, algebraic. Domain: $[1, \infty)$; range: $[0, \infty)$.

21. (a) $N(x) = [\![x/2]\!]$;
(b) $N(x) = [\![(x + 1)/2]\!]$.

3. Polynomial, rational, algebraic. Domain: R; range: R.

7. Rational, algebraic. Domain: $(-\infty, 1) \cup (1, \infty)$; range: $(-\infty, 2) \cup (2, \infty)$.

11. Irrational, algebraic. Domain: $(-3, \infty)$: range: $(0, \infty)$.

23. (a) Even;   (b) Odd;   (c) Odd;
(d) Neither;   (e) Odd;   (f) Even;
(g) Even;   (h) Even;   (i) Neither.

2.10 Page 71

1. (a) $\sqrt{x^2 - 1} + 2/x$;   (b) $\sqrt{x^2 - 1} - 2x$;
(c) $2\sqrt{x^2 - 1}/x$;   (d) $x\sqrt{x^2 - 1}/2$.
Domain of each is $[1, \infty)$.

5. $(f \circ g)(x) = 1/(2\sqrt{x - 1} - 5)$.   Domain is $[1, 29/4) \cup (29/4, \infty)$.

9. $\sqrt{x - 4}$. Domain is $[4, \infty)$.

13. Domain is R.

17. Even.

3. $(g \circ f)(x) = 2/\sqrt{x^2 - 1}$. Domain is $(-\infty, -1) \cup (1, \infty)$.

7. (a) $\sqrt{x - 4} + |x|$;   (b) $|x|\sqrt{x - 4}$;
(c) $\sqrt{x - 4}/|x|$. Domain of each is $[4, \infty)$.

11. $|[\![x]\!]|$. Domain is R.

15. (a) Neither;   (b) Even.

19. (a) Even;   (b) Odd;   (c) Even;
(d) Even;   (e) Odd.

2.11   Page 75

1. $f^{-1}(x) = (x + 1)/3$. Domain = range = R.

5. $f^{-1}(x) = 1/x + 5$. Domain: $(-\infty, 0) \cup (0, \infty)$;
   range: $(-\infty, 5) \cup (5, \infty)$.
9. $f^{-1}(x) = x^{1/3} + 4$. Domain = range = R.

3. $f^{-1}(x) = (1/2)(x^2 - 5)$. Domain: $[0, \infty)$; range:
   $[-5/2, \infty)$.
7. $f^{-1}(x) = x^{2/3}$. Domain: $[0, 8]$; range: $[0, 4]$.

2.12   Page 76

1. (c) and (d) define y as a function of x; (b) and (d)
   define x as a function of y; (d) defines a one-to-
   one function.
5. $x = 2/3$.
9. $F^{-1}(x) = -\sqrt{2x}$.
13. (a) Yes, domain = $\{x \mid x \neq 0, x \neq 5/2\}$;
    (b) No.
17. (a) $N(x) = [\![\sqrt{x}\,]\!]$;   (b) $N(x) = [\![\sqrt{x}/2]\!]$;
    (c) $N(x) = [\![(\sqrt[3]{x} + 1)/2]\!]$.

3. $9\sqrt{5}/5$.

7. $f^{-1}(x) = x + 3$.
11. $\phi^{-1}(w) = -\sqrt[3]{w}$.
15. Domain = R, range = $[0, \infty)$.

19. The points in the Cartesian plane that are out-
    side the parabola $y = -x^2$.

3.1   Page 82

1. 2.
5. -4.
9. $2.718 \cdots$.
13. (a) $(-3.4, -2.6)$;   (b) $(-3.1, -2.9)$;
    (c) $(-3.03, -2.97)$.

3. 4.
7. 1.
11. $I_2 = (1.7, 2.3)$.

3.2   Page 91

1. $0 < \delta < \epsilon/8$.
5. $0 < \delta < \epsilon/3$.
9. $\delta = 2\epsilon$ will do.

3. $0 < \delta \leq \epsilon$.
7. $\delta = 3\epsilon$ will do.
11. *Outline of proof.* $0 < |x - a| < \delta$
    $\implies |G(x) - 0| < \epsilon \iff -\epsilon < G(x) < \epsilon$
    $\implies -\epsilon < F(x) \leq G(x) < \epsilon \implies -\epsilon < F(x)$
    $< \epsilon \implies |F(x) - 0| < \epsilon$.

3.3   Page 97

11. 7.
15. 2.
19. 20.
23. $-1/9$.

13. 2.
17. 2.
21. 75.

3.4   Page 100

1. 0.
5. 0.
9. 1.
13. 2.

3. 2.
7. -1.
11. 0.

3.5   Page 105

1. Yes.

3. Yes.

5. No. $H(2)$ does not exist.

9. No; $g(3) = 4 \neq \lim\limits_{x \to 3} g(x) = 6$.

13. 1.
19. $(-\infty, \infty)$.

23. $x = 3/2$; $\lim\limits_{x \to 3/2} \Phi(x)$ does not exist.

7. No. Left-hand limit at 2 is 1, and right-hand limit at 2 is 2.

11. -4.

17. $[2, \infty)$.

21. $x = -2$ and $x = 3/2$; $g(-2)$ is not defined, $g(3/2)$ is not defined, and also $\lim\limits_{x \to 3/2} g(x)$ fails to exist.

3.6 Page 109

1. 15.
5. 1.748π square inches.
9. P_1:$(1, 0)$; P_2:$(1.6, 0.96)$; $\Delta y/\Delta x = 1.6$.
13. (a) $\Delta y = 2x_1 \, \Delta x + (\Delta x)^2$;
    (b) $\Delta y/\Delta x = 2x_1 + \Delta x$;
    (c) $\lim\limits_{\Delta x \leftarrow 3} \Delta y/\Delta x = 2x_1$;
    (d) $y - x_1^2 = 2x_1(x - x_1)$;
    (f) $2x - y - 1 = 0$. It is tangent to the curve at P_1:$(1, 1)$.

3. 0.05.
7. 323.21067π cubic inches.
11. P_1:$(2, 1/2)$; P_2:$(6/5, 25/18)$; $\Delta y/\Delta x = -10/9$.

3.7 Page 110

5. 3.
9. 0.
13. $(-\infty, \infty)$.
17. $(-\infty, \infty)$.
21. $\lim\limits_{x \to 3} f(x) = 1 \neq f(3) = 2$.
27. Yes. $\lim\limits_{x \to 0^+} H(x) = \lim\limits_{x \to 0^-} H(x) = 0 = H(0)$.

7. 1/4.
11. 4/3.
15. $(2, 5]$.
19. $(-\infty, -1)$, $(-1, 1)$, and $(1, \infty)$.
23. $(-\infty, -1)$, $(-1, 1)$, and $(1, \infty)$.
29. $[-2, 4]$.

4.1 Page 117

1. $m = 4$; $4x - y - 4 = 0$.
5. $m = -2$; $2x + y + 1 = 0$.
9. $m = -2/3$; $2x + 3y - 7 = 0$.
13. $m = 2$; $2x - y - 1 = 0$.

3. $m = 0$; $y - 1 = 0$.
7. $m = -2$; $2x + y - 4 = 0$.
11. $m = 1/2$; $x - 2y + 2 = 0$.

4.2 Page 121

1. 160 feet per second.
5. 24 feet per second.
9. 6.8 feet per second.
13. 0.25 foot per second.

3. 67.2 feet per second.
7. (a) 4;   (b) 2.25;   (c) 1.875.
11. -2 feet per second.
15. 5/3 seconds.

4.3 Page 124

1. $f'(x) = 5$.
5. $f'(x) = 9x^2$.
9. $f'(x) = 4x^3$.

3. $f'(x) = 16x$.
7. $f'(x) = 3x^2 - 2$.
11. $D_x F(x) = -3/(5x^2)$; $\mathfrak{D}_{F'} = \{x \mid x \in R, x \neq 0\}$.

13. $D_x F(x) = -24x/(2x^2 + 3)^2$; $\mathfrak{D}_{F'} = R$.
17. $dy/dx = -4/x^5$.
21. $F'(t) = t/\sqrt{t^2 + 4}$.
27. (a) At $x = 1$, the value of the right-hand derivative is 2, and the value of the left-hand derivative is -2.
    (b) $2x$.　　　　　　(c) $-2x$.

15. $dy/dx = -12/x^4$.
19. $g'(x) = -1/(2x\sqrt{3x})$.
23. $f'(u) = -u/(u^2 + 9)^{3/2}$.

4.4   Page 128

1. (a) 2.25;
   (b) 2.05. Instantaneous rate of change of y with respect to x at $x = 2$ is 2. Slope of tangent line at $(2, 1)$ is 2.
5. 144π.
9. $-\sqrt{2}/4$.
13. 8 feet per second.
17. At the end of $3\frac{1}{3}$ seconds.

3. $2x$.

7. $\sqrt{3}\,x/2$, where x is the length of a side.
11. $-2k/s^3$.
15. 128.8 feet per second.
19. $6/(5\pi)$ foot per minute.

4.5   Page 132

1. $f'(x) = 5$; R.
5. $g'(x) = 3x^2$; R.
9. $1/2$.
13. All integers.
17. The discontinuity in 9 can be removed by defining $f(1/2) = 7/2$; one of the discontinuities in 10 is removed by defining $\phi(-7/3) = -3/22$; and the discontinuity in 16 can be removed by defining $f(6) = 0$.

3. $G'(t) = 2t$; R.
7. $\Phi'(t) = 3t^2/(2\sqrt{t^3 - 1})$; $(1, \infty)$.
11. $(-\infty, -4) \cup (4, \infty)$.
15. All integers.
23. $\Phi'(0) = 0$.

4.6   Page 133

1. $f'(x) = a$.

5. $2x + y + 1 = 0$.
9. 100π.
13. (a) Parallel or identical;
    (b) Perpendicular.
17. (a) 64 feet per second; upward.
    (b) -32 feet per second; downward.
    (c) 256 feet.　(d) 8 seconds.

3. (a) $2x - 5$;　　　　(b) $-1/(x - 2)^2$;
   (c) $-1/(2\sqrt{9 - x})$.
7. P:(3, 1). Tangent line is $2x - y - 5 = 0$.
11. -2.
15. $-3/16$.

19. $v(c) = 3c^2 - 12c + 9 = 3(c - 1)(c - 3)$. Particle comes to a stop when $c = 1$ and when $c = 3$. It moves to the right during the first second, to the left during the second and third seconds, and to the right forever after 3 seconds.

5.1   Page 139

1. $3x^2$.
5. $8x$.
9. $44x^3 + 1$.
13. $22x^{10} - 24x^7 + 44x^3 + 6$.

3. $-18x^5$.
7. $12x^2 - 7$.
11. $49x^6 + 12x^5 - 12x$.
15. $20x^4 + 24x^3 - 9x^2 - 14x - 5$.

5.2 Page 142

1. $18x^2 - 26x - 30$.

3. $40x^3 + 90x^2 - 98x - 42$.

5. $4x^3 - 30x^2 + 54x - 10$.

7. $-6x/(3x^2 + 2)^2$.

9. $-(16x^3 - 9x^2)/(4x^4 - 3x^3 + 2)^2$.

11. $(6x^2 + 20x + 3)/(3x + 5)^2$.

13. $(2x^4 + 12x^3 - 29x^2 - 8x - 7)/(x^2 + 3x - 5)^2$.

15. $(-2t^4 - 18t^3 - 47t^2 + 44t + 15)/(t^3 + 6t^2 - t + 2)^2$.

17. (a) 23;   (b) $-17/9$.

5.3 Page 146

1. $-135(2 - 9x)^{14}$.

3. $10(5x + 1)(5x^2 + 2x - 8)^4$.

5. $8(x^3 - 4x^2 + 2x - 1)^7(3x^2 - 8x + 2)$.

7. $n(ax^2 + bx + c)^{n-1}(2ax + b)$.

9. $2(4x - 7)(12x + 5)$.

11. $2(x + 1)(7x - 9)(2x - 1)^2(x^2 - 3)$.

13. $-2/(x + 2)^3$.

15. $-6x^2/(x^3 - 4)^3$.

17. $-3(12x^3 + 1)/(3x^4 + x - 8)^4$.

19. $(x + 1)(3x - 11)/(3x - 4)^2$.

21. $4x(3x^2 + 2)(3x^2 - 17)/(2x^2 - 5)^2$.

23. $-44(2x + 3)/(4x - 5)^3$.

25. $4(t^2 + 4)^3(5t^2 - 18t - 20)/(5t - 9)^5$.

27. $[14s^2(2s^5 - 3)^6(8s^5 + 25s^2 + 18)]/(4s^3 + 5)^8$.

29. $132x^2(x^3 + 7)^3/[6 + (x^3 + 7)^4]^2$.

5.4 Page 149

1. $2/(3x^{1/3})$.

3. $1/(2x + 3)^{1/2}$.

5. $-2x/(x^2 + 4)^2$.

7. $3x^2/\sqrt{2x^3 + 7}$.

9. $1/[3\sqrt[3]{(x + 2)^2}]$.

11. $6/(5 - 2t)^4$.

13. $-9/[u^2\sqrt{9 + u^2}]$.

15. $6(4 - t^3)/(8 - t^3)^{2/3}$.

17. $(t - 1)/(2t^{3/2})$.

19. $2z/(2z - 1)^{3/2}$.

21. $-19/[2(7 - x)^{1/2}(12 + x)^{2/3}]$.

23. $10/[3(2p - 3)^{1/3}(3p - 2)^{5/3}]$.

25. $(4 - x)/[3x^{1/3}(2 + x)^{4/3}]$.

27. $-1/[2\sqrt{2u}\sqrt{5 - (2u)^{1/2}}]$.

29. Hint: Differentiate both members of the given identity.

5.5 Page 152

1. 4

3. $6(x + 1)$.

5. $42x^5 - 60x^4 + 30x$.

7. $2/x^3$.

9. $-1/(4x^{3/2})$.

11. $-26/(2x^2 - 13)^{2/3}$.

13. 6.

15. $24x$.

17. $24/(2x + 1)^4$.

19. $f'(x)$ is zero at $(0, 1)$ and at $(2, -3)$. $f''(1) = 0$. $f''(x) < 0$ on $(-\infty, 1)$ and the curve is concave downward there; $f''(x) > 0$ on $(1, \infty)$ and the curve is concave upward there.

21. $D_x y = -2/x^3$, $D_x^2 y = (-2)(-3)/x^4$, $D_x^3 y = (-2)(-3)(-4)/x^5$, $D_x^n y = (-1)^n(n + 1)!/x^{n+2}$.

25. (a) $y_1 L_1(x) + y_2 L_2(x) + y_3 L_3(x) + y_4 L_4(x)$, where $L_i(x) = A_i(x)/A'(x_i)$, $A(x) = (x - x_1)(x - x_2)(x - x_3)(x - x_4)$, and $A_i(x) = A(x)/(x - x_i)$ for $i = 1, 2, 3, 4$;

(b) $\dfrac{5(x - 2)(x - 4)(x - 7)}{-120} + \dfrac{-3(x + 1)(x - 4)(x - 7)}{30}$

$+ \dfrac{2(x + 1)(x - 2)(x - 7)}{-30} + \dfrac{(x + 1)(x - 2)(x - 4)}{120}$.

5.6 Page 156

1. x/y, $y \neq 0$.
3. $-y/x$.
5. $(1 - y^2)/(2xy)$.
7. $-(12x^2 + 11y^2)/(22xy - 6y^2)$.
9. $[(y^3 + 6)\sqrt{2xy} - y]/[y(2 - 3xy)\sqrt{2xy} + x]$.
11. $-(y/x)^{1/3}$.
13. $(24xy^2 - 9x^4)/(16y^3) = 3x/(8y)$.
15. $-2y^2/(2x^2y - 1)^3$.
17. $-x/y^5$.
19. $a^{1/2}/(2x^{3/2})$.

5.7 Page 158

1. $3x - y - 5 = 0$; $x + 3y - 15 = 0$.
3. $2x + y - 4 = 0$; $x - 2y + 8 = 0$.
5. $3x + 2y - 8 = 0$; $4x - 6y + 11 = 0$.
7. $\sqrt{2}x - y - 6 = 0$.
9. $(0, -2)$.
11. $8x - 16y - 3 = 0$.
13. $4x + y - 7 = 0$, $4x + y + 9 = 0$.
15. 2.

5.8 Page 163

1. 900 cubic inches per second.
3. $96\sqrt{5} \doteq 214.67$ miles per hour.
5. $4900/\sqrt{61} \doteq 627$ miles per hour.
7. $\sqrt{6}/6 \doteq 0.408$ foot per second.
9. $1/(4\pi) \doteq 0.08$ foot per second.
11. 1 inch per minute.
13. $0.324\pi \doteq 1.018$ square inches per second.
15. $12{,}100/\sqrt{122} \doteq 109.5$ feet per second.
17. $1/2$ inch per second.

5.9 Page 167

1. (a) $dy = 0.75$; (b) $dy = 2.25$.
3. $dy = (4x - 3)\,dx$.
5. $dy = -24x^2(3 + 2x^3)^{-5}\,dx$.
7. $dy = [(10x^4 + 4x^3)/\sqrt{4x^5 + 2x^4 - 5}]\,dx$.
9. $ds = (4t\,dt)/[5(t^2 - 3)^{3/5}]$.
11. $d\phi = -2(2x^2 - 3x + 14)(x - 3)^{-3}(x^2 + 14)^{-2}\,dx$.
13. $dy/dx = -(2x + 5)/(5x - 4y)$.
15. $dy/dx = -x^3/y^3$.

5.10 Page 170

1. (a) $\Delta y = 2.25$, $dy = 2$;
   (b) $\Delta y = 0.84$, $dy = 0.8$;
   (c) $\Delta y = 0.41$, $dy = 0.4$;
   (d) $\Delta y = 0.0401$, $dy = 0.04$.
3. $dy = 0.33$; $|\Delta y - dy| < 0.01211$.
5. 284.4π cubic feet.
7. 20.05.
9. 3.037.
11. 14.33.
13. 12.5663 feet.

5.11 Page 172

1. $dx/dy = 1/3$. $f^{-1}(y) = (y + 1)/3$, $y \in R$.
3. $dx/dy = 1/(18x)$. $f^{-1}(y) = \sqrt{y + 4}/3$, $y > -4$.
5. $dx/dy = 1/(3x^2)$. $f^{-1}(y) = (y + 2)^{1/3}$, $y > -2$.
7. $dx/dy = 2/(3\sqrt{x})$. $f^{-1}(y) = y^{2/3}$, $y > 0$.
9. $dx/dy = 1/[3(x - 4)^2]$. $f^{-1}(y) = y^{1/3} + 4$, $y \neq 0$.
11. $-1(2\sqrt{4 - y})$, $y < 4$.
13. $2/(y - 1)^2$, $y \neq 1$.
15. 3.
17. $11/2$.
19. $1/4$ and $1/7$.
21. $1/8$ and $1/64$.

5.12 Page 173

1. (a) $(7x + 13)(x - 2)^2(x + 7)^3$;
   (b) $(x - 3)^4(1 - 2x^2)(5 + 24x - 18x^2)$.
3. (a) $3(3t - 1)(9 - t)/(t + 4)^4$.
   (b) $10(1 + t^2)^4(2 + t - 3t^2)/(1 - 5t)^5$.

5. (a) $2x/\sqrt{2x^2 - 3}$;
   (b) $-1/[(1 - x)^{1/2}(1 + x)^{3/2}]$.
9. (a) 0;   (b) $(20!)/x^{21}$.
13. $2x - y - 1 = 0$ and $6x - 3y + 1 = 0$.
19. $dw = 8(4u^3 - 15u^2 - 4u + 16)\, du$.
23. $\Phi'(t) = 2t,\ t \in (-\infty, -2) \cup (2, \infty)$.
    $\Phi'(t) = -2t,\ t \in (-2, 2)$.
    Domain of Φ is R; domain of Φ' is $(-\infty, -2)$
    $\cup (-2, 2) \cup (2, \infty)$.

7. (a) $3/(16\pi)$ inch per second;
   (b) 3 square inches per second.
11. $-(y/x)^{1/3}$.
15. 1/6 foot per minute.
21. $dx/dy = -y^2/x^2,\ x \neq 0$.

6.1   Page 181

1. Maximum is $f(2) = 3$; minimum is $f(0) = -1$.
5. The only root of $F'(x) = 0$ is 2. Since F and F'
   exist on $(0, 3)$, the only possible point at which F
   can have an extreme value on $(0, 3)$ is $x = 2$ (by
   6.1.3). But $F(x) = (x - 2)^3$, and it is clear that
   $x < 2 \implies F(x) < 0$ and $x > 2 \implies F(x) > 0$.
   By the definition of maximum and minimum,
   therefore, $F(2)$ is not an extreme value of F on
   $(0, 3)$.
9. $z = 1/2$.
13. No; f does not satisfy (iii).

17. One. It is in $(-3, -2)$.

3. Maximum is $G(3) = 9$; minimum is $G(1) = -7/5$.
7. F is defined for all $x \in R$. Moreover, $F'(x) =$
   $1/(2\,|x|^{1/2}) \neq 0$ when $x \neq 0$, and $F'(0)$ fails to
   exist. So, by 6.1.3, the only point in *open* $(-2, 3)$
   at which F can have an extreme value is 0. Since
   $F(x) > 0$ for all $x \neq 0$, $F(0) = 0$ is the minimum
   of F on $(-2, 3)$.

11. No; f does not satisfy (i).
15. Two. One root in $(0, 1)$ and the other in
    $(-3, -2)$.

6.2   Page 184

1. $z = 0$.
5. No; because F is discontinuous at 3.
9. No; because ϕ is discontinuous at 0.

3. $z = \pm 2/\sqrt{3}$.
7. $z = 16/27$.

6.3   Page 190

1. Decreasing on $(-\infty, 2]$; increasing on $[2, \infty)$.
5. Decreasing on $(-\infty, -1]$; increasing on $[-1, \infty)$.

11. Increasing on $(-\infty, 0)$ and on $(0, \infty)$.

3. Increasing everywhere.
9. Increasing on $(-\infty, 0)$ and on $[4, \infty)$; decreasing
   on $(0, 4]$.

6.4   Page 191

1. (a) $s = 0$;   (b) $0 \leq t < 3$;   (c) $t = 3$;
   (d) $t > 3$;   (e) $t = 6$.
13. $t = 1$ and $t = 5/2$.

17. $4\frac{1}{6}$ seconds.

11. 10.4 feet per second when $t = 2$; 5 feet per
    second when $t = 5$.
15. (a) 36 feet;   (b) 16 feet per second upward;
    (c) 3 seconds.
19. $S = A(t - k)^3 + B$, where A and k are positive
    constants and B is any constant. The velocity is
    positive for all values of t except $t = k$, when it
    is zero.

6.5   Page 197

1. (a) 4;
   (b) $f(4) = -9$ is a local minimum.
5. (a) -1 and 1; (b) $f(-1) = 1$ is a local maximum,
   and $f(1) = -3$ is a local minimum.
9. (a) 0; (b) $\phi(0) = 1$ is a local minimum.
13. (a) -2; (b) $H(-2) = -4$ is a local minimum.

3. (a) 2;
   (b) $F(2) = 1$ is a local maximum.
7. (a) -2 and 2; (b) $G(-2) = 3$ is a local maximum, and $G(2) = -5$ is a local minimum.
11. (a) 1; (b) $g(1) = 2$ is a local maximum.
15. (a) 0; (b) $\Phi(0) = 0$ is a local minimum.

6.7   Page 200

1. The maximum is $f(0) = 4$, and the minimum is
   $f(3) = -5$.
5. The maximum is $g(0) = g(3) = 1$, and the minimum is $g(-1) = g(2) = -3$.
9. The maximum is $f(4) = 4$, and the minimum is
   $f(2) = 0$.

3. The maximum is $F(4) = 5$, and the minimum is
   $F(2) = -4$.
7. The maximum is $H(1) = 16$, and the minimum
   is $H(-1) = 0$.
11. *Outline of proof.* The only critical number for f
    is $x_1 = -B/(2A)$, $A \neq 0$. Since $f''(x) = 2A$ for
    all x, $f(x_1)$ will be the minimum if and only if
    $A > 0$. Now $f(x_1) = A[-B/(2A)]^2 + B[-B/(2A)]$
    $+ C = -(B^2 - 4AC)/(4A)$. Thus $f(x_1) \geq 0$ if
    and only if $B^2 - 4AC < 0$.

6.8   Page 205

1. 5 and 5.
5. 1024 cubic inches.
9. The piece for the square should be $64/(\pi + 4)$
   inches long.
13. 20 feet $\times$ 20 feet $\times$ 30 feet.
17. $32\pi r^3/81$ cubic units.

21. At a point $6\sqrt{7}/7 \doteq 2.2678$ miles from P.

25. Underwater from powerhouse to point
    $bw/\sqrt{a^2 - b^2}$ feet downstream from A, and the
    rest of the way on land.
29. Length of base = altitude = $\sqrt[3]{4}\,a/2$.
33. $2/(\pi + 1)$.

3. $2\sqrt{3}$ and $-2\sqrt{3}$.
7. $(4, \pm\sqrt{3})$.
11. A square.

15. $320\pi/3$ cubic inches.
19. Diameter of cylinder = length of cylinder =
    $\sqrt{2}$ times radius of sphere.
23. $s\sqrt[3]{I_1}/(\sqrt[3]{I_1} + \sqrt[3]{I_2})$ feet from the source
    whose intensity is I_1.
27. Depth = $1/2$ width.

31. Height = $27/20$ of length of an edge of base.

6.9   Page 209

1. Average cost is \$3.36, and the marginal cost is
   \$2.53.
5. (a) $R(x) = 20x + 4x^2 - x^3/3$; $R'(x) = 20 + 8x$
       $- x^2$.
   (b) $[0, 10]$.   (c) 4.
9. $p(x) = 6.00 - (0.15)\left(\dfrac{x - 4000}{250}\right)$, $x \geq 4000$;
   \$4.20 per yard.

3. 200.

7. The maximum revenue, \$639.29, is obtained
   when 25 units are sold. The marginal revenue
   at this level is $R'(25) = \$0.061$.
11. (a) For $0 \leq x \leq 4500$, $C_1(x) = 6000 + 1.40x$;
        and for $x < 4500$, $C_2(x) = 5100 + 1.60x$.
    (b) $p(x) = (11000 - x)/1000$, $x \geq 4000$;
    (c) 4700.

6.10 Page 215

1. Concave upward everywhere.

3. Concave upward for $x > 0$. Concave downward for $x < 0$. Point of inflection is $(0, 0)$.

5. Concave downward for $x < 1/2$. Concave upward for $x > 1/2$. Point of inflection is $(1/2, 3/2)$.

7. Concave upward for $x < -1$. Concave downward for $-1 < x < 0$. Concave downward for $0 < x < 1$. Concave upward for $1 < x$. Points of inflection are $(-1, 2)$ and $(1, 2)$.

9. Concave upward everywhere. No point of inflection.

11. Concave downward for $-3 < x < 3$. No point of inflection.

13. Concave downward on $(-\infty, 0)$. Concave upward on $(0, \infty)$. No point of inflection because f is not continuous at $x = 0$.

15. Concave downward on $(-\infty, 0)$. Concave upward on $(0, \infty)$. No point of inflection because G is discontinuous at 0.

6.11 Page 220

1. -2.

3. $2/5$.

5. $-\infty$.

7. 0.

11. $-\infty$.

15. $x = -1$ and $y = 0$.

17. $x = 3$ and $y = 2$.

19. $y = 2$ and $y = -2$.

21. $x = 0$ and $y = x$.

23. $x = 2$ and $y = x/2 + 1$.

6.12 Page 224

1. (b) Symmetric with respect to the origin; x intercepts are $-2, 0, 2$, and y intercept is 0.
   (c) Increasing on $(-\infty, -2\sqrt{3}/3)$, decreasing on $(-2\sqrt{3}/3, 2\sqrt{3}/3)$, and increasing on $(2\sqrt{3}/3, \infty)$.
   (d) Local maximum is $f(-2\sqrt{3}/3) = 16\sqrt{3}/9 \doteq 3$; local minimum is $f(2\sqrt{3}/3) = -16\sqrt{3}/9 \doteq -3$.
   (e) Concave downward on $(-\infty, 0)$; concave upward on $(0, \infty)$.
   (f) Point of inflection is $(0, 0)$.

3. (b) Symmetric with respect to the y axis; y intercept is 1.
   (c) Decreasing on $(-\infty, -3)$, increasing on $(-3, 0)$, decreasing on $(0, 3)$, and increasing on $(3, \infty)$.
   (d) $F(-3) = -3.05$ and $F(3) = -3.05$ are local minima; $F(0) = 1$ is a local maximum.
   (e) Concave upward on $(-\infty, -\sqrt{3})$, concave downward on $(-\sqrt{3}, \sqrt{3})$, and concave upward on $(\sqrt{3}, \infty)$.
   (f) Points of inflection are $(-\sqrt{3}, -5/4)$ and $(\sqrt{3}, -5/4)$.

5. (b) x intercept is -2.
   (c) Increasing on $(-\infty, -2)$, decreasing on $(-2, 0) \cup (0, 2)$, and increasing on $(2, \infty)$.
   (d) Local maximum is $g(-2) = 0$; local minimum is $g(2) = 8$.
   (e) Concave downward on $(-\infty, 0)$; concave upward on $(0, \infty)$.
   (g) The y axis is a vertical asymptote, and the line $y = x + 4$ is an oblique asymptote.

7. (a) Excluded region: $\{P{:}(x, y) \mid x < -3\}$.
   (b) x intercepts are -3 and 0; y intercept is 0.
   (c) Decreasing on $[-3, -2)$; increasing on $(-2, \infty)$.
   (d) $\phi(-2) = -4$ is a local minimum.
   (e) Concave upward on $(-3, \infty)$.

13. (a) Domain of H' is R.
$$H'(x) = \begin{cases} 3x^2 & \text{for } 0 \le x, \\ -3x^2 & \text{for } x < 0. \end{cases}$$
   (b) Domain of H'' is R.
$$H''(x) = \begin{cases} 6x & \text{for } 0 \le x, \\ -6x & \text{for } x < 0. \end{cases}$$

6.13 Page 229

1. 2.646.
5. -0.268.
9. (4.0608, 0.24626).

3. 0.241.
7. 1.213.
13. 1.2879.

6.14 Page 229

3. 11 feet per second when $t = 1$; -16 feet per second when $t = 4$.
7. (a) $s = -4$, $v = 9$.
   (b) At $t = 1$, $s = 0$; and at $t = 3$, $s = -4$.
   (c) $0 \le t < 1$, 4 feet; and for $t > 3$, without bound.
   (d) $1 < t < 3$, 4 feet.
   (e) $t = 3$.
   (f) $a < 0$ for $0 < t < 2$; $a > 0$ for $t > 2$.
13. 7.5 inches wide, 10 inches high.

5. 4.

9. (a) Yes; $z = \pm \sqrt{3}$.
   (b) No; $F'(0)$ does not exist.
   (c) No; G is not continuous at $x = \pm 1$.

15. Maximum is $f(0) = 2$; minimum is $f(-2) = f(2) = 0$. Concave upward for $-2 < x < 0$; concave downward for $0 < x < 2$.

7.2 Page 236

1. $2x^3 - 3x^2 + x + C$.
5. $-1/x^4 + C$.
9. $f(x) = -2x^{-2} - 2x^4 - 5x + C$.
13. $f(x) = 2x^3 - 3x^2 + C_1 x + C_2$.

3. $2x^9 - 5x^5 + x^3 + C$.
7. $f(x) = 5x^4 - 2x^3 + 17x + C$.
11. $f(x) = 2\sqrt{5x^5}/5 + C$.
15. $f(x) = 12x^{5/2} + x^3 + C_1 x + C_2$.

7.3 Page 239

1. $(3x + 1)^5/5 + C$.
5. $3(2x^5 + 9)^4/40 + C$.
9. $9(2x^2 - 11)^{4/3}/16 + C$.
13. $(3x^2/2) - 1/(x + 1) + C$.
17. $-\sqrt{a^2 - x^2}/a^2 + C$.

3. $(5x^3 - 18)^8/8 + C$.
7. $(5x^3 + 3x - 8)^7/21 + C$.
11. $-1/(x + 1) + C$.
15. $\sqrt{15x^2 + 35}/3 + C$.
19. (a) $f''(x) = -2/(x + 1)^3$; (b) $A_x(A_x f''(x)) = -1/(x + 1) + C_1 x + C_2$, where C_1 and C_2, the constants of antidifferentiation, are at our disposal; (c) Not necessarily. Our $f(x)$ is identically equal to $A_x(A_x f''(x))$ for all values of x except $x = -1$, if and only if we choose $C_1 = 0$ and $C_2 = 1$.

21. $x^2\sqrt{x - 1} + C$.

23. (a) $\frac{1}{2}x|x| + C$; (b) $\frac{1}{3}x^2|x| + C$; (c) $\frac{1}{4}x^3|x| + C$.

7.4 Page 241

1. $3x^2 + 12x - 2y = 0$.
5. $v = 39$ feet per second; $x = 112$ feet.
9. 289 feet.
13. 42 feet.

3. $3y = x^3 + 3x^2 - 9x - 12$.
7. $v = 106$ feet per second; $x = 124$ feet.
11. 352 feet.
15. $x(t) = \sqrt{3t + 1} + 2t - 1$.

8.1 Page 251

1. $a(3) = 6$, $A(3) = 9$.

3. $a(3) = 9$, $A(3) = 15$.

5. $a(10) = 16\frac{1}{4}$, $A(10) = 18\frac{3}{4}$.

9. $a(4) = 16\frac{2}{3}$, $A(4) = 22$.

7. $a(3) = 11$, $A(3) = 20$.

8.2   Page 253

1. 155.

5. 255.

9. $n(n + 1)(2n - 1)/2$.

3. 735/988.

7. $n(3n - 1)/2$.

11. $5n(n + 1)(3n^2 + 19n + 8)/12$.

8.3   Page 261

1. 15.6875.

5. $1729/576 \doteq 3.000$.

9. 50/3.

3. 4.75.

7. 12.

11. 36.

8.4   Page 267

1. 4.

5. 3/10.

9. $-15/2$.

13. $G(x) = x^4 + 2x^3 - 7x - 410$.

3. -560.

7. 52/3.

11. $G(x) = x^2 - x - 42$.

15. $G(x) = -1/(x - 3) - 1/13$.

8.5   Page 271

1. 122/9.

5. $5(\sqrt{2} - \sqrt{3})$.

9. $-3/2$.

15. 0.

3. $-16\frac{2}{3}$.

7. 0.

11. $31\frac{1}{2}$.

17. 128.

8.7   Page 277

1. 6.507.

5. 0.43; $n = 3$.

9. 3.29; $n = 12$.

3. 0.694.

7. 2.13; $n = 5$.

8.8   Page 278

1. 2.

5. 1/10.

9. $3^n - 1$.

13. $G(x) = (1/8)[(2x^2 - 1)^6 - 1]$.

17. 5.

3. 496/5.

7. 196.

11. 5/2.

15. 196/15.

19. No; f does not exist for $x = 0$.

9.1   Page 285

1. 9.

5. $4\frac{1}{4}$.

9. $5\frac{1}{3}$.

13. $2\frac{2}{3}$.

17. $4\frac{1}{2}$.

3. $7\frac{1}{3}$.

7. 10.

11. $4\frac{1}{2}$.

15. 2/3.

19. 8.

9.2   Page 291

1. $64\pi/5$.

5. 9π.

3. $3\pi/4$.

7. $32\pi/5$.

9. 8π.

13. $(4/3)\pi\ ab^2$.

17. $2\pi/3$.

11. 64π.

15. $64\pi/3$.

19. (a) $1024\pi/35$;   (b) $704\pi/5$.

9.3   Page 296

1. 24π.

5. (a) $128\pi/15$;   (b) $238\pi/15$.

9. $52\pi/3$.

3. (a) $128\pi/5$;   (b) $256\pi/15$.

7. (a) 8π;   (b) $8\pi/3$.

11. (a) $64\pi/5$;   (b) $208\pi/15$.

9.4   Page 298

1. $k = 16$; $W = 2$ foot-pounds.

9. $1600(2^{1.2} - 1) \doteq 2076$ inch-pounds.

13. $45{,}684\pi\delta \doteq 8{,}951{,}140$ foot-pounds.

5. 22 foot-pounds.

11. $1575\pi\delta/16 \doteq 19{,}287$ foot-pounds.

9.5   Page 301

1. $10{,}000\delta \doteq 624{,}000$ pounds.

5. $\delta(bh^2/6)$.

9. $640\delta \doteq 39{,}936$ pounds.

13. $603\delta \doteq 37{,}627$ pounds.

3. $6000\delta \doteq 374{,}400$ pounds.

7. $4000\delta/3 \doteq 83{,}200$ pounds.

11. $\delta(10\sqrt{3} + 2) \doteq 1204$ pounds.

9.6   Page 306

1. $3\sqrt{10}$.

5. 9.

9. $3\int_{-1}^{2} (9 - x^2)^{-1/2}\ dx$.

3. $37\frac{1}{3}$.

7. $4\frac{2}{3}$.

9.7   Page 308

1. πrs, where r is the radius of the base and s is the slant height.

5. 57π.

3. 50π.

7. $8429\pi/81$.

9.8   Page 319

1. $M_x = -3$; $M_y = 11$; $\bar{x} = 11/17$, $\bar{y} = -3/17$.

5. $M_x = 31$; $M_y = 9$; $\bar{x} = 9/16$, $\bar{y} = 31/16$.

9. $M_x = 128/5$, $M_y = 32$; $\bar{x} = 3$, $\bar{y} = 12/5$.

13. $M_x = -965/24$, $M_y = 35/8$; $\bar{x} = 21/92$,
    $\bar{y} = -193/92$.

17. $-279/20$.

3. $M_y = -3/2$; $M_x = 1/2$; $\bar{x} = -3/14$, $\bar{y} = 1/14$.

7. $\bar{x} = 0$, $\bar{y} = 8/5$.

11. $M_x = 64/7$, $M_y = 32/5$; $\bar{x} = 8/5$, $\bar{y} = 16/7$.

15. $M_x = 36/5$, $M_y = 9/4$; $\bar{x} = 1/2$, $\bar{y} = 8/5$.

19. $\bar{x} = -18/5$, $\bar{y} = 1$.

9.9   Page 325

1. $\bar{x} = 3h/4$.

5. $\bar{x} = -5/34$.

9. $\bar{y} = 2$.

3. $\bar{x} = 3/2$.

7. $\bar{x} = 12/5$.

11. $1575\pi\delta/16 \doteq 19{,}287$ foot-pounds.

9.10   Page 328

1. $x = 15$. Optimum output $= 1500$; total profit $=$
   \$1,125,000.

3. $x = 4$; total profit $= 16$.

5. $t = 8$; total net earnings $= \$512,000$.

7. $x = 9$; total net revenue is $\$108,000$.

9.11   Page 329

1. 32/3.

5. $\bar{x} = 2$, $\bar{y} = 8/5$.

9. $1200\pi\delta \doteq 235,123$ foot-pounds.

13. $162\pi/5$.

17. 53/6.

21. $\bar{x} = 5/4$, $\bar{y} = 0$.

3. $128\pi/3$.

7. $128\pi/3$.

11. 9/2.

15. $320\delta/3 \doteq 6,656$ pounds.

19. $485\delta \doteq 30,264$ pounds.

10.1   Page 339

1. Domain $= (-\infty, 5) \cup (5, \infty)$;
   $f'(x) = 4/(x - 5)$.

5. Domain $= (0, 2)$; $f'(x) = (2x - 6)/(2x - x^2)$.

9. $(8 - 3x)/(5x^2 - 10x)$.

13. $3 \ln |x - 3| + C = \ln |x - 3|^3 + C$.

17. $\ln \sqrt{3x + 11} + C$.

21. $\ln 10^{1/3}$.

25. $6 \ln 3$.

29. $-(x^3 + 33x^2 + 8)/[2(x^3 - 4)^{3/2}]$.

33. $x^{(x^2+1)}(1 + \ln x^2)$.

37. Since ln is increasing throughout its domain, $x > 4^K \implies \ln x > \ln 4^K$. Thus $\ln x > K$, by Exercise 36.

3. Domain $= (-\infty, 2) \cup (3, \infty)$;
   $f'(x) = (2x - 5)/(x^2 - 5x + 6)$.

7. $1 + \ln x$.

11. $(16 - 7x^3)/(8x - 2x^4)$.

15. $\ln (a^2 - x^2)^{-1/2} + C$.

19. $x \ln x - x + C$.

23. $2 - 2 \ln 2$.

27. $s = 4 \ln 5 - 6/5$.

31. $\dfrac{-10x^2 - 219x + 118}{6(x + 13)^{1/2}(x - 4)^2(2x + 1)^{4/3}}$.

35. $\displaystyle\int_1^2 (1/2)\, dt \le \int_1^2 (1/t)\, dt \le \int_1^2 1\, dt$.

10.3   Page 347

1. $2e^{2x+1}$.

5. $(e^x - e^{-x})/2$.

9. $(x^2 + 2x)e^x$.

13. $3x^2 e^{x^3}(x^3 + 1)$.

17. $\frac{1}{3}e^{3x+1} + C$.

21. $\ln (e^x - 1) + C$.

25. $F(t) = Ce^{kt}$.

29. 988.6 years.

3. $\dfrac{e^{\sqrt{x+1}}}{2\sqrt{x + 1}}$.

7. $e^{\ln x}/x$.

11. $[(\sqrt{x - 1} + 1)e^{\sqrt{x-1}}]/(2\sqrt{x - 1})$.

15. $e^x(x \ln x - 1)/[x(\ln x)^2]$.

19. $e^{\ln 2x} + C$.

23. $e^{(e^x)} + C$.

27. $k = \ln (1.04) \doteq 0.0392$.

31. *Outline.* Show that $f(0) = 0$ and that f is increasing on $(0, \infty)$, so that $e^x - x - 1 > 0$ for all $x > 0$. That is, $e^x > 1 + x$ for all $x > 0$.

10.4   Page 352

1. $3x^2(5^{x^3}) \ln 5$.

5. $y = 2^x(x \ln 2 + 1)$.

9. $(162x^2 \log_2 e)/(x^6 - 729)$.

15. $2e^{\sqrt{x-1}} + C$.

19. $5^{(\ln x)/x}/\ln 5 + C$.

23. $40/(81 \ln 3)$.

3. $7^{\sqrt{x-1}} \ln 7/(2\sqrt{x - 1})$.

7. $y = \log_{10} e/(2x\sqrt{\log_{10} x})$.

13. $2^{x^2}/\ln 4 + C$.

17. $8^{(x^2+2x)}/\ln 64 + C$.

21. $495/\ln 10$.

10.5   Page 360

1. (a) $7\pi/6$,   (b) $-\pi/3$,   (c) $7\pi/4$,
   (d) $-7\pi/2$,   (e) $5\pi/3$,   (f) -3π.

5. (a) $\tan 3\pi/2$ is undefined. (b) $\tan 3\pi/4 = -1$.
   (c) $\tan(-\pi/3) = -\sqrt{3}$. (d) $\tan(-\pi) = 0$.
   (e) $\tan 13\pi/6 = \sqrt{3}/3$.   (f) $\tan 7\pi/6 = \sqrt{3}/3$.

3. (a) $(0, -1)$, (b) $(-\sqrt{2}/2, \sqrt{2}/2)$,
   (c) $(1/2, -\sqrt{3}/2)$, (d) $(-1, 0)$, (e) $(\sqrt{3}/2, 1/2)$,
   (f) $(-\sqrt{3}/2, -1/2)$.

7. n is any integer; (a) $\pi/2 + n\pi$, (b) $2n\pi$,
   (c) $(2n + 1)\pi$, (d) $\pi/3 + 2n\pi$ and $5\pi/3 + 2n\pi$,
   (e) $2\pi/3 + 2n\pi$ and $4\pi/3 + 2n\pi$.

10.7   Page 370

1. $-4x \cos(x^2 - 2) \sin(x^2 - 2)$.
5. $\csc(1/t) \cot(1/t)/t^2$.
9. $2 \csc 2v$.
13. $-(1/2) \cos x^2 + C$.
17. $\tan x - x + c$.
21. $(-1/2)(\ln \cot x)^2 + C$.
25. $(1/2) \ln 2$.

3. $\csc x - 2 \csc^3 x$.
7. $-e^{\cot z} (\csc^2 z)$.
11. $-4[\csc^2 (\ln x - 2)]/x$.
15. $-\ln|\cos x| + C$.
19. $\sec x + C$.
23. $1/3$.
27. $f(\pi/4) = \sqrt{2}$ is maximum; $f(-3\pi/4) = -\sqrt{2}$ is minimum; $(-\pi/4, 0)$ and $(3\pi/4, 0)$ are points of inflection; concave upward on $(-\pi, -\pi/4) \cup (3\pi/4, \pi)$; concave downward on $(-\pi/4, 3\pi/4)$.

29. 2.
33. 2.
37. 64 feet.

31. 1.
35. 1.

10.8   Page 376

1. $2x/\sqrt{1 - x^4}$.
5. $e^x/(1 + e^{2x})$.
9. $\cos(\text{Tan}^{-1} t)/(1 + t^2) = (1 + t^2)^{-3/2}$
15. $\text{Tan}^{-1}(x + 2) + C$.
19. $(1/2) \text{Sec}^{-1} v^2 + C$.
23. $25\pi \text{Tan}^{-1} 4$.
27. 10 rpm.

3. $-7/\sqrt{2x(1 - 2x)}$.
7. $\text{Cot}^{-1} t - t(1 + t^2)^{-1}$.
13. $(1/2) \text{Tan}^{-1} 2x + C$.
17. $\text{Tan}^{-1}(\cos 2t) + C$.
21. π.
25. $1/13$ radian per second.

10.9   Page 381

13. $\sinh 2x$.
17. $2x \text{ sech } x - x^2 \text{ sech } x \tanh x$.
21. $2e^{2x} \sinh e^{2x}$.
25. $(1/2) \sinh(x^2 + 3) + C$.
29. $(1/2)(\ln \cosh x)^2 + C$.
33. $\ln \cosh 3 \doteq 2.3093$.

15. $2x \sinh(x^2 - 1)$.
19. $2 \sinh 4x \sinh 2x + 4 \cosh 2x \cosh 4x$.
23. $(x \sinh x - \cosh x)/x^2$.
27. $\ln \cosh x + C$.
31. $(e^2 - 1)/(2e)$.
35. $x = 0$; $y = 1$; $y = -1$.

10.10   Page 384

1. $3x^2/\sqrt{x^6 - 1}$.
5. $e^x/(1 - e^{2x})$.
9. $1/(x[1 - (\ln x)^2])$.
17. $-\tanh^{-1}(\cos x) + C$.

3. $\sinh^{-1} 2x + 2x/\sqrt{4x^2 + 1}$.
7. $-6/(x\sqrt{9 - x^4})$.
15. $(1/2) \sinh^{-1} 2x + C$.
19. $(1/15) \sinh^{-1} 5x^3 + C$.

10.11   Page 384

1. $1/x$.
5. $\sec^2 x$.
9. $\sec x$.
13. $15e^{5x}/(e^{5x} + 1)$.
17. $(x^2 - x)^{-1/2}$.
21. $20 \sec 5x(\tan^2 5x + \sec^2 5x)$.
25. $-\cos e^x + C$.
29. $3 \operatorname{Sin}^{-1} x + C$.
33. Decreasing on $[-\pi, -3\pi/4]$; $f'(-3\pi/4) = 0$; increasing on $(-3\pi/4, \pi/4)$; $f'(\pi/4) = 0$; decreasing on $(\pi/4, \pi)$. $f(-3\pi/4) = -\sqrt{2}$ is minimum, $f(\pi/4) = \sqrt{2}$ is maximum. Concave upward on $(-\pi, -\pi/4)$ and $(3\pi/4, \pi)$, concave downward on $(-\pi/4, 3\pi/4)$. Points of inflection are $(-\pi/4, 0)$ and $(3\pi/4, 0)$.

3. $2(x - 2)e^{x^2 - 4x}$.
7. $(\operatorname{sech}^2 \sqrt{x})/\sqrt{x}$.
11. $1/\sqrt{e^{2x} - 1}$.
15. $-e^{\sqrt{x}} \sin e^{\sqrt{x}}/(2\sqrt{x})$.
19. $-(\csc \sqrt{x} \cot \sqrt{x})/\sqrt{x}$.
23. $(1/3)e^{3x-1} + C$.
27. $e^{-1} \ln (e^{x+3} + 1) + C$.
31. $\operatorname{Cot}^{-1} (\ln x) + C = \operatorname{Tan}^{-1}(\ln x) + C$.
35. $\ln 5.25/\ln 2$.

11.2   Page 392

1. $(1/5)(x - 1)^5 + C$.
5. $(2 + t^2)^{3/2} + C$.
9. $\sec z + C$.
13. $e^{\sin x} + C$.
17. $(-1/2) \csc x^2 + C$.

3. $\ln |x + 1| + C$.
7. $\tan (s + 3) + C$.
11. $2 \sin \sqrt{x} + C$.
15. $(1/2) \operatorname{Sin}^{-1} x^2 + C$.
19. $(-1/5) \ln |\cos (5x - 1)| + C$.

11.3   Page 396

1. $(7x - 1)^{13}/91 + C$.
5. $(3/2) \sin^2 (x/3) + C$.
9. $(1/2)e^{x^2+2x-1} + C$.
13. $(1/20) (\ln t^2)^{10} + C$.
17. $(1/12) \ln (6x^2 - 19) + C$.
21. $(4/3)x^3 + (1/2)x^2 - (3/2) \ln (x^2 + 5) + C$.
25. $(5^{x^2-1})/\ln 25 + C$.

3. $(8x^6 - 2x^2 + 19)^{11}/44 + C$.
7. $(-1/4) \cos 2x^2 + C$.
11. $e^{\sqrt{2x+1}} + C$.
15. $1/(4 \cos^4 x) + C$.
19. $x^2 - x + \ln (x + 1) + C$.
23. $(2/9)(\sin^3 y + 4)^{3/2} + C$.
27. $e^{(\sin 2\theta)/2} + C$.

11.4   Page 400

1. $(-1/13) \cos (13x - 11) + C$.
5. $(-1/\pi) \cot \pi x + C$.
9. $(-1/2)(\cot e^{2y} + e^{2y}) + C$.
13. $x - \ln |\sin x| + C$.
17. $(1/10) \tan (5x^2 - 1) + C$.
21. $(-1/5) \ln |\cos (5u - 2)| + C$.
25. $(1/3) \tan 3y - y + C$.
29. $(1/4) \sec 4x + C$.
33. $(1/4) \sec (4t - 1) + C$.
37. $\tan x + e^{\sin x} + C$.
41. $(5/2) \operatorname{Sin}^{-1} (2x/3) + C$.
45. $2 \operatorname{Tan}^{-1} \sqrt{\theta} + C$.

3. $(2/\pi) \sin (\pi x/2) + C$.
7. $(-e^2/3) \cos e^{3t-2} + C$.
11. $(1/6) \sec (3w^2 + 1) + C$.
15. $(1/3) \sec 3\theta + C$.
19. $-\ln \cos t + C$.
23. $-\cot (\ln x) + C$.
27. $5 \operatorname{Sin}^{-1} e^x + C$.
31. $(-1/2) \csc \ln z^2 + C$.
35. $(1/3) \ln |\csc \sqrt{3z^2 - 5} - \cot \sqrt{3z^2 - 5}| + C$.
39. $-1/[3 \sin (t^2 - 2)] + C$.
43. $(1/2) \operatorname{Sec}^{-1} 2t + C$.
47. $\operatorname{Sin}^{-1} [(2x - 3)/3] + C$.

49. $(1/6) \, \mathrm{Sin}^{-1} \, (3y^2/4) + C.$

51. $\pi.$

53. $\ln (\sqrt{2} + 1).$

11.5   Page 404

1. $xe^x - e^x + C.$

3. $(-1/3)x \cos 3x + (1/9) \sin 3x + C.$

5. $x \, \mathrm{Tan}^{-1} \, x - (1/2) \ln (1 + x^2) + C.$

7. $(1/5)t \tan 5t + (1/25) \ln |\cos 5t| + C.$

9. $(2/3)x^{3/2} \ln x - (4/9)x^{3/2} + C.$

11. $(1/2)(x^2 + 1) \, \mathrm{Tan}^{-1} \, x - (1/2)x + C.$

13. $(1/2)w^2 \ln w - (1/4)w^2 + C.$

15. $(-1/2)(\sin x \cos x - x) + C.$

17. $(a^x/\ln a)(x - 1/\ln a) + C.$

19. $x^2 e^x - 2xe^x + 2e^x + C.$

21. $(e^t/2)(\sin t + \cos t) + C.$

23. $(x/2)[\sin (\ln x) - \cos (\ln x)] + C.$

25. $2.$

27. $(\pi/2, \pi/8).$

29. $243\pi(e^6 - 25)/(4e^6) \doteq 178.9.$

11.6   Page 410

1. $(1/2)x + (1/4) \sin 2x + C.$

3. $\sin x - (1/3) \sin^3 x + C.$

5. $-\cos t + (2/3) \cos^3 t - (1/5) \cos^5 t + C.$

7. $(1/2) \tan^2 y - \ln |\sec y| + C.$

9. $(1/27) \cos^9 3x - (1/7) \cos^7 3x + (1/5) \cos^5 3x - (1/9) \cos^3 3x + C.$

11. $-\csc x - \sin x + C.$

13. $(3/128)t - (1/256) \sin 8t + (1/2048) \sin 16t + C.$

15. $(1/15) \sec^5 3y - (1/9) \sec^3 3y + C.$

17. $(-1/3) \csc^3 x + C.$

19. $(-1/2) \cot^2 t + C.$

21. $(1/2) \cos y - (1/18) \cos 9y + C.$

23. $(-1/6) \cot^3 2x + (1/2) \cot 2x + x + C.$

25. $(1/7) \tan 7x + (1/21) \tan^3 7x + C.$

27. $(-1/2) \csc^2 x + \ln |\csc x| + C.$

29. $\tan x - \cot x + C.$

11.7   Page 416

1. $(2/5)(x + 3)^{3/2}(x - 2) + C.$

3. $(1/3)(t - 7)\sqrt{2t + 7} + C.$

5. $2\sqrt{x} - 4 \ln (\sqrt{x} + 2) + C.$

7. $(1/5)(5t - 1)(2t + 1)^{5/2} + C.$

9. $\sqrt{2y} \sin \sqrt{2y} + \cos \sqrt{2y} + C.$

11. $3x^{1/3} - 3 \ln |x^{1/3} + 1| + C.$

13. $2x^{1/2} - 3x^{1/3} + 6x^{1/6} - 6 \ln |x^{1/6} + 1| + C.$

15. $\sqrt{1 - x^2} - \ln |(1 + \sqrt{1 - x^2})/x| + C.$

17. $(1/3) \ln |(\sqrt{x^2 + 9} - 3)/x| + C.$

19. $\sqrt{x^2 - 16}/(16x) + C.$

21. $-(t + 1)/\sqrt{t^2 - 1} + C.$

23. $\mathrm{Sin}^{-1} \, [(x - 2)/4] + C.$

25. $\sqrt{y^2 + 4} + 4/\sqrt{y^2 + 4} + C.$

27. $-\sqrt{16 - t^2} + C.$

29. $(1/2)\sqrt{u^4 - 1}/u^2 + C.$

11.8   Page 421

1. $(1/4) \ln |(x - 3)/(x + 1)| + C.$

3. $\ln |x^2 - 6x + 18| + (7/3) \, \mathrm{Tan}^{-1} \, [(x - 3)/3] + C.$

5. $\mathrm{Sin}^{-1} \, [(x - 3)/5] + C.$

7. $\mathrm{Sin}^{-1} \, [(x - 2)/2] + C.$

9. $(x^2 - 4x + 5)^{1/2} + 2 \ln |(x^2 - 4x + 5)^{1/2} + x - 2| + C.$

11. $(1/2) \, \mathrm{Tan}^{-1} \, x + x/[2(x^2 + 1)] + C.$

13. $\dfrac{-1}{x^2 - 4x + 5} + \dfrac{7}{2} \, \mathrm{Tan}^{-1} \, (x - 2) + \dfrac{7(x - 2)}{2(x^2 - 4x + 5)} + C.$

11.9   Page 423

1. $18.$

3. $-9/88.$

5. $53/480.$

7. $5\pi/6 + \sqrt{3}/2 + 1.$

9. $274/5.$

11. $(4 - \pi)/12.$

13. $9952/315.$

15. $\ln [(\sqrt{2} + 1)/\sqrt{3}].$

17. 8/3.
21. $\bar{x} = 326/77$, $\bar{y} = 37/55$.
27. $\pi ab/2$.

19. 33/4.
23. $2 \ln [(\sqrt{2} - 1)(\sqrt{5} + 2)]$.

11.10   Page 430

1. $\ln |x| - \ln |x + 2| + C$.
5. $3 \ln |x + 4| - 2 \ln |x - 1| + C$.
9. $2 \ln |x| - \ln |x + 1| + \ln |x - 2| + C$.
13. $\ln |x - 3| - 4/(x - 3) + C$.
17. $2 \ln (x^2 + 4) - 2 \ln |x| + (1/2) \text{Tan}^{-1} (x/2) + C$.
21. $(1/2) [\ln (x^2 + 1) + 5/(x^2 + 1)] + C$.

3. $3 \ln |x - 3| + 2 \ln |x + 3| + C$.
7. $2 \ln |2x - 1| - \ln |x + 5| + C$.
11. $3x^2/2 - 3x + 8 \ln |x + 2| + \ln |x - 1| + C$.
15. $2 \ln |x| + \ln |x - 4| + 1/(x - 4) + C$.
19. $2 \ln |x + 3| - \ln |x - 2| - \text{Tan}^{-1} x + C$.
23. $\ln (125/576)$.

11.11   Page 432

1. $\dfrac{2}{1 - \tan (x/2)} + C$.
5. $(1/2) \ln [\tan (x/2)] - (1/4) \tan^2 (x/2) + C$.
9. $\dfrac{2}{\sqrt{15}} \text{Tan}^{-1} \left(\dfrac{4 \tan (x/2) + 1}{\sqrt{15}} \right) + C$.

3. $\dfrac{1}{5} \ln \left| \dfrac{3 \tan (x/2) + 1}{\tan (x/2) - 3} \right| + C$.
7. $-\ln |\tan (x/2) - 1| + C$.

11.12   Page 432

1. $3x^{1/3} - 3 \ln |x^{1/3} + 1| + C$.

5. $(1/6)(\ln y)^6 + C$.
9. $(\ln t)^2 + C$.
13. $2 \text{Tan}^{-1} e^x + C$.
17. $(-1/4) \cos 2x - (1/2) \cos x + C$.
21. $e^{4t}(8t^2 - 4t + 1)/32 + C$.
25. $(1/32) \sin 4\theta - (1/8)\theta \cos 4\theta + C$.

29. $\ln |t| - 6 \ln (t^{1/6} + 1) + C$.
33. $3 \sin x + C$.
37. $-(\text{Tan}^{-1} x)[(x^2 + 1)/2x^2] - 1/(2x) + C$.
41. $-2\sqrt{1 + \cos t} + C$.
45. $e^y/[a^2(a^2 - e^{2y})^{1/2}] + C$.

49. $x/(16\sqrt{x^2 + 16}) + C$.
53. $\ln 7 - 6/7 \doteq 1.09$.
57. $\bar{x} = 20/(3\pi)$, $\bar{y} = 8/(3\pi)$.
61. $\ln [(2\sqrt{3} + 3)/3] \doteq 0.77$.

3. $\ln |t - 1| - \ln (t^2 + t + 1)^{1/2}$
$\quad - \sqrt{3} \text{ Tan}^{-1} [(2t + 1)/\sqrt{3}] + C$.
7. $-2 \cos \sqrt{x} + C$.
11. $y \ln (y^2 + 9) - 2y + 6 \text{ Tan}^{-1} (y/3) + C$.
15. $\ln |t| - \ln (t^2 + 9)^{1/2} + (1/3) \text{ Tan}^{-1} (t/3) + C$.
19. $(1/4) \sinh 2x - (1/2)x + C$.
23. $\ln |x^{16}(x^2 - 9)/(x^2 - 1)^9| + C$.
27. $3x + \ln (x - 1)^2 - \ln (x^2 + x + 1)$
$\quad - 2\sqrt{3} \text{ Tan}^{-1} [(2x + 1)/\sqrt{3}] + C$.
31. $-\sqrt{9 - e^{2y}} + C$.
35. $\sqrt{7} \ln |(\sqrt{7} - \sqrt{7 - y^2})/y| + \sqrt{7 - y^2} + C$.
39. $2(w + 5)^{1/2}(w - 10)/3 + C$.
43. $\ln |\tanh (x/2)| + C$.
47. $\ln \left| \dfrac{x}{\sqrt{x^2 + 3}} \right| - \dfrac{2}{x} + \dfrac{2\sqrt{3}}{3} \text{ Tan}^{-1} \dfrac{x}{\sqrt{3}} + C$.
51. $4(3 - \sqrt{3})/3 \doteq 1.7$.
55. $3\pi(3 \ln 3 + 4)/2$.
59. $\bar{x} = 0$, $\bar{y} = (\sinh 4 + 4)/(8 \sinh 2)$.

11.14   Page 441

7. $xy = C$.
11. $\sqrt{1 + y^2} + \ln |\sec x + \tan x| = C$.
15. $\text{Tan}^{-1} y + (1/2) \ln |(1 + x)/(1 - x)| = C$.
19. Approximately 13 days.
23. 40.6.

9. $y = C \cos x$.
13. $y^2 = (2/x) - 2x + C$.
17. 4.2 hours.
21. 6.18%.
25. 5.4 miles per second.

11.15 Page 447

1. (a) 0.20052; (b) 0.2.
5. (a) 1.00068; (b) 1.
9. 0.7468.
13. 1.4028.

3. (a) 0.78539; (b) $\pi/4 \doteq 0.78540$.
7. $S_6 = 1.00003$; $-0.00060 \leq E_6 \leq 0$.
11. 1.111.
17. 3.8195.

12.2 Page 453

(a) Parabola; $e = 1$. (b) Ellipse; $e = 1/2$.

(c) Hyperbola; $e = 2$.

12.3 Page 456

1. $(4, 0)$; $x = -4$.
5. $(1/2, 0)$; $x = -1/2$.
9. $y^2 = 12x$.
13. $y^2 = -20x$.
17. $5x^2 + 36y = 0$.
21. $2\sqrt{3}\,x - y - 6 = 0$; $\sqrt{3}\,x + 6y - 42 = 0$.
25. $\sqrt{2}\,x + y - 3 = 0$; $\sqrt{2}\,x - 2y - 12 = 0$.
29. $3x - 2y - 6 = 0$.

3. $(0, -4)$; $y = 4$.
7. $(0, 3/2)$; $y = -3/2$.
11. $x^2 = 16y$.
15. $3y^2 - x = 0$.
19. $2x + y + 2 = 0$; $x - 2y - 9 = 0$.
23. $5x - 2\sqrt{5}\,y - 15 = 0$; $2x + \sqrt{5}\,y + 21 = 0$.
27. $(4, 2\sqrt{5})$.

12.5 Page 461

1. $e = \sqrt{5}/3$; foci: $(\pm\sqrt{5}, 0)$; vertices: $(\pm 3, 0)$; directrices: $x = \pm 9/\sqrt{5}$; lengths of major and minor axes: 6 and 4.
5. $e = \sqrt{3}/2$; foci: $(0, \pm 3\sqrt{3})$; vertices: $(0, \pm 6)$; directrices: $y = \pm 4\sqrt{3}$; lengths of major and minor axes: 12 and 6.
9. $225x^2 + 200y^2 = 45,000$.
13. $3x^2 + 9y^2 = 84$.

17. Tangent: $8x - 25y - 133 = 0$; normal: $25x + 8y + 15 = 0$.
23. 1664 tons.

3. Vertices: $(0, \pm 4)$; $e = \sqrt{7}/4$; foci: $(0, \pm\sqrt{7})$; directrices: $y = \pm 16/\sqrt{7}$; lengths of major and minor axes: 8 and 6.
7. $27x^2 + 36y^2 = 972$.

11. $12x^2 + 8y^2 = 96$.
15. Tangent: $2x + 3y - 12 = 0$; normal: $3x - 2y - 5 = 0$.
21. $x - \sqrt{2}\,y \pm 2 = 0$.

25. $a\sqrt{2}$ by $b\sqrt{2}$

12.6 Page 465

1. $e = \sqrt{41}/5$; vertices: $(\pm 5, 0)$; foci: $(\pm\sqrt{41}, 0)$; directrices: $x = \pm 25/\sqrt{41}$; asymptotes: $4x \pm 5y = 0$.
5. $e = \sqrt{13}/3$; vertices: $(0, \pm 3)$; foci: $(0, \pm\sqrt{13})$; directrices: $y = \pm 9/\sqrt{13}$; $2a = 6$, $2b = 4$; asymptotes: $2y \pm 3x = 0$.
9. $4y^2 - 8x^2 = 32$.
13. Asymptotes: $4x \pm 3y = 0$.
17. $64x - 45y - 31 = 0$, $45x + 64y - 500 = 0$.
21. $x - y - 1 = 0$.

3. $e = \sqrt{13}/3$; vertices: $(\pm 3, 0)$; foci: $(\pm\sqrt{3}, 0)$; directrices: $x = \pm 9/\sqrt{13}$; asymptotes: $2x \pm 3y = 0$.
7. $9x^2 - 16y^2 = 144$.

11. $x^2 - 4y^2 = 16$.
15. $x^2 - 4y^2 = 12$.
19. $(7, -3)$ and $(-7, 3)$.

12.7 Page 468

1. $16x^2 + 25y^2 - 400 = 0$.

3. $2x^2 + y^2 = 8$.

5. $21x^2 - 4y^2 - 84 = 0$.
9. $b^2x^2 + a^2y^2 = a^2b^2$.

7. $x^2/a^2 + y^2/[a^2(1 - e^2)] = 1$.

12.8 Page 473

1. $x'^2 + y'^2 = 1$.
5. $y'^2 = 10x'$.
9. $x'^2 = 16y'$.
13. $x'^2/16 - y'^2/9 = 1$.
17. $8x'^3 - y' = 0$.

3. $9x'^2 - 16y'^2 = 144$.
7. $16x'^2 + 5y'^2 = 80$.
11. $x'^2/16 + y'^2/4 = 1$.
15. $x'y' = 8$.
19. $x = x' - D/(2A), y = y' - E/(2C)$.

12.9 Page 479

1. $(\sqrt{2}, -2\sqrt{2}), (0, 2\sqrt{2}), (1, 1), (1, -3)$.
5. $x'^2 + y'^2 = 36$.
9. $31x'^2 + y'^2 + 10 = 0$.

13. $250x'^2 + 7x' + y' + 5 = 0$.

3. $y' + 2 = 0$.
7. $17x'^2 - 9y'^2 + 4 = 0$.
11. $130x'^2 - 39y'^2 + 11\sqrt{13}x' + 16\sqrt{13}y' + 13 = 0$.
15. $55x'^2 + 29y'^2 - 4 = 0$.

12.10 Page 484

1. $5x''^2 - 2y''^2 - 10 = 0$.
5. $2x''^2 - 3y''^2 - 6 = 0$.
9. $4x''^2 + 9y''^2 - 36 = 0$.

3. $y''^2 + 3x'' = 0$.
7. $x''^2 - 6y'' = 0$.
11. $3x''^2 + 16y''^2 - 48 = 0$.

12.11 Page 491

7. (a) $(2, 2\sqrt{3})$;
   (b) $(3\sqrt{2}/2, 3\sqrt{2}/2)$;
   (c) $(-5\sqrt{3}/2, -5/2)$;
   (d) $(-7/2, -7\sqrt{3}/2)$.
11. $r(\cos\theta - 4\sin\theta) + 2 = 0$.
15. $r = 4$.
19. $x + 6 = 0$.

9. (a) $(4, 7\pi/6)$;   (b) $(2, \pi/3)$;   (c) $(2, -\pi/4)$;
   (d) $(0, \theta)$, where θ is any real number.

13. $r\sin\theta + 5 = 0$.
17. $\sqrt{3}x - y = 0$.
21. $y - 4 = 0$.

12.13 Page 506

1. (a) $r\cos\theta - 8 = 0$;   (b) $r\sin\theta + 3 = 0$;
   (c) $\theta = \pi/3$;   (d) $\theta = 0$.
5. (a) $r - 8 = 0$;   (b) $r - 10\cos\theta = 0$;
   (c) $r - 12\sin\theta = 0$;   (d) $r + 8\cos\theta = 0$;
   (e) $r - 6\sin\theta = 0$.
9. Ellipse; vertices: $(4, 0), (4/3, \pi)$; center: $(4/3, 0)$; foci: $(0, 0), (8/3, 0)$; directrices: $r\cos\theta + 4 = 0$, $3r\cos\theta - 20 = 0$.
13. Ellipse; vertices: $(2, \pi/2), (10/9, 3\pi/2)$; center: $(4/9, \pi/2)$; foci: $(0, 0), (8/9, \pi/2)$; directrices: $9r\sin\theta - 53 = 0, r\sin\theta + 5 = 0$.

3. $r\cos\theta - 2 = 0$.

7. Parabola; vertex: $(-1, 0)$; directrix: $r\cos\theta + 2 = 0$.

11. Hyperbola; vertices: $(3, 0), (-15, \pi)$; center: $(9, 0)$; foci: $(0, 0), (18, 0)$; directrices: $r\cos\theta - 5 = 0, r\cos\theta - 13 = 0$.
15. Hyperbola; vertices: $(1, \pi/2), (-7, 3\pi/2)$; center: $(4, \pi/2)$; foci: $(0, \pi/2), (8, \pi/2)$; directrices: $4r\sin\theta - 7 = 0, 4r\sin\theta - 25 = 0$.

12.15 Page 512

1. $-\sqrt{3}/2; \sqrt{3}/5$.
5. $\sqrt{3} - 1; -(4 + 3\sqrt{3})$.

3. $-5\sqrt{3}/3; 2\sqrt{3}/3$.
7. $\sqrt{3} - 2; -1$.

9. $5\pi/4$; $(4 + 5\pi)/(4 - 5\pi)$.
15. $\theta = \pi/2$.
19. $(3\sqrt{3}/2, \pi/3)$, $(0, 0)$.
25. $-\sqrt{3}$.
29. $\tan \alpha = \tan \frac{1}{2}\theta$.

13. $\theta = 0$, $\theta = \pi/2$.
17. $\theta = 0$, $\theta = \pi/3$, $\theta = 2\pi/3$.
21. $(3, \pi/6)$, $(3, 5\pi/6)$, $(-6, 3\pi/2)$.
27. $\tan \psi_1 = -\cot \theta$ and $\tan \psi_2 = \tan \theta$.

12.16   Page 518

1. πa^2.
5. 24π.
9. 4.
13. $8\pi + 6\sqrt{3}$.
17. 51π.
21. $9(\sqrt{2} - 1)$.

3. $19\pi/2$.
7. $3a^2\pi/2$.
11. $4\pi - 6\sqrt{3}$.
15. 4π.
19. $4\sqrt{3} - 4\pi/3$.

12.17   Page 519

1. Parabola. Vertex: $(0, 0)$; focus: $(3/2, 0)$.

5. Parabola. Vertex: $(0, 0)$; focus: $(0, -9/4)$.
9. Ellipse. Vertices: $(\pm 5, 0)$; foci: $(\pm 4, 0)$.

13. $x^2/16 + y^2/12 = 1$.
17. $x^2/4 - y^2 = 1$.

21. Four-leaved rose. Symmetric to polar axis and $\pi/2$ − axis; domain: $[0, 2\pi]$; tangents at origin: $\theta = \pi/4$, $\theta = 3\pi/4$.
25. Limaçon. Symmetric to polar axis; domain: $[0, 2\pi]$.
29. Lemniscate. Symmetric to the line $\theta = \pi/4$ and to the origin; domain: $[0, \pi/2] \cup [\pi, 3\pi/2]$; tangents at the origin: $\theta = 0$, $\theta = \pi/2$.
33. $x'^2/9 + y'^2/4 = 1$.
37. $(x - 3)^2 + (y - 3)^2 = 9$.
41. $9\pi/2$.

3. Hyperbola. Vertices: $(2, 0)$, $(-6, \pi)$; foci: $(0, 0)$, $(8, 0)$.
7. Parabola. Vertex: $(2, \pi)$; focus: $(0, 0)$.
11. Ellipse. Vertices: $(1, 0)$, $(3, \pi)$; foci: $(0, 0)$, $(2, \pi)$.
15. $y^2 + 9x = 0$.
19. Circle. Symmetric with respect to the polar axis; domain: $[0, \pi]$; tangent at origin: $\theta = \pi$.
23. Circle. Symmetric with respect to polar axis and to $\pi/2$ axis; domain: $[0, 2\pi]$.

27. Line through the origin.

31. Spiral of Archimedes. Domain: $[0, \infty)$; tangent at the origin: $\theta = 0$.

35. $x'^2/10 - y'^2/3 = 1$.
39. -1.

13.3   Page 527

1. -1.
5. 0.
9. $1/3$.
13. 0.
17. $-9/4$.
21. $-1/2$.
25. -2.
29. Does not exist.
33. $-1/24$.

3. 0.
7. $11/10$.
11. 0.
15. Does not exist.
19. -1.
23. 2.
27. 0.
31. -1.
35. 2.

13.4   Page 529

1. 1.

3. 0.

5. 1.
9. 1.
13. 0.
17. $1/e$.
21. 1.
25. $-3/2$.
29. 0.

7. e^3.
11. -1.
15. 1.
19. e^2.
23. 0.
27. 1.

13.5 Page 534

1. Divergent.
5. $\pi/2$.
9. $1/3$.
13. Divergent.
17. $(\sin 2 + \cos 2)/(2e^2)$.
21. $1/2$.
25. 4.
29. $2 \ln (7/5)$.

3. $-1/64$.
7. $1/3$.
11. 100.
15. Divergent.
19. Divergent.
23. e^4.
27. π.
31. 1.

13.6 Page 535

(a) 3; yes.   (c) $\sqrt{3}$; no.   (e) 1; no.

13.7 Page 540

1. On $[1, \infty)$, g dominates f and h, and f dominates h.
7. Converges.
11. Converges.
15. Converges.

5. Converges.
9. Diverges.
13. Diverges.

13.8 Page 544

1. $3/2$.
5. π.
9. Divergent.
  6.

3. Divergent.
7. 3.
11. 12.
17. (a) 3;

   (b) the improper integral $\pi \int_0^1 (x^{-2/3})^2 \, dx$ diverges.

23. *Hint:* By repeated use of the recurrence formula in Exercise 22, show that when n is a positive integer

$$\Gamma(n + 1) = n(n - 1)(n - 2) \cdots 3 \cdot 2 \cdot 1 \cdot \Gamma(1).$$

Then show that $\Gamma(1) = 1$.

Outline of proof. Let $n > 0$. Then

$$\int_1^\infty x^{n-1} e^{-x} \, dx = \int_1^M x^{n-1} e^{-x} \, dx + \int_M^\infty x^{n-1} e^{-x} \, dx,$$

where $M > 1$ is the positive number corresponding to n whose existence was established in Exercise 18. Since $f(x) = x^{n-1} e^{-x}$ is continuous for all $x > 0$, $\int_1^M x^{n-1} e^{-x} \, dx$ exists; call its value I_M. Show that $\int_M^\infty x^{-2} \, dx = 1/M$ and use Exercise 18 to show that x^{-2} dominates $x^{n-1} e^{-x}$

on $[M, \infty)$. Use the comparison test (13.7.3) to show that $\int_M^\infty x^{n-1}e^{-x}\,dx$ converges and hence that

$$\int_1^\infty x^{n-1}e^{-x}\,dx = I_M + \int_M^\infty x^{n-1}e^{-x}\,dx$$

converges.

13.9   Page 545

1. 4.
5. 0.
9. e^2.
13. 1/3.
17. $e^2/2$.
21. $\pi/4$.
25. $1/\ln 2$.
29. $3(1 + \sqrt[3]{3})/2$.
33. Converges.

3. $-1/2$.
7. 0.
11. 1.
15. 1.
19. Diverges.
23. Divergent.
27. 6.
31. Converges.
35. Converges.

14.1   Page 552

1. $\{1, \frac{2}{3}, \frac{3}{5}, \frac{4}{7}, \frac{5}{9}, \cdots\}$. Converges. Limit $= \frac{1}{2}$.

5. $\left\{\dfrac{1}{\ln 2}, \dfrac{2}{\ln 3}, \dfrac{3}{\ln 4}, \dfrac{4}{\ln 5}, \dfrac{5}{\ln 6}, \cdots\right\}$. Diverges.

9. $\left\{e, \dfrac{e^2}{4}, \dfrac{e^3}{9}, \dfrac{e^4}{16}, \dfrac{e^5}{25}, \cdots\right\}$. Diverges.
13. $u_n = n$. Diverges.
17. $u_n = n/(2n - 1)$. Converges. Limit $= 1/2$.

3. $\{1, \frac{5}{3}, \frac{5}{3}, \frac{17}{11}, \frac{13}{9}, \cdots\}$. Converges. Limit $= 1$.

7. $\left\{0, \dfrac{\ln 2}{4}, \dfrac{\ln 3}{9}, \dfrac{\ln 4}{16}, \dfrac{\ln 5}{25}, \cdots\right\}$. Converges. Limit $= 0$.
11. $u_n = n/(n + 1)$. Converges. Limit $= 1$.

15. $u_n = \ln n/n$. Converges. Limit $= 0$.
19. $u_n = 2n^n/n^2$. Diverges.

14.2   Page 558

1. (b) $(1 - 1/5^n)/4$;   (c) converges;   (d) 1/4.
9. 2/9.

3. (b) $n/(n + 1)$;   (c) converges;   (d) 1.
11. 3010/999.

14.3   Page 565

1. Converges.

5. Divergent. $\displaystyle\lim_{n\to\infty}\left[\dfrac{n+1}{n(n + 2)}\Big/\dfrac{1}{n}\right] = 1$.

9. Diverges.
13. $Error < \displaystyle\int_5^\infty dx/(x^2 + 2x) < 0.169$.
17. Converges.

21. Converges.

3. Divergent. $\displaystyle\lim_{n\to\infty}\left[\dfrac{n}{(n + 1)^2}\Big/\dfrac{1}{n}\right] = 1$.
7. Converges.

11. Converges.

15. $Error < \displaystyle\int_5^\infty x^2e^{-x}\,dx < 0.248$.
19. No test. But note that this is a convergent p-series.

14.4   Page 570

1. Absolutely convergent.

3. Absolutely convergent.

5. Absolutely convergent.

7. Conditionally convergent.

9. Absolutely convergent.

14.5 Page 574

1. $[-1, 1]$.

3. $(-\infty, \infty)$.

5. $(-1, 1)$.

7. $(-1, 1]$.

9. $[-1, 1]$.

11. $(-2, 2)$.

13. $(-\infty, \infty)$.

15. $[0, 2)$.

17. $(-3, 1)$.

19. $[-6, -3]$.

14.6 Page 579

1. (a) $1/(1 + x) = 1 - x + x^2 - x^3 + x^4 - x^5 + \cdots$. (b) $(-1, 1)$.

3. (a) $\ln [(1 + x)/(1 - x)] = 2(x + x^3/3 + x^5/5 + \cdots)$. (b) $(-1, 1)$.

7. (a) $\operatorname{Tan}^{-1} x = x - x^3/3 + x^5/5 - x^7/7 + \cdots$. (b) $(-1, 1)$.

9. $\int \ln (1 + x) \, dx = x^2/2 - x^3/(2 \cdot 3) + x^4/(3 \cdot 4) - x^5/(4 \cdot 5) + \cdots$.

14.7 Page 583

1. $\ln (1 + x) = x - x^2/2 + x^3/3 - x^4/4 + \cdots + (-1)^{n+2} \int_0^x [(x - t)^n/(1 + t)^{n+1}] \, dt$.

3. $f^n(0) = 0$ if n is even; $f^n(0) = (-1)^{(n-1)/2}(n - 1)!$ if n is odd.

5. $1 - x + x^2/2! - x^3/3! + x^4/4! - \cdots + (-1)^{n+1}/n! \int_0^x (x - t)^n e^{-t} \, dt$.

7. $\sec x = 1 + x^2/2 + 5x^4/24 + 61x^6/720 + \cdots$.

9. $\operatorname{Sin}^{-1} x = x + x^3/6 + 3x^5/40 + 5x^7/112 + \cdots$.

11. $e^{-x} = 1 - x + x^2/2! - x^3/3! + x^4/4! + \cdots$.

13. $\tan x = x + x^3/3 + 2x^5/15 + 17x^7/315 + \cdots$.

15. $(\sin x)/x = 1 - x^2/3! + x^4/5! - x^6/7! + \cdots$.

17. $\sin x = \dfrac{1}{2} + \dfrac{\sqrt{3}}{2}\left(x - \dfrac{\pi}{6}\right) - \dfrac{1}{2 \cdot 2!}\left(x - \dfrac{\pi}{6}\right)^2 - \dfrac{\sqrt{3}}{2 \cdot 3!}\left(x - \dfrac{\pi}{6}\right)^3 + \dfrac{1}{3!} \int_{\pi/6}^x (x - t)^3 \sin t \, dt$.

19. $\tan x = 1 + 2\left(x - \dfrac{\pi}{4}\right) + \left(\dfrac{4}{2!}\right)\left(x - \dfrac{\pi}{4}\right)^2 + \left(\dfrac{16}{3!}\right)\left(x - \dfrac{\pi}{4}\right)^3 + \cdots$.

14.8 Page 589

1. $\sin 1°$ can be computed to five-decimal-place accuracy using the approximation $\sin x = x$.

3. Four nonvanishing terms of the Maclaurin expansion of $\sin x$ yield $\sin 28°$ accurate to eight decimal places.

5. Eight terms of the series for e yield three-decimal-place accuracy.

7. $e^x = e + e(x - 1) + (e/2!)(x - 1)^2 + \cdots + [f^{n+1}(\mu)/(n + 1)!](x - 1)^{n+1}, 1 < \mu < x$. Five terms of this series yield $e^{1.1}$ accurate to six decimal places.

9. $e^x \doteq 1 + x$, for small $|x|$.

11. $\sin (\pi/6 + x) \doteq 1/2 + (\sqrt{3}/2)x$ for small $|x|$.

15. $-0.1 < x < 0.1$.

17. $e^{.693} = 2.04$.

19. $\sin 0.201 = 0.200$.

23. $x - x^3/3(3!) + x^5/5(5!) - x^7/7(7!) + \cdots$. Its interval of convergence is $(-\infty, \infty)$.

25. $x + \dfrac{x^4}{8} + \displaystyle\sum_{i=1}^{\infty} \dfrac{(-1)^i(2i-1)! \, x^{3i+4}}{(i-1)! \, (i+1)! \, (2^{2i})(3i+4)}$. The interval of convergence is $(-1, 1)$.

14.9   Page 591

1. 3.
5. Converges absolutely.
9. Converges absolutely.
13. $[-1, 1]$.
17. $[2, 4)$.
21. $1/(1 - x)^2 = 1 + 2x + 3x^2 + \cdots + nx^{n-1} + \cdots$; $(-1, 1)$.

25. $\sin x + \cos x = 1 + x - \dfrac{x^2}{2!} - \dfrac{x^3}{3!} + \dfrac{x^4}{4!} + \dfrac{x^5}{5!} - \cdots$ for all real numbers x.
29. 0.

3. 0.
7. Diverges.
11. Converges conditionally.
15. $(3, 5]$.
19. $[4, 8)$.
23. $\sin^2 x = \sum \left[\dfrac{(-1)^{n-1}2^{2n-1}}{(2n)!} \right] x^{2n}$ for all real numbers x.
27. $\sum \dfrac{1}{n(n!)}$

15.2   Page 603

1. A simple, smooth curve. $y = 3x/2$.
5. A simple arc. $x^3 - y^2 = 0$, $-1 \le y \le 8$.
9. A simple, smooth, closed arc. $x^2/9 + y^2/25 = 1$.
13. A simple, smooth arc. $x^{1/2} + y^{1/2} - 2 = 0$, $0 \le x$, $0 \le y$.
17. $(2t^2 - 3)/(2t^2 + 3)$, $t \ne 0$.
21. $3x - y - 4 = 0$; $x + 3y - 28 = 0$.
25. $2x + y - 2\sqrt{3} = 0$; $3x - 6y - 8\sqrt{3} = 0$.
29. $(1/3)x^3 + C$.
33. $(e^{\pi/2} + e^{-\pi/2})/2$.
37. $x = (a - b) \cos \phi + b \cos [(a - b)/b] \phi$, $y = (a - b) \sin \phi - b \sin [(a - b)/b] \phi$.

3. A simple arc. $y = \sqrt{x + 4}$, $-4 \le x \le 0$.
7. A smooth arc but not simple. $x^2 = y^2(y + 4)$, $-4 \le y \le 5$.
11. A simple, smooth curve. $x^2/16 - y^2/9 = 1$, $x \ge 4$.
15. t, $t \ne 0$.
19. $(5 \sin t)/3$, $t \ne (2n + 1)\pi/2$, $n \in Z$.
23. $x + \sqrt{3}y - 6 = 0$; $\sqrt{3}x - y = 0$.
27. $-(x - 1)^4 + \dfrac{(x - 1)^3}{3} + (x - 1)^2 - 15(x - 1) + C$.
31. 8.
35. $3\pi a^2$.

15.3   Page 609

1. $8(37^{3/2} - 1)$ 27.
5. $\sqrt{2}(e^{\pi/2} - 1)$.
9. $4\sqrt{5} + 8 \ln(\sqrt{5} + 2)$.
13. $4a$.

3. 6π.
7. $2\sqrt{3} + \ln(2 + \sqrt{3})$.
11. $6a$.
15. $\sqrt{5}(e^\pi - 1)$.

15.4   Page 614

3. $[-3, -1]$; $[-5, 3]$; $[0, -4]$; $[3, -6]$; $[5, 3]$; $[4, 0]$; $[-3, \sqrt{2}]$; $[-\pi, -e]$; $[-4, 3\pi/2]$; $[-2, \sqrt{3}]$.
7. (a) $[6, 5]$;   (b) $[-8, -5]$;   (c) $[-5, 7]$; (d) $[3\sqrt{3}, -6]$.
15. $(-20 + 20\sqrt{17}, -20 - 20\sqrt{17}) \doteq (62.46, -102.46)$.

5. (a) $[6, 2]$;   (b) $[5, -9]$;   (c) $[-3, 2]$; (d) $[-4 - \sqrt{3}, -6]$.
13. $[400, 0] + [140\sqrt{2}, -140\sqrt{2}] = [400 + 140\sqrt{2}, -140\sqrt{2}]$.
17. $[2\sqrt{21}, 4]$.

15.5   Page 620

1. (a) $-14\mathbf{i} + 20\mathbf{j}$; (b) -18; (c) -38; (d) 425;
   (e) -100; (f) $13 - \sqrt{13}$.
5. (a) $-5\mathbf{i} + 5\mathbf{j}$; (b) $-6\mathbf{i} - 5\mathbf{j}$; (c) $-\sqrt{2}\,\mathbf{i} + e\mathbf{j}$;
   (d) $3\mathbf{i} - \frac{5}{3}\mathbf{j}$.
17. $k = (b_2 r_1 - b_1 r_2)/(a_1 b_2 - a_2 b_1)$,
    $m = (a_1 r_2 - a_2 r_1)/(a_1 b_2 - a_2 b_1)$.

3. (a) $-14/\sqrt{221}$; (b) $-5/\sqrt{29}$; (c) $\sqrt{2}/2$;
   (d) -1.
15. $k = 2$, $m = -1$.

19. $\mathbf{a} + \mathbf{b} + \mathbf{c} = \mathbf{0}$.

15.6   Page 623

1. (a) $\mathfrak{D} = \{t \mid t \in R,\ t \geq -4,\ t \neq 2\}$; (b) R;
   (c) $0 \leq t < 1$; (d) $\varnothing$.

5. (a) $\mathbf{r}'(t) = 4(2t + 3)\mathbf{i} - 2e^{2t}\mathbf{j}$, $\mathbf{r}''(t) = 8\mathbf{i} - 4e^{2t}\mathbf{j}$;
   (b) $\mathbf{r}'(t) = -2 \sin 2t\mathbf{i} - 3 \sin^2 t \cos t\mathbf{j}$, $\mathbf{r}''(t) = -4 \cos 2t\mathbf{i} + 3(\sin^3 t - 2 \sin t \cos^2 t)\mathbf{j}$;
   (c) $\mathbf{r}'(t) = -e^{-t}\mathbf{i} + 6t^{-1} \ln (t^3)\mathbf{j}$, $\mathbf{r}''(t) = e^{-t}\mathbf{i} + 6t^{-2}\{3 - \ln (t^3)\}\mathbf{j}$;
   (d) $\mathbf{r}'(t) = \frac{1}{2}(t + 1)^{-1/2}\mathbf{i} - (1 - t^2)^{-1/2}\mathbf{j}$, $\mathbf{r}''(t) = -\frac{1}{4}(t + 1)^{-3/2}\mathbf{i} - t(1 - t^2)^{-3/2}\mathbf{j}$.
7. $15t^4 + 1$.
9. $16 \cos 4t - 81 \cos 6t$.

13. $-6t(3t^2 - 4)^{-1/2} \sin \sqrt{3t^2 - 4} \cos \sqrt{3t^2 - 4}\mathbf{i}$
    $+ 9t(3t^2 - 4)^{-1/2}e^{1-3\sqrt{3t^2-4}}\mathbf{j}$.

3. (a) Continuous on $\mathfrak{D}$;
   (b) continuous on $\{t \mid t \in R,\ t \notin I\}$;
   (c) $0 < t \leq 1$; (d) none.

11. $e^{-3t}\{\frac{1}{2}(t - 1)^{-1/2} - 3(t - 1)^{1/2}\}\mathbf{i}$
    $+ e^{-3t}\{2t^{-1} - 3 \ln (2t^2)\}\mathbf{j}$.

15.7   Page 627

1. $\mathbf{v}(t) = -e^{-t}\mathbf{i} + e^t\mathbf{j}$; $\mathbf{a}(t) = e^{-t}\mathbf{i} + e^t\mathbf{j}$;
   $\mathbf{v}(1) = -e^{-1}\mathbf{i} + e\mathbf{j}$; $\mathbf{a}(1) = e^{-1}\mathbf{i} + e\mathbf{j}$;
   $|\mathbf{v}(1)| = e^{-1}(e + 1)$.
5. $\mathbf{v}(t) = 6t\mathbf{i} + 3t^2\mathbf{j}$; $\mathbf{a}(t) = 6\mathbf{i} + 6t\mathbf{j}$;
   $\mathbf{v}(2) = 12\mathbf{i} + 12\mathbf{j}$; $\mathbf{a}(2) = 6\mathbf{i} + 12\mathbf{j}$;
   $|\mathbf{v}(2)| = 12\sqrt{2}$.
9. $\mathbf{v}(t) = 3(1 - \cos t)\mathbf{i} + 3 \sin t\mathbf{j}$; $\mathbf{a}(t) = 3 \sin t\mathbf{i}$
   $+ 3 \cos t\mathbf{j}$; $\mathbf{v}(\pi/3) = (3/2)\mathbf{i} + 3\sqrt{3}/2\mathbf{j}$;
   $\mathbf{a}(\pi/3) = (3\sqrt{3}/2)\mathbf{i} + (3/2)\mathbf{j}$; $|\mathbf{v}(\pi/3)| = 3$.
13. 114.

3. $\mathbf{v}(t) = (-2 \sin t)\mathbf{i} - (3 \sin 2t)\mathbf{j}$; $\mathbf{a}(t) = (-2 \cos t)\mathbf{i}$
   $- (6 \cos 2t)\mathbf{j}$; $\mathbf{v}(\pi/3) = -\sqrt{3}\,\mathbf{i} - (3/2)\sqrt{3}\,\mathbf{j}$;
   $\mathbf{a}(\pi/3) = -\mathbf{i} + 3\mathbf{j}$; $|\mathbf{v}(\pi/3)| = \sqrt{39}/2$.
7. $\mathbf{v}(t) = -\sin t\mathbf{i} - 2 \sec^2 t\mathbf{j}$; $\mathbf{a}(t) = -\cos t\mathbf{i}$
   $- 4 \sec^2 t \tan t\mathbf{j}$; $\mathbf{v}(-\pi/4) = (\sqrt{2}/2)\mathbf{i} - 4\mathbf{j}$;
   $\mathbf{a}(-\pi/4) = (-\sqrt{2}/2)\mathbf{i} + 8\mathbf{j}$;
   $|\mathbf{v}(-\pi/4)| = \sqrt{66}/2$.
11. $\mathbf{v}(t) = ab \cosh bt\mathbf{i} + ab \sinh bt\mathbf{j}$;
    $\mathbf{a}(t) = ab^2 \sinh bt\mathbf{i} + ab^2 \cosh bt\mathbf{j}$; $\mathbf{v}(0) = ab\mathbf{i}$;
    $\mathbf{a}(0) = ab^2\mathbf{j}$; $|\mathbf{v}(0)| = |ab|$.
15. $15\pi^2/8$.

15.8   Page 632

1. (a) $\mathbf{T}(1/2) = (\sqrt{2}/2)\mathbf{i} + (\sqrt{2}/2)\mathbf{j}$.
   (b) $K(1/2) = \sqrt{2}/8$.
5. (a) $\mathbf{T}(\pi/3) = -(1/2)\mathbf{i} + (\sqrt{3}/2)\mathbf{j}$.
   (b) $K(\pi/3) = 4/9$.
9. $K(-3) = 2\sqrt{5}/25$; $q_1 = -1/2$, $q_2 = 4$.
13. $K(0) = 1/4$; $q_1 = 0$, $q_2 = 5$.

17. $(\sqrt{2}/2, -\ln \sqrt{2})$.

3. (a) $\mathbf{T}(\pi/4) = -(4/5)\mathbf{i} + (3/5)\mathbf{j}$.
   (b) $K(\pi/4) = 24\sqrt{2}/125$.
7. $K(1) = 2\sqrt{5}/25$; $q_1 = -4$, $q_2 = 7/2$.

11. $K(0) = 1$; $q_1 = 0$, $q_2 = 2$.
15. $K(\pi/4) = 2\sqrt{3}/9$; $q_1 = (\pi + 6)/4$,
    $q_2 = -\sqrt{2}$.
19. $(-\pi/2, -1)$ and $(\pi/2, 1)$.

15.9 Page 634

1. (a) $8t^2(4t^2 + 1)^{-1}\mathbf{i} + 4t(4t^2 + 1)^{-1}\mathbf{j}$.
   $2(4t^2 + 1)^{-1}\mathbf{i} - 4t(4t^2 + 1)^{-1}\mathbf{j}$.
   (b) $(8/5)\mathbf{i} + (4/5)\mathbf{j}$. $(2/5)\mathbf{i} - (4/5)\mathbf{j}$.
5. (a) $2t(t^2 - 1)(t^2 + 1)^{-1}\mathbf{i} + 4t^2(t^2 + 1)^{-1}\mathbf{j}$.
   $4t(t^2 + 1)^{-1}\mathbf{i} - 2(t^2 - 1)(t^2 + 1)^{-1}\mathbf{j}$.
   (b) $(12/5)\mathbf{i} + (16/5)\mathbf{j}$. $(8/5)\mathbf{i} - (6/5)\mathbf{j}$.

3. (a) $\mathbf{0}$. $-a\cos t\mathbf{i} - a\sin t\mathbf{j}$.
   (b) $\mathbf{0}$. $-(\sqrt{3}/2)a\mathbf{i} - (1/2)a\mathbf{j}$.

15.10 Page 635

1. Simple. $x - 3y - 2 = 0$.
5. A simple, closed arc. $(x + 2)^2/16 + (y + 2)^2/9 = 1$.
9. $2\pi^2$.

13. (a) $4\mathbf{i} - 17\mathbf{j}$; (b) -3; (c) -15;
    (d) -234; (e) -36; (f) 30.
17. (a) $t^{-1}\mathbf{i} - 6t\mathbf{j}$, $-t^{-2}\mathbf{i} - 6\mathbf{j}$;
    (b) $\cos t\mathbf{i} - 2\sin 2t\mathbf{j}$, $-\sin t\mathbf{i} - 4\cos 2t\mathbf{j}$;
    (c) $\sec^2 t\mathbf{i} - 4t^3\mathbf{j}$, $2(\sec^2 t\tan t)\mathbf{i} - 12t^2\mathbf{j}$;
    (d) $(e^t + 2e^{-2t})\mathbf{j}$, $(e^t - 4e^{-2t})\mathbf{j}$.
21. $\mathbf{v}(t) = -4\cos t\mathbf{i} + (4 + 4\sin t)\mathbf{j}$; $\mathbf{a}(t) = 4\sin t\mathbf{i}$
    $+ 4\cos t\mathbf{j}$; $\mathbf{v}(2\pi/3) = 2\mathbf{i} + (4 + 2\sqrt{3})\mathbf{j}$; $\mathbf{a}(2\pi/3)$
    $= 2\sqrt{3}\,\mathbf{i} - 2\mathbf{j}$; $|\mathbf{v}(2\pi/3)| = 4(2 + \sqrt{3})^{1/2}$.

3. A simple arc. $y = x^3/8$.
7. Tangent: $2x + 3y + (6\sqrt{2} - 7) = 0$; normal:
   $3x - 2y + (22 + 5\sqrt{2}/2) = 0$.
11. (b) $-2/3, 6/5$. (c) $[2, 8]$.
    (d) $[-8, -4]$. (e) $\sqrt{13}, \sqrt{61}$.
15. (a) $10\mathbf{i} + 2\mathbf{j}$; (b) $-6\mathbf{i} - 6\mathbf{j}$;
    (c) $(-15/2)\mathbf{i} + 4\sqrt{2}\mathbf{j}$; (d) $3\mathbf{i} - \pi\mathbf{j}$.
19. $\mathbf{v}(t) = 4t\mathbf{i} + 4\mathbf{j}$; $\mathbf{a}(t) = 4\mathbf{i}$; $\mathbf{v}(-1) = -4\mathbf{i} + 4\mathbf{j}$;
    $\mathbf{a}(-1) = 4\mathbf{i}$; $|\mathbf{v}(-1)| = 4\sqrt{2}$.

16.1 Page 642

3. Their y coordinates are zero; their z coordinates are zero.
7. Yes, the x axis; $|PQ| = 5$.
11. (a) $x^2 + y^2 + z^2 - 4x - 8y - 2z + 12 = 0$;
    (b) $x^2 + y^2 + z^2 + 8x - 12y + 3 = 0$;
    (c) $x^2 + y^2 + z^2 + 12x - 4y + 6z + 48 = 0$;
    (d) $x^2 + y^2 + z^2 - 16x + 2z + 61 = 0$.
15. $(2/3, -1/2, 5)$; 8.
19. $x^2 + y^2 + z^2 - 12x - 12y - 12z + 72 + 0$.

5. (a) The yz plane; (b) the y axis; (c) the negative x axis.
9. (a) $\sqrt{83}$; (b) $3\sqrt{10}$; (c) $\sqrt{14}$; (d) 3.
13. $(6, -7, 4)$; 10.

17. $x^2 + y^2 + z^2 - 4x - 8y - 10z + 20 = 0$.
21. $x^2 + y^2 + z^2 + 6x - 2y - 4z - 10 = 0$, and
    $x^2 + y^2 + z^2 - 10x + 6y - 12z + 46 = 0$.

16.2 Page 650

1. (a) $[3, 3, -2]$; (b) $[6, 2, 4]$;
   (c) $[5, -2, 5]$; (d) $[-3, 1, 6]$.
5. (a) $(3\mathbf{i} + \mathbf{j} - 7\mathbf{k})/\sqrt{59}$;
   (b) $(-2\mathbf{i} + 5\mathbf{j} - 3\mathbf{k})/\sqrt{38}$.
9. π.
13. $\mathbf{a} = [4, -17, 10]$, $\mathbf{b} = [7, 4, 4]$. (There are many other correct answers.)
17. $8(2\mathbf{i} + 4\mathbf{j} - 5\mathbf{k})/45$.
21. $\mathbf{b} - \mathbf{a} = \mathbf{c} - \mathbf{d}$.

3. (a) $\sqrt{21}$; (b) $5\sqrt{2}$; (c) 6; (d) 3.

7. $-4/9$.

11. $k[3, -1, 3]$, $k \neq 0$.
15. $5\sqrt{19}/19$.

19. $(1/2)(\mathbf{a} + \mathbf{b})$.
23. $\mathbf{c} = |\mathbf{b}|\mathbf{a} + |\mathbf{a}|\mathbf{b}$.

16.3   Page 654

1. (a) $-2\mathbf{i} + 7\mathbf{j} + 10\mathbf{k}$;   (b) $4\mathbf{i} + 15\mathbf{j} + 9\mathbf{k}$;
   (c) 29;   (d) $18\mathbf{i} + 47\mathbf{j} + 20\mathbf{k}$.
5. $\pm(4\mathbf{i} + 9\mathbf{j} + \mathbf{k})/(7\sqrt{2})$.
21. 60.

3. $k(-11\mathbf{i} - 2\mathbf{j} + 5\mathbf{k})$, $k \in R$, $k \neq 0$.

7. $3\sqrt{5}$.

16.4   Page 662

1. (a) $\mathbf{n} \cdot \mathbf{r} = 0$, where $\mathbf{n} = [3, -1, 2]$ and
   $\mathbf{r} = [x - 3, y + 1, z - 2]$;
   (b) $3x - y + 2z - 14 = 0$.
5. (a) $\mathbf{n} \cdot \mathbf{r} = 0$, where $\mathbf{n} = [5, -5, 7]$ and
   $\mathbf{r} = [x + 3, y - 2, z]$;
   (b) $5x - 5y + 7z + 25 = 0$.
9. Identical: (a) and (d); parallel: (b) and (c); perpendicular: (a) and (b), (a) and (c), (b) and (d).
13. $5x - 7y + 12z = 0$.
17. $B = 5$.
21. $7x + 5y + 4z + 5 = 0$.
27. $(\mathbf{a} \times \mathbf{b}) \times \mathbf{c}$ is perpendicular to $\mathbf{c}$.

3. (a) $\mathbf{n} \cdot \mathbf{r} = 0$,   where $\mathbf{n} = [9, -7, -5]$ and
   $\mathbf{r} = [x - 4, y - 2, z - 5]$;
   (b) $9x - 7y - 5z + 3 = 0$.
7. $\mathbf{n} \cdot \mathbf{r} = 0$,   where $\mathbf{n} = [2, -6, 1]$ and
   $\mathbf{r} = [x - 1, y - 1, z - 1]$. (There are many
   other correct answers.)
11. (a) $x = 4$;   (b) $y = -3$;   (c) $z = 6$.

15. $2x - 3y + z + 2 = 0$.
19. $2x - 4y + z - 5 = 0$.
23. $3x + y + 8 = 0$.

16.5   Page 668

1. (a) $x = -2t + 4$, $y = t - 6$, $z = 5t + 3$;
   (b) $(x - 4)/-2 = y + 6 = (z - 3)/5$;
   (c) $x + 2y + 8 = 0$, $5y - z + 33 = 0$,
   $5x + 2z - 26 = 0$.
5. (a) $x = -2t + 4$, $y = 4t + 2$, $z = 6t + 7$;
   (b) $(x - 4)/-2 = (y - 2)/4 = (z - 7)/6$;
   (c) $2x + y - 10 = 0$, $3y - 2z + 8 = 0$,
   $3x + z - 19 = 0$.
9. (a) $(4, 0, 3)$, $(0, 5, 1)$;
   (b) $x = 4t + 4$, $y = -5t$, $z = 2t + 3$;
   (d) $(x - 4)/4 = y/-5 = (z - 3)/2$.
13. (a) $(-4, 0, 6)$, $(2, 5, 0)$;
   (c) $x = 6t + 2$, $y = 5t + 5$, $z = -6t$;
   (d) $(x - 2)/6 = (y - 5)/5 = z/-6$.
17. $(x - 4)/1 = y/-5 = (z - 6)/2$.
21. $x/5 = y/-3$, $z - 4 = 0$.
25. $x = -t + 2$, $y = t - 1$, $z = t + 5$.
27. *Outline of proof.* If $P_1:(x_1, y_1, z_1)$ is any point on l, then

$$A_1 x_1 + B_1 y_1 + C_1 z_1 + D_1 = 0$$

and

$$A_2 x_1 + B_2 y_1 + C_2 z_1 + D_2 = 0.$$

Thus

$$(A_1 x_1 + B_1 y_1 + C_1 z_1 + D_1) + \lambda(A_2 x_1 + B_2 y_1 + C_2 z_1 + D_2) = 0 + \lambda 0 = 0$$

for each value of λ. Thus every point of l is on the graph of (1). Moreover, since (1) is a linear equation

3. (a) $x = -3t + 2$, $y = 4t + 5$, $z = 2t - 4$;
   (b) $(x - 2)/-3 = (y - 5)/4 = (z + 4)/2$;
   (c) $4x + 3y - 23 = 0$, $y - 2z - 13 = 0$,
   $2x + 3z + 8 = 0$.
7. (a) $x = 3t + 6$, $y = -2$, $z = 5$;
   (b) $y + 2 = 0$, $z - 5 = 0$;
   (c) $y + 2 = 0$, $z - 5 = 0$.

11. (a) $(-3, 4, 0)$, $(1, 0, 6)$;
    (c) $x = 2t + 1$, $y = -2t$, $z = 3t + 6$;
    (d) $(x - 1)/2 = y/-2 = (z - 6)/3$.
15. (a) $(0, 5, -1)$, $(7, 0, 4)$;
    (c) $x = 7t + 7$, $y = -5t$, $z = 5t + 4$;
    (d) $(x - 7)/7 = y/-5 = (z - 4)/5$.
19. $(x - 7)/8 = y - 4 = (z - 6)/11$.
23. $x = 6t$, $y = 5t$, $z = -8t$.

in x, y, and z for each value of λ, the graph of (1) is a plane for every λ. So the graph of (1) is a plane containing the line l, for each number λ.

29. $6x - 3y + 13 = 0$, $9x - 3z + 14 = 0$, $9y - 6z - 11 = 0$.

31. $31x + 29y - 34 = 0$, $8x + 29z + 9 = 0$, $8y - 31z - 19 = 0$.

16.7 Page 674

1. (a) Elliptic cylinder. (b) Circular cylinder. (c) Plane. (d) Parabolic cylinder. (e) Circular cylinder. (f) Cubic cylinder.

3. $y^2 = 4px$.

5. $y^2 + z^2 - 8y - 6z + 16 = 0$. ·

9. Circle in the plane $z = 3$, with center $(0, 0, 3)$ and radius $3\sqrt{3}$.

7. $z^2 = -12y$, $x = 0$.

11. $x^2 + y^2 = 27$.

13. $x^2 + y^2 = 4z$.

17. $y^3 - x^2 - z^2 = 0$.

21. $x^2 + y^2 = e^{2z}$, $z \geq 0$.

15. $25x^2 + 16y^2 + 16z^2 = 400$.

19. $x^2y^2 + x^2z^2 = 1$, $x > 0$.

23. (a) $y = z^3$, $x = 0$, about the z axis; (b) $x = 1/y$, $z = 0$, about the y axis. (c) $3y - 2z + 6 = 0$, $x = 0$, about the z axis.

16.10 Page 685

1. (1) Ellipsoid. (2) Hyperboloid of one sheet. (3) Hyperboloid of two sheets. (5) Elliptic paraboloid. (6) Quadric cone.

3. Quadric cone.

5. Hyperbolic paraboloid.

9. Ellipsoid.

13. Quadric cone.

7. Hyperbolic paraboloid.

11. Elliptic paraboloid.

16.11 Page 686

1. (a) 3 and $\sqrt{35}$.

(b) The direction cosines of **a** are $2/3$, $-1/3$, $2/3$, and of **b** are $5/\sqrt{35}$, $1/\sqrt{35}$, $-3/\sqrt{35}$

(c) $[2/3, -1/3, 2/3]$.   (d) $\cos \theta = 1/\sqrt{35}$, $\theta \doteq 80.2°$.

3. $k(10\mathbf{i} - 11\mathbf{j} - 3\mathbf{k})$, $k \in R$, $k \neq 0$.

5. (a) $y = 7$;   (b) $x = -5$;   (c) $z = -2$.

7. (a) $\mathbf{n} \cdot \mathbf{r} = 0$, where $\mathbf{n} = [5, -4, 8]$ and $\mathbf{r} = [x - 2, y + 4, z + 5]$. (b) $5x - 4y + 8z + 14 = 0$.

9. (a) $\mathbf{n} \cdot \mathbf{r} = 0$, where $\mathbf{n} = [6, 3, -1]$ and $\mathbf{r} = [x - 1, y + 2, z - 4]$. (b) $6x + 3y - z + 4 = 0$.

11. $C = 1$.

13. $65x + 9y - 31z - 107 = 0$.

15. (a) $(0, 4, -4)$ and $(-3, 2, 0)$. (c) $x = -3t$, $y = -2t + 4$, $z = 4t - 4$. (d) $(x/-3) = (y - 4)/-2 = (z + 4)/4$.

17. $(x - 4)/3 = (y - 5)/5 = (z - 8)/2$.

19. (a) Circular cylinder.   (b) Sphere. (c) Parabolic cylinder.   (d) Plane. (e) Hyperbolic cylinder.   (f) No graph.

21. (a) $z^4 - 36x^2 - 36y^2 = 0$; (b) $x^2 + z^2 - 6y = 0$.

17.1 Page 693

5. $(x - 1)/-\sqrt{3} = (y - 3\sqrt{3})/3$
$= [z - (1/3)\pi]/1.$

7. $(x + 3)/3 = (y - 2)/-4 = (z + 1)/5.$

11. (a) $t(3t - 2) \sin t + (t^3 + 2) \cos t + 60t^2$;
(b) $4t(5t^2 - 2 \sin t - t \cos t)\mathbf{i} + 4t(5 + t \sin t - 2 \cos t)\mathbf{j} + \{(t^3 + 2) \sin t + t(2 - 3t) \cos t\}\mathbf{k}.$

17.2 Page 696

1. $\mathbf{v}(t) = 3\mathbf{i} + 8t\mathbf{j} + 6t^2\mathbf{k}$; $\mathbf{a}(t) = 8\mathbf{j} + 12t\mathbf{k}$;
$|\mathbf{v}(1)| = \sqrt{109}.$

3. $\mathbf{v}(t) = (2 - 2t)\mathbf{i} + 3\mathbf{j} + 3t^2\mathbf{k}$; $\mathbf{a}(t) = -2\mathbf{i} + 6t\mathbf{k}$;
$|\mathbf{v}(1)| = 3\sqrt{2}.$

5. $\mathbf{v}(t) = -2 \sin 2t\mathbf{i} - 2e^{-t}\mathbf{j} + 3 \cos t\mathbf{k}$;
$\mathbf{a}(t) = -4 \cos 2t\mathbf{i} + 2e^{-t}\mathbf{j} - 3 \sin t\mathbf{k}$;
$|\mathbf{v}(0)| = \sqrt{13}.$

7. $\mathbf{v}(t) = -\sin (1/2)t\mathbf{i} + \cos (1/2)t\mathbf{j}$
$- (3/2) \tan (1/2)t\mathbf{k}$;
$\mathbf{a}(t) = (-1/2) \cos (1/2)t\mathbf{i} - (1/2) \sin (1/2)t\mathbf{j}$
$- (3/4) \sec^2 (1/2)t\mathbf{k}$; $|\mathbf{v}(\pi/3)| = \sqrt{7}/2.$

11. $10 + 9 \ln \sqrt{3}.$

13. $976/27.$

15. $\sqrt{2} (\sinh 2 + \sinh 1).$

17.3 Page 701

1. $K(2) = \sqrt{30}/50$; $\mathbf{T}(2) = [2/\sqrt{5}, 1/\sqrt{5}, 0]$;
$\mathbf{N}(2) = [1/\sqrt{30}, -2/\sqrt{30}, 5/\sqrt{30}]$;
$\mathbf{B}(2) = [1/\sqrt{6}, -2/\sqrt{6}, -1/\sqrt{6}].$

3. $K(\pi/9) = 9/10$;
$\mathbf{T}(\pi/9) = [3/(2\sqrt{10}), -3\sqrt{3}/(2\sqrt{10}), 1/\sqrt{10}]$;
$\mathbf{N}(\pi/9) = [-\sqrt{3}/2, -1/2, 0]$;
$\mathbf{B}(\pi/9) = [1/(2\sqrt{10}), -\sqrt{3}/(2\sqrt{10}), -3/\sqrt{10}].$

5. $K(\pi/2) = \sqrt{2}/(3e^{\pi/2})$;
$\mathbf{T}(\pi/2) = [1/\sqrt{3}, -1/\sqrt{3}, 1/\sqrt{3}]$;
$\mathbf{N}(\pi/2) = [-1/\sqrt{2}, -1/\sqrt{2}, 0]$;
$\mathbf{B}(\pi/2) = [1/\sqrt{6}, -1/\sqrt{6}, -2/\sqrt{6}].$

7. $K(0) = 2\sqrt{30}/25$; $\mathbf{T}(0) = [-2/\sqrt{5}, 1/\sqrt{5}, 0]$;
$\mathbf{N}(0) = [1/\sqrt{30}, 2/\sqrt{30}\ 5/\sqrt{30}]$;
$\mathbf{B}(0) = [1/\sqrt{6}, 2/\sqrt{6}, -1/\sqrt{6}].$

11. $\left[\dfrac{-e^t + e^{-3t}}{e^{2t} + e^{-2t} + 4}, \dfrac{2e^{2t} - 2e^{-2t}}{e^{2t} + e^{-2t} + 4}, \dfrac{e^{3t} - e^{-t}}{e^{2t} + e^{-2t} + 4}\right]$;
$\left[\dfrac{2e^t + 4e^{-t}}{e^{2t} + e^{-2t} + 4}, \dfrac{-2e^{2t} + 2e^{-2t}}{e^{2t} + e^{-2t} + 4}, \dfrac{4e^t + 2e^{-t}}{e^{2t} + e^{-2t} + 4}\right].$

13. $\left[\dfrac{-\cot^2 t \csc^2 t + \sec^2 t}{\cot^2 t + \sec^2 t}, \dfrac{\csc^2 t - \sec^2 t \tan^2 t}{\cot^2 t + \sec^2 t}, \dfrac{-\cot t \csc^2 t + \sec^2 t \tan t}{\cot^2 t + \sec^2 t}\right]$;
$\left[\dfrac{-\sec^2 t - \sec^2 t \csc^2 t}{\cot^2 t + \sec^2 t}, \dfrac{-\sec^2 t \cot^2 t - \sec^4 t - \csc^2 t + \sec^2 t \tan^2 t}{\cot^2 t + \sec^2 t}, \dfrac{\cot t \csc^2 t - \sec^2 t \tan t}{\cot^2 t + \sec^2 t}\right].$

17.5 Page 706

11. $\mathbf{v}(t) = 2\mathbf{i} + 3t^2\mathbf{j} + 2t\mathbf{k}$; $\mathbf{a}(t) = 6t\mathbf{j} + 2\mathbf{k}$;
$|\mathbf{v}(2)| = 2\sqrt{41}.$

15. $K(1) = \sqrt{266}/98$; $\mathbf{T}(1) = (\mathbf{i} + 2\mathbf{j} + 3\mathbf{k})/\sqrt{14}$;
$\mathbf{N}(1) = (-11\mathbf{i} - 8\mathbf{j} + 9\mathbf{k})/\sqrt{266}.$

18.1 Page 711

1. All points in the xy plane that are not on the line $x = y$.

3. The xy plane.

5. All points in the xy plane that are inside or on the ellipse $4x^2 + y^2 = 16$.

7. All points inside or on the ellipse $4x^2 + 9y^2 = 36$.

9. All points that are outside or on the ellipse $x^2/36 + y^2/49 = 1$.

11. $(0, \infty)$; $[0, 6).$

25. All points outside or on the sphere with radius 4 and center at the origin.

27. All points inside or on the ellipsoid $16x^2 + 9y^2 + 144z^2 - 144 = 0$.

29. All spheres with centers at the origin.

31. Hyperboloids of one sheet when $k > 0$; quadric cone if $k = 0$; and hyperboloids of two sheets when $k < 0$.

33. Hyperbolic cylinders when $k \neq 0$; two distinct planes containing the z axis when $k = 0$.

18.2   Page 716

1. $f_x(x, y) = 8(2x - y)^3$; $f_y(x, y) = -4(2x - y)^3$.

5. $f_x(x, y) = e^y \cos x$; $f_y(x, y) = e^y \sin x$.

9. $g_x(x, y) = -ye^{-xy}$; $g_y(x, y) = -xe^{-xy}$.

13. $f_x(x, y) = -2xy \sin (x^2 + y^2)$;
$f_y(x, y) = -2y^2 \sin (x^2 + y^2) + \cos (x^2 + y^2)$.

17. $f_{xy}(x, y) = 12xy^2 - 15x^2y^2 = f_{yx}(x, y)$.

21. $f_{xy}(x, y) = 0 = f_{yx}(x, y)$.

25. $f_x(\sqrt{5}, -2) = -4/21$;
$f_y(\sqrt{5}, -2) = -4\sqrt{5}/21$.

29. 3.

37. $f_x(x, y, z) = 6xy - yz$; $f_y(x, y, z) = 3x^2 - xz$
$+ 2yz^2$; $f_z(x, y, z) = -xy + 2y^2z$.

3. $f_x(x, y) = (x^2 + y^2)/(x^2 y)$;
$f_y(x, y) = -(x^2 + y^2)/(xy^2)$.

7. $f_x(x, y) = x(x^2 - y^2)^{-1/2}$;
$f_y(x, y) = -y(x^2 - y^2)^{-1/2}$.

11. $f_x(x, y) = 4/[1 + (4x - 7y)^2]$;
$f_y(x, y) = -7/[1 + (4x - 7y)^2]$.

15. $F_x(x, y) = 2 \cos x \cos y$;
$F_y(x, y) = -2 \sin x \sin y$.

19. $f_{xy}(x, y) = -6e^{2x} \sin y = f_{yx}(x, y)$.

23. $F_x(3, -2) = 1/9$; $F_y(3, -2) = -1/2$.

27. 1.

35. $180x^4y^2 - 12x^2$.

39. $f_x(x, y, z) = -yze^{-xyz} - y(xy - z^2)^{-1}$;
$f_y(x, y, z) = -xze^{-xyz} - x(xy - z^2)^{-1}$;
$f_z(x, y, z) = -xye^{-xyz} + 2z(xy - z^2)^{-1}$.

18.3   Page 722

1. At all points in the xy plane.

5. All points below the line $x - 2y + 1 = 0$.

9. No.

3. -1.

7. No. The function is not defined on the line $x - 2y = 0$.

11. The function does not exist on the line $y + 2 = 0$.

18.4   Page 725

1. $(3x^2 - y^2)\, dx + (3 - 2xy)\, dy$.

5. $e^x e^{-y}\, dx - e^x e^{-y}\, dy$.

9. $dw = -ste^{-rst}\, dr - rte^{-rst}\, ds - rse^{-rst}\, dt$.

13. $\Delta w = 0.044\ 186\ 84$; $dw = 0.045$.

17. $x\, \Delta y + y\, \Delta x + \Delta x\, \Delta y$.

3. $\dfrac{y}{1 + x^2y^2}\, dx + \dfrac{x}{1 + x^2y^2}\, dy$.

7. $(\sin \theta + 2r)\, dr + r \cos \theta\, d\theta$.

11. $\Delta z = 0.080\ 179\ 9216$; $dz = 0.08$.

15. $\Delta z \doteq 0.0397$; $dz = 0.04$.

19. 0.0044.

18.5   Page 729

1. $4(2u - 3v)x - (3u - 2v)(1 + 3y)$;
$(3v - 2u) - 3(3u - 2v)x$.

5. $(3u^2v \cos u^3)[-\sin (x - y)] + (\sin u^3)/(x + y)$;
$(3u^2v \cos u^3)[\sin (x - y)] + (\sin u^3)/(x + y)$.

9. $f_x(u, v) = f_y(u, v) = f_z(u, v) = 4uve^{x-y-z}$
$+ 2u^2/(x + y + z)$.

13. $8x \cos (r - \theta) + 14y \cos 3\theta$; $-8x \cos (r - \theta)$
$- 42yr \sin 3\theta$.

3. $(6u^2 + v)2x + (u - 9v^2)5y$;
$(6u^2 + v) + (u - 9v^2)(5x - 2y)$.

7. $v(x + y + z)^{-1} + ue^{x-y-z}$; $v(x + y + z)^{-1}$
$-ue^{x-y-z}$; $v(x + y + z)^{-1} - ue^{x-y-z}$.

11. $6x \sin 2\theta - 10y \sin (\theta - r)$; $12xr \cos 2\theta$
$+ 10y \sin (\theta - r)$.

15. $2(6x + 5y) \cos 2t - 2(5x + 2y) \sin 2t$.

17. $3[yz^2 \cos (xyz^2)]t^2 + 2[xz^2 \cos (xyz^2)]t$
    $+ 2xyz \cos (xyz^2)$.
21. $(y \sin x - \sin y)/(x \cos y + \cos x)$.
25. 2625π cubic inches per second.

19. $(3x^2 + 4xy)/(3y^2 - 2x^2)$.

23. $(y + 2yz)/(2z - 2xy)$; $(x + 2xz)/(2z - 2xy)$.
27. $\partial w/\partial x' = (\partial w/\partial x) \cos \theta + (\partial w/\partial y) \sin \theta$;
    $\partial w/\partial y' = -(\partial w/\partial x) \sin \theta + (\partial w/\partial y) \cos \theta$.

18.6   Page 738

1. $(2x - 3y) \cos \theta - (3x - 4y) \sin \theta$;
   $(1/2)(11 - 8\sqrt{3})$.
5. $-(y \cos \theta + x \sin \theta)/e^{xy}$; $-(\sqrt{3} + 1)e/2$.
9. $-3(2e^{-6} + 1) \cos \alpha + (3e^{-6} - 4) \cos \beta$
   $- (2e^{-6} + 1) \cos \gamma$.
13. $(1/6)(\mathbf{i} + 7\mathbf{j})$.
17. $\text{Tan}^{-1} (5/8)$.
21. $\theta = \pi - \text{Tan}^{-1} (6/5)$. Maximum is $\sqrt{61}$.

3. $(3x^2 + 8xy) \cos \theta + 4(x^2 - y^2) \sin \theta$;
   $91\sqrt{2}/2$.
7. $2 \cos \alpha - 2 \cos \beta + 4 \cos \gamma$.
11. $-21\mathbf{i} + 16\mathbf{j}$.

15. $6\mathbf{i} - 10\mathbf{j} - 14\mathbf{k}$.
19. $\theta = \text{Tan}^{-1} (-1/4)$. Maximum is $3\sqrt{17}$.
23. (a) $x^2/45 + y^2/20 = 1$, $z = 0$;
    (b) $(4/3)\mathbf{i} + \mathbf{j}$.

18.7   Page 744

1. $(16\mathbf{i} + 4\mathbf{j} + 8\sqrt{2}\,\mathbf{k}) \cdot [(x - 1)\mathbf{i} + (y - 2)\mathbf{j}$
   $+ (z - \sqrt{2}/2)\mathbf{k}] = 0$; $4(x - 1) + (y - 2)$
   $+ 2\sqrt{2}(z - \sqrt{2}/2) = 0$.
5. $(6\mathbf{i} - 12\mathbf{j} + 6\sqrt{2}\,\mathbf{k}) \cdot [(x - 3)\mathbf{i} + (y - 6)\mathbf{j}$
   $+ (z - 3\sqrt{2})\mathbf{k}] = 0$; $(x - 3) - 2(y - 6)$
   $+ \sqrt{2}(z - 3\sqrt{2}) = 0$.
11. Rel. max. is 0 at $(0, 0)$.
15. None.
19. $3\sqrt{3}\,\mathbf{i} + 3\sqrt{3}\,\mathbf{j} + 3\sqrt{3}\,\mathbf{k}$.
23. Hint: Use 18.7.5.

27. Approximately $y = 0.564x + 2.215$.

3. $(2\mathbf{i} - 6\mathbf{j} + 2\sqrt{7}\,\mathbf{k}) \cdot [(x - 1)\mathbf{i} + (y - 3)\mathbf{j}$
   $+ (z - \sqrt{7})\mathbf{k}] = 0$; $(x - 1) - 3(y - 3)$
   $+ \sqrt{7}(z - \sqrt{7}) = 0$.
7. $32, -19, -17$. $\dfrac{x - 1}{32} = \dfrac{y - 2}{-19} = \dfrac{z - 2}{-17}$.

13. Rel. min. is -8 at $(2, 2)$.
17. Base 8 feet by 8 feet; depth 4 feet.
21. $2\pi/3$; 4 inches.
25. Hint: Show that the second equation of (6) can
    be written $\bar{y} = m\bar{x} + b$, where $(\bar{x}, \bar{y})$ is the
    centroid.

18.8   Page 750

1. 6.
5. $72/7$.
9. 2 cubic feet.

3. 5.
7. 32.

18.9   Page 753

1. $f(x, y) = 5x^2 - 7xy + y^2 + C$.
5. $f(x, y) = (2x^3/5y^2) + C$.
9. $f(x, y) = x^3 \cos (y^2) + C$.

3. Not an exact differential.
7. $f(x, y) = 2xe^y - ye^x + C$.
11. $f(x, y, z) = x^3 + 2y^3 + 3z^3 + C$.

18.10   Page 759

1. (a) $100/3$;   (b) 60.
5. $815/14$.
9. $\dfrac{\partial P}{\partial y} = 2xy = \dfrac{\partial Q}{\partial x}$; 10.

3. (a) $828/35$;   (b) 144.
7. $(8 \ln 3)/5 + e^4(e^{32} - 1)/8$.
11. $\dfrac{\partial P}{\partial y} = -1 = \dfrac{\partial Q}{\partial x}$; $\dfrac{13}{6}$.

13. $\dfrac{\partial P}{\partial y} = \dfrac{2}{y} = \dfrac{\partial Q}{\partial x}$; 16 ln 2.

19. -23.

15. $\dfrac{\partial P}{\partial y} = 12x^2y + 4y^3 - 3 = \dfrac{\partial Q}{\partial x}$; -324.

18.11   Page 765

1. $-7/44$.
5. $2e^2(e^4 - 1) - 2\ln 3$.
9. $5k/104$, where k is the (positive) constant of proportionality in the force vector.

3. $618/5$.
7. $-(a^2 + b^2)/2$.

18.13   Page 767

1. (a) All points outside or on the ellipse $x^2 + 4y^2 - 100 = 0$.

   (b) All points to the right of or on the line $2x - y - 1 = 0$.

7. $f_x(x, y) = e^{-y}\sec^2 x$; $f_y(x, y) = -e^{-y}\tan x$;
   $f_{xx}(x, y) = 2e^{-y}\sec^2 x \tan x$;
   $f_{yy}(x, y) = e^{-y}\tan x$; $f_{xy}(x, y) = -e^{-y}\sec^2 x$;
   $f_{yx}(x, y) = -e^{-y}\sec^2 x$.

11. 1.

5. $f_x(x, y) = 12x^3y^2 + 14xy^7$; $f_y(x, y) = 6x^4y + 49x^2y^6$; $f_{xx}(x, y) = 36x^2y^2 + 14y^7$; $f_{yy}(x, y) = 6x^4 + 294x^2y^5$; $f_{xy}(x, y) = 24x^3y + 98xy^6$; $f_{yx}(x, y) = 24x^3y + 98xy^6$.

9. $450x^2y^4 - 42y^5$.

15. $\dfrac{1}{2}\left(\dfrac{v}{1 + u^2v^2}\right)\dfrac{y}{\sqrt{xy}} + \dfrac{1}{2}\left(\dfrac{u}{1 + u^2v^2}\right)\dfrac{1}{\sqrt{x}}$;
$\dfrac{1}{2}\left(\dfrac{v}{1 + u^2v^2}\right)\dfrac{x}{\sqrt{xy}} - \dfrac{1}{2}\left(\dfrac{u}{1 + u^2v^2}\right)\dfrac{1}{\sqrt{y}}$.

19. $5xz^2(3yzt^{1/2} + xyt^{-1} - 9xye^{3t})/(25x^4y^2 + z^6)$.

23. $\text{Tan}^{-1}(4/9)$.

27. $16\sqrt{3}/3$.

31. $\dfrac{\partial P}{\partial y} = \sec^2 x = \dfrac{\partial Q}{\partial x}$; 9.

13. No. The given function is not defined on the line $x + y = 0$.

17. $\dfrac{-v}{e^u}\left(\dfrac{yz}{1 + (xyz)^2}\right) + \dfrac{1}{e^u}\left(\dfrac{y}{xy + yz}\right)$;
$\dfrac{-v}{e^u}\left(\dfrac{zx}{1 + (xyz)^2}\right) + \dfrac{1}{e^u}\left(\dfrac{x + z}{xy + yz}\right)$;
$\dfrac{-v}{e^u}\left(\dfrac{xy}{1 + (xyz)^2}\right) + \dfrac{1}{e^u}\left(\dfrac{y}{xy + yz}\right)$.

21. $3(y\cos\theta + x\sin\theta)/(1 + 9x^2y^2)$; $(3\sqrt{3} - 6)/577$.

25. $(18\mathbf{i} + 16\mathbf{j} + 6\sqrt{11}\mathbf{k}) \cdot [(x - 1)\mathbf{i} + (y - 2)\mathbf{j} + (z - \sqrt{11}/3)\mathbf{k}] = 0$; $9(x - 1) + 8(y - 2) + 3\sqrt{11}(z - \sqrt{11}/3) = 0$.

29. Altitude: 4; radius of base: 2.

33. $-\pi/2$.

19.1   Page 774

1. 120.
5. 148.
9. 34.

3. $60\frac{2}{3}$.
7. $129\frac{1}{3}$.

19.2   Page 781

1. $-3/5$.
5. $(e^{27} - e)/2$.
9. $-\ln\sqrt{2}$.
13. 10.
17. 10.

3. 240.
7. $-\sqrt{2}/(2\pi)$.
11. $\pi^2/4 - \pi + 2$.
15. 20.
19. 144.

21. 28.

25. 128/3.

23. $(e + e^{-1} - 2)/2.$

19.3 Page 787

1. 139/80.

5. $\sqrt{3} - 1/2.$

9. $\pi/6 - 2(3^{-5/8})/5.$

3. $\ln \cos 1 + 4/3.$

7. $\sqrt{2} - \ln \sqrt{2}.$

11. $(\bar{x}, \bar{y}) = (2, 9/5); I_x = 117, I_y = 160; I_0 = 277.$

13. (a) $\bar{x} = \dfrac{e^2 - 8e + 33}{2(e^2 + 8e - 13)}, \bar{y} = \dfrac{8e^3 + 27e^2 - 53}{18(e^2 + 8e - 13)};$

   (b) $I_x = (27e^4 + 64e^3 - 139)/432, I_y = (e^2 + 32e - 81)/8;$

   (c) $I_0 = (27e^4 + 64e^3 + 54e^2 + 1728e - 4513)/432.$

15. *Outline of proof.* The mass, Δm, of the rectangular region S, having $P{:}(x, y)$ and $Q{:}(x + \Delta x, y + \Delta y)$ as opposite vertices, was defined by

(1)
$$\Delta m = \iint\limits_{S} \rho(x, y) \, dA.$$

By the mean value theorem for double integrals, there is a point (λ, μ) in the interior of S such that

(2)
$$\iint\limits_{S} \rho(x, y) \, dA = \rho(\lambda, \mu) \, \Delta A.$$

From (1) and (2), $\Delta m = \rho(\lambda, \mu) \, \Delta A$, or

$$\frac{\Delta m}{\Delta A} = \rho(\lambda, \mu), \qquad \Delta A \neq 0.$$

Thus

(3)
$$\lim_{d \to 0} \frac{\Delta m}{\Delta A} = \lim_{d \to 0} \rho(\lambda, \mu).$$

Since (λ, μ) is always inside the rectangular region S and ρ is continuous at $P{:}(x, y)$, the limit in the right-hand member of (3) exists and is equal to $\rho(x, y)$. Therefore

$$\lim_{d \to 0} \frac{\Delta m}{\Delta A} = \rho(x, y).$$

19.4 Page 793

1. $-64/15.$

5. 0.

9. 8/3.

3. 72/35.

7. $\pi.$

19.5 Page 799

1. $2\sqrt{3} + 4\pi/3.$

5. $\pi(b^2 - a^2).$

9. $(a^2 + b^2/2) \operatorname{Sin}^{-1}(a/b) + 3a\sqrt{b^2 - a^2}/2 + \pi(2a^2 + b^2)/4.$

13. $\bar{x} = 0, \bar{y} = 6/5.$

17. $I_x = 33k\pi a^5/40; I_y = 93k\pi a^5/40.$

3. $\pi a^2/8.$

7. $4\pi a^2.$

11. $\pi(1 - e^{-9})/4.$

15. $\bar{x} = 45/(4\pi), \bar{y} = 0.$

19. $I_x = I_y = k\pi(b^5 - a^5)/5.$

19.6 Page 803

1. $\sqrt{61}/3.$

5. $9 \operatorname{Sin}^{-1}(2/3).$

9. $5\pi.$

3. $\pi/3.$

7. $8\sqrt{2}.$

19.7 Page 809

1. -40.

3. $189/2$.

5. $2/3$.

7. 156.

9. $9/16$.

11. $\int_0^3 \int_0^2 \int_0^{(4-y^2)^{1/2}} (x + 2y - 3z)\, dx\, dy\, dz$
    $= 24 - 27\pi/2$.

13. $\int_0^4 \int_0^{3x/2} \int_0^{y^{1/2}} (x^2 + y^2 + z^2)\, dx\, dz\, dy = 62$.

15. $\int_0^1 \int_0^{z^{1/2}} \int_0^{y^2} (x^{1/2} + y^{1/2} + z^{1/2})\, dx\, dy\, dz$
    $= 125/462$.

17. $\int_0^3 \int_0^{(9-x^2)^{1/2}} \int_0^{9-x^2-y^2} dz\, dy\, dx = 81\pi/8$.

19. $\int_0^1 \int_0^{(2y-y^2)^{1/2}} \int_0^3 (xy - yz + zx)\, dz\, dx\, dy = (32 - 9\pi)/8$.

19.8 Page 812

1. $v = 4\int_0^{3\sqrt{2}} \int_0^{(2/3)(18-x^2)^{1/2}} \int_0^{(72-4x^2-9y^2)/36} dz\, dy\, dx$
   $= 12\pi$.

3. $v = \int_0^2 \int_0^{4-y^2} \int_0^{4-2y} dz\, dx\, dy = 40/3$.

5. $v = 8\int_0^3 \int_0^{(9-x^2)^{1/2}} \int_0^{(9-x^2)^{1/2}} dz\, dy\, dx = 144$.

7. $v = 4\int_0^2 \int_0^{(16-2z)^{1/2}} \int_0^{(16-2z-y^2)^{1/2}} dx\, dy\, dz$
   $= 36\pi$.

9. $v = \int_0^3 \int_0^{(9-x^2)^{1/2}} \int_{(x^2+y^2)/3}^{12-x^2-y^2} dz\, dy\, dx = 27\pi/2$.

11. $v = 4\int_0^1 \int_{x^2}^1 \int_0^{y^{1/2}} dz\, dy\, dx = 2$.

13. $v = 8\int_0^a \int_0^{(b/a)(a^2-x^2)^{1/2}} \int_0^{(c/ab)(a^2b^2-b^2x^2-a^2y^2)^{1/2}} dz\, dy\, dx = 4\pi\, abc/3$.

15. $\bar{x} = \bar{y} = \bar{z} = 4/15$.

17. $\bar{x} = 0,\ \bar{y} = 10/3,\ \bar{z} = 0$.

19. $\bar{x} = 0,\ \bar{y} = 615/196,\ \bar{z} = 0$.

21. $\bar{x} = 0,\ \bar{y} = 9/2,\ \bar{z} = 36/7$.

23. $\bar{x} = \bar{y} = 0,\ \bar{z} = 8/5$.

25. $k\int_0^1 \int_0^{1-x} \int_0^{1-x-y} (xy^2 + xz^2)\, dz\, dy\, dx$.

19.9 Page 818

1. (a) $(2\sqrt{2}, \pi/4, 3)$;   (b) $(2, \pi/3, -5)$;
   (c) $(8, -\pi/6, 6)$;   (d) $(5, \pi, -2)$.

3. $r = 3$.

5. $r^2 \cos 2\theta = 25$.

7. $r^2 + 4z^2 = 10$.

9. $x^2 + y^2 + z^2 = 9$.

11. $2x - 3y + z = 6$.

13. 8π.

15. $I_z = \pi a^4 h^2 k/4$, where a = radius of cylinder.

17. $I_z = 1024k\pi/15$.

19. $20\sqrt{5}\,\pi/3$.

19.10 Page 822

1. (a) $(4\sqrt{2}, 5\pi/3, \pi/4)$;   (b) $(12, \pi/6, \pi/4)$;
   (c) $(6, 5\pi/3, \pi/2)$;   (d) $(4, 3\pi/4, \pi/6)$.

3. (a) $(2\sqrt{2}, 2\sqrt{2}, 4\sqrt{3})$;
   (b) $(-\sqrt{3}, 1, 2)$;
   (c) $(13, 0, 0)$;
   (d) $(\sqrt{2}, \sqrt{6}, -2\sqrt{2})$.

5. $\rho^2(3 \sin^2 \phi - 2) = 0$.

7. $\rho^2(5 \sin^2 \phi \sin^2 \theta - 40 \cos^2 \phi + 4) = 36$.

9. $\rho \sin^2 \phi - \cos \phi = 0$.

11. $x^2 + z^2 = 1$.

13. $x^2 + y^2 + z^2 - 5z = 0$.

15. $x^2 + y^2 - z^2 = 1$.

17. $(0, 0, \bar{z})$, where $\bar{z} = (3/4)(1 + \sqrt{2}/2)$.

19. $8\pi c^7 k/21$, where c is the radius of the sphere.

21. $\pi kh(3s^5 - 5h^2 s + 2h^5)/45$, where s is the slant height.

20.1   Page 826

15. $x/y = C.$

17. $y = (x + C)/(1 - Cx).$

19. $e^x(x - 1) + (y^2/2) \ln |y| + C = 0$

20.2   Page 827

1. $y/x = C.$

3. $x^2 - y^2 = Cx.$

5. $xy = Ce^{x/y}.$

7. $x^2 + 2xy - y^2 = C.$

9. $3x^2 + xy - y^2 = C.$

11. $\sqrt{x^2 + y^2}/(xy)^2 = C.$

13. $x + \sin (y/x) = C.$

20.3   Page 830

1. $x^2y = C.$

3. $2xy + x^2 = C.$

5. $y^3 + 3xy - x^3 = C.$

7. $3x \ln y + x^3 = C.$

9. $\ln (1 + y^2) + 2 \operatorname{Tan}^{-1} x = C.$

20.4   Page 832

1. $ye^x = x + C.$

3. $y = a + C\sqrt{1 - x^2}.$

5. $y = xe^x + Cx.$

7. $y = 1 + C/x.$

9. $y = (1/2)x + Ce^{-2x} - 1/4.$

11. $(y - 1)e^{\int f(x)\,dx} = C.$

20.5   Page 836

1. $s = (1/2)gt^2 + C_1t + C_2.$

3. $y = C_1e^{ax} + C_2e^{-ax}$

5. $y = C_1e^{2x} + C_2 - e^x.$

7. $y = C_1(C_2e^{C_1x} - 1)/(C_2e^{C_1x} + 1).$

9. $y = C_1x^2 + C_2x + C_3 + xe^x - 3e^x.$

20.8   Page 842

1. $y = C_1e^{2x} + C_2e^{3x}.$

3. $y = C_1e^x + C_2e^{-7x}.$

5. $y = C_1e^{-3x/2} + C_2e^{x/3}.$

7. $y = C_1 \sin 3x + C_2 \cos 3x.$

9. $y = e^{-x/2}[C_1 \sin (\sqrt{3}/2)x + C_2 \cos (\sqrt{3}/2)x].$

20.9   Page 845

1. $y = C_1 \sin x + C_2 \cos x + 5.$

3. $y = C_1 \sin x + C_2 \cos x - \cos x \ln |\sin x| - x \sin x.$

5. $y = C_1 \sin x + C_2 \cos x - \cos x \ln |\sec x + \tan x|.$

7. $y = C_1e^{2x} + C_2e^{3x} + e^x.$

9. $y = C_1e^x + C_2e^{2x} + (e^x + e^{2x}) \ln (1 + e^{-x}).$

20.10   Page 850

1. $x = \cos 8t;$ frequency $= 4/\pi;$ period $= \pi/4.$

3. $v = dx/dt = -8 \sin 8t;$ maximum speed $=$ 8 feet per second.

5. $x = e^{-t/20} \cos 8t.$

7. Approximately 46 seconds.

9. $x = e^{-t/20} [\cos 8t - (19/160) \sin 8t].$

20.11   Page 854

1. $q = 10^{-6}(1 - e^{-t}).$

3. $i = 10^{-6}(1 - e^{-10^6t}).$

5. (a) $i = 2.25 \times 10^{-2} \cos 377t$ amperes;
   (b) $i = -2.25 \times 10^{-2} \cos 377t$ amperes;
   (c) $i = 12 \times 10^{-2} \sin 377t$ amperes.
9. $i = 12 \times 10^{-2} \sin 377t$ amperes.

7. $i = 4.4 \times 10^{-2} \sin 377t - 5.8 \times 10^{-2} \cos 377t$ amperes.

20.12   Page 855

1. $xy = C$.
5. $y = (e^x/4) + C_1 e^{-3x} + C_2$.
9. $y = -e^x + Ce^{2x}$.
13. $y = C_1 e^x + C_2 e^{-x} - 1$.
17. $x = -\cos 4t$; amplitude $= 1$ foot; period $= \pi/2$.

3. $4x^2 y - x^4 = C$.
7. $x \sin y + \sin^2 y = C$.
11. $y = C_1 e^x + C_2 e^{2x}$.
15. $y = C_1 \cos 2x + C_2 \sin 2x$.
19. $i = e^{-t} \sin t$.

index